SYSTEM IDENTIFICATION

(SYSID'03)

A Proceedings volume from the 13th IFAC Symposium on System Identification, Rotterdam, The Netherlands, 27 – 29 August 2003

Edited by

P.M.J. Van den HOF
Delft Center for Systems and Control,
Delft University of Technology,
Delft, The Netherlands

B. WAHLBERG
Royal Institute of Technology,
Stockholm, Sweden

S. WEILAND
Department of Electrical Engineering
Eindhoven University of Technology,
Eindhoven, The Netherlands

(In four volumes)

Volume 3

Published for the

INTERNATIONAL FEDERATION OF AUTOMATIC CONTROL

by

ELSEVIER LTD

ELSEVIER Ltd
The Boulevard, Langford Lane
Kidlington, Oxford OX5 1GB, UK

Elsevier Internet Homepage
http://www.elsevier.com

Consult the Elsevier Homepage for full catalogue information on all books, journals and electronic products and services.

IFAC Publications Internet Homepage
http://www.elsevier.com/locate/ifac

Consult the IFAC Publications Homepage for full details on the preparation of IFAC meeting papers, published/forthcoming IFAC books, and information about the IFAC Journals and affiliated journals.

First edition 2004

Library of Congress Cataloging in Publication Data

A catalogue record for this book is available from the Library of Congress

British Library Cataloguing in Publication Data

A catalogue record for this book is available from the British Library

ISBN 0-08-043709 5
ISSN 1474-6670

These proceedings were reproduced from manuscripts supplied by the authors, therefore the reproduction is not completely uniform but neither the format nor the language have been changed in the interests of rapid publication. Whilst every effort is made by the publishers to see that no inaccurate or misleading data, opinion or statement appears in this publication, they wish to make it clear that the data and opinions appearing in the articles herein are the sole responsibility of the contributor concerned. Accordingly, the publisher, editors and their respective employers, officers and agents accept no responsibility or liability whatsoever for the onsequences of any such inaccurate or misleading data, opinion or statement.

To Contact the Publisher

Elsevier welcomes enquiries concerning publishing proposals: books, journal special issues, conference proceedings, etc. All formats and media can be considered. Should you have a publishing proposal you wish to discuss, please contact, without obligation, the publisher responsible for Elsevier's industrial and control engineering publishing programme:

Christopher Greenwell
Publishing Editor
Elsevier Ltd
The Boulevard, Langford Lane
Kidlington, Oxford
OX5 1GB, UK

Phone: +44 1865 843230
Fax: +44 1865 843920
E.mail: c.greenwell@elsevier.com

General enquiries, including placing orders, should be directed to Elsevier's Regional Sales Offices – please access the Elsevier homepage for full contact details (homepage details at the top of this page).

Printed and bound in the United Kingdom
Transferred to digital print 2010

13th IFAC SYMPOSIUM ON SYSTEM IDENTIFICATION (SYSID 2003)

Sponsored by
International Federation of Automatic Control (IFAC)
IFAC Technical Committees on:
- Modeling, Identification and Signal Processing (MISP)
- Adaptive Control and Tuning (ACT)

Co-sponsored by
IEEE Control Systems Society
Division of Automatic Control (MRBT) of the Royal Institution of Engineers in The Netherlands (KIVI)
The Netherlands Organisation of Scientific Research (NWO)
Royal Netherlands Academy of Arts and Sciences (KNAW)
Dutch Institute of Systems and Control (DISC)
Faculty of Applied Sciences, Delft University of Technology (TUD)
Delft Center for Systems and Control (TUD)
Department of Electrical Engineering, Eindhoven University of Technology, The Netherlands (TU/e)
Stichting Meten en Regelen ER-THE, The Netherlands

Organizing Committee
P.M.J. Van den Hof – Delft University of Technology, Delft, The Netherlands
B. Wahlberg – Royal Institute of Technology, Stockholm, Sweden
S. Weiland – Eindhoven University of Technology, Eindhoven, The Netherlands

IPC Task Force
M. Deistler
M. Gevers
L. Ljung
M. Morari
J. Schoukens
P.M.J. Van den Hof
M. Viberg
B. Wahlberg

International Programme Committee (IPC)
P.M.J. Van den Hof; The Netherlands (Co-Chair)
B. Wahlberg, Sweden (Co-Chair)

P. Albertos; Spain
B. Anderson; Australia
E. Bai; USA
M. Basseville; France
R. Bitmead; USA
S. Bittanti* **; Italy
M. Blanke; Denmark
J. Bokor; Hungary
M. Campi; Italy
H.F. Chen; P.R. China
J. Chen; USA
R. de Callafon; USA
M. Deistler; Austria
B. de Moor; Belgium
J.J. Fuchs; France
K. Godfrey; UK

G. Goodwin; Australia
M. Gevers; Belgium
P. Guillaume; Belgium
L. Guo; P.R. China
H. Hjalmarsson; Sweden
H. Kimura; Japan
R. Kosut; USA
V. Krishnamurty; Australia
K. Kumamaru; Japan
I. Landau; France
J.H. Lee; USA
L. Ljung; Sweden
P. Mäkilä; Finland
T. McKelvey; Sweden
M. Milanese; Italy
M. Morari; Switzerland
B. Ninness; Australia
R. Ortega**; France
G. Picci; Italy
R. Pintelon; Belgium
B. Polyak; Russia
P. Regalia; France
D. Rivera; USA
W. Scherrer; Austria
J. Schoukens; Belgium
R. Schumann; Germany
R. Smith*; USA
T. Söderström*; Sweden
T. Sugie; Japan
R. Tempo; Italy
J. van Schuppen; The Netherlands
M. Verhaegen; The Netherlands
S. Veres; UK
M. Viberg; Sweden
A. Vicino; Italy
E. Walter; France

** Appointed by IFAC Technical Committee MISP*
*** Appointed by IFAC Technical Committee ACT*

National Organizing Committee (NOC)

P.M.J. Van den Hof (Finances, contacts NMO, Public Relations)
S. Weiland (PC Secretariat, Paper handling, Website)
A.C.P.M. Backx (Industrial participation, Sponsors)
M.H.G. Verhaegen (Publications)
Y. Zhu (Exhibitions)
T. Van der Weiden (Local arrangements, Technical and Social Events)

PREFACE

These Proceedings contain all the technical material presented at the 13th IFAC Symposium on System Identification (SYSID 2003), held in the Conference Center "De Doelen", Rotterdam, The Netherlands from 27 – 29 August 2003.

The SYSID symposium is organized every three years and is among the most successful symposia organized by IFAC. This has been the first SYSID symposium in the 3rd millennium and the second SYSID symposium to take place in The Netherlands, following The Hague symposium in 1973.

Being the only worldwide symposium that is fully directed towards system identification, it is the ideal opportunity for researchers and industrial engineers from very many disciplines to present and discuss the developments, the results and the future challenges in all aspects of modelling dynamical systems on the basis of experimental data.

The symposium covered all major aspects of system identification, experimental modelling, signal processing and adaptive control from theoretical and methodological developments to practical applications in a wide range of application areas. For the 13th edition of this symposium, the International Program Committee has taken steps to position SYSID 2003 as a meeting place where scientists and engineers from several research communities can meet to discuss issues related to these areas.

A total of 350 delegates from 40 different countries attended the conference. 100 of the participants were PhD students, showing that system identification is a very vital field of research. Out of a total of 422 papers that were submitted to SYSID 2003, the IPC selected 333 papers and these were incorporated in the final program. The selection was based on two referee reports per paper. The final program of the symposium was composed of 3 plenary papers, 6 semi-plenary papers, 232 papers in oral sessions, 82 posters and 10 software demonstrations. The Preprints of this Symposium appeared on CD-ROM and were distributed among the participants of the symposium. The Proceedings of SYSID-2003 contain 321 papers.

We hope that you, as reader or as researcher in the area of System Identification, will find the contents of these Proceedings useful and informative for your professional work.

We would like to thank all members of the International Program Committee (IPC), members of the IPC Taskforce and members of the National Organizing Committee for their work in the organization of this symposium and in the preparation of these Proceedings. We would also like to thank many friends and colleagues for their help and support in many practical matters related to SYSID 2003.

The editors,

Paul Van den Hof
Bo Wahlberg
Siep Weiland.

CONTENTS

VOLUME 1

PLENARY PAPER
FROM EXPERIMENTS TO CLOSED-LOOP CONTROL

IDENTIFICATION FOR CONTROL

NONLINEAR IDENTIFICATION

IDENTIFICATION OF MIMO COMMUNICATION CHANNELS

ESTIMATION IN PHYSICAL AND MEDICAL SYSTEMS

STOCHASTIC SYSTEMS

APPLICATIONS OF SYSTEM IDENTIFICATION

FINANCIAL ECONOMETRICS

SEMI-PLENARY
SNIPPETS OF IDENTIFICATION THEORY IN COMPUTER VISION

SEMI-PLENARY
INTERVAL ANALYSIS FOR GUARANTEED NONLINEAR PARAMETER ESTIMATION

IDENTIFICATION IN AUTOMOTIVE SYSTEMS

SENSOR IDENTIFICATION AND MONITORING

IDENTIFICATION OF NONLINEAR SYSTEMS I

MECHANICAL AND AEROSPACE APPLICATIONS

CLOSED-LOOP IDENTIFICATION

INDUSTRIAL APPLICATION OF IDENTIFICATION

PROCESS CONTROL SYSTEMS

CLOSED LOOP AND PERFORMANCE ISSUES

REPRODUCING KERNELS I

VOLUME 2

BLIND ESTIMATION AND EQUALIZATION

CONTINUOUS TIME IDENTIFICATION

INPUT DESIGN

IDENTIFICATION FOR FLIGHT TEST EXPLORATION

IDENTIFIABILITY

PLENARY PAPER
SYSTEM IDENTIFICATION FOR STRUCTURAL DYNAMICS AND VIBROACOUSTICS DESIGN ENGINEERING

SELECTED TOPICS IN IDENTIFICATION

REPRODUCING KERNELS II

IDENTIFICATION OF NONLINEAR BLOCK MODELS

NEW RESULTS IN SUBSPACE IDENTIFICATION

IDENTIFICATION FOR PROCESS CONTROL: INPUT DESIGN

IDENTIFICATION OF MECHANICAL SYSTEMS

SOFTWARE SESSION I

SEMI-PLENARY
DATA-BASED METHODS IN PROCESS CONTROL

SEMI-PLENARY
SUBSPACE ALGORITHMS

FILTERING AND ESTIMATION

DIAGNOSIS, DETECTION AND TRACKING

VOLUME 3

IDENTIFICATION OF NONLINEAR SYSTEMS II

IDENTIFICATION METHODS

CONTROLLER TUNING AND IDENTIFICATION

APPLICATIONS OF IDENTIFICATION

BIOENGINEERING SYSTEMS

PARTICLE FILTERS

WIENER HAMMERSTEIN MODELS

IDENTIFICATION USING BASIS FUNCTIONS

SUBSPACE IDENTIFICATION AND APPLICATIONS

IDENTIFICATION IN LARGE SCALE SYSTEMS

INDUSTRIAL APPLICATIONS OF IDENTIFICATION

SOFTWARE SESSION II

PLENARY PAPER
PREDICTION ALGORITHMS: COMPLEXITY, CONCENTRATION AND CONVEXITY

IDENTIFICATION AND PHYSICAL MODELING

IDENTIFICATION OF NONLINEAR SYSTEMS

VOLUME 4

EDUCATION AND TRAINING

RECURSIVE AND SUBSPACE IDENTIFICATION

PROCESS CONTROL: THEORY

APPLICATION OF SYSTEM IDENTIFICATION

OPTIMAL FILTERING

SEMI-PLENARY
IDENTIFICATION OF LINEAR SYSTEMS WITH NONLINEAR DISTORTIONS

SEMI-PLENARY
SOME PROBLEMS IN STATISTICAL INFERENCE FOLLOWING MODEL SELECTION

USER CHOICES IN SUBSPACE IDENTIFICATION

IDENTIFICATION OF STATIC AND DYNAMICAL NONLINEAR SYSTEMS

IDENTIFICATION AND MODEL VALIDATION

MODEL APPROXIMATION

PARAMETER ESTIMATION AND CONVERGENCE

IDENTIFICATION OF HYDROLOGIC SYSTEMS

ERRORS IN VARIABLE IDENTIFICATION

IFAC Publications
www.elsevier.com/locate/ifac

A PRUNING METHOD FOR THE IDENTIFICATION OF POLYNOMIAL NARMAX MODELS

L. Piroddi and W. Spinelli

Dipartimento di Elettronica e Informazione, Politecnico di Milano, Piazza Leonardo da Vinci 32, 20133, Milano (Italy), tel. ++39-2-23993556, fax. ++39-2-23993412, e-mail: piroddi@elet.polimi.it

Abstract: A pruning mechanism is developed for the identification of polynomial NARX/NARMAX models, in order to systematically delete redundant terms inserted in the structure selection phase. The proposed approach is capable of overcoming some limitations met by classical regressor deletion criteria, especially when the data available for identification are not adequately exciting or the model family does not include all the correct regressors. The effectiveness of the algorithm is shown by comparison analysis on some simulation examples. *Copyright © 2003 IFAC*

Keywords: identification algorithms, nonlinear models, NARMAX models, regressor elimination, pruning

1. INTRODUCTION

The problem of model structure selection in nonlinear black-box identification and, in particular, for the identification of polynomial NARX/ NARMAX models (Leontaritis and Billings 1985a, b), has received much attention in the recent years. Several methods, procedures and criteria have been introduced with the aim of finding the most appropriate structure given a model family in terms of set of possible regressors. In particular, both constructive and eliminating approaches have been devised. Constructive techniques start form an empty model and increment the model structure at each step, based on a specific criterion for the evaluation of the significance of each regressor. Conversely, eliminating methods start with the definition of a trial model, which is progressively reduced, by deleting the less important terms. Examples of both approaches are provided by the *backward-* and *forward-regression* versions of the orthogonal least squares estimator (Korenberg *et al.*, 1988; Billings *et al.*, 1989). Here, the significance of a regressor is assessed by means of the error reduction ratio (ERR), which measures the reduction in the variance of the residuals associated to the inclusion of the regressor in the model.

The downside with eliminating approaches is that a large enough initial model must be estimated, to include sufficient flexibility in the model structure. This typically causes computational and numerical problems to arise in the first stages of the identification procedure. On the other hand, constructive algorithms like the forward-regression orthogonal estimator (FROE) (Billings *et al.*, 1989) may occasionally select unnecessary terms, especially when the available data are not adequately exciting for the system to be identified (e.g. insufficient amplitude range, low frequency data, oversampled data, noisy data). For example, autoregressive terms are typically included in the first stages of the identification process (Billings and Aguirre, 1995), and other spurious regressors may be inserted at later stages, resulting in a redundant model. On the other hand, it is a well known fact that model robustness decreases as the model size increases, and that model over-parameterization can have various spurious dynamical effects and even cause model instability (Aguirre, 1994; Aguirre and Billings, 1995a). This is particularly critical in applications where the ultimate use of the model is for interpretation, simulation or long-range prediction purposes: while a significant degree of approximation in the model structure can be acceptable in the framework of short-term prediction, it appears that structural model errors severely affect the simulation performance of identified models (Palumbo and Piroddi, 2000; Palumbo *et al.*, 2001; Piroddi *et al.*, 2001; Leva and Piroddi, 2002).

An adequate structure selection method should thus complement a constructive approach with a procedure to delete the regressors contributing less to the model quality (Aguirre, 1994). This can be done *a posteriori* on the basis of different criteria, such as the statistical significance of the identified parameters, information criteria (Mendes and Billings, 2001), term clustering (Aguirre and Billings, 1995b), etc. However, these criteria generally yield controversial results and require significant experience on the part of the user to be applied correctly.

In this paper a pruning procedure is developed for consistent model reduction which operates at each iteration of the identification algorithm: after a new regressor is added to the current model, the real significance of each regressor term is assessed with respect to the adopted performance index, and redundant terms are eliminated. The effectiveness of the algorithm is shown by comparison analysis on some simulation examples.

The paper is organized as follows. The polynomial NARX representation and the FROE are briefly recalled in Section 2. The limitations of traditional deletion criteria are also discussed in Section 2. The pruning algorithm is developed in Section 3, and its behavior compared to that of the FROE by means of a simulation example in Section 4. Finally, Section 5 summarizes the main points of the paper.

2. NARMAX MODEL IDENTIFICATION

2.1. The NARMAX model representation

A well known representation of a wide class of nonlinear models is constituted by the general nonlinear difference equation model, known as the Nonlinear AutoRegressive Moving Average with eXogenous variables (NARMAX) model (Leontaritis and Billings 1985a, b):

$$y(t) = f(y(t-1), \dots, y(t-n_y), u(t-1), \dots, u(t-n_u), \xi(t-1), \dots, \xi(t-n_\xi)) + \xi(t), \quad (1)$$

where $u(\cdot)$, $y(\cdot)$ and $\xi(\cdot)$ are the model input, model output and white noise, respectively, n_y, n_u and n_ξ are the respective maximum lags, and $f(\cdot)$ is a suitable nonlinear function. In the following, $f(\cdot)$ is assumed to be a polynomial function with nonlinearity degree $l \in Z^+$. Polynomial expansions are linear-in-the-parameters models, so that algorithms of the Least Squares (LS) family can be employed for parameter estimation.

Model (1) can be expressed as follows

$$y(t) = \psi_p^T(t-1)\vartheta_p + \psi_n^T(t-1)\vartheta_n + \xi(t), \quad (2)$$

where $\psi_p^T(t-1)$ includes all the polynomial terms involving only $u(\cdot)$ and $y(\cdot)$ up to degree l and time $t-1$ (*process terms*), and $\psi_n^T(t-1)$ includes all the remaining terms (*noise terms*). Vectors ϑ_p and ϑ_n are the corresponding coefficients. If noise terms are absent, expression (2) can be used directly for estimation with ordinary LS. Otherwise, since noise terms are not measurable, sequence $\xi(\cdot)$ is estimated iteratively. In particular, in the Extended Least Squares (ELS) algorithm, the process terms are estimated first and the residual afterwards as follows:

$$\hat{\xi}(t) = y(t) - \hat{y}(t).$$

At this point the noise terms are included in the model and parameters ϑ_p and ϑ_n estimated. The process is iterated until convergence of the residual sequence.

2.2. The Forward-Regression Orthogonal Estimator and the Error Reduction Ratio

In the Forward-Regression Orthogonal Estimator (FROE) (Billings *et al.* 1989), the ELS estimation scheme is enhanced by means of an orthogonalization technique, to decouple the estimation of individual parameters. In this approach, the original model (2) is rephrased as:

$$y(t) = w^T(t)g + \xi(t),$$

where $w^T(t)$ are orthogonal regressors and g the corresponding parameters, constructed from (2). The model structure is iteratively incremented, starting from an empty model and including regressors in order of importance. Thanks to the orthogonality of the $w^T(t)$ regressors, the significance of each candidate regressor can be evaluated independently. This is done by computing the Error Reduction Ratio (ERR) associated to the regressor, a criterion which measures the increment to the overall output variance obtained by adding the regressor to the model, expressed as a fraction of the total variance. The ERR associated to the i-th term is computed as:

$$[\text{ERR}]_i = [\hat{g}_i^2 \sum_{t=1}^{N} w_i^2(t)] / [\sum_{t=1}^{N} y^2(t)], \quad (3)$$

where w_i is the i-th auxiliary orthogonal regressor and $\hat{g}_i$ is the corresponding estimated parameter. At each step, the regressor with the highest ERR value is added to the model. At the end of the procedure the estimates of the original parameters ϑ are obtained from $\hat{g}$.

In general terms, the FROE identification algorithm is sub-optimal (Billings *et al.*, 1989), since what is optimized is the *increment* to the explained variance and not the variance itself. Therefore, there is no guarantee of finding the optimal model within the family, i.e. the model with the minimum mean square prediction error (MSPE) and with the minimum number of regressors, and the algorithm may include redundant regressors in the model.

This is caused mainly by the use of the ERR criterion for model selection, which can partially fail to give a consistent measure of the importance of regressors. In fact, given a selected model structure, the ERR of each regressor can vary enormously depending on the *order* in which the regressors are considered, so that the actual significance of a term is difficult to assess. This has a number of practical consequences:

- Autoregressive terms are typically included in the first stages of the identification process (Billings and Aguirre, 1995), largely independently of the real model structure. In fact, such terms yield a big increment to the explained variance with respect to the empty model, especially with over-sampled or noisy data, though their real importance in the overall model may be much smaller.
- After the first few regressors are selected, which usually amount for most of the explained variance, the ERR values associated to additional regressors rapidly decrease, though their real importance in the model may be much greater. Therefore, the choice of additional regressors gets uncertain and sensitive to noise in the data.
- If an essential term is missing from the set of possible model regressors, it will be typically compensated by several spurious ones. Spurious terms may also be included because of noisy data (Aguirre, 1994).

2.3. *Traditional criteria for the detection and elimination of spurious regressors*

To obtain more compact and robust models and to improve the regressor selection process, it is important to develop a procedure for the detection and elimination of spurious or redundant regressors, *i.e.* those regressors which give the lowest contribution to the quality of the model.

For the reasons explained in the previous section, the ERR, as computed during the iterations of the FROE, does not constitute a reliable criterion for regressor deletion, since it is greatly affected by the order in which the terms are considered. In other words, the "worst" regressors (i.e. the regressors which contribute less to the model prediction performance) are not necessarily the regressors with the lowest ERR.

Other criteria often used in practice to reduce the model size evaluate the statistical significance of the identified parameters, or of entire clusters (Aguirre and Billings, 1995b). Regressors whose estimated parameters have a high standard deviation are deleted from the model. Similarly, terms belonging to the same cluster, *i.e.* regressors with the same type of polynomial nonlinearity, can be eliminated if the *cluster coefficient* (the sum of the parameters associated to all the terms of a cluster) is comparably small. Unfortunately, this analysis gets harder when the model size increases and may not lead to conclusive elements for regressor deletion, due to compensation effects. As noted in (Aguirre, 1994), relevant terms may have small coefficients, and therefore possibly be statistically ill-conditioned and candidates for elimination. Similar limitations of cluster analysis are pointed out in (Aguirre and Jácome, 1998).

Classical information criteria, such as the final prediction error (FPE), Akaike's information criterion (AIC), etc., can also be used to determine the correct size of the model (Akaike, 1974). Limitations of these criteria for model selection are discussed in (Aguirre and Billings, 1994).

3. THE PRUNING ALGORITHM

In view of the discussion in the previous section, a more reliable and systematic criterion for regressor deletion is needed. The elimination procedure should be included in the identification loop to control the growth in the model dimension and to make the regressor selection easier. Moreover, it should not be based on any of the criteria examined in the previous section, but it should be linked more directly to the performance index, *i.e.* the MSPE.

The proposed deletion procedure is called "pruning", a term borrowed from the neural networks' literature to indicate a systematic method for model dimension reduction. It is based on the following two simple observations:

- the least significant regressor in a model (*i.e.* a possible candidate for elimination) is the one whose elimination minimally increases the performance index (and yields the best sub-model);
- given two models of equal accuracy, the smaller one is to be preferred.

In the proposed approach, regressor selection and elimination are coupled at each iteration step of the identification algorithm as follows. At each iteration, a new regressor is added to the model structure, based on the ERR criterion. The following iterative pruning procedure is then applied:

1) For each selected regressor, the sub-model obtained after its elimination from the model is considered, its parameters re-estimated and the corresponding performance index computed.
2) The least significant regressor is identified as the one corresponding to the sub-model with the better value of the performance index.
3) If the same sub-model is also better than the model obtained at the previous iteration, the regressor is actually eliminated.
4) If a regressor elimination has actually taken place, a new check for eliminable regressors is effected (step 1), otherwise the iteration ends since no further regressor can be eliminated.

Briefly, a complete iteration of the identification algorithm either adds a new regressor to the current model or substitutes one or more of its terms with it, provided that this exchange of terms improves the model accuracy. In other words, the MSPE is always decreasing between iterations. The proposed algorithm combines the reduction in the MSPE with the reduction of the number of regressors, thereby improving the model robustness.

Notice that the identification of the worst regressor can also be performed by computing the ERR corresponding to each coefficient, provided the model terms are previously *reordered* so that the examined regressor is selected for *last*. In fact, the ERR associated to the i-th term selected for last is equal to:

$$[\mathrm{ERR}]_i = \frac{\mathrm{MSPE}(M_i) - \mathrm{MSPE}(M)}{\sum_{t=1}^{N} y^2(t)}, \quad (4)$$

where M is the current model and M_i the corresponding submodel without the i-th regressor. Then, the accuracy of each submodel amounts to:

$$\mathrm{MSPE}(M_i) = \mathrm{MSPE}(M) + [\mathrm{ERR}]_i \cdot \sum_{t=1}^{N} y^2(t),$$

and the best submodel is that obtained by elimination of the regressor with the lowest ERR.

A flow diagram of the pruning process is provided in figure 1, where J denotes the performance index. Since the selection criterion is based on the minimization of the MSPE, the natural choice is to use the same criterion for the elimination of the regressors. However, the explained procedure is independent of the actual performance index employed.

The pruning mechanism is particularly effective when a correct regressor is inserted: typically, a number of spurious compensating regressors is eliminated. Notice that the pruning mechanism may occasionally eliminate a correct regressor, but if this is a really significant term, it will be typically reintroduced in a later stage.

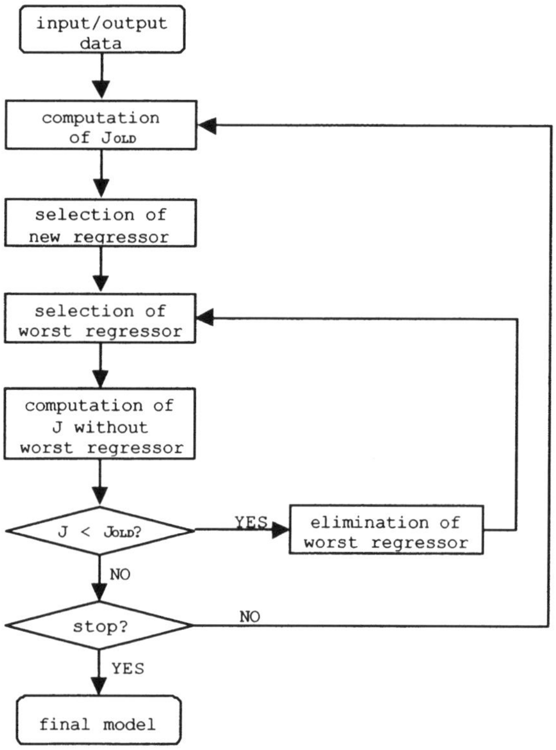

Fig. 1. Flow diagram of the pruning algorithm

The introduction of a systematic pruning step in the identification algorithm yields, in general, models not worst than those obtained with the FROE. At most, should the FROE always select the best terms, the pruning phase never takes place. On the contrary, if non-fundamental regressors are selected, the pruning may delete them.

Finally, notice that the introduction of the pruning algorithm may also provide a simple means for the detection of the optimal model size and therefore be used to devise a stopping criterion for the identification algorithm, as opposed to empirical criteria, like the FPE, the AIC, etc. A good compromise between model size and prediction accuracy is in general achieved after a significant pruning operation, in which many regressors have been deleted, and after which the performance index remains almost constant. On the other hand, if the performance index is still significantly decreasing after a pruning operation, the current model is probably not optimal.

4. A SIMULATION EXAMPLE

An identification experiment is performed generating a 500 data realization of system:

$$S: \begin{cases} y(t) = 0.5\, y(t-1) + 0.8\, u(t-2) + u(t-1)^2 + \\ -\ 0.05 y(t-2)^2 + 0.5 + \xi(t) \\ \xi(\cdot) \sim \mathrm{WN}(0,\, 0.05) \end{cases} \quad (5)$$

To represent the situation in which a fully exciting input signal cannot be employed to generate the identification data, the input is a zero mean, low frequency AR(2) process with unit variance.

The FROE is then applied to the data, operating on the family of NARX models of order 2 and degree 2 (which contains the exact model structure), and the identified model is reported in table 1. Here, regressors represented in italic are candidates for *a posteriori* elimination from the model, since statistically indistinguishable from 0 (the associated standard deviation is comparatively high).

Table 1 Model identified with the FROE

Iter.	Regressors	Estimated parameters	3·STD %	ERR
1	y(t-1)	5.068 E–01	13.8%	9.170 E–01
2	*y(t-2)*	6.654 E–03	338.4%	2.134 E–02
3	$u(t-1)^2$	1.004 E+00	6.1%	1.176 E–02
4	*$u(t-2)^2$*	-2.753 E–03	353.4%	2.489 E–02
5	*u(t-1)*	-5.296 E–02	202.1%	2.568 E–03
6	$y(t-2)^2$	-5.079 E–02	14.3%	8.708 E–03
7	1	4.851 E–01	13.9%	3.063 E–03
8	u(t-2)	8.631 E–01	15.4%	4.644 E–03

In this example, the FROE has identified the correct system structure *plus* some redundant terms, which can be spotted out and eliminated by standard deviation analysis. Notice that the deleted terms were included at early stages of the identification algorithm, with relatively high values of the ERR. This clearly depends on the regressor order within the model: if the spurious terms were selected *after* the right ones, their ERR would have been, respectively, 1.213 E–06, 2.015 E–06 and 2.677 E–05.

Consider now an identification experiment carried out on the same system after deleting the regressor $u(t-2)$ from the family of candidate regressors. This is representative of actual identification problems, where the model family is typically under-parameterized. Of course, the identification algorithm cannot obtain the correct structure (though it is still possible to obtain a satisfactory model). The identified model is reported in table 2.

This time, the information given by the standard deviation gets fuzzier and there are no elements for the deletion of spurious terms. However, the reader can easily verify that after the inclusion of the last regressors, the importance of regressors 2 and 4 is in fact greatly diminished: the total amount of ERR for the model in table 2 is 0.98928, while the more compact model without regressors 2 and 4 has a corresponding value of 0.9736. Therefore, the loss in explained variance due to the elimination of the two terms is equal to 0.01568 and not 0.04623 (sum of the corresponding ERR values). Compare also the two model structures of equal size obtained at the 5th iteration of the FROE (regressors 1-5) and the "pruned" model with regressors 1, 3, 5, 6, 7. The performance of the two estimated models in identification is extremely different: the un-pruned model misses some essential terms and incorporates some spurious ones, resulting in a mean square simulation error (MSSE) of 0.89242. The pruned model is far more efficient in modeling the system dynamics and obtains 0.1167. Hopefully, the pruning algorithm should automatically operate this term elimination.

Table 2 Model identified with the FROE with an under-parameterized model family

Iter.	Regressors	Estimated parameters	3·STD %	ERR
1	y(t–1)	7.377 E–01	10.9%	9.170 E–01
2	y(t–2)	-9.528 E–02	55.1%	2.134 E–02
3	$u(t-1)^2$	1.044 E+00	7.8%	1.176 E–02
4	$u(t-2)^2$	-3.598 E–01	32.3%	2.489 E–02
5	u(t–1)	5.556 E–01	12.5%	2.568 E–03
6	$y(t-2)^2$	-3.711 E–02	24.8%	8.708 E–03
7	1	3.365 E–01	25.1%	3.063 E–03

To verify this, the same experiment is repeated with the pruning algorithm. Table 3 reports the list of all selected and deleted regressors during an identification algorithm run of 10 iterations: the corresponding model is reported in table 4.

Table 3 First 10 iterations of the FROE with pruning with an under-parameterized model family: selected and deleted regressors

Iter.	Selected regressors	Deleted regressors	ERR
1	y(t-1)		9.170 E–01
2	y(t-2)		2.134 E–02
3	$u(t-1)^2$		1.176 E–02
4	$u(t-2)^2$	y(t-2)	3.253 E–02
5	$y(t-2)^2$		5.283 E–03
6	u(t-1)	$u(t-2)^2$	2.581 E–02
7	1		6.123 E–03
8	y(t-2)u(t-2)		2.050 E–03
9	y(t-2)u(t-1)		1.602 E–03
10	$u(t-2)^2$		6.268 E–04

Table 4 Model identified with the FROE with pruning with an under-parameterized model family

Iter.	Regressors	Estimated parameters	3·STD %	ERR
1	y(t–1)	6.002 E–01	9.9%	9.170 E–01
2	$u(t-1)^2$	9.011 E–01	7.1%	2.069 E–02
3	$y(t-2)^2$	-4.995 E–02	12.1%	1.778 E–02
4	u(t–1)	6.930 E–01	8.4%	2.581 E–02
5	1	3.884 E–01	19.3%	6.123 E–03

Figure 2 shows the corresponding evolution of the

MSPE and number of regressors during the identification process. The MSPE is always decreasing, as expected. The number of regressors increases always by 1, except at iterations 4 and 6, where a regressor substitution takes place. Notice that although $y(t-2)$ is inserted in the 2nd iteration, its real importance is much less, since it becomes inessential after the two regressors $u(t-1)^2$ and $u(t-2)^2$ are added. The final model is selected at iteration 7: the MSPE does not decrease significantly afterwards, while the model size increases steadily.

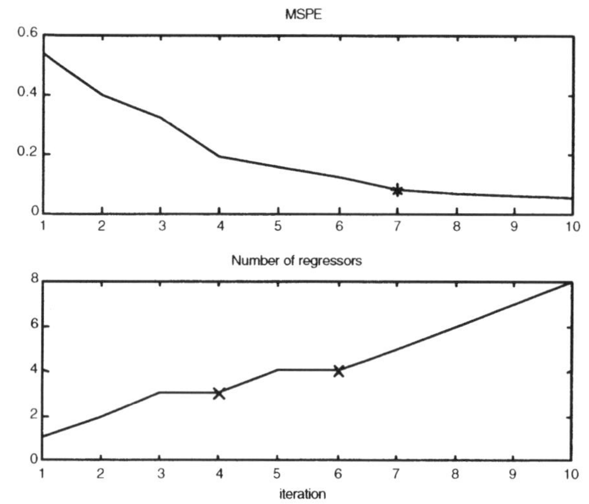

Fig. 2. Mean Square Prediction Error and number of regressors during the identification with pruning

5. CONCLUSIONS

A pruning procedure has been proposed to complement the forward-regression orthogonal estimator (FROE) for NARMAX model identification. Each iteration of the identification algorithm introduces a new regressor in the model, but may also eliminate other regressors included previously and now classified as redundant. Deletable terms are detected on the basis of an objective criterion, which measures the actual contribution of each term to the overall model quality. The proposed procedure increases the robustness of models identified with the FROE, especially when poorly exciting or noisy data are used or when the model family is under-parameterized. This features makes the version of the FROE with pruning more suitable for model identification intended for interpretation, simulation or long-range prediction purposes.

ACKNOWLEDGEMENTS

Paper supported by MURST project "Identification and Control of Industrial systems".

REFERENCES

Aguirre, L.A. (1994). Some remarks on structure selection for nonlinear models. *Int. J. of Bifurcation and Chaos*, **4**, 6, 1707-1714.

Aguirre, L.A. and S.A. Billings (1994). Validating identified nonlinear models with chaotic dynamics. *Int. J. of Bif. and Chaos*, **4**, 1, 109-125.

Aguirre, L.A. and S.A. Billings (1995a). Dynamical effects of overparametrization in nonlinear models. *Physica D*, **80**, 26-40.

Aguirre, L.A. and S.A. Billings (1995b). Improved structure selection for nonlinear models based on term clustering. *Int. J. of Control*, **62**, 569-587.

Aguirre, L.A. and C.R.F. Jácome (1998). Cluster analysis of NARMAX models for signal-dependent systems. *IEE Proc. – Control Theory Appl.*, **145** (4), 409-414.

Akaike, H. (1974). A new look at the statistical model identification. *IEEE Trans. on Aut. Control*, **19** (6), 716-723.

Billings, S.A. and L.A. Aguirre (1995). Effects of the sampling time on the dynamics and identification of nonlinear models. *Int. J. of Bifurcation and Chaos*, **5** (6), 1541-1556.

Billings, S.A., S. Chen and M.J. Korenberg (1989). Identification of MIMO non-linear systems using a forward-regression orthogonal estimator. *Int. J. of Control*, **49**, 2157-2189.

Korenberg, M., S.A. Billings, Y.P. Liu and P.J. McIlroy (1988). Orthogonal parameter estimation algorithm for non-linear stochastic systems. *Int. J. of Control*, **48**, 193-210.

Leontaritis, I.J., and S.A. Billings (1985a). Input-output parametric models for non-linear systems - Part I: deterministic non-linear systems. *Int. J. of Control*, **41**, 303-328.

Leontaritis, I.J., and S.A. Billings (1985b). Input-output parametric models for non-linear systems - Part II: stochastic non-linear systems. *Int. J. of Control*, **41**, 329-344.

Leva, A. and L. Piroddi (2002). NARX-based technique for the modeling of magneto-rheological damping devices. *Smart Materials and Structures*, **11**, 79-88.

Mendes, E.M.A.M. and S.A. Billings (2001). An Alternative Solution to the Model Structure Selection Problem. *IEEE Trans. on Sys., Man, and Cyb.*, **31** (6), 597-608.

Palumbo, P. and L. Piroddi (2000). Seismic behaviour of buttress dams: nonlinear modelling of a damaged buttress based on ARX/NARX models. *J. of Sound and Vibration*, **239**, 405-422.

Palumbo, P., L. Piroddi, S. Lancini and F. Lozza (2001). NARX Modelling of Radial Crest Displacements of the Schlegeis Arch Dam. *6th Benchmark Workshop on Numerical Analysis of Dams*, Salzburg (Austria), 17-19 October.

Piroddi, L., A. Feriani and F. Lozza (2001). Black-box modelling of a an anti-seismic isolator. *5th World Congress on Joints, Bearings and Seismic Systems for Concrete Structures, JBSS '01*, Rome (Italy), 7-11 October.

IFAC Publications
www.elsevier.com/locate/ifac

GENERALIZED ORTHONORMAL BASIS SELECTION FOR EXPANDING QUADRATIC VOLTERRA FILTERS

Alain Y. Kibangou *,** **, Gérard Favier** * **Moha M. Hassani** **

* *Laboratoire I3S/CNRS/UNSA*
B.P. 121 - 06903 Sophia Antipolis Cedex, France
** *Laboratoire d'Electronique et Instrumentation,*
FSSM/UCAM,B.P. 2390 - 40000 Marrakesh, Morocco

Abstract: Volterra models are very useful for signal and system representation due to their general non-linear structure and their property of linearity with respect to the kernels coefficients. However, when using such models we are confronted with a complexity problem that results from the very large number of the kernels coefficients. Expanding the kernels on a generalized orthonormal basis allows to significantly reduce this parametric complexity. In the present paper, a new procedure is proposed for selecting such a generalized orthonormal basis so that a least squares criterion is minimized in the case of a homogeneous quadratic model. *Copyright © 2003 IFAC*

Keywords: Volterra filters, non-linear systems, generalized orthonormal basis, identification.

1. INTRODUCTION

Truncated Volterra filters constitute a class of nonrecursive polynomial models, i.e. models without output feedback which guarantees their stability. Such models can approximate any time-invariant nonlinear system with fading memory (Boyd and Chua, 1985). However, the main drawback of these models is their over-parameterization. During the last decade, the issue of Volterra model complexity reduction has been addressed following three main different approaches. A first one consists in approximating a Volterra model by means of a parallel-cascade structure composed of linear filters and memoryless nonlinearities. An example of such an approximation to the quadratic filter is the Multi Memory Description (MMD) model composed of three linear FIR filters connected by a multiplier(Franck, 1995). (Korenberg, 1991) showed that any discrete-time finite memory nonlinear system having a finite order Volterra series representation can be exactly represented by a finite number of such cascade structures in parallel. A second approach is based on the use of SVD or LU type decompositions of the kernels matrices. This approach also leads to parallel-cascade approximations of Volterra models (Panicker and Mathews, 1998). A third approach consists in expanding the Volterra kernels on an orthonormal basis (OB) such as the Laguerre functions basis (Dumont and Fu, 1993; Campello *et al.*, 2001), or Generalized Orthonormal Basis (GOB) (Hacioglu and Williamson, 2001). The complexity reduction is then very depending on the choice of the GOB, i.e. the poles that characterize the GOB. So, when using this approach, an important problem to be solved is the determination of the GOB poles. Two solutions were recently proposed for this problem: A first one is based on an optimization by means of a gradient type algorithm (Malti, 1999; Hacioglu and Williamson, 2001). This solution presents the drawback that the cost function to minimize is non-linear with respect to the poles, so that the algorithm can converge to local minima. A second solution consists in minimizing an upper bound for the quadratic modelling error. This approach delivers an analytic optimal solution for the Laguerre pole associated with a linear model (Fu and Dumont, 1993) as well as quadratic or any order

Volterra models (Campello *et al.*, 2001; Campello *et al.*, 2003). In the present paper, a new procedure is proposed for building a GOB from a finite set of first order all-pass transfer functions associated with a set of stable discrete poles obtained by discretizing the interval]-1,1[.

The organization of the paper is as follows. In the next section, the quadratic Volterra kernel expansion on a GOB is presented. The coefficients of this expansion are estimated by means of the orthogonal least squares method as discussed in section 3. Then, the proposed procedure for building a GOB is described in section 4. An illustrative example is given in section 5 before concluding the paper in section 6.

2. QUADRATIC VOLTERRA KERNEL EXPANSION ON A GOB

In recent years, there has been a large interest in the use of OB for modelling linear as well as non-linear systems. Laguerre and Kautz bases are certainly the most used OBs for linear system identification and control (Wahlberg, 1991; Wahlberg, 1994; Oliveira *et al.*, 2000). Such OBs are well suited to model systems having kernels with first or second order dominant dynamics. In order to model more complex dynamics, several new GOBs have been proposed (Heuberger *et al.*, 1995; Ninness and Gustafsson, 1997; Bodin *et al.*, 1996). In this paper, we consider the GOB functions defined by their z-transform as in (Ninness and Gustafsson, 1997):

$$B_0(z) = \frac{\sqrt{1-|\xi_0|^2}}{z-\xi_0} \qquad (1)$$

$$B_k(z) = \frac{\sqrt{1-|\xi_k|^2}}{z-\xi_k}\prod_{i=0}^{k-1}\frac{1-\xi_i^* z}{z-\xi_i} \qquad (2)$$

$$k = 1,2,\cdots$$

The basis functions b_k are given by inverse z-transform of (1)-(2) and verify the orthonormality property

$$< b_k, b_l >= \sum_{n=0}^{\infty} b_k(n) b_l^*(n) = \delta_{k,l}$$

$\delta_{k,l}$ being the Kronecker symbol. This set of basis functions is complete in the signal space of causal ℓ_2 sequences (Ninness and Gustafsson, 1997) if and only if

$$\sum_{k=0}^{\infty}(1-|\xi_k|) = \infty$$

In the sequel, quadratic, homogeneous, discrete-time, time invariant, truncated Volterra filters with memory M are considered. The output of such a filter is given by

$$y(n) = \sum_{n_1=0}^{M-1}\sum_{n_2=0}^{M-1} h(n_1,n_2)\prod_{j=1}^{2} u(n-n_j) \qquad (3)$$

where u(n) designs the filter input. The number of the kernel coefficients is M^2.

The kernel of such a quadratic filter can be expanded on the GOB defined by equations (1)-(2) by means of a Kth-order truncated expansion:

$$h(n_1,n_2) = \sum_{k_1=0}^{K-1}\sum_{k_2=0}^{K-1} g^0_{k_1,k_2}\prod_{j=1}^{2} b_{k_j}(n_j) \qquad (4)$$

Assuming that the kernel is symmetric, which is not a restrictive assumption because any non-symmetric kernel can always be transformed into a symmetric one, we have

$$g^0_{k_1,k_2} = g^0_{k_2,k_1},\quad 0 \le k_1,k_2 \le K-1$$

and therefore a non-redundant expansion results:

$$h(n_1,n_2) = \sum_{k_1=0}^{K-1}\sum_{k_2=k_1}^{K-1} g_{k_1,k_2}\prod_{j=1}^{2} b_{k_j}(n_j) \qquad (5)$$

where

$$g_{k_1,k_2} = g^0_{k_1,k_2},\quad k_1 = k_2$$
$$g_{k_1,k_2} = 2g^0_{k_1,k_2},\quad k_1 < k_2$$

From equation (5), the input-output relation (3) becomes

$$y(n) = \sum_{k_1=0}^{K-1}\sum_{k_2=k_1}^{K-1} g_{k_1,k_2} s_{k_1,k_2}(n) \qquad (6)$$

where:

$$s_{k_1,k_2}(n) = s_{k_1}(n)s_{k_2}(n)$$

$$s_k(n) = \sum_{n_1=0}^{M-1} b_k(n_1)u(n-n_1)$$

In order to have a more tractable data organization, we use the following change of coordinates (Raz and Veen, 1998): $k_1 = i, k_2 = i+q$, which modifies equation (6) as

$$y(n) = \sum_{i=0}^{K-1}\sum_{q=0}^{K-1-i} g_{i,i+q}s_{i,i+q}(n) \qquad (7)$$

or equivalently

$$y(n) = \sum_{q=0}^{K-1}\sum_{i=0}^{K-1-q} g_{i,i+q}s_{i,i+q}(n) \qquad (8)$$

This model is characterized by $C(K,2) = \frac{(K+1)K}{2}$ parameters. The complexity reduction with respect to the standard form (3) is measured by the ratio $\frac{K(K+1)}{2M^2}$. When $K << M$, the Volterra model complexity is appreciably reduced. As it was previously mentioned, the value of K is depending on the choice of the poles ξ_k of the GOB. The nearest are the poles to the system's dynamics the smaller will be K. The problem of the poles determination will be addressed in section 4. In the next section, we present a method for estimating the expansion coefficients by means of the orthogonal least squares (OLS) algorithm.

3. EXPANSION COEFFICIENTS ESTIMATION

Let us assume that a record of N couples of input-desired output signals is available. Defining the vectors G of the expansion coefficients and φ_{k_1,k_2} of quadratic filtered inputs products:

$$\begin{aligned} G &= [g_{0,0}\ g_{1,1} \cdots g_{K-1,K-1} \\ &\quad g_{0,1} \cdots g_{K-2,K-1} \cdots g_{0,K-1}]^T \\ \varphi_{k_1,k_2} &= [s_{k_1,k_2}(0), \cdots, s_{k_1,k_2}(N-1)]^T \end{aligned}$$

and the matrix $\Phi \in \Re^{N \times C(K,2)}$ the columns of which are the vectors φ_{k_1,k_2}.

$$\begin{aligned} \Phi = [&\varphi_{0,0}\ \varphi_{1,1} \cdots \varphi_{K-1,K-1} \\ &\varphi_{0,1} \cdots \varphi_{K-2,K-1} \cdots \varphi_{0,K-1}] \end{aligned} \tag{9}$$

the vector $Y = [y(0)\ y(1) \cdots y(N-1)]^T$ of the Volterra model outputs can be written in the following matrix form:

$$Y = \Phi G \tag{10}$$

The expansion coefficients vector G is obtained by minimizing the least squares (LS) criterion:

$$\hat{G} = argmin_G \{E(G)\} \tag{11}$$

$$E(G) = \sum_{n=0}^{N-1} (d(n) - y(n))^2 = \|d - \Phi G\|^2$$

which leads to the well known LS solution :

$$\hat{G} = \left(\Phi^T \Phi\right)^{-1} \Phi^T d \tag{12}$$

where $d = [d(0)\ d(1) \cdots d(N-1)]^T$ is the vector of desired outputs. The vector of output errors is then given by:

$$d - \hat{Y} = d - \Phi \hat{G} = Pd \tag{13}$$

where P is the projection matrix defined as:

$$P = I - \Phi \left(\Phi^T \Phi\right)^{-1} \Phi^T \tag{14}$$

Using the idempotence property of the projection matrix ($P^T P = P^2 = P$), the minimum LS error E_{min} can be written as:

$$\begin{aligned} E_{min} &= \left\|d - \hat{Y}\right\|^2 \\ &= d^T P d \end{aligned} \tag{15}$$

Using the OLS algorithm allows to reduce the computational effort. The Gram-Schmidt orthogonalization procedure can be applied to transform the column vectors of Φ into a set of orthonormal vectors, so that Φ is decomposed into:

$$\Phi = \check{\Phi} U \tag{16}$$

where $\check{\Phi} \in \Re^{N \times C(K,2)}$ is a matrix with orthonormal columns $\check{\varphi}_{k_1,k_2}$, and $U \in \Re^{C(K,2) \times C(K,2)}$ is an upper triangular matrix.

For carrying this orthogonalization, Φ is considered in the following block structure:

$$\Phi = [D_0\ D_1 \cdots D_{K-1}] \tag{17}$$

where $D_q \in \Re^{N \times (K-q)}$, $q = 0, 1, \cdots, K-1$, is defined as:

$$D_q = \left[\varphi_{0,q}, \cdots, \varphi_{i,i+q}, \cdots, \varphi_{K-q-1,K-1}\right]$$

Then

$$\check{\Phi} = \left[\check{D}_0\ \check{D}_1 \cdots \check{D}_{K-1}\right] \tag{18}$$

The Φ orthonormalization simplifies the calculation of the projection matrix as follows:

$$\begin{aligned} P &= I - \Phi \left(\Phi^T \Phi\right)^{-1} \Phi^T \\ &= I - \check{\Phi} U \left(U^T \check{\Phi}^T \check{\Phi} U\right)^{-1} U^T \check{\Phi}^T \\ &= I - \check{\Phi} U U^{-1} \left(\check{\Phi}^T \check{\Phi}\right)^{-1} U^{-T} U^T \check{\Phi}^T \\ &= I - \check{\Phi} \check{\Phi}^T \end{aligned} \tag{19}$$

The LS solution can then be rewritten as

$$\hat{G} = U^{-1} \check{\Phi}^T d \tag{20}$$

4. OPTIMAL SELECTION OF THE GOB POLES

For selecting the poles of a GOB, we assume that N couples of measurements $\{u(n), d(n)\}_{n=0}^{N-1}$ and Q candidate poles grouped in the set $\Xi = \{\xi_0, \cdots, \xi_{Q-1}\}$ are available. Let us consider the real poles case. The candidate poles are selected in the segment]-1,1[to guarantee the model stability, and they are obtained by sampling this segment with a fixed step size. The computational complexity is obviously depending on the chosen step size. The GOBFs are built with these candidate poles. At the first step of the procedure, the set of the candidate GOBFs, given by their z-transforms, is:

$$\begin{aligned} F_0 = \{B_m(z) &= \frac{\sqrt{1-\xi_m^2}}{z - \xi_m}, \\ m &= 0, 1, \cdots, Q-1\} \end{aligned}$$

At the $(k+1)^{th}$ step, when k GOBFs have already been determined, the set F becomes :

$$F_k = \{B_{m,k}(z) = \frac{\sqrt{1-\xi_m^2}}{z-\xi_m}\prod_{i=0}^{k-1}\frac{1-\tau_i z}{z-\tau_i}$$
$$m = 0,1,\cdots,Q-1\}$$

where the poles $\tau_i, i \in \{0,\cdots,k-1\}$, correspond to the poles determined in the previous steps of the procedure.

According to (19), $P_k = I - \breve{\Phi}_k\breve{\Phi}_k^T$ designs the projection matrix associated with a GOB which contains k functions. The Φ_k matrix is partitioned as:

$$\breve{\Phi}_k = \left[\breve{D}_{0,k}\ \breve{D}_{1,k}\cdots\breve{D}_{k-1,k}\right]$$

where each block $\breve{D}_{q,k}$ is composed of $(k-q)$ column vectors $\breve{\varphi}_{i,i+q}; i=0,1,\cdots,k-1-q; q=0,1,\cdots,k-1$.

Incorporating a new function into the basis results in the incorporation of a supplementary column $\breve{\varphi}_{k-q,k}$ into matrix $\breve{D}_{q,k}$, for $q=0,1,\cdots,k-1$ and a new block $\breve{D}_{k,k+1}$ so that:

$$\begin{aligned}\breve{D}_{q,k+1} &= [\breve{D}_{q,k}\ \breve{\varphi}_{k-q,k}],\ q=0,1,\cdots,k-1\\ \breve{D}_{k,k+1} &= [\breve{\varphi}_{0,k}]\end{aligned} \tag{21}$$

Then

$$\begin{aligned}\breve{\Phi}_{k+1} &= \left[\breve{D}_{0,k+1}\cdots\breve{D}_{k-1,k+1}\ \breve{D}_{k,k+1}\right]\\ &= \left[\left[\breve{D}_{0,k}\ \breve{\varphi}_{k,k}\right]\cdots\left[\breve{D}_{k-1,k}\ \breve{\varphi}_{1,k}\right]\ \breve{\varphi}_{0,k}\right]\end{aligned} \tag{22}$$

This modification of $\breve{\Phi}$ implies the following update relation for the projection matrix

$$\begin{aligned}P_{k+1} &= I - \breve{\Phi}_{k+1}\breve{\Phi}_{k+1}^T\\ &= I - \sum_{i=0}^{k-1}\breve{D}_{i,k}\breve{D}_{i,k}^T - \sum_{i=0}^{k}\breve{\varphi}_{i,k}\breve{\varphi}_{i,k}^T\end{aligned}$$

or in a recursive form

$$P_{k+1} = P_k - \sum_{i=0}^{k}\breve{\varphi}_{i,k}\breve{\varphi}_{i,k}^T \tag{23}$$

The new function to be incorporated into the basis is chosen so that the output LS criterion be minimized, or equivalently the incremental variation $\Delta E_{k+1} = E_k - E_{k+1}$ be maximized.

$$\begin{aligned}\Delta E_{k+1} &= d^T\left(P_k - P_{k+1}\right)d\\ &= d^T\left(\sum_{i=0}^{k}\breve{\varphi}_{i,k}\breve{\varphi}_{i,k}^T\right)d\\ &= \sum_{i=0}^{k}\left(d^T\breve{\varphi}_{i,k}\right)^2\end{aligned} \tag{24}$$

So, the selected function $b_k(.)$ is the one associated with the pole $\tau_k = \xi_m$ that maximizes (24).

The selection procedure is stopped in using the GCV (Generalized Cross Validation) criterion (Orr, 1996), defined by:

$$\sigma_k^2 = N\frac{d^T P_k^2 d}{\left(tr\left(P_k\right)\right)^2} \tag{25}$$

where $tr(.)$ denotes the trace operator that, due to the orthonormality property of $\breve{\Phi}$, is such that:

$$\begin{aligned}tr(P_k) &= tr\left(I - \breve{\Phi}_k\breve{\Phi}_k^T\right)\\ &= N - \sum_{q=0}^{k-1} tr\left(\breve{D}_{q,k}\breve{D}_{q,k}^T\right)\\ &= N - \left(\sum_{q=0}^{k-1}\sum_{i=0}^{k-1-q}\breve{\varphi}_{i,i+q}^T\breve{\varphi}_{i,i+q}\right)\\ &= N - \frac{k(k+1)}{2}\end{aligned}$$

The optimal truncation order K is obtained when σ_k^2 is minimal or when it ceases to decrease significantly. The proposed algorithm for identifying the quadratic Volterra model expanded on an optimal GOB is summarized as follows:

(1) Discretization of the interval]-1,1[with a fixed step-size to get a set of poles.
(2) Construction of the set F_0 of candidate GOB functions.
(3) Iterative construction of an optimal GOB through the following calculations carried out at each iteration k:
   (a) Computation of the filtered input $s_k(n)$ by means of each candidate GOB function $b_{m,k}(.)$ and formation of the new Φ_k matrix.
   (b) Orthonormalization of the Φ_k matrix and selection of the GOB function $b_k(.)$ that maximizes the incremental variation (24) of the LS criterion.
   (c) Calculation of the truncation order selection criterion (25).
(4) If the GCV criterion ceases to decrease significantly, the Fourier coefficients are estimated by means of the OLS algorithm (20) and the identification algorithm is stopped. Otherwise, construction of the set F_{k+1} and return in (3.a).

5. ILLUSTRATIVE EXAMPLE

A simple example illustrating identification of a quadratic homogenous Volterra filter is presented in this section. The quadratic kernel of the simulated system is depicted in figure 2.a. It is a symmetric separable kernel such as $h(i,j) = h_1(i)h_1(j)$ where $h_1(.)$ is given by its z-transform:

$$H_1(z) = \frac{z^3(z+0.5)}{den(z)}$$

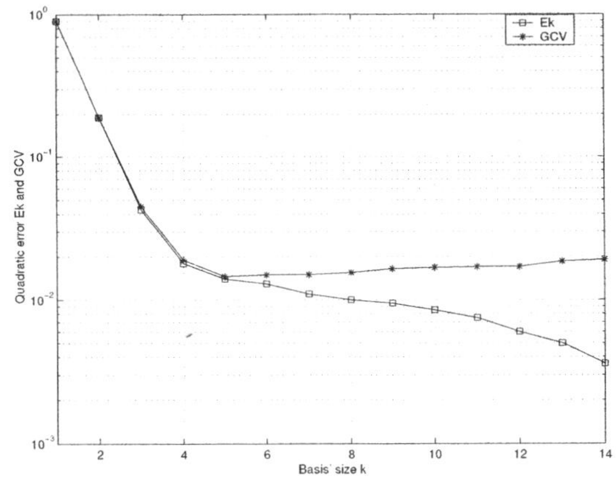

Fig. 1. Output QE and GCV criterion

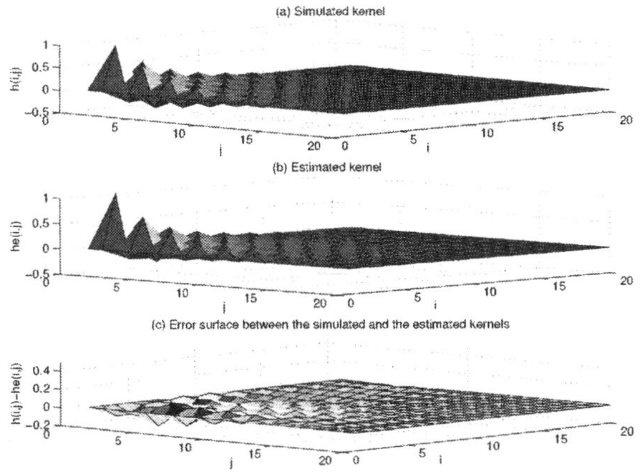

Fig. 2. Quadratic system identification

where $den(z)$ is a fifth order polynomial in z, the poles of which are $-0.8, -0.65, 0.5, 0.2$ and 0.01.

The associated truncated Volterra model with a memory $M = 20$, has 400 coefficients. Optimal selection of the GOBF poles is carried out from a record of 150 input/output data. The signal to noise ratio is 30 dB.

As shown in figure 1, when the basis' size, i.e. the truncation order, increases the output quadratic error (QE) E_k continuously decreases as expected from the theory. Unlike the QE criterion, the GCV one ceases to decrease for a truncation order greater than 5. This optimal truncation order provides a compromise between the basis' size and the model precision. The Volterra kernel is then represented by means of only 15 parameters. So the complexity reduction ratio is $\frac{3}{40}$. The selected poles are: $-0.8, 0.675, -0.75, 0.675, -0.675$. Three of these poles are close to those of the tranfer function H_1 that was used to generate the quadratic kernel.

The selected GOB is used for identifying the quadratic kernel coefficients from 3000 input-output data. Precision of estimation is plotted in figure 3 in terms of the Normalized Quadratic Error (NQE) with respect to the estimated kernel coefficients, defined as:

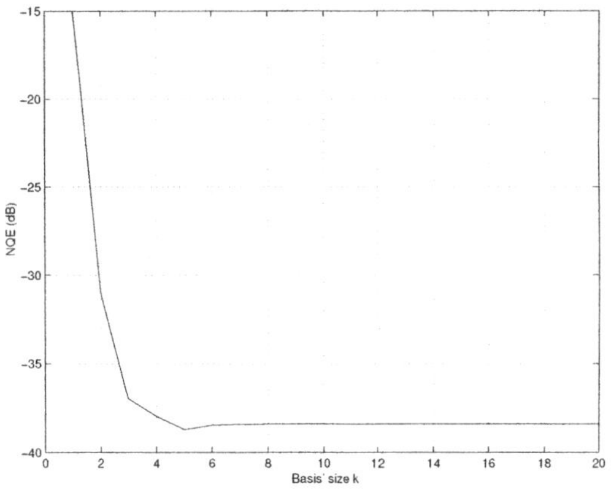

Fig. 3. NQE with respect to the kernel coefficients

$$NQE = 10Log \frac{\sum_{i=0}^{M-1} \sum_{j=0}^{M-1} (h(i,j) - h_e(i,j))^2}{\sum_{i=0}^{M-1} \sum_{j=0}^{M-1} (h(i,j))^2}$$

As shown in figure 3, the model precision (NQE) varies only slightly for truncation order higher than five, value which corresponds to the optimal order delivered by the GCV criterion.

6. CONCLUSION

During the last decade, several generalizations of the Laguerre functions basis have been proposed, which has lead to the notion of GOB. In this paper, a new method for selecting the best GOB to expand a given quadratic Volterra system has been presented. At each step of the proposed procedure, a new function is selected from the set of GOB functions obtained in using the construction method of (Ninness and Gustafsson, 1997) in which the basis functions are generated by multiplying first order all-pass transfer functions that can have different poles at each multiplication step. The pole associated with each new function is chosen as an element of a set of candidate poles that result from the discretization of the interval]-1,1[. The pole selection is depending on a LS criterion minimization that is carried out in using the OLS method. The GCV test is used to stop the procedure, i.e. to determine the number of GOB functions to take into account for the specific system to be modelled. Some simulation results have shown that the proposed method provides a good GOB selection allowing to significantly reduce the complexity of the Volterra model structure. This method allows both identifying the best GOB functions and the corresponding expansion coefficients, and determining an optimal truncation order. An extension of the proposed identification method to higher order Volterra systems is given in (Kibangou *et al.*, 2003).

REFERENCES

Bodin, P., T.O. E Silva and B. Wahlberg (1996). On the construction of orthonormal basis functions for system identification. In: *Proc. 13th IFAC World Congress*. Vol. I. San-Fransisco. pp. 369–374.

Boyd, S. and L.O. Chua (1985). Fading memory and the problem of approximating nonlinear operators with Volterra series. *IEEE Trans. Circuits and Systems* **CAS-32**(11), 1150–1161.

Campello, R. J. G. B., G. Favier and W. C. Amaral (2003). Optimal expansions of discrete-time Volterra models using Laguerre functions. In: *Proc. IFAC SYSID 2003*. Rotterdam, The Netherlands. To appear.

Campello, R. J. G. B., W. C. Amaral and G. Favier (2001). Optimal Laguerre series expansion of discrete Volterra models. In: *Proc. European Control Conference, ECC'2001*. Porto, Portugal. pp. 372–377.

Dumont, G.A. and Y. Fu (1993). Non-linear adaptive control via Laguerre expansion of Volterra kernels. *Intern. Journal of Adaptive Control and Signal Proces.* **7**, 367–382.

Franck, W.A. (1995). An efficient approximation to the quadratic Volterra filter and its application in real-time loudspeaker linearization. *Signal Processing* **45**(1), 97–113.

Fu, Y. and G.A. Dumont (1993). An optimum time scale for discrete Laguerre network. *IEEE Trans. Automatic Control* **38**(6), 934–938.

Hacioglu, R. and G.A. Williamson (2001). Reduced complexity Volterra models for nonlinear system identification. *Eurasip Journal on Applied Signal Processing* **4**, 257–265.

Heuberger, P.S.C., P.M.J. Van den Hof and O.H. Bosgra (1995). A generalized orthonormal basis for linear dynamical systems. *IEEE Trans. Automatic Control* **40**(3), 451–465.

Kibangou, A.Y., G. Favier and M.M. Hassani (2003). A growing approach for selecting generalized orthonormal basis functions in the context of system modelling. In: *Proc. IEEE-EURASIP Workshop on Nonlinear Signal and Image Processing, NSIP'03*. Grado, Italy. To appear.

Korenberg, M.J. (1991). Parallel cascade identification and kernel estimation for nonlinear systems. *Annals of Biomedical Engineering* **19**, 429–455.

Malti, R. (1999). Représentation de systèmes discrets sur la base des filtres orthogonaux: Application à la modélisation des systèmes dynamiques mutivariables. PhD thesis. Institut National Polytechnique de Loraine (INPL), Nancy, France.

Ninness, B.M. and F. Gustafsson (1997). A unifying construction of orthonormal bases for system identification. *IEEE Trans. Automatic Control* **42**(4), 515–521.

Oliveira, G.H.C., W.C. Amaral, G. Favier and G.A. Dumont (2000). Constrained robust predictive controller for uncertain processes modelled by orthonormal series functions. *Automatica* **36**(4), 563–571.

Orr, M.J.L. (1996). Introduction to radial basis function networks. Technical report. Centre for Cognitive Science, University of Edinburgh. Scotland, UK.

Panicker, T.M. and V.J. Mathews (1998). Parallel-cascade realizations and approximations of truncated Volterra systems. *IEEE Trans. Signal Processing* **46**(10), 2829–2832.

Raz, G.M. and B.D. Van Veen (1998). Baseband Volterra filters for implementing carrier based nonlinearities. *IEEE Trans. Signal Processing* **46**(1), 103–114.

Wahlberg, B. (1991). System identification using Laguerre models. *IEEE Trans. Automatic Control* **36**, 551–532.

Wahlberg, B. (1994). Laguerre and Kautz models. In: *Proc. IFAC SYSID'94*. Vol. 3. Copenhagen, Danemark. pp. 1–12.

IFAC
Publications
www.elsevier.com/locate/ifac

A LOCALISED FORGETTING METHOD FOR ON-LINE ADAPTATION OF GAUSSION RBFN MODELS

D.L. Yu, J.B. Gomm, D.W. Yu and D. Williams

Control Systems Research Group, School of Engineering
Liverpool John Moores University, Byrom Street, Liverpool L3 3AF, U.K.
E-mail: **d.yu@livjm.ac.uk**

Abstract: A localised forgetting method is proposed for on-line adaptation of Gaussion Radial Basis Function Network (RBFN) models. It is realised that the typically used forgetting factor is uniformly applied to the past data in entire operating space and is not correct for nonlinear systems where dynamics are different in different operating regions. The new method sets different regions with different forgetting factor according to the response of the local centre to the current measurement data. The method is based on the Recursive Orthogonal Least Squares (ROLS) algorithm and is simple. Application of the new method to the modelling of dissolved oxygen in a chemical reactor rig shows a smaller mean squared error (MSE) for one-step-ahead prediction than using the uniform forgetting, and indicates the effectiveness of the method. *Copyright © 2003 IFAC*

Keywords: radial basis function networks, model adaptation, time varying processes, adaptive control, non-linear systems.

1. INTRODUCTION

Although model-based control of nonlinear dynamic systems has been studied using neural network models since last decade, there are not a large number of real industrial applications that have been reported. One reason is that the stability analysis of these systems is difficult therefore a stable control is not guaranteed, another is that these controls are mainly based on fixed neural models which cannot model time varying effects that exist in many industrial systems, especially chemical and biotechnological processes. To maintain robust control of these plants and utilise the well-developed model-based control techniques, adaptive neural models are needed.

In various kinds of neural networks, radial basis function network (RBFN) is found to be suitable to on-line adaptation because its linear feature of the weights relating to the network output allows linear optimisation methods, e.g. least squares, to be used. Algorithms for recursive adaptation of the parameters (centres and weights) of an RBFN with a fixed structure have been investigated for process identification (Chen, *et al.*, 1992). Moreover, adaptation of both structure and weights were expected to enhance RBFN's modelling ability for time-varying and severely non-linear processes with reasonable network size. Some off-line adaptation techniques have been proposed for RBFNs to achieve an optimal structure (number and location of centres) for a certain training data set (e.g. Chen, *et al.*, 1991; Karayiannis and Mi, 1997). However, the structure will not be optimal when the model is used on-line and is subject to process dynamics or environment change. On-line structure adaptation of an RBFN was studied by several researchers. Platt (1991) proposed a resource allocating network and Liu et al. (1999) developed this as a variable neural network. However, both structure and parameter on-line adaptation will inevitably increase computing load and complexity of the software, which is not desirable for on-line use.

Orthogonal decomposition is a well known numerically robust technique to reduce ill-conditioning effects in solving least squares problems. Chen, *et al.* (1991) showed how batch orthogonal least squares (OLS) could be used to develop RBFNs by sequentially selecting centres based on evaluating an error reduction ratio in a forward selection algorithm. Yu, *et al.* (1997) developed a recursive OLS (ROLS) training algorithm for output layer weights in multi-input multi-output (MIMO) RBFNs. Results showed that the ROLS algorithm can be applied to a large training data set with a low requirement on computer memory, and can eliminate ill-conditioning in the data set, so that the accuracy of the RBFN model is greatly improved. When the ROLS is used on-line to update the RBFN's weights using new measurement data, a forgetting factor is usually used in the way for linear models. It is noted that dynamics of a nonlinear system is depend on operating points and is different from region to region. Forgetting the dynamics with a fixed rate for all regions while the new measurements from a specific region being used to update the weights will deteriorate the nonlinear model.

This paper proposes a novel localised forgetting method for RBFN with Gaussion basis function, which is capable of overcoming the problem described in the above, so as to maintain the model accuracy while it learns on-line.

2. THE LOCALLISED FORGETTING METHOD

This section introduces the proposed localised forgetting method applied to the Gaussion RBFNs. The ROLS algorithm is used and therefore will be briefly introduced. For the details of the training, readers are refereed to Yu, *et al.* (1997).

The nonlinear system to be modelled is represented by the following equation,

$$y(k) = f[y(k-1),\cdots,y(k-n_y),u(k),\cdots,u(k-n_u)]+e(k) \quad (1)$$

where $u \in \Re^m$, $y, e \in \Re^p$ e are system input, output and noise vectors, n_u and n_y are input and output orders respectively, An RBFN with Gaussian basis function used to model the nonlinear function in (1) is described by the following equation.

$$\hat{y}_j(k) = \sum_{i=1}^{n_h} w_{ji}(k)\exp\left(-\frac{\|x(k)-c_i\|_2^2}{\sigma_i^2}\right), \quad j = 1,\cdots,p \quad (2)$$

where n_h is the number of hidden layer nodes, $W(k) = [w_{ij}(k)]_{n \times n_h}$ is the weight matrix, $x(k) = [y(k-1) \ \cdots \ y(k-n_y) \ u(k-1) \ \cdots \ u(k-n_u)]^T$ is the network input vector, $c_i \in \Re^n$, $i = 1,\cdots,n_h$ is the ith centre and is represented as the ith column vector of the centre matrix C, σ_i, $i = 1,\cdots,n_h$ is the ith width and $\|\bullet\|_2$ is a second order norm to generate a distance from the input vector to the corresponding centre vector. Let the hidden layer output of the RBF network be denoted by $\phi(k) = [\varphi_1(k) \ \cdots \ \varphi_{n_h}(k)]^T$ with the ith entry given by the exponential function in (2).

The weights in the RBFN can be trained on-line with the ROLS algorithm. The least squares problem is formed as follows. Considering (2) for a set of N input-output training data, we have

$$Y = \hat{Y} + E = \Phi W + E \quad (3)$$

where $Y \in \Re^{N \times p}$ is the desired output matrix, $\hat{Y} \in \Re^{N \times p}$ is the neural network output matrix, $\Phi \in \Re^{N \times n_h}$ is the hidden layer output matrix, $E \in \Re^{N \times p}$ is the error matrix and

$$Y^T = [y(1), \ \cdots, \ y(N)], \ \hat{Y}^T = [\hat{y}(1), \ \cdots, \ \hat{y}(N)]$$
$$\Phi^T = [\phi(1), \ \cdots, \ \phi(N)], \ E^T = [e(1), \ \cdots, \ e(N)].$$

Equation (3) is to be solved for $W(t)$ to minimise the following time-varying cost function,

$$J(k) = \left\| \begin{bmatrix} \lambda Y(k-1) \\ \cdots \\ y^T(k) \end{bmatrix} - \begin{bmatrix} \lambda \Phi(k-1) \\ \cdots \\ \phi^T(k) \end{bmatrix} W(k) \right\|_F \quad (4)$$

where the F-norm of a matrix is defined as $\|A\|_F^2 = trace(A^T A)$ and $\lambda < 1$ is a forgetting factor to introduce exponential forgetting for the past data. It has been shown (Yu, *et al.*, 1997) that minimising (4) is equivalent to minimising the following cost function,

$$J(k) = \left\| \begin{bmatrix} \lambda \bar{Y}(k-1) \\ \cdots \\ y^T(k) \end{bmatrix} - \begin{bmatrix} \lambda R(k-1) \\ \cdots \\ \phi^T(k) \end{bmatrix} W(k) \right\|_F \quad (5)$$

where R is an $n_h \times n_h$ upper triangular matrix obtained from an orthogonal decomposition of Φ, and $\bar{Y}$ is obtained by applying the same decomposition to Y.

The objective function in (5) uses a constant forgetting factor to all past measurement data and consequently introduces an exponential forgetting to the past dynamics over the entire operating space. This is correct for a linear system, as the dynamics of a linear system is the same in the entire operating space. However, it is not true for nonlinear systems where the dynamics is dependant on the operating point. Therefore, the dynamics change represented by the current measurement data is only for a local area around the current operating point. If the past data over the whole operating space is forgotten, the dynamics for the regions other than that around the current operating point will be deteriorated. In the light of this consideration, a novel localised forgetting method is proposed here by applying different forgetting factors to different operating regions. For a Gaussion RBFN each hidden layer node output represents the dynamics that is valid only for the local region defined by the corresponding width. Furthermore, each centre output is represented by the corresponding column in the $\Phi(k-1)$ matrix in (4) or in the $R(k-1)$ in (5) of the ROLS algorithm. These lead to a simple modification of the objective function in (5) to enable the localised forgetting. A diagonal matrix with each non-zero entry being a different forgetting factor for each centre area is left-multiplied to $R(k-1)$ in (5).

$$J(k)=\left\| \begin{bmatrix} \bar{Y}(k-1) \\ \cdots \\ y^T(k) \end{bmatrix} - \begin{bmatrix} R(k-1)\Lambda(k) \\ \cdots \\ \phi^T(k) \end{bmatrix} W(k) \right\|_F \quad (6)$$

where

$$\Lambda(k)=\begin{bmatrix} \lambda_1(k) & & \\ & \ddots & \\ & & \lambda_{n_h}(k) \end{bmatrix} \quad (7)$$

and

$$\bar{Y}(k-1)=R(k-1)\Lambda(k)W(k-1) \quad (8)$$

if $\hat{Y}(k-1)$ in (5) can be decomposed into

$$\hat{Y}(k-1)=R(k-1)W(k-1) \quad (9)$$

Applying the following orthogonal decomposition,

$$\begin{bmatrix} R(k-1)\Lambda(k) \\ \cdots \\ \phi^T(k) \end{bmatrix} = Q(k) \begin{bmatrix} R(k) \\ \cdots \\ 0 \end{bmatrix} \quad (10)$$

$$\begin{bmatrix} \hat{Y}(k) \\ \cdots \\ \eta^T(k) \end{bmatrix} = Q^T(k) \begin{bmatrix} \bar{Y}(k-1) \\ \cdots \\ y^T(k) \end{bmatrix} \quad (11)$$

where Q is an orthogonal matrix and considering that the F-norm is preserved by orthogonal transformations, the following revised cost function is obtained,

$$J(t)=\left\| \begin{bmatrix} \hat{Y}(k)-R(k)W(k) \\ \cdots \\ \eta^T(k) \end{bmatrix} \right\|_F \quad (12)$$

which allows the optimal solution of $W(k)$ to be solved from

$$R(k)W(k)=\hat{Y}(k) \quad (13)$$

and leaves the residual at stage k as $\left\|\eta^T(k)\right\|_F$. It is noted that equation (13) satisfies assumption (9) and provides the calculation in (8) for the next sample time.

Each $\lambda_i(k), i=1,\cdots,n_h$ in (7) is a specific forgetting factor corresponding to the ith column in matrix $R(k)$ or the ith centre output $z_i(k)$ and should reflect the localization feature of each centre to current measurement data. It is considered that the hidden layer node output reflects this feature and therefore can be used as a measure of the feature for the current data. Thus, the normalised value of the hidden layer output is used as the different forgetting factor.

$$\lambda_i(k)=1-\beta\frac{\phi_i(k)}{\|\phi(k)\|_2}(1-\lambda_0), i=1,\cdots,n_h \quad (14)$$

where β is a real scalar coefficient used to adjust the average level of the forgetting factor.

Since $R(k)$ is an upper triangular matrix therefore $W(t)$ can be easily solved from (13) by backward substitution. The decomposition in (10)-(11) can be achieved efficiently by applying Givens rotations to an extended matrix to obtain the following transformation (Barry and Yu, 2000),

$$\begin{bmatrix} R(k-1)\Lambda(k) & R(k-1)\Lambda(k)W(k-1) \\ \phi^T(k) & y^T(k) \end{bmatrix} \rightarrow \begin{bmatrix} R(k) & \hat{Y}(k) \\ 0 & \eta^T(k) \end{bmatrix} \quad (15)$$

The procedure of the ROLS algorithm is therefore the following: at stage k, calculate forgetting matrix by (14) and (7), then form the left-hand side of the transformation in (15) and calculate $R(k)$ and $\hat{Y}(k)$, finally solve $W(k)$ by (13). Initial values for $R(t)$ and $\hat{Y}(t)$ can be assigned as $R(0)=\alpha I$ and $\hat{Y}(0)=0$, where α is a small positive number.

3. APPLICATION EXAMPLE

The localised forgetting method is applied to modelling an model predictive control of a chemical reactor rig using three multi-input single-output Gaussion RBFNs. To demonstrate the operation and performance of the proposed method, modelling of the dissolved oxygen using three different forgetting is presented here.

The reactor used in this research is a pilot system established in the laboratory to generally represent the dynamic behaviour of real chemical processes in industry. The schematic of the chemical reactor is shown in Fig.1. It consists of a continuously stirred tank (15 litres) to which the chemical solutions, NH_4OH, CH_3COOH and $N_{a2}SO_3$, and air are added. The liquid level in the tank is maintained at a pre-specified constant level by an outflow pump system. The concentrations and flow rates of solutions, CH_3COOH and $N_{a2}SO_3$, are constant except for some manual changes to mimic process disturbances. The concentration of NH_4OH is constant but the flow rate is adjustable by a servo-pump to regulate the pH value in the tank. The air-flow rate is also adjustable by a mass- flow meter connected to a compressing air network to regulate the percentage of the dissolved oxygen (pO_2) in the liquid in the tank. The tank is also equipped with an electric heating system to adjust the liquid temperature. The liquid in the tank is stirred continuously to make sure the pH, the dissolved oxygen and the temperature are consistent throughout the tank. All three variables are measured and displayed. A personal computer with analogue I/O is connected to the process to sample the measurements and issue the control outputs.

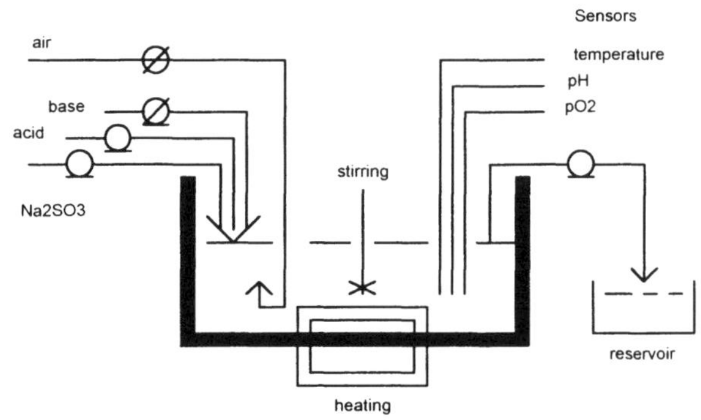

Fig.1 Schematic of the chemical reactor rig

With the three inputs, heating power, flow rate of ammonium hydroxide and flow rate of air, and the three outputs, liquid temperature, pH and percentage of dissolved oxygen, the process constitutes a MIMO, non-linear dynamic system. It has been shown in the experiments that the coupling between variables is very significant. The rate of absorption of oxygen into the liquid and the reaction of the sodium sulphite, for example, significantly depend on the liquid temperature. The process also suffers from many external disturbances, apart from those introduced manually, such as changes in the room temperature, perturbations in the concentrations of the inflow chemical solutions and air pressure in the compressing air network, different concentrations of H^+ and OH^- ions in the liquid at different times. In addition, the response times for the three variables are significantly different. The rise time for the temperature is very long due to the available heating power whereas the dissolved oxygen is quite short. All these effects cause the process to be non-linear in both dynamic and static behaviour, time varying and uncertain in parameters, multivariable with significant coupling, complex without a known mathematical model, suffering from unpredictable large disturbances.

Process inputs and outputs are chosen as

$$u=\begin{bmatrix} Q \\ f_b \\ f_a \end{bmatrix}, \quad y=\begin{bmatrix} T \\ pH \\ pO_2 \end{bmatrix} \tag{16}$$

where Q, f_b and f_a denote the heating power, the flow rate of the base and the flow rate of air respectively.

One possibility for choosing a sample time is based on examining the hold-up time (volume/total flow rate) of the process, which is approximately 10 minutes for this process. A suitable sample time could then be selected as, say, one minute if only the dynamics of the fluid flow rates are considered. However, since the dissolved oxygen responds to changes in the air flow rate quickly, especially at high temperature, therefore, the rise times for different variables should also be considered. From process step responses, approximate rise times to reach steady state are recognised as 45 minutes for temperature, 25 minutes for *pH*, 5 minutes for pO_2 when temperature is 30^0C and 2 minutes for pO_2 at 50^0C. Thus, a suitable sample interval for all variables was selected to be 10 seconds.

The input vector to the dissolved oxygen model is chosen as

$$x(k)=[pO_2(k-1) \quad pO_2(k-2) \quad T(k-1) \quad f_a(k-1)]$$

and the model output is $p\hat{O}_2(k)$. A set of process data with 1800 samples are collected when the process is activated by a special designed feedback control signals superimposed with a PRBS. The input output data is scaled linearly to [0 1] before it is used in model training and is scaled back after the prediction. Eight centres are chosen to be

$$c_i = x(200i), i = 1,\cdots,8$$

and the corresponding width is chosen to be the square root of sum-squared distances from the centre to the three nearest centres. The initial values of *R* and *W* are chosen as $R(0)=0.0001*I_{nh}$ and $W(0)=0$. The uniform forgetting factor is $\lambda_0 = 0.985$. It is noted that there is not uncertainty caused by random choice in the model structure and parameters and this will form a basis for a fire comparison.

Three models are developed off-line using the three different forgetting methods by simulating the on-line situation, which is considered the same as on-line.

(i) The first model is trained for the first 600 samples using the ROLS without forgetting, then stop updating at 601th sample.
(ii) The second model is trained every sample with a constant and uniform forgetting factor $\lambda_0 = 0.985$.
(iii) The third model is trained every sample with the developed method given by equations (14), (7) and (15) with $\beta = 3$.

The three models have the same centres, widths and model input data vectors, and also the same initial values for *R* and *W*. The total 1800 samples are used in the modelling, and the three models are used to predict the process output for one-step-ahead (before this data is used for training). The prediction results of the model 1 and 3 are displayed in the top and bottom graphs in Fig.2 together with the process output for comparison. The prediction by the second model is not displayed, as it is similar to that in the bottom graph. The prediction errors of the three models are calculated. It is noted in Fig.2 that there is not a significant time varying effects before sample 1200. Therefore, prediction errors for the samples 1201-1800 only are displayed in Fig.3 with the error by model 1,2 and 3 in the top, middle and bottom graphs. The models are also assessed by the Mean Squared Error (MSE) index as

$$MSE = \frac{1}{N}\sum_{k=1}^{N}[pO_2(k) - p\hat{O}_2(k)]^2 \quad (17)$$

for the prediction error from k=1201 to k=1800, to match displayed in Fig.3. The results of the three models are given as the follows.

MSE1 = 13.6762
MSE2 = 1.4674
MSE3 = 1.0164

It can be clearly seen in Fig.3 and the MSEs above that the second model is superior to the first model and the third one is the best in terms of smallest prediction error. From the top graph we notice that there exist a constant prediction error for the flat part of the process output. This is due to the process dynamics change and particularly the static gain change. Therefore, the non-adapted model will not follow these changes.

4. CONCLUSION

A localised forgetting method is proposed for Gaussion RBFN model and used in model on-line adaptation. Based on the ROLS the proposed method is simple to implement without increasing computing effort. Application of the method to dissolved oxygen modelling of a chemical reactor rig indicates that the method improves the weight on-line updating algorithm so that the model predicts much more accurately than the uniformly forgetting. The method can also be used for other forms of nonlinear model on-line adaptation.

ACKOWLEDGEMENT

The project is supported by the U.K. EPSRC with the grant No. GR/N18697.

REFERENCES

Chen, S., C.F.N. Cowan and P.M. Grant (1991), Orthogonal least squares learning algorithm for radial basis function networks, *IEEE Trans. On Neural Networks*, Vol.2, pp. 302-309.

Chen, S., S.A. Billings and P.M. Grant (1992), Recursive hybrid algorithm for non-linear system identification using radial basis function networks, *Int. J. Control*, Vol.55, pp. 1051-1070.

Gomm, J.B. and D.L. Yu (2000a), Selecting radial basis function network centres with recursive orthogonal least squares training, *IEEE Trans. on Neural Networks*, Vol. 11, No.3, pp. 306-314.

Karayiannis, N.B. and G.W. Mi (1997), Growing radial basis networks: merging supervised and unsupervised learning with network growth techniques, *IEEE Trans. on Neural Networks*, Vol.8, No.6, pp. 1492-1506.

Liu, G.P., Kadirkamanathan, V. and Billings, S.A. (1999), Variable neural networks for adaptive control of nonlinear systems, *IEEE Trans. on Systems, Man, and Cybernetics _Part C: Applications and Reviews*, Vol.29, No.1, pp. 34-43.

Luo, W., M.N. Karim, A.J. Morris and E.B. Martin (1996), Control relevant identification of a pH waste water neutralisation process using adaptive radial basis function networks, *Computers Chem. Engineering*. Vol.20, pp. 1017-1022.

Platt, J. (1991), A resource allocating network for function interpolation, *Neural Computation*, Vol.4, No.2, pp. 213-225.

Yu, D.L., J.B. Gomm and D. Williams (1997), A recursive orthogonal least squares algorithm for training RBF networks, *Neural Processing Letters*, Vol.5, No.3, pp. 167-176.

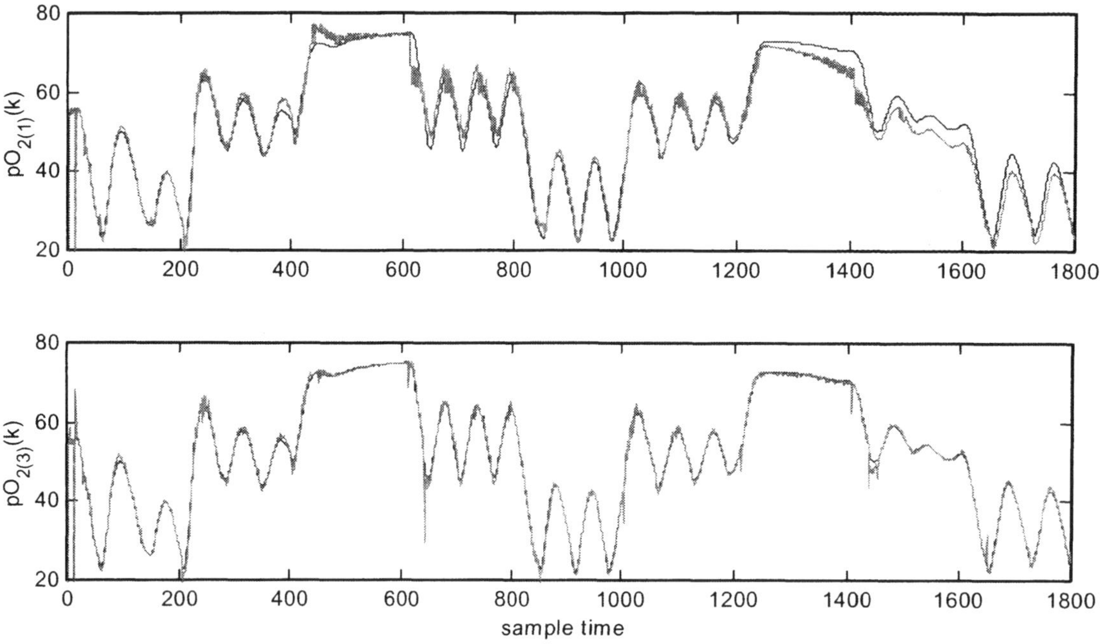

Fig. 2 Process output and predicted output by model 1 and 3

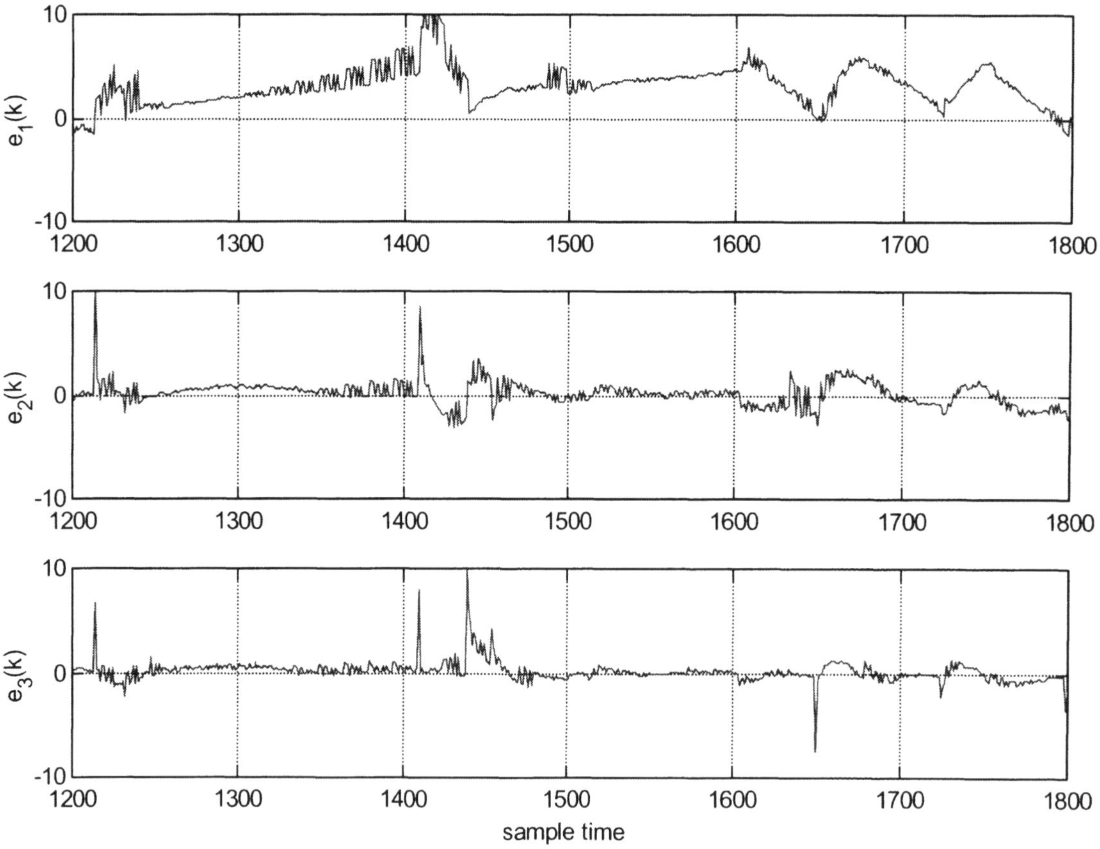

Fig.3. Prediction error for the dissolved oxygen by model 1, 2 and 3

IFAC Publications
www.elsevier.com/locate/ifac

Subspace identification of switching model

Komi Midzodzi PEKPE, Komi GASSO♯, Gilles MOUROT, José RAGOT
Centre de Recherche en Automatique de Nancy-CNRS UMR 7039
2, Avenue de la Forêt de Haye 54516 Vandoeuvre-lès-Nancy Cedex France
Tel: (33) 3 83 59 57 04, Fax: (33) 3 83 59 56 44
E-mail: {kpekpe, gmourot, jragot}@ensem.inpl-nancy.fr
♯Laboratoire PSI - CNRS FRE N° 2645
Place Emile Blondel - BP 08 F-76131 Mont-Saint-Aignan Cedex
Tel : (33) 2 32 95 98 75, Fax: (33) 2 32 95 97 08
E-mail: komi.gasso@insa-rouen.fr

Abstract— **Subspace identification of switching model is considered in this paper. Here the switching model is supposed to be a sum of weighted linear models. The method established uses recursive subspace identification to estimate the switching function and least squares method for local model Markov parameters estimation. To perform the computation of the weighting functions a two-steps algorithm (switching times determination and model merging) is given. Finally the local model parameter estimation is based on the estimation of the Markov parameters.** *Copyright © 2003 IFAC*

Keywords— **switching model, Markov parameters, least squares, weighting function, recursive subspace identification.**

I. Introduction

The objective of this paper is to identify switching models. Here a switching model structure is considered as a sum of weighted linear systems (local models), but only one model is active at any time. In this paper, we use a weighting function which determines the active model. Our goal is to estimate the weighting function and identify the local model, using the knowledge of the inputs and outputs only. Thus, knowing the weighting function allows to determine which local model is active. The estimation of the switching times is made by a recursive subspace identification method. The proposed method is not sensible to the change in input dynamics because the recursive subspace method proposed in [4] is not sensible to this fact. Then the switching times correspond only to a change in the system dynamic. To identify the local models, we use least squares method. The parameter estimation is done by the extraction of the Markov parameter matrix. The advantage of the proposed method results in fact that there is not any stage of nonlinear optimization, and the parameter estimation is unbiased.

After the formulation of the problem, the notations are introduced and the estimation of the Markov parameters is achieved. In section 4 the determination of the weighting function is performed. The identification of the local models procedure is given. Finally an example illustrates the performance of the proposed method.

II. Formulation of the model

The output of the switching model can be modeled as a weighted sum of outputs of h linear models as follows:

$$y_k = \sum_{s=1}^{h} \omega_{s,k} y_{s,k}, \tag{1}$$

where $y_k \in \mathbb{R}^\ell$, any weight $\omega_{s,k} \in \{0,1\}$ and $\sum_{s=1}^{h} \omega_{s,k} = 1$, $\forall\ k \in [1,q]$ (if we have q measurements of the inputs and outputs). $\omega_{s,k}$ represents the weighting function and $y_{s,k}$ the output of the sth local model.

Any local model is supposed linear of order n_s and can be described by the equation:

$$\begin{aligned} x_{s,k+1} &= A_s x_{s,k} + B_s u_k, \\ y_{s,k} &= C_s x_{s,k} + D_s u_k + e_k \end{aligned} \tag{2}$$

Here the output error $e_k \in \mathbb{R}^\ell$ is assumed to be a zero mean white noise sequence and uncorrelated with $u_k \in \mathbb{R}^m$ and has covariance matrix

$$\mathbf{E e_k e_t} = \begin{cases} R > 0, & \text{k=t} \\ 0, & \text{otherwise.} \end{cases}$$

We suppose each local model is stable.

The inputs u_k and outputs y_k of the switching model are supposed to be known. The object is to determine:

-the weighting function $\omega_{s,k}$ for each model,

-the order n_s and the h linear local model parameters. The parameters to be determined are A_s, B_s, C_s and D_s.

Note that, to obtain the gth weighting outputs, it is sufficient to make the product of the global outputs y_k by the gth weighting function:

$$\omega_{g,k} y_k = \omega_{g,k} \sum_{s=1}^{h} \omega_{s,k} y_{s,k} = \omega_{g,k} y_{g,k}, \ \forall k \in [1,q] \tag{3}$$

because

$$(\omega_{g,k})^2 = \omega_{g,k} \text{ and } \omega_{g,k} \times \omega_{s,k} = 0, \ (if\ g \neq s), \forall k \in [1,q] \tag{4}$$

In the following we use a recursive subspace identification technique to obtain the weighting function $\omega_{s,k}$. In order to obtain local Markov parameters, least squares tools will be used. In the next section the matrices used later are defined.

III. THE SYSTEM MATRICES

For the sth model, we can set the following definitions (note that, for simplicity we use i instead of i_s in this article).

Output weighted block Hankel matrix of the sth model is defined as ($i>n_s$):

$$Y^{\mathrm{i}}_{s,\omega} =$$

$$\begin{pmatrix} \omega_{s,i}y_{s,i} & \omega_{s,i+1}y_{s,i+1} & \dots & \omega_{s,i+j-1}y_{s,i+j-1} \end{pmatrix} \quad (5)$$

Input block Hankel matrix is defined as:

$$U = \begin{pmatrix} u_1 & u_2 & \dots & u_j \\ u_2 & u_3 & \dots & u_{j+1} \\ \vdots & \vdots & \vdots & \vdots \\ u_i & u_{i+1} & \dots & u_{i+j-1} \end{pmatrix} \quad (6)$$

The same definition is made for the Hankel block matrix E for the noise e_k.

The state sequence matrix X_s and the weighting matrix of the sth model Ω_s are defined as:

$$X_s = \begin{pmatrix} x^T_{s,1} & x^T_{s,2} & \dots & x^T_{s,j} \end{pmatrix}^T \quad (7)$$

and

$$\Omega_s = \begin{pmatrix} \omega_{s,i} & 0 & 0 & 0 \\ 0 & \omega_{s,i+1} & 0 & 0 \\ \vdots & \vdots & \vdots & \vdots \\ 0 & 0 & 0 & \omega_{s,i+j-1} \end{pmatrix} \quad (8)$$

Then we have:

$$Y^{\mathrm{i}}_{s,\omega} = \underbrace{\begin{pmatrix} y_{s,i} & y_{s,i+1} & \dots & y_{s,i+j-1} \end{pmatrix}}_{Y^i_s}\Omega_s \quad (9)$$

The extended observability matrix $\Gamma_{i,s}$ of the sth model is defined as:

$$\Gamma_{i,s} = \begin{pmatrix} C_s \\ C_sA_s \\ \vdots \\ C_sA^{i-1}_s \end{pmatrix} \in \mathbb{R}^{\ell i \times n_s} \quad (10)$$

The extended controllability matrix $\mathcal{C}_{i,s}$ of the sth model is defined as :

$$\mathcal{C}_{i,s} = \begin{pmatrix} B_s & A_sB_s & \cdots & A^{i-1}_s B_s \end{pmatrix} \quad (11)$$

The Markov parameter matrix for the deterministic part H^d_i and for the stochastic part H^{st}_i of the system are defined as[1]:

$$H^d_{s,i} = \begin{pmatrix} C_sA^{i-2}_sB_s & C_sA^{i-3}_sB_s & \dots & C_sB_s & D_s \end{pmatrix} \quad (12)$$

$$H^{st}_{s,i} = \begin{pmatrix} 0 & 0 & \dots & 0 & I \end{pmatrix} \quad (13)$$

IV. ESTIMATION OF THE MARKOV PARAMETER MATRIX FOR THE LOCAL MODEL

Local matrix input-output equation

It is essential to write the matrix input-output relation, that allows the extraction of the Markov parameter matrix. The following theorem summarizes this relation.

Theorem 1:

$$Y^{\mathrm{i}}_{s,\omega} = C_sA^{i-1}_sX_s\Omega_s + H^d_{s,i}U\Omega_s + H^{st}_{s,i}E\Omega_s \quad (14)$$

The proof of the theorem is established in appendix 1.

Now, the objective is to eliminate the term depending on the state X_s and the noise E, in order to obtain the Toeplitz matrix

$H^d_{s,i}$. The proposed method uses least squares and may be summarize in the next theorem.

Theorem 2: Under the assumptions that:
1. the sth local model is stable,
2. the matrix $U\Omega_s$ has full rank,
3. with an adequate value of i, A^{i-1} is neglected,

we have:

$$Y^{\mathrm{i}}_{s,\omega}(U\Omega_s)^T(U\Omega_s(U\Omega_s)^T)^{-1} \underset{j\to\infty}{\longrightarrow} H^d_{s,i} \quad (15)$$

The proof of the theorem can be found in appendix 2.

Now, from equation 15, we have an estimation of the Markov parameter matrix $H^d_{s,i}$.

Remark 1: to estimate the Markov parameter matrix $H^d_{s,i}$ by equation 15, we need the knowledge of the weighting function Ω_s. We use recursive subspace identification to compute this weighting function in the next section.

V. DETERMINATION OF THE WEIGHTING FUNCTION

It is important to note that the previous result involves the state matrix of the system which depends on the active local model.

The recursive subspace approach is used to determine the switching time between local models. As the local models are stable, the recursive subspace identification can be applied. This identification method does not allow to detect the change in inputs dynamics; therefore if a change

[1] The superscript 'd' and 'st' stand for 'deterministic' and 'stochastic' respectively.

is detected it corresponds only to change in the system dynamics. We recommend to the reader, to refer to [4] for the details of the recursive subspace identification procedure.

To begin with, we make a first computation of the weighting function with the recursive subspace algorithm. Then we merge the models which are closed to each other, and we recompute the new weighting functions.

A. The first computation of the weighting functions

The switching time is determined by the recursive subspace identification algorithm. Here we summarize the main stages of the algorithm of Oku and al.[4].

Let us define:

$$\Phi_j = \begin{pmatrix} \varphi_i(1) & \varphi_i(2) & \dots & \varphi_i(j-i) \end{pmatrix} \quad (16)$$

$$\varphi_i(k) = u_i(k) = \begin{pmatrix} u_k \\ \vdots \\ u_{i+k-1} \end{pmatrix}, k \in [1, j-i]. \quad (17)$$

$$U_j = \begin{pmatrix} u_i(i) & u_i(i+1) & \dots & u_i(j) \end{pmatrix} \quad (18)$$

Y_j is defined similarly to (18).

Θ_j which is the product of the extended observability matrix and the extended controllability matrix, can be estimated by the formula:

$\hat{\Theta}_j = Y_j \Pi_{U_j^\perp} \Phi_j^T \Psi_j$ with $\Psi_j = \left(\Phi_j \Pi_{U_j^\perp} \Phi_j^T\right)^{-1}$ and $\Pi_{F^\perp} = I - F^T(FF^T)^{(-)}F$, where I is the identity matrix with the appropriate size and F is a matrix.

If we define P_j as $P_j = (U_j U_j^T)^{-1}$, $\hat{\Theta}_j$ can be obtained also by a usual formulation of the recursive least squares method with a forgetting factor γ ($\gamma<1$, see [4]):

$$\hat{\Theta}_j = \hat{\Theta}_{j-1} - \beta_j(e_j + \hat{\Theta}_{j-1}q_j)q_j^T\Psi_{j-1}, \quad (19)$$

$$\Psi_j = \frac{1}{\gamma}(\Psi_{j-1} - \beta_j\Psi_{j-1}q_jq_j^T\Psi_{j-1}), \quad (20)$$

$$P_j = \frac{1}{\gamma}(P_{j-1} - \alpha_jP_{j-1}u_i(j)u_i^T(j)P_{j-1}), \quad (21)$$

$$Y_jU_j^T = \gamma Y_{j-1}U_{j-1}^T + y_i(j)u_i(j)^T, \quad (22)$$

$$\Phi_jU_j^T = \gamma\Phi_{j-1}U_{j-1}^T + \varphi_i(j-i)u_i(j)^T, \quad (23)$$

$$\alpha_j = \left(\gamma + u_i(j)^TP_{j-1}u_i(j)\right)^{-1}, \quad (24)$$

$$\beta_j = \left(\frac{1}{\alpha_j} + q_j^T\Psi_{j-1}q_j\right)^{-1}, \quad (25)$$

$$e_j = y_i(j) - Y_{j-1}U_{j-1}^TP_{j-1}u_i(j)^T, \quad (26)$$

$$q_j = \Phi_{j-1}U_{j-1}^TP_{j-1}u_i(j)^T - \varphi_i(j-i), \quad (27)$$

Let λ be a threshold designed according to the χ^2-distribution with ℓi degrees of freedom. The distance from the estimated parameter $\hat{\Theta}_j$ to the true parameter Θ_j, $D(\hat{\Theta}_j, \Theta_j)$ is defined as:

$$D(\hat{\Theta}_j, \Theta_j) = Trace\left(\hat{\Theta}_j - \bar{\Theta}_j\right)\frac{1}{\sigma_{\in\ i}^2}\left(\Phi_j\Pi_{U_j^\perp}\Phi_j^T\right)\left(\hat{\Theta}_j - \bar{\Theta}_j\right)^T, \quad (28)$$

where $\in_j = Y_j - \hat{Y}_j$, $\hat{Y}_j$ is an estimation of Y_j. An estimation of the variance of the modeling error $\sigma_\in^2$ is given by ([4]):

$$\hat{\sigma}_{\in\ j}^2 = \frac{1}{i[(j-i+1) - 2mi)]} \times$$
$$Trace\ (Y_j - \hat{\Theta}_j\Phi_j)\Pi_{U_j^\perp}(Y_j - \hat{\Theta}_j\Phi_j)^T \quad (29)$$

The *change test* is defined as (see [4]):

if $D(\hat{\Theta}_j, \Theta_j)<\lambda$: no change has occurred,

if $D(\hat{\Theta}_j, \Theta_j)>\lambda$: a change has occurred.

Note that in the implementation, the true parameter Θ_j is replaced by an approximation: $\bar{\Theta}_j$ (see [4]) computed as a least squares estimation of Θ_j over a sliding window:

$$\bar{\Theta}_j = \underset{min}{arg}\ Trace\ (\bar{Y} - \Theta\bar{\Phi}_j)\Pi_{U_j^\perp}(\bar{Y} - \Theta\bar{\Phi}_j)^T, \quad (30)$$

with

$$\bar{Y}_j = \begin{pmatrix} y_i(j-L+2i) & \dots & y_i(j) \end{pmatrix} \quad (31)$$

$$\bar{\Phi}_j = \begin{pmatrix} \varphi_i(j-L+i) & \dots & \varphi_i(j-i) \end{pmatrix} \quad (32)$$

Once a change at time instant t is detected, the recursive update equation are re-initialized at time instant t, the re-initialization technique is proposed in [4].

The first computation of the weighting functions is described by the algorithm 1 exposed below.

Algorithm 1 (first computation of the weighting functions):

Step1: *(initial conditions)*

set s←1 and k←0, where s is the local model index and k is the time index[2].

Step 2: *(change has not occurred)*

if change doesn't occur (i.e. $D(\hat{\Theta}_j, \bar{\Theta}_j)<\lambda$) then: k←k+1, go to step 4.

Step 3: *(change has occurred)*

if change occurs (i.e. $D(\hat{\Theta}_j, \bar{\Theta}_j)>\lambda$) then: s←s+1 and k←k+1.

Step 4: *(computation of the weighting function)*

model "s" is active: $\omega_{s,k} \leftarrow 1$ and $\omega_{r,k} \leftarrow 0$, $\forall\ r \neq s$.

Step 5: *(test of stop)*

h← s (h is the number of identified models)

if k<q then go to step 2,

if k=q then stop.

B. The fusion of model

After the first computation of the weighting function by the preceding algorithm, we estimate the Markov parameter matrix $H_{s,i}^d$ (for the h local models). Then, we seek the models which are 'similar'. Firstly, we estimate the covariance of the Markov parameter matrix $H_{s,i}^d$.

[2]The left arrow "← " denote the replacement of the value of the left hand side, by the right hand side.

B.1 Covariance estimate

The variance of the estimate error σ_y can be unbiasely estimated by:

$$\hat{\sigma}_y^2 = \frac{1}{i\,(j-im)} trace\ (\varepsilon\varepsilon^T), \tag{33}$$

where:
$\varepsilon = Y^i_{s,\omega} - \hat{Y}^i_{s,\omega}$ (see 9), $\hat{Y}^i_{s,\omega}$ is an estimation of $Y^i_{s,\omega}$.
The variance $\Sigma_{H^d_{\mathrm{s,i}}}$ of the output estimated Markov parameter matrix $H^d_{\mathrm{s,i}}$ is estimated by the following lemma.

Lemma 1:
Unbiased estimation of $\Sigma_{H^d_{\mathrm{s,i}}}$ is given by the formula:

$$\hat{\Sigma}_{H^d_{\mathrm{s,i}}} = i\hat{\sigma}_y^2[U\Omega_s(U\Omega_s)^T]^{-1}, \tag{34}$$

Now we can make a test on the models to find the models which are closed to each other.

B.2 The 'similar' models

The aim is to find the models close to each other (or "similar"). For that purpose, we define the distance from the estimated Markov parameters of model s to the parameters of the model r as:

$$d(\hat{H}^d_{s,i}, \hat{H}^d_{r,i}) = Trace\ \left((\hat{H}^d_{\mathrm{s,i}} - \hat{H}^d_{\mathrm{r,i}})\hat{\Sigma}^{-1}_{H_{\mathrm{s,r}}}(\hat{H}^d_{\mathrm{s,i}} - \hat{H}^d_{\mathrm{r,i}})^T\right). \tag{35}$$

with:

$$\hat{\Sigma}_{H_{\mathrm{s,r}}} = \frac{N_s}{N_s+N_r}\hat{\Sigma}_{H^d_{\mathrm{s,i}}} + \frac{N_r}{N_s+N_r}\hat{\Sigma}_{H^d_{\mathrm{r,i}}},$$

where $N_s = \sum_{k=1}^{q}\omega_{s,k}$ and $N_r = \sum_{k=1}^{q}\omega_{r,k}$.
If δ is a threshold designed according to the χ^2-distribution with ℓi degrees of freedom, then we perform the following 'similarity test'

B.21 *'similarity test'*

The models s and r are similar if:
if $d(\hat{H}^d_{\mathrm{s,i}}, \hat{H}^d_{\mathrm{r,i}}) \geq \delta$: the model s and the model r are not 'similar',
if $d(\hat{H}^d_{\mathrm{s,i}}, \hat{H}^d_{\mathrm{r,i}}) \leq \delta$: the model s and the model r are 'similar'.

B.3 Model merging

If $\mathbb{E}_{s0}$ is the *set of models which are "similar"* to s0, the models belonging to this set are replaced by the new model s0*. The new weighting function for the new model s0* is: $\omega^*_{s0,k} = \sum_{s\in\mathbb{E}_{s0}} \omega_{s,k}$ for k=1,...,q; moreover we estimate the Markov parameters of the new model s0*.

Remark 2: It is necessary to make a new "*similarity test*" *(see* B.21*)* for the merging of the new model (after a first merging), because it can exist two "similar" local models which have not been detected in the first step, because of the insufficiency of data used to identify the local models. But, by the new computation of the weighting functions and Markov parameters, the number of measurements used to identify the new local models is greater than those which are used to estimate the previous local models.

The merging of the local models is described by algorithm 2.

Algorithm 2 (models merging):
Step 1: *(initialization)*
st←0;
Step 2: (computation of Markov parameters)
compute the Markov parameters (by theorem 2)
Step 3: *(test)*
Check the "similarity *test*" (see B.21), and construct the *set of models which are "similar".*
Step 4: *(replacement of the weighting functions)*
For each *set of models* $\mathbb{E}_s$, if cardinal of $\mathbb{E}_s$ is greater than 1 then:
1) st←1;
2) $\omega_{s,k} \leftarrow \omega^*_{s,k}$, and cancel the model r, $r \in \mathbb{E}_s$ and $r \neq s$.
Step 5: *(renumber the local models)*
Reorganize local model index.
step 6: *(stop test)*
If st=1 then go to 1,
If st=0 then stop.

Now we have the final weighting function, we estimate the order and the parameters of each local model. The estimation procedure is described in next section.

VI. Estimation of the system parameters

The goal of this section is to estimate the system order n_s and matrices A_s, B_s, C_s and D_s for each local model.
• The system matrix D_s is directly obtained from the Markov parameter matrix $H^d_{\mathrm{s,i}}$ (see 12).
Having the Markov parameters (by theorem 2), one can use the algorithm of Kung [3], Ho and Kalman [1], Era [2], Zeiger and McEwen [6] to estimate the system matrices A_s, B_s and C_s.
In this paper we set $i>2(n_s+1)$. Let[3] ν =integer$(i/2)$.
Following we summarize the Era algorithm (*minimal and balanced realization* [2]).
• Build the Hankel matrices $\mathcal{H}^0_{\nu,\mathrm{s}}$ and $\mathcal{H}^1_{\nu,\mathrm{s}}$ which contains the Markov parameters and are defined by:

$$\mathcal{H}^{\mathrm{k}}_{\nu,\mathrm{s}} = \begin{pmatrix} C_sA_s^kB_s & C_sA_s^{k+1}B_s & \cdots & C_sA_s^{k+\nu-1}B_s \\ C_sA_s^{k+1}B_s & C_sA_s^{k+2}B_s & \cdots & C_sA_s^{k+\nu}B_s \\ \vdots & \vdots & \vdots & \vdots \\ C_sA_s^{k+\nu-1}B_s & C_sA_s^{k+\nu+1}B_s & \cdots & C_sA_s^{k+2\nu-2}B_s \end{pmatrix} \tag{36}$$

• Make a singular values decomposition of the matrix $\mathcal{H}^0_{\nu,\mathrm{s}}$:

[3]Integer(i) is the integer part of i.

$$\mathcal{H}^0_{\nu,s} = \begin{pmatrix} U_1 \\ U_2 \end{pmatrix} \begin{pmatrix} S_1 & 0 \\ 0 & S_2 \end{pmatrix} \begin{pmatrix} V_1^T \\ V_2^T \end{pmatrix} \simeq U_1 S_1 V_1^T \quad (37)$$

where S_2 contains the neglected singular values.

- The system order is equal to the number of singular values in S_1.
- The extended observability matrix $\Gamma_{\nu,s} = U_1 S_1^{1/2}$, and the controllability matrix $\mathcal{C}_{\nu,s} = S_1^{1/2} V_1^T$ are computed. Note that:

$\mathcal{H}^0_{\nu,s} = \Gamma_{\nu,s}\mathcal{C}_{\nu,s}$,
The matrix C_s can be determined from the first ℓ rows of $\Gamma_{\nu,s}$.

- The matrix B_s is equal to the first m columns of $\mathcal{C}_{\nu,s}$.
- The matrix A_s is given by the formula:

$A_s = S_1^{-1/2} U_1^T \mathcal{H}^1_{\nu,s} V_1 S_1^{-1/2}$.

VII. The simulation example

We consider a third order system described as:

$$x_{k+1} = A_s x_k + B_s u_k$$

$$y_k = C_s x_k + D_s u_k + K e_k$$

where s take the value 1 on the intervals [1, 999] and [1800, 2499] and the value 2 on the intervals [1000, 1799] and [2500, 3500]; and we have:

$$A_1 = \begin{pmatrix} 0.32 & 0.31 & 0 \\ -0.32 & 0.31 & 0 \\ 0 & 0 & -0.18 \end{pmatrix}, B_1 = \begin{pmatrix} 0.9 & -0.7 \\ 0.71 & -0.5 \\ 0.8 & 0.47 \end{pmatrix}$$

$$C_1 = \begin{pmatrix} -0.55 & 0.2 & 0.8 \\ 0.45 & 0.3 & 0.58 \end{pmatrix}, D_1 = \begin{pmatrix} 0.97 & 0.63 \\ -0.32 & 0.95 \end{pmatrix}$$

$$A_2 = \begin{pmatrix} -0.1 & -0.4 & 0 \\ 0.5 & -0.4 & 0 \\ 0 & 0 & 0.26 \end{pmatrix}, B_2 = \begin{pmatrix} 0.1 & -0.6 \\ 0.32 & -0.66 \\ 0.3 & 0.82 \end{pmatrix}$$

$$C_2 = \begin{pmatrix} -0.8 & -0.1 & 0.7 \\ 0.3 & 0.48 & 0.9 \end{pmatrix}, D_2 = \begin{pmatrix} 0.5 & 0.3 \\ -0.2 & -0.5 \end{pmatrix}.$$

Moreover $u_k \in \mathbb{R}^2$, and $y_k \in \mathbb{R}^2$, the noise $e_k \in \mathbb{R}^2$ is a zero mean white noise and $K = \frac{1}{\sqrt{5}} \times I_2$.

We suppose that we have i+j-1=3500 measurements of the inputs and outputs. The inputs u_k are white noises.

A. The first computation of the weights (use algorithm 1)

For the implementation of the recursive subspace method, we adopt i=15. Since the dimension of the outputs is ℓ=2, the degree of freedom of the χ^2 distribution is ℓi=30. The exponential forgetting factor γ is taken as 0.98 (see [4]).

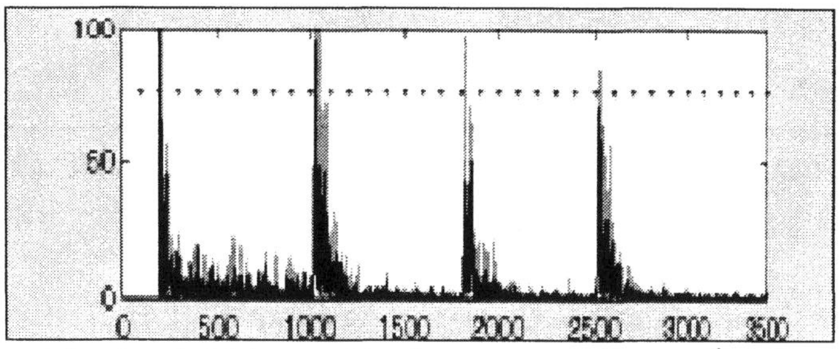

♦♦: threshold (99.999%, χ^2 distribution); –: $D(\hat{\Omega}_j, \bar{\Omega}_j)$

Figure 1: the switching times estimated by the recursive subspace algorithm

The recursive subspace algorithm determines the switching times and shows four local models. The first is active in time window [1:999], the second in time window [1000:1799], the third in time window [1800:2499] and the fourth in time window [2500:3500]. The switching time is correctly detected.

To compute the parameters for each local model, we set the index i equal to 15 (for each local model identification).

From the parameters obtained with the proposed method, we now estimate the poles for each local model (see figure 2).

100 Monte Carlo realizations are done in each case.

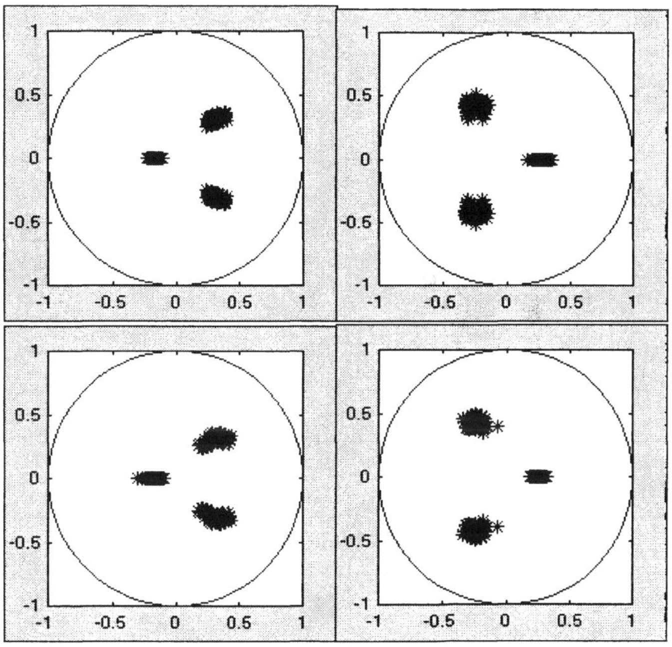

Figure 2: the poles of the four local models according to $\omega_{1,k}$, $\omega_{2,k}$, $\omega_{3,k}$ and $\omega_{4,k}$ respectively.

B. The fusion of the weighting functions (algorithm 2)

The estimated poles show that the local models 1 and 3 are "similar" (figure 2). The same remark hold for local models 2 and 4. That is confirmed by the "*similarity test*". The distance $d(\hat{H}^d_{s,i}, \hat{H}^d_{r,i})$ of 'similarity test' allows to merge the models without ambiguity. For, hundred experiments we carried out the distance and we have:

▶ $d(\hat{H}^d_{1,i}, \hat{H}^d_{2,i}) \in [2300, 2800]$,

► $d(\hat{H}^d_{1,i}, \hat{H}^d_{3,i}) \in [0.5, 4]$,

► $d(\hat{H}^d_{1,i}, \hat{H}^d_{4,i}) \in [2000, 2800]$,

► $d(\hat{H}^d_{2,i}, \hat{H}^d_{4,i}) \in [1, 4]$,

-The threshold (99.999%, χ^2 distribution), δ =59. Then we propose to merge models 1 and 3, that allow to defined a new weight $\omega^*_{1,k} = \omega_{1,k} + \omega_{3,k}$. We also merge the models 2 and 4 and define the new weight: $\omega^*_{2,k} = \omega_{2,k} + \omega_{4,k}$.

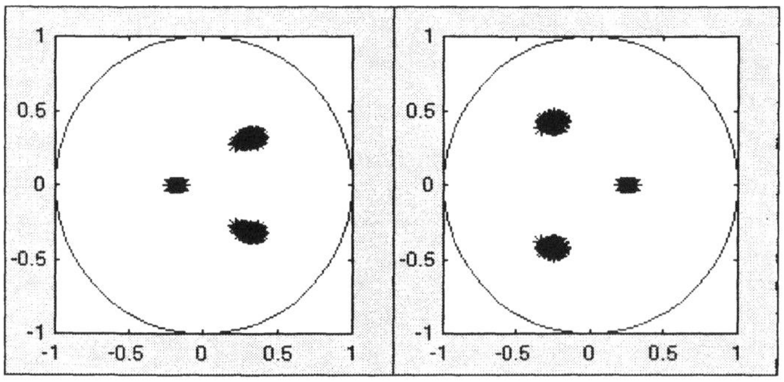

Figure 3: the estimated poles of the 1st and 2nd new local models according to $\omega^*_{1,k}$ and $\omega^*_{2,k}$ respectively.

The figure 3 shows the 100 Monte Carlo realizations for the estimated poles of the new models obtained with $\omega^*_{1,k}$ and $\omega^*_{2,k}$. As we can see, the new weighting functions $\omega^*_{1,k}$ and $\omega^*_{2,k}$ improve the variance of the estimated poles in each case. The estimated poles are unbiased.

VIII. Conclusion

Switching model identification is considered in this paper. The technique is based on subspace formulation and uses least squares method for the parameter estimation. The switching model is supposed to be a weighted sum of a local models. A recursive subspace identification technique is used to determine the switching time and a merging algorithm is given to estimate the final weighting function. Finally we estimate the local systems order and parameters from the new estimation of the local models Markov parameters. An illustrating example shows the application of the method.

References

[1] Ho, B. and Kalman, R. E., *Effective construction of linear state-variate models from input/output functions.* Proceedings. of 3rd Annual Allerton Conf. Circuit and System Theory, pp. 449-459, 1966.

[2] Juang, C., *Applied system identification.* Prentice -Hall, Englewood Cliffs, NJ, 1994.

[3] Kung, S. Y., *A new Identification and model reduction algorithm via singular value decompositions.* Proceedings. Twelth Asilomar Conf. on Circuits, Systems and Computers, pp. 705-714, 1978.

[4] Oku, H., G. Nijsse, M.Verhaegen and V. Verdult. *Change detection in the dynamics with recursive subspace.* Proceedings of the 40th CDC. Orlando, Florida. pp. 2297-2302, 2001.

[5] Van Overschee, P. and De Moor, B., *Subspace identification for linear systems-Theory, implementation, applications.* Kluwer Academic Publishers, 1996.

[6] Zeiger, H. P. and McEwen, A. J., *Approximate linear realisations of given dimension via Ho's algorithm.* IEEE Trans. on Auto. Control, pp. 153-155, 1974.

IX. Appendix 1

Proof: The proof of theorem 1 is established in free noise case, the stochastic case is obtained easily.

Form equation 5 we have:

$$Y^{i}_{s,\omega} =$$

$$\begin{bmatrix} \omega_{s,i}y_{s,i} & \omega_{s,i+1}y_{s,i+1} & \dots & \omega_{s,i+j-1}y_{s,i+j-1} \end{bmatrix}$$

$$\Rightarrow Y^{i}_{s,\omega} =$$

$$\Big[\ \omega_{s,i}(C_sA_s^{i-1}x_{s,1} + C_sA_s^{i-2}B_su_1 + \dots + D_su_i) \quad \dots$$
$$\omega_{s,i+j-1}(C_sA_s^{i-1}x_{s,j} + C_sA_s^{i-2}B_su_j + \dots + D_su_{i+j-1}) \Big]$$

$$\Rightarrow Y^{i}_{s,\omega} =$$

$$\begin{bmatrix} \omega_{s,i}C_sA_s^{i-1}x_{s,1} & \dots & \omega_{s,i+j-1}C_sA_s^{i-1}x_{s,j} \end{bmatrix} +$$
$$\Big[\ \omega_{s,i}(C_sA_s^{i-2}B_su_1 + \dots + D_su_i) \quad \dots$$
$$\omega_{s,i+j-1}(C_sA_s^{i-2}B_su_j + \dots + D_su_{i+j-1}) \ \Big]$$

$$\Rightarrow Y^{i}_{s,\omega} = C_sA_s^{i-1}X_s\Omega_s +$$

$$H^d_{s,i}\begin{pmatrix} \omega_{s,i}u_1 & \dots & \omega_{s,i+j-1}u_j \\ \omega_{s,i}u_2 & \dots & \omega_{s,i+j-1}u_{j+1} \\ \vdots & \vdots & \vdots \\ \omega_{s,i}u_i & \dots & \omega_{s,i+j-1}u_{i+j-1} \end{pmatrix}$$

Finally we obtain :

$$Y^{i}_{s,\omega} = C_sA_s^{i-1}X_s\Omega_s + H^d_{s,i}U\Omega_s$$

■

X. Appendix 2

Proof:

$$Y^{i}_{s,\omega} = C_sA_s^{i-1}X_s\Omega_s + H^d_{s,i}U\Omega_s + H^{st}_{s,i}E\Omega_s$$

if A^{i-1} is neglected then:

$$Y^{i}_{s,\omega} \simeq H^d_{s,i}U\Omega_s + H^{st}_{s,i}E\Omega_s$$

since A_s is assumed to be asymptotically stable, the covariance matrix of the state sequence $\left(\lim_{j\to\infty}\left(\frac{1}{j}(X_s\Omega_s)(X_s\Omega_s)^T\right)\right)$ is bounded. Note that, the elements of the weighting matrix Ω_s take the values 0 and 1.

By the least squares method we can estimate the Markov parameter matrix $H^d_{s,i}$:

$$Y^{i}_{s,\omega}(U\Omega_s)^T(U\Omega_s(U\Omega_s)^T)^{-1} \simeq H^d_{s,i}$$

■

IFAC
Publications
www.elsevier.com/locate/ifac

APPLICATION-ORIENTED NEURAL MODELLING

Kang Li and George Irwin

Intelligent Systems and Control Group
School of Electrical and Electronic Engineering
Queen's University Belfast
Ashby Building, Stranmillis Rd., Belfast BT9 5AH, UK

Abstract: The application-oriented neural modelling of non-linear dynamic systems, where 'a priori' knowledge is used in the choice of activation functions for constructing a feed-forward neural network, is introduced. A framework for constructing this new type of neural model using genetic algorithms is proposed, and is applied to identify a pollutant emission model for a coal-fired power generation plant. Comparison with a conventional MLP model shows that this new type of neural model produces better generalization performance. *Copyright © 2003 IFAC*

Keyword: Neural Networks, System Modelling, Genetic Algorithm, Power Systems and Power Plants, Pollutant Emissions

1. INTRODUCTION

Various types of neural networks, e.g. Multi-layer perceptrons (MLPs), Radial Basis Functions (RBFs), and wavelet networks have been proposed and widely applied to modelling and control of industrial processes and non-liner dynamic systems (Irwin, Warwick, and Hunt 1995). It has been shown that these networks can approximate a wide range of functions to an arbitrary degree of accuracy under certain conditions (Barron, 1993; Park and Sandberg, 1993; Girosi and Poggio, 1990). As an example, a one-hidden-layer neural network with single output can be formulated as follows

$$y_t = \boldsymbol{\Theta} \bullet \begin{bmatrix} \boldsymbol{h}_t \\ 1 \end{bmatrix}, \boldsymbol{h}_{ti} = \psi_i \left(\boldsymbol{C}^{(i)} \bullet \begin{bmatrix} \boldsymbol{u}_t \\ 1 \end{bmatrix} \right), i = 1,2,\ldots,p \quad (1)$$

where t is the time sequence, $y_t \in \Re^1$ is the network output, $\boldsymbol{h}_t^T = [h_{t1} \quad h_{t2} \quad \cdots \quad h_{tp}]$ are the output vector of the hidden layer, $\boldsymbol{u}_t^T = [u_{t1} \quad u_{t2} \quad \cdots \quad u_{tp}]$ are the inputs to the neural network, $\boldsymbol{\Theta}^T \in \Re^{p+1}$, $(\boldsymbol{C}^{(i)})^T \in \Re^{p+1}$, $i = 1,2,\ldots, p+1$ are the weights and bias for the output and hidden nodes respectively, and $\psi_i, i = 1,2,\ldots, p$, are the activation functions for hidden nodes.

The computational power of neural networks originates mainly from the non-linear activation function $\psi_i, i = 1,2,\ldots, p$ in (1). However, it is this very non-linearity that makes deeper analysis of the general properties and performance of the network difficult. Neural network research has concentrated on the development of neural networks with limited classes of non-linear functions, for which theoretic bounds for the approximation properties can be established (Barron, 1993). However, these results only show what is possible in terms of functional approximation and provide little application guidance. For example, it is impossible to continuously parameterise neural networks and approximate all continuous bounded functions uniformly, because the set of continuous bounded functions is not sup norm separable, unlike the parameter sets for neural networks (Stinchcombe, 1999). A good example is that no single hidden layer feed-forward neural network based on a sigmoid function can approximate the sine function over the complete real space $\Re^1$. This observation is important since generalization over unseen data is vital in system modelling and identification. The fact is that for many engineering systems, the data sets that can be used for modelling are always relatively small in size and unable to cover whole system dynamics. For example, most test data sets are acquired within short test period using restricted small perturbation signals due to safety and production concerns. Also operating conditions can vary from time to time and processes are always corrupted with noise and unexpected disturbances.

Although researches have been carried out on network construction to improve the generalisation performance (Rudolph, 1997), the activation functions are restricted to a limited class of non-linear functions, such as Volterra polynomial functions, radial functions, B-spline functions, or wavelets, etc. (Liu, Kadirkamanathan, Billings 1998). These approaches are based on the function approximation theory, very little system information is utilised. It is expectable that neural modelling using limited data samples and no system information may have the difficulty to produce a good generalisation performance.

The authors have recently conducted a research within the grey-box modelling framework, where a priori system knowledge has been used to extract model terms, which are then combined to produce a non-linear regression model. Since these terms are linked to physical reality the resultant model is interpretable (Li and Thompson, 2001; Li, et al, 2002). This is important, as a meaningful model may not only have the potential to achieve better generalisation performance, it can be used for plant operation and help operators to gain some physical insight into the system under control based on the model information. The overall man-machine system performance could therefore be significantly improved. Inspired by this research, a natural extension to neural modelling is to select specific activation functions for individual application. This method has however been neglected to date according to our literature survey and now is termed application-oriented neural modelling. The difficulties for this modelling method are twofold, i.e. how to choose the activation function and how to train the network. The choice of activation function plays a vital role in determining the subsequent learning approach. Almost all neural network architectures involve simple homogenous non-linear functions to ensure that the learning algorithm is analytically simple and computationally efficient to implement (McLoone, et al, 1998). In this new modelling approach, it is likely that heterogeneous functions will be used as the activation functions for nodes in the hidden layer. If this is the case, then the existing learning algorithms may be inappropriate. In addition, there is no apparent mechanism on how, and to what extent, 'a priori' knowledge about the specific application can be used to extract the activation function.

In this paper, genetic algorithm is used for activation function selection as well as for network training. The content is organized as follows. Section 2 gives the framework for building an application-oriented neural model where a genetic algorithm is used for constructing and training. This is then applied to modelling the NOx emission for operation and control of a coal-fired power generation plant. The result confirms the efficacy of the approach compared to a conventional MLP network. Section 4 contains the conclusions.

2. FRAMEWORK FOR APPLICATION ORIENTED NEURAL MODEL CONSTRUCTION

In application-oriented neural model construction, the first task is to select the activation functions based on 'a priori' knowledge of the specific application, since each application will exhibit its own particular distinctive non-linearity and dynamics. For example, a periodic function $f(t)$ of period $T = 2\pi/\omega$ can be better approximated by a Fourier series, in which the basis is a set of sine functions. This concept can be expressed more generally as follows. Suppose an unknown time-series function $f(\boldsymbol{u}_t)$ has the following form:

$$f(\boldsymbol{u}_t) = f(\psi_1(\boldsymbol{u}_t;\boldsymbol{c}_1),\ldots,\psi_q(\boldsymbol{u}_t;\boldsymbol{c}_q)) \qquad (2)$$

where $\boldsymbol{u}_t$ is the vector of input variables, $\psi_1(\boldsymbol{u}_t;\boldsymbol{c}_1),\ldots,\psi_q(\boldsymbol{u}_t;\boldsymbol{c}_q)$ are fundamental functions such as $sin(\boldsymbol{u}_t;c)$, $exp(\boldsymbol{u}_t;c)$, t is the time sequence and $\mathbf{c}_i, i = 1,2,\ldots,q$ are parameter vectors. It is assumed that these fundamental functions can be readily acquired from an analysis of the system dynamics and/or from fundamental physical system knowledge. Then, instead of approximating $f(\mathbf{u}_t)$ by a set of arbitrary expansion basis over independent variables $\mathbf{u}_t$, it can be approximated using one-order Tailor's expansion over $\psi_i(\boldsymbol{u}_t;\boldsymbol{c}_i), i=1,\cdots q$:

$$f(\boldsymbol{u}_t) = f(\boldsymbol{\psi}_1(\boldsymbol{u}_t;\boldsymbol{c}_1),\ldots,\boldsymbol{\psi}_q(\boldsymbol{u}_t;\boldsymbol{c}_q)) \approx$$

$$f(\boldsymbol{u}_t)\Big|_{\boldsymbol{u}_t=\boldsymbol{u}_0} + \sum_{i=1}^{q} b_i(\boldsymbol{\psi}_i(\boldsymbol{u}_0+\sigma\boldsymbol{u}_t;\boldsymbol{c}_i) - \boldsymbol{\psi}_i(\boldsymbol{u}_0;\boldsymbol{c}_i)) \qquad (3)$$

where $b_i = \dfrac{f'(\boldsymbol{u}_t)}{\psi_i'(\boldsymbol{u}_t;\boldsymbol{c}_i)}\Big|_{\boldsymbol{u}_t=\boldsymbol{u}_0}$. However, for a particular application, only a few distinctive types of fundamental function might be identified, so the number q in (3) can very small. In this case (3) is not sufficient to produce a better approximation performance. Instead, an application-oriented neural model can be used to approximate $f(\boldsymbol{u}_t)$, where $\psi_1(\boldsymbol{u}_t;\boldsymbol{c}_1),...,\psi_q(\boldsymbol{u}_t;\boldsymbol{c}_q)$ are used as the possible activation functions for the hidden nodes. That is, the activation functions in (1), $\psi_j, j=1,2,...,q$, are chosen as the fundamental functions defined in (2).

Eqn. (1) can be re-written as follows

$$y_t = \theta_0 + \sum_{i=1}^{p} \theta_i \varphi_i(\mathbf{u}_t;\mathbf{c}_i) \tag{4}$$

where

$$\varphi_i = \begin{cases} \psi_1(\boldsymbol{u}_t;\boldsymbol{c}_i) & 1=m_0 < i \le m_1 \\ \psi_2(\boldsymbol{u}_t;\boldsymbol{c}_i) & m_1 < i \le m_2 \\ \vdots & \vdots \\ \vdots & \vdots \\ \psi_q(\boldsymbol{u}_t;\boldsymbol{c}_i) & m_{q-1} < i \le m_q = p \end{cases}$$

and $(m_i - m_{i-1})$ $i=1,2,...,q$ are the numbers of hidden nodes with the same type of activation function, p is the total number of hidden nodes, $\theta_i, i=1,2,...,p$ are the weights for output node in (4), and θ_0 is the bias for the output node. Now assume that a data set of N samples is used for neural model construction. Applying the data set to (4) gives:

$$\boldsymbol{Y} = \boldsymbol{\Phi\Theta} + \boldsymbol{E} \tag{5}$$

where $\boldsymbol{Y} = [y_1, y_2, \cdots, y_N]^{\boldsymbol{T}}$ are the targets for neural network training, $\boldsymbol{E} = [\varepsilon_1, \varepsilon_2, \cdots, \varepsilon_N]^{\boldsymbol{T}}$ are the residues between the targets and outputs of the neural network model, and

$$\boldsymbol{\Phi} = [\boldsymbol{\varphi}_0, \boldsymbol{\varphi}_1, \boldsymbol{\varphi}_2, \cdots, \boldsymbol{\varphi}_p]$$

$$\boldsymbol{\varphi}_i = \begin{cases} [1,1,\cdots,1]^{\boldsymbol{T}}, i=0 \\ [\phi_i(\boldsymbol{u}_1), \phi_i(\boldsymbol{u}_2), \cdots, \phi_i(\boldsymbol{u}_N)]^{\boldsymbol{T}} \\ i=1,2,\cdots,p \end{cases}$$

$$\boldsymbol{\Theta} = [\theta_0, \theta_1, \cdots, \theta_q]^{\boldsymbol{T}}$$

The cost function can now be expressed as

$$\begin{aligned} &V(\boldsymbol{\Theta}; \boldsymbol{c}_1, \boldsymbol{c}_2, \cdots, \boldsymbol{c}_q) \\ &= \boldsymbol{E}^{\boldsymbol{T}} \boldsymbol{E} = (\boldsymbol{Y} - \boldsymbol{\Phi\Theta})^{\boldsymbol{T}} (\boldsymbol{Y} - \boldsymbol{\Phi\Theta}) \end{aligned} \tag{6}$$

In application-oriented neural model construction, the initial number of hidden nodes for each type of activation function can be relatively large. The optimal model after construction, however, will only contain a small number (say m) of hidden nodes with activation functions selected from $\{\varphi_i, i=1,2,\cdots,p\}$. In the construction process, let the coefficients $\theta_i, \forall i \in \{0,1,2,\cdots,p\}$ of the hidden nodes in $\{\varphi_i, i=1,2,\cdots,p\}$ that are not selected, be to zero. The cost function for the constructed models can then be simply represented by introducing a selection matrix $\boldsymbol{H}$:

$$\begin{aligned} &\boldsymbol{H} = diag(1, \delta_1, \delta_2, \cdots, \delta_p) \\ &\delta_i = 0 \text{ or } 1, \; i=1,2,...,p \end{aligned} \tag{7}$$

Applying $\boldsymbol{H}$ to (6) gives

$$\boldsymbol{Y} = \boldsymbol{\Phi H \Theta} + \boldsymbol{E} \tag{8}$$

and the cost function becomes

$$\begin{aligned} &V(\boldsymbol{H}; \boldsymbol{\Theta}; \boldsymbol{c}_1, \boldsymbol{c}_2, \cdots, \boldsymbol{c}_p) \\ &= (\boldsymbol{Y} - \boldsymbol{\Phi H \Theta})^{\boldsymbol{T}} (\boldsymbol{Y} - \boldsymbol{\Phi H \Theta}) \end{aligned} \tag{9}$$

Let $\boldsymbol{C} = \{\boldsymbol{c}_1, \boldsymbol{c}_2, \cdots, \boldsymbol{c}_p\}$, then the neural modelling then be expressed as:

$$\min_{\boldsymbol{H}, \boldsymbol{\Theta}, \boldsymbol{C}} V(\boldsymbol{H}; \boldsymbol{\Theta}; \boldsymbol{C}) \tag{10}$$

where $trace(\boldsymbol{H}) = m+1$, and if $(\boldsymbol{H})_{i,i} = 0$, then $\theta_i = 0$ and $\boldsymbol{c}_i = \boldsymbol{0}$. Neural model construction is therefore a minimisation problem with the objective function defined in (10). In (10), $\boldsymbol{H}$, $\boldsymbol{\Theta}$, $\boldsymbol{C}$ have to be optimised and it is a mixed integer non-linear optimisation problem. In this paper, genetic algorithms will be used as the optimisation tool.

Since GA (genetic algorithm) was first proposed by Holland in 1975, it has become well recognized as a powerful optimisation technique. The application of GAs to optimise neural structures is also not new and falls into two categories (Arifovic and Gencay, 2001; Blanco, et al 2001). One is to use GA as a means of learning artificial neural network connection weights that are coded, as binary or real numbers, in a genetic algorithm string (chromosome). The other is to use the genetic algorithm to evolve and select the artificial neural network architecture, together or independently from the evolution of weights. In the chromosome representation, an integer gene is used to represent the connections while the floating gene is used to represent weights and biases. However, for the application-oriented neural modelling proposed in this study, neural network construction is not applied to select the connections and weights, but rather the number of hidden nodes for each type of activation function, as well as the weights and bias. The GA for constructing the application-oriented neural model has therefore to be reformulated.

Position index

1	2		p	p+1	p+2		2p	2p+n₁	2p+n₂		2p+nₚ
δ_1	δ_2	...	δ_p	θ_1	θ_2	...	θ_p	$\mathbf{c}_1$	$\mathbf{c}_2$	...	$\mathbf{c}_p$

Integer Genes | Floating-point Genes

Fig. 1 Chromosome representation scheme

Position Index	1	2		p	p+1	p+2		2p	2p+n_1	2p+n_2		2p+n_p
Parent 1	$\delta_1(1)$	$\delta_2(1)$	...	$\delta_p(1)$	$\theta_1(1)$	$\theta_2(1)$	...	$\theta_p(1)$	$\mathbf{c}_1(1)$	$\mathbf{c}_2(1)$	...	$\mathbf{c}_p(1)$
Parent 2	$\delta_1(2)$	$\delta_2(2)$	...	$\delta_p(2)$	$\theta_1(2)$	$\theta_2(2)$	...	$\theta_p(2)$	$\mathbf{c}_1(2)$	$\mathbf{c}_2(2)$	...	$\mathbf{c}_p(2)$
Mask Code	1	0	...	1	1	1	...	0	0	1	...	1

Crossover ⇓ Operation

Child 1	$\delta_1(2)$	$\delta_2(1)$	...	$\delta_p(2)$	$\theta_1(1')$	$\theta_2(1')$	...	$\theta_p(1)$	$\mathbf{c}_1(1)$	$\mathbf{c}_2(1')$	...	$\mathbf{c}_p(1')$
Child 2	$\delta_1(1)$	$\delta_2(2)$	...	$\delta_p(1)$	$\theta_1(2')$	$\theta_2(2')$	...	$\theta_p(2)$	$\mathbf{c}_1(2)$	$\mathbf{c}_2(2')$	...	$\mathbf{c}_p(2')$

Fig. 2. The crossover operation

- *Fitness function* – Let *Vmax* = $\boldsymbol{Y}^T\boldsymbol{Y}$, the fitness function takes the following form:

$$fitness = V_{max} - V(\boldsymbol{H};\boldsymbol{\Theta};\boldsymbol{C}) \qquad (11)$$

- *Chromosome representation* – As the optimisation problem formulated in (10) is a typical mixed integer one, a mixed coding scheme is used. Fig. 1 illustrates this chromosome representation scheme, where $n_i = \sum_{k=1}^{i} n_{ck}, i = 1,2,\cdots,p$, and $n_{c1}, n_{c2},\ldots, n_{cp}$ are dimensions of $\mathbf{c}_1,\mathbf{c}_2,\ldots,\mathbf{c}_p$, respectively.
- *Selection Scheme* -The selection procedure in GAs is based on the "survival-of-the-fittest" mechanism. In this paper, the Roulette Wheel selection scheme is used.
- *Genetic Operations*- Due to the mixed gene-encoding scheme within the chromosomes, separated crossover operations on both the integer and floating-point parts of the two chromosomes (parents) are paralleled. For each operation, a mask code Mask with $2p+n_p$ bits is randomly generated, noting that the number of bits in the mask code equals the number of genes in a chromosome as shown in Fig. 2. According to Fig. 2, if a bit in the Mask is '1', exchange happens between the two genes in the corresponding position. For different genes, the exchange can be different:

$$gene^{(1')} = gene^{(2)}$$
$$gene^{(2')} = gene^{(1)} \qquad \text{(For integer genes)}$$

$$gene^{(1')} = \xi_c \cdot gene^{(1)} + (1-\xi_c)\cdot gene^{(2)}$$
$$gene^{(2')} = (1-\xi_c)\cdot gene^{(1)} + \xi_c \cdot gene^{(2)}$$
(For floating-point genes)

where, *gene*(1) and *gene*(2) refer to the two selected genes in parent 1 and parent 2; *gene*(1') and *gene*(2') are the two new genes in child 1 and child 2. Also, ξ_c is a random number, with a uniform distribution in [0,1].

The mutation operations also differ for integer and floating-point genes. Mutation of an integer gene is to change from 0 to 1 or from 1 to 0. Mutation of a floating-point gene simply replaces the value of the gene with a random number. After the genetic operations, a collection of offspring is generated with size *Sm*, the same size as the mating pool. The offspring are then used to replace part of the current population, producing a new generation.

- *Replacement Scheme*- To ensure the survival of the fittest and to prevent deterioration by either crossover or mutation or both, an elite preserving strategy is employed, whereby a few of the fittest chromosomes are duplicated directly in the next generation. The other chromosomes are replaced with the offspring.

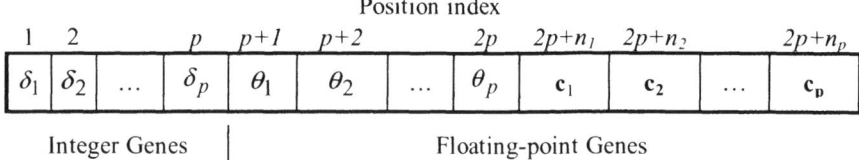

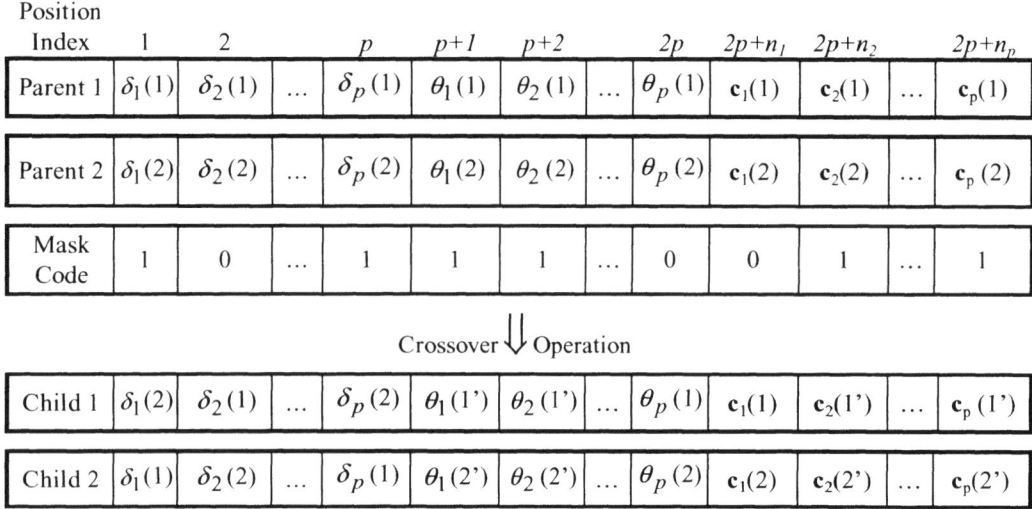

3. APPLICATION ORIENTED NEURAL MODEL FOR NO_X EMISSION IN A POWER PLANT

NOx emission in coal-fired power generation plants has been the subject of many researches over the last ten years due to its environmental impact. In order to achieve NOx reduction through the operational aspects of power station emission control, an essential step is to develop an appropriate NOx emission model for plant operation and control. Neural Networks have been used for NOx emission modelling and control (Li and Thompson, 2000, 2001, 2002), but generalisation capacity of models produced is quite restricted. In this paper the application-oriented neural model has been built. All data sets used in this study are real plant data.

In application oriented neural modelling, the first step is to extract fundamental functions from the NOx formation mechanisms. The ones that govern the NO formation is summarised as follows:

- Thermal NOx

$$\frac{d[NO]_T}{dt} = 2k_1[N_2][O_2] \times \frac{1-[NO]^2/k[O_2][N_2]}{1+k_{-1}[NO]/(k_2[O]+K_3[OH])} \quad (12)$$

where k_1, k_{-1} are the forward and backward reaction rates of N_2 and O_2 to NO and N. Also k_2 and k_{-2} are the forward and backward reaction rates of N and O_2 to NO and O, while k_3 is the forward reaction rate of N and OH to NO and H. Further, $k=(k_1/k_{-1})(k_2/k_{-2})$ is the equilibrium constant for the reaction between N_2 and O_2.

- Prompt NOx

$$\frac{d[NO]_p}{dt} = fT^{\beta} A_{pr}[O_2]^a[N_2][fuel]^b \exp(E_a/RT) \quad (13)$$

where f is a correction factor applicable for all aliphatic Alkane hydrocarbon fuels. T^{β} represents the non-Arrhenius behaviour of the equation at conditions where the maximum flame temperature is exceptionally high or low. A_{Pr} is the pre-exponential factor, a and b are reaction order constants for oxygen and fuel. And E_a is the activation energy.

- Fuel NOx

$$\frac{d[NO]_f}{dt}\Big|_{HCN\to NO} = 10^{10}\rho X_{CN} X_{O_2}^b \exp(-33700/T)$$
$$\frac{d[N_2]}{dt}\Big|_{NO\to N_2} = 3\times10^{12}\rho X_{CN} X_{NO}^b \exp(-30000/T) \quad (14)$$

where, X is the mole fractions of the chemical species, b is the reaction order for molecular oxygen (which is a function of oxygen concentration), and ρ is the density. The two reaction rates are included in the transport equations for HCN and NO and form the basis for the fuel NO post processor, which allows the calculation of NO formation for a pulverised coal flame.

Operation analysis shows the following operation variables are linked with the NOx formation, mass flow of primary air u_1 (Kg/s); mass flow of secondary air u_2 (Kg/s); mass flow of fuel u_3 (Kg/s); tilting position of burners firing port u_4 (degree). In addition, the past NOx outputs are also used as an input to the neural network model, u_5. Based on (12) to (14), two types of fundamental functions are identified.

$$\psi_1 = (\sum_{i=1}^{5}\omega_i u_i(t-k-d_i)+b_j)^{c_j} \quad (15)$$
$$j=1,2,\ldots,n_1,\ k=1,2,\ldots,n_d$$

$$\psi_2 = \exp(1/(\sum_{i=1}^{5}\omega_i u_i(t-k-d_i)+b_j), \quad (16)$$
$$j=1,2,\ldots,n_2,\ k=1,2,\ldots,n_d$$

where t is time, ω_i is the weight, b_j is the bias, c_j is the function parameter, n_1 and n_2 are the numbers of hidden nodes of the same type of activation function, n_d is the maximal order for variables, and d_i is the delay for input u_i.

A data set of 2300 samples is used for neural modelling. The GA evolves for 50 generations and the size of population pool is set to 60. The crossover probability is chosen to be 0.85, and the mutation probability is set to 0.02. Of 16 hidden nodes in the neural network, the GAs selected 8 hidden nodes for each type of the two activation functions. The resultant specialised neural network model has then been used to simulate the NOx emission in the power plant over another three different periods of operation time.

Finally, a conventional MLP model with one hidden layer of 16 sigmoid functions was built to model NOx emissions using the same training data, then tested over the same three periods of system operation. To improve the generalisation performance, early stopping technique is used in training the MLP, i.e. the training data set is split into two halves and training will stop as the validation error increases.

Table 1 compares the performance of these two neural network models using the index:

$$MP = \left(\sum_{i=1}^{N}(y(k)-\hat{y}(k))^2 \Big/ \sum_{i=1}^{N} y(k)^2\right)^{\frac{1}{2}} \times 100\% \quad (17)$$

According to table 1, the application-oriented neural network model is able to produce better generalization performance as expected. The prediction performance of the two network models over two plant operation periods is illustrated in Fig. 3 and 4.

4. CONCLUSION

A novel neural network, called the application-oriented neural model has been introduced, and a framework for constructing such a model of a non-linear dynamic system has been proposed. This has been applied to modelling and identification of the NOx emission of a coal-fired power generation plant. Comparison with a conventional MLP dynamic model, based on a sigmoid activation function, shows that the application oriented neural network model is able to give a better prediction performance over unseen data.

REFERENCES

Arifovic, J., R. Gencay (2001). Using genetic algorithms to select architecture of a feedforward artificial neural network. *Physica A*, **289**, 574-594

Barron, A.R. (1993). Universal Approximation bounds for superposition of a sigmoidal function, *IEEE Trans. Inform. Theory,* **39**, 930-945.

Blanco, A., M. Delgado and M. C. Pegalajar (2001). A real-coded genetic algorithm for training recurrent neural networks. *Neural Networks,* **14**, 93-105.

Girosi, F. and T. Poggio (1990). Neural networks and the best approximation property. *Biol. Cybernetics,* **63**: 169-176.

Irwin, G.W., K. Warwick, and K.J. Hunt (1995). Neural network applications in control. The Institute of Electrical Engineers, London.

Li, K., S. Thompson, G. R. Duan, J. Peng (2002). A case study of fundamental grey-box modelling, 15th IFAC World Congress on Automatic Control, Barcelona, July 2002.

Li, K., S. Thompson, (2001). Fundamental grey-box modelling. Proceedings of the European Control Conference 2001, Sept. 3 to 7, 2001, Oporto, Portugal, pp. 3648-3653.

Li, K., S. Thompson (2000). Developing NOx Emission Model for a Coal-fired Power Generation Plant Using Artificial Neural Networks. UKACC International Conference on CONTROL 2000, Cambridge, 4-7 Sept, 2000.

Liu, G., V. Kadirkamanathan, S. A. Billings (1998). On-line identification of nonlinear systems using Volterra polynomial basis function neural networks. *Neural Networks,* **11**, 1645–1657.

Park, J., I.W. Sandberg (1993). Approximation and radial-basis-function networks, *Neural Comput.* **5**, 305–316.

Rudolph, S (1997). On topology, size and generalization of nonlinear feed-forward neural networks, *Neurocomputing*, **16**, 1-22.

McLoone, S., Brown, M.D., Irwin, G.W. and Lightbody, G. (1998). A hybrid linear/nonlinear training algorithm for feedforward neural networks. *IEEE Transactions on Neural Networks*, **9**(4), 669-684.

Stinchcombe, M.B. (1999). Neural network approximation of continuous functionals and continuous functions on compactifications. *Neural Networks*, **12**, 467–477.

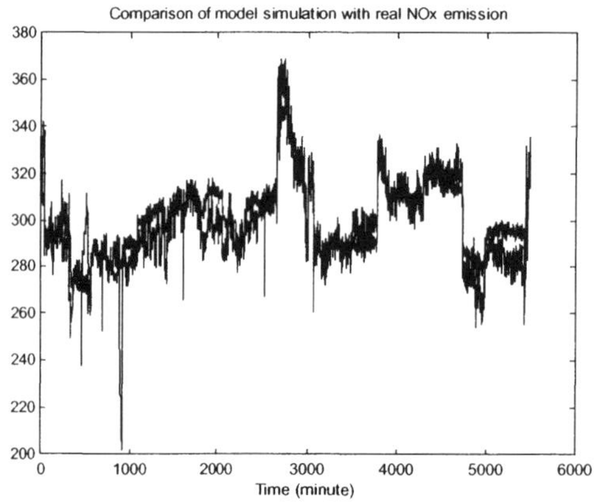

Fig. 3 Predictions of application-oriented neural model via real NOx emissions

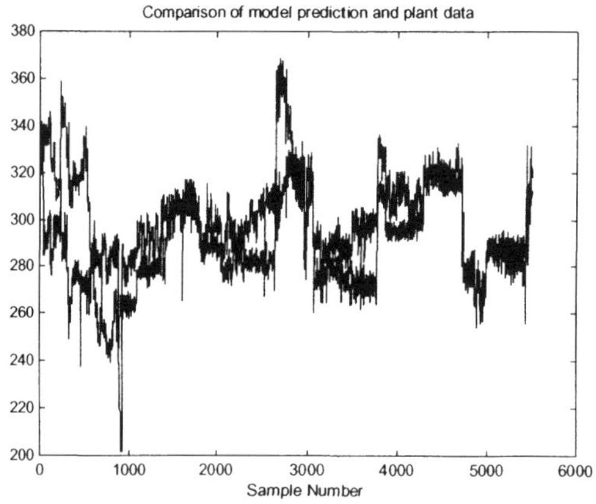

Fig. 4 Predictions of conventional MLP neural model via real NOx emissions

Table 1 Comparison of application oriented neural model with MLP model

	Training	Validation		
	Period 1 (2300 samples)	Period 2 (3000 samples)	Period 3 (2500 Samples)	Period 4 (1499 Samples)
Application oriented neural model	6.07%	4.93%	3.82%	8.24%
MLP model	6.44%	7.83%	6.19%	11.24%

IFAC
Publications
www.elsevier.com/locate/ifac

CLOSED-FORM FREQUENCY ESTIMATION USING SECOND-ORDER NOTCH FILTERS

S.M. Savaresi[†], S. Bittanti[†], H.C. So*

[†] *Dipartimento di Elettronica e Informazione, Politecnico di Milano, Piazza L. da Vinci, 32, 20133 Milano, ITALY.*
[*] *Department of Computer Eng. and Information Tech., City University of Hong Kong, Tat Chee Avenue, HONG KONG.*

Abstract. In this paper the problem of the frequency estimation of a sinusoid embedded in white noise is considered. The approach used herein is the minimization of the sample variance of the output of constrained notch filters fed by the noisy sinusoid. In particular, this paper focuses on closed-form expressions of the frequency estimate, which can be obtained using notch filters having an all-zeros FIR structure. In this paper it is shown that the FIR notch filters obtained from standard 2nd-order IIR filters are inadequate, and an alternate 2nd-order IIR notch filter is proposed, which provides an unbiased estimate of the frequency. The FIR filter obtained from the new IIR filter provides a closed-form unbiased frequency estimate. *Copyright © 2003 IFAC*

Keywords. Frequency estimation; harmonic analysis; notch filters; unbiased parameter identification.

1. INTRODUCTION

This paper deals with the estimation of the frequency of a harmonic signal $s(t) = A\cos(\Omega_0 t + \varphi)$, given its noisy measurement $y(t) = s(t) + n(t)$, $t = 1,2,...N$, where $n(t)$ is a zero-mean white Gaussian noise ($n \sim WGN(0,\sigma^2)$). This problem is frequently encountered in real-world applications, especially in the fields of adaptive control and signal processing, and numerous techniques have been developed for its treatment (see e.g. Bittanti and Picci, 1996, Bittanti and Savaresi, 2000, Hsu *et al.*, 1999, Kay, 1988, La Scala and Bitmead, 1996, Quinn and Fernandes, 1991, Renders *et al.*, 1984, Savaresi *et al.*, 2001, Schoukens *et al.*, 1992, Stoica, 1992). This paper focuses on the class of estimation methods based on constrained notch filters (see e.g. Händel and Nehorai, 1994, and references cited therein).

The basic idea underlying notch-filters-based estimation techniques is the minimization, with respect to Ω, of the loss function

$$J(\Omega) = \sum_{t=1}^{N} \varepsilon(t,\Omega)^2 , \tag{1}$$

where $\varepsilon(t,\Omega) = G(z^{-1},\Omega)y(t)$ is the output of a notch filter with transfer function $G(z^{-1},\Omega)$, fed by the measured signal $y(t)$. The notch of $G(z^{-1},\Omega)$ is centered around the frequency Ω.

In general, the dependence of $J(\Omega)$ on Ω is non-linear and non-convex; hence, iterative quasi-Newton minimization methods must be used. If the unknown frequency Ω_0 is time-varying, and the minimization of (1) is made recursively, the estimation algorithm usually is called *frequency tracker* (see e.g. Boashash, 1992).

Obviously, the most crucial design choice in a notch-based estimation technique is the selection of the structure and of the parameterization of the filter

$G(z^{-1},\Omega)$. Usually, 2nd-order IIR filter with a strongly constrained parameterization are used. Starting from 2nd-order filters, simple FIR filters or more sophisticated higher-order IIR filters have been developed and proposed (Händel *et al.*, 1998, Savaresi, 1997).

Two slightly different 2nd-order IIR notch filters are typically used in practice. They have the following expressions:

$$G_1(z^{-1},\Omega,\rho)=\frac{1-2\cos(\Omega)z^{-1}+z^{-2}}{1-2\rho\cos(\Omega)z^{-1}+\rho^2z^{-2}}, \quad (2)$$

$$G_2(z^{-1},\Omega,\rho)=\frac{1-2\cos(\Omega)z^{-1}+z^{-2}}{1-(1+\rho^2)\cos(\Omega)z^{-1}+\rho^2z^{-2}}. \quad (3)$$

In (2) and (3) the parameter ρ ($0\le\rho<1$) is known as the *de-biasing parameter* or the *poles-contraction factor* (note that ρ only affects the position of the poles). The difference of the poles position (when ρ varies) of (2) and (3) can be easily appreciated from Fig.1.

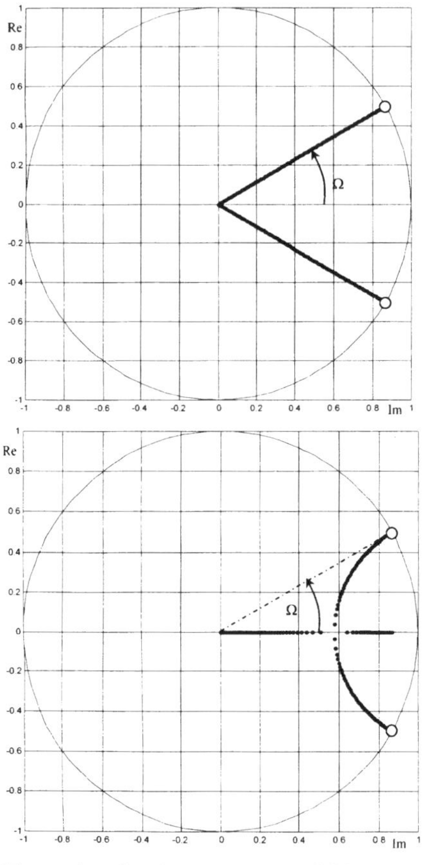

Fig.1. Example of pole placement of filters (2) (left) and (3) (right). The white bullets indicate the position of the zeros.

In the literature, filters of this type are also known as *constrained notch filter*, where the term *constrained* refers to the fact that their structure is strongly under-parameterized: the 5 parameters of a fully-parameterized 2nd-order digital IIR filter are reduced to one parameter only. As a matter of fact, since ρ is regarded as a design parameter, the only unknown parameter of (2) and (3) is the angular frequency Ω. The main difference between (2) and (3) is that (3) provides a *rigorously unbiased estimation of the frequency of a pure tone embedded in white noise*, whereas (2) provides a biased estimate. It is easy to see that such bias is negligible if $\rho\approx 1$; the problem of the bias becomes severe if $\rho<<1$. The properties of such filters have been discussed and analyzed in a large number of works (see e.g. Bittanti et al., 1997, and references cited therein).

The goal of this paper is to develop closed-form frequency estimators based on notch filters. The starting point of this work can be summarized in the following observations:

- closed-form expressions of the frequency estimator cannot be obtained if the notch filter has a IIR structure, due to the auto-regressive part of the filter;
- a constrained FIR notch filter can be obtained from $G_1(z^{-1},\Omega,\rho)$ by setting $\rho=0$; this is not possible using $G_2(z^{-1},\Omega,\rho)$ (note that $G_2(z^{-1},\Omega,0)$ is not a FIR);
- the closed-form frequency estimate obtained from $G_1(z^{-1},\Omega,0)$ is severely affected by a bias error.

Starting from these observations, the main results and original contributions of this paper are the following: a new 2nd-order IIR unbiasing constrained notch filter $G_3(z^{-1},\Omega,\rho)$ is developed and analyzed (section 2); it is shown, in Section 3, that a closed-form frequency estimate can be obtained using the FIR filters $G_1(z^{-1},\Omega,0)$ and $G_3(z^{-1},\Omega,0)$ (the major advantage of $G_3(z^{-1},\Omega,0)$ over $G_1(z^{-1},\Omega,0)$ is that it provides a rigorously unbiased estimate of Ω_0); it is shown that the closed-form frequency estimate provided by the FIR notch filter $G_3(z^{-1},\Omega,0)$, if the number N of data snapshots is large, tends to the frequency estimators provided by the "Pisarenko Harmonic Decomposition" approach, and by the "Youle-Walker" approach (Section 4).

2. A NEW UNBIASING NOTCH FILTER

As already remarked in the Introduction, one of the major drawbacks of the notch filter (2) (the most widely used in practice) is that it provides a biased estimation of Ω_0. This bias is particularly severe when $\rho<<1$. Starting from the cost function (1), a new unbiasing 2nd-order IIR notch filter can be obtained as follows.

Consider the long-run (asymptotic) version of the cost function (1), namely:

$$\bar{J}(\Omega)=\lim_{N\to\infty}\frac{1}{N}\left(\sum_{t=1}^{N}\varepsilon(t,\Omega)^2\right),$$

where $\varepsilon(t,\Omega)=G(z^{-1},\Omega)y(t)$.

It is easy to see that $\bar{J}(\Omega)$ can be given the following expression (see Bittanti *et al.*, 1997):

$$\bar{J}(\Omega) = \frac{1}{2\pi} \int_{-\pi}^{+\pi} \left| G(e^{j\omega}, \Omega) \right|^2 S_y(\omega) d\omega \,, \tag{4}$$

where $S_y(\omega)$ is the power spectrum of $y(t)$, which can be split into the power spectra of $s(t)$ and $n(t)$, namely: $S_y(\omega) = S_s(\omega) + S_n(\omega)$; $S_n(\omega) = \sigma^2$ and $S_s(\omega) = \frac{A^2}{2}\left(\frac{1}{2}\delta(\omega + \Omega_0) + \frac{1}{2}\delta(\omega - \Omega_0)\right)$.

Compute the asymptotic cost function $\bar{J}_1(\Omega)$ associated with the notch filter $G_1(z^{-1}, \Omega, \rho)$, by plugging in (4) the expression of the notch filter (2) and the expressions of $S_s(\omega)$ and $S_n(\omega)$:

$$\bar{J}_1(\Omega) = \bar{J}_1^{(s)}(\Omega) + \bar{J}_1^{(n)}(\Omega);$$

$$\bar{J}_1^{(n)}(\Omega) = \frac{1}{2\pi} \int_{-\pi}^{+\pi} \left| G_1(e^{j\omega}, \Omega, \rho) \right|^2 \sigma^2 d\omega \,,$$

$$\bar{J}_1^{(s)}(\Omega) = \frac{1}{2\pi}\left(\frac{A^2}{2} G_1(e^{j\omega}, \Omega_0, \rho)\right).$$

For the computation of $\bar{J}_1^{(n)}(\Omega)$ (the contribution to $\bar{J}_1(\Omega)$ due to the noise) we have resorted to the *Rugizka* algorithm (see Åström, 1970). The calculus of $\bar{J}_1^{(s)}(\Omega)$ calls for cumbersome but easier computations. The expressions obtained for $\bar{J}_1^{(s)}(\Omega)$ and $\bar{J}_1^{(n)}(\Omega)$ are:

$$\bar{J}_1^{(s)}(\Omega) = \frac{2A^2(\cos(\Omega) - \cos(\Omega_0))^2}{DEN}$$

$$\begin{aligned} DEN &= 1 + \rho^4 + 4\rho^2\cos^2(\Omega_0) + \\ &+ 4\rho^2\cos^2(\Omega) - 4\rho^3\cos(\Omega)\cos(\Omega_0) + \\ &- 4\rho\cos(\Omega)\cos(\Omega_0) - 2\rho^2 \end{aligned} \tag{5a}$$

$$\bar{J}_1^{(n)}(\Omega) = \frac{\sigma^2}{\pi} \frac{\rho^3 + \rho^2 - 6\rho\cos^2(\Omega) + \rho + 2\cos^2(\Omega) + 1}{(1+\rho)(\rho^2 + 2\rho\cos(\Omega) + 1)(\rho^2 - 2\rho\cos(\Omega) + 1)} \,. \tag{5b}$$

The bias in the frequency estimate obtained using $G_1(z^{-1}, \Omega, \rho)$ is due to the fact that $\bar{J}_1^{(n)}(\Omega)$ is a function of Ω. This dependence of $\bar{J}_1^{(n)}(\Omega)$ on Ω has the effect of moving the minimum of $\bar{J}_1(\Omega)$ away from Ω_0 (whereas Ω_0 is the minimum of $\bar{J}_1^{(s)}(\Omega)$). Now observe that the minimum of $\bar{J}_1^{(s)}(\Omega)$ does not change if $\bar{J}_1^{(s)}(\Omega)$ is multiplied by a strictly positive function of Ω and ρ, say $\eta(\Omega, \rho)$ (obviously for $\Omega \in [0, \pi)$ and $\rho \in [0, 1)$). This is due by the presence of the factor $(\cos(\Omega) - \cos(\Omega_0))$ in $\bar{J}_1^{(s)}(\Omega)$, which is null if $\Omega = \Omega_0$. Consider then the following function $\eta(\Omega, \rho)$:

$$\eta(\Omega, \rho) = \sqrt{\frac{(1+\rho)(\rho^2 + 2\rho\cos(\Omega) + 1)(\rho^2 - 2\rho\cos(\Omega) + 1)}{(1+\rho^2)(\rho^3 + \rho^2 - 6\rho\cos^2(\Omega) + \rho + 1 + 2\cos^2(\Omega))}} \tag{6}$$

Note that such function is the square-root of the inverse of $\bar{J}_1^{(n)}(\Omega)$ (but for the coefficient σ^2/π), multiplied by $(1+\rho^2)$).

A new unbiasing filter $G_3(z^{-1}, \Omega, \rho)$ can be obtained from $G_1(z^{-1}, \Omega, \rho)$ and (6) as follows:

$$\begin{aligned} G_3(z^{-1}, \Omega, \rho) &= \eta(\Omega, \rho) G_1(z^{-1}, \Omega, \rho) = \\ &= \eta(\Omega, \rho) \frac{1 - 2\cos(\Omega) z^{-1} + z^{-2}}{1 - 2\rho\cos(\Omega) z^{-1} + \rho^2 z^{-2}} \end{aligned} \tag{7}$$

Due to the fact that $G_3(z^{-1}, \Omega, \rho)$ is simply obtained by multiplying $G_1(z^{-1}, \Omega, \rho)$ by $\eta(\Omega, \rho)$, some remarks on the shape of $\eta(\Omega, \rho)$ are due (see Fig.2 where $\eta(\Omega, \rho)$ is plotted in the ranges $\Omega \in [0, \pi]$ and $\rho \in [0, 1]$).

- $\eta(\Omega, \rho)$ is not-null in the ranges $\Omega \in [0, \pi]$ and $\rho \in [0, 1]$; this can be easily seen from (6). This guarantees the well-posedness of the optimization problem based on the cost function (1).
- Note that $\eta(\Omega, 1) = 1$, whereas $\eta(\Omega, 0)$ strongly differs from 1; this is expected since $\eta(\Omega, \rho)$ is a sort of "de-biasing factor" of $G_1(z^{-1}, \Omega, \rho)$. Therefore $\eta(\Omega, \rho)$ leaves $G_1(z^{-1}, \Omega, \rho)$ almost unchanged if ρ is close to 1, whereas $\eta(\Omega, \rho)$ provides a strong correction to $G_1(z^{-1}, \Omega, \rho)$ for small values of ρ.
- Note that $\eta(\pi/2, \rho) = 1 \;\; \forall \rho \in [0, 1]$, and that $\eta(\Omega, \rho)$ is symmetric with respect to $\Omega = \pi/2$ in the range $\Omega \in [0, \pi]$ (Bittanti *et al.*, 1997).

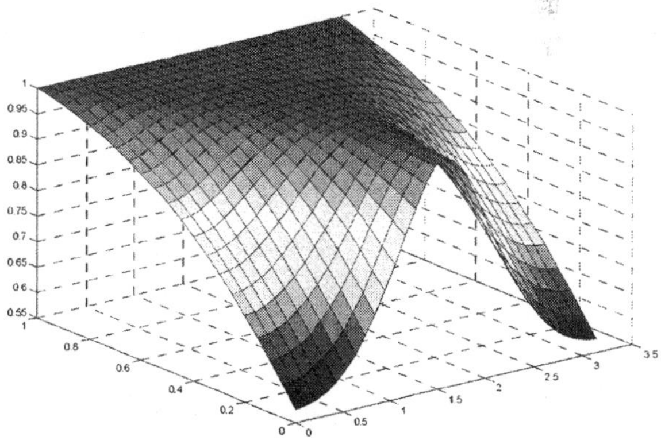

Fig.2. Shape of the function $\eta(\Omega, \rho)$ in the ranges $\Omega \in [0, \pi]$ and $\rho \in [0, 1]$.

In order to get a complete understanding of the differences between the three 2nd-order constrained notch filters $G_1(z^{-1}, \Omega, \rho)$, $G_2(z^{-1}, \Omega, \rho)$, and $G_3(z^{-1}, \Omega, \rho)$, it is interesting to compare the corresponding asymptotic cost functions $\bar{J}_1(\Omega)$, $\bar{J}_2(\Omega)$, and $\bar{J}_3(\Omega)$, respectively.

The closed-form expressions of $\bar{J}_1(\Omega)$ has already been computed in (5). Following the same procedure, $\bar{J}_2(\Omega)=\bar{J}_2^{(s)}(\Omega)+\bar{J}_2^{(n)}(\Omega)$ and $\bar{J}_3(\Omega)=\bar{J}_3^{(s)}(\Omega)+\bar{J}_3^{(n)}(\Omega)$ can be obtained as:

$$\bar{J}_2^{(s)}(\Omega)=\frac{2A^2(\cos(\Omega)-\cos(\Omega_0))^2}{DEN}$$

$$\begin{aligned}DEN=&1+\rho^4\cos^2(\Omega)-2\rho^4\cos(\Omega)\cos(\Omega_0)-2\rho^2+\\&+\rho^4+\cos^2(\Omega)+2\rho^2\cos^2(\Omega)-4\rho^2\cos(\Omega)\cos(\Omega_0)+\\&-2\cos(\Omega)\cos(\Omega_0)+4\rho^2\cos^2(\Omega_0)\end{aligned} \qquad (8a)$$

$$\bar{J}_2^{(n)}(\Omega)=\frac{\sigma^2}{(1+\rho^2)\pi} \qquad (8b)$$

$$\bar{J}_3^{(s)}(\Omega)=\frac{2A^2(\cos(\Omega)-\cos(\Omega_0))^2\eta^2(\Omega,\rho)}{DEN}$$

$$\begin{aligned}DEN=&1+\rho^4+4\rho^2\cos^2(\Omega_0)+4\rho^2\cos^2(\Omega)+\\&-4\rho^3\cos(\Omega)\cos(\Omega_0)-4\rho\cos(\Omega)\cos(\Omega_0)-2\rho^2\end{aligned} \qquad (9a)$$

$$\bar{J}_3^{(n)}(\Omega)=\frac{\sigma^2}{(1+\rho^2)\pi}. \qquad (9b)$$

By comparing $\bar{J}_1(\Omega)$, $\bar{J}_2(\Omega)$, and $\bar{J}_3(\Omega)$ it is apparent that $\bar{J}_2(\Omega)$ and $\bar{J}_3(\Omega)$ have the minimum exactly at Ω_0 (unbiased estimate), since $\bar{J}_2^{(n)}(\Omega)$ and $\bar{J}_3^{(n)}(\Omega)$ do not depend on Ω. On the contrary, as already observed, $\bar{J}_1(\Omega)$ provides a biased estimate since $\bar{J}_1^{(n)}(\Omega)$ is Ω-dependent.

To conclude this section, it is worth remarking that the new filter $G_3(z^{-1},\Omega,\rho)$ merges the two main appealing features of $G_1(z^{-1},\Omega,\rho)$ and $G_2(z^{-1},\Omega,\rho)$:

- similarly to $G_1(z^{-1},\Omega,\rho)$, a FIR filter can be obtained from $G_3(z^{-1},\Omega,\rho)$ by using $\rho=0$;
- similarly to $G_2(z^{-1},\Omega,\rho)$, $G_3(z^{-1},\Omega,\rho)$ provides an unbiased estimate of Ω_0 $\forall\rho\in[0,1)$.

These features will be fully exploited in the following section, in order to obtain closed-form frequency estimates based on FIR notch filters.

3. CLOSED-FORM FREQUENCY ESTIMATION

A closed-form notch-based frequency estimate cannot be obtained if the filter has a IIR structure. Consider the FIR filters obtained by simply setting $\rho=0$ in (2) and in (7) (it has been already observed that setting $\rho=0$ in $G_2(z^{-1},\Omega,\rho)$ does not yield a FIR filter), namely:

$$G_1(z^{-1},\Omega,0)=1-2\cos(\Omega)z^{-1}+z^{-2},$$

$$G_3(z^{-1},\Omega,0)=\sqrt{\frac{1}{2\cos^2(\Omega)+1}}(1-2\cos(\Omega)z^{-1}+z^{-2}).$$

Using such filters, closed-form frequency estimators from the data can be obtained as follows.

Closed form frequency estimator obtained using $G_1(z^{-1},\Omega,0)$.

Consider the following cost function, obtained by plugging in (1) the FIR notch filter $G_1(z^{-1},\Omega,0)$:

$$J_1(\Omega)=\sum_{t=1}^{N}\left(y(t)-2\cos(\Omega)y(t-1)+y(t-2)\right)^2,$$

and differentiate $J_1(\Omega)$ with respect to Ω:

$$\frac{dJ_1(\Omega)}{d\Omega}=2\sum_{t=1}^{N}\begin{bmatrix}\left(y(t)-2\cos(\Omega)y(t-1)+y(t-2)\right)\cdot\\\left(2\sin(\Omega)y(t-1)\right)\end{bmatrix}. \qquad (10)$$

Note that (10) is quadratic with respect to $\cos(\Omega)$; hence by solving $dJ_1(\Omega)/d\Omega=0$ with respect to $\cos(\Omega)$, it is easy to see that the following holds:

$$\cos(\Omega)=\frac{\sum_{t=1}^{N}y(t-1)(y(t)+y(t-2))}{\sum_{t=1}^{N}y(t-1)^2}.$$

The closed-form frequency estimator therefore is given by:

$$\hat{\Omega}_1=\arccos\left(\frac{\sum_{t=1}^{N}y(t-1)(y(t)+y(t-2))}{\sum_{t=1}^{N}y(t-1)^2}\right).\blacksquare \qquad (11)$$

As the number N of data grows, $\hat{\Omega}_1$ tends to the minimum of the asymptotic cost function (5) (in the special case of $\rho=0$). After some cumbersome computation, the following asymptotic expression of (11) is obtained:

$$\hat{\Omega}_1 \underset{N\to\infty}{=\to} \arccos\left(\frac{\pi A^2}{\pi A^2+\sigma^2}\cos(\Omega_0)\right). \qquad (12)$$

From (12), it is apparent that the frequency estimate is affected by a severe bias; note that the bias is null in the (trivial and unrealistic) case of zero noise ($\sigma^2=0$); it grows as the SNR decreases.

Closed form frequency estimator obtained using $G_3(z^{-1},\Omega,0)$.

Consider the following cost function, obtained by plugging in (1) the FIR notch filter $G_3(z^{-1},\Omega,0)$:

$$J_3(\Omega)=\sum_{t=1}^{N}\left(\frac{\left(y(t)-2\cos(\Omega)y(t-1)+y(t-2)\right)^2}{2\cos^2(\Omega)+1}\right),$$

and differentiate $J_3(\Omega)$ with respect to Ω:

$$\frac{dJ_3(\Omega)}{d\Omega}=\sum_{t=1}^{N}\left(\frac{NUM}{(2+\cos(2\Omega))^2}\right)$$

$$\begin{aligned}NUM=&2\left(y(t)-2\cos(\Omega)y(t-1)+y(t-2)\right)\cdot\\&\cdot\left(\begin{aligned}&(2+\cos(2\Omega))2\sin(\Omega)y(t-1)+\\&+\sin(2\Omega)\left(y(t)-2\cos(\Omega)y(t-1)+y(t-2)\right)\end{aligned}\right)\end{aligned} \qquad (13)$$

Consider now the problem of solving $dJ_3(\Omega)/d\Omega=0$ with respect to Ω. After some manipulation the following expression is obtained:

$$\sum_{t=1}^{N}\begin{pmatrix}\sin(\Omega)\left(y(t)-2\cos(\Omega)y(t-1)+y(t-2)\right)\cdot\\ \cdot\left(y(t-1)+\cos(\Omega)y(t)+\cos(\Omega)y(t-2)\right)\end{pmatrix}=0\text{.(14)}$$

Equation (14) admits a trivial solution: $\sin(\Omega)=0$. Assuming that $\Omega_0\neq\{0,\pi\}$, the following quadratic form (with respect to $\cos(\Omega)$) can be obtained from (14):

$$\begin{aligned}&2\left[\sum_{t=1}^{N}y(t-1)\left(y(t)+y(t-2)\right)\right]\cos^2(\Omega)+\\&+\left[\sum_{t=1}^{N}\left(2y(t-1)^2-\left(y(t)+y(t-2)\right)^2\right)\right]\cos(\Omega)+\\&-\left[\sum_{t=1}^{N}y(t-1)\left(y(t)+y(t-2)\right)\right]=0\end{aligned}\qquad(15)$$

From (15), a closed-form frequency estimator can be computed. It has the following expression:

$$\hat{\Omega}_3=\arccos\left(\frac{NUM}{4\left[\sum_{t=1}^{N}y(t-1)\left(y(t)+y(t-2)\right)\right]}\right).\blacksquare$$

$$\begin{aligned}NUM=&-\left[\sum_{t=1}^{N}\left(2y(t-1)^2-\left(y(t)+y(t-2)\right)^2\right)\right]+\\&+\sqrt{\left[\sum_{t=1}^{N}\left(2y(t-1)^2-\left(y(t)+y(t-2)\right)^2\right)\right]^2+8\left[\sum_{t=1}^{N}y(t-1)\left(y(t)+y(t-2)\right)\right]^2}\end{aligned}\qquad(16)$$

As the number N of data grows, $\hat{\Omega}_3$ tends to the minimum of the asymptotic cost function (8) (in the special case of $\rho=0$):

$$\hat{\Omega}_3\underset{N\to\infty}{\to}\Omega_0.$$

Thus, the new filter $G_3(z^{-1},\Omega,0)$ provides a simple closed-form unbiased estimate of Ω_0. Interestingly, (16) is closely related to the method given in So, 2002 and in So and Ip, 2002, even if the derivation of this result is completely different.

4. RELATED METHODS

In the literature, other closed-form frequency estimators for harmonic signals in white noise have been proposed and analyzed. Two celebrated estimators are the "Youle-Walker" estimator, and the "Pisarenko Harmonic Decomposition" (PHD) estimator (see e.g. Pisarenko, 1973, Xiao and Takodoro, 1994 and 1995). In this section they will be briefly recalled and compared with the asymptotic version of the notch-based estimator (16).

Youle-Walker approach.

Given $y(t)=s(t)+n(t)$, $s(t)=A\cos(\Omega_0 t+\varphi)$, $n\sim WGN(0,\sigma^2)$, the autocorrelation coefficients of order 1 and 2, say r_1 and r_2 respectively, are given by:

$$\begin{cases}r_1=E\left[y(t)y(t-1)\right]=\dfrac{A^2}{2}\cos(\Omega_0)\\ r_2=E\left[y(t)y(t-2)\right]=\dfrac{A^2}{2}\cos(2\Omega_0)\end{cases}\qquad(17)$$

By eliminating the parameter A in (17), the following equation is obtained:

$2r_1\cos^2(\Omega_0)-r_2\cos(\Omega_0)-r_1=0$.

Its solution with respect to Ω_0 provides the Youle-Walker frequency estimator, given by:

$$\hat{\Omega}_{YW}=\arccos\left(\frac{r_2+\sqrt{r_2^2+8r_1^2}}{4r_1}\right).\blacksquare\qquad(18)$$

PHD approach.

Given a zero-mean stationary signal $y(t)$, its autocorrelation matrix of order 3 is given by:

$$R=\begin{bmatrix}r_0&r_1&r_2\\r_1&r_0&r_1\\r_2&r_1&r_0\end{bmatrix},$$

$r_0=E\left[y(t)^2\right], r_1=E\left[y(t)y(t-1)\right], r_2=E\left[y(t)y(t-2)\right]$.

The eigenvector associated with the smallest eigenvalue of R has the following form:

$$\begin{bmatrix}1&-\dfrac{r_2+\sqrt{r_2^2+8r_1^2}}{2r_1}&1\end{bmatrix}^T.\qquad(19)$$

Pisarenko (Pisarenko, 1973) has proven that, if $y(t)=s(t)+n(t)$, $s(t)=A\cos(\Omega_0 t+\varphi)$, $n\sim WGN(0,\sigma^2)$, the smallest eigenvalue of R must have the following simple expression:

$$\begin{bmatrix}1&-2\cos(\Omega_0)&1\end{bmatrix}^T.\qquad(20)$$

By comparing (19) and (20), the PHD frequency estimator is obtained:

$$\hat{\Omega}_{PHD}=\arccos\left(\frac{r_2+\sqrt{r_2^2+8r_1^2}}{4r_1}\right).\blacksquare\qquad(21)$$

Interestingly enough, the Youle-Walker and PHD approaches provide exactly the same results. This has been recently proven and discussed in Xiao and Takodoro, 1994 and 1995.

Consider now the notch-based closed-form estimator (16). The following result holds.

Proposition 1.

Given a signal $y(t)=s(t)+n(t)$, where $s(t)=A\cos(\Omega_0 t+\varphi)$, $n\sim WGN(0,\sigma^2)$, the notch-filter based estimator (16) asymptotically converges towards $\hat{\Omega}_{YW}$ and $\hat{\Omega}_{PHD}$, namely:

$$\lim_{N\to\infty}\hat{\Omega}_3=\arccos\left(\frac{r_2+\sqrt{r_2^2+8r_1^2}}{4r_1}\right),$$

$r_1 = E[y(t)y(t-1)],\quad r_2 = E[y(t)y(t-2)]$.

Proof.

If N is large, the following hold:

$$\lim_{N\to\infty}\left[\frac{1}{N}\sum_{t=1}^{N} y(t-1)\left(y(t)+y(t-2)\right)\right]= \\ = \lim_{N\to\infty}\left[\frac{1}{N}\sum_{t=1}^{N} 2y(t)y(t-1)\right] = 2E[y(t)y(t-1)] \quad (22a)$$

$$\lim_{N\to\infty}\left[\frac{1}{N}\sum_{t=1}^{N}\left(2y(t-1)^2-\left(y(t)+y(t-2)\right)^2\right)\right]= \\ = -2E[y(t)y(t-2)] \quad (22b)$$

By plugging in (16) the asymptotic expressions (22), it is easy to see that:

$$\lim_{N\to\infty}\hat{\Omega}_3 = \arccos\left(\frac{r_2+\sqrt{r_2^2+8r_1^2}}{4r_1}\right),$$

$r_1 = E[y(t)y(t-1)],\quad r_2 = E[y(t)y(t-2)]$. ■

From a theoretical point of view the fact that (asymptotically) $\hat{\Omega}_{YW}$, $\hat{\Omega}_{PHD}$ and $\hat{\Omega}_3$ are exactly the same, is particularly interesting: it shows the equivalence of three classical approaches which have been independently conceived and developed following three completely different paths.

ACKNOWLEDGMENTS

This work has been supported by MIUR project "*New Methods for Identification and Adaptive Control for Industrial Systems*", and by the EU project "*Nonlinear and Adaptive Control*".

REFERENCES

Åström, K.J. (1970). Introduction to stochastic control theory. Academic Press.

Bittanti S., Campi M., Savaresi S.M. (1997). Unbiased estimation of a sinusoid in noise via Adapted Notch Filters. Automatica, vol.33, n.2, pp.209-215.

Bittanti, S., G. Picci (Eds.) (1996). Identification, Adaptation, Learning - The Science of Learning Models from Data. Computer and Systems Sciences Series, Springer-Verlag, Berlin.

Bittanti S., Savaresi S.M. (2000). On the parametrization and design of an Extended Kalman Filter Frequency Tracker. IEEE Transactions on Automatic Control, vol.45, n.9, pp.1718-1724.

Boashash B. (1992). Estimating and Interpreting the Instantaneous Frequency of a Signal. Proceedings of the IEEE, vol.80, n.4, pp.520-568.

Händel P., and Nehorai A. (1994). Tracking Analysis of an Adaptive Notch Filter with Constrained Poles and Zeros. IEEE Transactions on Signal Processing, vol.42, n.2, pp.281-291.

Händel P., Tichavsky P., Savaresi S.M. (1998). Large error recovery for a class of frequency tracking algorithms. International Journal on Adaptive Control and Signal Processing, Vol.12, pp.417-436.

Hsu L., Ortega R., Damm G. (1999). A Globally Convergent Frequency Estimator. IEEE Transactions on Automatic Control, Vol.44, No.4, pp.698-713.

Kay S.M. (1988). Modern Spectral Estimation: Theory and Applications. Prentice-Hall.

La Scala B., Bitmead R. (1996). Design of an Extended Kalman Filter Frequency Tracker. IEEE Transactions on Signal Processing, Vol.44, n.3, pp.739-742.

Quinn B.G, Fernandes J.M. (1991). A Fast Efficient Technique for the Estimation of Frequency. Biometrika, Vol.78, pp.489-497.

Pisarenko V.F. (1973). The retrieval of harmonics from a covariance function. J.Roy.Astr., Vol.33, pp.374-376.

Renders H., Schoukens J., Vilain G. (1984). High-Accuracy Spectrum Analysis of Sampled Discrete Frequency Signals by Analytical Leakage Compensation. IEEE Trans. on Instrum. and Meas., vol.33, n.4, pp.287-292.

Savaresi S.M. (1997). Funnel Filters: a new class of filters for frequency estimation of harmonic signals. Automatica, Vol.33, n.9, pp.1711-1718.

Savaresi S.M., R. Bitmead, W. Dunstan (2001). Nonlinear system identification using closed-loop data with no external excitation: the case of a lean combustion process. International Journal of Control, vol.74, n.18, pp.1796-1806.

Schoukens J., Pintelon R., Van hamme H. (1992). The Interpolated Fast Fourier Transform : A Comparative Study. IEEE Trans. on Instrum. and Meas., vol.41, n.2, pp.226–232.

So H.C. (2002). "A closed form frequency estimator for a noisy sinusoid". Proceedings of 45th IEEE Midwest Symposium on Circuits and Systems, Tulsa, Oklahoma, USA.

So H.C., Ip S.K. (2002). A novel frequency estimator and its comparative performances for short record lengths. Proceedings of 11th European Signal Processing Conference, Toulouse, France.

Stoica P. (1992). List of references on spectral line analysis. Signal Processing, vol.31, pp.329-340.

Xiao Y., Tadokoro Y. (1994). On Pisarenko and constrained Yule-Walker estimators of tone frequency. IEICE Transactions, vol.E77-A, n.8, pp. 1404-1406.

Xiao Y., Tadokoro Y. (1995). Statistical analysis of a simple constrained high-order Yule-Walker tone frequency estimator. IEICE Transactions, vol.E78-A, n.10, pp. 1415-1418.

IFAC Publications
www.elsevier.com/locate/ifac

L_1 PREDICTION ERROR SYSTEM IDENTIFICATION: A MODIFIED AIC RULE

J. C. Carmona, M. Ouladsine and M. El Adel *

* *LSIS CNRS UMR 6168, Marseille, France*

Abstract: In this paper we shall present our recent work on model estimation and model validation. An attractive expression of the AIC rule in the framework of L_1 prediction error is presented. This result based on an original formulation of the FPE criterion in the L_1 context, tends to improve either the task of model structure choice or the validation of the estimated models. Owing to the fact that the analytical properties of the LSAD criterion (least sum of absolute deviations) it was not evident at all that results comparable to the classical L_2 expressions would be found. *Copyright © 2003 IFAC*

Keywords: Identification. Model validation. L_1, L_2 estimation.

1. INTRODUCTION

Model identification and model validation for robust controller design still remains a delicate task (Ljung, 1999), (Smith and Doyle, 1992). In particular the system identification must deliver not only a relevant nominal model but also a reliable estimate of the uncertainty associated. To describe uncertainty classically two main approaches are used. The first one is developed in the statistical framework such as stochastic embedding (G.C. Goodwin and Seron, 1999) and model error modelling (W. Reinelt W. and Ljung, 2002)while the second one relies on deterministic assumptions such as the identification error is "unknown but bounded" like set membership identification techniques(see (A. Garulli and Zappa, 2000)). In fact in the model selection step, the model errors, also addressed as "bias" error, are balanced by the noise influence, also addressed as "variance" term. It is well-known that underestimated number of model parameters leads to a model not sufficiently "flexible" to describe the data, while an overestimation is used to "follow" the noise. A good solution for balancing both errors is to evaluate the model candidate on "fresh" data (validation data). In the case rather frequent where these data are not available the previous validation step is replaced by a prediction of its behavior. Cost function extensions are proposed to evaluate the cost function if validation data would be applied to the model. In the community AIC ((Akaike, 1974)) and MDL ((Rissanen, 1978)) rules are frequently encountered. A complexity term is introduced in the cost function in order to balance the quality of the fit versus the "complexity" of the model estimated. Even though recent relevant works ((J. Schoukens and Pintelon, 2002)) have been developed based on a non-parametric estimation of the noise model from "raw" data prior to the identification of the plant model, classically the prediction error framework is used, based on a parametric noise model estimated simultaneously with the plant where the residuals of the fit are used to scale the noise model. The *model error model* techniques illustrate this approach (see (Ljung, 1999)). Most of the time least squares estimation techniques are used in this prediction error framework. One of the main reason is based on the fact that a pseudo-gaussian noise favors least squares identification, despite its lack of robustness again the so-called "tail problem" (see (Rice and White, 1964)) and its statistical sensitivity to the data. Alternative solutions based on LSAD (least sum of absolute deviation) techniques leading to LP (linear programming) minimization problems with or without constraints are cleverly proposed by T.K. Gustafsson and al. (see (Gustafsson and Mäkilä, 1996)). The authors analyzed in particular a family of robustly convergent algorithms based on a "smoothed" LSAD criterion which seems to be more efficient than the classical "least squares" one in the case of a noise with Laplacian distribution. Despite these relevant results, L_1 identification approach strongly suffer from the lack of decision tools either in the model identification stage or in the validation's one: on the one hand rules like Akaike's FPE (final prediction error) criterion, and AIC (Akaike) or MDL (Rissanen) rules on the other end. In a previous paper ((Carmona and Alvarado, 2002)) we have presented a L_1 FPE criterion in order to facilitate the model order choice in the L_1 identification step. Our purpose in this paper is to propose an AIC rule version for L_1 model validation step in order to fill in the set of identification tolls in L_1 framework. This paper is organized as follows: in section 2we present some preliminary results like the convergence properties and the main asymptotic features of the L_1 estimate in the prediction error framework leading to the L_1 FPE criterion formulation. The main result is presented in section 3: the AIC rule in the case of Laplacian innovations and the complexity term associated. Then some conclusions and perspectives will be given.

2. PRELIMINARY RESULTS

2.1 The Prediction Error Framework

Let the system $\mathcal{S}$ be described as follows

$$y(t) = G(q)u(t) + v(t) = G(q)u(t) + H(q)e(t) \quad (1)$$

with $G(q) = B(q)/A(q)$ the plant model and $H(q) = C(q)/D(q)$ the monic noise model where $v(t)$ represents both the model error and the noise influence, and $e(t)$ are identical independent distributed (iid) sequences of random variables with means zero and variances λ. Let us then consider the *parameterized model structure* $\mathcal{M}$

$$y(t,\theta) = G(q,\theta)u(t) + H(q,\theta)e(t) \quad (2)$$

The (one step ahead) predictor associated

$$\widehat{y}(t,\theta) = H^{-1}(q,\theta)G(q,\theta)u(t) + [1 - H^{-1}(q,\theta)]y(t) \quad (3)$$

leads to the prediction errors

$$\begin{aligned} \varepsilon(t,\theta) &= y(t) - \widehat{y}(t,\theta) \\ &= H^{-1}(q,\theta)[y(t) - G(q,\theta)u(t)] \end{aligned} \quad (4)$$

Even though this assumption seems to be somehow unrealistic, it can be however convenient to suppose that the data Z^N are exactly generated by a "true system" $\mathcal{S}_0 \equiv \{G_0(q,\theta_0), H_0(q,\theta_0)\}$ belonging to the model set $\mathcal{M}$, so as:

$$\begin{aligned} y(t) &= G_0(q,\theta_0)u(t) + v_0(t) \\ &= G_0(q,\theta_0)u(t) + H_0(q,\theta_0)e_0(t) \end{aligned} \quad (5)$$

where the "true noise" $e_0(t)$ is a sequence of random iid variables with zero mean values and finite variances λ_0.

In the prediction error framework the parameter estimate $\widehat{\theta}_N \in \mathcal{D}_{\mathcal{M}} \subset R^d$ on a batch of data $Z^N = \{u(0), y(0), \cdots u(N), y(N)\}$ minimizes an estimation criterion or cost function of the model parameter θ

$$\widehat{\theta}_N = arg \min_{\theta \in \mathcal{D}_{\mathcal{M}}} V_N(\theta, Z^N) \quad (6)$$

where generally V_N is a well-defined scalar valued function based on a "norm" or positive scalar valued function $\ell(\varepsilon)$ such that

$$V_N(\theta, Z^N) = \frac{1}{N}\sum_{t=1}^{N} \ell(\varepsilon(t,\theta),\theta,t) \quad (7)$$

Classically, but without any limitation, one uses:

$$\ell(x) = \tfrac{1}{2}x^2 \quad , \text{ in the } L_2 \text{ framework,}$$
$$\ell(x) = |x| \quad , \text{ in the } L_1 \text{ framework.}$$

More precisely

$$V_N(\theta, Z^N) = \frac{1}{N}\sum_{t=1}^{N} \varepsilon(\theta,t)^2 \text{ or } \frac{1}{N}\sum_{t=1}^{N} |\varepsilon(\theta,t)| \quad (8)$$

Remark The function $|x|$ does not grow as fast as $|x|^2$ as $|x| \nearrow \infty$. This is what makes the L_1 regression less sensitive to large outlying residuals. It is important to emphasize that the lack of differentiability at the origin is not a serious hindrance. Indeed, the absolute value function can be smoothed in a small neighborhood of the origin. This practice is standard in statistics where robust regression/estimation is done by minimizing an objective function of the form:

$$\sum_{t=1}^{N} \psi(\varepsilon(t)) \quad (9)$$

where $\varepsilon(t)$ denotes the t-th residual, and where ψ is a convex function.

As we already pointed out, choosing the function $\psi(x) = x^2$ gives the ordinary least squares L_2 method, while the function $\psi(x) = |x|$ gives the least absolute deviations L_1 method. Choosing a function of the form:

$$\psi_\eta(x) = \begin{cases} x^2/(2\eta) & \text{if } |x| \le \eta \\ |x| - \eta/2 & \text{otherwise} \end{cases}$$

for some $\eta > 0$ will guarantee the same behaviour as the L_1 loss function for large residuals ($|\varepsilon(t)| > \eta$) and will guarantee the same behaviour as the L_2-loss function for small residuals ($|\varepsilon(t)| \le \eta$.) Moreover the function ψ_ε is continuously differentiable and it has a second derivative everywhere except at $x = \pm\eta$, and it converges toward the function $|x|$ when $\eta \searrow 0$. This remark implies that the mathematical proofs requiring continuous differentiability of the loss function and existence of the second derivative almost everywhere, can still be used with the L_1 loss function. The interested reader is referred for example to (Huber, 1981) and (Wilcox, 1997) for details and complements.

This idea has been already suggested in the identification literature (see for example (H. Akçay and Ljung, 1996) and more extensively (Poljak and Tsypkin, 1980)).

2.2 *Final Prediction Error criteria*

In the case of only estimation data are available, Akaike showed how to modify the cost (lost) function V_N to get a reasonable estimate of the model validation: the FPE criterion. In the L_2 approach, we have

$$\mathcal{J}_{L2}(\mathcal{M}) \approx \frac{1}{N}\sum_{t=1}^{N}\varepsilon(\theta,t)^2 + \lambda_0\frac{2d_{\mathcal{M}}}{N} \quad (10)$$

or

$$\mathcal{J}_{L2} \approx \frac{1+\frac{d}{N}}{1-\frac{d}{N}}\frac{1}{N}\sum_{t=1}^{N}\varepsilon^2(t,\widehat{\theta}_N) \quad (11)$$

In a previous paper (see (Carmona and Alvarado, 2002)) we proposed for the L_1 case the following L_1FPE criterion:

$$\mathcal{J}_{L1} \approx \frac{1}{N}\sum_{t=1}^{N}|\varepsilon(t,\widehat{\theta}_N)| + \frac{d_{\mathcal{M}}}{N} \quad (12)$$

The reader will certainly appreciate the fact that the criterium no longer depends on the noise variance λ_0 in the L_1 case.

3. AIC RULE IN THE L_1 CASE

The FPE criterion gives interesting results in the case the choice of the model structure within which the estimation is made $\mathcal{M}$ is not too delicate. But in certain applications, for example in propagative phenomenon, one can have some difficulties to choose the appropriate model structure and the compute the best candidate. In this case one can appreciate to have at his disposal a criterion allowing the estimation for a given data set Z^N in the same time of the best model structure $\hat{\mathcal{M}}$ and the best candidate within this structure $\hat{\theta}_N^{\mathcal{M}}$.

In the L_2 case, the Akaike ((Akaike, 1974)) proposed an AIC rule, and Rissanen ((Rissanen, 1978)) proposed an MDL rule given by, for $N \gg dim\theta$

$$W_N^0(\theta,\mathcal{M},Z^N) = V_N(\theta,Z^N)(1+U_N(\mathcal{M})) \quad (13)$$

with

$$U_N(\mathcal{M})) = 2\frac{dim\theta}{N} \quad (14)$$

for AIC, and:

$$U_N(\mathcal{M})) = dim\theta\frac{logN}{N} \quad (15)$$

for MDL, where $\mathcal{M}$ is a certain model structure depending on the parameters θ.

More precisely, Ljung (see (Ljung, 1999)) gives p. 506 the following expression of the AIC criterion :

$$\{\widehat{\theta}_N^{\mathcal{M}},\widehat{\mathcal{M}}\} \equiv arg\ \min_{\mathcal{M}\in M}\ \min_{\theta^{\mathcal{M}}\in\mathcal{D}_{\mathcal{M}}} \frac{1}{N}\left[-L_N(\theta^{\mathcal{M}},Z^N)+d_{\mathcal{M}}\right] \quad (16)$$

where $L_N(.)$ is the log-likelihood function, i.e. the log function of the residual PDF:

$$L_N(\theta,Z^N) = logf_e(\varepsilon(t,\theta),t,\theta) \quad (17)$$

Here the reader can clearly see that the minimization can be performed with respect to different model structures therefore simplifying the identification procedure. Moreover, in the L_2 case where the process innovations are suppose to be Gaussian distributed L. Ljung (see (Ljung, 1999)) showed in example 16.2 P. 506 that the above estimate minimize the *joint criterion* for the determination of the model structure (in fact its dimension $d_{\mathcal{M}}$) and the parameter values $\theta_N^{\mathcal{M}}$ within this structure $\mathcal{M}$ defined by expression (13) with the *complexity* term (14) as already mentioned.

In the L_1 case, i.e. assuming that the process innovations are Laplacian with mean values zero and variances λ unknown, we can show the following result

$$W_N^0(\theta,\mathcal{M},Z^N) = V_N(\theta,Z^N)(1+U_N(\mathcal{M})) \quad (18)$$

with the *complexity* term:

$$U_N(\mathcal{M}) = \frac{d_{\mathcal{M}}}{N} = \frac{dim\theta}{N} \quad (19)$$

to be compared with the corresponding expression (14). Here again he number 2 appearing both in L_2FPE and L_2AIC expressions and not in the corresponding L_1 expressions, seems to be specific to the L_2 estimation framework.

The proof of this result is given in Appendix bellow.

4. CONCLUSION

In this paper we have underlined the interest of identification in the L_1 framework. An interesting theoretical result about the asymptotic properties of the L_1 estimated model have been established. This result extend the classical Akaike's AIC to the L_1identification approach. It is remarkable that we conclude on results comparable to the L_2 case despite the quite different analytical properties of L_1 criterium. The reader may think about the first and second derivatives of this criterium with respect to the parameters. Moreover in the FPE criterion the problem of analytically assessing an approximation of the noise variance is no longer necessary that represent an interesting improvement of the method.

This result is to be confirmed by an experiment on a real process. Together with other results, we hope this paper shall contribute to the development of research directions of model estimation using various norm in the criterion related to.

5. APPENDIX

The residual probability density function is

$$f_e(\theta, Z^N) = \prod_{t=1}^{N} \frac{\exp - \sqrt{\frac{2}{\lambda}} |\varepsilon(t, \theta')|}{\sqrt{2\lambda}}$$

$$= \frac{1}{(\sqrt{2\lambda})^N} \exp - \sqrt{\frac{2}{\lambda}} \sum_{t=1}^{N} |\varepsilon(t, \theta')|$$

Where $\theta = [\theta', \lambda]$. Furthermore using the log-likelihood function

$$L_N(\theta, Z^N) = -\frac{N}{2} \ln 2\lambda - \sqrt{\frac{2}{\lambda}} \sum_{t=1}^{N} |\varepsilon(t, \theta')|$$

Therefore the "inner" minimization, i.e. with respect to θ within the structure $\mathcal{M}$, we have: $\widehat{\theta}^N = [\widehat{\theta}'_N, \widehat{\lambda}_N]$ where $\widehat{\lambda}_N$ is the experimental variance

$$\widehat{\lambda}_N = \frac{1}{N} \sum_{t=1}^{N} \varepsilon^2(t, \widehat{\theta}'_N)$$

with

$$\widehat{\theta}'_N = arg \min_{\theta \in \mathcal{D}_\mathcal{M}} \frac{1}{N} \sum_{t=1}^{N} |\varepsilon(t, \theta)|$$

leading approximately for N sufficiently large to

$$\sqrt{\widehat{\lambda}_N} \approx \frac{1}{\sqrt{N}} \sum_{t=1}^{N} |\varepsilon(t, \widehat{\theta}'_N)|$$

therefore

$$\sqrt{\frac{2}{\widehat{\lambda}_N}} \sum_{t=1}^{N} |\varepsilon(t, \widehat{\theta}'_N)| = \sqrt{2N}$$

Hence the empirical estimate of the log-likelihood function is

$$L_N(\widehat{\theta}_N, Z^N) = -\frac{N}{2} \ln 2 - \frac{N}{2} \ln \widehat{\lambda}_N - \sqrt{2N}$$

and the "outer minimization", i.e. with respect to the structure $\mathcal{M}$ becomes

$$\widehat{\mathcal{M}} = arg \min_{\mathcal{M} \in M} \frac{1}{N} \left[-L_N(\widehat{\theta}_N, Z^N) + d_{\}} \right]$$

$$= arg \min_{\mathcal{M} \in M} \left[\frac{1}{2} \ln 2 + \frac{1}{2} \ln \widehat{\lambda}_N + \sqrt{\frac{2}{N}} + \frac{d_\mathcal{M}}{N} \right]$$

In addition using the experimental values of the residuals it comes

$$\ln \widehat{\lambda}_N = 2 \ln \sqrt{\widehat{\lambda}_N} = 2 \ln \frac{1}{\sqrt{N}} \sum_{t=1}^{N} |\varepsilon(t, \widehat{\theta}'_N)|$$

$$\ln \widehat{\lambda}_N = 2 \ln \sqrt{N} \left(\frac{1}{N} \sum_{t=1}^{N} |\varepsilon(t, \widehat{\theta}'_N)| \right)$$

then

$$\frac{1}{2} \ln \widehat{\lambda}_N = \frac{1}{2} \ln N + \ln \left(\frac{1}{N} \sum_{t=1}^{N} |\varepsilon(t, \widehat{\theta}'_N)| \right)$$

Hence

$$\widehat{\mathcal{M}} = arg \min_{\mathcal{M} \in M} \left\{ \frac{1}{2} \ln 2 + \frac{1}{2} \ln N + \ln \left[\frac{1}{N} \sum_{t=1}^{N} |\varepsilon(t, \widehat{\theta}_N^\mathcal{M})| \right] + \frac{d_\mathcal{M}}{N} \right\} \quad (20)$$

the term to minimize, i.e. depending on $\mathcal{M}$, is

$$\ln \left[\frac{1}{N} \sum_{t=1}^{N} |\varepsilon(t, \widehat{\theta}_N^\mathcal{M})| \right] + \frac{d_\mathcal{M}}{N} \approx \ln \left[\frac{1}{N} \sum_{t=1}^{N} |\varepsilon(t, \widehat{\theta}_N^\mathcal{M})| \right] \left(1 + \frac{d_\mathcal{M}}{N} \right)$$

as $d_\mathcal{M} \ll N$, that ends the proof.

REFERENCES

A. Garulli, B. Kacewicz, A. Vicino and G. Zappa (2000). Error bounds for conditional algorithms in restricted complexity set membership identification. *IEEE Trans. Automat. Contr.* **45-1**, 160–164.

Akaike, H. (1974). A new look at the statistical model identification. *IEEE Trans. Automat. Contr.* **AC-19**, 716–723.

Carmona, J.C. and V.M. Alvarado (2002). L1 prediction error approach in system identification. *Proc. ACC02-IEEE, Anchorage, USA, AK May 8-10* pp. 3219–3223.

G.C. Goodwin, J.H. Brslavsky and M.M. Seron (1999). Non-stationary stochastic embedding for transfer function estimation. *Proc. of the 14th IFAC World Congress, Beijing, China.*

Gustafsson, T.K. and P.M. Mäkilä (1996). Modelling of uncertain systems via linear programming. *Automatica* **32-3**, 319–335.

H. Akçay, H. Hjalmarson and L. Ljung (1996). On the choice of norms in system identification. *IEEE Trans . on Aut. Contr* **41-9**, 1367–1372.

Huber, P.J. (1981). *Robust statistics.* John Wiley & sons. New York, NY.

J. Schoukens, Y. Rolain and R. Pintelon (2002). odified aic rule for model selection in combination with prior estimated noise models. *Automatica* **38**, 903–906.

Ljung, L. (1999). *System identification: theory for the user.* Prentice Hall PTR. NewYork.

Poljak, B.T. and J.A.Z. Tsypkin (1980). Robust identification. *Automatica* **16**, 53–61.

Rice, J.R. and J.S. White (1964). Norms for smoothing and estimation. *SIAM Rev.* **6**, 243–256.

Rissanen, J. (1978). Modelling by shortest data description. *Automatica* **14**, 465–471.

Smith, R. and J.C. Doyle (1992). Model validation: a connection between robust control and identification. *IEEE Trans. Automat. Contr.* **AC-37**, 942–952.

W. Reinelt W., A. Garulli A. and L. Ljung (2002). Comparing different approaches to model error modeling in robust identification. *Automatica* **38**, 787–803.

Wilcox, R.R. (1997). *Introduction to Robust Estimation and Hypothesis Testing.* Academic Press. San Diegi, Ca.

IFAC Publications
www.elsevier.com/locate/ifac

ON PARAMETER ESTIMATION OF ARMAX MODEL VIA BCLS METHOD

Li-Juan Jia *,[1] **Shunshoku Kanae** ** **Zi-Jiang Yang** **
Kiyoshi Wada **

* *Venture Business Laboratory (VBL), Kyushu University*
** *Dept. of Electrical and Electronic System Eng., Kyushu University*
6-10-1 Hakozaki, Higashiku, Fukuoka, 812-8581 Japan

Abstract: This paper studies the problem of parameter estimation of ARMAX model from a novel point of view. An efficient bias compensation least squares algorithm is proposed to provide consistent parameter estimate for ARMAX model. The main feature of our proposed algorithm is to introduce the auxiliary least squares linear backward predictors to construct the cross-correlations of least-squares (LS) error and backward prediction (BWP) errors. And with the help of the cross-correlations of LS error and BWP errors, estimate of the bias resulted from LS solution can be obtained. Consequently the consistent estimate for ARMAX model can be obtained via compensating the estimated bias of LS estimate. The batch-processing approach and the recursive processing approach for the proposed method are given. Theoretical analysis that compares the proposed method with the other existing methods such as bias-eliminated least-squares (BELS) method proposed by Zheng and instrumental variables (IV) method is carried out. Simulation results are presented to illustrate the effectiveness of the proposed algorithm. *Copyright © 2003 IFAC*

Keywords: ARMAX model, parameter estimation, bias compensation, least-squares method, backward prediction error

1. INTRODUCTION

ARMAX (AutoRegressive Moving Average with eX-ogenous variables) model has been used frequently in identification. Any linear finite-order system with stationary disturbances having a rational spectral density can be described by ARMAX model. In this paper the problem of identification for ARMAX model is considered.

It is well known that least squares (LS) method is the most widely used method for parameter estimation. Unfortunately, LS estimator is biasd when directly apply to ARMAX model estimation. To avoid the bias one possibility is to use PE method (Ljung, L., 1987). This leads to a more complex method and requires a numerical optimization of a nonlinear function that depends on the recorded data. Another choice to cope with the bias problem is the instrumental variables (IV) method (Söderström, T. and P. Stoica, 1989). The IV method, which is commonly used to estimate discrete time transfer functions requires construction of the so-called instrumental time series, which are causally related to the observed input and output, but independent of the noise on these observations.

On the other hand, there are some estimation methods via the bias-compensation principle appeared in the control and signal processing literature such as modified LS (MLS) method (Sagara, S. and K. Wada, 1977), correction LS (CLS) method (Stoica, P. and T. Söderström, 1982), bias-eliminated LS (BELS) method (Feng, C. B. and W. X. Zheng, 1991), and bias compensated LS (BCLS) method (Jia, L.J. *et al*, 2001). It has been indicated that BELS method belongs to the class of IV methods under certain conditions (T. Söderström, W.X. Zheng and P. Stoica, 1999; Jia, L.J., M. Ikenoue, C.Z. Jin and K. Wada, 2001). The key to bias compensation method is how to estimate the noise-induced bias. Some devices for estimation of bias have been proposed in these bias compensating methods.

[1] Corresponding author. Tel: +81-92-642-3958; Fax: +81-92-642-3939
Email addresses: jia@dickie.ees.kyushu-u.ac.jp(L.-J. Jia), {jin; yoh; wada}@ees.kyushu-u.ac.jp (C.-Z. Jin; Z.-J. Yang; K. Wada)

Recently, BELS method has been developed to apply for consistent parameter estimation of ARMAX model (Zheng, W. X., 2002). In order to estimate the bias of LS estimate, in BELS method, an auxiliary parameter vector is introduced to construct an augmented ARMAX model and the bias estimate is derived from this auxiliary augmented model. Being different from BELS method, this paper will propose a new approach to estimation of noise-induced bias. Moreover not only the batch-processing approach but also the recursive processing approach for the proposed method are given, but only the batch form of BELS method is described in Zheng (2002).

It is found that ARMAX model has an important property that the stochastic disturbance is orthogonal to the subspace spanned by the outputs. In this paper, an efficient algorithm is proposed by means of introducing the auxiliary estimates to provide consistent estimate for noise-induced bias. The introduced auxiliary estimates are used to construct one-step to n-step backward least-squares predictors. Then the cross-correlation functions of LS error and backward prediction (BWP) errors can be obtained. It is found out that the introduced backward prediction errors and input-output data vector are uncorrelated asymptotically or, in a least-squares estimation language, orthogonal. As we shall see, this is the key to obtain a consistent estimate for bias of LS estimate. With the help of these orthogonal properties, the bias of LS solution can be obtained via computing linear simultaneous equations which show the relationships between the cross-correlation functions of LS error and BWP errors and the correlation of output and stochastic noise. Consequently the consistent estimate for ARMAX model can be obtained via compensating the estimated bias of LS estimate. For the reason that the proposed method is established based on combining with the bias compensation principle and introduction of backward predictors, it can be called BCLS (bias compensated least squares)-B (backward) method for short.

This paper is organized as follows. Problem formulation and description of ARMAX model are presented in Section 2. The proposed BCLS-B method is shown in Section 3. Analysis of the relationship among the proposed BCLS-B method, Zheng's BELS method and IV method is given in Section 4. Section 5 illustrates simulation results, and conclusions are drawn in Section 6.

2. PROBLEM FORMULATION

A general single-input single-output (SISO) discrete-time system can be described by an ARMAX (AutoRegressive Moving Average with eXogenous variables) model

$$A(q^{-1})y(k) = B(q^{-1})u(k) + C(q^{-1})w(k) \quad (1)$$

where $u(k)$ is input, $y(k)$ is output, and q^{-1} is the backward shift operator, i.e., $q^{-1}u(k) = u(k-1)$,

$$\begin{aligned} A(q^{-1}) &= 1 + a_1 q^{-1} + \cdots + a_n q^{-n} \\ B(q^{-1}) &= b_1 q^{-1} + \cdots + b_m q^{-m} \\ C(q^{-1}) &= c_0 + c_1 q^{-1} + \cdots + c_n q^{-n}. \end{aligned}$$

Let

$$\begin{aligned} \boldsymbol{\theta} &= \left[a_1 \cdots a_n \; b_1 \ldots b_m\right]^T \\ \boldsymbol{y}_k &= \left[y(k-1) \cdots y(k-n)\right]^T \\ \boldsymbol{u}_k &= \left[u(k-1) \ldots u(k-m)\right]^T \\ \boldsymbol{\phi}_k^T &= \left[-\boldsymbol{y}_k^T \; \boldsymbol{u}_k^T\right] \end{aligned}$$

The ARMAX model can be further written as a linear regression form

$$y(k) = \boldsymbol{\phi}_k^T \boldsymbol{\theta} + e(k) \quad (2)$$

where

$$e(k) = C(q^{-1})w(k) \quad (3)$$

represents the stochastic noise acting on the system, and $w(k)$ stands for the source of the disturbance.

The following standard assumptions are made.

- (A1) $A(\cdot)$ has all zeros strictly outside the unit disc.
- (A2) The input $u(t)$ is stationary and persistently exciting of a sufficient order.
- (A3) The disturbance source $w(k)$ is a zero-mean white noise and statistically uncorrelated with $u(k)$.
- (A4) The orders of model (n, m) are known.

The problem under study is, using input-output data $\{u(k), y(k)\}$ to make a consistent estimation for the ARMAX model parameter $\boldsymbol{\theta}$.

Based on the least-squares criterion, the LS estimate $\widehat{\boldsymbol{\theta}}_{LS}$ can be obtained as

$$\begin{aligned} \widehat{\boldsymbol{\theta}}_{LS} &= \Big(\sum_{k=1}^{N} \boldsymbol{\phi}_k \boldsymbol{\phi}_k^T\Big)^{-1} \sum_{k=1}^{N} \boldsymbol{\phi}_k y(k) \quad (4) \\ &= \widehat{R}_{\phi\phi}^{-1} \widehat{\boldsymbol{r}}_{\phi y} \quad (5) \end{aligned}$$

where

$$\widehat{R}_{\phi\phi} = \frac{1}{N}\sum_{k=1}^{N} \boldsymbol{\phi}_k \boldsymbol{\phi}_k^T, \quad \widehat{\boldsymbol{r}}_{\phi y} = \frac{1}{N}\sum_{k=1}^{N} \boldsymbol{\phi}_k y(k).$$

and the LS error $\xi(k)$ is described as follows.

$$\begin{aligned} \xi(k) &= y(k) - \boldsymbol{\phi}_k^T \widehat{\boldsymbol{\theta}}_{LS} \\ &= \boldsymbol{\phi}_k^T(\boldsymbol{\theta} - \widehat{\boldsymbol{\theta}}_{LS}) + e(k) \quad (6) \end{aligned}$$

By eqn. (4), we found the orthogonal property between the LS error $\xi(k)$ and data vector $\boldsymbol{\phi}_k$ as

Property 1

$$\sum_{k=1}^{N} \boldsymbol{\phi}_k (y(k) - \boldsymbol{\phi}_k^T \widehat{\boldsymbol{\theta}}_{LS}) = \sum_{k=1}^{N} \boldsymbol{\phi}_k \xi(k) = \mathbf{0}$$

which will be useful for derivation of the proposed method.

Substituting eqn. (2) into eqn. (4) yields

$$\widehat{\boldsymbol{\theta}}_{LS} = \boldsymbol{\theta} + \Big(\sum_{k=1}^{N} \boldsymbol{\phi}_k \boldsymbol{\phi}_k^T\Big)^{-1} \sum_{k=1}^{N} \boldsymbol{\phi}_k e(k) \quad (7)$$

Taking probability limit, we have

$$\operatorname*{plim}_{N\to\infty} \widehat{\boldsymbol{\theta}}_{LS} = \boldsymbol{\theta} + E[\boldsymbol{\phi}_k \boldsymbol{\phi}_k^T]^{-1} E[\boldsymbol{\phi}_k e_k]$$

This equation indicates that LS method can not give consistent estimate for ARMAX model due to the asymptotical bias. The bias of LS estimate is

$$\boldsymbol{h} = \operatorname*{plim}_{N\to\infty} \widehat{\boldsymbol{\theta}}_{LS} - \boldsymbol{\theta} = E[\boldsymbol{\phi}_k \boldsymbol{\phi}_k^T]^{-1} E[\boldsymbol{\phi}_k e(k)].$$

3. BCLS METHOD

3.1 *Bias compensation*

Bias compensation principle proposed by Sagara and Wada can be described as

$$\widehat{\boldsymbol{\theta}}_{BCLS} = \widehat{\boldsymbol{\theta}}_{LS} - \widehat{\boldsymbol{h}}$$

It means that if the noise-induced bias in the LS parameter estimate is obtained, compensating the LS estimate for the estimated bias can generate consistent parameter estimate for ARMAX model.

By assumption (A3), the input $u(k)$ is orthogonal to the stochastic disturbance $e(k)$, the bias estimate becomes

$$\widehat{\boldsymbol{h}} = -\widehat{R}_{\phi\phi}^{-1} \begin{bmatrix} I_n \\ O \end{bmatrix} \widehat{\boldsymbol{r}}_{ye}.$$

where

$$\operatorname*{plim}_{N\to\infty} \widehat{\boldsymbol{r}}_{ye} = E[\boldsymbol{y}_k e(k)].$$

then the consistent BCLS estimate for ARMAX model can be obtained as

$$\widehat{\boldsymbol{\theta}}_{BCLS} = \widehat{\boldsymbol{\theta}}_{LS} + \widehat{R}_{\phi\phi}^{-1} \begin{bmatrix} I_n \\ O \end{bmatrix} \widehat{\boldsymbol{r}}_{ye} \quad (8)$$

It is clear that estimation of $\widehat{\boldsymbol{r}}_{ye}$ becomes the key to BCLS method and our purpose is to find an efficient approach for estimation of $\widehat{\boldsymbol{r}}_{ye}$.

3.2 *Backward predictor and prediction error*

By assumption (A3), it is also known that ARMAX model has an important feature that the stochastic disturbance $e(k)$ is orthogonal to the subspace spanned by the outputs $y(k-n-1), y(k-n-2), \cdots, y(k-2n)$. This property can be described as

Property 2.

$$E[e(k)y(k-n-i)] = 0, \quad i = 1\cdots n.$$

i.e.

$$\operatorname*{plim}_{N\to\infty} \frac{1}{N}\sum_{k=1}^{N} e(k)y(k-n-i) = 0, \quad i = 1\cdots n.$$

In this paper, utilizing this special property of ARMAX model helps us to find an ingenious approach to solution of the key problem to estimate the bias resulted from LS estimate.

Introduce the auxiliary estimates $\widehat{\boldsymbol{\beta}}_i$ $(i = 1\cdots n)$ defined as

$$\widehat{\boldsymbol{\beta}}_i = \Big[\sum_{k=1}^{N} \boldsymbol{\phi}_k \boldsymbol{\phi}_k^T\Big]^{-1} \sum_{k=1}^{N} \boldsymbol{\phi}_k y(k-n-i) \quad (9)$$

to construct i-step backward predictor

$$\widehat{y}(k-n-i) = \boldsymbol{\phi}_k^T \widehat{\boldsymbol{\beta}}_i \quad (10)$$

and i-step backward prediction error

$$\begin{aligned}\epsilon_i(k) &= y(k-n-i) - \widehat{y}(k-n-i) \\ &= y(k-n-i) - \boldsymbol{\phi}_k^T \widehat{\boldsymbol{\beta}}_i.\end{aligned}$$

From eqn. (9), it follows that

$$\sum_{k=1}^{N} \boldsymbol{\phi}_k (y(k-n-i) - \boldsymbol{\phi}_k^T \widehat{\boldsymbol{\beta}}_i) = \mathbf{0} \quad (11)$$

which shows the orthogonal property between data vector $\boldsymbol{\phi}_k$ and the backward prediction error $\epsilon_i(k)$.

Property 3.

$$\operatorname*{plim}_{N\to\infty} \frac{1}{N}\sum_{k=1}^{N} \boldsymbol{\phi}_k \epsilon_i(k) = \mathbf{0} \quad (i = 1\cdots n).$$

3.3 *Cross-correlation function of LS error and BWP errors*

For clarity and simplicity of presentation, we consider the one-step prediction error.

Denote f_1 be a cross-correlation function of the LS error $\xi(t)$ and one-step backward prediction error $\epsilon_1(t)$ described as

$$f_1 = \sum_{k=1}^{N} \xi(k)\epsilon_1(k) =$$

$$\sum_{k=1}^{N} (y(k) - \boldsymbol{\phi}_k^T \widehat{\boldsymbol{\theta}}_{LS})(y(k-n-1) - \boldsymbol{\phi}_k^T \widehat{\boldsymbol{\beta}}_1)$$

By eqn. (6) and the orthogonal property between data vector $\boldsymbol{\phi}_k$ and the backward prediction error $\epsilon_1(k)$, f_1 can be rewritten as

$$f_1 = \sum_{k=1}^{N} e(k)(y(k-n-1) - \boldsymbol{\phi}_k^T \widehat{\boldsymbol{\beta}}_1) \quad (12)$$

Taking probability limit, we have

$$\operatorname*{plim}_{N\to\infty}\frac{1}{N}f_1 = E[e(k)y(k-n-1)] - \operatorname*{plim}_{N\to\infty}\widehat{\boldsymbol{\beta}}_1^T\begin{bmatrix} I_n \\ \mathbf{0}\end{bmatrix}E[\boldsymbol{y}_k e(k)].$$

From **Property 2**, it follows that

$$\operatorname*{plim}_{N\to\infty}\frac{1}{N}f_1 = -\operatorname*{plim}_{N\to\infty}\widehat{\boldsymbol{\beta}}_1^T\begin{bmatrix} I_n \\ \mathbf{0}\end{bmatrix}E[\boldsymbol{y}_k e(k)]. \quad (13)$$

It is known that computation of n-dimension vector $\widehat{\boldsymbol{r}}_{ye}$ is a problem involving linear simultaneous equations in n unknowns. Eqn. (13) is only one linear equation which describes the relationship between $\widehat{\boldsymbol{r}}_{ye}$ and cross-correlation function f_1. In order to obtain the estimate $\widehat{\boldsymbol{r}}_{ye}$, we need n linear simultaneous equations.

Define $f_i, (i = 1\cdots n)$ be a cross-correlation function of the LS error $\xi(k)$ and i-step backward prediction error $\epsilon_i(k)$ described as

$$f_i = \sum_{k=1}^{N}\xi(k)\epsilon_i(k) = \sum_{k=1}^{N}(y(k) - \boldsymbol{\phi}_k^T\widehat{\boldsymbol{\theta}}_{LS,N})(y(k-n-i) - \boldsymbol{\phi}_k^T\widehat{\boldsymbol{\beta}}_i),$$

Taking probability limit, we have

$$\operatorname*{plim}_{N\to\infty}\frac{1}{N}f_i = E[e(k)y(k-n-i)] - \operatorname*{plim}_{N\to\infty}\widehat{\boldsymbol{\beta}}_i^T\begin{bmatrix} I_n \\ \mathbf{0}\end{bmatrix}E[\boldsymbol{y}_k e(k)].$$

From **Property 2**, it follows that

$$\operatorname*{plim}_{N\to\infty}\frac{1}{N}f_i = -\operatorname*{plim}_{N\to\infty}\widehat{\boldsymbol{\beta}}_i^T\begin{bmatrix} I_n \\ \mathbf{0}\end{bmatrix}E[\boldsymbol{y}_k e(k)]. \quad (14)$$

Let

$$\boldsymbol{f}^T = [f_1\dots f_n],$$

we can summarize the above results be

$$\boldsymbol{f} = \sum_{k=1}^{N}\begin{bmatrix} y(k-n-1) - \widehat{\boldsymbol{\beta}}_1^T\boldsymbol{\phi}_k \\ y(k-n-2) - \widehat{\boldsymbol{\beta}}_2^T\boldsymbol{\phi}_k \\ \vdots \\ y(k-2n) - \widehat{\boldsymbol{\beta}}_n^T\boldsymbol{\phi}_k\end{bmatrix}e(k) = \sum_{k=1}^{N}\boldsymbol{\rho}_k e(k) - \widehat{\mathcal{B}}\sum_{k=1}^{N}\boldsymbol{\phi}_k e(k),$$

and the solution of $\widehat{\boldsymbol{r}}_{ye}$ is obtained as

$$\widehat{\boldsymbol{r}}_{ye} = \left(\widehat{\mathcal{B}}\begin{bmatrix} I_n \\ O\end{bmatrix}\right)^{-1}\frac{1}{N}\boldsymbol{f} \quad (15)$$

where

$$\boldsymbol{\rho}_k = \begin{bmatrix} y(k-n-1) \\ y(k-n-2) \\ \vdots \\ y(k-2n)\end{bmatrix}, \quad \widehat{\mathcal{B}} = \begin{bmatrix} \widehat{\boldsymbol{\beta}}_1^T \\ \widehat{\boldsymbol{\beta}}_2^T \\ \vdots \\ \widehat{\boldsymbol{\beta}}_n^T\end{bmatrix}.$$

3.4 *Computation of Cross-correlation function* $\boldsymbol{f}$

From the definition of vector $\boldsymbol{f}$, **Property 1** and **Property 3**, we can derive the two forms for computing the cross-correlation function of LS error and BWP errors.

$$\begin{aligned}\frac{1}{N}\boldsymbol{f} &= \frac{1}{N}\sum_{k=1}^{N}\begin{bmatrix} y(k-n-1) \\ y(k-n-2) \\ \vdots \\ y(k-2n)\end{bmatrix}(y(k) - \boldsymbol{\phi}_k^T\widehat{\boldsymbol{\theta}}_{LS}) \\ &= \frac{1}{N}\sum_{k=1}^{N}\boldsymbol{\rho}_k y(k) - \frac{1}{N}\sum_{k=1}^{N}\boldsymbol{\rho}_k\boldsymbol{\phi}_k^T\widehat{\boldsymbol{\theta}}_{LS} \\ &= \widehat{\boldsymbol{r}}_{\rho y} - \widehat{R}_{\phi\rho}^T\widehat{\boldsymbol{\theta}}_{LS}\end{aligned} \quad (16)$$

or

$$\begin{aligned}\frac{1}{N}\boldsymbol{f} &= \frac{1}{N}\sum_{k=1}^{N}\left\{\begin{bmatrix} y(k-n-1) \\ y(k-n-2) \\ \vdots \\ y(k-2n)\end{bmatrix} - \begin{bmatrix} \widehat{\boldsymbol{\beta}}_1^T \\ \widehat{\boldsymbol{\beta}}_2^T \\ \vdots \\ \widehat{\boldsymbol{\beta}}_n^T\end{bmatrix}\boldsymbol{\phi}_k\right\}y(k) \\ &= \frac{1}{N}\sum_{k=1}^{N}\boldsymbol{\rho}_k y(k) - \widehat{\mathcal{B}}\frac{1}{N}\sum_{k=1}^{N}\boldsymbol{\phi}_k y(k) \\ &= \widehat{\boldsymbol{r}}_{\rho y} - \widehat{\mathcal{B}}\widehat{\boldsymbol{r}}_{\phi y}\end{aligned} \quad (17)$$

3.5 *Implementation of the BCLS-B Algorithm*

Based on the above discussions, the recursive form of BCLS-B algorithm is summarized as follows:

***Step* 1.** Calculate the LS estimate $\widehat{\boldsymbol{\theta}}_{LS,N}$ and the auxiliary estimate $\widehat{\mathcal{B}}_N$ by the conventional RLS algorithm:

$$\begin{aligned}\widehat{\boldsymbol{\theta}}_{LS,N} &= \widehat{\boldsymbol{\theta}}_{LS,N-1} + \frac{P_{N-1}\boldsymbol{\phi}_N\left(y_N - \boldsymbol{\phi}_N^T\widehat{\boldsymbol{\theta}}_{LS,N-1}\right)}{1 + \boldsymbol{\phi}_N^T P_{N-1}\boldsymbol{\phi}_N} \\ \widehat{\mathcal{B}}_N &= \widehat{\mathcal{B}}_{N-1} + \frac{P_{N-1}\boldsymbol{\phi}_N\left(\boldsymbol{\rho}_N - \boldsymbol{\phi}_N^T\widehat{\mathcal{B}}_{N-1}\right)}{1 + \boldsymbol{\phi}_N^T P_{N-1}\boldsymbol{\phi}_N} \\ P_N &= P_{N-1} - \frac{P_{N-1}\boldsymbol{\phi}_N\boldsymbol{\phi}_N^T P_{N-1}}{1 + \boldsymbol{\phi}_N^T P_{N-1}\boldsymbol{\phi}_N}\end{aligned}$$

***Step* 2.** Calculate $\boldsymbol{f}_N$:

$$\begin{aligned}\boldsymbol{r}_{\rho y,N} &= \boldsymbol{r}_{\rho y,N-1} + \boldsymbol{\rho}_N y_N \\ R_{\phi\rho,N} &= R_{\phi\rho,N-1} + \boldsymbol{\phi}_N\boldsymbol{\rho}_N^T \\ \boldsymbol{r}_{\phi y,N} &= \boldsymbol{r}_{\phi y,N-1} + \boldsymbol{\phi}_N y_N \\ \boldsymbol{f}_N &= \boldsymbol{r}_{\rho y,N} - R_{\phi\rho,N}^T\widehat{\boldsymbol{\theta}}_{LS,N} \quad \text{or} \\ &= \boldsymbol{r}_{\rho y,N} - \widehat{\mathcal{B}}_N\boldsymbol{r}_{\phi y,N}\end{aligned}$$

***Step* 3.** Calculate $\widehat{\boldsymbol{r}}_{ye}$

$$\widehat{\boldsymbol{r}}_{ye} = (\widehat{\mathcal{B}}_N \begin{bmatrix} I_n \\ O \end{bmatrix})^{-1} \frac{\boldsymbol{f}_N}{N}$$

***Step* 4.** Calculate the consistent BCLS-B parameter estimate via

$$\widehat{\boldsymbol{\theta}}_{BCLS,N} = \widehat{\boldsymbol{\theta}}_{LS,N} + P_N \begin{bmatrix} I_n \\ O \end{bmatrix} (\widehat{\mathcal{B}}_N \begin{bmatrix} I_n \\ O \end{bmatrix})^{-1} \boldsymbol{f}_N$$

***Step* 5.** N=N+1 and return to ***Step* 1** until convergence.

Initial values at $N = 0$ are given as follows:

$$P_0 = \alpha \boldsymbol{I}, \widehat{\boldsymbol{\theta}}_{LS,0} = \boldsymbol{0}, \widehat{\mathcal{B}}_0 = O,$$
$$\boldsymbol{r}_{\rho y,0} = \boldsymbol{0}, \boldsymbol{r}_{\phi y,0} = \boldsymbol{0}, R_{\phi\rho,0} = O.$$

The batch form is often applied for theoretical analysis and discussion, so we also summarize the batch form of BCLS-B algorithm as follows:

$$\widehat{\boldsymbol{\theta}}_{BCLS} = \widehat{\boldsymbol{\theta}}_{LS} + \widehat{R}_{\phi\phi}^{-1} \begin{bmatrix} I_n \\ O \end{bmatrix} \widehat{\boldsymbol{r}}_{ye} \tag{18}$$

$$\widehat{\mathcal{B}} = \sum_{k=1}^{N} \begin{bmatrix} y(k-n-1) \\ y(k-n-2) \\ \vdots \\ y(k-2n) \end{bmatrix} \boldsymbol{\phi}_k^T \widehat{R}_{\phi\phi}^{-1} = \widehat{R}_{\phi\rho}^T \widehat{R}_{\phi\phi}^{-1} \tag{19}$$

$$\widehat{\boldsymbol{r}}_{ye} = \left(\widehat{\mathcal{B}} \begin{bmatrix} I_n \\ O \end{bmatrix}\right)^{-1} \left(\widehat{\boldsymbol{r}}_{\rho y} - \widehat{R}_{\phi\rho}^T \widehat{\boldsymbol{\theta}}_{LS}\right) \tag{20}$$

or

$$\widehat{\boldsymbol{r}}_{ye} = \left(\widehat{\mathcal{B}} \begin{bmatrix} I_n \\ O \end{bmatrix}\right)^{-1} \left(\widehat{\boldsymbol{r}}_{\rho y} - \widehat{\mathcal{B}} \widehat{\boldsymbol{r}}_{\phi y}\right) \tag{21}$$

4. COMPARATIVE ANALYSIS

In this section, theoretical analysis is performed that compares the proposed BCLS-B method with Zheng's BELS method and IV method respectively.

4.1 *Relation to Zheng's BELS method*

Zheng's BELS method is revisited as follows.

$$\widehat{\boldsymbol{\theta}}_{BELS} = \widehat{\boldsymbol{\theta}}_{LS} + \widehat{R}_{\phi\phi}^{-1} \begin{bmatrix} I_n \\ O \end{bmatrix} \widehat{\boldsymbol{r}}_{ye}$$

$$\widehat{\boldsymbol{r}}_{ye} = \left(\widehat{R}_{\phi\rho}^T \widehat{R}_{\phi\phi}^{-1} \begin{bmatrix} I_n \\ O \end{bmatrix}\right)^{-1} \left(\widehat{\boldsymbol{r}}_{\rho y} - \widehat{R}_{\phi\rho}^T \widehat{\boldsymbol{\theta}}_{LS}\right)$$

In the BCLS-B method, substituting eqn. (19) and eqn. (5) into eqn. (20) and eqn. (21), then $\widehat{\boldsymbol{r}}_{ye}$ in eqn. (15) becomes

$$\begin{aligned}
\widehat{\boldsymbol{r}}_{ye} &= \left(\widehat{R}_{\phi\rho}^T \widehat{R}_{\phi\phi}^{-1} \begin{bmatrix} I_n \\ O \end{bmatrix}\right)^{-1} \left(\widehat{\boldsymbol{r}}_{\rho y} - \widehat{R}_{\phi\rho}^T \widehat{\boldsymbol{\theta}}_{LS}\right) \\
&= \left(\widehat{R}_{\phi\rho}^T \widehat{R}_{\phi\phi}^{-1} \begin{bmatrix} I_n \\ O \end{bmatrix}\right)^{-1} \left(\widehat{\boldsymbol{r}}_{\rho y} - \widehat{\mathcal{B}} \widehat{\boldsymbol{r}}_{\phi y}\right) \\
&= \left(\widehat{R}_{\phi\rho}^T \widehat{R}_{\phi\phi}^{-1} \begin{bmatrix} I_n \\ O \end{bmatrix}\right)^{-1} \left(\widehat{\boldsymbol{r}}_{\rho y} - \widehat{R}_{\phi\rho}^T \widehat{R}_{\phi\phi}^{-1} \widehat{\boldsymbol{r}}_{\phi y}\right).
\end{aligned}$$

It indicates that the two forms of $\widehat{\boldsymbol{r}}_{ye}$ are equivalent and the obtianed BCLS estimate is the same as the BELS's one.

4.2 *Relation to IV method*

Defining a vector $\boldsymbol{\eta}_k$ of dimension $n+m$ as

$$\boldsymbol{\eta}_k = \begin{bmatrix} \boldsymbol{u}_k \\ \boldsymbol{\rho}_k \end{bmatrix}.$$

By assumption (A3), we have

$$\operatorname*{plim}_{N\to\infty} \frac{1}{N} \sum_{k=1}^{N} \boldsymbol{\eta}_k e(k) = 0,$$

so we can choose the variable $\boldsymbol{\eta}_k$ be the instrumental variable, and the IV estimator is obtained by

$$\begin{aligned}
\widehat{\boldsymbol{\theta}}_{IV} &= \left(\sum_{k=1}^{N} \boldsymbol{\eta}_k \boldsymbol{\phi}_k^T\right)^{-1} \sum_{k=1}^{N} \boldsymbol{\eta}_k y(k) \\
&= \widehat{R}_{\eta\phi}^{-1} \widehat{\boldsymbol{r}}_{\eta y}
\end{aligned} \tag{22}$$

Let matrix $\widehat{\mathcal{B}}$ be partitioned into $\widehat{\mathcal{B}} = \begin{bmatrix} \widehat{\mathcal{B}}_1 & \widehat{\mathcal{B}}_2 \end{bmatrix}$, it follows that

$$\widehat{R}_{\phi\rho}^T \widehat{R}_{\phi\phi}^{-1} \begin{bmatrix} I_n \\ O \end{bmatrix} = \widehat{\mathcal{B}}_1.$$

Then we can obtain

$$\begin{aligned}
\widehat{\boldsymbol{\theta}}_{BCLS} &= \widehat{\boldsymbol{\theta}}_{LS} + \widehat{R}_{\phi\phi}^{-1} \begin{bmatrix} I_n \\ O \end{bmatrix} \widehat{\boldsymbol{r}}_{ye} \\
&= \widehat{\boldsymbol{\theta}}_{LS} + \widehat{R}_{\phi\phi}^{-1} \begin{bmatrix} I_n \\ O \end{bmatrix} \widehat{\mathcal{B}}_1^{-1} \left(\widehat{\boldsymbol{r}}_{\rho y} - \widehat{R}_{\phi\rho}^T \widehat{\boldsymbol{\theta}}_{LS}\right) \\
&= \widehat{R}_{\phi\phi}^{-1} \left\{ \widehat{\boldsymbol{r}}_{\phi y} + \begin{bmatrix} I_n \\ O \end{bmatrix} \widehat{\mathcal{B}}_1^{-1} \widehat{\boldsymbol{r}}_{\rho y} - \begin{bmatrix} I_n \\ O \end{bmatrix} \widehat{\mathcal{B}}_1^{-1} \widehat{\mathcal{B}} \widehat{\boldsymbol{r}}_{\phi y} \right\} \\
&= \widehat{R}_{\phi\phi}^{-1} \Big\{ \widehat{\boldsymbol{r}}_{\phi y} + \begin{bmatrix} I_n \\ O \end{bmatrix} \widehat{\mathcal{B}}_1^{-1} \widehat{\boldsymbol{r}}_{\rho y} + \begin{bmatrix} I_n \\ O \end{bmatrix} \widehat{\boldsymbol{r}}_{yy} \\
&\qquad - \begin{bmatrix} I_n \\ O \end{bmatrix} \widehat{\mathcal{B}}_1^{-1} \widehat{\mathcal{B}}_2 \widehat{\boldsymbol{r}}_{uy} \Big\} \\
&= \widehat{R}_{\phi\phi}^{-1} \left\{ \begin{bmatrix} 0 \\ \widehat{\boldsymbol{r}}_{uy} \end{bmatrix} + \begin{bmatrix} I_n \\ O \end{bmatrix} \begin{bmatrix} -\widehat{\mathcal{B}}_1^{-1} \widehat{\mathcal{B}}_2 & \widehat{\mathcal{B}}_1^{-1} \end{bmatrix} \begin{bmatrix} \widehat{\boldsymbol{r}}_{uy} \\ \widehat{\boldsymbol{r}}_{\rho y} \end{bmatrix} \right\} \\
&= \widehat{R}_{\phi\phi}^{-1} \begin{bmatrix} -\widehat{\mathcal{B}}_1^{-1} \widehat{\mathcal{B}}_2 & \widehat{\mathcal{B}}_1^{-1} \\ I_m & O \end{bmatrix} \widehat{\boldsymbol{r}}_{\eta y}.
\end{aligned}$$

Moreover due to

$$\begin{aligned}
\begin{bmatrix} -\widehat{\mathcal{B}}_1^{-1} \widehat{\mathcal{B}}_2 & \widehat{\mathcal{B}}_1^{-1} \\ I_m & O \end{bmatrix} &= \begin{bmatrix} \widehat{\mathcal{B}}_1^{-1} & -\widehat{\mathcal{B}}_1^{-1} \widehat{\mathcal{B}}_2 \\ O & I_m \end{bmatrix} \begin{bmatrix} O & I_n \\ I_m & O \end{bmatrix} \\
&= \left\{ \begin{bmatrix} O & I_n \\ I_m & O \end{bmatrix} \begin{bmatrix} \widehat{\mathcal{B}}_1 & \widehat{\mathcal{B}}_2 \\ O & I_m \end{bmatrix} \right\}^{-1}
\end{aligned}$$

and

$$\widehat{\mathcal{B}} \widehat{R}_{\phi\phi} = \widehat{R}_{\phi\rho}^T = \widehat{R}_{\rho\phi},$$

one obtains that

$$\widehat{\boldsymbol{\theta}}_{BCLS} = \widehat{R}_{\phi\phi}^{-1}\left\{\begin{bmatrix} O & I_n \\ I_m & O \end{bmatrix}\begin{bmatrix} \widehat{\mathcal{B}}_1 & \widehat{\mathcal{B}}_2 \\ O & I_m \end{bmatrix}\right\}^{-1}\widehat{\boldsymbol{r}}_{\eta y}$$
$$= \left\{\begin{bmatrix} O & I_n \\ I_m & O \end{bmatrix}\begin{bmatrix} \widehat{\mathcal{B}}_1 & \widehat{\mathcal{B}}_2 \\ O & I_m \end{bmatrix}\widehat{R}_{\phi\phi}\right\}^{-1}\widehat{\boldsymbol{r}}_{\eta y}$$
$$= \begin{bmatrix} \widehat{R}_{u\phi} \\ \widehat{\mathcal{B}}\widehat{R}_{\phi\phi} \end{bmatrix}^{-1}\widehat{\boldsymbol{r}}_{\eta y} = \begin{bmatrix} \widehat{R}_{u\phi} \\ \widehat{R}_{\rho\phi} \end{bmatrix}^{-1}\widehat{\boldsymbol{r}}_{\eta y}$$
$$= \widehat{R}_{\eta\phi}^{-1}\widehat{\boldsymbol{r}}_{\eta y} \quad (23)$$

is equivalent to the IV estimate in eqn. (22).

On the basis of the above comparative analysis, it is shown that the BCLS estimate, BELS estimate and IV estimate are identical. Hence the proposed BCLS-B method will inherit attractive properties of the IV method and BELS method as described in Zheng (2002) such as favorable estimation accuracy and low computational performance.

5. SIMULATION RESULTS

Consider a second-order ARMAX model given by

$$A(q^{-1}) = 1 - 1.5q^{-1} + 0.7q^{-2}$$
$$B(q^{-1}) = 1.0q^{-1} + 0.5q^{-2}$$
$$C(q^{-1}) = 1 + 1.0q^{-1} + 0.2q^{-2}.$$

The input $u(k)$ is taken as white noise with unit variance. The $w(k)$ is a white noise with zero mean and variance σ_w^2 is chosen as 0.09 and 0.81 respectively.

Table 1. Simulation results in $\sigma_w^2 = 0.09$

	a_1(-1.5)	a_2(0.7)	b_1(1.0)	b_2(0.5)
LS	-1.4635 ±0.00218	0.6663 ±0.0020	0.9993 ±0.0062	0.5372 ±0.0075
BELS	-1.5003 ±0.0024	0.7002 ±0.0018	0.9996 ±0.0063	0.5008 ±0.0061
IV	-1.5003 ±0.0024	0.7002 ±0.0018	0.9996 ±0.0063	0.5008 ±0.0061
BCLS-B	-1.5003 ±0.0024	0.7002 ±0.0018	0.9996 ±0.0063	0.5008 ±0.0061

Table 2. Simulation results in $\sigma_w^2 = 0.81$

	a_1(-1.5)	a_2(0.7)	b_1(1.0)	b_2(0.5)
LS	-1.2640 ±0.0108	0.4857 ±0.0098	0.9976 ±0.0138	0.7433 ±0.0169
BELS	-1.4988 ±0.0091	0.6992 ±0.0061	0.9964 ±0.0155	0.5087 ±0.0160
IV	-1.4988 ±0.0091	0.6992 ±0.0061	0.9964 ±0.0155	0.5087 ±0.0160
BCLS-B	-1.4988 ±0.0091	0.6992 ±0.0061	0.9964 ±0.0155	0.5087 ±0.0160

The mean and standard deviation of parameter estimates obtained by LS method, the BELS method, IV method and the proposed BCLS-B method for 20 runs are listed in Table 1and Table 2. Data length is chosen as 5000. Simulation results demonstrate that the LS method gives biased results. On the contrary, the BCLS-B method can give consistent estimates via compensating the asymptotical bias and estimation accuracy is satisfactory. The equivalence of BCLS-B method, BELS method and IV method is validated via simulation results.

6. CONCLUSIONS

The contribution of this paper is that a new approach to estimation of noise-induced bias in LS estimate is proposed from a novel point of view. It is worthy to find some useful properties that the stochastic noise is orthogonal to the subspace spanned by the outputs, the LS error and the introduced backward prediction errors are also (asymptotically) orthogonal to input-output data vector. With the help of these orthogonal properties, the efficient bias compensation method– BCLS-B method has been established to provide consistent parameter estimation for ARMAX model. It is seen that the proposed algorithm is straightforward. For the purpose of practical application and theoretical analysis, the recursive processing form and the batch-processing form of the proposed method are given. The equivalence of the proposed BCLS-B method, the BELS method and IV method are also studied. Simulation results have been performed to verify the theoretical discussions.

7. REFERENCES

Ljung, L. (1987). *System Identification: Theory for the User.*, Englewwod Cliffs, NJ: Prentice-Hall

Sagara, S. and K. Wada (1977). On-line modified least-squares parameter estimation on linear discrete dynamic systems. *Int.J.Control*,Vol.25,No.3 329-343,

Söderström, T. and P. Stoica (1989). *System Identification.* Prentice-Hall.

Stoica, P. and T. Söderström (1982). Bias correction in least-sqares identification. *Int.J.Control*, Vol.35, 449-457,

Feng, C. B. and W. X. Zheng (1991). Robust identification of stochastic linear systems with correlated output noise. *IEE Proc.-Control Theory and Applications*,Vol.138,No.5,484-492,

T. Söderström, W.X. Zheng and P. Stoica (1999). Comments on "On a Least-Squares-Based Algorithm for Identification of Stochastic Linear Systems". *IEEE Tans. on Signal Processing*, Vol. 47, No.5, 1395-1396

Wada, K., L. J. Jia, T. Hanada and J. Imai (2001). On BELS parameter estimation method of transfer functions. *Tans.IEE of Japan*, Vol.121-C, No.4, 795-799

Jia, L.J., M. Ikenoue, C.Z. Jin and K. Wada (2001). On bias compensated least squares method for noisy input-output system identification. *Proc. of the 40th IEEE Conference on Decision and Control, Orlando, Florida USA*, 3332-3337.

Zheng, W. X., (2002). On least-squares identification of ARMAX models *Proc. of 2002 IFAC 15th Triennial World Congress, Barcelona, Spain.*

IFAC Publications
www.elsevier.com/locate/ifac

ESTIMATION IN THE PRESENCE OF INTERFERENCES

Jean Jacques Fuchs

Irisa-Université de Rennes, Campus de Beaulieu,
35042 Rennes Cedex, France.
e-mail : fuchs@irisa.fr

Abstract: If, in a linear regression model, outliers are present on some components of the observation vector, one uses robust estimation schemes that downweight their influence. If interferences that perturb the whole observation vector are present, one generally assumes to know the interfering subspace and performs the parameter estimation on its orthogonal complement. A more general interference model is considered. The interferences belong to a known parametrized family that spans the whole space. The number and the range space of the interferences that are indeed present is unknown and a robust estimation scheme selects the interferences that are present and downweights their influence. *Copyright © 2003 IFAC*

Keywords: Robust estimation, Expectation-Maximization, Monte Carlo Markov Chains.

1. INTRODUCTION

A standard model for signal+interference+noise (Scharf, 1991) is:

$$y = a\theta + S\phi + \mathbf{n} \tag{1}$$

where y is the n-dimensional measurement vector, $a\theta$ is the signal model with known vector a and unknown amplitude θ, $S\phi$ represents the interferences that belong to the m-dimensional range of the (n,m) dimensional known matrix S and $\mathbf{n}$ is the additive white gaussian noise with a covariance matrix equal to the identity matrix. This is a more general model than the simpler signal+colored noise model, but it requires a precise knowledge of the interference subspace, an information that is seldom available. The proposed solution amounts in general to identify both θ and ϕ while keeping only the estimate of θ, (Scharf et al., 1994), (Lupas, 1989).

A different and more general model is as follows:

$$y = a\theta + Ae \tag{2}$$

with A an (n, m) full row rank known matrix and e a unknown random vector. The m columns in A that outnumber now n the number of rows model all the interferences potentially present of which only a very limited number are indeed active.

For Ae to model both the interferences and the white noise we take the components of the e vector to be random variables with a Huber density, i.e. a contaminated gaussian density:

$$p(x) = (1-\epsilon)\zeta(x) + \epsilon q(x)$$

with $\zeta(x) = \frac{1}{\sqrt{2\pi}}\, e^{-\frac{x^2}{2}}$ the standard gaussian density N(0,1) and $q(x)$ a symmetric density that minimizes the Fisher information contained in the samples. The real $\epsilon \in (0,1)$ allows to tune the contamination. This density has been designed and is used to model noise samples presenting outliers i.e. samples that are abnormally large. In our model this means that some few columns in A will have large weights while the remaining ones will be weighted by gaussian variables. Since A is full row rank for the model to be of interest, it can further be assumed orthonormal, the global gaussian contribution to y has then essentially an identity covariance matrix similarly to the $\mathbf{n}$ vector in (1) and the few columns in A with large weights will play the role of the interferences, the $S\phi$ vector in (1). The number and the structure of the interferences is now far less constrained than in (1), making (2) particularly attractive.

As we shall see below the price to pay for this flexibility is a more difficult estimation problem. In section 2, we further specify and justify the model. The estimation scheme is developped in section 3. It is based on a stochastic approximation version of an expectation-maximization algorithm that uses samples from a Monte Carlo markov chain. Some preliminary simulation results and conclusions are presented in section 4 and 5.

2. FORMULATION OF THE PROBLEM

2.1 *The Huber density*

The Huber density is the density that among those of the form: $p(x) = (1-\epsilon)\zeta(x) + \epsilon q(x)$ with $\zeta = N(0,1)$ the standard gaussian, minimizes the Fisher information with respect to the mean (see (Huber, 1981) for details). It has an interesting closed form:

$$p(x) = (1-\epsilon)\ \frac{1}{\sqrt{2\pi}}\ e^{-\frac{x^2}{2}} \qquad if\ |x| \leq h$$
$$p(x) = (1-\epsilon)\ \frac{1}{\sqrt{2\pi}}\ e^{\frac{h^2}{2}-h|x|} \qquad if\ |x| > h \quad (3)$$

where h is such that $\int_{-\infty}^{\infty} p(x)\ dx = 1$. This leads to the following implicit relation between the threshold h and the contamination factor ϵ:

$$1 = (1-\epsilon)\int_{-h}^{h}\zeta(x)dx + 2\frac{1-\epsilon}{h}\zeta(h).$$

For $\epsilon = 0$, $h = \infty$ and as ϵ increases towards one, h decreases to zero. For $\epsilon = .1$, one gets $h = 1.14$. One can note that the contaminating density $q(x) = 0$ for $|x| < h$, so that its contribution increases the number of large samples (outliers).

This density has been designed as a mean to develop estimation algorithms that are robust to such abnormal errors (Poljak, 1980), (Rousseeuw, 1987). Instead of modifying heuristically existing estimation schemes (such as least squares) to try to make them immune to outliers, the idea is to build a robust model of such spiky noises in order to develop corresponding optimal estimates, such as the maximum likelihood estimate associated with the density (3). In the sequel this density is used in a new signal+interference+noise model (2) to add some flexibility in the interference representation part.

2.2 *The signal + interference model*

We focus on the new model (2):

$$y = a\theta + Ae$$

with e a n-dimensional vector of independent samples of the *Huber* density (3) where we assume h or equivalently ϵ to be known.

If the (n,m) dimensional known matrix A is not full row rank θ can be estimated without errors and the problem is of little interest.

If A is square and non singular, one premultiplies both sides by A^{-1}, to get the standard robust linear regression model:

$$\bar{y} = \bar{a}\theta + e \qquad (4)$$

with $\bar{y} = A^{-1}y$ and $\bar{a} = A^{-1}a$. An estimate of θ is then obtained by maximizing the likelihood of the observations. It is the solution of the following optimization problem:

$$\min_{\theta}\sum_{1}^{n} f(r_i) \qquad \text{with} \qquad r_i = \bar{y}_i - \bar{a}_i\theta \qquad (5)$$

with f -known as Huber's function- defined by :

$$f(r) = r^2/2 \qquad |r| \leq h$$
$$= h|r| - h^2/2 \qquad |r| > h \qquad (6)$$

The parabola r^2 of the standard least-squares approach that is optimal if e in (4) is gaussian, is replaced in (5) by $f(r)$ that is a parabola in the vicinity of zero and increases linearly for $|r| > h$. The minimization in (5) is non trivial and one has to resort to iterative procedures such as the so-called iteratively reweighted least squares scheme (IRLS) (Huber, 1981), (Holland, 1977) or by iterating:

$$\theta^k = \bar{a}^+(\bar{y} - u^{k-1}) \qquad (7)$$
$$u^k = \min(\bar{y} - \bar{a}\theta^k + h,\ 0) + \max(\bar{y} - \bar{a}\theta^k - h,\ 0)$$

starting from $u^0 = 0$, see (Fuchs, 1999). The extrema are taken componentwise and $\bar{a}^+ = \bar{a}^T/\bar{a}^T\bar{a}$ is the pseudo-inverse of $\bar{a}$.

The case we are interested in is actually the one where A has more columns than rows: $m > n$. By premultiplying then both sides of (2) by an adequate square matrix A can be transformed into an orthogonal matrix denoted V^T. From now on we therefore assume without loss of generality that we observe y:

$$y = a\theta + V^Te \qquad \text{with} \qquad V^TV = I \qquad (8)$$

The idea being that the interferences belong to a known, possibly parametrized, family (the columns of V^T) but that we ignore how many and which interferences are present. Hence this model where only a few columns in V^T will be used to represent the interferences while most of them will model the additive white gaussian noise with (close to) unit covariance matrix, since $V^T V = I$.

The maximum likelihood estimate of θ is now quite difficult to get since the joint density of $V^T e$ is unknown. It is only under the pure gaussian assumption that the contribution of $V^T e$ remains gaussian and is simple to handle. Intuitively one can however expect that if the number of outliers in e is much smaller than n, the number of observations, one is still in a reasonable situation with a few outliers amid gaussian noise that globally may admit an n-dimensional representation.

3. THE ESTIMATION ALGORITHM

3.1 *The EM approach*

If one had access to additional observations say $z = W^T e$ with $U = [V\ W]$ a square orthogonal matrix, the complete model for $x = (y, z)$ would be:

$$x = \begin{bmatrix} y \\ z \end{bmatrix} = \begin{bmatrix} a \\ 0 \end{bmatrix} \theta + \begin{bmatrix} V^T \\ W^T \end{bmatrix} e \qquad (9)$$

Premultiplying then both sides by U:

$$Ux = Vy + Wz = Va\theta + e$$

would take us back to the usual linear regression situation (4) that is easy to handle. Since z is unknown we will try to use an Expectation Maximisation (EM) approach to solve this difficulty (Laird et al., 1977).

The ML estimate of θ in (8) is the solution to:

$$\theta_{ML} = \arg\max_{\theta}\ \ell_y(\theta) \qquad (10)$$

where y is an observed sample of the random vector Y and $\ell_y(\theta) = \log p_\theta(y)$ is the loglikelihood function. Let the random vector $X = (Y, Z)$ represent the complete data with density $p_\theta(x) = p_\theta(y, z)$, that is more informative than Y the observed data. If one had access to X, one would take as estimate $\hat{\theta}_{ML} = \arg\max_\theta\ \ell_x(\theta)$ with $\ell_x(\theta) = \log p_\theta(x)$. Since the complete data x corresponding to the observed sample y is unknown, $\ell_y(\theta)$ can only be estimated.

By Bayes rule, we have:

$$\log p_\theta(Y) = \log p_\theta(Y, Z) - \log p_\theta(Z|Y)$$

since the left-hand side does not depend on Z, one can integrate both sides with respect to a density $p_{\theta_-}(Z|y)$ for any value θ_- and y and preserve its original value to get:

$$\log p_\theta(y) = \int \log p_\theta(y, z) p_{\theta_-}(z|y) dz - \int \log p_\theta(z|y) p_{\theta_-}(z|y) dz$$

This last expression is generally written as:

$$\log p_\theta(y) = Q(\theta, \theta_-) - I(\theta, \theta_-)$$

One then shows that $I(\theta, \theta_-) \leq I(\theta_-, \theta_-)$ for any θ. In order to increase the likelihood $\ell_y(\theta) = \log p_\theta(y)$, it is therefore sufficient to maximise $Q(\theta, \theta_-)$ with respect to θ. This is the general approach taken by the EM algorithm that generates thus a sequence of estimates:

$$\theta_{k+1} = \arg\max_{\theta} \int \log p_\theta(y, z) p_{\theta_k}(z|y) dz \qquad (11)$$

that quite generally converges to a (local) maximum of $\ell_y(\theta)$.

There is however still a difficulty, while $p_\theta(y, Z)$ can indeed be evaluated for any Z and θ, no analytical expression of $Q(\theta, \theta_k)$ can be obtained.

In fact since an exact maximization of $Q(\theta, \theta_k)$ in (11) at each iteration is not necessary, it is sufficient to be able to evaluate the gradient (here the derivative) of $Q(\theta, \theta_k)$ with respect to θ at θ_k in order to perform a step that increases the likelihood. But this also is unfeasible and in order to get an estimate of this gradient one has to use Monte Carlo methods, i.e. to perform a simulation that generates samples z from the density $p_{\theta_k}(Z|y)$.

3.2 *A stochastic approximation algorithm*

Since the iteration in (11) is unfeasible, one can think of using a gradient algorithm to get an optimizing sequence:

$$\theta_{k+1} = \theta_k + \mu_k \int \frac{d \log p_\theta(y, z)}{d\theta} p_{\theta_k}(z|y) dz \qquad (12)$$

but this also is difficult to achieve since in order to get an estimate of the gradient one has to resort to a Metropolis-Hastings algorithm to get samples z from the density $p_{\theta_k}(Z|y)$ and evaluate the integral using an importance sampling approach.

It may therefore be preferable to go a step further and actually use a stochastic gradient algorithm. This stochastic EM algorithm (Celeux, 1992), (Delyon, 1999) then generates:

$$\theta_{k+1} = \theta_k + \gamma_k \frac{d \log p_{\theta_k}(y, z_k)}{d\theta} \qquad (13)$$

where z_k is a sample to be built at each step from the density $p_{\theta_k}(Z|y)$ and γ_k is a non-increasing small gain. This is a standard procedure in s-tochastic gradient or stochastic approximation algorithms: one replaces the true gradient or an estimate of it by just one sample and reduces drastically the size of the step made in this direction. One can further check that the expectation of the gradient in (13), is indeed the desired one:

$$\int \frac{d\log p_\theta(y,z)}{d\theta} p_\theta(z|y)dz$$
$$= \int \frac{dp_\theta(y,z)}{d\theta} \frac{1}{p_\theta(y,z)} \frac{p_\theta(y,z)}{p_\theta(y)} dz$$
$$= \frac{1}{p_\theta(y)} \frac{d}{d\theta} \int p_\theta(y,z)dz = \frac{d}{d\theta} \log p_\theta(y)$$

which is indeed the desired gradient since our aim is to maximize $\log p_\theta(y)$ (10).

3.3 *The Metropolis Hastings algorithm*

It remains to explain how to build samples from $p_{\theta_k}(Z|y)$ for the model (9) with e a vector of samples with density (3). Since the density $p_{\theta_k}(Z|y)$ is unknown, there is no mean to directly build the sought-for samples and the idea is to build a Markov Chain with invariant density $p_{\theta_k}(Z|y)$.

Indeed the only quantity that one can evaluate is $p_{\theta_k}(y, Z)$ which is linked to $p_{\theta_k}(Z|y)$ by Bayes rule and as we will see below this is sufficient to be able to apply a Monte Carlo Markow Chain (MCMC) algorithm (Meyn et al., 1993), (Robert, 1999) and more precisely the Metropolis-Hastings (MH) algorithm. The idea is to build a Markov chain that generates a sequence $\{z_j\}$ of random variables whose density converges to the invariant measure of the chain that is built to be $p_{\theta_k}(Z|y)$. Though the samples are of course not independent and convergence may be slow, it is a helpful tool for which some theoretical results are available. It is applicable to the present situation because only the ratio of the probabilities of two potential successive state has to be evaluated By Bayes rule one has:

$$\frac{p_{\theta_k}(z\ |y)}{p_{\theta_k}(z_k|y)} = \frac{p_{\theta_k}(y,z)}{p_{\theta_k}(y,z_k)}$$

It requires the definition of a proposal distribution say $q(Z)$ and proceeds as follows:

Given that the chain is in state z_k, propose a transition to z drawn from $q(Z)$, then:

$z_{k+1} = z$ with probability $\alpha(z, z_k)$ and
$z_{k+1} = z_k$ with probability $1-\alpha(z, z_k)$ where:

$$\alpha(z, z_k) = \min\ (1, \frac{p_{\theta_k}(y,\ z)q(z_k)}{p_{\theta_k}(y, z_k)q(z)}) \qquad (14)$$

The details of the procedure that is implemented in the simulations are presented below. Following again the stochastic approximation philosophy we only draw one sample from the Markov chain for each θ_k i.e. between two steps of the estimation algorithm in (13). Since the gain γ_k in (13) are either extremely small or decreasing this should not prevent convergence, though of course a precise analysis would be complex.

3.4 *The implementation details*

Let us summarize the procedure described so far. One observes a random n-dimensional vector y that is modeled as (8):

$$y = a\theta + V^T e$$

where a is a known vector, V^T is a (n,m) orthogonal matrix with $m > n$ and e an m-dimensional vector whose components are independent samples from Huber's density (3). The aim is to estimate the scalar θ in an efficient way. Since the likelihood function of y is unknown, this seems unfeasible. One therefore introduces so-called missing data $z = W^T e$ with W such that $U = [V\ W]$ is a square orthogonal matrix. The so completed model can be transformed into:

$$Ux = Vy + Wz = Va\theta + e$$

If one would observe $x = (y, z)$, the estimation of θ would be easy using IRLS or (7) for instance. Since z is unknown and that one can at best get samples from its conditional density one replaces the deterministic optimization algorithm (12) by a stochastic gradient algorithm (13) :

$$\theta_{k+1} = \theta_k + \gamma_k \frac{d\log p_{\theta_k}(y, z_k)}{d\theta}$$

where z_k is a sample from the density $p_{\theta_k}(Z|y)$ and γ_k is a non-increasing small gain.

In order to implement this procedure, one has to evaluate $p_\theta(y, z)$ in (14). From:

$$Ux = Va\theta + e$$

with U a square orthogonal matrix and thus with determinant equal to unity, it follows, with a slight change in notations:

$$p_X(x) = p_E(Ux - Va\theta)|U| = p_E(Ux - Va\theta)$$
$$\Rightarrow p_X(x) \equiv p_\theta(y,z) = \Pi_1^m p_E(e_i(\theta))$$

with $e_i(\theta)$ the i-th component of $Ux - Va\theta$. One further has:

$$\frac{d\log p_\theta(y,z)}{d\theta} = \sum \frac{d\log p_E(e_i)}{de_i} \frac{de_i}{d\theta}$$

From (3) one has:

$$\frac{d\log p_E(x)}{dx} = -\min(|x|, h)\text{sign}(x)$$

and thus:

$$\frac{d\log p_\theta(y,z)}{d\theta} = d^T Va$$

with d an m-dimensional vector with i-th component $d_i = \min(|e_i|, h)\ \text{sign}(e_i)$. It remains to describe the MH algorithm that is applied. The initial state of the $(m\text{-}n)$ dimensional MC z_0 is taken as a sample from a standard gaussian $N(0, I)$ and at each step, we propose a change to only one randomly chosen component in z_k

with the density $q(.)$ again the standard gaussian $N(0,1)$. We run (13,14) for 6000 steps and take $\gamma_k = .1/k$ in (13).

4. SIMULATION RESULTS

The observation y is obtained by simulating

$$y = a\theta + V^T e \qquad \text{with} \qquad V^T V = I$$

with V^T an arbitrary but known (25, 36) orthogonal matrix. The samples in vector e are drawn from (3) with $\epsilon = .2$ and thus $h = .862$ assumed to be known. The vector $a = \mathbf{1}$, a vector of ones and θ the parameter to be estimated is taken equal to one also. To assess how far we are from complete information we compare the obtained estimates to those, one gets under complete information when z also is observed. The Fisher information is then $a^T V^T V a = 25$ times the one contained in one observation and the Cramer Rao bound is then equal to $\sigma_{cr}^2 = .0818$ (Huber, 1981). The variance of the estimate obtained by the least squares approach (which is optimal if e were gaussian) is also known and equal to $\sigma_{ls}^2 = .114$. It is 40% higher than σ_{cr}^2. Indeed the components of $V^T e$ which are orthogonal linear combinations of $m = 36$ independent Huber random variables are close to independent gaussian random variables by the central limit theorem. The difference between the efficiency of these estimates and those obtained by ML, which we try to attain, is decreasing as m increases. Both estimates are unbiased since the models are linear and the densities even functions.

The results over 10000 independent realisations of e and V are presented in Table 1. The estimate of the correlation between the ML estimates under complete and uncomplete observation is .914.

	y obs.	y and z obs.	LS est.
mean	.997	.995	.993
std. dev.	.319	.295	.338
expected std. dev.	?	.286	.338

Table 1. The means and variances observed over 10000 independent realizations under partial and full observation, $a = \mathbf{1}$, $n = 25$, $m = 36$ and $\epsilon = .2$.

The corresponding results for $a = .5\ \mathbf{1}$ are presented in Table 2.

	y obs.	y and z obs.	LS est.
mean	.997	.995	1.0017
std. dev.	.624	.586	.673
expected std. dev.	?	.572	.676

Table 2. The means and variances observed over 10000 independent realizations under partial and full observation, $a = .5\ \mathbf{1}$, $n = 25$, $m = 36$ and $\epsilon = .2$.

5. PRELIMINARY CONCLUSIONS

Further simulations will be presented in the final paper. Many investigations remain to be done to validate this approach and the claim that it is indeed a model that allows to represent both the interferences and the additive noise. Simpler estimation schemes, along the lines of the algorithm presented in (Fuchs, 1999) can also be considered. But their link with the maximum likelihood approach is unclear.

6. REFERENCES

G. Celeux and J. Diebolt. (1992) A stochastic approximation type EM algorithm for the mixture problem. *Stochastics.*, 41, 127-146.

B. Delyon, M. Lavielle and E. Moulines. (1999) On a stochastic approximation version of the EM algorithm. *Annals of Statistics.*, vol. 27, 1, 94-128.

J.J. Fuchs. (1999) A new approach to robust regression. *14th World congress IFAC, vol. H, pp. 427-432, july 99, Beijing*

P.W. Holland and R.E. Welsh. (1977) Robust regression using iteratively reweighted least squares. *Comm. Stat.* A6, 813-828.

P.J. Huber. (1981) Robust Statistics. *John Wiley and sons.*, New York.

N.M. Laird, A.P. Dempster and D.B. Rubin. (1977) Maximum likelihood from incomplete dat via the E.M. algorithm. *Ann; Roy. Stat. Soc.*, B., 39, pp. 1-38, Dec. 1977.

R. Lupas and S. Verdu. (1973) Linear multiuser detectors for synchronous CDMA channels. *IEEE Trans. on I.T.*, 35, 123-126.

S.P. Meyn and R.L. Tweedie. (1993) Markov Chains and Stochastic Stability. *Springer Verlag* London, 1993.

B.T. Poljak and Y. Z. Tsypkin. (1980) Robust identification. *Automatica* vol. 16, 1, 53-63.

C. Robert and G. Casella. (1999) Monte Carlo Statistical Method. *Springer Verlag.*, New York.

P.J. Rousseeuw and A.M. Leroy. (1987) Robust regression and outlier detection. *John Wiley and sons.*, New York.

L.L. Scharf. (1973) Statistical Signal Processing. *Addison Wesley*

L.L. Scharf and B. Friedlander. (1994) Matched subspace detectors. *IEEE Trans. on S.P.*, 42, 8, pp. 2146-2156.

IFAC Publications
www.elsevier.com/locate/ifac

AUTOREGRESSIVE SPECTRAL ANALYSIS with RANDOMLY MISSING DATA

Piet M.T. Broersen, Stijn de Waele and Robert Bos

Signals and Systems Group, Department of Applied Physics
Delft University of Technology

Abstract: The joint data covariance matrix determines the likelihood of an arbitrary Gaussian process. Missing data influence the structure of the covariance. However, for stationary random processes it can still be characterized by the parameters of an autoregressive (AR) model. The best AR predictor includes all previous observations if data are incomplete. The missing data likelihood will here be approximated with only those observations that fall within a finite time interval. The resulting non-linear estimation algorithm requires no user provided initial solution. In various simulations, the spectral accuracy of likelihood methods was better than the accuracy of other spectral estimates for missing data. *Copyright © 2003 IFAC*

Keywords: conditional density, covariance estimation, maximum likelihood, missing observations, parameter estimation, spectral analysis.

1. INTRODUCTION

In many scientific and industrial measurements, records with missing observations are usual. In controlled experiments, sensor failure or outliers may cause missing data. In meteorological observations, the weather conditions may disturb the equidistant sampling scheme. In astronomical or satellite observations, data may be lost in the transmission channel. In paleoclimatic data, the relation between time and the physical depth causes an observed time series with missing observations on an equidistant time grid (Petit, *et al.*, 1999; Yiou, *et al.*, 1996).

A Fourier estimator that uses only the remaining observations is the method of Lomb (1976), which has been improved and analyzed by Scargle (1989). This computes Fourier coefficients as the least squares fit of sines and cosines to the available observations. This is a good estimator if the true process consists of a periodic function with additive white noise. The Lomb-Scargle spectrum is accurate in detecting strong spectral peaks but rather poor in describing slopes in the spectrum (Bos, *et al.*, 2002).

A group of methods uses algorithms that have been developed for spectral estimation from consecutive equidistant data. Three methods can be distinguished in this category. The first is interpolation between the remaining observations, linear, cubic, spline, nearest neighbor or sample and hold. The performance depends on the data characteristics and on the gaps between. No interpolation method gives good results for all types of data (de Waele and Broersen, 2000a). The second idea (Rosen and Porat, 1989) is to find an equidistant covariance estimate for the data and to use that for further analysis. As shown by de Waele and Broersen (1999), this technique does not guarantee that the estimated covariance function is positive semi-definite. Moreover, the sample covariances are known to be inefficient estimators for the covariance structure and will generally not produce accurate spectral estimates, not even for equidistant data (Broersen, 2002). The third method dynamically reconstructs the missing data, followed by the spectral estimation from the reconstructed consecutive signal. Isaksson (1993) compared several variants of this reconstruction method to ARX input-output data, including Kalman filtering, maximum-likelihood estimation and iterative reconstruction. The best method was derived from the iterative EM algorithm for missing data. The E step finds the conditional expectation of the missing data given the observed data together with the currently estimated model; the M step computes maximum likelihood estimates for the parameters of the model from all consecutive data, observed and reconstructed (Little and Rubin, 1987). A simplified iteration of data reconstruction and ARX estimation is described by Wallin, *et al.*, (2000). Another simplified AR reconstruction variant of Mirsaidi *et al.* (1997) does not use the conditional expectation principle but the simpler prediction with a fixed order AR(p) model. It substitutes the prediction for missing data based on only the past and it keeps the available data undisturbed.

A final group of estimators fits a time series model directly to the available observations. The Burg for segments algorithm of de Waele and Broersen (2000b) selects consecutive segments in the observed data and a variant of the Burg algorithm computes the parameters of an AR model. This has been applied to irregularly sampled data, which are made *equidistant with missing data* with slotted nearest neighbor resampling (Broersen, *et al.*, 2002). Jones (1980) described an exact maximum likelihood approach with Kalman filtering. This paper describes a new spectral estimator where an AR model is fitted with an approximate likelihood optimization. The specific choices for a good numerical performance of the algorithm are described and the results are in simulations compared with some existing methods. Furthermore, the new method is applied to paleoclimatic data to obtain an improved spectral estimate.

2. AR BURG for SEGMENTS

Autoregressive (AR) models describe stationary stochastic processes (Priestley, 1981). The power spectrum and the covariance are determined by the parameters of the AR model. An AR(p) process with consecutive observations can be written as:

$$x_n + a_1 x_{n-1} + \cdots + a_p x_{n-p} = \varepsilon_n , \qquad (1)$$

where ε_n is a purely random process. Almost any *stationary* stochastic process can theoretically be written as a unique AR(∞) process. Broersen (2002) argued that in practice finite order AR(p) models, that are estimated and selected from finite data records, will be accurate approximations. Hence, this model type can be applied to random physical, meteorological or astronomical phenomena. The covariance function and the power spectral density of the data can be computed from the estimated parameters of an AR model. The power spectrum of the estimated AR(p') model $\hat{A}_{p'}(z)$, defined as

$$\hat{A}_{p'}(z) = 1 + \hat{a}_1 z^{-1} + \cdots + \hat{a}_{p'} z^{-p'} , \qquad (2)$$

is, for arbitrary p', given by :

$$h(\omega) = \sigma_\varepsilon^2 / \left| \hat{A}_{p'}(e^{j\omega}) \right|^2 . \qquad (3)$$

The original algorithm of Burg (1967) for AR parameter estimation recursively estimates L reflection coefficients k_i from consecutive equidistant data. Reflection coefficients are used to recursively determine the parameters $\hat{A}_q(z)$ of all model orders q between 1 and L, with the Levinson-Durbin formulas (Kay and Marple, 1981):

$$\begin{aligned} &\hat{a}_1^1 = k_1 \\ &\hat{a}_i^q = \hat{a}_i^{q-1} + k_q \hat{a}_{q-i}^{q-1}, && 1 \le i < q \\ &\hat{a}_q^q = k_q && 1 \le q \le L \end{aligned} \qquad (4)$$

The Burg algorithm for segments has been developed for simultaneous AR estimation from S segments by de Waele and Broersen (2000b). It can be applied to missing data by considering consecutive parts between the gaps as separate segments. The number of contributing products diminishes sharply with the AR model order (Bos, *et al.*, 2002). The total number of contributions from the different segments for the calculation of k_L should at least be 10 for sufficient statistical reliability. This gives a practical limit for the maximum order L. This Burg for segments method can give good results if the missing fraction is small and the required AR order is low. Otherwise, too few segments are long enough to allow for orders higher than 1 or 2. For those cases, it is interesting to develop an algorithm that can use all available remaining data. Using only consecutive segments in the Burg estimate throws away a lot of information.

3. PROBABILITY DENSITY OF AR PROCESS

The normal or Gaussian probability density function of a variable x with mean μ and variance σ^2 is:

$$\mathbb{N}(x,\mu,\sigma^2) = \left(\frac{1}{2\pi\sigma^2} \right)^{1/2} \exp\left(-\frac{(x-\mu)^2}{2\sigma^2} \right). \qquad (5)$$

The joint probability density function (pdf) of N observations $\boldsymbol{X}$, where $\boldsymbol{X}$ is a jointly normally distributed vector stochastic variable

$$X = (x_1 \; x_2 \cdots x_N)^T \qquad (6)$$

with

$$\begin{aligned} &E\{X\} = \mu_x \\ &E\{(X-\mu_x)(X-\mu_x)^T\} = R_{xx} \end{aligned} \qquad (7)$$

is given by

$$f(X) = \frac{1}{2\pi^{\frac{N}{2}} \left| R_{xx} \right|^{\frac{1}{2}}} exp\left[-\frac{1}{2} \left\{ (X-\mu_x)^T R_{xx}^{-1} (X-\mu_x) \right\} \right] \qquad (8)$$

This is a general expression for arbitrary normal variables of any source. For equidistant observations of a stationary stochastic process, R_{XX} becomes a Toeplitz matrix with constant elements on all lines parallel to the diagonal. For AR processes, the elements of the autocovariance matrix R_{XX} can be expressed in the AR parameters. For AR processes with missing data, it is still possible to express the matrix R_{XX} in the AR parameters. However, the lines parallel to the diagonal do no longer contain the same numerical value in the missing data case.

With the general definition of the conditional density function, the joint probability density function of N arbitrarily distributed variables can always be written as a conditional product of the last N-k observations as one event, given the first k as a second event:

$$f(X) = f(x_{k+1},\cdots,x_N \mid x_1,\cdots,x_k) f(x_1,\cdots,x_k) . \quad (9)$$

For the event of a single variable with index k:

$$f(x_1,x_2,\cdots,x_k) = f(x_k \mid x_1,\cdots,x_{k-1}) f(x_1,\cdots,x_{k-1}) . (10)$$

With those intermediate results for k, the joint density $f(X)$ in (8) can for arbitrary distributions be written as a product of conditional density functions:

$$f(X) = \left(\prod_{k=2}^{N} f(x_k \mid x_1,\cdots,x_{k-1}) \right) f(x_1) . \quad (11)$$

The usual AR(p) derivations like in Priestley (1981) and in Brockwell and Davis (1991) use this property (11) to separate the likelihood expression in a part for the first p observations and a simple expression for the final N-p observations. Each of these conditional densities follows with (1) from the p preceding observations and the purely random ε_n. With missing data, an AR(1) process is still best predicted by the single closest previous observation. However, higher order AR(p) processes require all previous observations for the best prediction if data are missing. Moreover, the accuracy of prediction is no longer a constant for missing data. Therefore, the exact expressions for the conditional density in (11) become more complicated for the missing data case.

4. ARFIL: AR FINITE INTERVAL LIKELIHOOD ALGORITHM

The general theory of Gaussian variables gives a possibility to use AR(p) models in the conditional density of randomly available observations. Suppose that an arbitrary observation U with dimension 1, mean μ_U and variance σ_U^2 has a joint multivariate normal distribution with a k-dimensional vector Y, with mean μ_Y and covariance matrix R_{YY} or $\sigma_Y^2 \rho_{YY}$. The cross covariance matrix is denoted R_{UY} or $\sigma_U \sigma_Y \rho_{UY}$. Then the conditional density of U for a given vector Y is the normal distribution

$$f(U \mid Y) = \left(\frac{1}{2\pi\sigma_{U|Y}^2} \right)^{1/2} \exp\left(-\frac{(U-\mu_{U|Y})^2}{2\sigma_{U|Y}^2} \right), \quad (12)$$

where the mean and the variance are given by (Anderson and Moore, 1979, p. 25)

$$\begin{aligned} \mu_{U|Y} &= \mu_U + R_{UY} R_{YY}^{-1} (Y-\mu_Y) \\ &= \mu_U + (\sigma_U / \sigma_Y) \rho_{UY} \rho_{YY}^{-1} (Y-\mu_Y) \\ \sigma_{U|Y}^2 &= \sigma_U^2 s_{U|Y}^2 = \sigma_U^2 \left[1 - \rho_{UY} \rho_{YY}^{-1} \rho_{YU} \right] \end{aligned} \quad (13)$$

Consider a vector of N observations $\boldsymbol{X}$, consisting of $x_{t(1)}, \ldots, x_{t(N)}$, where $t(i)$ is a multiple of the sampling time T. They are samples of a stationary stochastic process with mean value zero and variance σ_X^2, sampled at kT and where data are missing. The conditional density of (12) will be used to predict $\hat{x}_{t(i)}$ from previous observations $x_{t(i-1)}, \ldots, x_{t(1)}$ which are not consecutive if data are missing. The probability density for the first observation $x_{t(1)}$ is:

$$f(x_{t(1)}) = \mathrm{N}(x_{t(1)}, 0, \sigma_x^2) = \left(\frac{1}{2\pi\sigma_x^2} \right)^{1/2} \exp\left(-\frac{(x_{t(1)})^2}{2\sigma_x^2} \right). \quad (14)$$

The conditional density $f(x_{t(k+1)}|x_{t(1)}, x_{t(2)}, \ldots, x_{t(k)})$ follows from (12) by substituting $x_{t(k+1)}$ for U and $(x_{t(1)}\, x_{t(2)} \ldots x_{t(k)})$ for Y. All elements of the covariance matrices required in (13) are for an AR(p) model completely determined by the AR parameters with the Yule-Walker relations (Kay and Marple 1981):

$$\rho_n + a_1 \rho_{n-1} + \cdots + a_p \rho_{n-p} = 0,\ n > 0;\ \rho_{-n} = \rho_n , \quad (15)$$

where ρ_n denotes $E(x_{iT} x_{(i+n)T})/\sigma_X^2$. The probability density function (11) of the total vector $\boldsymbol{X}$ can now exactly be written as the simple expression

$$f(X) = (2\pi\sigma_x^2)^{-N/2} \left(\prod_{j=1}^{N} s_{t(j)|t(j-1),\ldots,t(1)}^2 \right)^{-1/2} \exp\left\{ -\frac{1}{2} \sum_{j=1}^{N} (x_{t(j)} - \hat{x}_{t(j)|t(j-1),\ldots,t(1)})^2 / \sigma_x^2 s_{t(j)|t(j-1),\ldots,t(1)}^2 \right\} . (16)$$

The conditional variances s^2 are defined implicitly by (13), with $s^2_{t(1)}=1$. The AR(p) parameter vector $\underline{\hat{\boldsymbol{a}}}$ with elements $\hat{a}_1 \ldots \hat{a}_p$ is estimated in the case of missing data by minimizing $L(X,\underline{\hat{a}}) = -2 \log f(X) - N \log 2\pi$ with respect to $\underline{\hat{a}}$:

$$\begin{aligned} L(X;\underline{\hat{a}}) &= N \log \hat{\sigma}_x^2 + \sum_{j=1}^{N} \log \hat{s}_{t(j)|t(j-1),\ldots,t(1)}^2 \\ &+ \sum_{j=1}^{N} (x_j - \hat{x}_{t(j)|t(j-1),\ldots,t(1)})^2 / \hat{\sigma}_x^2 \hat{s}_{t(j)|t(j-1),\ldots,t(1)}^2 \end{aligned} \quad (17)$$

The usual variance estimate from N observations can be substituted in (17). However, by using the maximum likelihood estimate for σ_x^2,

$$\hat{\sigma}_x^2 = \frac{1}{N} \sum_{j=1}^{N} (x_{t(j)} - \hat{x}_{t(j)|t(j-1),\ldots,t(1)})^2 / \hat{s}_{t(j)|t(j-1),\ldots,t(1)}^2 \quad (18)$$

it is easily seen that substitution of this variance estimate in (17) yields the constant N for the last term and the first two terms of (17) are minimized together. This maximum likelihood choice (18) for the variance is implemented further.

The algorithm with (17) for the likelihood would still require too much computing time in practice, even for moderate N. The size of the predictor (12) would become N-1 for the last observation and a N-$1 \times N$-1 matrix has to be inverted. Generally, the prediction accuracy improves most with the nearest previous observations and further observations have almost no influence. Therefore, the maximum length of the predictor for an AR(p) model in missing data is limited in the algorithm to the finite time interval:

$$t(n) - t(n-K) \le 2pT / \gamma \quad (19)$$

where γ is the remaining fraction of the data. This algorithm is denoted **ARfil**, ***AR*** for a *f*inite *i*nterval *l*ikelihood. The number of observations within the interval varies with the index n; the average number is $2p$. It has been verified that using a larger interval in (19) would not have a noticeable influence on the estimated parameters. Estimated parameters are close to the exact likelihood results of Jones (1980). However, taking a much smaller interval will reduce the accuracy considerably. Of course, only the first p predecessors contribute to the prediction accuracy if observations are consecutive. Substituting more than p consecutive AR(p) observations in (13) just gives coefficients equal to zero in $R_{UY}R_{YY}^{-1}$ for the lags greater than p. It turns out that the prediction with only past observations conserves the symmetry property of the likelihood (17) in ARfil. This means that reversing the sequence of observations gives exactly the same value for the likelihood.

In principle, including the variance estimation with (18) makes sure that poles of the estimated AR model $\hat{A}_p(z)$ can only be inside the unit circle. That is a requirement for models to be stationary. A pole on the unit circle would theoretically give the value zero for the conditional variance s^2. This would have serious consequences for (18) where σ_x^2 would become infinity. Unfortunately, in practice non-linear unconstrained optimization algorithms may produce unusable estimates with a pole radius greater than 1. The use of constrained optimization will too often give a solution that is determined by the implied constraints instead of by the data.

A solution to the stationarity problem is to replace the estimation of parameters by the estimation of reflection coefficients with (4), like in Jones (1980). These are estimated in the Yule-Walker and the Burg methods for AR models (Kay and Marple 1981). Processes are stationary if and only if all reflection coefficients are smaller than 1. The ARfil algorithm uses unconstrained optimization of $\kappa_i = tan(\pi/2 * k_i)$. This guarantees that the estimated k_i is always in the range $-1 < k_i < 1$ for $-\infty < \kappa_i < \infty$. Hence, all estimated models are stationary. In contrast with the recursive Burg algorithm, all k_i, $i=1,\ldots,q$ are optimized simultaneously in ARfil to estimate an AR(q) model. All k_i have to be re-estimated and have different values for every model order q. It is easily shown in the way of construction of the AR(1), AR(2),..., covariance matrices for missing observations that low order reflection estimates are biased as long as the order of the estimated model is lower than the true process order. Automatic initial solutions for the non-linear optimization at order q are the reflection coefficients of the AR($q-1$) model with an additional zero for k_q, starting with $k_1=0$ for the AR(1) model. The automatic initial values require that a non-linear solution has to be determined for all intermediate orders from 1 until p to estimate the AR(p) model. This seeming disadvantage turns into an advantage for the use of order selection criteria. Their successful operation is outside the scope of this paper.

5. SIMULATIONS

The model error ME is a normalized squared error of prediction (Broersen, 2002), with p as the minimal expectation for AR(p). In simulations where the true process is known, it can be used as an accuracy measure for estimated spectral models. ME has relevance in the time domain, in the frequency domain and if no data are missing also statistically as an asymptotical equivalent of the Kullback-Leibler discrepancy. The first remarkable simulation result is that the ARfil algorithm as described never failed to converge to a stationary AR model. ARfil always produces a useful model. If enough observations remain, the ARfil model is reasonably accurate if the process can be approximated by an AR model and very accurate if the true process is low order AR.

The AR reconstruction of Mirsaidi *et al.* (1997) substitutes the unilateral AR prediction for missing data and keeps the available data undisturbed. In a gap of missing data, this reconstruction links up nicely at the front, like a causal impulse response. However, it neglects completely the smoothness of the fit at the backside of the gap, which can cause jumps in the reconstructed signal. Simulations show that even when using the true generating process parameters to predict the gaps, the performance is very poor in some examples. Furthermore, this true prediction was never better than the many variants that reconstruct a gap taking into account the transition to the gap at both sides. Therefore, unilateral AR prediction is not discussed further.

Isaksson (1993) and also Wallin, *et al.*, (2000) describe several methods to reconstruct missing data in both input and output data for the identification of ARX processes. The spectrum is estimated afterwards with an algorithm developed for consecutive data. In the simulations with AR data in this paper, the "misdata" algorithm of the Matlab System Identification Toolbox (Ljung, 1999) has been used for the reconstruction. The automatic built-in iteration method has been used for the reconstruction, as well as various estimated AR models: the AR model of Burg on segments and the AR model estimated after linear interpolation. Furthermore, the reconstruction with the true process parameters has been investigated as a limiting case for the achievable accuracy of this type of reconstruction. Also iterations of AR estimation and using that AR model for a new reconstruction have been investigated. As expected, the best results in reconstruction simulations are obtained with the true values for the AR parameters. Therefore, only the accuracy of those true reconstruction results are reported as an indication of the accuracy. In all but one simulation example, the estimated average ARfil quality was better than the quality that can be obtained with reconstruction with the true process parameters. The reported true order ARfil quality is always better than using estimated models of the true order for reconstruction. The difference between

ARfil and the other methods becomes still greater if models of selected orders are used.

The Lomb-Scargle spectrum converges to the raw periodogram for complete data if nothing is missing. Broersen (2002) showed that periodograms are always less accurate as spectral estimates for stochastic data than estimated and automatically selected time series models. The quality for Lomb-Scargle spectra is computed by transforming the spectrum into a covariance function, compute an AR(p) estimate from that covariance and calculate the model error of that AR(p) model, with the true process order p. Also a long AR(250) model is computed and evaluated with the model error.

The accuracy of the model estimated with ARfil will be compared to that of Burg on segments, to Lomb-Scargle spectra and to three reconstruction methods:

- using a linear interpolation for missing data. This is in most examples more accurate than nearest neighbor, cubic or spline interpolation. However, the overall accuracy of this simple method is poor, certainly if many data are missing.
- using the true process parameters to reconstruct the missing observations with "misdata" from the Matlab System Identification toolbox (Ljung, 1999).
- using the automatic iterative reconstruction of the "misdata" algorithm with a selected model.

The simulations in Table 1 present results for three processes, with a remaining fraction $\gamma = 0.75$ and N=500. N/γ equidistant observations are generated and a fraction $1-\gamma$ is discarded by giving each original observation the probability γ to be included. The process Ex-A is AR(5) with parameters [1 -1.98 1.76 -1.78 1.75 -0.73]. This gives a spectrum with two slopes and a peak at f=0.3. The second process Ex-B is an AR(5) process with a strong spectral peak on a weak background with [1 .1231 .3223 -.3223 -0.1231 0.3984]. The third process Ex-C is a low order AR(2) process with parameters [1 –1.2 0.5].

Table 1 clearly shows that the most accurate spectral estimates are obtained with Arfil, followed by Burg on segments. Reconstruction, interpolation and Lomb-Scargle perform less. The Cramér-Rao bound for the ME for consecutive data would be 5 for Ex-A and Ex-B and 2 for Ex-C. Reconstruction with the *true* parameters looks very good for Ex-B, but this

Table 1 The model error ME for estimated true order models, N=500 remaining observations, γ= 0.75.

Method	Ex-A	Ex-B	Ex-C
ARfil	4.1	14.6	1.8
Burg on segments	24.9	103.8	3.1
Linear interpolated	513.7	3756.4	3.3
AR *true* reconstructed	8.6	5.5	10.2
automatic reconstructed	92.6	207.9	32.8
Lomb-Scargle	3740.8	589.1	201.0
Lomb-Scargle, long AR	4691.5	613.9	464.6

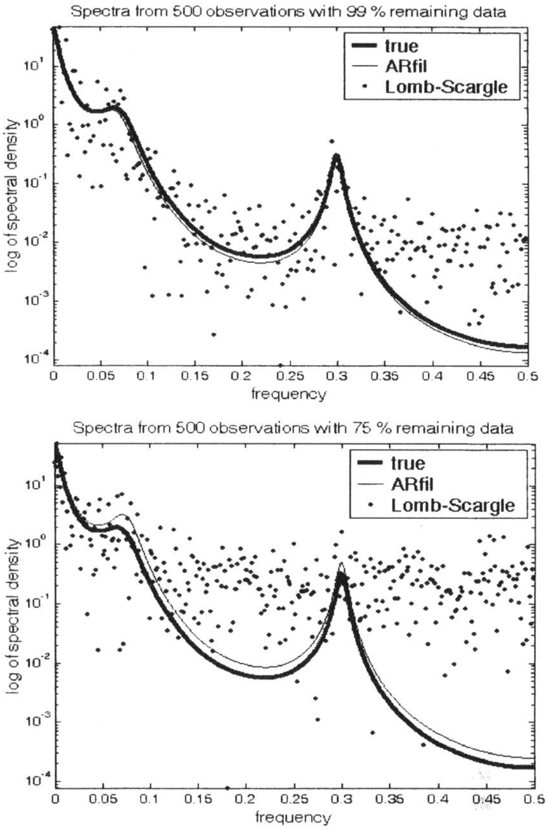

Fig.1. Estimated ARfil and Lomb-Scargle spectra for Ex-A, N=500, for γ=0.99 and 0.75, demonstrating bias of Lomb-Scargle, and problems in detecting slopes as well as peaks on spectral slopes.

Table 2 The model error ME for estimated AR(2) models for Ex-C as a function of γ, N = 1000.

γ	.9	.3	.1	.03	.01
ARfil	2.2	2.8	7.2	26.5	314
Burg	2.5	12.3	98.2	483	480

unattainable accuracy is lost completely in Table 1 in the automatic reconstruction method with estimated parameters. Fig.1 shows the bias and variance of the raw Lomb-Scargle spectra, causing its poor performance in ME in comparison with Arfil in Table 1. Fig.1 demonstrates that windowing cannot possibly make Lomb-Scargle spectra comparable to ARfil. Lomb-Scargle only detects the peak for f=0.3 on the spectral slope for γ greater than about 0.95.

Table 2 shows that the AR(2) process Ex-C is estimated accurately for γ=.03 from N=1000, with 97 % missing data. A benchmark number for the ME is 2.7N, which is obtained for an AR(0) white noise model. In other words, about 99% of the spectral power is explained if the ME is 26.5. As expected, Burg looses its accuracy for higher γ. Taking γ as low as 0.001 with N=10000 gives ME=0.155N for ARfil, which still explains 94 % of the spectrum.

The ARfil algorithm for missing data is applied to an interesting practical example of paleoclimatic data in Fig. 2. The search is for periodicities and other

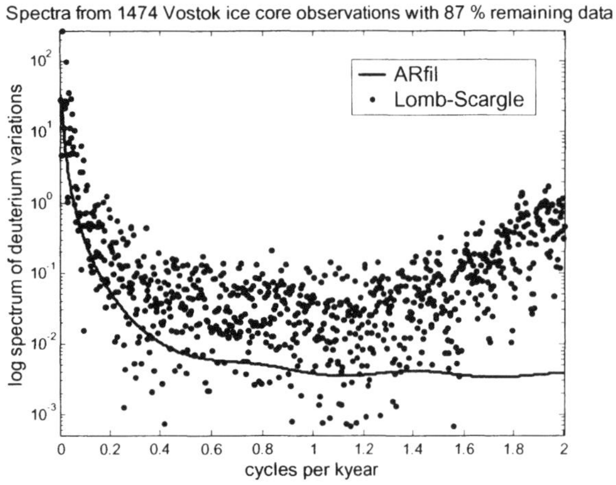

Fig.2. Spectra of paleoclimatic deuterium variations from the Antarctica Vostok 3310 m ice core data.

predictable climate variations in the past. The selected order was 7 for ARfil. Petit, *et al.* (1999) studied the occurrence of spectral peaks in the region below 0.06 cycles per kyear with Fourier spectra and Yiou, *et al.* (1996) produced spectra up to 2.5 cycles per kyear with AR models of orders up to 80. Their results for the climate differ from each other and also from the ARfil spectra. A new problem, yet unsolved, for this type of data is that recent data are sampled more densely than earlier data. The non-random distribution may cause extra uncertainties that can be studied or possibly solved with ARfil as a new and accurate candidate for the estimation.

6. CONCLUDING REMARKS

The finite interval likelihood maximization algorithm ARfil is numerically stable in estimating AR models from incomplete data. For few missing data, the performance of likelihood methods is better than that of methods that reconstruct the data. Moreover, likelihood methods can also be used for very sparse data where other methods fail. The new algorithm combines a spectrum that is guaranteed to be positive with accurate results. The quality of the estimated model is very good in simulations where the true process is a low order AR process. For few missing data, it is often comparable to the Cramér-Rao lower bound that could be obtained if the given number of observations were consecutive.

REFERENCES

Anderson B.D.O. and J.B. Moore (1979). *Optimal Filtering*, Englewood Cliffs, Prentice-Hall.

Bos, R., S. de Waele and P. M. T. Broersen (2002). Autoregressive spectral estimation by application of the Burg algorithm to irregularly sampled data. *IEEE Trans. on Instrumentation and Measurement,* **51**, no.6, pp. 1289-1294.

Brockwell P.J. and R.A. Davis (1991). *Time Series: Theory and Methods*, New York, Springer-Verlag.

Broersen, P.M.T. (2002). Automatic spectral analysis with time series models. *IEEE Trans. on Instrumentation and Measurement*, **51**, pp. 211-216.

Broersen, P.M.T., R. Bos and S. de Waele (2002). Spectral analysis of irregularly sampled data with autoregressive models. *Proc. 15th IFAC World Conference*, paper 651, 6 pp.

Burg, J. P. (1967). Maximum entropy spectral analysis. *Proc 37th Meeting Soc. Of Exploration Geophysicists,* 6 pp.

Isaksson, A.J. (1993). Identification of ARX-models subject to missing data. *IEEE Trans. On Automatic Control*, **38**, pp. 813-819.

Jones, R.H. (1980). Maximum likelihood fitting of ARMA models to time series with missing observations. *Technometrics*, **22**, pp. 389-395.

Kay S.M. and S.L. Marple (1981). Spectrum analysis-a modern perspective. *Proc. IEEE*, **69**, pp. 1380-1419.

Little, R.J.A. and D. B. Rubin (1987). *Statistical analysis with missing data*. New York: Wiley.

Ljung, L. (1999). *System Identification: Theory for the user (2nd ed.)*. Upper Saddle River NJ, Prentice Hall.

Lomb, N.R. (1976). Least squares frequency analysis of unequally spaced data. *Astrophysics and Space Science*, **39**, pp. 447-462.

Mirsaidi, S., G.A. Fleury and J. Oksman (1997). LMS-like AR modeling in the case of missing observations. *IEEE Trans. On Signal Processing*, **45**, pp. 1574-1583.

Petit, J. R., et al. (1999). Climate and atmospheric history of the past 420.00 years from the Vostok ice core, Antarctica. *Nature*, **399**, pp. 429-436.

Priestley, M.B. (1981). *Spectral Analysis and Time Series*. London, U.K.: Academic Press.

Rosen, Y. and B. Porat (1989). Optimal ARMA parameter estimation based on the sample covariances for data with missing observations. *IEEE Trans. On Information Theory*, **35**, pp. 342-349.

Scargle, J.D. (1989). Studies in astronomical time series analysis III. Fourier transforms, auto-correlation functions, and cross-correlation functions of unevenly spaced data. *The Astrophysics Journal*, **343**, pp. 874-887.

Waele, S. de and P. M. T. Broersen (1999). Reliable LDA-spectra by resampling and ARMA-modeling. *IEEE Trans. on Instrumentation and Measurement*, **48**, pp. 1117-1121.

Waele, S. de and P. M. T. Broersen (2000a). Error measures for resampled irregular data. *IEEE Trans. on Instrumentation and Measurement,* **49**, pp. 216-222.

Waele, S. de and P. M. T. Broersen (2000b). The Burg algorithm for segments. *IEEE Trans. on Signal Processing*, **48**, pp. 2876-2880.

Wallin, R., A.J. Isaksson and L. Ljung (2000). An iterative method for identification of ARX models from incomplete data. *Proc. CDC/IEEE Conference*, Sydney, Australia, pp. 203-208.

Yiou, P., E. Baert and M.F. Loutre (1996). Spectral analysis of climate data. *Surveys in Geophysics*, **17**, pp. 619-663.

IFAC Publications
www.elsevier.com/locate/ifac

ESTIMATING UNKNOWN PROBABILITY DENSITY FUNCTIONS FOR RANDOM PARAMETERS OF STOCHASTIC ARMAX SYSTEMS

Hong Wang and Yongji Wang *,[1]

* *Control Systems Centre,*
Department of Electrical Engineering and Electronics,
UMIST, Manchester, M60 1QD, The United Kingdom

Abstract: Different from existing parameter estimation algorithms where the values of parameters are required to be estimated, this paper presents a new method to estimate the unknown probability density functions of random parameters for non-Gaussian dynamic stochastic systems. The system is represented by an ARMAX model, where the parameters and the system noise term are random processes that are characterized by their unknown probability density functions. Under the assumption that each random parameter and the noise term are independent and are identically distributed sequence, a simple mathematical relationship is established between the measured output probability density function of the system and the unknown probability density functions of the random parameters and noise term. The mement generating function in probability theory has been used to transfer the multiple convolution integration into a simple algebraic operation. An identification algorithm is then established that estimates these unknown probability density functions of the parameters and the noise term by using the measured output probability density functions and the system input. *Copyright © 2003 IFAC*

Keywords: Dynamic stochastic systems; probability density function; ARMAX models; moment generating functions.

1. INTRODUCTION

Parameter estimation has long been recognized as an important research topic in control theory and applications. It forms a basis in constructing adaptive control and examples are seen in the well-known works (Astrom, 1970; Goodwin and Sin, 1984; Wellstead, 1995). In general, for time invariant parameter linear systems, the well-known least squares algorithm can be used. For parameters in nonlinear systems, extended Kalman filtering algorithms can be applied (Kalman, 1961). As for time-varying systems, modified least square algorithms can be used. However, in practice, there are some systems whose unknown parameters are random. Examples are some wet end chemical systems in paper-making (Smook, 1992; Wang, 1997, 2000) and wheat particle breakage systems in the food processing industry (Campbell and Webb, 2001). This does not include a large number of published research papers that look into the systems subjected to Markovian jumping parameters.

[1] Professor Yongji Wang is with the Automatic Control Department, HUST, China

Under the assumption that the system is subjected to a Gaussian input, in general parameter estimation using the least squares and the extended Kalman filtering can provide the information on the mean and the variance of the unknown parameters. However, this is also based upon the assumptions that the random parameters are Gaussian. When the system is subjected to non-Gaussian random parameters, the probability density function of the unknown random parameters cannot be easily estimated. This constitutes the main purpose of this paper, where a simple on-line identification algorithm is developed for the estimation of unknown probability density functions of random parameters and the noise term for linear ARMAX stochastic systems.

Indeed, the estimation of the arbitrary unknown probability density functions of random parameters in general non-Gaussian stochastic systems is important as it can provide global information on random parameters for closed loop control design. This will also help to construct control algorithms for the shape control of the output probability density functions, the subject that has been recently studied in (Karny, 1996; Wang, 1999; 2000, 2001, 2002).

In this paper, we consider a group of stochastic systems whose models are represented in ARMAX format. However, the parameters of the model are assumed random and are characterized by their unknown probability density functions. The task is therefore to use the measured output probability density function and the system inputs and outputs to estimate online these unknown probability density functions for the random parameters in the ARMAX model. With the development of online sensors, the output probability density function of some systems can now be measured (Wang, 2000).

2. MODEL REPRESENTATION FOR RANDOM PARAMETER SYSTEMS

In this paper, we consider the single-input and single-output physical model that relates the input sequence u_k, the output sequence of a system y_k and a stochastic noise term ω_k through the following ARMAX model (Wang,2002)

$$y_k = \sum_{i=1}^{n} a_i(k) y_{k-i} + \sum_{j=1}^{m} b_j(k) u_{k-j} + \omega_k \qquad (1)$$

where $y_k \in R^1$ and $u_k \in R^1$ are one dimensional output and input of the system, respectively, and $a_i(k) \in [\alpha, \beta]$, $(i = 1, 2, \cdots, n)$, $b_j(k) \in [\alpha, \beta]$, $(j = 1, 2, \cdots, m)$, and $\omega_k \in [\alpha, \beta]$ are Independent and Identically Distributed (IDD) bounded random parameters and external noise term. This means that for each of the indexes i and j, sequences $\{a_i(k)\}_{k=0,1,...}$, $\{b_j(k)\}_{k=0,1,...}$ and $\{\omega_k\}_{k=0,1,...}$ are independent random parameters and noise whose *unknown* probability density functions are defined from

$$P\{\alpha \le a_i(k) < \xi\} = \int_{\alpha}^{\xi} \gamma_{ai}(x)dx \qquad (2)$$

$$P\{\alpha \le b_j(k) < \xi\} = \int_{\alpha}^{\xi} \gamma_{bj}(x)dx \qquad (3)$$

$$P\{\alpha \le \omega_k < \xi\} = \int_{\alpha}^{\xi} \gamma_{\omega}(x)dx \qquad (4)$$

$$\xi \in [\alpha, \beta] \qquad (5)$$

where $\gamma_{ai}(x)$ and $\gamma_{bj}(x)$ are the unknown probability density functions for random parameters $a_i(k)$ and $b_j(k)$, respectively, $\gamma_{\omega}(x)$ is the unknown probability density function of ω_k. In equation (1), n and m are known structure orders of the system. Without loss of generality, we assume that

A1) all the parameters and ω_k in equation (1) are bounded. This means that α and β are finite numbers defined on $(-\infty, +\infty)$.

A2) there exists a known positive number M such that for $i = 1, 2, ..., n$ and $j = 1, 2, ..., m$

$$|\gamma_{ai}(x)| \le M; |\gamma_{bj}(x)| \le M; |\gamma_{\omega}(x)| \le M, \qquad (6)$$

for $\forall x \in [\alpha, \beta]$

A3) $\{u_k\}$ is bounded by M_u and the output sequence $\{y_k\}$ of equation (1) is also a bounded stochastic process at sample time k whose probability density function can also be defined on the bounded interval $[\alpha, \beta]$.

Assuming the current sample time is k, then the past inputs and outputs grouped by

$$\phi(k) = \{y_{k-1}, y_{k-2}, ..., y_{k-n}, u_{k-1}, ..., u_{k-m}\} \quad (7)$$

are measured values ready to be used to construct the required estimation algorithm. This indicates that at sample time k, output y_k is a linear combination of independent random parameters and the noise term through the measured values of $\phi(k)$. *The output probability density function at sample time k is in fact a conditional probability density function under available* $\phi(k)$. In this context, assumption **A3)** means that, given $\phi(k)$, the probability of y_k being inside the interval $[\alpha, \xi]$ is given by

$$P\{\alpha \le y_k < \xi|\phi(k)\} = \int_a^{\xi} \gamma_y(x|\phi(k))dx \quad (8)$$

where it can be seen that the conditional output probability density function $\gamma_y(x|\phi(k))$ of y_k is controlled by the control input in system (1). Since in many practical systems (e.g., papermaking, food processing, etc) the conditional output probability density function of the system is measurable (Wang, et, al, 1997, 2000), throughout this paper we assume that

A4) the triple $\{\gamma_y(x|\phi(k)), y_k, u_k\}$ are measurable at sample time k.

The purpose is to use the measured triple $\{\gamma_y(x|\phi(k)), y_k, u_k\}$ to estimate online the unknown probability density functions of the random parameters and the noise term, $\gamma_{ai}(x)$, $\gamma_{bj}(x)$, and $\gamma_\omega(x)$.

To develop an estimation algorithm for the probability density functions of the random parameters and the noise term, it is important to formulate a mathematical relationship that links the measured $\gamma_y(x|\phi(k))$ with unknown $\{\gamma_{ai}(x), \gamma_{bj}(x), \gamma_\omega(x)\}$ for all $i = 1, 2, ..., n$ and $j = 1, 2, ..., m$.

Under the assumption that all the random parameters and the random noise are independent, then from the results in probability theory, it can be shown that the probability density function $\gamma_y(x|\phi(k))$ is of an $(n+m+1)$-fold convolution of the form (Wang,2002)

$$\begin{aligned}
&\gamma_y(x|\phi(k)) = \\
&\int_{-\infty}^{x}\int_{-\infty}^{+\infty}\cdots\int_{-\infty}^{+\infty} \gamma_\omega(x - z_{n+m+1}) \\
&\prod_{j=0}^{m-1} \gamma_{b(m-j)}(z_{n+m-j} - z_{n+m-j-1}, u_{k-m+j}) \times \\
&\times \prod_{i=0}^{n-1} \gamma_{a(n-i)}(z_{n-i} - z_{n-i-1}, y_{k-n+i}) \\
&\times \gamma_{a1}(z_1, y_{k-1})dz_1dz_2\cdots dz_{n+m+1} \qquad (9)
\end{aligned}$$

where $\gamma_{ai}(x, y_{k-i})$ and $\gamma_{bj}(x, u_{k-j})$ are the conditional probability density function of random variables $a_i(k)y_{k-i}$, and $b_j(k)u_{k-j}$ for a given $\phi(k)$ at sample time k, respectively.

Since there are $(n+m+1)$-multiple integrations involved in expressing the output probability density function in (9), it is generally difficult to use equation (9) to construct an effective algorithm to estimate the unknown $\{\gamma_{ai}(x), \gamma_{bj}(x), \gamma_\omega(x)\}$.

This leads to the re-consideration of the expression for the output probability density function as shown in equation (9). The idea comes from the moment generating functions of probability density functions that will be described in the next section.

3. THE MOMENT GENERATING FUNCTIONS

In this section, we apply the moment generating functions (Melsa and Andrew, 1973) for general probability density functions that are defined on $[\alpha, \beta]$ interval. If we denote such a probability density function as $\gamma(x)$ with $x \in [\alpha, \beta]$, then its moment generating function is defined as

$$\Gamma(t) = \int_\alpha^\beta \gamma(x)e^{-tx}dx \qquad (10)$$

where $t \in [0, +\infty)$ is the real variable that defines the moment generating function $\Gamma(t)$ (Silverman, 1986).

Different from the standard moment generating function defined in (Silverman, 1986), "$-t$" is used in equation (10). However, this will not affect the formulations in the following text. Also, equation (10) is related to the Laplace transform where the variable t is taken as a real-valued.

Indeed, it can be shown that in terms of such a defined moment generating function, the convolution in the time domain variables can be transferred into an algebraic multiplication in the t-domain. As such, if we can apply the moment generating functions to all the probability density functions of the random parameters and the noise, the $(n + m + 1)$-fold convolutions in equation (9) can then be expressed as an algebraic multiplications of the transformed probability density functions (Wang, 2002). Since at sample time k, y_{k-i} and u_{k-j} are past outputs and inputs and are given, one can denote the moment generating functions of the **conditional** probability density functions of $a_i(k)y_{k-i}$ and $b_j(k)u_{k-j}$ as $\Gamma_{ai}(t, y_{k-i})$ and $\Gamma_{bj}(t, u_{k-j})$, respectively. Then by applying the moment generating operations to both sides of equation (9), it can be shown that

$$\Gamma_y(t|\phi(k)) = \Gamma_\omega(t)\prod_{j=1}^{m}\Gamma_{bj}(t, u_{k-j})\prod_{i=1}^{n}\Gamma_{ai}(t, y_{k-i}) \qquad (11)$$

where

$$\Gamma_y(t|\phi(k)) = \int_\alpha^\beta \gamma_y(x|\phi(k))e^{-tx}dx \quad (12)$$

$$\Gamma_{ai}(t, y_{k-i}) = \int_\alpha^\beta \gamma_{ai}(x, y_{k-i})e^{-tx}dx \quad (13)$$

$$\Gamma_{bj}(t, u_{k-j}) = \int_\alpha^\beta \gamma_{bj}(x, u_{k-j})e^{-tx}dx \quad (14)$$

where $\gamma_{ai}(x, y_{k-i})$ and $\gamma_{bj}(x, u_{k-j})$ are the **conditional** probability density functions of random variables $a_i(k)y_{k-i}$ and $b_j(k)u_{k-j}$, respectively. In probability theory, there are the following relationships between $\gamma_{ai}(x)$, $\gamma_{ai}(x, y_{k-i})$, $\gamma_{bj}(x)$ and $\gamma_{bj}(x, u_{k-j})$ as

$$\gamma_{ai}(x, y_{k-i}) = \frac{1}{|y_{k-i}|}\gamma_{ai}(\frac{x}{y_{k-i}}) \quad (15)$$

$$\gamma_{bj}(x, u_{k-j}) = \frac{1}{|u_{k-j}|}\gamma_{bj}(\frac{x}{u_{k-j}}) \quad (16)$$

We have, in terms of the t-variable, that

$$\Gamma_{ai}(t, y_{k-i}) = \Gamma_{ai}(y_{k-i}t) \quad (17)$$

and

$$\Gamma_{ai}(t, u_{k-j}) = \Gamma_{bj}(u_{k-j}t) \quad (18)$$

where $\Gamma_{ai}(t)$ and $\Gamma_{bj}(t)$ are the moment generating functions of the unknown probability density functions $\gamma_{ai}(x)$ and $\gamma_{bj}(x)$ in equations (2)-(3), respectively. As a result, equation (11) can be further expressed, in terms of the moment generating functions of the unknown probability density functions of random parameters $(a_i(k), b_j(k))$ and $\gamma_\omega(x)$, as

$$\Gamma_y(t|\phi(k)) = \Gamma_\omega(t)\prod_{j=1}^m \Gamma_{bj}(u_{k-j}t)\prod_{i=1}^n \Gamma_{ai}(y_{k-i}t) \quad (19)$$

Comparing equation (19) with equation (9), it can be seen that the use of the moment generating functions in equations (12)-(14) effectively simplifies the calculation of the output probability density functions from $\gamma_{ai}(x)$, $\gamma_{bj}(x)$ and $\gamma_\omega(x)$. This reveals the advantage of using the moment generating function to probability density functions. To further reduce the calculation of the multiplication, we introduce a logarithm operation to equation (19), this leads to

$$log(\Gamma_y(t|\phi(k))) = log(\Gamma_\omega(t)) + \sum_{j=1}^m log(\Gamma_{bj}(u_{k-j}t)) + \sum_{i=1}^n log(\Gamma_{ai}(y_{k-i}t)) \quad . \quad (20)$$

4. THE ESTIMATION OF PROBABILITY DENSITIES

To estimate $\{\gamma_{ai}(x), \gamma_{bj}(x), \gamma_\omega(x)\}$ using the measured past inputs, outputs and $\gamma_y(x|\phi(k))$, we just need to estimate $log(\Gamma_{ai}(y_{k-i}t))$, $log(\Gamma_{bj}(u_{k-j}t))$ and $log(\Gamma_\omega(t))$. Since variable t is defined on $[0, +\infty)$, we need to truncate the definition domain of functions $log(\Gamma_{ai}(y_{k-i}t))$, $log(\Gamma_{bj}(u_{k-j}t))$ and $log(\Gamma_\omega(t))$. Indeed, using the assumption **A2)**, it can be shown that

$$\Gamma_{ai}(t) = \int_\alpha^\beta \gamma_{ai}(x)e^{-tx}dx \le (\beta - \alpha)Me^{-\alpha t}$$
$$\Gamma_{bj}(t) = \int_\alpha^\beta \gamma_{bj}(x)e^{-tx}dx \le (\beta - \alpha)Me^{-\alpha t}$$
$$\Gamma_\omega(t) = \int_{-\alpha}^\beta \gamma_\omega(x)e^{-tx}dx \le (\beta - \alpha)Me^{-\alpha t} \quad . \quad (21)$$

As a result, it can be seen that when $t \to +\infty$, the following limitations should hold by considering equation (21).

$$lim_{t\to+\infty}\Gamma_{ai}(t) = lim_{t\to+\infty}\Gamma_{bj}(t) = 0$$
$$lim_{t\to+\infty}\Gamma_\omega(t) = 0 \quad (22)$$

This means that for any arbitrary small number $\epsilon > 0$, there is a large positive number $T > 0$ so that for all $\forall t > T$ the following inequalities hold for $i(= 1, 2, ..., n)$ and $j(= 1, 2, ..., m)$

$$|\Gamma_{ai}(t)| \le \epsilon; |\Gamma_{bj}(t)| \le \epsilon; |\Gamma_\omega(t)| \le \epsilon; \quad (23)$$

As such, we only need to estimate functions $\{\Gamma_{ai}(t), \Gamma_{bj}(t), \Gamma_\omega(t)\}$ for a fixed definition domain as specified by interval $t \in [0, T]$. Since the truncated functions $\{\Gamma_{ai}(t), \Gamma_{bj}(t), \Gamma_\omega(t)\}$ are continuous and defined on a bounded interval, B-spline approximations to these functions can be applied. In this context, there exists a big integer $N > 0$ and a set of basis functions $B_p(t), (p = 1, 2, ..., N)$, defined on $t \in [0, T]$, such that the following B-spline approximations (Girosi and Poggio, 1990)

$$log(\Gamma_{ai}(y_{k-i}t)) = \sum_{p=1}^N w_{ip}B_p(y_{k-i}t) + e_{ai}(t)$$
$$log(\Gamma_{bj}(u_{k-j}t)) = \sum_{p=1}^N v_{jp}B_p(u_{k-j}t) + e_{bj}(t),$$
$$log(\Gamma_\omega(t)) = \sum_{p=1}^N z_pB_p(t) + e_\omega(t)$$
$$\forall t \in [0, T] \quad (24)$$

hold, where $\{w_{ip}, v_{jp}, z_p\}$ are the unknown approximation coefficients to be estimated, and $\{e_{ai}(t), e_{bj}(t), e_\omega(t)\}$ are the approximation errors. At this stage, it can be seen that once the estimation for $\{w_{ip}, v_{jp}, z_p\}$ is made, the estimates for $\{\Gamma_{ai}(t), \Gamma_{bj}(t), \Gamma_\omega(t)\}$ can be readily obtained as all the basis functions are pre-specified. By selecting a proper set of basis functions, terms $\{e_{ai}(t), e_{bj}(t), e_\omega(t)\}$ can also be made to satisfy

$$max\{|e_{ai}(t)|, |e_{bj}(t)|, |e_\omega(t)|\} \le \frac{\epsilon}{(n+m+1)N} \quad (25)$$

where $\epsilon > 0$ is a small number. By substituting equation (25) into equation (21), it can be obtained that

$$log(\Gamma_y(t|\phi(k))) = \theta^T \eta(\phi(k), t) + e(t)$$
$$\forall t \in [0, T] \quad (26)$$
$$\theta^T = (z_1, z_2, ..., z_N, w_{11}, w_{12}, ..., w_{1N}, ..., w_{21}, w_{22}, ..., w_{2N}, ..., w_{n1}, w_{n2}, ..., w_{nN}, v_{11}, v_{12}, ..., v_{1N}, v_{21}, v_{22}, ..., v_{2N} v_{m1}, v_{m2}, ..., v_{mN}) \in R^{(n+m+1)N} \quad (27)$$
$$\eta(\phi(k), t)^T = (B_1(t), B_2(t), ..., B_N(t) B_1(y_{k-1}t), B_2(y_{k-1}t), ..., B_N(y_{k-1}t), ..., B_1(y_{k-n}t), B_2(y_{k-n}t), ..., B_N(y_{k-n}t), B_1(u_{k-1}t), B_2(u_{k-1}t), ..., B_N(u_{k-1}t), ..., B_1(u_{k-m}t), B_2(u_{k-m}t), ..., B_N(u_{k-m}t)) \in R^{(n+m+1)N} \quad (28)$$

where $e(t)$ is the combined approximation error obtained from

$$e(t) = e_\omega(t) + \sum_{i=1}^{n} e_{ai}(t) + \sum_{j=1}^{m} e_{bj}(t)$$

It can be shown using inequality (25) that such an error term should satisfy the following inequality

$$|e(t)| \le \epsilon, \ \forall t \in [0, T] \quad (29)$$

As a result, to estimate $\{\Gamma_{ai}(t), \Gamma_{bj}(t), \Gamma_\omega(t)\}$ for a fixed interval $t \in [0, T]$, it is sufficient to estimate the B-spline approximation coefficients grouped in vector θ. At this stage, the scaling least square estimation algorithm (Wang, 2000) can be readily used to estimate θ at sample time k. For this purpose, denote $\hat{\theta}(k)$ as the estimate of θ at sample time k, then by choosing a set of t-values in the interval $[0, T]$ as

$$\{t_1, t_2, ..., t_Q\}$$

where Q is another pre-specified integer, equation (26) leads to

$$log(\Gamma_y(t_q|\phi(k))) = \theta^T \eta(\phi(k), t_q) + e(t_q) \quad (30)$$

For equation (30), the following recursive least squares algorithm for index $q(= 1, 2, ..., Q)$ can be employed

$$\hat{\theta}_q = \hat{\theta}_{q-1} + \frac{P(q)\eta(\phi(k), t_q)\delta_{q,k}}{1 + \eta(\phi(k), t_q)^T P(q)\eta(\phi(k), t_q)}$$
$$\delta_{q,k} = log(\Gamma_y(t_q|\phi(k))) - \hat{\theta}^T_{q-1}\eta(\phi(k), t_q)$$
$$P^{-1}(q) = P^{-1}(q-1) + \eta(\phi(k), t_q)^T \eta(\phi(k), t_q) \quad (31)$$

where the initial values for $\hat{\theta}_q$ is $\hat{\theta}_1 = \hat{\theta}(k-1)$, and the resulting estimate for θ at sample time k is given by

$$\hat{\theta}(k) = \hat{\theta}_Q \quad (32)$$

Denote the estimated $\hat{\theta}(k)$ as

$$\hat{\theta}^T(k) = (z_1(k), z_2(k), ..., z_N(k), w_{1N}(k), ..., w_{21}(k), w_{22}(k), ..., w_{2N}(k), ..., w_{n1}(k), w_{n2}(k), ..., w_{nN}(k), v_{11}(k), v_{12}(k), ..., v_{1N}(k), v_{21}(k), v_{22}(k), ..., v_{2N}(k) v_{m1}(k), v_{m2}(k), ..., v_{mN}(k)) \in R^{(n+m+1)N} \quad (33)$$

then the estimated $log(\Gamma_{ai}(y_{k-i}t))$, $log(\Gamma_{bj}(u_{k-j}t))$, and $log(\Gamma_\omega(t))$ can be calculated from

$$log(\hat{\Gamma}_{ai}(y_{k-i}t)) = \sum_{p=1}^{N} w_{ip}(k)B_p(y_{k-i}t)$$
$$log(\hat{\Gamma}_{bj}(u_{k-j}t)) = \sum_{p=1}^{N} v_{jp}(k)B_p(u_{k-j}t)$$
$$log(\hat{\Gamma}_\omega(t)) = \sum_{p=1}^{N} z_p(k)B_p(t) \quad (34)$$

where the estimated moment generating functions of $\{\gamma_{ai}(x), \gamma_{bj}(x), \gamma_\omega(x)\}$ can be readily obtained. To summarize, the following algorithm can be obtained:-

(1) select a sufficient large T, an initial $\hat{\theta}(0)$ and $P(0)$,
(2) at sample time k, collect the measured $\{\gamma_y(x|\phi(k)), y_k\}$ and the past inputs and outputs $\{y_{k-i}, u_{k-j}\}$ for $i = 1, 2, ..., n$ and $j = 1, 2, ..., m$;
(3) calculate the moment generating function of $\gamma_y(x|\phi(k))$ as $\Gamma_y(t|\phi(k))$;
(4) select $\{t_1, t_2, ..., t_Q\}$ from the interval $[0, T]$;
(5) use the scaling least square algorithm (31) to estimate $\hat{\theta}(k)$;
(6) increase k by 1 and go back to step 2).

5. CONCLUSIONS

In this paper, an online identification algorithm has been developed for the estimation of the unknown probability density functions of the system random parameters and the noise term. Using the ARMAX representation and the assumption that the random parameters are independent and identically distributed sequences, a simple form of the transformed output probability density function can be obtained by using the moment generating function to all the probability density functions. A linear B-spline approximation is used to represent the logarithms of these transformed probability density functions and the weights of the B-spline expansion are estimated on-line via the scaling least squares algorithm similar to the one developed in (Wang, 2000).

The algorithm described in this paper can be readily applied to form an adaptive control framework for the shape and entropy control (Yue and Wang, 2003) of system output probability density functions (Wang, 2002). For example, when a target probability density function, $g(x)$ is given as defined on $x \in [\alpha, \beta]$, then to control the shape of $\gamma_y(x|\phi(k))$, one has to use equation (11) with estimated Γ_{ai} and Γ_{bj} to construct an estimate for $\gamma_y(x|\phi(k))$. This leads to

$$\hat{\Gamma}_y(t|\phi(k)) = \hat{\Gamma}_\omega(t) \prod_{j=1}^{m} \hat{\Gamma}_{bj}(t, u_{k-j}) \prod_{i=1}^{n} \hat{\Gamma}_{ai}(t, y_{k-i}) \tag{35}$$

If we denote $\Gamma_g(t)$ as the moment generating function of $g(x)$, then the following performance function can be used to measure the difference between $\hat{\Gamma}_y(t|\phi(k))$ and $\Gamma_g(t)$

$$J = \sum_{k=1}^{+\infty} \int_0^T (\hat{\Gamma}_y(t|\phi(k)) - \Gamma_g(t))^2 dt \tag{36}$$

As a result, by minimizing the performance function control input can be selected. This constitutes an effective closed loop adaptive control for the shape control of the conditional output probability density functions. Of course, the closed loop stability in probability sense should be established so as to guarantee its performance.

6. ACKNOWLEDGEMENTS

The authors would like to thank the financial support from the Leverhulme Trust under grant reference number F/00038D, and the Chinese NSF grants under (60128303, 69974017, 60274020).

REFERENCES

Astrom, K. J., (1970). *Introduction to Stochastic Control Theory*, Academic Press.

Campbell. G. M. , and Webb, C. (2001) On predicting roller milling performance,part I: the breakage equations, *Powder Technology*, Vol. 115,pp. 234 - 255.

Girosi, F. and Poggio, T. (1990) . Networks and the best approximation property, *Biol . Cybern.*, Vol. 63, pp. 169-176.

Karny, M. (1996) Towards fully probabilistic control design, *Automatica* Vol . 12, pp . 1719-1722.

Melsa J.L., Andrew P.S. (1973) *An Introduction to Probability and Stochastic Processes*, ed. Kailath T. Prentice Hall, Englewood Cliffs.

Silverman, B. W., (1986). *Density estimation of statistics and data analysis*, Chapman, Hall.

Smook, G. A., (1992). *Handbook for pulp and paper technologists*, Angus wilde Publication.

Wang, H. , (1999). Robust control of the output probability density functions for multivariable stochastic systems with guaranteed stability, *IEEE Trans on Automatic Control*, Vol. 44, pp. 2103-2107.

Wang, H., Wang, A. P., and Duncan, S. (1997). *Advanced Process Control for Paper, and Board Making*, PIRA International.

Wang, H. , (2000). *Bounded Dynamic Stochastic Distributions: Modelling and Control* , Springer-Verlag, London.

Wang, H. Kabore, P., and Baki, H . (2001) Lyapunov based design for bounded dynamic stochastic distribution control", *IEE Proc Control Theory and Applications*, Vol. 148, pp. 245 - 250.

Wang, H. and Zhang, J. H. , (2001) Bounded stochastic distribution control for pseudo ARMAX systems, *IEEE Transactions on Automatic Control*, Vol. 46, pp. 486-490.

Wang, H. (2002), Minimum entropy control of non -Gaussian dynamic stochastic systems *IEEE Transactions on Automatic Control*, Vol. 47, pp. 483-489.

Wang, Y. , and Wang, H. (2002) Output Probability density function control of linear stochastic systems with arbitrarily bounded random parameters a new application of the Laplace transforms, *Proc of the 2002 American Contr. Conf., pp. 4262 - 4267*

Yue, H. , and Wang, H. A., (2003) Minimum entropy control of closed loop tracking error, for dynamic Stochastic systems *IEEE Trans. , Automatic Control*, vol. 48, pp. 118 - 122.

IFAC Publications
www.elsevier.com/locate/ifac

ITERATIVE CONTROLLER TUNING BY MINIMIZATION OF A GENERALIZED DECORRELATION CRITERION [1]

Ljubiša Mišković Alireza Karimi Dominique Bonvin

Laboratoire d'Automatique
Ecole Polytechnique Fédérale de Lausanne
CH–1015 Lausanne, Switzerland
e-mail: ljubisa.miskovic@epfl.ch

Abstract: A controller tuning method based on the correlation approach is considered. A new, generalized, decorrelation criterion is proposed that allows tuning the controller parameters such that the reference signal be as little correlated as possible with both the input and output closed-loop errors. A frequency-domain analysis of the proposed criterion shows that the discrepancy between the true closed-loop system and the designed one is minimized in terms of the output *and* input sensitivity functions. Furthermore, it is shown that the noise has asymptotically no effect on the controller parameters. The theoretical results are illustrated via a simulation example.

Keywords: Data-driven control, iterative controller tuning, correlation approach

1. INTRODUCTION

Reliable mathematical descriptions of industrial plants are often difficult or impossible to obtain mainly due to the high complexity of the plants and/or the excessive cost of modelling. In these situations, the design of controllers using process information in the form of the experimental data collected under closed-loop operation seems to be a promising alternative to model-based design. Direct adaptive control (Åström and Wittenmark, 1989), iterative feedback tuning (Hjalmarsson, 2002), controller unfalsification (Safonov and Tsao, 1997) and control design based on simultaneous perturbation stochastic approximation (Spall and Cristion, 1998) are but a few examples of such data-driven methods.

In this line of research, a so-called iterative correlation-based tuning method has recently been proposed to address the model-following problem (Karimi *et al.*, 2002*a*; Karimi *et al.*, 2002*c*). The idea behind this approach is to tune the controller parameters to the extent that some external excitation signal be uncorrelated with the closed-loop output error between the true plant and the designed one. This way, the closed-loop output error is not affected by the model mismatch, and the output of the controlled plant tends towards the designed closed-loop output independently of the disturbance characteristics.

The correlation-based tuning approach has been applied to a magnetic suspension system in (Karimi *et al.*, 2002*b*), where the controller parameters are calculated as the solution of a correlation equation involving instrumental variables. Convergence and consistency of the controller parameters in the presence of disturbances and modeling errors has been analyzed in (Karimi *et al.*, 2002*a*). In (Karimi *et al.*, 2002*c*), the design objective is reformulated as the minimization of the 2-norm of the cross-correlation function between the closed-loop output error and the reference signal. Analysis of the proposed criterion in the

[1] This work is supported by Swiss National Science Foundation under grant No 2100-66876.01

frequency-domain shows that the algorithm, for a special case of instrumental variables, tries to minimize the integral of the difference between the achieved and designed output sensitivity functions weighted by the square of the spectrum of the reference signal. An adaptation of this approach to the regulation problem and its application to a benchmark problem posed for a special issue of European Journal of Control on the design and optimization of restricted-complexity controllers is treated in (Mišković *et al.*, 2002).

The tuning objective proposed in (Karimi *et al.*, 2002*c*) allows the achieved closed-loop system to approach the designed one in terms of the output sensitivity function. However, one could also make demand on the input sensitivity function. In order to handle mixed sensitivity specifications, this paper extends the criterion for controller tuning by adding the 2-norm of the cross-correlation function between the closed-loop input error and the reference signal. This way, the desired closed-loop output can be attained while taking into account some penalty on the control action, i.e. it is possible to make a trade-off between the specifications given in terms of the output sensitivity and those given in terms of the input sensitivity function. Analysis of the proposed generalized criterion in the frequency domain reveals the benefit of incorporating the new term.

The paper is organized as follows. Preliminary material and notations are given in Section 2. Section 3 briefly presents the correlation-based tuning approach. A generalization of the tuning criterion is developed in Section 4. In Section 5, controller tuning using the proposed criterion is illustrated via a numerical example. Finally, some concluding remarks are given in the last section.

2. PRELIMINARIES

Let the output of some unknown true plant be described by the discrete-time model:

$$y(t) = G(q^{-1})u(t) + v(t) \qquad (1)$$

where q^{-1} is the backward-shift operator, $G(q^{-1})$ is a linear time-invariant SISO discrete-time transfer operator, u(t) the input signal to the plant and $v(t)$ a disturbance signal. It is assumed that $v(t)$ is a zero-mean weakly stationary random process.

Consider the closed-loop system depicted in Fig.1, where $K(q^{-1}, \rho)$ is a linear time-invariant transfer function parametrized by the vector $\rho \in \mathcal{R}^{n_\rho}$, and $r(t)$ is an external excitation signal. It is assumed that measurements of $r(t)$ and $y(t)$ are available. The excitation signal $r(t)$ is assumed to be uncorrelated with the disturbance signal $v(t)$.

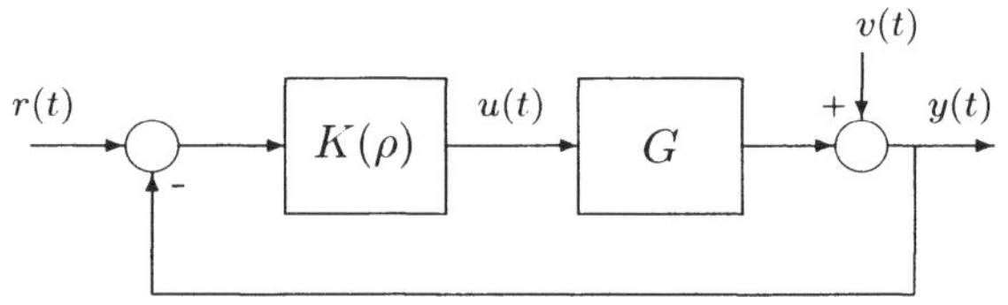

Fig. 1. Controlled plant

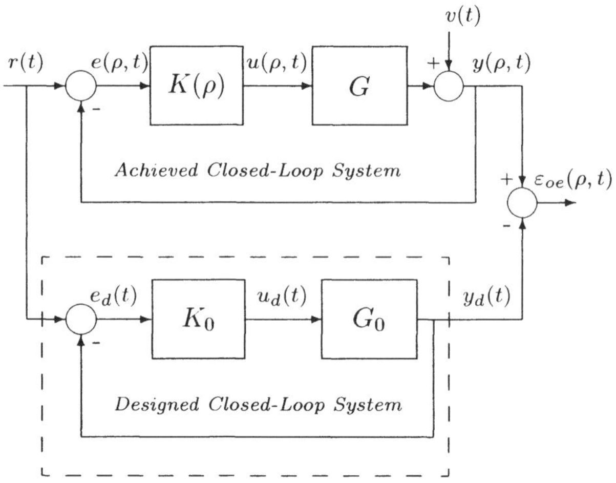

Fig. 2. Closed-loop output error resulting from a comparison of the achieved and designed closed-loop systems

As far as the notations are concerned, the signals collected under closed-loop operation using the controller $K(q^{-1}, \rho)$ will carry the argument ρ. Furthermore, to ease the notation, the backward-shift operator q^{-1} will be omitted in the sequel.

Let us define the following sensitivity functions:

- Output sensitivity function:

$$S(K, G) = (1 + KG)^{-1} \qquad (2)$$

- Input sensitivity function:

$$U(K, G) = K(1 + KG)^{-1} \qquad (3)$$

- Complementary sensitivity function:

$$T(K, G) = KG(1 + KG)^{-1} \qquad (4)$$

3. CORRELATION-BASED TUNING

A block diagram of the model-following problem is represented in Fig. 2. The upper part shows the achieved closed-loop system with the true plant, while the lower part is a realization of the designed closed-loop system that includes the plant model (G_0) and the initial controller (K_0). It is assumed that the initial controller is capable of meeting the specifications of the designed closed-loop system.

The closed-loop output error is defined as:

$$\varepsilon_{oe}(\rho, t) = y(\rho, t) - y_d(t) \qquad (5)$$

where $y(\rho, t)$ is the output of the achieved closed-loop system, and $y_d(t)$ the output of the designed closed-loop system.

Let the initial controller K_0 be applied to the true plant excited by the reference signal $r(t)$. Then, the closed-loop output error contains a contribution due to the difference between G and G_0 (modeling errors) and another contribution stemming from the disturbance $v(t)$. The effect of modeling errors is correlated with the reference signal, whereas that of disturbance is not. Thus, a reasonable way to tune the controller is to make the closed-loop output error $\varepsilon_{oe}(\rho, t)$ uncorrelated with the excitation signal $r(t)$. This way, the improved controller compensates the effect of modeling errors to the extent that the closed-loop output error contains only the filtered disturbance. However, since in practice perfect decorrelation between these two signals cannot be achieved, it is natural to define the tuning objective as the minimization of some norm of the cross-correlation function between $\varepsilon_{oe}(\rho, t)$ and $r(t)$.

Let define the correlation function $f_{oe}(\rho)$:

$$f_{oe}(\rho) = E\{\zeta(t)\varepsilon_{oe}(\rho, t)\} \qquad (6)$$

where $E\{\cdot\}$ is the mathematical expectation and $\zeta(t)$ a vector of instrumental variables that are correlated with the reference signal $r(t)$ and independent of the disturbance $v(t)$. Then, the tuning objective can be defined as the minimization of the following criterion:

$$J_{oe}(\rho) = ||f_{oe}(\rho)||_2^2 = f_{oe}^T(\rho) f_{oe}(\rho) \qquad (7)$$

where $||\cdot||_2$ represents the 2-norm. The control parameter vector ρ^* is given by:

$$\rho^* = \arg\min_{\rho} J_{oe}(\rho) \qquad (8)$$

Since this problem cannot be solved analytically, a numerical method is considered. The vector ρ^* is solution of the following gradient equation:

$$J'_{oe}(\rho) = f_{oe}^T(\rho)\frac{\partial f_{oe}(\rho)}{\partial \rho} = 0 \qquad (9)$$

This problem can be solved by the Robbins-Monro procedure using the following iterative formula (Robbins and Monro, 1951):

$$\rho_{i+1} = \rho_i - \gamma_i \left[Q(\rho_i)\right]^{-1} \left[J'_{oe}(\rho_i)\right]^T \qquad (10)$$

where γ_i is a scalar step size and $Q(\rho_i)$ a positive definite matrix. Under the assumption of boundedness of the signals in the loop, and with a step size tending to zero appropriately fast, this scheme converges to a local minimum of the criterion as the number of iterations tends to infinity (Karimi *et al.*, 2002*c*).

The gradient of the criterion involves the expectation of signals that are unknown and should be replaced by their estimates from closed-loop data. Let the correlation function be estimated by $\hat{f}_{oe}(\rho)$:

$$\hat{f}_{oe}(\rho) = \frac{1}{N}\sum_{t=1}^{N} \zeta(t)\varepsilon_{oe}(\rho, t) \qquad (11)$$

where N is the number of data points. Then, the derivative of the criterion is determined as follows:

$$J'_{oe}(\rho_i) = \hat{f}_{oe}^T(\rho_i)\,\frac{1}{N}\sum_{t=1}^{N} \zeta(t)\left.\frac{\partial \varepsilon_{oe}(\rho, t)}{\partial \rho}\right|_{\rho_i} \qquad (12)$$

An accurate value of this gradient cannot be computed because the derivative of $\varepsilon_{oe}(\rho, t)$ with respect to ρ is unknown. However, an unbiased model-free estimation can be obtained using two extra closed-loop experiments as is done in the IFT approach (Hjalmarsson, 2002). Note that the gradient could also be obtained from a plant model that is identified, for example, using closed-loop data (Karimi *et al.*, 2002*b*).

In order to improve the convergence speed, $Q(\rho_i)$ can be chosen as an approximation of the Hessian of the criterion (Gauss-Newton direction):

$$Q(\rho_i) = \left(\left.\frac{\partial \hat{f}_{oe}(\rho)}{\partial \rho}\right|_{\rho_i}\right)^T \left.\frac{\partial \hat{f}_{oe}(\rho)}{\partial \rho}\right|_{\rho_i} + \lambda I \qquad (13)$$

where the parameter λ should be chosen so as to ensure positive definiteness of the matrix $Q(\rho_i)$.

4. GENERALIZING THE CRITERION

In (Karimi *et al.*, 2002*c*), the frequency characteristics of the achieved closed-loop system have been compared with those of the designed closed-loop system for the following choice of instrumental variables:

$$\zeta^T(t) = [r(t + n_z), \ldots, r(t), \ldots, r(t - n_z)] \qquad (14)$$

where n_z is a sufficiently large integer number w.r.t. the order of the closed-loop system. On the other hand, the value of n_z should be much smaller than the number of data N in order to have an accurate estimation of the cross-correlation function. Analysis of the criterion of Eq. 7 has shown that the algorithm minimizes the integral of the difference between the achieved and the designed complementarity sensitivity functions weighted by the square of the reference signal spectrum $\Phi_r(\omega)$:

$$\lim_{n_z \to \infty} J_{oe}(\rho) = \frac{1}{2\pi}\int_{-\pi}^{\pi} |H_{oe}(e^{-j\omega}, \rho)|^2 \Phi_r^2(\omega) \qquad (15)$$

where $H_{oe}(\rho) = T(K(\rho), G) - T_0$ with $T_0 = T(K_0, G_0)$ being the designed complementary sensitivity function. If $r(t)$ is white noise with variance 1, and n_z tends to infinity, one has:

$$\rho^* = \arg\min_{\rho} \int_{-\pi}^{\pi} |T(e^{-j\omega}, \rho) - T_0(e^{-j\omega})|^2 d\omega$$
$$= \arg\min_{\rho} \int_{-\pi}^{\pi} |S(e^{-j\omega}, \rho) - S_0(e^{-j\omega})|^2 d\omega \quad (16)$$

where $S_0 = S(K_0, G_0)$ is the designed output sensitivity function. These relations show that both the achieved complementary sensitivity function $T(e^{-j\omega}, \rho)$ and the achieved output sensitivity function $S(e^{-j\omega}, \rho)$ tend to their respective designed functions. Thus, the tuned controller ensures the designed performance for the true plant with respect to tracking and output disturbance rejection.

However, when minimizing the criterion of Eq. 7, the achieved input sensitivity function $U(e^{-j\omega}, \rho)$ does not necessarily approach U_0. At some frequencies, $U(e^{-j\omega}, \rho)$ obtained by controller tuning may grow large, thus affecting robust stability. In addition, the controlled input $u(t)$ may exert a substantial effort on the actuators. To overcome this difficulty, the criterion can be generalized so as to incorporate the new term containing the 2-norm of the cross-correlation function between the closed-loop input error and the reference signal. This way, not only the output but also the input of the achieved closed-loop system will follow respectively the output and the input of the designed closed-loop system independently of the disturbance dynamics. Thus, let us modify the criterion of Eq. 7 as follows:

$$J(\rho) = k_{oe}||f_{oe}(\rho)||_2^2 + k_{ie}||f_{ie}(\rho)||_2^2 \quad (17)$$

where k_{oe} and k_{ie} are positive scalar weighting factors, and $f_{ie}(\rho)$ is the correlation function:

$$f_{ie}(\rho) = E\{\zeta(t)\varepsilon_{ie}(\rho, t)\} \quad (18)$$

The closed-loop input error $\varepsilon_{ie}(\rho, t)$ is given by:

$$\varepsilon_{ie}(\rho, t) = u(\rho, t) - u_d(t) \quad (19)$$

where $u_d(t)$ is the control input of the designed closed-loop system (see Fig. 3).

From Figs. 2 and 3, $\varepsilon_{oe}(\rho, t)$ and $\varepsilon_{ie}(\rho, t)$ can be written as:

$$\varepsilon_{oe}(\rho, t) = (T(\rho) - T_0)\, r(t) + S(\rho)v(t)$$
$$= H_{oe}(\rho)r(t) + S(\rho)v(t) \quad (20)$$

and

$$\varepsilon_{ie}(\rho, t) = H_{ie}(\rho)r(t) - U(\rho)v(t) \quad (21)$$

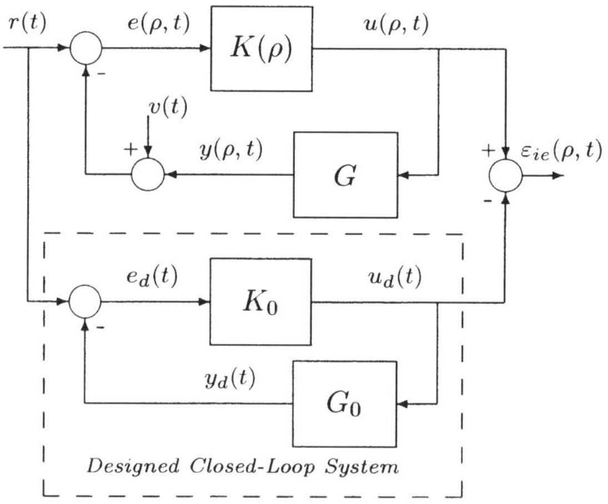

Fig. 3. Closed-loop input error resulting from a comparison of the achieved and designed closed-loop systems

where $H_{ie}(\rho) = U(\rho) - U_0$ with $U_0 = U(K_0, G_0)$ being the designed input sensitivity function.

Considering the vector of instrumental variables given in Eq. 14, and letting n_z tend to infinity, gives asymptotically (after straightforward calculations similar to those in (Karimi *et al.*, 2002*c*)):

$$J(\rho) = \frac{1}{2\pi} \int_{-\pi}^{\pi} \{k_{oe}|H_{oe}(e^{-j\omega}, \rho)|^2 +$$
$$k_{ie}|H_{ie}(e^{-j\omega}, \rho)|^2\} \, \Phi_r^2(\omega) d\omega \quad (22)$$

If $r(t)$ is white noise with variance 1, one has:

$$\rho^* = \arg\min_{\rho} \int_{-\pi}^{\pi} \{k_{oe}|T(e^{-j\omega}, \rho) - T_0(e^{-j\omega})|^2$$
$$+ k_{ie}|U(e^{-j\omega}, \rho) - U_0(e^{-j\omega})|^2\} \, d\omega$$
$$= \arg\min_{\rho} \int_{-\pi}^{\pi} \{k_{oe}|S(e^{-j\omega}, \rho) - S_0(e^{-j\omega})|^2$$
$$+ k_{ie}|U(e^{-j\omega}, \rho) - U_0(e^{-j\omega})|^2\} \, d\omega \quad (23)$$

This relation shows that there is a trade-off between the minimization of $||S(\rho) - S_0||_2$ and that of $||U(\rho) - U_0||_2$. By minimizing this criterion, the mixed sensitivity specifications are satisfied, and the achieved closed-loop system tries to preserve the robustness properties of the designed one. Furthermore, it is easy to see that the criterion of Eq. 17 is not influenced by the disturbance signal $v(t)$. With regard to this criterion, two extreme cases can be considered: (i) When $(k_{oe}, k_{ie}) = (1, 0)$, Eq. 23 reduces to Eq. 16 and $S(\rho)$ is forced towards S_0; (ii) when $(k_{oe}, k_{ie}) = (0, 1)$, $U(\rho)$ is pushed towards its designed function U_0.

5. SIMULATION EXAMPLE

In this section, the properties of the proposed tuning method are illustrated via an example.

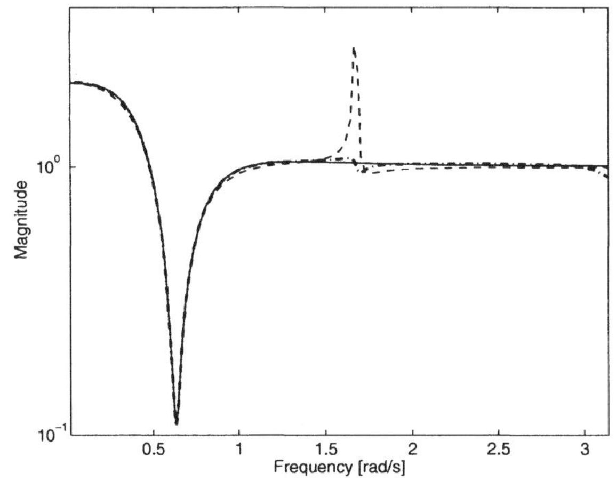

Fig. 4. Output sensitivity functions $S_0(e^{j\omega})$ (solid), $S_{init}(e^{j\omega})$ (dash) and $S(e^{j\omega})$ (dash-dot) for $(k_{oe}, k_{ie}) = (1, 0)$

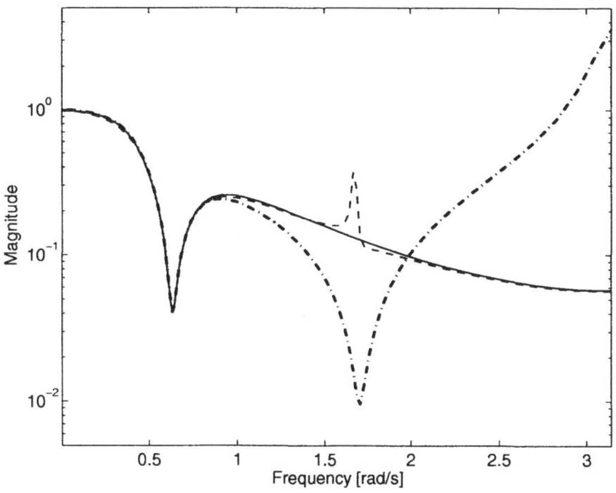

Fig. 5. Input sensitivity functions $U_0(e^{j\omega})$ (solid), $U_{init}(e^{j\omega})$ (dash) and $U(e^{j\omega})$ (dash-dot) for $(k_{oe}, k_{ie}) = (1, 0)$

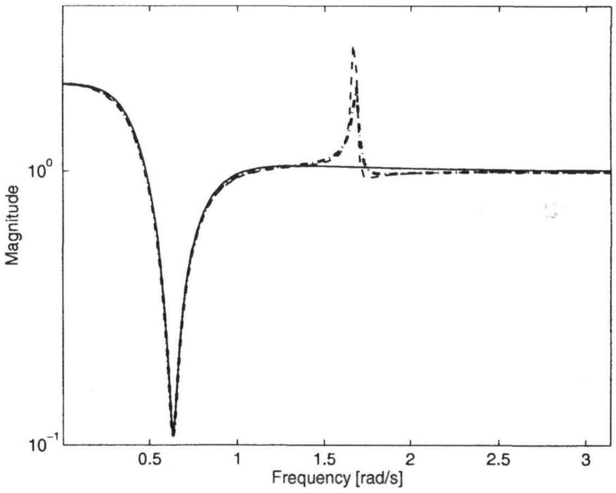

Fig. 6. Output sensitivity functions $S_0(e^{j\omega})$ (solid), $S_{init}(e^{j\omega})$ (dash) and $S(e^{j\omega})$ (dash-dot) for $(k_{oe}, k_{ie}) = (0, 1)$

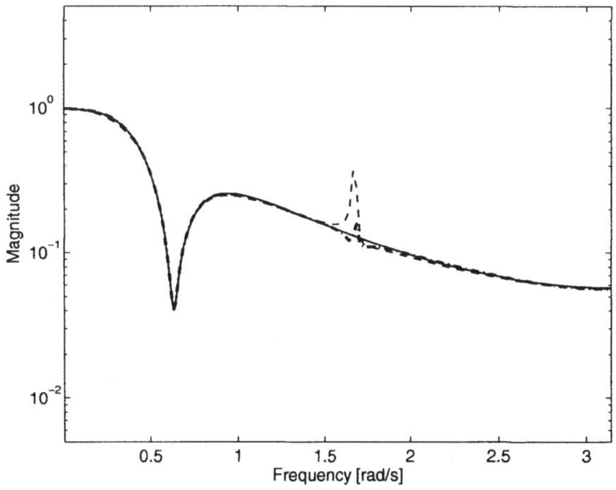

Fig. 7. Input sensitivity functions $U_0(e^{j\omega})$ (solid), $U_{init}(e^{j\omega})$ (dash) and $U(e^{j\omega})$ (dash-dot) for $(k_{oe}, k_{ie}) = (0, 1)$

Consider the following 4th-order true plant:

$$G = \frac{0.385q^{-2} + 0.525q^{-3}}{1 - 1.353q^{-1} + 1.55q^{-2} - 1.282q^{-3} + 0.915q^{-4}}$$

The system has two very lightly damped resonant modes and one unstable zero. The following second-order model G_0 has been identified:

$$G_0 = \frac{0.6043q^{-2} - 0.1562q^{-3} - 0.0306q^{-4}}{1 - 1.5822q^{-1} + 0.9629q^{-2}}$$

Let the initial 3rd-order controller K_0 be:

$$K_0 = \frac{-0.1530q^{-1} - 0.038q^{-2}}{1 - 0.8093q^{-1} + 0.2141q^{-2} - 0.012q^{-3}}$$

When K_0 is applied to the true plant G, there is significant deterioration of the performance due to model mismatch (see dashed line in Figs. 4 and 5). To improve the behaviour of the closed-loop system, a 4th-order controller K is to be tuned on the true plant for three different choices of the weighting factors k_{oe} and k_{ie}. The tuning procedure is carried out in 8 iterations, with each iteration being performed using a different realization of the disturbance signal $v(t)$ with a noise-to-signal ratio of 7% in terms of variance. The vector of instrumental variables is chosen as in Eq. 14 with $n_z = 72$, and the reference signal $r(t)$ is a PRBS generated by a 7-bit shift register with data length $N = 2048$. In all iterations, the initial step size $\gamma_i = 0.5$ is used. If the algorithm provides a controller that destabilizes the closed-loop system, the step-size is then divided by 2.

The first choice of weighting factors $(k_{oe}, k_{ie}) = (1, 0)$ corresponds to the minimization of $||H_{oe}||_2$. Fig. 4 shows the output sensitivity functions S_0, $S_{init} = S(K_0, G)$ and S for the designed, initial and final closed-loop systems, respectively. It can be seen that S_0 and S are almost superposed, i.e. the tuning algorithm has succeeded in minimizing H_{oe} to a large extent. However, comparing the corresponding input sensitivity functions U_0, $U_{init} = K_0 S_{init}$ and U shown in Fig. 5, it is easy to see that U gets large at high frequencies which can significantly deteriorate the robustness of the closed-loop system.

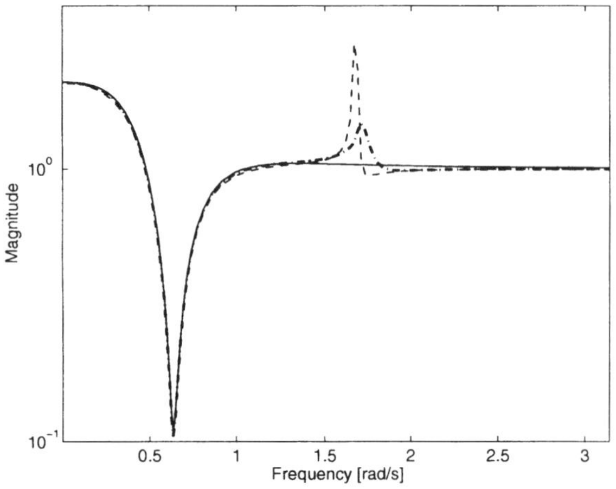

Fig. 8. Output sensitivity functions $S_0(e^{j\omega})$ (solid), $S_{init}(e^{j\omega})$ (dash) and $S(e^{j\omega})$ (dash-dot) for $(k_{oe}, k_{ie}) = (0.5, 0.5)$

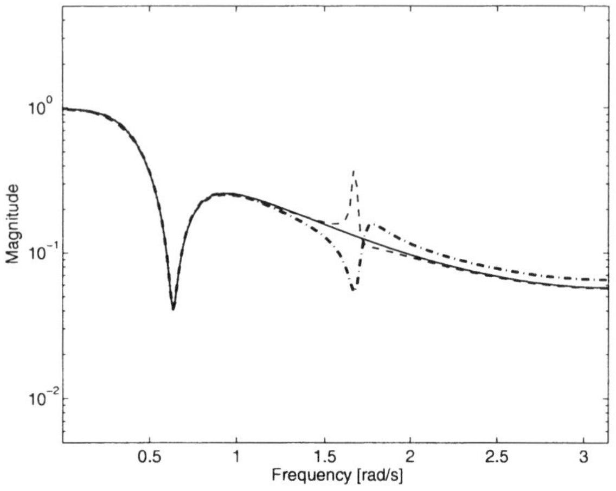

Fig. 9. Input sensitivity functions $U_0(e^{j\omega})$ (solid), $U_{init}(e^{j\omega})$ (dash) and $U(e^{j\omega})$ (dash-dot) for $(k_{oe}, k_{ie}) = (0.5, 0.5)$

For $(k_{oe}, k_{ie}) = (0, 1)$, $||H_{ie}||_2$ is minimized. Figs. 6 and 7 depict the corresponding output sensitivities S_0, S_{init} and S, and input sensitivities U_0, U_{init} and U. A comparison of the curves shows that, though the resulting controller K has not succeeded in reducing the peak of the output sensitivity function S, the final input sensitivity function U is very similar to U_0.

Finally, for the case $(k_{oe}, k_{ie}) = (0.5, 0.5)$, there is a trade-off in minimizing $||H_{oe}||_2$ and $||H_{ie}||_2$. Figs. 8 and 9 shows that the resulting controller K has reduced the peak of the output sensitivity function S and, at the same time, the discrepancy between U_0 and U remains small.

Table 1 gives the performance of the tuning procedure in function of the weighting factors k_{oe} and k_{ie}. These numerical results confirm the qualitative shapes seen in Figs. 4-9. The minima of $||H_{oe}||_2$ and $||H_{ie}||_2$ are achieved for $(k_{oe}, k_{ie}) = (1, 0)$ and $(k_{oe}, k_{ie}) = (0, 1)$, respectively. However, when minimizing only $||H_{oe}||_2$ or $||H_{ie}||_2$, the deviation of the other sensitivity does increase. In contrast, the controller obtained with $(k_{oe}, k_{ie}) = (0.5, 0.5)$ reduces both $||H_{oe}||_2$ and $||H_{ie}||_2$.

Table 1. Results of tuning

	Iteration	$\|\|H_{oe}\|\|_2$	$\|\|H_{ie}\|\|_2$
	1st	0.3002	0.0388
$k_{oe} = 1, k_{ie} = 0$	8th	0.0284	0.4209
$k_{oe} = 0, k_{ie} = 1$	8th	0.1493	0.0091
$k_{oe} = k_{ie} = 0.5$	8th	0.1018	0.0284

6. CONCLUSIONS

An extension of the controller-tuning criterion based on the correlation approach has been proposed. The new criterion is defined as the weighted sum of the 2-norms of the cross-correlation functions between a reference signal and the output and input closed-loop errors. If the assumption of independence between the reference signal and the disturbance holds, the criterion remains asymptotically unaffected by the disturbance characteristics. A frequency-domain analysis of the proposed criterion has shown that, depending on the values of the weighting factors k_{oe} and k_{ie}, there is a trade-off in meeting the designed output and input sensitivities. Simulation results illustrate the features and the applicability of the new tuning approach.

7. REFERENCES

Åström, K. J. and B. Wittenmark (1989). *Adaptive Control.* Addison-Wesley.

Hjalmarsson, H. (2002). Iterative feedback tuning - an overview. *Int. Journal of Adaptive Control and Signal Processing* **16**, 373–395.

Karimi, A., L. Mišković and D. Bonvin (2002*a*). Convergence analysis of an iterative correlation based controller tuning method. In: *15th IFAC World Congress, Barcelona, Spain.*

Karimi, A., L. Mišković and D. Bonvin (2002*b*). Iterative correlation-based controller tuning: Application to a magnetic suspension system. *Control Engineering Practice, to appear.*

Karimi, A., L. Mišković and D. Bonvin (2002*c*). Iterative correlation-based controller tuning: Frequency-domain analysis. In: *41st IEEE-CDC, Las Vegas, USA.*

Mišković, L., A. Karimi and D. Bonvin (2002). Correlation-based tuning of a restricted complexity controller for an active suspension system. *Submitted to European J. of Control.*

Robbins, H. and S. Monro (1951). A stochastic approximation method. *Ann. Math. Stat.* **22**, 400–407.

Safonov, M. G. and T-C. Tsao (1997). The unfalsified control concept and learning. *IEEE Trans. on Automatic Control* **42**(6), 843–847.

Spall, J. C. and J. A. Cristion (1998). Model-free control of nonlinear stochastic systems with discrete-time measurements. *IEEE Trans. on Automatic Control* **43**(9), 1198–1210.

IFAC Publications
www.elsevier.com/locate/ifac

SUBSPACE IDENTIFICATION BASED PID CONTROL TUNING

A.Sanchez [*,1], **M.R. Katebi** [*] **M.A. Johnson** [*]

[*] *Industrial Control Centre, University of Strathclyde*
Glasgow G1 1QE, Scotland

Abstract: This paper introduces a new method to design a PID controller using a data driven approach for identification. The tuning algorithm is developed within the subspace identification framework, which is used to identify an open loop model from closed-loop data. The PID parameters are calculated by minimising a quadratic performance index over a finite future horizon. The inclusion of a closed-loop condition to guarantee stability results in a constrained nonlinear optimisation problem. The paper includes simulation results for the case of a dissolved oxygen control loop in an activated sludge wastewater treatment plant. *Copyright © 2003 IFAC*

Keywords: subspace identification, LQG, PID control, tuning

1. INTRODUCTION

This paper reports a new data driven algorithm to design a PID controller. The algorithm is presented within the subspace identification framework, and has been developed for a single loop case; however, it could be extended for a multivariable system with mutiple loops.

The design method conveys the identification of an open loop model for the plant and a nonlinear constrained optimisation. The first step is to identify a model for the plant. In most practical cases control loops are in operation in the process, so collected data is of closed-loop nature. Due to this circumstance, it is more realistic to use a closed-loop identification routine to obtain a plant model. The second step is the calculation of the controller parameters. This is achieved by minimising a parameterised cost function. The design assures that the closed-loop response will approach the reference asymptotically over a finite future horizon while keeping the PID parameters constant.

PID controllers have become the industry preferred process controller due to its simplicity of use and ease of understanding. Over the years several algorithms to tune and design this controller has been developed, begining from the traditional Ziegler-Nichols rules introduced in 1942. More advanced tuning methods have appeared periodically over time with more or less complexity and each trying to address different problems in the determination of adequate PID parameters. For example Hjalmarsson *et al.* (1994) address the problem of a model free tuning, by developing an iterative algorithm. Other recent approaches use the idea of trying to make the PID response as close as possible to that obtainable from more sophisticated methods such as presented in (Katebi and Moradi 2001), where an MPC framework is employed. Other methods approach the design of PID controllers by using the method of optimal reduced order controllers, which consists in restricting the optimal control design to one of limited parameters. A different approach is to tune the controller dynamically by minimising a GPC criterion in each time instant as reported in (Uduehi *et al.* 2002). The resulting controller is said to be optimal in the GPC sense. The approach followed in this paper

[1] Author for correspondence: asanchez@eee.strath.ac.uk

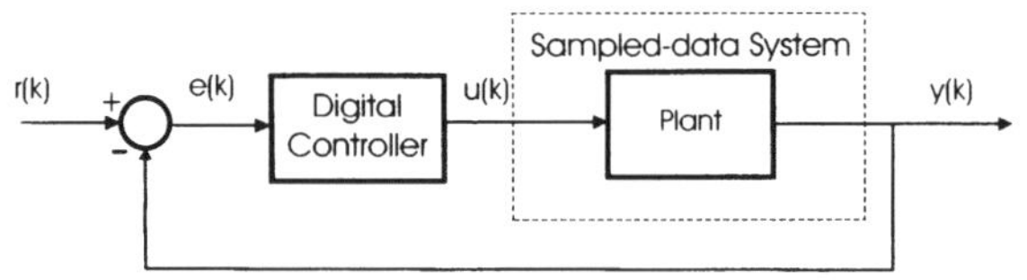

Figure 1. Closed-Loop System

exploits the principle of restricting the solution of the LQG problem to a three-term controller.

The paper is organized in the following way: the algorithm to identify the plant is presented in section 2. Subsequently the parameterisation and solution of the cost function is given in section 3. Simulation results for a simple application case are presented in section 4, and finally at the end of the paper, conclusions are drawn.

2. CLOSED-LOOP SUBSPACE IDENTIFICATION

The initial part of this section is based on (Favoreel *et al.* 1998) and (van Overschee and De Moor 1996), where an algorithm to identify a linear model for a plant from closed-loop measurements using subspace identification is developed. The second part of this section is based on the results from (Kadali *et al.* 2003), where the subspace model is modified to include control increments.

2.1 Subspace model representation

For this paper, it is not necessary to obtain the full state-space system representation but a simplified model as in equation (1) for the closed-loop discrete system depicted in Figure (1).

Consider the single-input single-output system operating in closed-loop shown in Figure (1). Let the closed-loop system be represented by the model in equation (1).

$$Y_f = L_w^c \cdot W_p + L_u^c \cdot M_f \tag{1}$$

Consider then a sufficiently large amount of data {u(k)} and {y(k)} and knowledge of the PID controller parameters so the past and future block Hankel matrices for u(k) and y(k) can be constructed by considering a backward horizon of dimension M and a future horizon of dimension N, as in equations (2) and (3) for the case of u(k).

$$U_p = \begin{bmatrix} u_0 & u_1 & \cdots & u_{j-1} \\ u_1 & u_2 & \cdots & u_j \\ \vdots & \vdots & \ddots & \vdots \\ u_{M-1} & u_M & \cdots & u_{M+j-2} \end{bmatrix} \tag{2}$$

$$U_f = \begin{bmatrix} u_M & u_{M+1} & \cdots & u_{M+j-1} \\ u_{M+1} & u_{M+2} & \cdots & u_{M+j} \\ \vdots & \vdots & \ddots & \vdots \\ u_{N+M-1} & u_{N+M} & \cdots & u_{N+M+j-2} \end{bmatrix} \tag{3}$$

Define also the matrix W_p as in equation (4) and M_f as in equation (5), where the matrix H_N^c is the block Toeplitz matrix of the controller and is defined by equation (6).

$$W_p = \begin{bmatrix} Y_p \\ U_p \end{bmatrix} \tag{4}$$

$$M_f = U_f + H_N^c Y_f \tag{5}$$

$$H_N^c = \begin{bmatrix} D_c & 0 & \cdots & 0 \\ C_c B_c & D_c & \cdots & 0 \\ \vdots & \vdots & \ddots & \vdots \\ C_c A_c^{N-2} B_c & C_c A_c^{N-3} B_c & \cdots & D_c \end{bmatrix} \tag{6}$$

It is then possible to find L_w^c and L_u^c by minimising (7) using a *QR-factorisation*.

$$\min_{L_w^c, L_u^c} \left\| Y_f - \begin{pmatrix} L_w^c & L_u^c \end{pmatrix} \begin{pmatrix} W_p \\ M_f \end{pmatrix} \right\|_F^2 \tag{7}$$

Finally, the open-loop matrices L_w and L_u can be found by using equations (8) and (9). L_w must be approximated to a rank deficient matrix with rank-n, where 'n' is found by a *Singular Value Decomposition*.

$$L_u = L_u^c (I - H_N^c L_u^c)^{-1} \tag{8}$$

$$L_w = (I + L_u H_N^c L_u) L_w^c \tag{9}$$

The plant model is given as a function of the future input vector $\hat{u}_f$ and the past input-output vector w_p as in equation (10),where $\hat{u}_f$ and w_p are defined in equations (11) and (12). In particular, $\hat{u}_f$ is a vector of future control inputs of length *N*.

$$\hat{y}_f = L_w \cdot w_p + L_u \cdot \hat{u}_f \tag{10}$$

$$\hat{u}_f = \begin{bmatrix} u_1 \\ \vdots \\ u_N \end{bmatrix} \tag{11}$$

$$w_p = \begin{bmatrix} y_{-M+1} \\ \vdots \\ y_0 \\ --- \\ u_{-M+1} \\ \vdots \\ u_0 \end{bmatrix} \quad (12)$$

2.2 *Incremental subspace representation*

Equation (10) gives the best prediction of the output $\hat{y}_f$ given the future inputs $\hat{u}_f$ and past output-inputs w_p; however, in this case and in many control approaches it is sometimes more useful to have a model defined in terms of the changes in the control signal rather than the signal itself. To do such a modification several approaches have been suggested in the literature as in (Ruscio 1997*a*, Ruscio 1997*b*) and (Kadali *et al.* 2003). In particular the method described in (Kadali *et al.* 2003) uses an integrated noise model. This approach is convenient for this application since it involves a stochastic model, which as will be described later, simplifes the solution.

As described in (Kadali *et al.* 2003), by considering an integrating white noise model as in equations (13) and (14), the plant described by equation (10) can be written in incremental form as in equation (15) for a single input - single output case.

$$e_{k+1} = e_k + a_k \quad (13)$$

$$e_k = \frac{a_k}{\Delta} \quad (14)$$

$$\hat{y}_f = y_t + L_w^{\Delta} \cdot \Delta w_p + L_u^{\Delta} \cdot \Delta \hat{u}_f \quad (15)$$

where L_w^{Δ} , L_u^{Δ} and y_t are defined in equations (16), (17) and (18) respectively.

$$L_u^{\Delta} = L_u \begin{bmatrix} 1 & 0 & \cdots & 0 \\ 1 & 1 & \cdots & 0 \\ \vdots & \vdots & \ddots & \vdots \\ 1 & 1 & \cdots & 1 \end{bmatrix} \quad (16)$$

$$L_w^{\Delta} = \begin{bmatrix} l_{w_1} \\ l_{w_2} + l_{w_1} \\ \vdots \\ \sum_{i=1}^{N} l_{w_i} \end{bmatrix} \quad (17)$$

$$y_t = \begin{bmatrix} y_0 \\ y_0 \\ \vdots \\ y_0 \end{bmatrix} \quad (18)$$

with l_{w_i} being the ith row vector of L_w.

3. PID CONTROL DESIGN

The tuning and design of PID controllers has been a subject of major research for many years now. Efforts are now concentrated on trying to improve the performance of these controllers using several innovative techniques like incorporating more sophisticated controllers in cascade with the PID loop like in (Katebi and Grimble 1999). Optimal control theory has also been used for the design of reduced order controllers as in (Johnson and Sánchez 2003). The approach developed in this paper calculates a PID controller which is optimal in the sense of minimi-sing a cost index over a finite horizon. The algorithm uses the plant model described by equation (15) and the PID controller structure in order to minimise the parameterised cost index. The cost index is a parametrised LQG criterion, which is minimised over a finite forward horizon by considering a backward horizon of measurements and assuring stability, in the BIBO sense, by constraining the future error trajectory.

Consider the closed-loop system in Figure (1), with plant model described by equation (15). Assume as well sufficient knowledge of the signals r, y and u. The problem is to find a set of controller parameters such the cost function of equation (19) is minimised.

$$J = \left(r_f - \hat{y}_f\right)^T \mathrm{Q} \left(r_f - \hat{y}_f\right) + \Delta\hat{u}_f^T \mathrm{R}\, \Delta\hat{u}_f \quad (19)$$

A discrete PID controller can be defined by equation (20), and the incremental control action by equation (21).

$$u_k = kp \cdot e_k + ki \sum_{n=1}^{k} e_n + kd\,(e_k - e_{k-1}) \quad (20)$$

$$\Delta\hat{u}_k = u_k - u_{k-1} \quad (21)$$

From equations (20) and (21), it is easy to find the time expression for the control action increment as in equation (22).

$$\Delta\hat{u}_0 = \rho_1 e_0 + \rho_2 e_{-1} + \rho_3 e_{-2} \quad (22)$$

where: $\rho_1 = kp + ki + kd$, $\rho_2 = -kp - 2kd$ and $\rho_3 = kd$.

Equation (22) can be parameterised as in equation (23).

$$\Delta\hat{u}_k = \begin{bmatrix} e_0 & e_{-1} & e_{-2} \end{bmatrix} \begin{bmatrix} \rho_1 \\ \rho_2 \\ \rho_3 \end{bmatrix} \quad (23)$$

Which in simplified notation can be written as:

$$\Delta\hat{u}_0 = \begin{bmatrix} e_0 & e_{-1} & e_{-2} \end{bmatrix} \cdot \rho \quad (24)$$

Notice however, that in order to comply with a digital PID structure the controller parameter vector ρ must comply with the following linear constraints:

$$\begin{bmatrix} -1 & 0 & 0 \\ 0 & 1 & 0 \\ 0 & 0 & -1 \end{bmatrix} \rho \leq \begin{bmatrix} 0 \\ 0 \\ 0 \end{bmatrix} \quad (25)$$

or equivalently:

$$\varphi \cdot \rho \leq 0 \quad (26)$$

The future increment control action can then be written as in equation (27) and in simplified form in equations (28):

$$\Delta\hat{u}_f = \begin{bmatrix} e_1 & | & e_0 & | & e_{-1} \\ e_2 & | & e_1 & | & e_0 \\ \vdots & | & \vdots & | & \vdots \\ e_N & | & e_{N-1} & | & e_{N-2} \end{bmatrix} \rho \quad (27)$$

$$\Delta\hat{u}_f = \varepsilon(\rho) \cdot \rho \quad (28)$$

where

$$\varepsilon(\rho) = \begin{bmatrix} \xi_1 & | & \xi_2 & | & \xi_3 \end{bmatrix} \quad (29)$$

By replacing equation (28) in (15) and in (19) and defining $\Phi = r_f - y_t - L_w^\Delta \Delta w_p$ it is possible to deduce an equivalent expression for the cost function given by (30), where r_f has been set to 0.

$$\begin{aligned} J = {} & \rho^T \left(\varepsilon^T L_u^{\Delta^T} Q L_u^\Delta \varepsilon + \varepsilon^T R \varepsilon\right) \rho \\ & -2\rho^T \left(\varepsilon^T L_u^{\Delta^T} Q \Phi\right) \\ & +\Phi^T Q \Phi \end{aligned} \quad (30)$$

Equation (30) has the quadratic form $x^T Ax + x^T B + C$, which can be minimised subject to the linear constraints of equation (25). Alternatively the cost index of equation (30) can be formulated as a *least-squares* problem, which improves numerical stability.

3.1 Formulation as a Least-Squares Problem

The matrices ε and L_u^Δ are usually ill-conditioned, therefore the numerical algorithm used to calculate the minimum of (30) is of vital importance. A good solution to this problem is by using a *least-squares* approach. Since $\mathtt{Q}$ and $\mathtt{R}$ are positive-definite or at least positive-semidefinite, it is possible to find matrices $S_\mathtt{R}$ and $S_\mathtt{Q}$ such that:

$$\begin{aligned} S_\mathtt{Q}^T S_\mathtt{Q} &= \mathtt{Q} \\ S_\mathtt{R}^T S_\mathtt{R} &= \mathtt{R} \end{aligned} \quad (31)$$

It is relatively simple to prove that the minimisation of (30) is equivalent to the minimisation of the norm of (32).

$$\min_\rho \left\| \begin{bmatrix} S_\mathtt{Q} \cdot L_u^\Delta \cdot \varepsilon(\rho) \\ S_\mathtt{R} \cdot \varepsilon(\rho) \end{bmatrix} \rho - \begin{bmatrix} S_\mathtt{Q} \Phi \\ 0 \end{bmatrix} \right\|^2 \quad (32)$$

3.2 Closed-loop condition

In equation (32) it remains how to calculate the Hankel matrix of errors $\varepsilon(\rho)$ such that the system is stable. A solution to this problem is to calculate the future errors based on past data by complying with the closed-loop equations. Since $r_f$has been set to 0, it is evident that the matrix ε is equal to:

$$\begin{aligned} \varepsilon &= \begin{bmatrix} \xi_1 & | & \xi_2 & | & \xi_3 \end{bmatrix} \\ &= -\begin{bmatrix} T_{f1} \cdot \hat{y}_f & | & T_{f2} \cdot \hat{y}_f & | & T_{f3} \cdot \hat{y}_f \end{bmatrix} \\ &\quad - \begin{bmatrix} T_{p1} \cdot y_p & | & T_{p2} \cdot y_p & | & T_{p3} \cdot y_p \end{bmatrix} \end{aligned} \quad (33)$$

where:

$$T_{f1} = I_N \quad (34)$$

$$T_{f2} = \begin{bmatrix} 0 & \cdots & 0 & 0 \\ 1 & \cdots & 0 & 0 \\ \vdots & \ddots & \vdots & \vdots \\ 0 & \cdots & 1 & 0 \end{bmatrix} \quad (35)$$

$$T_{f3} = \begin{bmatrix} 0 & \cdots & 0 & 0 & 0 \\ 0 & \cdots & 0 & 0 & 0 \\ 1 & \cdots & 0 & 0 & 0 \\ \vdots & \ddots & \vdots & \vdots & \vdots \\ 0 & \cdots & 1 & 0 & 0 \end{bmatrix} \quad (36)$$

$$T_{p1} = 0_N \quad (37)$$

$$T_{p2} = \begin{bmatrix} 0 & 0 & \cdots & 1 \\ 0 & 0 & \cdots & 0 \\ \vdots & \vdots & \ddots & \vdots \\ 0 & 0 & \cdots & 0 \end{bmatrix} \quad (38)$$

$$T_{p3} = \begin{bmatrix} 0 & 0 & \cdots & 1 & 0 \\ 0 & 0 & \cdots & 0 & 1 \\ \vdots & \vdots & \ddots & \vdots & \vdots \\ 0 & 0 & \cdots & 0 & 0 \end{bmatrix} \quad (39)$$

Then by using equations (15) , (28) and (33), the follwing set of equations are obtained:

$$\Omega(\rho)\begin{bmatrix}\xi_1\\ \xi_2\\ \xi_3\end{bmatrix} = \omega \tag{40}$$

where

$$\Omega = \begin{bmatrix} I_N + q_0 L_u^\Delta & q_1 L_u^\Delta & q_2 L_u^\Delta \\ q_0 T_{f2} L_u^\Delta & I_N + q_1 T_{f2} L_u^\Delta & q_2 T_{f2} L_u^\Delta \\ q_0 T_{f3} L_u^\Delta & q_1 T_{f3} L_u^\Delta & I_N + q_2 T_{f3} L_u^\Delta \end{bmatrix} \tag{41}$$

$$\omega = \begin{bmatrix} \Phi \\ T_{f2}\Phi - T_{p2} y_p \\ T_{f3}\Phi - T_{p3} y_p \end{bmatrix} \tag{42}$$

It is interesting to see that the left hand side of (40) are signals to be predicted over the forward horizon (i.e. future), while the right hand side are signals previously recorded in the past.

3.3 Stability condition

The problem of stability can be addressed by using a result presented in (Giovanini and Marchetti 1999). In this paper it is demonstrated that to assure exponential stability of reduced order digital controllers it is sufficient to comply with the following condition:

$$|\Delta\hat{u}(N)| \leq \sigma \tag{43}$$

This condition transforms into the constraint of equation (44), which can be written in simplified notation as in equation (45). The constraint in (45) is nonlinear since it has to be solved simultaneously with the closed-loop condition of equation (40).

$$\begin{bmatrix} e_N & e_{N-1} & e_{N-2} \\ -e_N & -e_{N-1} & -e_{N-2} \end{bmatrix}\rho - \begin{bmatrix}\sigma\\ \sigma\end{bmatrix} \leq \begin{bmatrix}0\\ 0\end{bmatrix} \tag{44}$$

$$\Theta \cdot \rho - \Psi \leq 0 \tag{45}$$

The use of this condition in the optmisation will produce a controller with which the system is closed-loop stable; however, it might be possible that this condition cannot be met and in that case the optimisation will be infeasible. Some suggestions to avoid infeasible optimisations are to enlarge the horizon or relax the stability constraints. It is also important to note, that this condition assures exponential stability for a step change in the input.

4. SIMULATION RESULTS

The following simulation exemplifies the use of this tuning algorithm. The example comprises the control of dissolved oxygen in a simulation benchmark for an activated sludge wastewater treatment plant

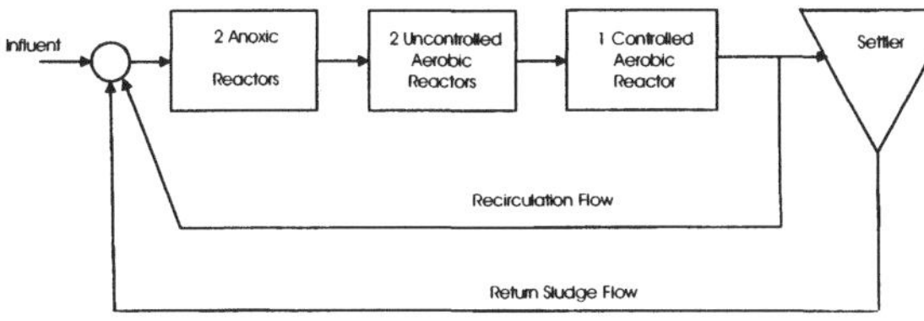

Figure 2. The COST Simulation Benchmark.

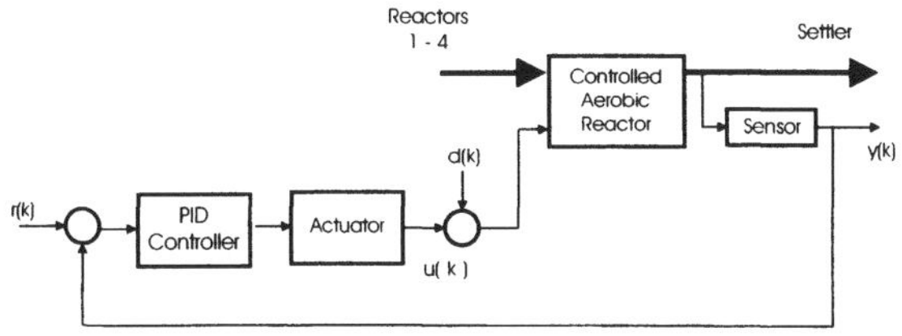

Figure 3. Dissolved Oxygen control loop.

Table 1. Optimisation Specifications

N	M	R	Q	σ
100	100	8	5	0.01

(Copp 2002). The wastewater treatment facility is composed by five bioreactors, each described by the ASM1 model (Henze *et al.* 1987), two internal recirculation flows and one non-reactive settler as shown in Figure (2). The first two bioreactors are anoxic and the remaining three are aerated. From the aerated reactors, only the last one is controlled. The input to the model is the airflow rate scaled to a base of 10 and the output is the oxygen concentration in the reactor. The airflow is pumped into the reactor through blowers (actuators) which are commanded by a PID controller. Figure (3) shows details of the control loop in the fifth aerobic reactor. The control loop also accounts for unmeassured disturbances as changes in the plant load. These disturbances are included in the form of the signal d(k). The identification considers a 1 minute sampling rate with an oxygen sensor with 1 minute time delay. The initial controller, with which the plant was identified in closed-loop, is a PI with parameters shown in Table (2) .

For the identification, 1200 points of data have been collected with the system excited by a pseudo random binary signal (PRBS) of zero mean and 0.5 [mg/l] amplitude around a 1 [mg/l] setpoint. The forward and backward horizon have been set to a length of 100. The system has been aproximated to a fifth order system by setting n=5 in the singular value decomposition of the identification algorithm. The algorithm was implemented in MATLAB, and the solution takes around 10 to 20 iterations depending on how stringent the stability constraints are, as shown in Figure (5). Figure (4) shows the response of the obtained controller for a unit step when designed with the specifications in Table (1) , while Table (2) shows the initial and optimal controller parameters.

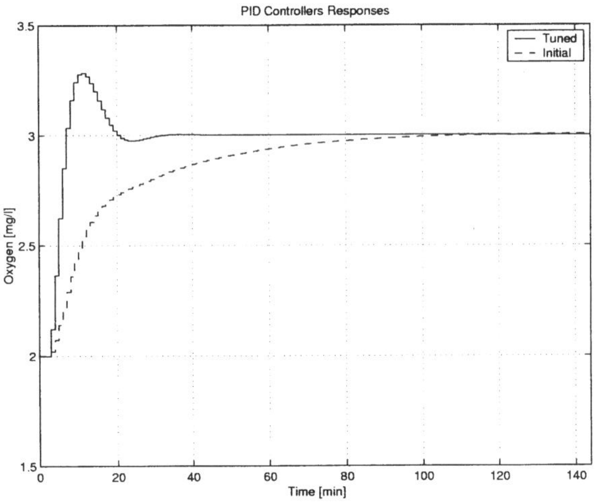

Figure 4. Comparison between initial PID response and tuned PID response.

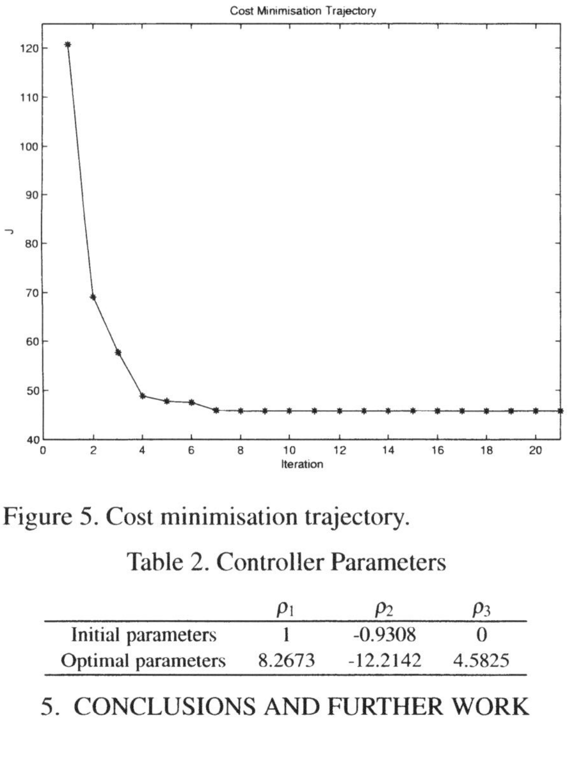

Figure 5. Cost minimisation trajectory.

Table 2. Controller Parameters

	ρ_1	ρ_2	ρ_3
Initial parameters	1	-0.9308	0
Optimal parameters	8.2673	-12.2142	4.5825

5. CONCLUSIONS AND FURTHER WORK

This paper presents an algorithm which allows the design of a PID controller within a subspace identification framework. The algorithm can be divided in two steps: identification of the plant model using closed-loop data followed by the parameter calculation. The parameters are calculated as the result of minimising a LQG criterion over a finite forward horizon. The method considers that the system response will asymptotically approach the desired value over the horizon.

ACKNOWLEDGEMENTS

The authors express their gratitude to the European Commission under whose contract EVK1-CT-2000-00056 the SMAC project and this work has been performed. Thanks are also given to Dr. Leonardo Givannini for his comments on stability.

REFERENCES

Copp, J., (Ed.) (2002). *COST Action 624 - The COST Simulation Benchmark: Description and Simulation Manual.* European Comission - European cooperation in the field of scientific and technical research.

Favoreel, W., B. De Moor, M. Gevers and P. Van Overschee (1998). Closed loop model-free subspace-based LQG -design. Technical Report ESAT-SISTA/TR 1998-108. Departement Elecktrotechniek - Katholieke Universiteit Leuven.

Giovanini, L. and J. Marchetti (1999). Shaping time-domain responses with discrete controllers. *Ind. Eng. Chem. Res.* **38**, 4777–4789.

Henze, M., C.P.L. Grady, W. Gujer, G.v.R. Marais and T. Matsuo (1987). Activated sludge model no.1. Technical report. IAWQ Scientific and Technical Report No.1.

Hjalmarsson, H., S. Gunnarsson and M. Gevers (1994). A convergent iterative restricted complexity control design scheme. In: *Conference on Decision and Control.* Lake Buena Vista, Florida, USA.

Johnson, M.A. and A. Sánchez (2003). Process control loop tuning and monitoring using LQG optimality with applications in wastewater treatment plant. In: *IEEE 4th International Conference on Control & Automation.* Montreal, Canada.

Kadali, R., B. Huang and A. Rossiter (2003). A data driven approach to predictive controller design. *Control Engineering Practice* **11**(3), 261–278.

Katebi, M.R. and M.H. Moradi (2001). Predictive PID controllers. *IEE Proc.-Control Theory Appl.* **148**(6), 478–487.

Katebi, M.R. and M.J. Grimble (1999). Intgrated control, guidance and diagnosis for reconfigurable autonomous underwater vehicle control. *International Journal of Systems Science* **30**(9), 1021–1032.

Ruscio, D.D. (1997*a*). Model based predictive control: An extended state-space approach. In: *Proc. of the* 36th *Conference on Decision and Control.* San Diego, CA. pp. 3210–3217.

Ruscio, D.D. (1997*b*). Model predictive control and identification: A linear state-space approach. In: *Proc. of the* 36th *Conference on Decision and Control.* San Diego, CA. pp. 3202–3209.

Uduehi, D., A. Ordys and M.J. Grimble (2002). Multivariable PID controller design using online generalised predictive control optimisation. In: *Proc. of the 2002 IEEE International Conference on Control Applications.* Glasgow.

van Overschee, P. and B. De Moor (1996). Closed-loop subspace system identification. Technical Report ESAT-SISTA/TR 1996-521. Departement Elecktrotechniek - Katholieke Universiteit Leuven.

IFAC
Publications
www.elsevier.com/locate/ifac

EVOLUTIONARY TUNING OF PID PARAMETERS

Toru Yamamoto

Dept. of Technology & Information Education,
Graduate School of Education,
Hiroshima University
1-1-1 Kagamiyama, Higashi-Hiroshima, 739-8524 Japan

Abstract: PID control schemes have been widely used in most process control systems represented by chemical processes for a long time. However, it is still a very important problem how to determine or tune the PID parameters, because these parameters have a great influence on the stability and the performance of the control system. In this paper, a new evolutionary tuning algorithm of PID parameters is proposed. A suitable set of PID parameters is calculated based on the relationship between the generalized predictive control(GPC) and the PID control. Then, the GPC includes some user-specified parameters. The suitable value of the user-specified parameter is sought by the genetic algorithm(GA). According to the proposed scheme, the search area of PID parameters is sharply reduced, and the computational burden is drastically shortened. *Copyright © 2003 IFAC*

Keywords: PID control, Genetic algorithm, Generalized predictive control, Process control, Self-adjustment

1. INTRODUCTION

Lots of studies on advanced complex control schemes have been carried out vigorously due to the marvelous development of the computer technology. However, PID control schemes still continue to be widely used for most industrial control systems, particularly in the chemical process industry. This is mainly because PID controllers have simple control structures, and are simple to maintain and tune. Therefore, it is still attractive to design PID controllers. Since it plays a great role in determining the behavior of the control system how to tune the PID parameters, a lot of tuning methods have been studied up to now(Zieglar and Nichols, 1942; Chien, *et al.*, 1952; Åström and Hägglund, 1988; Åström, *et al.*, 1992). However, the effective technique is not still established.

On the other hand, the genetic algorithm(GA), which simulated the mechanism of heredity or evolution in the living things, has attracted our attentions as a method which gives us fairly quickly suitable or suboptimal answers for search, optimization or machine learning problems(Goldberg, 1992). Moreover, some searching methods of PID parameters using GA have also been reported(Porter and Jones, 1992; Jones, *et al.*, 1996; Omatu and Deris, 1996). However, since three PID parameters are directly searched in those methods, it costs fairly long time to get a suitable set of values due to the wide search area.

The main motivation in this study is to present a new tuning method of PID parameters using the GA. First, a tuning scheme of PID parameters based on the relation with a generalized predictive control(GPC)(Clarke, Mohtadi and Tuffs, 1987) is discussed. The GPC is the control technique based on a multi-step prediction, and is effective for systems with large time-delays. Moreover, it has robustness in the case where the time-delay is changed on the way of controlling. Although GPC is one of the advanced control and has been widely employed to chemical processes, it is necessary to build the exclusive control device for implementing the GPC. Therefore, the GPC is employed only for the plant where the effect corresponding to investment is expected. On the other hand, since PID controllers are supervised by the existing DCS, it is quit easy to change PID parameters. According to the PID tuning scheme to be considered in this paper, PID controller can be designed whose performance is almost equivalent to the GPC without any economic load. However, the GPC includes two user-specified parameters, *i.e.*, a weighting coefficient for the control input and a prediction horizon. Since the control performance strongly depends on the user-specified parameter, it is necessary to further consider the suitable scheme of determining these parameter.

In this paper, a new tuning method of PID parameters is proposed, in which the user-specified parameters are sought using GA, and are transformed into the corresponding PID parameters. That is, it means that a suitable set of the user-specified parameters in the performance criterion of the GPC law is firstly sought by using GA, and then a set of PID parameters is tuned using the above set as a mediator. Therefore, according to the newly proposed PID tuning scheme, since search areas of the user-specified parameters are sharply reduced, the time to get a suitable set of PID parameters can be drastically shorten. This paper is organized as follows. First, the descriptive model of the controlled object is derived, and a tuning scheme of PID parameters based on the relation with the GPC is explained. Next, an evolutionary computation to find the suitable user-specified parameter is considered. Finally, the effectiveness of the proposed scheme is evaluated on a numerical simulation example, and the extension of the proposed scheme in an adaptive manner is discussed.

2. PROBLEM FORMULATION

2.1 System description

Let z^{-1} be the backward shift operator, and the following discrete-time model describes the process:

$$\alpha(z^{-1})y(t) = \beta(z^{-1})u(t-1) + \eta(t)/\Delta \quad (1)$$

where

$$\left.\begin{aligned} \alpha(z^{-1}) &= 1 + \alpha_1 z^{-1} + \cdots + \alpha_{n_1} z^{-n_1} \\ \beta(z^{-1}) &= \beta_0 + \beta_1 z^{-1} + \cdots + \beta_{n_2} z^{-n_2}. \end{aligned}\right\} (2)$$

Moreover, $u(t)$, $y(t)$ and $\eta(t)$ are the input, the output and the modeling error in (1) with zero mean and covariance σ_η^2, respectively. $f(u,y)$ denotes the nonlinear function. In this paper, the system model (1) is called as the 'full model', because it explains the behavior of the system as exactly as possible. In this paper, it is assumed that that the full model (1) can be obtained. If it is not so, the full model may be replaced by an appropriate model, for example, a neural network which simulates approximately the behavior of the controlled object. It is used only for calculating the fitness function in GA as shown in the section 4.4.

On the other hand, consider the second-order linear model which approximates the full model (1) by the following equation:

$$A(z^{-1})y(t) = B(z^{-1})u(t-1) + \xi(t)/\Delta \quad (3)$$

where

$$\left.\begin{aligned} A(z^{-1}) &= 1 + a_1 z^{-1} + a_2 z^{-2} \\ B(z^{-1}) &= b_0 + b_1 z^{-1} + ... + b_m z^{-m}. \end{aligned}\right\} (4)$$

$\xi(t)$ denotes the modeling error in (3) with zero mean and covariance σ_ξ^2. In general, since the control law based on the full model (1) seems to be complex and very difficult in implementation, the second-order linear model (3) is often used in designing the process control system. Therefore, the model (3) is called as a 'design-oriented model'. Note that a_2 is set to 0 in (4) if the controlled object is approximately expressed as the first-order system. Moreover, the degree of $B(z^{-1})$, m, should be set large enough in order to cope with the higher components which cannot be approximated as the second-order linear system.

2.2 Brief review of PID control law

Consider the velocity-type PID control law(Cameron and Seborg, 1983) described as

$$\Delta u(t) = \frac{k_c \cdot T_s}{T_I} e(t) - k_c\{\Delta + \frac{T_D}{T_s}\Delta^2\}y(t) \quad (5)$$

where

$$e(t) := w(t) - y(t). \quad (6)$$

Δ denotes the differential operator defined as $\Delta := 1 - z^{-1}$. $w(t)$ is the reference signal and is given by piecewise constant components. k_c , T_I and T_D are the proportional gain, the reset time and the derivative time, respectively. Moreover, T_s denotes the sampling interval.

For convenience, let $L(z^{-1})$ be

$$L(z^{-1}) := k_c\{\Delta + \frac{T_s}{T_I} + \frac{T_D}{T_s}\Delta^2\} \quad (7)$$

then, (5) can be rewritten by

$$L(z^{-1})y(t) + \Delta u(t) - L(1)w(t) = 0. \quad (8)$$

Since it plays a great role in determining the behavior of the closed-loop system how to tune the PID parameters in (5) or (8), lots of works for PID tuning schemes have been reported up to now. A tuning scheme of PID parameters is considered in the following section, which is derived based on the relation with GPC.

3. GPC BASED PID CONTROLLER DESIGN

3.1 Generalized predictive control

The GPC is one of model predictive control schemes, and since it is based on multi-step prediction, it is an effective technique for systems with ambiguous time-delays and/or time-variant time-delays. In chemical processes, it is usually difficult to estimate the time-delays exactly. Therefore, the GPC is considered in this paper.

First, consider the following cost function of the GPC:

$$\begin{aligned} J = E\{ &\sum_{j=N_1}^{N_2} [y(t+j) - w(t)]^2 \\ &+\lambda\sum_{j=1}^{NU}[\Delta u(t+j-1)]^2\}, \end{aligned} \quad (9)$$

where λ denotes the user-specified parameter which means the weighting factor for the control input. $w(t)$ denotes the reference signal given by piecewise constants. Furthermore, the period from N_1 thru N_2 denotes the prediction horizon, and NU denotes the control horizon. For simplicity, they are respectively set as $N_1 = 1$, $N_2 = N$ and $NU = N$, where N is designed in consideration of the time constant and the time-delay of the controlled object.

The control law based on minimizing the cost function (9) is given by

$$\begin{aligned} &\sum_{j=1}^{N} p_j F_j(z^{-1})y(t) + \{1 + z^{-1}\sum_{j=1}^{N} p_j S_j(z^{-1})\} \\ &\cdot\Delta u(t) - \sum_{j=1}^{N} p_j w(t) = 0. \end{aligned} \quad (10)$$

$F_j(z^{-1})$ and $S_j(z^{-1})$ are calculated by the following Diophantine equations:

$$1 = \Delta A(z^{-1})E_j(z^{-1}) + z^{-j}F_j(z^{-1}) \quad (11)$$

$$E_j(z^{-1})B(z^{-1}) = R_j(z^{-1}) + z^{-j}S_j(z^{-1}), \quad (12)$$

where

$$\left.\begin{aligned} E_j(z^{-1}) &= 1 + e_1 z^{-1} + \cdots + e_{j-1}z^{-(j-1)} \\ F_j(z^{-1}) &= f_{j,0} + f_{j,1}z^{-1} + f_{j,2}z^{-2}. \end{aligned}\right\} (13)$$

$$\left.\begin{aligned} R_j(z^{-1}) &= r_0 + r_1 z^{-1} + \cdots + r_{j-1}z^{-(j-1)} \\ S_j(z^{-1}) &= s_{j,0} + \cdots + s_{j,m-1}z^{-(m-1)}. \end{aligned}\right\} (14)$$

Moreover, $P(z^{-1})$ is defined by

$$P(z^{-1}) := \sum_{j=1}^{N} p_{N-j+1} z^{-(j-1)}, \quad (15)$$

where p_j is calculated by

$$\begin{aligned} [p_1, p_2, \cdots, p_N] &:= [1, 0, \cdots, 0] \\ &\cdot(\mathbf{R}^T\mathbf{R} + \mathbf{\Lambda})^{-1}\mathbf{R}^T. \end{aligned} \quad (16)$$

In (16), the matrix $\mathbf{R}$ which consists of coefficients of $R_j(z^{-1})$ is defined by

$$\mathbf{R} := \begin{bmatrix} r_0 & & & \\ r_1 & r_0 & & 0 \\ \cdot & \cdot & \cdot & \\ \cdot & & \cdot & \cdot \\ \cdot & & & \cdot \\ r_N & r_{N-1} & \cdots & r_0 \end{bmatrix}, \quad (17)$$

and λ is as follows:

$$\mathbf{\Lambda} := \text{diag}\{\lambda\}. \quad (18)$$

3.2 PID parameter tuning

By replacing the coefficient polynomial of the second term in (10) into the static gain, the following equation can be obtained:

$$\begin{aligned} &\frac{1}{\nu}\sum_{j=1}^{N} p_j F_j(z^{-1})y(t) \\ &+\Delta u(t) - \sum_{j=1}^{N} p_j/\nu \cdot w(t) = 0 \end{aligned} \quad (19)$$

where ν is defined as

$$\nu := 1 + \sum_{j=1}^{N} p_j S_j(1). \quad (20)$$

If the following relationship is satisfied:

$$L(z^{-1}) = \frac{1}{\nu}\sum_{j=1}^{N} p_j F_j(z^{-1}) \quad (21)$$

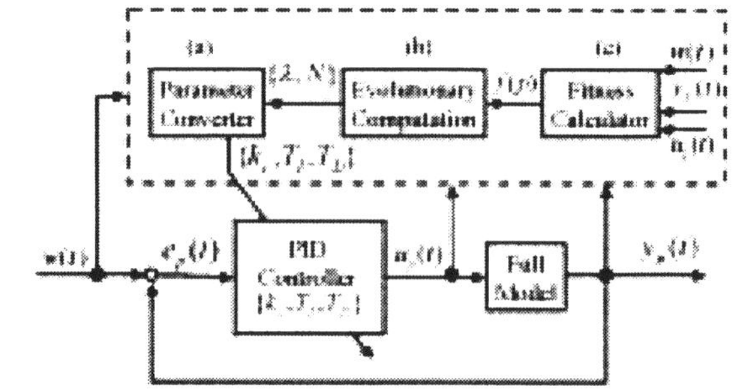

Fig. 1. Schematic diagram of the evolutionary computation of PID parameters.

then, (19) becomes identical to (8). Because the relation $\sum_{j=1}^{N} p_j F_j(1) = \sum_{j=1}^{N} p_j$ can be immediately obtained from (11). Therefore, PID parameters can be calculated based on (8) and (21) as follows:

$$\left.\begin{aligned} k_c &= -\frac{1}{\nu}(\tilde{f}_1 + 2\tilde{f}_2) \\ T_I &= -\frac{\tilde{f}_1 + 2\tilde{f}_2}{\tilde{f}_0 + \tilde{f}_1 + \tilde{f}_2} T_s \\ T_D &= -\frac{\tilde{f}_2}{\tilde{f}_1 + 2\tilde{f}_2} T_s, \end{aligned}\right\} \quad (22)$$

where $\tilde{f}_i$ is defined as

$$\sum_{j=1}^{N} p_j F_j(z^{-1}) := \tilde{f}_0 + \tilde{f}_1 z^{-1} + \tilde{f}_2 z^{-2}. \quad (23)$$

However, it is necessary to choose a suitable set of the user-specified parameters, *i.e.*, λ and N in order to calculate PID parameters based on (22). Then, a new scheme in which a suitable set of them is sought by using the GA, is considered in the following section.

4. EVOLUTIONARY COMPUTATION

4.1 Outline of evolutionary computation

Fig.1 shows the schematic diagram of the evolutionary computation of PID parameters. In Fig.1, The part surrounded by the dash line is the tuning part of PID parameters every one generation in GA. It is composed of (a) the parameter conversion part in which the user-specified parameters $\{\lambda, N\}$ sought by GA are converted to the corresponding PID parameters, (b) the evolutionary computation part of the user-specified parameters based on the fitness values, and (c) the calculation part of the fitness function for each individual.

The feature of the proposed method can be summarized as follows:

1) As mentioned in the section 3, PID parameters are not directly sought by GA, but they are calculated using the user-specified parameters $\{\lambda, N\}$ in the performance criterion of the GPC, which are sought by GA. Here, since the search areas of the parameters $\{\lambda, N\}$ are restricted by taking the control performance of the closed-loop system into consideration, the time to get a suitable set of PID parameters can be shorten drastically.
2) By means of selecting the fitness function appropriately, the response of the closed-loop system can be shaped fairly freely. For example, the rise property, the overshoot and the fluctuation of the control input can be specified fairly in detail.
3) The full model does not have to be an analytical formulation, but it may be only the model whose input-output property is closely similar to that of the controlled object in a sense. Therefore, as mentioned in the section 2, various neural network models can be utilized, and it is possible to take any nonlinear property in the model freely.

Here, consider more precisely the tuning part of PID parameters (a), (b) and (c) mentioned above.

4.2 Parameter converter

The significant search spans of the parameters λ and N in the control engineering sense can be considered as follows.

On λ, investigating the stability of the closed-loop system using an appropriate approach, for example, the Bode diagram, the search span of λ can be determined so as to give a suitable gain or phase margin. N relates to the rise-time and the time-delay of the controlled object. Therefore, based on *a priori* information, the search span of N can be determined in consideration of the sum of the time constant and the time-delay.

From these control engineering restrictions, the seek span of the corresponding PID parameters can be limited to fairly small area in practice.

4.3 Evolutionary computation

The evolutionary computation of the parameters $\{\lambda, N\}$ is discussed in this section. These parameters are calculated by using a real-coded GA(Ono, *et al.* , 2000).

First, the user-specified parameters λ and N are arranged as cells included in the l−th string $s_l(n_3)(l = 1, 2, \cdots, n_4; n_3 = 1, 2, 3)$, that is, $s_l(1) = \lambda$, $s_l(2) = \sigma$ and $s_l(3) = \delta$. Furthermore, n_4 denotes the number of population. Therefore, $s_l(\cdot)$ denotes the l−th individual. After the above preparation, a real-coded GA is employed, which is explained as follows.

i) **Initialization**
Determine the search spans of the user-specified parameters by taking account of the control engineering sense, and product the initial individuals with random real numbers within the search spans.

ii) **Selection**
Calculate the fitness value $f(l)(l = 1, 2, \cdots, n_4)$ corresponding to the individual s_l, and select γ percent individuals with superior fitness values. Here, the fitness function $f(\cdot)$ can be freely designed by taking account of the desired control performance, which is discussed in the following section. γ percent superior individuals are saved in the next generation.

iii) **Crossover**
The $(100 - \gamma)$ percent remains are generated by the crossover. Choose two individuals s_a and S_b among γ percent superiors. The new individual s_c is generated by

$$s_c(n_3) = \frac{I(X) + Y}{2} \tag{24}$$

where

$$X := \max\{f(s_a(n_3)), f(s_b(n_3))\} \tag{25}$$

$$Y := \frac{s_a(n_3) + s_b(n_3)}{2}. \tag{26}$$

$I(X)$ denotes the individual with the fitness value x, that is, $I(\max\{f(s_a(n_3)), f(s_b(n_3))\})$ means the individual with the superior fitness value within $s_a(n_3)$ and $s_b(n_3)$. Note that this procedure is employed for every cell $n_3(= 1, 2, 3)$ included in s_a and s_b.

iv) **Mutation**

- ω percent cells for all new individuals are replaced with

$$s_l(n_3) \leftarrow s_{\max}(n_3) - s_l(n_3), \tag{27}$$

where $s_{\max}(n_3)$ denotes the maximum value of the search range.

The procedure from ii) thru iv) is repeated for generations which are designed in advance.

4.4 Fitness function

As mentioned above, the response of the closed-loop system can be shaped fairly freely by using a suitable fitness function. For example, consider the following as a fitness function:

$$F(p) = 1/(1 + f(p)), \tag{28}$$

where

$$f(P) = \sum_{t=1}^{N} \{y_m(t) - y_P(t)\}^2 \tag{29}$$

or

$$f(P) = \sum_{t=1}^{N} |y_m(t) - y_P(t)|. \tag{30}$$

$y_m(t)$ denotes the reference model output excited by the reference signal $w(t)$. $y_P(t)$ is the output when the PID controller is employed for the full model, whose PID parameters correspond to each individual in the real-coded GA.

Usually, a kind of quadratic performance criteria is mainly used so that it may be convenient to obtain a control law analytically. However, in parameter tuning by the GA, it is a feature that the fitness function equivalent to the performance criterion can be freely chosen for the control purpose as mentioned above.

5. SIMULATION EXAMPLE

In this section, the newly proposed scheme is evaluated on a numerical simulation example.

Consider the following nonlinear system expressed by a Hammerstein model:

$$\begin{aligned} y(t) &= 0.8y(t-1) - 0.3y(t-2) \\ &\quad +0.4x(t-1) + 0.2x(t-2) + \eta(t)/\Delta, \end{aligned} \tag{31}$$

where $x(t-1)$ is defined by

$$x(t-1) := u(t-1) + 0.5u^2(t-1). \tag{32}$$

$eta(t)$ denotes a white Gaussian noise with zero mean and covariance 0.01. Suppose that the full model is also given by (31) and (32).

Carrying out the parameter estimation by a recursive least squares method using the input-output data of (31) and (32), the design-oriented model was obtained as follows:

$$\begin{aligned} y(t) &= 1.11y(t-1) - 0.47y(t-2) \\ &\quad +0.58u(t-1) + 0.26u(t-2) \\ &\quad -0.47u(t-3). \end{aligned} \tag{33}$$

First, for comparison purposes, the control result by Chien, Hrones & Reswick(CHR) method(1952), which is well-known as a PID parameter tuning method, is shown in Fig.2. Here, PID parameters were chosen as $k_c = 0.36$, $T_I = 4.07$ and $T_D = 0.5$. Fig.2 illustrates that there is a some oscillation in the transient state, and the settling time becomes large.

Next, the newly proposed scheme was employed, and the control result is shown in Fig.3, where the parameters used in GA were set as follows.

- **Individual**: The string is composed of λ and N with real numbers, which are randomly

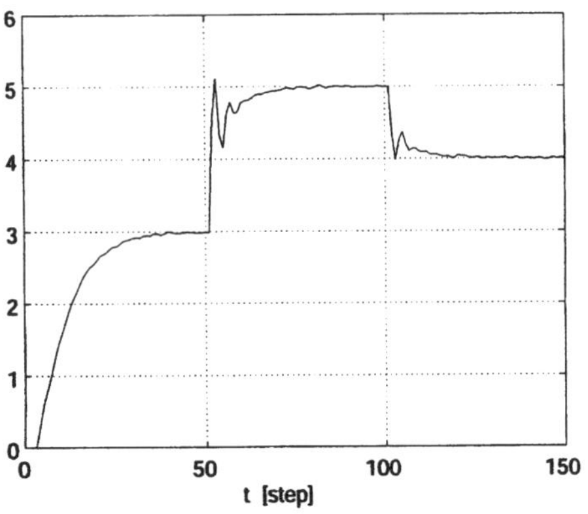

Fig. 2. Control result using CHR type PID control scheme.

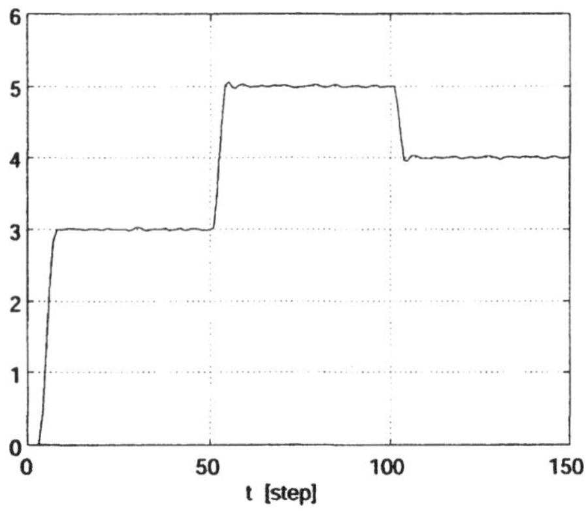

Fig. 3. Control result using the proposed control scheme.

generated. The number of the population was set to 100.

- **Generation**: The generation number was set to 50.
- **Fitness function**: The fitness function is given by (28) and (30).
- **Selection**: The conventional Elite strategy(Goldberg, 1992) was employed, that is, 20 percent superior individuals were saved in the next generation.
- **Crossover**: The 80 percent remains were generated by the crossover. The crossover was performed by (24), (25) and (26).
- **Mutation**: The ratio of the mutation was set to 20 percent.
- **Seek spans**: $0.0 \leq \lambda \leq 20.0, \quad 1 \leq N \leq 30$

Moreover, the reference model was designed as:

$$y_m(t) = \frac{z^{-1}0.40}{1 - 0.74z^{-1} + 0.14z^{-2}} w(t).. \quad (34)$$

λ and N were sought as 0.523 and 4, respectively. Then, PID parameters were given by $k_c = 0.174$, $T_I = 0.726$ and $T_D = 0.848$, and the sampling interval T_s was set to 1.0[sec].

From Fig.3, it is clear that the oscillation is considerably removed, and that the newly proposed scheme works well.

6. CONCLUSIONS

In this paper, a new evolutionary tuning scheme of PID parameters has been proposed. Moreover, the effectiveness of the newly proposed scheme has been verified by the numerical simulation example. The procedure of the newly proposed method is summarized as follows.

1) A suitable set of the user-specified parameters in the performance criterion of the GPC law is firstly sought by the GA.
2) PID parameters are calculated by using the suitable set of the user-specified parameters.

Therefore, since the seek spans of the user-specified parameters are restricted to a fairly small area from the viewpoint of the control engineering restrictions, the time required to seek a suitable set of the parameters can be drastically shortened.

REFERENCES

Åström, K.J. and T.Hägglund (1988). Automatic Tuning of PID Controllers. *Instrument Society of America.*

Åström,K.J., CC.Hang, P.Persson and W.K.Ho (1992). Towards Intelligent PID Control. *Automatica.* **28**, 1–9.

Åström,K.J., CC.Hang, P.Persson and W.K.Ho (1983). A Self-Tuning Controller with a PID Structure. *Int. J. of Control.* **38**, 401–417.

Chien,K.L., J.A.Hrones and J.B.Reswick (1952). On the Automatic Control of Generalized Passive Systems. *Trans. ASME.* **74**, 175–185.

Clarke,D.W., C.Mohtadi and P.S.Tuffs (1987). Generalized Predictive Control. *Automatica.* **23**, 856–875.

Goldberg, D.E. (1992). Genetic Algorithm in Search, Optimization & Machine Learning. *Addison-Wesley.*

Jones,A.H., N.Ajlouni and M.Uzam (1987). On-Line Frequency Domain Identification and Genetic Tuning of PID Controllers. *Proc. of IEEE Conference on Emerging Technologies and Factory Automation.* , 261–266.

Omatu,S. and S.Deris (1996). Stabilization of Inverted Pendulum by the Genetic Algorithm. *Proc. of IEEE International Conference on Evolutionary Computation.* , 700–705.

Ono,I., M.Yamamura and H.Kita (2000). Real-Coded Genetic Algorithms and Their Applications (in Japanese). *J. of Japanese Society for Artificial Intelligence.* **15**, 259–266.

Porter,B. and A.H.Jones (1992). Genetic Tuning of Digital PID Controllers. *Electronics Letter.* **28**, 843–844.

Ziegler,G.J. and N.B.Nichols (1942). Optimum Settings for Automatic Controllers. *Trans. ASME.* **64**, 759–768.

IFAC Publications
www.elsevier.com/locate/ifac

ADAPTIVE, CAUTIOUS, PREDICTIVE CONTROL WITH GAUSSIAN PROCESS PRIORS

Roderick Murray-Smith * **Daniel Sbarbaro** **
Carl Edward Rasmussen *** **Agathe Girard** ****

* *Department of Computing Science, University of Glasgow, Glasgow G12 8QQ, Scotland, UK. & Hamilton Institute, NUI Maynooth, Ireland E-mail: rod@dcs.gla.ac.uk*
** *Departamento de Ingeniería Eléctrica, Universidad de Concepción, Chile. E-mail: dsbarbar@die.udec.cl*
*** *Max Planck Institute, Tübingen, Germany. E-mail: carl@tuebingen.mpg.de*
**** *Department of Computing Science, University of Glasgow, Glasgow G12 8QQ, Scotland, UK. E-mail: agathe@dcs.gla.ac.uk*

Abstract: Nonparametric Gaussian Process models, a Bayesian statistics approach, are used to implement a nonlinear adaptive control law. Predictions, including propagation of the state uncertainty are made over a k-step horizon. The expected value of a quadratic cost function is minimised, over this prediction horizon, without ignoring the variance of the model predictions. The general method and its main features are illustrated on a simulation example.

Keywords: Cautious control, Gaussian process priors, nonparametric models, nonlinear model-based predictive control, propagation of uncertainty.

1. INTRODUCTION

Gaussian process priors provide a flexible, nonparametric approach to modelling nonlinear systems. In Murray-Smith and Sbarbaro (2002), the use of Gaussian process priors allowed us to analytically obtain a control law which perfectly minimises the expected value of a quadratic cost function, without disregarding the variance of the model prediction as an element to be minimised. This led naturally and automatically to a regularising *caution* in control behaviour in following the reference trajectory, depending on model accuracy close to the current state. This paper expands on the previous work by making the cost function more flexible, using our recent results on inference with uncertain states Girard et al. (2003) and allowing control based on multistep predictions.

1.1 *Background*

Several authors have proposed the use of non-linear models as a base to build nonlinear adaptive controllers. Agarwal and Seborg (1987), for instance, have proposed the use of known nonlinearities, capturing the main characteristic of the process, to design a Generalized Minimum Variance type of self-tuning controller. In many applications, however, these nonlinearities are not known, and non-linear parameterision must be used instead. A popular choice has been the use of Artificial Neural Networks for estimating the nonlinearities of the system Narendra and Parthasarathy (1990); Bittanti and Piroddi (1997). These researchers adopted the *certainty equivalence principle* for designing the controllers, where the model is used in the control law as if it were the true system.

In order to improve the performance of nonlinear adaptive controllers based on nonlinear models, the accuracy of the model predictions should also be taken into account. A common approach to consider the uncertainty in the parameters, is to add an extra term in the cost function of a Minimum Variance controller, which penalizes the uncertainty in the parameters of the nonlinear approximation Fabri and Kadirkamanathan (1998). Another similar approach, based on the minimization of two separate cost functions, has been proposed in Filatov et al. (1997). The first one is used to improve the parameter estimation and the second one to drive the system output to follow a given reference signal. This approach is called *bicriterial* and it has also be extended to deal with nonlinear systems Sbarbaro et al. (1998).

The above ideas are closely related to the work done on dual adaptive control Fel'dbaum (1960), where the main effort has been concentrated on the analysis and design of adaptive controllers based on the use of the uncertainty associated with parameters of models with fixed structure. The dual control literature has struggled with the challenges of numerical integration over high-dimensional spaces. Most approaches to dual control have pursued approximations which lead to sub-optimal solutions Wittenmark (1995); Filatov and Unbehauen (2000).

1.2 *Model structure – parametric vs. nonparametric?*

Most control engineering applications are still based on parametric models, where the functional form is fully described by a finite number of parameters, often a linear function of the parameters. Even in the cases where flexible parametric models are used, such as neural networks, spline-based models, multiple models etc, the uncertainty is usually expressed as uncertainty of parameters (even though the parameters often have no physical interpretation), and do not take into account uncertainty about model structure, or distance of current prediction point from training data used to estimate parameters.

Non-parametric models retain the available data and perform inference conditional on the current state and local data (called 'smoothing' in some frameworks). As the data are used directly in prediction, unlike the parametric methods more commonly used in control contexts, non-parametric methods have advantages for off-equilibrium regions, since normally in these regions the amount of data available for identification is much smaller than that available in steady state. The uncertainty of model predictions can be made dependent on local data density, and the model complexity automatically related to the amount and distribution of available data (more complex models need more evidence to make them likely). Both aspects are very useful in sparsely-populated transient regimes.

2. GAUSSIAN PROCESS PRIORS

In a Bayesian framework the model is based on a prior distribution over the infinite-dimensional space of functions. As illustrated in O'Hagan (1978), such priors can be defined as Gaussian processes. These models have attracted a great deal of interest recently – see for example reviews such as Williams (1998). ? showed empirically that Gaussian processes were extremely competitive with leading nonlinear identification methods on a range of benchmark examples. The further advantage that they provide analytic predictions of model uncertainty makes them very interesting for control applications. Use of GPs in a control systems context is discussed in Murray-Smith et al. (1999); Leith et al. (2000). Integration of prior information in the form of state or control linearisations is presented in Solak et al. (2003). A simulation of Model Predictive Control with GPs is presented in Kocijan et al. (2003).

In the following, let $\mathbf{x}$ and $\mathbf{y}$ be the N input and output pairs used for identification and $\mathbf{x}^*, y^*$ the pair used for prediction. Instead of parameterising $y^i = f(\mathbf{x}^i)$ as a parametric model, we can place a prior directly on the space of functions where f is assumed to belong. A Gaussian process represents the simplest form of prior over functions – we assume that any p points have a p-dimensional multivariate Normal distribution. We will assume zero mean, so for the case with partitioned data $\mathbf{y}$ and y^* we will have the multivariate Normal distribution

$$\begin{bmatrix} \mathbf{y} \\ \mathbf{y}^* \end{bmatrix} \sim \mathcal{N}(0, \mathbf{\Sigma}_F), \quad \mathbf{\Sigma}_F = \begin{bmatrix} \mathbf{\Sigma} & \mathbf{k}(\mathbf{x}^*) \\ \mathbf{k}(\mathbf{x}^*)^T & C(\mathbf{x}^*, \mathbf{x}^*) \end{bmatrix}. \tag{1}$$

where $\mathbf{\Sigma}_F$ is the full covariance matrix. Like the Gaussian distribution, the Gaussian Process is fully specified by a mean and its covariance function, so we denote the distribution $GP(\mu, C)$. The covariance function $C(\mathbf{x}_i, \mathbf{x}_j)$ expresses the covariance between y_i and y_j.

2.1 *The covariance function*

The model's prior expectations can be adapted to a given application by altering the covariance function. In this paper, we use a straightforward covariance function,

$$C(\mathbf{x}_i, \mathbf{x}_j; \Theta) = v_0 \rho(|\mathbf{x}_i - \mathbf{x}_j|, \alpha),$$

so that the hyperparameter vector is $\Theta = [v_0, \alpha_{1,..p}]^T$ and p is the dimension of vector $\mathbf{x}$. The function $\rho(d)$ is a distance measure, which should be one at $d = 0$ and which should be a monotonically decreasing function of d. The one used here was

$$\rho(|\mathbf{x}_i - \mathbf{x}_j|, \alpha) = e^{-\frac{1}{2}\sum_{k=1}^{p} \alpha_k (x_{ik} - x_{jk})^2}. \tag{2}$$

In most cases we will only have uncertain knowledge of Θ. With unknown hyperparameters, we can use

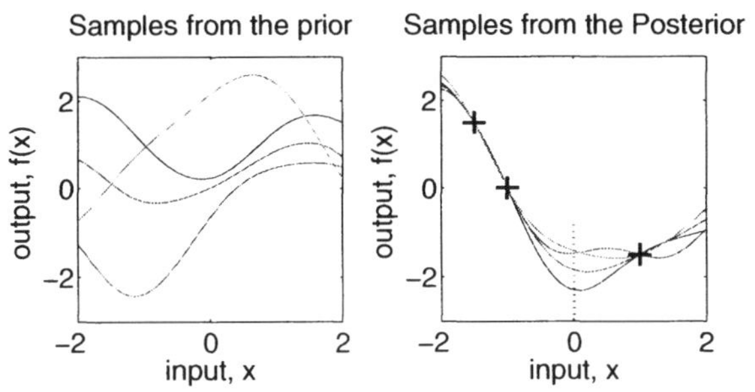

Fig. 1. Sampling from a Gaussian Process. Sample realisations are drawn from the Gaussian process prior (left hand figure), and then sample functions are drawn from prior conditioned on three training points (right hand figure).

maximum likelihood methods, with standard gradient-based optimisation tools to optimise the log-likelihood by adapting the hyperparameters. The hyperparameters are usually given a vague prior distribution, such as a gamma prior Neal (1997). Each hyperparameter of the covariance function can be given an independent prior distribution. The use of Gamma priors does not add significant complexity to the optimisation, and if used appropriately makes the model behaviour more robust with small training sets, leading to increased robustness and higher performance in the early stages of adaptation, but the relative advantage decreases with the amount of initial data available, as would be expected.

2.2 *Prediction*

As in the multinormal case, we can divide the joint probability into a marginal Gaussian process and a conditional Gaussian process. The marginal term gives us the probability of the training data,

$$P(\mathbf{y}|\mathbf{x}) = (2\pi)^{-\frac{N}{2}}|\Sigma|^{-\frac{1}{2}}e^{-\frac{1}{2}\mathbf{y}^T\Sigma^{-1}\mathbf{y}}. \quad (3)$$

The conditional part of the model, which best relates to a traditional regression model, is therefore the Gaussian process which gives us the output posterior density function conditional on the training data $\mathbf{x}, \mathbf{y}$ and the test points $\mathbf{x}^*$.

We have $p(y^*|\mathbf{x},\mathbf{y},\mathbf{x}^*) = \frac{p(y^*,\mathbf{y})}{p(\mathbf{y})}$, that is

$$p(y^*|\mathbf{x},\mathbf{y},\mathbf{x}^*) = \frac{1}{(2\pi)^{\frac{1}{2}}|\sigma^2(\mathbf{x}^*)|^{\frac{1}{2}}}e^{-\frac{1}{2}\frac{(y^*-\mu(\mathbf{x}^*))^2}{\sigma^2(\mathbf{x}^*)}}$$

where

$$\mu(\mathbf{x}^*) = \mathbf{k}(\mathbf{x}^*)^T\Sigma^{-1}\mathbf{y} \quad (4)$$
$$\sigma^2(\mathbf{x}^*) = C(\mathbf{x}^*,\mathbf{x}^*) - \mathbf{k}(\mathbf{x}^*)^T\Sigma^{-1}\mathbf{k}(\mathbf{x}^*), \quad (5)$$

with $k^i(\mathbf{x}^*) = C(\mathbf{x}^*,\mathbf{x}^i)$. So we can use $\mu(x^*)$ as the expected model output, with a variance of $\sigma^2(\mathbf{x}^*)$.

2.2.1. *Prediction at a random input* If we now assume that the test input $\mathbf{x}^*$ has a Gaussian distribution, $\mathbf{x}^* \sim \mathcal{N}(\boldsymbol{\mu}_{\mathbf{x}^*},\Sigma_{\mathbf{x}^*})$, the predictive distribution is now obtained by integrating over $\mathbf{x}^*$:

$$p(y^*|\boldsymbol{\mu}_{\mathbf{x}^*},\Sigma_{\mathbf{x}^*}) = \int p(y^*|\mathbf{x},\mathbf{y},\mathbf{x}^*)p(\mathbf{x}^*)d\mathbf{x}^*,$$

where $p(y^*|\mathbf{x},\mathbf{y},\mathbf{x}^*)$ is as specified by (4) and (5).

In Girard et al. (2002, 2003); Quinonero-Candela et al. (2003), we suggest an analytical Gaussian approximation to solve this integral, computing only the mean and variance of $p(y^*|\boldsymbol{\mu}_{\mathbf{x}^*},\Sigma_{\mathbf{x}^*})$. Assuming a covariance function such as (2), we arrive at

$$m(\boldsymbol{\mu}_{\mathbf{x}^*},\Sigma_{\mathbf{x}^*}) = v_0\mathbf{q}^T\boldsymbol{\beta} \quad (6)$$
$$v(\boldsymbol{\mu}_{\mathbf{x}^*},\Sigma_{\mathbf{x}^*}) = C(\boldsymbol{\mu}_{\mathbf{x}^*},\boldsymbol{\mu}_{\mathbf{x}^*}) + v_0^2\mathrm{Tr}\left[(\beta\beta^{\mathbf{T}} - \Sigma^{-1})\mathbf{Q}\right] - m(\boldsymbol{\mu}_{\mathbf{x}^*},\Sigma_{\mathbf{x}^*})^2 \quad (7)$$

where $\boldsymbol{\beta} = \Sigma^{-1}\mathbf{y}$, $\mathbf{W} = \mathrm{diag}[\alpha_1,\ldots,\alpha_p]$, I is the $p \times p$ identity matrix and

$$q_i = |\mathbf{W}^{-1}\Sigma_{\mathbf{x}^*}+I|^{-\frac{1}{2}}e^{-\frac{1}{2}(\boldsymbol{\mu}_{\mathbf{x}^*}-\mathbf{x}_i)^T(\Sigma_{\mathbf{x}^*}+\mathbf{W})^{-1}(\boldsymbol{\mu}_{\mathbf{x}^*}-\mathbf{x}_i)}$$

$$Q_{ij} = |2\mathbf{W}^{-1}\Sigma_{\mathbf{x}^*}+I|^{-\frac{1}{2}}e^{-\frac{1}{2}(\mathbf{x}_i-\mathbf{x}_j)^T(2\mathbf{W})^{-1}(\mathbf{x}_i-\mathbf{x}_j)} e^{-\frac{1}{2}(\mathbf{xb}-\boldsymbol{\mu}_{\mathbf{x}^*})^T(\frac{1}{2}\mathbf{W}+\Sigma_{\mathbf{x}^*})^{-1}(\mathbf{xb}-\boldsymbol{\mu}_{\mathbf{x}^*})},$$

where $\mathbf{xb} = (\mathbf{x}_i + \mathbf{x}_j)/2$ (see Girard et al. (2002, 2003); Quinonero-Candela et al. (2003) for the detailed calculations).

Note that as $\Sigma_{\mathbf{x}^*}$ tends to zero we have $m(\boldsymbol{\mu}_{\mathbf{x}^*},\Sigma_{\mathbf{x}^*}) \rightarrow \mu(\boldsymbol{\mu}_{\mathbf{x}^*})$ and $v(\boldsymbol{\mu}_{\mathbf{x}^*},\Sigma_{\mathbf{x}^*}) \rightarrow \sigma^2(\boldsymbol{\mu}_{\mathbf{x}^*})$, as we would expect.

3. DERIVATION OF CONTROL LAW

The objective of this paper is to control a multi-input, single-output, nonlinear system of the form,

$$y(t+1) = f(\mathbf{x}(t),u(t)) + e(t+1) \quad (8)$$

where $\mathbf{x}(t)$ is the state vector at time t, which in this paper will be defined as $\mathbf{x}(t) = [y(t),\ldots,y(t-n),u(t-1),\ldots,u(t-m),v_1(t),\ldots,v_l(t)]$, $y(t+1)$ the output, $u(t)$ the current control vector, $v_i(t), i = 1,\ldots,l$ are external known signals, f is a smooth nonlinear function, bounded away from zero. For notational simplicity we consider single control input systems, but extending the presentation to vector $u(t)$ is trivial. The noise term $e(t)$ is assumed zero mean Gaussian, with unknown variance σ^2.

The cost function proposed is:

$$J = E\{(y_d(t+1) - y(t+1))^2$$

where $y_d(t)$ is a bounded reference signal. [1]

Using the fact that $\text{Var}\{y\} = E\{y^2\} - \mu_y^2$, where $\mu_y = E\{y\}$, the cost function can be written as:

$$J = (y_d(t+1) - E\{y(t+1)\})^2 + \text{Var}\{y(t+1)\}.$$

With most models, estimation of $\text{Var}\{y\}$, and (as we shall see below) $\frac{\partial \text{Var}\{y\}}{\partial u(t)}$ would be difficult, forcing the use of certainty equivalence assumptions, adding extra terms to the cost function,or pursuing other sub-optimal solutions Wittenmark (1995); Filatov and Unbehauen (2000). With the Gaussian process prior (assuming smooth, differentiable covariance functions – see O'Hagan (1992)) straightforward analytic solutions can be obtained.

3.1 *Multistep optimisation*

In Murray-Smith and Sbarbaro (2002) we presented the one-step ahead version of the controller, but at $T = 1$ we do not get any 'probing' or 'active learning' benefit from the one-step controller–it only minimises the loss at the next time-step, and we limited ourselves to non minimum-phase systems. We now wish to use multistep optimisation. The cost function proposed is:

$$J_T = \frac{1}{T} \sum_{k=1}^{T} E\{(y_d(t+k) - y(t+k))^2\} \quad (10)$$

3.1.1. *Multistep prediction* We use the framework described in section 2.2.1 to propagate model predictions *and* uncertainties as we predict ahead in time. That is, we compute $E\{y(t+k)\}$ and $\text{Var}\{y(t+k)\}$ using equations (6) and (7) respectively. [2] This enables us to incorporate the uncertainty about intermediate estimates $\hat{y}(t+k-1), \ldots \hat{y}(t+k-n)$, which affects both the expected value, and its variance at $t+k$. See Girard et al. (2002) for more details on the application of the prediction at a Gaussian random input to time-series iterative k-step ahead prediction.

3.1.2. *Sensitivity equations for GPs* Given the cost function (10), and observations to time t, if we wish to find the optimal $u(t)$, we need the derivative of J_T,

$$\frac{\partial J_T}{\partial u(j)} = \sum_{k=j+1}^{t+T} \frac{\partial J_k}{\partial u(j)} \quad (11)$$

$$\frac{\partial J_k}{\partial u(j)} = -2\,(y_d(k+1) - \mu_y(k+1))\,\frac{\partial \mu_y(k+1)}{\partial u(j)} + \frac{\partial \text{Var}\{y(k+1)\}}{\partial u(j)} \quad (12)$$

where

$$\frac{\partial \mu_y(k+1)}{\partial u(j)} = \frac{\partial \mu_y(k+1)}{\partial \mu_y(k)}\frac{\partial \mu_y(k)}{\partial u(j)} + \frac{\partial \mu_y(k+1)}{\partial \text{Var}\{y(k)\}}\frac{\partial \text{Var}\{y(k)\}}{\partial u(j)} \quad (13)$$

$$\frac{\partial \text{Var}\{y(k+1)\}}{\partial u(j)} = \frac{\partial \text{Var}\{y(k+1)\}}{\partial \mu_y(k)}\frac{\partial \mu_y(k)}{\partial u(j)} + \frac{\partial \text{Var}\{y(k+1)\}}{\partial \text{Var}\{y(k)\}}\frac{\partial \text{Var}\{y(k)\}}{\partial u(j)} \quad (14)$$

The calculation of these terms requires use of the sensitivity equations for a change to $u(j)$, i.e. the derivatives of model predictions of mean and variance from $k = j, \ldots, T$. This is relatively straightforward for GPs with differentiable covariance functions, with hyperparameters fixed during the T-step prediction. In other models the sensitivity of the variance term would be much more difficult to estimate than in the GP case.

Given the derivative $\partial J_T / \partial u(t)$ we can use standard algorithms to optimise the values of $u(t), \ldots, u(t+T-1)$ to maximise J_T, i.e. we find

$$J_{opt} = \min_{u(t),\ldots,u(t+T-1)} \frac{1}{T} \sum_{k=1}^{T} E\{(y_d(t+k) - y(t+k))^2\}$$

This can then be repeated in a step-wise fashion to produce a T-step-ahead model predictive controller. An obvious issue is the computational effort needed in each iteration of the optimisation, because of the need to recalculate the inverse of the covariance matrix of the training data, with each new training point. [3]

At $\frac{\partial J_T}{\partial u(t)} = 0$, the resulting optimal control signal is significantly affected by the derivative of the variance of the model over the prediction horizon. If we had not included the variance term in cost function (10), or if we were in a region of the state-space where the variance was zero, the optimal control law would be different, and designers typically have to tune control effort damping parameters. The regularisation inherent to the GP approach make a control effort penalty constant, or regulariser unnecessary in many applications.

[1] It is straightforward to generalise this to more practical cost function, such as $J = E\{(y_d(t+1) - y(t+1))^2\} + (R(q^{-1})u(t))^2$, where the polynomial $R(q^{-1})$ is defined as:

$$R(q^{-1}) = r_0 + r_1 q^{-1} + \ldots + r_{n_r} q^{-n_r} \quad (9)$$

and its coefficients can be used as tuning parameters.

[2] Note that for $k = 1$, $\mathbf{x}^*$ will be composed of *known* lagged outputs so that $\mathbf{\Sigma}_{\mathbf{x}^*} = \mathbf{0}$ and we can use (4) and (5).

[3] An alternative to optimisation is to integrate over the variables of interest, by performing numerical integration of the joint density of future y and u using Monte Carlo techniques. Possibly the simplest strategy is the Metropolis method, which may be slow due to very small possible proposal widths. A more promising method is Hybrid Monte Carlo (HMC) Duane et al. (1987); MacKay (1998), which takes the derivative of the density into account and avoids slow random walk behaviour. HMC would be easilly implemented since all the required derivatives are readily available, although the computational cost will still be significant.

4. SIMULATION

The example considers the following non-linear functions:

$$y(t+1) = \frac{y(t)y(t-1)y(t-2)u(t-1)(y(t-2)-1)+u(t)}{1+y(t-1)^2+y(t-2)^2}$$

where $\mathbf{x} = \left[y(t)\ y(t-1)\ y(t-2)\ u(t-1)\right]^T$, from Narendra and Parthasarathy (1990). The observation noise has a variance $\sigma^2 = 0.001$, and we had 5 initial data points. We simulated a tracking experiment, where the model had to follow the reference trajectory (blue), while learning the behaviour of the system. We used a standard optimisation routine (MATLAB's fminsearch) to find the optimal control values at each point, and did not use explicit sensitivity equations to estimate the gradient. At each timestep, after $u(t)$ has been calculated, applied, and the output observed, we add the information $\mathbf{x}(t), u(t), y(t+1)$ to the training set, and the new Σ_1 increases in size to $N+1 \times N+1$. We then optimise the hyperparameters of the covariance function to further refine the model. The results are shown in Figure 2.

In simulation, including the variance term in the cost function led to more robust control and also led to a reduction in tracking error over the sample indicated in the figure, when compared to an identical model which had ignored the variance term in the cost function (the certainty equivalence approach).

T-step ahead predictions tend to show more caution for larger values of T. When the model state moves towards higher uncertainty regions (due to system complexity, or lack of training data), the control signal will be appropriately damped, and provides a smoother control signal. This leads to poorer immediate tracking performance, but greater robustness in the face of uncertainty. The choice of an appropriate prediction horizon T is obviously very important. A feature of the predictions is that as T increases, the mean prediction eventually tends towards zero (for a zero mean GP), and the variance of the prediction will also saturate at prior levels.

The hyperparameters in Figure 2(b) make few rapid changes, seeming well-behaved during learning.

5. CONCLUSIONS

Gaussian process priors can provide a flexible, general modelling tool for adaptive nonlinear control problems. The framework provides analytic estimates of model uncertainty and of derivatives of both model mean and uncertainty. In this paper we presented an approximate method for propagation of uncertainty for multistep predictions, which also allows you to calculate an analytic derivative of the cost function with respect to control actions. The experimental comparisons presented here used numerical optimisation over the prediction horizon, and found that the inclusion

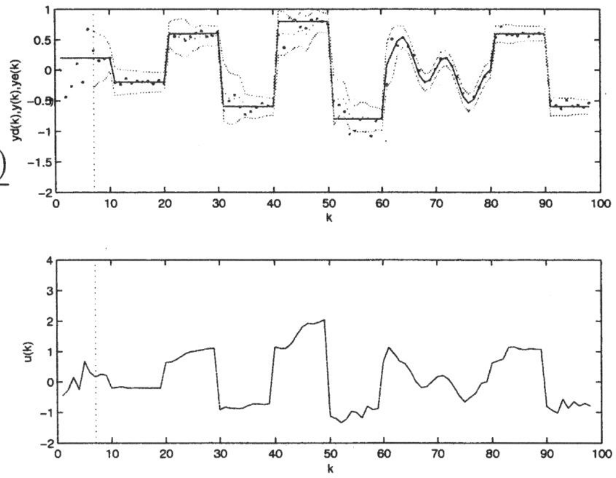

(a) Simulation of nonlinear GP-based controller, showing it controlling the target system in an adaptive manner from an initial condition of 5 training points, prediction horizon $T = 1$

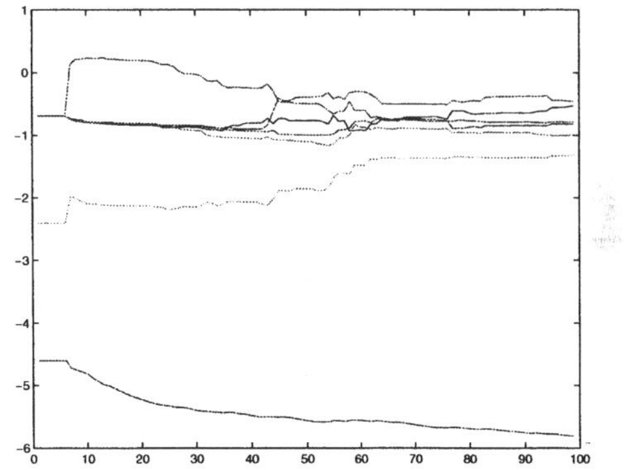

(b) Covariance function hyperparameters adapting for $T = 1$ case.

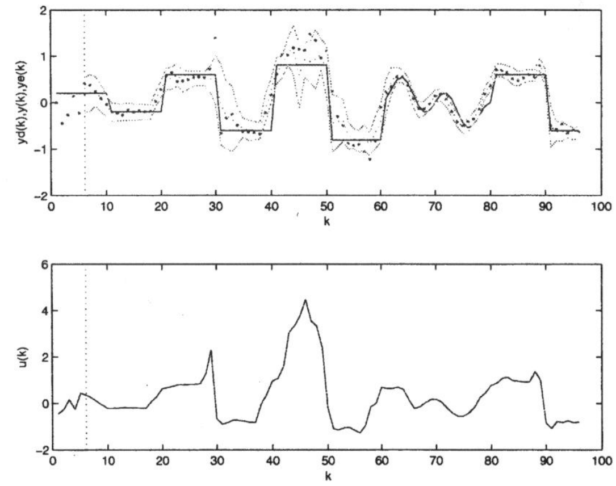

(c) Simulation of nonlinear GP-based controller, prediction horizon $T = 3$

Fig. 2. Simulation results for nonlinear system, showing modelling accuracy, control signals and development of the hyperparameters.

of the variance estimate, derived from estimates of propagated errors improve tracking performance, and robustness.

As a consequence of this multistep predictive controller, and the retention of the model variance in the cost function, we automatically get appropriate regularising behaviour *(caution)* which adapts to the uncertainty local to the current state.

ACKNOWLEDGEMENTS

RM-S & DS are grateful for support from FONDECYT Project 700397. RM-S, AG & CR gratefully acknowledge the support of the *Multi-Agent Control* Research Training Network (EC TMR grant HPRN-CT-1999-00107), and RM-S for the EPSRC grant *Modern statistical approaches to off-equilibrium modelling for nonlinear system control* GR/M76379/01.

References

M. Agarwal and D. E. Seborg. Self-tuning controllers for nonlinear systems. *Automatica*, (2):209–214, 1987.

S. Bittanti and L. Piroddi. Neural implementation of GMV control shemes based on affine input/output models. *Proc. IEE Control Theory Appl.*, (6):521–530, 1997.

S. Duane, A. D. Kennedy, and D. Roweth. Hybrid Monte Carlo. *Phys. Lett. B*, 195:216–222, 1987.

S. Fabri and V. Kadirkamanathan. Dual adaptive control of nonlinear stochastic systems using neural networks. *Automatica*, 34(2):245–253, june 1998.

A. A. Fel'dbaum. Dual-control theory I-IV. *Avtomatika i Telemekhanika*, 21 and 22(9,11 (vol. 21), 1, 2 (vol. 22)):1240–1249, 1453–1464, 3–16,129–142, 1960.

N.M. Filatov and H. Unbehauen. Survey of adaptive dual control methods. *Proc. IEE Control Theory Appl.*, (1):119–128, 2000.

N.M. Filatov, H. Unbehauen, and U. Keuchel. Dual pole placement controller with direct adaptation. *Automatica*, 33(1):113–117, 1997.

A. Girard, C. E. Rasmussen, and R. Murray-Smith. Gaussian process priors with uncertain inputs: multiple-step ahead prediction. Technical Report TR-2002-119, Department of Computing Science, University of Glasgow, 2002.

A. Girard, C. E. Rasmussen, J. Quinonero-Candela, and R. Murray-Smith. Gaussian process priors with uncertain inputs – application to multiple-step ahead time series forecasting. In S. Becker, S. Thrun, and K. Obermayer, editors, *Advances in Neural Information processing Systems 15*, 2003.

J. Kocijan, R. Murray-Smith, C.E. Rasmussen, and B. Likar. Predictive control and Gaussian process models. Submitted for publication, 2003.

D. J. Leith, R. Murray-Smith, and W. E. Leithead. Nonlinear structure identification: A Gaussian Process prior/Velocity-based approach. In *Control 2000, Cambridge*, 2000.

D.J.C. MacKay. Introduction to Monte Carlo methods. In M. I. Jordan, editor, *Learning and Inference in Graphical Models*, pages 175–204. Kluwer, 1998.

R. Murray-Smith, T. A. Johansen, and R. Shorten. On transient dynamics, off-equilibrium behaviour and identification in blended multiple model structures. In *European Control Conference, Karlsruhe, 1999*, pages BA–14, 1999.

R. Murray-Smith and D. Sbarbaro. Nonlinear adaptive control using non-parametric Gaussian process prior models. In *15th IFAC World Congress on Automatic Control, Barcelona*, 2002.

K. Narendra and K. Parthasarathy. Identification and Control of Dynamical Systems Using Neural Networks. *IEEE Trans. on Neural Networks*, 1(1):4–27, January/February 1990.

R. M. Neal. Monte Carlo implementation of Gaussian process models for Bayesian regression and classification. Technical Report 9702, Department of Statistics, University of Toronto, 1997.

A. O'Hagan. On curve fitting and optimal design for regression (with discussion). *Journal of the Royal Statistical Society B*, 40:1–42, 1978.

A. O'Hagan. Some Bayesian numerical analysis. In J. M. Bernardo, J. O. Berger, A. P. Dawid, and A. F. M. Smith, editors, *Bayesian Statistics 4*, pages 345–363. Oxford University Press, 1992.

J. Quinonero-Candela, A. Girard, J. Larsen, and C. E. Rasmussen. Propagation of uncertainty in Bayesian kernels models – application to multiple-step ahead forecasting. In *ICASSP, Hong Kong*, 2003.

D. Sbarbaro, N.M. Filatov, and H. Unbehauen. Adaptive dual controller for a class of nonlinear systems. In *Proceedings of the IFAC Workshop on Adaptive systems in Control and Signal Processing*, pages 28–33, Glasgow, U.K., 1998.

E. Solak, R. Murray-Smith, W. E. Leithead, D. J. Leith, and C. E. Rasmussen. Derivative observations in Gaussian process models of dynamic systems. In S. Becker, S. Thrun, and K. Obermayer, editors, *Advances in Neural Information processing Systems 15*, 2003.

C. K. I. Williams. Prediction with Gaussian processes: From linear regression to linear prediction and beyond. In M. I. Jordan, editor, *Learning and Inference in Graphical Models*, pages 599–621. Kluwer, 1998.

B. Wittenmark. Adaptive dual control methods: An overview. In *Preprints of 5th IFAC symposium on Adaptive Systems in Control and Signal processing, Hungary*, pages 67–72, 1995.

IFAC Publications
www.elsevier.com/locate/ifac

CONTROLLER DESIGN FOR SYSTEMS SUFFERING NONLINEAR DISTORTIONS

Michael Solomou (*), David Rees (*), Neophytos Chiras (**)

(*): School of Electronics, University of Glamorgan, Pontypridd, Rhondda Cynon Tâf, CF37 1DL, Wales, UK
(**): Praxis Critical Systems Ltd, 20 Manvers Street, Bath, BA1 1PX, UK
msolomo1@glam.ac.uk

Abstract: The aim of the study presented in this paper is to provide some initial results over the potential use of best linear approximation models in the design of model-based linear controllers for systems suffering nonlinear distortions. This is illustrated on a nonlinear mechanical resonating system. Parametric best linear approximation models are estimated from frequency response function measurements using random phase multisines. A full nonlinear model for the system is also estimated. This provides the basis for the design of simple optimal linear controllers, and the performance obtained from the controllers based on the nonlinear model is set as the benchmark. Parametric linear models are also estimated from measurements taken using Schroeder phase multisine signals and controllers based on these models are designed. It is shown that in the presence of nonlinear distortions the choice of input excitation in estimating models that describe the best linear approximation to the nonlinear system is crucial. It is also shown that this choice will have an effect on the performance of the model-based controllers. *Copyright © 2003 IFAC*

Keywords: identification, linear models, nonlinear distortion, nonlinear models, optimal control.

1. INTRODUCTION

The study presented in this paper primarily deals with the design of linear controllers based on *related linear dynamic system* (RLDS) measurements of a nonlinear system. The measurement of the RLDS of systems disturbed by nonlinearities has been extensively studied by Schoukens *et al.* (1998). They showed that by using random phase multisine signals and performing a number of experiments, it is possible to obtain improved frequency response function (FRF) measurements in the presence of nonlinear distortions. Such measurements can be regarded as the best linear approximation to the nonlinear system at a given operating point.

This approach, combined with frequency-domain techniques, is applied to estimate parametric linear models from measurements taken on an electrical simulator of a nonlinear mechanical resonating system at different operating points. These measurements are taken using random phase multisines and Schroeder phase multisines. Furthermore, a full Nonlinear Autoregressive Moving Average with eXogenous inputs (NARMAX) model that describes the system at all operating points is estimated.

This leads to the design of simple optimal linear controllers based on the linear models and the NARMAX model. The performance of the controllers obtained using the NARMAX model is considered to be the benchmark controller performance for the device under test (DUT). It is shown that the performance of the controllers obtained from the models estimated using the random phase multisines, lies very close to the benchmark performance.

The paper covers the following: Sections 2 and 3 provide the necessary theoretical background regarding FRF measurements in the presence of system nonlinearities, the measurement of the RLDS and the estimation of NARMAX models. Section 4 gives a description of the experimental setup, which leads to the estimation of linear and nonlinear parametric models of the DUT. Optimal controller design based on the estimated models is then described in Section 5 and conclusions are finally drawn in Section 6.

2. LINEAR FRF IN THE PRESENCE OF NONLINEARITIES

Consider a periodic multiharmonic signal applied to a time-invariant system. In steady state any nonlinearities present will generate an output contribution, which will be the same for each successive period of the signal. This will introduce a distortion into the estimated linear FRF which, in contrast to the error introduced by stochastic effects, will not reduce with averaging

$$H(j\omega_k)=\frac{\overline{Y}(j\omega_k)}{\overline{U}(j\omega_k)}=\frac{\overline{Y}_l(j\omega_k)+\overline{Y}_{nl}(j\omega_k)}{\overline{U}(j\omega_k)} \quad (1)$$

where $Y_l(j\omega_k)$ is the linear response and $Y_{nl}(j\omega_k)$ is the nonlinear distortion at the test frequencies.

The behaviour of linear FRF measurements in the presence of nonlinear distortions depends on the class of excitation signals. In this paper the excitation signals are multisines of F cosines with dc excluded, defined in the frequency domain as

$$\begin{aligned} U(j\omega) &= \sum_{\substack{k=-F\\k\neq 0}}^{F} a_k e^{j\phi_k}\delta(\omega - i_k\omega_0) \\ a_{-k} &= a_k \\ i_{-k} &= -i_k \qquad k=1,2,\ldots F. \\ \phi_{-k} &= -\phi_k \end{aligned} \quad (2)$$

where $\boldsymbol{a}$ is a vector of amplitudes, $\mathbf{i}$ a vector of harmonic numbers and $\boldsymbol{\phi}$ a vector of phases.

The distortions introduced by a power-series nonlinearity on multisine signals are the product of time-domain multiplication and hence convolution in the frequency domain. In order to clarify the influence of power-series nonlinearities on multisine signals, Evans (1998) proposed a methodology by which nonlinear distortions are divided into two types:

Harmonic Contributions: For the case of odd-order nonlinearities, harmonic contributions are generated by combinations of a test frequency with pairs of equal positive and negative test frequencies, and therefore they only fall at the test frequencies. Each harmonic contribution is in-phase to the test frequency it falls at. This introduces a systematic bias on the estimated FRF which is equal in magnitude for each test frequency and depends only on the amplitude of the input harmonics and the odd-order nonlinearities.

Interharmonic Contributions: These are generated by frequency combinations that do not follow the pattern of harmonic contributions and they depend entirely on the properties of the input harmonic frequencies. The phases of these interharmonic contributions will adopt the properties of the input harmonic phases (Solomou *et al.*, 2002).

It should be mentioned that harmonic and interharmonic contributions have been previously described in the literature as Type I and Type II contributions respectively. For their detailed description the reader is referred to the literature.

3. LINEAR AND NONLINEAR MODELS

3.1 The RLDS Model

Extensive work conducted by Schoukens *et al.* (1998) examined the influence of power-series nonlinearities on multisine signals with independent uniformly distributed random phases between [0, 2π], termed *random phase multisines*. By analysing the asymptotic properties of the FRF estimated using random phase multisines, they showed that the nonlinear distortions can be split in two classes, namely systematic and stochastic contributions. These have the same properties as harmonic and interharmonic contributions respectively for the case of random phase multisines as was shown by Solomou *et al.* (2002).

Schoukens *et al.* (1998) showed that for random excitations there exists an RLDS to which the expected value of the FRF estimate converges as a function of signal realisations or as a function of signal harmonics. It differs from the underlying linear system by the systematic bias introduced by the harmonic contributions. This is a deterministic component, independent of the random phase of the excitation. Moreover, the FRF estimate is scattered around its expected value and these deviations do not converge to zero. This effect is created by the action of the interharmonic contributions on the FRF measurements, and can be represented by a simple noise source that models the interharmonic contribution to the FRF.

Taking this into account, the estimated FRF $H(j\omega_k)$ can be expressed as

$$H(j\omega_k)=\frac{Y(j\omega_k)}{U(j\omega_k)}=H_0(j\omega_k)+H_B(j\omega_k)+\ldots \quad (3)$$
$$+H_S(j\omega_k)+N_G(k)$$

$U(j\omega_k)$ is the input signal to the system, $Y(j\omega_k)$ is the output signal to the system, $H_0(j\omega_k)$ is the underlying linear system, $H_B(j\omega_k)$ is the bias due to the harmonic contributions, $H_S(j\omega_k)$ is the interharmonic contributions, and $N_G(k)$ represents the output noise. Neither of the nonlinear components is decreasing if the number of frequencies increases. Therefore, the right hand side of (3) can be split in two parts: a first part that is linearly related with the input, leading to the related linear dynamic system $H_{RLDS}(j\omega_k) = H_0(j\omega_k) + H_B(j\omega_k)$ and a second part that is uncorrelated with the input, which leads to $H_S(j\omega_k) + N_G(k)$.

Measurements of the RLDS are obtained by eliminating the influence of $H_S(j\omega_k)$ and $N_G(k)$ from the FRF measurement. This is achieved by averaging the FRFs estimated through a number of different realisations of the random phase multisines. The RLDS can be considered as the best linear approximation to the overall nonlinear system at a given operating point, but it is clear that this approximation strongly depends on the class of excitation signals. Schoukens *et al.* (1998) have shown that by applying a well-chosen parametric identification scheme, the estimated linear models will describe the RLDS of the nonlinear system at a given operating point.

3.2 The NARMAX Model

The NARMAX approach was introduced by Leontaritis and Billings (1985) as a means of describing the input-output relationship of a nonlinear system. The model represents the extension of the well-known ARMAX model to the nonlinear case, and is defined as

$$y(k) = f\begin{bmatrix} y(k-1),...,y(k-n_y),u(k-1),..., \\ u(k-n_u),...e(k-1),...,e(k-n_e) \end{bmatrix} + \\ +e(k) \quad (4)$$

where f is a nonlinear function; $y(k)$, $u(k)$ and $e(k)$ represent the output, input and noise signals respectively; and n_y, n_u and n_e are their associate maximum lags. The NARMAX representation constitutes a powerful tool for nonlinear modelling and it includes a family of other nonlinear representations such as block-oriented models and Volterra functional series (Liu, 1988).

The selection of the NARMAX structure is based on the error reduction ratio (ERR), which is defined as:

$$ERR_i = \frac{g_i^2 \sum_{k=1}^{N} w_i^2(k)}{\sum_{k=1}^{N} y^2(k)} \quad (5)$$

where g_i are the coefficients and $w_i(k)$ are the terms of an auxiliary model constructed in such a way that the terms $w_i(k)$ are orthogonal over the data records. A forward-regression algorithm is employed to select at each step the term with the highest ERR. The procedure is usually stopped by using an information criterion such as the Akaike Information Criterion (AIC).

4. EXPERIMENTAL RESULTS

4.1 The Experimental Setup

A nonlinear mechanical resonating system (mass, viscous damping, nonlinear spring) was simulated using an electrical circuit. The displacement $y(t)$ (output) is related to the force $u(t)$ (input) by the following nonlinear, second order differential equation:

$$m\frac{d^2y(t)}{dt} + d\frac{dy(t)}{dt} + ay(t) + by^3(t) = u(t) \quad (6)$$

For small excitations the spring becomes almost linear so that the underlying linear system consists of a second-order resonance system. Special-odd random phase multisines (Vanhoenacker *et al.*, 2000) are used to measure the FRF of the system, where f_k=kf_0, k=1, 3, 9, 11, 17, 19... F, F=151 and f_0≈0.2980 Hz.

The behaviour of the system is illustrated in Fig. 1. It can be seen that at a small excitation level (10 mV_{rms}) the underlying linear system $H_0(j\omega_k)$ can be measured. The impact of the nonlinearity is made visible by increasing the excitation level 100 mV_{rms}. From Fig. 1 it can be seen that the resonance frequency is shifted to the right, the peak value is decreased and the FRF estimate becomes highly noisy, which can prove misleading during the model estimation phase.

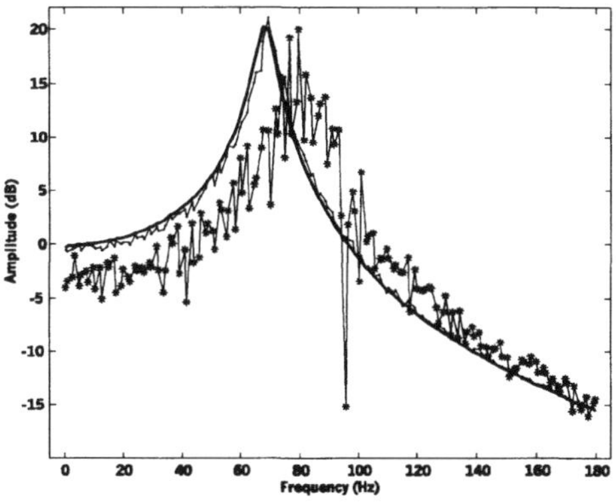

Fig. 1. Evolution of the estimated FRF of the nonlinear system at excitation levels 10 (Bold line), 30 (Thin line) and 100 (Stars) mV_{rms}.

Figure 2 illustrates that you can smooth the estimated FRFs by averaging over different realisations of the random phase multisines. This significant improvement in the resulting FRF measurement is achieved by the averaging procedure since the influence of the interharmonic contributions at the test frequencies is now reduced. The end product is an FRF measurement which differs from the underlying linear system only in terms of the bias term introduced by the action of the harmonic contributions at the test frequencies.

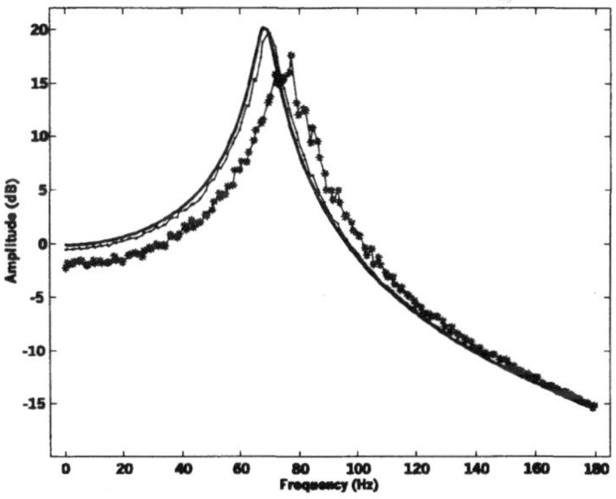

Fig. 2. RLDS measurement of the nonlinear system at excitation levels 10 (Bold line), 30 (Thin line) and 100 (Stars) mV_{rms}.

Such improvement however, cannot be obtained using Schroeder phase multisines (Schroeder, 1970), which are signals with a swept sine-like behaviour. Figure 3 shows the FRF measurement obtained for a Schroeder phase multisine with 601 consecutive harmonics. This is compared to the FRF measurement obtained after averaging through 5 different realisations of a random phase multisine with 601 consecutive harmonics. Both signals have an rms value of 30 mV.

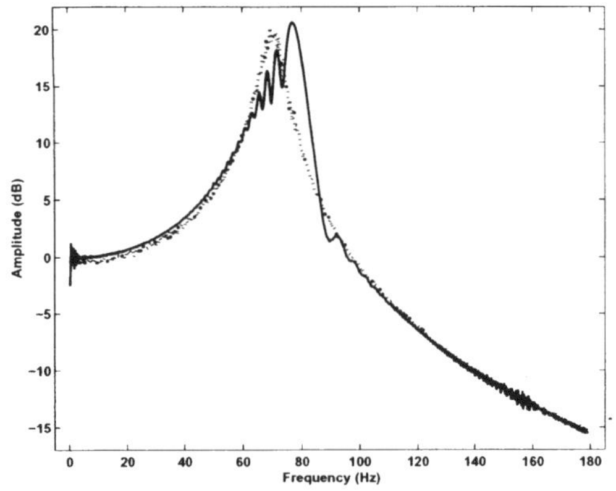

Fig. 3. Measured FRF of the nonlinear system at 30 mV_{rms} excitation level. Random phase multisine (Dots). Schroeder phase multisine (Solid line).

From Fig. 3 it can be seen that the FRF measured with the Schroeder phase multisine strongly deviates from the FRF measured with the random phase multisine. This is simply because the interharmonic components will interact in a different manner because of the Schroeder phases. This greatly distorts the FRF measurement and without prior knowledge no second order system is recognized any more. This illustrates that in the presence of nonlinear distortions the choice of input excitation is crucial if the aim is to measure the RLDS to the system at a given operating point. Such choice will have an immediate effect on the estimated linear models as will be discussed next.

4.2 Linear and nonlinear modelling of the system

Initially, linear transfer function models of the DUT were estimated by averaging the FRFs estimated from 20 different realisations of the special-odd random phase multisines by using the frequency-domain estimator described in Pintelon *et al.* (2001). These models were estimated at the operating points of 10, 30, 70 and 100mV_{rms}. Linear models were also estimated for the DUT from measurements taken using Schroeder phase multisines with 151 special-odd harmonics. Table 1 provides a comparison of the estimated models at the operating point of 30mV_{rms}.

Table 1 Estimated linear models at operating point of 30mV_{rms}

Signal Type	Model Parameters			
	Poles	f_{poles} (Hz)	*Zeros*	f_{zeros} (Hz)
Random	-21.3 + 439.9*j*	70.1		
	-21.3 – 439.9*j*	70.1		
	-84.6	13.5	-84.5	13.4
Schroeder	-30.95 +469.4*j*	74.9		
	-30.95 - 469.4*j*	74.9		
	-387.7	60.7	-455.3	72.4

From Table 1 it can be seen that the parameters of the two models differ. This is to be expected since the nonlinear influence on the two types of signals is different, as was shown in Fig. 3. According to the theory, the models estimated using the random phase multisines represent the best linear approximation of the nonlinear system at the operating point of test. This is not the case for the models estimated using the Schroeder phase signals. This argument will be further investigated in Section 5, where the models estimated from the two types of multisines will be used as a basis for model-based optimal controller design.

A NARMAX model for the DUT was also estimated using the methodology described in Section 3, which resulted in the following nonlinear difference equation.

$$y(k) = 1.9836y(k-1) - 0.9913y(k-2) + ...$$
$$+ 0.0077u(k-1) - 9.69 \cdot 10^{-5}u(k-2) - ... \quad (7)$$
$$- 0.0869y^3(k-1) - 3.47 \cdot 10^{-5}$$

The NARMAX model of (7) provides a good representation of the dynamic behaviour of the nonlinear system. This can be illustrated by examining Fig. 4 which shows the measured output along with the simulation error when the NARMAX model and the 70 mV_{rms} linear (RLDS) model estimated with the random phase multisines, are excited with a 70 mV_{rms} band-limited random noise signal.

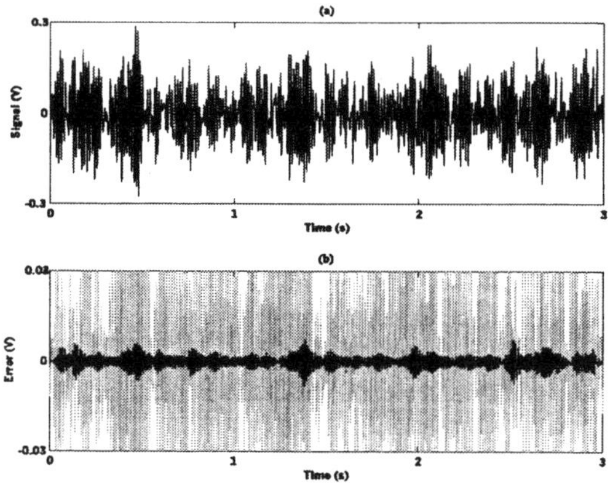

Fig. 4. Validation results for 70mV_{rms}, showing (a) the measured output and (b) the simulation error obtained with the RLDS (grey) and the NARMAX (black) models.

From Fig. 4 it is obvious that the simulation error of the NARMAX model is by far smaller than that of the RLDS model. Similar simulations were carried out for excitations of 10, 30, 70 and 100 mV_{rms}, which verified that the NARMAX model offers a good representation of the system global dynamics. For this reason the NARMAX model will be used to examine the performance of model-based controllers designed for the DUT without having to implement any hardware and thus avoiding the need to perform additional measurements.

5. CONTROLLER DESIGN

This section examines the performance of linear model-based controllers for the DUT, which are designed using the NARMAX model and the linear models obtained with the random phase and the Schroeder phase signal measurements. The analysis can be carried out through simulation by representing the DUT with the NARMAX model described in (7).

The structure employed to control the system is the simple proportional plus integral with derivative feedback (PI-D) controller structure shown in Fig. 5, where K_p is the proportional gain, K_i is the integral gain and K_d is the derivative gain. This configuration has been selected over the classic PID structure to reduce the effect of set-point differentiation, since it introduces very large changes to the control effort which is not desirable (Golten and Verwer, 1991).

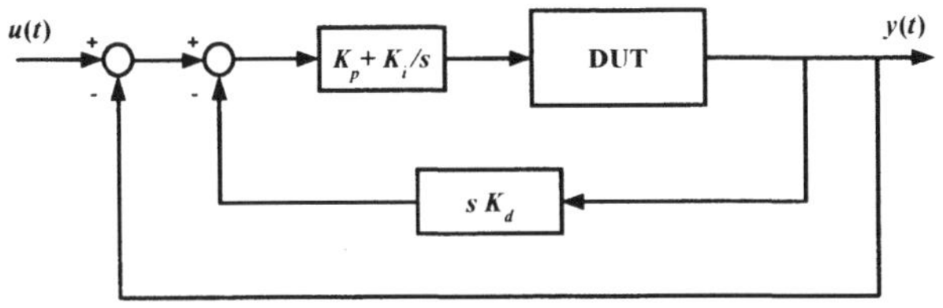

Fig. 5. Controller structure for the DUT.

Optimal controller design using the various models will be examined, which is achieved by minimising a performance index. In this study we use the integral of time multiplied by absolute error, ITAE, since it provides the best selectivity amongst other performance indices. A nonlinear optimisation algorithm is employed which searches to find the optimum values for K_p, K_i and K_d to minimise the ITAE performance index J. This procedure can be expressed in the continuous time as

$$J_{ITAE} = \int_{t=0}^{0.15} t\left|u(t) - y(t)\right| dt \qquad \min_{Kp,\, Ki,\, Kd} J_{ITAE} \tag{8}$$

where $0.01 \leq K_p \leq 9$, $0.01 \leq K_i \leq 1.5$ and $0.01 \leq K_d \leq 6$.

Table 2 shows the ITAE performance indices and controller parameters obtained from the NARMAX model at four step input levels of increasing amplitude. Step Inputs 1, 2, 3 and 4 represent step input amplitudes of 20, 60, 140 and 200 mV respectively. It is seen that at different step input levels the controller parameters will change in order to produce the optimum ITAE performance index. The ITAE indices obtained from the NARMAX model will be used as the benchmark performance indices.

The nonlinear optimisation algorithm is also employed to obtain optimum controller parameters for the same step inputs using the linear models estimated for the DUT. When the optimum controller parameters are obtained, new ITAE indices are calculated by applying these controller parameters on the closed loop system (Fig. 4), where the DUT is now represented by the NARMAX model. Table 3 shows the ITAE indices obtained for the increasing step input levels, when using the linear (RLDS) models estimated using the random phase multisine measurements.

Table 2 Controller parameters and ITAE indices for NARMAX model

Step Input	ITAE	K_p	K_i	K_d
1	**0.46**	9	0.21	5.46
2	**1.52**	9	0.22	5.29
3	**3.59**	9	0.26	5.39
4	**4.86**	9	0.31	5.52

Table 3 ITAE indices for RLDS models estimated at different operating points using random multisines

Step Input	Operating point			
	10mV$_{rms}$	**30mV$_{rms}$**	**70mV$_{rms}$**	**100mV$_{rms}$**
1	**0.63**	0.63	1.07	1.27
2	2.25	**1.58**	2.94	3.59
3	11.8	9.60	**4.01**	5.43
4	31.2	26.5	12.3	**7.37**

It can be seen that a given RLDS model will not produce good ITAE indices for all step inputs, as was the case for the NARMAX model. Instead, it will produce good ITAE indices for the step input that lies closer to the operating point at which the model was estimated. For example, while the RLDS model estimated at 100 mV$_{rms}$ gives the best ITAE index for Step Input 4, it fails to do so for Step Input 3, where the 70 mV$_{rms}$ RLDS model gives the best ITAE index. It can be seen that the optimum ITAE indices for the four different step input levels lie across the main diagonal of Table 3.

Table 4 shows the ITAE indices obtained for the increasing step input levels, when using the linear models estimated from the special-odd Schroeder phase multisine measurements.

Table 4 ITAE indices for linear models estimated at different operating points using Schroeder multisines

Step Input	Operating point			
	10mV$_{rms}$	**30mV$_{rms}$**	**70mV$_{rms}$**	**100mV$_{rms}$**
1	**0.62**	0.64	1.27	1.49
2	2.17	**1.95**	3.71	4.34
3	11.14	9.46	**8.17**	10.97
4	29.1	19.9	10.83	**12.01**

Once again it can be seen that the optimum ITAE indices for the four different step input levels lie across the main diagonal of Table 4, as was the case for Table 3. However, the optimal ITAE indices in the case of the models estimated from the Schroeder phase multisines are larger than those obtained from the models estimated using the random phase multisines. This difference is much more pronounced for the operating points of 70 and 100 mV_{rms}.

Fig. 6 shows the closed-loop step response of the system configuration shown in Fig. 5 for Step Input 3 (140 mV), using the controller parameters generated by the NARMAX model and the linear models estimated with the special-odd Schroeder phase multisines and the special-odd random phase multisines.

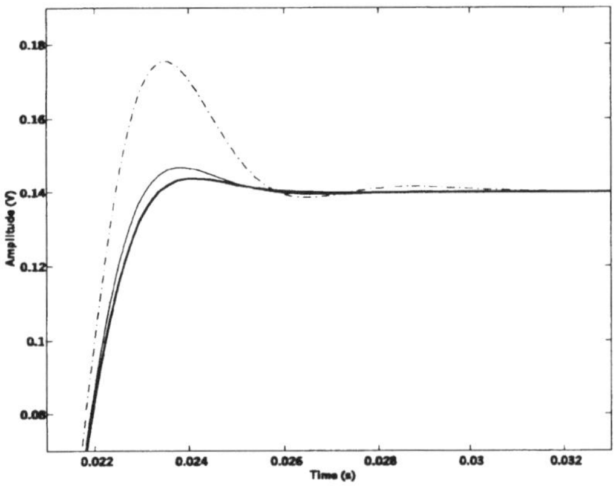

Fig. 6. Closed loop response at Step Input 3 for optimum controller parameters obtained from: NARMAX model (thick solid), and linear models estimated with Schroeder phase (dash-dot) and random phase (thin solid) multisines.

It is seen that in the case of the model estimated using the random phase multisines the step response is very similar to that of the NARMAX model. This is to be expected since the ITAE indices generated by the two models have very close values. However, the step response obtained using the model estimated from the Schroeder phase multisine measurements exhibits a large overshoot, which is unacceptable.

From the overall analysis conducted in this section the relationship between model quality and controller performance is obvious. In the presence of nonlinear distortions the user needs to make a careful choice of the type of excitation employed to measure the RLDS at a given operating point, since it will have a direct influence on the estimated models. If the approximation process is to be successful then test signals with random properties are strongly recommended over Schroeder phase signals.

6. CONCLUSIONS

This paper has demonstrated the use of best linear approximation models in the design of linear optimal controllers for systems suffering nonlinear distortions. This has been illustrated on a nonlinear mechanical resonating system. Emphasis was given on the choice of input excitation employed to estimate the RLDS models. It was shown that the controllers based on the model estimated with random phase multisines outperform those based on the models estimated with Schroeder phase multisines.

Moreover, the performance indices obtained from the controllers based on the random multisine models lie very close to those obtained from the controllers based on the NARMAX model. This by itself is an important result, since it indicates that for systems suffering nonlinear distortions it is possible to design good controllers without the need for a full nonlinear model, which is a more involved and complicated task.

7. ACKNOWLEDGEMENTS

Testing of the nonlinear mechanical resonating system was conducted at the Department ELEC of the Vrije Universiteit Brussel (VUB) with the permission of Professor Johan Schoukens. The authors would like to thank all the staff involved.

The authors would also like to express their deepest gratitude to the late Dr. Ceri Evans, who suddenly died in August 2002, for his invaluable contribution to the work presented in this paper. This paper is dedicated to his memory.

REFERENCES

Evans, C. (1998). *Identification of Linear and Nonlinear Systems Using Multisine Signals, with a Gas Turbine Application*. Ph.D. dissertation, School of Electronics, University of Glamorgan, U.K.

Golten, J. and A. Verwer (1991). *Control System Design and Simulation*, McGraw-Hill, London.

Leontaritis, I. J., and S.A. Billings (1985). Representations of non-linear systems: the NARMAX model. *Int. Jour. of Control*, vol. 49, no. 3, pp. 1013-1032.

Liu, Y. P. (1988). *Identification of Nonlinear Systems: The NARMAX Polynomial Model Approach*. Ph.D. dissertation, University of Sheffield, Department of Automatic Control & Systems Engineering, U.K.

Pintelon, R., J. Schoukens, W. Van Moer and Y. Rolain (2001). Identification of linear systems in the presence of nonlinear distortions. *IEEE Trans. Instrum. Meas.*, vol. 50, no. 4, pp. 855-863.

Schoukens, J., T. Dobrowiecki and R. Pintelon (1998). Parametric and nonparametric identification of linear systems in the presence of nonlinear distortions – A frequency domain approach. *IEEE Trans. Instrum. Meas.*, vol. 43, pp. 176-190.

Schroeder, M. R. (1970). Synthesis of low peak-factor signals and binary sequences of low autocorrelation, *IEEE Trans. Information Theory*, vol. 16, pp. 85-89.

Solomou, M., C. Evans, D. Rees and N. Chiras (2002). Frequency Domain Analysis of Nonlinear Systems Driven by Multiharmonic Signals. in *Proc. of the 19th IEEE Instrum. Meas. Tech. Conf. IMTC/2002*, Alaska, vol. 2, pp. 799-806.

Vanhoenacker, K., T. Dobrowiecki and J. Schoukens (2001). Design of multisine excitations to characterize the nonlinear distortions during FRF-measurements. *IEEE Trans. Instrum. Meas.*, vol. 50, no. 5, pp. 1097-1102.

IFAC Publications
www.elsevier.com/locate/ifac

HOW THE OUTPUT SATURATION OF A REGULATOR INFLUENCES THE REACHABLE PERFORMANCE AND ROBUSTNESS MEASURES

L. Keviczky and **Cs. Bányász**

Computer and Automation Research Institute, Hungarian Academy of Sciences
H-1111 Budapest, Kende u 13-17, HUNGARY
Phone: +361-466-5435; Fax: +361-466-7503
e-mail: keviczky@sztaki.hu ; banyasz@sztaki.hu

Abstract: The paper presents a simple procedure how to compute direct numerical relationships between performance, robustness measures and the output amplitude limit of a regulator for time delay control plants with higher order. *Copyright © 2003 IFAC*

Keywords: performance, robustness, amplitude limit

1. INTRODUCTION

A major control design paradigm is how to find compromise between the contradictory requirements of performance and robustness. This paradigm becomes even more sophisticated if in a practical application the always existing amplitude constraint at the output of the regulator should also be considered. Because of the "nonlinear" character of the problem algebraic solutions usually do not exist. Besides the usual "trial and error" approach another possible method is to "rescale" the original problem in order to remain in the linear operating range of the regulator (Keviczky and Bányász (1997)). As very useful design tool, a direct relationship (Keviczky and Bányász (2002))

$$Robustness = f\{Performance; Constraints; Plant\} \quad (1)$$

was developed formerly for a first order (or dominant pole) time-delay plant, where the variables have the meaning: *Robustness*= ρ=Nyquist stability margin; *Performance* = $\omega_{b,c}=1/T_w$ =Closed-loop bandwidth; *Constraints*= $p_s = T/T_w$ =Actuator's amplitude limit $\leq \overline{p}_s$; *Plant*=parameters. In this paper our goal is to extend these results for some higher order processes.

2. THE *GTDOF* CONTROLLER STRUCTURE

The *generic two-degree of freedom* (*GTDOF*) system (Keviczky (1995)) is used in this paper, which is based on the *Youla-parametrization* providing all realizable stabilizing regulators (*ARS*) for open-loop stable plants.

A *GTDOF* control system is shown in Fig. 1, where y_r, u, y and w are the reference, process input, output and disturbance signals, respectively. The optimal *ARS* regulator of the *GTDOF* scheme (Keviczky and Bányász (1999)) is given by

$$R_o = \frac{P_w K_w}{1 - P_w K_w S} = \frac{Q_o}{1 - Q_o S} = \frac{P_w G_w S_+^{-1}}{1 - P_w G_w S_- z^{-d}} \quad (2)$$

where

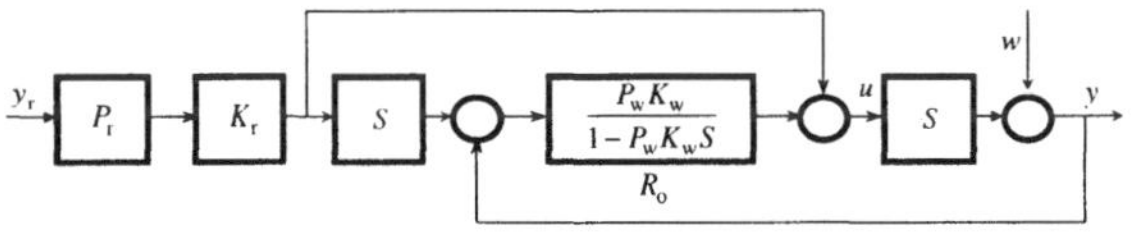

Fig. 1 The *generic TDOF* (*GTDOF*) control system

$$Q_o = Q_w = P_w K_w = P_w G_w S_+^{-1} \quad (3)$$

is the associated *Youla-parameter* (Maciejowski (1989)) furthermore

$$Q_r = P_r K_r = P_r G_r S_+^{-1} \; ; \; K_w = G_w S_+^{-1} \; ;$$
$$K_r = G_r S_+^{-1} \quad (4)$$

assuming that the process is factorable as

$$S = S_+ \overline{S}_- = S_+ S_- z^{-d} \quad (5)$$

where S_+ means the inverse stable (IS) and S_- does the inverse unstable (IU) factors, respectively. z^{-d} corresponds to the discrete time delay, which is the integer multiple of the sampling time. Here P_r and P_w are assumed stable and proper transfer functions (reference models). An interesting result presented in Keviczky and Bányász (1999) was that the optimization of the *GTDOF* scheme can be performed in $\mathcal{H}_2$ and $\mathcal{H}_\infty$ norm spaces by the proper selection of the serial G_r and G_w embedded filters.

3. THE SIMPLEST CASE: A FIRST ORDER TIME-DELAY PLANT

In our analysis the continuous time equivalent of the *GTDOF* is applied, where the following simple assumptions are used:

$$P_w = 1/(1+sT_w) \quad ; \quad S = e^{-s\tau}/(1+sT) \tag{6}$$

so the *IS* process is a first order time delay lag and the reference model is a first order lag. The continuous-time optimal *ARS* regulator based on (2) is now

$$R_o = \frac{1}{1-P_w e^{-s\tau}}\left(P_w S_+^{-1}\right) = \frac{1}{1-e^{-s\tau}/(1+sT_w)}\frac{1+sT}{1+T_w} \tag{7}$$

which can be easily realized, e.g., by a simple closed-loop. Note that R_o has a pole at $s=0$, so it is an integrating regulator.

The sensitivity function of the *GTDOF* system is

$$E = 1-P_w e^{-s\tau} = 1-\frac{1}{1+sT_w}e^{-s\tau} = \frac{1+sT_w-e^{-s\tau}}{1+sT_w} \tag{8}$$

The well-known Nyquist stability margin (the simplest robustness measure) is

$$\rho_m = \rho_{min}(R) = \min_\omega |\rho(\omega, R)| = \min_\omega |1+RS| = \frac{1}{\|E\|_\infty} \tag{9}$$

which is the distance between the point $(-1+0j)$ and the closest point of $Y(j\omega)$ and the reciprocal value of the norm $\|E\|_\infty$. ρ_m depends only on our design goal (T_w) and on the process behavior time delay (τ), more exactly on their relative value $x = T_w/\tau$. Unfortunately there is no simple analytical solution to obtain the relationship $\rho_m(x)$, only a numerical procedure can be applied following the graphical interpretation for $\|E\|_\infty$ using MATLAB.

In the first order example it is easy to see from the form of R_o that the initial peak for a unit step excitation is $R_o(\omega=\infty) = T/T_w$, so the forcing factor (power surplus) is

$$p_s = T/T_w \tag{10}$$

which comes from a simple physical interpretation: we should like to speed up the closed-loop from the original open-loop bandwidth $\omega_{b,o} = 1/T$ to the desired closed-loop reference bandwidth $\omega_{b,c} = 1/T_w$, therefore p_s is their ratio

$$p_s = \omega_{b,c}/\omega_{b,o} = T/T_w \tag{11}$$

Because

$$\nu = T/\tau = T/T_w \; T_w/\tau = p_s x \tag{12}$$

it is not difficult to draw a complex figure to show the relationships between ν, x, ρ_m parametrized by p_s as Fig. 2 shows.

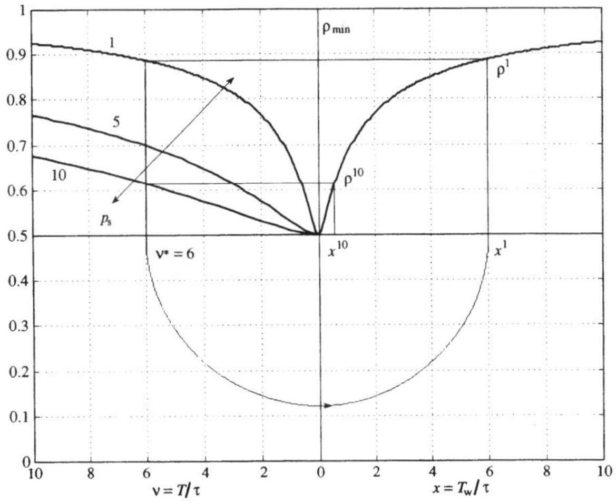

Fig. 2. Complex relationships between ν, x, ρ_m parametrized by p_s

The interpretation of $\rho_m(x)$ is very important, because this curve gives the theoretically best reachable robustness measure with any controller for an arbitrary *IS* time-delay plant. This measure is $\rho_m(0) \to 0.5$ for cases when the P_w reference model requires a very fast transient response from the time-delay process. The measure is $\rho_m(\infty) \to 1$, if τ is negligible comparing to the time lag of P_w.

Here the $\rho_m(x)$ curve is calculated for the most significant $p_s = 1, 5, 10$ parameters. The interpretation of Fig. 2 is very simple. Assume a plant behavior $\nu^* = T/\tau = 6$. Then one can get faster closed-loop response than the open-loop for all

$$x < x^1 = \nu^* = T/\tau = 6 \tag{13}$$

It is very rare that a practical actuator can apply $p_s > 10$ limits, so we have the lower limit x^{10} what corresponds to $p_s = 10$. So the reachable performance domain $x^{10} < x = T_w/\tau < x^1$ and the reachable robustness domain $\rho^{10} < \rho_m < \rho^1$ are obtained for the plant parameter $\nu^* = T/\tau = 6$. Note that x^1, x^{10}, ρ^1 and ρ^{10} depend on the given plant parameter ν^*. The calculations can be performed for another p_s, too.

4. EXAMPLES OF HIGHER ORDER TIME-DELAY PLANTS

Second order (complex) dominant poles

In some cases the dominant time lag is not first order, instead it is a second order term corresponding to a conjugate complex pole pair, so the time delay process is

$$S = \frac{e^{-s\tau}}{1+2\xi Ts+T^2s^2} = S_+ e^{-s\tau} \quad ; \quad 0<\xi<1 \quad (14)$$

It is not reasonable to use the usual first order reference model P_w, now, because the equivalent continuous-time optimal *ARS* regulator is not realizable in this case. A better selection is to use a second order reference model

$$P_w = 1/(1+sT_w)^2 \quad (15)$$

when the regulator is

$$R_o = \frac{1}{1-P_w e^{-s\tau}}\left(P_w S_+^{-1}\right) = \frac{1}{1-e^{-s\tau}/(1+sT_w)^2} \frac{1+2\xi Ts+T^2s^2}{(1+sT_w)^2} \quad (16)$$

It is easy to see from the form of R_o that the initial peak for a unit step excitation is $R_o(\omega=\infty) = T^2/T_w^2$, so the forcing factor is

$$p_s = T^2/T_w^2 \quad (17)$$

Note that p_s does not depend on ξ, which can be easily checked by simple MATLAB or SIMULINK simulation. However, p_s will not be equal exactly to the bandwidth improvement ratio $\omega_{b,c}/\omega_{b,o}$, because it depends on ξ. In the sequel the previously discussed relationships between $\nu = T/\tau$ and $x = T_w/\tau$ using p_s by (4) as parameter, will be investigated, because the meaning of p_s is the same as in the first order P_w case.

The demonstration of the change of the crossover frequency ω_c to ω_c' for the uncompensable process time delay $e^{-s\tau}$ is similar to the case of a first order P_w as the Fig. 3 shows.

The $\|E\|_\infty$ of the sensitivity function can also be determined graphically on Fig. 3, which is the farthest distance of $P_w(j\omega)e^{-j\omega\tau}$ from the point $(1+0j)$, see (9), and the Nyquist stability margin is obtained from (10).

The sensitivity function of the *GTDOF* system is now

$$E = 1 - P_w e^{-s\tau} = 1 - \frac{1}{(1+sT_w)^2} e^{-s\tau} = \frac{(1+sT_w)^2 - e^{-s\tau}}{(1+sT_w)^2} \quad (18)$$

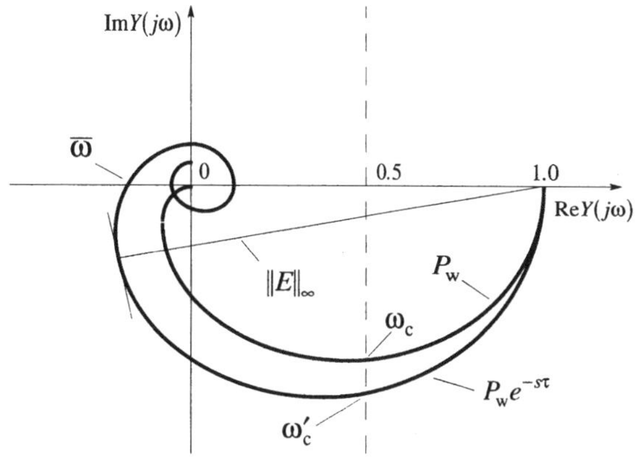

Fig. 3. Change of the crossover frequency ω_c to ω_c'

As it was mentioned earlier for the problem (14)-(16)

$$p_s = T^2/T_w^2 \neq \omega_{b,c}/\omega_{b,o} \quad (19)$$

stands now, i.e.

$$\nu = \frac{T}{\tau} = \frac{T}{T_w}\frac{T_w}{\tau} = \sqrt{p_s}\frac{T_w}{\tau} = \sqrt{p_s}\, x \quad (20)$$

It is not difficult to draw a complex figure to show the relationships between ν, x, ρ_m parametrized by p_s as Fig. 4 shows.

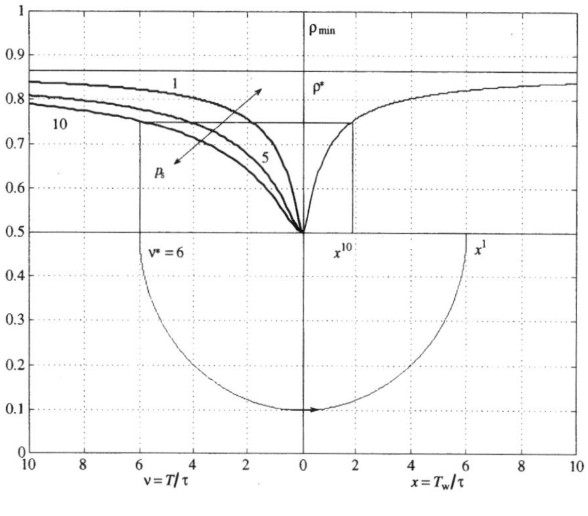

Fig. 4. Complex relationships between ν, x, ρ_m parametrized by p_s

It is interesting to observe, that the limiting performance x^{10} is greater than it was in the first order case for the same plant behavior $\nu^* = T/\tau = 6$. This is because both the upper and the lower part of the curve series in Fig. 2 is "pushed" for the second order case forming Fig. 4, so the controller design has a smaller selection range ($x^{10} < x < x^1$ and $\rho^{10} < \rho_m < \rho^1$) now.

The upper part of Fig. 4 is shown separately in Fig. 5, where one can observe the $\rho^* = 0.866$ limit, which is different from the plot for first order P_w. It is easy to calculate this limit considering $\|E\|_\infty$ for $\tau \to 0$.

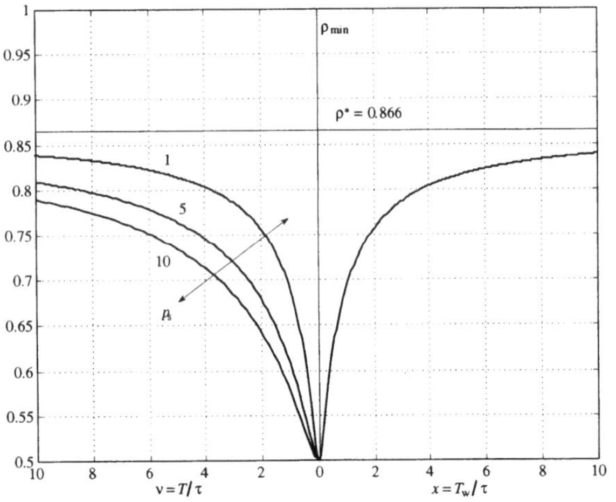

Fig. 5. Complex relationships between ν, x, ρ_m parametrized by p_s

Simple calculations give

$$|E(j\omega)|_{\tau=0} = \sqrt{\left(\frac{T_w \omega}{1+T_w^2\omega^2}\right)^2 \left(4+T_w^2\omega^2\right)} \tag{21}$$

which has a maximum (obtained by the symbolic MATLAB) at

$$\omega^* = \frac{\sqrt{2}}{T_w} \tag{22}$$

where

$$\|E\|_\infty = \left|E(j\omega^*)\right|_{\tau=0} = \frac{2}{\sqrt{3}} = 1.155 \tag{23}$$

so finally

$$\rho^* = \frac{1}{\|E\|_\infty} = \frac{\sqrt{3}}{2} = 0.866 \tag{24}$$

n-order dominant poles

It is possible to compute the above limit $\rho^*(n)$ in the function of the order n of P_w, if

$$P_w = \frac{1}{(1+sT_w)^n} \quad ; \quad S = \frac{1}{(1+sT)^n} e^{-s\tau} \tag{25}$$

This function is plotted in Fig. 6. Although, this plot shows only the considerable decrease of the reachable robustness limit $\rho^*(n)$ by n, but the computations give that the lower part of the curves move up, which means that the acceptable design region is smaller and smaller for higher order plants.

It is also possible to use the real part of $P_w(j\omega)e^{-j\omega\tau}$ at the first intersection with the real axis at $\bar{\omega}$ as a lower limit for $\|E\|_\infty$, so its reciprocal value is an appropriate upper limit for $\rho_m = \rho_{min}$:

$$\rho_m \le \frac{1}{1 - \mathrm{Re}\left\{P_w(j\bar{\omega})e^{-j\bar{\omega}\tau}\right\}} = \bar{\rho}_m \tag{26}$$

The $\rho_m(x)$ and $\bar{\rho}_m(x)$ curves obtained by MATLAB numerical calculations are plotted on Fig. 7 for the second order P_w. This upper limit is much worse approximation than what can be calculated for a first order P_w, shown in Fig. 8.

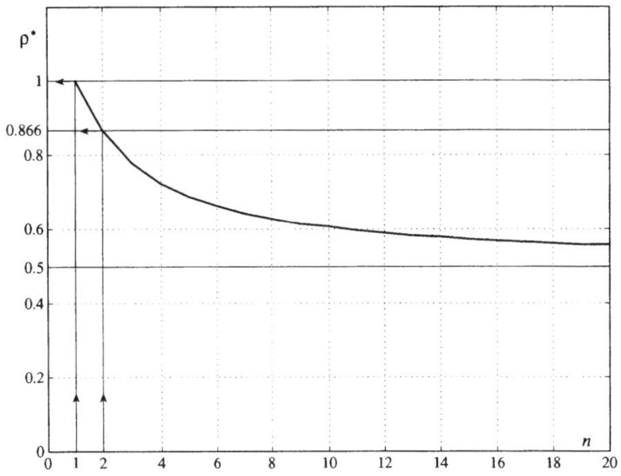

Fig. 6. The robustness limit $\rho^*(n)$ in the function of the plant order n

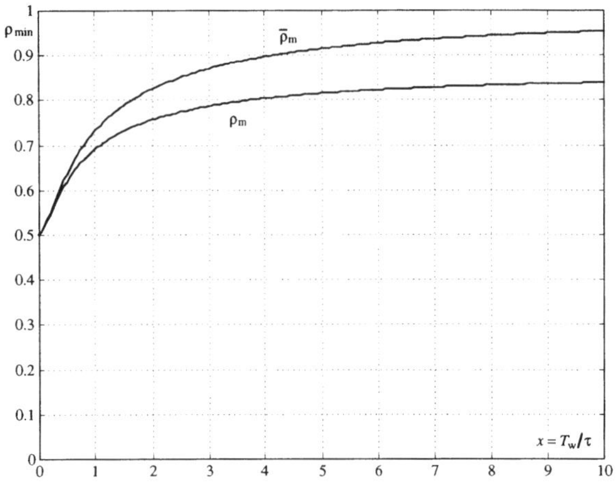

Fig. 7. The reachable robustness measure ρ_m and a practical upper bound $\bar{\rho}_m$ ($n = 2$)

In case of a higher order plant one can always use the dominant time constant as T in these investigations. All further lag terms (higher order denominator in S) make the situation worse lowering the $\rho_m(x)$ curve (decreasing the robustness of the closed-loop). The influence of minimum phase lead terms (higher order numerator in S) improves the situation by pushing up the curve $\rho_m(x)$.

The influence of IU non-minimum phase lead terms (higher order unstable numerator in S) has the same effects as further lags in the denominator of S. Consider a first order nonminimum phase time-delay process with

$$S = \frac{1 - sT_1}{1 + sT} e^{-s\tau} \qquad (27)$$

The numerical computations are shown in Fig. 9.

It is interesting to observe that a considerably large unstable zero has a much worse influence than a higher order plant.

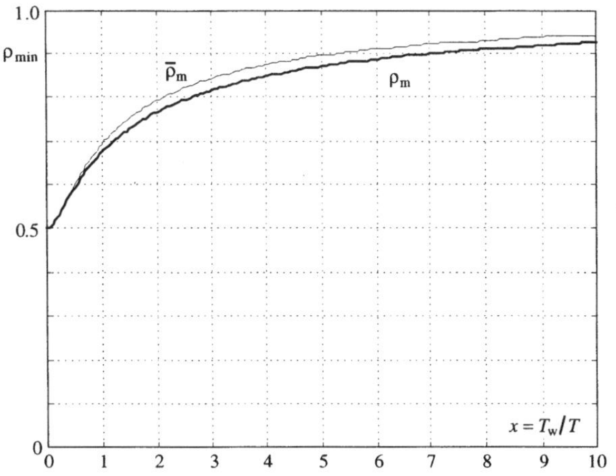

Fig. 8 The reachable robustness measure ρ_m and a practical upper bound ($n = 1$)

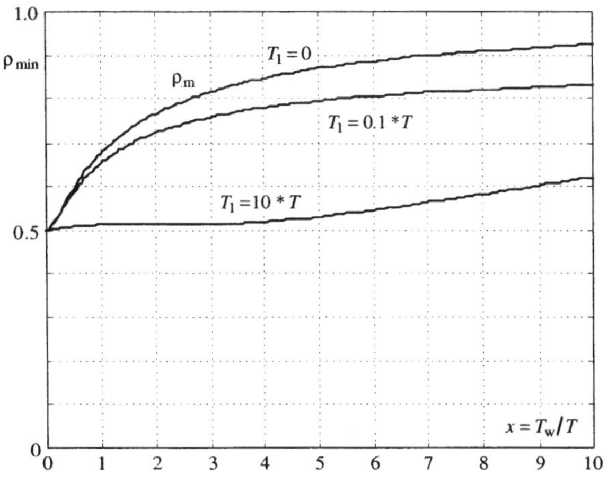

Fig. 9 The function $\rho_m(x)$ parametrized by T_1/T

5. CONCLUSIONS

A simple algorithm was presented to compute the direct relationships (1) between the most important actuator, process parameters and rubustness measures. The presented plots are simple and very important, because they give the ultimate control limits reachable by any regulator. In case of a higher order plant one can always use the dominant time constant as T in these investigations. At the same time the designer can expect a considerable decrease of the reachable performance/robustness region for higher order plants and for those having dominant unstable zeros. These results can be used in adaptive and/or off-line or iterative tuning of discrete-time controller in practical cases, when the output of the regulator has amplitude saturation.

REFERENCES

Keviczky, L. (1995). Combined identification and control: another way. (Invited plenary paper.) *5th IFAC Symp. on Adaptive Control and Signal Processing, ACASP'95*, 13-30, Budapest, H.

Keviczky, L. and Cs. Bányász (1997). An iterative redesign technique of reference models: How to reach the maximal bandwidth? *11th IFAC Symposium on System Identification SYSID'97*, Fukuoka, Japan, 619-624.

Keviczky, L. and Cs. Bányász (1999). Optimality of two-degree of freedom controllers in $\mathcal{H}_2$- and $\mathcal{H}_\infty$-norm space, their robustness and minimal sensitivity. *14th IFAC World Congress*, **F**, 331-336, Beijing, PRC.

Keviczky, L. and Cs. Bányász (2002). Direct relationships of performance, robustness measures and amplitude constraint. CDC'2002 Las Vegas, USA.

Maciejowski, J.M. (1989). *Multivariable Feedback Design*, Addison Wesley.

This work was supported by the Hungarian NSF (OTKA) and the Control Engineering Research Group of the Hungarian Academy of Sciences.

IFAC Publications
www.elsevier.com/locate/ifac

RANDOM LOADING IDENTIFICATION OF A PLASTIC GLASS CANTILEVER BEAM

DongSheng Li, XingLin Guo and HongNan Li

State Key Laboratory of Coastal and Offshore Engineering, School of Civil and Hydraulic Engineering Dalian University of Technology, Dalian 116023, China

Abstract: The loading power spectral density matrices of a cantilever beam of plastic glass subjected to stationary random excitations are identified by utilizing inverse pseudo excitation method using response measurements and structural frequency response functions. Numerical simulation is conducted for the purpose of optimal selection of sensor locations. Advantages of this method include its computational efficiency compared with other approaches and the benefit of alleviating the ill conditioning of frequency response function near some resonant frequencies. The identified loadings agree with actual ones perfectly well and this method is robust to random noise. *Copyright © 2003 IFAC*

Keywords: identification, inverse modelling, vibration, linear, mechanical

1. INTRODUCTION

The solution for the responses of a linear structure subjected to multiple stationary random excitations has been solved successfully by pseudo excitation method (PEM), see (Lin *et al.*,1992, 1994); this is the direct problem, i.e. the computation of the power spectral densities (PSDs) of the responses from given excitation PSD of a structure. In parallel to this direct problem, there are two types of inverse problems. The first is the system identification problem, for which the properties of a structure are identified from the known loading and response PSD functions. As viewed from modal testing, this is to extract structural dynamic properties (resonant frequencies, mode shapes and modal damping ratios) from the frequency response function (FRF), and the FRF is obtained by measuring appropriate input and output points. Many publications cover such problems and their successful applications in engineering (Ewins, 1984). The second is so-called loading identification problem, for which the responses and the system properties of a structure (FRF) are known and used to identify the loading information. This second type of inverse problem has received comparatively less attention in the technical literature, especially for identifying random excitations. In fact, many loads in engineering applications need to be identified. The identification of such loads is of great importance. For instance, the traffic loads of bridges; the seismic excitations of buildings; the interaction forces between moving machines and their bases, and so on. These excitations are difficult, or sometimes impossible, to measure directly, whereas the measurement of some of the responses are relatively easy. Therefore it is essential to develop indirect means to identify these excitations. The inverse pseudo excitation method (IPEM), a counterpart or an inversion form of PEM, is an efficient approach to deal with such problems (Lin *et al.*, 2001).

This paper addresses the loading identification problem of a structure subjected to stationary random excitations by IPEM. This algorithm first decomposes the known response PSD matrix to produce a pseudo response vector, which will be used to generate a pseudo excitation vector by multiplication with a corresponding FRF, at a given frequency point. The FRF is determined by measuring inputs and corresponding outputs in prior tests. Further the loading PSD matrix is reconstructed by the multiplication of appropriate forms of the pseudo excitation vectors. Above method yielded excellent identification results when applied in model tests for a plastic glass cantilever beam. The main advantages of IPEM are its efficiency to solve loading identification problem by saving computation time substantially and its accuracy to predict unknown excitations.

The earliest investigations of force determination were to determine the external vibratory forces exerted on the rotor hub of a dynamic model of a helicopter (Barlett and Flannelly, 1979). Subsequently, a real AH-1G helicopter was tested to determine the magnitudes and phase angles of in-flight rotor vibratory forces acting on the airframe using acceleration responses and mobility calibration matrix (Giansante *et al.*, 1982). Their method belongs to the direct inversion of FRF. Hillary and Ewins (Hillary and Ewins, 1984) tested a cantilever beam by using two sinusoidal forces of the same frequency but different magnitudes. Their experimental results shown that the poor identified forces were caused largely by the contamination of measurement noise, and that strain gauge measurement instead of accelerometers in the lower frequency range, will improve the reconstructed excitations. Three different types of real structures, a machine tool, an automobile engine and an air conditioner under operating condition were tested and the experiments resulted in a good agreement between reconstructed and actual forces by virtue of the inversion of FRFs (Okubo *et al.*, 1985). They also analysed the different influences of noises on the identified results using a synthesized model. Desanghere and Snoeys (Desanghere and Snoeys, 1985) conducted experiments on a real longitudinal beam of a car frame excited by three electro-magnetic shakers, and researched extensively different influences on the identification results, such as noise on dynamic responses, perturbation of modal parameters and limited number of modes from an analytical example by the modal coordinate transformation method (MCTM). But this method is weakened drastically by the lack of participating structural modes, especially by the lack of the higher order modes that are difficult to obtain accurately. In addition, most of these explorations dealt with deterministic forces, for instance impact forces, but not random excitations and the identification results are hardly satisfactory.

Several researchers also investigated other aspects of loading identification problem. Callahan and Piergentili (Callahan and Piergentili, 1994) discussed a situation in which the input locations were not known and found that when the assumed force locations are actual force inputs, the forces are adequately determined; otherwise, the forces are distributed evenly to all locations in the assumed set by using MCTM. Avitabile (Avitabile and Chandler, 2001) proposed a test reference identification procedure (TRIP) method for the selection of multiple reference locations by a pre-modal test with a very limited set of potential reference locations. TRIP was developed mainly due to the fact that a finite element model may be incorrect and if it is used to help identify reference locations for test, then it may not produce the best test or even miss modes. Two thorough reviews of loading identification literature concerning different methods and existing problems can be found in Karl (Karl, 1987) and Dobson (Dobson and Rider, 1990).

This paper describes the IPEM method and the associated theory. Computer simulation for the purpose of optimal selection of sensor locations from several candidate sets and experiment results of the cantilever beam for loading identification are also presented.

2. THEORY OF INVERSE PSEUDO EXCITATION METHOD

The fundamental formula relating the response PSD matrices to the excitation PSD matrices by FRFs for solving stationary random vibrations on a linear structure in the frequency domain is (Newland, 1984):

$$[S_{yy}] = [H]^*[S_{xx}][H]^T \tag{1}$$

In which $[S_{xx}]$ is the known excitation PSD matrix, $[H]$ is the FRF matrix, $[S_{yy}]$ is the response PSD matrix to be computed and the superscripts * and T represent complex conjugate and transpose respectively.

Eq.(1) is used for computing the responses of a structure subjected to known excitations in the direct problem. On the contrary, in the inverse problem of loading identification, the responses of a structure are already known and are used to compute the unknown excitations. The given response PSD matrix $[S_{yy}]$ can generally be assumed to be a p x p matrix with rank r ($\leqslant p$), and can be decomposed into the following form according to IPEM,

$$[S_{yy}] = \sum_{j=1}^{r} \{b\}_j^* \{b\}_j^T \tag{2}$$

This equation can generally be realized in terms of the spectral decomposition scheme of a Hermitian matrix. If all the elements of $[S_{yy}]$ are real quantities, then $\{b\}_j$ will take very simple forms. In such cases, $\{b\}_j$ is the product of the square root of an eigen-value and the corresponding normalized eigen-vector, which are easily obtained by matrix eigen-value decomposition method (Bathe and Wilson, 1976). Assume a structure is excited by a pseudo excitation $\{x\}_j = \{a\}_j e^{i\omega t}$, then the pseudo response and its relation to the pseudo excitation is:

$$\{b\}_j e^{i\omega t} = [H]\{x\}_j = [H]\{a\}_j e^{i\omega t} \tag{3}$$

$$\{a\}_j = [H]^+\{b\}_j \tag{4}$$

Where the superscript + represents Moore-Penrose generalized inversion. Hence, the excitation PSD matrix $[S_{xx}]$ is given in terms of $\{a\}_j$ as

$$[S_{xx}] = \sum_{j=1}^{r} \{a\}_j^* \{a\}_j^T \tag{5}$$

Therefore the excitation matrix $[S_{xx}]$ can be computed from the known response PSD $[S_{yy}]$ matrix by above algorithm.

An FRF matrix is required to be inversed once for any given discrete frequency in order to compute the excitations by above IPEM algorithm, therefore this method reduces computation time significantly compared to conventional approaches. Usually the determination of n external excitations requires at least m $(m \geqslant n)$ measured responses, and the inversion of the FRF matrix is pathological at some spectral regions as indicated by Dobson (Dobson and Rider, 1990). The singular value decomposition (SVD) scheme is employed in IPEM to moderate the severity of such rank deficiency in the inversion process of the rectangular FRF matrix (n x m).

$$[H] = [U][S][V]^T \qquad (7)$$

$$[H]^+ = [V][S]^+[U]^H \qquad (8)$$

Consequently the steps of the loading identification test procedure are set up as follows.
(1) Computer simulation for sensor displacement.
(2) System or FRF calibration. Obtain the FRF matrices by artificially exciting the structure.
(3) Responses measuring. Measure all the responses simultaneously under normal conditions.
(4) Loading reconstruction. Compute the loadings according to Eq.(2) to Eq.(8).

3. SELECTION OF SENSOR LOCATIONS

Selection of measurement locations by means of computer simulation is an efficient approach before conducting field tests (Lin *et al.*, 2001). First, assume several candidate sets of measurement locations and arrange them according to the descending order of their contribution to the operational deflection shape resulting from the mode shape products. Meanwhile the installation of sensors on such locations should be convenient. Then apply white noise excitations to the structure, and compute the corresponding FRF matrix by an FEM program for the selected combination of random responses to excitations. The response PSD matrices at the frequencies of interest are then computed by implementing PEM; These PSD matrices are further used to generate the excitation PSD matrices based on the above IPEM. Repeat this process for the preliminary selected combinations of random responses in sequence, and compare the identified excitation PSD matrices with the assumed white noise spectral as well as the condition numbers of the FRF matrices for each set, the best one set with smallest identification disparities and overall condition numbers will be used in later experimental identification of the excitation PSD matrices. Computer simulation can still be implemented even if analytical modal of a structure is not available. The reason for this situation is the inability of mathematical modelling techniques to adequately describe the structure because of inaccurate material parameters or impractical boundary conditions. In such cases, use FRF matrices measured in prior experiments instead in the simulation process, and other related procedures remain the same. In this paper, FRFs were obtained by actual tests.

The three dimensions of the plastic glass cantilever beam, which was vertically fixed, are 88.5x7.5x1.2 cm (L x W x T). Two horizontal forces are applied at nodes 4 and 5 (Fig.1) and the assumed excitation PSD matrix at all given discrete frequencies is:

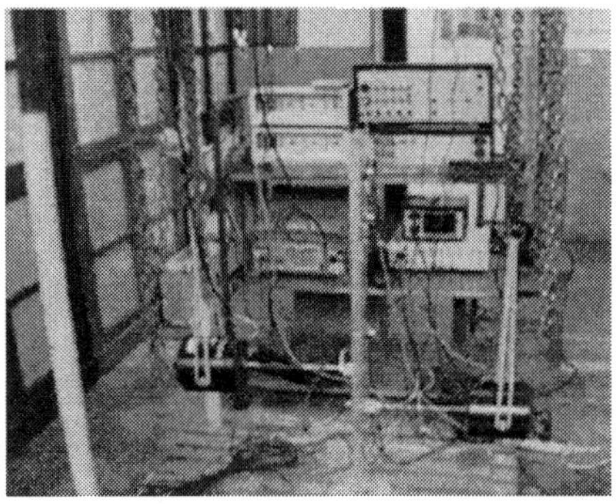

Figure 1-Plastic glass cantilever beam

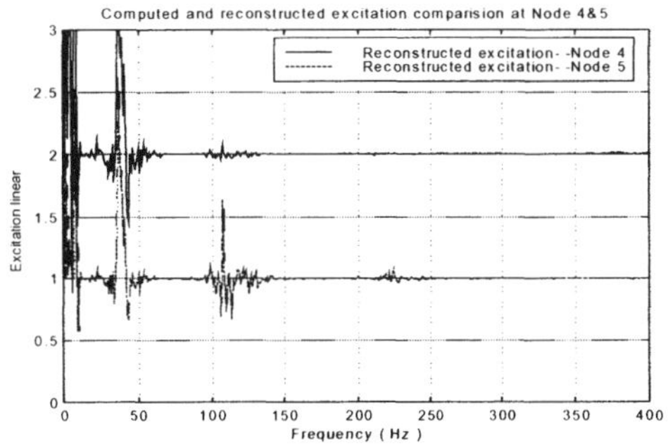

Figure 2-Measured points at 11,8,7,5

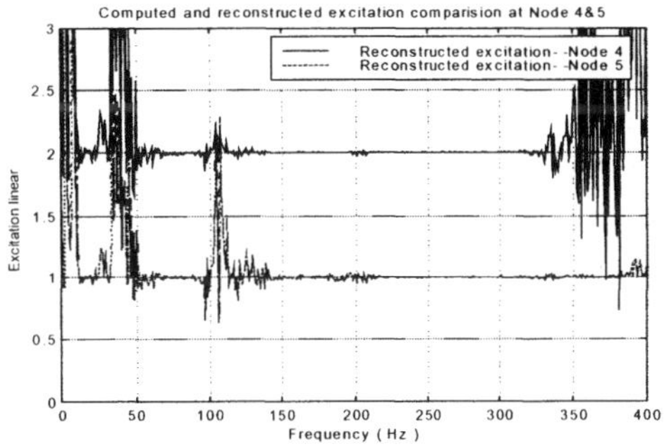

Figure 3-Measured points at 10,9,6,4

$$[S_{xx}] = \begin{bmatrix} 2.0 & 0 \\ 0 & 1.0 \end{bmatrix}$$

Fig.2 and Fig.3 show the simulation results of loading identification for two different sets of measured locations. The first set includes four responses at locations 5,8,9,11 (Fig.2) and the second at locations 3,5,9,11(Fig.3). It is obvious that the first set is better, which was then chosen for subsequent experiments.

4. EXPERIMENT IDENTIFICATION RESULTS

Two uncorrelated excitations were applied at points 4, 5 horizontally and four acceleration responses were measured at 5,8,9,11 simultaneously (Fig.1) as in the simulation process. The significant frequency range is from 0 to 400 Hz, which includes the first five natural frequencies. Fig. 4 and Fig. 5 show that the identified loading PSD matrices agree with the actual ones perfectly well.

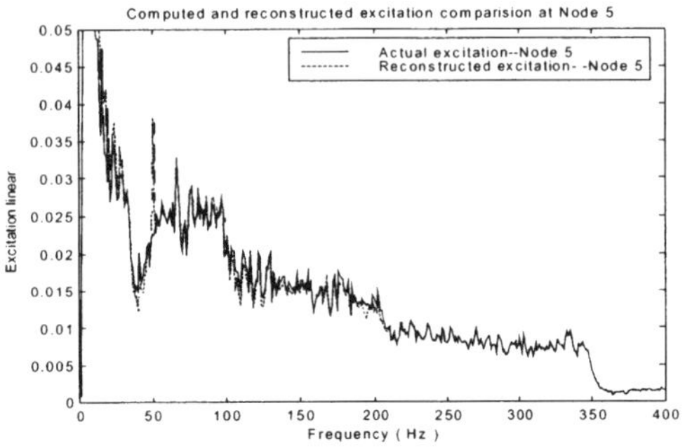

Fig.4-Point 5 identification result comparison

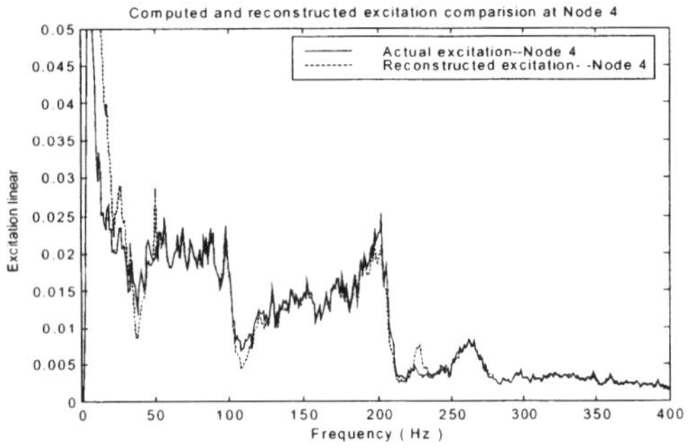

Fig.5-Point 4 identification result comparison

It should be noted that the identification results are very sensitive to the mounting condition of the shakers, especially misalignment of two ends of the drive rode has remarkable influence on the test results. To avoid such adverse stiffness attachment effects, following measures have been adopted to deal with such delicate installation problems. When measuring the FRFs due to excitation 1 (the 1st shaker) in step 2 of section 2, the drive rod of the 2nd shaker should always keep its connection as the in-situ tests, and vice versa. Otherwise, the measured FRFs are not correct and such faulty FRFs will never yield correct identification outputs. In other words, the test settings while measuring the FRFs must keep the same state strictly as that when measuring the responses used for loading identification in step 3. This can be easily attained by switching on or off corresponding power amplifiers according to what is going to be measured, and keep the settings of other related facilities unchanged. In addition, both forces and all the acceleration responses must be measured simultaneously in step 3. That means better FRFs and coherence functions between inputs and outputs will get better identification results as shown in Fig. 6 and Fig.7.

CONCLUSIONS AND DISCUSSIONS

Actual model tests based on IPEM are presented on a cantilever beam subjected to stationary random loadings. The promising identification results demonstrate its potential application in engineering practice. This method significantly overcomes the ill conditioning of the FRFs at frequencies close to some resonances and anti-resonances, and reduces computational time considerably.

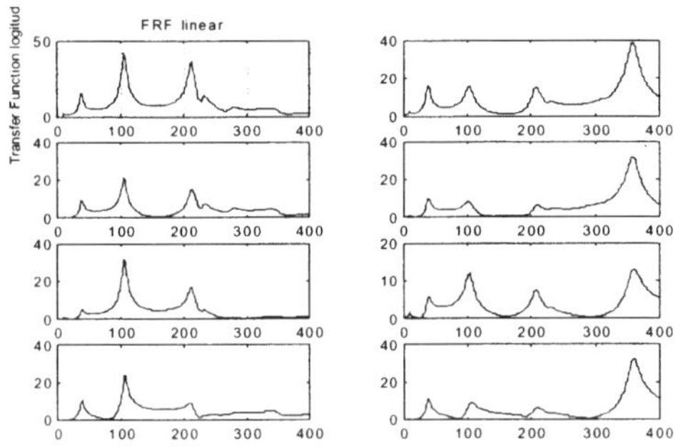

Fig.6-Frequency response function

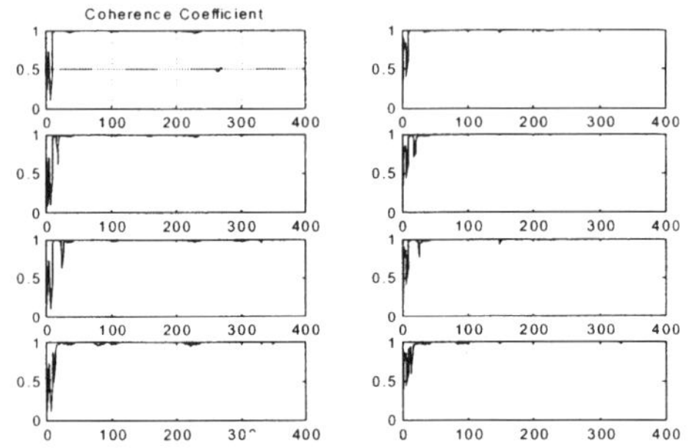

Fig.7- Input/Output coherence function

However, the singularity of the FRF matrices near some resonant frequencies is not completely solved even though SVD is introduced. Epsilon decomposition method, which deserves further research, is an appealing way to deal with this problem if epsilon could be effectively determined. Another problem to be solved is that sometimes the exact positions that the force actually apply on the structure are not known, or the FRFs (Step 2 of section 2) can not be measured in-situ accurately if the mounting places are not accessible or the working conditions offend in some situations.

REFERENCES

Avitabile P, Chandler D.(2001). Selection of Measurement References using the TRIP Method. *Nineteenth International Modal Analysis Conf.*, 125-131

Barlett FD, Flannelly WG(1979). Modal verification of force determination for measuring vibratory loads. *Journal of the American Helicopter Society*,19(4),10-18.

Bathe KJ, Wilson EL(1976). *Numerical methods in finite element analysis*. Pretentice-Hall, New Jersey.

Callahan OJ, Piergentili F(1994). Force estimation using operational data. *Proc. of 8th International Modal Analysis Conf.*, 1586-1592.

Desanghere G, Snoey R(1985). Indirect identification of excitation forces by modal coordinate transformation. *Proc. of 3rd International Modal Analysis Conf.*, 685-690.

Dobson BJ, Rider E(1990). A review of the indirect calculation of excitation forces from measured structural response data. *Journal of mechanical engineering science : Part C (Proc. Institute of mechanical engineers)*, 204,69-75.
Ewins DJ(1984). *Modal Testing: Theory and Practice*. Research Studies Press Ltd, London.
Giansante N, Jones R, Galapodas NJ(1982). Determination of in-flight helicopter loads. *Journal of the American Helicopter Society*, 27(3),58-64.
Hillary B, Ewins DJ(1984). The use of strain gages in force determination and frequency response measurements. *Proc. of 2nd International Modal Analysis Conf.*,627-634.
Karl SK(1987). Force Identification problems-an overview. *Proc. of the 1987 Society of Experimental Mechanics Spring Conf. on Experimental Mechanics*, 838-844.
Lin JH(1992). A fast CQC algorithm of PSD matrices for random seismic responses. *Computers and Structures*, 44,683-687.
Lin JH, Zhang WS, Li JJ(1994). Structural responses to arbitrarily coherent stationary random excitations. *Computers and Structures*, 50(5),629-633.
Lin JH, Guo XL, Zhi H, Howson WP, Williams FW(2001). Computer simulation of structural random loading identification. *Computers and Structures*,**79**,375-387.
Newland NE(1984). *An introduction to random vibration and spectral analysis*, Longman Press, London.
Okubo N, Tanabe S, Tatsuno T(1985). Identification of forces generated by a machine under operation condition. *Proc. of 3rd International Modal Analysis Conf.*, 920-927.

IFAC
Publications
www.elsevier.com/locate/ifac

ON SEQUENTIAL IDENTIFICATION OF A DIFFUSION TYPE PROCESS WITH MEMORY

Uwe Küchler and Vyacheslav Vasil'iev

Institute of Mathematics
Humboldt University Berlin
Unter den Linden 6, D-10099
Berlin, Germany
e-mail: kuechler@mathematik.hu-berlin.de

Department of Applied Mathematics
and Cybernetics
Tomsk State University
Lenina 36, 634050 Tomsk, Russia
e-mail: vas@vmm.tsu.ru

Abstract: The sequential estimation problem of dynamic parameters in stochastic linear systems with memory is solved. The estimation procedure is based on the maximum likelihood method and yields estimators with guaranteed accuracy in the mean square sense. The proposed procedure works for that cases, when the eigenvalues of the information matrix of the observed process have certain rates of increase. The asymptotic behaviour of the duration of observations is investigated. *Copyright © 2003 IFAC*

Keywords: stochastic differential equations with memory, time delay, maximum likelihood estimator, sequential analysis, guaranteed accuracy

1. PROBLEM SETTING[1]

Differential stochastic equations with time delay, or, general speaking, with memory, are widely used to model phenomena in economics and natural sciences. Often one has to determine some underlying parameters from the observations of the running process. This paper presents a sequential estimator for unknown dynamic parameters in stochastic linear systems with memory.

Consider the stochastic differential equation with time delay given by

$$dX(t) = \vartheta_0 X(t)dt + \vartheta_1 X(t-1)dt + dW(t), \; t \geq 0, \tag{1}$$

$$X(t) = X_0(t), t \in [-1, 0].$$

Assume for reasons of citation, that X_0 is continuous. The vector parameter $\vartheta = (\vartheta_0, \vartheta_1)' \in \mathcal{R}^2$ is supposed to be unknown.

[1] Research was supported by RFFI - DFG 02-01-04001 and RFFI 00-01-00880 Grants

The problem is to estimate ϑ based on continuous observations of $X(\cdot)$.

It is well known that (1) has a uniquely determined solution $(X(t), t \geq -1)$ which admits the representation

$$X(t) = \vartheta_1 \int_{-1}^{0} x_0(t-s-1)X_0(s)ds + x_0(t)X_0(0) + $$

$$+ \int_{0}^{t} x_0(t-s)dW(s), \; t > 0,$$

$X(t) = X_0(t)$ for $t \in [-1, 0]$ and satisfies

$$E_\vartheta \int_0^T |X(s)|^2 ds < \infty, \; 0 < T < \infty,$$

if $E_\vartheta \int_{-1}^0 |X_0(s)|^2 ds < \infty$, see e.g. (Mohammed and Scheutzow, 1990; Mao, 1997; Gushchin and

Küchler, 1999).

Here the function $x_0(\cdot)$ denotes the so-called fundamental solution of the corresponding to (1) linear deterministic equation

$$x_0(t) = 1 + \int_0^t (\vartheta_0 x_0(s) + \vartheta_1 x_0(s-1))ds, \qquad t \geq 0, \tag{2}$$

$$x_0(t) = 0, \ t \in [-1, 0).$$

The asymptotic properties of the maximum likelihood estimators (MLE) of the unknown parameter ϑ have been investigated by Gushchin and Küchler (1999).

Sequential parameter estimation problems for the drift of diffusions without time delay have been studied e.g. by Novikov (1971), Liptzer and Shiryaev (1977), Konev and Pergamenshchikov (1985, 1987, 1992) and under noisy observations by Vasiliev and Konev (1987, 1990).

For one-dimensional diffusions a sequential estimation procedure has been constructed by Novikov (1971), Liptzer and Shiryaev (1977). For processes arising from linear stochastic differential equations without time delay having multidimensional parameters special sequential estimation methods have been developed by Konev and Pergamenshchikov (1985, 1987, 1992). For time-delay systems of the type (1) a sequential estimation procedure for some special chosen set $\Theta \subset \mathcal{R}^2$ of vector parameter $\vartheta = (\vartheta_0, \vartheta_1)'$ has been constructed by Küchler and Vasiliev (2001).

We assume that the parameter ϑ belongs to some fixed $\Theta \subset \mathcal{R}^2$ which will be specified below and we shall construct a sequential estimator for ϑ having a preassigned accuracy in the mean square sense. To define Θ we introduce the following notations, see (Gushchin and Küchler, 1999) for details.

Let $s = u(r)$ $(r < 1)$ and $s = w(r)$ $(r \in \mathcal{R}^1)$ be functions given by the parametric representation $(r(\xi), s(\xi))$ in $\mathcal{R}^2$:

$$r(\xi) = \xi \cot \xi, \ s(\xi) = -\xi / \sin \xi$$

with $\xi \in (0, \pi)$ and $\xi \in (\pi, 2\pi)$ respectively.

Consider the set Λ of all (real or complex) roots of the so-called characteristic equation corresponding to (2)

$$\lambda - \vartheta_0 - \vartheta_1 e^{-\lambda} = 0$$

and put $v_0 = v_0(\vartheta) = \max\{Re\lambda | \lambda \in \Lambda\}$,

$$v_1 = v_1(\vartheta) = \max\{Re\lambda | \lambda \in \Lambda, \ Re\lambda < v_0\}.$$

By $m(\lambda)$ we denote the multiplicity of the solution $\lambda \in \Lambda$. It can be easily shown that $v_0 < \infty$. Note that $\lambda = \vartheta_0 - 1 \in \Lambda$ if and only if $\vartheta_1 = -e^{\vartheta_0 - 1}$.

The estimation procedure will be constructed for the region Θ defined as

$$\Theta = \Theta_1 \cup \Theta_2 \cup \Theta_3,$$

where

$$\Theta_1 = \{\vartheta \in \mathcal{R}^2 | \ v_0(\vartheta) < 0\},$$

$$\Theta_2 = \Theta_2' \cup \Theta_2'', \ \ \Theta_3 = \Theta_3' \cup \Theta_3'',$$

$$\Theta_2' = \{\vartheta \in \mathcal{R}^2 | \ 0 < v_0(\vartheta)/2 < v_1(\vartheta) < v_0(\vartheta),$$

$$m(v_0(\vartheta)) = 1 \text{ and } v_0(\vartheta) \in \Lambda\},$$

$$\Theta_2'' = \{\vartheta \in \mathcal{R}^2 | \ 0 < v_1(\vartheta) \leq v_0(\vartheta)/2,$$

$$m(v_0(\vartheta)) = 1 \text{ and } v_0(\vartheta) \in \Lambda\},$$

$$\Theta_3' = \{\vartheta \in \mathcal{R}^2 | \ v_0(\vartheta) > 0; \ v_0(\vartheta) \in \Lambda, \ m(v_0) = 2\},$$

$$\Theta_3'' = \{\vartheta \in \mathcal{R}^2 | \ v_0(\vartheta) > 0 \text{ and } v_0(\vartheta) \notin \Lambda\}.$$

Obviously, by all sets Θ_1, Θ_2', Θ_2'', Θ_3', Θ_3'' are pairwise disjoint.

Note, that this division is very related to a classification used in Gushchin and Küchler (1999), where can be found a figure giving an imagination of these sets. In particular, Θ_1 is the set of parameters ϑ for which there exists a stationary solution of (1).

The problem of sequential estimation for the case $\Theta_1 \cup \Theta_3''$ has been solved by Küchler and Vasiliev (2001). But their estimation procedure does not work for $\vartheta \in \Theta_2 \cup \Theta_3'$. In principle, the problem of sequential estimation of $\vartheta \in \Theta$ may be attacked with the help of an analogue to the procedure, constructed by Galtchouk and Konev (2001).

However, as we can see later, in some regions of Θ it needs a longer observation time than the procedure of Küchler and Vasil'iev (2001). Here we shall construct our estimators as a combination of the sequential estimators of both types.

2. SEQUENTIAL ESTIMATION PLANS

The just mentioned two estimators based on the MLE, which has the form:

$$\hat{\vartheta}(T) = G^{-1}(T)\Phi(T), \ \ G(T) = \int_0^T Y(t)Y'(t)dt,$$

$$Y(t) = \begin{pmatrix} X(t) \\ X(t-1) \end{pmatrix}, \ \ \Phi(t) = \int_0^T Y(t)dX(t)$$

see (Gushchin and Küchler, 1999).

Gushchin and Küchler (1999) completely determined the asymptotic behavior of the MLE $\hat{\vartheta}(T)$ for $T \to \infty$ and for that they had to distinguish eleven different cases for the parameter ϑ from $\mathcal{R}^2$.

Denote by $\lambda_{min}(T)$ and $\lambda_{max}(T)$ the minimal and maximal eigenvalues of the information matrix $G(T)$ respectively. According to Gushchin and Küchler (1999) and Küchler and Vasiliev (2001) these eigenvalues have the following rates of increase (in the almost sure sense) for increasing T in all considered regions:

Region	$\lambda_{min}(T)$	$\lambda_{max}(T)$
Θ_1	T	T
Θ_2	e^{2v_1T}	e^{2v_0T}
Θ_3'	$T^{-2}e^{2v_0T}$	$T^2e^{2v_0T}$
Θ_3''	e^{2v_0T}	e^{2v_0T}

Denote by $(\sigma_1(\varepsilon),\ \vartheta_1^*(\varepsilon))$ and $(\sigma_2(\varepsilon),\ \vartheta_2^*(\varepsilon))$ two sequential estimation plans, which will now be defined and which are modifications of sequential plans constructed by Küchler and Vasil'iev (2001) and Galtchouk and Konev (2001) respectively.

Put

$$\varrho_1 = \sum_{n\geq 1} 1/n^d \quad \text{where } d > 1 \text{ is fixed.}$$

Now choose two unbounded strictly increasing sequences of positive numbers $(c_k)_{k\geq 0}$ and $(d_k)_{k\geq 1}$, such that $c_1 > 1,\ c_2 > 2$,

$$\varrho_2 = \sum_{n\geq 1} 1/d_n < \infty, \tag{3}$$

as well as

$$\sum_{n\geq 1} \frac{c_n^\alpha}{d_n} = \infty \quad \text{for any } \alpha > 0. \tag{4}$$

Let ε be any positive number being fixed in the sequel. Moreover, assume c is a real number with $c > 1$.
Now for every $n \geq 1$ we define (with $\inf\{\emptyset\} = \infty$)

$$\tau_1(n,\varepsilon) = \inf\{t > 0:\ trG(t) = \varepsilon^{-1}c^n\},\ c > 1,$$

$$\tau_2(n,\varepsilon) = \inf\{t > 0:\ c_0^{-1}trG(T_G) +$$
$$+ \int_{T_G}^t Y'(s)G^{-1}(s)Y(s)ds = c_n \ln \varepsilon^{-1}\},\ n \geq 1$$

where $T_G = \inf\{t \geq 0:\ \lambda_{min}(t) \geq c_0\}$.

As the next step we introduce for any $k \geq 1$ the numbers $\beta_1(k)$ and $\beta_2(k)$ by

$$\beta_1(k) = \varepsilon(c^k k^d)^{-\frac{1}{2}}\|G^{-1}(\tau_1(k,\varepsilon)\|^{-1},$$

and

$$\beta_2(k) = \varepsilon^{\frac{1}{2}}(c_k d_k \ln \varepsilon^{-1})^{-\frac{1}{2}}\|G^{-\frac{1}{2}}(\tau_2(k,\varepsilon))\|^{-1}$$

respectively.

For all $k, n \geq 1$ and for $i = 1, 2$ put

$$S_i(n,\varepsilon) = \sum_{k=1}^n \beta_i^2(k),$$

$$\hat\vartheta_i(k) = \hat\vartheta(\tau_i(k,\varepsilon))$$

and

$$\nu_i = \nu_i(\varepsilon) = \inf\{n \geq 1:\ S_i(n,\varepsilon) \geq \varrho_i\}.$$

With these notations we define two sequential plans $(\sigma_1(\varepsilon),\ \vartheta_1^*(\varepsilon))$ and $(\sigma_2(\varepsilon),\ \vartheta_2^*(\varepsilon))$ as follows:

the durations of the observations are given by

$$\sigma_i(\varepsilon) = \tau_i(\nu_i,\varepsilon),\quad i = 1, 2$$

and the estimators of the parameter ϑ are defined by

$$\vartheta_i^*(\varepsilon) = S_i^{-1}(\nu_i,\varepsilon)\sum_{k=1}^{\nu_i}\beta_i^2(k)\hat\vartheta_i(k),\quad i = 1, 2.$$

3. THE PROPERTIES OF SEQUENTIAL ESTIMATION PLANS

In this section the basic properties of the considered sequential estimation plans $(\sigma_1(\varepsilon),\ \vartheta_1^*(\varepsilon))$ and $(\sigma_2(\varepsilon),\ \vartheta_2^*(\varepsilon))$ are given.

Under conditions (3) and (4) on the sequences (c_n) and (d_n) the following relationships for the duration of observations $\tau_1(\nu_1,\varepsilon)$ and $\tau_2(\nu_2,\varepsilon)$ hold with P_ϑ–probability one:

a) for $\vartheta \in \Theta_1$ (stationary case):

$$\overline{\lim_{\varepsilon\to 0}}\ \varepsilon\tau_1(\nu_1,\varepsilon) < \infty,$$

$$0 < \lim_{\varepsilon\to 0}\ \varepsilon^{c_1}\tau_2(\nu_2,\varepsilon) < \infty$$

(note that it holds $c_1 > 1$ by assumptions); in particular,

$$c^{\tilde\nu - 1}\cdot(trJ_1)^{-1} \leq \underline{\lim_{\varepsilon\to 0}}\ \varepsilon\cdot\tau_1(\nu_1,\varepsilon) \leq$$

$$\leq \overline{\lim_{\varepsilon\to 0}}\ \varepsilon\cdot\tau_1(\nu_1,\varepsilon) \leq c^{\tilde\nu}(trJ_1)^{-1},$$

where

$$\tilde\nu = \tilde\nu(\varrho_1) = \inf\{N \geq 1:\ \sum_{n=1}^N \frac{c^n}{n^d} >$$

$$> \varrho_1(trJ_1\cdot\|J_1^{-1}\|)^2\},$$

$$J_1 =$$

$$= \begin{pmatrix} \int_0^\infty x_0^2(t)dt & \int_0^\infty x_0(t)x_0(t+1)dt \\ \int_0^\infty x_0(t)x_0(t+1)dt & \int_0^\infty x_0^2(t)dt \end{pmatrix}$$

and

$$\lim_{\varepsilon \to 0} \varepsilon^{c_1} \tau_2(\nu_2, \varepsilon) = |J_1|^{-1} |G(T_G)| e^{-c_0^{-1} trG(T_G)}$$

where $|J|$ denotes the determinant of the matrix J;

b) for $\vartheta \in \Theta_2 \cup \Theta_3$:

$$\lim_{\varepsilon \to 0} (\ln \varepsilon^{-1})^{-1} \tau_1(\nu_1, \varepsilon) = \begin{cases} \frac{1}{2(2v_1 - v_0)}, & \vartheta \in \Theta_2', \\ \infty, & \vartheta \in \Theta_2'', \\ \frac{1}{2v_0}, & \vartheta \in \Theta_3 \end{cases}$$

and

$$\lim_{\varepsilon \to 0} (\ln \varepsilon^{-1})^{-1} \tau_2(\nu_2, \varepsilon) = \begin{cases} \frac{c_\mu}{2(v_0 + v_1)}, & \vartheta \in \Theta_2, \\ \frac{c_2}{4v_0}, & \vartheta \in \Theta_3 \end{cases}$$

where for $\vartheta \in \Theta_2$

$$\mu = \inf\{n \geq 1 \, : c_n > \frac{v_0 + v_1}{v_1}\}.$$

Note that $\frac{c_\mu}{2(v_0+v_1)} > \frac{1}{2v_1}$ and $\frac{c_2}{4v_0} > \frac{1}{2v_0}$.

With the notations

$$\Theta_2^* = \Theta_2' \cap \{\vartheta : \frac{v_0 + v_1}{2v_1 - v_0} < c_\mu\}$$

and

$$\Theta_2^{**} = \Theta_2' \cap \{\vartheta : \frac{v_0 + v_1}{2v_1 - v_0} > c_\mu\}$$

we have from a) and b) for sufficiently small ε :

- for $\vartheta \in \Theta_1 \cup \Theta_2^* \cup \Theta_3$:

$$\tau_1(\nu_1, \varepsilon) < \tau_2(\nu_2, \varepsilon) \;\; P_\vartheta - \text{a.s.},$$

- for $\vartheta \in \Theta_2^{**} \cup \Theta_2''$:

$$\tau_1(\nu_1, \varepsilon) > \tau_2(\nu_2, \varepsilon) \;\; P_\vartheta - \text{a.s.},$$

- for $\vartheta \in \Theta_2' \setminus [\Theta_2^* \cup \Theta_2^{**}]$:

$$\lim_{\varepsilon \to 0} \frac{\tau_1(\nu_1, \varepsilon)}{\tau_2(\nu_2, \varepsilon)} = 1 \;\; P_\vartheta - \text{a.s.}$$

Furthermore for any $0 < \varepsilon < 1$ the estimators $\vartheta_1^*(\varepsilon)$ and $\vartheta_2^*(\varepsilon)$ have the properties

$$\sup_{\Theta \setminus \Theta_2''} E_\vartheta ||\vartheta_1^*(\varepsilon) - \vartheta||^2 \leq \varepsilon$$

and

$$\sup_{\Theta} E_\vartheta ||\vartheta_2^*(\varepsilon) - \vartheta||^2 \leq \varepsilon.$$

We can see that under the same upper bound ε for the mean square deviation of the estimators $\vartheta_i^*(\varepsilon)$, $i = 1, 2$, the durations of observation $\tau_1(\nu_1, \varepsilon)$ and $\tau_2(\nu_2, \varepsilon)$ have various rates of increase in various regions of the unknown parameter ϑ. Moreover it should be pointed out that the estimation procedure of Galtchouk and Konev (2001) works better in the cases, when the rates of increase of $\lambda_{min}(T)$ and $\lambda_{max}(T)$ are essentially different (case $\Theta_2^{**} \cup \Theta_2''$). In all the other cases the procedure of Küchler and Vasil'iev (2001) has the shorter duration of observations asymptotically.

Note that the presented sequential estimation procedure $(\sigma_1(\varepsilon),\ \vartheta_1^*(\varepsilon))$ works in a more wide region $\Theta_1 \cup \Theta_2' \cup \Theta_3$ than the procedure constructed by Küchler and Vasil'iev (2001) for the case $\Theta_1 \cup \Theta_3''$.

The obtained properties of sequential plans lead to the following construction of the estimation procedure for all considered regions of ϑ.

4. GENERAL SEQUENTIAL ESTIMATION PROCEDURE

To minimize the duration of observation we define the sequential plan $(\sigma(\varepsilon),\ \vartheta(\varepsilon))$ of estimation $\vartheta \in \Theta$ as a combination of the two presented estimators by the formulae

$$\sigma(\varepsilon) = \min(\tilde{\sigma}_1(\varepsilon), \tilde{\sigma}_2(\varepsilon)),$$

$$\vartheta(\varepsilon) = \tilde{\chi}(\varepsilon)\tilde{\vartheta}_1(\varepsilon) + (1 - \tilde{\chi}(\varepsilon))\tilde{\vartheta}_2(\varepsilon),$$

$$\tilde{\chi}(\varepsilon) = \chi(\tilde{\sigma}_1(\varepsilon) \leq \tilde{\sigma}_2(\varepsilon)),$$

where $\chi(a \leq b) = 1,\ a \leq b;\ 0,\ a > b.$

Here we put for some $0 < \delta < 1$ the sequential plans $(\tilde{\sigma}_1(\varepsilon),\ \tilde{\vartheta}_1(\varepsilon)) = (\sigma_1(\varepsilon),\ \vartheta_1^*(\varepsilon))$ with $\delta^{-1}\varrho_1$ instead ϱ_1 in the definition of stopping times $\nu_1(\varepsilon)$ and $(\tilde{\sigma}_2(\varepsilon),\ \tilde{\vartheta}_2(\varepsilon)) = (\sigma_2(\varepsilon),\ \vartheta_2^*(\varepsilon))$ with $(1-\delta)^{-1}\varrho_2$ instead ϱ_2 in the definition of stopping times $\nu_2(\varepsilon)$.

We summarize the basic properties of considered plans in the following

Theorem. Assume that the underlying process $(X(t))$ satisfies the equation (1) and the conditions (3) and (4) hold. Then for any ε and δ from $(0, 1)$ and every $\vartheta \in \Theta$ the sequential estimation plan $(\sigma(\varepsilon),\ \vartheta(\varepsilon))$ of ϑ is closed $(\sigma(\varepsilon) < \infty \;\; P_\vartheta - \text{a.s.})$. They possess the following properties:

1°. for any $0 < \varepsilon < 1$

$$\sup_{\Theta} E_\vartheta ||\vartheta(\varepsilon) - \vartheta||^2 \leq \varepsilon$$

2°. the following relations hold with $P_\vartheta-$ probability one:

a) for $\vartheta \in \Theta_1$ (stationary case):

$$\overline{\lim_{\varepsilon \to 0}} \, \varepsilon \sigma(\varepsilon) \le c^{\tilde{\nu}(\delta^{-1} \varrho_1)} \cdot (tr J_1)^{-1},$$

b) for $\vartheta \in \Theta_2 \cup \Theta_3$:

$$\lim_{\varepsilon \to 0} (\ln \varepsilon^{-1})^{-1} \sigma(\varepsilon) =$$

$$= \begin{cases} \min\left(\frac{1}{2(2v_1 - v_0)}, \frac{c_\mu}{2(v_0 + v_1)}\right), & \vartheta \in \Theta_2', \\ \frac{c_\mu}{2(v_0 + v_1)}, & \vartheta \in \Theta_2'', \\ \frac{1}{2v_0}, & \vartheta \in \Theta_3; \end{cases}$$

3°. the estimator $\vartheta(\varepsilon)$ is strongly consistent:

$$\lim_{\varepsilon \to 0} \vartheta(\varepsilon) = \vartheta \quad P_\vartheta - a.s.$$

for all $\vartheta \in \Theta$.

In particular, it follows from this theorem that for every $\vartheta \in \Theta$ the proposed sequential plan $(\sigma(\varepsilon), \vartheta(\varepsilon))$ has the better rate of convergence of the duration $\sigma(\varepsilon)$ in the sence *a)* and *b)* than both procedures of Galtchouk and Konev (2001) and Küchler and Vasil'iev (2001), mentioned above.

Remark. In the Ornstein–Uhlenbeck case $\vartheta_1 = 0$ the sequential plan $(\sigma(\varepsilon), \vartheta(\varepsilon))$ coincides with the sequential plan $(\tilde{\sigma}_1(\varepsilon), \tilde{\vartheta}_1(\varepsilon))$ for suffitiently small ε. Moreover, in this case the set Θ is reduced to $\tilde{\Theta}_1 = \{\vartheta : v_0(\vartheta) < 0, \ v_1(\vartheta) = 0\} \subset \Theta_1$. The theorem remains true even for the case $v_0(\vartheta) > 0$, $v_1(\vartheta) = 0$ (nonstationary case), for details see (Küchler and Vasil'iev, 2001).

5. REFERENCES

Galtchouk, L. and Konev, V. (2001). 'On sequential estimation of parameters in semimartingale regression models with continuous time parameter'. *The Annals of Statistics*, **29**, 5, 1508-1536.

Gushchin, A. A. and U. Küchler (1999). 'Asymptotic inference for a linear stochastic differential equation with time delay'. *Bernoulli*, **5**, 6, 1059-1098.

Konev, V. V. and S. M. Pergamenshchikov (1985). 'Sequential estimation of the parameters of diffusion processes'. *Problems of Inform. Trans.*, **21**, 1, 48-62 (in Russian).

Konev, V. V. and S. M. Pergamenshchikov (1987). 'Sequential estimation of the parameters of unstable dynamical systems in continuous time'. In F. Tarasenko, editor, *Math. Stat. and Appl., Publishing House of Tomsk University*, Tomsk, 11, 85-94 (in Russian).

Konev, V.V. and S. M. Pergamenshchikov (1992). 'Sequential estimation of the parameters of linear unstable stochastic systems with guaranteed accuracy'. *Problems of Inform. Trans.*, **28**, 4, 35-48 (in Russian).

Küchler, U. and Vasiliev, V. (2001). 'On sequential parameter estimation for some linear stochastic differential equations with time delay'. *Sequential Analysis*, **20**, 3, 117-146.

Liptzer, R.S. and Shiryaev A.N. (1977). *Statistics of Random Processes.* Springer-Verlag, New York, Heidelberg.

Mao, X. (1997). *Stochastic Differential Equations and Application.* Harwood Publishing, Chichester.

Mohammed, S.E-A. and Scheutzow, M.K.R . (1990). 'Lyapunov exponents and stationary solutions for affine stochastic delay equations'. *Stochastics and Stochastic Reports*, 29, 259-283 (in Russian).

Novikov, A.A. (1971). 'The sequential parameter estimation in the process of diffusion type'. *Probab. Theory and its Appl.*, **16**, 2, 394-396 (in Russian).

Vasiliev, V. A. and V. V. Konev (1987). 'On sequential identification of linear dynamic systems in continuous time by noisy observations'. *Probl. of Contr. and Inform. Theory*, **16**, 2, 101-112.

Vasiliev, V. A. and V. V. Konev (1990). 'On sequential parameter estimation of continuous dynamic systems by discrete time observations'. *Probl. of Contr. and Inform. Theory*, **19**, 3, 197-207.

IFAC Publications
www.elsevier.com/locate/ifac

INCREMENTAL IDENTIFICATION OF TRANSPORT COEFFICIENTS IN DISTRIBUTED SYSTEMS

André Bardow, Wolfgang Marquardt [1]

Lehrstuhl für Prozesstechnik, RWTH Aachen,
D-52056 Aachen, Germany

Abstract: In this work model identification of state-dependent transport coefficients in distributed systems is considered. In many cases it is difficult to formulate a suitable candidate model based on physical insight or prior knowledge. General parameterizations have therefore to be employed leading to a large number of unknown parameters. This may be prohibitive for distributed systems due to the computational cost of parameter estimation. An incremental model identification procedure is therefore employed here. This approach reflects the model development process itself and splits the identification into a sequence of inverse problems. Thereby, uncertainty in each step is minimized and computational cost is reduced substantially. The implementation presented here uses results from inverse problems theory and is applied to the estimation of a concentration dependent diffusion coefficient.

Keywords: identification, regularization, distributed parameter systems, parameter estimation, discrimination

1. INTRODUCTION

In many physical systems the model structure for the kinetic phenomena is unknown and has to be inferred from experimental data. Often it is even difficult to formulate reasonable candidate models based on prior knowledge and general parameterizations have to be used instead. They are inserted into the balance equations and measured data is used to estimate the unknown parameters in these fully specified models. This approach is therefore called *simultaneous identification* here. Model discrimination techniques are finally employed to identify the most suitable model (Asprey and Macchietto, 2000).

During the course of model identification, different parameterizations and levels of detail may be tested leading to a large number of estimation problems. In addition, many unknown parameters may have to be estimated in each problem, since general approximations are used. Furthermore, these problems occur usually for very complex systems which are often described by partial differential equations making the estimation problems computationally expensive. Another drawback of the simultaneous approach is that by inserting the candidate model into the balance equations the model will be biased if there is a structured modeling error (Walter and Pronzato, 1997).

In this work, a general incremental approach to model identification is used which minimizes model uncertainty and reduces computational cost in estimation problems. The approach follows the steps common to model development in the identification procedure. It has been previously applied to cases where candidate models could be easily proposed (Bardow and Marquardt, 2003). The extension to the more general case will be illustrated with examples from diffusive mass transfer in distributed systems.

Diffusion is the rate-limiting step in many mass transfer operations, such as extraction or absorption and in (heterogeneous) chemical reactions. Therefore, constitutive equations describing diffusive transport are required to design mass transfer equipment. Fick diffusion coefficients are usually measured in experi-

[1] Correspondence should be addressed to W. Marquardt, marquardt@lfpt.rwth-aachen.de

ments. In concentrated liquid mixtures they are in general a strong function of composition. Today, there is still uncertainty about a valid model (Taylor and Krishna, 1993).

In the next section, the common approach to model development will be revisited and the incremental approach to model identification is derived from it. The method will then be applied to a numerical example and different features of the new approach will be analyzed. Finally, conclusions are given.

2. INCREMENTAL MODEL IDENTIFICATION

2.1 *Model Development*

The key idea of incremental model identification is to exploit model structure by following the steps of model development. Model development is commonly performed in incremental steps starting by formulation of the balance equations. For conciseness it will be presented for isothermal binary diffusion. The mole balance for component 1 can then be given as

$$\text{Model B:} \quad \frac{\partial c_1}{\partial t} + \frac{\partial (c_1 u^V)}{\partial z} = -\frac{\partial J_1^V}{\partial z}, \quad (1)$$

where u^V is the volume average velocity, c_1 the concentration and J_1^V the diffusive flux of component 1.

In a next step, constitutive models for the convective and diffusive flux have to be given. In diffusion experiments the volume average velocity is usually negligible, i.e. $u^V = 0$ (Tyrell and Harris, 1984). Still, the modeler has to specify the diffusive flux. For binary diffusion, all models can be related to Fick's law (Taylor and Krishna, 1993)

$$\text{Model T:} \quad J_1^V = -D_{12}^V \frac{\partial c_1}{\partial z}, \quad (2)$$

where D_{12}^V denotes the Fick diffusion coefficient.

A further constitutive relation is required to describe this diffusion coefficient. Since the functional form of its concentration dependence is today still subject to discussion a generic relationship is given

$$\text{Model D:} \quad D_{12}^V = f(x, \theta) \quad (3)$$

where x represents the mole fraction and θ collects all constant coefficients. If the model f and the value of its parameters θ are given, the model can be solved.

It should be stressed that the sequence in model development shown here is generic and not limited to mass transfer processes (Marquardt, 1995).

2.2 *Model Identification*

The simultaneous approach to identification is computationally expensive and may lead to biased estimates. It neglects the inherent model structure with its sequence of models, each containing further assumptions about the process. In contrast, the incremental approach follows these steps as shown in Fig. 1.

2.2.1. *Model B: Balances* The balance equation (1) contains the least uncertainty. Without introducing potentially uncertain constitutive equations, the diffusive flux itself is computed from this equation as a function of space and time,

$$\text{Model B:} \quad J_1^V(z,t) = -\int_0^z \frac{\partial c_1(z',t)}{\partial t}\, dz', \quad (4)$$

if concentration can be measured with high resolution as a function of space and time. In this equation, the lower boundary ($z = 0$) is assumed impermeable as common in diffusion experiments.

It should be noted that the estimation of the diffusive flux requires only the solution of the linear Eq. (4) independent of the number of candidate models. All following estimation problems are only algebraic. This decoupling of the problem reduces the computational expense substantially.

The main difficulty in the solution of Eq. (4) is the estimation of the time derivative of the measured concentration data. This is known to be an ill-posed problem, i.e. small errors in the data will be amplified (Hansen, 1998). Regularization techniques have to be employed to compute a stable approximation. Here, a smoothing spline technique is used (Reinsch, 1967). The estimate is the minimizer of the functional

$$\min_c \left|\left|c - c_{\text{measured}}\right|\right| + \lambda \left|\left|\frac{\partial^2 c}{\partial t^2}\right|\right| \quad (5)$$

which corresponds to the well-known Tikhonov regularization method. The time derivative is finally taken as the derivative of the smoothing spline.

The crucial step is the selection of the regularization parameter λ which balances data and regularization error. Different methods are available from the literature (Hansen, 1998). Heuristic methods will only be used here since they apply even if there is no a priori knowledge about the measurement error. Generalized cross-validation (GCV) is derived from leave-one-out cross-validation where one data point is dropped from the data set. The regularization parameter is chosen which predicts the missing point best on average (Craven and Wahba, 1979). The L-curve is a log-log-plot of the smoothing norm over the residual norm (Hansen, 1998). This graph usually has a typical L-shape since the residual norm will be large for large λ while the smoothing norm is minimized. For small λ the residual norm will be minimized but the smoothing norm is large due to the ill-posed nature of the problem leading to oscillations in the solution. At intermediate λ-values both error contributions are balanced. The optimal regularization parameter is therefore chosen

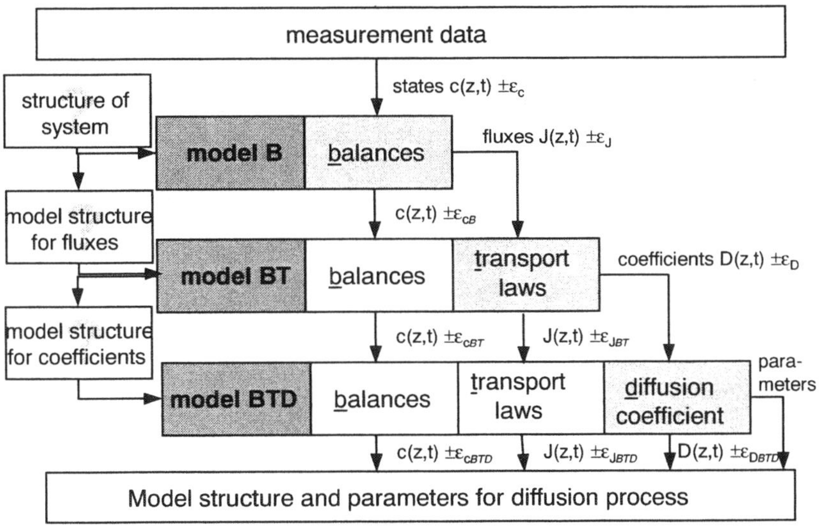

Fig. 1. Incremental approach to identification of diffusion models

as the corner point of the L-curve corresponding to the maximum curvature with respect to the regularization parameter. Computational routines for both methods are available in the Regularization Toolbox (Hansen, 1999).

With the estimated derivative $(\partial c_1/\partial t)$ the flux $J_1^V(z,t)$ can be computed from Eq. (4). Here, a simple trapezoidal rule is used for integration.

2.2.2. *Model BT: Transport laws* In general, several models for the fluxes may be considered. The unknown transport coefficients contained are then computed as function of space and time. In the binary case, the diffusion coefficients are calculated from Eq. (2) as

$$\text{Model T:} \quad D_{12}^V(z,t) = -\frac{J_1^V(z,t)}{\partial c_1(z,t)/\partial z}. \quad (6)$$

The spatial derivative is also calculated using the smoothing spline approach. It can be seen that the coefficient can only be estimated if the spatial gradient differs from zero and it will be very sensitive to noise if the gradient is small.

In the case of several candidate flux models they can be tested already at this level. Since transport coefficients have a physical interpretation which results in certain restrictions (e.g. positivity), those values violating any restriction could be discarded already at this stage.

2.2.3. *Model BTD: Diffusion coefficients* If reasonable candidates for the Fick diffusion coefficient can be formulated the incremental approach allows robust and efficient model identification as was shown previously (Bardow and Marquardt, 2003).

Here, the case when no model candidate is available is considered. A general parametrization for the diffusion coefficient is therefore introduced. The parametrization should be capable of approximating any function. Following closely Hanke and Scherzer (1999) the concentration range [0,1] is divided into m intervals. The diffusion coefficient is approximated by a piecewise constant function in each interval

$$D_{12}^V(x,\theta) = \theta_i \quad \text{for } x \in \left[\frac{i-1}{m}, \frac{i}{m}\right]. \quad (7)$$

The unknown parameter vector θ can now be estimated from the diffusion coefficients (6). This yields one equation for every measurement time and position. Collecting the estimated diffusion coefficients in a vector $\mathbf{D_{12}^V}$ the estimation problem can be given

$$\mathbf{D_{12}^V} = -\mathbf{A}\,\theta \quad (8)$$

with an extremely sparse matrix $\mathbf{A}$ which contains only one "1" per row. It turns out in practise that it is advantageous to insert the diffusion coefficient model (7) into the transport law (2) to avoid explicit division by the spatial concentration gradient (see Eq. (6)). The resulting equation is then

$$\mathbf{J_1^V} = -\tilde{\mathbf{A}}\,\theta\,. \quad (9)$$

where the "1" is replaced by the value of the estimated spatial derivative $\frac{\partial c}{\partial z}$ in matrix $\tilde{\mathbf{A}}$.

This discrete ill-posed problem can be solved using any technique known from literature (Hansen, 1998). Because of the large problem size and the sparsity of matrix $\tilde{\mathbf{A}}$ iterative regularization methods are the most appropriate choice for the solution (Hanke and Scherzer, 1999). Here, the conjugate gradient method (CG) is employed using the Regularization Toolbox (Hansen, 1999). A preconditioner enhancing smoothness is used.

The number of CG-iterations serves as the regularization parameter. The L-curve is used as a stopping rule.

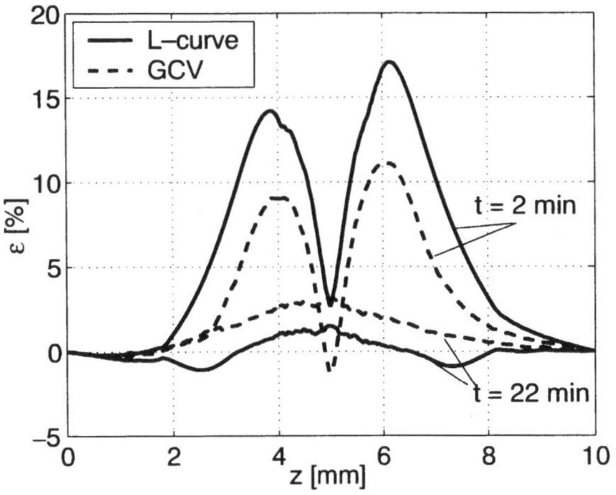

Fig. 2. Errors in estimated flux J_1 using GCV and L-curve ($\sigma = 10^{-6}$)

3. NUMERICAL EXAMPLE

In order to exemplify the incremental approach to model identification the following model for the diffusion coefficient (3) is considered

$$D_{12}^V(x_1) = \theta_1 x_1 + (1 - x_1)(\theta_2 + \theta_3 x_1). \quad (10)$$

The task is to estimate this constitutive equation from mole fraction measurements x_1. The considered example is challenging because of the non-monotonous diffusion coefficient (Cannon and DuChateau, 1980). Such models are used to design mass transfer equipment.

The assumed setup corresponds to Raman diffusion experiments (Bardow *et al.*, 2003). Here, two liquids of different composition are layered on top of each other in a diffusion cell ($L = 10$ mm). Then, one-dimensional mixing occurs by diffusion. Measurements are taken with a resolution of $\Delta z = 0.1$ mm and $\Delta t = 120$ s. The experiment runs for $t_{final} = 2$ hr. Ideal mixture behavior is assumed. The mole fraction x_1 can therefore computed from the molar concentration c_1 using

$$x_1 = c_1 / c_t \quad (11)$$

$$c_t = 1 / \left(V_2^0 + (V_1^0 - V_2^0) x_1 \right). \quad (12)$$

where V_i^0 is the known molar volume of pure component i. Gaussian noise is added to simulated data to represent noisy measurements. The noise level was chosen as $\sigma^2 = 10^{-5}$. This is a typical value for binary Raman experiments (Bardow *et al.*, 2003). Because of the use of simulated data it is possible to evaluate the different steps of the incremental algorithm.

3.1 *Flux estimation*

Following the formulation of the balance equations (1) the diffusive flux is estimated. The crucial step of derivative estimation can be done here either by differentiating the mole fraction data or by first computing the concentration (Eqns. (11)-(12)) and the computation of its derivative. The latter choice may seem more natural since the concentration derivative shows in Eq. (4). But the derivation of general cross validation assumes Gaussian noise (Craven and Wahba, 1979). This error distribution can be expected for the mole fractions but will be skewed for the concentration if the difference between the molar volumes is large. Both approaches were therefore tested. In practical situations the difference is negligible and both methods lead to very similar solutions (not shown here). In the following the derivative will be calculated from the concentration data.

Table 1. Flux RSS for GCV and L-curve

Method	$\sigma^2 = 0$	$\sigma^2 = 10^{-6}$	$\sigma^2 = 10^{-4}$
GCV	$0.896 \cdot 10^{-3}$	$0.907 \cdot 10^{-3}$	$2.21 \cdot 10^{-3}$
L-curve	$0.786 \cdot 10^{-3}$	$0.884 \cdot 10^{-3}$	$4.47 \cdot 10^{-3}$

The performance of the regularization parameter choice methods is now analyzed. Fig. 2 shows the errors in the estimated fluxes J_1^V. The error ε is defined by

$$\varepsilon(z,t) = \frac{J_{1,estimate}^V(z,t) - J_{1,true}^V(z,t)}{\max_z J_{1,true}^V(z,t)}. \quad (13)$$

The errors are large initially. Note that the flux is a Dirac impulse for time $t = 0$. The steep gradients at the beginning of the run cannot be easily distinguished from measurement noise. This is in accordance with observations by other authors who realized that their estimates improved if the first data points were omitted (Hanke and Scherzer, 1999). This suggestion is also followed here. For larger times the estimate is quite good.

It is difficult to decide from Fig. 2 which regularization parameter choice method performs better. Tab. 1 gives the residual sum of squares (RSS) for different noise levels covering the experimental range. The shown values are averaged over 32 samples.

Both methods perform very similar. The L-curve is slightly better if the measurement error is low but performs poorer for large measurement variances. The L-curve suffers from some objections from a theoretical viewpoint (Hansen, 1998). Even so, the results here show its practical value. Still, the GCV method will be used in the following to compute the diffusive flux.

3.2 *Diffusion coefficient*

The computed flux values, the estimated spatial derivatives and the measured mole fractions are used to estimate the diffusion coefficient as presented above. The optimal iteration number for the CG iterations is chosen by the L-curve as shown in Fig. 3. The estimated and the true concentration dependence of the diffusion coefficient D_{12}^V are compared in Fig. 4.

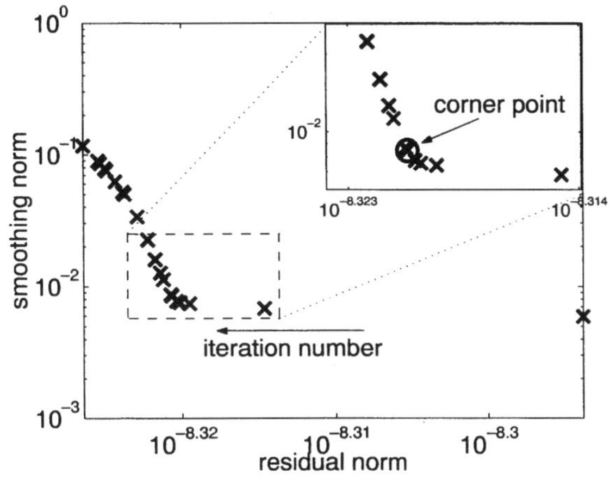

Fig. 3. L-curve for choice of iteration number

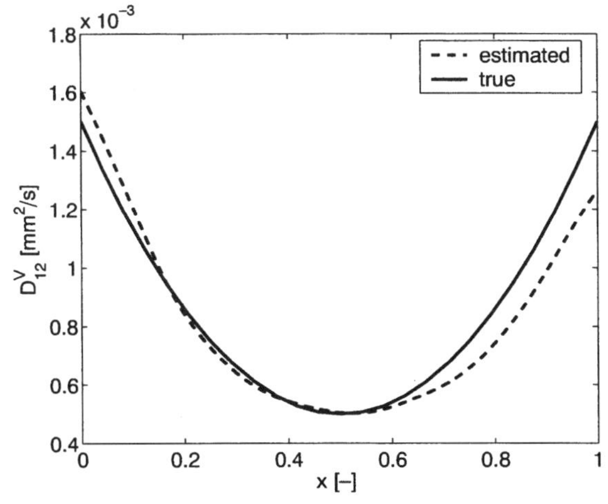

Fig. 4. Estimated and true diffusion coefficient

The shape of the concentration dependence is well captured. It should be noted that only data from one experiment was used. Commonly, more than 10 experiments are employed (Tyrell and Harris, 1984). The minimum is found accurately in location and value.

The discretization level of the diffusion coefficient had only minor influence on the final result. Here, the concentration range was split into $m = 500$ intervals, i.e. 500 parameters have to be estimated. This prohibits the use of the simultaneous approach whereas the incremental approach takes an average CPU time of only 8 sec on a standard desktop PC. This substantial reduction in computation time is mainly due to the decoupling of the problem and the use of an equation error scheme.

Even though a model for the diffusion coefficient has been found, it is desirable in practise to obtain a simpler expression than the piecewise constant representation. Such a simple form can now easily be found based on the previous investigation. Usually, polynomials of different order are used to represent the diffusion coefficient (Tyrell and Harris, 1984). It is obvious from Fig. 4 that a constant or linear function will not be suitable. Therefore, only polynomials of higher order have to be considered. As an example, a quadratic and a cubic polynomial are proposed. They can now simply be fitted to the piecewise constant diffusion coefficient found by the incremental approach. It seems advantageous though to use the simultaneous approach to re-adjust the parameters. Thereby, the influence of error propagation is reduced. The incremental approach here helps to generate a small number of suitable candidate models and good starting values for their parameters.

Table 2. Estimated diffusion coefficient models and residual sum of squares

model	θ_1	θ_2	θ_3	θ_4	$\hat{S}$
true	1.5	1.5	-4.0	-	-
quadratic_0	1.305	1.531	-3.731	-	53.3e3
quadratic_{sim}	1.503	1.500	-4.008	-	6.08e3
cubic_0	1.360	1.476	-3.731	-0.723	6.09e3
cubic_{sim}	1.499	1.503	-4.006	0.006	6.07e3

Tab. 2 gives the coefficients of the polynomials estimated from the piecewise constant representation. They serve as initial guesses to the simultaneous identification procedure (indicated by the subscript "0"). The final results from the simultaneous approach are also given.

The values obtained from the fit to the solution of the incremental approach is already close to the true solution. The residual sum of squares ($\hat{S}$), here scaled by the variance, is already within an order of magnitude of the optimal value. Note that the residual sum of squares is not explicitly minimized in the incremental approach. The larger difference for the quadratic case results from the deviation for larger concentrations (cf. Fig. 4). In the cubic case the $\hat{S}$ values are almost identical for the diffusion coefficient from the incremental and the simultaneous approach showing that the diffusion coefficient found by the incremental approach is an excellent approximation within measurement accuracy.

From simple inspection, the quadratic model would be favored since the objective decreases only by 0.3% adding the additional parameter in the cubic model. Model discrimination techniques such as Akaike's Information Criterion (AIC) (Walter and Pronzato, 1997) or a-posteriori model probabilities (Stewart *et al.*, 1998) are used to quantify this preference. They balance two terms: one representing lack-of-fit and a penalizing term for the number of parameters. For the example considered both methods favor the cubic model since the likelihood dominates due to the large number of measurements and the high accuracy.

AIC can be expressed as (Walter and Pronzato, 1997)

$$\phi_{aic} = \frac{1}{n_t}\left(-\ln L + n_p\right) \qquad (14)$$

where n_t is the total number of measurements, L the likelihood function and n_P the number of parameters in the model. For the example considered, the ϕ_{aic}-

values of the quadratic and cubic model differ by far less than 1%. Further experiments would be required to gain a higher certainty.

The a posteriori probability p of model M being based on data Y can be stated as (Stewart *et al.*, 1998)

$$p(M|Y) \propto p(M)2^{-n_P/2}\exp(-\hat{S}_M/2) \quad (15)$$

where $p(M)$ is the prior probability for the model, here chosen equal initially. A probability share $\pi_M =$ can now be calculated for each model (Stewart *et al.*, 1998)

$$\pi_M = p(M|Y)/\sum_I p(I|Y) \quad (16)$$

in order to rank the models. In the considered example the probability share for the quadratic model is only 30.9% due to the dominant influence of the sum of residuals over the parameter penalty term. After a total of six experiments experiments, the cubic model will be selected with a probability higher than 99%.

4. CONCLUSIONS

Estimation of transport coefficients when no reasonable candidate model can be proposed is very difficult and computational expensive using the simultaneous approach to model identification. Here, the incremental approach was therefore employed. The stepwise procedure decouples the estimation problem and thereby reduces computational cost substantially.

The approach is especially suited for measurements with high spatial and temporal resolution. These data are used to solve an infinite dimensional problem, the estimation of the diffusive flux. This step may be error-prone with low-resolution data. But it could be shown that the method compares well with results from simultaneous identification strategies even though the resolution assumed here was five times coarser than it may be obtained from real experiments today (Bardow *et al.*, 2003).

The method is robust with respect to the selection of the regularization parameter choice method. Both, general cross-validation and the L-curve performed well for the flux estimation problem.

The state dependence of the transport coefficient could be well established by the incremental approach from a single run even for non-monotonous cases. The resulting diffusion coefficient could be used directly or simpler relations may derived from the piecewise constant parameterization. Model discrimination techniques are finally used to select a suitable model. The results indicate that these methods tend to select overparmeterized models for the case of high-resolution measurements with low variance considered here.

Future work will concentrate on the extension of the incremental approach to multicomponent mixtures. The transport coefficients will then depend on all state variables leading to problems in more than one dimension. It can be expected that the gain from the use of the incremental approach will be even larger for these cases.

Acknowledgments – The authors gratefully acknowledge the financial support of the Deutsche Forschungsgemeinschaft (DFG) within the Collaborative Research Center (SFB) 540 "Model-based Experimental Analysis of Kinetic Phenomena in Fluid Multi-phase Reactive Systems".

5. REFERENCES

Asprey, S.P. and S. Macchietto (2000). Statistical tools for optimal model building. *Comp. Chem. Eng.* **24**(1), 831–834.

Bardow, A. and W. Marquardt (2003). Identification of multicomponent mass transfer by means of an incremental approach. Accepted for: *European Symposium on Computer Aided Process Engineering – ESCAPE 13.*

Bardow, A., W. Marquardt, V. Goeke, H.-J. Koss and K. Lucas (2003). Model-based measurement of diffusion using Raman spectroscopy. *AIChE. J.* **49**(2), 323–334.

Cannon, J.R. and P. DuChateau (1980). An inverse problem for a nonlinear diffusion equation. *SIAM J. Appl. Math.* **48**, 272–289.

Craven, P. and G. Wahba (1979). Smoothing noisy data with spline functions - estimating the correct degree of smoothing by the method of generalized cross-validation. *Numer. Math.* **31**, 377–403.

Hanke, M. and O. Scherzer (1999). Error analysis of an equation error method for the identification of the diffusion coefficient in a quasi-linear parabolic differential equation. *SIAM J. Appl. Math.* **59**(3), 1012–1027.

Hansen, P.C. (1998). *Rank-deficient and discrete ill-posed problems.* SIAM, Philadelphia.

Hansen, P.C. (1999). Regularization tools version 3.0 for Matlab 5.2. *Numer. Algorithms* **20**, 195–196.

Marquardt, W. (1995). Towards a process modeling methodology. In: *Methods of model-based control* (R. Berber, Ed.). pp. 3–41. Kluwer, Dordrecht.

Reinsch, C.H. (1967). Smoothing by spline functions. *Numer. Math.* **10**, 177–183.

Stewart, W.E., Y. Shon and G.E.P. Box (1998). Discrimination and goodness of fit of multiresponse mechanistic models. *AIChE. J.* **44**(6), 1404–1412.

Taylor, R. and R. Krishna (1993). *Multicomponent mass transfer*. John Wiley, New York.

Tyrell, H.J.V. and K.R. Harris (1984). *Diffusion in liquids.* Butterworths, London.

Walter, E. and L. Pronzato (1997). *Identification of parametric models: from experimental data.* Springer, Berlin.

IFAC Publications
www.elsevier.com/locate/ifac

ON THE STRUCTURE OF STATIC BALANCED FLOW SYSTEMS

E. Weyer[1] **A. Gleiß**[2] **M. Deistler**[3] **K. Gruber**[3] **T. Matyus**[3]

[1] *CSSIP, Department of Electrical and Electronic Engineering, The University of Melbourne, Parkville, VIC 3010, Australia.*
Email: e.weyer@ee.mu.oz.au
[2] *Institute of Medical Statistics, University of Vienna, Vienna, Austria*
[3] *Department of Econometrics and System Theory, Technical University of Vienna, Argentinerstraße 8, A-1040 Vienna, Austria*

Abstract: In this paper we consider identification of static balanced flow systems. Static balanced flow systems are characterised by balancing equations and other physical laws governing the systems. These equations are used for improving estimates of flows and transfer coefficients, and we examine the structure of the equations in the case where in addition to measured flows there are also measured transfer coefficients. It turns out that in most cases the constraints imposed by the equations are linear and bilinear. An extension of the reduced balance scheme which can be used with measured transfer coefficients in some special cases is also provided. *Copyright © 2003 IFAC*

Keywords: Static balanced flow systems, system identification, data reconciliation

1. INTRODUCTION

In this paper we consider identification of static balanced flow systems. Many environmental systems can be modelled as balanced flow systems (Baccini and Brunner (1991), van der Voet et al (1995)), and analysis and modelling of such systems have gained an increased importance. Typical features of such systems are that they are large scale and that they can be decomposed into subsystems which satisfy balance equations, e.g. mass or energy balances. Examples of application areas are accounting of phosphorus in an agricultural region (Gleiß et al (1998)), and waste management in a large city (Matyus et al (2002)). The uses of balanced flow models are many and include monitoring of flows of a particular substance (e.g. Stigliani et al (1993)) and early recognition of cumulative releases or accumulation of harmful substances in the environment (Baccini et al (1993)). Here we will mainly focus on the data reconciliation problem, i.e. we want to improve the estimates of the measured and unmeasured flows by utilising the fact that the balance equations for each subsystem must be satisfied.

This work is a continuation of Gleiß et al (1998). Here we consider systems with measured transfer coefficients in more detail and examine the structure of the governing equations. Moreover, multi level systems are also considered, e.g. systems where both a mass and an energy balance are satisfied. We also present an extension of the algorithm for flow estimation, which can be used in some simple cases involving measured transfer coefficients.

The paper is organised as follows. In the next section we give a short introduction to balanced flow systems before we present some preliminary problem formulations for the data reconciliation problem. The structure of the

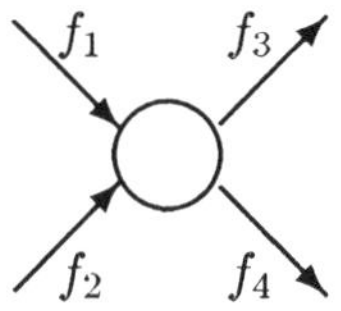

Fig. 1. Subsystem of balanced flow system

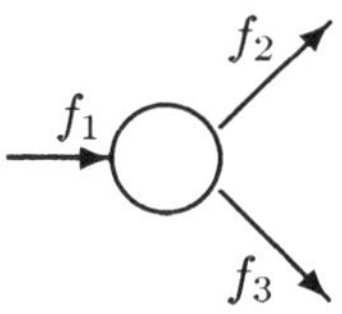

Fig. 2. Simple balanced flow system

problem is examined in section 4, and a simple algorithm is presented in 5.

2. BALANCED FLOW SYSTEMS

A balanced subsystem of a static flow system can be represented by a node where the sum of the accumulated inflows is equal to the sum of the accumulated outflows. By a flow we will understand an accumulated flow, and the time period it has been accumulated over is referred to as the balancing period. For the subsystem in Figure 1 we have $f_1 + f_2 = f_3 + f_4$. This means that over a balancing period there is no accumulation or depletion of the flow substance in the node.

For the linear case, the transfer coefficients are given as the ratio between an outflow and the sum of the inflows, e.g. $f_3 = \alpha_3(f_1 + f_2)$, $f_4 = \alpha_4(f_1 + f_2)$, where α_3 and α_4 are the transfer coefficients. Of course, $\alpha_3 + \alpha_4 = 1$. For simplicity we have assumed that to each outflow all inflows contribute the same fraction. In a certain sense this is no restriction since a subsystem can always be decomposed into finer subsystems such that each subsystem satisfies this condition as the following example shows.

Example 1. Assume that $f_3 = \alpha_{31}f_1 + \alpha_{32}f_2$ and $f_4 = \alpha_{41}f_1 + \alpha_{42}f_2$. The system in Figure 1 can be decomposed as follows: Subsystem 1: $f_1 = f_{11} + f_{12}$, $f_{11} = \alpha_{31}f_1$, $f_{12} = \alpha_{41}f_1$, Subsystem 2: $f_2 = f_{21}+f_{22}$, $f_{21} = \alpha_{32}f_2$, $f_{22} = \alpha_{42}f_2$, Subsystem 3: $f_3 = f_{11}+f_{21}$, Subsystem 4: $f_4 = f_{12} + f_{22}$.

By interconnecting subsystems we obtain the total system under consideration which typically will have external in- and outflows to and from the surrounding environment.

The balancing periods may be different, but we will assume that they are multiple of a smallest period which allows us to express all balancing equations for a node with respect to a common period.

Example 2. Consider the system in Figure 2, and assume that the balancing period for f_1 is two weeks and that the balancing period for f_2 and f_3 is one week. We denote the variables associated with the first and second week by a $'$ and a $''$ respectively. For the first week we have $f_1'-f_2'-f_3' = 0$, $f_2' = \alpha_2'f_1'$, $f_3' = \alpha_3'f_1'$, $\alpha_1'+\alpha_2' = 1$. For the second week we have $f_1''-f_2''-f_3'' = 0$, $f_2'' = \alpha_2''f_1''$, $f_3'' = \alpha_3''f_1''$, $\alpha_1''+\alpha_2'' = 1$. For f_1' and f_1'' we do not have individual measurements, but we have measurements of their sum $f_1 = f_1' + f_1''$. Also, if the transfer coefficients are equal for the two balancing periods it gives us $\alpha_1' = \alpha_1''$, and $\alpha_2' = \alpha_2''$. Note that some of the equations and variables are redundant.

The vector f of all flows is partitioned as $f^T = [z^T\ x^T\ y^T]$ where z are the exactly known flows, $z = z^0$, x are the measured flows, with measurement values x^*, and y are the unknown or unmeasured flows. Similarly the vector α of all transfer coefficients are partitioned as $\alpha^T = [\alpha_k^T, \alpha_m^T, \alpha_u^T]$, where α_k are the exactly known transfer coefficients with values α_k^0, α_m are the measured ones with measurement values α_m^* and α_u are the unknown transfer coefficients.

3. DATA RECONCILIATION.

The aim is to improve the estimates of the flows and transfer coefficients by utilising the prior physical knowledge provided by the balance equations and other physical relationships that may be available. The best way of mathematically formulating the problem will of course depend on the particular application, but it is common to formulate it as a constraint optimisation problem. We let $\hat{x}, \hat{y}, \hat{\alpha}_m$ and $\hat{\alpha}_u$ denote the improved estimates.

For the case where there were no measured transfer coefficients Gleiß et al (1998) formulated the problem as

$$\min_{\hat{x},\hat{y}}(\hat{x} - x^*)^T\Sigma^{-1}(\hat{x} - x^*)$$

subject to $g(z^0, \hat{x}, \hat{y}) = 0$, where Σ^{-1} is a weighting matrix reflecting the relative accuracy of the measurements in x^*. The constraints g were linear in z^0, $\hat{x}$ and $\hat{y}$.

In the case of measured transfer coefficients the problem becomes a lot harder. One straightforward extension of the above problem formulation is

$$\begin{aligned}\min_{\hat{x},\hat{y},\hat{\alpha}_m,\hat{\alpha}_u}\ & (\hat{x} - x^*)^T\Sigma_1^{-1}(\hat{x} - x^*)+ \\ & (\hat{\alpha} - \alpha^*)^T\Sigma_2^{-1}(\hat{\alpha} - \alpha^*)\end{aligned} \tag{1}$$

Subject to $g(z^0, \hat{x}, \hat{y}, \alpha_k^0, \hat{\alpha}_m, \hat{\alpha}_u) = 0$.

If the constraints containing measured transfer coefficients can be separated from those containing unknown and exactly known transfer coefficients, the following approach can be taken

$$\min_{\hat{x},\hat{y},\hat{\alpha}_u} (\hat{x} - x^*)^T \Sigma_1^{-1} (\hat{x} - x^*) + g_1^T(z^0, \hat{x}, \hat{y}, \alpha_m^*) \Sigma_2^{-1} g_1(z^0, \hat{x}, \hat{y}, \alpha_m^*) \tag{2}$$

subject to $g_2(z^0, \hat{x}, \hat{y}, \alpha_k^0, \hat{\alpha}_u) = 0$, where g_1 are the constraints involving the measured transfer coefficients. In section 5 we extend the reduced balance scheme to this case. The reduced balance scheme is a condensed representation of the system which does not contain any balance equations involving unmeasured flows. The problem formulation (2) leaves it open how to obtain the estimate $\hat{\alpha}_m$ since only the measured values α_m^* are used in the criterion function. One possibility is to use a weighted average of the measured values α_m^* and the corresponding ratios between the reconciled flow estimates.

Regardless of how the problem is posed, there is a need to examine the structure of the constraints. In particular, it is important to determine whether the constraints $g = 0$ have a feasible solution. If this is not the case, it means that there is a misspecification of the physical knowledge. Moreover, after the measured variables have been estimated by minimisation of the criterion function, it is common to solve the constraints for the unmeasured variables in order to obtain an estimate of them, e.g. $\hat{y} = \text{sol}_y g(z^0, \hat{x}, y) = 0$ in the case where there are no transfer coefficients involved. Clearly, the number of solutions to $g = 0$ and the dimension of the solution space become important issues in order to determine structural properties such as the identifiability of the system. In the next section we will therefore examine the structure of the constraints.

4. STRUCTURE OF CONSTRAINTS

The constraint equations introduced here represent typical features of balanced flow systems, and the assumptions are (approximately) satisfied in many cases. However, we have not tried to be as general as possible and there will be cases where the assumptions are not satisfied or we have constraints of a different type. Also, issues such as "minimal representations" of the constraints and the most useful or suitable form for expressing the constraints, are not touched upon.

Below subscript i is used for enumeration of the variables (flows and transfer coefficients).

A1 The flows and transfer coefficients are real.

This makes the problem of determining existence and uniqueness of solutions to the constrains more difficult since the field of real numbers is not algebraically closed.

A2 The flows are non-negative.

This assumption implies that we do know the direction of the flows. It is satisfied in many systems, e.g. gravity fed irrigation systems (the water does not run uphill). If the flow direction is not known a priori, the flow can be modelled as two non-negative flows, one in each direction, with the added constraint that one of the two flows must be zero.

A3 The balance equations are of the form $\sum_i \beta_i f_i = 0$. where β_i is either 1 or -1 depending on whether f_i is an inflow or outflow.

A4 The transfer coefficient equations are of the form

$$f_i = \alpha_i \left(\sum_j f_{i_j} \right) \tag{3}$$

where the f_{i_j}s are all the inflows to a node and f_i an outflow. As we saw in section 2, there is no restriction in assuming this particular form of the transfer coefficient equations since we can always rewrite the system such that this condition is satisfied. If α_i is exactly known (3) represent a linear relationship between flows. If α_i is unknown, and there is no additional information about α_i, (3) does not contain any useful information, and the constraint can be removed. If α_i is measured we have a bilinear constraint between the flows and the transfer coefficient with the structural property that each term on the right hand side is a product of a transfer coefficient and a flow.

A5 The transfer coefficients satisfy $0 \leq \alpha_i \leq 1$, and the sum of all transfer coefficients for a node sums up to 1.

This is a redundant constraint which follows from the other constraints above.

A6 Additional linear constrains between flows are of the form $\sum_i a_i f_i = 0$ where the a_is are real coefficients.

The right hand side being equal to zero rather than a constant, represents no restriction since we can introduce an artificial flow which is identical to some known constant. Constraints of this type are useful for expressing that a measured flow is an accumulation over a multiple of the smallest balancing period as illustrated in Example 2. Also note that these constraints have the same structural properties as the transfer coefficients equations in **A4** in that each term is a product of a flow and an a coefficient.

A7 For some flows and transfer coefficients we may have lower and upper bounds on their values, i.e. $f_{\min,i} \leq f_i \leq f_{\max,i}$ and $\alpha_{\min,i} \leq \alpha_i \leq \alpha_{\max,i}$.

A8 Additional relationships between transfer coefficients, e.g. $\sum_i \gamma_i \alpha_i = 0$ where the γ_is are real coefficients.

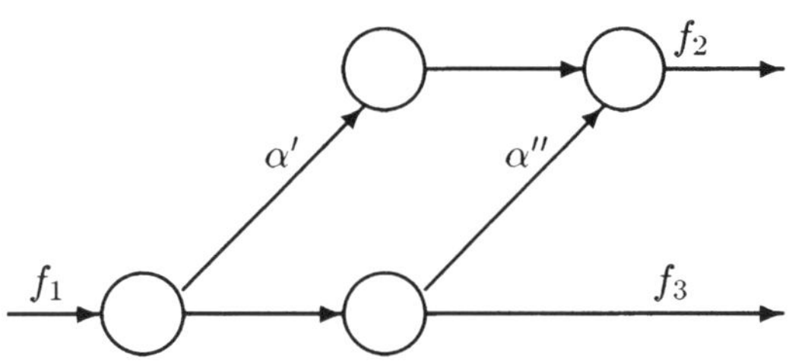

Fig. 3. Total system

These constraints are useful for expressing that some transfer coefficients are equal as illustrated in Example 2. As another example transfer coefficients on a mass balance level and an energy balance level may be the same, even though the transfer coefficients themselves may be unknown. Most relationships between transfer coefficients will be linear, but we can have polynomial relationships if some nodes are aggregated at one level as the following example will illustrate.

Example 3. Assume that the system in Figure 2 is an aggregation of the system in Figure 3, and that the transfer coefficients between f_1 and f_2 are the same in both cases which means that $\alpha_2 = \alpha' + (1 - \alpha')\alpha''$. Hence in general we can have polynomial relationship between transfer coefficients.

4.1 *Discussion*

With the exception of the polynomial constraints arising in **A8** all equality constraints are linear or bilinear. However, the bilinear case is not much simpler than the polynomial case, so from a mathematical point of view, we may as well regard the equality constraints to be a set of polynomial equations. The solution to such a system is known as an affine variety (Cox et al (1997)) which points us to the area of algebraic geometry for answers to questions regarding existence, identifiability and the dimension of the solution space to the equality constraints. Other issues whose solutions may partly be found using algebraic geometry are to determine whether some constraints are redundant and the existence of useful (for the intended purpose or task ahead) parametric representation of the affine variety.

Such questions are in general difficult to answer, and it is made particularly difficult here, since the field $\mathbb{R}$ is not algebraically closed (i.e. the solution to a polynomial equation with real coefficients may be complex), and in addition to the equality constraints we also have inequality constraints (e.g all flows are non-negative). Furthermore many results from geometric algebra are generic, that is, they hold in almost all cases, but they do not necessarily hold for the special situations occurring in balanced flow systems. This indicates that the structure of the constraints as exemplified in **A1-A8** should be exploited as much as possible when investigating the above mentioned questions for a particular system.

4.2 *Some practical issues*

In many large scale systems the data quality is highly varying. Many measurements are very uncertain and they are subject to gross errors. This calls for routines for outlier detection in order to remove or modify erroneous measurement values before the data set is used.

Moreover, robustness is usually much more important than optimality of the estimates, and criterion functions which are robust (Huber (1981)) against deviation from the assumptions should be employed. For example, the assumption that we do know the variance of the measurement errors is often unrealistic, likewise an assumption that they are normally distributed.

Finally, after one has obtained improved estimates of flows and transfer coefficients, there should be a validation (or rather invalidation) step in order to examine whether the obtained results are "in agreement" with the prior information. What is meant with "in agreement" is of course highly problem dependent, and it is a question of how much mismatch between reconciled data, measured data and prior information we can live with for a particular application.

5. REDUCED BALANCE SCHEME

The data reconciliation problems are constrained optimisation problems with both equality and inequality constraints, and general methods for solving such problems can be found in e.g. Polak (1997). Here we present a simple extension of the reduced balance scheme which can be used in some special cases. Admittedly, this approach can not take inequality constraints such as **A2** and **A7** into account, and the quadratic criterion used is not especially robust, but it is a simple scheme which extends a commonly used method in balanced flow analysis.

5.1 *Derivation of problem formulation*

We assume that there are no constraints of the type **A8**, and that the a_is are known for all constraints of type **A6**. Moreover **A2** and **A7** are not taken into account.

The first requirement excludes multi level systems where it is known that two or more levels have a transfer coefficient in common. It also excludes systems where a transfer coefficient is constant over several balancing periods.

Under these assumptions the equality constraints can be written as

$$A_p(\alpha_k^0, \alpha_m)f = 0$$

since constraints involving unknown transfer coefficients can be removed. Next we remove the constraints which involve measured transfer coefficients. This is equivalent to removing rows from $A_p(\alpha_k^0, \alpha_m)$. The remaining constraints can be rewritten as

$$A(\alpha_k^0)\begin{bmatrix} z^0 \\ \hat{x} \\ \hat{y} \end{bmatrix} = 0$$

where $A(\alpha_k^0)$ is known since α_k^0 is known. Using the procedure of the reduced balance scheme we apply a nonsingular transformation S such that

$$SA(\alpha_k^0) = \begin{bmatrix} A_{00} & A_{01} & 0 \\ A_{10} & A_{11} & A_{12} \end{bmatrix} \tag{4}$$

where the dependence on α_k^0 is not shown explicitly. In the removed constraints we replace α_m with the measured values α_m^*, such that they read

$$[A_{20}(\alpha_k^0, \alpha_m^*)\ A_{21}(\alpha_k^0, \alpha_m^*)\ A_{22}(\alpha_k^0, \alpha_m^*)]\begin{bmatrix} z^0 \\ \hat{x} \\ \hat{y} \end{bmatrix} = 0$$

and we formulate the data reconciliation problem as

$$\min_{\hat{x}\ \hat{y}} (\hat{x} - x^*)^T Q(\hat{x} - x^*) + [z^{0T}\ \hat{x}^T\ \hat{y}^T]\begin{bmatrix} A_{20}^T \\ A_{21}^T \\ A_{22}^T \end{bmatrix} Q_2[A_{20}\ A_{21}\ A_{22}]\begin{bmatrix} z^0 \\ \hat{x} \\ \hat{y} \end{bmatrix} \tag{5}$$

subject to (4). Q and Q_2 are positive definite matrices. The second term imposes a penalty if the ratio between reconciled flows are different from the corresponding measured transfer coefficients. The minimisation problem (5) may not have a unique solution and in order to guarantee a unique solution we introduce new flow variables. We can assume that the rows of $[A_{20}\ A_{21}\ A_{22}]$ are linearly independent, if not we can remove constraints until the rows become linearly independent. A condition for the minimisation problem to have a unique solution is that

$$\begin{bmatrix} Q & 0 \\ 0 & 0 \end{bmatrix} + \begin{bmatrix} A_{21}^T \\ A_{22}^T \end{bmatrix} Q_2[A_{21}\ A_{22}] \tag{6}$$

is positive definite, which is the case if and only if $A_{22}^T Q_2 A_{22}$ is positive definite, which again is the case if and only if the columns of A_{22} are linearly independent. The dimension of A_{22} is $m \times n_y$ where m is the number of constraints involving measured transfer coefficients and n_y is the number of elements in the y vector (number of unknown flows). Let $\bar{A}_{22}$ be a matrix whose rows form a basis for the rows in A_{22}, i.e. $A_{22} = T\bar{A}_{22}$ for some matrix T. Augment $\bar{A}_{22}$ with a $(n_y - m) \times n_y$ matrix B such that $\begin{bmatrix} \bar{A}_{22} \\ B \end{bmatrix}$ is nonsingular, and introduce the transformed flows

$$\bar{y} = \begin{bmatrix} \bar{y}_1 \\ \bar{y}_2 \end{bmatrix} = \begin{bmatrix} \bar{A}_{22}y \\ By \end{bmatrix}$$

Only the flows $\bar{y}_1$ occur in the minimisation criterion, and it can be rewritten as

$$\min_{\hat{x}\ \hat{\bar{y}}_1} (\hat{x} - x^*)^T Q(\hat{x} - x^*) + [z^{0T}\ \hat{x}^T\ \hat{\bar{y}}_1^T]\begin{bmatrix} A_{20}^T \\ A_{21}^T \\ T^T \end{bmatrix} Q_2\,[A_{20}\ A_{21}\ T]\begin{bmatrix} z^0 \\ \hat{x} \\ \hat{\bar{y}}_1 \end{bmatrix}$$

T has full column rank and the minimisation problem has a unique solution. If not we could have written $T = T_1P$ ($T_1(m \times m_2)$, $P(m_2 \times m_1)$) where T_1 has full column rank, and $A_{22} = T_1P\bar{A}_{22}$, implying that $P\bar{A}_{22}$ is a basis for the rows in A_{22} contradicting that $\bar{A}_{22}$ is a basis since $P\bar{A}_{22}$ has fewer rows than $\bar{A}_{22}$.

The constraints are given by

$$\begin{bmatrix} A_{00} & A_{01} & 0 \\ A_{10} & A_{11} & A_{12}\begin{bmatrix} \bar{A}_{22} \\ B \end{bmatrix}^{-1} \end{bmatrix}\begin{bmatrix} z^0 \\ \hat{x} \\ \hat{\bar{y}}_1 \\ \hat{\bar{y}}_2 \end{bmatrix} = 0$$

Next we apply a nonsingular transformation $\begin{bmatrix} I & 0 \\ 0 & S_2 \end{bmatrix}$ to obtain a reduced balance scheme of the form

$$\begin{bmatrix} A_{00} & A_{01} & 0 & 0 \\ \tilde{A}_{10} & \tilde{A}_{11} & \tilde{A}_{12} & 0 \\ \tilde{A}_{20} & \tilde{A}_{21} & \tilde{A}_{22} & \tilde{A}_{23} \end{bmatrix}\begin{bmatrix} z^0 \\ \hat{x} \\ \hat{\bar{y}}_1 \\ \hat{\bar{y}}_2 \end{bmatrix} = 0$$

By construction $\tilde{A}_{23}$ has full row rank, and hence the last constraints do not affect the minimisation problem.

5.2 *Problem formulation*

In summary: The data reconciliation problem can be formulated as

$$\min_{\hat{x}\ \hat{\bar{y}}_1} (\hat{x} - x^*)^T Q(\hat{x} - x^*) + [z^{0T}\ \hat{x}^T\ \hat{\bar{y}}_1^T]\begin{bmatrix} A_{20}^T \\ A_{21}^T \\ T^T \end{bmatrix} Q_2\,[A_{20}\ A_{21}T]\begin{bmatrix} z^0 \\ \hat{x} \\ \hat{\bar{y}}_1 \end{bmatrix} \tag{7}$$

subject to

$$\begin{bmatrix} A_{00} & A_{01} & 0 \\ \tilde{A}_{10} & \tilde{A}_{11} & \tilde{A}_{12} \end{bmatrix}\begin{bmatrix} z^0 \\ \hat{x} \\ \hat{\bar{y}}_1 \end{bmatrix} = 0$$

where Q and Q_2 are positive definite, T has full column rank and the constraints are linearly independent.

The solution to the constraint minimisation problem can be found using Lagrange multipliers, and it is given by

$$\begin{bmatrix} \hat{x} \\ \hat{\bar{y}}_1 \end{bmatrix} = \tilde{Q}^{-1} \begin{bmatrix} Q \\ 0 \end{bmatrix} x^* - \tilde{Q}^{-1} \begin{bmatrix} A_{21}^T Q_2 A_{20} \\ T^T Q_2 A_{20} \end{bmatrix} z^0 + \frac{1}{2}\tilde{Q}^{-1} \begin{bmatrix} A_{01}^T & \tilde{A}_{11}^T \\ 0 & \tilde{A}_{12}^T \end{bmatrix} C^{-1}(R_1 x^* + R_2 z^0)$$

where

$$R_1 = - \begin{bmatrix} A_{01} & 0 \\ \tilde{A}_{11} & \tilde{A}_{12} \end{bmatrix} \tilde{Q}^{-1} \begin{bmatrix} Q \\ 0 \end{bmatrix}$$

$$R_2 = \begin{bmatrix} A_{01} & 0 \\ \tilde{A}_{11} & \tilde{A}_{12} \end{bmatrix} \tilde{Q}^{-1} \begin{bmatrix} A_{21}^T Q_2 A_{20} \\ T^T Q_2 A_{20} \end{bmatrix} - \begin{bmatrix} A_{00} \\ \tilde{A}_{10} \end{bmatrix}$$

$$\tilde{Q} = \begin{bmatrix} Q + A_{21}^T Q_2 A_{21} & A_{21}^T Q_2 T \\ T^T Q_2 A_{21} & T^T Q_2 T \end{bmatrix}$$

$$C = \frac{1}{2} \begin{bmatrix} A_{01} & 0 \\ \tilde{A}_{11} & \tilde{A}_{12} \end{bmatrix} Q^{-1} \begin{bmatrix} A_{01}^T & \tilde{A}_{11}^T \\ 0 & \tilde{A}_{12}^T \end{bmatrix}$$

It can be shown that provided that there are no inconsistencies in the constraints involving exactly known flows the matrix C is invertible. The inequality constraints have not been used, and one must check if the obtained solution satisfies these constraints.

5.3 *Computation of unknown flows*

The constraints which were not used in the optimisation problem are given by

$$[\tilde{A}_{20}\ \tilde{A}_{21}\ \tilde{A}_{22}\ \tilde{A}_{23}] \begin{bmatrix} z^0 \\ \hat{x} \\ \hat{\bar{y}}_1 \\ \hat{\bar{y}}_2 \end{bmatrix} = 0$$

where by construction $\tilde{A}_{23}$ has full row rank. Some flows are of more interest than others, and a procedure for deciding whether a flow can be uniquely determined is as follows. Write the flow of interest as $y_i = a_i^T \hat{\bar{y}}_2$. Introduce an invertible linear transformation

$$\begin{bmatrix} y_i \\ y_r \end{bmatrix} = \begin{bmatrix} a_i^T \\ A_r \end{bmatrix} \hat{\bar{y}}_2 = \tilde{A}_r \hat{\bar{y}}_2$$

Let $\tilde{A}_{23r}$ be the matrix $\tilde{A}_{23}\tilde{A}_r^{-1}$ with the first column removed. Then y_i can be uniquely determined if $\tilde{A}_{23r}$ has reduced row rank.

5.4 *Estimates of measured transfer coefficients*

The procedure leaves it open how to calculate the estimates of the measured transfer coefficients. Obvious choices are a weighted average of the measured values and the ratio of the reconciled flows or just the ratio of the reconciled flows. In the latter case we have $\hat{\alpha}_i = \hat{f}_i / (\sum_j \hat{f}_{i_j})$. This means that the terms involving expressions of the type $(\hat{f}_i - \alpha_i^*(\sum_j \hat{f}_{i_j}))$ in the criterion function (7) can be rewritten using the expressions $\sum_j \hat{f}_{i_j}(\hat{\alpha}_i - \alpha^*)$, i.e. we have a criterion function of the type (1), but with flow dependent weights.

6. CONCLUSIONS

In this paper we have examined the constraint equations occurring in static balanced flow systems with measured transfer coefficients. In typical cases these constraints are polynomial or bilinear. However, mathematically, the treatment of bilinear systems is in general no simpler than the full polynomial case. It is therefore important to utilise the structure and prior information about a given system as much as possible.

7. REFERENCES

[1] Baccini P. and P.H. Brunner (1991). *The Metabolism of the Anthroposphere.* Springer

[2] Baccini P., H. Daxbeck, E. Glenck and G. Henseler (1993). METAPOLIS - Güterumsatz und Stoffwechselprozesse in den Privathaushalten einer Stadt. NFP 25 "Stadt und Land".

[3] Cox D., J. Little and D. O'Shea (1997). *Ideals, Varieties and Algorithms.* Springer

[4] Gleiß A., T. Matyus, G. Bauer, M. Deistler, E. Glenck, C. Lampert (1998). "Identification of Material Flow Systems - Extensions and Case Study" *Envir. Sci and Pollut. Res*, Vol 5, no 4, pp. 137-144.

[5] Gleiß A., M. Deistler and T. Matyus (2000). "Linear Material Flow Models" *Proceedings of 12th IFAC Symposium on System Identification*, Santa Barbara June 2000.

[6] Huber (1981). *Robust statistics.* John Wiley.

[7] Matyus T., A Gleiß, K. Gruber and G. Bauer (2002). "Data Reconciliation, Structure Analysis and Simulation of Waste Flows - Case Study Vienna" To appear in *Waste management and research.*

[8] Polak E (1997). *Optimization : algorithms and consistent approximations.* Springer.

[9] Stigliani W.M, P.R. Jaffe and S. Anderberg (1993). "Heavy Metal Pollution in the Rhine Basin" Env. Sci. and Technology, Vol. 27, no. 5, pp 785-793.

[10] Voet, E. van der, R. Heijungs, P. Mulder, R. Huele, R. Kleijn, L. van Oers, (1995). "Substance flows through the economy and environment of a region - part 2: Modelling." *Envir. Sci and Pollut. Res*, Vol 2, no. 3, pp. 137-144.

IFAC
Publications
www.elsevier.com/locate/ifac

ENDOGENEITY AND IDENTIFICATION IN TRANSPORTATION SYSTEMS: ECONOMETRIC RELATIONSHIPS TO PARTIAL OBSERVABILITY

Naveen Kumar Juvva
Venky N. Shankar
Songrit Chayanan

Department of Civil and Environmental Engineering
University of Washington, Box 352700
Seattle, WA 98195
Tel: (206) 616-1259
Fax: (206) 543-1543
vns@u.washington.edu

Abstract: This paper addresses key econometric issues related to partial observability and identification due to endogeneity in transportation systems. The purpose of this essay is to bring to the forefront the identification problem in transportation systems stemming from partial observability. Inferences from recent work in pavement condition and traffic safety models are used to build this essay. Given the highly empirical nature of transportation contexts, sources of endogeneity and forms of exogeneity and consequent impacts on identification and parameter estimation are discussed. *Copyright © 2003 IFAC*

Keywords: endogeneity, identification, partial observability, likelihoods

1. INTRODUCTION

Econometric models serve the decision makers in formulating policies in social, economic and political fields. Econometricians make use of the data and/or pre-policy issues to arrive at the modelling results. One significant aspect of the data and the ad hoc policy is that the observations/documentation is not entirely reliable and complete, often the case in the transportation systems (travel demand, safety, infrastructure) modelling, political decision making, health policies, etc. Fundamentally they are partial observability problems. The extent to which partial observability is fundamental in the context of transportation as a whole is illustrated in figure 1. The transportation infrastructure context is used as the empirical basis for outlining this discussion of the partial observability problem, since they present a comprehensive basis for econometric inquiry. According to figure 1, the partial observability arising from the historical behaviour and measurement are both more or less related to data. The fundamental way of looking at it would be that there is not the right data to be modelled, i.e. data is missing or variables are omitted. The bottom portion of the figure depicts the data issues primarily related to partial observability and the latent variable approach to this problem. In the case of measurement issue, as is observed in the popular example of pavement performance modelling, the actual pavement condition process is not observed and is hence the functional relationship between the observed and latent performance indicators is modelled. As for the historical behaviour issue, the actual process is modelled as a suitable combination of the observed and latent processes, as is done in modelling zero accident counts in highway safety through Zero-Inflated Poisson/Negative Binomial models. The top portion of the figure illustrates the issues that can aptly be related to model estimation techniques. These (endogeneity, strong and weak exogeneity and identification, for instance) are mostly associated with the latent variable modelling or other facets of modelling partial observability.

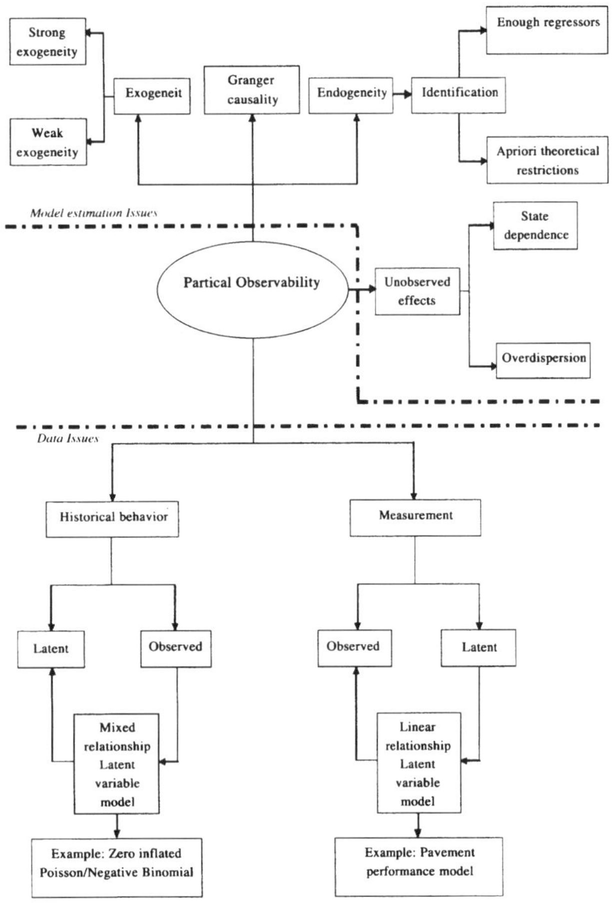

Fig. 1. Conceptual framework of partial observability in transportation modelling.

Often in cases of transportation systems modelling, the resultant of two latent processes is observed, and the inquiry is then guided by the vector of measured attributes associated with the product, and not factors affecting the latent processes. To parse this problem, one can employ the bivariate probit model (Poirier 1980), where the product is a discrete outcome, or adopt a sequential partial observability model depending on the nature of the measured dependent variates. When one considers the array of transportation infrastructure contexts where measurable outcomes are counts or discrete ordered or unordered variables, parsing the problem into estimable latent processes becomes a significant econometric challenge. The risk of not recognizing this challenge can result in simplistic models that can overlook data-related issues such as unobserved heterogeneity, endogeneity or selectivity bias. For example, if one were to examine accident counts for a highway corridor, the reported count could be the result of two latent processes, namely, the probability state of the highway being susceptible to an accident in its lifetime, and the probability of the highway having "n" accidents in any given year. A simple Poisson model applied to this case to model the count of accidents per year would result in inconsistent and inefficient estimates. The model would overlook the fact that the highway's limited accident history may provide spurious indication of its safety characteristic over its lifetime, and ignore overdispersion phenomena that arise from unobserved effects as well. In addition, if the safety characteristic of the highway is mis-estimated because of the Poisson model, it may create a selectivity bias issue, because the location may be sampled for high-expected risk when it is not truly a hazardous location. In addition, it may create endogeneity issues since road improvement policies would be driven by misestimated risk.

In general, the key to the partial observability problem is identifying the latent processes at work first. The partially observed variable could be then modeled as a linear or a mixed relationship between the observed and latent/two latent variables. Several econometric issues arise then. First and foremost is the issue of latent variable modelling itself. In addition, issues relating to endogeneity, including identification and related issues, weak exogeneity, strong exogeneity and granger causality, and unobserved effects, including overdispersion and state dependence arise. The relevance and impacts of such econometric issues in transportation models will be discussed in the following sections of this paper.

2. LATENT VARIABLE MODELLING

Before we discuss the issues mentioned above, such as endogeneity, identification, etc with regard to latent variable modelling, a brief explanation of the latent variable processes would be helpful in understanding the econometric relationships of these issues to partial observability. The latent variable approach helps in modelling the actual processes underlying the system performance. The observed random variables are modelled as functions of these latent variables. The exact functional relationships depend on the system behaviour. The simplest and most popular relationships are linear and mixed relationships. The mixed relationships consist of two dimensions: conditional likelihood and unconditional likelihood. An example of unconditional likelihood would be Zero-Inflated Poisson/Negative Binomial models employed in accident frequency modelling (Lambert 1992, Greene 1994 and Shankar, et al. 1997) and that of conditional likelihood would be bivariate ordered-response probit model employed in accident severity modelling (Yamamoto and Shankar 2002). The most common application of the linear relationship approach is in the area of pavement performance modelling.

3. ENDOGENEITY AND IDENTIFICATION

Consider, as an example for the linear relationship approach for latent variable, the latent variable model for pavement deterioration discussed in Ben-Akiva, et al., 1991. Traditional deterioration models link index of pavement condition to a set of explanatory variables. The problem with this kind of approach is that the pavement condition actually cannot be observed; so what is being modeled is a made up index of performance, which is subject to bias. In the latent variable approach, the dependent variable of the pavement deterioration relationship is unobserved. This relationship cannot be estimated by itself, but forms only one part (structural model) of the latent variable model (Ramaswamy and Ben-Akiva 1990). The second part is the measurement model, which defines the relationship between the underlying condition of the pavement and the measurements of condition indicators. The latent variable model simultaneously estimates the parameters of both the deterioration (structural) and measurement modes, thereby developing a deterioration model that is directly linked to the observed pavement data (Ben-Akiva and Ramaswamy 1990). This model gives better results than the traditional performance model. The significant variables have expected signs.

Structural model (deterioration model):

$$S = \beta_1 A + \boldsymbol{\beta}_2 \mathbf{X}_1 + \eta_1 \tag{1}$$

$$A = \gamma_1 S + \boldsymbol{\gamma}_2 \mathbf{X}_2 + \eta_2 \tag{2}$$

Measurement model:

$$f(I_1, I_2, ...) = \lambda_1 S + \varepsilon \tag{3}$$

where β_1, β_2, γ_1, γ_2, λ_1 are the estimable parameters; S is the latent variable of pavement performance; A is the maintenance variable; I_1, I_2, etc are the pavement condition indicator measurements; f(.) denotes a function of the condition indicators; X_1, X_2 are the vectors of explanatory variables affecting deterioration and maintenance respectively. It is to be noted here that the model specification contains the latent dependent variable, which requires us to use the calculated values for the performance indices in the model estimation. As a result, the propagation of error in a system of equations is not only theoretically possible but also empirically highly probable. This offsets any advantage a system of equations approach can provide in terms of statistical efficiency.

The system of simultaneous equations in the structural model is due to the argument that the pavement maintenance variable is endogenous, if the analysis is done on in-service pavements. This gives rise to two important issues:

- Is the maintenance variable contemporaneously endogenous?
- Is the system of equations identifiable?

In in-service pavements monitoring, maintenance is performed in response to the condition of the pavement. More maintenance has been performed over time on pavements that are in worse conditions or those that deteriorate faster. Typically, such pavements tend to carry higher loads. Because of increased maintenance on high-traffic pavements, it is conceivable that higher traffic pavements are in a better condition than the lower- traffic pavements. If a deterioration model is estimated with condition as the dependent variable and traffic as an independent variable, such a situation will clearly produce a counterintuitive sign for the parameter of the traffic variable. Thus, the simultaneous system of equations is specified. An issue that comes along with the system of equations is the identification issue, i.e., whether all the equations in the system can be identified. Models estimated previously (Ramaswamy and Ben-Akiva 1990) assumed zero correlation between the error terms in the structural model equations. One could also look at other types of a priori theoretical restrictions, such as restrictions on the parameters. Also, enough regressors could be included in the system of equations so that identification is ensured. This form of endogeneity is also possible in the accident rate modelling. In a model with the number of accidents as the dependent variable and, roadway geometrics and environmental factors as the explanatory variables, the speed variable could be considered as endogenous. Classical econometrics recommends diagnostics (Hausman 1978) for identifying endogeneity and also treatments such as instrumental variables when identification is a limitation (Hausman 1983). At the expense of inefficient parameter estimates, we are able to then estimate every equation in the system, albeit with significant multicollinearity. The instrumental variable approach is reliable in linear contexts, but in non-linear contexts can be more erroneous as a treatment than the original endogenous variable itself. Coming back to the pavement latent model, in the structural model, it could be argued that the maintenance variable is not contemporaneously endogenous. There could be a lagged effect of deterioration on maintenance. The pavement performance models that were previously estimated were cross-sectional in nature. The data collected was averaged over the period of time during which the data was observed. This gives rise to the following problems:

- This does not aid the process of decision-making regarding the maintenance and inspection activities for the pavement management. The decision making process requires the model estimates for regular intervals of time at which the decisions could be made. But the cross-sectional model gives the average condition of the pavement over a certain period of time.
- The actual form of correlation between the maintenance and deterioration variables is not incorporated into the model.
- Aggregation bias results from aggregating the data for the period of time for which the data was collected.

An appropriate model would be a time-series model, with lagged effects of the maintenance and deterioration variables. This along with the issues of weak exogeneity, strong exogeneity and granger causality will be discussed in the next section.

Endogeneity could be also be caused by the following:

- Omitted variables
- Self-selection

In the case of omitted variables, endogeneity can be considered as a case of partial observability. In a structural model of the latent performance model of infrastructure safety for example, the explanatory (exogenous) variables include ADT (Annual Daily Traffic) and some environmental variables. Variables like speed are omitted in the model specification. Speed is very likely correlated with inclusive variables like ADT and environmental variables such as rain, temperature, etc. So if,

$$S_i = \beta_1 A + \beta_2(ADT) + \beta_3(\text{rainfall}) + \beta_4(speed) + \eta_1 \quad (4)$$

and if speed is omitted, it can be added to the error term. So we have

$$S_i = \beta_1 A + \beta_2(ADT) + \beta_3(\text{rainfall}) + \eta_2$$
$$(\eta_2 \text{ incorporates speed}) \quad (5)$$

ADT and rainfall variables are correlated with speed, which is missing in the above equation and is incorporated into the error term. Hence, the ADT and rainfall variables are correlated with the error term η_2. Estimation using OLS gives biased and inconsistent estimates (Greene 2000). This forces us to consider ADT and rainfall as endogenous variables and employ appropriate estimation techniques. Endogeneity can also be caused by self-selection. Consider the example of ITS applications. In high accident locations (high accidents due to severe weather conditions), it would be considered helpful to install intelligent transportation systems (ITS) for advance warning to drivers based on real time weather information. The effect of inclement weather conditions on the accident probability for any site is determined using econometric models. If we wanted to find out the effectiveness of the ITS dummy in accident prevention, we would encounter the problem of endogeneity. The ITS dummy depends on whether the location in question is highly accident prone or not. Only high accident locations have the ITS dummy and hence, the variable is not constant in fixed in repeated samples (i.e., it is stochastic). This problem may also be viewed as a selectivity bias problem that could be solved by Heckman's two-step method (Heckman 1979).

4. WEAK EXOGENEITY, STRONG EXOGENEITY AND GRANGER CAUSALITY

In the previous section, we mentioned the time-series model as an appropriate model for the pavement deterioration model. It could be written as:

$$S_t = \alpha_0 + (\beta_1 A_{t-1} + \beta_2 A_{t-2} + \ldots) + (\gamma_1 S_{t-1} + \gamma_2 S_{t-2} + \ldots) + \delta_1 \mathbf{X}_1 + \varepsilon_1 \quad (6)$$
$$A_t = \rho_0 + (\lambda_1 A_{t-1} + \lambda_2 A_{t-2} + \ldots) + (\theta_1 S_{t-1} + \theta_2 S_{t-2} + \ldots) + \varphi_1 \mathbf{X}_2 + \varepsilon_2 \quad (7)$$

This brings up issues of weak exogeneity, strong exogeneity and Granger causality. Weak exogeneity (see Ericsson, et al. 1998 for example) is the requirement for the conditional estimation to be without loss of information from conditioning. Considering the factorization of the joint density into a conditional density and marginal density,

$$F_x(y_t, z_t \mid \mathbf{X}_{t-1}, \theta) = f_{y|z}(y_t \mid z_t, \mathbf{X}_{t-1}, \lambda_1) f_z(z_t \mid \mathbf{X}_{t-1}, \lambda_2) \quad (8)$$

z_t is weakly exogenous for the parameters of interest ψ if and only if

- $\psi = \psi(\lambda_1)$, that is ψ is a function of λ_1 alone; and
- λ_1 and λ_2 are variation free

Condition 1 ensures that ψ can be learned from λ_1. Together, conditions 1 and 2 exclude the possibility that ψ depends on λ_2, either directly (condition 1) or indirectly (condition 2). Hence, no information about the parameters of interest can be derived from the marginal model. Because ψ can be learned uniquely and completely from the conditional model, weak exogeneity is a sufficient condition for efficient inference on ψ from the conditional model. The marginal distribution of policy variables may be difficult to model empirically, due to changes in policy regime. In such a situation, valid conditioning on these variables can greatly assist empirical modelling. Failure of either condition 1 or 2 precludes inference without loss of information when using the conditional model alone. The following are the consequences:

- The parameters of interest cannot be obtained from the conditional model, as with errors-in-variables and simultaneity.
- Inference is distorted.
- Knowledge of λ_2 is required to identify ψ.
- Efficiency may be lost because λ_2 contains useful information about λ_1, as with classical cross equation restrictions.

Granger causality is one of the two conditions required for strong exogeneity, which bears on conditional impulse response analysis and conditional forecasting interalia. It can be defined as the presence of feedback from one variable to another, with Granger non-causality defined as the absence of such feedback. Suppose that the marginal density $f_z(.)$ does not depend on Y_{t-1}; that is

$$f_z(z_t \mid X_{t-1}, .) = f_z(z_t \mid Z_{t-1},) \quad (9)$$

Then, y does not Granger cause z. (Granger 1969 and Engle et al. 1983). Weak exogeneity and Granger causality combine to give strong exogeneity. z_t is strongly exogenous for the parameters of interest ψ if z_t is weakly exogenous for ψ and if y does not Granger cause z. Strong exogeneity allows conditional forecasting from the conditional model without loss of information. The forecasts of z over several periods may be constructed, and then forecasts of y are generated from the conditional model, conditional on that set of forecasts for z. If y did Granger forecast z, then forecasts of y and z would need to be constructed together, one period at a time or else risk losing valuable information.

In the context of the pavement performance modelling, if one considers the probability space to consist of variables of deterioration, maintenance, and other explanatory variables, then, then the policy instrument (z) is the maintenance variable and the target variable for the policy (y) is the deterioration variable. So, the maintenance variable is weakly exogenous if no information can be derived from the marginal model. The conditions for Granger noncausality are not likely to be satisfied in practice because past values of target variable, which is the deterioration, typically influence the present choice of policy instrument's (maintenance) values. So, the strong exogeneity of maintenance is not feasible in practice. So, for ensuring no loss of information, the forecasts of maintenance and deterioration will have to be constructed together, one period at a time.

5. HETEROGENEITY AND STATE DEPENDENCE

The other forms of partial observabilty commonly encountered in transportation are:

- Heterogeneity
- State dependence

It was mentioned that the ZIP model takes care of the overdispersion due to excess zeroes in the accident data. There is also the negative binomial model, which accounts for the overdispersion in the form of unobserved heterogeneity.
For the NB model,

$$\mathrm{var}[y_i] = E[y_i](1 + \alpha E[y_i]) \tag{10}$$

where α is the overdispersion parameter. For the ZIP model,

$$\mathrm{var}[y_i] = E[y_i]\left[1 + \left(\frac{q_i}{1-q_i}\right)E[y_i]\right] \tag{11}$$

Thus, $\left(q_i/1-q_i\right)$ is the counterpart to the α in the NB model. For the ZINB, which tries to incorporate both forms of overdispersion,

$$\mathrm{var}[y_i] = E[y_i]\left[1 + \left(\frac{q_i+\alpha}{1-q_i}\right)E[y_i]\right] \tag{12}$$

The overdispersion is considered to arise from these two independent sources. But in the estimation, the contributions are exactly mutually exclusive.

Another concept related to the unobserved heterogeneity is the concept of state dependence. State dependence creates potential estimation problems in discrete probability models (Ben-Akiva and Lerman 1985), if the information on previous outcomes is used to determine the current probability outcomes. It is an interesting situation in which the observed habitual behaviour or cyclical behaviour of infrastructure could be confused with the unobserved heterogeneity. In a model of infrastructure maintenance, maintenance policy choice in year 'i' could be considered to be an independent variable in modelling the policy choice in year 'i+1'. So,

$$y_t = \alpha + \boldsymbol{\beta}\mathbf{x} + \gamma y_{t-1} \tag{13}$$

where x is the vector of explanatory variables (infrastructure, vehicle and institutional characteristics). Conceptually, this may make sense since it could be capturing important cyclical behaviour, known as state dependence (Heckman 1981; Madanat, et al. 1997). Unfortunately, the inclusion of such a state could also pick up residual heterogeneity, which would lead us to observe spurious state dependence. If the disturbance term is considered to include unobserved characteristics (namely unobserved heterogeneity) that are commonly present among select highway sections, it is unclear whether the estimated coefficient of this variable is capturing true cyclical behaviour (state dependence) or is simply picking up some mean commonality in highway section disturbance terms (unobserved heterogeneity). Hence, extreme caution must be used when interpreting the coefficients of variables that are based on previous outcomes.

6. CONCLUSIONS

Partial observability is a fundamental issue in transportation modelling, and as illustrated through Figure 1 and key econometric issues, it leads to several opportunities for erroneous inferences if proper treatments or methodological approaches are not employed. The modelling of latent processes gives rise to other potential forms of partial observability. These potential forms of partial observability like endogeneity, exogeneity, causality and state dependence should not be overlooked, considering the dynamic nature of infrastructure modelling. Specification tests help in identifying forms of endogeneity and suggestions to care of identification issues are made. Linear and non-linear frameworks have to be treated differently in instrumental regressions. Extreme caution in modelling and identifying the processes can help estimation problems like state dependence and heterogeneity.

7. RECOMMENDATIONS

The econometric relationships of the endogeneity and identification problems to partial observability have been discussed. The identification of a generalised structural approach for the infrastructure approach remains an issue, and in the absence of a priori theoretical restrictions, estimation of all equations may not be possible, given the limited set of explanatory regressors that are usually available in infrastructure contexts. The alternative approach is to examine the system as a series of instrumented regressions, but with special attention to instrumentation in non-linear frameworks, to ensure consistency of parameters.

REFERENCES

Ben-Akiva, M. and Lerman, S. R. (1985). *Discrete Choice analysis:Theory and application to travel demand*, The MIT Press, Camb.idge, MA.

Ben Akiva, M. and Ramaswamy, R. (1990). Estimation of latent pavement performance from damage measurements. Proc., 3rd International Conference on bearing Capacity of Roads and Airfields, Trondheim, Norway.

Ben-Akiva, M., Humplick, F., Madanat, S. and Ramaswamy, R. (1991). Latent Performance Approach to Infrastructure Management. *Transportation Research Record* **1311**, TRB, National Research Council, Washington, D.C.

Engle, R. F., Hendry, D. F., and Richard, J. F. (1983). Exogeneity, *Econometrica*, **51**, 277-304.

Ericsson, N.R., Hendry, D.F., and Mizon, G.E. (1998). Exogeneity, Conintegration, and Economic Policy Analysis. *Journal of Business & Economic Statistics*, **16(4)**, 370-87.

Granger, C. W. J. (1969). Investigating casual relations by econometric methods and cross-spectral methods, *Econometrica*, **37**, 424-438.

Greene, W. (1994). Accounting for excess zeroes and sample selection in Poisson and negative binomial regression models. (Working Paper EC-94-10). Stern School of Business, New York University, New York.

Greene, W. (2000). *Econometric Analysis*, Fourth Edition, Prentice Hall.

Hausman, J. A. (1978). Specification tests in econometrics. *Econometrica,* **46**, 1251-1271.

Hausman, J. (1983). Specification and estimation of simultaneous equation models, In: *Handbook of Econometrics* (Griliches, Z. and Intriligator, M. Ed)), Amsterdam: North Holland.

Heckman, J. (1979). Sample selection bias as a specification error. *Econometrica*, **47 (1)**, 156-161.

Heckman, J. J. (1981) Heterogenity and State Dependence, in: Rosen, S., eds, *Studies in labour markets*, University of Chicago Press, Chicago, IL, pp. 91-139.

Lambert, D. (1992). Zero-Inflated Poisson regression, With an application to defects in manufacturing. *Technometrics*, **34, 1**, pp. 1-14.

Madanat, S. M., Karlaftis, M. G. and Mc Carthy, P. S. (1997). Probabilistic infrastructure deterioration models with panel data. *Journal of Infrastructure systems*, **3(1)**, 4-9.

Poirier, D. J. (1980). Partial observability in bivariate probit models. *J Economet* , **12**, 209-217.

Ramaswamy, R. and Ben Akiva, M. (1990). Estimation of highway pavement deterioration from in-service pavement data. *Transportation Research Record*, **1272**, TRB, National Research Council, Washington, D.C.

Shankar, V., Milton, J. and Mannering, F. L. (1997). Modelling accident frequencies as zero-altered probability processes: an empirical enquiry. *Accident Analysis and prevention*, **29, 6**, 829-837.

Yamamoto T. and Shankar V.N. (2003). Bivariate Ordered-Response Models of Occupant Severities in Traffic Accidents. Working Paper, Department of Civil and Environmental Engineering, University of Washington

IFAC Publications
www.elsevier.com/locate/ifac

TOOL FOR EQUAL OPPORTUNITY EVALUATION IN DYNAMIC ORGANIZATIONS

P. Albertos, I. Benítez, J. L. Díez and J. A. Lacort

Department of Systems Engineering and Control. (DISA)
Universidad Politécnica de Valencia , PO Box. 22012, E-46071 Valencia, Spain
E-mail: pedro@aii.upv.es

Abstract: in this paper, an attempt to model and evaluate the equal opportunity (EO) conditions of an organization is presented. The main goal is to develop tools to estimate this condition. For that, a measurement system based on clustering techniques is first proposed. In this way, and using numeric, logic and qualitative information, a degree of fulfillment of the gender-based EO is estimated. A rough model of the organization based on basic principles allows evaluating the EO condition of the organization and pointing out the effects of different strategies of hiring and promoting people. The tool is intended as an instrument to make decisions in public and private organizations sensitive to this issue. *Copyright © 2003 IFAC*

Keywords: Modeling, Social systems, Gender diversity, Fuzzy Clustering, Quantitative Socio-dynamics

1. INTRODUCTION

In modern structured organizations, there is a growing interest in taking advantage of the diversity of their members to get the best performances. The diversity concept can be applied to different characteristics, such as age, sex, race, nationality, religion, etc. The diversity criteria emphasizes the inter-individual variability, in such a way that each one is valued on his/her own, depending on his/her skill and knowledge but not on his/her origin. Equal Opportunities (EO) measures in the organizations through the application of the diversity criteria allow pointing out the organization's performance in using its human resources. The EO strategy can be a useful opportunity to improve the labor competitiveness and the better use of the available human resources. Gender diversity is a common human resource factor and is the subject of this work.

The university environment has been one of the most attractive sectors to study about the gender EO, mainly because the University is a complex structure with different activities and many levels of operation. There are students, administrative staff and technicians, teaching/research people and directors.

The EO issue has been investigated in some European projects. Probably one of the best known is the research project "Equal opportunities at Universities. Towards a gender mainstreaming approach" (Stevens and Van Lamoen, 2001). The main aim of the project was to develop a manual for gender EO based on the mainstreaming strategy, as a result of analyzing the information and experiences concerning equal opportunities policies and gender mainstreaming in universities and research institutions.

Similar to other social systems, the main problem in analyzing the system's behavior and formulating proposals for better performing, takes root on the qualitative and even subjective nature of this information. Not only the variables are measured or interpreted in a qualitative way, also the system's structure and relationships, and even the system's goals, present the same weakness, (Albertos and Barberá, 1996). For instance, information about the satisfaction degree, level of interaction or knowledge about how to perform a task is clearly qualitative and subjective, whereas data about the salaries, the number of people in a department or the working schedule can be easily quantified. In order to analyze the organization's behavior, the main features of all this information should be extracted.

In our study, the main purpose is to analyze the EO condition of an institution, by means of an automatic treatment procedure of a set of heterogeneous data. There have been different approaches to deal with these characteristics (Helbing, 1995) and fuzzy logic has been proposed (Barberá and Albertos, 1994) as a mathematical tool to represent the approximated knowledge implicit in qualitative measurements.

A further step in the analysis of an organization is to know about its dynamic behavior. Some human characteristics are time varying and their evolution can be strongly influenced by the external actions as well as the working conditions. The effects are neither instantaneous nor continuous. To model the interaction among the different agents in the system may be a good starting point.

Based on the dynamic model of a social system, once validated with historical data, the effect of different external actions can be analyzed. Thus, if properly used, it could be a tool to determine organization policies to achieve, in particular, the gender EO.

The above referred European project proposed a number of EO indicators, most of them of a qualitative nature, suitable to analyze the performance of an academic institution. These results are also used in the framework of the European project "Divers@: Gender and Diversity"[1] the authors are involved in. One of the goals of this project is to develop a methodology to revise and analyze the presence of the EO indicators and the role of diversity and gender perspective in these organizations. Afterwards, and based on the diagnosis results, corrective measures will be designed. The aim is to develop a general enough methodology able to be applied to other public and private institutions and organizations.

In this paper, the general structure of this methodology is presented. The proposed models and procedures are based on simulation results and should be the basis to develop such a practical tool. The paper is organized as follows: in the next section, a measurement system is defined to get a static picture of an organization. Collecting inputs, environmental and system variables, methods are explained to obtain the values of the proposed EO indicators. Then, suggesting the use of psychological and social principles, the characteristics of its members and the relationships among them are expressed and a dynamic model is presented, trying to capture the dynamic nature of the organization. For a better comprehension, a hypothetical organization is described and a simulation example is included in section 4, showing its evolution. A final conclusion section is also included.

[1] This research is partly funded by the European Social Fund Project ES296 Divers@: Gender and Diversity.

2. SYSTEM STATIC ANALYSIS

Classic modeling techniques based on differential equations and detailed variables relationships are not suitable for complex systems (Johansson, 1993) such as those in biochemical engineering, aerospace or social models. It is also well known the added complexity that the human factor implies when trying to determine and predict results (Pearson and Boudarel, 2001; Wander, *et al.*, 2001). Artificial Intelligence and Fuzzy Logic (Zadeh, 1965) techniques can be used to tackle the uncertainty of the system, because of its universal function approximation capabilities and the parallelism to human reasoning process (Wang, 1997).

The first goal of this work is to evaluate the degree of gender discrimination in an organization from structured data as provided by the organization at a particular time, T_a. This information is composed by different characteristics of its workers and the relationships among them considered relevant for the analysis of the gender EO condition (e.g. number of men and women, salary, satisfaction degree, motivation, etc).

Partial indices should be computed for different subsystems to evaluate the EO condition. The process is illustrated in figure 1. The data should be structured and segregated (S), clustered (C) and interpreted (EO E). Afterwards, a global EO condition is evaluated by weighted averaging of the partial results (F).

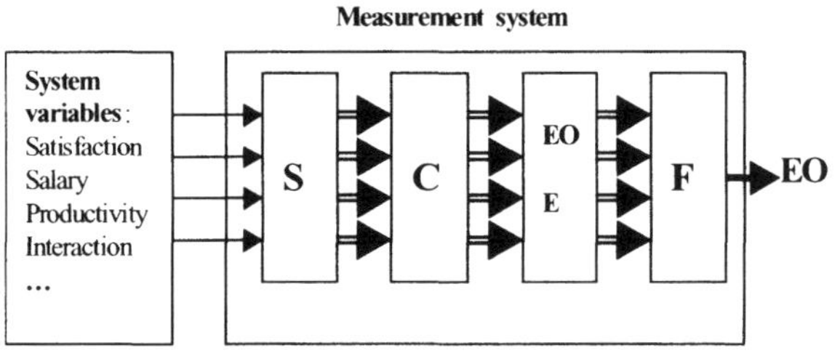

Fig. 1. Static analysis of an organization.

2.1 Data Structuring.

Initially, the proposed data structure consists on a matrix. Each column represents a worker and every worker has a fixed number of characteristics, each one appearing in the corresponding row, as shown in figure 2 (matrix **X**). Characteristics can be qualitative or quantitative.

It is assumed that every organization has a certain hierarchy, consisting of different levels, which can be defined by professions, departments, tasks, or other factors. Thus, it is clear that some characteristics, like the position or level in the organization structure, are more relevant, making difficult to compare individuals not having the same characteristic.

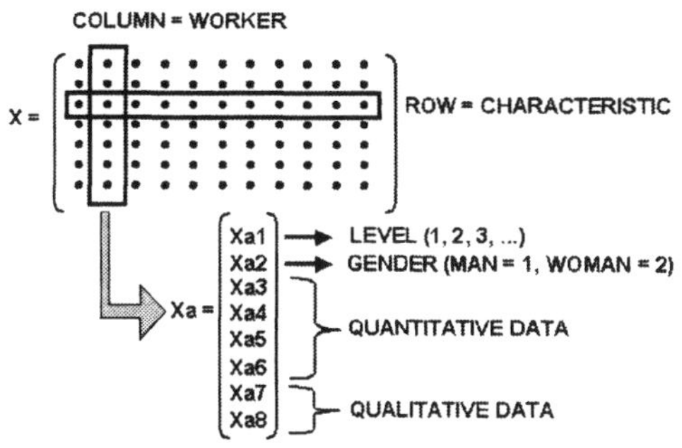

Fig. 2. Matrix of workers.

The algorithms use all the known values of every characteristic, so the system's accuracy will depend on how these values reflect the situation of the organization.

The first row of the organization or X matrix would represent the hierarchical level of the person attached to the corresponding column. The second row would indicate the worker's gender. Other relevant characteristics can be described in the following rows.

2.2 Data Segregation.

Several options can be taken to analyze the organization X. An interesting proposal would be to analyze the data grouping them by similar value of relevant characteristics, that is, to establish known differences among workers prior to apply any further data processing.

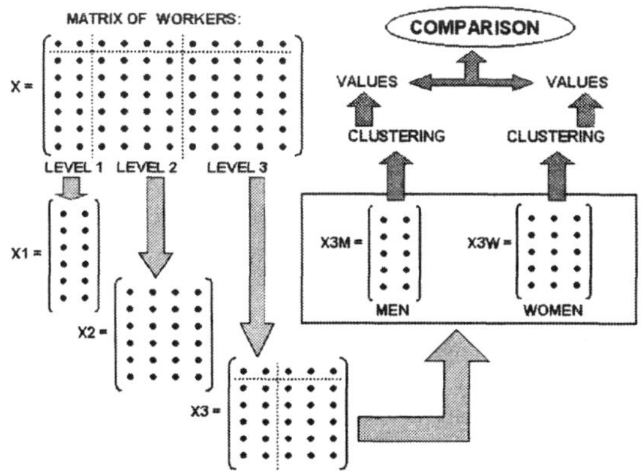

Fig. 3. Example of data segregation.

For instance, matrix X could be divided by level (or any other characteristic suitable of being segregated), or by level and gender. As our interest is to detect differences in opportunities according to gender, and the level influences most of the characteristics, the second option will be followed from now on.

A useful strategy consists in splitting this matrix into sections or submatrices (fig. 3), according to the level (fig. 3, X1, X2, X3); therefore, as many matrices as levels are created. The same can be done by the gender characteristic, (fig. 3, X3M, X3W). As a result, for each level two matrices are created. The next step is to analyze these matrices using unsupervised learning (optimization) methods as, for example, clustering algorithms.

2.3 Clustering Methods.

Classical clustering methods classify objects or individuals in clusters, according to a similarity, in a mathematical sense (Bezdek, 1987). In the simplest approach, if an object belongs to one cluster, then it does not belong to anyone else. Given a set of N objects (workers) $X = \{x_1, ..., x_N\}$, each object x_k having a number j of characteristics or dimensions, $x_k = \{x_{k1}, x_{k2}, ..., x_{kj}\}$, the classical partitioning space M_c in an expected number of c clusters is defined by a partition matrix $U \in R^{cxN}$ whose elements μ_{ik} (membership factor for cluster i of object k) are defined by:

$$\mu_{ik} \in \{0, 1\}, \quad 1 \le i \le c \quad 1 \le k \le N \qquad (1)$$

$$\sum_{i=1}^{c} \mu_{ik} = 1, \quad 1 \le k \le N \qquad (2)$$

$$0 < \sum_{k=1}^{N} \mu_{ik} < N, \quad 1 \le i \le c \qquad (3)$$

Undefined situations may appear when an object does not clearly fall within a cluster, but near two or more of them, or when two or more clusters overlap.

A much more realistic analysis can be done using fuzzy clustering algorithms. These algorithms extend the classical approach providing more flexibility. Objects being fuzzy clustered do not belong to a specific cluster exclusively, but to them all at different percentages. Thus, every object k has a different membership factor μ_{ik} for every cluster i, being the sum of all percentages again equal to 100%. The only difference, then, between classical M_c and fuzzy F_c partitioning spaces, is that the membership factor μ_i in M_c takes values 0 or 1, while μ_i in F_c varies from 0 to 1 (Bezdek, 1987). Thus, equation (1) becomes:

$$\mu_{ik} \in [0,1], \quad 1 \le i \le c \quad 1 \le k \le N \qquad (4)$$

Applying a fuzzy clustering algorithm, such as the fuzzy c-means (Bezdek, 1987), to any of the previously described submatrices of workers will give as a result two matrices. One contains the mathematical centers of the clusters (matrix of centers), having as many as clusters have been demanded, all with the same dimensions, one for each characteristic analyzed. The other matrix (fuzzy partition matrix) indicates, for each worker, the membership percentage (varying from 0 to 1) to each cluster. Thus, the elements of the fuzzy partition matrix represent scaled membership values of every worker (column) to every cluster (row). The projection of these membership values for each dimension or characteristic of the worker yields clouds of points that can be adjusted (using interpolation, for example) to membership functions.

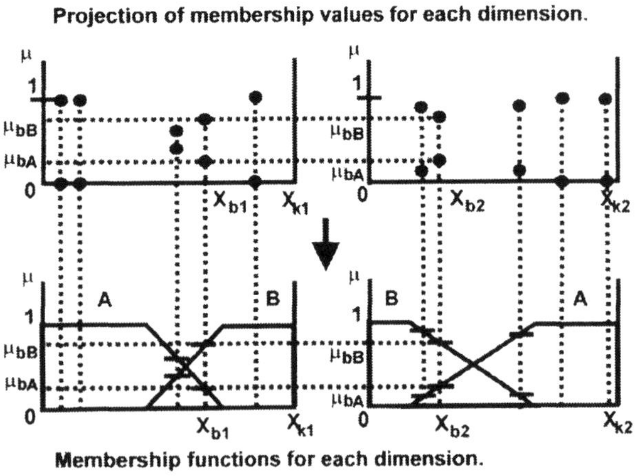

Fig. 4. Fuzzy Clustering. Membership functions.

An example of how membership functions can be obtained is shown in figure 4. Assume a set of N objects $X = \{x_1, ..., x_N\}$, each object x_k having 2 dimensions, x_{k1} and x_{k2}. Assume two clusters are requested, A and B. If a fuzzy clustering algorithm is applied on set X, a fuzzy partition matrix is obtained, having two rows (clusters) and N columns (objects). Each object x_k has, therefore, one membership value with respect to (w.r.t.) cluster A and one membership value w.r.t. cluster B, both of them varying between 0 and 1, their summation being 1. Two graphics are depicted, plotting, for each dimension (fig. 4, x_{b1} and x_{b2}), the membership values (fig. 4, μ_{bA}, μ_{bB}) of the object x_b w.r.t. the two clusters, A and B. If all the membership values of all the objects are plotted, and the resulting points are adjusted to a function separately, according to the cluster they belong to, membership functions are obtained, such as the four membership functions shown in figure 4 (2 dimensions x 2 clusters = 4 membership functions).

2.4 EO evaluation.

The so explained clustering process may be applied to the full matrix X (fig. 2) or to any subdivision of matrices. This strategy, chosen at the beginning of section 2.2, is detailed next. First, the data gathered from an organization are arranged in matrices per level and gender, as previously mentioned (fig. 3). All the matrices are then fuzzy clustered. The requested number of clusters per level is either defined by the expert or obtained empirically, applying a validity criterion. The resulting fuzzy partition matrices are represented on graphics relating membership factors μ_{ik} versus characteristics x_{kj}.

The obtained projected points are adjusted to predefined membership functions, obtaining a number of different membership functions equal to the product of the number of characteristics given and the number of clusters assigned for each characteristic. For two matrices corresponding to the same level and characteristic, that is, for men and women, the same number of linguistic variables should be assigned in order to allow comparison.

At this point, a set of specific information can be obtained from each membership function and stored, i.e., its center, its singular points, its amplitude, the number of workers pertaining to it, the density, etc. These values are compared between men and women, at the same level and characteristic, to obtain the value of an EO index (fig. 1, EO E). These indices tell about gender inequality regarding specific characteristics, such as salary, productivity, etc. If the values of the membership functions corresponding to men and women are equal, there will be no discrimination w.r.t. that specific characteristic. If they are not equal, there will exist discrimination. The farther men and women membership functions values are from one another, the higher the inequality index value will be.

2.5 Results.

Once the inequality indicators between men and women have been obtained for all levels and characteristics, a joint EO index can be evaluated either for level or for characteristic. Finally, a global measurement of the discrimination can be obtained by properly averaging all the partial EO indices. In this stage, a fuzzy model is used, setting membership functions for each indicator, inputs and outputs, and a set of rules (fig. 1, F).

The flexibility of the described system has permitted to evaluate as many indices as wanted or needed for further analysis and to obtain conclusions. Moreover, this static analysis, used to evaluate the EO condition, can be applied to the evaluation of other factors, considered relevant by the experts. The accuracy of the system will strongly depend on which characteristics are being used and the validity of the provided data.

3. DYNAMIC ANALYSIS

Analyzing the members of an organization and the organization itself at a given time can be very interesting to evaluate the current discrimination level. However, both the organization and its members are dynamic entities and are bound to change along the time. Thus, there is a justified interest in being able to forecast their evolution in a period of time and to consider the effects of external actions in order to avoid undesired results by implementing appropriate corrective actions.

It is evident that members of any organization undergo several and important changes as individuals as well as active members of the organization, and therefore, after a period of time the scenario could be potentially very different from the current one.

To begin with, it should be considered that some characteristics of an individual, such as capability, productivity or hierarchical level, are expected to change in the future. Among the different reasons for these changes, these are quite common: fulfillment of expectations, disappointment caused by failure in achieving personal goals or recognition, getting familiar with work environment or perceiving discrimination from the colleagues of the opposite gender.

Also, whilst some events will equally happen to all the members of a group or team (i.e. company philosophy, economic background, etc), some others will affect specific individuals more directly. Personal issues will affect people individually, and some of them will not find a corrective action from the organization. It should be also considered that some individuals are more susceptible to change than others, and the variability changes from a characteristic to another.

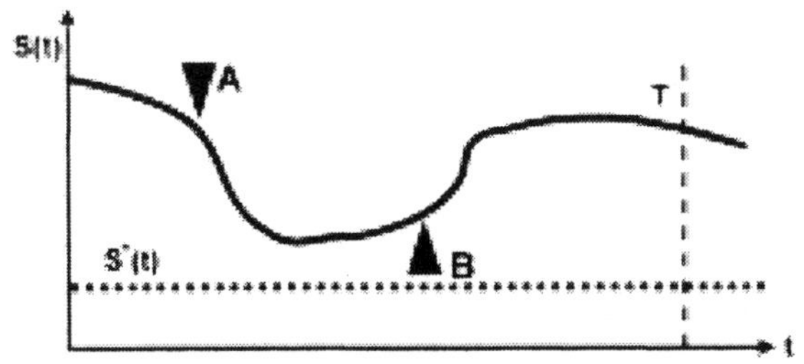

Fig. 5. Characteristics evolution along the time.

For instance, in figure 5, A and B represent two events, with opposite effect on the grade of satisfaction of an individual S(t) over a period of time T. There is a latency until the consequences of each event are noticeable. If the given characteristic S(t) reaches a lower limit, S*(t), that person will look for an alternative, leaving the organization.

Due to the subjectivity of the discussed parameters, it can be difficult to establish values for certain characteristics at an individual level. In this case this characteristic could be ignored and a projection can be done anyway. However, a better approach would be to make several assumptions to allocate the value within a range. If a set of historical data from the organization is available, the accuracy of the prediction can be tested and, incorporating the new information in subsequent simulations, the system's expertise can be enhanced.

The characteristics do not only depend on time, but also on other factors. Thus, the expert's knowledge (in particular that of the human resources staff) to consider the events occurred in the organization affecting the workforce is necessary to develop a dynamic model of the organization. The members relations, their characteristics and the organization policy, determining if the members are treated equally, will lead to a sketch of the organization.

For example, variations in wages (Δw) and satisfaction (Δs) could depend on several factors and characteristics, as shown by the equations (5) and (6):

$$\Delta w = f(\text{Level}, \text{Experience}, \text{Organization Policies}, \ldots) \quad (5)$$

$$\Delta s = f(\text{Level}, \text{Interaction}, \text{Wages}, \text{Expectations}, \ldots) \quad (6)$$

In the same way, interaction with individuals of the same or upper levels has to be considered as a factor to determine changes in the parameters defining a personal status.

3.1 Simulation.

The estimation must be reviewed and fed with current information in order to tune predictions going forward. The amount of historical information will in great measure determine the goodness of our predictions.

The basic information is composed by temporal series of data taken from the organization, including system variables (levels' composition, lists of salaries, questionnaires, interviews), external actions or inputs (promotion policy, internal communication measures, positive discrimination), or environment variables (labor market, organization success, etc). Based on information taken at a specific time T_a, and on the model, we can simulate the evolution of the characteristics going forward along the time.

Real data is valuable, since it will allow to enhance the model in three aspects: model validation, parameters adjustment and robustness improvement.

4. EXAMPLE

As an example, a hypothetical organization with six hierarchical levels has been simulated. Starting with random and standard values, the system evaluates workers characteristics and interactions. The gender distribution at the different levels will change according to the organization's evolution, being driven by the rewarding, hiring and promotion policies. At any time, a static analysis may be performed to evaluate the EO condition. Based on a set of dynamic equations, similar to (5) and (6), the individual characteristics are updated monthly. It is also considered that those with a satisfaction level below a lower limit leave the institution. Every year, vacancies are filled both with intern and extern individuals, according to the organization's philosophy. In the results collected in Table 1, where M = Men, W = Women and V = Vacancies, two opposite scenarios have been considered. In the first case (Table 1, shadowed area), the organization's promotion policy is in favor of men. As a result the EO index, expressed by the balance of men and women at the different levels is very poor.

In case 2 (Table 1, white area), a positive discrimination in favor of women is applied in the promotion policy. As a result, the balance is reversed and even some women reach the second level. It must be emphasized that the promotion policy is not the only available action. The parameters affecting the variation of satisfaction or interaction are also crucial to determine the organization's evolution.

Table 1. Cases 1 and 2. Considerations about discrimination.

Initial State			**Case 1**					**Case 2**				
Data			**Final Data**			**Gone**		**Final Data**			**Gone**	
M	**W**	**V**	**M**	**W**	**V**	**M**	**W**	**M**	**W**	**V**	**M**	**W**
4	0	0	4	0	0	2	0	4	0	0	1	0
8	0	0	8	0	0	1	0	5	3	0	0	0
12	2	2	14	2	0	2	0	2	14	0	1	3
25	4	3	30	2	0	2	2	10	22	0	3	2
52	8	4	56	8	0	0	0	3	61	0	2	1
107	16	5	107	21	0	10	3	35	93	0	19	3
208	**30**	**14**	**219**	**33**	**0**	**17**	**5**	**59**	**193**	**0**	**26**	**9**

Once the current and the projected states are known, suggested actions will correct the evolution of the organization, leading it towards an EO environment.

Likewise, the expected outcome of a corrective action can be analyzed, justifying the cost of implementing the actions required, and the time needed to see the benefits as, in general, they will not be immediate. Other simulations to determine the implications of a particular characteristic or group of characteristics may be done. In this way, specific causes of leaving the organization can be detected and, if so required, corrected. For instance, a dynamic analysis can point out the satisfaction degree of workers between 35 and 50 years old, or those with primary studies.

5. CONCLUSIONS

The modeling of social systems is always a difficult task due to the nature of the data and the processes themselves. But modeling is also a tool to better understand the system's behavior and to extract the knowledge of the experts.

In this paper, a framework to analyze the EO condition of an organization has been presented. The results obtained by the tool, at this moment, rely on hypothetical data and information provided by the experts and should be validated by actual data. Nevertheless, the structure of the proposed application is versatile enough to allow improvements and refinements and it is planned to be used in the development of a European project to determine the EO condition at the University and other public and private institutions.

6. REFERENCES

Albertos, P. and Barberá, E. (1996) Control Structures in Motivational Processes. *Proc. of the XIII IFAC World Congress*. S. Francisco. **Vol. J** pp 101-106.

Barberá E. and P. Albertos (1994). Fuzzy Logic Modeling of Social Behavior. *Cybernetics and Systems*. **Vol 25, No.2** pp 343-358.

Bezdek J. C. (1987). *Pattern recognition with Fuzzy Objective Function Algorithms*. Ed. Plenum Press.

Helbing, D. (1995). *Quantitative Sociodynamics*. Kluwer Academic Publishers.

Johansson R. (1993). *System Modeling and Identification*. Ed. Prentice Hall International. Series in Information and Systems Sciences.

Pearson, D. W. and Boudarel M-R. (2001). Pair Interactions: Real and Perceived Attitudes. *Journal of Artificial Societies and Social Simulation*. **Vol , no. 4**. http://www.soc.surrey.ac.uk/JASSS/4/4/4.html

Stevens, I. and Van Lamoen, I. (2001). *Manual on Gender Mainstreaming at Universities*. Garant, Leuven - Apeldoorn.

Wander, J. Popping, R. and Van de Sande, H. (2001). Clustering and fighting in two-party crowds: simulation the approach-avoidance Conflict. *Journal of Artificial Societies and Social Simulation*. **Vol 4, no.3**. http://www.soc.surrey.ac.uk/JASSS/4/3/7.html

Wang, L.-X. (1997). *A Course in Fuzzy Systems and Control*. Ed. Prentice Hall International

Zadeh, L. A. (1965). *Fuzzy Sets*. Information Control. **Vol 8**, p. 338-353

IFAC Publications
www.elsevier.com/locate/ifac

LINEARIZATION IN THE PARAMETERS VIA DIFFERENTIAL ALGEBRA TECHNIQUES

Maria Pia Saccomani *

* *Department of Information Engineering, University of Padova, Padova, Italy; Email:* pia@dei.unipd.it

Abstract: Recently, differential algebra tools have been applied to the study the identifiability of dynamic systems described by polynomial or rational equations. These methods all exploit the characteristic set of the differential ideal generated by the polynomials defining the system. In this paper it will be shown that the procedures based on differential algebra to test identifiability provide a very useful linear reparametrization of the input-output model of the system. This linear reparametrization can be used to derive explicit one-shot least squares estimates of the parameters, thus avoiding the usual bottleneck of nonlinear parameter optimization which has to be performed by iterative local search with no guarantee of reaching global minima. *Copyright © 2003 IFAC*

Keywords: Nonlinear parameter identification; Differential algebra;

1. INTRODUCTION

It is well-known that the only existing practical way to approach parameter identification of nonlinear system is through iterative optimization routines which are often unreliable, in the sense that they give no guarantee of converging to a true minimum, and hence require expensive and time consuming random search in the parameter space. Recently, a number of authors have applied differential algebra tools to study identifiability of non linear systems of the polynomial/rational function type (see (Ollivier, 1990; Ljung and Glad, 1994; Glad, 1990; Audoly *et al.*, 2001; Saccomani *et al.*, 1997)). These methods are based on elimination theory for algebraic differential systems, the main tool being the computation of the so-called *characteristic set* of a certain differential ideal associated to the dynamic equations of the system. This characteristic set can in principle be found by symbolic computation. In particular, there is good evidence that the approach introduced by Ollivier in his thesis (Ollivier, 1990) provides the right tool for testing global identifiability of the system, (Saccomani *et al.*, 1997). A reparametrization of the input-output relation of the system by the so-called *exhaustive summary* plays a major role in this approach. The reparametrization is in fact a *linear* reparametrization of the input-output relation of the model. It seems to have passed unnoticed that this reparametrization can be exploited and employed in a number of system theoretic problems. One of them is system parameter estimation for nonlinear systems. This is so since the reparametrization leads, in principle quite immediately, to one-shot least-squares estimates of the exhaustive summary, and hence to parameter estimates which do not require iterative optimization. Recovering the true parameters from the exhausive summary (which is possible if and only if the model is globally identifiable) is a problem of solving a system of (static) agebraic equations which is much easier and can be approached by standard algorithms.

2. BACKGROUND ON DIFFERENTIAL IDEALS AND CHARACTERISTIC SETS

For a formal treatment of differential algebra, the reader is referred to (Ritt, 1950; Forsman, 1991; Carrà Ferro, 1989). Here we shall just recall the definitions and notions which are necessary to set notations used in the rest of the paper.

Let $\mathbf{z} := [z_1, \ldots, z_n]$ be a vector of smooth functions of the variable t (time); the totality of polynomials in the variables z_i and their derivatives with coefficients in a field K, is a *differential polynomial ring* which will be denoted $K[\mathbf{z}]$.

Consider a set S of differential polynomials belonging to $K[\mathbf{z}]$. The *differential ideal* $I = I_S$ generated by S, is the smallest subset of $K[\mathbf{z}]$ containing S, which is closed with respect to addition, multiplication by arbitrary elements of $K[\mathbf{z}]$ and with respect to differentiation. The elements of S are *generators* of the ideal.

A differential ideal I is called *prime* if $A_iA_j \in I$ implies that $A_i \in I$ or $A_j \in I$ and *perfect* if $A \in I$ whenever A^k does (i.e. a perfect ideal coincides with its own radical).

In order to handle differential ideals, a *ranking*, i.e. a total ordering among the variables and their derivatives, must be introduced (Ritt, 1950). Let $z_i^{(\mu)}$ and $z_j^{(\nu)}$ be arbitrary derivatives. Then the ranking should be such that, for arbitrary positive integer k:

$$z_i^{(\nu)} < z_i^{(\nu+k)}; \quad z_i^{(\mu)} < z_j^{(\nu)} \Rightarrow z_i^{(\mu+k)} < z_j^{(\nu+k)}$$

The *leader* u_j of a polynomial A_j is the highest ranking derivative of the variables appearing in that polynomial (in particular it can be a derivative of order zero).

The polynomial A_i is said to be of *lower rank* than A_j if $u_i < u_j$ or, whenever $u_i = u_j$ and $\deg_{u_i}(A_i) < \deg_{u_j}(A_j)$, where $\deg_u(A)$ denotes the algebraic degree. A polynomial A_i will be said to be *reduced with respect to a polynomial* A_j if A_i contains neither the leader of A_j with equal or greater algebraic degree, nor its derivatives. If A_i is not reduced with respect to A_j it can be reduced by using the *pseudodivision algorithm* described below.

(1) if A_i contains the kth-derivative, $u_j^{(k)}$ (possibly $k = 0$), of the leader of A_j, A_j is differentiated k times so its leader becomes $u_j^{(k)}$;
(2) let be R the rest of the division between the two polynomials A_i and $A_j^{(k)}$ with respect to the variable $u_j^{(k)}$, then R is reduced with respect to $A_j^{(k)}$. The polynomial R is called the *pseudoremainder* of the pseudodivision;
(3) the polynomial A_i is replaced by the pseudoremainder R and the process is iterated using $A_j^{(k-1)}$ in place of $A_j^{(k)}$ and so on, until the pseudoremainder is reduced with respect to A_j.

A set of differential polynomials $A := \{A_1, A_2, ..., A_r\}$ that are all reduced with respect to each other, is called an *autoreduced set*.

Let π a differential polynomial. If we apply the pseudodivision algorithm to reduce π with respect to all $A_j, j = 1, ..., r$, the final reminder is called the *pseudoremainder of* π *with respect to the autoreduced set* A. Such a pseudoremainder is said to be reduced with respect to A (compare with (Ritt, 1950) where the autoreduced set corresponds to a *chain*).

Two autoreduced sets, $A = \{A_1, A_2, ..., A_r\}$ and $B = \{B_1, B_2, ...B_s\}$ ordered in increasing rank so that $A_1 < A_2 < ... < A_r, B_1 < B_2 < ... < B_s$, are ranked according to the following principle.

- If there is an integer k, $k \leq \min(s, r)$ such that $\text{rank}A_i = \text{rank}B_i$, $i = 1, ...k - 1$, $\text{rank}A_k < \text{rank}B_k$ then A is said to be of lower rank than B.
- If $r < s$ and $\text{rank}A_i = \text{rank}B_i$, $i = 1, ..r$, then A is also said to be of lower rank than B.

Definition 2.1. A lowest rank autoreduced set that can be formed with polynomials from a given set S of differential polynomials, is called a *characteristic set* of S.

The concept of characteristic set of a differential ideal has been introduced by Ritt (Ritt, 1950) who also proposed the pseudodivision algorithm to construct it. The important property of a characteristic set is that it can be used to generate a differential ideal by means of a finite number of polynomials. In particular, the characteristic set of a prime ideal spans the whole associated ideal (Forsman, 1991).

In principle, the characteristic set is not unique. It can however be normalized in such a way as to render it unique. One such normalization has been proposed by Rabinowitsch (Mishra, 1993).

3. THE BASIC MODEL

Consider a single-input single-output nonlinear dynamic system depending on a vector parameter $\mathbf{p}$, subject to random measurement errors

$$\dot{\mathbf{x}}(t) = \mathbf{f}(\mathbf{x}(t), u(t), \mathbf{p}) \quad (1)$$

$$y(t) = h(\mathbf{x}(t), u(t), \mathbf{p}) + w(t) \quad (2)$$

where the state variable $\mathbf{x}(t)$ evolves in an open set X of the n-dimensional space $\mathbb{R}^n$; and y is the scalar output. The constant unknown p-dimensional parameter vector $\mathbf{p}$ belongs to some open subset $\mathcal{P} \subset \mathbb{R}^p$.

We assume for simplicity that the state of the model is disturbance free; however at the price of some additional complications, we could accomodate also smooth unknown disturbance functions acting on the state. The admissible scalar input functions $t \to u(t)$, are assumed to be smooth (infinitely differentiable) functions of time. The output error term, $w(t)$, is an unobservable function (either a disturbance or a measurement error term) which affects additively the measured output. At no additional cost we can treat the case of say *multiplicative measurement noise* or, more generally, of noise entering in the output equation in a general *algebraic* fashion. The noise w is assumed "wideband" but not exactly "white" so that it can be differentiated a suitable number of times.

The system is assumed to be scalar-input/scalar-output for simplicity. The general procedure of this paper can be applied to a general MIMO system, but our scope here is mainly to convey the basic idea. The essential assumptions in this paper is that $\mathbf{f}$ and h are *polynomial functions* in the variables $\mathbf{x}, u$. The dependence on $\mathbf{p}$ may be rational. We shall assume that there is no feedback, i.e. u is a free variable, in particular not allowed to depend on $\mathbf{x}$.

We shall occasionally refer to the concept of, *a priori* parameter identifiability of the model. This concept deals with the (theoretical) uniqueness of solutions to the problem of recovering the model parameters from noise free input-output data. *A priori* identifiability analysis (Audoly *et al.*, 1998; Audoly *et al.*, 2001; Ljung and Glad, 1994; Vajda *et al.*, 1989; Walter and Lecourtier, 1982; Chappel and Godfrey, 1992; Saccomani *et al.*, 2003) refers to the "ideal model" (1, 2) where $\mathbf{w}$ is set equal to zero. It is obviously a necessary but not sufficient condition to ensure identification of the model from real input/output data.

3.1 *The Characteristic set*

We set
$$\bar{y}(t) := y(t) - w(t)$$
and associate to the dynamic system (1, 2) a corresponding system, Σ, of $n+1$ differential polynomials in the $n+1$ variables $\mathbf{x}, u$:

$$\dot{\mathbf{x}}_k - \mathbf{f}_k(\mathbf{x}, u, \mathbf{p}) \qquad k = 1, ..., n \tag{3}$$
$$\bar{y} - h(\mathbf{x}, u, \mathbf{p}) \tag{4}$$

We shall consider these differential polynomials in the ring $R(\mathbf{p})[\mathbf{u}, \mathbf{y}, \mathbf{x}]$, where $R(\mathbf{p})$ is the field of rational functions of the parameter vector $\mathbf{p}$ (Saccomani *et al.*, 1997; Audoly *et al.*, 2001). Hence the variables in this polynomial differential ring are the states, the input and the new output $\bar{y}$ and, possibly, their derivatives.

The polynomials (3, 4) can be looked upon as the generators of a differential ideal in a differential ring. We shall denote by I the ideal in $R(\mathbf{p})[\mathbf{u}, \mathbf{x}]$, generated by the differential polynomials (3) of the "state" equations, and by I_Σ the ideal generated by the full system (3, 4).

It has been proven by Diop (Diop, 1992) that the state-space description structure of the system ensures that the ideal I is prime. A further theoretical result (Ritt, 1950, par. 27, cap. 1) ensures that also I_Σ, which includes polynomials in the ring $R(\mathbf{p})[\mathbf{u}, \bar{\mathbf{y}}, \mathbf{x}]$, is prime.

As regards to the choice of the ranking, the ranking used in the literature declares the inputs as the lowest ranked components, followed by the outputs and the highest rank is given to the state variables. Normally, we shall choose the following order relation:

$$\begin{aligned} &u < \dot{u} < \ddot{u} < \ldots < \bar{y} < \dot{\bar{y}} < \ddot{\bar{y}} < \ldots \\ &< x_1 < x_2 < \ldots < \dot{x}_1 < \dot{x}_2 < \ldots \end{aligned} \tag{5}$$

A different ordering of the components of $\mathbf{x}$ may lead to a different characteristic sets.

Remark 3.1. If we choose the ranking: $u < \dot{u} < \ldots < x_1 < x_2 < \ldots < x_n < \dot{x}_1 < \dot{x}_2 < \ldots$ it turns out that the set of polynomials (3) is already an autoreduced set and hence a *characteristic set of* I (Ollivier, 1990; Glad, 1990).

Now consider the polynomials (3, 4). We calculate a characteristic set of I_Σ with respect to the ranking (5). The result is a family of differential polynomials of the following form:

$$\begin{aligned} &A_0(u, \bar{y}) \\ &A_1(u, \bar{y}, x_1) \\ &A_2(u, \bar{y}, x_1, x_2) \\ &\vdots \\ &A_n(u, \bar{y}, x_1, \ldots, x_n) \end{aligned} \tag{6}$$

This corresponds to a finite set of $n+1$ algebraic differential equations which represent the totality of functions $(u, \bar{y}, \mathbf{x})$ satisfying the equations of the original system (1, 2).

Note that the first differential polynomial A_0 of (6) does not depend on $\mathbf{x}$. In fact, A_0 is obtained after elimination of the state variables $\mathbf{x}$

from the set (3, 4). The corresponding polynomial differential equation

$$A_0(u, y+w) = 0 \tag{7}$$

is called the *(noisy) input-output relation* of the system, while $A_0(u,y) = 0$ is the *ideal (noise-free) input output relation*, considered e.g. in identifiability analisis. The latter describes all input-output pairs which satisfy the system equations (1, 2) once w is set equal to zero.

The input-output relation(s) of the system can always be be computed explicitly (at least in theory) by means of Ritt's pseudodivision algorithm (see (Ritt, 1950)), even if in practice this may not always be possible, depending on the complexity of the problem.

We do not show the explicit dependence on $\mathbf{p}$ in (7) since the polynomial on the left hand side belongs to the differential ring $\mathbf{R}(\mathbf{p})[\mathbf{u}, \bar{\mathbf{y}}]$ and the dependence on $\mathbf{p}$ is implicit in the choice of the field of coefficients. Also, since the coefficients are rational functions of the parameters it is possible to normalize the input-output relation by making the highest degree coefficient of the leader in A_0, equal to one[1]. Hence, after making the input-output polynomial "monic" in the above sense, (7) can be written

$$a_0(u,\bar{y}) + \sum_{i=1}^{\nu} \theta_i(\mathbf{p})\, a_i(u,\bar{y}) = 0 \tag{8}$$

where $a_i(u,\bar{y})$, $i = 0, 1, \ldots, \nu$, is the i-th monomial term in $A_0(u,\bar{y})$ (here i is an index running over the monomial indices of A_0, ordered, say, in a lexicographic ordering) and the rational functions $\theta_i(\mathbf{p})$ constitute a "canonical" set of coefficients of the input-output polynomial differential equation, uniquely attached to the dynamical system. We shall in fact assume that they form a *a minimal set* of parameters. In this case they will be called the *exhaustive summary* of the model (Ollivier, 1990), since the $\theta_i(\mathbf{p})$ embody in a minimal way the parameter dependence of the model, as seen from the input-output terminals.

Let now $\mathbf{w}$ be the column vector made by stacking the variables $w, \dot{w}, \ldots, w^{(n)}$, as they appear in the differential equation (7). We shall assume that the amplitudes $w(t), \dot{w}(t), \ldots, w^{(n)}(t)$ are statistically small and introduce a linearization with respect to $\mathbf{w}$ of the (normalized) input output model

$$A_0(u, y+w) \simeq A_0(u,y) + \frac{\partial A_0(u, y+w)}{\partial \mathbf{w}}_{|\mathbf{w}=0} \mathbf{w}$$

so that (8) takes the form

[1] This is only one possible normalization. Other choices are equally legitimate, provided the normalized coefficients turn out to be unique.

$$a_0(u,y) = -\sum_{i=1}^{\nu} \theta_i(\mathbf{p})\, a_i(u,y) + + \sum_{i=0}^{\nu}\sum_{j=0}^{n} \beta_i(\mathbf{p})\, b_{ij}(u,y) w^{(j)} \tag{9}$$

where the , $b_{ij}(u,y)$ are monomials. Introducing $\boldsymbol{\theta} := [\theta_1 \ldots \theta_\nu]^T$, $\boldsymbol{\beta} := [\beta_1 \ldots \beta_\nu]^T$, and

$\mathbf{a}(u,y) := -[a_1(u,y) \ldots a_\nu(u,y)]$ $\quad \mathbf{B}(u,y) := [b_{i,j}$

(8) can conveniently be written (to first order in $\mathbf{w}$) as

$$a_0(u,y) = \mathbf{a}(u,y)\boldsymbol{\theta} + \varepsilon \tag{10}$$

$$\varepsilon := \boldsymbol{\beta}^T \mathbf{B}(u,y)\, \mathbf{w} \tag{11}$$

where the noise term ε depends linearly on $\boldsymbol{\beta}$, a known functions of the θ_i's.

Remark 3.2. Under the reasonable assumption that w is a Gaussian process, (10), (11), is a conditionally Gaussian model, given the observations y, u. Hence assuming that the input-output data are given quantities (together with a suitable number of their derivatives) the parameter θ can be estimated by standard techniques. However, for reasons of space, below we shall discuss only a simple linear regression approach which subsumes parameter independent and known variance of the additive noise term. This method will in general provide a reasonable answer only in case of very low noise.

4. IDENTIFICATION BY LINEAR REGRESSION

Assume that input-output function measurements $\{u(t_k), y(t_k)\}$, are available for a suitably large set of time instants $\{t_1, ..., t_N\}$, and assume also that the derivatives appearing in the input-output relation $A_0(u,y)$, of the variables u, y can somehow be computed (or estimated) at the same instants of time. Then the exhaustive summary parameters $\theta_i = \theta_i(\mathbf{p})$ $\; i = 1, ..., \nu$ can be estimated by solving a linear regression problem. The regression equation can be written considering the linear relation (10) evaluated sequentially at the time instants $\{t_1, ..., t_N\}$. Let us denote for simplicity $a_i(u(t_k), y(t_k))$ by $a_i(t_k)$, $i = 1, \ldots \nu$; $k = 1, \ldots, N$. Then (10) implies that

$$\begin{bmatrix} a_0(t_1) \\ \vdots \\ a_0(t_N) \end{bmatrix} = \begin{bmatrix} -a_1(t_1) & \ldots & -a_\nu(t_1) \\ \vdots & \ldots & \vdots \\ -a_1(t_N) & \ldots & -a_\nu(t_N) \end{bmatrix} \begin{bmatrix} \theta_1 \\ \vdots \\ \theta_\nu \end{bmatrix} + \begin{bmatrix} \varepsilon_1 \\ \vdots \\ \varepsilon_N \end{bmatrix} \tag{12}$$

where the vector with components ε_k is an error term whose variance matrix is parameter dependent but in principle computable once the θ_i are given.

Naturally, a basic requirement of this identification scheme is that, for input functions which are "sufficiently exciting" and for a large enough data set, the parameters should be uniquely determinable from the input-output data measured from the dynamic system (1, 2). This means that the regression equation (12) should be uniquely solvable in the unknowns θ_i, $i = 1, ..., \nu$. In particular the regression matrix formed with the monomials $\{a_i(t_k)\}$ should be of *full column rank* ν. This solvability condition is a condition on the input of the system and on the sampling schedule used in the measurement. It is the natural analogue of the "persistence of excitation of the input" in linear system identification.

In order to recover the original parameters after the exhaustive summary is estimated, one needs to solve in $\mathbf{p}$ a system of ν algebraic equations of the form

$$\theta_i(\mathbf{p}) = \hat{\theta}_i \qquad i = 1, \ldots, \nu \qquad (13)$$

where the $\hat{\theta}_i$ are the estimates provided by the linear identification step. *Global input-output identifiability* of the system (1, 2) guarantees the injectivity of the map $\boldsymbol{\theta}$ from the p-dimensional parameter space $\mathcal{P}$ to its range (a subset of the ν-dimensional Euclidean space) and hence unique solvability of equation (13). Unfortunately however the range space of the map $\boldsymbol{\theta}$ is a thin set in $\mathbb{R}^{\nu}$ and because of the the noise in the estimates $\hat{\theta}_i$, the system (13) will in general be overdetermined (incompatible). Hence we are eventually confronted with solving a nonlinear estimation problem, but this time the problem is a static problem in the parameter space. One possible way to go is to use a (static) nonlinear least squares routine. Global iniectivity of $\boldsymbol{\theta}$ guarantees that there are no local minima and a unique solution exists.

Example 4.1. Consider the following system:

$$\begin{cases} \dot{x}_1 = -p_0 u - p_2 x_1 - p_3 x_2 \\ \dot{x}_2 = p_3 x_1 x_2 - p_1 x_1 \\ y = x_1 + w \end{cases} \qquad (14)$$

with parameter set $\mathcal{P} = \mathbb{R}^4_+ \setminus \{\mathbf{0}\}$. Using the standard ranking of the input, output and state variables defined by

$$u < \dot{u} < \ddot{u} < w < \dot{w} < \ddot{w} < y < \dot{y} < \ddot{y} < \\ < x_1 < x_2 < \dot{x}_1 < \dot{x}_2 \qquad (15)$$

we find the input-output differential equation

$$A_0 \equiv -\ddot{y} - \ddot{w} - p_0\dot{u} - p_2(\dot{y} + \dot{w}) + \\ p_3(\dot{y} + \dot{w})(y + w) + p_0 p_3 u(y + w) + \\ p_2 p_3 (y + w)^2 + p_1 p_3 (y + w) \qquad (16)$$

from which the following exhaustive summary is constructed

$$\begin{array}{ll} \theta_1(\mathbf{p}) = p_0 & \theta_2(\mathbf{p}) = p_2 \\ \theta_3(\mathbf{p}) = p_3 & \theta_4(\mathbf{p}) = p_0 p_3 \\ \theta_5(\mathbf{p}) = p_2 p_3 & \theta_6(\mathbf{p}) = p_1 p_3 \end{array} \qquad (17)$$

It is evident that all parameters p_0, p_1, p_2, p_3 can be recovered uniquely from the exhaustive summary. Thus the system is globally identifiable (from "generic" initial conditions).

Assume w and its derivatives are "small". Then the regression equation is derived by evaluating at the sampling instants the following equation

$$\ddot{y} = -\theta_1\dot{u} - \theta_2\dot{y} + \theta_3\dot{y}y + \theta_4 uy + \theta_5(y)^2 + \theta_6 y \\ -\ddot{w} + (-\theta_2 + \theta_3 y)\dot{w} + (\theta_3\dot{y} + \theta_4 u + 2\theta_5 y + \theta_6)w$$

which is manifestly linear in the new parameters.

5. CONCLUSIONS

In this paper it has been observed that, using methods based on differential algebra the estimation of the parameters of a nonlinear model can be reduced to the solution of a standard least squares problem. Difficulties can arise in evaluating various derivatives of the input-output functions required in setting up the regression equation. These difficulties may be addressed in various ways, e.g. by spline or exponential smoothing in biomedical applications, depending on the problem at hand.

6. REFERENCES

Audoly, Stefania, D'Angiò Leontina, Saccomani Mariapia and Cobelli Claudio (1998). Global identifiability of linear compartmental models. *IEEE Trans. Biomedical Eng.* **45**(3), 36–47.

Audoly, Stefania, D'Angiò Leontina, Saccomani Mariapia and Cobelli Claudio (2001). Global identifiability of nonlinear models of biological systems. *IEEE Trans. Biomedical Eng.* **48**(1), 55–65.

Carrà Ferro, Giuseppa (1989). Gröbner bases and differential algebra. In: *Lecture Notes in Computer Science, (Vol. 356)*. pp. 129–140.

Chappel, M.J. and Keith R. Godfrey (1992). Structural identifiability of the parameters of a nonlinear batch reactor model. *Math. Biosci* **108**, 245–251.

Diop, S. (1992). Differential algebraic decision methods and some applications to system theory. *Theoretical Comp. Sci.* **98**, 137–161.

Forsman, K. (1991). Constructive Commutative Algebra in Nonlinear Control Theory. PhD thesis. Linköping University, Sweden.

Glad, S.T. (1990). Differential algebraic modelling of nonlinear systems. In: *Realization and Modelling in System Theory, MTNS'89.* Vol. 1. Birkheuser. pp. 97–105.

Ljung, L. and S.T. Glad (1994). On global identifiability for arbitrary model parameterizations. *Automatica* **30**(2), 265–276.

Mishra, B. (1993). *Algorithmic Algebra.* Springer-Verlag texts and monographs in computer science. Springer. Berlin.

Ollivier, F. (1990). Le problème de l'identifiabilité structurelle globale: étude théorique, méthodes effectives et bornes de complexité. PhD thesis. École Polytéchnique.

Ritt, J.F. (1950). *Differential Algebra.* RI: American Mathematical Society. Providence.

Saccomani, Mariapia, Stefania Audoly and Leontina D'Angiò (2003). Parameter identifiability of nonlinear systems: the role of initial conditions. *Automatica* **39**, 619–632.

Saccomani, Mariapia, Stefania Audoly, Giuseppina Bellu, Leontina D'Angiò and Claudio Cobelli (1997). Global identifiability of nonlinear model parameters. In: *Proceedings of SYSID '97 11th IFAC Symp. System Identification, (vol. 3).* pp. 219–224.

Vajda, S., K. Godfrey and H. Rabitz (1989). Similarity transformation approach to identifiability analysis of nonlinear compartmental models. *Math. Biosci.* **93**, 217–248.

Walter, Eric and Y. Lecourtier (1982). Global approaches to identifiability testing for linear and nonlinear state space models. *Math. and Comput. in Simul.* **24**, 472–482.

IFAC Publications
www.elsevier.com/locate/ifac

A PENALTY FUNCTION APPROACH TO HIV/AIDS MODEL PARAMETER ESTIMATION

R. Filter [*,1] **X. Xia** [*]

[*] *Department of Electrical, Electronic and Computer Engineering, University of Pretoria, Pretoria 0002, South Africa*

Abstract: This paper proposes a procedure of parameter estimation for all parameters of the three dimensional HIV model. The least square based procedure uses standard optimization routines and aims to allow parameter extraction for individual patients. It is shown how additional information from outside a measurement dataset can be included in the estimation routine to increase the reliability and accuracy of parameter estimates. This procedure is also applied to a long-term dataset of the HIV/AIDS progression to find possible variations in parameters.

Keywords: Physical parameters, parameter estimation, parameter variation, bioengineering and medical systems

1. INTRODUCTION

A helpful tool to decide on dosages in treatment of HIV/AIDS would be a model that describes the disease and the influence of drugs on the virus. A basic model, which is also the model under consideration in this paper, is a three-dimensional model as described in (Nowak and May, 2000) and (Perelson and Nelson, 1999). This model has helped to reshape the perception of the disease (Ho *et al.*, 1995; Wei *et al.*, 1995; Fauci *et al.*, 1996). This was achieved by estimating key parameters, such as the half-lives of infected $CD4^+$ T cells and free virus. These published estimates are for a subset of the parameters only, and give an indication of parameter values that can be expected in general. In order to use this model as a tool for treatment decisions, it is necessary to determine all six parameters of the model for individual patients. Even though there are general observations that can be made from the model and its structure, it is only when the model is tailored to each patients individual parameters that clear benefits in the treatment strategy arise.

This paper presents a procedure that can be used to extract all six model parameters on a per-patient basis, even under less favorable conditions.

[1] Corresponding author. Phone: +27 (82) 442 3415 Fax: +27 (12) 362 5000 Email: s9801305@postino.up.ac.za

2. PROCEDURES

The three-dimensional model of HIV/AIDS, considered here, consists of three variables: the population sizes of uninfected cells (T), infected cells (T^*), and free virus particles (v). Free virus particles infect cells at a rate proportional to the product of their abundances, βvT. The rate constant, β, describes the efficacy of this process. Infected cells produce free virus particles at a rate proportional to their abundance, kT^*. Infected cells die at a rate δT^*, and free virus particles are removed from the system at a rate cv. By assuming a constant production rate, s and death rate dT for the uninfected cells, the three-dimensional model of virus dynamics is obtained (Nowak and May, 2000; Perelson and Nelson, 1999):

$$\begin{cases} \dot{T} = s - dT - \beta T v, \\ \dot{T^*} = \beta T v - \delta T^*, \\ \dot{v} = kT^* - cv. \end{cases} \tag{1}$$

Furthermore, for the purpose of estimating model parameters, it will be assumed that plasma viral load and $CD4^+$ T cell count are measured. That is, the measurement outputs are $y_1 = T$, and $y_2 = v$. This is in accordance with the current prevailing medical practice (Panel on Clinical Practices for Treatment of HIV Infection, 2001).

For ease of notation in the following sections $\chi = [\ s\ d\ \beta\ \delta\ c\ k\ ]^T$, is defined as the vector of model parameters and $\hat{\chi} = [\ \hat{s}\ \hat{d}\ \hat{\beta}\ \hat{\delta}\ \hat{c}\ \hat{k}\ ]^T$ as the estimate of these parameters. Also the state vector $\boldsymbol{x}(t)$ is defined at at time t as $\boldsymbol{x}(t) = [\ T(t)\ T^*(t)\ v(t)\ ]^T$, and the initial state vector $\boldsymbol{x}_0 = [\ T_0\ T_0^*\ v_0\ ]^T$.

As with the method in Xia (2002), parameter estimation in this paper is least square (LSQ) based, but with two important differences. Firstly, derivative estimation is only present when a nominal curve is generated by a numerical ordinary differential equation (ODE) solver. This estimation is not influenced by measurement noise. Secondly, the cost function is not limited to the LSQ distance, allowing it to be expanded to accommodate a diverse base of knowledge in order to increase the accuracy of parameter estimation.

A pre-existing implementation of the Nelder-Mead Simplex search method is used here as the optimization routine, to find a set of parameters that minimizes the cost function [2]. At each iteration of the search, the cost function is called by the optimization routine. A curve is generated with the current $\hat{\chi}$ to compute the square distance between data-points and the nominal curve. The cost is augmented with values pertaining to external information and returned to the main routine. When a pre-set tolerance is met by the optimization routine, it exits with the final parameter estimation. The basic steps of the cost function are as follows:

(1) The function receives a list of data points for the CD4$^+$ T cell and the virus count with their respective time points. Together with these, $\hat{\chi}$ is also passed to the function.
(2) When constraints are specified for χ, they are enforced at this point. (By default only positive values are allowed for the parameters.)
(3) The function uses $\hat{\chi}$ and solves the dynamic model of eq. (1). This is done within the framework of a pre-existing numerical ODE solver.
(4) The numerical solution is used to calculate the difference between each data point and its predicted value. The differences are squared and summed.
(5) Any additional penalty values are calculated and added to the total, which is then returned to the optimization function.

This method is not prone to the derivative estimation error, as is the LSQ method considered in (Xia, 2002). In the rest of this section, the cost function will be discussed in more detail.

[2] The estimation procedure is not dependent on the use of the Nelder-Mead search method; other search methods could be used. It is important that the chosen search method does not rely on a smooth cost function.

2.1 Pure LSQ with correction factor

Apart from the problem of derivative estimation, there is a second major drawback in the pure LSQ method. The equations that are fitted to the data, contain product terms of CD4$^+$ T cell and virus counts. This essentially forces the data vectors to be of equal length for proper estimation. If this is not the case, interpolation has to be employed, which degrades the results. Since the penalty function method does not require any product terms, there is no constraint on the length of CD4$^+$ T cell and virus data vectors. In fact none of the points of the two data vectors have to coincide in time. Thus, for a nominal set of parameters, $\hat{\chi}$, and initial conditions, $\hat{\boldsymbol{x}}_0$, a curve is generated with a numerical ODE solver to find $\hat{T}$ and $\hat{v}$. Together with N measurements of T and K measurements of v, at time $t_1, \ldots, t_N$, and $\tau_1, \ldots, \tau_K$ respectively, we define the basic cost function as,

$$J_w = \sum_{n=1}^{N} \frac{(\hat{T}(t_n) - T_n)^2}{N \operatorname{mean}(T_n)} + \sum_{k=1}^{K} \frac{(\hat{v}(\tau_k) - v_k)^2}{K \operatorname{mean}(v_k)}. \quad (2)$$

The differences in data points may not jeopardize the balance of the penalty function, thus both data vectors have to be weighted by their mean value and their length. For equal length data vectors, division by their length is not necessary.

2.1.1. Logarithmic distance From (CDC Working group, 2001) it is known that the tests used to determine viral load are log based. Even with the highest precision tests, a log variance of 0.6 can be expected in the measurements. For the *Roche® Amplicor HIV-1 Monitor™* test, the most commonly used test by the participating laboratories in (CDC Working group, 2001), a log difference of up to 2.2 was noted. For this case the cost function can be modified as follows,

$$J_l = \sum_{n=1}^{N} \frac{(\hat{T}(t_n) - T_n)^2}{\operatorname{mean}(T_n) N} + \sum_{k=1}^{K} \frac{(\log \hat{v}(\tau_k) - \log v_k)^2}{\operatorname{mean}(\log v_k) K}. \quad (3)$$

Thus, the logarithmic distance between virus data points is used in the least square calculation. In this example the CD4$^+$ T cell data term is still computed as a linear value. Similar changes can be made for this term if necessary.

2.1.2. Additional refinements It is often the case that a set of data on its own does not contain enough information to determine all parameters, but when the prevailing circumstances are known, this knowledge allows the extraction of key parameters. In these situations the custom penalty function is helpful, since outside knowledge of the dataset can be incorporated into the parameter estimation cycle.

The main points where refinements can be incorporated are at step 2 and 5 in the penalty function. Some of the common refinements that are used in this paper are described below.

Enforcing limits
This is usually done at step 2 in the penalty function, by checking the parameters against a predefined range and correcting any parameters that do not fall within the specifications. A second option would be to add these limits directly to the penalty value at step 5. This would allow for weighted penalties if any of the limits are violated.

Prior knowledge of parameters
When a parameter is known from another source (*e.g.*, experiment, assumption or literature), this knowledge is incorporated at step 2. At the same time the optimization routine has to be instructed not to search for the parameters that are already known.

Prevailing conditions
When prevailing conditions (*e.g.*, $c > \delta$) for a dataset are known from other sources, these conditions are usually added by means of an additional term in step 5. This term must be scaled to ensure its proper influence.

As an example, consider the experiment described in (Perelson and Nelson, 1999, pp. 16–19). In this experiment key assumptions were made in order to extract two of the six parameters. Firstly, each patient was assumed to be at steady state ("set-point" has been reached) before initiation of therapy. This is a prevailing condition for that experiment. Viral load data was available before the experiment, which indicated that the viral loads were in steady state. Secondly, Nowak and May (2000, p.32) state a prior knowledge that infected cells live longer than free virus. This information is reflected in J_r by adding two terms to the basic LSQ cost.

$$J_r = J + k_1 \max(\frac{d\hat{v}_s}{dt}, 0) + k_2 \max(\hat{\delta} - \hat{c}, 0), \quad (4)$$

where J is either J_w or J_l, and $\hat{v}_s$ is the vector of computed viral loads, truncated after a few days. In this case scaling constants k_1 and k_2 are chosen such that any violation of the prevailing conditions would result in a marked increase of the penalty function.

The first refinement term corresponds to the knowledge that the patient is in steady state before initiation of therapy. Thus, no positive derivative should be allowed initially, for the viral load. An intuitive way to see this is to note that therapy results in a decline of virions, thus, an increasing virion count could only be the result of fluctuations before therapy, which is not possible since the patient was in steady state at the start of therapy. The second refinement term corresponds to the statement that the average infected CD4$^+$ T cell lives longer than free virions.

3. RESULTS

3.1 Generated data

Before the described procedure is put to use on actual data, its efficacy with generated, well posed, data is checked. In order to do this, a dataset is generated by numerically solving (1). The underlying parameters for the test set are chosen to accommodate the range of parameters described in literature (Covert and Kirschner, 2000; Culshaw and Ruan, 2000; Jeffrey *et al.*, 2003; Jeffrey and Xia, 2002; Müller *et al.*, 2001; Nowak and May, 2000; Perelson and Nelson, 1999). Some parameter values vary notably among authors, as discussed by Müller *et al.* (2001). For this case, the test parameters are biased toward (Perelson and Nelson, 1999) and (Nowak and May, 2000).

3.1.1. Ideal data with a random component As a starting point, an ideal dataset is generated for $\chi = [\ 10\ 0.01\ 5 \times 10^{-6}\ 0.5\ 3.0\ 1000\ ]^T$, $\boldsymbol{x}_0 = [\ 1000\ 1\ 100\ ]^T$ and $t = 0,\ 1,\ 2, \ldots,\ 99,\ 100$.

Without random component, the estimation routine achieves near-perfect estimation of parameters. For the results in this section, a random component is added to the data in order to simulate inaccuracies in the measurements. The errors for the viral load are taken from a normal distribution with a log-variance of 0.6. This is based on measurement variances described by the CDC Working group (2001).

First an estimate of parameters is made without any assumptions added to the cost function. The result for an estimation of parameters with a normal log-variance of 0.6 in the virus and a normal variance of 50 in the CD4$^+$ T cell count is shown in figure 1, which corresponds to $\hat{\chi} = [\ 10.2\ 0.011\ 7.5 \times 10^{-6}\ 4.22\ 0.45\ 831.23]^T$. Note that $\hat{\delta} = 4.22$ and $\hat{c} = 0.45$ are not correctly estimated. Their values should be exchanged, since the original data has as basis $\delta = 0.5$ and $c = 3$. This difference is clearly seen in the number of infected cells, but not in the CD4$^+$ T cell or virus counts. In the language of Jacquez (1985) δ and c are not identifiable. Xia and Moog (2003) have shown that a transformed system with parameters s, λ, β, k, $\theta_1 = \delta c$, and $\theta_2 = \delta + c$ is completely identifiable. Since it is known that the transformed system is completely identifiable, additional knowledge of δ and c allows complete determination of the original system parameters. As described in section 2.1.2, the cost function has to be augmented with the knowledge that $\delta < c$ (Nowak and May, 2000) in order to correct their exchange.

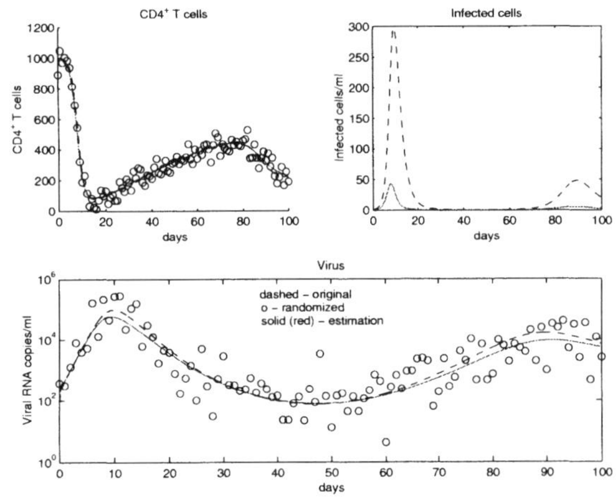

Fig. 1. Estimate of parameters on noisy data, without assumptions in cost function. Original parameter simulation – dashed; Estimated parameter simulation – solid (red).

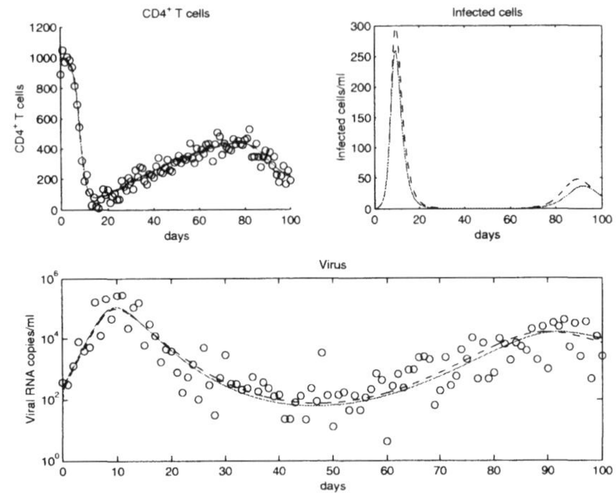

Fig. 2. Estimate of parameters on noisy data, assuming that $\delta < c$. Original parameter simulation – dashed; Estimated parameter simulation – solid (red).

After the cost function is updated, a second estimation is performed. Figure 2 shows the result, which is unchanged from the previous estimation, except that $\hat{\delta}$ and $\hat{c}$ are exchanged, i.e. $\hat{\chi} = [\ 10.7\ 0.015\ 4.5 \times 10^{-6}\ 0.58\ 2.05\ 896.49\ ]^T$. Note that the order of $\hat{\delta}$ and $\hat{c}$ is correct. The correct estimation is a direct result of the update in the cost function that $\delta < c$. No notable influence can be seen in the virus and CD4$^+$ T cell counts, but the difference in the number of infected cells is now absent. It is clear that the estimation procedure responds correctly when outside knowledge is added. The obvious question that follows now is, if the estimation is still reasonable with less data-points.

3.1.2. A dataset with reduced number of data points The theoretical limit for the minimum number of data points for an estimation of parameters was derived by Xia and Moog (2003). From the results with a pure LSQ method we know that

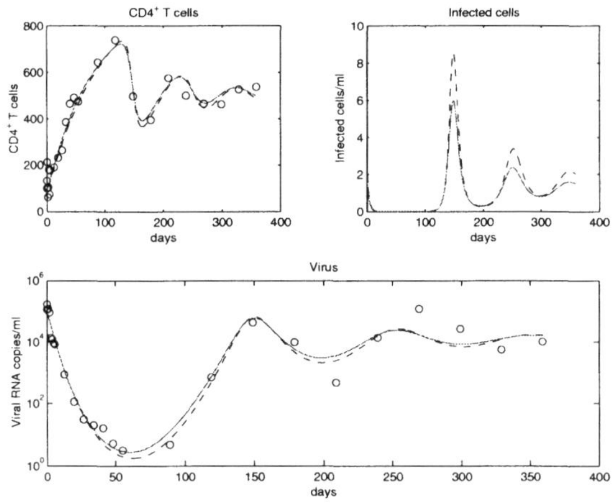

Fig. 3. Estimate for dataset with reduced number of points. Original parameter simulation – dashed; Estimated parameter simulation – solid (red).

the ideal case soon fails in reality when noise is added to the system, or when the dataset is not densely populated. It is not viable to take daily samples for patients living with HIV/AIDS, thus a compromise has to be made. This could be done by increasing the time between measurements as time progresses (Jeffrey and Xia, 2002; Jeffrey *et al.*, 2003). Before any time schedule for measurements can be finalized, its viability for model identification has to be determined. As an example, to show the applicability of the proposed procedures, a dataset is created with $\chi = [\ 10\ 0.01\ 8 \times 10^{-7}\ 3.2\ 0.5\ 4000\ ]^T$, $\boldsymbol{x}_0 = [\ 100\ 2.8\ 112500\ ]^T$ and the time points as in figure 3. As with the previous dataset, a random component, with normal log-variance of 0.6 in virus counts and normal variance of 50 in CD4$^+$ T cell counts, is added to both the CD4$^+$ T cell and virus counts. The influence of this random component can be seen when the markers are compared with the dashed curve, in figure 3.

The choice of parameters for this dataset is based on some informed assumptions about the scenario where measurements might be taken. The values for $\boldsymbol{x}_0$ are chosen by simulating for χ with nominal $\boldsymbol{x}_0$ and $\beta = 8 \times 10^{-6}$ and reading the steady state value from the graph. This steady state value now becomes the initial value to be used as $\boldsymbol{x}_0$ values for the test dataset. The choice of $\beta = 8 \times 10^{-7}$ corresponds to the assumption that the medication taken by a patient is 90% effective. Thus after the steady state is found with $\beta = 8 \times 10^{-6}$, it is lowered to 10% of its original value to simulate treatment. After a complete parameter estimation with this data, the results are plotted in figure 3. The estimate gives $\hat{\chi} = [\ 11.2\ 0.012\ 7.4 \times 10^{-7}\ 3.89\ 0.49951\ 5156\ ]^T$. Note that the assumption that the patient is in steady state is not added to the estimation procedure, since this assumption is only made to generate the dataset

Table 1. Comparison of results with a published experiment. The half-life for c is computed as $t_{c\frac{1}{2}} = \frac{\ln 2}{c}$ and similarly for δ.

Patient number	Method	Virus clearance $\hat{c}$ (day^{-1})	$t_{c\frac{1}{2}}$ (days)	Infected cell loss $\hat{\delta}$ (day^{-1})	$t_{\delta\frac{1}{2}}$ (days)
104	published	3.7	0.2	0.5	1.4
	penalty function	2.03	0.34	0.51	1.36
	$\boldsymbol{x}_0$ *fixed*	*3.40*	*0.20*	*0.50*	*1.38*
105	published	2.1	0.3	0.5	1.3
	penalty function	0.66	1.05	0.66	1.05
	$\boldsymbol{x}_0$ *fixed*	*2.22*	*0.31*	*0.45*	*1.54*
107	published	3.1	0.2	0.5	1.4
	penalty function	2.07	0.33	0.50	1.39
	$\boldsymbol{x}_0$ *fixed*	*3.07*	*0.23*	*0.49*	*1.41*

and might not be available in real life situations. If such information is available, it is possible to increase the accuracy of estimation.

3.2 Reproducing a published experiment

Some data sets do not contain enough information on their own to allow a complete parameter estimate. In this section, the parameter estimation for three patients in (Perelson and Nelson, 1999, pp. 16–19) is repeated using the custom penalty method. This allows the proposed procedure to be checked in a situation where many different assumptions have to be added to the identification routine. Results can be compared with the published data, revealing any discrepancies. In this way the proposed procedure can be seen in context. The method described in this paper is used to extract the same two parameters as in the experiment. Note that *all* the assumptions described in the experiment are included in this estimate by customizing the penalty function as in section 2.1.2. This is done to verify that the procedure described here correctly handles the addition of custom penalty terms.

After the basic assumptions are added, there is a distinct, and consistent difference in $\hat{c}$ between the published results and the estimation by custom penalty function. Since the estimation of c is dependent on the shoulder region of the virus count (Nowak and May, 2000), a dependence on $\boldsymbol{x}_0$ is to be expected. Initially no outside knowledge about the initial conditions, $\boldsymbol{x}_0$, was added to the estimation routine, since these values are not explicitly mentioned, *i.e.* the viral load and CD4$^+$ T cell steady states were not fixed. If the assumption is made that the steady state values were available at the start of the experiment, then one can estimate the parameters again with the same cost function, but with fixed $\boldsymbol{x}_0$. This results in the values given in table 1. The discrepancy between manually fixing and dynamically searching for $\boldsymbol{x}_0$ indicates that the small data window for this experiment does not allow clear information

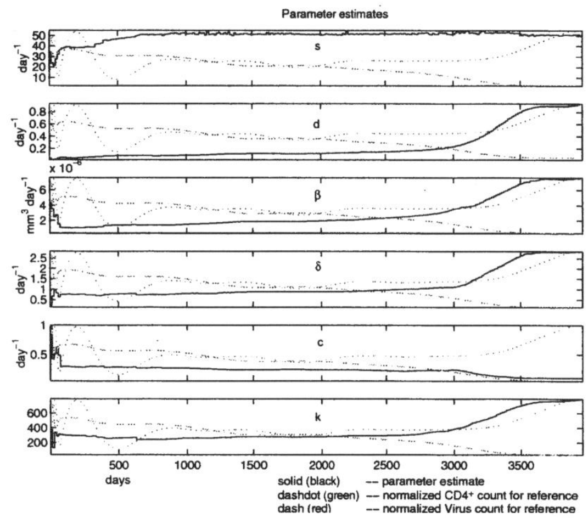

Fig. 4. Progressive estimate of parameters from HIV infection to the onset of AIDS. (s limited to 50).

to be found about the initial conditions. From the results it can be seen that the estimation of c, and, to a lesser degree, of δ, is dependent on information from outside the dataset, and that the addition of this information influences the results of the estimation routine.

3.3 Parameter variation from infection to the onset of AIDS

In order to get an indication of the variation of parameters over the course of the disease, a dataset from (Fauci *et al.*, 1996) is considered in this section. Parameter estimates are made from infection to the onset of AIDS. It should be noted that the original dataset is interpolated, before it is used to estimate parameters. The parameters are estimated for fixed data window, which is moved over the data at fixed increments. Because of the large time difference between measurements, combined with interpolation, results at each point might be erratic if a perfect fit of the curve is searched for. In a bid to overcome these problems, the estimation procedure is instructed to exit with a high tolerance for error. Figure 4 shows the result for a window size of 60 days, with estimates made 15 days apart. On each graph, the normalized CD4$^+$ T cell and virus loads are plotted to give a visual reference of the disease progression. These estimates are the initial steps taken to find possible parameter variations. No previous studies were found to give rigorous approaches in finding these variations. The approach was to increase the search tolerance of the optimization routine. Now each window's search is initialized with $\hat{\chi}$ from the previous window. This allows for continuity of parameters. Intuitively this corresponds to the assumption that parameters vary smoothly over time as the disease progresses. Thus the estimation procedure factors the results from the previous window into the next estimation, resulting in an estimate, $\hat{\chi}$ that is a modification of the previous window's $\hat{\chi}$. Each modification is just enough to ensure that the penalty value changes by less than the specified tolerance. In conclusion one can see that estimation stops as soon as a *possible* (as opposed to a locally *near optimal*) solution is found. An intuitive description of this procedure would be that the parameter estimate for each window is the nearest plausible neighbor to the estimate of the bordering windows.

Note, in figure 4, that the increase in $\hat{s}$ is halted by the assumption that $s \leq 50$. Also note that most of the estimations tend toward unreasonable values with the onset of AIDS at about 3000 days. The estimations $\hat{\delta}$, $\hat{c}$ and $\hat{k}$ stay fairly constant until the onset of AIDS, whereas the estimations for $\hat{s}$, $\hat{d}$ and $\hat{\beta}$ show a steady increase, even before the onset of AIDS.

4. CONCLUSION

By designing and implementing a custom penalty function for the extraction of model parameters, a diverse base of information from outside the basic dataset can be used to extract model parameters for the three dimensional HIV/AIDS model. From the results, it is clear that by customizing an LSQ based penalty function, one can identify parameters in situations where an orthodox pure LSQ method would fail. Together with the advantage of higher quality estimates in situations where the data is well-posed, the developed method can use outside information from the dataset and still provide reasonable estimates for data that cannot be used with a pure LSQ method. The presented method is a step forward in the effort to supply patients with an individual parameter estimation. The estimates made in literature were for at most two parameters per dataset, whereas the procedures described in this paper can estimate all six parameters. The conditions for successful estimation, as described by Xia and Moog (2003), still apply, if no external information is available to support the basic LSQ cost function.

The use of standard tables for data acquisition is a step forward in the right direction. One such table proposed for use in hospitals and clinics is considered. From the results it can be seen that the table would allow enough information to extract a good estimation for the parameters of the three-dimensional HIV/AIDS model.

Many of the experiments in literature have data windows that are too small to make definite conclusions about the parameters from the data alone. This is one of the main problems that are encountered when comparing results with those published previously. In order to compare results, external knowledge gained from articles, or the description of the experiment, is included in the cost function of the estimation routine.

Parameter variations during the course of the disease are still not well understood, and the basis of their variation needs to be found. One of the reasons for this poor understanding is the lack of high quality data that can be analyzed. From the results, one can see that the parameters may vary considerably over the course of HIV/AIDS. Even though seemingly tangible results are presented here, the reader is reminded that the underlying data points are separated by considerable lengths of time. It would be presumptuous to draw any definite conclusions from the results, other than the conclusions that parameters *do* vary over time and that the possibility should be considered that parameters vary in the manner of these results.

REFERENCES

CDC Working group (2001). "Guidelines for Laboratory Test Result Reporting of Human Immunodeficiency Virus Type 1 Ribonucleic Acid Determination". *http://www.cdc.gov/*.

Covert, Douglas J. and Denise Kirschner (2000). Revisiting early models of the host-pathogen interactions in HIV infection. *Cmnts. Theor. Biol.* **5**(6), 383–411.

Culshaw, Rebecca V. and Shingui Ruan (2000). A delay-differential equation model of HIV infection of CD4$^+$ T-cells. *Mathematical Biosciences* (165), 27–39.

Fauci, A. S., G. Pantaleo, S. Stanley and D. Weissman (1996). Immunopathogenic mechanisms of HIV infection. *Annals of Internal Medicine* **124**, 654–663.

Ho, David D., A. U. Neumann, Alan S. Perelson and *et al* (1995). Rapid turnover of plasma virions and CD4 lymphocytes in HIV-1 infection. *Nature* **273**, 123–126.

Jacquez, John A. (1985). *Compartmental Analysis in Biology and Medicine.* 2 ed.. Uni of Michigan Press.

Jeffrey, A. M. and X. Xia (2002). Estimating the viral load response time after HIV chemotherapy. Africon '02. George, South Africa,.

Jeffrey, A. M., X. Xia and I.K. Craig (2003). When to initiate HIV therapy: A control theoretic approach. *IEEE Transactions on Biomedical Engineering.* (to appear).

Müller, Viktor, Athanasius F. M. Marée and Rob J. De Boer (2001). Small variations in multiple parameters account for wide variations in HIV-1 set points: a novel modelling approach. *Proc. R. Soc. Lond. B. Biol. Sci.* **268**, 235–242.

Nowak, Martin A. and Robert M. May (2000). *Virus Dynamics: Mathematical Principles of Immunology and Virology.* Oxford University Press.

Panel on Clinical Practices for Treatment of HIV Infection (2001). "Guidelines for the Use of Antiretroviral Agents in HIV-Infected Adults and Adolescents". *http://www.hivatis.org.*

Perelson, Alan S and Patrick W. Nelson (1999). Mathematical analysis of HIV-1. *SIAM Review* **41**(1), 3–44.

Wei, X., S. K. Ghosh, M. E. Taylor, V. A. Johnson and *et al* (1995). Viral dynamics in HIV-1 infection. *Nature* **273**, 117–121.

Xia, X. (2002). Estimation of HIV/AIDS parameters. IFAC. 15TH IFAC World Congress, Barcelona, Spain.

Xia, X and C H Moog (2003). Identifiability of non-linear systems with applications to HIV/AIDS models. *IEEE Transactions on Automatic Control* **48**(2), 330–336.

IFAC
Publications
www.elsevier.com/locate/ifac

SENSITIVITY ANALYSIS AND PARAMETER IDENTIFICATION OF WASTEWATER TREATMENT SYSTEM BASED ON ACTIVATED SLUDGE MODELS

Jo Sato * Hiromitsu Ohmori **

* *Graduate School of Science and Technology, Keio University*
** *Department of System Design Engineering, Keio University*

Abstract: In this paper, the application of sensitivity analysis in order to determine the key parameters of wastewater treatment systems, along with identification method based on this result has been proposed. The effectiveness of the proposed method has been verified by using real influent data provided by a pilot plant. *Copyright © 2003 IFAC*

Keywords: Activated Sludge Models, Sensitivity Analysis, Parameter Identification

1. INTRODUCTION

Public awareness on environmental issues has enhanced dramatically as the influence of chemical substances on human body came to light. There are great hopes of quantum leap in the field of wastewater treatment, and at the same time, design, control, and the operational management of the system based on more rational thinking are becoming inevitable; regulations regarding discharges into the rivers are becoming intense.

Activated sludge process is a wastewater treatment process with the assistance of microorganisms. The inflow to the treatment plant consists of various substances, including those chemical substances which are assumed to cause environmental concerns. These hazardous substances must be removed from the final effluent, and this is accomplished by having the microorganisms react effectually during the treatment process.

One had to consider just organic carbon removal until the recent years. However, it is necessary to consider simultaneous removal of nitrogen and phosphorus from now on. In order to develop new wastewater treatment system that assures high quality, high reliability, and low operational cost, numerous proposals of precise mathematical models regarding wastewater quality and quantity has been made (Henze *et al.*, 2000).

Activated Sludge Model No.1–No.3 (ASM1 - ASM3), proposed by a task group from the International Water Association(IWA), are especially gaining popularity for its preciseness. These fine models may be excellent from physico-chemistry point of view, but that does not necessarily hold from the control engineering point of view.

In fact, the redundancy regarding number of parameters in the fine models has been pointed out (Julien *et al.*, 1998). Thus, the purpose of this paper is to select out the key parameters of the process by applying sensitivity analysis, and to propose an identification algorithm based on this result. It can be observed later that approximately the same state trajectory is achieved from the simplified model.

2. WASTEWATER TREATMENT PROCESS

2.1 *A2O Process*

Wastewater treatment process mainly consists of two processes: treatment of wastewater and treatment of sludge. From the treatment standpoint, the process can be divided to three parts: physical process, biological process, and chemical process. More specifically, for example, they are processes such as solid-liquid separation by precipitation, decomposition of organic matter by microorganisms, and chemical treatment such as coagulation sedimentation, respectively. This paper will focus mainly on biological process.

The so-called Anaerobic-Anoxic-Oxic (A2O) process is one of the wastewater treatment plant configurations that enables simultaneous removal of nitrogen and phosphorus (Yamanaka *et al.*, 2000). The plant consists of three biological reactors and one clarifier.

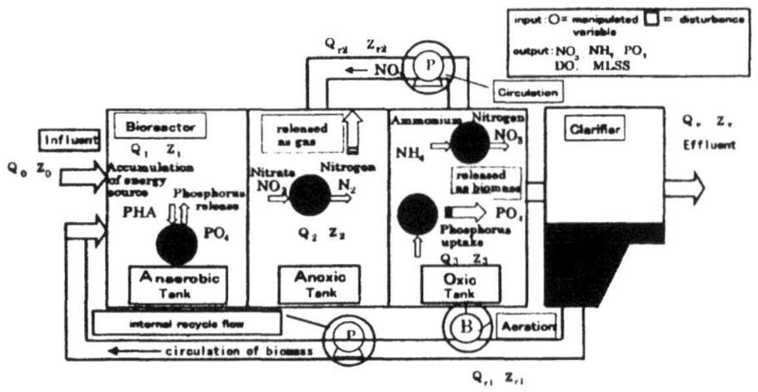

Fig. 1. A2O Process

The aim is to selectively vitalize through three reactors the nitrifying bacteria and denitrifying bacteria involved in nitrogen removal, and phosphorus accumulating organism involved in phosphorus removal process. Ultimately, nitrogen and phosphorus would be discharged from the system as gas and biomass, respectively. Conceptual diagram is shown in Fig.1.

2.2 *The Process Model*

Bioreactor model consists of three models. The first is the activated sludge model (ASM), which describes the dynamics of nitrogen / phosphorus removal process carried out by microorganisms such as heterotrophs, autotrophs, and phosphorus accumulating organisms. The second model represents the oxygen suppliance executed by aeration. The third model represents the hydraulics of the system.

Assuming that all tanks are well mixed, as usually is the case in ASM, the water quality vector of the kth tank $z_k[g/m^3] \in R^n$, can be formulated as the following based on the fine model.

$$\dot{z}_1 = (D_{r1}z_{r1} + D_0 z_0 - D_1 z_1) + r_1(z_1) \quad (1)$$
$$\dot{z}_2 = (D_{r2}z_{r2} + D_1 z_1 - D_2 z_2) + r_2(z_2) \quad (2)$$
$$\dot{z}_3 = (D_2 z_2 - D_{r2}z_{r2} - D_3 z_3) + r_3(z_3, q_B) \quad (3)$$

$D = \frac{Q}{V_k}[1/day]$ denotes the dilution rate where $V_k[m^3]$ is the volume of the kth tank, and $q_B[m^3/day]$ is the aeration rate. $r(z) \in R^n$ represents the reaction rate, and contains monod type nonlinearities. The dimension of $z_k \in R^n$ depends on which ASM to be used.

The control inputs are: aeration rate q_B, internal recycle flow rate $Q_{r1}[m^3/day]$, and circulation rate $Q_{r2}[m^3/day]$. The influent $z_0[g/m^3]$ is treated as disturbance. r_k is the reaction rate vector which takes

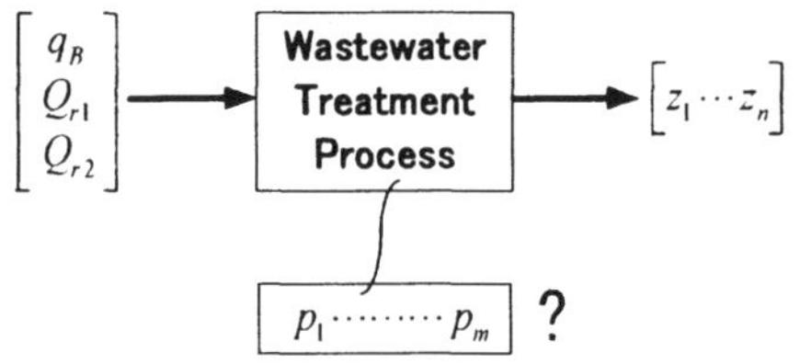

Fig. 2. Parameters to be Identified

into account the biological processes involved as described specifically in (Henze *et al.*, 2000). Aeration rate q_B is built into r_3, and adds an extra term $K_L a(DO_{sat} - DO)$ that represents oxygen transfer besides the ordinary reaction rate $r'(z_3)$ as shown in the equation below.

$$\frac{dDO_3(t)}{dt} = r'(z_3) + K_L a(DO_{sat} - DO) \quad (4)$$
$$K_L a = K q_B \quad (5)$$

Here, $DO_3(t)\,[g/m^3]$ represents the concentration of dissolved oxygen (DO) in the third reactor, $K_L a[1/day]$ the colligated oxygen transfer coefficient, a $K[1/m^3]$ a constant, and $r'(z_3)$ a component of z_3 which corresponds with DO concentration.

Considering identification, one must confront the problem with a fine model as shown in Fig.2.
There will be an extreme difficulty in estimating all m parameters. In fact, $m = 67$ where $n = 19$ for ASM2d. The situation is no better in ASM2 since $m = 65$ against $n = 19$. Even the simplest of the fine models, ASM1 is for sure to encounter severe identifiability problem since $m = 19$ where $n = 13$. The relationship between number of parameters m and dimension of the system n for all activated sludge models is listed in Table 1.

Table 1. Difficluty of Identification Among Different Models

	Number of Parameters m	Dimension of the System n (per bioreactor)
ASM1	19	13
ASM2	65	19
ASM2d	67	19
ASM3	49	13

Notice from Table 1 that dimension of the system n depends not only on the type of fine model to be used, but also on number of bioreactors present in the system. For example, focusing on ASM2d, there are hopes of identifying all 67 parameters from theoretical standpoint since there are 3 bioreactors plus a clarifier in A2O process, which adds up to $n = 76$. This, however, is unrealistic in terms of intense calculation and number of sensors that needs to be installed.

Thus, the approach would be to identify only the key parameters, those parameters that are essential to the process. Decreasing m, the number of parameters to be identified relieves the calculation intensity and installation of so many sensors, which both leads to cost efficiency.

3. SENSITIVITY ANALYSIS

Let the sensitivity function, z_p of the system

$$\dot{z}(p,t) = f(z(p,t), p, u) \tag{6}$$

be defined as

$$z_p = \frac{\partial z}{\partial p} \tag{7}$$

or in terms of components,

$$(z_p)_{ij} = \frac{\partial z_i}{\partial p_j} \tag{8}$$

In other words, the sensitivity function $(z_p)_{ij}$ is a mathematical description which expresses the influence of a small perturbation in parameter p_j on state variable z_i. Differentiating (6) by parameter vector $p \in R^m$ gives

$$\dot{z_p}(t) = \frac{\partial f}{\partial z} z_p(t) + \frac{\partial f(t)}{\partial p} \tag{9}$$

The sensitivity functions z_p can be obtained by integrating (6) and (9) simultaneously (Holmberg, 1982).

For the following $f_k (k = 1, 2, 3)$,

$$f_1 = (D_{r1} z_{r1} + D_0 z_0 - D_1 z_1) + r_1(z_1) \tag{10}$$
$$f_2 = (D_{r2} z_{r2} + D_1 z_1 - D_2 z_2) + r_2(z_2) \tag{11}$$
$$f_3 = (D_2 z_2 - D_{r2} z_{r2} - D_3 z_3) + r_3(z_3, q_B) \tag{12}$$

when

$$z_p = \begin{bmatrix} z_{1,p_1} & \cdots & z_{1,p_m} \\ \vdots & \ddots & \vdots \\ z_{n,p_1} & \cdots & z_{n,p_m} \end{bmatrix} \tag{13}$$

the right side terms of (9) can be written as

$$\frac{\partial f_k}{\partial z} = \begin{bmatrix} \frac{\partial f_{k,z_1}}{\partial z_1} & \cdots & \frac{\partial f_{k,z_1}}{\partial z_n} \\ \vdots & \ddots & \vdots \\ \frac{\partial f_{k,z_n}}{\partial z_1} & \cdots & \frac{\partial f_{k,z_n}}{\partial z_n} \end{bmatrix} \tag{14}$$

$$\frac{\partial f_k}{\partial p} = \begin{bmatrix} \frac{\partial f_{k,z_1}}{\partial p_1} & \cdots & \frac{\partial f_{k,z_1}}{\partial p_m} \\ \vdots & \ddots & \vdots \\ \frac{\partial f_{k,z_n}}{\partial p_1} & \cdots & \frac{\partial f_{k,z_n}}{\partial p_m} \end{bmatrix} \tag{15}$$

Fig.4 is an example of sensitivity curve for ammonium concentration utilizing ASM2d obtained by solving (9), when the process behaved as Fig.3.

Relative sensitivity values pz_p are plotted against time in order to enable comparisons of the influence of different parameters. The relative sensitivity $iNBM \cdot S_{nh4,iNBM}$, for instance, expresses how much the value of the state S_{nh4} will change if the value of the parameter $iNBM$ is perturbed by $iNBM$ units. The sensitivity functions express to which parameter a variable is most sensitive at a certain time intant.

Fig. 4 has $m = 61$ curves plotted. Here, $m = 61$ and $n = 18$ was chosen for calculation purpose.

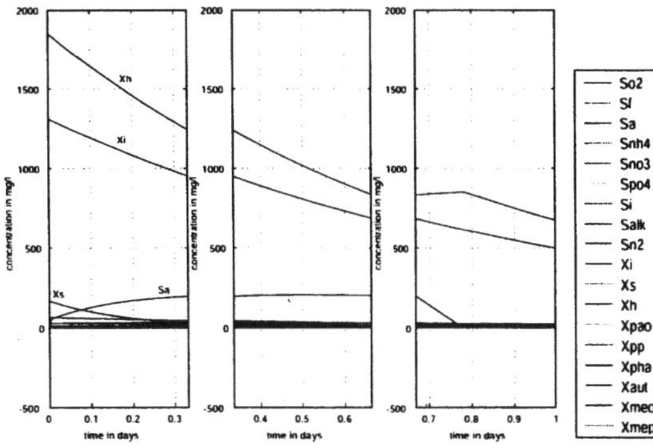

(a) anaerobic (b) anoxic (c) oxic

Fig. 3. Response of the A2O Process

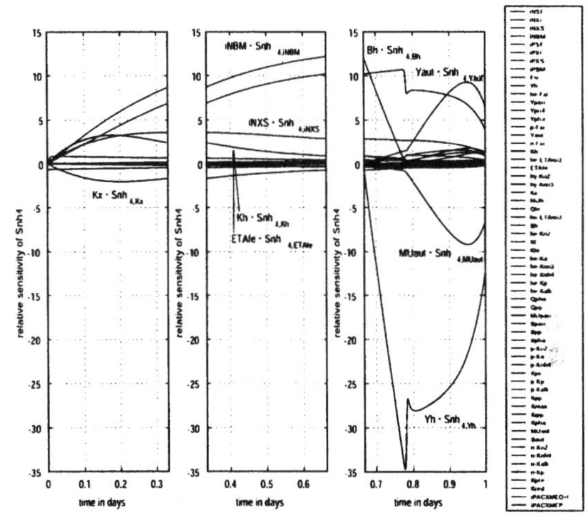

(a) anaerobic (b) anoxic (c) oxic

Fig. 4. Parameter Sensitivity of NH_4 Concentration

$n \neq 19$ and $m \neq 67$ as shown previously in Table 1 for ASM2d since the 19th component of the system just simply illustrates the dynamics of total substances involved in the system. 19th component does not provide any new information. Parameters associated with this 19th componenet has also been neglected, leaving a system of $m = 61$ and $n = 18$.

It can be observed that most curves stay near zero; effect that these parameters have on the concentration changes of S_{nh4} can be assumed to be very diminutive, thus not so sensitive. Here, assumption was made that if the maximum absolute value of specific sensitivity curve is less than 20% of the highest maximum absolute value of pz_p, the parameter has low sensitivity. In Fig. 4 (c) for example, parameter Yh is considered highly- sensitive. Parameter MU_{aut} is not as highly-sensitive, but still considered sensitive.

Conducting the same analysis for the rest of z_p components, Table 2 was obtained. Table 2 shows parameters that are likely to be key parameters of ASM2d. The check mark in Table 2 implies high-sensitiveness of a parameter to certain z component. Only the parameters that have been concluded to have high sensitivity for at least two water component, dominant in the process, are listed in the table for space-convenience.

Table 2. Candidates for Key Parameters

	Yh	he_Fxi	$Ypo4$	$Yaut$	Kh	η_{fe}	Kx	μ_h	Bh	he_Ko2	$Qpha$	Qpp	μ_{pao}	$Bpao$	μ_{aut}
$So2$	√			√					√						√
Sf					√	√	√		√						
Sa	√				√	√	√		√						
$Snh4$	√			√	√	√	√		√						√
$Sno3$	√			√					√						√
$Spo4$	√		√		√	√	√		√		√	√			
Si															
$Salk$															
$Sn2$	√			√				√	√	√					√
Xi	√	√							√						
Xs	√	√			√	√	√		√						
Xh	√							√	√						
$Xpao$	√							√	√	√			√	√	
Xpp	√		√						√		√	√	√	√	
$Xpha$	√		√						√		√	√	√		
$Xaut$	√			√				√	√	√					√
$Xmeoh$															
$Xmep$															

For low sensitive parameters, substituting literature values (Henze *et al.*, 2000) should be adequate; identification process is necessary only for highly-sensitive-parameters. Notice that the number of parameters to be identified reduced to 15 from the original 61. Validity of this choice of key parameters will be surveyed in Section 6. The same analysis can be conducted by any other activated sludge models.

4. IDENTIFICATION

Generally speaking, there lacks the information necessary to identify the parameters despite the aplenty data achieved in an activated sludge model-based-wastewater treatment process. For instance in case of online identification, the values of most z in (1)-(3) cannot be measured directly from existing sensors (Hiraoka and Tsumura, 2000). Focus will now move on to the identification of the wastewater treatment process by making the following assumptions:

Assumptions:

- The component values of the state variable in oxic tank z_3 are all available
- $z_{3,i} \neq 0, \forall t, (i = 5, 7, 9, 10, 12, 13, 14, 15, 16)$

Component values of oxic tank z_3 was chosen since this is where the most powerful input, the aeration, have direct effect on the system. Even from the control standpoint, the values should be measured in the last compartment of the bioreactors according to (Cadet and Béteau, 2002).

The following is the basic identification procedure. Recursive least squares method was used for simplicity.

(1) Select out the key parameters by conducting the sensitivity analysis
(2) Transform (3) so that they are linear-in-key parameters
(3) Divide both sides of the equation by $z_{3,i}$ and use the fact that $\dot{z}_{3,i}/z_{3,i} = \frac{d}{dt}log z_{3,i}$
(4) Use M-sequential signal for input signal
(5) Determine the parameter values by recursive least squares method for those that can be independently estimated
(6) Reduce the number of key parameters by imposing more strict selecting criteria if none can be independently estimated
(7) Determine the rest of the parameter values. Use the estimated values achieved in step 5 in case they have converged when necessary

5. CASE STUDY

Identification of μ_{pao}, b_{pao}

From the z_{15} component of (2),

$$\dot{X}_{pao} = \mu_{pao}(C_1 + C_2) X_{pao} - B_{pao}C_3X_{pao} + D_2X_{pao2} - (D_3 + D_{r2}) X_{pao} \quad (16)$$

$$C_1 = p\_So2Yes \cdot p\_Snh4Yes \cdot p\_SalkYes \cdot p\_Spo4Yes \cdot fxphaYes \quad (17)$$

$$C_2 = p\_\eta_{no3} \cdot p\_So2No \cdot p\_Sno3Yes \cdot fxphaYes \cdot p\_Snh4Yes \cdot p\_SalkYes \cdot p\_Spo4Yes \quad (18)$$

$$C_3 = p\_SalkYes \quad (19)$$

can be obtained when

$$p\_So2Yes = \frac{So2}{p\_Ko2 + So2} \quad (20)$$

$$p\_So2No = \frac{p\_Ko2}{p\_Ko2 + So2} \quad (21)$$

$$p\_Snh4Yes = \frac{Snh4}{p\_Knh4 + Snh4} \quad (22)$$

$$p\_Sno3Yes = \frac{Sno3}{p\_Kno3 + Sno3} \quad (23)$$

$$p\_Spo4Yes = \frac{Spo4}{p\_Kp + Spo4} \quad (24)$$

$$p\_SalkYes = \frac{Salk}{p\_Kalk + Salk} \quad (25)$$

$$fxphaYes = \frac{Xpha/Xpao}{Kpha + Xpha/Xpao} \quad (26)$$

Since this is already linear-in-the-parameters, rewriting the equation to matrix form yields

$$\left[\dot{X}_{pao} - D_2X_{pao2} + (D_3 + D_{r2}) X_{pao}\right] =$$

$$\left[\, (C_1 + C_2)\, X_{pao},\ -C_3 X_{pao} \,\right] \begin{bmatrix} \mu_{pao} \\ b_{pao} \end{bmatrix} \tag{27}$$

In order to alleviate identification result from the adverse effect of $\dot{X}_H$, the following is introduced by assuming that $X_H \neq 0$:

$$\left[\frac{d}{dt} log X_{pao} - D_2 \frac{X_{pao2}}{X_{pao}} + (D3 + D_{r2}) \right] =$$

$$\left[\, C_1 + C_2,\ -C_3 \,\right] \begin{bmatrix} \mu_{pao} \\ b_{pao} \end{bmatrix} \tag{28}$$

By letting

$$\begin{cases} y_N = \dfrac{d}{dt} log X_{pao,N} - D_2 \dfrac{X_{pao2,N}}{X_{pao,N}} + (D3 + D_{r2}) \\ \boldsymbol{z}_N^T = \left[\, C_{1,N} + C_{2,N},\ -C_{3,N} \,\right] \\ \hat{\boldsymbol{\theta}}_N = \begin{bmatrix} \hat{\boldsymbol{\theta}}_{1,N} \\ \hat{\boldsymbol{\theta}}_{2,N} \end{bmatrix} = \begin{bmatrix} \hat{\mu}_{pao,N} \\ \hat{b}_{pao,N} \end{bmatrix} \end{cases} \tag{29}$$

then (28) can be rewritten as

$$y_N = \boldsymbol{z}_N^T \hat{\boldsymbol{\theta}}_N \tag{30}$$

Applying the recursive least squares algorithm,

$$\begin{cases} \hat{\boldsymbol{\theta}}_N = \hat{\boldsymbol{\theta}}_{N-1} + \boldsymbol{P}_N \boldsymbol{z}_N \left(y_N - \boldsymbol{z}_N^T \hat{\boldsymbol{\theta}}_{N-1} \right) \\ \boldsymbol{P}_N = \boldsymbol{P}_{N-1} - \dfrac{\boldsymbol{P}_{N-1} \boldsymbol{z}_N \boldsymbol{z}_N^T \boldsymbol{P}_{N-1}}{1 + \boldsymbol{z}_N^T \boldsymbol{P}_{N-1} \boldsymbol{z}_N} \end{cases} \tag{31}$$

gives the value of $\hat{\boldsymbol{\theta}}$(Nakamizo, 1988), consequently, $\hat{\mu}_{pao}$ and $\hat{b}_{pao}$. The subscript N in equation (29) denotes the Nth concentration measurement of each wastewater components at time $t = N$, and $\hat{\boldsymbol{\theta}}_N$ is the estimated value $\hat{\boldsymbol{\theta}}$ at $t = N$.

6. SIMULATION

Since there exists no conventional method for identification of an activated sludge model, evaluatuation of the proposed method was carried out by comparing the output error under the following condition. To confirm the effectiveness of the proposed method, a 10% modeling error has been inflicted deliberately; key parameter values have been increased by 10%.

The proposed method will try to estimate new values for the key parameters that would compensate for the changes among low-sensitive parameters. The result, hereafter, will be referred to as *estimated parameters*. Substituting literature values, hereafter referred as the *correct parameters*, for insensitive parameters, the proposed identification algorithm was executed under disturbance of real influent data obtained from a pilot plant as shown in Fig. 5.

In this way, small output error of the generated state trajectories by both correct and estimated parameters would mean success in both the sensitivity analysis and the identification process.

Fig. 6 shows the output trajectory of the system generated by using (a) correct parameters, and (b) estimated parameters. Fig. 7 illustrates the error between the two models.

Despite the inflicted modeling error, the output error remained relatively small; the effectiveness of the proposed identification method based on sensitivity analysis can be confirmed. Approximately the same state trajectory has been duplicated by the estimated parameters.

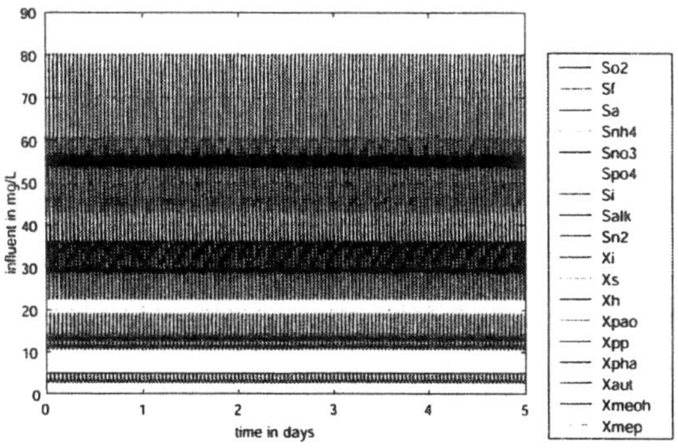

Fig. 5. Influent Data

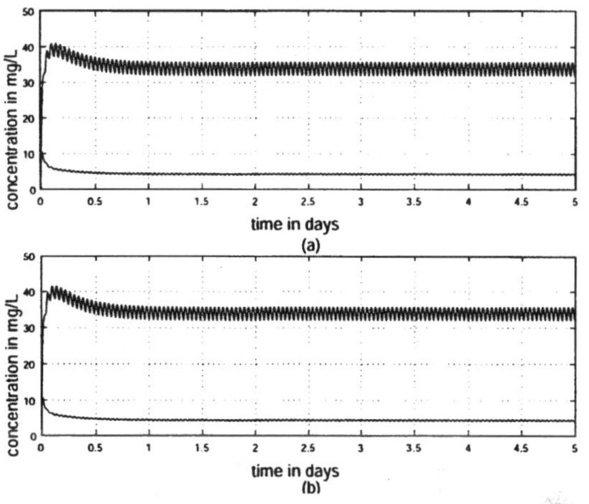

Fig. 6. Output Response (a) $\boldsymbol{Z}_{correct}$ (b) $\boldsymbol{Z}_{estimated}$

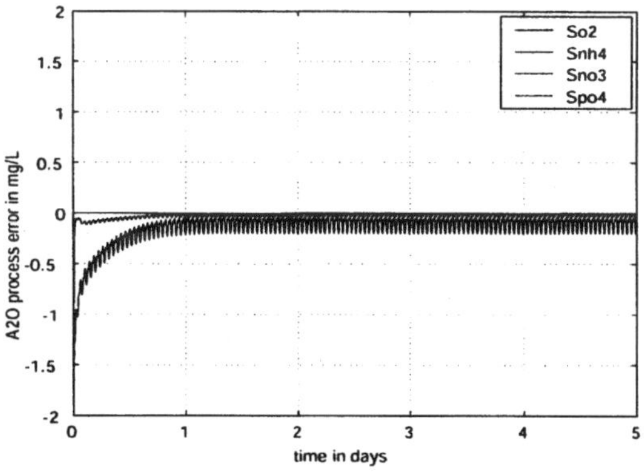

Fig. 7. Output Error

7. CONCLUSIONS

This paper demonstrated the powerfulness of sensitivity analysis in determining the key parameters for the activated sludge models. The effectiveness of the proposed method in achieving nearly the same state trajectory has been qualitatively confirmed.

8. ACKNOWLEDGEMENT

This work is supported by the Grant-in-Aid for Scientific Research (B) (Project Number:13555120) of Japan Society for the Promotion of Science, and also by the following people who have provided special research materials, intellectual stimulation, and other forms of assistance: Hiroshi Ito, Kyushu Institute of Technology; Koichi Hidaka, Tokyo Metropolitan College of Aeronautical Engineering; Akihiro Nagaiwa, Toshiba; and Osamu Yamanaka, Toshiba.

9. REFERENCES

Cadet, C. and J.F. Béteau (2002). New control strategy design to improve effluent quality in wastewater treatment plants.

Henze, M., W. Gujer, T. Mino and M. Loosdrecht (2000). *Activated Sludge Models ASM1, ASM2, ASM2d and ASM3.* IWA Publishing. London.

Hiraoka, M. and K. Tsumura (2000). *Analysis and Control of Environmental System Based on AR Model.* EICA.

Holmberg, A. (1982). On the practical identifiability of microbial growth models incorporating michaelis-menten type nonlinearities. *Mathematical Biosciences* **62**, 23–43.

Julien, S., J.P. Babary and P. Lessard (1998). Theoretical and practical identifiability of a reduced order model in an activated sludge process doing nitrification and denitrification. *Wat. Sci. Tech* **37**, 309–316.

Nakamizo, T. (1988). *Signal Analysis and System Identification.* Corona Publishing.

Yamanaka, O., A. Nagaiwa, M. Tsutsumi, Y. Nagamori and Y. Hatsushika (2000). A model predictive control of biological nitrogen and phosphorus removal in activated sludge process. *Proc. of 8th SICE Symposium on Control Technology* pp. 459–462.

IFAC
Publications
www.elsevier.com/locate/ifac

A METHODOLOGY FOR NONLINEAR SYSTEM IDENTIFICATION USING VOLTERRA SERIES. APPLICATION TO AN ANAEROBIC DIGESTOR.

G. Bibes * P. Coirault * R. Ouvrard * and J.P. Steyer **

* *Laboratoire d'Automatique et d'Informatique Industrielle - ESIP*
40 avenue du Recteur Pineau 86022 Poitiers FRANCE
e-mail: coirault@esip.univ-poitiers.fr
** *Laboratoire de Biotechnologie de l'Environnement-INRA*
Avenue des étangs 11100 Narbonne FRANCE
e-mail: steyer@ensam.inra.fr

Abstract: This paper introduces a Volterra series model for a continuous nonlinear system. Volterra kernels are expanded on generalized orthonormal functions around a continuous component. An application to validate this model is presented: a biological wastewater treatment process. The Volterra model presented is validated in simulation and with experimental results. *Copyright © 2003 IFAC*

Keywords: Nonlinear system, Volterra series, generalized orthonormal bases, wastewater biological treatment.

1. INTRODUCTION

Most of systems that exist in nature are nonlinear. It is generally difficult to develop models for processes whose dynamics exhibit significant nonlinearities. Some of the reasons for this difficulty lie in the large variety of nonlinearities and processes. One category of process, considered here, is industrial wastewater treatment. In this paper, we are going to consider anaerobic digestion, one of the oldest and most efficient wastewater treatment processes.

Different ways to model nonlinearities already exists such as NARMAX models (Chen, 1989), multi-model approaches (Boukhris *et al.*, 1999) or Volterra series (Doyle *et al.*, 2002). Many methods to estimate Volterra kernels are presented in literature (Billings and Fakhouri, 1980), (Nishiyama and Kashiwagi, 1997), (Hassouna *et al.*, 2001). In this paper, we propose to model the considered nonlinear biological system with Volterra series expanded on generalized orthonormal bases.

The first four parts of this paper are about Volterra series and their expansion. The third part explains how generalized orthonormal bases are used to expand Volterra kernels around a continuous component. At last, in the fifth part, we apply this proposed model to the anaerobic digestor.

2. FUNCTIONAL SERIES MODELS

One way to represent a large number of nonlinear systems is a functional series expansion, which provides an explicit relationship between the system input and output. In the single input-single output (SISO) case, a functional H may be written:

$$y(t) = H[u(t)] \qquad (1)$$

where $y(t) \in R$ is the system output and $u(t) \in R$ is the system input. Let us note that we assume that the system is initially at rest, meaning $y(0) = 0$ and $u(t) = 0$ for $t \leq 0$. For real systems, such a condition is common. By definition, the functional H may be expanded into an infinite sum of functionals:

$$y(t)=\sum_{k=1}^{\infty}H_k[u(t)] \quad (2)$$

In order to find a model that fit at best to a real nonlinear system, it may not be necessary to use an infinity of term in the sum (2). If nonlinearities of the system are too strong, the number of terms $H_k[u(t)]$ required for a close approximation becomes very large. In the other side, if nonlinearities are sufficiently smooth, the transient response of the system is determined by the first terms of the series.

This is why the functional H is truncated to its first N terms in order to correctly represent the real system. The value of the term N depends strongly of the kind of system nonlinearities. The model output $\hat{y}(t)$ is given by:

$$\hat{y}(t)=\sum_{k=1}^{N}H_k[u(t)] \quad (3)$$

3. VOLTERRA SERIES MODELS

One important class of nonlinear SISO models is the class of finite Volterra series models. Volterra series may be considered as a subset of functional series. Finite Volterra series models are defined by equation:

$$\hat{y}(t)=H\left[u(t)\right]= \sum_{k=1}^{N}\underbrace{\int_0^t\cdots\int_0^t}_{k} h_k\left(\tau_1,\ldots,\tau_k\right)\prod_{i=1}^{k}u(t-\tau_i)d\tau_i \quad (4)$$

where $u(t)$ is the input, $\hat{y}(t)$ is the model output. The term $h_k\left(t_1,\ldots,t_k\right)$ is the kth order Volterra kernel, a locally bounded and continuous piecewise function. We assume here that the real system is causal, i.e. the output at any given time does not depend on future inputs $u(t+\alpha)$, $\alpha>0$.

Let us note that if we take $N=1$, we recover the linear convolution model. Volterra series models (4) is a very general method to represent nonlinear systems.

However, even for simple systems, the form of Volterra kernels is rather complicated. In absence of model knowledge, a current approach is to expand the kernels on multidimensional orthonormal bases, which can be easily builded.

Proposition 1 *Let $(\nu_1,\ldots,\nu_n)$ be a multiindex of length n such that $\nu_1\geq\ldots\geq\nu_n$, and $\{\phi_\nu\}$ a monodimensional orthonormal basis of $\mathbf{L}^2(\mathbf{T})$, where $\mathbf{T}=(\mathbf{0},\infty)$ is the set of times, and $\mathbf{L}^2$ the Hardy space. A n-dimensional orthonormal basis of $\mathbf{L}^2(\mathbf{T}^n)$ is obtained from $\{\phi_\nu\}$ by forming the products:*

$$\{\mathbf{W}_{\nu_1,\ldots,\nu_n}\}=\{\phi_{\nu_1}\times\ldots\times\phi_{\nu_n}\} \quad (5)$$

Proof Consider the scalar product defined on the linear space $\mathbf{L}^2(\mathbf{T})$ by:

$$\left\langle\mathbf{W}_{\nu_1,\ldots,\nu_n},\mathbf{W}_{\mu_1,\ldots,\mu_n}\right\rangle= \left(\frac{1}{2\pi}\right)^n\int_{-\infty}^{+\infty}\cdots\int_{-\infty}^{+\infty}\mathbf{W}_{\nu_1,\ldots,\nu_n}(j\omega_1,\ldots,j\omega_n) \mathbf{W}_{\mu_1,\ldots,\mu_n}(-j\omega_1,\ldots,-j\omega_n)d\omega_1\ldots d\omega_n \quad (6)$$

From (5) and (6), we obtain:

$$\begin{aligned}&\left\langle\mathbf{W}_{\nu_1,\ldots,\nu_n},\mathbf{W}_{\mu_1,\ldots,\mu_n}\right\rangle\\ &=\left\langle\phi_{\nu_1}\times\ldots\times\phi_{\nu_n},\phi_{\mu_1}\times\ldots\times\phi_{\mu_n}\right\rangle\\ &=\left\langle\phi_{\nu_1},\phi_{\mu_1}\right\rangle\times\ldots\times\left\langle\phi_{\nu_n},\phi_{\mu_n}\right\rangle\\ &=\prod_{i=1}^{n}\delta_{\nu_i\mu_i}\end{aligned} \quad (7)$$

where $\delta_{\nu_i\mu_i}$ is the Kronecker symbol. ■

Proposition 2 *Consider the finite Volterra series model (4). It admits an expansion on the set of orthonormal bases $\{\{\mathbf{W}^{(1)}_{\nu_1}\},\{\mathbf{W}^{(2)}_{\nu_1\nu_2}\},\ldots,$ $\{\mathbf{W}^{(N)}_{\nu_1\ldots\nu_N}\}\}$ given by*

$$\hat{y}(t)=\sum_{k=1}^{N}\sum_{\nu_1=1}^{M_k}\cdots\sum_{\nu_k=1}^{\nu_{k-1}}\xi_{\nu_1\ldots\nu_k}\prod_{j=1}^{k}\mathcal{I}^{(k)}_{\nu_j}(t) \quad (8)$$

where $\mathcal{I}^{(k)}_{\nu_j}(t)=\int_0^t\phi^{(k)}_{\nu_j}(\tau)u(t-\tau)d\tau$.

Proof Consider the kth order Volterra kernel $h_k(t_1,\ldots,t_k)$. It admits a series expansion on the orthonormal basis $\left\{\mathbf{W}^{(k)}_{\nu_1\ldots\nu_k}\right\}$:

$$\begin{aligned}h_k(t_1,\ldots,t_k)=&\sum_{\nu_1=1}^{\infty}\cdots\sum_{\nu_k=1}^{\nu_{k-1}-1}\sigma_{\nu_1\ldots\nu_k}\\ &(\phi^{(k)}_{\nu_1}(t_1)\ldots\phi^{(k)}_{\nu_k}(t_k)+\ldots+\phi^{(k)}_{\nu_1}(t_k)\ldots\phi^{(k)}_{\nu_k}(t_1))\\ &+\sum_{\nu_1=1}^{\infty}\sigma_{\nu_1\ldots\nu_1}\phi^{(k)}_{\nu_1}(t_1)\ldots\phi^{(k)}_{\nu_k}(t_k)\end{aligned} \quad (9)$$

The series (9) converges on $\mathbf{L}^2(\mathbf{T}^n)$. It can be truncated to its first terms:

$$\hat{h}_k(t_1, ..., t_k) = \sum_{\nu_1=1}^{M_k} ... \sum_{\nu_k=1}^{\nu_{k-1}-1} \sigma_{\nu_1...\nu_k}$$
$$(\phi_{\nu_1}^{(k)}(t_1)...\phi_{\nu_k}^{(k)}(t_k) + ... + \phi_{\nu_1}^{(k)}(t_k)...\phi_{\nu_k}^{(k)}(t_1))$$
$$+ \sum_{\nu_1=1}^{M_k} \sigma_{\nu_1...\nu_1} \phi_{\nu_1}^{(k)}(t_1)...\phi_{\nu_k}^{(k)}(t_k) \tag{10}$$

By posing

$$\mathcal{I}_{\nu_i}^{(k)}(t) = \int_0^t \phi_{\nu_i}^{(k)}(\tau_i) u(t - \tau_i) d\tau_i \tag{11}$$

and

$$\xi_{\nu_1...\nu_k} = \begin{cases} \sigma_{\nu_1...\nu_k} \text{ if } \nu_p = \nu_q \ \forall \ p, q \\ \dfrac{1}{k!}\sigma_{\nu_1...\nu_k} \text{ else} \end{cases} \tag{12}$$

the Volterra model is given by (8).■

Kernel's structure may be very complex. Each kernel model's dimension depends on the selected functions $\phi_{\nu_i}^{(j)}$. Let us consider the set $\{\phi_n\}$ of generalized orthonormal functions described in (Ouvrard *et al.*, 1999) and (Akçay and Ninness, 1999):

$$\phi_n(s) = \frac{\sqrt{2\,\mathrm{Re}\{p_n\}}}{s + p_n} \prod_{i=1}^{n-1} \frac{p_i - s}{p_i + s} \tag{13}$$

This set forms a basis of $\mathbf{L}^2(\mathbf{T})$. This particular choice of ϕ_n allows to fit to the system dynamics by choosing the poles p_i suitably.

4. RESTRICTED COMPLEXITY MODEL: CASE OF SISO SYSTEM

4.1 *Proposed model definition*

Let us consider the following SISO nonlinear system:

$$\begin{cases} \dot{x}(t) = f(x(t)) + g(x(t))u(t) \\ y(t) = h(x(t)) \end{cases} \tag{14}$$

where $x(t) \in \mathbf{R}^n$ is the state vector, $u(t) \in \mathbf{R}$ the input and $y(t) \in \mathbf{R}$ the output. $f(x)$, $g(x)$ and $g(x)$ are sufficiently smooth nonlinear functions of $x(t)$. The input $u(t)$ is supposed to be of the form:

$$u(t) = u_0 + \Delta u(t) \tag{15}$$

where u_0 is constant (i.e. does not depend on time). Thus are separated the static part u_0 of the input from its dynamic part $\Delta u(t)$. From Volterra series model (4), one obtain:

$$\hat{y}(t) = \sum_{k=1}^{N} \underbrace{\int_0^t \cdots \int_0^t}_{k} h_k(\tau_1, \ldots, \tau_k) \prod_{i=1}^{k} [u_0 + \Delta u(t - \tau_i)] d\tau_i \tag{16}$$

The expansion of Volterra kernels $h_k(\tau_1, \ldots, \tau_k)$ on generalized orthonormal bases and some other mathematical manipulations lead to the following proposition, which gives an input/output model of system (14).

Proposition 3 *An input/output model of the system (14), expanded on Volterra series using generalized orthonormal bases, is given by*

$$\hat{y}(t) = \sum_{k=1}^{N} \left(\sum_{\nu_1=1}^{M_k} \cdots \sum_{\nu_k=1}^{\nu_{k-1}} \xi_{\nu_1...\nu_k} \left[u_0^k \prod_{i=1}^{k} \phi_{\nu_i}(0) + \sum_{i=1}^{k} u_0^{k-i} \mathcal{J}_{k,i}(\Delta u(t)) \right] \right) \tag{17}$$

where $\mathcal{J}_{k,i}(\Delta u(t)) =$

$\sum^{C_k^{k-i}} \left(\prod_{j=1}^{k-i} \phi_{\nu_j}(0)\right) \left(\prod_{n=1}^{i} \mathcal{I}_{\nu_n}(\Delta u(t))\right)$, *with,* $\forall\ i \in [1, k-i]$, $\forall\ n \in [1, i]$:

$$\begin{cases} \{\nu_j\} \cup \{\nu_n\} = \{1, ..., k+i\} \\ \{\nu_j\} \cap \{\nu_n\} = \emptyset \end{cases} \tag{18}$$

Proof Let us expand the Volterra kernel $h_k(\tau_1, \ldots, \tau_k)$ in the expression (16) on a generalized orthonormal basis. One obtain:

$$\hat{y}(t) = \sum_{k=1}^{N} \left(\sum_{\nu_1=1}^{M_k} \cdots \sum_{\nu_k=1}^{\nu_{k-1}} \xi_{\nu_1...\nu_k} \left[\prod_{i=1}^{k} \int_0^t \phi_{\nu_i}^{(k)}(\tau_i)[u_0 + \Delta u(t - \tau_i)] d\tau_i \right] \right) \tag{19}$$

with $\prod_{i=1}^{k} \int_0^t \phi_{\nu_i}^{(k)}(\tau_i)[u_0 + \Delta u(t - \tau_i)] d\tau_i) =$

$\prod_{i=1}^{k} \left(u_0 \int_0^t \phi_{\nu_i}^{(k)}(\tau_i) d\tau_i + \int_0^t \phi_{\nu_i}^{(k)}(\tau_i) \right.$

$\left. \Delta u(t - \tau_i) d\tau_i \right)$

Let us pose

$$\begin{cases} \int_0^t \phi_{\nu_i}^{(k)}(\tau_i)d\tau_i = \phi_{\nu_i}(0) = a_i \\ \int_0^t \phi_{\nu_i}^{(k)}(\tau_i)\Delta u(t-\tau_i)d\tau_i = \mathcal{I}_{\nu_i}^{(k)}(\Delta u(t)) \\ \qquad\qquad = b_i \end{cases} \quad (20)$$

A generalization of the binomial theorem gives:

$$\prod_{i=1}^{k}(u_0 a_i + b_i) = u_0^k \prod_{i=1}^{k} a_i + \sum_{i=1}^{k} u_0^{k-i} \left[\sum^{c_k^{k-i}} \left(\prod_{j=1}^{k-i} a_j \right) \left(\prod_{n=1}^{i} b_n \right) \right] \quad (21)$$

with $\{j\} \cup \{n\} = \{1, ..., k+i\}$ and $\{j\} \cap \{n\} = \emptyset$.

Replacing a_i and b_j in the equation (21) by their expression (20) leads to the result (17) of the proposition 3. ■

4.2 *Discussion*

Compared to a more "classic" Volterra model, this proposed Volterra model is expanded around the input $u(t)$ continuous component (or mean value) u_0. Such an expansion allows to separate the static part from the dynamic part. Consequently, it is only necessary to identify the dynamic part of the model and, in order to do so, the number of Volterra kernels will be less important than in the case of a "classic" Volterra model.

Such a result may also be developed for some multi input-single output (MISO) nonlinear systems. Each input $u_i(t)$ is written $u_i(t) = u_{0i} + \Delta u_i(t)$ and, instead of the term $g(x(t))u(t)$ in (14), the sum $\sum_{i=1}^{m} g_i(x(t))u_i(t)$ for m inputs must be considered.

4.3 *Identification method*

Having obtained a Volterra series model (17), the second stage is concerned with the estimation of expanded coefficients $\xi_{\nu_1 \dots \nu_k}$.

Define $\Xi = \{\xi_{\nu_1 \dots \nu_k}\}$ the set of parameters.

Since the model (17) is linear in the parameters $\xi_{\nu_1 \dots \nu_k}$, it turns out that the simplest way to address this problem is via the minimization of the least square criterion $\mathcal{V}(\Xi) = \|y(t) - \hat{y}(t, \Xi)\|_2^2$.

5. APPLICATION: ANAEROBIC DIGESTION

Wastewater treatment is currently an interesting and widely open research area. Different kinds of processes may be applied to industrial effluents in order to clean them from some useless (or dangerous) organic matter. There exists some biological methods such as anaerobic digestion, a process in which organic matter is degraded into a mixture of methane (CH_4) and carbon dioxide (CO_2). The proposed model is going to be applied to such a biological process, which can be considered as a nonlinear SISO phenomenon.

5.1 *Process description and modelling*

The pilot plant considered is a continuous anaerobic digester in which raw industrial wine distillery vinasses are mixed with water into a 1 m^3 upflow fixed bed reactor. For more details about the plant and its online instrumentation, see a complete description in (Steyer *et al.*, 2002). In order to correctly model the process, it is assumed that the anaerobic digestion can be described by a two-stage process including two groups of bacterial populations, each group having homogeneous characteristics. During the first step, the acidogenic bacteria (X_1) consume the organic substrate (S_1) and produce CO_2 and volatile fatty acids (S_2) following the acidogenesis reaction (with reaction rate $r_1 = \mu_1 X_1$) :

$$k_1 S_1 \rightarrow X_1 + k_2 S_2 + k_4 CO_2 \quad (22)$$

Through the second step, the methanogenic bacteria (X_2) degrades the volatile fatty acids, thus producing CO_2 and CH_4. The methanization reaction (rate $r_2 = \mu_2 X_2$) is:

$$k_3 S_2 \rightarrow X_2 + k_5 CO_2 + k_6 CH_4 \quad (23)$$

μ_1 and μ_2 represent the specific growth rate of acidogenesis and methanization respectively. All terms y_i ($i = 1, ..., 6$) are yield coefficients.

A model representing this two-stage process was developped, identified and experimentally validated (Bernard *et al.*, 2001):

$$\begin{cases} \dot{X}_1 = (\mu_1 - \alpha D)X_1 \\ \dot{X}_2 = (\mu_2 - \alpha D)X_2 \\ \dot{Z} = D(Z^i - Z) \\ \dot{S}_1 = D(S_1^i - S_1) - k_1\mu_1 X_1 \\ \dot{S}_2 = D(S_2^i - S_2) + k_2\mu_1 X_1 - k_3\mu_2 X_2 \\ \dot{C}_{TI} = D(C_{TI}^i - C_{TI}) + k_7(k_8 P_{CO_2} + Z \\ \qquad - C_{TI} - S_2) + k_4\mu_1 X_1 + k_5\mu_2 X_2 \end{cases} \quad (24)$$

where C_{TI} and Z are respectively the concentrations of total inorganic carbon and strong ions. The upper index "i" indicates the inputs of the process. The term α is a proportionality parameter experimentally determined. The variable D is the dilution rate at the process input. P_{CO_2} is the CO_2 partial pressure. Terms k_7 and k_8 are respectively liquid/gas transfer rate and Henry's constant.

Strong nonlinearities in the system behavior are due to the growth rates expressions:

$$\begin{cases} \mu_1 = \mu_{\max 1} \dfrac{S_1}{K_{S_1} + S_1} \\ \mu_2 = \mu_0 \dfrac{S_2}{K_{S_2} + S_2 + \left(\frac{S_2}{K_{I_2}}\right)^2} \end{cases} \tag{25}$$

where $\mu_{\max 1}$ and μ_0 are parameters associated with biomass growth rates, K_{S_1} and K_{S_2} are saturation parameters and K_{I_2} is an inhibition constant. Precise definitions and values of these parameters are given in (Bernard *et al.*, 2001).

5.2 *Simulation results*

Simulation input/output data obtained with the model (24) are used in order to find the model expanded on Volterra series using generalized orthonormal bases. Simulations runs were performed with a Simulink model of the two-stage process.

All inputs except dilution rate $D(t)$ remain constant. The considered input $D(t)$ in figure 1 is a pseudo-random sequence variable in amplitude.

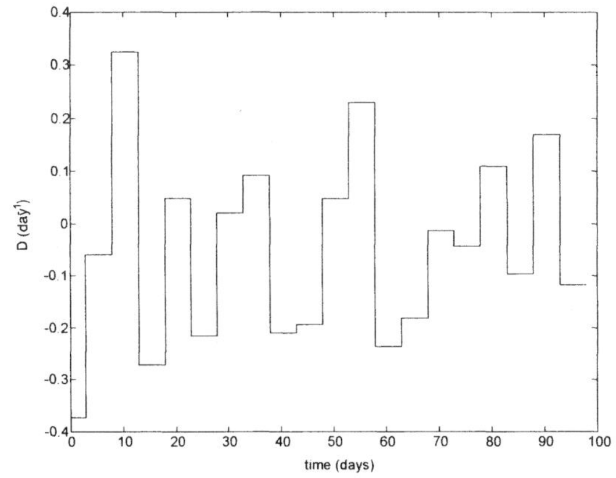

Figure 1: simulator input (dilution rate $D(t)$)

The sample period is 30 minutes. The process total time is 100 days. Such an input variation and amplitude is necessary to sensitize all Volterra kernels.

The considered output obtained in figure 2 with the Simulink model is $COD(t) = S_1(t) + S_2(t) * 64/1000$, the Chemical Oxygen Demand (in g/l), which represent the organic substrate concentration.

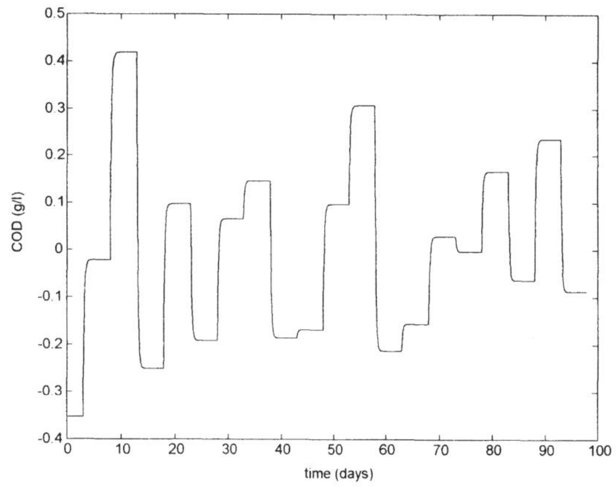

Figure 2: simulated output $COD(t)$

We are trying to find a Volterra model of the transfert between $D(t)$ and $COD(t)$. Such a transfert is strongly nonlinear.

In order to apply the model described in the first part of this paper, we are going to consider only the dynamic part of $D(t)$ and $COD(t)$. Let us pose:

$$\begin{cases} D(t) = D_0 + \Delta D(t) \quad \text{with } D_0 = 0.5658 \\ COD(t) = COD_0 + \Delta COD(t) \text{ with } COD_0 = 1.2589 \end{cases} \tag{26}$$

The global model comprises the first 4 Volterra kernels. Their order of decomposition on the orthonormal basis defined by (13) are respectively $M_1 = 4$, $M_2 = 2$, $M_3 = 2$ and $M_4 = 1$. The poles of generalized functions are $p_1 = -1$, $p_2 = -2$, $p_3 = -3$ and $p_4 = -4$. The total number of parameter is 9. This Volterra model is therefore relatively parsimonious. The figure 3 is a comparison between simulated output and estimated output of the Volterra model.

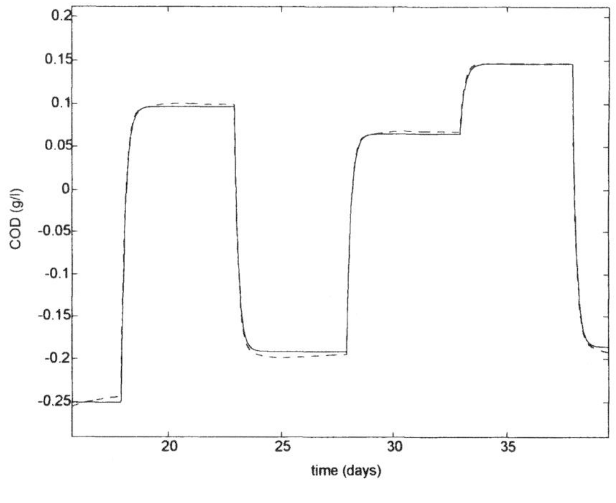

Figure 3: comparison between simulated output (clear line) and Volterra model output (dark dashed line)

5.3 *Validation*

Experimental datas used to validate the Volterra model expanded on generalized orthonormal functions come from on-line measurements obtained in September 2001. For more details, see (Steyer and Bernard, 2003). Experimental measurement and output model are compared in figure 4.

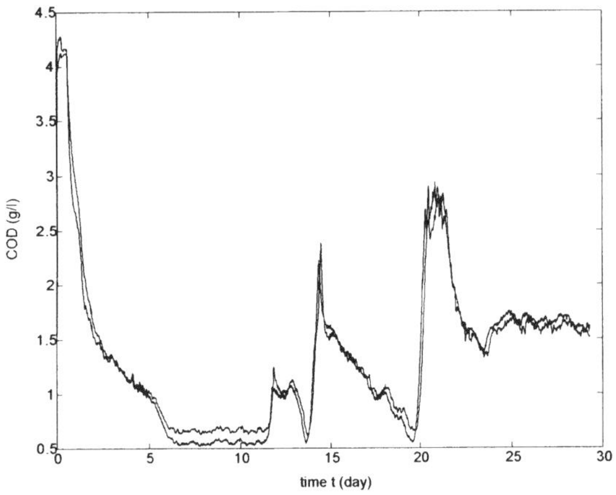

Figure 4: comparison between experimental data (dark line) and output model (clear line)

Over 30 days of experimentation, the output model follows quite well the data. Let us notice that the form of the experimental input is completely different from the one used to estimate the parameters in the previous section.

Therefore, the Volterra model proposed is experimentally validated and may be used to explain the system behavior in the case of a relatively large variety of inputs.

6. CONCLUSION

In this paper, Volterra series kernels are expanded on generalized orthonormal bases. The obtained model allows to represent the behavior of smooth nonlinear systems with a Volterra series truncated to its first terms.

The objective was to model the behavior of an anaerobic digestor into which an organic substrate is decomposed and converted into biogas, microbial biomass and residual organic matter. A single input-single output Volterra model is obtained, compared to another process model (a set of differential equations), and then expermimentally validated. The Volterra model is relatively parsimonious and able to predict the digestor behavior over a large period of time, without having to develop a model directly from experimental data.

TheVolterra model considered here is obtained and validated for the system output corresponding to COD. The next step would be to develop other Volterra models to represent the transfer function between the input and other possible system outputs such as volatile fatty acids or biogas flowrate.

7. REFERENCES

Akçay, H. and B. Ninness (1999). Orthonormal basis functions for modelling continuous-time systems. *Signal Processing* **77**(3), 261–274.

Bernard, O., Z. Hadj-Sadok, D. Dochain, A. Genovesi and J-Ph. Steyer (2001). Dynamical model development and parameter identification for an anaerobic wastewater treatment process. *Biotechnology and Bioengineering* **75**(4), 424–438.

Billings, S.A. and S.Y. Fakhouri (1980). Identification of nonlinear systems using correlation analysis and pseudorandom inputs. *International Journal of Control* **11**, 261–279.

Boukhris, A., G. Mourot and J. Ragot (1999). Non-linear dynamic system identification: A multi-model approach.. *International Journal of Control* **72**(7/8), 591–604.

Chen, S. (1989). Representation of nonlinear systems: The NARMAX model. *Int. J. Control* **49**(3), 1013–1032.

Doyle, F.J., R.K. Pearson and B.A. Ogunnaike (2002). *Identification and Control Using Volterra Models*. Springer.

Hassouna, S., P. Coirault and R. Ouvrard (2001). Continuous nonlinear system identification using volterra series expansion. *Proceedings of the ACC* pp. 4874–4879.

Nishiyama, E. and H. Kashiwagi (1997). Identification of volterra kernels of nonlinear systems by use of m-sequence correlation. *Proceedings of 11th IFAC Symposium SYSID'97, Japan* **2**, 739–743.

Ouvrard, R., P. Coirault, T. Poinot and J.C. Trigeassou (1999). Continuous time system identification with orthonormal functions: A new approach.. *Proceedings of the ECC'99.*

Steyer, J-Ph., J-C. Bouvier, T. Conte, P. Gras and P. Sousbie (2002). Evaluation of a four year experience with a fully instrumented anaerobic digestion process. *Water Science and Technology* **45**(4-5), 495–502.

Steyer, J.P. and O. Bernard (2003). An example of the benefits obtained from the long term use of mathematical models in wastewater biological treatment.. *Proceedings of the 4th Mathmod, Vienna* pp. –.

IFAC Publications
www.elsevier.com/locate/ifac

SOME RELATIONS OF SENSITIVITY FUNCTIONS IN BIO-REACTOR MODELS [1]

Julio A. Romero Pérez *
José Luis Navarro Herrero **

* *jromero@isa.upv.es*
** *joseluis@isa.upv.es*
Department of Systems Engineering and Control
Technical University of Valencia, Spain

Abstract
Sensitivity functions are a basic tool in the parametric identification of models. They can be used to obtain the gradiente for the optimization algorithms in the parameters estimation and for calculating the Fisher information matrix in experiment design. In this work it is shown that for a class of bio-reactor models, a fixed relation between some sensitivity functions exists. This property allows the reduction of the computation load in numeric calculation of the sensitivity trajectories. Simplified equations for gradient and Fisher information matrix calculation using this property are given. *Copyright © 2003 IFAC*

Keywords: bio-reactor, unstructured models, sensitivity functions

1. INTRODUCTION

Non structured models are commonly used in the design of bio-reactor controllers, (González *et al.*, 1998), (Zeng and Dahhou, 1995), (Roux and Dahhou, 1994), (Proell *et al.*, 1991), (VanImpe and Bastin, 1995),(Banga *et al.*, 1998), (Gauthier *et al.*, 1992), (McLain *et al.*, 1999). This kind of model approaches the external dynamics of this systems in a simple way, because it is assumed that the physiologic state of all micro-organisms in the vessel is the same, so the biomass growth in the bio-reactor is homogeneous. The behavior of the micro-organisms is considered like a black box, describing the characteristics of the biomass growth as functions of the concentration of several substances, as well as the capacity of the biomass to transform some substances in others.

The non-structured model is a non-linear state space representation with a set of parameters that must be calculated for the biomass at the growth conditions. These parameters can be obtained from micro-organism property tables, or they can be obtained with a parametric identification procedure. In this case, it is necessary to design some proper experiments for the particular model structure and perform a non-linear parametric identification, usually based on the minimization of a cost function using the gradient respect to the parameters.

Several techniques have been developed for the calculation of the gradient. A numeric approximation to the gradient can be obtained using finite differences method, where the gradient is obtained causing small perturbations in the model parameters. Other more efficient and more exact methods are based on the *sensitivity functions*,(Walter and Pronzato, 1997).

[1] Supported by the Science and Technology Ministry of Spain (CICYT DPI2002-0525)

Sensitivity functions are also used in experiment design, where it is looked for a sufficiently exciting condition for the process. The experiment design is solved as an optimization problem. Different approaches for experiment design have been proposed in the literature, (Walter and Pronzato, 1990). These approaches are based on the structure of Fisher information matrix, and the sensitivity functions are the key piece for the calculation of this matrix.

If a linear model is used, the sensitivity functions can be analytically obtained from the state space equations. For the non lineal in it's parameters model,

$$\dot{x} = f(x, \theta, u), \; x(0) = x_0, \quad y = C(\theta)x(t) \tag{1}$$

the sensitivity trajectories with respect to the parameters vector θ are given by:

$$\frac{\partial y}{\partial \theta_i} = \frac{\partial C}{\partial \theta_i} x + C \frac{\partial x}{\partial \theta_i} \tag{2}$$

where the values of $\frac{\partial x}{\partial \theta_i}$ are calculated by numeric integration of the extended state space model:

$$\begin{aligned}
&\frac{d}{dt} x = f(x, \theta, u) \\
&\frac{d}{dt}\frac{\partial x}{\partial \theta_1} = \frac{\partial f(x,\theta,u)}{\partial x^T}\frac{\partial x}{\partial \theta_1} + \frac{\partial f(x,\theta,u)}{\partial \theta_1} \\
&\quad\vdots \\
&\frac{d}{dt}\frac{\partial x}{\partial \theta_p} = \frac{\partial f(x,\theta,u)}{\partial x^T}\frac{\partial x}{\partial \theta_p} + \frac{\partial f(x,\theta,u)}{\partial \theta_p}
\end{aligned}$$

In this work an interesting property of the sensitivity functions for non structured models of bio-reactors is presented. The application of the property reduce the computational load during the model identification. The structure of the article is as follows: first, the structural property of the bio-reactor that establish a relationship between some sensitivity values is presented. Next, simplified equations for gradient and Fisher information matrix calculation using this property are given. The advantages are shown by means of two case studies. Finally the derived conclusions of this work are pointed out.

2. MAIN RESULT

The non-structured and non-segregate model for bio-reactor can be wrote as follow (Bastin and Dochain, 1990):

$$\dot{\xi} = K\Phi(\xi, \theta) - D\xi + D\xi_i \tag{3}$$

where:

- ξ: component concentrations in the bio-reactor (state variables).
- $K_{N\times M}$: yield coefficient matrix. The values k_{ij} is the yield coefficient of the component i in reaction j. If the component is reactant, then the coefficient is negative and if the component is a product the coefficient is positive. N is the number of the components and M is the number of reactions.
- $\Phi(\xi, \theta)_{M\times 1}$: Vector of reaction rates equations. The component Φ_j is the rate of the reaction j. θ is the vector of the parameters for this equations.
- ξ_i: inlet component concentration.
- $D = F/V$, dilution rate, F inlet flow rate, V volume.

For simplicity, it is supposed that it is an anaerobic growth or that the oxygen is in excess, so the gaseous components have been removed from (1).

The sensitivity functions of states with respect to parameter θ_i are given by:

$$\frac{d}{dt}\frac{\partial \xi}{\partial \theta_i} = K\left(\frac{\partial \Phi(\xi,\theta)}{\partial \xi}\frac{\partial \xi}{\partial \theta_i} + \frac{\partial \Phi(\xi,\theta)}{\partial \theta_i}\right) - D\frac{\partial \xi}{\partial \theta_i} \tag{4}$$

If $rank(K_{N\times M}) = M$ (the number of reactions), then (4) can be decoupled in two subsystems:

$$\begin{aligned}
\frac{d}{dt}\frac{\partial \xi_a}{\partial \theta_i} &= K_a\left(\frac{\partial \Phi(\xi,\theta)}{\partial \xi}\frac{\partial \xi}{\partial \theta_i} + \frac{\partial \Phi(\xi,\theta)}{\partial \theta_i}\right) - D\frac{\partial \xi_a}{\partial \theta_i} \\
\frac{d}{dt}\frac{\partial \xi_b}{\partial \theta_i} &= K_b\left(\frac{\partial \Phi(\xi,\theta)}{\partial \xi}\frac{\partial \xi}{\partial \theta_i} + \frac{\partial \Phi(\xi,\theta)}{\partial \theta_i}\right) - D\frac{\partial \xi_b}{\partial \theta_i}
\end{aligned} \tag{5}$$

where $K_{a_{(M\times M)}}$ is a full rank submatrix of K; K_b is the remaining coefficients of K, and the vector variables are induced by the decomposition. Then, a transformation can be defined by

$$Z(t) = A_0\frac{\partial \xi_a}{\partial \theta_i}(t) + \frac{\partial \xi_b}{\partial \theta_i}(t) \tag{6}$$

where $A_{0_{(N-M)\times M}}$ is the solution of the matrix equation:

$$A_0 K_a + K_b = 0 \tag{7}$$

and the sensitivity model (5) is equivalent to

$$\begin{aligned}
\frac{d}{dt}\frac{\partial \xi_a}{\partial \theta_i} &= K_a\left(\frac{\partial \Phi(\xi,\theta)}{\partial \xi}\frac{\partial \xi}{\partial \theta_i} + \frac{\partial \Phi(\xi,\theta)}{\partial \theta_i}\right) - D\frac{\partial \xi_a}{\partial \theta_i} \\
\dot{Z} &= -DZ
\end{aligned} \tag{8}$$

Starting from this property, an algebraic relationship among the vectors $\frac{\partial \xi_a}{\partial \theta_i}(t)$ and $\frac{\partial \xi_b}{\partial \theta_i}(t)$ can be obtained. The base of that algebraic equation is the transformation (6).

From equation (7), $A_0 = -K_b K_a^{-1}$, and replacing A_0 in equation (6)

$$Z(t) = -K_b K_a^{-1} \frac{\partial \xi_a}{\partial \theta_i}(t) + \frac{\partial \xi_b}{\partial \theta_i}(t) \qquad \forall t \qquad (9)$$

It is considered that the initial conditions $\xi(0)$ of the experiment are independent of parameters θ. This is a natural assumption for system identification, since the initial condition of the model are supposed to be known constants. Then

$$\frac{\partial \xi_a}{\partial \theta_i}(0) = \frac{\partial \xi_b}{\partial \theta_i}(0) = 0 \Rightarrow Z(0) = 0 \qquad (10)$$

Because Z=0 is an equilibrium point of equation (8), from the previous condition:

$$Z(t) = 0 \quad \forall t \qquad (11)$$

Therefore, from equation (9)

$$\frac{\partial \xi_b}{\partial \theta_i}(t) = K_b K_a^{-1} \frac{\partial \xi_a}{\partial \theta_i}(t) = -A_0 \frac{\partial \xi_a}{\partial \theta_i}(t) \; \forall t \quad (12)$$

The sensitivity functions

$$\frac{\partial \xi}{\partial \theta}(t) = \begin{bmatrix} \frac{\partial \xi}{\partial \theta_1}(t) & \frac{\partial \xi}{\partial \theta_2}(t) & \cdots & \frac{\partial \xi}{\partial \theta_p}(t) \end{bmatrix} \qquad (13)$$

where

$$\frac{\partial \xi}{\partial \theta_j}(t) = \begin{bmatrix} \frac{\partial \xi_a}{\partial \theta_j}(t) \\ \frac{\partial \xi_b}{\partial \theta_j}(t) \end{bmatrix} = \begin{bmatrix} \frac{\partial \xi_a}{\partial \theta_j}(t) \\ -A_0 \frac{\partial \xi_a}{\partial \theta_j}(t) \end{bmatrix} \qquad (14)$$

could be wrote as

$$\frac{\partial \xi}{\partial \theta}(t) = \begin{bmatrix} I_M \\ -A_0 \end{bmatrix} \frac{\partial \xi_a}{\partial \theta}(t) \qquad (15)$$

where

$$\frac{\partial \xi_a}{\partial \theta}(t) = \begin{bmatrix} \frac{\partial \xi_a}{\partial \theta_1}(t) & \frac{\partial \xi_a}{\partial \theta_2}(t) & \cdots & \frac{\partial \xi_a}{\partial \theta_p}(t) \end{bmatrix} \qquad (16)$$

The equation (15) means that it's posible to obtain the values for all sensitivity function from the sensitivity values of ξ_a.

Replacing (15) in the first equation of (5),

$$\frac{d}{dt}\frac{\partial \xi_a}{\partial \theta_i} = \left(K_a \frac{\partial \Phi(\xi, \theta)}{\partial \xi} \begin{bmatrix} I \\ -A_0 \end{bmatrix} - DI \right) \frac{\partial \xi_a}{\partial \theta_i} + K_a \frac{\partial \Phi(\xi, \theta)}{\partial \theta_i} \qquad (17)$$

The sensitivity functions of ξ_a with respect to the parameter θ_i can be calculated by numeric integration of the previous equation.

3. APPLICATIONS

Bio-reactor modelling requires the identification of the matrix K coefficients and the reaction rates equations parameters θ. Taking into account that in many cases provided experiments with full state measurements are available, the identification of K and θ can be completely decouple, (Bastin *et al.*, 1992),(Chen *et al.*, 1990); this means that K is known, therefore A_0 can be calculated. Then, the behavior of all the sensitivity functions can be obtained by means of the sensitivity functions of a state subset ξ_a, combining the equations 15 and 17. The total number of differential sensibility equations for a model with N states (number of substances) and n_p parameters is $N \times n_p$ equations. The values of the sensitivity function are obtained by the numeric integration of these equations. Applying the previous property, the number of equations to be solved decreases to $M \times n_p$, where M is the number of reactions, $M < N$.

Using this property the equations for Fisher information matrix and the gradient of the cost computation can be modified.

3.1 Calculation of the Fisher information matrix

The Fisher information matrix is defined as follow:

$$\Im = \frac{1}{n_m} \sum_{i=1}^{n_m} \left(\frac{\partial y}{\partial \theta}(t_i) \right)^T W^{-1} \left(\frac{\partial y}{\partial \theta}(t_i) \right) \qquad (18)$$

where n_m is the number of measures; W is the measurement noise covariance matrix, and $\frac{\partial y}{\partial \theta}(t_i)$ is the vector of output sensitivity functions. If all the states are measured, the output sensitivity equations are replaced by the equations of the state sensitivities.

$$\Im = \frac{1}{n_m} \sum_{i=1}^{n_m} \left(\frac{\partial \xi}{\partial \theta}(t_i) \right)^T W^{-1} \left(\frac{\partial \xi}{\partial \theta}(t_i) \right) \qquad (19)$$

Usually, the noise is assumed to be *white*, so that W is diagonal

$$W = \begin{bmatrix} W_{11} & \varnothing \\ \varnothing & W_{22} \end{bmatrix} \quad (20)$$

Replacing the relationship (15) in equation (19)

$$\Im = \frac{1}{n_m} \sum_{i=1}^{n_m} \left(\frac{\partial \xi_a}{\partial \theta}(t_i) \right)^T W' \left(\frac{\partial \xi_a}{\partial \theta}(t_i) \right) \quad (21)$$

where

$$W' = (W_{11}^{-1} + A_0^T W_{22}^{-1} A_0) \quad (22)$$

This result demonstrates that the Fisher information matrix can be obtained only using the sensitivity functions for the subset of states ξ_a.

3.2 Calculation of the gradient

Parameter estimation of non-linear systems, usually requires the calculation of the cost function gradient with respect the parameters.

If the error vector $\varepsilon_i(\theta)$ is defined as

$$\varepsilon_i = \xi(t_i) - \hat{\xi}(t_i, \theta) \quad (23)$$

and supposing full state measurement, $y(t) = I_N \xi(t)$, then a quadratic cost function of the error can be defined by

$$J(\theta) = \sum_{i=1}^{n_m} \varepsilon_i(\theta)^T Q \varepsilon_i(\theta) \quad (24)$$

Usually, Q is considered to be a diagonal matrix. so the $J(\theta)$ gradient with respect to θ is

$$\frac{\partial J(\theta)}{\partial \theta} = \frac{1}{2} \sum_{i=1}^{n_m} \varepsilon_i(\theta)^T Q \frac{\partial \varepsilon_i(\theta)}{\partial \theta} \quad (25)$$

Provided that the measured value $\xi(t_i)$ does not depend on θ, the error sensitivity is related to that of the model output by

$$\frac{\partial \varepsilon_i(\theta)}{\partial \theta} = -\frac{\partial \xi}{\partial \theta}(t_i) \quad (26)$$

therefore

$$\frac{\partial J(\theta)}{\partial \theta} = -\frac{1}{2} \sum_{i=1}^{n_m} \varepsilon_i(\theta)^T Q \frac{\partial \xi}{\partial \theta}(t_i) \quad (27)$$

substituting the relationship 15 in the previous equation

$$\frac{\partial J(\theta)}{\partial \theta} = -\frac{1}{2} \sum_{i=1}^{n_m} \varepsilon_i(\theta)^T Q \begin{bmatrix} I_M \\ -A_0 \end{bmatrix} \frac{\partial \xi_a}{\partial \theta}(t_i) \quad (28)$$

Here, like in the Fisher information matrix calculation, sensitivity values of the vector ξ_a are enough to obtain the gradient of the cost funcional with respect to θ.

4. CASE STUDIES

4.1 Simple microbial growth process driven by Haldane law

The model for this process is

$$\begin{bmatrix} \dot{x} \\ \dot{s} \end{bmatrix} = \begin{bmatrix} 1 \\ -k_1 \end{bmatrix} \frac{\mu_{max} s x}{\kappa + s + \frac{s^2}{K_i}} - D \begin{bmatrix} x \\ s \end{bmatrix} + D \begin{bmatrix} 0 \\ s_i \end{bmatrix} \quad (29)$$

where x is the concentration of the microorganisms, s is the substrate concentration.

Here $\xi = [x \;\; s]^T$ and $\theta = [\mu_{max} \;\; \kappa \;\; K_i]$.

If $\xi_a = x$ and $\xi_b = s$ then $K_a = 1$, $K_b = -k_1$. Applying the equation 12, the values of sensitivity function for x and s satisfy the following condition:

$$\begin{bmatrix} \frac{\partial s}{\partial \mu_{max}}(t) \\ \frac{\partial s}{\partial \kappa}(t) \\ \frac{\partial s}{\partial K_i}(t) \end{bmatrix} = -k_1 \begin{bmatrix} \frac{\partial x}{\partial \mu_{max}}(t) \\ \frac{\partial x}{\partial \kappa}(t) \\ \frac{\partial x}{\partial K_i}(t) \end{bmatrix}, \quad \forall t \quad (30)$$

Therefore it is necessary solve by numeric integration 3 sensitivity equations instead of 6.

4.2 Two reactions process driving by Monod law

For the process of two reaction

$$S_1 \longrightarrow X_1 + S_2$$
$$S_2 \longrightarrow X_2$$

the dynamic model is defined by:

$$\begin{bmatrix} \dot{s_1} \\ \dot{x_1} \\ \dot{s_2} \\ \dot{x_2} \end{bmatrix} = \begin{bmatrix} -k_1 & 0 \\ 1 & 0 \\ k_3 & -k_2 \\ 0 & 1 \end{bmatrix} \begin{bmatrix} \frac{\mu_{max1} s_1 x_1}{\kappa_1 + s_1} \\ \frac{\mu_{max2} s_2 x_2}{\kappa_2 + s_2} \end{bmatrix} - D \begin{bmatrix} s_1 - s_{1i} \\ x_1 \\ s_2 \\ x_2 \end{bmatrix} \quad (31)$$

Monod law is considered for both reactions. The vector of kinetic equation parameters is $\theta = [\mu_{max1} \;\; \kappa_1 \;\; \mu_{max2} \;\; \kappa_2]$

For the following selection of ξ_a and ξ_b

$$\xi_a = \begin{bmatrix} x_1 \\ s_2 \end{bmatrix} \quad \xi_b = \begin{bmatrix} s_1 \\ x_2 \end{bmatrix} \quad (32)$$

the decomposition of the matrix K is

$$K_a = \begin{bmatrix} 1 & o \\ k_3 & -k_2 \end{bmatrix}, \quad K_b = \begin{bmatrix} -k_1 & 0 \\ 0 & 1 \end{bmatrix} \quad (33)$$

Then

$$A_0 = -K_b K_a^{-1} = \begin{bmatrix} k_1 & 0 \\ -\frac{k_3}{k_2} & k_2^{-1} \end{bmatrix} \quad (34)$$

For the calculation of the Fisher information matrix or the cost function gradient, equations 21 and 28, is enough know the sensitivity functions of ξ_a. Then the number of sensitivity differential equations to be solved decreases from 16 to 8, the 8 sensitivity functions of ξ_a.

5. CONCLUSIONS

In the paper the relationship between sensitivity functions in bio-reactor models has been shown. This property allows the reduction of the number of sensitivity differential equation that must be integrated and it can reduced significantly the computational effort during the parameter identification and in experiments design. The equations for the computation of Fisher information matrix and gradient of the cost are restated and only depends explicitly on the sensitivities $\frac{\partial \xi_a}{\partial \theta}(t)$. Two case studies shown the utility of this property.

ACKNOWLEDGEMENTS

Julio A. Romero would like to acknowledge gratefully the support of the Agencia Española de Cooperación Internacional (AECI) during his stay at the Systems Engineering and Control Department of the Universidad Politécnica de Valencia.

REFERENCES

Banga, J. R., R. Irizarry-Rivera and W. D. Seider (1998). Stochastic optimization for optimal and model-predictive control. *Computers Chemical Engineering* **22**(4/5), 603–612.

Bastin, G. and D. Dochain (1990). *On-line estimation and Adaptive Control of Bioreactors.* Elsevier Science Plublishers.

Bastin, G., L. Chen and V. Chotteau (1992). Can we identify biotechnological processes?. In: *Proceedings on IFAC Symposium on Modeling and Control of Biotechonological Processes* (Pergamon, Ed.). pp. 83–88.

Chen, L., G. Bastin and D. Dochain (1990). Structural identifiability of the yield coefficients in nonlinear compartmental models for bioprocesses. In: *Proceedings on IEEE Conference of Desision and Control.* pp. 1074–1079.

Gauthier, J. P., H. Hammouri and S. Othman (1992). A simple observer for nonlinear systems application to bioreactors. *IEEE Transactions on Automatic Control* **37**(6), 875–879.

González, J., R. Aguilar, J. Álvarez Ramírez and M. A. Barrón (1998). Nonlinear regulation for a continuous bioreactor via a numerical uncertainty observer. *Chemical Engineering Journal* **69**, 105–110.

McLain, R. B., M. A. Henson and M. Pottmann (1999). Direct adaptive control of partially known nonlinear systems. *IEEE Transactions on Neural Networks* **10**(3), 714–721.

Proell, T., A. Hilaly, M. N. Karim and D. Guyre (1991). Comparison of different optimization and control schemes in an industrial scale microalgae fermentation. In: *Proc. ACC'91.* pp. 1323–1328.

Roux, G. and B. Dahhou (1994). Adaptive nonlinear control of a real live fermentation process.. In: *Proceedings of the Third IEEE Conference on Control Applications, Glasgow, UK.* pp. 909–914.

VanImpe, J. F. and G. Bastin (1995). Optimal adaptive control of fed-batch fermentation processes. *Control Engineering Practice* **3**(7), 939–954.

Walter, E. and L. Pronzato (1990). Qualitative and quantitative experimental design for phenomenological models. a survey.. *Automatica* **26**(2), 195–213.

Walter, E. and Luc Pronzato (1997). *Identification of parametric models from experimental data..* Springer.

Zeng, F. Y. and B. Dahhou (1995). Reference model adaptive estimation applied to a continuous flow fermentation process. *Control Engineering Practice* **3**(7), 939–954.

IFAC
Publications
www.elsevier.com/locate/ifac

AN EXPERIMENTAL OBJECT-ORIENTED MODELLING OF AN HYDRAULIC VALLEY

T. Bastogne[†], **A. Libaux**[‡]

[†] *Centre de Recherche en Automatique de Nancy (CRAN),*
CNRS UMR 7039
Université Henri Poincaré, Nancy 1,
BP 239, F-54506 Vandœuvre-lés-Nancy Cedex, France,
Phone: (33) 3 83 68 44 73 - Fax: (33) 3 83 68 44 62
`thierry.bastogne@cran.uhp-nancy.fr`

[‡] *EDF CIH FCC, 73 373 Le Bourget-du-Lac Cedex, France*

Abstract: This communication presents an object-oriented modelling methodology for the development of complex systems simulators and an application to an hydroelectric system. The model structure is described by a multiport object-oriented diagram. Its implementation and simulation are based on the object-oriented modelling language Modelica© and on a multiformalism simulation platform Dymola©. System identification techniques are used for both estimating the behavioural models of objects and for the calibration of the multiport diagram. The originality of this contribution is the introduction of the behavioural formalism of the systems theory in the object-oriented modelling framework. *Copyright © 2003 IFAC*

Keywords: object-modelling technique, system identification, hydroelectric systems

1. INTRODUCTION

Modelling and simulation are two important disciplines which are common to all fields of engineering and science. Since the advent of cybernetics (Wiener, 1948), a large amount of literature on modelling and simulation of complex systems has been published (Elmqvist, 1978), (Cellier, 1991), (Zeigler *et al.*, 1999). Despite all these advances, the development of a global, transparent and flexible modelling procedure is still a challenge, both for industry applications and engineering education (Otter and Elmqvist, 1997), (Isermann, 1999). This need is particularly necessary for large-scale industrial processes which are often composed of multi-domain components. Among the most popular simulation platforms, the majority is very user friendly but restricted to specific domains of engineering, e.g. *Flowmaster* devoted to network fluid flow analysis, *Pspice* for analog modelling of electrical circuits, *Adams* and *Simpack* for modelling of mechanical systems or *Simulink* for data-flow modelling. Another difficulty is the causal nature of some components. Indeed, a physical system is acausal by nature, but since the definition of the CSSL (Continuous System Simulation Language) standard in 1967, most of the modelling languages like *Simulink* are essentially block oriented with input and output variables. As a consequence these languages are inappropriate to correctly describe physical processes. Moreover, the *a priori* and theoretical knowledge about the process is never completely available and the experimental modelling is often a requirement. In a classical grey-box system identification approach, the structure of the system is defined through theoretical modelling, while the values of parameters are determined by the use of experimental techniques. The mathematical structure is generally defined after multiple combinations and substitutions of equations

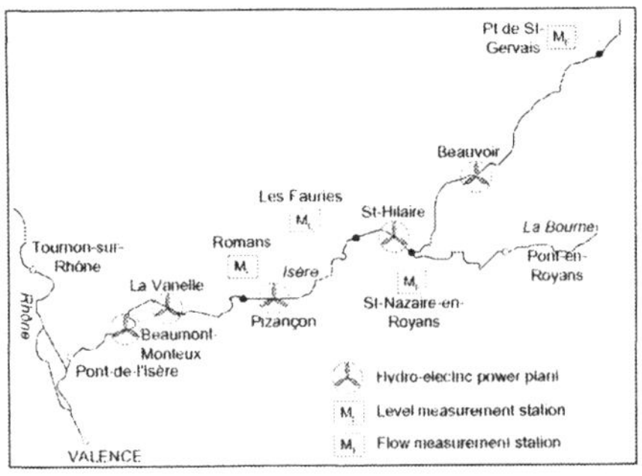

Figure 1. Hydro-electric power plants along the 'Basse-Isère' river

to finally lead to a compact model in which physical components of the process do not explicitly appear. An alternative between physical modelling and system identification, called semi-physical modelling, has been proposed by (Graebe and Bohlin, 1992). This approach is an application of system identification, where physical insight into the application is used to come up with suitable nonlinear transformations of the raw measurements, so as to allow for a good model structure. However in both cases, if only one component is changed, it is generally all the model structure which has to be modified. This sensitivity of the model to the least change in the process structure, points out the impossibility to perpetuate and capitalize the modelling effort with this usual experimental modelling approach.

However, efforts have converged to develop both multiformalism platforms of simulation (Zeigler *et al.*, 1999) and object-oriented languages for physical systems modelling like Modelica (Mattsson *et al.*, 1998). Furthermore, the behavioural formalism of systems theory (Willems, 1986), (Polderman and Willems, 1997) allows the mathematical modelling of dynamical systems and interconnected systems. Based on these recent advances, this communication proposes an object-modelling approach which allows the modeler to integrate black-box models and physical equations in a modular model structure, described by a multiport object-oriented diagram (MOOD) and implemented in Modelica. This approach federates existing modelling concepts in a procedure devoted to the development of 'High-fidelity' dynamic models (Maciejowski, 1997), i.e. models for simulation, training or safety certification purposes. The originality of this contribution is the introduction of the behavioural formalism in the object-modelling framework in order to develop a mathematical expression of objects. A practical assessment of this approach is presented herein via its application to a hydraulic valley.

2. THE 'BASSE-ISÈRE' RIVER VALLEY

2.1 General description of the industrial process

As described in figure 1, the 'Basse-Isère' river between Grenoble and Valence in the south-east of France, is formed by five managed reaches provided with $40/50$ MW hydro-electric plants at Beauvoir (BV), Saint-Hilaire (SH), Pizançon (PZ), La Vanelle (LV) and Beaumont-Monteux (BM). The river flow rate may vary from 70 to $3000m^3/s$ in times of flood. The 'Bourne' river, between Beauvoir and Saint-Hilaire, is a tributary river which adds to the main flow inducing large disturbances. Some reaches have also critical water level points located near dwellings and roads that must be secured. Note that the total ingoing and outgoing flows cannot be measured directly. However, they can be estimated from the measured aperture of the actuators (turbines and sluice gates) stored in local controllers.

As illustrated in figure 2, a typical managed reach is formed by a reach and its downstream barrage. In this application, the length of reaches varies between $8km$ and $15km$. The controlled variable is either the downstream water level or a critical water level at a given point of the reach. The manipulated variable is the outflow rate while the inflow rate and the eventual tributary flows are regarded as disturbances. Each barrage is composed of a dam (sluice gates) and a hydro-electric power plant (turbines).

3. SEMANTICS OF A MULTIPORT DIAGRAM

A multiport object-oriented diagram is a modular model, noted Σ, defined by :

$$\Sigma = (\mathcal{O}, \mathcal{L}) \quad (1)$$

where $\mathcal{O} = \{\mathcal{O}_1, \cdots, \mathcal{O}_m\}$ is a set of **objects** which compose the diagram and $\mathcal{L} = \{\mathcal{L}_1, \cdots, \mathcal{L}_n\}$ is a set of **links** (connections) which allow the objects to exchange power and information flows with other objects, m and n denote the number of objects and links of Σ.

3.1 Objects

As depicted in equation 2, an object $\mathcal{O}_i$ is composed of a behavioural model $\mathcal{B}_{\mathcal{O}_i}$ and a set of ports Π_i by which it communicates with its environment.

$$\mathcal{O}_i = (\mathcal{B}_{\mathcal{O}_i}, \Pi_i) \quad (2)$$

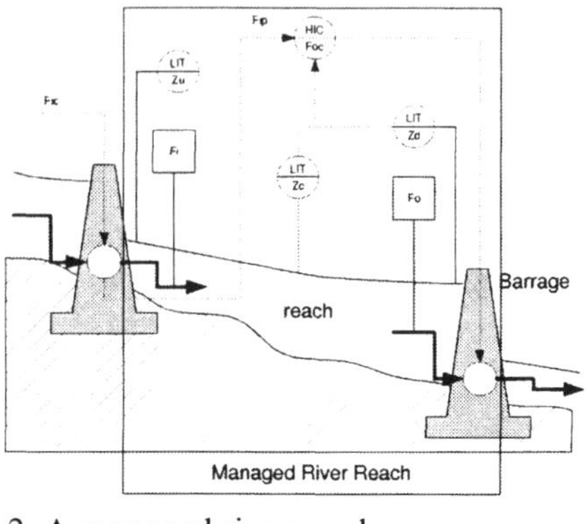

Figure 2. A managed river reach

Π_i is defined by :

$$\Pi_i = \{\mathcal{P}_k\} \qquad k \in \{1, \cdots, p\} \tag{3}$$

$\mathcal{P}_k$ is a port of communication where k is its number and p the number of ports in the diagram. $\mathcal{B}_{\mathcal{O}_i}$ is defined by :

$$\mathcal{B}_{\mathcal{O}_i} = \{t \in \mathbb{T}, \theta \in \Theta, w(t) \in \mathbb{W}, l(t) \in \mathbb{L}| \cdots \tag{4}$$
$$\cdots f_1(t, \theta, w, l) = f_2(t, \theta, w, l)\} \tag{5}$$

with apologies for the abusive notation. $f_1(\cdot)$, $f_2(\cdot)$ express the behavioural equations of the object. θ is a vector of parameters and $(w(t), l(t))$ are the variables of the model. They are divided into manifest (external) variables : $w(t)$ and latent (local) variables : $l(t)$. Θ denotes the parameter space, $\mathbb{W}$ the manifest signal space and $\mathbb{L}$ the latent variable space. t is the time variable and $\mathbb{T}$: the time axis.

Two other attributes complete the definition of $\mathcal{O}_i$. The first one is its **identity** composed of its name and its graphic icon. The second one is its **colour** defined according to the *a priori* knowledge about the object (see section 4.2).

3.2 Ports

Two types of ports are considered : the **physical** ports and the **signal** ports which allow the object to exchange power and information flows respectively. $\mathbb{P}_P$ and $\mathbb{P}_S$ denote the spaces of power and signal ports.

The oriented **physical** ports : $\mathcal{P}_P \in \mathbb{P}_P$ are symbolized by a black circle. The status at a physical port is defined by a couple of *across/through* variables : $\mathcal{P}_P = (\alpha(t), \varphi(t))$. The power flow $P(t)$ passing through the physical port is given by : $P(t) = \alpha(t) \cdot \varphi(t)$. In accordance with the Modelica convention, the positive flow of *through* variables is into the object.

The oriented **signal** ports : $\mathcal{P}_S \in \mathbb{P}_S$ are causal interfaces by which objects exchange input and output signals : $\mathcal{P}_{S+} = (u(t))$ or $\mathcal{P}_{S-} = (y(t))$ symbolized by a black and a white arrow respectively.

Finally, it appears that the vector $w(t)$ of manifest variables is composed of input, output, across or through variables.

3.3 Links

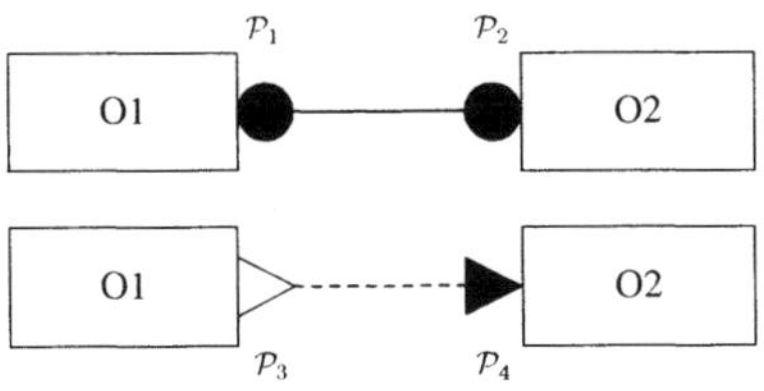

Figure 3. Types of connections : $(\mathcal{P}_1, \mathcal{P}_2) \in \mathcal{P}_P, (\mathcal{P}_3, \mathcal{P}_4) \in \mathcal{P}_S$

A link $\mathcal{L}_i$ between objects is defined as a connection of ports :

$$\mathcal{L}_i = \mathcal{C}(\mathcal{P}_{k_1,l_1}, \cdots, \mathcal{P}_{k_q,l_q}) \tag{6}$$

with $i \in \{1, \cdots, n\}$. $\mathcal{C}(\cdot)$ denotes the connection function and q the number of ports connected by $\mathcal{L}_i$. In the case of signal links, $q = 2$. As previously mentioned and shown in figure 3, two types of flows are considered : the information and power flows, which are graphically represented by dotted and solid lines respectively. Accordingly there are two types of links and two types of connection functions. The behavioural models of physical and signal links, $\mathcal{B}_{\mathcal{L}_P}$ and $\mathcal{B}_{\mathcal{L}_S}$ are given by :

$$\mathcal{B}_{\mathcal{L}_P} = \left\{ \begin{pmatrix} \alpha_1 \\ \vdots \\ \alpha_q \end{pmatrix} \in \mathbb{A}, \begin{pmatrix} \varphi_1 \\ \cdots \\ \varphi_q \end{pmatrix} \in \mathbb{F} \right| \cdots$$
$$\left. \begin{matrix} \alpha_1(t) = \cdots = \alpha_q(t) \\ \varphi_1(t) + \cdots + \varphi_q(t) = 0 \end{matrix} \right\} \tag{7}$$
$$\mathcal{B}_{\mathcal{L}_S} = \{u \in \mathbb{I}, y \in \mathbb{O} | u(t) := y(t)\} \tag{8}$$

where $\mathbb{A}$, $\mathbb{F}$, $\mathbb{I}$ and $\mathbb{O}$ denote the domains of across, through, input and output variables. Note in equation 8 the assignment operator in comparison with the equality operator in equation 7.

4. AN OBJECT-MODELLING APPROACH OF PHYSICAL SYSTEMS

As illustrated in figure 4, the object-modelling approach used in this application is made up of four main steps : the hierarchical decomposition of the process, the behavioural modelling of the objects, their interconnection in a multiport diagram and the model calibration. The diagram of figure 4 points out in dotted lines the correlation of system identification techniques with the proposed modelling procedure. System identification techniques are essentially used for the modelling of black- and grey-box objects and in the calibration step. The *hybrid* problem in interfacing theoretical with experimental knowledge is solved by the modularity property of the object-oriented approach which allows the modeler to connect white- and black-box objects independently of the nature of their behavioural models.

4.1 Hierarchical decomposition of the process

The first step of the proposed object-modelling approach can be regarded as a 'qualitative' modelling phase handling the complexity of process. The objective is to determine the number of objects : m and the resolution level of the model, i.e. a high-resolution model (microscopic model) or a low-resolution model (macroscopic model)? Off course, the answer mainly depends on the final application of the model. But it also depends on the *a priori* knowledge we have on

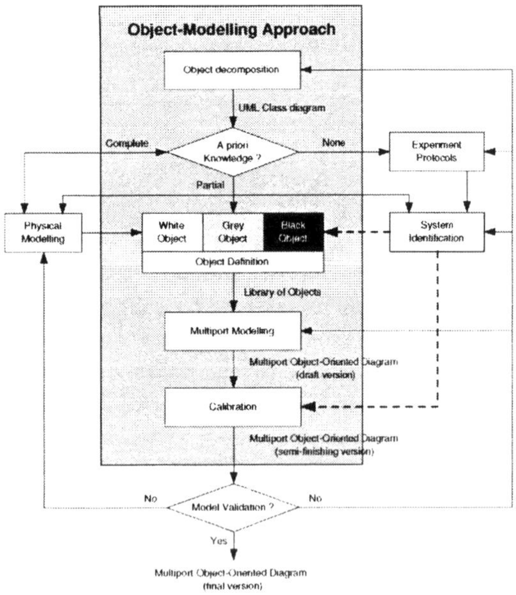

Figure 4. The object-modelling procedure

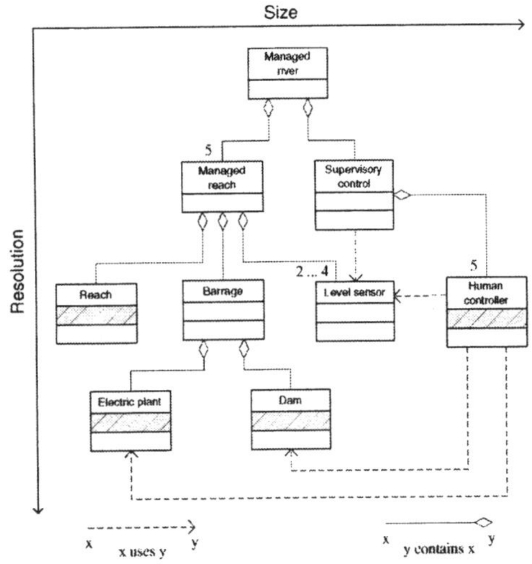

Figure 5. UML class diagram of the 'Basse-Isère' managed river

the process and on the available measurements. In this application, each power plant is composed of several turbine groups and each dam is composed of a number of sluice gates. However, the flow rates through those actuators are not measured. Moreover, the structure of the local flow controllers is not known. This lack of information prevents us to identify models of turbines and valves, and limits the resolution of the decomposition. Consequently, by abstraction, details of the plant like turbines, sluice gates and local flow controllers have been aggregated in two broader objects entitled : *Power plant* and *Dam* respectively. The result of the decomposition is represented in figure 5 by a UML (Unified Modelling Language) class diagram in which the composition and use links described by solid and dotted lines respectively. The complexity of the process is decomposed on two axis : size and resolution. This diagram shows a decomposition of the 'Basse-Isère' river in nine object classes : $m = 9$.

4.2 A priori *knowledge*

In a second step, it is proposed to define for each object class a colour attribute according to the *a priori* knowledge about its dynamical behaviour. This attribute is defined by analogy with the system identification terminology, i.e. :

- white box : if the theoretical laws or physical equations and the values of parameters are known,
- grey box : if there only exists a partial knowledge about the object, i.e. values of some parameters or the structure of physical equations.
- black box : if no *a priori* knowledge about the object is available.

In this application, as depicted in figure 5, most of the objects are assumed to be completely known and are then regarded as white-box objects. The other components, i.e. the reach, the power plant, the dam and the level controller, are described by grey-box models in which some parameters remain unknown. At this stage of the object-modelling procedure, the number of grey- and black-box elements in the UML class diagram informs the modeler about the experimental effort to perform for a complete modelling of the process.

5. OBJECTS MODELLING

In this third step, the behavioural formalism of dynamical systems theory (Willems, 1986) is borrowed to describe the ports Π_i and the dynamical models $\mathcal{B}_{\mathcal{O}_i}$ of the objects $\mathcal{O}_i$.

Let's examine the example of the river reach object class. $\mathcal{B}_{reach}$ is given by :

$$\dot{z}_d(t) = \frac{1}{A}[F_i(t-\tau_i) - F_o(t-\tau_o)] \quad (9)$$

$$P_i(t) = P_o(t) = 10^5 \quad (10)$$

$$\Delta z(t) = aF(t) + bF(t)^2 \quad (11)$$

with:

$$t \in \mathbb{T} = \mathbb{R}^+ \quad (12)$$

$$(\, A \;\, \tau_i \;\, \tau_o \;\, a \;\, b \,) \in \Theta = \mathbb{R}^{+3} \times \mathbb{R}^2 \quad (13)$$

$$\left.\begin{array}{l} (\, P_i(t) \;\, P_o(t) \,) \in \mathbb{A} \\ (\, F_i(t) \;\, F_o(t) \,) \in \mathbb{F} \\ (\, z_d(t) \;\, \Delta z(t) \,) \in \mathbb{O} \end{array}\right| \in \mathbb{W} = \mathbb{R}^{+6} \quad (14)$$

$$F(t) \in \mathbb{L} = \mathbb{R}^+ \quad (15)$$

where z is the water level, F_i and F_o are the inflow and outflow rates, P_i and P_o are the input and output pressures of the reach. F and Δz denote the steady-states values of the mean flow and of the altitude loss of the reach, i.e. the level difference of water between the upstream and downstream levels. A is the water surface, τ_i and τ_o denote the retention time-delays. If A, τ_i and τ_o are known for each reach of the river, this is not the case for a and b which are two

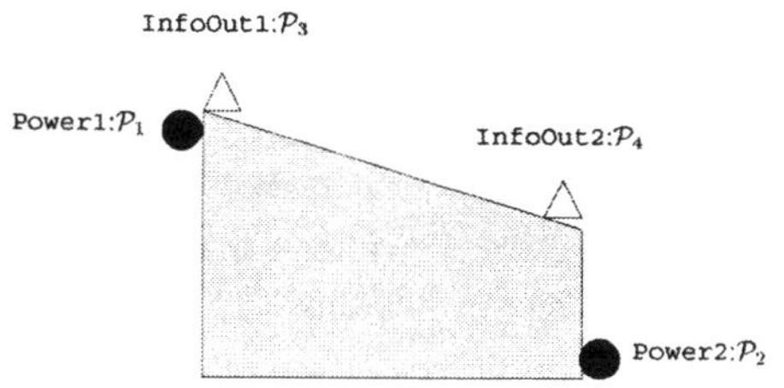

Figure 6. Icon of the reach object

unknown parameters. Equation 9 expresses the storage behaviour of the reach while equation 10 corresponds to the equality assumption of the pressures in an open-channel. Equation 11 shows the static model of the altitude loss of water. The communication of the reach object with its environment is modelled by four ports (two physical and two signal ports) defined by :

$$\left.\begin{array}{l}\mathcal{P}_1 = \left(P_i(t)\ F_i(t)\right) \\ \mathcal{P}_2 = \left(P_o(t)\ F_o(t)\right)\end{array}\right| \in \mathbb{P}_P \qquad (16)$$

$$\left.\begin{array}{l}\mathcal{P}_3 = \left(z_d(t)\right) \\ \mathcal{P}_4 = \left(\Delta z(t)\right)\end{array}\right| \in \mathbb{P}_S \qquad (17)$$

The icon of the reach object is presented in figure 6. Let's now examine the implementation in Modelica of $\mathcal{B}_{reach}$, presented in the algorithm 1. The first part of this algorithm is devoted to the definition of the ports where manifest variables, i.e. across/through and input/output variables are attached to the power and information ports. Ports labelled `Power1`, `Power2`, `InfoOut1`, `InfoOut2` correspond to $\mathcal{P}_1$, $\mathcal{P}_2$, $\mathcal{P}_3$ and $\mathcal{P}_4$. In a second part, two delay functions are used and applied to the flow variables $F_i(t)$ and $F_o(t)$. One filter bloc `TransferFunction1`, not detailed herein, has been used to estimate the steady-state values of the mean flow $F(t)$.

Algorithm 1. Modelica model of the reach object

```
equation
// PORTS (manifest variables)
     Power1.a =Pi;
     Powerl.f =Fi;
     Power2.a =Po;
     Power2.f =-Fo;
     InfoOut1.signal[1] = Deltaz;
     InfoOut2.signal[1] = zd;
// BEHAVIORAL MODEL
     Fid=delay(Fi,taui);
     Fod=delay(Fo,tauo);
     der(zd)=1/A*(Fid-Fod);
     Pi=105;
     Po=105;
     F=(Fi+Fo)/2;
     TransferFunction1.inPort.signal[1]=F;
     TransferFunction1.outPort.signal[1]=Ff;
```

Algorithms 2 and 3 show the implementation of two ports in Modelica. For each port, a connector class is defined. Connecting power ports means that across variables are equal while through variables (marked by the prefix *flow*) are sum to zero. Concerning the output information port, the output signals are marked with the keyword *output*.

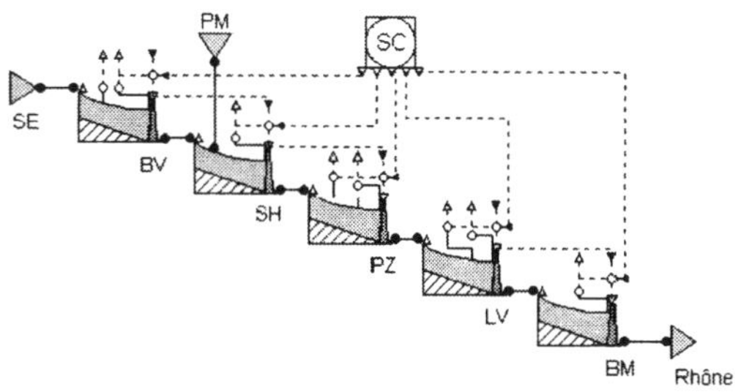

Figure 7. Multiport object-oriented diagram of the 'Basse-Isère' river implemented in Dymola

Algorithm 2. Modelica object class of physical ports

```
connector Power
     Real a;
     flow Real f;
end Power
```

Algorithm 3. Modelica object class of output information ports

```
connector InfoOut
   parameter Integer n=1;
   output Real signal[n];
end InfoOut
```

5.1 Multiport object-oriented diagram

All the objects which compose the managed river have been gathered in a library to be reused for other modelling tasks. The multiport object-oriented diagram is got by cutting, pasting and plugging together objects in the same way as an experimenter would plug together real components in a laboratory. The objects are connected by links corresponding to the various exchanges of power and information between the components of the system. The multiport diagram of the managed river is presented in figure 7. The behavioural model of this multiport diagram can be expressed as the intersection of the behavioural models of the objects and ports which compose it :

$$\mathcal{B}_\Sigma = \{\mathcal{B}_{\mathcal{O}_1} \cap \cdots \cap \mathcal{B}_{\mathcal{O}_m} \cap \mathcal{B}_{\mathcal{L}_1} \cap \cdots \cap \mathcal{B}_{\mathcal{L}_n}\} \quad (18)$$

Concretely, i.e. from a compilation point of view, $\mathcal{B}_\Sigma$ is a system of equations (link equations + object equations). Those equations are then sorted to constitute a another system which is reduced and solved by standard numerical procedures and finally compiled as C code.

6. MODEL CALIBRATION

It is generally necessary to adjust some parameters of the model in order to match some properties between the system and its model. Each reach of the managed river has been calibrated one by one. The altitude loss and the dynamical models have been estimated separately. Each of these estimation steps has been expressed as the minimization of a quadratic criterium :

$$\hat{\theta} = \arg\min_{\theta} \|e(\theta)\|^2, \qquad (19)$$

where $\hat{\theta}$ and θ denote the vectors of the estimated and unknown parameters respectively. In this application,

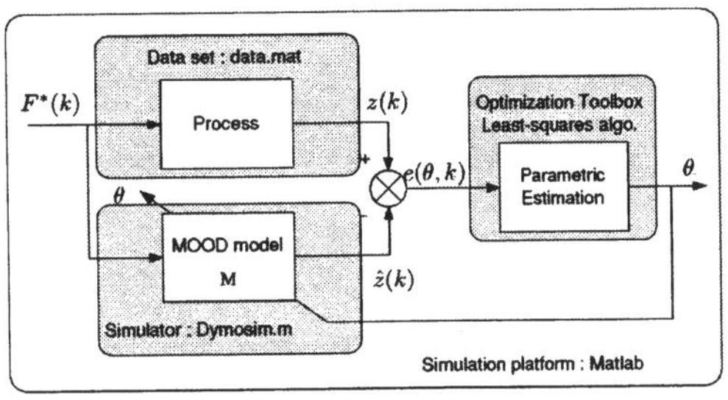

Figure 8. Calibration procedure based on the output error method

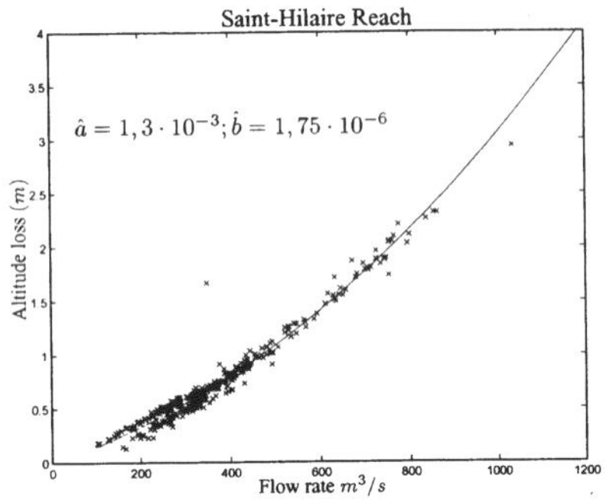

Figure 9. Estimation results of the altitude loss model

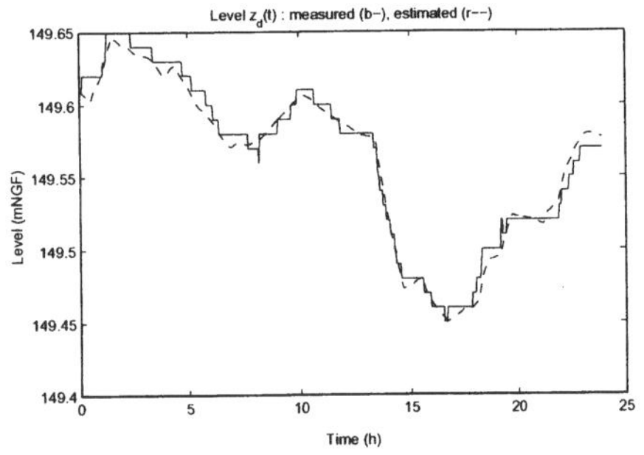

Figure 10. Estimated and measured water levels after the calibration step (Saint-Hilaire reach)

$e(\theta)$ is the output error. The implementation of this calibration procedure is shown in figure 8. The simulator of each model of the reaches has been supported by Dymosim in Dymola (Elmqvist *et al.*, 1999), a standalone program which can be used in several different environments. Dymosim is called from the Matlab environment in which algorithms of the Optimization toolbox (The MathWorks, 2002) are used to solve the equation (19). Estimation results of the altitude loss model for one reach are presented in figure 9. Figure 10 presents the calibration results of the dynamical model of the downstream level. These models have been separately cross-validated.

7. CONCLUSION

This communication presents the development of an object-modelling approach and its application to a managed river in view to develop a simulator of an hydraulic valley. The model is described by a multiport object-oriented diagram, implemented in the modelling language : Modelica and simulated in the multiformalism simulation platform : Dymola. The originality of this contribution consists in combining the flexibility of the object-oriented modelling approaches with the behavioural formalism of systems theory. This formalism provides a mathematical expression of the multiport diagram.

REFERENCES

Cellier, F. E. (1991). *Continuous System Modeling.* Springer-Verlag.

Elmqvist, H. (1978). A structured model language for large continuous systems. PhD thesis. Dept. of Automatic Control, Lund Institute of Technology. Sweden. Report CODEN: LUTFD2(/TFRT-1015).

Elmqvist, H., D. Brück and M. Otter (1999). *Dymola - Dynamic Modeling Laboratory. User's Manual.* Dynasim AB.

Graebe, S. F. and T. Bohlin (1992). Identification of nonlinear stochastic grey box models. In: *IFAC Int. Symp. on Adaptative Systems in Control and Signal Processing.* Laboratoire d'Automatique de Grenoble, Grenoble, France. pp. 401–406.

Isermann, R. (1999). Modeling, identification and simulation of mechatronic systems. In: *Proc of the 14th IFAC World Congress.* Beijing, P. R. China. pp. 395–406.

Maciejowski, J. M. (1997). Reconfigurable control using constrained optimization. In: *Proc. of the 4th European Control Conference* (G. Bastin and M. Gevers, Eds.). Vol. Plenary Lectures and Mini-Courses. pp. 107–130.

Mattsson, S.E., H. Elmqvist and M. Otter (1998). Physical system modeling with modelica. *Control Engineering Practice* **6**, 501–510.

Otter, O. and H. Elmqvist (1997). Energy flows modeling of mechatronic systems via object diagrams. In: *Proc. of the 2nd MATHMOD.* Vienna. pp. 705–710.

Polderman, J. W. and J. C. Willems (1997). *Introduction to Mathematical Systems Theory - A Behavioral Approach.* Texts in Applied Mathematics, 26. Springer.

The MathWorks, Inc. (2002). *Optimization Toolbox User's Guide.* 2 ed.

Wiener, N. (1948). *Cybernetics or Control and Communication in the Animal and the Machine.* MIT Press.

Willems, J. C. (1986). From time series to linear systems. *Automatica.* Part I: Vol. 22, No. 5, pp. 561-580, 1986, Part II: Vol. 22, No. 6, pp. 675-694, 1986, Part III: Vol. 23, No. 1, pp. 87-115, 1987.

Zeigler, B. P., H. Praehofer and T. G. Kim (1999). *Theory of Modeling and Simulation - Integrating Discrete Event and Continuous Complex Dynamic Systems.* Academic Press. Second Edition.

IFAC Publications
www.elsevier.com/locate/ifac

PARTICLE FILTERS FOR SYSTEM IDENTIFICATION WITH APPLICATION TO CHAOS PREDICTION

Fredrik Gustafsson and Paul Hriljac

Department of Electrical Engineering, Linköpings universitet,
SE-581 83 Linköping, Sweden
Tel: +46 13 282226; fax: +46 13 282622
e-mail: fredrik@isy.liu.se
College of Engineering, Embry-Riddle Aeronautical University,
3200 Willow Creek Road USA
e-mail: hriljap@erau.edu

Abstract: The theory of the particle filter, or sequential Monte Carlo methods, has made substantial progress the last decade. The number of applications has increased substantially the last three years, in particular in navigation and telecommunication areas. In this contribution, we will first point out how the particle filter can be used for system identification, using a quite general problem formulation, and it is pointed out in which kind of application the particle filter can be an attractive alternative to classical system identification methods.
This is then demonstrated on prediction of time series arising from chaotic dynamical systems. The specific dynamical system considered is the so called logistical map with an unknown parameter, which belongs to the chaotic regime. *Copyright © 2003 IFAC*

Keywords: system identification, particle filter, dynamical systems, non-linear systems, estimation, chaos, prediction, Cramer-Rao

1. INTRODUCTION

Since the seminal paper Gordon et al. (1993a), a large number of papers developing the theory of particle filtering have appeared. State of the art is summarized in the recent monograph Doucet et al. (2001a). During the five years following the seminal paper above, the theory was developed mainly by statisticians. When the theory was mature enough, the signal processing community quickly adopted the results, and developed practical algorithms for applications. Since 2000, it can be noted that the number of publications on applications has increased substantially, in particular in navigation, Gustafsson et al. (2002), and telcommunications areas. Special sessions on particle filter are now frequent on major signal processing conferences. However, so far the impact on the automatic control and system identification societies have been quite limited.

Anyhow, the particle filter offers a general tool for estimating unknown parameters in non-linear models of moderate complexity. A general parametric nonlinear state space model, see Chapter 5.3 in Ljung (1999), with additive noise processes is given by:

$$z_{t+1} = f(z_t;\theta_t) + v_t^z, \quad (1a)$$
$$y_t = h(z_t;\theta_t) + e_t, \quad (1b)$$

The functions $f(z_t;\theta)$ and $h(z_t;\theta)$ are given by the model and may contain unknown parameters as given by the vector θ. Joint state and parameter estimation aims at estimating both the state z_t and the parameters θ simultaneously.

The literature on system identification mainly describes black-box approaches to system identification for non-linear systems. Particular algorithms for special structures as Wiener and Hammerstein models are known. In the general case, with a completely known model structure, prediction error methods can be applied. These involve differentiation of the non-linearities, and often work quite well when the noise is well-behaved (uni-modal and symmetric) and the non-linearities are smooth. Maximum-likelihood approaches are possible for some particular structures, where the 'certainty equivalence' principle can be applied (iterate between estimating states and parameters) where for instance the EM-algorithm can work well. For general models, including arbitrary non-linearities (discontinuous ones for instance), hard constraints on the parameters and states, or multi-modal noise distributions (aircraft maneuver as state noise, radar lobes as noise distribution *etc.*), no general theory applies or standard methods give poor performance.

To put this in a framework where the particle filter can be applied, the state vector is augmented with the parameter vector. The new state vector is denoted

$$x_t = \begin{pmatrix} z_t \\ \theta_t \end{pmatrix} \tag{2}$$

which is governed by the relations

$$\begin{bmatrix} z_{t+1} \\ \theta_{t+1} \end{bmatrix} = \begin{bmatrix} f(z_t;\theta_t) \\ \theta_t \end{bmatrix} + \begin{bmatrix} v_t^z + w_t^z \\ v_t^\theta + w_t^\theta \end{bmatrix} \tag{3a}$$

$$y_t = h(z_t;\theta_t) + e_t, \tag{3b}$$

Here we have distinguished the physical noise v_t from the instrumental roughening noise w_t. As will be described later, this is something needed for the particle filter to explore the whole state space, which fills in the gaps between the finite number of particles. By introducing an additional noise to the samples the depletion problem can be reduced. This technique is called *jittering* Fearnhead (1998) or *roughening* Gordon et al. (1993a). To summarize the stochastic assumptions, the different noise processes in (3) are:

- Physical state noise v_t^z.
- Roughening state noise w_t^z, which has turned out to be beneficial for the particle filter performance. Loosely speaking, it helps the particles explore the whole state space.
- Random walk noise v_t^θ on the parameters, for making the algorithm adaptive to slow changes in the parameters. In system identification, $v_t^\theta = 0$, which we will mostly assume here.
- Roughening parameter noise w_t^θ, which makes the particles explore a small neighborhood. This noise should decay with time in system identification, to get a converging estimate.

The particle filter provides an appealing framework for this nonlinear and non-Gaussian estimation problems. The aim of the particle filter is to recursively estimate the posterior density function $p(X_t|Y_t)$, where $X_t = \{x_0, \dots, x_t\}$. According to the Bayesian philosophy, $p(X_t|Y_t)$ contains all there is to know about the process at time t. From this density, we can then obtain any point estimate we like. Loosely speaking the particle filter can be interpreted as a large number of simulations, where each simulation consists of a sample from the distribution we want to estimate. There is a weight accociated with each sample, which contains information on how *likely* the corresponding sample is. These samples together with the corresponding weights will constitute a discrete approximation of the posterior density. Hence, for a sufficiently large number of samples, the particle filter provides a tool to approximate $p(z_t, \theta_t|Y_t)$ arbitrarily well. However, the large state dimension might be prohibitive for the practical use of the particle filter, and this is its main drawback, besides the obvious demand on sufficent computational resources. As a coarse rule of thumb, the particle filter should not be applied to problems with more than five states.

In an accompanying paper, Schön and Gustafsson (2003), marginalization techniques are applied to possible linear sub-structures in the model, which makes it possible to apply the filter to larger problems and furthermore decreases the requirement on computational power.

2. THE PARTICLE FILTER

2.1 *The idea*

The particle filter provides an approximative solution for the problem of recursively estimating the *posterior* density function $p(X_t|Y_t)$, for a nonlinear discrete time system on the form (1). In this article we are interested in one of the marginals of the posterior density, namely the *filtering* density, $p(x_t|Y_t)$. Using this density function we can then compute an estimate of any inference function $g(x_t)$ we like, for instance the state estimator $g(x_t) = z_t$ or the output prediction error variance $g(x_t) = \text{Var}(y_t - h(z_t;\theta_t))$. We will use $I(g(x_t))$ to denote this estimate, according to

$$I(g(x_t)) = E_{p(x_t|Y_t)}[g(x_t)] = \int g(x_t)p(x_t|Y_t)dx_t \tag{4}$$

More specifically, the particle filter provides an approximative solution to the optimal recursive Bayesian filter given by the prediction density, $p(x_{t+1}|Y_t)$, and the filtering density, $p(x_t|Y_t)$, Jazwinski (1970)

$$p(x_{t+1}|Y_t) = \int p(x_{t+1}|x_t)p(x_t|Y_t)\,dx_t, \tag{5a}$$

$$p(x_t|Y_t) = \frac{p(y_t|x_t)p(x_t|Y_{t-1})}{p(y_t|Y_{t-1})}. \tag{5b}$$

These equations are in general very hard to solve analytically, except in a few special cases, *i.e.* when the model is linear and the noise is Gaussian. In that case the solution is given by the Kalman filter, see Anderson and Moore (1979). The particle filter provides us with an approximative solution to these integrals by using a large set of samples (also called particles, hence the name particle filter), $\{x_t^{(i)}\}_{i=1}^N$, which constitutes a discrete approximation of $p(x_t|Y_t)$, according to

$$\hat{p}_N(x_t|Y_t) = \sum_{i=1}^{N} \bar{q}_t^{(i)} \delta(x_t^{(i)} - x_t), \tag{6}$$

where $\bar{q}_t^{(i)} = q_t^{(i)} / \sum_{j=1}^{N} q_t^{(j)}$ are the normalized *importance* weights. These weights are introduced due to the fact that we cannot sample from the true density functions. For a more thorough discussion of these weights see *e.g.* Doucet et al. (2001b); Doucet (1998). These weights are updated using the likelihood function according to

$$q_{t+1}^{(i)} = p(y_{t+1}|x_{t+1}^{(i)})\bar{q}_t^{(i)}, \tag{7}$$

which means that the most *likely* samples, *i.e.* the samples that correspond to a large likelihood, will be assigned a large weight. There is still one problem that remains to be solved, and that is that the approach described above will lead to that the variance of the importance weights increases over time and thus the estimate will finally diverge, see *e.g.* Doucet (1998) for a formal proof of this fact. What happens is that the samples spread out and the weights will be almost zero for most of the samples. This can be avoided, using *resampling*. This key-step, which made the particle filter work in practice was introduced in Gordon et al. (1993b), based on the weighted bootstrap presented in Smith and Gelfand (1992). The resampling step consists of drawing N samples with replacement, where the probability of drawing $X_t^{(i)}$ is given by the corresponding importance weight, $\bar{q}_t^{(i)}$. This makes sense since the importance weight will be large if the corresponding sample is close to the true state.

Apart from the resampling step the basic ideas for the particle filter have been around since the 1940:s. The first article, known to the authors, introducing the overall ideas is Metropolis and Ulam (1949). In the automatic control community the ideas were introduced in the late 1960:s by Handschin and Mayne (1969); Handschin (1970), but then they were forgotten again until the late 1980:s, when more computer power became available, and since then there has been a lot of research activity in this area.

2.2 *Obtaining the estimates*

As described above the particle filter provides us with an estimate of the filtering density, $p(x_t|Y_t)$, from which we can deduce various point estimates according to (4). An estimate of the mean of the state, can now be obtained by combining (4), with $g(x_t) = x_t$, and (6),

$$\begin{aligned}\hat{x}_{t|t} &= E_{p(x_t|Y_t)}[x_t] = \int x_t p(x_t|Y_t)dx_t \\ &\approx \int x_t \hat{p}_N(x_t|Y_t)dx_t \\ &= \int x_t \sum_{i=1}^{N} \bar{q}_t^{(i)} \delta(x_t^{(i)} - x_t)dx_t = \sum_{i=1}^{N} \bar{q}_t^{(i)} x_t^{(i)}.\end{aligned}$$

Similarly an estimate of the variance can be obtained using $g(x_t) = (x_t - \hat{x}_{t|t})(x_t - \hat{x}_{t|t})^T$.

A numerical approximation to (5) is given in the following algorithm.

Algorithm 1. The Particle Filter

Given: A parametric state space model (3) with known parametric functions $f(z;\theta)$ and $h(z;\theta)$. Prior densities p_{z_0}, p_{θ_0} and noise densities $p_{v_t,z}$, p_{e_t} and possibly also $p_{v_t,\theta}$ for adaptive filtering.

Design parameters: Number of particles N. Roughening densities $p_{w_t,z}$ and $p_{w_t,\theta}$.

(1) *Initialization:* Generate $x_0^{(i)} \sim p_{x_0}, i = 1, \ldots, N$. Each sample of the state vector is referred to as a *particle*.

(2) *Measurement update:* Update the weights by the likelihood (more generally, any importance function, see Doucet et al. (2001a)):

$$\begin{aligned}q_t^{(i)} &= q_{t-1}^{(i)} p(y_t|x_t^{(i)}) \\ &= q_{t-1}^{(i)} p_{e_t}(y_t - h(x_t^{(i)}, \theta_t^{(i)}))\end{aligned}$$

for $i = 1, 2, \ldots, N$, and normalize to $\bar{q}_t^{(i)} = q_t^{(i)} / \sum_i q_t^{(i)}$. As an approximation of $\mathrm{E}(x_t|Y_t)$, take

$$\hat{x}_t \approx \sum_{i=1}^{N} \bar{q}_t^{(i)} x_t^{(i)}.$$

(3) *Re-sampling:*

(a) *Bayesian bootstrap.* Take N samples with replacement from the set $\{x_t^{(i)}, \theta_t^{(i)}\}_{i=1}^N$ where the probability to take sample i is $\bar{q}_t^{(i)}$. Let $q_t^{(i)} = 1/N$. This step is also called *Sampling Importance Re-sampling (SIR)*.

(b) *Importance sampling.* Only resample as above when the effective number of samples is less than a threshold N_{th},

$$N_{\text{eff}} = \frac{1}{\sum_i (\bar{q}_t^{(i)})^2} < N_{\text{th}},$$

see Bergman (1999); Doucet et al. (2000); Kong et al. (1994); Liu (1996). Here $1 \leq N_{\text{eff}} \leq N$, where the upper bound is attained when all particles have the same weight, and the lower bound when all probability mass is at one particle. The threshold can be chosen as $N_{\text{th}} = 2N/3$.

(4) Prediction: Take $v_t^{z,(i)} \sim p_{v,z}$, $v_t^{\theta,(i)} \sim p_{v,\theta}$, $w_t^{z,(i)} \sim p_{w,z}$, and $w_t^{\theta,(i)} \sim p_{w,\theta}$, and simulate

$$\begin{aligned}x_{t+1}^{(i)} &= f(x_t^{(i)}; \theta^{(i)}) + v_t^{z,(i)} + w_t^{z,(i)} \\ \theta_{t+1}^{(i)} &= \theta_t^{(i)} + v_t^{\theta,(i)} + w_t^{\theta,(i)}\end{aligned}$$

for $i = 1, 2, \ldots, N$.

(5) Let $t := t + 1$ and iterate to item 2.

The key point with re-sampling is to prevent high concentration of probability mass at a few particles. Without this step, some $\bar{q}_t^{(i)}$ will converge to 1 and the filter would brake down to a pure simulation. The re-sampling can be efficiently implemented using a classical algorithm for sampling N ordered independent identically distributed variables Bergman (1999); Ripley (1988).

The roughening state noise processes are just instrumental, which have no physical counterpart. The rationale is to fill in the gap in between the finite number particles to make sure that the whole state space will be covered. Furthermore, when the particle cloud becomes denser around its true value, the roughening can be decreased. A simple choice is a Gaussian distribution $p_{w,z} = \mathrm{N}(0, (\sigma^2_{w,z}/t)I)$ and $p_{w,\theta} = \mathrm{N}(0, (\sigma^2_{w,\theta}/t)I)$, whose variance decays to zero. One could here compare to recursive implementations of the least squares method, where the step size in the parameter update decreases as $1/t$.

As a generalization, in adaptive algorithms the parameter vector is assumed slowly (compared to the dynamics) time-varying. The step size in the recursive least squares algorithm with forgetting factor or the Kalman filter for parameter estimation decays as $1/t$ initially in the transient phase, and then converges/fluctuates around a constant value. Algorithm 1 can be made adaptive by having a non-zero parameter noise w_t^θ.

3. APPLICATION EXAMPLE: CHAOS

The prediction and control of chaotic dynamical systems is a problem relevant to a number of fields in science and engineering Schuster (1995); Kapitaniak (1998); Lorenz (1993); Devaney (1989). Chaotic behavior has been postulated and/or observed in classical mechanics Henon and Heiles (1964), quantum mechanics Gutzwiller (1990), chemical reactions Vidal and Pacault (1981), civil engineering Naschie (1990), electrical circuits Chua (1992), and climatology Vallis (1988). Prediction and understanding of chaotic phenomena are frequently based on approximations of the underlying attractor and its dynamics Farmer and Sidorowich (1987), Casdagli (1989), Sugihara and May (1990), Sauer (1993). This approximation is typically performed by using values of the time series from embedding this attractor in some phase-space. Then approximations of the dynamics on the attractor are obtained in various neighborhoods using various methods, some linear, some not. Two important issues encountered are dealing with noise in the observations and dealing with relatively small amounts of data. These are important issues because the underlying geometry of chaotic attractors is frequently very fine, with varying structure at all levels of magnification. Another method sometimes used is neural networks Lapedes and Farber (1987) and Casdagli and Eubank (1992). The point of this paper is to examine the possible use of a particle filter in this context. In this paper, it is assumed that one has a time series of observations of some potentially chaotic system, and also a model of the underlying system, but with unknown parameters. It is further assumed that both the system and the observations have a small amount of error.

The paper in particular studies one of the simplest known models which can produce chaotic behavior, the logistics map,

$$z_{t+1} = \theta z_t(1 - z_t). \tag{8}$$

For θ in the the range $[3.56994568, 4]$ chaotic behaviour is frequently observed Schuster (1995). To reflect the fact that one may have a model of a potentially chaotic system but not the underlying parameters, our paper assumes that θ is unknown and includes it as a state variable.

The logistic map can in our notation be formulized as

$$z_{t+1} = \theta z_t(1 - z_t) + v_t \tag{9a}$$

$$y_t = z_t + e_t. \tag{9b}$$

In such an application, the state noise may simply be quantization noise from a finite precision simulation, and the measurement noise may be quantization noise from the sensor. However, we will investigate the case of Gaussian noise in Section 6, to be able to compute the Cramer-Rao lower bound.

The state space form (3) with state $x_t = (z_t, \theta)^T$ now becomes

$$x_{t+1} = \begin{pmatrix} \theta_t z_t(1 - z_t) \\ \theta_t \end{pmatrix} + \begin{pmatrix} v_t^z + w_t^z \\ w_t^\theta \end{pmatrix} \tag{10a}$$

$$y_t = z_t + e_t \tag{10b}$$

The parameter random walk noise $v_t^\theta = 0$, since a time constant θ is assumed.

4. PREDICTION

The simplest idea to predict future states in (1) is to estimate the state x_t and parameter θ using the particle filter, then use these values for a pure simulation of the state equation (1a). However, we get no information of the uncertainty in the prediction in this way. This is a particularly pronounced drawback for a chaotic model as (10). What we want to do is to evolve $p(x_t|Y_t)$ to $p(x_{t+k}|Y_t)$ using (1a). One idea to approximate this distribution is to use the particle approximation $x_t^{(i)} \sim p(x_t|Y_t)$, and to simulate each particle according to (1a). If this is done without both state and roughening noise $v_t^z = v_t^\theta = w_t^z = w_t^\theta = 0$, then a the state prediction with the smallest variance is obtained. On the other hand, keeping the state noise v_t with no roughening $w_t^z = w_t^\theta = 0$, the *a posterior* distribution $x_{t+k}^{(i)} \sim p(x_{t+k}|Y_t)$ is approximated, which may be needed in decision theory, statistical approaches to control, and risk calculations.

5. CRAMER-RAO LOWER BOUND

The Cramer-Rao lower bound for non-linear filtering takes a quite simple form for Gaussian noise. Intuitively, the best one can hope for is to get the same performance as the Kalman filter on a linearized model, which happens when the estimation error is small enough so the non-linearities f, h are approximately linear. One can show formally, see N. Bergman and Gordon (2001), that

$$\text{Cov}(\hat{x}_{t|t}) \geq J^{-1} = P_{t|t},$$

where J is the Fisher information matrix, for any estimator $\hat{x}_{t|t}$. The Cramer-Rao lower bound $P_{t|t}$ for this case is given by the Ricatti equation

$$\begin{aligned} P_{t|t-1} &= F_t P_{t-1|t-1} F_t^T + Q_t \\ P_{t|t} &= P_{t|t-1} - \\ & P_{t|t-1} H_t^T (H_t P_{t|t-1} H_t^T + R)^{-1} H_t P_{t|t-1}. \end{aligned}$$

For the chaos model (9), we get

$$F_t = \left.\frac{df}{dx}\right|_{x=x_t} = \begin{pmatrix} \theta(1-2z_t) & z_t(1-z_t) \\ 0 & 1 \end{pmatrix}$$

$$H_t = \left.\frac{dh}{dx}\right|_{x=x_t} = (1,\ 0),$$

which is the associated linear system with (10), evaluated at the trajectory x_t.

6. SIMULATIONS

The chaotic system

$$z_{t+1} = \theta z_t(1-z_t) + v_t \tag{11a}$$

$$y_t = z_t + e_t \tag{11b}$$

is simulated with 100 samples, $\theta = 3.95$, $\text{Var}(v_t^z) = 7 \cdot 10^{-14}$, $\text{Var}(e_t) = 8 \cdot 10^{-5}$ and initial state $x(1) = 0.513$,

The particle filter with $N_p = 1000$, $\text{Var}(w_t^z) = \text{Var}(w_t^\theta) = 10^{-4}$ is applied initialized with uniformly distributed random numbers in $\theta \in [3.8,\ 4]$ and $x(1) \in [0,\ 1]$. Figure 1 shows the particle filter output and its resulting simulation. Figure 2 illustrates the estimation error. We see that the proposed filter provides a converging parameter estimate, good state estimates and also gives results comparable to the Cramer-Rao lower bound.

Finally, Figure 3 illustrates how the predictive ability of chaos quickly deteriorates with the horizon k. Still, better performance than just guessing is achieved up to five time steps ahead, after the initial transient (the 25 first samples are removed here) has vanished. Note that a simulation initialized at any $x(1)$, would give a state variance of 0.3. The standard deviation of such a difference is thus $\sqrt{2}$ larger than the variance of $x(t)$ itself. That is, since the model is chaotic, long term predictions, even using the true θ, tend to be approximately 0.45 in the limit $k \to \infty$, which is the value the curve in Figure 3 converges to.

7. CONCLUSIONS

The particle filter shows promise in the problem of simultaneous state and parameter estimation in general, which was exemplified in an application of prediction of chaotic time series. The filter is applicable when:

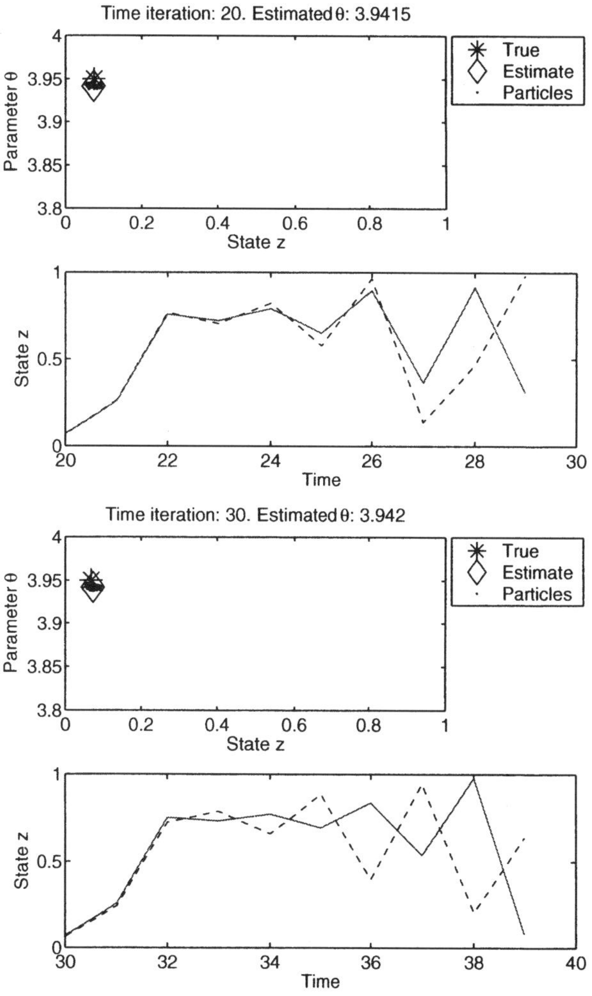

Fig. 1. Particle cloud and particle filter estimate of x, θ at time $i = 20$ and $i = 30$, respectively, compared to the true values. Lower plot shows the predictions based on the estimated parameter and state compared to the true state.

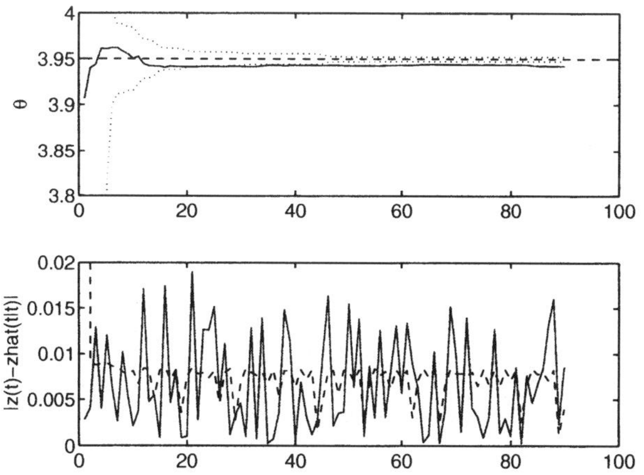

Fig. 2. Estimated $\hat{\theta}$ with Cramer-Rao confidence interval around true parameter, filter error $|x_t - \hat{x}_{t|t}|$ compared to Cramer-Rao lower bound.

- the problem is of moderate complexity (not more than, roughly, five states and parameters),
- the computational resources are good enough, compared to the sample interval,

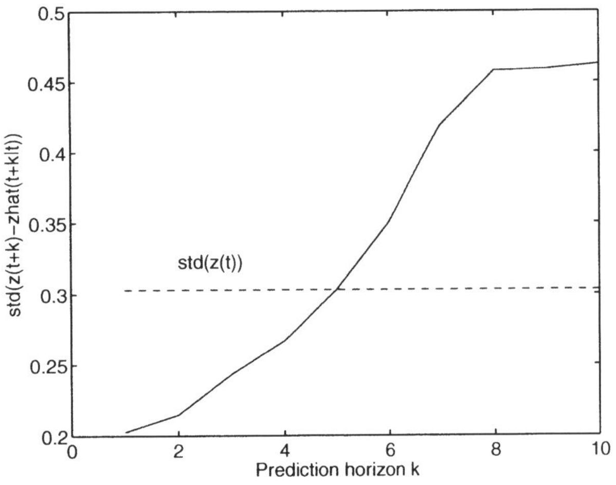

Fig. 3. Prediction error $\sqrt{\mathrm{E}(z_{t+k} - \hat{z}_{t+k|t})^2}$ (k=1,2,...,10). As a comparison, the state variance itself is 0.3.

- prior knowledge on a region containing the true initial state and parameter vector is available.

Having said these limitation, the particle filter has been demonstrated to be quite a flexible tool, handling all kind of non-linear systems, even including by hard constraints or geographical information systems. Any known noise distribution can be used to model *e.g.* sensor characteristic, state disturbances and prior knowledge on stability *etc.*. For system identification, where a part of the state vector corresponds to the time invariant parameter vector, the known special trick of adding artificial roughening noise is needed. With a carefully decaying roughening noise variance, a converging estimate is obtained.

The principle was demonstrated on a model for chaos, which is non-linear in the states and where the Cramer-Rao lower bound can be calculated.

References

B.D.O. Anderson and J.B. Moore. *Optimal Filtering*. Information and system science series. Prentice Hall, Englewood Cliffs, New Jersey, 1979.

N. Bergman. *Recursive Bayesian Estimation: Navigation and Tracking Applications*. Dissertation nr. 579, Linköping University, Sweden, 1999.

M. Casdagli. Nonlinear prediction of chaotic time series. *Physica D*, 35:335, 1989.

M. Casdagli and S. Eubank. *Nonlinear Modeling and Forecasting*. Addison-Wesley, 1992.

L.O. Chua. The genesis of chua's circuit. *Archiv fur Elektronik und Ubertragungstechnik*, 46:250–257, 1992.

Robert Devaney. *An Introduction to Chaotic Dynamical Systems, Second Edition*. Addison-Wesley, 1989.

A. Doucet. On sequential simulation-based methods for Bayesian filtering. Technical Report CUED/F-INFENG/TR.310, Signal Processing Group, Department of Engineering, University of Cambridge, 1998.

A. Doucet, N. de Freitas, and N. Gordon, editors. *Sequential Monte Carlo Methods in Practice*. Springer Verlag, 2001a.

A. Doucet, S.J. Godsill, and C. Andrieu. On sequential simulation-based methods for Bayesian filtering. *Statistics and Computing*, 10(3):197–208, 2000.

A. Doucet, N. Gordon, and V. Krishnamurthy. Particle Filters for State Estimation of Jump Markov Linear Systems. *IEEE Transactions on Signal Processing*, 49(3):613–624, 2001b.

J.D. Farmer and J. Sidorowich. Predicting chaotic time series. *Phys. Rev. Lett.*, 59:845, 1987.

P. Fearnhead. *Sequential Monte Carlo methods in filter theory*. PhD thesis, University of Oxford, 1998.

N.J. Gordon, D.J. Salmond, and A.F.M. Smith. A novel approach to nonlinear/non-Gaussian Bayesian state estimation. In *IEE Proceedings on Radar and Signal Processing*, volume 140, pages 107–113, 1993a.

N.J. Gordon, D.J. Salmond, and A.F.M. Smith. A novel approach to nonlinear/non-Gaussian Bayesian state estimation. In *IEE Proceedings on Radar and Signal Processing*, volume 140, pages 107–113, 1993b.

F. Gustafsson, F. Gunnarsson, N. Bergman, U. Forssell, J. Jansson, R. Karlsson, and P-J. Nordlund. Particle filters for positioning, navigation and tracking. *IEEE Transactions on Signal Processing*, 50(2), February 2002.

M.C. Gutzwiller. *Chaos in Classical and Quantom Mechanics*. Springer, 1990.

J.E. Handschin. Monte carlo techniques for prediction and filtering of non-linear stochastic processes. *a*, 6:555–563, 1970.

J.E. Handschin and D.Q. Mayne. Monte carlo techniques to estimate the conditional expectation in multi-stage non-linear filtering. *International journal of control*, 9:547–559, 1969.

M. Henon and C. Heiles. The applicability of the third integral of the motion: Some numerical results. *Astron, J.*, 69:73, 1964.

A.H. Jazwinski. *Stochastic processes and filtering theory*. Mathematics in science and engineering. Academic Press, New York, 1970.

Tomaz Kapitaniak. *Chaos for Engineers, Theory, Applications, and Control*. Springer-Verlag, 1998.

A. Kong, J. S. Liu, and W. H. Wong. Sequential imputations and Bayesian missing data problems. *J. Amer. Stat. Assoc.*, 89(425): 278–288, 1994.

A. Lapedes and R. Farber. Nonlinear signal processing using neural networks: Prediction and signal modeling. *preprint, Los Alamos*, 1987.

J.S. Liu. Metropolized independent sampling with comparison to rejection ampling and importance sampling. *Statistics and Computing*, 6:113–119, 1996.

L. Ljung. *System identification, Theory for the user*. Prentice Hall, Englewood Cliffs, NJ, second edition, 1999.

Edward Lorenz. *The Essence of Chaos*. University of Washington Press, 1993.

N. Metropolis and S. Ulam. The Monte Carlo Method. *Journal of the American Statistical Association*, 44(247):335–341, 1949.

A. Doucet N. Bergman and N.J. Gordon. Optimal estimation and cramer-rao bounds for partial non-gaussian state-space model. *Ann. Inst. Stat. Math*, 52(1):97–112, 2001.

M.S. El Naschie. *Stress Stability and Chaos*. McGraw-Hill, 1990.

B.D. Ripley. *Stochastic Simulation*. John Wiley, 1988.

T. Sauer. Time series prediction using delay coordinate embedding. *Time Series Prediction: Forecasting and Understanding the Past*, 1993.

T. Schön and F. Gustafsson. Particle filters for system identification of state-space models linear in either parameters or states. In *SYSID*, Rotterdam, NL, 2003.

Heinz Georg Schuster. *Deterministic Chaos: An Introduction, Third Edition*. VCH, 1995.

A.F.M. Smith and A.E. Gelfand. Bayesian statistics without tears: A sampling-resampling perspective. *The American Statistician*, 46(2):84–88, May 1992.

G. Sugihara and R.M. May. Nonlinear forecasting as a way of distinguishing chaos from measurement error in time series. *Nature*, 334:734, 1990.

G.K. Vallis. Conceptual models of el nino and the southern oscillation. *J. Geophys. Res.*, 93:13979–14019, 1988.

C. Vidal and A. Pacault. *Nonlinear Phenomena in Chemical Dynamics*. Springer, 1981.

IFAC
Publications
www.elsevier.com/locate/ifac

PARTICLE FILTERS FOR SYSTEM IDENTIFICATION OF STATE-SPACE MODELS LINEAR IN EITHER PARAMETERS OR STATES [1]

Thomas Schön and Fredrik Gustafsson

Division of Automatic Control and Communication Systems
Department of Electrical Engineering
Linköpings universitet, SE-581 83 Linköping, Sweden
Tel: +46 13 281373; fax: +46 13 282622
{schon, fredrik}@isy.liu.se

Abstract: The potential use of the marginalized particle filter for nonlinear system identification is investigated. The particle filter itself offers a general tool for estimating unknown parameters in non-linear models of moderate complexity, and the basic trick is to model the parameters as a random walk (so called roughening noise) with decaying variance. We derive algorithms for systems which are non-linear in either the parameters or the states, but not both generally. In these cases, marginalization applies to the linear part, which firstly significantly widens the scope of the particle filter to more complex systems, and secondly decreases the variance in the linear parameters/states for fixed filter complexity. This second property is illustrated on an example of chaotic model. The particular case of freely parametrized linear state space models, common in subspace identification approaches, is bi-linear in states and parameters, and thus both cases above are satisfied. One can then choose which one to marginalize. *Copyright © 2003 IFAC*

Keywords: System identification, Nonlinear estimation, Recursive estimation, Particle filters, Kalman filters, Bayesian estimation.

1. INTRODUCTION

In this contribution, the particle filter (Doucet *et al.*, 2001*a*) is applied to some classical system identification problems (Ljung, 1999) based on time-varying parametric state-space models

$$z_{t+1} = f(z_t; \theta_t) + v_t^z, \tag{1a}$$

$$y_t = h(z_t; \theta_t) + e_t, \tag{1b}$$

where $z \in \mathbb{R}^n$ is the state variable, $\theta_t \in \mathbb{R}^d$ is the parameter vector, and $y \in \mathbb{R}^m$ is the output variable. The additive noise terms are assumed to be independent and identically distributed (i.i.d.).

[1] This work was supported by the Swedish Research Council.

First, we briefly review the problem formulation in (Gustafsson and Hriljac, 2003). By augmenting the state vector with the parameters, $x_t^T = (z_t^T, \theta_t^T)$, and assuming a random walk parameter variation (of which constant parameters is a special case), we get

$$\begin{bmatrix} z_{t+1} \\ \theta_{t+1} \end{bmatrix} = \begin{bmatrix} f(z_t; \theta_t) \\ \theta_t \end{bmatrix} + \begin{bmatrix} v_t^z + w_t^z \\ v_t^\theta + w_t^\theta \end{bmatrix} \tag{2a}$$

$$y_t = h(z_t; \theta_t) + e_t, \tag{2b}$$

where the noises are physical state noise v_t^z, state roughening noise w_t^z, parameter random walk for time-varying parameters v_t^θ and parameter roughening noise w_t^θ. The roughening noise is instrumental in the particle filter to get good performance, and is a second level design parameter. For system identification, $v_t^\theta = 0$ and w_t^θ has a variance decaying to zero, which yields converging parameter estimates. The particle filter recursively approximates the posterior density function $p(X_t|Y_t)$, where $X_t = \{x_0, \ldots, x_t\}$, and the approximation converges to the true posterior when the number of particles tends to infinity. The only problem is that the practical limit for 'infinity' depends on the dimension of x_t, that is, the sum of number of parameters θ_t and states z_t. As a very coarse rule of thumb, do not try to use the particle filter for more than five unknowns.

Now, if there is linear substructure available in the model this can be exploited using *marginalization*. Conceptually, marginalization means that the linear states are marginalized out and then we can apply optimal filters for these states and the particle filter is only applied to the truly nonlinear states. In this way, the samples in the particle filter will live in a lower dimensional space. Hence, we will intuitively obtain *more accurate estimates* for a given number of samples, since we use the optimal filters for a part of the state vector. Alternatively, we can apply the particle filter on *more complex models*. These are the practical implications of our contribution.

We will in this contribution consider the two following special cases of (1a):

(1) The model is affine in the parameters and possibly nonlinear in the states, *i.e.*,

$$f(z_t;\theta_t) = A_t(z_t)\theta_t + f_t^z(z_t) \tag{3}$$
$$h(z_t;\theta_t) = C_t(z_t)\theta_t + h_t^z(z_t). \tag{4}$$

(2) The model is affine in the states and possibly nonlinear in the parameters, *i.e.*,

$$f(z_t;\theta_t) = A_t(\theta)z_t + f_t^{\theta_t}(\theta_t) \tag{5}$$
$$h(z_t;\theta_t) = C_t(\theta)z_t + h_t^{\theta_t}(\theta_t). \tag{6}$$

2. THE PARTICLE FILTER

We here briefly present the theory and main algorithm. For a more intuitive presentation, see the accompanying paper (Gustafsson and Hriljac, 2003).

2.1 *Recursive Bayesian estimation*

Consider systems that are described by the generic state space model (2). The optimal Bayesian filter in this case is given below. For further details, consult (Doucet *et al.*, 2001*b*; Bergman, 1999).

Denote the observations at time t by $Y_t = \{y_0, \ldots, y_t\}$. The Bayesian solution to compute the posterior distribution, $p(x_t|Y_t)$, of the state vector, given past observations, is given by (Bergman, 1999)

$$p(x_{t+1}|Y_t) = \int p(x_{t+1}|x_t)p(x_t|Y_t)\,dx_t, \tag{7a}$$
$$p(x_t|Y_t) = \frac{p(y_t|x_t)p(x_t|Y_{t-1})}{p(y_t|Y_{t-1})}. \tag{7b}$$

For expressions on $p(x_{t+1}|x_t)$ and $p(y_t|x_t)$ in (7) we use the known probability densities $p_{e_t}(x)$ and $p_{v_t+w_t}(x) = p_{v_t} * p_{w_t}(x)$, with all noises assumed independent,

$$p(x_{t+1}|x_t) = p_{v_t+w_t}\big(x_{t+1} - f(x_t)\big), \tag{8a}$$
$$p(y_t|x_t) = p_{e_t}\big(y_t - h(x_t)\big). \tag{8b}$$

2.2 *Implementation*

A numerical approximation to (7) is given by

$$p(x_t|Y_t) \approx \sum_{i=1}^{N} q_t^{(i)}\delta(x_t - x_t^{(i)}), \tag{9}$$

where $\delta(\cdot)$ is the Dirac delta function. The particles $x_t^{(i)}$ and the corresponding weights $q_t^{(i)}$ represent a sampled version of the posterior distribution $p(x_t|Y_t)$ (Doucet *et al.*, 2001*b*), and intuitively, the more samples the better approximation.

2.3 *The algorithm*

The discussion in the previous section is summarized in the algorithm below. This is the algorithm presented in (Gordon *et al.*, 1993) under the name, *Bayesian bootstrap filter*.

Algorithm 1. The Particle Filter

(1) Generate N samples $\{x_{0|-1}^{(i)}\}_{i=1}^N$ from $p(x_0)$.
(2) Calculate the weights $q_t^i = p(y_t|x_t^{(i)})$ and normalize, *i.e.*, $\bar{q}_t^{(i)} = \frac{q_t^{(i)}}{\sum_{j=1}^N q_t^{(j)}}$.
(3) Resample with replacement,
$$P(x_{t|t}^{(i)} = x_{t|t-1}^{(i)}) = \bar{q}_t^{(i)}.$$
(4) Predict (*i.e.*, simulate) new particles by
$$x_{t+1|t}^{(i)} = f(x_{t|t}^{(i)}) + v_t^{(i)} + w_t^{(i)}.$$
(5) Iterate from step (2).

The particle filter can be interpreted as a simulation-based method, *i.e.*, N possible state trajectories, $x_t^{(i)}$, $i = 1, \ldots, N$, are simulated. Based on the measurements each trajectory is assigned a weight, $\bar{q}_t^{(i)}$, representing the probability of that trajectory being the correct one.

3. MARGINALIZATION FOR VARIANCE REDUCTION

Consider the case where the model is linear in some of the states. Then the Kalman filter can be used to estimate the linear states, denoted x_t^l, and the particle filter can be used to estimate the nonlinear states, denoted x_t^n. To separate the problem of estimating $p(x_t^l, x_t^n|Y_t)$ into one linear and one nonlinear problem, Bayes' rule is used [2]

$$p(x_t^l, X_t^n|Y_t) = p(x_t^l|X_t^n, Y_t)p(X_t^n|Y_t). \tag{10}$$

Here the density $p(x_t^l|X_t^n, Y_t)$ is given by the Kalman filter and the particle filter is used to estimate $p(X_t^n|Y_t)$. This means that the particles live in a lower-dimensional space, and it can indeed be proven (Doucet *et al.*, 2001*c*; Nordlund, 2002) that the variance of any function of the state and parameter is decreased or remains constant when using marginalization for a given number of particles. This technique of marginalizing out the linear state is also referred to as Rao-Blackwellization (Doucet *et al.*, 2001*c*).

Before we state the theorem we have to introduce some notation. Let the estimate of any inference function of the state vector be given by

$$I(g(x_t)) = E_{p(x_t|Y_t)}[g(x_t)] = \int g(x_t)p(x_t|Y_t)dx_t, \tag{11}$$

and its estimate using N particles and the standard particle filter be denoted by $\hat{I}_N^s(g(x_t))$. When the marginalized particle filter is used the same estimate is denoted by $\hat{I}_N^m(g(x_t))$.

Theorem 1. Assume i.i.d. samples $\{x_t^{(i)}\}_{i=1}^N$ computed by Algorithm 1 and that the expected value and variance of the inference function $g(x_t)$ and the likelihood q_t exist and are finite. Then there is a central limit theorem stating that for large N,

$$\hat{I}_N^s(g(x_t)) \approx \mathcal{N}(I(g(x_t), R_s(N)),$$
$$\hat{I}_N^m(g(x_t)) \approx \mathcal{N}(I(g(x_t), R_m(N)),$$

where $R_s(N) \geq R_m(N)$.

See *e.g.*, (Doucet *et al.*, 2001*c*) for a proof.

Asymptotically as the number of particles tend to infinity there is nothing to gain in using marginalization, since then the particle filter will provide a perfect description of $p(x_t^l, x_t^n|Y_t)$. However, since we only can use a finite number of particles it is certainly useful to marginalize and use the optimal filter, *i.e.*, the Kalman filter, for the linear states. For details concerning the marginalized particle filter, the reader is referred to *e.g.*, (Chen and Liu, 2000), (Doucet *et al.*, 2001*c*), (Nordlund, 2002).

[2] We have to use all the old nonlinear states, X_t^n, in order to make the Kalman filter work. The density $p(x_t^l, x_t^n)$ is then obtained by integrating $p(x_t^l, X_t^n)$ over the old nonlinear states, X_{t-1}

4. MODELS

In this section it will be shown how the particle filter can be used to estimate the nonlinear states and the Kalman filter to estimate the linear states, using the marginalization technique discussed above. All noise terms associated with the linear states are here assumed to be Gaussian, which means that the optimal estimator for the linear states/parameters is given by the Kalman filter. For the details concerning the Kalman filter equations, the state transition densities, and the likelihood functions in Algorithms 2 and 3 the reader is referred to (Nordlund, 2002). First there will be a discussion on models that are linear in the states and nonlinear in the parameters. This is followed by the reversed case, *i.e.,* linear in the parameters and nonlinear in the states.

4.1 *State-space models linear in the states*

A state-space model linear in the states and possibly nonlinear in the parameters is written as

$$z_{t+1} = A_t(\theta_t)z_t + f^\theta(\theta_t) + v_t^z \tag{12a}$$
$$\theta_{t+1} = \theta_t + v_t^\theta + w_t^\theta \tag{12b}$$
$$y_t = C_t(\theta_t)z_t + h^\theta(\theta_t) + e_t, \tag{12c}$$

where $v_t^z \in \mathcal{N}(0, Q_t^{v,z})$, $v_{\theta,t} \in \mathcal{N}(0, Q_t^{v,\theta})$ and $w_{\theta,t} \in \mathcal{N}(0, Q_t^{w,\theta})$ [3]. Note that we can let the roughening noise w_t^z be zero when using marginalization. The posterior density will here be separated using Bayes' rule according to

$$p(z_t, \Theta_t|Y_t) = p(z_t|\Theta_t, Y_t)p(\Theta_t|Y_t). \tag{13}$$

Note that we here consider the posterior of the complete parameter trajectory Θ_t, but only the last state vector z_t. The first density on the right hand side in (13) is given by the Kalman filter, while the second one is approximated by the particle filter. That is, we randomize particles in the parameter space according to our prior, and then each particle trajectory will be associated with one Kalman filter. The exact algorithm is given below.

Algorithm 2. The particle filter for linear states
Below, we let $C_t^{(i)} = C_t(\theta_{t|t-1}^{(i)})$ and $A_t^{(i)} = A_t(\theta_{t|t}^{(i)})$ for ease of notation.

[3] The noise on the non-linear part, here w_t^θ and v_t^θ, can in fact have an arbitrary distribution. Similarly, The PDF $p(\theta_0)$ does not have any restrictions, since it is only used in the particle filter, the same goes for $p(e_t)$ if $C = 0$ in (12c).However, we leave these generalizations as a remark and assume Gaussian distributions.

(1) Sample $\theta_{0|-1}^{(i)} \sim p(\theta_0)$ and set, for $i = 1, \dots, N$, $\{z_{0|-1}^{(i)}, P_{0|-1}\}_{i=1}^{N} = \{z_0, \Pi_0\}$.

(2) Calculate the weights

$$\begin{aligned} q_t^{(i)} =& p(y_t|\Theta_t^{(i)}, Y_{t-1}) \\ =& \mathcal{N}(h^\theta(\theta_{t|t-1}^{(i)}) + C_t^{(i)}\hat{z}_{t|t-1}^{(i)}, \\ & R_t + C_t^{(i)}P_{t|t-1}(C_t^{(i)})^T \end{aligned}$$

and normalize, *i.e.*, $\bar{q}_t^{(i)} = \frac{q_t^{(i)}}{\sum_{j=1}^{N} q_t^{(j)}}$.

(3) Resample with replacement,

$$P(\theta_{t|t}^{(i)} = \theta_{t|t-1}^{(j)}) = \bar{q}_t^{(j)}$$

(4) Kalman filter measurement update:

$$\begin{aligned} S_t^{(i)} =& R_t + C_t^{(i)}P_{t|t-1}^{(i)}(C_t^{(i)})^T \\ L_t^{(i)} =& P_{t|t-1}^{(i)}(C_t^{(i)})^T(S_t^{(i)})^{-1} \\ \hat{z}_{t|t}^{(i)} =& \hat{z}_{t|t-1}^{(i)} + L_t^{(i)}(y_t - h^\theta(\theta_{t|t}^{(i)}) - C_t^{(i)}\hat{z}_{t|t-1}^{(i)}) \\ P_{t|t}^{(i)} =& P_{t|t-1}^{(i)} - L_t^{(i)}S_t^{(i)}(L_t^{(i)})^T \end{aligned}$$

(5) Predict (i.e., simulate) new particles by $p(\theta_{t+1|t}|\Theta_t, Y_t)$, where

$$\theta_{t+1|t}^{(i)} \sim p(\theta_{t+1|t}|\Theta_t^{(i)}, Y_t) = \mathcal{N}(\theta_{t|t}^{(i)}, Q_t^{v,\theta} + Q_t^{w,\theta}).$$

(6) Kalman filter time update:

$$\begin{aligned} \hat{z}_{t+1|t}^{(i)} &= A_t^{(i)}\hat{z}_{t|t}^{(i)} + f^\theta(\theta_{t|t}^{(i)}) \\ P_{t+1|1}^{(i)} &= A_t^{(i)}P_{t|t}^{(i)}A_t^T(\theta_{t|t}^{(i)}) + Q_t^{v,z} + Q_t^{w,z} \end{aligned}$$

(7) Compute relevant estimates from $p(z_t, \theta_t|Y_t)$.

$$\hat{\theta}_t = \sum_{i=1}^{N} \bar{q}_t^{(i)}\hat{\theta}_{t|t}^{k,(i)} \tag{15a}$$

$$\hat{z}_t = \frac{1}{N}\sum_{i=1}^{N} z_{t|t}^{(i)} \tag{15b}$$

(8) Iterate from step (2).

Comparing the algorithms 1 and 2 we see that the differences are in the prediction step, which now consists of a Kalman filter update stage (split into step 4 and 6) besides the prediction of the nonlinear states.

In some cases the same Riccati recursion can be used for all the particles, and hence a lot of computations can be saved. This occurs when the matrices A_t and C_t in (12) are independent of θ_t. In this case $P_{t|t}^{(i)} = P_{t|t}$. for all $i = 1, \dots, N$ and hence the covariance only has to be updated once for each t. More on this can be found in (Gustafsson *et al.*, 2002).

4.2 *State-space models linear in the parameters*

A state-space model that is linear in the parameters can be written as

$$z_{t+1} = A_t(z_t)\theta_t + f^z(z_t) + v_t^z + w_t^z \tag{16a}$$

$$\theta_{t+1} = \theta_t + v_t^\theta \tag{16b}$$

$$y_t = C_t(z_t)\theta_t + h^z(z_t) + e_t. \tag{16c}$$

In this case the posterior will be split the other way around, compared to the previous section, *i.e.*,

$$p(Z_t, \theta_t|Y_t) = p(\theta_t|Z_t, Y_t)p(Z_t|Y_t). \tag{17}$$

The last density is approximated by the particle filter, while the first one can be solved by a Kalman filter for a parameter estimation problem in a linear regression framework. The corresponding algorithm will thus be

Algorithm 3. The particle filter for linear parameters Below, we let $C_t^{(i)} = C_t(z_{t|t-1}^{(i)})$ and $A_t^{(i)} = A_t(z_{t|t}^{(i)})$ for ease of notation.

(1) Sample $z_{0|-1}^{(i)} \sim p(z_0)$ and set for $i = 1, \dots, N$ $\{\theta_{0|-1}^{(i)}, P_{0|-1}\}_{i=1}^{N} = \{\theta_0, \Pi_0\}$.

(2) Calculate the weights

$$\begin{aligned} q_t^{(i)} =& p(y_t|Z_t^{(i)}, Y_{t-1}) \\ =& \mathcal{N}(h(z_{t|t-1}^{(i)}) + C_t^{(i)}\hat{\theta}_{t|t-1}^{(i)}, \\ & R_t + C_t^{(i)}P_{t|t-1}^{(i)}(C_t^{(i)})^T) \end{aligned}$$

and normalize, *i.e.*, $\bar{q}_t^{(i)} = \frac{q_t^{(i)}}{\sum_{j=1}^{N} q_t^{(j)}}$.

(3) Resample with replacement,

$$P(z_{t|t}^{(i)} = z_{t|t-1}^{(j)}) = \bar{q}_t^{(j)} \tag{18}$$

(4) Kalman filter measurement update:

$$\begin{aligned} S_t^{(i)} &= (R_t + C_t^{(i)}P_{t|t-1}^{(i)}(C_t^{(i)})^T) \\ L_t^{(i)} &= P_{t|t-1}^{(i)}C_t^T(S_t^{(i)})^{-1} \\ \hat{\theta}_{t|t}^{(i)} &= \hat{\theta}_{t|t-1}^{(i)} + L_t^{(i)}(y_t - h(z_{t|t}^{(i)}) - C_t^{(i)}\hat{\theta}_{t|t-1}^{(i)}) \\ P_{t|t}^{(i)} &= P_{t|t-1}^{(i)} - L_t^{(i)}S_t^{(i)}(L_t^{(i)})^T \end{aligned} \tag{19}$$

(5) Predict (i.e., simulate) new particles by $p(z_{t+1|t}|Z_t, Y_t)$, where

$$\begin{aligned} z_{t+1|t}^{(i)} \sim& p(z_{t+1|t}|Z_t^{(i)}, Y_t) \\ =& \mathcal{N}(A_t^{(i)}\hat{\theta}_{t|t}^{(i)}, Q_t^{\theta,v} + Q_t^{\theta,w} + \\ & A_t^{(i)}P_{t|t}^{(i)}(A_t^{(i)})^T). \end{aligned} \tag{20}$$

(6) Kalman filter time and state update:

$$\begin{aligned} \hat{\theta}_{t+1|t}^{(i)} &= \hat{\theta}_{t|t}^{(i)} + K_t^{(i)}(z_{t+1|t}^{(i)} - A_t^{(i)}\hat{\theta}_{t|t}^{(i)}) \\ P_{t+1|t}^{(i)} &= P_{t|t}^{(i)} + Q_t^{v,z} + Q_t^{w,z} - \\ & \quad K_t^{(i)}(Q_t^n + A_t^{(i)})P_{t|t}^{(i)}(A_t^{(i)})^T)(K_t^{(i)})^T \\ K_t^{(i)} &= P_{t|t}^{(i)}(A_t^{(i)})^T(Q_t^n + A_t^{(i)}P_{t|t}^{(i)}(A_t^{(i)})^T)^{-1} \end{aligned}$$

(7) Compute relevant estimates from $p(z_t, \theta_t|Y_t)$.

$$\hat{z}_t = \frac{1}{N}\sum_{i=1}^{N} z_{t|t}^{(i)} \tag{22a}$$

$$\hat{\theta}_t = \sum_{i=1}^{N} \bar{q}_t^{(i)}\hat{\theta}_{t|t}^{k,(i)} \tag{22b}$$

(8) Iterate from step (2).

The measurements used in the Kalman filter are thus the "normal" measurements y_t and the predicted state trajectory $z_{t+1|t}$, *i.e.*, the samples from the particle filter. Step 6 in the current algorithm contains a measurement update, using the prediction (since this contains information about θ_t) from the particle filter, and a time update.

5. PARAMETRIC INNOVATION MODELS

An interesting special case of the two different model types discussed above is when we consider "the intersection" of the two types, *i.e.*, a model that is bilinear in the states, z_t, and in the parameters, θ_t.

A particular case of interest is a general state-space model in innovation form

$$z_{t+1} = A(\theta_t)z_t + K(\theta_t)e_t \quad (23a)$$
$$y_t = C(\theta_t)z_t + e_t, \quad (23b)$$

where the parameters enter linearly in A, K, and C. The posterior will here be according to (17). One popular approach here is so called subspace identification, see (?). This class of algorithms usually perform very well and gives consistent estimates. One limitation is that it is hard to give the *a posterior* distribution of the parameters, even in the Gaussian case, and this is perhaps where the particle filter can help. However, the particle filter has perhaps not so much to offer to this bi-linear model, and this case is mentioned more to show the relation to classical system identification problems.

Assume, to avoid ambiguities in the state coordinates, an observer canonical form and scalar output, where $C = (1, 0, \ldots 0)$ and that all parameters in A and K are unknown. Then, given the state trajectory and measurement, we have from (16) the linear regression $y_t = Az_t + K(y_t - z_t^{(1)})$. This regression problem has to be solved for each particle $z_t^{(i)}$.

In the case where there are more states to be estimated than parameters, *i.e.*, $\dim z > \dim \theta$ it is better to split the density $p(Z_t, \theta_t|Y_t)$ in (17) the other way around, *i.e.*, as in (13). This time, a Kalman filter estimating the states z_t for each particle $\theta_t^{(i)}$ is needed. In this way the dimension of the state estimated by the particle filter is kept as low as possible. An example where this situation typically occurs is in gray-box identification (Ljung, 1999).

6. CHAOS EXAMPLE

The ideas presented in this article will be illustrated using the following chaotic model

$$z_{t+1} = (1 - z_t)z_t\theta + v_t, \quad (24a)$$
$$y_t = z_t + e_t, \quad (24b)$$

where, z_t, is the state variable, y_t, is the measurement, θ is the unknown parameter, v_t is the process noise, and e_t is the measurement noise. Both these noise densities are Gaussian distributed. The aim is to recursively estimate both the state, z_t, and the parameter, θ. This model is linear in the time-invariant parameter θ and nonlinear in the state z_t. This fits our framework, according to Section 4.2 and hence Algorithm 3 can be applied. This problem has also been studied in (Gustafsson and Hriljac, 2003), where the particle filter was directly applied to the augmented state $x_t = (z_t, \theta_t)$. The model (24) can be written on the form (16), *i.e.*,

$$z_{t+1} = A_t(z_t)\theta_t + v_t + w_t^z, \quad (25a)$$
$$\theta_{t+1} = \theta_t + w_t^\theta, \quad (25b)$$
$$y_t = h_t(z_t) + e_t, \quad (25c)$$

where $A_t(z_t) = z_t(1 - z_t)$ and $h_t(z_t) = z_t$. The two noises $w_t^z \sim \mathcal{N}(0, Q_t^{w,z})$ and $w_t^\theta \sim \mathcal{N}(0, Q_t^{w,\theta})$ are roughening noises. Furthermore $e_t \sim \mathcal{N}(0, R_t)$.

In the simulations, two different particle filters were used, the standard particle filter, Algorithm 1, applied to the augmented state vector, x_t, and the marginalized particle filter according to Algorithm 3. The true value of θ is 3.92, and the initial guess is $\theta_{0|-1} \sim \mathcal{N}(3.83, 0.04)$. The initial state is $z_0 \sim \mathcal{N}(0, 1)$. We do not use any process noise, however we have roughening noises $Q_0^{w,z} = Q_0^{w,\theta} = 10^{-2}$, which is decreased at each time step, according to (Gustafsson and Hriljac, 2003). The measurement noise has variance $R_t = 10^{-5}$, and we have used 200 Monte Carlo simulations. In Fig. 1 the filtered estimates of θ are shown using these two algorithms for 150, 1000, and 10000 particles respectively. In order to make the dif-

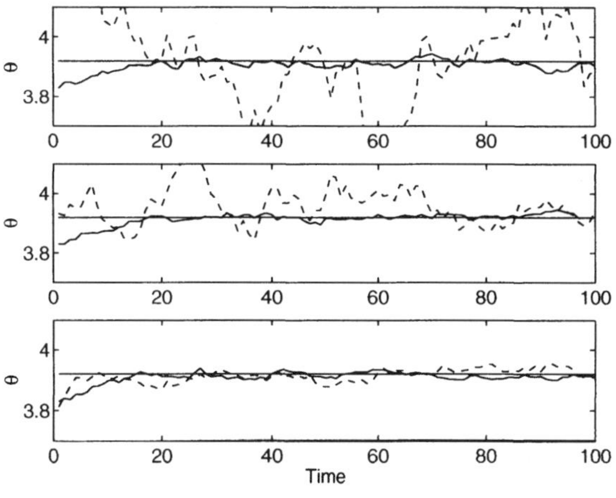

Fig. 1. Estimates of θ using the standard (dotted) and the marginalized (solid) particle filters. The true θ is shown using a dashed line. Top plot - 150 particles, middle - 1000 particles, bottom - 10000 particles.

ference more apparent the Root Mean Square Error (RMSE) is plotted in Fig. 2 as a function of the number of particles used in the simulations. Note that the RMSE values are calculated from time 50. In that way the transient effects are not included in the RMSE values. According to Theorem 1 the estimates should

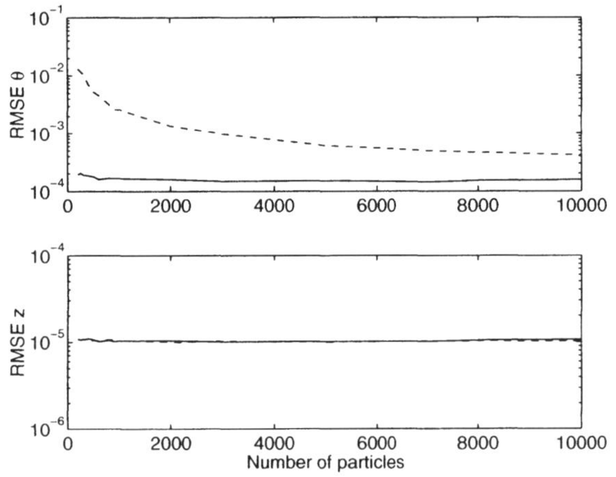

Fig. 2. RMSE values for $\hat{\theta}$ (top) and $\hat{z}$ (bottom) as a function of the number of particles used. Notice that a log-scale has been used in the plots, and that a dashed line has been used for the standard particle filter and a solid line for the marginalized particle filter.

be better or the same when we use the marginalized particle filter. From Fig. 2 we can see that this is indeed the case. It is only the estimate of the linear part, θ, that is improved, this is also consistent with the theory, see *e.g.*, (Nordlund, 2002) for the theoretical details. That this is true in the simulations is apparent by Fig. 2, from which it is clear that the estimate of the linear part (top) clearly is better using the marginalized particle filter. The estimate of the nonlinear part, z_t, has the same quality. Of course if we could use an infinite number of particles the results using the standard and the marginalized particle filter would have been the same, since then the particle filter would be able to provide an arbitrarily good estimate of $p(x_t|Y_t)$. We can see indications of this fact in the top plot in Fig. 2, since the more particles that are used the closer the estimates get.

7. CONCLUSIONS

The potential use of particle filtering for system idenfication of unknown parameters in non-linear systems was explained in the accompanying paper (Gustafsson and Hriljac, 2003). Here, we have proposed the use of marginalized particle filters. More specifically, we studied the cases where the model is either linear in the states and nonlinear in the parameters, or nonlinear in the states and linear in the parameters. The algorithms were given for these two cases. It is straightforward to give the algorithm for an arbitrary mix of linear and nonlinear states and parameters. The advantage of marginalization is that one can apply the filter to larger problems with more states and parameters, or that fewer particles and thus less filter complexity is needed for a given performance. Finally an example was given, which illustrates the improvement in estimation performance compared to using a standard particle filter.

REFERENCES

Bergman, N. (1999). Recursive Bayesian Estimation: Navigation and Tracking Applications. Dissertation nr. 579. Linköping University, Sweden.

Chen, R. and J.S. Liu (2000). Mixture Kalman filters. *Journal of the Royal Statistical Society* **62**(3), 493–508.

Doucet, A., de Freitas, N. and Gordon, N., Eds.) (2001*a*). *Sequential Monte Carlo Methods in Practice*. Springer Verlag.

Doucet, A., de Freitas, N. and Gordon, N., Eds.) (2001*b*). *Sequential Monte Carlo Methods in Practice*. Springer Verlag.

Doucet, A., N. Gordon and V. Krishnamurthy (2001*c*). Particle Filters for State Estimation of Jump Markov Linear Systems. *IEEE Transactions on Signal Processing* **49**(3), 613–624.

Gordon, N.J., D.J. Salmond and A.F.M. Smith (1993). A novel approach to nonlinear/non-Gaussian Bayesian state estimation. In: *IEE Proceedings on Radar and Signal Processing*. Vol. 140. pp. 107–113.

Gustafsson, F. and P. Hriljac (2003). Particle filters for prediction of chaos. Accepted for publication at the 13th IFAC Symposium on System Identification.

Gustafsson, F., F. Gunnarsson, N. Bergman, U. Forssell, J. Jansson, R. Karlsson and Nordlund P-J (2002). Particle filters for positioning, navigation and tracking. *IEEE Transactions on Signal Processing*.

Ljung, L. (1999). *System identification, Theory for the user*. System sciences series. 2 ed.. Prentice Hall. Upper Saddle River, NJ.

Nordlund, P.J. (2002). Sequential Monte Carlo Filters and Integrated Navigation. Licenciate thesis. Thesis No. 945.

IFAC
Publications
www.elsevier.com/locate/ifac

FAULT DETECTION, ISOLATION AND DIAGNOSIS WITH PARTICLE FILTERS FOR NONLINEAR STOCHASTIC SYSTEMS

Visakan Kadirkamanathan * **Ping Li** **

* *Automatic Control and Systems Engineering Dept, University of Sheffield, UK*
** *Electrical and Electronic Engineering Dept, Loughborough University, UK*

Abstract: In this paper, we provide a particle filtering based approach to the problems of fault detection isolation and diagnosis of general nonlinear stochastic systems. The algorithms use decision criteria based on the generalised likelihood ratios which are computed using the samples based probability distribution information of the state variables given by the particle filters. *Copyright © 2003 IFAC*

Keywords: Particle filter, Monte Carlo methods, Likelihood ratio, Nonlinear systems, Stochastic systems, Fault detection and isolation and Fault Diagnosis.

1. INTRODUCTION

The increasing complexity and growing demands for reliability of modern control system have stimulated the development of different fault detection and isolation (FDI) approaches, as can be seen from the survey papers (see e.g. (Isermann, 1984), (Basseville, 1988), (Frank, 1990), (Patton and Chen, 1996). In model-based approaches, the FDI is based on available input-output measurements and a mathematical model of the system to be monitored. One of the main difficulties in FDI of dynamic systems is due to the presence of unknown and unmeasured variables, typically state variables x. Two approaches are commonly used to deal with them: *estimation* and *elimination*. The estimation of x is usually performed with observers for deterministic systems, or filters for stochastic systems, which lead to observer-based and innovation-based FDI approaches respectively. The elimination of x directly explores the analytical redundancy embodied in the mathematic model. For the linear system, this leads to the well-known parity space-based FDI approach. For nonlinear deterministic systems, the nonlinear observer-based approaches have been reviewed in (Garcia and Frank, 1997). However, in comparison with linear systems, the literature addressing model-based FDI for nonlinear stochastic systems is not extensive, the main reason being that the estimation of the state vector of a nonlinear stochastic system is not easy.

Recently, the *particle filter*, a simulation-based method for nonlinear non-Gaussian state estimation, has attracted much attention (see (Gordon *et al.*, 1993), (Kitagawa, 1998), (Bolviken *et al.*, 2001), (Doucet *et al.*, 2001). This interest stems from the great advantage of the particle filter being able to handle any functional nonlinearity and system or measurement noise of any probability distribution.

The paper is organized as follows: In Section 2, the fault detection, isolation and diagnosis problem is formulated. This is followed by the description of the particle filter in Section 3. Then, its application to fault detection is developed in Section 4 followed by its extension to fault isolation. Fault diagnosis is addressed in section 6. This is followed by simulation results in section 7 and conclusions in section 8.

2. PROBLEM FORMULATION

The dynamical system of interest in this paper is nonlinear and stochastic, given in discrete-time state-space representation,

$$\begin{aligned} x_k &= f_k(x_{k-1}, \theta, w_{k-1}) \\ y_k &= h_k(x_k, \theta, v_k) \end{aligned} \tag{1}$$

where $x \in \Re^n$ is the system state, $y \in \Re^m$ is the output measurement, $\theta \in \Re^p$ is the system parameter, $w \in \Re^n$ is the system disturbance and $v \in \Re^m$ is the measurement noise. The functions $f_k(\cdot) : \Re^n \times \Re^n \mapsto \Re^n$ and $h_k(\cdot) : \Re^n \times \Re^m \mapsto \Re^m$ can be both nonlinear or linear and assumed known. The noise and disturbance are assumed to be additive and their characteristics known, generally taken to be zero mean Gaussian white noise.

The type of faults of interest here are the failure type where the system parameter values jump to a new value reflected in a change in the function $f(\cdot)$ and /or $h(\cdot)$. Such faults can be detected using the state observer approach or the filtering approach. The idea is to generate estimates of the states and the predicted outputs from these state estimates. The residuals or innovation from the output prediction are used in a measure which changes significantly under a failure type fault. Such a fault detection scheme facilitates on-line application since the state estimates and the predicted outputs can be generated on-line.

The problems of fault detection, isolation and diagnosis can be defined as follows:

Fault Detection: Let condition $S_0 : \theta \in \Theta_0$ denote the condition that the system is normal or fault free. Let condition $\bar{S}_0 : \theta \notin \Theta_0$, with $\Theta_0 \cap \bar{\Theta}_0 = \emptyset$, be the complementary condition that the system is not normal or that the system is faulty. Then, fault detection problem requires the detection of the transition event $E : S_0 \to \bar{S}_0$. The assumptions are that the system is initially in condition S_0 and that only one transition event takes place in the time interval of interest [0,KT], where T is the sampling interval.

Fault Isolation: Let conditions $S_i : \theta \in \Theta_i$ for $i = 1,2,\ldots,M$ be mutually exclusive so that $\Theta_i \cap \Theta_j = \emptyset$ for all $i \neq j$ and exhaustive so that $\Theta_1 \cup \Theta_2 \cup \ldots \cup \Theta_M = \bar{\Theta}_0$, where the conditions represent mutually exclusive and exhaustive set of known system fault conditions. The fault isolation problem requires the identification of the condition (amongst the M possible) following the transition event under the same assumptions.

Fault Diagnosis: The fault diagnosis problem requires the estimation of the parameters $\theta \notin \Theta_0$ so that the level of fault $\tilde{\theta} = \theta - \theta^0$ can be estimated. If is known that ith fault has taken place so that the system is in condition S_i, then $\theta \in \Theta_i$ requires estimation. To avoid the estimation of parameters lying in particular region in the parameter space, the estimation procedure can also be carried out without imposing additional constraints.

3. PARTICLE FILTER

The Bayesian approach to dynamic state estimation problems involves the construction of the probability density function (PDF) of the current state x_k, given the measurements up to time k. If Z_k is denoted to be the set of measurements up to time k, *ie.*, $Z_k = \{y_1, y_2, \cdots, y_k\}$, then the Bayesian solution would be to calculate the PDF $p(x_k|Z_k)$. This PDF will encapsulate all the information about the state x_k which is contained in the measurements Z_k and the prior PDF of x_0.

For linear Gaussian systems where the PDF can be summarised by means and covariances, the Kalman filter is used to propagate and update the means and covariances of the PDF. For general nonlinear, non-Gaussian systems, there is no simple way to proceed. The reason is that there is no general analytic expression for the required PDF. *Particle filter*, a sequential Monte Carlo filter, was proposed as a new way of representing and recursively generating an approximation to the conditional PDF $p(x_k|Z_k)$ (Gordon *et al.*, 1993). The key idea is to represent the PDF by a swarm of points called "particles", rather than by a function over the state space. As the number of particles increases, they effectively provide a good approximation to the required PDF. Following (Gordon *et al.*, 1993), the particle filter algorithm can be described as follows:

- Assume that there is a set of random samples (particles) $\{x_{k-1}(i) : i = 1,2,\cdots,N\}$ from the PDF $p(x_{k-1}|Z_{k-1})$.
- **Prediction:** Sample N values $\{w_{k-1}(i) : i = 1,2,\cdots,N\}$ from the PDF of system noise w_{k-1}. Use these to generate new swarm of points $\{x_k^*(i) : i = 1,2,\cdots,N\}$, where,

  $$x_k^*(i) = f_k(x_{k-1}(i), w_{k-1}(i)) \tag{2}$$

  based on equation (1).
- **Update:** Assign each $x_k^*(i)$ a weight $q_k(i)$ for $i = 1,2,\cdots,N$, after measurement y_k is received. The weights are given by,

  $$q_k(i) = \frac{\tilde{q}_k(i)}{\sum_{j=1}^{N} \tilde{q}_k(j)} \tag{3}$$

  where $\tilde{q}_k(i)$ are the un-normalised weights

  $$\tilde{q}_k(i) = p(y_k|x_k^*(i)) \tag{4}$$

  This defines a discrete distribution over $\{x_k^*(i) : i = 1,2,\cdots,N\}$, which assigns probability mass $q_k(i)$ to the element $x_k^*(i)$.
- **Resample:** Resample independently N times from the above discrete distribution. The resulting particles $\{x_k(i) : i = 1,2,\cdots,N\}$ satisfies

$$P\{x_k(i) = x_k^*(j)\} = q_k(j) \quad \text{for all } i \quad (5)$$

and forms an appropriate sample from the posterior PDF $p(x_k|Z_k)$.

- The prediction, update and resample steps form a single iteration and is recursively applied for each k.

The particle filter algorithm above is described as SIR (sampling importance resampling) particle filter due to the resampling method employed being SIR. Several alternatives with improved performances are available in the literature such as the regularised particle filter, auxillary particle filter and the use of Markov chain Monte Carlo sampling techniques (Doucet *et al.*, 2001).

4. FAULT DETECTION

The advantages of using the complete PDF of the system state in a fault detection scheme is bound to be superior than one which uses Gaussian approximations, such as in the extended Kalman filter (EKF). The problem of fault detection is to test for the hypothesis that the system is in condition S_0 (normal) against it being in condition $\bar{S}_0$ (faulty). The decision to accept or reject the normal hypothesis can be made based on the ratio of the conditional probabilities,

$$L_k = \frac{\Pr(S_0|Z_k)}{\Pr(\bar{S}_0|Z_k)} = \frac{p(Z_k|S_0)\Pr(S_0)}{p(Z_k|\bar{S}_0)\Pr(\bar{S}_0)} \quad (6)$$

If $d_k \geq 1$ then accept the normal hypothesis that system is in condition S_0 and if $d_k < 1$ then reject the normal hypothesis. If the prior probabilities are assumed equal, then d_k becomes the likelihood ratio test. The likelihoods can be determined if the system parameters under both possible conditions are known. In practice, only the nominal parameters θ^0 are likely to be known and hence LR cannot be computed. Fault detection then must rely only on the likelihood from the normal model which can be recursively computed for on-line implementation by decomposing the likelihood $p(Z_k|S_0) = \prod_{i=1}^{k} p(y_i|Z_{i-1},S_0)$. This is in fact the innovations likelihood that emerges from the state estimation stage (with particle filters) and can be computed through the following stages:

- **Output prediction:** The output prediction samples $\{y_k^*(i) : i = 1,2,\cdots,N\}$ are generated using the measurement equation in (1), where,

$$y_k^*(i) = h_k(x_k^*(i)) \quad (7)$$

- **Innovations likelihood:** With residuals or innovations defined as $r_k(i) = y_k - y_k^*$, the innovations likelihood is given by,

$$p(r_k|Z_k) = \frac{1}{N}\sum_{i=1}^{N} \tilde{q}_k(i) \quad (8)$$

The relevant likelihood based on all observations arise in the detection criteria under the assumption that the system is either normal or faulty throughout the time interval of interest. For change detection, a robust criteria is to compute the likelihood over a window of length κ. The windowed likelihood is

$$D(k) = \prod_{j=k-\kappa+1}^{k} p(r_k|Z_k) \quad (9)$$

or equivalently the negative log likelihood is

$$\mathscr{L}(k) = \sum_{j=k-\kappa+1}^{k} -\ln(p(r_k|Z_k)) \quad (10)$$

is computed and the condition $\mathscr{L}(k) > \varepsilon$ is tested for the presence of a fault.

5. FAULT ISOLATION

For the development of the particle filter based fault isolation scheme, we assume that the normal behavior and all possible faults of the physical system to be monitored can be described by the parameters taking values $\theta^0 \in \Theta_0$, $\theta^1 \in \Theta_1$, ..., $\theta^M \in \Theta_M$ associated with conditions $S_0, S_1, \ldots S_M$ respectively. Then $M+1$ models, each with the relevant fixed parameter values as above can be constructed. Fault isolation is then to determine which of the M possible faulty models the system has jumped to. In this scenario, the likelihood ratio test is the most powerful or optimal test (see e.g. (Basseville and Nikiforov, 1998),(Bar-Shalom and Li, 1993)), in the sense that it minimizes the probability of miss alarm (or maximizes the probability of detection subject to a given probability of false alarm.

Again, the idea is to compute the joint likelihood of the observations conditional on each hypothesized model through Monte Carlo estimation which uses the complete sample-based pdf information provided by the particle filter. Then we activate in parallel M log-likelihood ratio (LLR) tests for conditions $S_l(l = 1,2,\cdots,M)$ versus S_0. More specifically, the joint log-likelihood ratio to be computed in the present case is actually as follows:

$$L_j^k(h) = \sum_{t=j}^{k} \ln \frac{p(y_t|S_h, Z_{t-1})}{p(y_r|S_0, Z_{r-1})} \quad (11)$$

The calculation is performed *via* Monte Carlo integration using the complete sample based pdf information from the particle filter as in the previous section.

The typical behavior of the sum of LLR $L_1^k(l)$ defined by (11) shows, on average, a negative drift before the onset of a fault and a positive drift after the onset of a fault as k increases (see e.g. (Basseville and Nikiforov, 1998)). The decision function for fault detection which involves the double maximization is then given by:

$$d_k = \max_{1 \leq j \leq k} \max_{1 \leq l \leq M} L_j^k(l) \quad (12)$$

the value of d_k will, on average, be below zero if no fault takes place and drift positively away, as k increases, from zero after a fault occurs. Thus, the fault alarm is set at time t_a determined by:

$$t_a = \min\{k : d_k > \lambda\} \quad (13)$$

where the threshold $\lambda > 0$ is chosen to provide a reasonable tradeoff between false alarms and non detection. Following the fault detection, fault isolation is achieved by finding out the faulty model index l which, along with the maximum likelihood estimate (MLE) $\hat{t}_0$ of fault onset time, is given by:

$$(\hat{l}, \hat{t}_0) = \arg \max_{1 \leq j \leq t_a} \max_{1 \leq l \leq M} L_j^{t_a}(l) \quad (14)$$

The full implementation of the new Monte Carlo filtering-based LLR detector as described above requires a linearly growing number of calculations, as $L_j^k(l)$ must be calculated for $l = 1, \cdots, M$, and *all* possible fault onset times up to the present, i.e. $j = 1, \cdots, k$. The standard method to avoid this problem is to constrain the search in a fixed width (say κ) "sliding window" of the most recent past observations. In practice, the window size κ can be chosen as large as possible subject to the requirement for the computational time. If the window is sufficiently wide to insure detection and identification of all important faults, this approximation avoids the aforementioned difficulty. The detector proposed in this section can be considered as an extension of the Willsky-Jones' GLR detector to the general nonlinear non-Gaussian case through Monte Carlo filtering and Monte-Carlo integration.

The scheme outlined above requires the knowledge of all fault modes in the system so that the parameter values and hence all the possible models for the system are known. In practice, this is unrealistic and models for only a smaller subset of fault modes are available. Also, the level of fault in each of these cases is determined a priori and hence more than a single model is required to capture the various levels of even a single mode of failure. An alternative approach is to partition the parameter space into regions based on the subsets of the parameter vector which are typically associated with subsystems. Then, if parameters of the model are estimated, by determining the region into which the parameter vector is present, the fault can be isolated to the corresponding subsystem. However, parameter estimation is necessitated by this procedure and noting that fault diagnosis also requires parameter estimation, fault isolation and diagnosis can be performed simultaneously.

6. FAULT DIAGNOSIS

Fault diagnosis using the observer approach requires the simultaneous estimation of the state and the parameters, which can be achieved using the augmented states model (Anderson and Moore, 1979). The use of particle filter for estimating the states and the parameters has been demonstrated in (Kitagawa, 1998). The idea is to use an augmented state $\xi^T = [x^T \quad \theta^T]$ and re-write the state-space model in terms of ξ. The following set of equations result:

$$\begin{aligned} \begin{bmatrix} x_k \\ \theta_k \end{bmatrix} &= \begin{bmatrix} f_k(x_{k-1}, \theta_{k-1}, w_{k-1}) \\ \theta_{k-1} + w'_{k-1} \end{bmatrix} \\ y_k &= h_k(x_k, \theta_k, v_k) \end{aligned} \quad (15)$$

where the disturbance term w'_{k-1} is introduced by the use of a random walk model for parameter evolution to allow the exploration of the parameter space. The random walk model is an integral process and is marginally stable. Application of particle filters does not guarantee asymptotic convergence in this case (Crisan and Doucet, 2002) and thus a stable process with a pole just inside unit circle may be used in implementation.

Given the above state space representation, the particle filter, can be used to estimate the states and the parameters. Such an estimate is given by,

$$\hat{\xi}_k = \sum_{i=1}^{N} q_k(i) \xi_k^*(i) \quad (16)$$

The estimate is essentially a weighted average of the particles or samples representing the underlying distribution. The parameter estimates $\hat{\theta}_k$ can be compared to the nominal values θ^0 as a means for fault detection and its deviation $\tilde{\theta}_k = \theta^0 - \hat{\theta}_k$ be used for fault isolation.

The augmented state space model is attractive in principle. However, this increases the dimensionality of the model and thereby demands the increase in the number of particles for suffciently accurate results. Identifiability of the parameters and the states is also an issue. It is possible under particular restrictions to design a more efficient algorithm for fault isolation and diagnosis. If we can identify subsets of parameters associated with subsystems and further assume that system faults will always be restricted to only one subsystem, then, we can define M augmented models of reduced dimensionality along with the nominal system model:

Nominal particle filter M_0 :

$$\begin{aligned} x_k^{(0)} &= f(x_k^{(0)}, \theta_0, w_{k-1}) \\ y_k &= h(x_k^{(0)}, \theta_0, v_k) \end{aligned}$$

The set of M augmented state space models are:

Adaptive particle filter M_l :

$$\begin{bmatrix} x_k^{(l)} \\ \theta_{l,k} \end{bmatrix} = \begin{bmatrix} f(x_k^{(l)}, \theta_l^0, \theta_{l,k-1}, w_{k-1}) \\ \theta_{1,k-1} + w_1' \end{bmatrix}$$

$$y_k = h(x_k^{(l)}, \theta_l^0, \theta_{l,k}, v_k)$$

where, for $l = 1, \cdots, M$, θ_l^0 is the part of θ^0 complementary to its lth subsystem parameter set which at time instant k takes the values $\theta_{l,k}$. It can be seen that the state space descriptions for each model are essentially the same as that for nominal model, but the state vector is augmented by a different component of θ.

As we have M adaptive particle filters running in parallel and each estimates parameters of one subsystem it seems that the FDI could be achieved using these M parameter estimates. Unfortunately, because each adaptive particle filter estimates only a subset of parameters and assumes that the other parameters are known and constant, a change in a single parameter affects all the models and hence the parameter estimates. It is thus not straightforward to determine which parameter has really changed. This motivates the development of the following FDI method.

Prior to the occurrence of any fault, the output predictions given by the filters described above are all filtered versions of the actual measurement y, therefore, after an initial transition, they are identical up to filtering errors resulting only from the intrinsic uncertainty in the system model. Consequently, the likelihood of the innovations based on each particle filter are all close to each other. Thus, the joint LLRs $L_j^k(l)(l = 1, \cdots, M)$ are all close to zero after an initial period of transition.

In the presence of a fault, due to different adaptations of the above Monte Carlo filters, the output predictions behave differently, and the likelihood of the output predictions based on different adaptive particle filter and the nominal particle filter are no longer close to each other and the joint LLRs $L_j^k(l)(l = 1, \cdots, M)$ will drift away from zero. If $\theta^f (1 \le f \le M)$ is the changed parameter, then the likelihood $p(y_k|M_f, Z_{k-1})$ based on the fth adaptive particle filter will be greater than the likelihood based on any other adaptive particle filter and the nominal particle filter. Thus $L_j^k(f)$ will positively drift away from zero as k increases and take the greatest value among the M LLRs.

Based on the above argument, define the decision function for fault detection as:

$$d_k = \max_{1 \le l \le M} \max_{1 \le j \le k} L_j^k(l) \qquad (17)$$

then d_k is close to zero in the fault-free condition and increases positively after the occurrence of a fault. Fault detection could thus be achieved by thresholding d_k and a fault alarm is set at the time t_a determined by:

$$t_a = \min\{k : d_k > \lambda\} \qquad (18)$$

where $\lambda > 0$ is a threshold which depends on the noise level in the monitored system and can be determined by simulation. The fault isolation is achieved, after fault detection, by finding out the index f of the faulty parameter which is given by:

$$f = \arg \max_{1 \le m \le M} L_{t_a}^k(m) \qquad (19)$$

Once again, the implementation of the LLR is constrained in a sliding window with width κ to avoid the linearly growing computation.

7. SIMULATION RESULTS

The example is a univariate growth model taken from (Gordon *et al.*, 1993), where the system is is described by,

$$\begin{aligned} x_k &= 0.5x_{k-1} + a\frac{x_{k-1}}{(1+x_{k-1}^2)} \\ &\quad + b\cos(1.2(k-1)) + w_{k-1} \\ y_k &= cx_k^2 + v_k \end{aligned}$$

where w_k and v_k are uncorrelated zero mean Gaussian white noise with variance $Q_w = 1$ and $Q_v = 10$ respectively. The parameters to be monitored are collected in the vector $\theta = [a, b, c]^T$. The nominal value of θ is $\theta^0 = [a_0, b_0, c_0]^T = [25, 8, 0.05]^T$.

Two Monte Carlo simulation experiments have been carried out. In the first experiment, the component fault is simulated to occur at time $k = 201$ at which the parameter a jumps from the nominal value a_0 to $0.5a_0$. In the second experiment, the sensor fault is simulated to occur at time $k = 201$ at which the parameter c is shifted from c_0 to $0.5c_0$. The new method for FDI proposed in this paper is used to detect and isolate these faults.

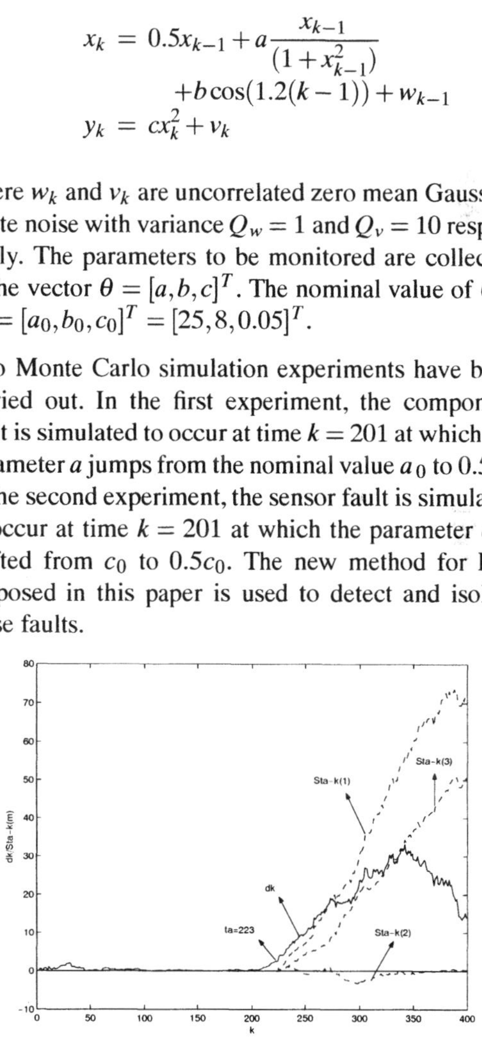

Fig. 1. d_k and $L_{t_a}^k(m)(m = 1, 2, 3)$ from the first experiment (parameter a changes).

In the two experiments, the sample sizes for nominal Monte Carlo filter and adaptive Monte Carlo filters are chosen as $N = 1000$, the width of sliding window for maximizing the joint LLRs is chosen as $\kappa = 50$, the threshold in (18) is chosen as $\lambda = 2.5$. The decision

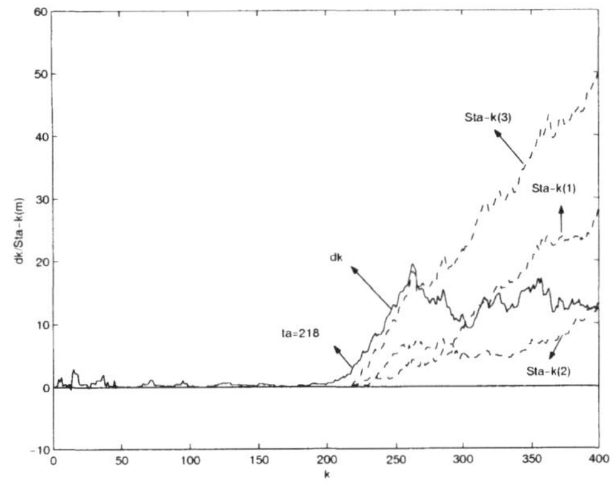

Fig. 2. d_k and $L^k_{t_a}(m)(m = 1,2,3)$ from the second experiment (parameter c changes).

function d_k defined and the cumulative LLRs $L^k_{t_a}(m)$ from these two experiments are shown in Fig.1 and Fig.2 respectively. The component fault is detected at time $t_a = 223$ and the sensor fault is detected at time $t_a = 218$. We can see, from these figures, that d_k remains steady around zero before the fault takes place, and drifts positively away from zero after the fault occurs and the cumulative LLR $L^k_{t_a}(m)$ corresponding to the changed parameter (e.g. $L^k_{t_a}(1)$ in Fig.1 and $L^k_{t_a}(3)$ in Fig.2) takes the greatest value. These results show that the proposed FDI method is able to detect the faults in time and to isolate the faults correctly.

8. CONCLUSIONS

In this paper, algorithms based on particle filtering for the problems of fault detection, fault isolation and fault diagnosis have been developed. Decision criteria are based on likelihood ratios which can be computed accurately due to the complete PDF information being provided by the particle filters.

REFERENCES

Anderson, B.D.O. and J.B. Moore (1979). *Optimal Filtering*. Prentice-hall. Englewood Cliffs, NJ.

Bar-Shalom, Yaakov and Xiao-Rong Li (1993). *Estimation and Tracking: principles, techniques, and software*. Artech House. Norwood, MA.

Basseville, M. (1988). Detecting changes in signals and systems — a survey. *Automatica* **24**(3), 309–326.

Basseville, M. and I. Nikiforov (1998). *Detection of abrupt changes — Theory and Application*. Internet. http://www.irisa.fr/sigma2/michele/publis.html.

Bolviken, E., P.J. Acklam, N. Christophersen and J-M. Stordal (2001). Monte Carlo filters for non-linear state estimation. *Automatica* **37**(2), 177–183.

Crisan, D. and A. Doucet (2002). A survey of convergence results on particle filtering methods for practitioners. *IEEE Transactions on Signal Processing* **50**(3), 736–746.

Doucet, A., Freitas, N.de and Gordon, N., Eds.) (2001). *Sequential Monte Carlo Methods in Practice*. Statistics for Engineering and Information Science. Springer-Verlag. New York.

Frank, P.M. (1990). Fault diagnosis in dynamic systems using analytical and knowledge based redundancy — a survey and new results. *Automatica* **26**, 459–474.

Garcia, E.Alcorta and P.M. Frank (1997). Deterministic nonlinear observer-based approaches to fault diagnosis: A survey. *IFAC J. of Control Eng. Practice* **5**(5), 663–670.

Gordon, N.J., D.J. Salmond and A.F.M. Smith (1993). Novel approach to nonlinear non-Gaussian Bayesian state estimation. *IEE Proceedings-F* **140**(2), 107–113.

Isermann, R. (1984). Process fault detection based on modeling and estimation methods — a survey. *Automatica* **20**, 387–404.

Kitagawa, G. (1998). A self-organizing state-space model. *Journal of the American Statistical Association* **93**(443), 1203–1215.

Patton, R.J. and J. Chen (1996). Robust fault detection and isolation (FDI) systems. In: *Control and Dynamic Systems*. Vol. 74. pp. 171–224. Academic Press. Inc.

IFAC Publications
www.elsevier.com/locate/ifac

MONTE CARLO MIXTURE KALMAN FILTER AND ITS APPLICATION TO SPACE-TIME INVERSION

Tomoyuki Higuchi [*,1] **Jun'ichi Fukuda** [**]

[*] *The Institute of Statistical Mathematics, 4-6-7 Minami-Azabu, Minato-ku, Tokyo 106-8569, Japan*
[**] *Earthquake Research Institute, University of Tokyo, Yayoi 1-1-1, Bunkyo-ku, Tokyo 113-0032, Japan*

Abstract: It is important to precisely know the whole time history of various types of fault slip events to understand the physics of earthquake generation. We develop a new time dependent inversion method for imaging transient fault slips from geodetic data. Past studies employed a linear Gaussian state space model and applied Kalman filter. The Kalman filter based methods, however, do not allow any variation to the temporal smoothness (or roughness) of fault slips. In the present study, we develop/apply a new filtering scheme, Monte Carlo mixture Kalman filter (MCMKF), to the time dependent inversion. MCMKF allows variation to the temporal smoothing of slips in the following scheme; (1) we prepare a finite number of competing state space models, each of which follows a different state space model, (2) we introduce a switching structure among these competing models. *Copyright © 2003 IFAC*

Keywords: Monte Carlo calculation, Kalman filter, Filtering techniques, State-space model, Inverse dynamic problem

1. INTRODUCTION

Recently continuous measurements of surface deformation with dense Global Positioning System (GPS) network have revealed that transient crustal deformations with a time scale of hours up to years play a very important role in seismic cycles. Accurate estimates of these spatio-temporal variation of such slow events provides us with an opportunity to understand an earthquake mechanism. From geodetic view point, it is therefore important to investigate detailed spatio-temporal process of slow events and dense GPS array record provides the most suitable data for this end (Heki *et al.*, 1997; Ozawa *et al.*, 2001).

Several studies have tried to image the spatio-temporal variation of transient fault slip (Segall and Matthews, 1997). One efficient way to retrieve slip distribution is that a space-time history of fault slip is modeled by using linear Gaussian state space model, i.e., state space model (Anderson and Moore, 1979; Kitagawa and Gersch, 1996) and estimated by Kalman filter (Kitagawa and Gersch, 1996). This method is referred to as the Network Inversion Filter (NIF) and recognized as one of the standard techniques in the application domain.

In the NIF framework, the temporal smoothness of the fault slip is controlled by a scaling parameter of the employed stochastic model. This scaling parameter, often referred to as hyper-parameter, is determined with maximum likelihood method. In the NIF framework, the scaling parameter is held fixed over the observation period. The constancy of the scaling parameter seems to obscure the causal relationship among multiple events, and hence motivated us to explore a new approach to the time dependent inversion such that the scaling parameter is variable in time.

[1] Partially supported by the MEXT, Japan

In the present study, we develop a new filtering algorithm, called as Monte Carlo mixture Kalman filter (MCMKF) for imaging time-dependent fault slip from geodetic data. In section 2 a new model to identify a temporal variation of the scaling parameter is proposed and formulated by the conditional dynamic linear model (CDLM). A basic idea of MCMKF is explained in section 3. In section 4, the recursive calculations, prediction and filtering, are derived. Section 5 describes a procedure for the state estimation based on the model averaging principle. Section 6 gives a summary of the MCMKF procedure and brief discussion on the numerical experiments.

2. CONDITIONAL DYNAMIC LINEAR MODEL (CDLM)

In order to allow temporal variation of the hyper-parameter for slip acceleration, we first prepare a finite number of competing different state space models, each of which has different hyper-parameter value. Then, we realize the temporal change of the value by introducing a switching structure among them. For the limited space, we cannot give a detailed explanation for the system and the observation models in the NIF framework (Fukuda *et al.*, 2003). In stead, we remark that a departure of our model from the NIF is a time dependency of system noise variances. Namely, our model can be rewritten as follows:

$$\boldsymbol{x}_n = F_n \boldsymbol{x}_{n-1} + \boldsymbol{v}_n, \qquad \boldsymbol{v}_n \sim N(\mathbf{0}, Q_n(I_n)) \qquad (1)$$

$$\boldsymbol{y}_n = H \boldsymbol{x}_n + \boldsymbol{w}_n, \qquad \boldsymbol{w}_n \sim N(\mathbf{0}, R_n). \qquad (2)$$

In this case only Q_n among the four matrices is dependent on I_n. Thus a value of I_n specifies a system model with a certain value of the hyper-parameter. The dimension of the state vector depends on a way of how a displacement region of interest is numerically represented. It usually exceeds over one hundred, and then a direct approach of the particle filter (Doucet *et al.*, 2001) to combine many generalized state space model (Kitagawa, 1998; Higuchi and Kitagawa, 2000; Higuchi, 2001) cannot deal with this problem.

This generalization to incorporate a time dependency of hyperparameters can be formulated in the conditional dynamic linear model (CDLM) (Chen and Liu, 2000; Chen *et al.*, 2000; Liu *et al.*, 2001). The CDLM can be defined as:

System model

$$\boldsymbol{x}_n = F_n(I_n)\boldsymbol{x}_{n-1} + \boldsymbol{v}_n, \qquad (3)$$

Observation model

$$\boldsymbol{y}_n = H_n(I_n)\boldsymbol{x}_n + \boldsymbol{w}_n \qquad (4)$$

where $\boldsymbol{v}_n \sim N(\mathbf{0}, Q_n(I_n))$ and $\boldsymbol{w}_n \sim N(\mathbf{0}, R_n(I_n))$. The indicator vector I_n is a discrete latent variable which takes an integer value between $1 \sim M$. Usually a number of models treated in the Mixture Kalman filter is about $2 \sim 3$, but we consider a problem of dealing with a large number of models, $M \simeq 100$. Given I_n, F_n, H_n, Q_n, and R_n are known matrices of appropriate dimension. The CDLM is a direct generalization of the dynamic linear model (DLM) and retain a capability of dealing with outliers, sudden jumps, clutters, and other nonlinear features (Liu *et al.*, 2001). The CDLM includes other types of generalization of DLM, e.g., Partial non-Gaussian state space model (Shephard, 1994; Bergman *et al.*, 2001), Markov switching state space model, (Kim and Nelson, 1999; Frühwirth-Schnatter, 2001) and Dynamic liner models with switching (Shumway and Stoffer, 1991).

3. BASIC IDEA OF MONTE CARLO MIXTURE KALMAN FILTER (MCMKF)

In this section, we introduce a new filtering scheme and call it as Monte Carlo mixture Kalman filter (MCMKF) that allows us to choose the optimal model from many candidates or to average over many models.

3.1 Model switching structure

The MCMKF algorithm requires a stochastic model which describes a time-dependent structure for I_n. In this study, I_n is assumed to follow a stationary Markov process, i.e.,

$$p(I_n|\boldsymbol{I}_{1:n-1}) = p(I_n|I_{n-1}) \qquad (5)$$

where $\boldsymbol{I}_{i:j} = (I_i, I_{i+1}, \cdots, I_j)$ and $p()$ denotes probability density function. An evolution of I_n is realized by Markov switching model with transition probability given by

$$\pi_{ij} = \Pr(I_n = j | I_{n-1} = i) \qquad (6)$$

where Pr denotes realization probability. In the following subsections, we present an algorithm that determines time evolution of I_n.

3.2 Monte Carlo mixture Kalman filter

The MCMKF algorithm consists of two steps. First, temporal variation of the probability distribution of indicator variable I_n is determined. Second, temporal variation of the probability distribution of the state vector $\boldsymbol{x}_n$ is estimated following the history of I_n.

Let $\boldsymbol{y}_{i:j}$ and $\boldsymbol{I}_{i:j}$ be a set of data vectors and indicator variable from time t_i to time t_j, respectively, i.e., $\boldsymbol{y}_{i:j} = (\boldsymbol{y}_i, \boldsymbol{y}_{i+1}, \cdots, \boldsymbol{y}_j)$ and $\boldsymbol{I}_{i:j} = (I_i, I_{i+1}, \cdots, I_j)$. In MCMKF, two conditional joint distributions of $\boldsymbol{I}_{1:n}$: (i) predictive distribution $p(\boldsymbol{I}_{1:n}|\boldsymbol{y}_{1:n-1})$ and (ii) filter distribution $p(\boldsymbol{I}_{1:n}|\boldsymbol{y}_{1:n})$, are approximated by many "particles" that can be considered as independent realizations from each distribution. Let $\boldsymbol{I}^{(j)}_{1:i|k} =$

$(I^{(j)}_{1|k}, I^{(j)}_{2|k}, \cdots I^{(j)}_{i|k})$ be the jth realization of the conditional distribution $p(\boldsymbol{I}_{1:i}|\boldsymbol{y}_{1:k})$. Each distribution is approximated by N_p ($N_p \gg 1$) realizations as follows:

$$\left\{\boldsymbol{I}^{(1)}_{1:n|n-1}, \boldsymbol{I}^{(2)}_{1:n|n-1}, \cdots, \boldsymbol{I}^{(N_p)}_{1:n|n-1}\right\} \sim p(\boldsymbol{I}_{1:n}|\boldsymbol{y}_{1:n-1}) \tag{7}$$

$$\left\{\boldsymbol{I}^{(1)}_{1:n|n}, \boldsymbol{I}^{(2)}_{1:n|n}, \cdots, \boldsymbol{I}^{(N_p)}_{1:n|n}\right\} \sim p(\boldsymbol{I}_{1:n}|\boldsymbol{y}_{1:n}) \tag{8}$$

where

$$\Pr\left(\boldsymbol{I}_{1:n} = \boldsymbol{I}^{(j)}_{1:n|n-1}|\boldsymbol{y}_{1:n-1}\right) = \frac{1}{N_p}, \tag{9}$$

$$\Pr\left(\boldsymbol{I}_{1:n} = \boldsymbol{I}^{(j)}_{1:n|n}|\boldsymbol{y}_{1:n}\right) = \frac{1}{N_p}. \tag{10}$$

In this study, we refer to $\{\boldsymbol{I}^{(1)}_{1:n|n-1}, \boldsymbol{I}^{(2)}_{1:n|n-1}, \cdots, \boldsymbol{I}^{(N_p)}_{1:n|n-1}\}$ and $\{\boldsymbol{I}^{(1)}_{1:n|n}, \boldsymbol{I}^{(2)}_{1:n|n}, \cdots, \boldsymbol{I}^{(N_p)}_{1:n|n}\}$ as "approximated predictive distribution" and "approximated filter distribution", respectively. Given realizations of I_n, $I^{(j)}_n = I^{(j)}_{n|k}$, following CDLM is satisfied for each $I^{(j)}_n$ ($j = 1, \cdots, N_p$):

$$\boldsymbol{x}^{(j)}_n = F_n(I^{(j)}_n)\boldsymbol{x}^{(j)}_{n-1} + \boldsymbol{v}^{(j)}_n, \quad \boldsymbol{v}^{(j)}_n \sim N(\boldsymbol{0}, Q_n(I^{(j)}_n)) \tag{11}$$

$$\boldsymbol{y}_n = H_n(I^{(j)}_n)\boldsymbol{x}^{(j)}_n + \boldsymbol{w}^{(j)}_n, \quad \boldsymbol{w}^{(j)}_n \sim N(\boldsymbol{0}, R_n(I^{(j)}_n)). \tag{12}$$

Using (11) and (12), we will later show that a set of particles approximating the predictive distribution and the filter distribution is obtained recursively by two steps:

$$\text{Prediction:} \quad \left\{\boldsymbol{I}^{(1)}_{1:n-1|n-1}, \cdots, \boldsymbol{I}^{(N_p)}_{1:n-1|n-1}\right\} \longrightarrow \left\{\boldsymbol{I}^{(1)}_{1:n|n-1}, \cdots, \boldsymbol{I}^{(N_p)}_{1:n|n-1}\right\}, \tag{13}$$

$$\text{Filtering:} \quad \left\{\boldsymbol{I}^{(1)}_{1:n|n-1}, \cdots, \boldsymbol{I}^{(N_p)}_{1:n|n-1}\right\} \longrightarrow \left\{\boldsymbol{I}^{(1)}_{1:n|n}, \cdots, \boldsymbol{I}^{(N_p)}_{1:n|n}\right\}. \tag{14}$$

4. RECURSIVE CALCULATION

4.1 Prediction

In this subsection, we show that an approximated predictive distribution at time t_n $\{\boldsymbol{I}^{(1)}_{1:n|n-1}, \cdots, \boldsymbol{I}^{(N_p)}_{1:n|n-1}\}$ is obtained from an approximated filter distribution at time t_{n-1} $\{\boldsymbol{I}^{(1)}_{1:n-1|n-1}, \cdots, \boldsymbol{I}^{(N_p)}_{1:n-1|n-1}\}$. We assume that $\{\boldsymbol{I}^{(1)}_{1:n-1|n-1}, \cdots, \boldsymbol{I}^{(N_p)}_{1:n-1|n-1}\}$ and $\boldsymbol{y}_{1:n-1}$ are given. Then the probability $\Pr(\boldsymbol{I}_{1:n} = \boldsymbol{I}^{(j)}_{1:n|n-1}|\boldsymbol{y}_{1:n-1})$ is manipulated as

$$\begin{aligned}
&\Pr(\boldsymbol{I}_{1:n} = \boldsymbol{I}^{(j)}_{1:n|n-1}|\boldsymbol{y}_{1:n-1}) \\
&= \Pr(I_n = I^{(j)}_{n|n-1}, \boldsymbol{I}_{1:n-1} = \boldsymbol{I}^{(j)}_{1:n-1|n-1}|\boldsymbol{y}_{1:n-1}) \\
&= \Pr(I_n = I^{(j)}_{n|n-1}|\boldsymbol{I}_{1:n-1} = \boldsymbol{I}^{(j)}_{1:n-1|n-1}, \boldsymbol{y}_{1:n-1}) \\
&\cdot \Pr(\boldsymbol{I}_{1:n-1} = \boldsymbol{I}^{(j)}_{1:n-1|n-1}|\boldsymbol{y}_{1:n-1}) \\
&= \Pr(I_n = I^{(j)}_{n|n-1}|\boldsymbol{I}_{1:n-1} = \boldsymbol{I}^{(j)}_{1:n-1|n-1}) \\
&\cdot \Pr(\boldsymbol{I}_{1:n-1} = \boldsymbol{I}^{(j)}_{1:n-1|n-1}|\boldsymbol{y}_{1:n-1}) \\
&= \Pr(I_n = I^{(j)}_{n|n-1}|I_{n-1} = I^{(j)}_{n-1|n-1})\frac{1}{N_p}.
\end{aligned} \tag{15}$$

(15) indicates that $\left\{\boldsymbol{I}^{(1)}_{1:n|n-1}, \cdots, \boldsymbol{I}^{(N_p)}_{1:n|n-1}\right\}$ is obtained by sampling a realization $I^{(j)}_{n|n-1}$ with probability or weight $\Pr(I_n = I^{(j)}_{n|n-1}|I_{n-1} = I^{(j)}_{n-1|n-1})$, and setting $\boldsymbol{I}^{(j)}_{1:n|n-1} = (\boldsymbol{I}^{(j)}_{1:n-1|n-1}, I^{(j)}_{n|n-1})$. Note that $\Pr(I_n = I^{(j)}_{n|n-1}|I_{n-1} = I^{(j)}_{n-1|n-1})$ is given by the Markovian transition probability defined by (6).

4.2 Filtering

In this subsection, we show that an approximated filter distribution at time t_n $\{\boldsymbol{I}^{(1)}_{1:n|n}, \cdots, \boldsymbol{I}^{(N_p)}_{1:n|n}\}$ is obtained from an approximated predictive distribution at time t_n $\left\{\boldsymbol{I}^{(1)}_{1:n|n-1}, \cdots, \boldsymbol{I}^{(N_p)}_{1:n|n-1}\right\}$. Given the observation $\boldsymbol{y}_n$, the probability $\Pr(\boldsymbol{I}_{1:n} = \boldsymbol{I}^{(j)}_{1:n|n-1}| \boldsymbol{y}_{1:n-1})$ is updated as follows:

$$\begin{aligned}
&\Pr(\boldsymbol{I}_{1:n} = \boldsymbol{I}^{(j)}_{1:n|n-1}|\boldsymbol{y}_{1:n}) \\
&= \Pr(\boldsymbol{I}_{1:n} = \boldsymbol{I}^{(j)}_{1:n|n-1}|\boldsymbol{y}_{1:n-1}, \boldsymbol{y}_n) \\
&= p(\boldsymbol{y}_n|\boldsymbol{I}_{1:n} = \boldsymbol{I}^{(j)}_{1:n|n-1}, \boldsymbol{y}_{1:n-1}) \\
&\cdot \Pr(\boldsymbol{I}_{1:n} = \boldsymbol{I}^{(j)}_{1:n|n-1}|\boldsymbol{y}_{1:n-1})/p(\boldsymbol{y}_n|\boldsymbol{y}_{1:n-1}) \\
&= \{p(\boldsymbol{y}_n|\boldsymbol{I}_{1:n} = \boldsymbol{I}^{(j)}_{1:n|n-1}, \boldsymbol{y}_{1:n-1}) \\
&\cdot \Pr(\boldsymbol{I}_{1:n} = \boldsymbol{I}^{(j)}_{1:n|n-1}|\boldsymbol{y}_{1:n-1})\}/ \\
&\{\sum_{j=1}^{N_p} p(\boldsymbol{y}_n|\boldsymbol{I}_{1:n} = \boldsymbol{I}^{(j)}_{1:n|n-1}, \boldsymbol{y}_{1:n-1}) \\
&\cdot \Pr(\boldsymbol{I}_{1:n} = \boldsymbol{I}^{(j)}_{1:n|n-1}|\boldsymbol{y}_{1:n-1})\} \\
&= \frac{w^{(j)}_n \frac{1}{N_p}}{\sum_{j=1}^{N_p} w^{(j)}_n \frac{1}{N_p}} = \frac{w^{(j)}_n}{\sum_{j=1}^{N_p} w^{(j)}_n}
\end{aligned} \tag{16}$$

where

$$w^{(j)}_n = p(\boldsymbol{y}_n|\boldsymbol{I}_{1:n} = \boldsymbol{I}^{(j)}_{1:n|n-1}, \boldsymbol{y}_{1:n-1}). \tag{17}$$

Equation (16) means that the filter distribution $p(\boldsymbol{I}_{1:n}|\boldsymbol{y}_{1:n})$ is approximated by giving weight proportional to $w^{(j)}_n$ to the jth particle of approximated predictive distribution. For the next prediction step, it is necessary to represent the approximated filter distribution with equally weighted particles $\left\{\boldsymbol{I}^{(1)}_{1:n|n}, \cdots, \boldsymbol{I}^{(N_p)}_{1:n|n}\right\}$. This is achieved by generating N_p particles $\left\{\boldsymbol{I}_{1:n|n}^{(1)}, \cdots, \boldsymbol{I}_{1:n|n}^{(N_p)}\right\}$ by resampling $\left\{\boldsymbol{I}_{1:n|n-1}^{(1)}, \cdots, \boldsymbol{I}_{1:n|n-1}^{(N_p)}\right\}$ with probability proportional to $\left\{w_n^{(1)}, \cdots, w_n^{(N_p)}\right\}$.

4.3 Recursive calculation for the state vector estimation

From (11) and (12), $p(\boldsymbol{x}_{n-1}|\boldsymbol{I}_{1:n-1} = \boldsymbol{I}_{1:n-1|n-1}^{(j)}, \boldsymbol{y}_{1:n-1})$ and $p(\boldsymbol{x}_n|\boldsymbol{I}_{1:n} = \boldsymbol{I}_{1:n|n-1}^{(j)}, \boldsymbol{y}_{1:n-1})$ becomes Gaussian distributions. Let us define mean vectors and covariance matrices of the two distributions as follows:

$$p(\boldsymbol{x}_{n-1}|\boldsymbol{I}_{1:n-1} = \boldsymbol{I}_{1:n-1|n-1}^{(j)}, \boldsymbol{y}_{1:n-1}) \sim N(\boldsymbol{x}_{n-1|n-1}^{(j)}, V_{n-1|n-1}^{(j)}) \quad (18)$$

$$p(\boldsymbol{x}_n|\boldsymbol{I}_{1:n} = \boldsymbol{I}_{1:n|n-1}^{(j)}, \boldsymbol{y}_{1:n-1}) \sim N(\boldsymbol{x}_{n|n-1}^{(j)}, V_{n|n-1}^{(j)}) \quad (19)$$

Since $\boldsymbol{I}_{1:n-1|n-1}^{(j)}$ is assumed to be given, the CDLM (11) and (12) reduces to a linear Gaussian state space model and thus $\boldsymbol{x}_{n-1|n-1}^{(j)}$ and $V_{n-1|n-1}^{(j)}$ are calculated by Kalman filter. $\boldsymbol{x}_{n|n-1}^{(j)}$ and $V_{n|n-1}^{(j)}$ are also calculated by Kalman filter using $\boldsymbol{x}_{n-1|n-1}^{(j)}$, $V_{n-1|n-1}^{(j)}$ and $I_{n|n-1}^{(j)}$ as follows:

$$\boldsymbol{x}_{n|n-1}^{(j)} = F_n(I_{n|n-1}^{(j)})\boldsymbol{x}_{n-1|n-1}^{(j)} \quad (20)$$

$$V_{n|n-1}^{(j)} = F_n(I_{n|n-1}^{(j)})V_{n-1|n-1}^{(j)}F_n^T(I_{n|n-1}^{(j)}) + Q_n(I_{n|n-1}^{(j)}). \quad (21)$$

Note that $I_{n|n-1}^{(j)}$ is obtained by the prediction scheme of the MCMKF.

From (12), the predictive distribution of data also becomes a Gaussian:

$$p(\boldsymbol{y}_n|\boldsymbol{I}_{1:n} = \boldsymbol{I}_{1:n|n-1}^{(j)}, \boldsymbol{y}_{1:n-1}) \sim N(\boldsymbol{y}_{n|n-1}^{(j)}, W_{n|n-1}^{(j)}) \quad (22)$$

where

$$\boldsymbol{y}_{n|n-1}^{(j)} = H_n(I_{n|n-1}^{(j)})\boldsymbol{x}_{n|n-1}^{(j)} \quad (23)$$

$$W_{n|n-1}^{(j)} = H_n(I_{n|n-1}^{(j)})V_{n|n-1}^{(j)}H_n^T(I_{n|n-1}^{(j)}) + R_n(I_{n|n-1}^{(j)}). \quad (24)$$

The left hand side of (22) is the weight $w_n^{(j)}$ defined in (17). Thus $w_n^{(j)}$ follows the Gaussian distribution with mean $\boldsymbol{y}_{n|n-1}^{(j)}$ and covariance matrix $W_{n|n-1}^{(j)}$ as follows:

$$w_n^{(j)} = (2\pi)^{-N_d/2}\left|W_{n|n-1}^{(j)}\right|^{-1/2} \exp\left[-\frac{1}{2}(\boldsymbol{y}_n - \boldsymbol{y}_{n|n-1}^{(j)})^T W_{n|n-1}^{(j)-1}(\boldsymbol{y}_n - \boldsymbol{y}_{n|n-1}^{(j)})\right] \quad (25)$$

where $\left|W_{n|n-1}^{(j)}\right|$ is the absolute value of the determinant of $W_{n|n-1}^{(j)}$. Because $\boldsymbol{x}_{n|n-1}^{(j)}$ and $V_{n|n-1}^{(j)}$ are given in (20) and (21), the weight $w_n^{(j)}$ is obtained using (23), (24) and (25).

By using the prediction and the filtering algorithm recursively, we finally obtain N_p particles $\{\boldsymbol{I}_{1:N_e|N_e}^{(1)}, \cdots, \boldsymbol{I}_{1:N_e|N_e}^{(N_p)}\}$ that approximate $p(\boldsymbol{I}_{1:N_e}|\boldsymbol{y}_{1:N_e})$, the posterior distribution of $\boldsymbol{I}_{1:N_e}$ conditioned on all of available data. Here, N_e is the number of observation epochs. $p(\boldsymbol{I}_{1:N_e}|\boldsymbol{y}_{1:N_e})$ is called smoother distribution of $\boldsymbol{I}_{1:N_e}$. A sequence of each particle, $\boldsymbol{I}_{1:N_e|N_e}^{(j)} = [I_{1|N_e}^{(j)}, I_{2|N_e}^{(j)}, \cdots I_{N_e|N_e}^{(j)}]$, is called the trajectory.

This filtering algorithm is conceptually similar to the storing state vector algorithm in the Monte Carlo filter proposed by Kitagawa (1996). He applied the Monte Carlo approximation directly to the distribution of the state, whereas we apply the approximation to the distribution of the indicator variable. He showed that in the Monte Carlo filter the repetition of resampling gradually decreases the number of different realizations of state vector as time passes because the number of realizations is finite. Therefore the shape of the distribution of the state deteriorates as time passes. Kitagawa (1996) showed that this difficulty can be eliminated by employing fixed L-lag smoother rather than fixed interval smoother (Anderson and Moore, 1979). Although we apply the Monte Carlo approximation to the indicator variable instead of the state, this situation also applies to the MCMKF. Thus following Kitagawa (1996), we modify the MCMKF filtering algorithm as follows:

For fixed L, generate N_p particles $\{\boldsymbol{I}_{n-L:n|n}^{(1)}, \boldsymbol{I}_{n-L:n|n}^{(2)}, \cdots, \boldsymbol{I}_{n-L:n|n}^{(N_p)}\}$ by the resampling of $\{\boldsymbol{I}_{n-L:n|n-1}^{(1)}, \boldsymbol{I}_{n-L:n|n-1}^{(2)}, \cdots, \boldsymbol{I}_{n-L:n|n-1}^{(N_p)}\}$ with probability proportional to $\{w_n^{(1)}, \cdots, w_n^{(N_p)}\}$ defined in (17).

It is recommended to take L not so large (say, 10 or 20 at the largest 50) (Kitagawa, 1996; Higuchi and Kitagawa, 2000). We adopt $L = 20$ in our application study shown in Section 6.

5. MODEL AVERAGING

5.1 State Estimation

We present here an algorithm to estimate the state using all the N_p trajectories $\{\boldsymbol{I}_{1:N_e|N_e}^{(1)}, \cdots, \boldsymbol{I}_{1:N_e|N_e}^{(N_p)}\}$. In this case, $F_n(I_n^{(j)})$, $Q_n(I_n^{(j)})$, $H_n(I_n^{(j)})$ and $R_n(I_n^{(j)})$ $(j = 1, \cdots, N_p)$ in (11) and (12) reduce to sets of known matrices which have different time evolutions corresponding to trajectories. Thus the CDLM defined by (11) and (12) reduces to the conventional linear Gaussian state space model to which Kalman filter is applicable for state estimation:

$$x_n^{(j)} = F_n^{(j)} x_{n-1}^{(j)} + v_n^{(j)}, \quad v_n^{(j)} \sim N(\mathbf{0}, Q_n^{(j)}) \quad (26)$$

$$y_n = H_n^{(j)} x_n^{(j)} + w_n^{(j)}, \quad w_n^{(j)} \sim N(\mathbf{0}, R_n^{(j)}) \quad (27)$$

where $F_n^{(j)} = F_n(I_{n|N_e}^{(j)})$, $Q_n^{(j)} = Q_n(I_{n|N_e}^{(j)})$, $H_n^{(j)} = H_n(I_{n|N_e}^{(j)})$ and $R_n^{(j)} = R_n(I_{n|N_e}^{(j)})$. Let

$$x_{i|k}^{(j)} = \mathrm{E}(x_i | y_{1:k}, I_{1:N_e} = I_{1:N_e|N_e}^{(j)}) \quad (28)$$

$$V_{i|k}^{(j)} = \mathrm{Cov}(x_i | y_{1:k}, I_{1:N_e} = I_{1:N_e|N_e}^{(j)}) \quad (29)$$

be the conditional mean and the covariance matrix of the state at time t_i given the data $y_{1:k}$ for jth trajectory. $\left\{x_{n+1|n}^{(j)}, V_{n+1|n}^{(j)}\right\}_{j=1}^{N_p}$, $\left\{x_{n|n}^{(j)}, V_{n|n}^{(j)}\right\}_{j=1}^{N_p}$ and $\left\{x_{n|N_e}^{(j)}, V_{n|N_e}^{(j)}\right\}_{j=1}^{N_p}$ are recursively obtained by Kalman filter. Given $\left\{x_{n|N_e}^{(j)}, V_{n|N_e}^{(j)}\right\}_{j=1}^{N_p}$, distribution of the final estimate for x_n, $p(x_n | y_{1:N_e})$, is written as

$$\begin{aligned} p(x_n | y_{1:N_e}) &= \sum_{j=1}^{N_p} p(x_n, I_{1:N_e} = I_{1:N_e|N_e}^{(j)} | y_{1:N_e}) \\ &= \sum_{j=1}^{N_p} p(x_n | I_{1:N_e} = I_{1:N_e|N_e}^{(j)}, y_{1:N_e}) \\ &\quad \Pr(I_{1:N_e} = I_{1:N_e|N_e}^{(j)} | y_{1:N_e}) \\ &= \frac{1}{N_p} \sum_{j=1}^{N_p} N(x_{n|N_e}^{(j)}, V_{n|N_e}^{(j)}). \end{aligned} \quad (30)$$

In the 3-rd equality, (10), (28) and (29) are used. Therefore $p(x_n | y_{1:N_e})$ is non-Gaussian distribution with mean

$$x_{n|N_e} = \frac{1}{N_p} \sum_{j=1}^{N_p} x_{n|N_e}^{(j)}. \quad (31)$$

Estimation of standard deviation error bounds for $x_{n|N_e}$ is not straightforward because $p(x_n | y_{1:N_e})$ is non-Gaussian distribution. In this study, error bounds for $x_{n|N_e}$ are approximately obtained as follows:

(1) Generate N_s realizations of $N(x_{n|N_e}^{(j)}, V_{n|N_e}^{(j)})$, $\mathscr{X}_{n,1}^{(j)}$, $\mathscr{X}_{n,2}^{(j)}$, $\cdots$, $\mathscr{X}_{n,N_s}^{(j)}$, for each trajectory, $j = 1, 2, \cdots, N_p$.
(2) Estimate covariance matrix of $p(x_n | y_{1:N_e})$, $V_{n|N_e}$, by

$$\begin{aligned} V_{n|N_e} &= \frac{1}{N_p N_s - 1} \sum_{j=1}^{N_p} \sum_{k=1}^{N_s} \\ &\quad \left[\mathscr{X}_{n,k}^{(j)} - \bar{\mathscr{X}}_{n,k}^{(j)}\right] \left[\mathscr{X}_{n,k}^{(j)} - \bar{\mathscr{X}}_{n,k}^{(j)}\right]^T \end{aligned} \quad (32)$$

where

$$\bar{\mathscr{X}}_{n,k}^{(j)} = \frac{1}{N_p N_s} \sum_{j=1}^{N_p} \sum_{k=1}^{N_s} \mathscr{X}_{n,k}^{(j)}. \quad (33)$$

The procedure for state estimation using all trajectories described above is computationally massive both in calculation time and in memory. More efficient algorithm is implemented by reducing number of trajectories to which Kalman filter is applied. This is done by sampling N_p' ($N_p' < N_p$) trajectories randomly from N_p trajectories $\{I_{1:N_e|N_e}^{(1)}, \cdots, I_{1:N_e|N_e}^{(N_p)}\}$. Once N_p' trajectories are selected, the procedure for state estimation is identical to the case using all N_p trajectories. The distribution of the final estimate for x_n, $p(x_n | y_{1:N_e})$, and its mean vector, $x_{n|N_e}$, are obtained by replacing N_p in (30) and (31) with N_p', respectively.

5.2 *Likelihood of the meta-model*

In this subsection, we present a formula for the log-likelihood of the model. Let $\boldsymbol{\theta}$ be a vector that contains temporally invariable hyper-parameters. Given $\boldsymbol{\theta}$, the likelihood of the model is expressed by

$$\begin{aligned} L(\boldsymbol{\theta}) &= p(y_{1:N_e} | \boldsymbol{\theta}) \\ &= \prod_{n=1}^{N_e} p(y_n | y_{1:n-1}, \boldsymbol{\theta}). \end{aligned} \quad (34)$$

If we use all N_p trajectories for state estimation, $p(y_n | y_{1:n-1}, \boldsymbol{\theta})$ in (34) is given by

$$\begin{aligned} &p(y_n | y_{1:n-1}, \boldsymbol{\theta}) = \\ &\sum_{j=1}^{N_p} p(y_n, I_{1:n} = I_{1:n|N_e}^{(j)} | y_{1:n-1}, \boldsymbol{\theta}) \\ &= \sum_{j=1}^{N_p} p(y_n | y_{1:n-1}, I_{1:n} = I_{1:n|N_e}^{(j)}, \boldsymbol{\theta}) \\ &\quad \Pr(I_{1:n} = I_{1:n|N_e}^{(j)} | y_{1:n-1}, \boldsymbol{\theta}) \\ &= \frac{1}{N_p} \sum_{j=1}^{N_p} p(y_n | y_{1:n-1}, I_{1:n} = I_{1:n|N_e}^{(j)}, \boldsymbol{\theta}). \end{aligned} \quad (35)$$

Combining (34) and (35) yields the following formula for the log-likelihood of the model $l(\boldsymbol{\theta})$. If we use N_p' trajectories randomly sampled from N_p trajectories, the log-likelihood of the model is obtained by replacing N_p with N_p'.

The goodness of the model is evaluated by the Akaike information criterion (AIC) (Akaike, 1974). The AIC is defined as

$$\mathrm{AIC} = -2l(\boldsymbol{\theta}) + 2(\text{number of unknown parameters}). \quad (36)$$

6. SUMMARY

6.1 *An algorithm for MCMKF*

The MCMKF algorithm that we propose in this study is summarized as follows:

(1) Initialization: For $j = 1, \ldots, N_p$,
 (a) Sample $I_{0|0}^{(j)} \sim p(I_{0|0})$.
 (b) Set $(x_{1|0}^{(j)}, V_{1|0}^{(j)})$.
(2) For $n = 1, \ldots, N_e$,
 (a) For $j = 1, \ldots, N_p$,

(i) Sample $I^{(j)}_{n|n-1} \sim \Pr(I_n = I^{(j)}_{n|n-1} | I_{n-1} = I^{(j)}_{n-1|n-1})$.

(ii) Set $\boldsymbol{I}^{(j)}_{1:n|n-1} = (\boldsymbol{I}^{(j)}_{1:n-1|n-1}, I^{(j)}_{n|n-1})$.

(iii) Compute $w^{(j)}_n = p(\boldsymbol{y}_n | \boldsymbol{I}_{1:n} = \boldsymbol{I}^{(j)}_{1:n|n-1}, \boldsymbol{y}_{1:n-1})$.

(iv) Update $(\boldsymbol{x}^{(j)}_{n-1|n-1}, V^{(j)}_{n-1|n-1})$ to obtain $(\tilde{\boldsymbol{x}}^{(j)}_{n|n}, \tilde{V}^{(j)}_{n|n})$ using Kalman filter.

(b) Obtain $\{(\boldsymbol{I}^{(j)}_{1:n|n}, \boldsymbol{x}^{(j)}_{n|n}, V^{(j)}_{n|n})\}^{N_p}_{j=1}$ by the resampling of $\{(\boldsymbol{I}^{(j)}_{1:n|n-1}, \tilde{\boldsymbol{x}}^{(j)}_{n|n}, \tilde{V}^{(j)}_{n|n})\}^{N_p}_{j=1}$ with probability proportional to $w^{(j)}_n$.

(3) Obtain the distribution of the final estimate for $\boldsymbol{x}_n$, $p(\boldsymbol{x}_n | \boldsymbol{y}_{1:N_e})$, based on N_p trajectories $\left\{\boldsymbol{I}^{(1)}_{1:N_e|N_e}, \cdots, \boldsymbol{I}^{(N_p)}_{1:N_e|N_e}\right\}$.

6.2 Application Result

A temporally invariable scaling parameter as in the NIF could not trace abrupt changes because optimized scaling parameter would be too small to allow such a sudden change of fault slip, and vice versa. As a result, estimated slip evolution would be flattened during the event and oscillatory in steady-state period, and hence, it would be hardly possible to identify the initiation of events. In order to overcome this difficulties, we propose the CDLM and apply the MCMKF for its state estimation. We apply this space-time inversion method to simulated data which are generated by an infinitely long strike slip fault. Results show that the proposed method can reproduce rapidly accelerating and decelerating fault slip and coseismic slip as well as slow variation of fault slip rate, even in a case that noise level is so high that signal is invisible. We confirmed that a benefit of applying our approach is maximized when deformation rate varies rapidly or coseismic deformation exists, and signal-to-noise ratio is low. In addition we address that the MCMKF is designed to deal with the CDLM and then can be applicable to a wide variety of the nonlinear non-Gaussian state space models. The MCMKF allows us to integrate various type of time series models and to generate a flexible time series model automatically.

REFERENCES

Akaike, H. (1974). A new look at the statistical model identification. *IEEE Trans. Autom. Control* **AC-19**, 716–723.

Anderson, B.D.O. and J.B. Moore (1979). *Optimal Filtering*. Prentice-Hall. New Jersey.

Bergman, N., A. Doucet and N. Gordon (2001). Optimal estimation and Cramér-Rao bounds for partial non-Gaussian state space models. *Annals of Institute of Statistical Mathematics* **53**, 97–122.

Chen, R. and J.S. Liu (2000). Mixture Kalman filters. *J. R. Statist. Soc. B.* **62**, 493–508.

Chen, R., X. X. Wang and J.S. Liu (2000). Adaptive joint detection and decoding in flat-fading channels via mixture Kalman filtering. *IEEE Trans. on Information Theory* **46**, 2079–2094.

Doucet, A., J.F.G. de Freitas and N.J. Gordon (2001). *Sequential Monte Carlo Methods in Practice*. Springer-Verlag. New York.

Frühwirth-Schnatter, S. (2001). Fully Bayesian analysis of switching Gaussian state space models. *Annals of Institute of Statistical Mathematics* **53**, 31–49.

Fukuda, J., T. Higuchi, S. Miyazaki and T. Kato (2003). A new approach to time dependent inversion of geodetic data using Monte Carlo mixture Kalman filter. *Geophysical Journal International (submitted)*.

Heki, K., S. Miyazaki and H. Tsuji (1997). Silent fault slip following an interplate thrust earthquake at the Japan trench. *Nature* **386**, 595–598.

Higuchi, T. (2001). Evolutionary time series model with parallel computing. In: *The Third JAPAN-US Joint Seminar on Statistical Time Series Analysis*. pp. 183–190.

Higuchi, T. and G. Kitagawa (2000). Knowledge discovery and self-organizing state space model. *IEICE Transactions on Information and Systems* **E83-D**, 36–43.

Kim, C.-J. and Ch.R. Nelson (1999). *State space models with regime switching*. MIT press. Cambridge.

Kitagawa, G. (1996). Monte Carlo filter and smoother for non-Gaussian nonlinear state space model. *Journal of Computational and Graphical Statistics* **5**, 1–25.

Kitagawa, G. (1998). Self-organizing state space model. *Journal of the Americal Statistical Association* **93**, 1203–1215.

Kitagawa, G. and W. Gersch (1996). *Smoothness Priors Analysis of Time Series*. Springer-Verlag. New York.

Liu, J.S., R. Chen and T. Logvinenko (2001). A theoretical framework for sequential importance sampling with resampling. In: *Sequential Monte Carlo Methods in Practice* (A. Doucet, J.F.G. de Freitas and N.J. Gordon, Eds.). pp. 225–246. Springer-Verlag. New York.

Ozawa, S., M. Murakami, M. Kaidzu, T. Tada, T. Sagiya, Y. Hatanaka, H. Yarai and T. Nishimura (2001). Detection and monitoring of ongoing aseismic slip in the tokai region, central Japan. *Science* **298**, 1009–1012.

Segall, P. and M. Matthews (1997). Time dependent inversion of geodetic data. *Journal of Geophysical Research* **102**, 22391–22409.

Shephard, N. (1994). Partial non-Gaussian time series models. *Biometrika* **81**, 115–131.

Shumway, R.H. and D.S. Stoffer (1991). Dynamic liner models with switching. *Journal of the American Statistical Association* **86**, 763–769.

IFAC Publications
www.elsevier.com/locate/ifac

A PARTICLE IMPLEMENTATION OF THE RECURSIVE MLE FOR PARTIALLY OBSERVED DIFFUSIONS

Arnaud Guyader * **François Le Gland** ** **Nadia Oudjane** ***

* *Université de Haute Bretagne, 35043 RENNES Cédex, France*
** *IRISA / INRIA, 35042 RENNES Cédex, France*
*** *EDF, Division R&D, 92141 CLAMART Cédex, France*

Abstract In this paper, the problem of identifying a hidden Markov model (HMM) with general state space, e.g. a partially observed diffusion process, is considered. A particle implementation of the recursive maximum likelihood estimator for a parameter in the transition kernel of the Markov chain is presented. The key assumption is that the derivative of the transition kernel w.r.t. the parameter has a probabilistic interpretation, suitable for Monte Carlo simulation. Examples are given to show that this assumption is satisfied in quite general situations. As a result, the linear tangent filter, i.e. the derivative of the filter w.r.t. the parameter, is absolutely continuous w.r.t. the filter and the idea is to jointly approximate the (prediction) filter and its derivative with the empirical probability distribution and with a weighted empirical distribution associated with the same and unique particle system. Application to the identification of a stochastic volatility model is presented. *Copyright © 2003 IFAC*

Keywords: hidden Markov model, stochastic volatility model, nonlinear filter, linear tangent filter, particle filter, recursive MLE.

1. HIDDEN MARKOV MODEL

The state sequence $\{X_k, k \geq 0\}$ is a Markov chain taking values in the space E, with transition kernel $Q(x,dx')$, i.e.

$$\mathbb{P}[X_{k+1} \in dx' \mid X_k = x] = Q(x,dx') .$$

The kernel $Q(x,dx')$ could depend on a parameter, that should be either estimated, or monitored (i.e. changes w.r.t. a nominal value should be detected), however the dependence w.r.t. the parameter is not written explicitly, so as to avoid intricated notations. The following assumption is made

It is *easy* to *simulate* a r.v. X with probability distribution $Q(x,dx')$, even though the analytical expression of the kernel $Q(x,dx')$ is not known, or is so complicated that it is *pratically impossible* to *compute* such integrals as

$$Q\,\phi(x) = \int_E Q(x,dx')\,\phi(x') ,$$

or

$$Q\,\mu(dx') = \int_E \mu(dx)\,Q(x,dx') .$$

This is the case for instance if the Markov chain $\{X_k, k \geq 0\}$ is obtained by sampling a diffusion process $\{X'_t, t \geq 0\}$ at discrete time instants $\{t_k, k \geq 0\}$, i.e. if $X_k = X'_{t_k}$, with

$$dX'_t = b(X'_t)\,dt + \sigma(X'_t)\,dW_t , \qquad (1)$$

where $\{W_t, t \geq 0\}$ is a Brownian motion. In this case, to *simulate* a r.v. with probability distribution $Q(x,dx')$ simply reduces to simulate (with an appropriate numerical discretisation scheme) the solution at time t_{k+1} of the stochastic differential equation (1) starting from the initial condition $X'_{t_k} = x$ at time t_k.

[1] This work was partially supported by the CNRS, under the *MathSTIC* project *Chaînes de Markov Cachées et Filtrage Particulaire*, and under the *AS–STIC* project *Méthodes Particulaires* (AS 67), and by the A.M. Liapunov French–Russian Institute project *Mathématiques Financières*.

The state sequence $\{X_k, k \geq 0\}$ is not observed, but instead an observation sequence $\{Y_k, k \geq 0\}$ is available, which has the following property : given the hidden states $\{X_k, k \geq 0\}$, the observations $\{Y_k, k \geq 0\}$ are mutually independent, and the conditional probability distribution of Y_k depends only on the hidden state X_k at the same time instant, and by definition

$$\mathbb{P}[Y_k \in dy \mid X_k = x] = g(x,y)\,\lambda(dy) ,$$

and

$$\Psi_k(x) = g(x, Y_k) .$$

This *memoryless channel* assumption is satisfied for instance in the case where

$$Y_k = h(X_k) + V_k ,$$

where $\{V_k, k \geq 0\}$ is a white noise sequence (i.e. a sequence of mutually independent r.v.'s) with probability distribution $q(v)\,dv$, independent of $\{X_k, k \geq 0\}$. In this case

$$\mathbb{P}[Y_k \in dy \mid X_k = x] = q(y - h(x))\,dy ,$$

and

$$\Psi_k(x) = q(Y_k - h(x)) .$$

The *memoryless channel* assumption is also satisfied in the case where the covariance of the observation noise depends on the hidden state, i.e. in the case where

$$Y_k = r(X_k)\,V_k ,$$

where $\{V_k, k \geq 0\}$ is a white noise sequence with probability distribution $q(v)\,dv$, independent of $\{X_k, k \geq 0\}$. In this case, provided the matrix $r(x)$ is invertible for any $x \in \mathbb{R}^m$, it holds

$$\mathbb{P}[Y_k \in dy \mid X_k = x] = \frac{q([r(x)]^{-1}y)}{\det r(x)}\,dy ,$$

and

$$\Psi_k(x) = \frac{q([r(x)]^{-1}Y_k)}{\det r(x)} .$$

Given observations, the objective of nonlinear filtering is to estimate the hidden states, and to this effect the probability distributions

$$\mu_k(dx) = \mathbb{P}[X_k \in dx \mid Y_0, \cdots, Y_k] ,$$

and

$$\mu_{k|k-1}(dx) = \mathbb{P}[X_k \in dx \mid Y_0, \cdots, Y_{k-1}] ,$$

are introduced. The evolution of the sequence $\{\mu_k, k \geq 0\}$ taking values in the space of probability distributions on E, is very easily described by the following steps

$$\mu_{k-1} \longrightarrow \mu_{k|k-1} = Q\,\mu_{k-1}$$

$$\longrightarrow \mu_k = \Psi_k \cdot \mu_{k|k-1} ,$$

where

$$\mu_{k|k-1}(dx') = Q\,\mu_{k-1}(dx') = \int_E \mu_{k-1}(dx)\,Q(x,dx') ,$$

can happen to be *difficult* (if not just impossible) to *compute*, and where $\cdot$ denotes the projective product, i.e.

$$\mu_k(dx) = \Psi_k \cdot \mu_{k|k-1}(dx) = \frac{\Psi_k(x)\,\mu_{k|k-1}(dx)}{\langle \mu_{k|k-1}, \Psi_k \rangle} .$$

2. LINEAR TANGENT KERNEL / EXTENDED KERNEL, ETC.

If the transition kernel $Q(x,dx')$ depends on a parameter, then the filter μ_k depends also on the parameter, and one would like to compute the linear tangent filter w_k, i.e. the derivative of the filter μ_k w.r.t. the parameter. To this end, one needs first to study the linear tangent kernel $\Gamma(x,dx')$, i.e. the derivative of the transition kernel $Q(x,dx')$ w.r.t. the parameter, and the following assumption is made

Assumprion AC : The following probabilistic representation holds for the linear tangent kernel $\Gamma(x,dx')$

$$\Gamma\phi(x) = \int_E \Gamma(x,dx')\,\phi(x')$$

$$= \mathbb{E}[\phi(X_{k+1})\,\Xi_{k+1} \mid X_k = x] ,$$

where $\{(X_k, \Xi_k), k \geq 0\}$ is a Markov chain taking values in the product space $E \times F$, such that

$$\mathbb{P}[X_{k+1} \in dx', \Xi_{k+1} \in ds' \mid X_k = x, \Xi_k = s]$$

$$= K(x,dx',ds') .$$

The following assumption, which extends the similar assumption introduced in Section 1, is made

It is *easy* to *simulate* a r.v. (X,Ξ) with probability distribution $K(x,dx',ds')$, even though the analytical expression of the kernel $K(x,dx',ds')$ is not known, or is so complicated that it is *pratically impossible* to *compute* such integrals as

$$\Gamma\phi(x) = \int_{E\times F} s'\,\phi(x')\,K(x,dx',ds') ,$$

or

$$\Gamma\mu(dx') = \int_{E\times F} \mu(dx)\,s'\,K(x,dx',ds') .$$

Example 2.1. Let the Markov chain $\{X_k, k \geq 0\}$ taking values in $E = \mathbb{R}^m$, be defined by

$$X_{k+1} = f(X_k) + g(X_k)\,W_k ,$$

where only the function f depends on the parameter, and where $\{W_k, k \geq 0\}$ is a sequence of independent r.v.'s taking values in $\mathbb{R}^q$ with probability distribution $p(w)\,dw$. In the simple case where $g(x) = I$ for any $x \in \mathbb{R}^m$, the transition kernel $Q(x,dx')$ is given by

$$Q(x,dx') = p(x' - f(x))\,dx' ,$$

and one can show directly that

$$\Gamma(x,dx') = \frac{-p'}{p}(x' - f(x))\, \partial f(x)\, Q(x,dx')\ ,$$

where ∂f denotes the derivative of the function f w.r.t. the parameter, i.e. Assumption AC is satisfied, with

$$\Xi_{k+1} = \frac{-p'}{p}(W_k)\, \partial f(X_k)$$

This result generalizes to the case where for any $x \in \mathbb{R}^m$ the matrix $g(x)$ has full rank, and the vector $\partial f(x)$ belongs to the range of $g(x)$.

Example 2.2. Let the Markov chain $\{X_k, k \geq 0\}$ taking values in $E = \mathbb{R}^m$, be defined by

$$X_{k+1} = f(X_k) + c\, g(X_k)\, W_k\ ,$$

where $\{W_k, k \geq 0\}$ is a sequence of independent r.v.'s taking values in $\mathbb{R}^q$ with probability distribution $p(w)\, dw$. In the simple case where $g(x) = I$ for any $x \in \mathbb{R}^m$, the transition kernel $Q(x,dx')$ is given by

$$Q(x,dx') = p(\frac{x' - f(x)}{c})\, \frac{dx'}{c^q}\ ,$$

and one can show directly that

$$\Gamma(x,dx') =$$

$$= [\frac{-p'}{p}(\frac{x' - f(x)}{c})\, \frac{x' - f(x)}{c^2} - \frac{q}{c}]\, Q(x,dx')\ ,$$

i.e. Assumption AC is satisfied, with

$$\Xi_{k+1} = \frac{1}{c}[\frac{-p'}{p}(W_k)\, W_k - q]\ ,$$

and

$$\Xi_{k+1} = \frac{1}{c}\,(\,|W_k|^2 - q)\ ,$$

in the special case where W_k is a zero mean Gaussian r.v. with identity covariance matrix. This results actually holds without any assumption on the matrix g.

Notice that in the above two examples, the r.v. Ξ_{k+1} depends only on (X_k, W_k), in which case it does not seem necessary to simulate Ξ_{k+1} in addition to W_k. In some cases, it is even possible to express W_k in terms of (X_k, X_{k+1}), and finally the r.v. $\Xi_{k+1} = I(X_k, X_{k+1})$ depends only on (X_k, X_{k+1}). This apparently very particular situation is actually very general, as the following result shows.

Lemma 2.3. Under Assumption AC

$$\Gamma(x,dx') = I(x,x')\, Q(x,dx')\ ,$$

with

$$I(x,x') = \mathbb{E}[\Xi_{k+1} \mid X_k = x, X_{k+1} = x']\ ,$$

for any $x, x' \in E$.

However, and as the following example shows, there exist situations where the existence of the function I does not imply that an easy–to–compute explicit expression exists, whereas in opposition the joint simulation of (X_{k+1}, Ξ_{k+1}) is easy.

Example 2.4. Let the Markov chain $\{X_k, k \geq 0\}$ be defined by sampling at discrete time instants $\{t_k, k \geq 0\}$ a diffusion process $\{X'_t, t \geq 0\}$, i.e. $X_k = X'_{t_k}$, with

$$dX'_t = b(X'_t)\, dt + \sigma(X'_t)\, dW_t\ ,$$

where only the drift function b depends on the parameter, and where $\{W_t, t \geq 0\}$ is a Brownian motion. If for any $x \in \mathbb{R}^m$, the matrix $\sigma(x)$ has full rank, and the vector $\partial b(x)$ belongs to the range of $\sigma(x)$, then Assumption AC is satisfied, with

$$\Xi_{k+1} = \int_{t_k}^{t_{k+1}} [\partial b(X'_t)]^*$$

$$\sigma(X'_t)\, [\sigma^*(X'_t)\, \sigma(X'_t)]^{-1}\, dW_t\ ,$$

where ∂b denotes the derivative of the drift function b w.r.t. the parameter, see (Cérou *et al.*, 2001) or (Fournié *et al.*, 1999) for the simpler case where for any $x \in \mathbb{R}^m$ the matrix $\sigma(x)$ is invertible. It is easy (with an appropriate numerical discretization scheme) to jointly simulate (X_{k+1}, Ξ_{k+1}), but in opposition there does not exist in general a simple analytical expression for

$$I(x,x') = \mathbb{E}[\Xi_{k+1} \mid X_k = x, X_{k+1} = x']\ .$$

By definition

$$\Gamma(x,dx') = \int_F s'\, K(x,dx',ds')\ ,$$

and

$$Q(x,dx') = \int_F K(x,dx',ds')\ .$$

On the product space $E \times E \times F$, define the projection $\pi_0\ : (x,x',s') \longmapsto x$ on the (first) space E, the projection $\pi\ : (x,x',s') \longmapsto x'$ on the (second) space E and the projection $\pi_F\ : (x,x',s') \longmapsto s'$ on the auxiliary space F. For any probability distribution μ on the space E, the probability distribution $\mu \otimes K$ is defined on the product space $E \times E \times F$ by

$$(\mu \otimes K)(dx,dx',ds') = \mu(dx)\, K(x,dx',ds')\ .$$

It follows that

$$Q\mu = (\mu \otimes K) \circ \pi^{-1}\ ,$$

and

$$\Gamma\mu = (\pi_F\, (\mu \otimes K)) \circ \pi^{-1}\ ,$$

and

$$(w \ll \mu \quad \Longrightarrow \quad Qw = ((\frac{dw}{d\mu} \circ \pi_0)\, (\mu \otimes K)) \circ \pi^{-1})\ .$$

Lemma 2.5. Under Assumption AC, $\Gamma\mu \ll Q\mu$ for any probability distribution μ on E, with Radon–Nikodym derivative (which depends on μ)

$$\frac{d(\Gamma\mu)}{d(Q\mu)}(x') = \mathbb{E}_\mu[\Xi_{n+1} \mid X_{n+1} = x']\ .$$

Lemma 2.6. If the finite signed measure w is absolutely continuous w.r.t. the probability distribution μ, then $Qw \ll Q\mu$, with Radon–Nikodym derivative

$$\frac{d(Qw)}{d(Q\mu)}(x') = \mathbb{E}_\mu[\frac{dw}{d\mu}(X_k) \mid X_{k+1} = x'] .$$

By definition

$$F_k(\mu)\, w = \frac{\Psi_k w}{\langle \mu, \Psi_k \rangle} - \frac{\langle w, \Psi_k \rangle}{\langle \mu, \Psi_k \rangle} \frac{\Psi_k \mu}{\langle \mu, \Psi_k \rangle} ,$$

is the derivative at point μ and in the direction w, of the mapping $\mu \longmapsto \Psi_k \cdot \mu$.

Lemma 2.7. If the finite signed measure w is absolutely continuous w.r.t. the probability distribution μ, then $F_k(\mu)\, w \ll \Psi_k \cdot \mu$, with Radon–Nikodym derivative

$$\frac{d(F_k(\mu)\, w)}{d(\Psi_k \cdot \mu)}(x) = \frac{dw}{d\mu}(x) - \langle \Psi_k \cdot \mu, \frac{dw}{d\mu} \rangle .$$

3. PARTICLE APPROXIMATION OF SOME FINITE SIGNED MEASURES

With the notations of the previous section, it easily seen that the probability distribution $Q\mu$ and the finite signed measures $\Gamma\mu$ and Qw can be put in the general form $(r\,(\mu \otimes K)) \circ \pi^{-1}$ for some appropriate choice of the weight function r, namely $r \equiv 1$, $r = \pi_F$ and $r = \dfrac{dw}{d\mu} \circ \pi_0$ respectively. The weighted particle approximation of a finite signed measure of the general form $r\,(\mu \otimes K)$ is defined by

$$r\,(\mu \otimes K) \approx r S^N (\mu \otimes K)$$

$$= \frac{1}{N} \sum_{i=1}^{N} r(\xi_0^i, \xi^i, \Xi^i)\, \delta_{(\xi_0^i, \xi^i, \Xi^i)} ,$$

where the r.v.'s $\{\xi_0^i, \xi^i, \Xi^i, i = 1, \cdots, N\}$ form an N–sample with probability distribution $\mu \otimes K$, which can be easily achieved in the following manner : independently for any $i = 1, \cdots, N$

$$\xi_0^i \sim \mu(dx) ,$$

and

$$(\xi^i, \Xi^i) \sim K(\xi_0^i, dx', ds') ,$$

and the corresponding particle approximation for the marginal measure $(r\,(K \otimes \mu)) \circ \pi^{-1}$ is defined by

$$(r\,(\mu \otimes K)) \circ \pi^{-1} \approx (r S^N (\mu \otimes K)) \circ \pi^{-1}$$

$$= \frac{1}{N} \sum_{i=1}^{N} r(\xi_0^i, \xi^i, \Xi^i)\, \delta_{\xi^i} .$$

In particular for the weight functions $r \equiv 1$, $r = \pi_F$ and $r = \dfrac{dw}{d\mu} \circ \pi_0$, it holds

$$Q\mu = (\mu \otimes K) \circ \pi^{-1} \approx \frac{1}{N} \sum_{i=1}^{N} \delta_{\xi^i} ,$$

$$\Gamma\mu = (\pi_F\,(\mu \otimes K)) \circ \pi^{-1} \approx \frac{1}{N} \sum_{i=1}^{N} \Xi^i\, \delta_{\xi^i} ,$$

and

$$Qw = ((\frac{dw}{d\mu} \circ \pi_0)\,(\mu \otimes K)) \circ \pi^{-1} \approx \frac{1}{N} \sum_{i=1}^{N} \frac{dw}{d\mu}(\xi_0^i)\, \delta_{\xi^i} ,$$

respectively. For any test function ϕ defined on E, it holds

$$\sup_{\|\phi\|=1} \mathbb{E}| \frac{1}{N} \sum_{i=1}^{N} \phi(\xi^i) - \langle Q\mu, \phi \rangle | \leq \frac{1}{\sqrt{N}} ,$$

$$\sup_{\|\phi\|=1} \mathbb{E}| \frac{1}{N} \sum_{i=1}^{N} \Xi^i\, \phi(\xi^i) - \langle \Gamma\mu, \phi \rangle |$$

$$\leq \frac{1}{\sqrt{N}} \{\sup_{x \in E} \int_{E \times F} |s'|^2\, K(x, dx', ds')\}^{1/2} ,$$

and

$$\sup_{\|\phi\|=1} \mathbb{E}| \frac{1}{N} \sum_{i=1}^{N} \frac{dw}{d\mu}(\xi_0^i)\, \phi(\xi^i) - \langle Qw, \phi \rangle |$$

$$\leq \frac{1}{\sqrt{N}} \{\int_E |\frac{dw}{d\mu}(x)|^2\, \mu(dx)\}^{1/2} .$$

4. JOINT PARTICLE APPROXIMATION OF THE FILTER AND THE LINEAR TANGENT FILTER

Recall that the evolution of the sequence $\{\mu_k, k \geq 0\}$ taking values in the space of probability distributions on E, is described by the following two steps

$$\mu_{k-1} \longrightarrow \mu_{k|k-1} = Q\mu_{k-1}$$

$$\longrightarrow \mu_k = \Psi_k \cdot \mu_{k|k-1} .$$

If w_k denotes at each time instant the linear tangent filter, i.e. the derivative of the filter μ_k w.r.t. the parameter, then the evolution of the sequence $\{w_k, k \geq 0\}$ taking values in the linear tangent space to the space of probability distributions on E, i.e. taking values in the space of finite signed measures on E with zero total mass, is described by the following two steps, which are linear tangent versions of the prediction step and correction step respectively

$$w_{k-1} \longrightarrow w_{k|k-1} = Qw_{k-1} + \Gamma\mu_{k-1}$$

$$\longrightarrow w_k = F_k(\mu_{k|k-1})\, w_{k|k-1} .$$

Under Assumption AC, it is easily seen by induction, and using Lemmas 2.5, 2.6 and 2.7, that at each time instant $w_{k|k-1} \ll \mu_{k|k-1}$ and $w_k \ll \mu_k$. In view of this absolute continuity property, and of the key assumption that it is *easy* to *simulate* r.v.'s with probability distribution $K(x,dx',ds')$, the idea is to jointly approximate the predictor $\mu_{k|k-1}$ and its derivative $w_{k|k-1}$ w.r.t. the parameter with the empirical probability distribution and a weighted empirical distribution associated with the same and unique N–sample, i.e.

$$\mu_{k|k-1} \approx \mu^N_{k|k-1} = \frac{1}{N} \sum_{i=1}^{N} \delta_{\xi^i_{k|k-1}} ,$$

and

$$w_{k|k-1} \approx w^N_{k|k-1} = \frac{1}{N} \sum_{i=1}^{N} \rho^i_{k|k-1} \, \delta_{\xi^i_{k|k-1}} .$$

With this definition $w^N_{k|k-1} \ll \mu^N_{k|k-1}$, with Radon–Nikodym derivative

$$r^N_{k|k-1}(x) = \frac{1}{|I^N_{k|k-1}(x)|} \sum_{i \in I^N_{k|k-1}(x)} \rho^i_{k|k-1} ,$$

where $I^N_{k|k-1}(x) = \{i = 1, \cdots, N : \xi^i_{k|k-1} = x\}$, for any x in the support supp $\mu^N_{k|k-1}$ of the discrete probability distribution $\mu^N_{k|k-1}$. Notice that in most cases, the particle locations $\{\xi^i_{k|k-1}, i = 1, \cdots, N\}$ happen to be all distinct, and the much simpler relation

$$r^N_{k|k-1}(\xi^i_{k|k-1}) = \rho^i_{k|k-1} ,$$

holds for any $i = 1, \cdots, N$. This approximation is completely characterized by the set $\{\xi^i_{k|k-1}, \rho^i_{k|k-1}, i = 1, \cdots, N\}$ of particles and weights, and the algorithm is completely described by the mechanism which builds $\{\xi^i_{k+1|k}, \rho^i_{k+1|k}, i = 1, \cdots, N\}$ from $\{\xi^i_{k|k-1}, \rho^i_{k|k-1}, i = 1, \cdots, N\}$. This mechanism is as follows :

(i) the correction step is applied *exactly* to $\mu^N_{k|k-1}$, which results in

$$\mu^N_k = \Psi_k \cdot \mu^N_{k|k-1} = \sum_{i=1}^{N} \frac{\Psi_k(\xi^i_{k|k-1}) \, \delta_{\xi^i_{k|k-1}}}{\sum_{j=1}^{N} \Psi_k(\xi^j_{k|k-1})}$$

$$= \sum_{i=1}^{N} \omega^i_k \, \delta_{\xi^i_{k|k-1}} ,$$

and the linear tangent correction step is applied *exactly* to $w^N_{k|k-1}$, which results in

$$w^N_k = F_k(\mu^N_{k|k-1}) \, w^N_{k|k-1}$$

$$= [r^N_{k|k-1} - \langle \mu^N_k, r^N_{k|k-1} \rangle] \, \mu^N_k ,$$

(ii) instead of trying to *compute* $Q\,\mu^N_k$, the following particle approximation

$$\mu^N_{k+1|k} = S^N(Q\,\mu^N_k) = \frac{1}{N} \sum_{i=1}^{N} \delta_{\xi^i_{k+1|k}} ,$$

is used, instead of trying to *compute*

$$Q w^N_k = Q(r^N_{k|k-1} \, \mu^N_k) - \langle \mu^N_k, r^N_{k|k-1} \rangle \, Q\,\mu^N_k$$

$$= ((r^N_{k|k-1} \circ \pi_0) \, (\mu^N_k \otimes Q)) \circ \pi^{-1}$$

$$- \langle \mu^N_k, r^N_{k|k-1} \rangle \, Q\,\mu^N_k ,$$

the following weighted particle approximation

$$((r^N_{k|k-1} \circ \pi_0) \, S^N(\mu^N_k \otimes Q)) \circ \pi^{-1}$$

$$- \langle S^N(\mu^N_k), r^N_{k|k-1} \rangle \, S^N(Q\,\mu^N_k)$$

$$= \frac{1}{N} \sum_{i=1}^{N} [r^N_{k|k-1}(\xi^i_k)$$

$$- \frac{1}{N} \sum_{j=1}^{N} r^N_{k|k-1}(\xi^j_k)] \, \delta_{\xi^i_{k+1|k}} ,$$

is used, and instead of trying to *compute*

$$\Gamma \mu^N_k = (\pi_F \, (\mu^N_k \otimes K)) \circ \pi^{-1} ,$$

the following weighted particle approximation

$$(\pi_F \, S^N(\mu^N_k \otimes K)) \circ \pi^{-1} = \frac{1}{N} \sum_{i=1}^{N} \Xi^i_{k+1} \, \delta_{\xi^i_{k+1|k}} ,$$

is used, hence finally the weighted particle approximation

$$w^N_{k+1|k} = \frac{1}{N} \sum_{i=1}^{N} [r^N_{k|k-1}(\xi^i_k) + \Xi^i_{k+1}$$

$$- \frac{1}{N} \sum_{j=1}^{N} r^N_{k|k-1}(\xi^j_k)] \, \delta_{\xi^i_{k+1|k}}$$

$$= \frac{1}{N} \sum_{i=1}^{N} \rho^i_{k+1|k} \, \delta_{\xi^i_{k+1|k}} ,$$

where the r.v.'s $\{\xi^i_k, \xi^i_{k+1|k}, \Xi^i_{k+1}, i = 1, \cdots, N\}$ form an N–sample with probability distribution $\mu^N_k \otimes K$, which can be easily achieved in the following manner : independently for any $i = 1, \cdots, N$

$$\xi^i_k \sim \mu^N_k(dx) ,$$

which is easy, since the probability distribution μ^N_k is discrete, and

$$(\xi^i_{k+1|k}, \Xi^i_{k+1}) \sim K(\xi^i_k, dx', ds') ,$$

which is easy, by assumption.

An alternate redistribution scheme has been proposed recently by (Doucet and Tadić, n.d.)

5. PARTICLE FILTER IMPLEMENTATION OF THE RECURSIVE MLE

In this section, the parameter is denoted by θ and dependence w.r.t. the parameter appears explicitly in the notation for the transition kernel $Q^\theta(x,dx')$, and for the linear tangent kernel $K^\theta(x,dx',ds')$. Monitoring the parametric model, i.e. detecting a small change from a nominal value, corresponding to the normal behaviour of the system, has been addressed in (Cérou and Le Gland, 2000). Another question is to identify the parametric model, and it is natural to consider the recursive MLE, which is defined by the following relation

$$\widehat{\theta}_k = \widehat{\theta}_{k-1} + \gamma_k \frac{\langle \widehat{w}_{k|k-1}, \Psi_k \rangle}{\langle \widehat{\mu}_{k|k-1}, \Psi_k \rangle} , \qquad (2)$$

where typically $\gamma_k \simeq k^{-2/3}$, and the averaged estimator (which achieves the minimum variance of the estimation error) is obtained by post–processing

$$\overline{\theta}_k = \overline{\theta}_{k-1} + \frac{1}{k} (\widehat{\theta}_k - \overline{\theta}_{k-1}) .$$

Here, the adaptive filter $\{\widehat{\mu}_k, k \geq 0\}$ and the adaptive linear tangent filter $\{\widehat{w}_k, , k \geq 0\}$ satisfy the same equations as the filter and the linear tangent filter respectively, in which the value of the parameter is adapted at each time instant according to equation (2), i.e.

$$\widehat{\mu}_{k-1} \longrightarrow \widehat{\mu}_{k|k-1} = Q^{\widehat{\theta}_{k-1}} \widehat{\mu}_{k-1}$$

$$\longrightarrow \widehat{\mu}_k = \Psi_k \cdot \widehat{\mu}_{k|k-1} ,$$

and

$$\widehat{w}_{k-1} \longrightarrow \widehat{w}_{k|k-1} = Q^{\widehat{\theta}_{k-1}} \widehat{w}_{k-1} + \Gamma^{\widehat{\theta}_{k-1}} \widehat{\mu}_{k-1}$$

$$\longrightarrow \widehat{w}_k = F_k(\widehat{\mu}_{k|k-1}) \widehat{w}_{k|k-1} ,$$

respectively. The particle implementation of the recursive MLE is

$$\widehat{\theta}^N_k = \widehat{\theta}^N_{k-1} + \gamma_k \, [\sum_{i=1}^{N} \rho^i_{k|k-1} \, \omega^i_k] ,$$

and

$$\overline{\theta}^N_k = \overline{\theta}^N_{k-1} + \frac{1}{k} (\widehat{\theta}^N_k - \overline{\theta}^N_{k-1}) .$$

The mathematical analysis of the asymptotic properties of the estimator $\widehat{\theta}^N_k$ as $k \to \infty$ and $N \to \infty$ is far beyond the scope of this paper, and would rely on joint stability properties of the filter and the linear tangent filter, which is a very difficult question. Even the asymptotic properties of the estimator $\widehat{\theta}_k$ as $k \to \infty$ are difficult to prove, unless some mixing assumption holds for the transition kernels $Q^\theta(x,dx')$ and the linear tangent kernels $\Gamma^\theta(x,dx')$, which practically implies that the state–space E should be compact, see e.g. (Douc and Matias, 2001) where only the non–recursive MLE is studied.

6. APPLICATION TO A STOCHASTIC VOLATILITY MODEL

The following stochastic volatility model, with mean–reverting hidden diffusion

$$dX'_t = a(b - X'_t)\, dt + c X'_t \, dW'_t , \qquad X'_0 > 0$$

$$dY'_t = \sqrt{X'_t} \, dV'_t ,$$

where $\{W'_t, t \geq 0\}$ and $\{V'_t, t \geq 0\}$ are independent Brownian motions, has been considered by (Genon-Catalot *et al.*, 2000). An alternate discrete–time observation model is considered in the present paper, in which an approximate Markov chain $\{X_k, k \geq 0\}$ is used, based on a Euler (or an alternate splitting–up) scheme, i.e.

$$X_k = (1 - a\Delta_k) X_{k-1} + ab\Delta_k + cX_{k-1} W_k ,$$

where $\Delta_k = t_{k+1} - t_k$ and where $\{W_k, k \geq 0\}$ is a Gaussian white noise sequence with variance Δ_k, and

$$Y_k = \sqrt{X_k} V_k ,$$

where $\{V_k, k \geq 0\}$ is a Gaussian white noise sequence independent of $\{W_k, k \geq 0\}$, hence

$$\Psi_k(x) = \frac{1}{\sqrt{x}} \exp\{-\frac{Y_k^2}{2x}\} , \qquad x > 0 .$$

In this case, it follows from Examples 2.1 and 2.2 that Assumption AC is satisfied, with

$$\Xi_{k+1} = (\frac{b - X_k}{cX_k} W_k , \frac{a}{cX_k} W_k , \frac{1}{c} (|W_k|^2 - 1)) .$$

7. REFERENCES

Cérou, Frédéric and François Le Gland (2000). Efficient particle filters for residual generation in partially observed SDE's. In: *Proceedings of the 39th Conference on Decision and Control, Sydney 2000*. IEEE–CSS. pp. 1200–1205.

Cérou, Frédéric, François Le Gland and Nigel J. Newton (2001). Stochastic particle methods for linear tangent filtering equations. In: *Optimal Control and Partial Differential Equations. In honour of professor Alain Bensoussan's 60th birthday* (José-Luis Menaldi, Edmundo Rofman and Agnès Sulem, Eds.). pp. 231–240. IOS Press. Amsterdam.

Douc, Randal and Catherine Matias (2001). Asymptotics of the maximum likelihood estimator for general hidden Markov models. *Bernoulli* **7**(3), 381–420.

Doucet, Arnaud and Vladislav B. Tadić (n.d.). Parameter estimation in general state–space models using particle methods. *Annals of the Institute of Statistical Mathematics*. To appear.

Fournié, Éric, Jean-Michel Lasry, Jérôme Lebuchoux, Pierre-Louis Lions and Nizar Touzi (1999). Applications of Malliavin calculus to Monte Carlo methods in finance. *Finance and Stochastics* **3**(4), 391–412.

Genon-Catalot, Valentine, Thierry Jeantheau and Catherine Laredo (2000). Stochastic volatility models as hidden Markov models and statistical applications. *Bernoulli* **6**(6), 1051–1079.

IFAC Publications
www.elsevier.com/locate/ifac

ONLINE SAMPLING FOR PARAMETER ESTIMATION IN GENERAL STATE SPACE MODELS

Christope Andrieu * **Arnaud Doucet** ** **Vladislav B. Tadić** ***

* *Department of Mathematics, Bristol University, BS8 1TW Bristol, UK*
** *Department of Engineering, Cambridge University, CB2 1PZ Cambridge, UK*
*** *Department of Electrical Engineering, Melbourne University, 3010 Melbourne, Australia*

Abstract: We consider the class of stationary nonlinear non Gaussian state space models with unknown static parameters. We propose original online stochastic gradient type algorithms to estimate these parameters. These algorithms rely on the simulation of artificial observations. Contrary to all the methods we are aware of in this framework, optimal state estimation is not required by our methods and the proposed algorithms are computationally efficient. Their efficiency is assessed through simulation. *Copyright © 2003 IFAC*

Keywords: Parameter Estimation, Sequential Importance Sampling, State Space Models, Stochastic Approximation

1. INTRODUCTION

1.1 State Space Models and Problem Statement

Let $\{X_n\}_{n\geq 0}$ and $\{Y_n\}_{n\geq 1}$ be R^p and R^q-valued stochastic processes defined on a measurable space $(\Omega, \mathscr{F})$ while $\theta \in \Theta$ where Θ is an open subset of R^d. The process $\{X_n\}_{n\geq 0}$ is an unobserved (hidden) Markov process of initial density μ, i.e. $X_0 \sim \mu$, and Markov transition density $f_\theta(\cdot,\cdot)$; i.e.

$$X_{n+1} | X_n = x \sim f_\theta(\cdot|x). \quad (1)$$

One observes the process $\{Y_n\}_{n\geq 1}$ and it is assumed that the observations are conditionally independent upon $\{X_n\}_{n\geq 0}$ of marginal density $g_\theta(\cdot,\cdot)$; i.e.

$$Y_n | X_n = x \sim g_\theta(\cdot|x). \quad (2)$$

All densities are defined with respect to some appropriate dominating measures; e.g. Lebesgue. This class of models include many nonlinear and non-Gaussian time series models such as

$$X_{n+1} = \varphi_\theta(X_n, V_{n+1}),\ Y_n = \psi_\theta(X_n, W_n)$$

where $\{V_n\}_{n\geq 0}$ and $\{W_n\}_{n\geq 1}$ are independent sequences and φ_θ, ψ_θ are deterministic functions.

We will restrict ourselves to stationarity state space models where, for any $\theta \in \Theta$, the Markov process $\{X_n\}_{n\geq 0}$ is ergodic and admits an invariant distribution π_θ.

Let us assume that the true value of the parameter θ is θ^* and that only the process $\{Y_n\}_{n\geq 1}$ is observed. We are interested in deriving recursive algorithms to estimate θ^*. This complex problem has numerous applications in electrical engineering, econometrics, statistics, etc.

Further on we will denote for any sequence z_k/random process Z_k $z_{i:j} = (z_i, z_{i+1}, \ldots, z_j)$ and $Z_{i:j} = (Z_i, Z_{i+1}, \ldots, Z_j)$.

1.2 A Brief Literature Review

Following the introduction of Sequential Monte Carlo (SMC) methods (Doucet, *et al.*, 2001), i.e. particle filters, many methods have been recently proposed to address this problem. There are roughly three categories of methods.

• *Filtering methods.* A standard approach followed in the literature consists of setting a prior distribution on the unknown parameter θ and then considering the extended state $S_n = (X_n, \theta)$. This converts the parameter estimation into an optimal filtering problem. One can then apply, at least theoretically, standard particle filtering techniques (Doucet, *et al.*, 2001) to estimate the joint posterior density $p\left(x_n, \theta \middle| Y_{1:n}\right)$ and thus $p\left(\theta \middle| Y_{1:n}\right)$. In this approach, the parameter space is only explored at the initialization of the algorithm. Consequently the algorithm is inefficient; after a few iterations the marginal posterior distribution of the parameter is approximated by a single delta Dirac function. To limit this problem, several authors have proposed to use kernel density estimation methods. However, this has the effect of transforming the fixed parameter into a slowly time-varying one. A pragmatic approach consists of introducing explicitly an artificial dynamic model on the parameter of interest; see (Higuchi, 1997), (Kitagawa, 1998). To avoid the introduction of such a model, an approach proposed in (Gilks and Berzuini, 2001) consists of adding Markov chain Monte Carlo (MCMC) steps so as to add "diversity" among the particles. However, this approach does not really solve the fixed-parameter estimation problem. More precisely, the addition of MCMC steps does not make the dynamic model ergodic. Thus, there is an accumulation of errors over time and the algorithm can diverge.

• *Recursive Maximum Likelihood*, RML is a stochastic gradient type algorithm to maximize the average log-likelihood. This approach requires the computation of the optimal filter $p_\theta\left(x_n \middle| Y_{1:n}\right)$ and its derivative with respect to θ. This is the approach followed in (LeGland and Mevel, 1997) for finite state space HMM. Algorithms to compute numerically the derivative for general state space models have been proposed in (Cérou, et al., 2001) and (Doucet and Tadić, 2003).

• *Online Expectation-Maximization (EM).* Online EM is an alternative stochastic gradient type algorithm. Such algorithms have been proposed for finite state space HMM and linear Gaussian state space models. It is formally possible to come up with a similar algorithm for general state space models. However it requires the computation of an (online) approximation of the joint density $p_\theta\left(x_{1:n} \middle| Y_{1:n}\right)$ whose dimension increases with time. To avoid this problem, one can use the split data likelihood (Rydén, 1994); see (Andrieu and Doucet, 2003).

All these methods rely on computing non standard posterior distributions and thus require the use of numerical techniques such as SMC methods. For some applications, SMC methods are still too computationally intensive. This has motivated the development of computationally cheaper methods.

1.3 Contributions

We propose here alternative algorithms to address the problem of recursive parameter estimation in general state space models. These algorithms rely on the introduction of a non standard cost function (i.e. different from the classical Kullback-Leibler cost function) which can be shown to be minimized for θ^*. The main difference between our algorithms and all other algorithms we are aware of is that state estimation is not required in our framework. The algorithms are computationally several orders of magnitude cheaper than algorithms based on SMC methods.

The rest of this paper is organized as follows. In Section 2, we introduce the cost function to maximize. In Section 3, we present recursive algorithms to optimize this cost function and discuss the implementation issues. Finally in Section 4, we present an application to a stochastic volatility model arising in econometrics.

2. STATIONARY DISTRIBUTION AND COST FUNCTION

In the stationary regime, the distribution of the random variables $\left(X_{(k-1)L+1:kL}, Y_{(k-1)L+1:kL}\right)$ satisfies

$$P_\theta\left(\left(X_{(k-1)L+1:kL}, Y_{(k-1)L+1:kL}\right) \in d\left(x_{1:L}, y_{1:L}\right)\right) \\ = \pi_\theta\left(x_{1:L}, y_{1:L}\right) dx_{1:L} dy_{1:L}$$

where $d(\cdot)$ is the dominating measure and

$$\pi_\theta\left(x_{1:L}, y_{1:L}\right) = \pi_\theta\left(x_1\right) g_\theta\left(y_1 \middle| x_1\right) \\ \times \prod_{i=2}^{L} f_\theta\left(x_i \middle| x_{i-1}\right) g_\theta\left(y_i \middle| x_i\right)$$

where we recall that π_θ corresponds to the invariant density of the latent Markov process.

We will consider the following cost function

$$J(\theta) = \int \left(\pi_\theta\left(y_{1:L}\right) - \pi_{\theta^*}\left(y_{1:L}\right)\right)^2 dy_{1:L} \quad (3)$$

where

$$\pi_\theta\left(y_{1:L}\right) = \int \pi_\theta\left(x_{1:L}, y_{1:L}\right) dx_{1:L}.$$

The cost function $J(\theta)$ is defined under weak assumptions and it is clear that θ^* minimizes it. The parameter L is chosen large enough to ensure identifiability of θ using $\pi_\theta\left(y_{1:L}\right)$.

3. ONLINE SAMPLING ALGORITHMS

We will distinguish here two cases. If the invariant distribution π_θ is known, then one can devise a standard stochastic gradient type algorithm to minimize $J(\theta)$. If this distribution is unknown, then it is necessary to use a (randomized) finite difference scheme.

3.1 Invariant distribution known

We describe briefly in this section a stochastic gradient algorithm to minimize $J(\theta)$. One has

$$J(\theta)=\int \pi_\theta(y_{1:L})(\pi_\theta(y_{1:L})-2\pi_{\theta^*}(y_{1:L}))\,dy_{1:L}$$
$$+\text{terms independent of }\theta.$$

The derivative $\frac{1}{2}\nabla J(\theta)$ of the cost function is given by

$$\int \nabla\pi_\theta(y_{1:L})(\pi_\theta(y_{1:L})-\pi_{\theta^*}(y_{1:L}))\,dy_{1:L}.$$

Under regularity conditions, one has

$$\begin{aligned}\nabla\pi_\theta(y_{1:L}) &= \nabla\int\pi_\theta(x_{1:L},y_{1:L})\,dx_{1:L}\\ &=\int[\nabla\log\pi_\theta(x_{1:L},y_{1:L})]\,\pi_\theta(x_{1:L},y_{1:L})\,dx_{1:L}\\ &=\int[\nabla\log\pi_\theta(x_{1:L},y_{1:L})]\,g_\theta(y_{1:L}|x_{1:L})\,\pi_\theta(x_{1:L})\,dx_{1:L}\end{aligned}$$

where one uses the following notation $\pi_\theta(x_{1:L},y_{1:L}) = g_\theta(y_{1:L}|x_{1:L})\,\pi_\theta(x_{1:L})$ with

$$g_\theta(y_{1:L}|x_{1:L})=\prod_{i=1}^{L}g_\theta(y_i|x_i)$$

to emphasize that this density does not depend on on the invariant density of $\{X_n\}_{n\geq 0}$.

To sum up, one obtains for $\frac{1}{2}\nabla J(\theta)$

$$\begin{aligned}&\int\nabla\log\pi_\theta(x_{1:L},y_{1:L})\times g_\theta(y_{1:L}|x_{1:L})\,\pi_\theta(x_{1:L})\\ &\times(\pi_\theta(y_{1:L})-\pi_{\theta^*}(y_{1:L}))\,dx_{1:L}dy_{1:L}.\end{aligned}\tag{4}$$

Let $X^*_{(k-1)L+1:kL}$ denote a realization of $\pi_{\theta_k}(x_{1:L})$, $Y^*_{(k-1)L+1:kL}$ a realization from $\pi_{\theta_k}(y_{1:L})$ (the variables $X^*_{(k-1)L+1:kL}$ and $Y^*_{(k-1)L+1:kL}$ being statistically independent) and $Y_{(k-1)L+1:kL}$ a realization from $\pi_{\theta^*}(y_{1:L})$ (i.e. the available data), an asymptotically [1] unbiased gradient estimate of (4) for $\theta=\theta_k$ is given by

$$\begin{aligned}&\tfrac{1}{2}\widehat{\nabla}J(\theta_k)=\\ &\nabla\log\pi_{\theta_k}\left(X^*_{(k-1)L+1:kL},Y^*_{(k-1)L+1:kL}\right)\\ &\times g_{\theta_k}\left(Y^*_{(k-1)L+1:kL}\middle|X^*_{(k-1)L+1:kL}\right)\\ &-\nabla\log\pi_{\theta_k}\left(X^*_{(k-1)L+1:kL},Y_{(k-1)L+1:kL}\right)\\ &\times g_{\theta_k}\left(Y_{(k-1)L+1:kL}\middle|X^*_{(k-1)L+1:kL}\right)\end{aligned}\tag{5}$$

The stochastic gradient algorithm to minimize $J(\theta)$ follows. It relies on a non-increasing positive stepsize sequence $\{\gamma_k\}_{k\geq 0}$ satisfying $\sum\gamma_k=\infty$, $\sum\gamma_k^2<\infty$; one usually selects $\gamma_k\propto k^{-\alpha}$ with $\alpha\in(\frac{1}{2},1]$.

Sampling step

- Sample $\widetilde{X}_{(k-1)L+1:kL}\sim\pi_{\theta_k}(\cdot)$,

$Y^*_{(k-1)L+1:kL}\sim g_{\theta_k}\left(\cdot|\widetilde{X}_{(k-1)L+1:kL}\right)$.

- Sample $X^*_{(k-1)L+1:kL}\sim\pi_{\theta_k}(\cdot)$.

Gradient estimation step

- Compute $\frac{1}{2}\widehat{\nabla}J(\theta_k)$ using (5).

Parameter updating step

$$\theta_{k+1}=\theta_k-\gamma_{k+1}\widehat{\nabla}J(\theta_k).$$

Remark. A truly recursive algorithm, i.e. updating the parameter estimate at each time step, can be derived easily by using a "sliding" window instead of partitioning the data in separated blocks.

Note that $\widetilde{X}_{(k-1)L+1:kL}$ and $X^*_{(k-1)L+1:kL}$ need to be statistically independent to ensure that $X^*_{(k-1)L+1:kL}$ and $Y^*_{(k-1)L+1:kL}$ are independent. Contrary to all algorithms we are aware of, this algorithm only requires the simulation of variables according to the prior distribution.

It is possible to reduce the variance of the gradient estimate by using Sequential Importance Sampling (SIS) techniques (Doucet, *et al.*, 2001); i.e. instead of imputing the latent process according to its (stationary) prior distribution one can come up with "clever" importance distributions to reduce the variance of the gradient. One can also avoid sampling from the invariant distribution and use instead the following algorithm.

Sampling step

- Sample $\widetilde{X}_{(k-1)L+1:kL}\sim f_{\theta_k}\left(\cdot|\widetilde{X}_{(k-1)L}\right)$,

$Y^*_{(k-1)L+1:kL}\sim g_{\theta_k}\left(\cdot|\widetilde{X}_{(k-1)L+1:kL}\right)$.

- Sample $X^*_{(k-1)L+1:kL}\sim f_{\theta_k}\left(\cdot|X^*_{(k-1)L}\right)$.

Gradient estimation step

- Compute $\frac{1}{2}\widehat{\nabla}J(\theta_k)$ using (5).

Parameter updating step

$$\theta_{k+1}=\theta_k-\gamma_{k+1}\widehat{\nabla}J(\theta_k).$$

In the above algorithm, one uses the following notation

$$\begin{aligned}&f_\theta\left(x_{(k-1)L+1:kL}\middle|x_{(k-1)L}\right)\\ &=\prod_{i=(k-1)L+1}^{kL}f_\theta(x_i|x_{i-1}).\end{aligned}$$

It is important to remark that even if one can avoid sampling directly from the invariant distribution, one needs to know it analytically to compute the gradient estimate. This is a restriction even if this density is known in many important applications; e.g. state space models with a linear Gaussian evolution equation. In cases where this density is not known, one needs to use an alternative algorithm developed in the following subsection.

[1] The true system needs to reach its stationary regime.

3.2 Invariant distribution unknown

We propose here another stochastic approximation method namely SPSA (Simultaneous Perturbation Stochastic Approximation) due to Spall (Spall, 1998) as an alternative way to optimize the cost function. With SPSA, the gradient is approximated via a finite difference method using only the estimates of the cost function of interest. This technique has proved successful among other finite difference methods as it only requires two estimates of the cost function regardless of the dimension d of the parameter to obtain a gradient estimate. The SPSA technique requires all elements of θ to be varied randomly simultaneously to obtain two estimates of the cost function. The two estimates required are of the form $J(\theta \pm \text{perturbation})$ for a two-sided gradient approximation. In this case, the gradient estimate $\widehat{\nabla} J(\theta_k) = \left(\widehat{\nabla} J_1(\theta_k), \ldots, \widehat{\nabla} J_d(\theta_k)\right)^{\mathrm{T}}$ is given by

$$\widehat{\nabla} J_i(\theta_k) = \frac{\widehat{J}(\theta_k + c_k \Delta_k) - \widehat{J}(\theta_k - c_k \Delta_k)}{2 c_k \Delta_{k,i}}$$

where $\{c_k\}_{k \geq 1}$ denotes a sequence of positive scalars such that $c_k \to 0$ and

$$\Delta_k = (\Delta_{k,1}, \Delta_{k,2}, \cdots, \Delta_{k,d})$$

is a d-dimensional random perturbation vector. Careful selection of algorithm parameters γ_n, c_n and Δ_n is required to ensure convergence. The $\{\gamma_k\}_{k \geq 1}$ and $\{c_k\}_{k \geq 1}$ sequences generally take the form of $\gamma_k \propto k^{-\alpha}$ and $c_k \propto k^{-\beta}$ respectively with non-negative coefficients α and $\beta \in \left(\frac{1}{2}, 1\right]$. Each component of Δ_k is usually generated from a symmetric Bernoulli ± 1 distribution. See (Spall, 1998) for guidelines on coefficient selection.

To obtain an estimate of $J(\theta)$ for a given value, one recalls that

$$J(\theta) = \int \pi_\theta(y_{1:L}) \left(\pi_\theta(y_{1:L}) - \pi_{\theta^*}(y_{1:L})\right) dy_{1:L}$$
$$+ \text{terms independent of } \theta.$$

and for any θ', θ''

$$\int \pi_{\theta'}(y_{1:L}) \pi_{\theta''}(y_{1:L}) dy_{1:L} =$$
$$\int\int g_{\theta'}(y_{1:L} | x_{1:L}) \pi_{\theta'}(x_{1:L}) \pi_{\theta''}(y_{1:L}) dx_{1:L} dy_{1:L}.$$

It follows that by sampling a realization $X^+_{(k-1)L+1:kL}$ (resp. $X^-_{(k-1)L+1:kL}$) from $\pi_{\theta_k + c_k\Delta_k}(x_{1:L})$ (resp. from $\pi_{\theta_k - c_k\Delta_k}(x_{1:L})$), a realization $Y^+_{(k-1)L+1:kL}$ (resp. $Y^-_{(k-1)L+1:kL}$) from $\pi_{\theta_k + c_k\Delta_k}(y_{1:L})$ (resp. from $\pi_{\theta_k + c_k\Delta_k}(y_{1:L})$) and $Y_{(k-1)L+1:kL}$ a realization from $\pi_{\theta^*}(y_{1:L})$ (i.e. the data available), one obtains an (asymptotically) unbiased estimate of

$$\frac{1}{2}\left(J(\theta_k + c_k\Delta_k) - J(\theta_k - c_k\Delta_k)\right)$$

using

$$\begin{aligned}
&\tfrac{1}{2}\left(\widehat{J}(\theta_k + c_k\Delta_k) - \widehat{J}(\theta_k - c_k\Delta_k)\right) = \\
&g_{\theta_k + c_k\Delta_k}\left(Y^+_{(k-1)L+1:kL} \middle| X^+_{(k-1)L+1:kL}\right) \\
&-g_{\theta_k + c_k\Delta_k}\left(Y_{(k-1)L+1:kL} \middle| X^+_{(k-1)L+1:kL}\right) \\
&-g_{\theta_k - c_k\Delta_k}\left(Y^-_{(k-1)L+1:kL} \middle| X^-_{(k-1)L+1:kL}\right) \\
&+g_{\theta_k - c_k\Delta_k}\left(Y_{(k-1)L+1:kL} \middle| X^-_{(k-1)L+1:kL}\right).
\end{aligned} \tag{6}$$

Note that one needs $X^+_{(k-1)L+1:kL}$ (resp. $X^-_{(k-1)L+1:kL}$) to be statistically independent from $Y^+_{(k-1)L+1:kL}$ (resp. $Y^-_{(k-1)L+1:kL}$). However $X^+_{(k-1)L+1:kL}$ (resp. $X^-_{(k-1)L+1:kL}$) can be statistically dependent from $X^-_{(k-1)L+1:kL}$ (resp. $X^+_{(k-1)L+1:kL}$) and $Y^-_{(k-1)L+1:kL}$ (resp. $Y^+_{(k-1)L+1:kL}$).

The algorithm to minimize $J(\theta)$ follows.

Sampling step

- Sample $\Delta_k = \left(\Delta_{k,1}, \ldots, \Delta_{k,d}\right)$.
- Sample $X^+_{(k-1)L+1:kL} \sim \pi_{\theta_k + c_k\Delta_k}(\cdot)$
- Sample $X^-_{(k-1)L+1:kL} \sim \pi_{\theta_k - c_k\Delta_k}(\cdot)$.
- Sample $\widetilde{X}^+_{(k-1)L+1:kL} \sim \pi_{\theta_k + c_k\Delta_k}(\cdot)$,

$Y^+_{(k-1)L+1:kL} \sim g_{\theta_k + c_k\Delta_k}\left(\cdot \middle| \widetilde{X}^+_{(k-1)L+1:kL}\right)$.

- Sample $\widetilde{X}^-_{(k-1)L+1:kL} \sim \pi_{\theta_k - c_k\Delta_k}(\cdot)$,

$Y^-_{(k-1)L+1:kL} \sim g_{\theta_k - c_k\Delta_k}\left(\cdot \middle| \widetilde{X}^-_{(k-1)L+1:kL}\right)$.

Gradient estimation step

- Compute $\frac{1}{2}\left(\widehat{J}(\theta_k + c_k\Delta_k) - \widehat{J}(\theta_k - c_k\Delta_k)\right)$ using (6).
- For $i = 1$ to d, evaluate gradient components as

$$\widehat{\nabla} J_i(\theta_k) = \frac{\widehat{J}(\theta_k + c_k\Delta_k) - \widehat{J}(\theta_k - c_k\Delta_k)}{2 c_k \Delta_{k,i}}$$

Parameter updating step

$$\theta_{k+1} = \theta_k - \gamma_{k+1} \widehat{\nabla} J(\theta_k).$$

The problem is that it is typically impossible to sample exactly from the distributions $\pi_{\theta_k + c_k\Delta_k}$ and $\pi_{\theta_k - c_k\Delta_k}$. However one can use the following algorithm instead.

Sampling step

- Sample $\Delta_k = \left(\Delta_{k,1}, \ldots, \Delta_{k,d}\right)$.
- Sample $X^+_{(k-1)L+1:kL} \sim f_{\theta_k + c_k\Delta_k}\left(\cdot \middle| X^+_{(k-1)L}\right)$
- Sample $X^-_{(k-1)L+1:kL} \sim f_{\theta_k - c_k\Delta_k}\left(\cdot \middle| X^-_{(k-1)L}\right)$.

• Sample $\widetilde{X}^{+}_{(k-1)L+1:kL} \sim f_{\theta_k+c_k\Delta_k}\left(\cdot|\widetilde{X}^{+}_{(k-1)L}\right)$, $Y^{+}_{(k-1)L+1:kL} \sim g_{\theta_k+c_k\Delta_k}\left(\cdot|\widetilde{X}^{+}_{(k-1)L+1:kL}\right)$.

• Sample $\widetilde{X}^{-}_{(k-1)L+1:kL} \sim f_{\theta_k-c_k\Delta_k}\left(\cdot|\widetilde{X}^{-}_{(k-1)L}\right)$, $Y^{-}_{(k-1)L+1:kL} \sim g_{\theta_k-c_k\Delta_k}\left(\cdot|\widetilde{X}^{-}_{(k-1)L+1:kL}\right)$.

Gradient estimation step

• Compute $\frac{1}{2}\left(\widehat{J}(\theta_k+c_k\Delta_k)-\widehat{J}(\theta_k-c_k\Delta_k)\right)$ using (6).

• For $i=1$ to d, evaluate gradient components as

$$\widehat{\nabla} J_i(\theta_k) = \frac{\widehat{J}(\theta_k+c_k\Delta_k)-\widehat{J}(\theta_k-c_k\Delta_k)}{2c_k\Delta_{k,i}}$$

Parameter updating step

$$\theta_{k+1} = \theta_k - \gamma_{k+1}\widehat{\nabla} J(\theta_k).$$

To reduce the variance of the gradient, SIS techniques can also be used in this case. Further variance reduction can be achieved by using common random numbers so as to introduce correlations between $X^{+}_{(k-1)L+1:kL}$ and $X^{-}_{(k-1)L+1:kL}$ or/and $Y^{+}_{(k-1)L+1:kL}$ and $Y^{-}_{(k-1)L+1:kL}$ (Kleinman, *et al.*, 1999).

4. APPLICATION

We demonstrate our methodology on a non linear state space model. Let us consider the following stochastic volatility model arising in econometrics

$$X_{n+1} = \phi X_n + \sigma V_{n+1},$$
$$Y_n = \beta \exp(X_n/2)\, W_n,$$

where $V_n \overset{\text{i.i.d.}}{\sim} \mathcal{N}(0,1)$ and $W_n \overset{\text{i.i.d.}}{\sim} \mathcal{N}(0,1)$ are two mutually independent sequences of independent identically distributed (i.i.d.) Gaussian random variables, independent of the initial state X_0. We are interested in estimating the parameter $\theta = (\beta, \phi, \sigma)$ where $\Theta = (0,1)\times(0,M)\times(0,M)$ with $M = 100$. In this case, the stationary distribution of the hidden process is $\mathcal{N}\left(0, \frac{\sigma^2}{1-\phi^2}\right)$. We simulate 10000 observations with $\theta^* = (1, 0.8, 1)$. The algorithm is using $L = 2$.

5. DISCUSSION

We have proposed new algorithms to perform parameter estimation in general state space models. These algorithms do not perform state estimation and are computationally efficient. Algorithmically, variance reduction techniques can be developed to improve their performance. Theoretically, our algorithms are "standard" stochastic gradient algorithms for which convergence results will be reported elsewhere.

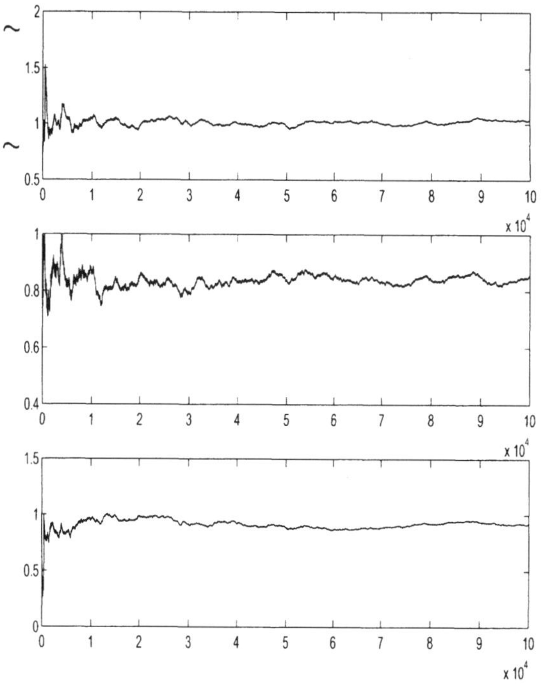

Fig. 1. Sequence of parameter estimates $\theta_n = (\beta_n, \phi_n, \sigma_n)$ for $N = 10000$. From top to bottom: β_n, ϕ_n and σ_n.

6. ACKNOWLEDGMENTS

This work was supported by the Engineering and Physical Sciences Research Council, UK.

7. REFERENCES

Andrieu C., and Doucet A. (2003). On-line Expectation-Maximization type algorithms for parameter estimation in general state space models. *Proc. IEEE Conf. ICASSP.*

Cérou F., LeGland F. and Newton N.J. (2001) Stochastic particle methods for linear tangent equations. in *Optimal Control and PDE's - Innovations and Applications* (eds. J. Menaldi, E. Rofman and A. Sulem), pp. 231-240, IOS Press, Amsterdam.

Doucet A., de Freitas J.F.G. and Gordon N.J. (eds.) (2001) *Sequential Monte Carlo Methods in Practice.* New York: Springer-Verlag.

Doucet A. and Tadić V.B. (2003) Parameter estimation in general state space models using particle methods. *Ann. Inst. Stat. Math.*, to appear.

Gilks W.R. and Berzuini C. (2001). Following a moving target - Monte Carlo inference for dynamic Bayesian models. *J. R. Statist. Soc.* B, **63**, pp. 127-146.

Higuchi T. (1997) Monte Carlo filter using the genetic algorithm operators. *J. Statist. Comp. Simul.*, **59**, 1-23.

Kitagawa G. (1998). A self-organizing state space model. *J. Am. Statist. Ass.*, **93**, 1203-1215.

Kleinman N.L., Spall J.C. and Naiman D.Q. (1999) Simulation-based optimisation with stochastic approximation using common random numbers. *Management Science*, **45**, pp. 1570-1578.

LeGland F. and Mevel L. (1997) Recursive identification in hidden Markov models, *Proc. 36th IEEE Conf. Decision and Control*, pp. 3468-3473.

Rydén T. (1994) Consistent and asymptotically normal parameter estimates for hidden Markov models. *Annals of Statistics*, **22**, 1884-1895.

Spall J.C. (1998) Implementation of the Simultaneous Perturbation Algorithm for Stochastic Optimisation. *IEEE Trans. Aerospace Elect. Sys.*, **34**, 817-823.

IFAC Publications
www.elsevier.com/locate/ifac

Nonlinear Structure Identification With Application to Wiener-Hammerstein Systems

D.J. Leith[1], W.E. Leithead[1,2], R. Murray-Smith[1,3]

[1]Hamilton Institute, National University of Ireland, Maynooth, Co. Kildare, Ireland
[2]Dept. of Electronics & Electrical Engineering, University of Strathclyde, Glasgow G1 1QE, U.K.
[3]Dept. of Computing Science, University of Glasgow, Glasgow G12 8QQ, U.K.

Abstract

While there exists a substantial literature on the identification of Hammerstein and Wiener models, the identification of Wiener-Hammerstein models has received considerably less attention yet this is a model class of very great practical importance. This paper proposes an elegant approach to estimating Wiener-Hammerstein systems from measured data. *Copyright © 2003 IFAC*

1. Introduction

While there exists a substantial literature on the identification of Hammerstein and Wiener models, the identification of Wiener-Hammerstein models has received considerably less attention (recent papers include Bai 1998, 2002; see also the survey of Haber & Unbehauen 1990) yet this is a model class of very great practical importance. In this paper we consider the identification of transversal Wiener-Hammerstein systems

$$x_i = a_n r_{i-n} + \ldots + a_o r_i$$
$$z_i = f(x_i) \qquad (1)$$
$$y_i = b_m z_{i-m} + \ldots + b_o z_i$$

Assume that N>n+m noisy input-output measurements $\{(r_i, \hat{y}_i)\}_{i=1}^{N}$ are available, where $\hat{y}_i = y_i + n_i$, n_i zero mean Gaussian noise. One common strategy is to seek to directly fit the parameters in (1) to the measured data. An alternative strategy, however, is to adopt an indirect approach as follows. Let $\mathbf{R}_i$ denote the delayed input vector

$$\mathbf{R}_i = \begin{bmatrix} r_{i-(n-m)} \\ r_{i-(n-m-1)} \\ \vdots \\ r_i \end{bmatrix} \qquad (2)$$

and D={R_i, i=n+m,...N} be the set of measured delayed input vectors. Reformulating the dynamics in terms of $\mathbf{R}_i$ and y_i yields

$$y_i = F(\mathbf{R}_i) \qquad (3)$$

where

$$F(\mathbf{R}_i) = b_m f(\mathbf{M}_{m+1}\mathbf{R}_i) + \ldots + b_o f(\mathbf{M}_1 \mathbf{R}_i)$$
$$\mathbf{M} = \begin{bmatrix} a_n & \cdots & a_o & 0 & \cdots & 0 \\ 0 & a_n & \cdots & a_o & 0 \cdots & 0 \\ & & & \ddots & & \\ 0 & & \cdots & 0 & a_n \cdots & a_o \end{bmatrix} \qquad (4)$$

and $\mathbf{M}_k$, k=1..m+1 denotes the kth row of **M**. Without loss it is assumed that f(0)=0, $\nabla f(0)=1$. We also assume that $\nabla^2 f(\mathbf{R}_i)$ is *not* identically zero[1] $\forall$ $\mathbf{R}_i \in D$, $a_n=1$ and that at least one of the coefficients b_k, k=0..m is non-zero. Since the input measurements are noise free, inference of the map F in (3) relating output y_i to input $\mathbf{R}_i$ may be formulated as a regression problem. Many methods exist which might be employed to estimate F. In this paper, we employ a non-parametric Gaussian process prior approach; see Appendix. Our objective is then to determine a minimal realisation (1) of the system from this non-minimal input-output formulation. The derivation of minimal realisations of nonlinear systems is well known to be a difficult task in general. However, the minimal realisation task for Wiener-Hammerstein systems is potentially more tractable. Indeed, the key issue is to determine the matrix **M** and it is this which is the focus of the present paper. This matrix effectively partitions the linear dynamics into the input filter and the output filter elements. Once **M** is known, the identification task reduces to the estimation of the Hammerstein model relating the output y_i to the transformed input $\mathbf{MR}_i$. This is a relatively straightforward problem for which standard techniques exist.

Notation: The notation used is standard. For a matrix $\mathbf{M} \in \Re^p$, null(**M**) denotes the null space of **M**, *i.e.* null(**M**) $=\{\mathbf{v} \in \Re^p : \mathbf{Mv}=0\}$, and range(**M**) denotes the range space of **M**, *i.e.* range(**M**)= $\{\mathbf{v} \in \Re^p : \mathbf{Mv} \neq 0\}$. We let basis{V} denote any basis of the subspace $V \subseteq \Re^p$ *i.e.* any orthonormal set of vectors spanning V. For a twice differentiable map **F**: $\Delta \subseteq \Re^p \rightarrow P \subseteq \Re^q$, $\mathbf{H_F}(\mathbf{z})$ denotes the Hessian $\mathbf{H_F}(\mathbf{z}) = [\nabla(\nabla F_1(\mathbf{z}))^T \ldots (\nabla F_q(\mathbf{z}))^T]^T$ with F_i denoting the ith element of the map **F**.

2. Deterministic Case

Assume, initially, that the nonlinear map F in (3) is known. We assume also that the associated Hessian map, $\mathbf{H_F}$, is

[1] This is a form of identifiability condition: if $\nabla^2 f(\mathbf{R}_i)=0$ $\forall$ $\mathbf{R}_i \in D$, then on the basis of the measured data it is readily verified that f cannot be distinguished from a linear function.

known. The latter assumption is little more than a smoothness assumption on F as the Hessian may be immediately derived provided F is specified in a closed-form manner. For example, in a classical regression context where F might be expressed as a sum of basis functions,

$$F(\mathbf{R}) = \sum_{k=1}^{n_\varphi} \theta_k \varphi_k(\mathbf{R}) \qquad (5)$$

we have that

$$\mathbf{H}_F(\mathbf{R}) = \sum_{k=1}^{n_\varphi} \theta_k \nabla\left(\nabla\varphi_k(\mathbf{R})\right)^T \qquad (6)$$

Here, the basis functions φ_k might for example be Gaussian, sigmoidal, polynomial *etc*. Notice that for a given model structure, once the parameters θ_k of the map F are specified the Hessian $\mathbf{H}_F$ is also completely specified.

With regard to the Wiener-Hammerstein system (3), it is readily verified that

$$\mathbf{H}_F(\mathbf{R}) = \mathbf{M}^T\nabla^2 f(\mathbf{R})\begin{bmatrix} b_m\mathbf{M} \\ \vdots \\ b_o\mathbf{M}\end{bmatrix} \qquad (7)$$

where $\mathbf{R}\in\Re^{n+m+1}$. Hence, provided $\nabla^2 f(\mathbf{R})\neq 0$ it follows immediately that

$$H_F(\mathbf{R})\mathbf{v} = 0 \Leftrightarrow \mathbf{M}\mathbf{v} = 0 \ \forall \mathbf{v}\in\Re^{n+m} \qquad (8)$$

i.e. null($H_F(\mathbf{R})$)=null($\mathbf{M}$). Since $\nabla^2 f(\mathbf{R})$ is not everywhere zero in D (or else the system is trivially linear), this property may be expressed as

$$\bigcap_{\mathbf{R}\in D} \text{null}(\mathbf{H}_F(\mathbf{R})) = \text{null}(\mathbf{M}) \qquad (9)$$

This observation is key importance in the sequel as it implies that the matrix **M** may be determined by direct inspection of the Hessian of the input-output map F.

Remarks

(i) The matrix **M** has a specific block diagonal structure. By inspection, the coefficients, a_i, of the input filter and the delay taps of the output filter can be directly inferred via (4). Identification of the remaining system elements is now relatively straightforward. For example, once the input filter is known, the output filter can be inferred from the transfer function of the linearised dynamics about any equilibrium point. With the input and output linear dynamics known (*i.e.* the coefficients a_k and b_k) the scalar nonlinear function f might be determined directly from the curve

$$\bar{y} = \left(\sum_{k=0}^{m} b_k\right) f(\bar{x}) \qquad (10)$$

relating the equilibrium output $\bar{y}$ to $\bar{x} = \sum_{k=0}^{n} a_k \bar{r}$, where $\bar{r}$ is the associated equilibrium input value.

(ii) In general, the matrix **M** is not uniquely determined by the input-output properties of the system: we cannot distinguish between structured matrices **M** with identical null spaces. This arises because scalings in the coefficients a_k can be absorbed in the function f. This degree of freedom is removed here through the constraint that a_n is unity.

3. Nonlinear System Identification

The structural analysis in section 2 is deterministic. In this section we consider the extension to a probabilistic context. We assume that probabilistic (stochastic process) descriptions are available of F(**R**) and its directional derivative $\mathbf{H}_F(\mathbf{R})\mathbf{v}$ (for any given direction **v**). The stochastic process description of the directional derivative $H_F(\mathbf{R})\mathbf{v}$ can be determined from the stochastic process description of F(**R**) (*e.g.* in the case of a Gaussian process, the mean and covariance of[2] $H_F(\mathbf{R})\mathbf{v}$ are appropriate derivatives of the mean and covariance of F(**R**) – see Appendix). It is perhaps worth emphasising that this certainly does not require differentiation of the raw, noisy data. The latter is, of course, highly inadvisable.

Remark It is important to note that, while most previous work on the identification of Wiener-Hammerstein systems has assumed that the nonlinear function f has a specific parametric structure (typically polynomial), stochastic process descriptions do not necessarily require the imposition of a parametric structure. Non-parametric descriptions (*e.g.* Green & Silverman 1994, Neal 1996, Williams 1998) are characterised by drawing inferences directly from the measured data using smoothness information but *without* assuming an underlying parameterisation. (Various forms of smoothness assumption are typically employed: any specific assumption may of course be more or less appropriate in a particular application context). An example of a non-parametric non-linear description is a Gaussian process prior model used here: see Appendix.

3.1 Estimating M

The condition $\bigcap_{\mathbf{R}\in D} \text{null}(\mathbf{H}_F(\mathbf{R})) = \text{null}(\mathbf{M})$ is equivalent to

(i) $\mathbf{H}_F(\mathbf{R})\mathbf{v} = 0 \ \forall \ \mathbf{v} \in \text{basis}\{\Psi_l\}, \mathbf{R}\in D$

(ii) $\mathbf{H}_F(\mathbf{R})\mathbf{v} \neq 0 \forall \mathbf{v} \in \text{basis}\{\Psi_{nl}\}, \mathbf{R}\in D$

where Ψ_l denotes null(**M**) and Ψ_{nl} denotes range(**M**). Assuming that the dimension of Ψ_l is known and that $\mathbf{H}_F(\mathbf{R})\mathbf{v}$ is now described by a probability distribution[3], the probability distribution corresponding to *(i)* is

$$p_{\mathbf{M}} = p\left(\mathbf{H}_F(\mathbf{R})\mathbf{v} = 0 : \mathbf{v} \in \text{basis}\{\Psi_l\}, \mathbf{R}\in D\right) \qquad (11)$$

Exploiting the block diagonal structure of **M** for Wiener-Hammerstein systems, we have that basis{Ψ_l} may be expressed as

$$\text{basis}\{\Psi_l\}=\{M_1, M_2, \ldots, M_{m+1}\} \qquad (12)$$

where

[2] Note that, with an abuse of notation, we use $\mathbf{H}_F(\mathbf{R})$ to also refer to the stochastic process but the meaning should be clear.

[3] While $\mathbf{H}_F(\mathbf{R})\mathbf{v}$ is a matrix, the elements are stacked into a vector for working with standard multivariate probability descriptions.

$$M_1 = [a_n \ \cdots \ a_o \ 0 \ \cdots \ 0]^T$$
$$M_2 = [0 \ a_n \ \cdots \ a_o \ 0 \ \cdots \ 0]^T \quad (13)$$
$$\vdots$$
$$M_{m+1} = [0 \ \cdots \ 0 \ a_n \ \cdots \ a_o]^T$$

Coefficients a_k, k=1..n for which the probability

$$p_M = p(\mathbf{H}_F(\mathbf{R})\mathbf{v} = 0 : \mathbf{v} \in \{M_1, M_2, \ldots, M_{m+1}\}, \mathbf{R} \in D) \quad (14)$$

is maximal therefore provide an estimate of **M**.

The foregoing assumes that the dimension of Ψ_l is known *i.e.* the order m of the output filter. In using (13) we also assume knowledge of the order n of the input filter. This may be relaxed to an assumption only on the overall model order n+m (thereby avoiding assumptions as to partitioning of the dynamics between input and output filters). We proceed as follows. We sequentially increase the dimension of Ψ_l, re-estimating the basis. Let Ψ_l^i denote the estimated sub-space of dimension *i* and let Ψ_l and Ψ_{nl} denote the true sub-spaces. We must have that $\Psi_l^i \cap \Psi_{nl} \neq \emptyset$ for *i>m*. Consequently, as *i* is increased beyond the true dimension *m*, we can expect the value of p_M to abruptly decrease (since $\mathbf{H}_F(\mathbf{z})\mathbf{v} \neq 0 \ \forall \ \mathbf{v} \in \text{basis}\{\Psi_{nl}\}$) and can thereby infer the dimension from the data.

4. Example

Consider the Wiener-Hammerstein nonlinear system illustrated in Figure 1. Reformulating the dynamics in terms of the measured variables (input r and output y) yields

$$y(t_n) = 0.3(\mathbf{M}_1\mathbf{R})^2 + 0.165(\mathbf{M}_3\mathbf{R})^2 \quad (15)$$

where $\mathbf{R} = [r(t_i) \ \ r(t_{i-1}) \ \ r(t_{i-2}) \ \ r(t_{i-3})]^T$ with

$$\mathbf{M} = \begin{bmatrix} 0.9184 & 0.3674 & 0 & 0 \\ 0 & 0.9184 & 0.3674 & 0 \\ 0 & 0 & 0.9184 & 0.3674 \end{bmatrix} \quad (16)$$

The plant output in response to a Gaussian input is measured: data is collected for 15 seconds with a sampling interval of 0.1 seconds (150 data points). Gaussian white noise of standard deviation 0.1 units is added to the output measurement (the underlying signal has a peak magnitude of 0.5, so this represents a substantial level of noise). A non-parametric Gaussian Process prior model is used with explanatory variables/regressors $[r(t_i) \ r(t_{i-1}) \ r(t_{i-2}) \ r(t_{i-3})]^T$ and model output $y(t_i)$. The change in p_M as the dimension of Ψ_l (the dimension of the output filter m) is varied indicates a dimension of two as expected. The associated estimate of **M** is

$$\hat{\mathbf{M}} = \begin{bmatrix} 0.9292 & 0.3694 & 0 & 0 \\ 0 & 0.9292 & 0.3694 & 0 \\ 0 & 0 & 0.9292 & 0.3694 \end{bmatrix} \quad (17)$$

The estimate evidently agrees well with the true **M**, particularly in view of the low signal to noise ratio and small number of data points on which it is based (150 points from a four dimensional map).

5. Conclusions

The is paper investigates new ways of estimating Wiener-Hammerstein models from measured data. A constructive algorithm is proposed and its application is illustrated in a simple example.

Acknowledgements

This work was supported by Science Foundation Ireland grant 00/PI.1/C067, by the EC through EC TMR grant HPRNCT-1999-00107, and EPSRC grants GR/M76379/01 and GR/R15863/01.

Appendix – Non-parametric Gaussian process priors

Consider a smooth function f(.) dependent on $\mathbf{z} \in D \subseteq \Re^P$. To avoid cumbersome notation, f is scalar. Suppose N measurements, $\{(\mathbf{z}_i, y_i)\}_{i=1}^N$, of the value of the function with additive Gaussian white measurement noise, i.e. $y_i = f(\mathbf{z}_i) + n_i$, are available and denote them by M. It is of interest here to use this data to learn the mapping f(**z**) or, more precisely, to determine a probabilistic description of f(**z**). Note that this is a regression formulation and it is assumed the input **z** is noise free[4]. The probabilistic description of the function, f(**z**), adopted is the stochastic process, $f_\mathbf{z}$. By necessity, the probability distributions of $f_\mathbf{z}$ for every choice of value of $\mathbf{z} \in D$ are required together with the joint probability distributions of $f_{\mathbf{z}_i}$ for every choice of finite sample, $\{\mathbf{z}_1, \ldots, \mathbf{z}_k\}$, from D, for all k>1. Of course, the joint probability distributions of lower dimensionality must be the marginal distributions of those of higher dimensionality. The $E[f_\mathbf{z}]$ as **z** varies is interpreted as a fit to f(**z**).

In the Bayesian probability context, the prior belief is placed directly on the probability distributions describing $f_\mathbf{z}$ which are then conditioned on the observations, M, to determine the posterior probability distributions. In Gaussian Process prior model it is assumed that the prior probability distributions for the $f_\mathbf{z}$ are all Gaussian with zero mean (in the absence of any evidence the value of f(**z**) is as likely to be positive as negative). To complete the statistical description, requires only a definition of the covariance function $C(f_{\mathbf{z}_i}, f_{\mathbf{z}_j}) = E[f_{\mathbf{z}_i}, f_{\mathbf{z}_j}]$, for all $\mathbf{z}_i$ and $\mathbf{z}_j$. The resulting posterior probability distributions are also Gaussian. The Gaussian assumption may seem strangely restrictive initially, but recall that this is simply a prior on the relevant stochastic process space and so places few inherent restrictions on the class of nonlinear functions that can be modelled. Indeed, it can be shown that the result is, in fact, a Bayesian form of kernel regression model (Green & Silverman 1994) subsuming RBF, spline and many neural network models (Williams 1998).

[4]No attempt to being made here to propagate a Gaussian or other distribution through a nonlinear function.

This model is used to carry out inference as follows. We have that $p(f_z|M) = p(f_z,M)/p(M)$ where $p(M)$ acts as a normalising constant. Hence,

$$p(f_z|M)\ \alpha\ \exp\left[-\frac{1}{2}\begin{bmatrix} f_z & \mathbf{F}^T \end{bmatrix}\begin{bmatrix} \Lambda_{11} & \Lambda_{21}^T \\ \Lambda_{21} & \Lambda_{22} \end{bmatrix}^{-1}\begin{bmatrix} f_z \\ \mathbf{F} \end{bmatrix}\right]$$

where $\mathbf{F} = [y_1,\cdots,y_N]^T$, Λ_{11} is $C(f_z,f_z)$, the ijth element of the covariance matrix Λ_{22} is $C(f_{z_i},f_{z_j})$ and the ith element of vector Λ_{21} is $C(f_{z_i},f_z)$. Both Λ_{11} and Λ_{21} depend on $\mathbf{z}$. Applying the partitioned matrix inversion lemma, it follows that

$$p(f_z|M)\ \alpha\ \exp\left[-\frac{1}{2}(f_z-\hat{f}_z)\Lambda_z^{-1}(f_z-\hat{f}_z)\right] \quad (18)$$

with mean $\hat{f}_z = \Lambda_{21}^T\Lambda_{22}^{-1}\mathbf{F}$ and variance $\Lambda_z = \Lambda_{11} - \Lambda_{21}^T\Lambda_{22}^{-1}\Lambda_{21}$. Note that $\hat{f}_z$ is simply a z-dependent weighted linear combination of the measured data points.

Furthermore, assume that the related stochastic process, $f_z^{\delta e_i}$, where $f_z^{\delta e_i} = (f_{(z+\delta e_i)} - f_z)/\delta$ and $\mathbf{e}_i$ is a unit basis vector, is well-defined in the limit as $\delta \to 0$, i.e. all the necessary probability distributions for a complete description exist. Denote the derivative stochastic process, i.e. the limiting random process, by $f_z^{e_i}$. The $E[f_z^{e_i}]$ as $\mathbf{z}$ varies is interpreted as a fit to $\frac{\partial f}{\partial z_i}(\mathbf{z})$ when the partial derivative of $f(\mathbf{z})$ in the direction $\mathbf{e}_i$ exists. Provided the covariance $C(f_{z_i},f_{z_j})$ is sufficiently differentiable, it is known (O'Hagan 1992) that $f_z^{e_i}$ is itself Gaussian and that

$$E[f_z^{e_i}] = \frac{\partial}{\partial z_i}h_f(\mathbf{z}) \quad ; \quad h_f(\mathbf{z}) = E[f_z] \quad (19)$$

where z_i denotes the ith element of $\mathbf{z}$; that is, the expected value of the derivative stochastic process is just the derivative of the expected value of the stochastic process. In addition,

$$E[f_{z_0}^{e_i}f_{z_1}^{e_j}] = \nabla_i^1\nabla_j^2 C_f(\mathbf{z}_0,\mathbf{z}_1)\,;C_f(\mathbf{z}_0,\mathbf{z}_1) = E[f_{z_0}f_{z_1}] \quad (20)$$

where $\nabla_i^1 Q(\mathbf{z}_0,\mathbf{z}_1)$ denotes the partial derivative of $Q(\mathbf{z}_0,\mathbf{z}_1)$ with respect to the ith element of its first argument, *etc*.

The above procedure can be repeated to construct second derivative stochastic processes. The means and covariances can be determined by recursive application of (19) and (20).

In the examples discussed in this paper, a straightforward smoothness prior covariance function is used which ensures that outputs associated with nearby inputs should have higher covariance than more widely separated inputs; specifically,

$$C(f_{z_i},f_{z_j}) = \gamma\exp\left[-\sum_k\left((z_i)_k-(z_j)_k\right)^2/2\alpha_k\right]+\beta\delta_{ij} \quad (21)$$

where $(z_i)_k$ denotes the kth element of vector $\mathbf{z}_i$. The value of a_k characterises the rate of variation of the function in dimension k, thereby, estimating the relative smoothness of different input dimensions. The parameter β is the variance of the measurement noise, n, on the output. The hyperparameters (β, a_k, γ) are adapted to maximise the likelihood $p(M|(\beta,\alpha_k,\gamma))$. The covariance function, (13), is sufficiently smooth for the derivative and second derivative stochastic processes to be well-defined and the relations (11) and (12) to apply (O'Hagan 1992).

References

BAI, E.W., 1998, An Optimal Two-Stage Identification Algorithm for Hammerstein-Wiener Systems. *Automatica*, **34**, pp333-338.

BAI, E.W., 2002, A Blind Approach to the Hammerstein-Wiener Model Identification. *Automatica*, **38**, pp967-979.

GREEN, P.J., SILVERMAN, B.W., 1994, *Nonparametric Regression and Generalised Linear Models*. (Chapman & Hall, London).

HABER,R., UNBEHAUEN, H., 1990, Structure Identification of Nonlinear Dynamic Systems – A Survey of Input/Ouput Approaches. *Automatica*, **26**, pp651-677.

O'HAGAN, A., 1978, On curve fitting and optimal design for regression, *J. Royal Stat Soc. B*, **40**, 1-42.

WILLIAMS, C. K. I., 1998, Prediction with Gaussian Processes: From linear regression to linear prediction and beyond. *In Learning and Inference in Graphical Models* (M. I. Jordan, Ed.), Kluwer.

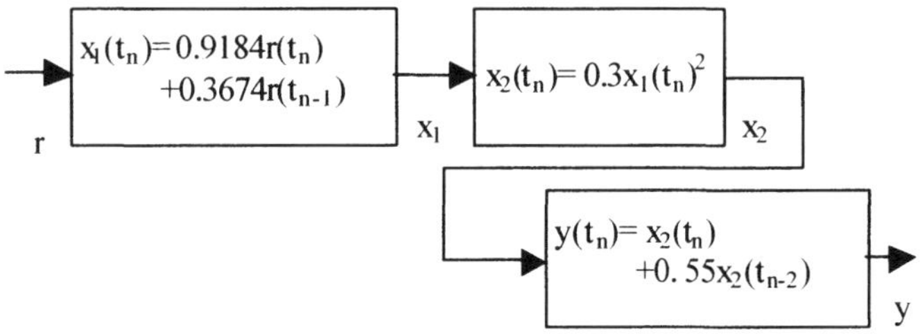

Figure 1 Block diagram representation of example system.

IFAC
Publications
www.elsevier.com/locate/ifac

IDENTIFICATION OF A WIENER SYSTEM WITH SOME GENERAL DISCONTINUOUS NONLINEARITIES

Fen Guo, Georg Bretthauer

Forschungszentrum Karlsruhe GmbH
Institute for Applied Computer Science
76344 Karlsruhe, Germany
email: fenguo@iai.fzk.de; bretthauer@iai.fzk.de

Abstract: A nonlinear dynamic system can be described by a Wiener system which is a linear dynamic subsystem combined with a nonlinear static subsystem. And general discontinuous nonlinearities are very popular in a real system. This paper presents a new recursive identification method to a Wiener system with different general discontinuous nonlinearities. It is assumed that the linear subsystem model structure and the type of the discontinuous nonlinearities are known in prior. By using the key term separation principle and constructing intermediate variables, such a Wiener system can be approximately transformed into a pseudo-linear MISO system. Using the adaptive recursive pseudo-linear regressions (RPLR) for a linear MISO dynamic system and smoothing and filtering techniques to estimate the intermediate variables, satisfied parameter estimates of the nonlinear dynamic system can be obtained in the presence of a colored measurement noise without parameter redundancy. Simulation examples are also given to illustrate the correctness of the developed method. *Copyright © 2003 IFAC*

Keywords: Nonlinear system identification; Wiener system; Discontinuous nonlinearities.

1. INTRODUCTION

A Wiener system (Fig. 1), in which a nonlinear static subsystem follows a linear dynamic subsystem, is widely used to simulate nonlinear dynamic systems in practice. General discontinuous nonlinearities appear often in real systems and processes, like the direction-dependent nonlinearity, the preload nonlinearity, the dead-zone nonlinearity and the saturation nonlinearity (Fig. 2). Therefore, the study of identification method for a Wiener system with such a general discontinuous nonlinearity has important significance.

Some identification methods for a Wiener system with a continuous nonlinearity were introduced and different linear and nonlinear subsystem structures have been applied in literatures, see (Narendra & Gallman, 1966; Kortmann & Unbehauen, 1987; Boutayeb & Darouach, 1994; Vörös, 1995). In some literatures a strong assumption was made to suppose the nonlinear block to be invertible which aims to minimize the intermediate error between the output of the linear subsystem and the input of the nonlinear subsystem, see (Wigren, 1993; Kalafatis, 1997; Bruls et al., 1997; Hagenblad, 1999; Zhu, 1999). Pearson and Pottmann (2000) introduced a identification method but with the known nonlinear block. Several researchers studied identification methods with discontinuous nonlinearities. Vörös (1997) described and directly identified these nonlinearities in a Hammerstein system. Bai (2002) used the same identification principle to identify the simple discontinuous nonlinearities in a Hammerstein system. Zeng (2000) identified these discontinuous nonlinearities in a Wiener and a Hammerstein system, but with the inverse of the nonlinearities.

Usually, it is desirable to find a model as simple as possible for the data. Actually, there is an implicit (sometimes explicit) tradeoff between the acceptable model complexity and how well it matches the data. However, to write out the description of the whole model analytically, it presents usually nonlinearities in parameters or redundancy of parameters in identification because of the substitutions of the intermediate variable and the relevant identification algorithm would be also complicated and inefficient. It is clear that a complicated continuous polynomial can also be used to approximate some discontinuous nonlinearities. But direct identification of discontinuous nonlinearities is no doubt more efficient. It can be concluded from the literatures that the most important attempt is trying to reduce parameter redundancy by using special linear and nonlinear model structures and defining some switch functions and intermediate variables to write the discontinuous nonlinearities in a continuous form. Some other attempts are to select a parameterization and approximation or relax algorithm to simplify the computation procedure to fit the individual nonlinear model to process data.

The purpose of this paper is under the prediction error method and model scheme to develop a new recursive identification method and some strategies to identify a Wiener system with different general discontinuous nonlinearities in the presence of a colored measurement noise without parameter redundancy. It will be seen that from the known linear subsystem model structure and a prior knowledge about the behavior of discontinuous nonlinearities, the intermediate variables could be separated, defined and estimated recursively only based on the observed input and output data. Then such a Wiener system can be approximately transformed into a pseudo-linear MISO system and identified with RPLR method. The rest of this paper is organized as follows: Section 2 introduces a Wiener system and the some general discontinuous nonlinearities. The identification method and strategies will also be established. Section 3 derives the identification algorithms for a Wiener system with different general discontinuous nonlinearities. Section 4 presents the simulation examples. Finally, section 5 gives a brief summary of the major results and conclusions.

2. SYSTEM STATEMENT AND IDENTIFICAION STRATEGIES

A Wiener system can be presented as Figure 1.

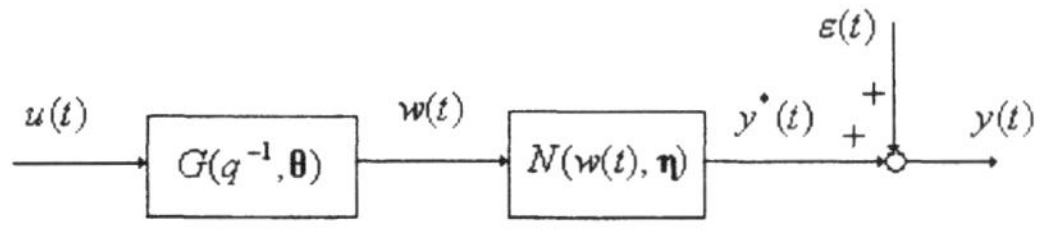

Figure 1 A Wiener system

Where, $u(t)$ is system input. $y(t)$ is system output. $y^*(t)$ is unmeasurable system output without measurement noise. $w(t)$ is a unmeasurable intermediate variable. $\boldsymbol{\theta}$ is the parameter vector determining the linear dynamic subsystem $G(q^{-1},\boldsymbol{\theta})$. The symbol q^{-1} denotes the discrete shift operator. $\boldsymbol{\eta}$ is the parameter vector determining the nonlinear static subsystem $N(\bullet,\boldsymbol{\eta})$. The colored measurement noise $\varepsilon(t)$ is a zero mean white noise $e(t)$ through a linear filter $H(q^{-1},\xi)$:

$$H(q^{-1},\xi)=\frac{C(q^{-1})}{D(q^{-1})} \tag{1}$$

where, $C(q^{-1})=1+c_1q^{-1}+\cdots+c_{n_c}q^{-n_c}$ and $D(q^{-1})=1+d_1q^{-1}+\cdots+d_{n_d}q^{-n_d}$. ξ is the noise filter parameter vector. We assume that $\varepsilon(t)$ is independent of $u(t)$. The dynamic linear subsystem $G(q^{-1},\boldsymbol{\theta})$ is described as:

$$G(q^{-1},\boldsymbol{\theta})=\frac{B(q^{-1})}{F(q^{-1})} \tag{2}$$

where, $B(q^{-1})=b_1q^{-1}+b_2q^{-2}+\cdots+b_nq^{-n}$ and $F(q^{-1})=1+f_1q^{-1}+\cdots+f_nq^{-n}$. It is assumed that the system orders n_c, n_d, n are known. Without losing generality and to simplify the derivation, it is assumed here that the delay of linear block is one and $b_1=1$. Some general discontinuous nonlinearities, as the nonlinear static subsystem $N(\bullet,\boldsymbol{\eta})$, are described as Figure 2.

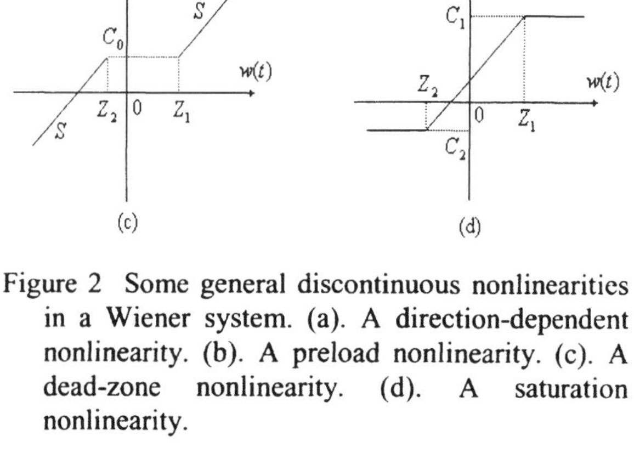

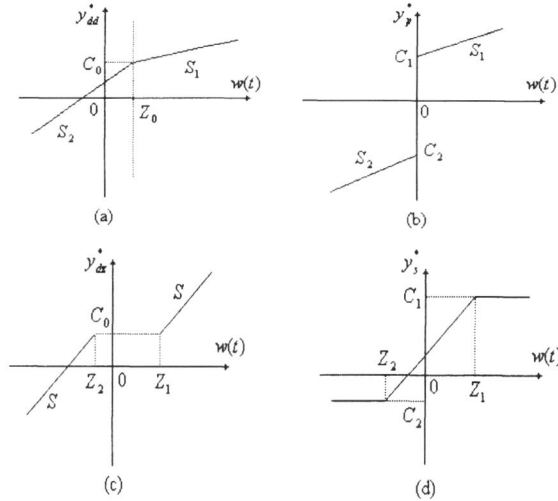

Figure 2 Some general discontinuous nonlinearities in a Wiener system. (a). A direction-dependent nonlinearity. (b). A preload nonlinearity. (c). A dead-zone nonlinearity. (d). A saturation nonlinearity.

Where, the unknown constant nonzero slopes S, S_1, S_2 and the boundaries, cross-points and center-points, $Z_2 < Z_0 < Z_1$, $C_2 < C_0 < C_1$, are within $-\infty < S, S_1, S_2, Z_0, Z_1, Z_2, C_0, C_1, C_2 < \infty$. Therefore, the discontinuous nonlinear relationships are:

$$y_{dd}^{\bullet}(t) = \begin{cases} S_1 \cdot [w(t) - Z_0] + C_0 & w(t) \geq Z_0 \\ S_2 \cdot [w(t) - Z_0] + C_0 & w(t) < Z_0 \end{cases} \quad (3)$$

$$y_p^{\bullet}(t) = \begin{cases} S_1 \cdot w(t) + C_1 & w(t) > 0 \\ 0 & w(t) = 0 \\ S_2 \cdot w(t) + C_2 & w(t) < 0 \end{cases} \quad (4)$$

$$y_{dz}^{\bullet}(t) = \begin{cases} S \cdot [w(t) - Z_1] + C_0 & w(t) > Z_1 \\ 0 & Z_2 \leq w(t) \leq Z_1 \\ S \cdot [w(t) + Z_2] + C_0 & w(t) < Z_2 \end{cases} \quad (5)$$

$$y_s^{\bullet}(t) = \begin{cases} C_1 & w(t) > Z_1 \\ \dfrac{C_1 - C_2}{Z_1 - Z_2}(w(t) - Z_0) + C_0 & Z_2 \leq w(t) \leq Z_1 \\ C_2 & w(t) < Z_2 \end{cases} \quad (6)$$

in eq. (6), $Z_0 = \frac{1}{2}(Z_1 - Z_2)$ and $C_0 = \frac{1}{2}(C_1 - C_2)$.

We consider the identification problem in a prediction error method and model scheme. It is known that a Wiener system is over-parameterization if the linear and nonlinear subsystems are parameterized separately, that is, a constant gain can be distributed arbitrarily between the linear and nonlinear subsystems. In order to get a unique solution, the gain of one subsystem must be fixed. A simpler solution is just to fix one of the parameters of the linear or nonlinear subsystem constant during the minimization. To identify the other parameters, we compare the predicted output $\hat{y}(t/\vartheta)$ with the measured output $y(t)$ in the following prediction error criterion:

$$\vartheta = \arg\min_{\vartheta} \sum \frac{1}{2}[(y(t) - \hat{y}(t/\vartheta)]^2 \quad (7)$$

where, $\vartheta = [\boldsymbol{\theta}, \boldsymbol{\eta}, \boldsymbol{\xi}]$. A possible prediction model could be written as:

$$\hat{y}(t/\vartheta) = \boldsymbol{\varphi}^{\mathrm{T}}(t, \vartheta) \cdot \vartheta \quad (8)$$

The pseudo-regression vector $\boldsymbol{\varphi}(t, \vartheta)$ contains relevant past data, partly reconstructed using the current model which has two meanings: the past date in $\boldsymbol{\varphi}(t, \vartheta)$ could be nonlinear but the parameter vector ϑ is linear; on the other hand the parameter-dependent reconstructed elements (intermediate variables) are determined in some recursive fashion and arrive at the recursive pseudo-linear regressions (RPLR) estimates. As Ljung (1987) pointed out: no matter how $\boldsymbol{\varphi}(t, \vartheta)$ is formed, it is the known data at time t and it can contain arbitrary transformations of measured data. We could rewrite eq. (8) as:

$$\hat{y}(t/\vartheta) = \boldsymbol{\varphi}_1^{\mathrm{T}}(t, \vartheta_1) \cdot \vartheta_1 + \boldsymbol{\varphi}_2^{\mathrm{T}}(t, \vartheta_2) \cdot \vartheta_2 + \cdots + \boldsymbol{\varphi}_s^{\mathrm{T}}(t, \vartheta_s) \cdot \vartheta_s \quad (9)$$

with arbitrary functions $\boldsymbol{\varphi}_i(t, \vartheta_i)$ of past data for $i = 1 \cdots s$. Eq. (9) could be regarded as a finite-dimensional parameterisation of a general, unknown nonlinear predictor. The key is how to choose the functions $\boldsymbol{\varphi}_i(t, \vartheta_i)$, for $i = 1 \cdots s$, and this is where physical insight into the system is required. Eq. (9) can also be seen as a transforming result from a s pseudo multiple inputs pseudo-linear system,

$$y(t) = G_1(q^{-1}, \boldsymbol{\theta}_1) \cdot u_1(t) + G_2(q^{-1}, \boldsymbol{\theta}_2) \cdot u_2(t) + \cdots + G_s(q^{-1}, \boldsymbol{\theta}_s) \cdot u_s(t) + H(q^{-1}, \boldsymbol{\xi}) \cdot e(t) \quad (10)$$

Direct substitutions of eq. (2) into eqs. (3)-(6) can cancel the intermediate variable $w(t)$ but produce redundant parameters. Vörös (1995) introduced the key term separation principle. The basic idea is a form of half-substitution, that is, only the separated key term will be substituted with their front expression, in order to reduce the appearance of nonlinearity tied with parameters in the system structure. Because of the separation of linear block and nonlinear block in a Wiener system and according to the characteristics of the discontinuous nonlinearities, some other possible intermediate variables will be defined. By introducing these intermediate variables, one can compress or merge the redundant terms. To form a predictor model, these intermediate variables will be independently constructed and will be approximated as pseudo-inputs of the system.

The unmeasurable intermediate variables can only be estimated recursively by some estimated system parameters. In a RPLR identification the parameter variation will strongly affect identification quality. Therefore, smoothing and filtering techniques in parameter estimation are necessary which guarantee the original convergence and improve the stability of the parameter identification process, to mitigate the estimate errors of system parameters and to avoid the possible oscillations for better convergence. Some possible smoothing techniques are exponentially average smoother using a moving window with fixed length, a piecewise line of polynomial fitting approach using a moving window with another fixed window length, Kalman filter and fuzzy sets. And we use the forgetting factor approach with a variant forgetting factor, $\lambda(t) = \lambda(t-1) + (1 - \lambda(t-1))\Delta\lambda$ to adaptation and to identify the parameters of the system. With the system input, output data and the estimated intermediate variables, subsystems can be reconstructed in other type structure or in plot curve.

Based on the discussed above, new identification algorithms can be derived respectively. Under a colored measurement noise, the parameter identification is carried out only with measured system input and output data. The model parameters are explicitly given or recalculated without redundancy. The identifiable conditions and identification strategies are as follows:

1) The Wiener system is stable and the orders of subsystems and type of general discontinuous nonlinearities are known a priori.
2) Defining the intermediate variables and transforming the Wiener system into a pseudo-linear MISO system with explicit parameters without parameters redundancy.
3) Fix one parameter, e.g., $b_1 = 1$, to obtain a unique parameterization. Using the key term principle to produce a prediction error model.
4) The persistent system input $u(t)$, the multiple pseudo-inputs and the interference signal $\varepsilon(t)$ are independent.
5) With the smoothing techniques for estimation of the corresponding parameters and using adaptive identification methods, the original parameters of the Wiener system can be identified and sole recalculated.

3. ALGORITHMS DERIVATION

3.1. *With a direction-dependent nonlinearity or a preload nonlinearity*

By introducing a switching sequence:

$$h(t) = \begin{cases} 0 & w(t) \geq Z_0 \\ 1 & w(t) < Z_0 \end{cases} \tag{11}$$

the direction-dependent nonlinearity (Figure 2(a)), eq. (3) can be rewritten as:

$$\begin{aligned} y^*(t) &= S_1 \cdot w(t) + (S_2 - S_1) \cdot w(t) \cdot h(t) \\ &- S_1 \cdot Z_0 + C_0 + (S_1 - S_2) \cdot Z_0 \cdot h(t) \end{aligned} \tag{12}$$

where, the $w(t)$ in the first term is the key term. Half-substituting eq. (2) into the key term $w(t)$, the Wiener system output $y(t)$ is:

$$\begin{aligned} y(t) &= y^*(t) + H(q^{-1},\xi) \cdot e(t) \\ &= \frac{S_1 \cdot B_1(q^{-1})}{F_1(q^{-1})} \cdot u(t) + (S_2 - S_1) \cdot w(t) \cdot h(t) \\ &- S_1 \cdot Z_0 + C_0 + (S_1 - S_2) \cdot Z_0 \cdot h(t) + \frac{C(q^{-1})}{D(q^{-1})} \cdot e(t) \end{aligned} \tag{13}$$

With the same switching sequence but $Z_0 = 0$, the preload nonlinearity (Figure 2(b)), eq. (4) can be rewritten as:

$$\begin{aligned} y^*(t) &= S_1 \cdot w(t) + (S_2 - S_1) \cdot w(t) \cdot h(t) \\ &+ C_1 + (C_2 - C_1) \cdot h(t) \end{aligned} \tag{14}$$

where, the $w(t)$ in the first term in eq. (14) is the key term. Half-substituting eq. (2) into the key term $w(t)$, the Wiener system output $y(t)$ is:

$$\begin{aligned} y(t) &= y^*(t) + H(q^{-1},\xi) \cdot e(t) \\ &= \frac{S_1 \cdot B_1(q^{-1})}{F_1(q^{-1})} \cdot u(t) + (S_2 - S_1) \cdot w(t) \cdot h(t) \\ &+ C_1 + (C_2 - C_1) \cdot h(t) + \frac{C(q^{-1})}{D(q^{-1})} \cdot e(t) \end{aligned} \tag{15}$$

Eqs. (13) and (15) showed that a Wiener system with a direction-dependent nonlinearity or a preload nonlinearity has been transformed into a pseudo-linear MISO system with four independent pseudo-inputs, $u(t)$, $w(t) \cdot h(t)$, 1 and $h(t)$. The intermediate variable $w(t)$ can be recursively estimated with eq. (2).

3.2. *With a dead-zone nonlinearity or a saturation nonlinearity*

In the dead-zone nonlinearity (Figure 2(c)), we assume the width of the dead-zone is $B_Z = Z_1 - Z_2 > 0$ and the center point of it is Z_0. Then eq. (5) can be rewritten as:

$$\begin{aligned} y^*(t) &= C_0 + S \cdot w(t) - S \cdot \frac{1 + \mathrm{sgn}(\frac{B_Z}{2} - |w(t) - Z_0|)}{2} \cdot w(t) \\ &- S \cdot Z_0 \cdot \frac{1 - \mathrm{sgn}(\frac{B_Z}{2} - |w(t) - Z_0|)}{2} \\ &+ S \cdot \frac{B_Z}{2} \cdot \frac{\mathrm{sgn}(\frac{B_Z}{2} - |w(t) - Z_0|) - 1}{2} \cdot \mathrm{sgn}(w(t) - Z_0) \end{aligned} \tag{16}$$

where, $\mathrm{sgn}(\cdot)$ is the standard sign function. We define the following intermediate variables:

$$sw(t) = -0.5 - 0.5 \cdot \mathrm{sgn}(\frac{B_Z}{2} - |w(t) - Z_0|) \cdot w(t) \tag{17}$$

$$ss_1(t) = -0.5 + 0.5 \cdot \mathrm{sgn}(\frac{B_Z}{2} - |w(t) - Z_0|) \tag{18}$$

$$ss_2(t) = \frac{\mathrm{sgn}(\frac{B_Z}{2} - |w(t) - Z_0|) - 1}{4} \cdot \mathrm{sgn}(w(t) - Z_0) \tag{19}$$

Therefore, eq. (16) can be simplified as:

$$\begin{aligned} y^*(t) &= S \cdot w(t) + S \cdot sw(t) + C_0 \\ &+ S \cdot Z_0 \cdot ss_1(t) + S \cdot B_Z \cdot ss_2(t) \end{aligned} \tag{20}$$

Where, the $w(t)$ is the key term. Half-substituting eq. (2) into the key term $w(t)$, the Wiener system output $y(t)$ is:

$$\begin{aligned} y(t) &= y^{*}(t) + H(q^{-1},\xi)\cdot e(t) \\ &= S\cdot\frac{B(q^{-1})}{F(q^{-1})}u(t) + S\cdot sw(t) + C_0 \\ &+ S\cdot Z_0\cdot ss_1(t) + S\cdot B_Z\cdot ss_2(t) + \frac{C(q^{-1})}{D(q^{-1})}e(t) \end{aligned} \tag{21}$$

For the saturation nonlinearity (Figure 2(d)), we assume the two square widths of linear part between the two saturation boundaries are $B_Z = Z_1 - Z_2 > 0$ and $B_C = C_1 - C_2 > 0$, and the center point is (Z_0, C_0). Then, eq. (6) can be rewritten as:

$$\begin{aligned} y^{*}(t) &= \frac{1+\operatorname{sgn}(\frac{B_Z}{2} - |w(t)-Z_0|)}{2}\cdot\frac{B_C}{B_Z}\cdot(w(t)-Z_0) \\ &+ \frac{1+\operatorname{sgn}(|w(t)-Z_0| - \frac{B_Z}{2})}{2}\cdot\frac{B_C}{2}\cdot\operatorname{sgn}(w(t)-Z_0) + C_0 \end{aligned} \tag{22}$$

We define the following intermediate variables:

$$sw(t) = 0.5\cdot\operatorname{sgn}(\frac{B_Z}{2} - |w(t)-Z_0|)\cdot w(t) \tag{23}$$

$$ss_1(t) = -0.5 - 0.5\cdot\operatorname{sgn}(\frac{B_Z}{2} - |w(t)-Z_0|) \tag{24}$$

$$ss_2(t) = \frac{1+\operatorname{sgn}(|w(t)-Z_0| - \frac{B_Z}{2})}{4}\cdot\operatorname{sgn}(w(t)-Z_0) \tag{25}$$

Therefore, eq. (22) can be simplified as:

$$\begin{aligned} y^{*}(t) &= \frac{B_C}{2B_Z}\cdot w(t) + \frac{B_C}{B_Z}\cdot sw(t) \\ &+ \frac{B_C}{B_Z}\cdot Z_0\cdot ss_1(t) + B_C\cdot ss_2(t) + C_0 \end{aligned} \tag{26}$$

Where, the $w(t)$ is the key term. Half-substituting eq. (2) into the key term $w(t)$, the Wiener system output $y(t)$ is:

$$\begin{aligned} y(t) &= y^{*}(t) + H(q^{-1},\xi)\cdot e(t) \\ &= \frac{B_C}{2B_Z}\cdot\frac{B(q^{-1})}{F(q^{-1})}u(t) + \frac{B_C}{B_Z}\cdot sw(t) \\ &+ \frac{B_C}{B_Z}\cdot Z_0\cdot ss_1(t) + B_C\cdot ss_2(t) + C_0 + \frac{C(q^{-1})}{D(q^{-1})}e(t) \end{aligned} \tag{27}$$

Eqs. (21) and (27) showed that a Wiener system with a dead-zone nonlinearity or a saturation nonlinearity has been transformed into a pseudo-linear MISO system with four independent corresponding pseudo-inputs, $u(t)$, $sw(t)$, $ss_1(t)$ and $ss_2(t)$. The unmeasurable intermediate variables $w(t)$, $sw(t)$, $ss_1(t)$ and $ss_2(t)$ can be recursively estimated with corresponding eq. (2) and eqs. (17)-(19) or eqs. (23)-(25), respectively.

4. SIMULATION EXAMPLES

Using the identification algorithms and strategies discussed above, we'll consider identification of a Wiener system under a colored measurement noise, with the same linear dynamic subsystem following with the different discontinuous nonlinearities respectively. A exponentially average smoother using a moving window with fixed length $Mov = 4$ will be used to filter the estimated parameters to calculate the intermediate variables.

A basic Wiener system consists of a linear block,

$$w(t) = \frac{q^{-1} + 0.5004q^{-2}}{1 - 1.5q^{-1} + 0.7q^{-2}}u(t) \tag{28}$$

and the following discontinuous nonlinearities, respectively.

$$y_{dd}^{*}(t) = \begin{cases} 0.45(w(t)-2)+1 & w(t) \geq 2 \\ 1.2(w(t)-2)+1 & w(t) < 2 \end{cases} \tag{29}$$

$$y_{p}^{*}(t) = \begin{cases} 0.45w(t)+1.55 & w(t) > 0 \\ 0 & w(t) = 0 \\ 1.2w(t)-2.7 & w(t) < 0 \end{cases} \tag{30}$$

$$y_{dz}^{*}(t) = \begin{cases} [w(t)-3]+1 & w(t) > 3 \\ 0 & -1 \leq w(t) \leq 3 \\ [w(t)+1]+1 & w(t) < -1 \end{cases} \tag{31}$$

$$y_{s}^{*}(t) = \begin{cases} 3 & w(t) > 1 \\ [w(t)-1]+1 & -3 \leq w(t) \leq 1 \\ -1 & w(t) < -3 \end{cases} \tag{32}$$

A standard random numbers is used as input $u(t)$ which has zero mean and unit variance. A Noise/Signal=10% colored noise with the noise filter:

$$H(q^{-1},\xi) = \frac{1 + 0.2q^{-1} + 0.1q^{-2}}{1 - 1.2q^{-1} + 0.5q^{-2}} \tag{33}$$

is added to the system output. The algorithm variable settings are $\lambda(0) = 0.7$, $\Delta\lambda = 0.01$. The initial estimates of the unknown parameters are also taken as zero. 2000 data points are collected. By applying the standard RPEM function in MATLAB with forgetting factor algorithm for linear MISO system, we get the satisfying identification results shown in Tables 1-4, respectively.

Table 1 Identification results with direction dependent nonlinearity

2000	b_2	f_1	f_2	S_1	S_2
10%	0.5485	-1.519	0.7262	0.5331	1.1607
Real	0.5004	-1.500	0.7000	0.4500	1.2000
Z_0	C_0	c_1	c_2	d_1	d_2
2.1309	1.1159	0.2728	0.0988	-1.120	0.4371
2.0000	1.0000	0.2000	0.1000	-1.200	0.5000

Table 2 Identification results with preload nonlinearity

2000	b_2	f_1	f_2	S_1	S_2
10%	0.4914	-1.509	0.7024	0.4989	1.2480
Real	0.5004	-1.500	0.7000	0.4500	1.2000
C_1	C_2	c_1	c_2	d_1	d_2
1.1628	-2.287	-0.265	0.2086	-1.127	0.4777
1.5500	-2.700	0.2000	0.1000	-1.200	0.5000

Table 3 Identification results with dead-zone nonlinearity

2000	b_2	f_1	f_2	S	C_0
10%	0.4941	-1.519	0.7163	0.9613	1.1288
Real	0.5004	-1.500	0.7000	1.0000	1.0000
Z_1	Z_2	c_1	c_2	d_1	d_2
3.0278	-0.939	0.2252	0.1556	-1.139	0.4378
3.0000	-1.000	0.2000	0.1000	-1.200	0.5000

Table 4 Identification results with saturation nonlinearity

2000	b_2	f_1	f_2	C_1	C_2
10%	0.6326	-1.481	0.6883	2.7964	-0.820
Real	0.5004	-1.500	0.7000	3.0000	-1.000
Z_1	Z_2	c_1	c_2	d_1	d_2
0.9656	-3.009	0.0249	0.0750	-1.217	0.5216
1.0000	-3.000	0.2000	0.1000	-1.200	0.5000

5. CONCLUSIONS

A new recursive identification method for a Wiener system with different general discontinuous nonlinearities are proposed under some common assumptions. The intermediate variables will be defined and estimated recursively. Then the Wiener system is transformed approximately into a pseudo-linear MISO system. It is based only on the observed input and output data and the recursively estimated intermediate variable data. From the derivations and simulations, it is seen that the identification algorithms and strategies are clear and efficient. It can be easily extended to other discontinuous nonlinearities or their combinations and to other block-oriented nonlinear dynamic systems.

REFERENCES

Bai, E. W. (2002), Identification of linear systems with hard input nonlinearities of known structure, *Automatica*, Vol. 38, pp. 853-860,.

Boutayeb, M. and Darouach, M. (1994), Recursive identification method for the Hammerstein model: Extension to nonlinear MISO systems, *Control Theory and Advanced Technology*, Vol. 10, No. 1.

Greblicki, W. (1999), Recursive identification of continuous-time Wiener systems, *International Journal Control*, Vol. 72. No. 11, pp. 981-989.

Hagenblad, A. and Ljung, L. (1998), Maximum likelihood identification of Wiener models with a linear regression initialization, *In Proceedings of the 37th IEEE Conference on Decision and Control*, pp. 712-713, Tampa, Florida, USA.

Kalafatis, A. D., Wang, L., and Cluett, W. R. (1997), Identification of Wiener-type nonlinear systems in a noisy environment, *International Journal of Control*, Vol. 66, No. 6, pp. 923-941.

Kortmann, M. (1989), Die Identifikation nichtlinearer Ein- und Mehrgrößensysteme auf der Basis nichtlinearer Modelansatze, *Dissertation, Fortschr.-Ber. VDI Reihe 8*, Nr. 177, Düsseldorf, VDI-Verlag.

Ljung, L. (1987), *System Identification: Theory for the User*, Prentice-Hall, Inc., Englewood Cliffs, New Jersey.

Narendra, K. S. and Gallman, P. G. (1966), An iterative method for the identification of nonlinear systems using a Hammerstein model, *IEEE Transactions on Automatic Control*, pp. 546-550.

Pearson, R. K. and Pottmann, M. (2000), Gray-box identification of block-oriented nonlinear models, *Journal of Process Control*, Vol. 10, pp. 301-315.

Vörös, J. (1995), Identification of nonlinear dynamic systems using extended Hammerstein and Wiener models, *Control-Theory and Advanced Technology*, Vol. 10, No. 4, Part 2, pp. 1203-1212.

Vörös, J. (1997), Parameter identification of discontinuous Hammerstein systems, *Automatica*, Vol. 33, No. 6, pp. 1141-1146.

Wigren, T. (1993), Recursive prediction error identification using the nonlinear Wiener model, *Automatica*, Vol. 29, No. 4, pp. 1011-1025.

Zhu, Y. (1999b), Parametric Wiener model identification for control, *In the 14th World Congress of IFAC*, pp. 37-42, Beijing, China.

IFAC Publications
www.elsevier.com/locate/ifac

NONLINEAR MODEL IDENTIFICATION USING WORKING POINT VARIABLES

Yucai Zhu

Section CS, Faculty of Electrical Engineering
Eindhoven University of Technology
P.O. Box 513, 5600 MB Eindhoven, The Netherlands
Tel. +31.40.2473246, y.zhu@tue.nl

Also at: Tai-Ji Control
Grensheuvel 10, 5685 AG Best, The Netherlands

Abstract: The identification of dynamic models with nonlinear gains is studied. The nonlinear gain is modelled as a function of so-called working-point variable (WPV). The working-point variable is introduced to represent the changes of process operation range. Parameter estimation is solved for the models. A relaxation iteration scheme is used to estimate the initial high order model and then a model reduction is used to obtain the reduced model. The convergence and consistency of the method are studied. Simulation study will be used to illustrate the estimation method. *Copyright © 2003 IFAC*

Keywords: Identification, nonlinear models, working-point variable (WPV), parameter estimation, relaxation algorithm

1. INTRODUCTION

Nonlinear system identification has received more attention in both research institutions and industries. One way to specify model structure is to combine linear dynamic models with static (memoryless) nonlinear functions. The most popular models are Hammerstein model, Wiener model and their combinations. Many researchers have studied the parametric identification of Hammerstein model; see e.g., Narendra and Gallman (1966), Stoica and Söderström (1982) and Zhu, (2000). Some work has been done on the identification of Wiener models; see, e.g., Billings and Fakhouri (1982), Greblicki (1992), Wigren (1993), Verhaegen (1998) and Zhu (1999). One extension of the Hammerstein and Wiener models is in the form of an L-N-L model where a nonlinear block is embedded between two linear blocks and it is called a Wiener-Hammerstein model; see Billings and Fakhouri (1982). Often correlation analysis are used to identify this kind of models and they belong to nonparametric model identification. A recursive estimation of a parametric Wiener-Hammerstein model is proposed in Boutayeb and Darouach (1995). Another combination of Hammerstein and Wiener models consists of a linear block embedded between two static nonlinear gains, hence the so-called N-L-N Hammerstein-Wiener model (Zhu, 2002).

In this work, the concept of working-point variable will be introduced and used to model process nonlinearity. A working-point variable (WPV) is a measured process variable, or, a known function of measured process variables, that reflects the change of process operation range. As a first step in this direction, only dynamic models with nonlinear gains will be studied. The advantages of this model strudcture are: 1) it is simple to comprehend and simple to identify; 2) it is simple to used in control and 3) it is more general than than Hammerstein models and Wiener models. In Section 2 we will introduec the model structure. Section 3 will treat parameter estimation, convergence and consistency analysis of the model. Simulation studies are performed in Section 4. Section 5 contains the conclusions.

2. NONLINEAR GAIN AS A FUNCTION OF A WPV

The process model changes can be seen as the consequences of the changes of operation range or working point. From this viewpoint, the concept of working-point variable can be introduced to model process nonlinearity. A working-point variable (WPV) of a process is defined as a measured variable, or, a known function of measured variables, that represents the change of process operation range. For an electrical motor, a WPV can be the load of the motor, or, its speed; for a distillation column, a working-point variable can be a tray temperature, the feed rate or a product quality. A working-point variable should be chosen so that it can catch process nonlinearity. In general, a WPV is neither an input variable, nor an output variable. However, the the input or the output can also be used as a WPV.

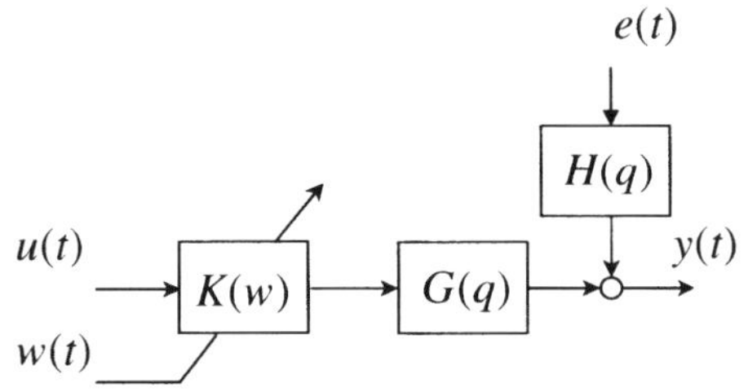

Figure 1. Model nonlinear gain using working-point variable

Denote $u(t)$, $y(t)$ as the process input and output at time t, and $w(t)$ as the chosen WPV. To model the nonlinear gain using the WPV $w(t)$, one can write the process model as (Figure 1)

$$\begin{aligned} y(t) &= G(q)K(w(t))u(t) + H(q)e(t) \\ &= \frac{B(q)}{A(q)}K(w(t))u(t) + \frac{C(q)}{D(q)}e(t) \end{aligned} \quad (1)$$

where $A(q)$, $B(q)$, $C(q)$ and $D(q)$ are polynomials of delay operator q^{-1}

$$\begin{aligned} A(q) &= 1 + a_1 q^{-1} + ... + a_n q^{-l} \\ B(q) &= b_0 + b_1 q^{-1} + ... + b_n q^{-l} \\ C(q) &= 1 + c_1 q^{-1} + ... + c_n q^{-l} \\ D(q) &= 1 + d_1 q^{-1} + ... + d_n q^{-l} \end{aligned}$$

l is called the order of the linear part of the model, $K(w(t)) \neq 0$ is a static nonlinear function of WPV $w(t)$, and $\{e(t)\}$ is a white noise sequence with zero mean value and variance λ^2. It is assumed that the nonlinear gain $K(w(t))$ is continuous and will not change sign in the operation range of WPV $w(t)$.

Model (1) will be called an input WPV model, because the nonlinear gain is at the input side. The linear part of the model is the well known Box-Jenkins model.

Similarly, one can propose an output WPV model with the nonlinear gain at the output side as (Figure 2)

$$\begin{aligned} y(t) &= K(w(t))[G(q)u(t) + H(q)e(t)] \\ &= K(w(t))[\frac{B(q)}{A(q)}u(t) + \frac{C(q)}{D(q)}e(t)] \end{aligned} \quad (2)$$

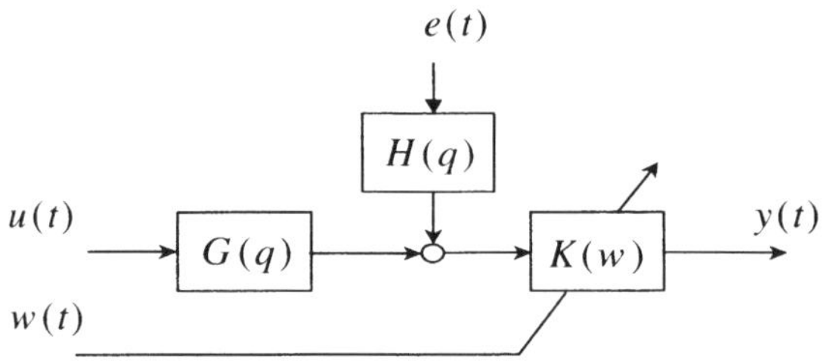

Figure 2 WPV model with output nonlinearity

Note that, in the output WPV model, the disturbance term $H(q)e(t)$ is placed before the output nonlinear block, which is different from the normal assumption that a stationary disturbance acts at the process output. This disturbance model implies *nonlinear* output disturbance: the output disturbance is large when the process gain is large (process is sensitive) and is small when the gain is small (process is insensitive). This assumption is realistic from a process operation point of view. A disadvantage of this model is that the output measurement noise can not be modelled properly. In practice, however, the influence of unmeasured disturbance is, in general, much greater than that of the measurement noise due to the advances in sensor technologies.

We will use cubic splines to model the nonlinear gain $K(w(t))$. Let K denote a set of knots $\{w_1, w_2, ..., w_m\}$ which are real numbers and satisfy

$$w_1 = w_{\min} < w_2 < w_3 < ... < w_m = w_{\max}$$

A cubic spline function defined for all real numbers w is given as

$$K(w) = \sum_{k=2}^{m-1} \beta_k |w - w_k|^3 + \beta_m + \beta_{m+1} w \quad (3)$$

where $[\beta_2, \beta_3, ..., \beta_{m+1}]$ are fixed real numbers, namely, the parameters to be estimated. Here m is called the number of knots which can be seen as the "degree" or "order" of the cubic splines. Note that the number of parameters of the cubic splines is m. It is easy to verify that the first and the second derivatives of the function are continuous and hence the function is smooth. The parameters of the model cannot be uniquely determined without introduce further constraint. A solution to this problem is to fix the gain of the nonlinear block. For example one can let $\beta_{m+1} = 1$.

When the WPV is the input, $w(t) = u(t)$, the input WPV model (1) is equivalent to a Hammestein model studied in Zhu (2000).

When the WPV is the output, $w(t) = y(t)$, the output WPV model (2) is equivalent to a Wiener model. The advantage of the output WPV model over Wiener models is that the nonlinear function need not be monotone and invertable. The nonlinear function in a Wiener model is normally assumed monotone and invertable, because nonlinear inversion is often involved in Wiener model estimation (Greblicki, 1992 and Zhu, 1999).

It is very simple to compensate for the nonlinear gain when using a WPV model in control. Assume that $K(w(t))$ does not change its sign, just devide the linear controller gain by $K(w(t))$. No inversion is needed.

One can further combine the input WPV model and the output WPV model to obtain a model with nonlinear gains at both the input and the output, the so-called N-L-N WPV model. In this case, generally, two WPV's will be needed.

The identification test is done in closed-loop and the closed-loop system is assumed to be stable. To assure persistent excitation, a suitable test signal is applied at the setpoint and/or at the input. An open loop test is a special case of the closed-loop test.

3. PARAMETER ESTIMATION

The parameter estimation is solved for input WPV model (1). But the results of this section hold equally for the output WPV model (2). Assume that an identification test has been performed, possibly in a closed-loop operation. Denote the input-output and WPV data as

$$Z^N = [u(1),\, y(1), w(1),\, \ldots,\, u(N),\; y(N),\; w(N)] \tag{4}$$

For given orders l, m, determine the parameters of the model (1) and (3) using the input-output and WPV data from the test by minimizing the loss function

$$V(\theta,\, Z^N) = \frac{1}{N}\sum_{t=1}^{N} \varepsilon^2(t,\theta) \tag{5}$$

where

$$\varepsilon(t,\theta) = H^{-1}(q)[y(t) - G(q)K(w(t))u(t)] \tag{6}$$

$$= \frac{D(q)}{C(q)}[y(t) - \frac{B(q)}{A(q)}K(w(t))u(t)] \tag{7}$$

The error defined in (6) is highly nonlinear in model parameters. Direct minimization of the loss function (5) is difficult and can run into numerical problems. It desirable to reduce this complexity by looking for some simpler numerical schemes. It is well know that any linear prediction error model structure can be approximated arbitrarily well by a ARX (AutoRgressive with eXternal input), or, equation error model with sufficiently high order. Based on this fact, approximate the Box-Jenkins model in (1) by a high order ARX model

$$y(t) = \frac{B^n(q)}{A^n(q)}K(w(t))u(t) + \frac{1}{A^n(q)}e(t)$$

or

$$A^n(q)y(t) = B^n(q)K(w(t))u(t) + e(t) \tag{8}$$

where $A^n(q)$ and $B^n(q)$ are polynomials of q^{-1}, n is the order of the ARX model and $\{e(t)\}$ is white noise with zero mean and variance R. Then, the loss function for parameter estimation for model (8) becomes

$$V_{ARX}(\theta,\, Z^N) = \frac{1}{N}\sum_{t=1}^{N} \varepsilon^2(t,\, \theta) \tag{9}$$

where

$$\varepsilon(t,\, \theta) = A^n(q)y(t) - B^n(q)K(w(t))u(t) \tag{10}$$

and

$$\theta = [a_1,\, \cdots,\, a_n,\, b_0,\, \cdots,\, b_n,\, \beta_2,\, \cdots,\, \beta_{m+1}]^T \tag{11}$$

is the parameter vector of model (6). This vector varies over a set D_M which is a compact subset of R^{n_θ} where n_θ is the number of parameters.

Note that the error $\varepsilon(t,\, \theta)$ is linear in the parameters of $A^n(q)$ and bilinear in the parameters of $K(w(t))$ and $B^n(q)$. Hence one can use the following relaxation algorithm for parameter estimation.

3.1 *Relaxation Algorithm*

Initialization.

Set $K(w) = 1$ and estimate $A^n(q)$ and $B^n(q)$ using linear least-squares.

Iteration. Denote $\hat{A}^n_{(i)}(q)$, $\hat{B}^n_{(i)}(q)$, and $\hat{K}_{(i)}(w)$ as the estimates from iteration i, then

1) Calculate the parameters of $\hat{K}_{(i+1)}((u(t))$ for fixed $\hat{A}^n_{(i)}(q)$ and $\hat{B}^n_{(i)}(q)$ by minimizing

$$\frac{1}{N}\sum_{t=1}^{N}[\hat{A}^n_{(i)}(q)y(t) - \hat{B}^n_{(i)}(q)K_{(i+1)}(w(t))u(t)]^2 \tag{12}$$

2) Calculate $\hat{A}^n_{(i+1)}(q)$ and $\hat{B}^n_{(i+1)}(q)$ for fixed $\hat{K}_{(i+1)}(w)$ by minimizing

$$\frac{1}{N}\sum_{t=1}^{N}[A^n_{(i+1)}(q)y(t) - B^n_{(i+1)}(q)\hat{K}_{(i+1)}(w(t))u(t)]^2 \tag{13}$$

Go back to 1). Stop when convergence occurs.

Both steps are linear least-squares problems.

3.2 Convergence and Consistence

Now, we will show that this relaxation scheme is not an ad hoc trick but a theoretically sound optimization method. To this end, we need to define persistent excitation condition on the test input and WPV. Denote

$$\Phi_w = \begin{bmatrix} |w(1)-w_1|^3 & \cdots & 1 & w(1) \\ |w(2)-w_1|^3 & \cdots & 1 & w(2) \\ \vdots & & \vdots \ \vdots & \vdots \\ |w(N-n)-w_1|^3 & \cdots & 1 & w(N-n) \end{bmatrix} \tag{14}$$

as the WPV data matrix over knots $\{w_1, ..., w_m\}$, and denote the input data matrix as

$$\Phi_u = \begin{bmatrix} u(1) & u(2) & \cdots & u(n) \\ u(2) & u(3) & \cdots & u(n+1) \\ \vdots & \vdots & \vdots & \vdots \\ u(N-n) & \cdots & \cdots & u(N) \end{bmatrix} \tag{15}$$

We say that the input signal $u(t)$ and the WPV $w(t)$ are *jointly persistently exciting* with orders (n, m) over the knots $\{w_1, w_2, ..., w_m\}$, if matrices Φ_w and Φ_u both have full column ranks for all $N >> \max(n, m_1)$.

First, some assumptions on the process, the model and test conditions are given.

A1 The data set Z^N is generated by the process

$$y(t) = f_s[u^{t-1}, y^{t-1}, w(t)] + e_o(t) \tag{16}$$

Here $f_s[u^{t-1}, y^{t-1}, w(t)]$ is a deterministic function of past input and past output up to time $t-1$ and WPV up to time t. The noise $\{e_o(t)\}$ is zero mean white noise with bounded moments of order $4+\delta$ for some $\delta > 0$. The nonlinear function $f_s[u^{t-1}, y^{t-1}, w(t)]$ is continuous. The WPV signal $w(t)$ is noise-free.

A2 The input-output and WPV data are bounded. This means that, if the data are generated in open loop test, the process is stable; if the process is unstable, the data are generated under feedback control and the closed-loop system is stable.

A3 The true process (16) has the same parametrization as the model (8) and (3)

$$A^o(q)y(t) = B^o(q)K^o(w(t))u(t) + e_o(t) \tag{17}$$

with the same orders. $A^o(q)$ and $B^o(q)$ are coprime.

A4 The input $u(t)$ and WPV $w(t)$ are jointly persistently exciting with orders $(2n, m)$ over knots $\{w_1, w_2, ..., w_m\}$. If the input-output data is generated in closed-loop, the input is not determined purely by linear output feedback, namely, for any linear filters $L(q)$ and $C(q)$ such that $L(q)$ is non zero

$$\lim_{N\to\infty} \inf \frac{1}{N}\sum_{t=1}^{N} E|L(q)u(t) + C(q)y(t)|^2 > 0 \tag{18}$$

Theorem 1. *Assume that A1 - A4 hold. Then the relaxation algorithm (12) and (13) will converge to a local minimum of the loss function (9).*

The proof is omitted due to the paper size. It is the same as proof of Theorem 1 in Zhu and Weiland (2003). No global convergence result is known yet for this kind of algorithm. Therefore, good initial values of the estimate is important for finding a good estimate.

In the following we will establish the convergence of criterion $V_{ARX}(\theta, Z^N)$ and consistency of parameter estimate $\hat{\theta}$.

Theorem 2. *Assume that conditions A1–A2 are true, then the following convergence results hold:*

$$\sup_{\theta\in D_M} |V_{ARX}(\theta, Z^N) - E\varepsilon^2(t, \theta)| \to 0 \text{ w.p. 1 as } N \to \infty \tag{19}$$

where w. p. 1 means with probability 1 and

$$\hat{\theta}^N \to \theta^* \quad \text{w. p. 1 as } N \to \infty \tag{20}$$

where

$$\hat{\theta}^N = \arg\min_{\theta\in D_M} V_{ARX}^N(\theta, Z^N)$$

$$\theta^* = \arg\min_{\theta\in D_M} E\varepsilon^2(t, \theta)$$

Moreover, assume that conditions A3 and A4 also hold and denote the true parameter vector as θ^o. Then the estimate $\hat{\theta}^N$ is consistent, namely,

$$\hat{\theta}^N \to \theta^o \quad \text{w. p. 1 as } N \to \infty \tag{21}$$

Proof. See Zhu (2001)..

The convergence result holds under very general and weak conditions. The true nonlinear functions need not be cubic splines and the model orders need not to be correct. The limiting model will always minimize the loss function of the prediction error of $y(t)$. Unstable process can be treated without problems, because the predictor is stable for equation error models.

3.3 Model Reduction for Linear ARX Model

The obtained WPV model with high order ARX model is unbiased, provided that the process can be modelled by an WPV model, but the variance error of the ARX model is high due to its high order. Model reduction can be used to reduce the variance error. The following asymptotic result is used for developing an frequency domain criterion for model reduction.

Assume that nonlinear gain $K(w(t))$ is known exactly and $K(w(t)) \neq 0$. Denote the compensated input

$$w_1(t) = \frac{u(t)}{K(w(t))}$$

and use the data set

$$[w_1(1),\ y(1),\ w_1(2),\ y(2),\ ...,\ w_1(N),\ y(N)]$$

to estimate a high order ARX model. Then it can be shown (see Ljung, 1985) that the estimated frequency response of the high order model is unbiased and its error follows a Gaussian distribution with variance given as

$$var[\hat{G}^n(e^{i\omega})] \approx \frac{n}{N}\frac{\Phi_v(\omega)R}{\Phi_{w_1}(\omega)R - |\Phi_{w_1e}(\omega)|^2} \qquad (22)$$

where $\hat{G}^n(e^{i\omega})$ is the frequency response of the estimated high order ARX model, n is the order of the ARX model, N is the number of data points, $\Phi_v(\omega)$ is the power spectrum of the disturbance, R is the variance of white noise $\{e(t)\}$ that generated the disturbance, $\Phi_{w_1}(\omega)$ is the power spectrum of input to the linear block $w_1(t)$ and $\Phi_{w_1e}(\omega)$ is the cross-spectrum between the white noise $\{e(t)\}$ and $w_1(t)$ (due to feedback). Note that this result hold for closed-loop tests.

If we view the frequency response of the high order estimates as the noisy observations of the true transfer function, we can then apply the maximum likelihood principle. Based on (22), it can be shown that when $N \to \infty$, the asymptotic negative log-likelihood function for the linear model is given by (Wahlberg, 1989):

$$\frac{1}{2\pi}\int_{-\pi}^{\pi}\left|\hat{G}^n - G^l\right|^2 \frac{\Phi_{w_1}\lambda^2 - |\Phi_{w_1e}|^2}{\Phi_v\lambda^2}d\omega \qquad (23)$$

where $G^l(e^{i\omega})$ is the frequency response of the reduced model to be calculated.

In practice, however, the signals $w_1(t)$ is not known exactly and it is calculated using the nonlinear function estimates $K(w(t))$. This implies that the frequency weighting in model reduction will deviate from the theoretical one. But experience has shown that the model reduction is robust to variations in the weighting function. The minimization of the loss function (23) needs nonlinear search. It can be done directly in the frequency domain; it can also be converted to a time-domain parameter estimation problem. If disturbance model is needed in control, it can be obtained by a model reduction on $1/\hat{A}^n(q)$ using the same idea as for the process model.

Remark on Model Validation

Denote $G^o(e^{i\omega})$ as the frequency response of the true model of the linear part. Because the error of $\hat{G}^n(e^{i\omega})$ is Gaussian with its variance given in (22), we can define a 3σ bound for the the error in the frequency domain

$$|G^o - \hat{G}^n| \le 3\sqrt{\frac{n}{N}\frac{\Phi_v R}{\Phi_{w_1}R - |\Phi_{w_1e}|^2}} \quad \text{w.p. } 99.9\% \qquad (24)$$

Assume that the nonlinearity are cancelled completely, which implies that the nonlinear blocks are known perfectly, then, for a given controller, robust stability and performance analysis can be applied using linear robust control theory; see, e.g., Skogestad and Postlethwaite (1996).

Remark on estimating the output WPV model. The output WPV model (2) can be rewritten as

$$\frac{1}{K(w(t))}y(t) = \frac{B(q)}{A(q)}u(t) + \frac{C(q)}{D(q)}e(t) \qquad (25)$$

Parametrizing $K_2(w(t)) = \frac{1}{K(w(t))}$ using cubic splines and approximating the right hand side of the equation using a high order ARX will result in

$$K_2(w(t))y(t) = \frac{B^n(q)}{A^n(q)}u(t) + \frac{1}{A^n(q)}e(t) \qquad (26)$$

Then, with slight modification, the relaxation algorithm can be used to estimate $A^n(q)$, $B^n(q)$ and $K_2(w(t))$. It is easy to see that the convergence and consistent results hold equally for the estimation scheme. The same model reduction method can be used to obtain the Box-Jenkins model of the linear part. An estimate of the original nonlinear function $K(w(t))$ can be obtained from the estimate $\frac{1}{\hat{K}_2(w(t))}$ using least-squares method.

4. SIMULATION

A Wiener process. Simulated process:

$$x(t) = \frac{q^{-1} + 0.5q^{-2}}{1 - 1.5q^{-1} + 0.7q^{-2}}u(t) + v(t) \qquad (27)$$

$$y(t) = f(x(t)) = x(t) + 1.2x^3(t)$$

and

$$v(t) = \frac{\alpha}{1 - 0.9q^{-1}}e(t)$$

where $\{e(t)\}$ is zero mean white Gaussian noise.

The input signal used is multi level random noise with average switch time of 10 samples in the range of $[0.0, 0.4]$. The number of samples is 1000. First noise-free data are used to test the correctness of the algorithm. The linear model has the same order as the process, the degree of the cubic splines is 15. The relaxation algorithm converged in 5 iterations. The fit error of the output WPA model is 0.003% at the output. The fit error of a linear output error model is 30%. The nonlinear gain identified from the noise-free data will be used as the "true nonlinear gain".

Then 10% disturbance is added before the output nonlinear block to generate noisy data. The degree of the nonlinear function is 18 and the rest remains the same. The relaxation algorithm converged in 5 iterations. The input-output data are shown in Figure 3. The fit of the nonlinear function and of the step response of the linear block (after gain correction).are shown in Figure 3. The result is very satisfactory, considering the level of the disturbance.

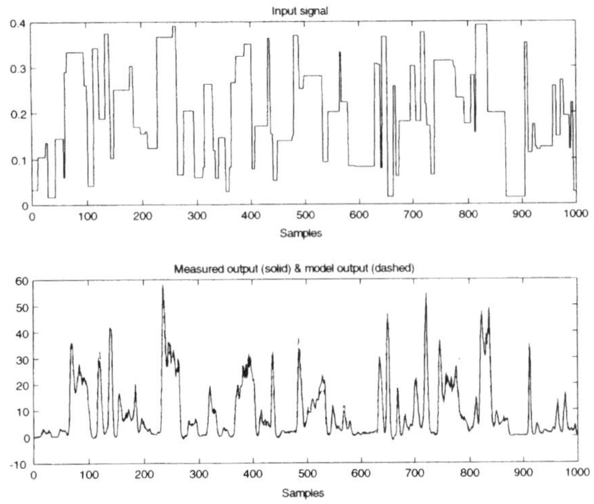

Figure 3. Input-output data of the Wiener process and simulated output. 10% disturbance before the nonlinear gain

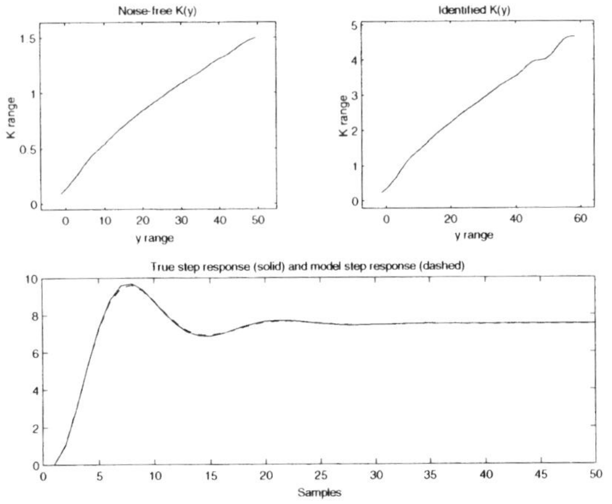

Figure 4. The "true" (noise-free) and identified nonlinear gains of the Wiener process and the fit of the step responses of the linear part. Note that the difference in the K ranges is caused by the nonuniqueness of the gain distributions

5. CONCLUSION AND DISCUSSION

Working-point variable (WPV) is introduced and used to model nonlinear gains. Input WPV model and output WPV model have been proposed. In parameter estimation, the bilinear-in-parameters property of the model is used to derive the relaxation algorithm. The convergence and consistency are proved. Then, the linear model is reduced using a frequency domain criterion. The effectiveness of the method have been shown in a simulation example. The advantage of the two WPV models over Hammerstein and Wiener models is twofold: 1) it is more general, meaning that it can model the nonlinear gains that is not a function of the input or the output; 2) it can treat non monotone output nonlinearity.

REFERENCES

[1] Billings, S.A. and S.Y. Fakhouri (1982). Identification of systems containing linear dynamics and static nonlinear elements. *Automatica*, Vol. 18, No. 1, pp. 15-26.

[2] Boutayeb, M. and M. Darouach (1995). Recursive identification method for MISO Wiener-Hammerstein model. *IEEE Trans. Autom. Control*, Vol. AC-40, pp.287–291.

[3] Greblicki, W. (1992). Nonparametric identification of Wiener systems. *IEEE Trans. Information Theory*, Vol. 38, No. 5, pp. 1487-1493.

[4] Lancaster P. and K. Šalkauskas (1986). *Curve and Surface Fitting: An Introduction.* Academic Press, London.

[5] Ljung, L. (1978). Convergence analysis of parametric identification methods. *IEEE Trans. Autom. Control*, Vol. AC-23, pp. 770-783.

[6] Ljung, L. (1985). Asymptotic variance expressions for identified black-box transfer function models. *IEEE Trans. Autom. Control*, Vol. AC-30, pp. 834-844.

[7] Narendra, K.S. and P.G. Gallman (1966). An iterative method for the identification of nonlinear systems using Hammerstein model. *IEEE Trans. Autom. Control*, Vol. AC-11, No. 31, pp. 546-550.

[8] Skogestad, S. and I. Postlethwaite (1996). *Multivariable Feeadback Control.* Wiley, Chichester.

[9] Stoica, P. and T. Söderström (1982). Instrumental-variable methods for identification of Hammerstein systems. *Int. J. Control*, Vol. 35, No. 3, pp. 459-476.

[10] Verhaegen, M. (1998). Identification of the temperature-product quality relationship in a multi-component distillation column. *Chemical Engineering Communications*, Vol. 163, pp. 111-132

[11] Wigren, T. (1993). Recursive prediction error identification using the nonlinear Wiener model. *Automatica*, Vol. 29, No. 4, pp. 1011-1025.

Zhu, Y.C. (1999). Parametric Wiener model identification for control. *Proceedings of IFAC Congress*, July 5-9, 1999, Beijing.

[12] Zhu, Y.C. (2000). Hammerstein model identification for control using ASYM. *International Journal of Control*, Vol. 73, No.18 pp. 1692-1702.

[13] Zhu, Y.C. (2002). Estimation of an N-L-N Hammerstein-Wiener model. *Automatica*. Vol. 38, No. 9, pp 1607-1614.

[14] Zhu, Y.C. and S. Weiland (2003). Identification of a class of nonlinear ARX models. Submitted to *IEEE Trans. Autom. Control.*

IFAC Publications
www.elsevier.com/locate/ifac

IDENTIFICATION OF WIENER-HAMMERSTEIN MODELS WITH CUBIC NONLINEARITY USING LIFRED

A. H. Tan[1] and K. R. Godfrey[2]

*1: Faculty of Engineering, Multimedia University, 63100 Cyberjaya, Malaysia;
Email: htai@mmu.edu.my
2: School of Engineering, University of Warwick, Coventry, CV4 7AL, U.K.;
Email: K.Godfrey@warwick.ac.uk*

Abstract: The identification of Wiener-Hammerstein models using linear interpolation in the frequency domain (LIFRED) is extended from models with quadratic nonlinearity to models with cubic nonlinearity. The modifications to the algorithm are discussed and a simulation example is presented. A further technique is proposed which enables the estimation of the gain response of the first linear subsystem from estimation lines in the output which are distorted due to contributions of several combinations of the input harmonics. This new technique is less susceptible than the first approach to the effects of noise and a possible reason for this is discussed. *Copyright © 2003 IFAC*

Keywords: frequency methods, nonlinearity, non-parametric identification, signal processing, Wiener-Hammerstein models.

1. INTRODUCTION

Most processes exhibit some form of nonlinearity. For the purpose of modelling and control, such processes have often been linearised around their operating points. However, this is unsatisfactory if the nonlinear distortions are relatively large. Even though a dedicated nonlinear model may be used, the solution often comes with a high cost of complexity. Due to these reasons, simple block-oriented models provide excellent alternatives to using completely linear models or dedicated nonlinear models.

The identification of block-oriented models has received considerable attention in the literature. Such models include the Wiener model (where the static nonlinearity succeeds the linear dynamics) – see, for example Billings & Fakhouri (1977) and Westwick & Verhaegen (1996); the Hammerstein model (where the static nonlinearity precedes the linear dynamics) – see, for example Narendra & Gallman (1966); and the Wiener-Hammerstein model (where the static nonlinearity is sandwiched between two blocks of linear dynamics, as shown in Figure 1) - see, for example, Vandersteen, *et al.* (1997) and Bershad, *et al.* (2000). However, the identification of third order Volterra kernels of block-oriented models is a subject which has not been extensively studied. There are papers on the subject, for example Evans, *et al.* (1996) and Lawrence (1981), but these are far fewer in number than those on the identification of second order kernels of these models. Often, a technique proposed for the identification of a quadratic nonlinearity is simply assumed to work for higher order nonlinearities. Even though this may be true in the ideal, noiseless situation, the technique cannot be guaranteed to work in practical situations, where imperfections such as noise and distortions have to be taken into account. Possible reasons for the lack of explicit results for models with third order kernel are that the added dimension makes analysis much more complicated and it is also difficult to present information visually. In addition, the total power of the third order kernel may be low, resulting in it being ignored particularly if a long testing time is required to obtain results associated with this kernel. In this paper, the identification technique LIFRED (Linear Interpolation in the FREquency Domain), proposed in Tan & Godfrey (2002) for Wiener-Hammerstein models with quadratic nonlinearity, is extended to identify such models consisting of a cubic nonlinearity. The modifications to the algorithm are discussed and a simulation example is presented to illustrate the applicability of the method.

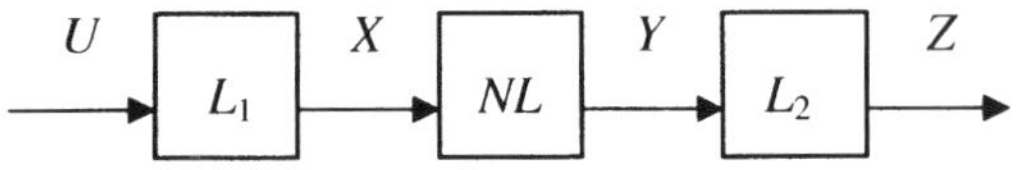

Fig. 1. Wiener-Hammerstein model. *L*: Linear dynamics; *NL*: Static nonlinearity.

In Section 4 of the paper, a second technique is proposed for identifying the gain response of the first linear subsystem. This makes use of the estimation lines in the output which are distorted due to contributions of several combinations of the input harmonics. It is shown that this new technique yields more accurate estimates in the presence of noise than the first method. The results obtained are significant considering that these distorted lines are normally discarded and not used in the identification process.

2. MODIFICATIONS TO THE LIFRED ALGORITHM

2.1. Signal Design

The LIFRED algorithm is based on the separation of the dynamics of the two linear subsystems using the symmetry properties of the Volterra kernel. After the contributions of the two subsystems are separated, their frequency responses are estimated using linear interpolation. It is important that the distortions at the test frequencies should be avoided. These can be classified into Type I and Type II (Evans, *et al.*, 1994). Type I distortions are caused by the combination of a test frequency with a pair of equally positive and negative frequencies, resulting in a contribution falling at that particular test frequency. Type II distortions are caused by any other combinations which are not classified as Type I, which result in a contribution at a test frequency.

It is possible to design signals which do not suffer from any Type II distortions, for a given order of nonlinearity, by proper selection of their harmonic components. These signals are termed *no interharmonic distortion* (NID) multisines, and they possess a sparse spectrum with a log-tone appearance (Evans, *et al.*, 1994). For the identification of the third order Volterra kernel, the signal is theoretically more sparse than that used to identify the second order kernel since the number of possible contributions of the input frequencies is larger when identifying a third order kernel. Also Type I contributions from points in the third order kernel where the frequencies $f_1 = -f_2$ or $f_1 = -f_3$ or $f_2 = -f_3$ will fall at the input frequencies and cannot be separately measured (Evans, *et al.*, 1996). (However, they can still be used in the identification process, as will be shown later in Section 4.)

In this paper, a signal which consists of only odd harmonics is used. This ensures that the measurement points of the third order kernel will not be corrupted by even order nonlinearities. The signal used, which was designed in Evans, *et al.*, (1996), is

$$fv = [241\ \ 451\ \ 663\ \ 877\ \ 1095\ \ 1319\ \ 1581\ \ 1817\ \ 2109\ \ 2347] \tag{1}$$

It should be noted that the measurement of the third order Volterra kernel is much more sensitive to the effects of input signal distortion compared with the second order kernel. This is due to the fact that the number of possible combinations of the input frequencies increases approximately at the rate F^n, where F is the number of harmonics in the input signal and n is the order of the kernel. It should be stressed that this is not a weakness of the LIFRED method. Rather, it is a problem associated with the use of multisines for the measurement of the third order Volterra kernel (although the problem is less serious with NID multisines), if input signal distortion could not be avoided. However, such problems are not encountered in a purely digital system, since the NID multisines can be exactly realised.

It is assumed that the approximate bandwidths of the linear subsystems are known *a priori*, and that the power at each of the estimation lines is large enough to negate the effects of noise.

2.2. Calculation of the Frequency Response Gain of the Second Linear Subsystem

The third order Volterra kernel of a Wiener-Hammerstein model is given by

$$|H_3(s_1,s_2,s_3)| = |L_1(s_1)||L_1(s_2)||L_1(s_3)||L_2(s_1+s_2+s_3)| \tag{2}$$

where H_3 denotes the Laplace transform of the third order Volterra kernel; and L_1 and L_2 are the linear kernels of the first and second linear subsystems respectively. The gain at the output of the system is

$$|Z(s_1+s_2+s_3)| = |L_1(s_1)||L_1(s_2)||L_1(s_3)||L_2(s_1+s_2+s_3)||U(s_1)||U(s_2)||U(s_3)| \tag{3A}$$

$$|Z(s_1+s_2-s_3)| = |L_1(s_1)||L_1(s_2)||L_1(-s_3)||L_2(s_1+s_2-s_3)||U(s_1)||U(s_2)||U(-s_3)| \tag{3B}$$

$$|Z(s_1-s_2+s_3)| = |L_1(s_1)||L_1(-s_2)||L_1(s_3)||L_2(s_1-s_2+s_3)||U(s_1)||U(-s_2)||U(s_3)| \quad (3C)$$

$$|Z(s_1-s_2-s_3)| = |L_1(s_1)||L_1(-s_2)||L_1(-s_3)||L_2(s_1-s_2-s_3)||U(s_1)||U(-s_2)||U(-s_3)| \quad (3D)$$

Using the symmetry property, $|L_1(s_2)| = |L_1(-s_2)|$, $|L_1(s_3)| = |L_1(-s_3)|$, $|U(s_2)| = |U(-s_2)|$ and $|U(s_3)| = |U(-s_3)|$. Hence,

$$\frac{|Z(s_1+s_2+s_3)|}{|Z(s_1-s_2-s_3)|} = \frac{|L_2(s_1+s_2+s_3)|}{|L_2(s_1-s_2-s_3)|} = \alpha_1 \quad (4A)$$

$$\frac{|Z(s_1+s_2-s_3)|}{|Z(s_1-s_2-s_3)|} = \frac{|L_2(s_1+s_2-s_3)|}{|L_2(s_1-s_2-s_3)|} = \alpha_2 \quad (4B)$$

$$\frac{|Z(s_1-s_2+s_3)|}{|Z(s_1-s_2-s_3)|} = \frac{|L_2(s_1-s_2+s_3)|}{|L_2(s_1-s_2-s_3)|} = \alpha_3 \quad (4C)$$

In the program written, the smallest of the harmonics $|s_1 + s_2 - s_3|$, $|s_1 - s_2 + s_3|$ and $|s_1 - s_2 - s_3|$ is used in the denominators of equations (4); this harmonic is stored in a vector called *minus*. (Note that $|s_1 + s_2 + s_3|$ is definitely larger than the above three harmonics.) The values of α_1, α_2 and α_3 are calculated for all possible combinations of the input frequencies where $f_1 \neq f_2 \neq f_3$. For an input signal with F harmonics, there are F different choices for the first harmonic, F-1 choices for the second harmonic and F-2 choices for the third harmonic, while there are also six different ways to combine these three harmonics. This results in $(F^3 - 3F^2 + 2F)/6$ sets of α_1, α_2 and α_3. All the entries in *minus* are then arranged in an increasing order.

In the interpolation algorithm, the first entry in *minus* according to the new order is arbitrarily set to unity. Next, the gains at the corresponding three harmonics (which are linked to *minus* by α_1, α_2 and α_3) are calculated using (4). The harmonics in *minus* are considered in increasing order, and are estimated using linear interpolation. After each harmonic in *minus* is estimated, equations (4) are used to calculate the gains at the three corresponding harmonics. This continues until the gains at all the estimation lines are calculated. The program flow is similar to that in Tan & Godfrey (2002), except that there are now three different attenuation ratios being considered instead of only one.

It should be noted that the algorithm attempts to estimate the lower frequencies first because the estimation lines are closer at these frequencies. Furthermore, most practical systems are lowpass which means that the measurement at lower frequencies is likely to be more accurate.

It should be stressed here that the complete gain response of L_2 is obtained in a single experiment and the computational burden is very low. These advantages are evident throughout the entire LIFRED procedure.

2.3. *Calculation of the Frequency Response Gain of the First Linear Subsystem (Method 1)*

When $s_1 = s_2 = s_3 = s$, (3A) can be written as

$$|Z(3s)| = |L_1(s)|^3 |L_2(3s)| |U(s)|^3 \quad (5)$$

The values of $|L_2(3s)|$ can be easily calculated by linear interpolation on the frequency response gain curve of L_2. The gain of L_1 is estimated using

$$|L_1(s)| = \frac{1}{|U(s)|}\left(\frac{|Z(3s)|}{|L_2(3s)|}\right)^{1/3} \quad (6)$$

The advantage of using this method is its simplicity. The gain of L_1 is obtained directly from (6). However, this may not be the best in the presence of output noise. An alternative method (Method 2) is proposed in Section 4. This utilizes the information present in the largest frequency lines in the output, in order to increase the signal-to-noise ratios at the estimation lines.

2.4. *Simultaneous Calculation of the Frequency Response Phases of the First and Second Linear Subsystems*

For a third order kernel, the phase relationship is

$$\angle H_3(s_1,s_2,s_3) = \angle L_1(s_1) + \angle L_1(s_2) + \angle L_1(s_3) + \angle L_2(s_1+s_2+s_3) \quad (7)$$

and

$$\angle Z(s_1+s_2+s_3) = \angle L_1(s_1) + \angle L_1(s_2) + \angle L_1(s_3) + \angle L_2(s_1+s_2+s_3) + \angle U(s_1) + \angle U(s_2) + \angle U(s_3) \quad (8A)$$

$$\angle Z(s_1+s_2-s_3) = \angle L_1(s_1) + \angle L_1(s_2) - \angle L_1(s_3) + \angle L_2(s_1+s_2-s_3) + \angle U(s_1) + \angle U(s_2) - \angle U(s_3) \quad (8B)$$

$$\angle Z(s_1-s_2+s_3) = \angle L_1(s_1) - \angle L_1(s_2) + \angle L_1(s_3) + \angle L_2(s_1-s_2+s_3) + \angle U(s_1) - \angle U(s_2) + \angle U(s_3) \quad (8C)$$

$$\angle Z(s_1-s_2-s_3) = \angle L_1(s_1) - \angle L_1(s_2) - \angle L_1(s_3) + \angle L_2(s_1-s_2-s_3) + \angle U(s_1) - \angle U(s_2) - \angle U(s_3) \quad (8D)$$

Subtracting (8B) from (8A), and (8D) from (8C)

$$\angle L_2(s_1+s_2+s_3) - \angle L_2(s_1+s_2-s_3) = \angle Z(s_1+s_2+s_3) - \angle Z(s_1+s_2-s_3) - 2\angle L_1(s_3) - \angle 2U(s_3) = \gamma_1 \quad (9A)$$

$$\angle L_2(s_1-s_2+s_3) - \angle L_2(s_1-s_2-s_3) = \angle Z(s_1-s_2+s_3) - \angle Z(s_1-s_2-s_3) - 2\angle L_1(s_3) - \angle 2U(s_3) = \gamma_2 \quad (9B)$$

where γ_1 and γ_2 are defined as the difference terms.

The program used is similar to that in Tan & Godfrey (2002) and only the main modifications will be described here. The term $\angle L_1(s_3)$ with s_3 set to the lowest harmonic in the input is assigned an arbitrary value ϕ since the phase contribution due to the pure time delay in the Wiener-Hammerstein model can be equally well attributed to the linear subsystems L_1 and L_2. This is in contrast to first assigning the phase at the third lowest harmonic in Tan & Godfrey (2002). The different approach is necessary because the third harmonic in (1) is quite large and it is preferable not to interpolate between this and the phase at dc. The values of $\angle L_2(|s_1 + s_2 + s_3|)$, $\angle L_2(|s_1 + s_2 - s_3|)$, $\angle L_2(|s_1 - s_2 + s_3|)$ and $\angle L_2(|s_1 - s_2 - s_3|)$ for all combinations of s_1 and s_2 (with s_3 still set to the lowest harmonic) are estimated using a combination of linear interpolation and solving (9). Next, the rest of the phases of L_1 are estimated using (8A). The phases at the remaining estimation lines of L_2 can also be calculated directly using (8).

3. SIMULATION EXAMPLE

To illustrate the technique, a Wiener-Hammerstein system was simulated, with L_1 a second order Chebyshev filter with 3dB of ripple in the passband and a cut-off frequency of 0.1 Hz, and L_2 a third order Chebyshev filter with 2dB of ripple in the passband and a cut-off frequency of 0.3 Hz. The nonlinearity was $Y = X^3$. The input signal was a multisine of length $N = 16384$ with the harmonics given in (1). The sampling frequency was set at 1 Hz. The highest estimation lines used in estimating L_1 and L_2 were placed at 0.143 Hz and 0.430 Hz respectively. This gave sufficient excitation in the passbands of both L_1 and L_2. The experiment was repeated with band-limited white noise added to the system output. The signal-to-noise ratio was approximately 30dB at the frequency band of interest. Twenty five experiments were carried out for the noisy case.

It is worth noting that only an approximate knowledge of the bandwidths of the linear subsystems is required *a priori*. This may pose a problem if the bandwidth of L_2 is very different from three times that of L_1 because then it is difficult to set a sampling frequency which gives sufficient excitation in the passbands of both L_1 and L_2. In particular, if the bandwidth of L_1 is much smaller than that of L_2, the signal power will be very much attenuated after passing through L_1, hence resulting in a low accuracy in the estimates of L_2 (and subsequently those of L_1). One way to overcome this is to increase the signal power. However, this may or may not be practical depending on the specific application.

The frequency response of the linear subsystems L_1 and L_2 are illustrated in Figures 2 and 3 respectively. (For the noisy case, the results of only one of the 25 experiments are shown.) These were plotted assuming that the phase contribution due to the pure time delays of L_1 and L_2 are known exactly, in order to facilitate comparison.

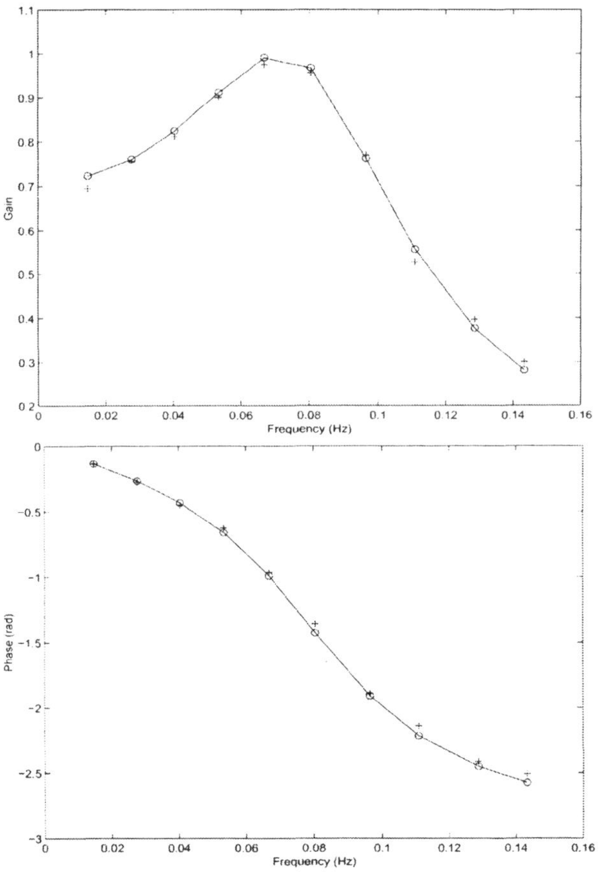

Fig. 2. Frequency response gain (top) and phase (bottom) of L_1. Solid line: Actual values; Circles: Estimates obtained without noise; Plusses: Estimates obtained with noise.

It can be seen from Figures 2 and 3 that the estimates obtained without noise are extremely accurate, while those obtained in the presence of noise are still very good. This confirms the usefulness of the LIFRED technique. It is also clear from the above figures that the frequency response estimate of L_1 is more accurate than that of L_2. This is consistent with the statistical values of the complex errors E and E_S as defined by

$$E = \frac{1}{F}\sum_{k=1}^{F}\left|H(j\omega_k) - \hat{H}(j\omega_k)\right| \qquad (10A)$$

$$E_S = \frac{1}{F}\sum_{k=1}^{F}\left(H(j\omega_k) - \hat{H}(j\omega_k)\right)^2 \qquad (10B)$$

where H and $\hat{H}$ represent the actual and estimated frequency responses at angular frequency ω_k respectively; and F is the number of harmonics in the input multisine. These errors are shown in Table 1. For the experiments with noise, the mean values of the errors are tabulated with their variance values given in brackets.

Table 1. Values of E and E_S for L_1 and L_2.

Errors	No noise	With noise
E for L_1	$8.97 * 10^{-4}$	$2.94 * 10^{-2}$ $(1.53*10^{-4})$
E_S for L_1	$1.04 * 10^{-6}$	$1.76 * 10^{-3}$ $(3.58*10^{-6})$
E for L_2	$2.22 * 10^{-3}$	$4.55 * 10^{-2}$ $(2.31*10^{-4})$
E_S for L_2	$5.61 * 10^{-6}$	$3.39 * 10^{-3}$ $(7.89*10^{-6})$

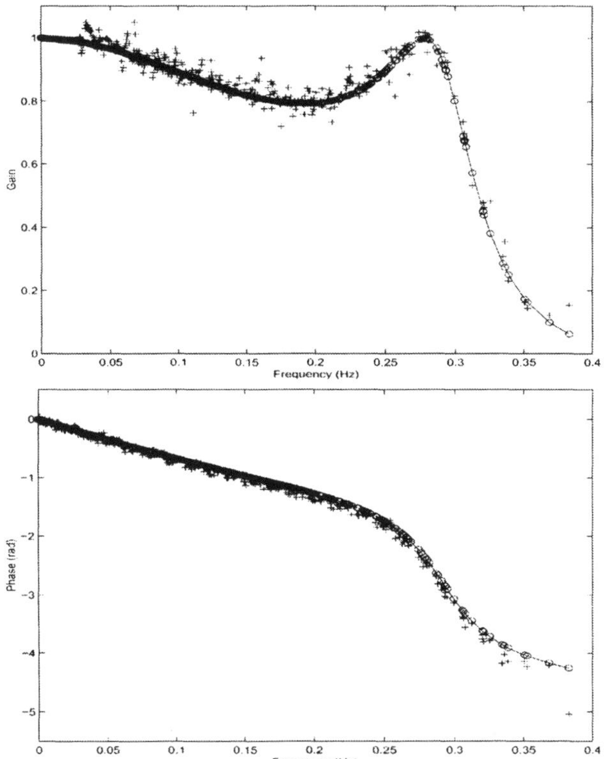

Fig. 3. Frequency response gain (top) and phase (bottom) of L_2. Solid line: Actual values; Circles: Estimates obtained without noise; Plusses: Estimates obtained with noise.

4. ESTIMATION USING LINES DISTORTED BY TYPE I DISTORTION (METHOD 2)

In Evans, *et al.*, (1996), it is shown that Type I contributions cause distortions in the third order kernel where the frequencies $f_1 = -f_2$ or $f_1 = -f_3$ or $f_2 = -f_3$ fall at the input frequencies and cannot be separately measured. These points are normally discarded and not used in the extraction of the frequency response of the linear subsystems. However, it will be shown here that these points can contribute to the estimation of $|L_1|$. To avoid confusion with the previous sections in the paper, the harmonics used will be represented using fv_k, with k denoting the kth element of fv given in (1). Using this notation, and considering all combinations of the input which result in contributions at the input frequencies in the output,

$$|Z(fv_1)| = |L_1(fv_1)||L_2(fv_1)||L_1(fv_1)^2 + L_1(fv_2)^2 + \ldots + L_1(fv_{10})^2||U(fv)|^3$$

$$|Z(fv_2)| = |L_1(fv_2)||L_2(fv_2)||L_1(fv_1)^2 + L_1(fv_2)^2 + \ldots + L_1(fv_{10})^2||U(fv)|^3$$

$$\ldots$$

$$|Z(fv_{10})| = |L_1(fv_{10})||L_2(fv_{10})||L_1(fv_1)^2 + L_1(fv_2)^2 + \ldots + L_1(fv_{10})^2||U(fv)|^3 \quad (11)$$

$|U(fv)|$ is in this case equal to $|U(fv_1)| = |U(fv_2)| = \ldots = |U(fv_{10})|$. From (11)

$$|L_{init}(fv_1)| = |L_1(fv_1)||L_1(fv_1)^2 + L_1(fv_2)^2 + \ldots + L_1(fv_{10})^2| = |Z(fv_1)|/(|L_2(fv_1)||U(fv)|^3)$$

$$|L_{init}(fv_2)| = |L_1(fv_2)||L_1(fv_1)^2 + L_1(fv_2)^2 + \ldots + L_1(fv_{10})^2| = |Z(fv_2)|/(|L_2(fv_2)||U(fv)|^3)$$

$$\ldots$$

$$|L_{init}(fv_{10})| = |L_1(fv_{10})||L_1(fv_1)^2 + L_1(fv_2)^2 + \ldots + L_1(fv_{10})^2| = |Z(fv_{10})|/(|L_2(fv_{10})||U(fv)|^3) \quad (12)$$

From (12), $|L_1(fv_n)|$ is directly proportional to $|L_{init}(fv_n)|$, where n = 1 to 10. ($|L_1(fv_n)| = |L_{init}(fv_n)|/a$.)

$$\frac{|L_{init}(fv_n)|}{a}\left|\left(\frac{L_{init}(fv_1)}{a}\right)^2 + \left(\frac{L_{init}(fv_2)}{a}\right)^2 + \ldots + \left(\frac{L_{init}(fv_{10})}{a}\right)^2\right| = |L_{init}(fv_n)| \quad (13)$$

The value of a (and subsequently $|L_1|$) can be solved using

$$\frac{1}{a}\left|\left(\frac{L_{init}(fv_1)}{a}\right)^2 + \left(\frac{L_{init}(fv_2)}{a}\right)^2 + \ldots + \left(\frac{L_{init}(fv_{10})}{a}\right)^2\right| - 1 = 0 \quad (14)$$

The values of $|L_1|$ obtained using the above method for the system considered in Section 3 are plotted in Figure 4. Unfortunately, a similar method could not be applied to calculate the phase because the input multisine signal does not have a constant phase. The values of the errors obtained are tabulated in Table 2.

Table 2. Values of E and E_S for L_1 with the gain calculated from estimation lines distorted by Type I distortion.

Errors	No noise	With noise
E for L_1	$1.45 * 10^{-2}$	$2.76 * 10^{-2}$ $(8.23*10^{-5})$
E_S for L_1	$2.90 * 10^{-4}$	$1.13 * 10^{-3}$ $(8.46*10^{-7})$

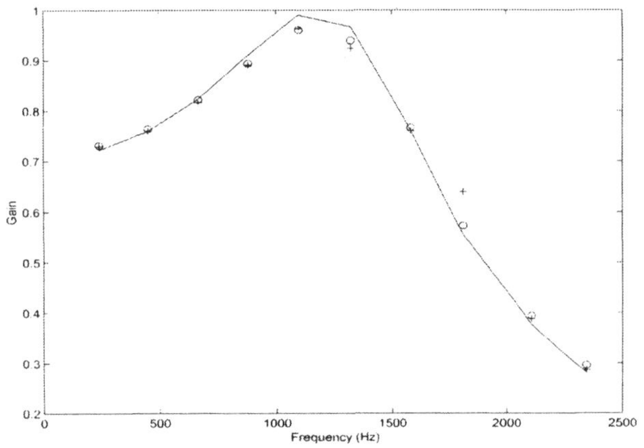

Fig. 4. Frequency response gain of L_1 obtained from the distorted estimation lines. Solid line: Actual values; Circles: Estimates obtained without noise; Plusses: Estimates obtained with noise.

Comparing Tables 1 and 2, it can be seen that in the noiseless situation, the errors are larger using Method 2. However, in the presence of noise, the corresponding errors are smaller. A possible explanation is that the estimation lines have a much larger amplitude than those used in Section 3, due to several different combinations of the input harmonics. This explains both the lower accuracy in the noiseless case (since the distortions cause errors in the estimation process) and the higher accuracy when output noise is present (since the larger amplitude at these estimation lines gives a better signal-to-noise ratio). The discrete Fourier transform magnitude of the output (Z in Figure 1) is illustrated in Figure 5. The larger amplitude of the estimation lines described above can be clearly seen in this figure.

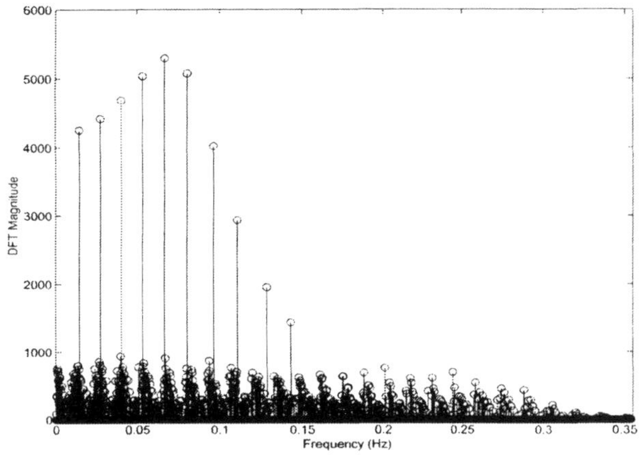

Fig. 5. Discrete Fourier transform magnitude at the output of the Wiener-Hammerstein process.

It should be noted that the nonlinear block will introduce a scaling gain factor into the system. In the simulations conducted in this paper, the scaling factor has been set to unity. This is also assumed in all the equations given, in order to reduce the complexity involved. However, a non-unity scaling factor can be easily incorporated provided its value is known. This does not pose any problem since the information on the nonlinear block is assumed to be known *a priori*.

5. CONCLUSIONS

In this paper, the technique of LIFRED has been extended to identifying the dynamics of the linear subsystems of a Wiener-Hammerstein model from the third order Volterra kernel. The basic ideas are similar to those for the second order case but the analysis is, of necessity, somewhat more complicated. A simulation example was conducted to illustrate the applicability of the method. High accuracy was achieved even in the presence of noise. However, unlike in the second order case, input distortion will cause a greater problem in the measurement of the third order kernel.

It was shown in this paper that the estimation lines which are normally discarded due to Type I distortion could be used to estimate the gain of the first linear subsystem. This is advantageous if the system is noisy. A possible reason for this is that these estimation lines have a larger magnitude due to the contributions of several different combinations of the input harmonics. Hence, they have a larger signal-to-noise ratio and are less susceptible to the effects of noise.

REFERENCES

Bershad, N. J., S. Bouchired & F. Castanie (2000). Stochastic analysis of adaptive gradient identification of Wiener-Hammerstein systems for Gaussian inputs, *IEEE Trans. Signal Processing*, **48**, 557-560.

Billings, S. A. & S. Y. Fakhouri (1977). Identification of nonlinear systems using the Wiener model, *Electron. Lett.*, **13**, 502-504.

Evans, C., D. Rees & L. Jones (1994). Nonlinear disturbance errors in system identification using multisine test signals, *IEEE Trans. Instrum. Meas.*, **43**, 238-244.

Evans, C., D. Rees, L. Jones & M. Weiss (1996). Periodic signals for measuring nonlinear Volterra kernels, *IEEE Trans. Instrum. Meas.*, **45**, 362-371.

Lawrence, P. J. (1981). Estimation of a Volterra functional series of a nonlinear system using frequency response data, *IEE Proc. D*, **128**, 206-210.

Narendra, K. S. & P. G. Gallman (1966). An iterative method for the identification of nonlinear systems using the Hammerstein model, *IEEE Trans. Automat. Contr.*, **11**, 546-550.

Tan, A. H. & K. Godfrey (2002). Identification of Wiener-Hammerstein models using linear interpolation in the frequency domain (LIFRED), *IEEE Trans. Instrum. Meas.*, **51**, 509-521.

Vandersteen, G., Y. Rolain & J. Schoukens (1997). Non-parametric estimation of the frequency-response functions of the linear blocks of a Wiener-Hammerstein model, *Automatica*, **33**, 1351-1355.

Westwick, D. & M. Verhaegen (1996). Identifying MIMO Wiener systems using subspace model identification methods, *Signal Processing*, **52**, 235-258.

IFAC
Publications
www.elsevier.com/locate/ifac

PERFORMANCE INVESTIGATION OF SLICOT WIENER SYSTEMS IDENTIFICATION TOOLBOX

Vasile Sima *,[1]

* *National Institute for Research & Development in Informatics*
Bd. Mareşal Averescu, Nr. 8–10, 71316 Bucharest 1, Romania

Abstract: This paper summarizes the results obtained through a systematic and extensive investigation of the performance of the new system identification toolbox for Wiener-type multivariable discrete-time systems, incorporated in the freely available Fortran 77 **S**ubroutine **L**ibrary **in Co**ntrol **T**heory (SLICOT). This toolbox provides drivers, computational routines, and MATLAB or Scilab interfaces, implementing several algorithmic approaches. The input-output data sets considered in this investigation include those in the DAISY collection, freely available at the site `www.esat.kuleuven.ac.be/sista/daisy`. The results show that SLICOT is reliable, efficient, and able to solve large identification problems. *Copyright © 2003 IFAC*

Keywords: Estimation, Identification for Control, Identification Methods, Signals and Systems, Subspace Methods

1. INTRODUCTION

Finding mathematical models of dynamic systems from measured data, by system identification techniques, is the first, essential step for both system analysis and control system design. The increasing complexity of analysis and synthesis problems requires very reliable and efficient algorithms and associated software for system identification. Reliability, efficiency, and ability to solve large, industrial identification problems received a special consideration in the development of the new, multivariable system identification toolbox—SLIDENT—incorporated in the Fortran 77 **S**ubroutine **L**ibrary **in Co**ntrol **T**heory (SLICOT) (Benner *et al.*, 1999). SLIDENT provides drivers, computational routines, and MATLAB (MathWorks, 1999) or Scilab (Gomez, 1999) interfaces, which implement several algorithmic approaches, as well as standard or fast techniques for data compression and problem solving. Initially, the SLIDENT toolbox addressed linear systems only (Sima, 2000), but it has been recently extended to Wiener-type systems (Schneider *et al.*, 2002), which consist of a linear part and a static nonlinearity.

Linear time-invariant discrete-time systems are identified by SLIDENT tools using subspace-based techniques. These techniques are also used to initialize the algorithms for estimating the linear part of Wiener systems. Both MOESP (Multivariable Output Error state SPace) (Verhaegen, 1993), and N4SID (Numerical algorithm for Subspace State Space System IDentification) (Van Overschee and De Moor, 1994; Van Overschee and De Moor, 1996) approaches are covered, with special emphasis on reliability and efficiency. The block-Hankel structure is exploited using either covariance calculations (Sima, 1999), or displacement rank techniques (Mastronardi *et al.*, 2001). For Wiener systems, the linear part is parameterized using the output normal form (Peeters *et al.*, 1999), and the nonlinearity is modelled by a neural network with one hidden layer. The parameters are estimated by a Levenberg-Marquardt algorithm (Kelley, 1999; Press *et al.*, 1992) tailored for the problem at hand. By a suitable ordering

[1] Partially supported by the European Community BRITE-EURAM III *Thematic Networks Programme NICONET* (project BRRT–CT97-5040)

of the parameters, the Jacobian matrix is put in a block diagonal form with an additional right block column. This structure is exploited. Two implementations of the Levenberg-Marquardt algorithm are provided: a standard implementation, which uses Cholesky factorization, or a conjugate gradients (CG) algorithm, for solving the symmetric positive-definite linear systems involved, and a MINPACK-like (Moré *et al.*, 1980), but LAPACK-based (Anderson *et al.*, 1999), structure exploiting implementation, which uses QR factorization. The second implementation is scaling invariant.

The numerical results obtained by extensive comparisons with the available techniques show that SLICOT system identification toolbox is highly efficient. The included fast QR or Cholesky factorization algorithms enabled to significantly increase the computational efficiency for estimating linear systems (or the linear part of Wiener systems), while preserving the same accuracy as for the standard algorithms. In comparison with MATLAB codes, speed-up factors between 15 and 240 have been obtained (Sima *et al.*, 2002). Exploiting the Jacobian structure in the Levenberg-Marquardt algorithm for Wiener systems identification also produced a remarkable efficiency improvement.

2. WIENER SYSTEMS IDENTIFICATION

A state space representation of a discrete-time Wiener system is given as

$$\begin{aligned} x_{k+1} &= Ax_k + Bu_k, \\ z_k &= Cx_k + Du_k, \\ y_k &= f(z_k) + v_k, \end{aligned} \tag{1}$$

where x_k is the n-dimensional state vector at time k, u_k is the m-dimensional input (control) vector, y_k is the ℓ-dimensional output vector, z_k is the output of the linear part, $\{v_k\}$ is an output disturbance or noise sequence, $f(\cdot)$ is a square nonlinear vector function from $\mathbb{R}^\ell$ to $\mathbb{R}^\ell$, and A, B, C, and D are real matrices of appropriate dimensions. The *identification problem* for (1) asks for finding the system order, n, and the quadruple of system matrices (A, B, C, D) for the linear part, and an approximation of $f(\cdot)$, using an upper bound, s, on n, and the available input and output data sequences, $\{u_k\}$ and $\{y_k\}$, $k = 1{:}N$. It is assumed that the input sequence $\{u_k\}$ is sufficiently persistently exciting as defined, e.g., in (Verhaegen, 1994) and statistically independent from the perturbation $\{v_k\}$.

A systematic three-step approach to solve the Wiener system identification problem was given in (Verhaegen, 1998). The first step identifies the linear part in (1), using a subspace-based approach. The second step finds initial values of the weighting coefficients of a neural network parameterizing the function $f(\cdot)$ in (1). Finally, the parameters of the linear and nonlinear parts are refined by optimization calculations, starting with values corresponding to the results of the first two steps. This approach has been implemented in the SLICOT Library. Specifically, the linear part of the Wiener model (1) is first approximated by subspace techniques, temporarily assuming that $f(\cdot)$ is the identity function. The estimated matrices (A, B, C, D) are used to compute the estimated sequence $\{\widehat{z}_k\}_{k=1}^N$, and this, in turn, is used to start the approximation of the nonlinear part. This part is modelled as a set of ℓ single layer neural networks,

$$\begin{aligned} f_r(z_k) = \sum_{i=1}^{\nu} \Big(\alpha(r,i)\phi\Big(\sum_{j=1}^{\ell} \beta(r,i,j) z_j(k) + b(r,i) \Big) \Big) \\ + b(r,\nu+1) + \varepsilon_r(k), \qquad r = 1{:}\ell, \end{aligned} \tag{2}$$

where $f_r(\cdot)$ and $z_r(k)$ denote the r-th entry of the vector function $f(\cdot)$ and the vector $z(k) := z_k$, respectively, the coefficients $\alpha(r,i)$, $\beta(r,i,j)$, $b(r,i)$, and $b(r,\nu+1)$ are real numbers to be estimated, the integer ν represents the number of neurons, and $\varepsilon(k)$ is the approximation error. The hyperbolic tangent is used as a function $\phi(\cdot)$. All constants α, β, and b in (2) are stacked in the parameter vector θ, where $\theta = \left(\theta_1^T \,|\, \theta_2^T \,|\, \cdots \,|\, \theta_\ell^T \right)^T \in \mathbb{R}^{\ell((\ell+2)\nu+1)}$, with

$$\begin{aligned} \theta_r := \big(\beta(r,1,1), \ldots, \beta(r,\nu,\ell), \alpha(r,1), \ldots, \alpha(r,\nu), \\ b(r,1), \ldots, b(r,\nu+1) \big)^T, \qquad r = 1{:}\ell, \end{aligned}$$

and are estimated by solving the following nonlinear least squares problem

$$\min_\theta \sum_{k=1}^{N} \left\| \begin{bmatrix} y_1(k) - \widehat{y}_1(k) \\ \vdots \\ y_\ell(k) - \widehat{y}_\ell(k) \end{bmatrix} \right\|^2, \tag{3}$$

$$\begin{aligned} \widehat{y}_r(k) := \sum_{i=1}^{\nu} \Big(\alpha(r,i)\phi\Big(\sum_{j=1}^{\ell} \beta(r,i,j) \widehat{z}_j(k) + b(r,i) \Big) \Big) \\ + b(r,\nu+1), \qquad r = 1{:}\ell. \end{aligned} \tag{4}$$

Note that (3) is equivalent to ℓ independent nonlinear least squares problems, which are solved separately. The estimated system matrices of the linear part and the estimated parameters in the vector θ are then used to initialize the optimization of the parameters in the fully parameterized Wiener system with a fixed order of the state vector. To this aim, the linear part is converted to the output normal form (Peeters *et al.*, 1999), and the corresponding $l := n(\ell+m+1) + \ell m$ parameters (including the initial state vector, $x(1)$) are added at the end of the vector θ, getting $\Theta \in \mathbb{R}^c$, $c := \ell((\ell+2)\nu+1) + l$. Define $e := \mathrm{vec}\left([e_1 \cdots e_N]^T \right)$, with $e_k := y_k - \widehat{y}_k$. The Jacobian matrix J of the resulting nonlinear least squares problem has the following structure

$$J = \left[\begin{array}{cccc|c} J_1 & 0 & \cdots & 0 & L_1 \\ 0 & J_2 & \cdots & 0 & L_2 \\ \vdots & \vdots & \ddots & \vdots & \vdots \\ 0 & 0 & \cdots & J_\ell & L_\ell \end{array} \right], \quad J_c = \begin{bmatrix} J_1 & L_1 \\ J_2 & L_2 \\ \vdots & \vdots \\ J_\ell & L_\ell \end{bmatrix}, \tag{5}$$

where $J_r \in \mathbb{R}^{N\times((\ell+2)v+1)}$ and $L_r \in \mathbb{R}^{N\times l}$ are full matrices, corresponding to the nonlinear and linear part, respectively, $r = 1{:}\ell$. The submatrices J_r, $r = 1{:}\ell$, are computed analytically, and the block-matrix $\left[L_1^T \cdots L_\ell^T\right]^T$ is computed by a forward-difference approximation. In implementation, the Jacobian J is stored in a compressed form, J_C in (5); specifically, the matrices $J_1, \ldots, J_\ell$ are concatenated in the first block column of an array, and the matrices $L_1, \ldots, L_\ell$ are stored in the second block column of that array.

The full nonlinear least squares problem is written as (3), with θ replaced by Θ, and it is no longer separable. This problem, as well as the ℓ separate problems in (3), are solved by a Levenberg-Marquardt algorithm. The version of the algorithm described in (Moré, 1978; Moré *et al.*, 1980), optimizes the variables Θ, using the values of the *error functions*, e, and of the Jacobian matrix J. A "trust region" method is used. The algorithm tries to update Θ by the formula $\Theta \leftarrow \Theta - p$, using an approximate solution of the system of linear equations

$$(J^T J + \lambda DD)p = J^T e, \tag{6}$$

where $\lambda \geq 0$ is the Levenberg-Marquardt parameter, p is the correction vector, $D \in \mathbb{R}^{c\times c}$ is a diagonal nonsingular scaling matrix, and either

$$\lambda = 0, \quad \text{and} \quad \left(\|Dp\| - \delta\right) \leq 0.1\delta, \quad \text{or} \tag{7}$$

$$\lambda > 0, \quad \text{and} \quad \left|\,\|Dp\| - \delta\,\right| \leq 0.1\delta, \tag{8}$$

with δ the radius of the trust region. The MINPACK implementation in (Moré *et al.*, 1980), which uses the QR factorization (with column pivoting) of J, cannot generally be afforded for Wiener systems identification, due to the large computational effort involved. In the SLICOT implementation of the QR-based algorithm, a special QR factorization with block column pivoting is used. The procedure consists in two phases. In the first phase, the algorithm uses standard QR factorizations with column pivoting for each block J_r in (5), $J_r P_r = Q_r\left[R_r^T\ 0\right]^T$, $r = 1{:}\ell$, and applies the orthogonal matrix Q_r^T to the matrix L_r of the last block column and to part of e. After all block rows have been processed, suitable block rows of J are interchanged so that the zeroed submatrices in the first ℓ block columns are moved to the bottom part. The same block row permutation Z is also applied to the updated vector e. At the end of the first phase, the structure of the processed matrix J is

$$\left[\begin{array}{cccc|c} R_1 & 0 & \cdots & 0 & L_1^{(1)} \\ 0 & R_2 & \cdots & 0 & L_2^{(1)} \\ \vdots & \vdots & \ddots & \vdots & \vdots \\ 0 & 0 & \cdots & R_\ell & L_\ell^{(1)} \\ 0 & 0 & \cdots & 0 & L_1^{(2)} \\ \vdots & \vdots & \ddots & \vdots & \vdots \\ 0 & 0 & \cdots & 0 & L_\ell^{(2)} \end{array}\right],$$

where R_r are square upper triangular, $r = 1{:}\ell$. In the second phase, the submatrix $L_{1:\ell}^{(2)}$ is triangularized using an additional QR factorization with pivoting, $L_{1:\ell}^{(2)} P_{\ell+1} = Q_{\ell+1}\left[R_{\ell+1}^T\ 0\right]^T$, and the last part of e is updated; the columns of the submatrix $L_{1:\ell}^{(1)}$ are also permuted accordingly. Therefore, the column pivoting is restricted to each such local block column. In implementation, both J and R are stored in the compressed form mentioned above, and R overwrites J; an additional block, corresponding to $R_{\ell+1}$, is needed in the second block-column. The returned matrix R is

$$R = \begin{bmatrix} R_1 & \widetilde{L}_1 \\ R_2 & \widetilde{L}_2 \\ \vdots & \vdots \\ R_\ell & \widetilde{L}_\ell \\ 0 & R_{\ell+1} \end{bmatrix}, \tag{9}$$

where $\widetilde{L}_r = L_r^{(1)} P_{\ell+1}$, $r = 1{:}\ell$, but the zero submatrix is not set. For efficiency of the later calculations, matrix R is delivered in a two-dimensional array with the leading dimension c, possibly much smaller than ℓN, the number of rows of J in (5). If $\ell = 1$, the matrix J is triangularized in a single phase, by one QR factorization with pivoting. In this case, the column pivoting is global. All calculations are based on this special factorization and storage scheme. The corresponding MEX-file is called `wident`.

More details about the algorithm are given below. The implemented Levenberg-Marquardt algorithm has an outer loop; in each iteration, the Jacobian matrix is evaluated and its QR factorization is computed, that is, $JP = QR$, where P is a block permutation matrix, Q has orthogonal columns, and R is an upper triangular matrix with diagonal elements of nonincreasing magnitude for each diagonal block of $J^T J$. On the first iteration, the initial approximation of the solution is scaled with some factors d_i, hence $\Theta_i \leftarrow d_i\Theta_i$, $d_i > 0$, $i = 1{:}c$, where d_i are either given, or are set internally to the column norms of the initial Jacobian matrix. (In the next iterations, d_i is updated with the maximum norm of the i-th column found so far, if internal scaling was chosen.) Also, the step bound δ is initialized as a constant multiple of the norm of the scaled Θ. Then, an inner loop is started, which updates at each iteration the Levenberg-Marquardt parameter λ, the direction p, and the vector $RP^T p$. The norm of the scaled correction p, the scaled predicted reduction of the sum of squares (found using $RP^T p$), and the scaled directional derivative are computed. The error functions are evaluated at $\Theta - p$ and the scaled actual reduction is calculated. The ratio of the actual to the predicted reduction is used to assess the progress of the iterations, and to decide how to proceed, by updating λ, the step bound δ, the approximate parameter vector Θ, the error functions e, and their norms.

Consider now the computation of the Levenberg-Marquardt parameter. Briefly speaking, a non-negative value for λ is sought such that if p solves the system

$$Jp = e, \quad \sqrt{\lambda} D p = 0, \tag{10}$$

in the least squares sense, then either (7) or (8) are satisfied. Based on the block QR factorization of J, $JP = QR$, an upper triangular matrix S, such that

$$P^T(J^T J + \lambda DD)P = S^T S, \tag{11}$$

is computed and used in the solution process. The matrix S has the same structure as R.

The algorithm for solving the special optimization problem corresponding to (10) first computes the Gauss-Newton direction, i.e., the solution p of (10) for $\lambda = 0$. An approximate basic least squares solution is found if the Jacobian is rank deficient. If the Gauss-Newton direction is not acceptable, then an iterative Newton algorithm updates the given λ using improved lower and upper bounds for λ. Only a few iterations are generally needed for convergence of the algorithm. (Since these iterations use the same matrix R, the full upper triangles of R_k, $k = 1:\ell+1$, and the submatrices $\widetilde{L}_k$, $k = 1:\ell$, are preserved, the transposed of the strict upper triangles of the blocks S_k are stored in the corresponding strict lower triangles of R_k, $k = 1:\ell+1$, and the diagonals are stored elsewhere.) If, however, the limit of 10 iterations is reached, then the returned λ will contain the best value obtained so far. If the Gauss-Newton step is acceptable, it is stored in p, and λ is set to zero, hence $S = R$.

Let Q_+ denote the orthogonal matrix in the QR factorization of J (with Q having the first c columns of Q_+), and let $\widetilde{e}_1$ denote the first c elements of $Q_+^T e$. Then, solving the system (10) in the least squares sense is equivalent to solving

$$Rq = \widetilde{e}_1, \qquad P^T \widetilde{D} P q = 0, \tag{12}$$

where $\widetilde{D} = \sqrt{\lambda} D$, and $p = Pq$. If this system has not full rank, then an approximate least squares solution is computed. Using (11), S can equivalently be obtained from a special QR factorization of the matrix $\left[R^T \;\; P^T \widetilde{D} P\right]^T$, hence the system (12) is equivalent to $Sq = \bar{e}_1$, where $\bar{e}_1$ contains the first c components of the vector resulted by applying to $\left[\widetilde{e}_1^T \;\; 0\right]^T$ the transformations which produced S. Specifically, standard plane rotations are used to annihilate the elements of the matrix $P^T \widetilde{D} P$, updating the upper triangular matrix R and $\left[\widetilde{e}_1^T \;\; 0\right]^T$. A basic least squares solution is computed. If one or more of the submatrices S_k, $k = 1:\ell+1$, is singular, then the computed result is not the basic least squares solution for the whole problem, but a concatenation of (least squares) solutions of the individual subproblems involving R_k, $k = 1:\ell+1$.

The standard Levenberg-Marquardt algorithm solves at each iteration the positive definite linear system of equations in (6) with $D = I_c$. The matrix $J^T J + \lambda I_c$ has an additional last nonzero block-row, comparing to J in (5), but the block-diagonal part has normally much smaller blocks than (5) (since $(\ell+2)\nu + 1 \ll N$). The conjugate gradients algorithm does not build the matrix $J^T J$, but uses matrix-vector multiplications, with matrices J and J^T. The Cholesky-based algorithm computes the matrix $J^T J + \lambda I_c$, exploiting its structure, and then factorizes this matrix, which usually has relatively small order. These calculations are implemented in several Fortran routines and the MEX-file `widentc`. Parameter scaling was not included, and the algorithm structure is simpler than for the block QR factorization-based algorithm described above.

Major advantages of the SLIDENT toolbox include increased reliability and efficiency. These follow from the use of structure-exploiting algorithms and certified linear algebra codes. Condition number estimates are provided for essential intermediate results. The fast algorithms used enable to speed-up the calculations by one to three orders of magnitude.

3. PERFORMANCE INVESTIGATION OF WIENER SYSTEMS IDENTIFICATION TOOLBOX

Some results of a systematic and extensive performance investigation of the SLICOT Wiener systems identification toolbox are summarized in this section. The input-output data sets considered in this investigation include the sets from the DAISY collection, available from the site `www.esat.kuleuven.ac.be/sista/daisy`. DAISY contains several large data sets from various domains. Performance results for SLICOT linear systems identification tools have been presented in (Sima and Van Huffel, 2001), where 22 applications have been dealt with. The same applications are considered here, but their description is omitted, due to the space limitations. The numerical results shown below have been obtained on an IBM PC computer at 500 MHz, with 128 Mb memory, using Compaq Visual Fortran V6.5, and non-optimized BLAS. MATLAB 6.1 has been employed.

SLICOT combination of MOESP and N4SID techniques (`slmoen4`) with fast QR factorization (Mastronardi *et al.*, 2001) was used to identify the linear part. For most applications, the number of neurons for the hidden layer was chosen as $\nu = 12$. The initial values for the parameters of the nonlinear part were generated randomly using the LAPACK uniform $(-1,1)$ random number generator with the initial seed set to $[1998, 1999, 2000, 2001]$, to guarantee the use of the same starting point for various algorithms. Tolerances for the Levenberg-Marquardt algorithm were normally chosen to be 10^{-4}. (Two tolerances are used, one for the initialization of the parameters of the nonlinear part, and the other one for the whole optimization.) Since the data sets contain noise, smaller tolerances are rarely useful, and their effect will be an unreasonable number of iterations, without sensible improvement of the accuracy. The number of samples in the estimation set, N, was chosen either $t/2$ or t, where t denotes the total number of available samples; the largest value was $N = 49998$, for Applications 21 (with the application numbering in (Sima and Van Huffel, 2001)). The number of estimated parameters varied between 50, for Applications 12 and 15, and 977, for Application 16, for which only the first 7 outputs (from 28) have been considered. When several experiments have been performed for the same application (e.g., with different tolerances), the mean values of the results were plotted.

Figure 1 shows the execution times for the MEX-files `widentc` and `widentc` with options Cholesky or CG. Since the execution times vary with several orders of magnitude, their logarithms (base 10) were plotted. The size of the estimation set for Applications 1 and 3 equals the corresponding total number of samples, t.

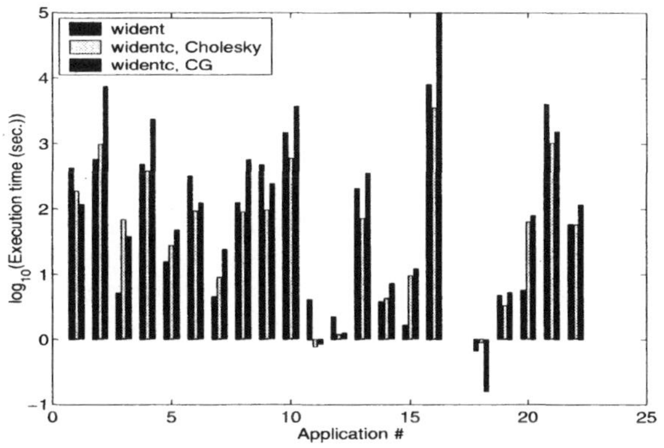

Fig. 1. Decimal logarithms of the execution times in seconds for solving the Wiener system identification problem for various applications.

Figure 2 shows the sums of squares of the prediction errors for both linear and Wiener system identification. The errors are computed for all available samples, t, even if only half of them were used for parameter estimation (except for Applications 1 and 3).

For most applications, the MINPACK-like SLIDENT implementation was highly efficient, and produced more accurate results (for the same tolerances) than the standard Levenberg-Marquardt algorithm, using Cholesky- or conjugate gradients-based linear systems solvers. But most often, the Cholesky-based implementation was faster. For instance, the execution times needed for Wiener system identification for Application 16, corresponding to a steel subframe flexible structure ($t = 8523$, $n = 20$, $m = 2$, $\ell = 28$, but the first 7 outputs only have been used), were 7956.51, 3481.84, and 98595.72 seconds, for `wident`, `widentc` with Cholesky, and `widentc` with CG, respectively. The corresponding optimization problem has 977 variables, and $7 \times 4261 = 29827$ nonlinear error functions, since $N = t/2$. The error

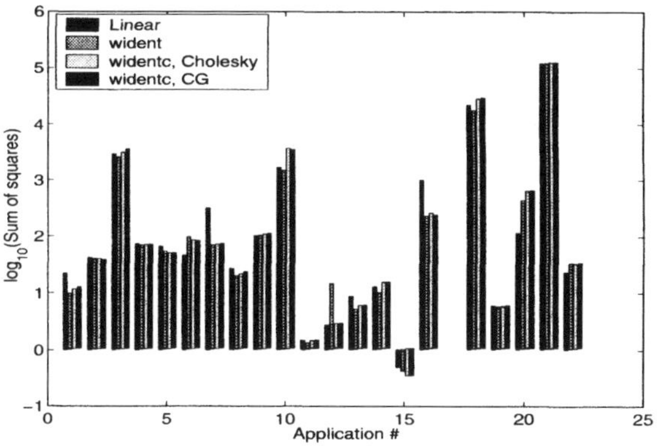

Fig. 2. Decimal logarithms of the sums of squares of the prediction errors for solving the linear and Wiener system identification problem for various applications.

norms for the whole time horizon were `2.25e+2`, `2.49e+2`, and `2.26e+2`, respectively, compared to `1.00e+3`, for the linear model. Hence, the faster Cholesky implementation was less accurate.

Figures 3 and 4 show the trajectories of prediction error norms and their mean values, respectively, for both linear and Wiener systems identification, for Application 16. The mean values have been computed on a moving window with a length of 40 samples. The plotted trajectories correspond to the MEX-file `wident`. Clearly, the Wiener model significantly reduces the prediction error, and has a smoothing effect.

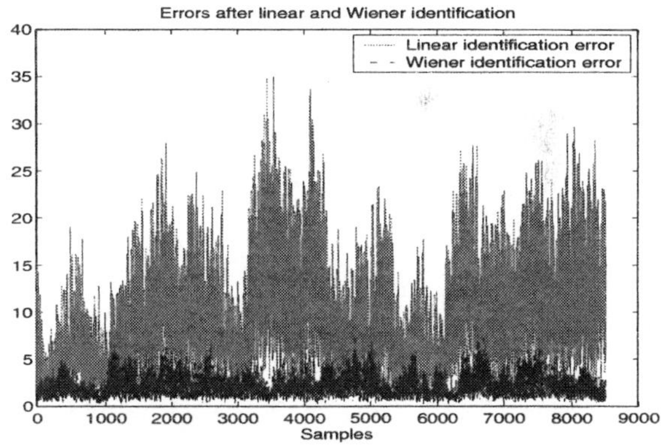

Fig. 3. Prediction error norms for Application 16 for linear and Wiener system identification ($t = 8523$, $N = t/2$, $c = 977$, the first 7 outputs only).

4. CONCLUSIONS

Algorithmic and numerical issues related to Wiener multivariable system identification have been described. The techniques are implemented in the new system identification toolbox for the SLICOT Library. The results show that this toolbox is reliable, efficient, and powerful enough to solve large identification problems.

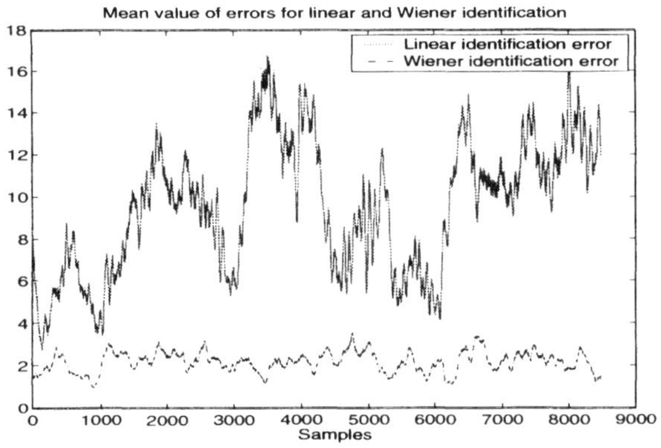

Fig. 4. Mean values of prediction error norms for Application 16 for linear and Wiener system identification.

5. ACKNOWLEDGEMENTS

The development of the SLIDENT toolbox benefited by the support of S. Van Huffel, M. Verhaegen, and A. Varga. R. Schneider and A. Riedel wrote the initial version of the CG-based implementation.

REFERENCES

Anderson, E., Z. Bai, C. Bischof, S. Blackford, J. Demmel, J. Dongarra, J. Du Croz, A. Greenbaum, S. Hammarling, A. McKenney and D. Sorensen (1999). *LAPACK Users' Guide: Third Edition*. Software · Environments · Tools. SIAM. Philadelphia.

Benner, P., V. Mehrmann, V. Sima, S. Van Huffel and A. Varga (1999). SLICOT — A subroutine library in systems and control theory. In: *Applied and Computational Control, Signals, and Circuits* (B. N. Datta, Ed.). Vol. 1, chapter 10. pp. 499–539. Birkhäuser, Boston.

Gomez, C. (Ed.) (1999). *Engineering and Scientific Computing with Scilab*. Birkhäuser, Boston.

Kelley, C. T. (1999). *Iterative Methods for Optimization*. SIAM. Philadelphia.

Mastronardi, N., D. Kressner, V. Sima, P. Van Dooren and S. Van Huffel (2001). A fast algorithm for subspace state-space system identification via exploitation of the displacement structure. *J. Comput. Appl. Math.* **132**(1), 71–81.

MathWorks (1999). *Using MATLAB. Version 5.*

Moré, J. J. (1978). The Levenberg-Marquardt algorithm: Implementation and theory. In: *Numerical Analysis* (G. A. Watson, Ed.). Vol. 630 of *Lecture Notes in Mathematics*. pp. 105–116. Springer-Verlag. Berlin, Heidelberg and New York.

Moré, J. J., B. S. Garbow and K. E. Hillstrom (1980). User's guide for MINPACK-1. Report ANL-80-74. Applied Math. Division, Argonne National Laboratory. Argonne, Illinois.

Peeters, R., B. Hanzon and M. Olivi (1999). Balanced realizations of discrete-time stable all-pass systems and the tangential Schur algorithm. In: *Proceedings of the European Control Conference 31 August–3 September 1999, Karlsruhe, Germany.*

Press, W. H., S. A. Teukolsky, W. T. Wetterling and B. P. Flannery (1992). *Numerical Recipes. The Art of Scientific Computing*. Second ed. Cambridge University Press. New York.

Schneider, R., A. Riedel, V. Verdult, M. Verhaegen and V. Sima (2002). SLICOT system identification toolbox for nonlinear Wiener systems. SLICOT Working Note 2002-6. Katholieke Universiteit Leuven (ESAT/SISTA). Leuven, Belgium. URL `ftp://wgs.esat.kuleuven.ac.be/pub/WGS/REPORTS/SLWN2002-6.ps.Z`, 26 pages.

Sima, V. (1999). Cholesky or QR factorization for data compression in subspace-based identification ? In: *Proceedings of the Second NICONET Workshop on "Numerical Control Software: SLICOT, a Useful Tool in Industry", December 3, 1999, INRIA Rocquencourt, France*. pp. 75–80.

Sima, V. (2000). SLICOT linear systems identification toolbox. SLICOT Working Note 2000-4. Katholieke Universiteit Leuven (ESAT/SISTA). Leuven, Belgium. URL `ftp://wgs.esat.kuleuven.ac.be/pub/WGS/REPORTS/SLWN2000-4.ps.Z`, 30 pages.

Sima, V., D. M. Sima and S. Van Huffel (2002). SLICOT System Identification Software and Applications. In: *Proceedings of the 2002 IEEE International Conference on Control Applications and IEEE International Symposium on Computer Aided Control System Design, CCA/CACSD 2002, September 18–20, 2002, Glasgow, Scotland, U.K.* pp. 45–50.

Sima, V. and S. Van Huffel (2001). Performance investigation of SLICOT system identification toolbox. In: *Proceedings of the European Control Conference, ECC 2001, 4–7 September, 2001, Porto, Portugal*. pp. 3586–3591.

Van Overschee, P. and B. De Moor (1994). N4SID: Two subspace algorithms for the identification of combined deterministic-stochastic systems. *Automatica* **30**(1), 75–93.

Van Overschee, P. and B. De Moor (1996). *Subspace Identification for Linear Systems : Theory – Implementation – Applications*. Kluwer Academic Publishers. Boston/London/Dordrecht.

Verhaegen, M. (1993). Subspace model identification. Part 3: Analysis of the ordinary output-error state-space model identification algorithm. *Int. J. Control* **58**(3), 555–586.

Verhaegen, M. (1994). Identification of the deterministic part of MIMO state space models given in innovation form from input-output data. *Automatica*, **30**(1), 61–74.

Verhaegen, M. (1998). Identification of the temperature-product quality relationship in a multi-component distillation column. *Chemical Engineering Communications* **163**, 111–132.

IFAC Publications
www.elsevier.com/locate/ifac

RATIONAL BASES GENERATED BY BLASCHKE PRODUCT SYSTEMS

F. Schipp * **and J. Bokor** **

* *Department of Numerical Analysis*
Eötvös Loránd University
H-1117 Budapest, Pázmány Péter sétány 1/D

** *Computer and Automation Research Institute*
Hungarian Academy of Sciences
H-1111 Budapest, Kende u. 13–17

Abstract: This paper presents an approach to generate rational orthogonal bases in $\mathcal{H}_2$ that can be used in signal modelling and identification. This approach is based on defining Blaschke product systems and the bases are generated by function compositions that act as dilation on $\mathcal{H}_2$ when compared to previous approaches where shift operators are used. The coefficients of a finite order approximating model can be computed by FFT. *Copyright © 2003 IFAC*

Keywords: Rational bases, Blaschke functions, Fourier representation

1. INTRODUCTION

In signal processing, control, communication or system identification signals are usually represented as elements of a certain space like ℓ_2 for discrete time energy constrained or ℓ_∞ for amplitude constrained signals. A signal representation depends on the choice of a particular basis in these spaces. Bases that are constructed by using a unit element e_0 of a function space and a shift operator $\mathcal{S}$ acting on the basis elements as $e_1 = \mathcal{S}e_0, e_2 = \mathcal{S}e_1, \dots$ are typical construction leading to the standard or trigonometric bases. In $\ell_2[0,\infty)$ this corresponds to fix a unit sequence $e_0 = (1,0,0\dots)^T$, $e_1 = (0,1,0,\dots)^T$ and to specify the shift operator with matrix representation $\mathcal{S} = (e_1, e_2, \dots)$. It is known that using this signal representation, a bounded linear operator G that maps an input signals to an output signal space can be characterized as those that commute with the shift, i.e. $G\mathcal{S} = \mathcal{S}G$ and its matrix representation is Toeplitz. The corresponding results in $\mathcal{H}_2$ can be obtained by using the isomorphism generated by the $\mathcal{Z}$-transform. In the above situation signals are represented as $\mathcal{Z}$-transforms of their ℓ_2 representations and bounded linear operators appear as element of the Hardy space $\mathcal{H}_\infty$ acting by multiplication.

In the years of nineties several authors examined the construction and use of rational bases in $\mathcal{H}_2$. The simplest case is to choose a first order rational function and use the Blaschke-shift represented by first order inner functions. All elements of this so-called discrete Laguerre-basis are generated by multiplying the a basis element by the shift. The generalization to this idea appeared in the papers (Heuberger, 1991; Heuberger et al., 1990, 1991, 1995), and its use in system identification in (Akcay ,2002; van den Hof et al., 1995; Schipp and Bokor, 1997, 1998; Wahlberg, 1991, 1994; Milanese and Taragna, 1999) in relation to using linearly parameterized rational models. Important results are published in (Ninness and Gustafsson, 1997; Ninness et al., 1998; Casini et al., 2001;

Fisher and Medvegyev, 1998). An associated system theory has been elaborated by (Szabó and Bokor, 1997; de Hoog et al., 2001, 2002) and others.

The above approaches are based on the use of shift operator to generate rational bases. This paper offers an alternative for generating orthogonal bases in $\mathcal{H}_2$. This is based on defining product systems using Blaschke function compositions that can be interpreted as using dilation instead of the shift operators. The resulting rational orthogonal basis have several good properties, e.g. the Fourier coefficients of a function represented in this basis can be computed by FFT, and the Fourier series interpolate the function at the measurement points.

The paper defines a product systems first and then describes the representation of $\mathcal{H}_2$ signals in the resulting orthogonal basis. Results for construction of the associated FFT algorithm is are given, too.

2. PRODUCT SYSTEMS IN $\mathcal{H}_2$

For the definition of product systems lets start with the basis in $\mathcal{H}_2$ given by $e_m(z) = z^m$ $(m \in \mathbb{N} := \{0, 1, \dots\},\ z \in \mathbb{C})$.

Namely, the multiplication by $e_1(z) = z$ can be identified as the shift on $\mathcal{H}_2$. The power functions e_m and the powers of the shift $\mathcal{S}^m$ can be generated by the functions $\phi_n := e_{2^n}$ $(n \in \mathbb{N})$ and by the corresponding subsystem $\mathcal{T}_n := \mathcal{S}^{2^n}$ $(n \in \mathbb{N})$. If the numbers $m \in \mathbb{N}$ are represented in binary form

$$m = \sum_{n=0}^{\infty} m_n 2^n \quad (m_n \in \{0, 1\}) \tag{1}$$

then $\mathcal{S}^m$ can be written as

$$e_m = \prod_{n=0}^{\infty} \phi_n^{m_n}, \quad \mathcal{S}^m = \prod_{n=0}^{\infty} \mathcal{T}_n^{m_n}. \tag{2}$$

The system of function $(e_m, m \in \mathbb{N})$ is usually called as *the product system of* $(\phi_n, n \in \mathbb{N})$.

There is another important property utilized in a subsequent FFT-algorithm of the system $(e_m, m \in \mathbb{N})$. This is that the generating functions $(\phi_n, n \in \mathbb{N})$ can be obtained by function compositions starting from $A(z) := \phi_1(z) = z^2$ $(z \in \mathbb{C})$ as

$$\begin{aligned} \phi_0(z) &= z, \\ \phi_n &= \phi_{n-1} \circ A = A \circ A \circ \cdots \circ A, \end{aligned} \tag{3}$$

here $n \in \mathbb{N}^* := \{1, 2, \dots\}$. Concerning the Blaschke-shift operator, consider the function

$$\begin{aligned} B_b(z) &:= \epsilon(b) \frac{z - b}{1 - \bar{b}z}, \\ \epsilon(b) &:= \frac{1 - \bar{b}}{1 - b} \quad (z, b \in \mathbb{C}). \end{aligned} \tag{4}$$

The scaling factor $\epsilon(b)$ results in normalizing condition $B_b(1) = 1$. It is known that in case of $b \in \mathbb{D} := \{z \in \mathbb{C} : |z| < 1\}$ the functions B_b map the unit disc and its boundary $\mathbb{T} := \{z \in \mathbb{C} : |z| = 1\}$ onto itself, and the class of functions

$$\mathcal{B}_1 := \{B_b : b \in \mathbb{D}\}$$

form a non commutative group with respect to the operation of function compositions. The unit element of this group is the identity map $B_0(z) = z$ $(z \in \overline{\mathbb{D}} := \mathbb{D} \cup \mathbb{T})$ that corresponds to the parameter $b = 0$ and the inverse of an element B_b is the inverse of the function B_b corresponding to the parameter $-\epsilon(b)b$.

The most frequently used rational orthogonal bases like Laguerre, Kautz, Takenaka systems can be obtained as finite products of the above defined functions. Most specifically, defining all the parameters as $b = 0$ one obtains the standard basis $(e_m, m \in \mathbb{N})$.

The class of system is closed under function composition but it is not closed under function multiplication. To achieve both one introduce the set of n-times products

$$\mathcal{B}_n := \{f_1 f_2 \cdots f_n : f_j \in \mathcal{B}_1 \ (j = 1, \dots, n)\}.$$

Denote by $\mathcal{B}_0$ the set of constant functions having values in $\mathbb{T}$. Then it can be proved that the class of functions

$$\mathcal{B} := \cup_{n=0}^{\infty} \mathcal{B}_n$$

will be closed under both operations. This class represents a *subclass of inner functions that are closed under function composition and multiplication.*

Generalizing the above constructions let start with second order Blaschke products $A_n := B_{a_n} B_{-a_n}$ $(n \in \mathbb{N})$ with parameters $a_n \in \mathbb{D}$ $(n \in \mathbb{N})$. Since $A_n(z) = B_{b_n}(z^2)$, these functions represent a twofold map of the boundary of the unit circle $\mathbb{T}$ onto itself, i.e. for every $z \in \mathbb{T}$ there exist two elements $z_1, z_2 = -z_1 \in \mathbb{T}$ such that $A_n(z_1) = A_n(z_2) = z$. Consider now the Blaschke function $A_0 := B_{a_0}$ and the functions ϕ_n obtained by function compositions:

$$\begin{aligned} \phi_0 &:= A_a, \\ \phi_n &:= A_n \circ A_{n-1} \circ \cdots \circ A_1 \circ A_0 \\ &= A_n \circ \phi_{n-1} \quad (n \in \mathbb{N}^*). \end{aligned} \tag{5}$$

Then the generators ϕ_n of the product system are the above Blaschke products consisting 2^n Blaschke factors. It is obvious that the standard $e_m(z) = z^m$ $(m \in \mathbb{N},\ z \in \mathbb{C})$ system can be obtained by choosing $a_n = 0$.

In the general case the number of points used for the computation of the coefficients in the generated basis is 2^n and these are given by

$$\begin{aligned} X_n &:= \phi_n^{-1}(1) \\ &= \{z \in \mathbb{T} : \phi_n(z) = 1\} \quad (n \in \mathbb{N}) \end{aligned} \tag{6}$$

The number of elements in X_n is 2^n and $A_n(1) = 1$ $(n \in \mathbb{N})$ implies

$$X_n = \phi_{n-1}(A_n^{-1}(1)) = \qquad (7)$$
$$= \phi_{n-1}^{-1}(1) \cup \phi_{n-1}^{-1}(x_n) = X_{n-1} \cup X'_{n-1},$$

where $x_n \in \mathbb{T}$ is the complex number satisfying $x_n \neq 1$ and $A_n(x_n) = 1$. In this case set X_n can be derived from X_{n-1} by adding a 2^{n-1}-element point set. Denote the product system of $\phi = (\phi_n, n \in \mathbb{N})$ by

$$\psi_m := \prod_{j=0}^{\infty} \phi_j^{m_j} \quad \Big(m = \sum_{j=0}^{\infty} m_j 2^j \in \mathbb{N}\Big). \qquad (8)$$

We will show that product system $(\psi_m,\ 0 \leqq m < 2^N)$ is orthonormal with respect to scalar products

$$[f,g]_N := \int_{X_N} f\overline{g}\, d\mu_N = 2^{-N} \sum_{x \in X_N} f(x)\overline{g(x)}, \qquad (9)$$

where μ_N is a discrete measure on set X_N defined as $\mu_N(\{x\}) := 2^{-N}$.

Several rational systems applied in control theory can be derived on the basis of the above procedure. With choices $A_0(z) = z$, $A_n(z) := z^2$, $(z \in \mathbb{T},\ n \in \mathbb{N}^*)$ we get the *trigonometric system* $\psi_m(z) = z^m$ $(m \in \mathbb{N}, z \in \mathbb{T})$ as a product system.

The *discrete Laguerre system* corresponding to parameter $a \in \mathbb{D}$ – not considering factor $\sqrt{1-|a|^2}/(1-\bar{a}z)$ – is similar to the product system with choices $A_0 := B_a$, $A_n(z) := z^2$ $(z \in \mathbb{T},\ n \in \mathbb{N}^*)$, namely

$$\psi_m(z) = (z-a)^m/(1-\bar{a}z)^m \quad (m \in \mathbb{N}, z \in \mathbb{T}).$$

Similarly, the *Malmquist–Takenaka system* generated by the periodic zeros $a_n \in \mathbb{D}$, $a_{n+N} = a_n$ $(n \in \mathbb{N})$ can be decomposed into N product systems that can be generated separately using the above procedure. Specifically, this also holds for *Kautz systems*.

Using the definition of product systems and applying induction the kernel functions

$$D_{2^N}(x,t) := \sum_{k=0}^{2^N-1} \psi_k(x)\overline{\psi_k(t)} \quad (x,t \in X, N \in \mathbb{N}) \qquad (10)$$

of the product system can be written as

$$D_{2^N}(x,t) = \prod_{j=0}^{N-1} (1 + \phi_j(x)\overline{\phi_j(t)}). \qquad (11)$$

This is a generalization of *Paley*'s identity concerning Dirichlet kernels of Walsh system. Orthogonality of discrete product system can also be derived from the product representation (11) of the kernel function.

In the following theorem we summarize the basic properties of product systems.

Theorem 1. Formula (11) holds for the kernel functions of index 2^N of any product systems. The product system (8) is orthogonal with respect to the scalar product (9), further, the kernel functions can be written as

$$D_{2^N}(x,t) = \begin{cases} 2^N & (x = t, x,t \in X_N), \\ 0 & (x \neq t, x,t \in X_N). \end{cases} \qquad (12)$$

Partial sums corresponding to indices 2^N of the expansion of an arbitrary function $f : \mathbb{T} \to \mathbb{C}$ reproduces the function f in points of X_N:

$$(S_{2^N} f)(x) = f(x) \quad (x \in X_N). \qquad (13)$$

Equation (13) expresses the fact that partial sums $S_{2^N} f$ interpolates the function f in the points of X_N.

3. FAST FOURIER TRANSFORM FOR RATIONAL PRODUCT SYSTEMS

For computing Fourier coefficients

$$c_n := \Big[f, \psi_n\Big]_N = 2^{-N} \sum_{x \in X_N} f(x)\overline{\psi_n(x)}$$
$$(n = 0, 1, \ldots, 2^N - 1)$$

one need to perform $2^N 2^N = 2^{2N}$ multiplications and $2^N(2^N - 1)$ additions (neglecting the computing of function values and the normalization by 2^N). Function f can be reconstructed in the points of X_N based on Fourier coefficients and using the following formula that is equivalent to (13)

$$f(x) = \sum_{n=0}^{2^N-1} c_n \psi_n(x) \quad (x \in X_N).$$

The number of necessary calculations is just the same as in the preceding case.

Using the two special properties of ψ_n $(0 \leqq n < 2^N)$ mentioned in the introduction one can construct an algorithm for computing Fourier coefficients and reconstructing function f in $O(N2^N)$ steps.

Denote $X_N^n := T_n(X_N)$ $(n = 0, 1, \ldots, N)$ the image of X_N by the mapping T_n. Then

$$X_N^N = \{1\},$$
$$X_N^n := A_n(X_N^{n-1}) \quad (n = 1, 2, \ldots, N)$$

and the number of elements in X_N^n is 2^{N-n}. Elements of $X_N^n := \{x_k^n : 0 \leqq k < 2^{N-n}\}$ are indexed such a way that

$$A_n(x_{2k}^{n-1}) = A_n(x_{2k+1}^{n-1}) = x_k^n$$
$$(0 \leqq k < 2^{N-n}, n = 1, 2, \ldots, N)$$

is fulfilled. Easy to prove that using such indexing the following equality holds

$$T_n(x_k^0) = x^n_{[k2^{-n}]} \quad (0 \leqq k < 2^N, 0 \leqq n \leqq N), \tag{14}$$

where $[r]$ denotes the integer part of real number r. Using the above notations and relationships we define a recursion for $Y^n_{k,\ell}$ $(0 \leqq k < 2^{N-n}$, $0 \leqq \ell < 2^n)$ with which the Fourier coefficients corresponding to the product system can be computed in $O(N2^N)$ steps. This algorithm – that contains the classical FFT algorithm as a special case – is described in the following theorem.

Theorem 2. Starting from the sequence $A_n : X \to X$ $(n \in \mathbb{N}^*)$ of twofold mappings let us introduce the discrete orthogonal product system ϕ_n using equations (5) and (8). For the initial values corresponding to the parameter $n = 0$ set

$$Y^0_{k,0} := f(x^0_k) \quad (0 \leqq k < 2^N), \tag{15}$$

and define the sequence $Y^n_{k,\ell}$ recursively with respect to n by

$$Y^n_{k,j2^{n-1}+\ell} := \frac{x^{n-1}_{2k}}{2}\left(Y^{n-1}_{2k,\ell} + (-1)^j Y^{n-1}_{2k+1,\ell}\right)$$
$$(0 \leqq k < 2^{N-n}, 0 \leqq \ell < 2^{n-1}, \quad j = 0, 1 \; n = 1, 2, \ldots, N). \tag{16}$$

In the N-th step we arrive at the Fourier coefficients

$$Y^N_{0,m} = \left[f, \psi_m\right]_N \quad (m = 0, 1, \ldots, 2^N - 1). \tag{17}$$

Since the number of computing in one step is $O(2^{N-n}2^{n-1}2) = O(2^N)$, therefore the total amount of necessary computation is $O(N2^N)$. The inputs of the algorithm (of a number 2^N) are the function values $f(x^0_k)$ $(0 \leqq k < 2^N)$, the outputs (also of a number 2^N) are the Fourier coefficients $[f, \psi_m]$ $(0 \leqq m < 2^N)$. The algorithm can be performed using a buffer capable of containing 2^N complex numbers (plus some extra space). Easy to see that if numbers $Y^n_{k,\ell}$ are stored step-by-step in array Z of length 2^N according to rule

$$Z(2^n k + \ell) := Y^n_{k,\ell}$$
$$(0 \leqq k < 2^{N-n}, 0 \leqq \ell < 2^n, n = 0, 1, \ldots, N),$$

then we have a so-called *goose-step* algorithm consuming a buffer space of 2^N.

We remark that in the above described algorithm, we only utilized properties (5) and (8) of the product system. It follows that the algorithm can also be utilized for calculating coefficients corresponding to biorthogonal product systems of any systems that are in form (5).

The function ϕ_k is the product of 2^k Blaschke functions. Denote

$$Z_k := \{z \in \mathbb{C} : \phi_k(z) = 0\} \quad (k \in \mathbb{N})$$

the set of root of the function ϕ_k. Then the set of poles of ϕ_k are of the form

$$P_k := \{z^* := \frac{1}{\bar{z}} : z \in Z_k\} \quad (k \in \mathbb{N}).$$

For true rational functions R we have

Theorem 3. Let $R : \mathbb{C} \to \mathbb{C}$ a true rational function with poles P satisfying

$$P \subseteq \cup_{k=0}^{n-1} P_k. \tag{18}$$

Then $S_{2^n} R = R$ on the whole complex plane.

If we know the poles $P = \{p_k : 0 \leqq k < n\}$ of the true rational function R, we can construct a system ϕ_j $(0 \leqq j < n)$ such a way that (18) is satisfied.

Since ϕ_k is a Blaschke product with 2^k factors, therefore b is a pole of ϕ_k if and only if the mirror $a := (b^*)^{-1}$ with respect to the unit circle is a zero of ϕ_k, i.e. if $\phi_k(a) = 0$. Starting from the numbers $b_k := 1/\overline{p_k}$ $(0 \leqq k < n)$ we define the sequence a_k and the functions $A_k = B_{a_k} B_{-a_k}$ and $\phi_k := A_k \circ \cdots \circ A_1 \circ A_0$ recursively:

$$a_0 := b_0, \; a_k := \phi_{k-1}(b_k) \quad (1 \leqq k < n).$$

Then b_k is a root of ϕ_k for $0 \leqq k < n$. Indeed $\phi_0(b_0) = B_{a_0}(a_0) = 0$ and by definition

$$\phi_k(b_k) = A_{a_k}(\phi_{k-1}(b_k)) = A_{a_k}(a_k) = 0.$$

Consequently $p_k = b^*_k$ is a pole of ϕ_k and (18) is satisfied.

4. A SIMULATION EXAMPLE

In order to illustrate the above results consider the 4th order system with complex conjugate poles: $p_{12} = -0.9 \pm i0.2$, $p_{34} = .8 \pm i0.5$. The poles of the rational basis generated by a simple choice of $b = 0.3 + i0.25$ in Eq. 4 are shown on Figure 1 for $m = 4, 6$. I can be seen that the poles concentrate close the the interior of the unit circle as the m increases. This can be an advantage e.g. when identifying lightly damped vibrating or oscillating systems.

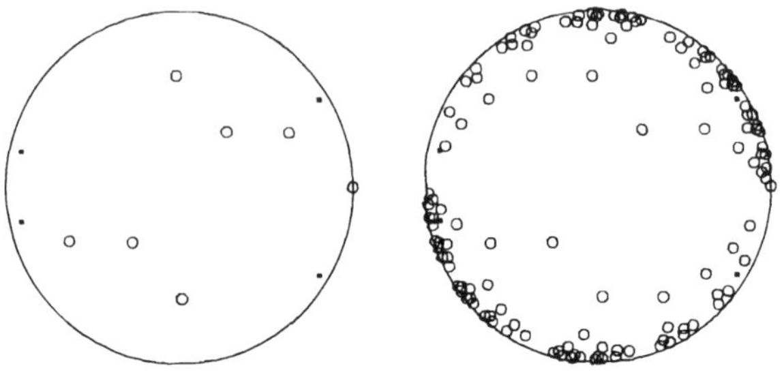

Fig. 1. Pole locations of the basis functions for $m = 4, 6$

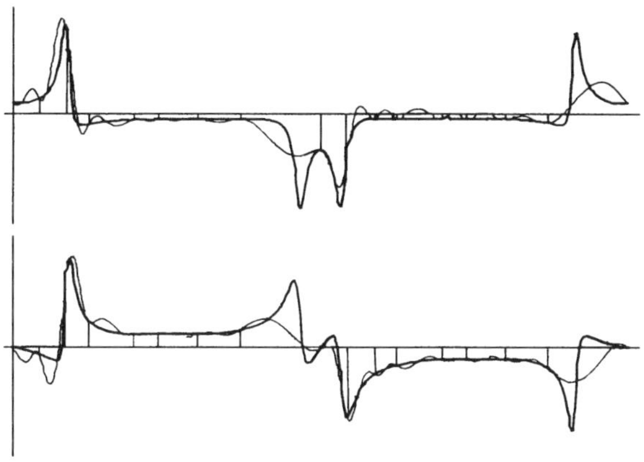

Fig. 2. Fourier synthesis, Re and Im parts of the true and fitted transfer functions, $m = 4$

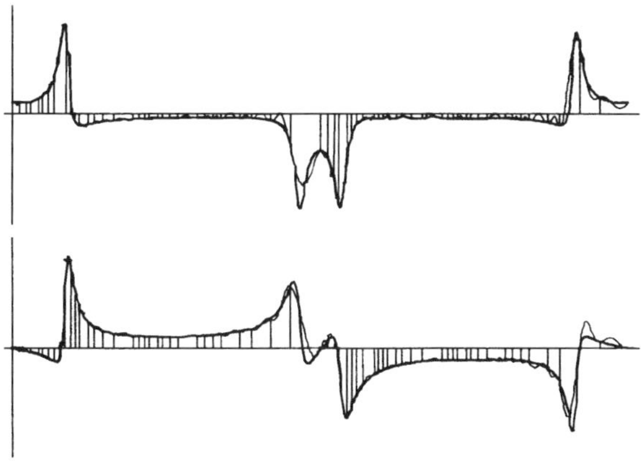

Fig. 3. Fourier synthesis, Re and Im parts of the true and fitted transfer functions, $m = 6$

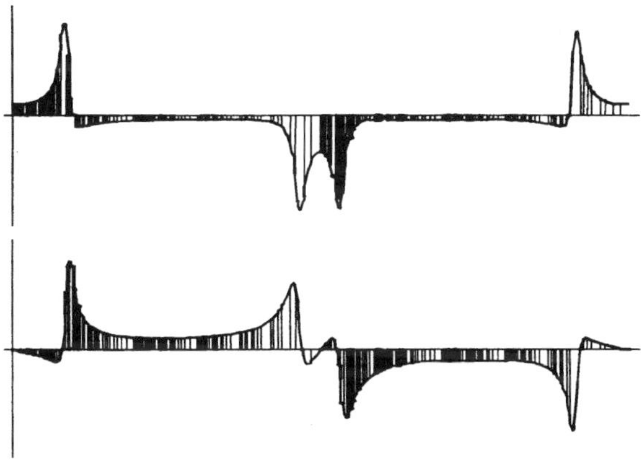

Fig. 4. Fourier synthesis, Re and Im parts of the true and fitted transfer functions, $m = 8$

The location of the measurement points, i.e. the elements of X_n and the plot of the real and imaginary parts of the transfer function of the system generating the data are shown on Figures 2–4. The plot of the reconstructed approximate models obtained from the computed Fourier coefficients in the rational basis for $m = 4, 6, 8$ is plotted, too. The mean square errors of the reconstruction are $e_4 = 0.713$, $e_6 = 0.323$, $e_8 = 0.0621$, respectively.

The plot of the basis functions ψ_m, $m = 1, \ldots, 4$ are shown on Figure 5 for illustration.

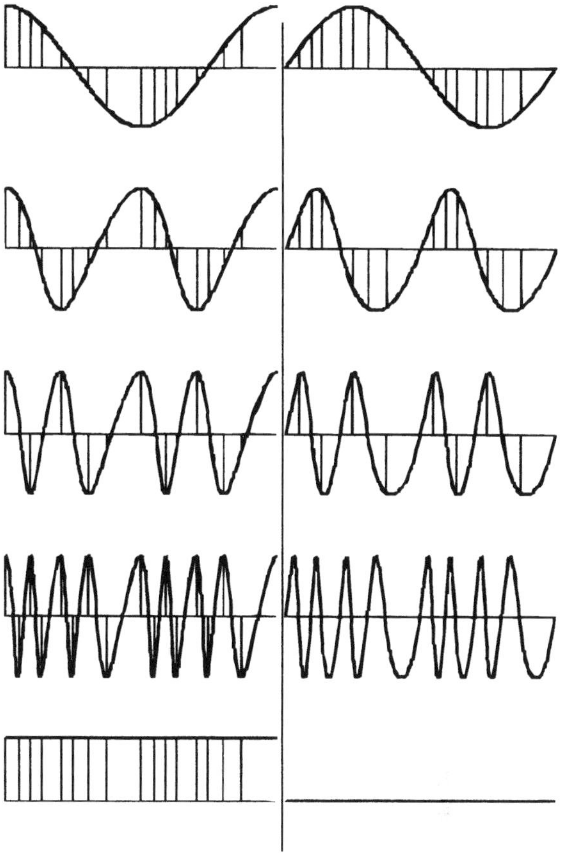

Fig. 5. Re and Im parts of the basis functions, $m = 4$

5. CONCLUSIONS

This paper presented an approach for generation of rational orthogonal bases in $\mathcal{H}_2$ that can be used in signal modelling and identification. This approach is based on defining Blaschke product systems and function compositions that act as dilation on $\mathcal{H}_2$ when compared to previous approaches where shift operators are used. Results on signal representation using their Fourier series in the generated bases and on the construction of FFT algorithms for computing the coefficients are given, too.

REFERENCES

Akcay, H. (2002). General orthonormal bases for robust identificcation in $\mathcal{H}_\infty$. 41st IEEE Conference on Decision and Control. Las Vegas, Nevada, USA, pp. 2619-2624.

Bodin, P., Oliveira e Silva, T. and B. Wahlberg (1996). On the construction of orthonormal basis functions for system identification. *Prepr. 13th Triennial IFAC World Congress*, San Francisco, 369-374.

Casini, M., Garulli, A. and Vicino, A. (2001). On worst-case approximation of feasible system sets via orthonormal basis functions. In: Proc of the 40th IEEE Conference on Decision and Control. Orlando, Florida, USA.

Fischer, B.R. and A. Medvedev (1998). Laguerre shift identification of a pressurized process. Proceedings of the American Control Conference, Philadelphia, Pennsylvania, Vol. 3, pp. 1933-1937.

Heuberger, P.S.C. and O.H. Bosgra (1990). Approximate system identification using system based orthonormal functions. *Proc. 29th IEEE Conf. Decis. Control*, Honolulu, HI, 1086–1092.

Heuberger, P.S.C. (1991). *On Approximate System Identification with System Based Orthonormal Functions.* Dr. Dissertation, Delft University of Technology, The Netherlands.

Heuberger, P.S.C., P.M.J. Van den Hof and O.H. Bosgra (1995). A generalized orthonormal basis for linear dynamical systems. *IEEE Trans. Autom. Control* **40**, 451–465.

de Hoog, T.J., P.S.C. Heuberger and P.M.J Van den Hof (2000). Partial realization in generalized bases: Algorithm and Example. *Prepr. of SYSID'2000*, Santa Barbara, CA.

de Hoog, T.J., Heuberger, P.S.C., Van den Hof, P.M.J., Szabó, Z. and Bokor, J. (2001) Minimal partial realization from orthonormal basis function expansions *Prooceedings of 40th IEEE Conference on Decision and Control*, Orlando, Florida.

T.J. de Hoog, Szabó Z., Heuberger P.S.C., Van den Hof P.M.J. and J. Bokor (2002). Minimal partial realization from generalized orthonormal basis function expansions. Automatica, 38(4):655-669.

Milanese, M. and Taragna M. (1999). H-infinity identification of 'soft' uncertainty models. System Control Letters 37(4):217-228.

Ninness, B.M. and F. Gustafsson (1997). A unifying construction of orthonormal basis for system identification. *IEEE Trans. Automatic Control* **42**, 515–521.

Ninness, B.M., H. Hjalmarsson and F. Gustafsson (1998). Generalized Fourier and Toeplitz results for rational orthonormal bases. *SIAM Journal on Control and Optimization* **37**, 429–460.

Oliveira e Silva, T. (1995). Optimality conditions for truncated Kautz networks with two periodically repeating complex conjugate poles. *IEEE Trans. Autom. Control* **40**, 342-346.

Schipp, F. and J. Bokor (1997). L_∞ system approximation algorithms generated by ϕ-summations. *Automatica* **33**, 2019-2024.

Schipp, F. and J. Bokor (1998). Approximate identification in Laguerre and Kautz bases. *Automatica* **34**, 463–468.

Szabó, Z. and J. Bokor (1997), Minimal state space realization for transfer functions represented by coefficients using generalized orthonormal basis. Proceedings of the 36th IEEE Conference on Decision and Control, **1**, 169-174.

Szabó, Z., J. Bokor and F. Schipp (1999). Identification of rational approximate models in H_∞ using generalized orthonormal basis. *IEEE Trans. Autom. Control* **44**, 153–154.

Van den Hof, P.M.J., P.S.C. Heuberger and J. Bokor (1995). System identification with generalized orthonormal basis functions. *Automatica* **31**, 1821-1834.

Wahlberg, B. (1991). System identification using Laguerre models. *IEEE Trans. Autom. Control* **36**, 551–562.

Wahlberg, B. (1994a). System identification using Kautz models. *IEEE Trans. Autom. Control* **39**, 1276–1282.

Wahlberg, B. and P.M. Mäkilä (1996). On approximation of stable linear dynamical systems using Laguerre and Kautz functions. *Automatica* **32**, 693-708.

IFAC Publications
www.elsevier.com/locate/ifac

MORE ON SPARSE REPRESENTATIONS IN ARBITRARY BASES.

Jean Jacques Fuchs

Irisa-Université de Rennes, Campus de Beaulieu,
35042 Rennes Cedex, France.
e-mail : fuchs@irisa.fr

Abstract: The purpose of this contribution is to generalize some recent results on sparse representations of signals in redundant bases. The question that is considered is the following : let A be a known (n, m) matrix with $m > n$, one observes $b = AX$ where X is known to have $p < n$ nonzero components, under which conditions on A and p is it possible to recover X by solving a convex optimization problem such as a linear or quadratic program? The solution is known when A is the concatenation of two unitary matrices, we extend it to arbitrary matrices. *Copyright © 2003 IFAC*

Keywords: Sparse representations, linear programming, quadratic programming..

1. INTRODUCTION

Let us consider a set of m n-dimensional vectors a_j with $m > n$ and let us denote A the (n, m) matrix having these vectors as columns. Any linear combinations b of these m vectors can then be written as: $b = AX$ with X a m dimensional vector of weights.

If X has just a few non-zero components it may be possible to recover the value of X from the knowledge of b and the aim of this contribution is to give condition under which this is feasible.

If the true solution is known to be sparse, one seeks a mean allowing to single out among the infinity of possible solutions the or a solution having the smallest possible number of nonzero components i.e. the solution of the following optimization problem :

$$\min_X \|X\|_0 \qquad \text{subject to :} \quad AX = b \qquad (P_o)$$

where $\|X\|_0$ denotes the number of non-zero components in X. This is a difficult problem that can only be solved using a combinatorial approach i.e. testing systematically all the potential combinations of columns. This approach is thus unfeasible and it is usual to consider instead the following much simpler optimization problem :

$$\min_X \|X\|_1 \qquad \text{subject to :} \quad AX = b \qquad (LP)$$

where $\|X\|_1 = \sum |x_i|$ denotes the ℓ_1 norm of X. The problem (LP) is easily transformed into a linear program whose solution is straightforward to obtain.

Donoho and Huo (Donoho, 2001) investigated the case where A is the concatenation of two square orthogonal matrices U_1 and U_2 : $A = [U_1\ U_2]$, and proved that if :

$$\|X\|_0 \leq \frac{1}{M} \qquad \text{with :} \quad M = \sup_{i \neq j} |a_i^T a_j| \qquad (1)$$

the sparsest solution is unique. They further established that under the stronger condition

$$\|X\|_0 \leq \frac{1}{2}(1 + \frac{1}{M}) \qquad \text{with :} \quad M \text{ as above} \qquad (2)$$

the sparsest solution is also the unique solution of (LP). This result has later been refined in (Elad, 2002) but still in the case where A is the concatenation of two orthogonal bases.

Below we prove that condition (2) is indeed sufficient for arbitrary A matrices built upon vectors with unit euclidean norm. This normalization is obviously satisfied for the concatenation of orthogonal matrices.

A similar result has independently been obtained in (Gribonval, 2002) using a completely different approach.

Instead of (LP) we consider a parametrized quadratic program (QP) that encompasses (LP) as a special case and show that if $X_0 \in \{X|AX = b\}$ satisfies (2) then it is the unique minimum point of both (QP), in a sense to be defined later, and (LP). This establishes the result since if a sparser representation existed, the same reasoning would hold and one would arrive at a contradiction since both (LP) and (QP) are convex optimization problems for which all minimum points are located in a convex set that reduces to an unique point under (2).

To establish this result we merely apply more general but non explicit results presented in (Fuchs, 1997), (Fuchs, 1998) to this very specific problem. The proof we present goes through if the vectors a_j are not normalized or the weights known to be greater or equal to zero but its last part would be more intricate.

2. THE CRITERION

Let us consider the following optimization problem:

$$\min_X \frac{1}{2}\|AX - b\|_2^2 + h\|X\|_1 \ , \quad h > 0 \qquad (QP)$$

If one introduces new variables $x_i^+ = \max(x_i, 0)$, $x_i^- = \max(-x_i, 0)$ and replaces x_i by $x_i^+ - x_i^-$ and $|x_i|$ by $x_i^+ + x_i^-$, this unconstrained nonsmooth optimization problem is converted into a quadratic program where these new variables x_i^+ and x_i^- are constrained to be greater or equal to zero (Luenberger, 1973).

This optimization problem has thus an unique global minimum that can be obtained using standard algorithms available from any scientific program library. This criterion and similar ones have been considered for a while now (Chen, 1999),(Moal, 1998), (Fuchs, 1998), (Fuchs, 1999).

To assess the role played by h in this criterion, one can make the following remarks about the optimum X^* of (QP) as h goes from 0 to $+\infty$.

◇ For $h = 0$, one is left with $\min_X \|b - AX\|_2^2$, and since there are more unknowns than equations ($m > n$), the value of the minimum is zero and it is attained for all points in a convex set (a linear manifold), some of them having at most n nonzero components.

◇ For $h = 0^+$ i.e. for h positive and arbitrarely close to zero, the solution is attained at the point(s) in the previous set having least ℓ_1-norm. Again there is a solution having at most n nonzero components.

◇ For $h \geq \| A^T b \|_\infty$, the optimum is attained at $X^* = 0$.

If X is sparse, one can thus expect that for h small enough, it may be possible to recover the columns that were used to build b with however a biased value of X. This bias converges to zero with h and can be corrected.

Further insight on the role of h can be gained from the dual of (QP), which is (Fuchs, 2001) :

$$\min_X \|AX\|_2^2$$
$$\text{subject to} \quad \|A^T(AX - b)\|_\infty \leq h \qquad (DQP)$$

The constraint of the dual says that -at the optimum- the residues or reconstruction errors in r defined as $r = b - AX$ are such that their correlations with the columns of A never exceeds h. Note that these correlations are the outputs of the matched filter applied to the reconstruction errors. Since the ℓ_2 norm of the columns of A is equal to one, this also says that this criterion allows for reconstruction errors that are of order h. Their size can thus be fixed by tuning h.

3. OPTIMALITY CONDITIONS FOR (QP)

In order to find the conditions under which the true value of X is also the location of the unique optimum of (QP) we propose to seek the conditions under which this point satisfies the necessary and sufficient condition (NSC) for a strict global minimum of (QP).

The results we present in this section are established in the appendix. The criterion (QP) is unsconstrained, convex but not continuously differentiable. A necessary and sufficient condition (NSC) for X^* to be a global minimum of (QP) is that the vector zero be a sub-gradient of the criterion at X^* (Fletcher, 1991). To write the NSC, it is helpful to distinguish between the nonzero components and the zero components of X^*. We denote $\bar{X}^*$ the reduced dimensional vector of dimension $\|X^*\|_0$ built upon the nonzero components of X^*. Similarly $\bar{A}$ denotes the associated columns in A. One then has e.g. $AX^* = \bar{A}\bar{X}^*$. The NSC characterizing X^* are then :

$$\bar{A}^T(b - \bar{A}\bar{X}^*) = h \ \text{sign}\bar{X}^* \qquad (NSC_1)$$
$$|a_j^T(b - A\ X^*)| \leq h \ \text{ for } \ a_j \notin \bar{A} \qquad (NSC_2)$$

where $\text{sign}(x_j) = 1,\ 0, -1$ when x_j is respectively $< 0,\ = 0,\ > 0$. The first condition that concerns the nonzero components of X^* collected in $\bar{X}^*$, yields :

$$\bar{X}^* = \bar{A}^+ \, b - h \, (\bar{A}^T \bar{A})^{-1} \, \text{sgn}\bar{X}^* \quad (3)$$

with $\bar{A}^+ = (\bar{A}^T \bar{A})^{-1} \bar{A}^T$ the pseudo-inverse of $\bar{A}$. This is only an implicit relation that does not allow to compute $\bar{X}^*$ since $\bar{X}^*$ appears on both sides. The optimum of (QP) can only be obtained through an iterative search. The second term (3) is a bias term induced by the regularization term in (QP). Once $\bar{X}^*$, and thus $\bar{A}^*$, are known, it is easy to remove the bias.

The second condition (NSC_2) concerns the zero components in X^* those associated with the columns that are not in $\bar{A}$. If the inequalities in (NSC_2) are satisfied strictly, X^* is a strict unique minimum of (QP). Both conditions will play a central role in the sequel where (NSC_2) will appear as an identifiability or separability condition,

4. THE SEPARABILITY CONDITION

We are now ready to answer the question posed in the introduction : given an $(n,\ m)$ matrix A with $m > n$ and a vector b linear combination of $p < n$ columns of A, under which condition can this linear combination be retrieved by solving (QP)?

Let us denote X_0 the exact solution, $\bar{X}_0$ the p-dimensional vector built upon the nonzero component in X_0 and $\bar{A}_0$ the (n, p) dimensional matrix built accordingly with the associated columns of A so that $b = AX_0 = \bar{A}_0 \bar{X}_0$.

For a nonzero h, the best we can expect is that X_0 and X^*, the solution to (QP), have their nonzero components at the same locations and with the same signs. From section 3 we know that this fully characterizes the optimum of (QP), which is then equal to, (see (3)) :

$$\begin{aligned} \bar{X}^* &= \bar{A}_0^+ \, b - h \, (\bar{A}_0^T \bar{A}_0)^{-1} \, \text{sign}\bar{X}_0 \\ &= \bar{X}_0 - h \, (\bar{A}_0^T \bar{A}_0)^{-1} \, \text{sign}\bar{X}_0 \quad (4) \end{aligned}$$

It remains to check if this solution completed by zeroes satisfies (NSC_2), the second part of the NS conditions :

$$|a_j^T \, (b - \bar{A}_0 \bar{X}^*)| < h \quad \forall a_j \notin \bar{A}_0$$

Replacing $\bar{A}_0 \bar{X}^*$ by its value and introducing the vector d_0 :

$$d_0 = \bar{A}_0^{+T} \, \text{sign}\bar{X}_0 \quad (5)$$

these conditions become :

$$|a_j^T d_0| < 1 \quad \forall a_j \notin \bar{A}_0 \quad (6)$$

To summarize, the following conditions have thus to be satisfied for (QP) to retrieve the *true* solution :

- h has to be small enough for : $\text{sign}\bar{X}^* = \text{sign}\bar{X}_0$, in (4)
- the scenario-dependent vector $d_0 = \bar{A}_0^{+T} \text{sign}\bar{X}_0$ must be such :

$$|a_j^T d_0| < 1 \quad \forall a_j \notin \bar{A}_0$$

The first of these two conditions depends upon the magnitude of the components while the second is independent of it. The second says that there must exist two *separating* hyperplanes $H_\pm$, associated with a single vector d_0 : $H_\pm = \{a | a^T d_0 = \pm 1\}$ such that the true columns lie in these hyperplanes $\bar{A}_0^T d_0 = \text{sign}\bar{X}_0$ and the wrong ones lie in between $|a_j^T d_0| < 1$.

This is why we call the conditions (6), which are already presented in (Fuchs, 1997), (Fuchs, 1998), the separability conditions.

5. THE SPARSITY CONDITION

Let us transform the separability condition (5, 6):

$$|a_j^T \bar{A} (\bar{A}^T \bar{A})^{-1} sign(\bar{X}| < 1 \ \ \forall a_j \notin \bar{A}$$

into the more useable but also more conservative condition (2).

Remember that :

$$M = \sup_{i \neq j} |a_i^T a_j| \qquad \text{for} \quad 1 \leq i, j \leq m$$

Let us rewritte : $\bar{A}^T \bar{A} = I - H$ where H is built with the non-diagonal elements of $\bar{A}^T \bar{A}$. The diagonal elements of $\bar{A}^T \bar{A}$ are equal to one since we assume $a_j^T a_j = 1, \ \forall j$.

Since all the non-zero components in the order p square matrix H are smaller than M, its spectral radius $\rho(H)$ satisfies : $\rho(H) \leq (p-1)M$ by Gersgorin theorem (Householder, 1964).

If $p < 1 + (1/M)$, a condition implied by (2) holds, the spectral radius verifies $\rho(H) < 1$ and one can apply Neuman's lemma (Householder, 1964) to get :

$$(\bar{A}^T \bar{A})^{-1} = (I - H)^{-1} = I + \sum_{k>0} H^k$$

For any matrix B, we denote $|B|$ the matrix with entries $|b_{i,j}|$ and write $B < C$ if $b_{i,j} < c_{i,j}, \forall i, j$, one then has :

$$\begin{aligned} |(\bar{A}^T \bar{A})^{-1}| &= |I + \sum_{k>0} H^k| \leq I + \sum_{k>0} |H^k| \\ &\leq I + \sum_{k>0} |H|^k \leq I + \sum_{k>0} ME^k = (I - ME)^{-1} \end{aligned}$$

where E is a square matrix whose elements are equal to one except those on the diagonal. We also write $E = \mathbf{1} - I$, where $\mathbf{1}$ denotes here and below a matrix of ones with adequate dimension. It is then easy to check that :

$$I - ME = (1+M)I - M\mathbf{1}$$
$$\Rightarrow (I-ME)^{-1} = \alpha I + \beta \mathbf{1}$$

with : $\alpha = \frac{1}{1+M}$ and $\beta = \frac{M}{(1+M)(1+M-Mp)}$ which in turn leads to : $|(\bar{A}^T\bar{A})^{-1}| < (\alpha+\beta)I + \beta E$.

Since : $|a_j^T \bar{A}| < M\mathbf{1^T}$, condition (6) yields :

$$|a_j^T \bar{A}(\bar{A}^T\bar{A})^{-1} sign(\bar{X})|$$
$$< M\mathbf{1^T}|(\mathbf{\bar{A}^T\bar{A}})^{-1}|\mathbf{1}$$

substituting the bound of $|(\bar{A}^T\bar{A})^{-1}|$ one gets :

$$|a_j^T \bar{A}(\bar{A}^T\bar{A})^{-1} sign(\bar{X})|$$
$$< M(p(\alpha+\beta) + (p^2-p)\beta) < 1$$

which after some easy computations becomes condition (2) if one replaces p by $\|X\|_0$.

This completes the proof of the result announced in the introduction and establishes that if h is taken small enough, the true value X_0 and the solution X^* of (QP) have their nonzero components at the same locations and with the same signs. The relation (4) indicates precisely what is meant by "small enough" and taking $h = 0^+$ garanties that $X_0 = X^*$ as is the case for the (LP) criterion.

Remark : It may be of interest to distinguish between the vectors belonging to a potential selection : $a_i \in \bar{A}_0$ and the others and define :

$$M_o = \sup_{i \neq j} |a_i^T a_j| \qquad \text{for} \quad a_i, a_j \in \bar{A}_0.$$

The condition then becomes :

$$\|X\|_0 \leq \frac{1+M_o}{M+M_o}$$

which is weaker than (2) if $M_o < M$. In a detection-estimation context (Fuchs, 1999) where the aim is to detect which components are present in b, M_o is linked to the resolution one expects to be able to attain while M is linked to the precision with which one expects to be able to locate the components. It is then quite natural to have $M_o \ll M$ and this condition is thus less conservative than (2).

6. THE SPECIAL CASE OF THE ℓ_1 NORM

As indicated in section 2, if h in (QP) decreases to zero, the optimum of (QP) converges to the solution of (LP) and relation (4) :

$$\bar{X}^* = \bar{X}_0 - h\ (\bar{A}_0^T\bar{A}_0)^{-1}\ \text{sign}\bar{X}_0$$

indicates clearly how the tradeoff between both terms in (QP) affects the optimum.

A direct analysis of the condition under which (LP) retrieves the true solution can indeed be achieved quite easily along the lines used for the (QP) criterion. Let us do so in this section using the notations introduced above.

Let us recall the (LP) criterion:

$$\min_X \|X\|_1 \qquad \text{subject to :} \quad AX = b \qquad (LP)$$

The principal difficulty arises from the fact that the expected solution to (LP) is degenerate i.e. it has less than n nonzero components and in order to characterize the unicity of the solution one has to introduce the dual linear program :

$$\min_d d^T b \quad \text{subject to :} \quad \|d^T A\|_\infty \leq 1 \quad (DLP)$$

which as (LP) is a linear program that is not in standard form. For the sought-for vector X_0, that is entirely characterized by the knowledge of $\bar{A}_0$, to be the unique solution of (LP) one needs to be able to associate with it a solution d_0 of (DLP). The conditions for X_0, d_0 to be optimal are simply the equality of the criterions and primal and dual feasibility, i.e.:

$$\|X_0\|_1 = d_0^T b, \quad AX_0 = b, \quad \|d_0^T A\|_\infty \leq 1$$

The first of these conditions can be transformed into : $\bar{X}_0^T sign(\bar{X}_0) = d_0^T A X_0$, the second is trivialy satisfied and the last is simialr to (6).

The condition for the vector X_0 to be the unique solution of (LP) are actually :

$$\exists\ d_0 \ni \bar{A}_0^T d_0 = sign(\bar{X}_0)$$
$$|a_j^T d_0| < 1 \ \ \forall a_j \notin \bar{A}_0 \qquad (7)$$

This condition is similar to (6) which has to hold for (QP), but the vector d_0 for which it has to hold is now less constrained since it belongs to a (n-p) dimensional linear manifold. The vector d_0 defined in (5) is just one possible candidate, the one of least ℓ_2 norm.

This is not a minor difference and it is quite easy to build a (toy) example where (LP) works and (QP) does not, i.e. for which d_0 in (5) does not satisfy (7) while there exists a d_0 that does.

7. CONCLUSIONS

The results proposed in (Donoho, 2001) for the case where the redundant bases is limited to the concatenation of two orthonormal bases has been extended to an arbitrary set of vectors.

Using (QP) instead of (LP) further allows to handle the case where noise is present in the observations i.e. if

$$b = AX + e$$

with e a vector of white gaussian noise, for instance. In this situation, the relative magnitudes of the nonzero components in X and standard deviation of the noise model plays a crucial role and the problem of the recovery of the true decomposition of X is of a different nature. These topics have been investigated with the (QP) criterion in (Fuchs, 1998), (Fuchs, 1999) (Fuchs, 2001) and in the case of the following extended (LP) criterion :

$$\min_X \|X\|_1 \quad \text{subject to :} \quad \|AX - b\|_\infty \leq \rho$$

in (Fuchs, 1996), (Fuchs, 1997), (Fuchs, 2000).

8. APPENDIX

In this appendix we establish the optimality condition for (QP) :

$$\min_X \frac{1}{2}\|b - AX\|_2^2 + h \|X\|_1 \qquad \text{(QP)}$$

The criterion (QP) is convex but not continuously differentiable. It can be transformed into a quadratic program and the optimality conditions can be obtained by writing the Kuhn Tucker conditions which are necessary and sufficient for convex problems.

We use here a more direct way in writing that a necessary and sufficient condition (NSC) for X^* to be a global minimum of (QP) is that the vector 0 is a sub-gradient of the criterion at X^* (Fletcher, 1991). A vector γ is a sub-gradient of f at X^* if $f(X) \geq f(X^*) + \gamma^T(X - X^*)$. Since (QP) is non-smooth at zero only, it is worthwhile to distinguish the zero components from the non-zero components in X^*. We collect the non zero components of X^* in $\bar{X}^*$ and the associated columns of A in $\bar{A}$ so that $AX^* = \bar{A}\bar{X}^*$ and introduce the notation $\text{sign}(X)$ which is such that $\|X\|_1 = X\text{sign}(X)$.

• The sub-gradient is unique and equal to the gradient for the non-zero components in $\bar{X}^*$. By nulling the gradient of (QP) with respect to $\bar{X}$ at $\bar{X}^*$, one gets :

$$-\bar{A}^T(b - \bar{A}\bar{X}^*) + h \;\text{sign}\bar{X}^* = 0$$

which leads to the following implicit expression for the non-zero components of X^* :

$$\bar{X}^* = \bar{A}^+ b - h\;(\bar{A}^T\bar{A})^{-1}\;\text{sign}\bar{X}^*$$

where $\bar{A}^+ = (\bar{A}^T\bar{A})^{-1}\bar{A}^T$ denotes the pseudo-inverse of $\bar{A}$.

• The vector 0 is a sub-gradient for the zero components in X^*, if the criterion increases when they are taken non zero. This is the case if the absolute value of the partial derivative of the quadratic term in (QP) is smaller than h. For the j-th component of X one gets :

$$|a_j^T(b - A\,X^*)| \leq h \qquad \forall a_j \notin \bar{A}$$

The NSC for X^* to be a minimum of (QP) are thus :

$$a_j^T(b - AX^*) = h\;\text{sign}\bar{x}_j{}^* \quad \text{for } x_j^* \neq 0 \quad (NSC_1)$$
$$|a_j^T(b - A\,X^*)| \leq h \quad \text{for } x_j^* = 0 \quad (NSC_2)$$

Introducing the ℓ_∞-norm, these conditions can be rewritten more compactlty as :

$$\|\; A^T\;(\hat{b} - AX^*)\;\|_\infty \leq h$$

They are the constraints of the following dual problem :

$$\min_X \|AX\|_2^2$$
$$\text{s.t.} \quad \|A^T(AX - b)\|_\infty \leq h \qquad (DQP)$$

that is equivalent to (QP) as is easily shown by using the equivalent quadratic programs (Fuchs, 2001).

For convex optimisation problems, the minimum is unique and potentially attained for all points in a convex set. A sufficient condition for the minimum of (QP) to be strict, i.e. to be attained at an unique point, is that the inequalities in (NSC_2) are strict. This follows for instance from the second order sufficient conditions to be satisfied at a strict minimum (Luenberger, 1973).

9. REFERENCES

D.L. Donoho and X. Huo. (2001) Uncertainty principles and ideal atomic decomposition. *IEEE Trans. on I.T.*, 47, 11, 2845-2862.

M. Elad an A.M. Bruckstein. (2002) A generalized uncertainty principle and sparse representation in pairs of bases. *IEEE Trans. on I.T.*, 48, 9, 2558-256;

R. Gribonval and M. Nielsen. (2002) Sparse representations in unions of bases. *Technical Report INRIA*, 1499, nov. 2002.

J.J. Fuchs. 1997 Une approche l'estimation et l'identification simultanées. *XVI Colloque GRETSI*, ,vol. 2, pp. 1273-1276, Grenoble.

J.J. Fuchs. 1998 Detection and estimation of superimposed signals. *IEEE ICASSP*, vol. III, pp. 1649-1652, Seattle.

D. G. Luenberger.(1973) Introduction to linear and nonlinear programming. *Addison Wesley.*

S. Chen, D. Donoho and M. Saunders (1999) Atomic Decomposition by basis pursuit. *SIAM J. Scientific Computing*, 20, 1, 33-61.

N. Moal and J.J. Fuchs. 1998 Sinusoids in white noise : a quadratic programming approach. *IEEE ICASSP*, vol. IV, pp. 2221-2224, Seattle, 1998.

J.J. Fuchs. 1999 Multipath time-delay detection and estimation. *IEEE Trans. S.P.*, 47, 237–243, jan. 1999.

J.J. Fuchs. 2001 On the application of the global matched filter to DOA estimation with uniform circular arrays. *IEEE-T-SP*, vol. 49, p. 702–709, avr. 2001.

A.S. Householder. 1964 The theory of matrices in numerical analysis. *Blaisdell.* N.Y..

R. Fletcher. (1991) Practical methods of optimization. *Wiley.*

J.J. Fuchs. 1996 Linear programming in spectral estimation. Application to array processing. *IEEE ICASSP*, vol.6, pp. 3161-3164, Atlanta.

J.J. Fuchs. 1997 Extension of the Pisarenko method to sparse linear arrays *IEEE-T-SP*, 45: 2413–2421, Oct. 1997.

J.J. Fuchs and B. Delyon Minimum L_1-norm reconstruction function for oversampled signals: Application to time-delay estimation. *IEEE Trans. on I.T.*, 46, 7, . 1666–1673.

IFAC
Publications
www.elsevier.com/locate/ifac

ON SPECTRAL ANALYSIS USING MODELS WITH PRE-SPECIFIED ZEROS [1]

Bo Wahlberg *

* *S3 - Automatic Control, KTH, SE-100 44 Stockholm, Sweden.*
(bo.wahlberg@s3.kth.se)

Abstract: The fundamental theory of Lindquist and co-workers on the rational covariance extension problem provides a very elegant framework for ARMA spectral estimation. Here the choice of zeros is completely arbitrary, and can be used to tune the estimator. An alternative approach to ARMA model estimation with pre-specified zeros is to use a prediction error method based on generalizing autoregressive (AR) modeling using orthogonal rational filters. Here the motivation is to reduce the number of parameters needed to obtain useful approximate models of stochastic processes by suitable choice of zeros, without increasing the computational complexity.
The objective of this contribution is to discuss similarities and differences between these two approaches to spectral estimation. *Copyright © 2003 IFAC*

Keywords: ARMA estimation, time-series modelling, covariance extension,

1. INTRODUCTION AND PROBLEM FORMULATION

The concept of representing complex systems by simple models is fundamental in science. The aim is to reduce a complicated process to a simpler one involving a smaller number of parameters. The quality of the approximation is determined by its usefulness, e.g. its predictive ability. Autoregressive (AR) and autoregressive moving-average (ARMA) models are the dominating parametric models in spectral analysis, since they give useful approximations of many stochastic processes of interest. The ARMA model leads to non-linear optimization problems to be solved for best approximation, while the special case of AR modeling only involves a quadratic least squares optimization problem. Hence, AR models are of great importance in applications where fast and reliable computations are necessary.

The fact that the true system is bound to be more complex than a fixed order AR model has motivated the analysis of high-order AR approximations, where the model order is allowed to tend to infinity as the number of observations tends to infinity. However, aspects such as the number of observations, computational limitations and numerical sensitivity set bounds on how high an AR order can be tolerated in practice. Herein, we shall study discrete orthogonal rational function model structures, which reduce the number of parameters to be estimated without increasing the numerical complexity of the estimation algorithm.

Suppose that $\{y(t),\, t = \ldots -1, 0, 1, \ldots\}$ is a stationary linear regular random process with Wold representation,

$$y(t) = \sum_{k=0}^{\infty} h_k^0 e(t-k), \quad h_k^0 \in \mathbf{R}, \quad h_0^0 = 1. \qquad (1)$$

Here $\{e(t)\}$ is a sequence of random variables with the properties

$$\mathrm{E}\{e(t)|\mathscr{F}_{t-1}\} = 0, \mathrm{E}\{e(t)^2|\mathscr{F}_{t-1}\} = \sigma_0^2,$$
$$\mathrm{E}\{e(t)^4\} < \infty, \qquad (2)$$

[1] This work was partially supported by The Swedish Science Foundation

The transfer function, often called the noise filter or the shaping filter,

$$H_0(q) = \sum_{k=0}^{\infty} h_k^0 q^{-k}, \quad H_0(\infty) = 1, \tag{3}$$

is a function of the shift operator q, $qe(t) = e(t+1)$. By q^{-1} we mean the corresponding delay operator $q^{-1}e(t) = e(t-1)$. The power spectral density of $\{y(t)\}$ equals

$$\Phi(e^{i\omega}) = \sigma_0^2 |H_0(e^{i\omega})|^2. \tag{4}$$

We shall assume that the complex function $[H_0(z)]^{-1}$, $z \in \mathbb{C}$, is analytic in $|z| > 1$ and continuous in $|z| \geq 1$. Then

$$[H_0(z)]^{-1} = \sum_{k=0}^{\infty} a_k^0 z^{-k}, \quad |z| \geq 1. \tag{5}$$

We shall impose a further smoothness condition on $[H_0(z)]^{-1}$, namely

$$\sum_{k=0}^{\infty} k|a_k^0| < \infty. \tag{6}$$

By truncating the expansion (5) at $k = n$, we obtain a nth order autoregressive (AR) approximation of (1),

$$A_n^0(q)y(t) = e(t), \quad A_n^0(q) = 1 + \sum_{k=1}^{n} a_k^0 q^{-k}. \tag{7}$$

A crucial question is how large an order n must be chosen to obtain a useful AR approximation. From (5) and (6) we know that the process (1) can be arbitrarily well approximated by an AR model (in the mean square sense) by taking the order n large enough. However, nothing is said about the rate of convergence. Assume that $H_0(z)$ is a rational function with zeros $\{z_i\}$, $|z_i| < 1$. The error in the AR approximation (7) is then of order δ^n, where $\delta = \max_j |z_j|$. Hence, zeros close to the unit circle imply a slow rate of convergence and consequently a high model order n.

This motivates the investigation of alternative approximations which are less sensitive to the location of the zeros. As discussed above the AR approximation corresponds to a truncated series expansion of $[H_0(z)]^{-1}$ in the basis functions $\{z^{-k}\}$. An natural extension is to replace $\{z^{-k}\}$ by more general orthonormal rational functions $\{F_k(z)\}$ in order to get more efficient representations.

Over the last decades, there has been considerable interest in the systems, signal processing and control literature in representing linear time-invariant dynamic systems using an expansion in a rational orthonormal basis. The recent monograph (Bultheel *et al.*, 1999) gives an excellent overview of the field of orthogonal rational functions, which can be viewed as generalizations of orthogonal polynomials. In parallel to the efforts in applied mathematics a very similar theory has been developed in the fields of signals, systems and control. See (Bokor *et al.*, 1999) for a thorough presentation of this theory. The paper (Wahlberg, 2002) provides an overview of orthogonal rational functions using a transformation approach.

We will make use of the following mathematical notation. Let P^T denote the transpose of a matrix, and P^* the complex conjugate transpose. Let $\mathbb{E}$ denote the exterior of the unit disc: $\{z \in \mathbb{C} : |z| > 1\}$, and $\mathbb{T}$ the unit circle: $\{z \in \mathbb{C} : |z| = 1\}$. By $\mathscr{H}_2(\mathbb{E})$ we mean the Hardy space of square integrable functions on $\mathbb{T}$, analytic in the region $\mathbb{E}$. We denote the corresponding inner product for $X(z), Y(z) \in \mathscr{H}_2(\mathbb{E})$ by

$$\begin{aligned} \langle X, Y \rangle &= \frac{1}{2\pi} \int_{-\pi}^{\pi} X(e^{i\omega})^* Y(e^{i\omega}) \omega \\ &= \frac{1}{2\pi i} \oint_{\mathbb{T}} X^T(1/z) Y(z) \frac{dz}{z}. \end{aligned} \tag{8}$$

Two functions $F_1(z), F_2(z) \in \mathscr{H}_2(\mathbb{E})$ are called orthonormal if $\langle F_1, F_2 \rangle = 0$ and $\langle F_1, F_1 \rangle = \langle F_2, F_2 \rangle = 1$.

2. ORTHONORMAL RATIONAL FUNCTIONS

First, we will review some basic facts for orthogonal all-pass transfer functions and corresponding state space realizations. Consider a real, single-input single-output, exponentially stable, all-pass transfer function $H_b(z)$, $H_b(z)H_b(1/z) = 1$, of order m, specified by its poles $\{\xi_j; |\xi_j| < 1, j = 1 \dots m\}$. Such a transfer function, often called an inner function, can be represented by a Blaschke product

$$H_b(z) = \prod_{j=1}^{m} \frac{1 - \xi_j^* z}{z - \xi_j}, \quad |\xi_j| < 1. \tag{9}$$

It can also be represented by an orthogonal state-space realization

$$\begin{bmatrix} x(t+1) \\ y(t) \end{bmatrix} = \begin{bmatrix} A & B \\ C & D \end{bmatrix} \begin{bmatrix} x(t) \\ u(t) \end{bmatrix},$$

$$\begin{bmatrix} A & B \\ C & D \end{bmatrix} \begin{bmatrix} A & B \\ C & D \end{bmatrix}^T = I. \tag{10}$$

Such a realization can easily be obtained by change of state variables, and is by no means unique. The special cases of a first order system and a second order system are, however, of special importance:

$$\begin{bmatrix} A & B \\ C & D \end{bmatrix} = \begin{bmatrix} a & \sqrt{1-a^2} \\ \sqrt{1-a^2} & -a \end{bmatrix}, \tag{11}$$

$$H_b(z) = \frac{1 - az}{z - a}, \quad -1 < a < 1,$$

$$\begin{bmatrix} A & B \\ C & D \end{bmatrix} = \begin{bmatrix} b & \sqrt{1-b^2} & 0 \\ c\sqrt{1-b^2} & -bc & \sqrt{1-c^2} \\ \sqrt{1-b^2}\sqrt{1-c^2} & -b\sqrt{1-c^2} & -c \end{bmatrix} \tag{12}$$

$$H_b(z) = \frac{-cz^2 + b(c-1)z + 1}{z^2 + b(c-1)z - c}, \quad -1 < b, c < 1.$$

An all-pass transfer function can be factorized as $H_b(z) = H_{b1}(z)H_{b2}(z)$, where $H_{b1}(z)$ and $H_{b2}(z)$ are lower order all-pass transfer functions with orthogonal state-space realizations

$$(A_1,B_1,C_1,D_1), \quad \text{and} \quad (A_2,B_2,C_2,D_2), \qquad (13)$$

respectively. Denote the corresponding state vectors by $x_1(t)$ and $x_2(t)$. It turns out that an orthogonal state space realization of $H_b(z)$ can be directly obtained by using the lumped state vector $x(t) = [x_1^T(t)\, x_2^T(t)]^T$. This observation is due to Mullis and Roberts, see (Roberts and Mullis, 1987). By recursively using the factorization results, an orthogonal state space realization (A,B,C,D) of $H_b(z)$ can be constructed by cascading orthogonal state space realization of its first order all-pass factors with a real pole (11), and its second order all-pass factors with two complex conjugated poles (12).

Study the input-to-state transfer function

$$V(z) = (zI - A)^{-1}B, \qquad (14)$$

for an orthogonal state space realization of an all-pass transfer function. Denote the kth canonical unit vector, $e_k = (0\ldots1\ldots0)^T$, where 1 is in position k. We then have the following fundamental result: Assume that (9) and (10) hold. The transfer functions $\{F_k(z) = e_k^T V(z),\, k = 1\ldots m\}$ are then orthonormal in $\mathcal{H}_2(\mathbb{E})$. The cascade realization gives

$$F_k(z) = \frac{\sqrt{1-|\xi_k|^2}}{z-\xi_k}\prod_{j=1}^{k-1}\frac{1-\xi_j^* z}{z-\xi_j}, \qquad (15)$$

which forms a complete set in $\mathcal{H}_2(\mathbb{E})$ if

$$\sum_{j=1}^{\infty} 1-|\xi_j| = \infty. \qquad (16)$$

3. LEAST SQUARES ESTIMATION

A natural extension of a AR model is to use the model structure

$$[H(z)]^{-1} = 1 + \sum_{k=1}^{n} f_k F_k(z), \qquad (17)$$

where $\{F_k(z)\}$ is a set of orthonormal rational functions with pre-specified poles. This is nothing but a way to represent an ARMA model with fixed zeros, i.e.

$$H(z) = \frac{C_*(z)}{A(z)}, \qquad (18)$$

where $C_*(z)$ is specified by the poles of $\{F_k(z)\}$, i.e. the zeros of $H(z)$. Consider the filtered process

$$y_c(t) = \frac{1}{C_*(q)}y(t). \qquad (19)$$

The problem of estimating a rational orthogonal model is completely equivalent to estimating an AR model for the process $\{y_c(t)\}$. The obvious question is why one should study a more complex model structure then the simple AR. There are several answers. One is that this structure leads to much more robust filter implementations, and numerically better scaled estimation problems. This structure also allows for simpler analysis as will be shown below.

Define the regression vector

$$\varphi(t) = [-F_1(q)y(t)\ldots - F_n(q)y(t)]^T. \qquad (20)$$

Given observations $\{y(1)\ldots y(N)\}$, the least squares estimate of the parameter vector $\theta = (f_1\ldots f_n)^T$ is given by

$$\hat{\theta} = R_N^{-1} f_N,$$

$$R_N = \frac{1}{N}\sum_{t=1}^{N}\varphi(t)\varphi^T(t),\; f_N = \frac{1}{N}\sum_{t=1}^{N}\varphi(t)y(t). \qquad (21)$$

Let $R = \mathrm{E}\{R_N\}$, i.e. the covariance matrix of the regression vector.

The numerical properties of the least squares problem depend on the condition number of R. The optimal case would be if R equals the identity matrix. This is the case for orthonormal rations functions models if $y(t)$ equals white noise, which of course is not of interest. For the AR case, $F_k(z) = z^{-k}$, the covariance matrix has a Toeplitz structure and a classical result is

$$\min_{\omega}\Phi(e^{i\omega}) \le \mathrm{eig}\{R\} \le \max_{\omega}\Phi(e^{i\omega}), \qquad (22)$$

where $\Phi(e^{i\omega})$ denotes the power spectral density of $y(t)$. It can be shown that the same result holds for a general orthonormal rational function model, see (Ninness *et al.*, 1999). The covariance matrix is also of interest in order to determine the statistical properties of the estimate since

$$\mathrm{Var}\{\hat{\theta}\} \approx \frac{\sigma_0^2}{N}R^{-1}, \qquad (23)$$

where the expression is asymptotic in the number of data N, and the bias due to model errors is neglected.

The sensitivity of the parameter vector θ as such is often of secondary interest. A more invariant measure is the variance of the estimated frequency function

$$\mathrm{Var}\{[\hat{H}(e^{i\omega})]^{-1}\} \approx$$
$$\frac{\sigma_0^2}{N}[F_1(e^{i\omega})\ldots F_n(e^{i\omega})]R^{-1}[F_1(e^{i\omega})\ldots F_n(e^{i\omega})]^*. \qquad (24)$$

It is possible to derive a more explicit expression for the following cases:

For large model orders n we have

$$\mathrm{Var}\{[\hat{H}(e^{i\omega})]^{-1}\} \approx \frac{1}{N}\frac{\sum_{k=1}^{n}|F_k(e^{i\omega})|^2}{|H_0(e^{i\omega})|^2}. \qquad (25)$$

From (15)

$$|F_k(e^{i\omega})|^2 = \frac{1-|\xi_i|^2}{|e^{i\omega}-\xi_k|^2}, \qquad (26)$$

where $\{\xi_k\}$ are the zeros of $H(z)$. See e.g. (Ninness *et al.*, 1999).

If the zeros of $H_0(z)$ is known to belong to a given set, we have the following result. Assume that

$$H_0(z) = \frac{C_0(z)}{A_0(z)}, \quad H(z) = \frac{C_1(z)C_0(z)}{A(z)}, \tag{27}$$

where degree$[H(z)] = n \geq$ degree$[H_0(z)] = n_0$. Then

$$\text{Var}\{[\hat{H}(e^{i\omega})]^{-1}\} \approx \frac{1}{N}\frac{\sum_{k=1}^{n_0}|F_k^0(e^{i\omega})|^2 + \sum_{j=1}^{n-n_0}|F_j^1(e^{i\omega})|^2}{|H_0(e^{i\omega})|^2}. \tag{28}$$

Here $\{F_k^0(z)\}$ are the orthogonal basis functions constructed using the poles of $H_0(z)$, and the $n - n_0$ basis functions $\{F_k^1(z)\}$ are constructed from the zeros $C_1(z)$. This result is a slight generalization of the AR variance expression given in (Lie and Ljung, 2001). The idea of proof is rather simple. Rewrite the prediction error problem as

$$y(t) = (1 - [H(q)]^{-1})y(t) + \varepsilon(t) \tag{29}$$

to obtain the artificial problem of estimating the transfer function of the system

$$y(t) = \frac{(C(q) - A(q))C_0(z)}{C(q)A^0(q)}e(t) + \varepsilon(t), \tag{30}$$

where $e(t)$ is a given white noise input signal with variance σ_0^2, $\varepsilon(t)$ is white measurement noise with variance σ_0^2, and $y(t)$ is the output signal. The transfer function $G(z) = (C(z) - A(z))/C_1(z)A^0(z)$ will have relative degree one with pre-specified poles. The corresponding orthogonal basis functions can be chosen as $\{F_k(z) = F_k^0(z)\}$, $k = 1\ldots n_0$ corresponding the the roots of $A_0(z) = 0$, and $F_{j+n_0}(z) = F_j^1(z)H_b(z)$, $j = 1\ldots n - n_0$, where $H_b(z)$ is the all-pass transfer function with the same poles as $H_0(z)$ and $\{F_k^1(z)\}$ are constructed from $C_1(z)$. Since $e(t)$ is white noise with variance σ_0^2, the covariance matrix of the regression vector for this estimation problem equals $R = \sigma_0^2 I$ and

$$\text{Var}\{\hat{G}(e^{i\omega})\} \approx \frac{\sum_{k=1}^{n_0}|F_k^0(e^{i\omega})|^2 + \sum_{j=1}^{n-n_0}|F_j^1(e^{i\omega})|^2}{N}. \tag{31}$$

Now $[H(e^{i\omega})]^{-1} = [1 - G(e^{i\omega})][H_0(e^{i\omega})]^{-1}$ and thus

$$\text{Var}\{[\hat{H}(e^{i\omega})]^{-1}\} = \frac{\text{Var}\{\hat{G}(e^{i\omega}\}}{|H_0(e^{i\omega})|^2}, \tag{32}$$

which gives (28). By setting $C_1(z) = z^{n-n_0}$ we obtain Theorem 5.1 in (Lie and Ljung, 2001).

If the spectral density of $y(t)$ was known one could use basis functions that are orthonormal for the weighted scalar product

$$< F_1^w, F_2^w >_w = \frac{1}{2\pi i}\oint_{\mathbb{T}} F_1^w(1/z)F_2^w(z)\Phi(z)\frac{dz}{z}. \tag{33}$$

This is very closely related to using a feedforward lattice model. Since R then equals the unity matrix, we directly obtain

$$\text{Var}\{[\hat{H}(e^{i\omega})]^{-1}\} \approx \frac{\sigma_0^2}{N}\sum_{k=1}^{n}|F_k^w(e^{i\omega})|^2. \tag{34}$$

Here we have neglected bias errors. Even if its easy to calculate the basis functions $F_k^w(z)$, it is more difficult to explicitly relate them to the poles and zeros of the model and the system.

To conclude: We shown how the choice of zeros will influence the variance of the estimator. The best choice is of course to take the true zeros of the process, i.e. $C(z) = C_0(z)$. The other extreme is to use a high order AR model with all zeros at $z = 0$. Orthogonal rational function models give a compromise between these extremes.

Next, we will list how some more results for AR estimation are generalized to spectral estimation using orthogonal rational functions models. The asymptotic, as the number of data tends to infinity, prediction error method estimate minimizes the cost function

$$\frac{1}{2\pi}\int_{-\pi}^{\pi}|H(e^{i\omega})|^{-2}\Phi(e^{i\omega})d\omega. \tag{35}$$

Denote the corresponding minimum by σ^2. As we have shown the orthogonal rational function model approach is theoretically equivalent to AR estimation of the process

$$y_c(t) = \frac{1}{C_*(q)}y(t),$$
$$[H(z)]^{-1} = 1 + \sum_{k=1}^{n} f_k F_k(z) = \frac{A(z)}{C_*(z)}. \tag{36}$$

It is well known that the AR estimate will converge to an stable filter as the number of data tends to infinity. This means that $[\hat{H}(z)]^{-1} = 1 + \hat{a}_k F_k(z)$ will converge to a stable and minimum phase transfer function.

It is also well known that the covariance function of the limiting AR estimate perfectly fits the n first covariance values of the underlying process. Let $r_c(\tau) = \text{E}\{y_c(t)y_c(t+\tau)\}$, $\tau = 0,\ldots n$. Hence, we solve the following covariance matching problem

$$r_c(\tau) = \frac{1}{2\pi}\int_{-\pi}^{\pi} e^{i\omega\tau}\frac{\sigma^2}{|A(e^{i\omega})|^2}d\omega,$$
$$= \frac{1}{2\pi}\int_{-\pi}^{\pi}\frac{e^{i\omega\tau}}{|C_*(e^{i\omega}|^2}\sigma^2|H(e^{i\omega})|^2 d\omega,$$
$$\tau = 0\ldots n. \tag{37}$$

4. THE COVARIANCE EXTENSION PROBLEM

Consider the covariance lags

$$r(\tau) = \text{E}\{y(t+\tau)y(t)\}, \tag{38}$$

and the corresponding spectral density

$$\Phi(e^{i\omega}) = \sum_{k=-\infty}^{\infty} r(\tau)e^{-i\omega\tau},$$

$$r(\tau) = \frac{1}{2\pi}\int_{-\pi}^{\pi} e^{i\omega\tau}\Phi(e^{i\omega})d\omega. \qquad (39)$$

Assume an ARMA model $H(z) = C(z)/A(z)$ with innovation variance σ^2, and let

$$Q(z) = A(z)A(1/z)/\sigma^2 = q_0 + q_1(z+z^{-1})\ldots + q_n(z^n + z^{-n}). \qquad (40)$$

Notice that $Q(z) > 0$ on the unit circle, i.e. a positive function. An ARMA process with degree constraint and covariances $r(\tau)$, $\tau = 0\ldots n$ must satisfy

$$r(\tau) - \frac{1}{2\pi}\int_{-\pi}^{\pi} e^{i\omega\tau}\frac{|C(e^{i\omega})|^2}{Q(e^{i\omega})}d\omega = 0, \quad \tau = 0,\ldots n \qquad (41)$$

But this is just the derivative of the cost-function:

$$V(q) = [r(0)\ldots r(n)]q - \frac{1}{2\pi}\int_{-\pi}^{\pi} \log Q(e^{i\omega})|C(e^{i\omega})|^2 d\omega \qquad (42)$$

with respect to $q = (q_0 \ldots q_n)^T$. The main result of Byrnes, Lindquist and co-workers is that $V(q)$ is a convex function. Hence we have only need to solve a convex optimization problem to find the solution of the covariance realization problem! The first term

$$[r(0)\ldots r(n)]q = \frac{1}{2\pi}\int_{-\pi}^{\pi} Q(e^{i\omega})\Phi(e^{i\omega})d\omega \qquad (43)$$

is just the AR prediction error cost function (which is quadric in A). The second term

$$-\frac{1}{2\pi}\int_{-\pi}^{\pi} \log Q(e^{i\omega})|C(e^{i\omega})|^2 d\omega \qquad (44)$$

can be viewed a barrier function to impose the positivity constraint. We refer to (Byrnes *et al.*, 2001*a*) for an excellent overview of the convex optimization approach to the rational covariance extension problem.

A natural question is why should one be interested in perfect fit of the n first covariances? It is well known that a fixed number of covariances is not a sufficient statistics for an ARMA model. As shown in Byrnes et. al. (Byrnes *et al.*, 2001*b*) it is possible to extend the covariance fitting problem to certain filter banks. Using a similar idea we will finally sketch on an extensions of this problem to the prediction error framework. As discussed in the previous section, the prediction error method approach gives perfect fit to the first $n+1$ covariances of $y_c(t)$, and corresponds to what is called the central solution. Using the method of Lindquist et. al. one could as well determine an ARMA model $C(z)/A(z)$ of order n which fits $r_c(\tau)$, by solving

$$r_c(\tau) - \frac{1}{2\pi}\int_{-\pi}^{\pi} e^{i\omega\tau}\frac{|C(e^{i\omega})|^2}{|C_*(e^{i\omega})|^2 Q(e^{i\omega})}d\omega = 0, \quad \tau = 0,\ldots n \qquad (45)$$

The same trick applies here, i.e. integration w.r.t. q gives the convex cost function

$$V_c(q) = [r_c(0)\ldots r_c(n)]q - \frac{1}{2\pi}\int_{-\pi}^{\pi} \log Q(e^{i\omega})\frac{|C(e^{i\omega})|^2}{|C_*(e^{i\omega})|^2}d\omega, \qquad (46)$$

which can be used to calculate the model $H(z) = C(z)/A(z)$. The choice $C(z) = C_*(z)$ gives back the prediction error solution, while using other $C(z)$ allows for different extensions. The first term is nothing but the standard prediction error, and it correspond to

$$[r_c(0)\ldots r_c(n)]q = \mathrm{E}\{([H(q)]^{-1}y(t))^2\},$$
$$[H(z)]^{-1} = 1 + \sum_{k=1}^{n} f_k F_k(z) \qquad (47)$$

with respect to $(f_1 \ldots f_n)$. It seems more difficult to express the second term using $\{F_k(z)\}$. It is, however, possible to represent

$$\frac{C(z)}{C_*(z)} = 1 + \sum_{k=1}^{n} \bar{f}_k F_k(z). \qquad (48)$$

It would be of interest to investigate if this representation leads to better numerical properties.

5. CONCLUSION AND FUTURE WORK

Models are always approximations of true data generating process. The quality of a model depends heavily on its intended use. If the objective is prediction, the prediction error approach is optimal. If one is interested in spectral properties in certain frequency band the answer is more difficult. The covariance extension approach and its generalizations provide promising methods for this case. The objective of this paper has been to discuss some relation between these two approaches to spectral estimation, and to show that pre-specified zeros indeed are a valuable tool to tune a spectral estimators.

6. REFERENCES

Bokor, J., P. Heuberger, B. Ninness, T. Oliveira e Silva, P. Van den Hof and B. Wahlberg (1999). Modelling and identification with orthogonal basis functions. In: *Workshop Notes, 14:th IFAC World Congress, Workshop nr 6*. Beijing, PRC.

Bultheel, A., P. González-Vera and O. Njåstad (1999). *Orthogonal Rational Functions*. Cambridge Monographs on Applied and Computational Mathematics. Cambridge University Press. Cambridge.

Byrnes, C.I., S.V. Gusev and A. Lindquist (2001*a*). From finite windows to modeling filters: A convex optimization approach. *SIAM Review* **43**(4), 645–675.

Byrnes, C.I., T. T. Geeorgiou and A. Lindquist (2001*b*). A new approach to spectral estimation: A tunable high-resolution spectral estimator. *IEEE Transactions on Signal Processing* **48**(11), 3189–3205.

Lie, L.-L. and L. Ljung (2001). Asymptotic variance expressions for estimated frequency functions. *IEEE Trans. Autom. Control* **46**(12), 1887–1899.

Ninness, B.M., H. Hjalmarsson and F. Gustafsson (1999). Generalised Fourier and Toeplitz results for rational orthonormal bases. *SIAM Journal on Control and Optimization* **37**(2), 429–460.

Roberts, R.A. and C.T. Mullis (1987). *Digital Signal Processing*. Addison-Wesley Publishing Company. Reading, Massachusetts.

Wahlberg, B. (2002). Orthonormal rational functions: A transformation analysis. *SIAM Review*. Accepted for publication.

IFAC Publications
www.elsevier.com/locate/ifac

IDENTIFICATION OF RATIONAL SPECTRAL DENSITIES USING ORTHONORMAL BASIS FUNCTIONS

A. Blomqvist [*,1] **G. Fanizza** [*,1,2]

[*] *Division of Optimization and Systems Theory, Department of Mathematics, Royal Institute of Technology, SE-100 44 Stockholm, Sweden*

Abstract: This paper gives an algorithm for identifying spectral densities using orthonormal basis functions. Mathematically, this amounts to identifying a time-invariant linear SISO system with the additional constraint that the transfer function should be positive-real. Thus, we solve the long-standing problem of how to incorporate this positivity constraint while using orthonormal basis functions. The procedure is a variant of the THREE algorithm introduced by Byrnes, Georgiou and Lindquist. The relation between and numerical properties of the proposed and the THREE algorithms are discussed. The orthonormal basis functions are better scaled for a concentrated pole selection in the basis, which increases the accuracy of the estimates. A numerical example which highlights this phenomenon and illustrates the algorithm is given. *Copyright © 2003 IFAC*

Keywords: System identification, Spectral estimation, Filter banks, Spectral density function, Linear filters

1. INTRODUCTION

Estimating rational spectral densities is equivalent to identifying time-invariant linear spectral factors under the additional constraint that spectral density should be real and positive on the unit circle. Then the spectral density can be written as the real part of a positive-real function. Therefore, the identification problem is an instance of identification of *passive* systems (see Caines (1988)). The nontrivial positivity constraint has (so far) not been possible to include in the identification procedures of Ho-Kalman type.

Different types of orthonormal basis functions for identification of general time-invariant linear systems have been thoroughly studied. They provide the means to incorporate a priori information of the system dynamics in the basis. But they also have advantages in terms of computational complexity and numerical conditioning. de Hoog et al. (2002) considers basis functions similar to the ones here for identifying general time-invariant linear systems; also see Ninness and Gustafsson (1997). These also give an introduction to orthonormal basis functions and their use in system identification.

Here the problem is treated from a quite different starting point leading up to an analytic interpolation problem. The algorithm uses the framework of Georgiou (2001) and is therefore closely related to the THREE algorithm proposed in Byrnes et al. (2000). The mathematical foundation is given in Georgiou and Lindquist (2002). The main contribution is the demonstration of how to identify passive systems using orthonormal basis functions. Furthermore, the paper contains a comparison to THREE basis functions, in particular for concentrated pole sets.

First the algorithm is described, then a comparison to the THREE algorithm is made and finally a numerical example is given. The notation is standard.

[1] This research was supported by the Swedish Research Council (VR)
[2] This research was supported by TMR

2. IDENTIFICATION ALGORITHM

The proposed identification procedure has four components: an orthonormal basis, a filter bank, an analytic interpolation problem and a homotopy continuation method. In the following each component is discussed and it is shown how to bring the components together.

2.1 *The Generalized Orthonormal Basis*

The orthonormal basis considered in this paper is generated from a finite set of points in the open unit disc. These points will be the poles of the basis functions. The basis is generalized in the sense that Laguerre, Kautz and other bases are special cases.

Let $\{\xi_k\}_{k=1}^{n} \subset \mathbb{D}$ be given. Then define the functions

$$G_k(z) := \frac{\sqrt{1-|\xi_k|^2}}{z-\xi_k} \prod_{j=1}^{k-1} \left(\frac{1-\xi_j^* z}{z-\xi_j} \right) \quad \forall k. \tag{1}$$

These basis functions date back to early work in the 1920's by Takenaka and Malmquist. They are constructed by all-pass factors with the balanced state-space realization:

$$\frac{1-\xi_k^* z}{z-\xi_k} \sim \left[\begin{array}{c|c} \xi_k & \sqrt{1-|\xi_k|^2} \\ \hline \sqrt{1-|\xi_k|^2} & -\xi_k \end{array} \right], \tag{2}$$

Due to the well-known recursive relationship between two such function $G_k(z)$ and $G_l(z)$:

$$G_k(z)G_l(z) \sim \left[\begin{array}{cc|c} A_k & 0 & B_k \\ B_l C_k & A_l & B_l D_k \\ \hline D_l C_k & C_l & D_l D_k \end{array} \right], \tag{3}$$

the finite product of all-pass functions allows a balanced, minimal state-space realization. This will prove to be very useful in the filter bank construction in the next section.

Remark 1. If the poles are complex-conjugated, i.e., $\bar{\xi}_k$ is in the pole set whenever $\xi_k \notin \mathbb{R}$ is, we can get a real-valued state-space realization using e.g. a 2-parameter Kautz model.

2.2 *The Filter Bank Construction*

A filter bank constituted of the basis functions will put the problem into the desired form. Define the $(n+1) \times 1$ vector-valued analytic function

$$G(z) := \begin{bmatrix} G_0(z) \\ G_1(z) \\ \vdots \\ G_n(z) \end{bmatrix}, \tag{4}$$

where $G_0(z) \equiv 1$. It can be viewed as a bank of filters as showed in Figure 1. Now, construct a minimal balanced state-space realization $\{A, B, C, D\}$ as in Section 2.1 for the last basis function, $G_n(z)$. Then, a minimal, i.e., $(n+1)$-dimensional, Input-to-State (IS) realization, $\{A, B\}$, for the filter bank is given by the same A and B. The pair (A, B) will be a controllable pair.

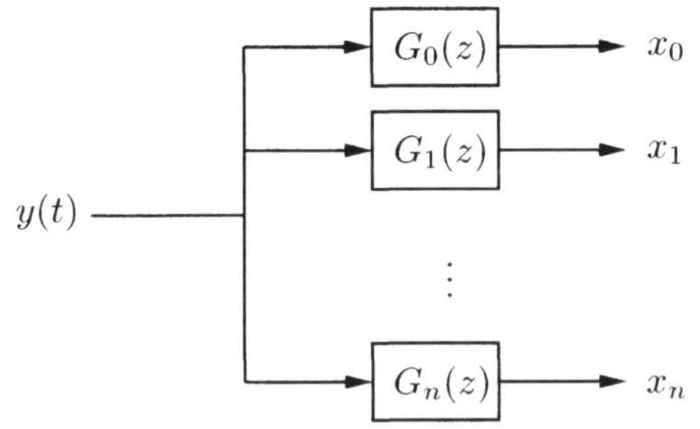

Fig. 1. The filter bank

The Linear Predictive Filter, a maximum entropy filter, is one example that falls under this framework. As pointed out in Georgiou (2001) the corresponding basis functions are $\{z^{-k}\}_{k=0}^{n}$. Thus, they can be interpreted as being generated as in (1) from a set with all poles at origin.

2.3 *The Analytic Interpolation Problem*

Given the IS realization of the filter bank, the identification problem can be stated as an analytic interpolation problem; see Georgiou (2001). The interpolant whose spectral density is closest to an a priori estimate $\Psi(z)$ in terms of the Kullback-Leibler distance is given by Georgiou and Lindquist (2002) and is shown to be the unique minimizer of a certain convex optimization problem.

The analytic interpolation problem is derived in Georgiou (2001). First, a characterization of the state-covariance matrices $P := \epsilon\{x^k x^{k*}\}$ (where ϵ denotes the expectation) for an IS filter is given. A state-covariance matrix should be positive definite and fulfill the algebraic condition

$$P = \frac{1}{2}(WE + EW^*), \tag{5}$$

where E is the reachability Gramian and thus the unique positive definite solution to the Lyaponov equation $E - AEA^* = BB^*$. For the orthonormal basis, E is the identity matrix, so P is a Toeplitz matrix. Furthermore, W and A commutes and W admits the representation

$$W = w(A) = w_0 I + w_1 A + \cdots + w_n A^n. \tag{6}$$

With this notation, the positive-real part of the spectral density, $f_y(z) : \Phi(z) = f_y(z) + f_y^*(z)$, should fulfill the interpolation condition

$$f_y(A^*) = W^*. \tag{7}$$

This leads to a generalization of the Nevanlinna-Pick interpolation problem. The state covariance matrix plays the role of a generalized Pick matrix.

Now assume that given data we estimate W consistent with (5) and want to minimize the so-called Kullback-Leibler distance to a given spectral density while meeting the interpolation condition (7). The solution is given by the following theorem of Georgiou and Lindquist (2002):

Theorem 2. Given $P \succ 0$ and (A, B) as earlier and a spectral density $\Psi(z)$, there is a unique $\Phi(z) = f_y(z) + f_y^*(z)$ that minimizes Kullback-Leibler distance

$$\mathbb{S}(\Psi \| \Phi) := \int \Psi \log \frac{\Psi}{\Phi} \tag{8}$$

subjected to $f_y(A^*) = W^*$. The minimizer takes the form

$$\hat{\Phi} = \frac{\Psi}{G^* \hat{\Lambda} G}$$

where $\hat{\Lambda}$ is the unique interior minimizer to the convex functional

$$\mathbb{J}_\Psi(\Lambda) := \text{trace}(\Lambda P) - \int \Psi \log(G^* \Lambda G). \tag{9}$$

Here and later the integration limits and variables are suppressed:

$$\int g := \frac{1}{2\pi} \int_{-\pi}^{\pi} g(e^{i\theta}) d\theta.$$

If $\Psi(z) \equiv 1$, that is white noise with unit covariance, the solution is called the *maximum entropy solution*.

Remark 3. In de Hoog et al. (2002) a corresponding interpolation problem is also derived. That interpolation problem apparently look the same as the one of Theorem 2 but they are fundamentally different: here the interpolant is the positive-real part of the spectral density while it is the spectral factor itself in de Hoog et al. (2002).

2.4 *The Homotopy Continuation Method*

The optimization problem (9) in Theorem 2 may suffer from numerical problems, as pointed out in Georgiou and Lindquist (2002). In Blomqvist and Nagamune (2002) a numerical algorithm which seems to have better numerical properties is developed along the lines of Nagamune (2001).

The original, convex, functional $\mathbb{J}_\Psi(\Lambda)$ has the property that the gradient is infinite on the boundary of the feasible region. This causes numerical problems when the minimizer is close to the boundary. To avoid this, change variables to $\alpha(z) := C(z)\tau(z)$, where $C^*G^*(z)G(z)C = G^*(z)\Lambda G(z)$, and let $\boldsymbol{\alpha}$ be a vector with the n coefficients of $\alpha(z)$. The functional can then be written

$$J_\Psi(\alpha) = \boldsymbol{\alpha}^T K \boldsymbol{\alpha} - 2 \int \Psi \log \alpha, \tag{10}$$

where

$$K := L_n^{-T} \Gamma^{-1} P \Gamma^{-T} L_n^{-1}, \tag{11}$$

with L_n nonsingular and given from the poles along with Γ as the controllability matrix of the IS realization of the filter bank. Thus, the problem is of the type in Nagamune (2001) and the same homotopy continuation method can be applied. The condition number of the matrix K will be of importance and will be discussed in Section 4.

3. ESTIMATING P AND W FROM DATA

The state covariance matrix from an IS filter as in Section 2.2 is necessarily positive semi-definite but it also fulfills the algebraic condition (5). However, experimental data typically does not. This section deals with the problem how to enforce the algebraic condition.

Given measurements $\{x^k\}_{k=1}^N$ (each x^k is a vector with components $x_0^k \dots x_n^k$) an estimate of the state covariance matrix P can be computed, e.g., as

$$\hat{P} = \frac{1}{N} \sum_{k=1}^{N} x^k x^{k*}.$$

For this estimate there is typically no solution to (5). Georgiou proposes a least-squares solution in terms of the coefficients $\{w_k\}_{k=0}^n$ in (6). However, this will not guarantee the modified P to be positive semi-definite.

To circumvent the problem of non-positive semi-definite state covariances, the least-squares program can be stated as a semi-definite program. This will guarantee the modified P to be positive at the cost of computational effort. More precisely, P is taken to be the solution of

$$(P) \quad \begin{aligned} &\min \ \|\hat{P} - P\|_{\text{Frob}} \\ &\text{s.t.} \ \ P = \frac{1}{2}(WE + EW) \succeq 0\,, \\ &\qquad AW = WA \end{aligned} \tag{12}$$

where E is as in (5). This can be rewritten as the semi-definite program

$$(SDP) \quad \begin{aligned} &\min \ t \\ &\text{s.t.} \ \|\operatorname{Vec}(\hat{P}) - \operatorname{Vec}(P)\|_2 \leq t\,, \\ &\qquad -P \preceq 0 \end{aligned} \tag{13}$$

where feasible matrices are parameterized by (5) and (6). The program (SDP) is on standard form and there are several software packages available.

Remark 4. The program (SDP) is computationally considerably more expensive than a least-squares solution. Thus, an implementation should first compute the least-squares solution and check the positivity before proceeding to solve the (SDP).

Remark 5. This also allows the possibility, by requiring $P \succeq \varepsilon I$ for some ε, to avoid a high condition number in the optimization problem (10) caused by inverting a nearly singular matrix.

4. COMPARISON TO THE THREE ALGORITHM

The proposed algorithm resembles the THREE algorithm of Byrnes et al. (2000) and they will be compared in the subsequent. Firstly, the convergence rate of the expansion coefficients is discussed. Secondly, the numerical properties of the optimization problem (10) are analyzed. They turn out to be essentially the same, even though the orthonormal approach is internally better scaled. Finally, the accuracies of the state covariance estimates are considered. Generally, the orthonormal construction will be less noise-sensitive when the poles are concentrated, which is an interesting case when the system dynamics is concentrated to one frequency region. This will cause the main difference in the performance for the methods.

4.1 *Convergence rate of expansion coefficients*

A fast rate of convergence in terms of the expansion coefficients for the system to be identified has been a major reason for using orthonormal basis functions in identification, see for instance Wahlberg (1991). The same arguments can be used for the subclass of systems studied in this paper. If the poles of the basis functions are close to the dominant ones of the spectral density to be identified, it will require fewer coefficients to catch most of the dynamics. Thus this motivates both algorithms.

4.2 *Numerical properties of the optimization problem*

The THREE and orthonormal basis functions span the same space, so there is a non-singular coordinate transformation, T, between the basis functions:

$$G^{\text{THREE}}(z) = TG^{\text{Orth}}(z).$$

This gives a relation between the IS realizations

$$\begin{aligned} G^{\text{THREE}}(z) &= T(I - zA^{\text{Orth}})^{-1}B^{\text{Orth}}, \\ &= (I - zTA^{\text{Orth}}T^{-1})^{-1}TB^{\text{Orth}}, \\ &= (I - zA^{\text{THREE}})^{-1}B^{\text{THREE}}, \end{aligned}$$

which in turn gives relations between the controllability matrices and the true state covariances matrices respectively:

$$\Gamma^{\text{THREE}} = T\Gamma^{\text{Orth}}, \tag{14}$$

$$\begin{aligned} P^{\text{THREE}} &= \int G^{\text{THREE}}\Phi G^{\text{THREE}\,*}, \\ &= \int TG^{\text{Orth}}\Phi G^{\text{Orth}\,*}T^T, \\ &= TP^{\text{Orth}}T^T. \end{aligned} \tag{15}$$

In particular the accuracy of the estimates and condition number of the matrix K in (11) are important; for the maximum entropy solution the K matrix is actually inverted. For the true state covariance matrices, i.e., not noise-corrupt, (15) hold. Then K is invariant under coordinate changes:

$$\begin{aligned} K^{\text{THREE}} &= L_n^{-T}\Gamma^{\text{THREE}-1}P^{\text{THREE}}\Gamma^{\text{THREE}-T}L_n^{-1}, \\ &= L_n^{-T}\Gamma^{\text{Orth}-1}TT^{-1}P^{\text{Orth}}T^{-T}T^T\Gamma^{\text{Orth}-T}L_n^{-1}, \\ &= K^{\text{Orth}} := K. \end{aligned}$$

Therefore the matrices K, and thus the optimization problems, will converge when the data sequence goes to infinity.

Thus we can conclude that for reasonably long data sequences the numerical properties are essentially the same for the two problems. Even so, it is instructive to consider the internal conditioning of the K matrices. Firstly, consider the controllability matrix Γ. If the pole set is concentrated, i.e., if many poles are closely located, the condition number of Γ will be higher for the THREE filter bank. A two-pole example illustrates this:

Example 6. Given the poles $\{a, a+\varepsilon\}$ the IS realization of the filter banks for the THREE algorithm and the orthonormal basis functions are easily determined. The corresponding controllability matrices are

$$\begin{aligned} \Gamma^{\text{THREE}} &= \begin{bmatrix} 1 & a \\ 1 & a+\varepsilon \end{bmatrix}, \\ \Gamma^{\text{Orth}} &= \begin{bmatrix} \tilde{b} & a\tilde{b} \\ -a\hat{b} & \hat{b}(1-|a|^2-a^2-\varepsilon a) \end{bmatrix}, \end{aligned}$$

respectively and where $\hat{b} = \sqrt{1-|a+\varepsilon|^2}$ and $\tilde{b} = \sqrt{1-|a|^2}$. If the absolute value of ε is small compared to a, the condition numbers under the Euclidean norm can be computed as

$$\begin{aligned} \kappa_2(\Gamma^{\text{THREE}}) &= \kappa_2\left(\begin{bmatrix} 1 & 0 \\ 1 & 1 \end{bmatrix}\begin{bmatrix} 1 & a \\ 0 & \varepsilon \end{bmatrix}\right), \\ &= \kappa_2\left(\begin{bmatrix} 1 & a \\ 0 & \varepsilon \end{bmatrix}\right) = \frac{1}{\varepsilon}, \\ \kappa_2(\Gamma^{\text{Orth}}) &= \kappa_2\left(\begin{bmatrix} \tilde{b} & 0 \\ -a\hat{b} & \hat{b}(1-|a|^2-\varepsilon a) \end{bmatrix}\begin{bmatrix} 1 & a \\ 0 & 1 \end{bmatrix}\right), \\ &= \kappa_2\left(\begin{bmatrix} \tilde{b} & 0 \\ -a\hat{b} & \hat{b}(1-|a|^2-\varepsilon a) \end{bmatrix}\right), \\ &= \frac{\sqrt{1-|a|^2}}{\sqrt{1-|a+\varepsilon|^2}(1-|a|^2-\varepsilon a)}, \end{aligned}$$

respectively. Here the condition number for the THREE controllability matrix is considerably larger. Having several poles close to each other enforces this behavior.

Secondly, consider state covariance estimate $\hat{P}$. Its condition number will depend on the condition number of the corresponding controllability matrix. For infinite data sequences we have the bounds:

$$\frac{1}{\kappa_2(\Gamma)^2} \le \frac{\kappa_2(P)}{\kappa_2(L_n^T K L_n)} \le \kappa_2(\Gamma)^2. \qquad (16)$$

This gives a relation for the condition number of the state-covariance matrices:

$$\kappa_2(P^{\mathrm{Orth}}) \le \frac{\kappa_2(\Gamma^{\mathrm{Orth}})^2}{\kappa_2(\Gamma^{\mathrm{THREE}})^2} \kappa_2(P^{\mathrm{THREE}}). \qquad (17)$$

Thus the condition number of the state covariance matrix will generally be higher when the corresponding controllability matrix is. The conclusion is that the optimization problem (10) will have approximately the same conditioning for the two sets of basis functions. However, will the problem with the orthonormal basis function be internally better scaled.

Here it is interesting to consider the basis functions z^{-k}. They are generically well-scaled since the corresponding controllability matrix is the identity matrix.

4.3 *Accuracy in estimating the state covariance*

For relatively short data sequences there are two effects that will determine the accuracy. Firstly, the orthonormal basis functions span the space so that each state contain information that is not in the prior states. From (17) we see that this will reduce the condition number of the state covariance matrix. This makes it less sensitive to noise. Secondly, the time lags in the filter banks will decrease the accuracy. As pointed out in Byrnes et al. (2000) one of the merits of the THREE algorithm is the time lag is one for all filters in the filter bank. These two effects work in different directions and which that is dominant will depend on the choice of poles for the filter banks.

5. A NUMERICAL EXAMPLE

Consider the shaping filter

$$W(z) = \frac{z - 0.9}{z - 0.8}.$$

Driven by Gaussian white noise, the output of the system is measured as depicted in Figure 2. The task is to recover the original system as accurately as possible. The shaping filter $W(z)$ is chosen to a simple first order filter. The zero relatively close to the pole will make it fairly hard to identify.

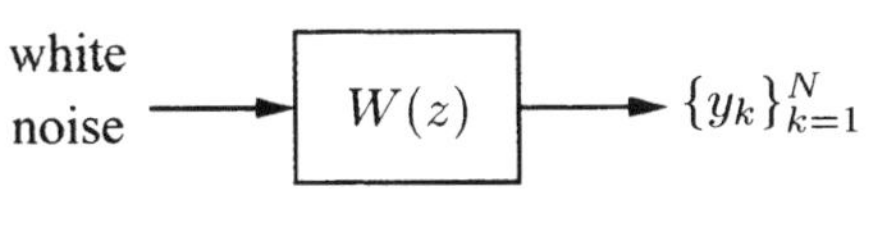

Fig. 2. The shaping filter in the numerical example

Given the artificial measurements $\{y_k\}_{k=1}^N$, the shaping filter (and correspondingly the spectral density) is identified using a Linear Prediction Filter (LPC), the THREE algorithm (THREE) and the orthonormal basis function procedure of this paper (Orth). The poles are chosen to be the same for the THREE and orthonormal algorithms and they are chosen to be in the vicinity of the pole at 0.80, but not necessarily identically to it. Since the THREE algorithm does not allow for repeated poles in the basis the case of distinct poles is consider. Also note that since the proposed algorithm is developed in the same framework as the THREE, the optimization problem (10) is also solved using the THREE algorithm.

In order to compare two estimates of the shaping filters, we compute the Kullback-Leibler distance, defined in (8), between the true and the estimated normalized spectral densities: $\mathbb{S}(\hat{\Phi}^{true} \| \hat{\Phi}^{estimate})$. The normalization guarantees that the distance will be nonnegative and that it is zero exactly when the normalized densities coincide.

The system is driven for a while so that the stationary assumption of the filter is at least approximatively valid. For each parameter set the 100 Monte Carlo simulations are performed and the average Kullback-Leibler distance is computed. Here the maximum entropy solution is computed, for simpler comparison. This means that the accuracy of the estimates can be significantly increased if a good initial estimate is given.

As discussed in Section 4 both the number of basis functions in the filter bank, this is the order of the identified system, and the length of the data sequence affects procedure differently. In the Table 1 the results are given for all combinations of the data length $N = 20, 200$ & 2000 and the number of basis functions $n = 1, 3$ & 5. In addition a few estimate for the case $n = 1$ and $N = 2000$ is plotted in Figure 3. The basis function poles are chose to be $\{0.75\}$, $\{0.75, 0.70$ & $0.80\}$, and $\{0.75, 0.70, 0.80, 0.72$ & $0.78\}$ for the different values of n. Note that for the orthonormal basis functions the ordering of the poles is vital (see Bodin et al. (2000) for a discussion in the case without positivity constraint).

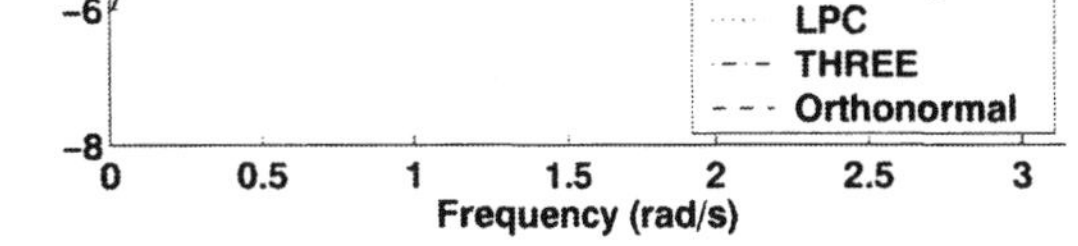

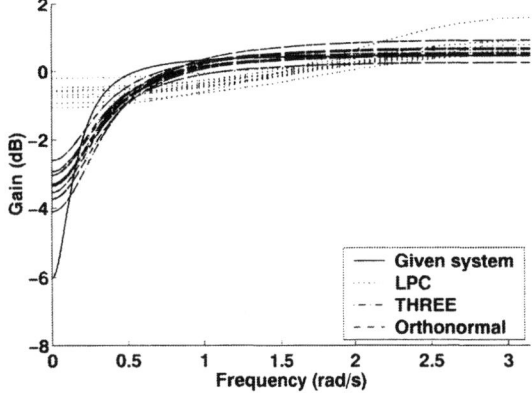

Fig. 3. The true and some identified spectral density for $n = 1$ and $N = 2000$

Table 1. Kullback-Leibler distances between true and estimated spectral densities

N	20			200			2000		
n	LPC	THREE	Orth	LPC	THREE	Orth	LPC	THREE	Orth
1	0.0633	0.0350	0.0392	0.0189	0.0052	0.0053	0.0153	0.0032	0.0032
3	0.1779	0.0354	0.1367	0.0235	0.0311	0.0097	0.0099	0.0288	0.0010
5	0.2704	0.0308	0.2136	0.0300	0.0324	0.0177	0.0076	0.0321	0.0025

The results are as expected from the discussion in Section 4. It is clear that the non-default basis choice can increase the accuracy in the estimates. The THREE algorithm gets problems in estimating the state covariance matrix in more cases than the orthonormal algorithm. For very short data sequences the THREE algorithm produces a better result when i works. Both the THREE and the orthonormal algorithms seem to perform worse when the number of basis functions increases. Thus, they are better suited for directly identifying a low-order model rather than first identifying a high-order model that is to be model-order reduced.

Remark 7. There is no comparison to the approximative identification algorithm of de Hoog et al. (2002) made since that algorithm is designed for deterministic identification, that is when the input signal is available. However, it would be possible to replace the least square estimates of Van den Hof et al. (1995) with estimate given from the interpolation problem and translated to the spectral factor. Applying the method directly to the transfer function $f_y(z)$ would not guarantee positive-realness.

Remark 8. In Byrnes et al. (2000) and Georgiou (2001) the increased precision in certain frequency ranges is emphasized, but in this example the whole frequency region is considered.

6. CONCLUSIONS

This paper provides a procedure for spectral estimation using orthonormal basis functions. The filter bank framework allow for combination of different types of basis functions including the orthonormal of this paper. It is discussed and illustrated how the orthonormality in the basis can be important; in particular this seems to be important for concentrated poles in the basis. An extension to vector processes seems fairly straight forward in the light of Blomqvist et al. (2002).

ACKNOWLEDGMENT

The authors want to thank their advisor Anders Lindquist and Ryozo Nagamune for suggestion of research topic and guidance.

References

A. Blomqvist, A. Lindquist, and R. Nagamune. Matrix-valued Nevanlinna-Pick interpolation with complexity constraint: An optimization approach. Preprint, 2002.

A. Blomqvist and R. Nagamune. A numerical algorithm for Kullback-Leibler approximations. Preprint, 2002.

P. Bodin, L. Villemoes, and B. Wahlberg. Selection of best orthonormal rational basis. *SIAM J. Contr. and Optimiz.*, 38(4):995–1032, 2000.

C. I. Byrnes, T. T. Georgiou, and A. Lindquist. A new approach to spectral estimation: A Tunable High-Resolution Spectral Estimator. *IEEE Trans. Signal Processing*, 48(11):3189–3205, November 2000.

P. E. Caines. *Linear Stochastic Systems*. John Wiley & Sons, New York, 1988.

T. de Hoog, Z. Szabó, P. Heuberger, P. Van den Hof, and J. Bokor. Minimal partial realization from generalized orthonormal basis function expansions. *Automatica*, 38:655–669, 2002.

T. T. Georgiou. Spectral estimation via selective harmonic amplification. *IEEE Trans. Automat. Control*, 46(1):29–42, January 2001.

T. T. Georgiou and A. Lindquist. Kullback-Leibler approximation of spectral density functions. Preprint, 2002.

R. Nagamune. A robust solver using a continuation method for Nevanlinna-Pick interpolation with degree constraint. In *Proceeding of the 40th IEEE Conference on Decision and Control*, pages 1119–1124, Orlando, Florida, December 2001.

B. Ninness and F. Gustafsson. A unifying construction of orthonormal bases for system identification. *IEEE Trans. Automat. Control*, 42(4):515–521, 1997.

P. Van den Hof, P. Heuberger, and J. Bokor. System identification with generalized orthonormal basis functions. *Automatica*, 31:1821–1834, 1995.

B Wahlberg. System identification using Laguerre models. *IEEE Trans. Automat. Control*, 36(5):551–562, 1991.

IFAC Publications
www.elsevier.com/locate/ifac

ORTHONORMAL BASIS FUNCTIONS FOR MODELING CONTINUOUS-TIME FRACTIONAL SYSTEMS [1]

Mohamed AOUN[+], Rachid MALTI[*], Francois LEVRON[□] and Alain OUSTALOUP[+]

[+] *LAP – UMR 5131 CNRS – Université Bordeaux I – ENSEIRB*
351 cours de la Libération, 33405 Talence cedex, France
Tel : +33 (0)556 842 418 Fax : +33 (0)556 846 644
aoun@lap.u-bordeaux.fr

[*] *I²S – IUT de Sénart – Université Paris 12*
Avenue Pierre Point, 77127 Lieusaint, France
Tél: +33 (0)164 135 183 Fax: +33 (0)164 134 503
malti@univ-paris12.fr

[□] *Institut de Mathématique de Bordeaux*
Université de Bordeaux I
levron@math.u-bordeaux.fr

Abstract: The classical Laguerre functions are known to be divergent as soon as their differentiation order is non-integer. They are hence inappropriate for representing fractional differentiation systems. A complete orthogonal basis, having fractional differentiation orders and a single pole, is synthesized. It extends the well-known definition of Laguerre functions to fractional systems. Hence a new class of fixed denominator models is provided for system identification. Fourier coefficients are computed using a least squares method. The least squares error is plotted versus the differentiation order and the pole, in an example, and shows that an optimal differentiation order may be located away from an integer number. Hence, the use of the synthesized basis is fully justified. *Copyright © 2003 IFAC*

Keywords: Orthogonal functions, fractional differentiation, Laguerre filters, System identification.

1. INTRODUCTION

Over the last twelve years, identification and control of linear stable dynamic systems using orthonormal functions have widely been used; see for instance (Wahlberg, 1991), (Van den Hof *et al.*, 1995), (Ninness and Gustafsson, 1997), (Malti *et al.*, 1998) and all references therein. The most popular orthogonal functions used in control engineering are: Laguerre functions (having a single pole), Kautz functions (having two complex conjugate poles) and the Generalized Orthogonal Basis (GOB) functions which extend the two former definitions to any number of real or complex conjugate poles.

El-Sayed (1999) has proposed to extend the definition of Laguerre functions by simply allowing their differentiation orders to be real. However, Abbot (2000) has proven that Laguerre functions are divergent as soon as their differentiation order is non-integer commenting on El-Sayed's work.

In this paper, a fractional orthonormal basis will be synthesized by applying Gram-Schmidt orthogonalization procedure. Hence a fixed denominator model will be proposed for the class of fractional differentiation systems. The new orthonormal basis will have a single pole and differentiation orders multiples of a single number, $n \in]1/2, 2[$, known as commensurate order. Laguerre functions will be considered as a special case (when $n = 1$) of this basis. Up to our knowledge, this is the first fractional orthonormal basis ever proposed.

The interest of fractional differentiation systems comes from the fact that studies on real systems such as thermal (Battaglia *et al.* 2000) or electrochemical (Darling and Newman, 1997), reveal inherent fractional differentiation behavior. The use of classical methods (based on integer order differentiation) is thus inappropriate in identifying these fractional systems. Thus, models using fractional differentiation, have been developed (Le Lay, 1998; Trigeassou et al., 1999; Cois, 2002; Aoun et *al.*, 2002).

The paper is organized as follows. The second section presents some mathematical background on fractional differentiation and its use in dynamical system representation and defines orthogonal basis functions and more precisely the classical Laguerre functions. Section 3 focuses on the synthesis of the new fractional orthonormal basis. In section 4 a numerical example of system identification using the new basis is developed.

2. MATHEMATICAL BACKGROUND

2.1. Definition of fractional differentiation

The concept of differentiation to an arbitrary order (also called fractional differentiation) was defined in the 19th century. The main contribution to the establishment of the definition is due to Riemann and Liouville. Their main concern was to extend differentiation by using not only integer but also non-integer orders. The nth fractional order differentiation of a continuous real

[1] This work is a part of the '*Mobilité*' research project: '*identification de systèmes linéaires et non linéaires à derivées fractionnaires*' financially supported by STIC – CNRS, gratefully acknowledged.

function $f(t)$ is defined as (Miller and Ross, 1993):

$$D^n f(t) \overset{\Delta}{=} \left(\frac{d}{dt}\right)^{\lfloor n+1 \rfloor} \left(I^{1-n} f(t)\right) \quad (1)$$

where $\lfloor n+1 \rfloor$ is the floor of $n+1$ and $I^{1-n} f(t)$ is the fractional integral of order $1-n$ of $f(t)$:

$$\left(I^{1-n} f\right)(t) \overset{\Delta}{=} \frac{1}{\Gamma(1-n)} \int_0^t \frac{f(\tau)}{(t-\tau)^n} d\tau , \quad (2)$$

Second definition of fractional differentiation is due to Grünwald:

$$D^n f(t) = \lim_{h \to 0} \frac{1}{h^n} \sum_j^{t/h} (-1)^j \binom{n}{j} f(t - jh) \quad (3)$$

The two definitions (1) and (3) are equivalent when $f(t)$ is relaxed at $t = 0$.

From equation (3), it can be noted that fractional differentiation is not a local operator. The value of the fractional derivative function at t depends on the whole past of the function. However, in the case where the differentiation order is an integer value, the derivative function depends only on some local points.

The Laplace transform of $D^n f(t)$, when $f(t)$ is relaxed at $t = 0$, is given by (Oldham and Spanier, 1974):

$$\mathcal{L}\{D^n f(t)\} = s^n F(s) \quad (4).$$

This result is coherent with the classical case where n is an integer. Consequently, it is easy to define a symbolic representation of a fractional dynamic system, using a differential equation, a transfer function, or a state space representation.

* *Differential equation*

$$\begin{aligned} & a_1 s^{n_{a1}} Y(s) + \cdots + a_L s^{n_{aL}} Y(s) \\ & = b_1 s^{n_{b1}} U(s) + \cdots + b_J s^{n_{bJ}} U(s) \end{aligned} \quad (5)$$

* *Transfer function*

$$\frac{Y(s)}{U(s)} = \frac{b_1 s^{n_{b1}} + \ldots + b_J s^{n_{aJ}}}{a_1 s^{n_{a1}} + \ldots + a_L s^{n_{aL}}} = \frac{B(s)}{A(s)} \quad (6)$$

* *State space representation* (Matignon, 1994):

$$\begin{cases} \left(D^{(n)} \mathbf{x}\right)(t) = \mathbf{A}\, \mathbf{x}(t) + \mathbf{B}\, u(t) \\ y(t) = \mathbf{C}\, \mathbf{x}(t) \end{cases} . \quad (7)$$

with n a real number known as commensurate order.

2.2. Scalar product, orthogonality and Laguerre functions

Laguerre, Kautz and GOB functions form a complete orthonormal basis in $L_2[0, \infty[$, according to the usual definition of the scalar product:

$$\langle l_n(t), l_m(t) \rangle = \int_0^\infty l_n(t) l_m(t) dt = \delta_{nm} \quad (8)$$

which reciprocal in the frequency domain is obtained by Plancherel's theorem:

$$\langle L_n(s), L_m(s) \rangle = \frac{1}{2\pi j} \int_{-j\omega}^{j\infty} L_n(j\omega) \overline{L_m(j\omega)}\, d\omega = \delta_{nm} \quad (9)$$

Any function $f(t) \in L_2[0, \infty[$, thus satisfying:

$$\langle f(t), f(t) \rangle^{\frac{1}{2}} = \|f\|_2 < \infty ,$$

can be written as a linear combination of these functions:

$$F(s) = \sum_{n=0}^{\infty} a_n L_n(s) \quad (10)$$

Usually, (10) is truncated to a given order N which is justified by the fact that Fourier coefficients are convergent as n tends to infinity. $F(s)$ is hence approximated by the finite sum:

$$F(s) \approx F_N(s) = \sum_{n=0}^{N} a_n L_n(s) \quad (11)$$

The Fourier coefficients are computed by minimizing the least squares criterion:

$$J = \int_0^\infty \left(f(t) - f_N(t)\right)^2 dt \quad (12)$$

Equation (12) corresponds to the L_2 norm of the approximation error, according to the definition of the scalar product (8):

$$J = \|f(t) - f_N(t)\|^2 \quad (13)$$

Minimizing J and taking advantage of the orthogonality of Laguerre functions, the Fourier coefficients are obtained by computing the scalar product:

$$a_n = \langle f(t), l_n(t) \rangle \quad (14)$$

either in the time or the frequency domain.

As stated previously $l_n(t)$ can represent any set of orthonormal functions in $L_2[0, \infty[$ (even the new basis which will be synthesized in the next section).

In the case where Laguerre functions are used, $l_n(t)$ is defined as:

$$l_n(t) = \sqrt{2\lambda} \frac{e^{\lambda t}}{n!} \frac{d^n\left(t^n e^{-2\lambda t}\right)}{dt^n} \quad (15)$$

They have the following Laplace transform:

$$L_n(s) = \sqrt{2\lambda} \frac{(s-\lambda)^n}{(s+\lambda)^{n+1}} \quad (16)$$

Laguerre functions are well suited for modelling systems having a dominant dynamics, because they have a single pole.

However, if $F(s)$ is a fractional system, Laguerre functions will fail to capture the fractional dynamics of the system with a small number of functions. As a consequence, N will be large in (11).

On the other hand, as soon as n becomes non-integer, (Abott, 2000) has shown that Laguerre functions (15) are divergent:

$$\int_0^\infty \left(l_n(t)\right)^2 dt = \infty \qquad \forall\, n \in \mathbb{R}^+ - \mathbb{N}^+ \quad (17)$$

Therefore, the generalization of Laguerre function by simply allowing to the differentiation order to be real is not possible.

The main contribution of this paper is to build a new basis function which has, as Laguerre functions, a single pole. Moreover, it will have an additional degree of freedom: differentiation order.

3. SYNTHESIS OF A FRACTIONAL ORTHONORMAL BASIS

Gram-Schmidt orthogonalization procedure is applied on function series:

$$F_1(s)=\frac{1}{s^n+\lambda},\ F_2(s)=\frac{1}{\left(s^n+\lambda\right)^2},\ldots,F_m(s)=\frac{1}{\left(s^n+\lambda\right)^m}$$

where: $m=1,2,3...$ (18)

These generating functions have two degrees of freedom: an eigenvalue $-\lambda$; $\lambda\in\mathbb{R}^{+*}$ as in the case of Laguerre filters and differentiation orders multiples of a single commensurate order $n\in\mathbb{R}^{+*}$.

These functions are defined in the time domain by their inverse Laplace-transform.

Important remarks:

1. Orthogonalization procedure is possible iff all functions $\|F_m\|\in L_2[0,\infty[\ \forall\ m=1,2,\ldots$ In other words the following inequality must be satisfied:

$$\|F_m\|^2=\int_0^\infty f_m^2(t)dt<\infty \qquad \forall\ m=1,2,3,\ldots$$

2. It is important to note that this condition is more restrictive than the L_2 stability. Indeed, $F_m(s)$ must be $L_2[0,\infty[$ convergent although the impulse dirac delta is not $L_2[0,\infty[$ convergent.

It is easy to check that stability conditions of $F_1(s)$ (where $\mathcal{L}^{-1}\{F_1(s)\}\in\mathbb{R}$ i.e. time-response of $F_1(s)$ is a real signal), is: $\lambda>0$ and $0<n<2$ (19)

It was shown in (Malti et al., 2002) that a necessary and sufficient condition for $\mathcal{L}^{-1}\{F_1(s)\}$ to belong to $L_2[0,\infty[$ is: $\lambda>0$ and $0.5<n<2$ (20)

As it can be noticed, $F_1(s)$ may have an $L_2[0,\infty[$ stable transfer function. However its impulse response is not $L_2[0,\infty[$. This is due to the fact that a dirac delta distribution $\notin L_2[0,\infty[$ (it is not a bounded input signal in the sense of L_2). Hence, when $0<n\le 0.5$ and $\lambda>0$, $F_1(s)$ is stable but $\mathcal{L}^{-1}\{F_1(s)\}\notin L_2[0,\infty[$.

As noted previously an orthogonalization procedure is possible if $\mathcal{L}^{-1}\{F_1(s)\}\in L_2[0,\infty[$. Consequently, $F_m(s)$ series can be orthogonalized if conditions (20) are satisfied.

By applying Gram-Schmidt orthogonalization procedure on $F_m(s)$, the orthonormal functions, named $G_m(s)$ $m=1, 2, \ldots \infty$, are obtained. The first step in this procedure consists of normalizing the first function $F_1(s)$.

$$G_1(s)=\frac{\alpha_{1,1}}{s^n+\lambda} \qquad (21)$$

Hence, $\alpha_{1,1}=\|F_1\|^{-1}$

The second function of the orthogonal basis:

$$G_2(s)=\alpha_{2,1}G_2(s)+\alpha_{2,2}F_2(s) \qquad (22)$$

is computed so that:

$$\langle G_1(s),G_2(s)\rangle=0 \text{ and } \langle G_2(s),G_2(s)\rangle=1.$$

These two conditions permit obtaining the $\alpha_{1,2}$ and $\alpha_{2,2}$. Generalizing this procedure to any m will permit to compute all functions of the orthogonal basis $G_m(s)$ $m=1, 2, 3, 4, \ldots$ by applying:

$$G_m(s)=\alpha_{m,1}G_1(s)+\cdots+\alpha_{m,m-1}G_{m-1}(s)+\alpha_{m,m}F_m(s) \qquad (23)$$

where all coefficients $\alpha_{m_1,m_2}, (m_1,m_2)\in\mathbb{N}^2$ are computed according to the orthonormality conditions:

$$\langle G_{m_1},G_{m_2}\rangle=\begin{cases}0 & \text{if } m_1\ne m_2\\ 1 & \text{if } m_1=m_2\end{cases} \qquad (24)$$

One can check that all coefficients α_{m_1,m_2} must satisfy:

$\forall m_2=1,2\cdots,\infty$ and $\forall m_1=1,2,\cdots,(m_2-1)$

$$\alpha_{m_1,m_2}=-\alpha_{m_2,m_2}\langle G_{m_1},F_{m_2}\rangle \qquad (25)$$

$$\alpha^2_{m_2,m_2}=\left(\|F_{m_2}\|^2-\sum_{i=1}^{m_2-1}\langle G_i,F_{m_2}\rangle^2\right)^{-1} \qquad (26)$$

Computation of all α_{m_1,m_2} is presented in appendix.

For giving the M orthogonal functions basis, one can write the following matrix notation

$$G=\begin{bmatrix}1 & 0 & \cdots & 0\\ -\alpha_{2,1} & 1 & & \\ \vdots & \ddots & \ddots & \\ -\alpha_{M,1} & \cdots & -\alpha_{M,M-1} & 1\end{bmatrix}^{-1}\begin{bmatrix}\alpha_{1,1} & 0 & \cdots & 0\\ 0 & \alpha_{2,1} & \ddots & \vdots\\ \vdots & \ddots & \ddots & 0\\ 0 & \cdots & 0 & \alpha_{M,M}\end{bmatrix}F$$

(27)

This allows to express all orthogonal functions $G^T=[G_1(s)\ \ G_2(s)\ \ \cdots\ \ G_M(s)]$ in terms of the initial functions $F^T=[F_1(s)\ \ F_2(s)\ \ \cdots\ \ F_M(s)]$ as defined in (18).

Example

When the fractional order $n=1.5$ and the eigenvalue $\lambda=1.5$ the first three functions of the base are:

$$G_1(s)=\frac{1.49}{s^{1.5}+1.5}$$

$$G_2(s)=\frac{0.7}{s^{1.5}+1.5}+\frac{-2.37}{\left(s^{1.5}+1.5\right)^2} \qquad (28)$$

$$G_3(s)=\frac{-0.12}{s^{1.5}+1.5}+\frac{-1.17}{\left(s^{1.5}+1.5\right)^2}+\frac{3.56}{\left(s^{1.5}+1.5\right)^3}$$

Impulse responses of these three functions are plotted on Figure (1).

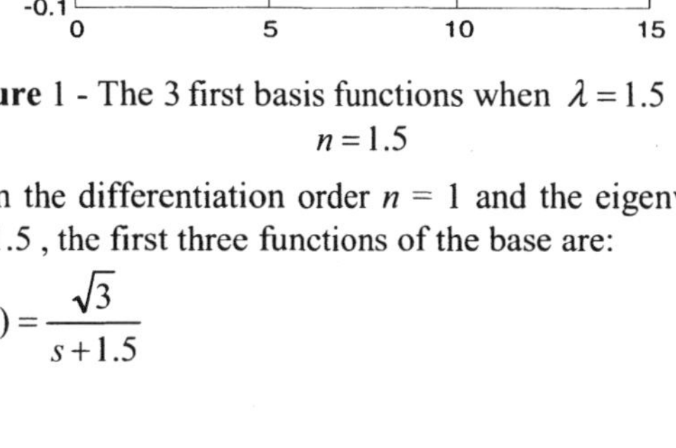

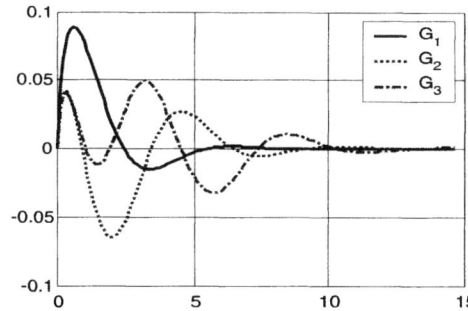

Figure 1 - The 3 first basis functions when $\lambda=1.5$ and $n=1.5$

When the differentiation order $n=1$ and the eigenvalue $\lambda=1.5$, the first three functions of the base are:

$$G_1(s)=\frac{\sqrt{3}}{s+1.5}$$

$$G_2(s)=\sqrt{3}\left(\frac{1}{s+1.5}+\frac{-3}{(s+1.5)^2}\right) \qquad (29)$$

$$G_3(s)=\sqrt{3}\left(\frac{1}{s+1.5}+\frac{-6}{(s+1.5)^2}+\frac{9}{(s+1.5)^3}\right)$$

It is easy to check that (29) correspond to the definition of Laguerre functions (16). Hence, Laguerre functions can be considered as a special case, corresponding to $n = 1$, of the new set of fractional orthonormal functions. The three Laguerre functions are plotted in figure (2).

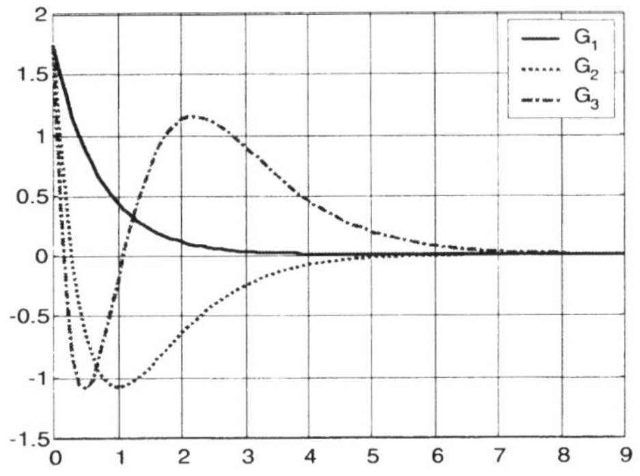

Figure 2 - The 3 first Laguerre base filters when $\lambda = 1.5$.

Comparing figures 1 and 2, one can notice that fractional orthonormal functions are more or less oscillatory depending on the differentiation order n. An additional degree of freedom is provided. It would be interesting for the reader to plot some other orthonormal functions for $0.5 < n < 1$, in order to check their behavior. Moreover, by applying initial value theorem on $G_m(s)$ of (23) or (27), shows that the impulse response at $t = 0$ is ($\forall m = 1, 2, 3, \ldots$):

- $g_m(0) = \infty$ if $½ < n < 1$
- $g_m(0) = \sqrt{2\lambda}$ if $n = 1$ (Laguerre filters)
- $g_m(0) = 0$ if $1 < n < 2$

Remark: Laguerre functions are known to form a complete basis in $L_2[0, \infty[$. Proof of completeness is yet to be generalized to any commensurate differentiation order $n \in\ ]1/2, 2[$ of the new orthonormal basis. ■

4. SYSTEM IDENTIFICATION USING THE FRACTIONAL ORTHONORMAL BASIS

The fractional orthonormal basis is used for system identification with fixed denominator models. An *a priori* knowledge or a rough estimation of a fractional ARX model with a single pole and a single differentiation order can be used to fix n and λ. These two parameters are then plugged in formulae (25), (26), (27) and the orthonormal basis synthesized as explained in the appendix. Then, Fourier coefficients are computed using a least squares method.

Assume $u(t), y(t), t \in [0,T]$ input and output data generated using a linear fractional model H. Then $H(s)$ can be approximated as a linear combination of M orthonormal functions $\{G_m(s), m = 1 \ldots M\}$ with:

$$H(s) \approx \sum_{m=1}^{M} g_m G_m(s) \qquad (30)$$

$$H(s) \approx g^T G(s) \qquad (31)$$

where $g = [g_1, g_2, \ldots g_M]^T$ and $G(s) = [G_1(s)\ G_2(s) \cdots G_M(s)]$.

The truncation order M is fixed so as to obtain a satisfactory approximation. Akaike criterion information can be used. Then the identification procedure consists of computing optimal coefficient vector $g = [g_1, g_2, \ldots g_M]^T$ minimizing the least square error:

$$J = \frac{1}{T}\int_0^T (\varepsilon(t))^2\, dt = \langle \varepsilon(t), \varepsilon(t) \rangle \qquad (32)$$

where

$$\varepsilon(t) = \sum_{m=1}^{M} g_m u_{G_m}(t) - y(t). \qquad (33)$$

$u_{G_m}(t)$ and $u_G(t)$ are defined respectively as:

$$u_{G_m}(t) = G_m(t) \otimes u(t)$$

$$u_G(t) = [u_{G_1}(t)\ u_{G_2}(t)\ \cdots\ u_{G_M}(t)]$$

The optimum values of Fourier coefficients g are given by the least squares formula:

$$\hat{g} = \left[\int_0^T \left(u_G(t)^T u_G(t)\right) dt\right]^{-1} \int_0^T u_G(t)^T y(t) dt \qquad (34)$$

or after a numerical discretization, by defining Y as a column vector of system's output and X as a matrix which columns are filters' outputs, (34) can be approximated by:

$$\hat{g} = (X^T X)^{-1} X^T Y$$

4.1. Example

To illustrate the use of fractional differentiation orthonormal basis in system identification, a dynamical system is identified using simulated data in a noisy context. Consider a simple example where the following fractional transfer function is simulated.

$$H(s) = \frac{1}{s^{1.5}+1.5} + \frac{1}{s^{1.5}+1.6} + \frac{1}{s^{1.5}+1.8} \qquad (35)$$

Output data are corrupted by a stationary zero mean white noise having a noise to signal ratio:

$$10 \log_{10}\left(\frac{\text{Energy of signal}}{\text{Energy of noise}}\right) = 13\text{dB} \qquad (36)$$

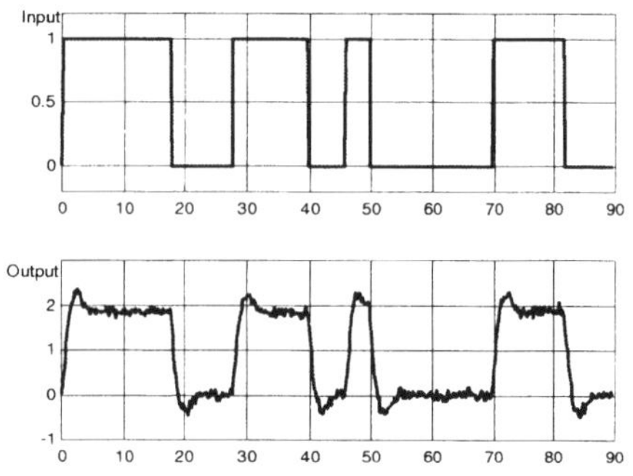

Figure 3 Input and output identification data.

H is approximated using the fractional orthonormal base G. Fourier coefficients are computed by minimizing the least squares error (32). It was noticed that a convenient approximation can be obtained for a truncation order equal to two.

In order to observe the evolution of the least squares error in terms of differentiation order n and eigenvalue λ iso-contours of J are plotted in figure (4) for a truncation of $M = 2$. n varies from 0.6 to 1.9 with a step of 0.2 and λ from 0.2 to 10 with a step of 0.1.

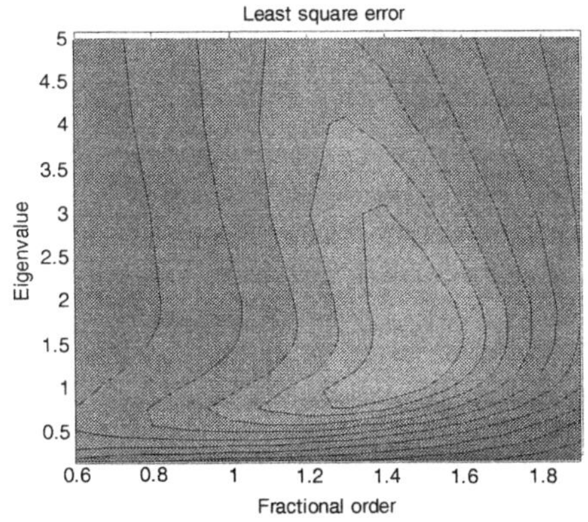

Figure 4 Isocontour of the least square error.

As it can be noticed the optimal values are around $(n, \lambda) = (1.5, 1.6)$. The optimal differentiation order is far from an integer value as expected since the initial system (35) has a commensurate order of 0.5.

To compute the final approximation, n and λ were chosen to be $(n, \lambda) = (1.5, 1.6)$. Computing Fourier coefficients, the following is obtained:

$$\widehat{H}(s) = 1.89\,G_1(s) - 0.01G_2(s)$$

where $G_1(s)$ and $G_2(s)$ are orthogonal functions and correspond to the definition (28).

The value of the normalized criterion is:

$$J = 4\times10^{-3}$$

System's and model's outputs are plotted on validation data on figure (5).

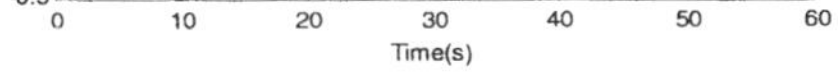

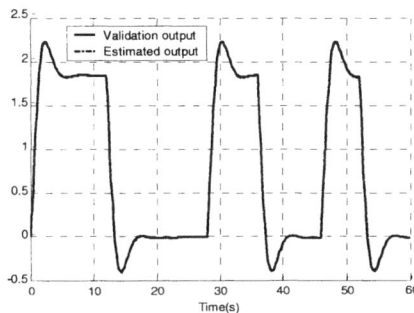

Figure 5 plot of validation output data.

For the sake of comparison, H was also approximated with the classical Laguerre filters.

The Laguerre model leading to the same criterion value as fractional base model is obtained for an optimal pole equal to 1.6. and a length of base 5. It needs more parameters than a fractional base model.

$$H(s) \cong 1.48L_1(s) - 0.94L_2(s) - 0.54L_3(s) + 0.44L_4(s) + 0.22L_5(s)$$

5. CONCLUSION

In this paper we have build an orthonormal basis for fractional systems. It has two degrees of freedom: an eigenvalue (as the pole of Laguerre functions) and a commensurate differentiation order n which can vary in]1/2, 2[. Laguerre functions are considered as a special case corresponding to $n = 1$. As far as we know, this is the first basis ever developed for fractional systems. It was used in the context of system identification with fixed denominator models.

6. APPENDIX

The objective of the appendix is to give an algorithm for computing α_{m_1,m_2}, the coefficients which will ensure orthogonality as described in (27). The computation of the scalar product $\langle F_h, F_m\rangle$, assumed known in the first part, is detailed in the second part of the appendix.

6.1. Computation of α_{m_1,m_2}

The algorithm is iterative. For a better comprehension let's develop (25) and (26) up to $m_2 = 3$.

1. start by computing $\alpha_{1,1} = \|F_1\|^{-1}$.
2. compute all coefficients $\alpha_{2,k}$ in the following order:
   - $k = 2$: $\alpha_{2,2} = \left(\langle F_2, F_2\rangle - \langle G_1, F_2\rangle^2\right)^{-1/2}$

   Replace G_1 by its equivalent:

   $$\alpha_{2,2}^2 = \left(\langle F_2, F_2\rangle - \left(\alpha_{1,1}\langle F_1, F_2\rangle\right)^2\right)^{-1}$$

   - $k = 1$: $\alpha_{1,2} = -\alpha_{2,2}\langle G_1, F_2\rangle, \alpha_{1,2} = -\alpha_{2,2}\alpha_{1,1}\langle F_1, F_2\rangle$

   As shown, $\alpha_{1,2}$ is computed using previously computed $\alpha_{1,1}$ and $\alpha_{2,2}$.
3. compute all coefficients $\alpha_{3,k}$ in the following order:
   - $k = 3$ (all needed coef. are already computed):

   $$\alpha_{3,3} = -\left|\left(\langle F_3, F_3\rangle - \left(\alpha_{1,1}\langle F_1, F_3\rangle\right)^2 - \left(\alpha_{2,1}\alpha_{1,1}\langle F_1, F_3\rangle + \alpha_{2,2}\langle F_2, F_3\rangle\right)^2\right)\right|^{-1/2}$$

   - $k = 2$: $\alpha_{2,3} = -\alpha_{3,3}\left(\alpha_{2,1}\alpha_{1,1}\langle F_1, F_3\rangle + \alpha_{2,2}\langle F_2, F_3\rangle\right)$
   - $k = 1$: $\alpha_{1,3} = -\alpha_{3,3}\alpha_{1,1}\langle F_1, F_3\rangle$

At each step start always by computing $\alpha_{k,k}$, and then compute $\alpha_{j,k}$ for $j = k$-1, k-2, ... 1. Computing each of these coefficients will require some already computed ones.

6.2. Computing $\langle F_h, F_m\rangle$

Recall that F_h and F_m are the orthogonalized functions defined by (18).

$$\langle F_h, F_m\rangle = \left\langle \frac{1}{\left(s^n + \lambda\right)^h}, \frac{1}{\left(s^n + \lambda\right)^m} \right\rangle \tag{37}$$

where $\lambda \in \mathbb{R}^{+*}$ and $n \in \mathbb{R}^{+*}$.

$$\langle F_h, F_m\rangle = \frac{1}{2\pi}\int_{-\infty}^{+\infty} \frac{1}{\left((j\omega)^n + \lambda\right)^h} \left(\overline{\frac{1}{\left((j\omega)^n + \lambda\right)^m}}\right) d\omega \tag{38}$$

then

$$\langle F_h, F_m\rangle = \frac{1}{2\pi}\int_0^{\infty} \frac{d\omega}{\left((j\omega)^n + \lambda\right)^h \overline{\left(j^n\omega^n + \lambda\right)}^m} + \frac{1}{2\pi}\int_0^{\infty} \frac{d\omega}{\left((-j\omega)^n + \lambda\right)^h \overline{\left((-j)^n\omega^n + \lambda\right)}^m} \tag{39}$$

The following change of variable is applied: $\omega^n = x$, $d\omega = \frac{1}{n}x^{\frac{1}{n}-1}dx$. Moreover, by defining:

$$I\left(\alpha, \beta, \frac{1}{n}, h, m\right) = \int_0^{\infty} \frac{x^{\frac{1}{n}-1}dx}{(\alpha x + 1)^h (\beta x + 1)^m} \tag{40}$$

(39) can be rewritten as

$$\langle F_h,F_m\rangle=\frac{1}{2\pi\lambda^{h+m}}\left[I\left(\lambda^{-1}e^{j\frac{\pi}{2}n},\lambda^{-1}e^{-j\frac{\pi}{2}n},\frac{1}{n},h,m\right)+I\left(\lambda^{-1}e^{-j\frac{\pi}{2}n},\lambda^{-1}e^{j\frac{\pi}{2}n},\frac{1}{n},h,m\right)\right] \tag{41}$$

Note that $\alpha=\overline{\beta}$.

The integral I can be decomposed as follow:

$$I\left(\alpha,\beta,\frac{1}{n},h,m\right)=\int_0^{\infty}\left[\frac{x^{\frac{1}{n}-1}}{\alpha x+1}\left(\sum_{l=1}^{h-1}\frac{A_l}{(\alpha x+1)^l}+\sum_{l=1}^{m}\frac{B_l}{(\beta x+1)^l}\right)\right] \tag{42}$$

The coefficients A_l and B_l are given according to:

$$A_l=\left(-\frac{\beta}{\alpha}\right)^{(h-l)}\left(-\frac{\beta}{\alpha}+1\right)^{-m-h+l}\binom{m+h-l-1}{m-1}$$

$$B_l=\left(-\frac{\alpha}{\beta}\right)^{(m-l)}\left(-\frac{\alpha}{\beta}+1\right)^{-m-h+l}\binom{m+h-l-1}{h-1}$$

Then, the last term is rewritten so as to obtain:

$$I\left(\alpha,\beta,\frac{1}{n},h,m\right)=\sum_{l=1}^{h-1}\int_0^{\infty}\frac{x^{\frac{1}{n}-1}A_l}{(\alpha x+1)^{l+1}}dx+\sum_{l=1}^{m}B_l I(\alpha,\beta,\frac{1}{n},1,l) \tag{43}$$

The left integral of (43) was computed in [Gradshteyn *et al.*, 1980 integral 3.194.4 p 285]. The solution depends on n:

1. $$\sum_{l=1}^{h-1}\int_0^{\infty}\frac{x^{\frac{1}{n}-1}A_l dx}{(\alpha x+1)^{l+1}}=\frac{\pi}{\alpha^{\frac{1}{n}}}\frac{1}{\sin\left(\frac{\pi}{n}\right)}\sum_{l=1}^{h-1}\left(A_l(-1)^l\binom{\frac{1}{n}-1}{l}\right)$$

if $2>n>0.5$ and $n\neq 1$ (44)

2. $$\sum_{l=1}^{h-1}\int_0^{\infty}\frac{x^{\frac{1}{n}-1}dx}{(\alpha x+1)^{l+1}}=\frac{1}{\alpha}\sum_{l=1}^{h-1}\frac{1}{l} \qquad \text{if } n=1 \tag{45}$$

The right part in (43) is now computed. First of all, note that:

$$I\left(\alpha,\beta,\frac{1}{n},1,l\right)=I\left(\beta,\alpha,\frac{1}{n},l,1\right) \tag{46}$$

Using (43) one can write:

$$I\left(\beta,\alpha,\frac{1}{n},l,1\right)=\sum_{k=1}^{l-1}\int_0^{\infty}\frac{x^{\frac{1}{n}-1}A'_k}{(\beta x+1)^{k+1}}+B'_1\, I(\beta,\alpha,\frac{1}{n},1,1) \tag{47}$$

with $A'_k=\left(-\frac{\alpha}{\beta}\right)^{(l-k)}\left(-\frac{\alpha}{\beta}+1\right)^{-1-l+k}$

and $B'_1=\left(-\frac{\beta}{\alpha}\right)^{(1-l)}\left(-\frac{\beta}{\alpha}+1\right)^{-l}$

From (41) $I\left(\alpha,\beta,\frac{1}{n},1,1\right)=\pi\lambda^2\langle F_1,F_1\rangle$

Then, using the energy expression of F_l (Malti et *al.* 2002) (47) can be written

$$I\left(\beta,\alpha,\frac{1}{n},l,1\right)=\sum_{k=1}^{l-1}\int_0^{\infty}\frac{x^{\frac{1}{n}-1}A'_l}{(\beta x+1)^{k+1}}+B'_1\pi\left(\frac{\lambda^{\left(\frac{1}{n}-1\right)}\cot\left(\frac{\pi}{2}n\right)}{n\sin\left(\frac{\pi}{n}\right)}\right)^2$$

if $2>n>0.5$ and $n\neq 1$ (48)

$$I\left(\beta,\alpha,\frac{1}{n},l,1\right)=\sum_{k=1}^{l-1}\int_0^{\infty}\frac{x^{\frac{1}{n}-1}A'_l}{(\beta x+1)^{k+1}}+B'_1\frac{\pi}{4} \qquad \text{if } n=1 \tag{49}$$

(48) and (49) are deduced from (44). This completes the computations.

7. REFERENCES

Abbott P.C. (2000). *Generalized Laguerre polynomials and quantum mechanics. J. Phys. A: Math.* Gen. 33 No 42 pp. 7659-7660.

Aoun M., Malti R., Cois O., Oustaloup A. (2002). *System identification using fractional hammerstein models.* 15th IFAC World Congress 2002, Spain.

Battaglia, J.-L. Le Lay L., Batsale J.-C., Oustaloup A., Cois O. (2000). *Heat flux estimation through inverted non integer identification models.* Int J. of Thermal Science, Vol. 39 n°3, pp. 374-389.

Cois O. (2002). *Systèmes linéaires non entiers et identification par modèle non entier : application en thermique.* Ph. D. Thesis, Univ. of Bordeaux I. Fra.

Darling R. and Newman J. 1997. *On the short behavior of porous intercalation electrodes.* J. Electrochem. Soc. Vol 144 n° 9, pp. 3057-3063.

El-Sayed A.M. (1999). *On the generalized Laguerre polynomials of arbitrary (fractional) orders and quantum mechanics.* J. Phys. A: Math. Gen. 32. pp. 8647-8654.

Gradshteyn I.S. & Ryshik I.M. (1980). *Table of integrals, series, and products.* Academic press.

Le Lay L. (1998). *Identification fréquentielle et temporelle par modèle non entier.* Ph. D. Thesis, Université de Bordeaux I. France.

Ljung, L. (1999). *System identification: Theory for the user.* Prentice-hall.

Malti R. Ekongolo S.B. and Ragot J. 1998. *Dynamic SISO and MIMO system approximations based on optimal Laguerre models.* IEEE TAC. V.43(9).

Malti R., Cois O., Aoun M., Levron F. and Oustaloup A. (2002). *Energy of fractional order transfer functions.* 15th world IFAC Congress. b'02, Spain

Matignon D. (1994) *Représentations en variables d'état de modèles de guides d'ondes avec dérivation fractionnaire*, Ph. D. Thesis, Univ. Paris-Sud, Orsay.

Ninness B. and Gustafsson F. (1997). *A unifying construction of orthonormal bases for system identification.* IEEE TAC. V. 42, No.4, pp. 515-521.

Miller K.S. and Ross B. (1993). *An introduction to the fractional calculus and fractional differential equations.* A Wiley-Interscience publication.

Oldham K. B. and J. Spanier (1974). *The fractional calculus.* Academic Press, New York and London.

Oustaloup, A. (1983) *Systèmes asservis linéaires d'ordre fractionnaire.* Masson

Trigeassou J.-C., T. Poinot, J. Lin, A. Oustaloup, F. Levron (1999). *Modeling and identification of a non integer order system.* ECC'99 Karlsruhe, Germany.

Van den Hof P.M.J., Heuberger P.S.C. and Bokor J. (1995). *System identification with generalized orthonormal basis functions.* Automatica. 31(12).

Wahlberg B. (1991). *System Identification Using Laguerre Models.* IEEE TAC, Vol 36, p. 551-562.

IFAC
Publications
www.elsevier.com/locate/ifac

Adaptive Laguerre Time Scaling Factor in Predictive Control

M. El Adel, M. Ouladsine and J.C Carmona

LSIS CNRS UMR 6168
Domaine Universitaire de Saint Jérôme
Avenue Escadrille Normandie Niémen 13397
Marseille Cedex 20
Tel : 0491056062. Fax : 0491056033
Email : mostafa.eladel@lsis.org

Abstract : *This paper proposes a design method of estimation of the Laguerre time scaling factor in the predictive control context. The presented approach is based on the minimization of a cost function in the frequency domain. By discretizing the continuous Laguerre network, a simple predictive control law is presented using only a system matrices of such network. This makes possible to design an adaptive predictive control for unstructured unstable systems whose priori knowledges are not available. The simulation results show that the proposed scheme provides satisfactory performances.*

Key words: *Adaptive predictive control. Laguerre network. Adaptive Laguerre time scaling factor.*

1. INTRODUCTION

The behaviour of the adaptive controller in the presence of unmodelled dynamics as well as the lack of priori knowledges about the process have led to abandon the usual ARMA transfert function representation for a representation by anorthonormal series [1]. Practically, the the orthogonal Laguerre basis of fuctions have been used often (Zervos et al., 1988), (Wang L. et al., 1994), (Tangy N. et al, 1995), (Clowes G.J. et al, 1965) and (Yoon Sang K. Et al, 1999). Such basis however, involves one of the most important parameter called the *Laguerre time scaling factor*. It is well known that if this parameter is chosen suitably, then the Laguerre series can efficiently approximate any stable plant), (Wang L. et al., 1994). The optimal choice of this time scaling factor has been a topic of concern in many works. For system approximation, the problem was first investigated in the time continuous case by Clowes (Clowes G.J. et al, 1965) who established for a qth order truncated series, the well-known general optimality equation: $C_{q-1}(p)C_q(p)=0$ (C is the Laguerre parameter vector, p is the Laguerre time scaling factor and q is the projection order). He illustrated how to select the optimal time scaling factor p for approximation of a rational transfer function using a series expansion of Laguerre functions. Silva (Silva O.E, 1994) extended the problem and generalized the optimality equation for linear systems with any input signal. Masnadi et al (Masnadi et al, 1991) have resolved analytically the above equation for the discrete time systems with known impulse responses that are of the kind of sum of exponentials. However, except for this restrictive case, this leads moreover to a computational difficulty. In practice, Wang and Cluett (Wang L. et al., 1994) have extended numerically the optimality equation for general $\mathbf{L_2}$ stable linear systems. They have obtained as direct application of these extended results, the approximation of irrational transfer functions, which minimize the frequency domain error in the $\mathbf{L_2}$ sense. They have also given empirical solution for the optimal time scaling factor for first order plus delay systems. These approaches however, are all limited to the case of known parameters. Based on the knowledge of the input-output signals, Malti et al (Malti et al, 1998) have proposed a general procedure to approximate linear stable time-invariant (LTI) systems, where the Laguerre parameters are estimated using the least squares method. They have applied the Newton Raphson's iterative technique to compute the optimal Laguerre time scaling factor from the estimated values of the Laguerre parameter vector obtained using an open loop global least squares method. Although this procedure can be applied in system identification, model reduction and noisy modeling, the method requires however to consider that the input signals are persistently exciting and jointly correlated. Further in control theory, the persistently exciting condition is not always satisfied under the closed-loop operation, where the input signals are computed according to the control laws that achieve the desired tracking and performances. Motivated by the need for reduced a priori constraints to be imposed on the considered system (namely the persistently exciting condition and the plant stability), our objective in this paper is to provide a simple adaptive algorithm giving the optimal Laguerre time scaling factor without need for hypothesis about the input signals. The presented approach consists on modeling the system output by an orthonormal Laguerre network put in the stable, observable and controllable state space form. Such a modelisation can be seen as the projection of the considered process's output signal onto the basis whose components are the outputs of each block of fig.1. This permits to derive a simple predictive control law depending on the estimated Laguerre projection parameters. We will show that these parameters may be time varying and depending on the time scaling factor. This most important parameter is simultaneously updated during the parameter estimation phase. For this updating, we use the normalized $\sigma-$ modification algorithm (Ouaaline N. et al, 1998). This permits us to ovoid the convergence problem which often arises from the poor excitations and to deal with even an unstable plant. The paper is organized as follows: In section II, the fundamental Laguerre functions and the problem formulation are presented. The estimation algorithm giving the parameter vector and the time scaling factor of the Laguerre network is presented in section III. The predictive control law is stated in section IV. Finally, the potential of the proposed approach is illustrated by numerical examples and a real application in section V.

2. FUNDAMENTAL LAGUERRE SERIES

The well-known continuous Laguerre functions form an orthonormal set in $L_2[0,+\infty)$ and they are defined by:

$$f_n(t,p)=\sqrt{2p}\frac{\exp(pt)d^{n-1}}{(n-1)!dt^{n-1}}[t^{n-1}\exp(-2pt)] \qquad (1)$$

where $n=1,2,\cdots$ is the order of the function $f_n(t,p)$, and $p>0$ is the time scaling factor.

The Laplace transform of (1) is given by:

$$F_n(s,p)=\sqrt{2p}\frac{(s-p)^{n-1}}{(s+p)^n} \qquad (2)$$

These functions form an orthonormal set in the s-domain and they are represented by the simple and convenient Laguerre network of figure 1 [1],[4].

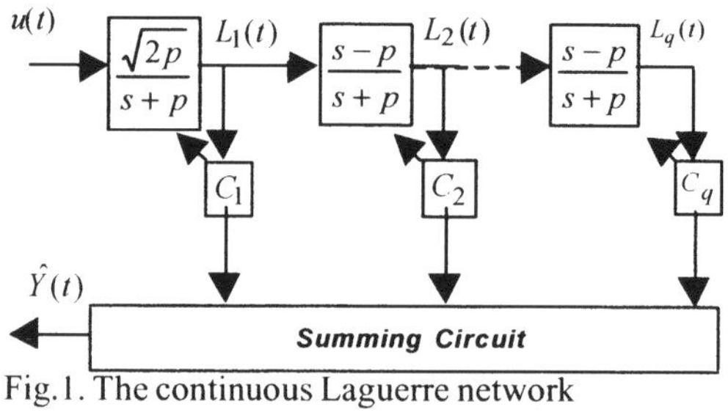

Fig.1. The continuous Laguerre network

Moreover, the Laguerre filter truncated series capture the behavior of a large class of stable linear systems (Malti et al, 1998). This permits to consider that for any stable transfer function $G(s)$ that is assumed to be strictly proper, analytical and continuous in $R(s)\geq 0$ ($R(s)$ is the real part of the complex s) and by letting $p>0$, there exists a sequence C_k such that:

$$G(s)=\sum_{n=1}^{+\infty}C_nF_n(s,p) \qquad (3)$$

For practical use, the infinite sum (3) must be reduced to a q-order expansion terms which are known to be the best q-terms approximation to $G(s)$ i.e.

$$G(s)=\sum_{n=1}^{q}C_nF_n(s,p) \qquad (4)$$

The Laguerre parameter vector C_n may be obtained by using the (RLS) algorithm (Zervos et al., 1988). The regression data used to determine this vector is considered to be the discrete state space vector : $L(t)=[L_1(t)\cdots L_q(t)]^T$ obtained by discretizing the continuous Laguerre network of figure1. It has been demonstrated (El Adel M. et al, 2000), (El Adel M. et al, 1999) and (El Adel M. et al, 1998) that the discrete Laguerre network can be given in the discrete state space form as :

$$L(t+1)=AL(t)+bu(t) \qquad (5)$$

where $u(t)$ is the system's input and $L(t)$ is the q dimensional state vector. A and b are respectively the lower triangular $(q\times q)$ matrix and the q dimensional vector defined as:

if T is the sampling period of the Laguerre network, and

$$a_1=\exp(-pT)\ ;\ a_3=-Ta_1-\frac{2}{p}(a_1-1)$$
$$a_2=T+\frac{2}{p}(a_1-1)\ ;\ a_4=\sqrt{2p}\frac{(a_1-1)}{p} \qquad (6)$$

then

$$A=\begin{bmatrix} a_1 & 0 & \cdots & 0 \\ \frac{-a_1a_2-a_3}{T} & a_1 & \cdots & 0 \\ \vdots & \cdots & \ddots & \vdots \\ \frac{(-1)^{q-1}a_2^{q-2}(a_1a_2+a_3)}{T^{q-1}} & \cdots & \frac{-a_1a_2-a_3}{T} & a_1 \end{bmatrix} \qquad (7)$$

and

$$b=[a_4\ (-a_2/T)a_4\ \cdots\ (-a_2/T)^{q-1}a_4]^T \qquad (8)$$

The above state space is stable $(p>0)$ observable and controllable [1].

It has been demonstrated [11-13] that the output of the process to be modeled is approximated by the weighted sum as:

$$Y(t)=\hat{C}^T(t)L(t)+e(t) \qquad (9)$$

where $\hat{C}(t)$ is the unknown time varying Laguerre parameter [11-13] and $e(t)$ is the error term resulting form the truncated series and all the uncertainties about the plant.

3. ESTIMATION ALGORITHM

3.1 Adaptive Laguerre parameter vector

In adaptive control, the basic recursive least squares algorithm (RLS) is known to have optimal properties when the parameters are time-invariant. However, it is unsuitable for tracking time-varying parameters, since the algorithm gain converges to zero. In addition, the results of the use of the (RLS) algorithm for parameter estimation is that the covariance matrix may grow which causes large changes in the parameter estimates and the system output. This often occurs in the presence of poor excitation and leads to covariance wind-up problem. Thus many modified versions have been proposed. Here, we are inspired by the σ-modification (Ouaaline N. et al, 1998) to ovoid the wind-up of the covariance matrix in the presence of a poor excitation, and by the normalization procedure (Goodwing G.C. et al, 1977) to deal with unbounded unmodelled dynamics and / or unstable systems. The modified normalized recursive least squares algorithm ($N\sigma$-RLS) is introduced as follows:

Let us investigate the equation (9). For many reasons, the error term in this equation may be unbounded and the model is not suitable yet from an estimation point of view. Further, consider a bounded and non zero signal defined as follows:

$$Z(t)=\mu Z(t-1)+\max\left\{\left\|L(t)\right\|;Z_o\right\} \qquad (10)$$

$$0<\mu<1,\ Z(0)>0 \ \textit{and}\ Z_o>0$$

Dividing both sides of equation (9) by $Z(t)$, one has:

$$\bar{Y}(t)=\hat{C}^T(t)\bar{L}(t)+\bar{e}(t) \qquad (11)$$

where :

$$\bar{Y}(t)=\frac{Y(t)}{Z(t)},\ \bar{L}(t)=\frac{L(t)}{Z(t)} \text{ and } \bar{e}(t)=\frac{e(t)}{Z(t)} \qquad (12)$$

Thus, the normalized signals (12) become bounded.

Using the results in (Ouaaline N. et al, 1998) and (Goodwing G.C. et al, 1977), the ($N\sigma-RLS$) algorithm is given by:

$$\hat{C}(t)=\hat{C}(t-1)+\frac{P(t-1)L(t)e(t)}{Z^2(t)+L^T(t)P(t-1)L(t)} \qquad (13)$$

$$P(t)=\frac{1}{\lambda(t)}\left[P(t-1)-\frac{P(t-1)L(t)L^T(t)P(t-1)}{Z^2(t)+L^T(t)P(t-1)L(t)}\right]-\sigma P^2(t-1) \qquad (14)$$

$$0<\sigma<1 \qquad (15)$$

The forgetting sequence $\lambda(t)$ is introduced because the parameters may be time-varying. Such a sequence is initialized as:

$$\lambda(t)=\lambda_o\lambda(t-1)+(1-\lambda_o),\ \ddot{e}_o<0.98,\ \ddot{e}(0)\geq 0.98 \qquad (16)$$

The covariance matrix $P(t)$ satisfies the following properties [2]:

- $\eta I_q<P(t)<\rho I_q$ with $\eta>0$ and $0<\rho<+\infty$
- If $L(t)\to 0$, then $P(t)\to\gamma I_q$ as t tends to infinity, where I_q is the identity matrix and γ is a scalar.

3.2 Adaptive Laguerre time-scaling factor

In the s-domain, the true output to be modeled $Y(t)$ can be represented by a stable transfer function $G(s)$ i.e.

$$\int_{-\infty}^{+\infty}\left|G(j\omega)\right|^2 d\omega<+\infty \qquad (17)$$

In the same away from (2) and (4), $\hat{Y}(t)$ can be given by:

$$\hat{G}(s)=\sum_{n=1}^{q}\hat{C}_nF_n(s,p) \qquad (18)$$

The standard calculation of $\hat{C}_n$ is given by [3]:

$$\hat{C}_n=\frac{1}{2\pi}\int_{-\infty}^{+\infty}\hat{G}^*(j\omega)F_n(j\omega,p)d\omega \qquad (19)$$

($*$) denotes the conjugate of the complex transfer function $G(j\omega)$, and $F_n(j\omega,p)$ is deduced from (2) by letting $s=j\omega$. From (19), it is obvious that the parameter $\hat{C}_n$ depends only of the time scaling factor p. One can rewrite it as $\hat{C}_n=\hat{C}_n(p)$.

Then, consider the cost function to be minimized as:

$$\Lambda(p)=\frac{1}{2\pi}\int_{-\infty}^{+\infty}\left|G(j\omega)-\sum_{n=1}^{q}\hat{C}_n(p)F_n(j\omega,p)\right|^2 d\omega \qquad (20)$$

substituting the coefficients $\hat{C}_n(p)$ from (19) into (20) leads to:

$$\Lambda(p)=\frac{1}{2\pi}\int_{-\infty}^{+\infty}\left|G(j\omega)\right|^2 d\omega-\sum_{n=1}^{q}\hat{C}_n^2(p) \qquad (21)$$

Then, the minimum of $\Lambda(p)$ with respect to p corresponds to the maximum of

$$\sum_{n=1}^{q}\hat{C}_n^2(p) \qquad (22)$$

Here, the main problem is to find the time scaling factor p which maximizes (22). In other words, we must look for the maximum of (22) as a function of p. This maximum is simply obtained by using the first derivative of (22), from which the variation of p is deduced. As the stability of the Laguerre network depends essentially on the time-scaling factor p then, this factor must be constrained to a positive and real interval such that the network's stability is always ensured. This constraint is obtained by means of the second derivative of (22) which must be strictly positive in order to have maximal values with respect to p. This gives conditions that must be achieved to keep the time scaling factor in a real and positive set. So, the simple adaptive recursive algorithm giving the optimal value of p from the estimated parameter vector $\hat{C}_q(p)$ is deduced.

Before proving the main theorem giving the optimal value of p from the parameter vector $\hat{C}_n(p)$, we first require the following lemma (Wang L. et al., 1994).

Lemma 1

For some $p>0$, the Laplace transform of the Laguerre functions in equation (2) satisfies the equality:

$$2p\frac{dF_n(s,p)}{dp}=nF_{n+1}(s,p)-(n-1)F_{n-1}(s,p) \qquad (23)$$

for the proof, see [3].

Theorem 1

Given that the Laguerre parameters are obtained from (19) with respect to the estimate values of $\hat{C}_n(p)$, and that the plant output is stable in the sense of (17), then

$$\frac{d}{dp}\sum_{n=1}^{q}\hat{C}_k^2(p)=\frac{q}{p}\hat{C}_q(p)\hat{C}_{q+1}(p) \qquad (24)$$

Proof

From (19), one has:

$$2p\frac{d\hat{C}_n(p)}{dp}=\frac{2p}{2\pi}\int_{-\infty}^{+\infty}\hat{G}^*(j\omega)\frac{d}{dp}F_n(j\omega,p)d\omega \qquad (25)$$

Applying lemma 1 gives:

$$2p\frac{d\hat{C}_n(p)}{dp}=n\hat{C}_{n+1}(p)-(n-1)\hat{C}_{n-1}(p) \qquad (26)$$

Multiplying both sides of equation (25) by $\hat{C}_n(p)$ and using (23), one has:

$$2\hat{C}_n(p)\frac{d\hat{C}_n(p)}{dp}=\frac{1}{p}[n\hat{C}_n(p)\hat{C}_{n+1}(p)-(n-1)\hat{C}_n(p)\hat{C}_{n-1}(p)] \quad (27)$$

considering that

$$\frac{d}{dp}\sum_{n=1}^{q}\hat{C}_n^2(p)=2\sum_{n=1}^{q}\hat{C}_n(p)\frac{d}{dp}\hat{C}_n(p) \quad (28)$$

then, the summation of both sides of (27) gives:

$$\frac{d}{dp}\sum_{n=1}^{q}\hat{C}_n^2(p)=\frac{q}{p}\hat{C}_q(p)\hat{C}_{q+1}(p) \quad (29)$$

which completes the proof.

Remark 1

For a given model order q, the equation (29) tells us that we are looking for the value of p which corresponds to zeros of $\hat{C}_{q+1}(p)$. Otherwise, the model order can be reduced to $q-1$ without any change in the model accuracy (i.e. $\hat{C}_q(p)=0$) [3]. In other words, the maximum of (22) tends to reduce the model order.

In the following, we will change the indices q by $q-1$ and $q+1$ by q. The second derivative of (22) gives:

$$\frac{d^2}{dp^2}\sum_{n=1}^{q}\hat{C}_n^2(p)=\frac{q}{p}\left[\frac{d}{dp}\hat{C}_{q-1}(p)\hat{C}_q(p)+\hat{C}_{q-1}(p)\frac{d}{dp}\hat{C}_q(p)-\frac{1}{p}\hat{C}_{q-1}(p)\hat{C}_q(p)\right] \quad (30)$$

From (30), one can deduce that the behavior of the equation (29) as a function of p and $\hat{C}_n(p)$ tends to have maximal values with respect to p if $\frac{d^2}{dp^2}\sum_{n=1}^{q}\hat{C}_n^2(p)>0$. This is satisfied if the following inequality holds:

$$\frac{dp}{p}<\left[\frac{d\hat{C}_{q-1}(p)}{\hat{C}_{q-1}(p)}+\frac{d\hat{C}_q(p)}{\hat{C}_q(p)}\right] \quad (31)$$

Let consider the following variation of p as: $dp=p(t+1)-p(t)$, one has the condition that must be achieved to keep the values of $\hat{p}(t)$ in a real and positive set.

$$\hat{p}(t+1)<\hat{p}(t)\left[1+\frac{\Delta\hat{C}_{q-1}(p)}{\hat{C}_{q-1}(p)}+\frac{\Delta\hat{C}_q(p)}{\hat{C}_q(p)}\right] \quad (32)$$

where:

$$\begin{cases}\Delta\hat{C}_{q-1}(p)=\hat{C}_{q-1}(\hat{p}(t+1))-\hat{C}_{q-1}(\hat{p}(t))\\ and\ \ \Delta\hat{C}_q(p)=\hat{C}_q(\hat{p}(t+1))-\hat{C}_q(\hat{p}(t))\end{cases} \quad (33)$$

A suitable manipulation of equation (29) and by integrating it, one has the recursive adaptive algorithm giving $\hat{p}(t)$ as:

$$\hat{p}(t+1)=\hat{p}(t)\left[1+\frac{\Delta\left\|\hat{C}(\hat{p}(t))\right\|^2}{q\cdot\hat{C}_q(\hat{p}(t))\cdot\hat{C}_{q-1}(\hat{p}(t))}\right] \quad (34)$$

where:

$$\Delta\left\|\hat{C}(\hat{p}(t))\right\|^2=\sum_{n=1}^{q}\hat{C}_n^2(\hat{p}(t+1))-\sum_{n=1}^{q}\hat{C}_n^2(\hat{p}(t))=\left\|\hat{C}(\hat{p}(t+1))\right\|^2-\left\|\hat{C}(\hat{p}(t))\right\|^2 \quad (35)$$

Definition 1

A nonnegative real sequence $S(t)$ is said to be δ–asymptotically small in the mean (δ– ASM) if

$$\lim_{k\to\infty}\sup\lim_{l\to\infty}\sup\left[\frac{1}{k}\sum_{t=l+1}^{t=k+l}S(t)\right]\leq\delta$$

Lemma 2

The combined adaptive estimation algorithm (12-15), and (32-35) has the following properties:

P_1 - There exists a positive scalar R_c such that, for all $t\in IN$ one has: $\left\|\hat{C}(t)\right\|\leq R_c$

P_2 - There exists a nonnegative scalar K_δ such that the normalized adaptation error sequences

$$\left|\bar{Y}(t)-\hat{\bar{Y}}(t)\right| \text{ is } K_\delta\text{-ASM}$$

P_3 - There exists a nonnegative scalar K_c such that

$$\left\|\hat{C}(t)-\hat{C}(t-1)\right\| \text{ is } K_c-\text{ASM}$$

P_4 - $\lim_{t\to+\infty}\hat{p}(t)=p^*$

where p^*=Constant is the optimal value of p in the sense of **P_3**.

The proof of **P_1 - P_3** may be carried out along the same lines as in reference (M'saad M. et al, 1991). The proof of **P_4** is trivial: from (34) and (35) when the projecting parameter $\hat{C}(p)$ converges (**P_3**), it is clear that $\hat{p}(t)$ remains constant and is equal to it optimal value p^*, which completes the proof.

Remark 2

From equation (34) and property **P_3** of lemma 2 $\hat{p}(t)$ is bounded.

4. PREDICTIVE CONTROL

For the predictive control, it can be easily derived as follows:

From (9), one can predict the plant output for d steps as :

$$\hat{Y}(t+d)=Y(t)+\hat{C}^T(t)[L(t+d)-L(t)]+\Omega(t)-e(t) \quad (36)$$

where: $\Omega(t)=[\hat{C}(t+d)-\hat{C}(t)]^T L(t+d)$ and $e(t)=Y(t)-\hat{Y}(t)$ (37)

Assuming that $u(t)=u(t+1)=\cdots=u(t+d-1)$ i.e. $u(t)$ remains constant along the horizon d, the recursive use of (6) gives:

$$L(t+d)=A^d L(t)+[A^{d-1}+\cdots+I_q]\,bu(t) \quad (38)$$

Substituting (38) into (36), one obtains:

$$\hat{Y}(t+d)=Y(t)+\hat{K}^T(t)L(t)+\hat{\beta}(t)u(t)+\Omega(t)-e(t) \quad (39)$$

where: $\hat{K}^T(t)=\hat{C}^T(t)[A^d-I_q]$ and $\hat{\beta}(t)=\hat{C}^T(t)[A^{d-1}+\cdot\cdot+I_q]b$ (40)

According to [1], the first order set point reference trajectory is given by:

$$Y_r(t+1)=\alpha Y(t)+(1-\alpha)Y_{sp}(t) \quad (41)$$

where $0<\alpha<1$ and $Y_{sp}(t)$ are the desired set points. According to the same reference (Zervos et al., 1988) and for d steps ahead reference trajectory one has:

$$Y_r(t+d)=\alpha Y_r(t+d-1)+(1-\alpha)Y_{sp}(t) \quad (42)$$

By recursive substitution of the equation (42), the later becomes:

$$Y_r(t+d)=\alpha^d Y(t)+(1-\alpha^d)Y_{sp}(t) \quad (43)$$

Setting $\hat{Y}(t+d)-\Omega(t)+e(t)=Y_r(t+d)$, one can find the required control input as follows:

$$u(t)=\frac{1}{\hat{\beta}(t)}[Y_r(t+d)-Y(t)-\hat{K}^T(t)L(t)] \quad (44)$$

Remark 3

1- From the expression of $u(t)$, $\hat{\beta}(t)$ must be non zero. This is true only if $k_d>d$, where k_d is the plant time delay and d is the prediction horizon (Zervos et al., 1988).

2- Using the expression of $\hat{K}^T(t)$ defined in (40), the equation (44) can be expressed as a function of the error term as:

$$u(t)=\frac{1}{\hat{\beta}(t)}[Y_r(t+d)-\hat{C}^T(t)A^d L(t)-e(t)] \qquad (45)$$

The convergence of the proposed predictive control law is shown in light of the following theorem:

Theorem 2:

Assume that the plant described in (9) is controlled by (44), and provided that the $(N\sigma-RLS)$ algorithm (12-15), (32-35), is used to find the estimates $\hat{C}(t)$ and $\hat{p}(t)$ with $\hat{\beta}(t)\neq 0$ then, the indirect adaptive control scheme described above is globally convergent in the sense that:

i) $u(t)$ and $Y(t)$ remain bounded

ii) $\left|Y(t)-Y_{sp}(t)\right|$ is ϕ-ASM

Proof:

i) Substituting (44) into (6) and using (40), yields the closed loop equation:

$$\begin{aligned} L(t+1) &= [A-b\hat{\beta}^{-1}(t)\hat{C}^T(t)\alpha^d-b\hat{\beta}^{-1}(t)\hat{C}^T(t)A^d]L(t) \\ &\quad +b\hat{\beta}^{-1}(t)(1-\alpha^d)[Y_{sp}(t)-Y(t)+\hat{Y}(t)] \end{aligned} \qquad (46)$$

For stability, we examine the matrix:

$$[A-b\hat{\beta}^{-1}(t)\hat{C}^T(t)\alpha^d-b\hat{\beta}^{-1}(t)\hat{C}^T(t)A^d] \qquad (47)$$

As long as the eigenvalues of the above matrix for some value of d are inside the unit disk in the z-plane, then the closed-loop system is stable. Now, for sufficiently large d and while $\hat{\beta}(t)\neq 0$, the second term in expression (47) approaches zero because the definition of α $(0<\alpha<1)$.

The third term in expression (47) also approaches zero under the same conditions as above because the square matrix A is lower triangular and its eigenvalues appear along the main diagonal. It is straightforward then to show that the powers A^d approach zero because all the eigenvalues of A are less than one in modulus. Finally, the first term in expression (47) is always a stable matrix and the closed loop system is stable.

ii) The equation (46) can be rewritten in the reduced form as :

$$L(t+1)=G(t)L(t)+\nu(t) \qquad (48)$$

From **i)**, it is easy to show that the free system $L(t+1)=G(t)L(t)$ is exponentially stable,

with: $\left\|\nu(t)\right\|\leq\left\|b\right\|\left|\hat{\beta}^{-1}(t)\right|(1-\alpha^d)\left(\left|Y_{sp}(t)\right|+\left|\hat{Y}(t)-Y(t)\right|\right)$ (49)

After suitable manipulation of this expression, and using lemma.1, one has:

$$\left\|\nu(t)\right\|\leq\left\|b\right\|\left|\hat{\beta}^{-1}(t)\right|(1-\alpha^d)\left(\left|Y_{sp}(t)\right|+\left|Z(t)\right|\left|\tilde{Y}(t)-\tilde{Y}(t)\right|\right) \qquad (50)$$

Since $Y_{sp}(t)$ and $\left|Z(t)\right|$ are both bounded, then from part $\mathbf{P_2}$ of lemma 2, one can deduce that $\left\|\nu(t)\right\|$ is bounded. Moreover, the expression (49) can be rewritten as:

$$\left\|\nu(t)\right\|\leq\left\|b\right\|\cdot\left|\hat{\beta}^{-1}(t)\right|\cdot(1-\alpha^d)\left(\left|\hat{Y}(t)\right|+\left|Y_{sp}(t)-Y(t)\right|\right) \qquad (51)$$

From the expression of $Y(t)=\hat{C}^T(t)L(t)+e(t)$, and since $\left\|\hat{C}(t)\right\|$ (part $\mathbf{P_1}$ of lemma.2) and $\left\|L(t)\right\|$ (part i) of theorem.2) are both bounded, then $\hat{Y}(t)$ is bounded.

From lemma 2, it follows that:

$$\left|\hat{Y}(t)\right|\leq R_c\left\|L(t)\right\| \ \ and \ \ \left|e(t)\right|\leq\left|Z(t)\right|K_\delta \qquad (52)$$

on the other hand:

$$\begin{aligned}\left|Y(t)-Y_{sp}(t)\right| &\leq\left|Y_{sp}(t)\right|+R_c\left\|L(t)\right\|+\left|Z(t)\right|K_\delta \\ &\leq \sup\left|Y_{sp}(t)\right|+R_c\max\left\|L(t)\right\|+\max\left|Z(t)\right|K_\delta\end{aligned}$$

In light of definition 1, one can deduce that $\left|Y(t)-Y_{sp}(t)\right|$ is bounded and converges to ϕ as t tends to infinity with:

$$\phi=\sup\left|Y_{sp}(t)\right|+R_c\max\left\|L(t)\right\|+\max\left|Z(t)\right|K_\delta$$

Thus $\left|Y(t)-Y_{sp}(t)\right|$ is $\phi-$ ASM, which completes the proof.

5. SIMULATION RESULTS

Example 1

We consider the following minimum phase plant [16] and we assume that it is given as unstructured system.

$$H(z)=\frac{B(z)}{A(z)}=\frac{b_0 z+b_1}{z^2+a_1 z+a_2}$$

where $a_1=-1.6065$, $a_2=0.6065$ $b_0=0.1065$ $b_1=0.0902$

The plots in figures (2-4) show respectively the behavior of the output, the adaptive time scaling factor and the input of the considered plant.

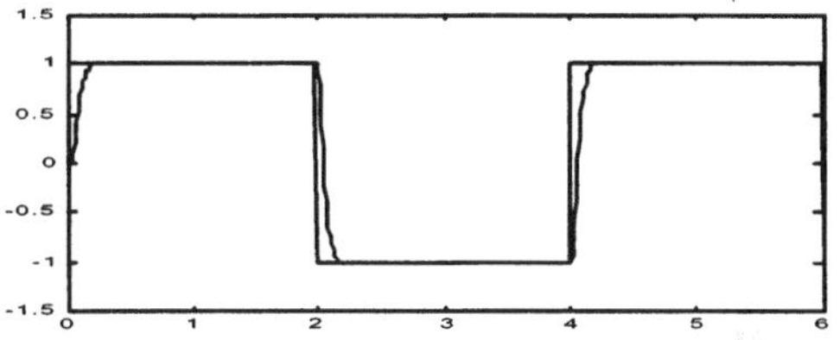

Fig.2- The output $Y(t)$

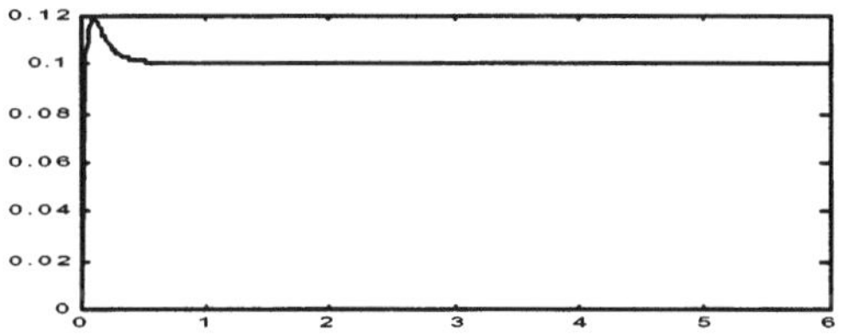

Fig.3- The time scaling $\hat{p}(t)$

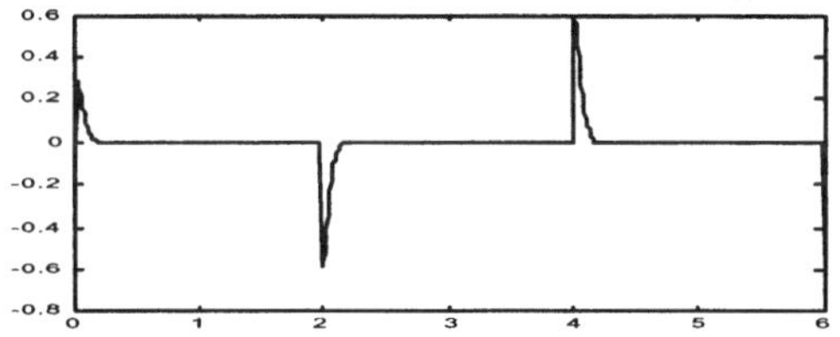

Fig.4- The input $u(t)$

Example2

Here, we consider the same non minimum phase plant (Yoon sang K. et al, 1999), and we assume that it is given as unstructured system.

$$H(z)=\frac{B(z)}{A(z)}=\frac{b_0 z+b_1}{z^2+a_1 z+a_2}$$

where $a_1=-1.7$, $a_2=0.72$ $b_0=0.4$ $b_1=0.8$

The plots in figures (5-7) show respectively the behaviour of the output, the adaptive time scaling factor and the input of the considered plant.

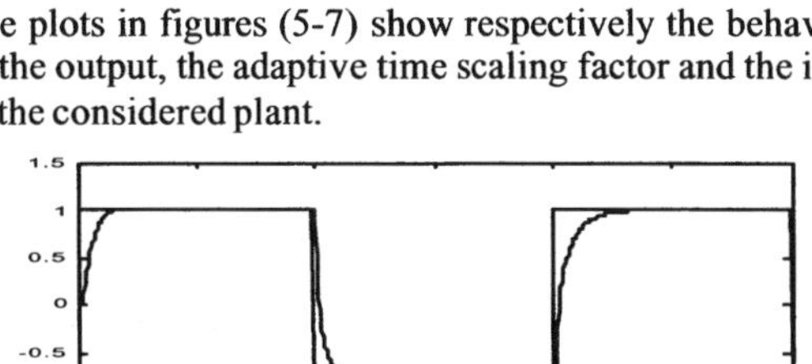

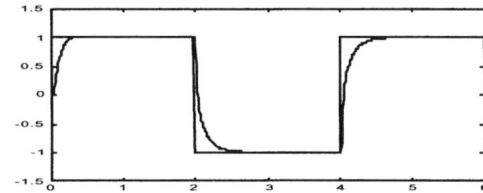

Fig.5- The output $Y(t)$

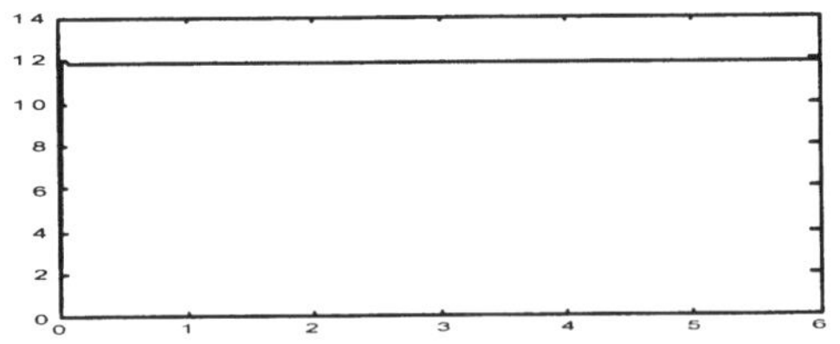

Fig.6- The time scaling $\hat{p}(t)$

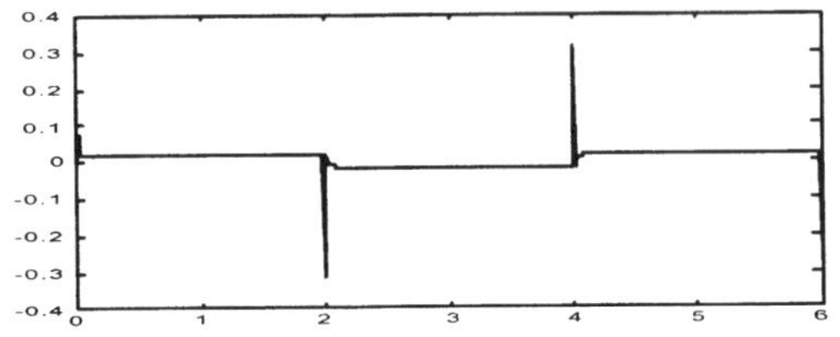

Fig.7- The input $u(t)$

Robustness

Because this method doses not relay on predifined model structure, and because the normalization procedure in the updating phase for determining the projecting parameters can deal even with the unstable plant, we expect it to be more robust when the pant is unstable. For this we consider the continuous time unstable state space model given by :

$$\dot{X}(t)=\begin{bmatrix} 0 & 1 \\ 2 & -0.5 \end{bmatrix} X(t)+\begin{bmatrix} 0 \\ 1 \end{bmatrix} u(t)$$

$$Y(t)=[1 \quad 0.5]X(t)$$

The eigenvalues are $\lambda_1 = 1.1861$; $\lambda_2 = -1.6861$

It is clear that this system is unstable $(\lambda_1 > 0)$. The proposed approach is applied with the normalization procedure (10-16). The obtained results are shown in figures (8-10).

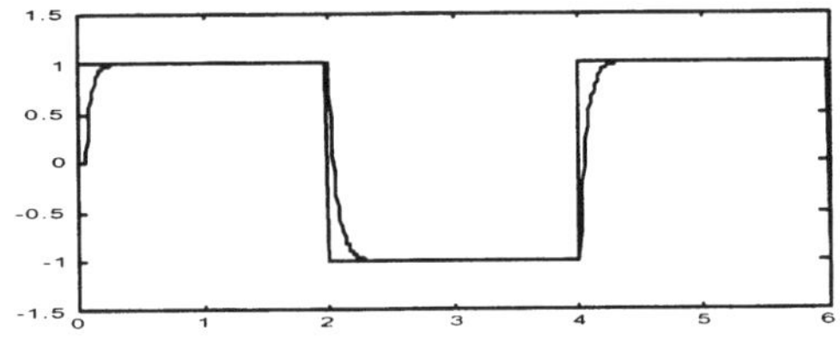

Fig.8- The output $Y(t)$

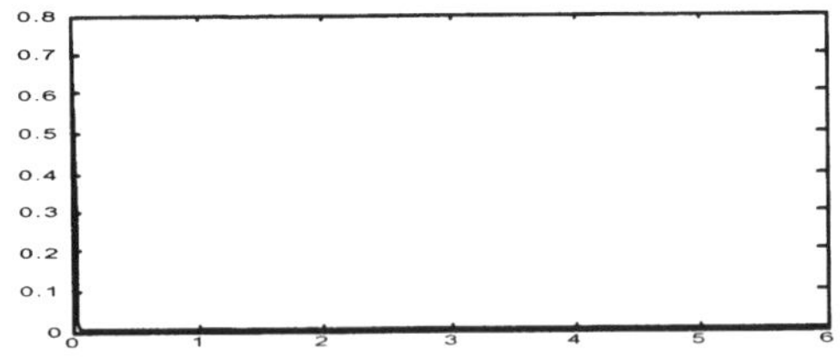

Fig.9- The time scaling $\hat{p}(t)$

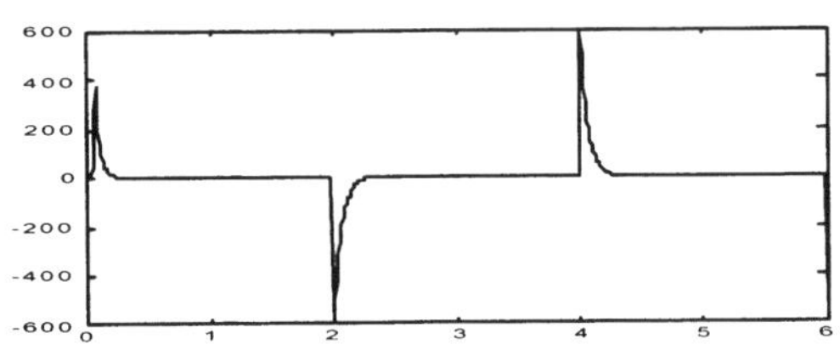

Fig.10- The input $u(t)$

Conclusion

In this paper, we have proposed a new approach to adapt the Laguerre time scaling factor from the estimated Laguerre parameters for unstructured and eventually unstable systems. The obtained results are used to construct a simple predictive control law depending only on a system matrices of Laguerre network. This makes possible to design an adaptive predictive control for unstructured unstable systems whose priori knowledges are not available. The results in simulations show the good behavior of the output, the time scaling factor and the control input in both the minimum, nonminimum phase and unstable systems.

REFERENCES

Clowes G.J," Choice of the time-scaling factor for linear systems approximations using orthonormal Laguerre functions," IEEE Trans. Automat. Contr., vol. AC-10, pp.487-489, 1965.

El Adel M., M. Makoudi and L. Radouane," Decentralized adaptive control for Large-scale continuous systems based on Laguerre model," SAMS Systems Analysis Modelling Simulations,vol.37, pp.261-276. 2000.

El Adel M., M. Makoudi and L. Radouane," Decentralized adaptive control of linear interconnected systems based on Laguerre series representation. Automatica, vol. 35, No. 11, pp. 1873-1881, 1999.

El Adel M., M. Makoudi and L. Radouane," Decentralized adaptive control using Laguerre model with adaptation of time scaling factor," 3ème conférence Maghrébine d'Automatique, d'Electrotechnique et d'Electronique Indusrtielle 8 et 9 décembre, Alger, Algérie, 1998.

Goodwin G.C and R. L. Payne, "Dynamics system identification: Experiment design and data analysis ". Academic Press, New York, 1977.

Malti R., S. B. Ekongolo and J. Ragot," Dynamic SISO and MISO system approximations based on optimal Laguerre models," IEEE. Trans. On Automatic control, vol.43, no.9, pp. 1318-1328. Sept. 1998.

M'saad M. and I. D. Landau," Adaptive control: an overview," IFAC ADCHEM. Toulouse, France, 1991.

Ouaaline N. and L. Radouane," Pole-zero estimation of speech signal based on Zero-tracking algorithm," Int. J. of adaptive control and signal processing, vol.12, pp.1-12, 1998.

Silva O.E," Optimality conditions for truncated Laguerre networks," IEEE Trans. Signal Process., vol.42, pp. 2528-2530, Sept. 1994.

Tanguy N., P. Vilbé, and L. C. Calvez," Optimum choice of the free parameter in orthonormal approximations," IEEE. Trans. On Automatic control, vol.40, no.10, pp. 1811-1813, Oct. 1995.

Wang L. and W. R. Cluett," Optimal choice of time-scaling factor of linear systems approximation using Laguerre models," IEEE. Trans. On Automatic control, vol.39, no.7, pp. 1463-1467, July, 1994.

Yoon Sang KIM. " A STC Design Based on D-L Network for Discrete Unstructured Systems". IEICE. Trans. Fundamentals, Vol. E82-A, No.12, December 1999.

Zervos C. and G. A. Dumont," Deterministic adaptive control based on Laguerre series representation," Int. J. Control, vol.48, no.6, pp. 2333-2359, 1988.

IFAC Publications
www.elsevier.com/locate/ifac

IDENTIFICATION OF MIMO STATE SPACE MODELS FOR HELICOPTER DYNAMICS

Marco Lovera

Dipartimento di Elettronica e Informazione, Politecnico di Milano,
Piazza Leonardo da Vinci 32, 20133 Milano, Italy,
Tel. +39-02-23993592, Fax +39-02-23993412,
E-mail: lovera@elet.polimi.it

Abstract: A considerable amount of work has been dedicated in the past to the problem of identification of helicopter flight dynamics, both in the time and frequency domain, however, limited attention has been devoted so far to the application of subspace methods to the problem. The aim of this paper is to show that subspace based identification techniques can be used to determine accurate discrete-time and continuous-time linear models for helicopter dynamics. The identification techniques are applied to simulated data generated by a physical model that describes coupled rotor-fuselage dynamics for a realistic rotorcraft.

Keywords: Helicopter dynamics, Subspace methods, Identification algorithms, State space models.

1. INTRODUCTION

The derivation of accurate dynamic models for helicopter aeromechanics is becoming more and more important, as progressively stringent requirements are being imposed on rotorcraft control systems: as the required control bandwidth increases, model accuracy becomes a vital part of the design problem. System identification has been known for a long time as a viable approach to the derivation of control oriented dynamic models in the rotorcraft field. In particular, a number of contributions are available on the experimental characterization of flight dynamics, see for example the recent review paper (Hamel and Kaletka, 1997) and the references therein. In recent years, most of the research activity was performed in the frequency domain, see for example (Tischler and Cauffman, 1992; Jones and Celi, 1997) and (La Civita *et al.*, 2002) and the references therein for related results available in the field of small-scale helicopters. On the other hand, much less activity has been carried out in the development of time domain methods for helicopter identification. Specifically, very little work has been done to explore the applicability of subspace methods to this problem and a limited number of contributions is available (see, e.g., (Verhaegen and Varga, 1994; Bittanti and Lovera, 1997)). Subspace methods could provide a viable approach to the problem, for a number of reasons. First of all, the subspace approach can deal in a very natural way with MIMO problems; in addition, all the operations performed by subspace algorithms can be implemented with numerically stable and efficient tools from numerical linear algebra. Finally, information from separate data sets (such as generated during different experiments on the system) can be merged in a very simple way into a single state space model.

In the light of the above considerations, the aims of the present paper are the following:

- To show that subspace identification provides useful tools for the identification of helicopter dynamics, both in discrete and continuous time;
- To demonstrate, on the basis of simulation experiments, that accurate models can be obtained by this approach;
- To assess the role of the periodicity of helicopter dynamics in determining the prediction performance of time-invariant helicopter models.

2. HELICOPTER MODEL

The baseline simulation model used in this study is a nonreal-time, blade element type, coupled rotor-fuselage simulation model (see (Theodore and Celi, 2000) for details). The fuselage is assumed to be rigid and dynamically coupled with the rotor. A total of nine states describe fuselage motion through the nonlinear Euler equations. Fuselage and blades aerodynamics are described through tables of aerodynamic coefficients, and no small angle assumption is required. A coupled flap-lag-torsion elastic rotor model is used. Blades are modeled as Bernoulli-Euler beams. The rotor is discretized using finite elements, with a modal coordinate transformation to reduce the number of degrees of freedom. The elastic deflections are not required to be small. Blade element theory is used to obtain the aerodynamic characteristics on each blade section. Quasi-steady aerodynamics is used, with a 3-state dynamic inflow model.

The trim procedure is the same as in (Celi, 1991). The rotor equations of motion are transformed into a system of nonlinear algebraic equations using a Galerkin method. The algebraic equations enforcing force and moment equilibrium, the Euler kinematic equations, the inflow equations and the rotor equations are combined in a single coupled system. The solution yields the harmonics of a Fourier expansion of the rotor degrees of freedom, the pitch control settings, trim attitudes and rates of the entire helicopter, and main and tail rotor inflow. Linearized models (of order 53) are extracted numerically, by perturbing rotor, fuselage, and inflow states about a trimmed equilibrium position. In particular, whenever forward flight conditions are considered, the linearised models turn out to be time-periodic ones, with a period given by $T = \frac{2\pi}{N\Omega}$, where N is the number of rotor blades and Ω is the rotor angular frequency.

3. SUBSPACE IDENTIFICATION ALGORITHMS

The aim of this work is to assess the practical applicability of subspace identification techniques to the helicopter identification problem. In this paper, subspace algorithms of the MOESP class are used. In particular, for discrete time identification the so-called PI MOESP algorithm is applied (see (Verhaegen, 1994)), while the continuous time identification is carried out using the results in (Haverkamp, 2001).

The MOESP class of identification algorithms operates in two steps: first an orthonormal basis for the observability subspace of the system is extracted from the measured input/output data, and is used to compute an estimate of the model order n and of the A and C matrices of the state space description. Subsequently, the B and D matrices are obtained, usually by solving the least squares problem associated with the minimisation of the simulation/prediction error for the identified model. The interested reader is referred to the cited works for details about the operation and the implementation of the identification algorithms.

4. SIMULATION STUDY

The model described in Section 2 is used to generate data for black-box identification of linear models describing the response of the helicopter to perturbations in the control inputs. In particular, this study focuses on the analysis of the response of the body linear and angular rates to perturbations in the pilot inputs. Only the results obtained in identifying a linear, discrete time model for the response of the rigid body modes (linear velocities u, v, w and angular rates p, q and r) of the helicopter to perturbations on the collective inputs are presented for brevity, but similar results have been obtained by applying analogous input perturbations to the longitudinal and lateral cyclic channels.

In the identification experiments, multiharmonic input signals have been used, i.e., signals of the form:

$$u(t) = \sum_{k=1}^{N} A\cos(k\omega_f t + \phi_k) \qquad (1)$$

The design of this signal depends on the choice of the amplitude A, the main frequency ω_f, the total number of harmonics N and the phases ϕ_k. For the experiment illustrated in this abstract, the following choices have been made:

- $A = 0.1$;
- $N = 250$;
- $\omega_f = 0.15 rad/s$.

With these choices the highest frequency of u becomes $N\omega_f = 37.5$ rad/s, i.e., a frequency range which allows for the accurate extraction of the response of the rigid body modes without exciting in a significant way the rotor modes. Note that having defined the phase ϕ_1 of the first harmonic, the phases ϕ_k of the higher harmonics have been selected according to the Schroeder formula (see (Schroeder, 1970)): this particular phasing allows to obtain multiharmonic signals with a reduced value for the peak amplitude, so that the system remains in its linear operating region around the selected equilibrium (trim) point. From Figure 1 one can see that the time history of the designed input signal is very similar to the sine sweep inputs used in actual flight testing (see (Hamel and Kaletka, 1997)). This input sequence has been applied to the collective input channel of the linearised helicopter models corresponding to forward flight conditions ranging from $V = 40$ kn to $V = 120$ kn and the response of the linear and angular rates of the fuselage have been collected, using a sampling interval $T_s = 0.05s$. Note that measurement noise has not been taken into account in the simulated experiment illustrated herein.

The collected input/output data have been used to compute an estimate of the observability matrix (with

$i = 40$ block rows) for the system; Figure 2 shows a logarithmic plot of the singular values of the estimate of the observability matrix computed in the case $V =$ 100 kn, from which one can see that a reasonable estimate for the rank of the matrix is 10, which leads to choice of a model order $n = 10$, on the basis of the available input/output data.

A set of 10th order models have been then identified, and the frequency response of the identified models has been subsequently compared with the frequency response of the full linearised helicopter model. Such a comparison is illustrated in Figures 3, 6, 9,12 and 15 for the case of the linear velocities and in Figures 4, 7, 10,13 and 16 for the case of the angular velocities. As can be seen from the Figures, the identified model provides an accurate fit of the actual response over the range of frequency corresponding to the input excitation. Note, in passing, the strong dependence of the linearised dynamics of the system from the considered flight condition.

The validation of the identified model has been also carried out in the time domain, by measuring the accuracy of the model in response to a doublet input signal (along the lines of (Hamel and Kaletka, 1997), see Figure 18), which consists of a (low pass filtered) double step. The relative error norm of the validation experiment is below 5% on all the considered output variables (the time histories of the estimated outputs are omitted for brevity). This result can be considered more than satisfactory, keeping in mind that the doublet input excites also high frequency modes which are not included in the reduced order identified model.

Another interesting issue is whether there is some correspondence between the eigenvalues of the (low order) identified model and the eigenvalues of the linearised helicopter model. This issue is a relevant one if the identified models obtained via the application of subspace algorithms are to be used as initial estimates for a subsequent model optimisation step based on physical model parameterisations (as in (Hamel and Kaletka, 1997; Jones and Celi, 1997)). The comparison is shown in Figures 5, 8, 11,14 and 17 (in which only the low frequency modes of the linearised model are shown), from which one can see that the eigenvalues of the identified model match very accurately the low frequency modes of the linearised model, and therefore they can be readily given a physical interpretation in terms of the actual rigid body modes of the helicopter.

Finally, a characterisation of the frequency domain uncertainty of the identified model could be obtained using, e.g., the approach described in (Bittanti and Lovera, 2000).

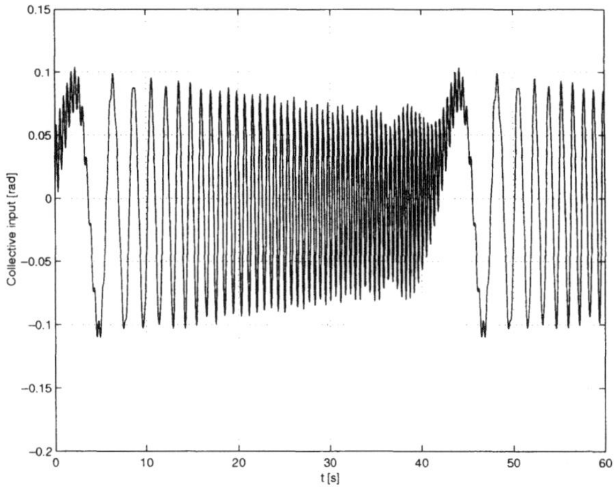

Fig. 1. Time history of applied collective input perturbation.

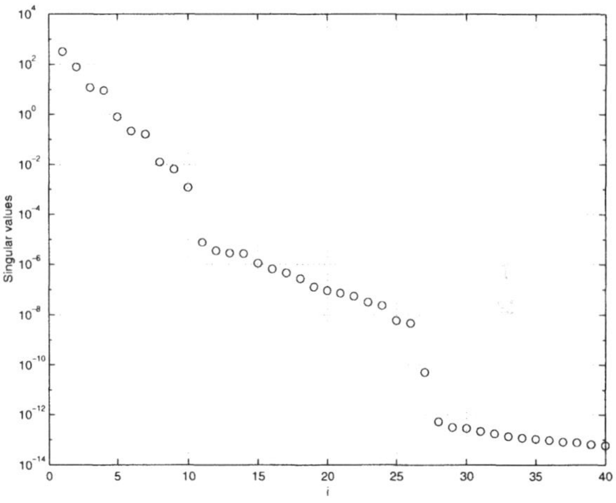

Fig. 2. Singular values of the observability matrix ($i = 40, V = 100$).

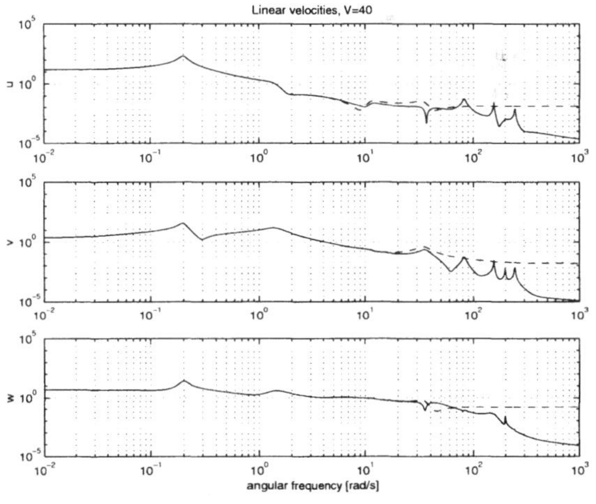

Fig. 3. Frequency response from collective input to linear velocities, 40 kn (real: solid line; estimated: dashed line).

ACKNOWLEDGMENTS

This research is supported by the MIUR project "Innovative Techniques for Identification and Adaptive Control of Industrial systems" and by the US Army Research Office project "Control of systems with pe-

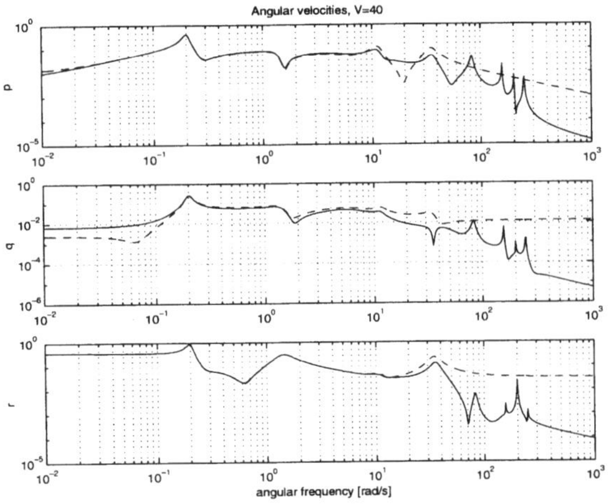

Fig. 4. Frequency response from collective input to angular velocities, 40 kn (real: solid line; estimated: dashed line).

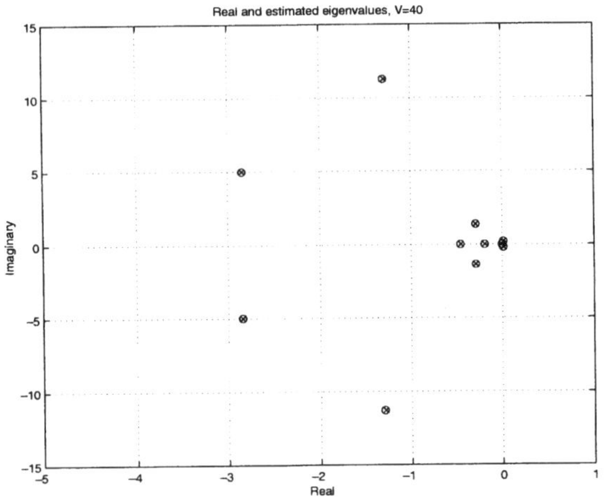

Fig. 5. Eigenvalues of the identified model (x) compared with low frequency modes of the linearised helicopter model (o), , 40 kn.

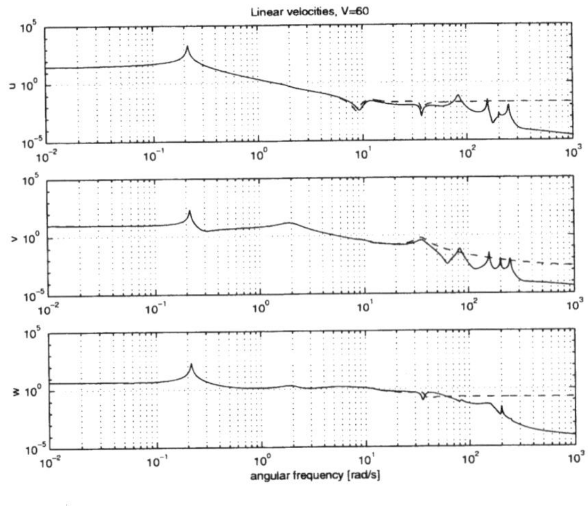

Fig. 6. Frequency response from collective input to linear velocities, 60 kn (real: solid line; estimated: dashed line).

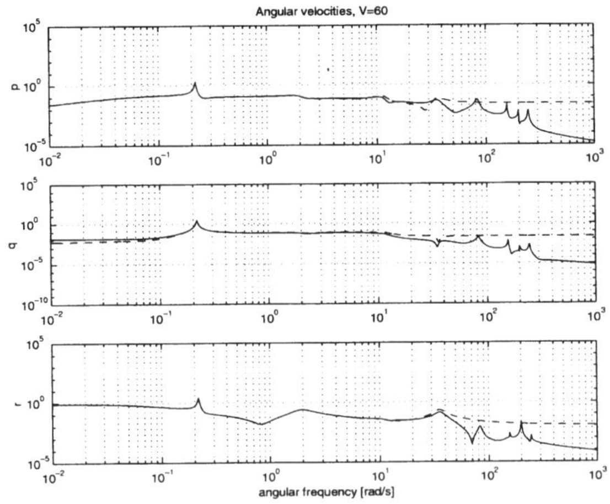

Fig. 7. Frequency response from collective input to angular velocities, 60 kn (real: solid line; estimated: dashed line).

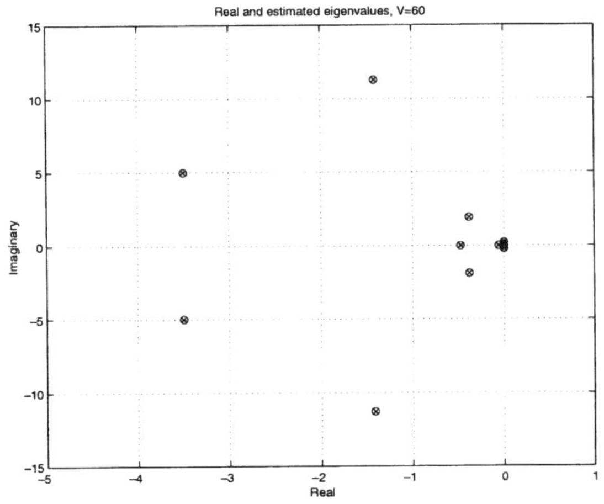

Fig. 8. Eigenvalues of the identified model (x) compared with low frequency modes of the linearised helicopter model (o), 60 kn.

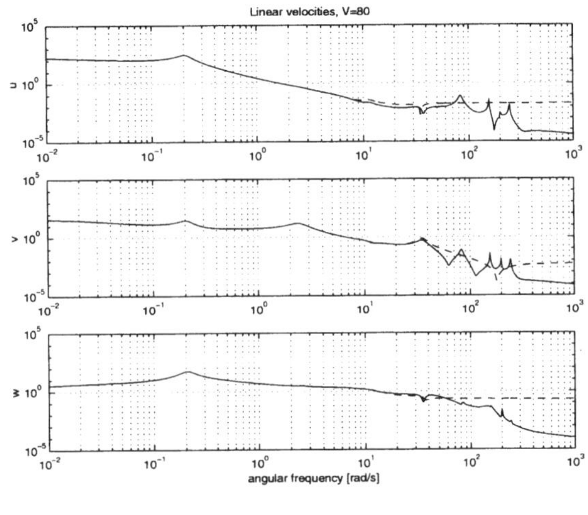

Fig. 9. Frequency response from collective input to linear velocities, 80 kn (real: solid line; estimated: dashed line).

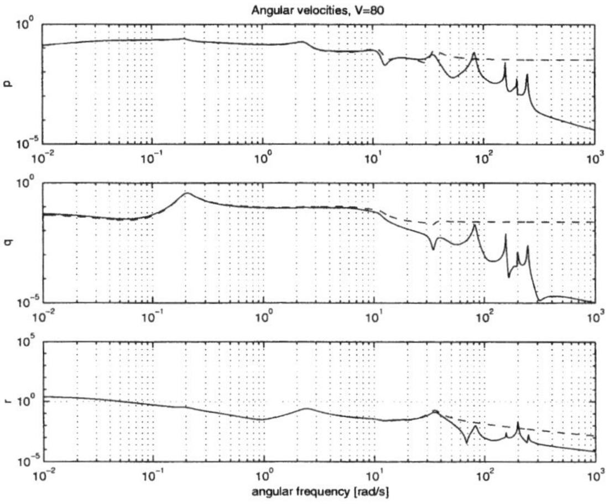

Fig. 10. Frequency response from collective input to angular velocities, 80 kn (real: solid line; estimated: dashed line).

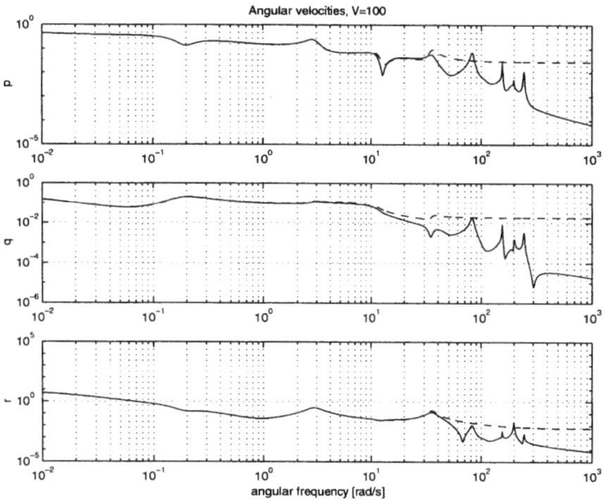

Fig. 13. Frequency response from collective input to angular velocities, 100 kn (real: solid line; estimated: dashed line).

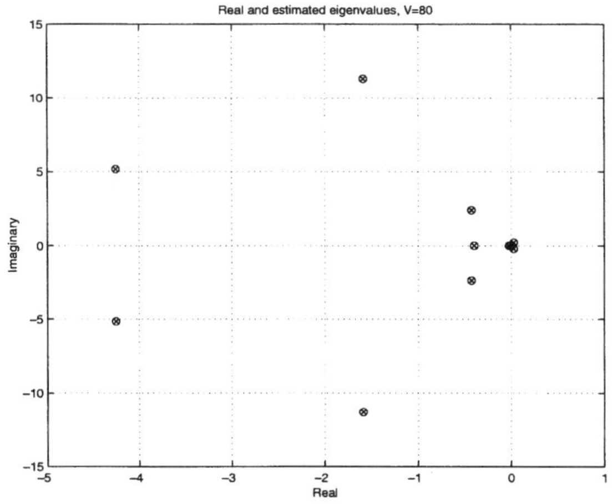

Fig. 11. Eigenvalues of the identified model (x) compared with low frequency modes of the linearised helicopter model (o), 80 kn.

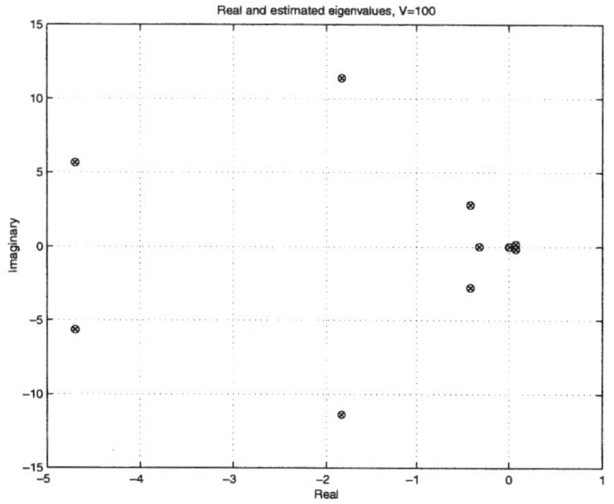

Fig. 14. Eigenvalues of the identified model (x) compared with low frequency modes of the linearised helicopter model (o), 100 kn.

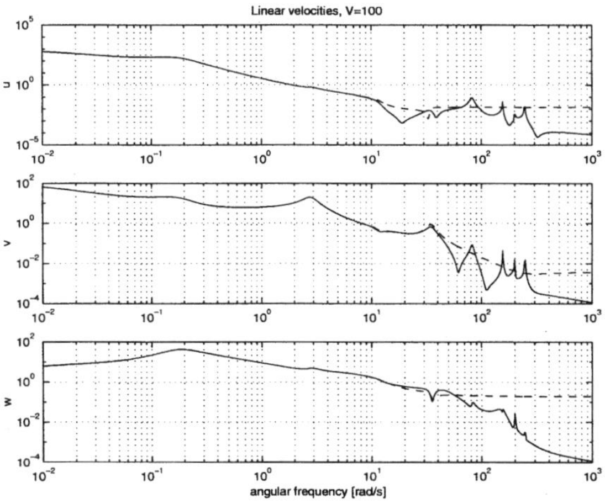

Fig. 12. Frequency response from collective input to linear velocities, 100 kn (real: solid line; estimated: dashed line).

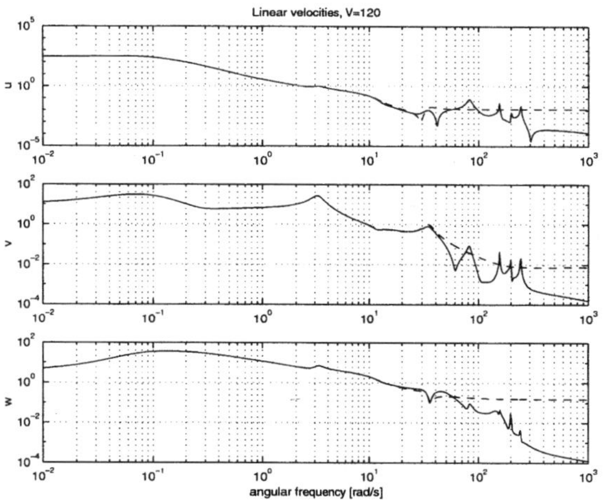

Fig. 15. Frequency response from collective input to linear velocities, 120 kn (real: solid line; estimated: dashed line).

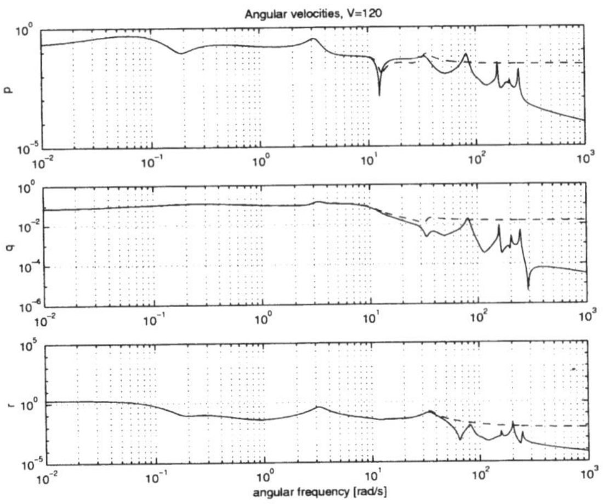

Fig. 16. Frequency response from collective input to angular velocities, 120 kn (real: solid line; estimated: dashed line).

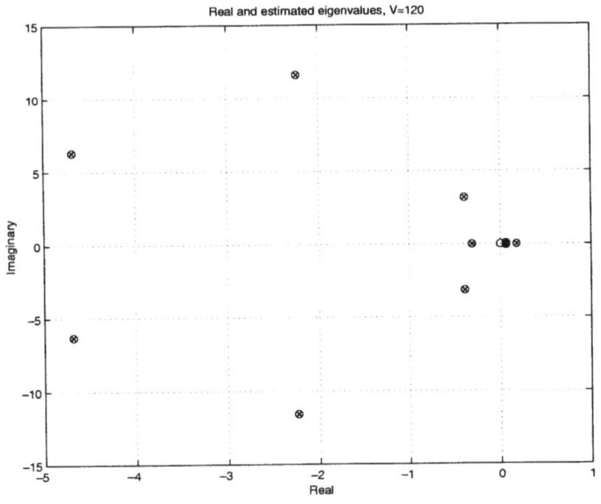

Fig. 17. Eigenvalues of the identified model (x) compared with low frequency modes of the linearised helicopter model (o), 120 kn.

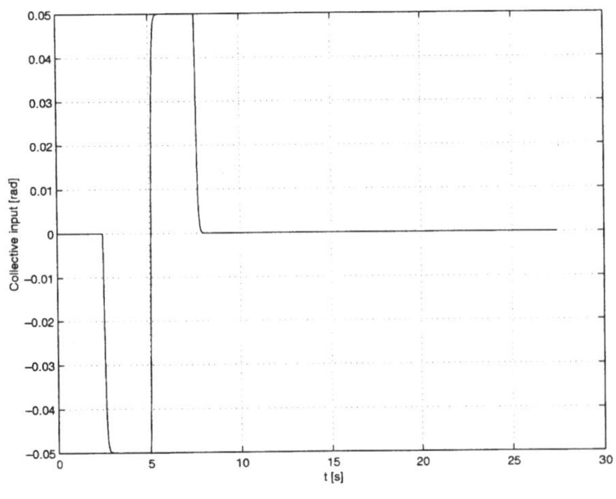

Fig. 18. Doublet input signal used for model validation (real: solid line; estimated: dashed line).

riodic coefficients with application to active rotor control".

5. REFERENCES

Bittanti, S. and M. Lovera (1997). Identification of linear models for a hovering helicopter rotor. In: *Proceedings of the 11th IFAC Symposium on system identification*. Fukuoka, Japan.

Bittanti, S. and M. Lovera (2000). Bootstrap-based estimates of uncertainty in subspace identification methods. *Automatica* **36**(11), 1605–1615.

Celi, R. (1991). Hingeless rotor dynamics in coordinated turns. *Journal of the American Helicopter Society* **36**, 39–47.

Hamel, P. and J. Kaletka (1997). Advances in rotorcraft system identification. *Progress in Aerospace Sciences* **33**(3-4), 259–284.

Haverkamp, B. (2001). State space identification-theory and practice. PhD thesis. Delft University of Technology.

Jones, C. and R. Celi (1997). Frequency response sensitivity functions for helicopter system identification. *Journal of the American Helicopter Society* **42**(3), 244–253.

La Civita, M., W. Messner and T. Kanade (2002). Modelling of small-scale helicopters with integrated first-principles ans system identification techniques. In: *Proceedings of the 58th American Helicopter Society Annual Forum*.

Schroeder, M.R. (1970). Synthesis of low-peak-factor signals and binary sequences with low autocorrelation. *IEEE Transactions on Information Theory* **16**(1), 85–89.

Theodore, C. and R. Celi (2000). Flight dynamic simulation with refined aerodynamic and flexible blade modeling. In: *Proceedings of the 56th Annual Forum of the American Helicopter Society, Virginia Beach, VA*.

Tischler, M. and M. Cauffman (1992). Frequency-response method for rotorcraft system identification: Flight applications to BO-105 coupled rotor/fuselage dynamics. *Journal of the American Helicopter Society* **37**(3), 3–17.

Verhaegen, M. (1994). Identification of the deterministic part of mimo state space models given in innovations form from input-output data. *Automatica* **30**(1), 61–74.

Verhaegen, M. and A. Varga (1994). Some experience with the MOESP class of subspace model identification methods in identifying the BO105 helicopter. Technical Report TR R165-94. DLR.

IFAC Publications
www.elsevier.com/locate/ifac

ESTIMATION OF DAMPED AND UNDAMPED SINUSOIDS WITH APPLICATION TO ANALYSIS OF ELECTROMAGNETIC FDTD SIMULATION DATA [1]

Tomas McKelvey *, **Thomas Rylander** ** and **Mats Viberg** *

* *Dept. of Signals and Systems, Chalmers University of Technology, SE-412 96 Gothenburg, Sweden, {mckelvey,viberg}@s2.chalmers.se*
** *Center for Computational Electromagnetics, Department of Electrical and Computer Engineering, University of Illinois at Urbana-Champaign, Urbana, Illinois 61801-2991, USA, tryl@uiuc.edu*

Abstract: Computational electromagnetics deals with the problem of finding efficient and reliable solutions of Maxwell's equations. This is important in a variety of practical applications, such as antenna design and automatic target recognition, and time-domain computations are attractive for a large class of these problems. However, one is more often interested in the frequency characteristics of the solution. Using the Fourier transform to determine the resonance frequencies and damping factors requires long time series, implying severe computational complexity. An attractive alternative is to apply high-resolution frequency estimation, developed in the signal processing society over the past few decades. This is a challenging problem, though, as the number of interesting frequency components can be as large as several hundreds. To overcome the difficulties, we propose a frequency-domain subspace method that yields accurate frequency and damping estimates in a selected frequency band. The required electromagnetic simulation time can thereby be reduced by several orders of magnitude. *Copyright © 2003 IFAC*

Keywords: spectral estimation, system identification, estimation of harmonics, resonance frequency, linear multivariable systems, identification algorithms, state-space realization, Maxwell's equations

1. INTRODUCTION

The problem of estimating damped and undamped sinusoids from measurements has received considerable attention over the years, and a wide range of methods have been developed. The books by Kay (Kay, 1988) and Stoica and Moses (Stoica and Moses, 1997) give good overviews of existing techniques. However, the case when the frequencies are closely spaced is still regarded as challenging, particularly when the number of data samples are scarce. In this case the non-parametric techniques, such as the periodogram and its derivatives, often cannot resolve the frequencies. Some of the classical parametric techniques, such as AR modeling, work fine if the number of frequencies are low or moderate. If the number of frequencies is large and also closely spaced, only the so called high resolution techniques such as Kung's method (Kung, 1978) ESPRIT (Roy and Kailath, 1989) or MUSIC can yield satisfactory results. Common for both techniques is that they use geometrical properties of the sample covariance matrix from time domain samples to derive the parameters.

In this paper the focus is on data where the frequencies to be estimated reside in a relatively narrow frequency band, and where considerable disturbances exist out-

[1] Tomas McKelvey and Mats Viberg were in part supported by the Swedish Research Council (VR). Thomas Rylander was in part supported by grants from the National Graduate School in Scientific Computing (NGSSC) and the Swedish Research Council on Engineering Science (TFR).

side the band of interest. The number of sinusoids are also large within the band. Hence, a high resolution parametric technique must be applied. However, since the signal has considerable energy in the other frequency bands, a straightforward application of these techniques will fail, since the methods cannot be tuned to focus on a particular frequency band. A possibility is to apply prefiltering prior to the construction of the sample covariance matrices. Such filters will inevitably introduce transient effects, which will perturb the signal unless ample amount of data is available so the transient regions are removed and only a subset of the time domain signal is used. This is often not desired, since fewer data will lead to less accurate results.

Recently, a frequency domain method has been developed which overcomes several of the aforementioned problems. In (McKelvey and Viberg, 2001), the method is developed for the undamped case. The method can be viewed as a frequency domain version of the well known ESPRIT algorithm. The algorithm is based on using the discrete Fourier transform (DFT) on the time domain data, and hence move the problem to the frequency domain. A subset of the DFT data, corresponding to the frequencies of interest, is then used in the estimation. The technique has two main advantages:

Data compression The first step of the method is to calculate the DFT from the time domain data. This step is computationally inexpensive when using the fast Fourier transform (FFT). Now, if only a narrow subband is of interest only the corresponding DFT points are retained for further processing. Working on a subset of data has positive effects regarding both algorithmic speed and memory requirements.

Filtering The selection of DFT points from a subband corresponds to a band-pass filtering operation. Signal energy outside the band is suppressed according to the sinc formula from the rectangular windowing function. Furthermore, the proposed algorithm is completely insensitive to this filtering operation with regard to the normally troublesome spectral leakage phenomenon. In other words, the frequency domain filtering gives no transient effects which are inevitable using standard time-domain filtering.

First an overview of the estimation technique is given. Then an application is studied where data from a numeric electromagnetic simulation of Maxwell's equations are analyzed.

1.1 *Problem formulation and signal model*

Consider the signal model

$$y(t) = \sum_{k=1}^{n} \alpha_k e^{\beta_k t} + w(t) \tag{1}$$

where $\beta_k \in \mathbb{C}$ and $\alpha_k \in \mathbb{C}^p$ are the unknown complex frequencies and gains respectively, and $w(t)$ denotes disturbances. Normally the complex frequency is represented as $\beta_k = -\gamma_k + i2\pi f_k$, where γ_k is the damping and f_k is the relative frequency. To make the signal model (1) unique we assume $f_k \in (-.5, .5)$, $\beta_k \neq \beta_i$ for $k \neq i$ and $\alpha_k \neq 0$ for all k. The Q-value is a measure of the level of damping for each exponential, and is here defined as

$$Q_k = \frac{\sqrt{\gamma_k^2 + (2\pi f_k)^2}}{2\gamma_k}$$

The signal $y(t)$ is sampled at N points at $t = 0, \ldots, N-1$, and the unknown frequencies and gains are to be estimated. The model includes vector valued signals if $p > 1$, and real valued signals is a special case of the model, where α_k and β_k appear in complex conjugate pairs. Undamped exponentials have $\gamma_k = 0$ and consequently an infinite Q-value.

As a preparation for development of the estimation method a matrix notation is introduced.

$$\begin{aligned} A &= \text{diag}\left[e^{\beta_1}, e^{\beta_2}, \cdots, e^{\beta_n}\right] \in \mathbb{C}^{n\times n} \\ C &= \left[\alpha_1 \ \alpha_2 \ \cdots \ \alpha_n\right] \in \mathbb{C}^{p\times n} \\ x_0 &= \left[1 \ 1 \ \cdots \ 1\right]^T \in \mathbb{R}^{n\times 1} \end{aligned} \tag{2}$$

With the matrix notation we can compactly write $y(t) = CA^t x_0$. Note that the A matrix is diagonal, and hence the n eigenvalues of A are e^{β_k}, $k = 1, \ldots, n$. The representation of $y(t)$ given by the matrix triple (A, C, x_0) defined in (2) is called a *realization* and is *not* unique. Take an arbitrary non-singular matrix $T \in \mathbb{C}^{n\times n}$. Then the matrix triple $(\tilde{A}, \tilde{C}, \tilde{x}_0) \triangleq (T^{-1}AT, CT, T^{-1}x_0)$ is also a valid realization of $y(t)$, i.e. $y(t) = \tilde{C}\tilde{A}^t\tilde{x}_0$. Note that eigenvalues of $\tilde{A}$ are still e^{β_k}.

With the matrix notation a simple recursive formula for calculating $y(t)$ exists, and is called a *state-space model*

$$\begin{aligned} x(t+1) &= Ax(t), \quad x(0) = x_0 \\ y(t) &= Cx(t) \end{aligned} \tag{3}$$

where the state $x(t) \in \mathbb{C}^{n\times 1}$ is the memory vector in the recursion. Associated with a state-space model is the *observability matrix*

$$\mathscr{O}_s = \begin{bmatrix} C \\ CA \\ \vdots \\ CA^{s-1} \end{bmatrix} \in \mathbb{C}^{ps\times n} \tag{4}$$

If the number of block rows s is larger than n, it is called the extended observability matrix. Note that if $\mathscr{O}_s$ is the observability matrix of the realization (A, C, x_0), then $\mathscr{O}_s T$ is the observability matrix of the realization $(\tilde{A}, \tilde{C}, \tilde{x}_0)$. This implies that the *range space* of the observability matrix is invariant w.r.t. the realization and is a property of the signal $y(t)$. If the complex frequencies β_k are all distinct and the relative frequencies belong to the interval $(-.5, .5)$, $s \geq n$ and for all k, $\alpha_k \neq 0$, then $\mathscr{O}_s$ has full rank n. The result is well known (Stoica and Moses, 1997) and follows by noting that $\mathscr{O}_s$ has a Vandermonde type structure.

2. FREQUENCY DOMAIN REPRESENTATION

As the DFT is an integral part of the method, it is vital to transfer the signal model (3) to the DFT domain. A well-known property of the DFT of a finite signal is that the resulting DFT coefficients are samples of the discrete time Fourier transform (DTFT) if the signal periodically repeats itself outside the measurement interval $t = 0,\ldots,N-1$. The signal model (1), in general, does not imply such a periodic repetition except for a very restrictive set of parameters, i.e. the undamped case, and where the product between frequencies f_k and the measurement time N is an integer. By the introduction of an artificial input signal we can, for every measurement interval N and signal model of the form (1), create an observationally identical model which is periodic. We call this *periodic embedding*.

2.1 *Periodic embedding*

A slight modification of the original signal model (3) is here introduced. Consider the modified model

$$\begin{aligned} x(t+1) &= Ax(t)+Bu(t), \quad x(0)=x_0 \\ y(t) &= Cx(t) \end{aligned} \tag{5}$$

where the introduced "input" signal $u(t)$ is defined as

$$u(t) \triangleq \begin{cases} 1, & t = kN-1, \quad k = 1,2,\ldots \\ 0 & \text{otherwise} \end{cases} \tag{6}$$

Note that $u(t)$ is an $N-$periodic signal since $u(t) = u(t+kN)$, $k = 1,2,\ldots$. If B is chosen as

$$B \triangleq (I-A^N)x_0 \tag{7}$$

it directly follows that $x(N) = x_0$ and also $y(t)$ in (5) is $N-$periodic. Note that in the measurement interval ($t = 0,\ldots,N-1$) the output $y(t)$ of the modified signal model (5) equals the output of the original signal model (3), and consequently both models are valid for the observed signal. The column vector B is identically zero when $(I-A^N) = 0$. This happens in the undamped case, when the observation interval encompasses exactly an integer number of periods for all sinusoids, and the modification by an input is then not necessary to make the signal $N-$periodic.

2.2 *Frequency domain model*

The N-point Discrete Fourier Transform of a signal $x(t)$ is for $k = 0,\ldots,N-1$ given by

$$X_k \triangleq \mathrm{DFT}\{x(t)\}_k \triangleq \sum_{t=0}^{N-1} x(t)W_N^{-kt} \tag{8}$$

where $W_N^k \triangleq e^{i\frac{2\pi k}{N}}$. Consider the modified model (5). Due to the $N-$periodicity of $x(t)$ it follows that $x(N) = x(0)$ and hence

$$\begin{aligned} \mathrm{DFT}\{x(t+1)\}_k &= \sum_{t=0}^{N-1} x(t+1)W_N^{-kt} \\ &= W_N^k \sum_{t=1}^{N} x(t+1)W_N^{-k(t+1)} = W_N^k X_k. \end{aligned}$$

Therefore, the modified signal model can be expressed in the DFT-domain as

$$\begin{aligned} W_N^k X_k &= AX_k + BU_k \\ Y_k &= CX_k \end{aligned} \tag{9}$$

where $Y_k \triangleq \mathrm{DFT}\{y(t)\}_k$ and $U_k \triangleq \mathrm{DFT}\{u(t)\}_k = W_N^k$.

3. ESTIMATION ALGORITHM

The derivation starts by forming a vector relation by repeatedly using (9). It is easy to show that for DFT frequency k we have

$$\mathbf{Y}_k = \mathcal{O}_s X_k + \Gamma_s \mathbf{U}_k \tag{10}$$

where

$$\mathbf{Y}_k \triangleq \begin{bmatrix} Y_k \\ W_N^k Y_k \\ W_N^{2k} Y_k \\ \vdots \\ W_N^{(s-1)k} Y_k \end{bmatrix}, \quad \mathbf{U}_k \triangleq \begin{bmatrix} U_k \\ W_N^k U_k \\ W_N^{2k} U_k \\ \vdots \\ W_N^{(s-1)k} U_k \end{bmatrix}$$

and the lower triangular Toeplitz matrix Γ_s is

$$\Gamma_s \triangleq \begin{bmatrix} 0 & & & & \\ CB & 0 & & & \\ CAB & \ddots & \ddots & \ddots & \\ \vdots & & & \ddots & \\ CA^{s-2}B & CA^{s-3}B & \cdots & CB & 0 \end{bmatrix}$$

Now, assume a subset of the frequencies on the DFT-grid is selected. Denote the M selected frequency indices by $k_1,\ldots,k_M$; and we will use these M points of the DFT for estimating the frequencies. The reason for considering a subset of the DFT data is to provide suppression of interfering signal components residing outside the frequency band of interest. The method works for any size $M > 2n$, and it is possible to use all DFT points, i.e. $M = N$. The final data matrices are formed by assembling the data vectors as

$$\mathbf{Y} \triangleq \begin{bmatrix} \mathbf{Y}_{k_1} & \mathbf{Y}_{k_2} & \cdots & \mathbf{Y}_{k_M} \end{bmatrix} \tag{11}$$

$$\mathbf{U} \triangleq \begin{bmatrix} \mathbf{U}_{k_1} & \mathbf{U}_{k_2} & \cdots & \mathbf{U}_{k_M} \end{bmatrix} \tag{12}$$

$$\mathbf{X} \triangleq \begin{bmatrix} X_{k_1} & X_{k_2} & \cdots & X_{k_M} \end{bmatrix} \tag{13}$$

and the vector relation (10) is expanded into a matrix one

$$\mathbf{Y} = \mathcal{O}_s \mathbf{X} + \Gamma_s \mathbf{U} \tag{14}$$

Hence the DFT-data in $\mathbf{Y}$ is a sum of two matrix components. The rank of the matrix product $\mathcal{O}_s\mathbf{X}$ equals n, and the second term is the product of a known matrix ($\mathbf{U}$) and the unknown Γ_s. In order to retrieve the frequencies, it suffices to get an estimate of the range space of the extended-observability matrix $\mathcal{O}_s$. It is important to note that the second term in (14) is due to the non-periodicity of the data. It's removal, which is necessary to recover $\mathcal{O}_s$, has the effect of canceling the spectral leakage introduced by the DFT. Employing the periodic embedding is just a simple way to explain this phenomenon, and does not introduce any artificial assumptions of the signal outside the observation interval.

The matrix equation (14) has a structure which is well known in the area of subspace based system identification methods (Verhaegen and Dewilde, 1992; Viberg, 1995; McKelvey *et al.*, 1996).

3.1 *Subspace based method*

Starting from the derived identity (14) the identification method consists of the following steps:

(1) Remove the influence of the term $\Gamma_s \mathbf{U}$. This is done by a projection.
(2) Find a matrix Z_s which is an estimate of the range space of $\mathcal{O}_s$ using the singular value decomposition.
(3) From Z_s estimate a matrix $\hat{A}$ which is similar to the original matrix A.
(4) Determine the eigenvalues $\hat{\lambda}_k$ of the matrix $\hat{A}$ and let the complex frequency estimates be $\hat{\beta}_k = \log \hat{\lambda}_k$.
(5) Estimate the complex gains α_k of the sinusoids using linear regression.

For more details we refer the reader to (McKelvey and Viberg, 2001) which deals with the undamped case.

4. TEST CASE

This test case is based on a metal box of sides 60 mm × 60 mm × 20 mm, from which the square lid is removed. The bottom of the metal box is covered by a substrate with $\varepsilon_r = 2$ and thickness 2 mm. There are two metal strip lines on the substrate as shown in Figure 1. The strip line to the left is electrically connected to the walls of the box. We solve for the electromagnetic fields by the Finite-Difference Time-Domain scheme (FDTD) (Yee, 1966; Taflove, 1995), with the cell size h = 2 mm, given a line source at point A in Figure 1. The line source is implemented by assigning the value $E_i(t)$ to the electric field along the FDTD edge, which connects the strip with the box, and we choose the time dependence to be a sinusoidal which is amplitude modulated with a Gaussian, i.e. $E_i(t) = E_0 \exp\left(-(t-t_0)^2/d_0^2\right) \sin\left(2\pi f_0(t-t_0)\right)$, where f_0 = 4 GHz, $t_0 = 6/f_0$ and $d_0 = 1/f_0$. The electric field is recorded at point B in Figure 1, again along the FDTD edge which connects the strip with the box. The recording is started after the power of the exciting source has diminished. The output can then be regarded as a sum of damped sinusoids. The time step used in the simulations is $\Delta t = h/\sqrt{3}c_0$, where c_0 is the speed of light.

The goal of the simulation is to establish the resonance frequencies of the conducting structure and their associated Q-values and amplitudes. Such knowledge is important in the design phase of microwave electronic systems in order to, e.g., achieve high electromagnetic compatibility (EMC). Based on the recorded time signal it is therefore desirable to estimate a parametric model from which the relevant frequencies, dampings etc. can be calculated. As the test case lacks an analytical solution, the following procedure was applied to validate the estimates. The FDTD simulation was calculated for $2^{20} \simeq 10^6$ time steps in order to generate a time domain signal such that the DFT and the DTFT will be very close. In Figure 2 a graph of the 8192 first time domain samples of the signal is shown. In Figure 3 the DFT of the entire signal is shown for the frequency band 0-0.035 $/\Delta t$ (Hz) as a solid line. Based on these data a parametric model with 11 frequencies (22 states) is estimated using the algorithm in the paper. A total of 3667 frequency points are used. A very good agreement is noted compared with the original data sequence, so this model is regarded as the correct one. To solve the FDTD for such a massive amount of times steps requires prohibitive amount of computational resources. Hence, similar results are desired using only a fraction of time steps. In a second estimation, again a 22 states model is estimated now using only 2^{13} = 8192 FDTD time steps. This time 287 DFT points reside in the 0-0.035 band and are used in the estimation. The computational time for the estimation was 2.7 seconds on a 930 MHz Intel PIII using Matlab R12.1.

The DTFT of the estimated model is given by

$$\hat{Y}_k = \hat{C}(W_N^k I - \hat{A})^{-1} \hat{B} U_k \tag{15}$$

The DFT estimate $\hat{Y}_k$ is calculated on a grid corresponding to 0-0.035 in frequencies with $N = 2^{20}$. In Figure 3 the estimate is shown together with the DFT calculated from the long time sequence. The dash-dotted line shows the magnitude error between the two curves. The agreement is very good.

In Table 1 a comparison between the two parametric models is given. The quantities f, α, Q corresponds to the model derived from 2^{20} time steps while the quantities given by $\hat{f}, \hat{\alpha}, \hat{Q}$ are the estimated values from using 8192 time steps. The agreement is striking. Particularly the frequencies agree very well. Only the last frequency has a higher deviation. This is probably due to the very closely spaced neighboring frequency with a much higher Q value.

5. CONCLUSIONS

A parametric technique for the estimation of damped and undamped signals based on DFT data has been presented. The application of the algorithm to an electromagnetic test problem shows promising results. We find that it can reproduce damping rates and eigenfrequencies to high accuracy, given a relatively small number of time steps. In particular, this is useful for weakly damped modes where the standard DFT would require simulations with an unrealistic number of time steps for results of reasonable accuracy.

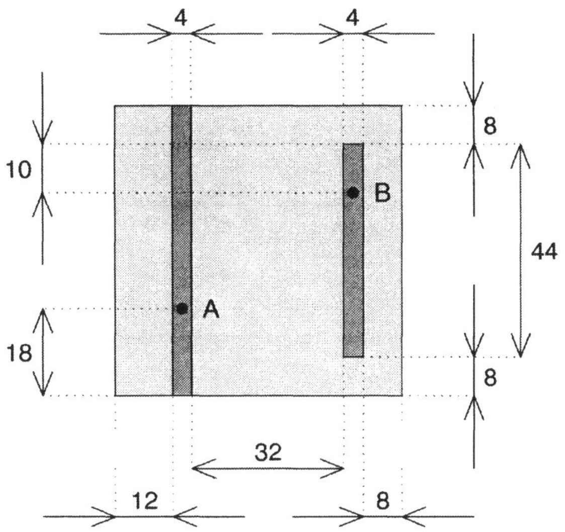

Fig. 1. The metal box viewed from above. The substrate and the metal strips are shown by light and dark gray areas, respectively. All measures of length are given in millimeters.

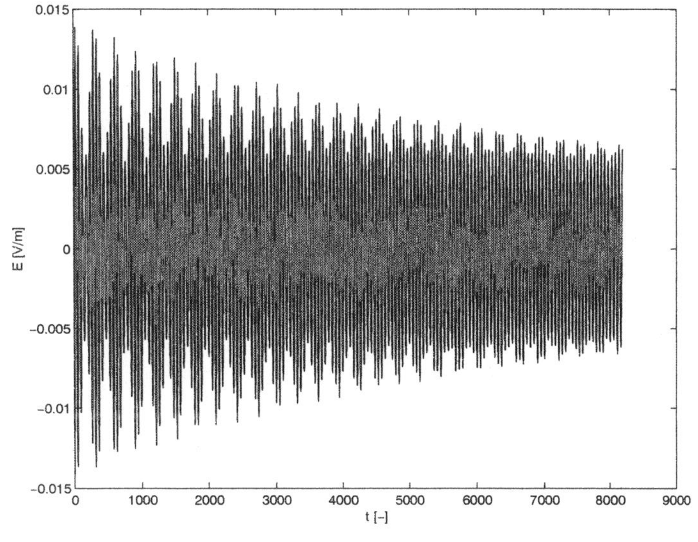

Fig. 2. Time domain data from FDTD simulation

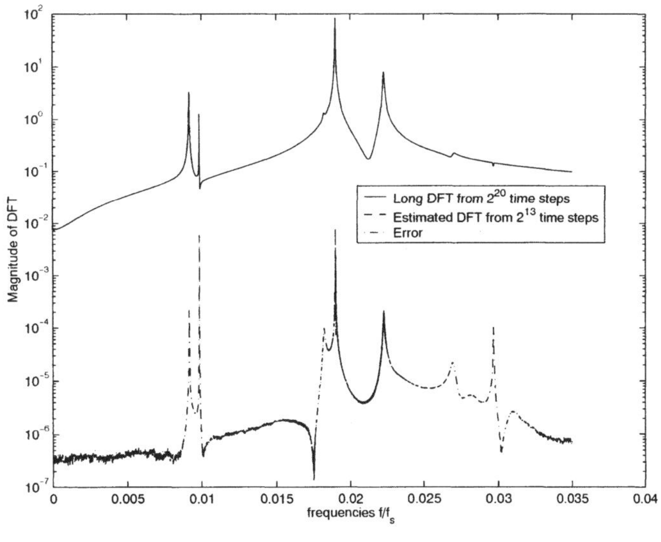

Fig. 3. DFT of data (2^{20} samples) and DTFT of estimated parametric model

f	$\hat{f}-f$	$\|\alpha\|$	$\|\hat{\alpha}\|-\|\alpha_0\|$	Q	$\hat{Q}-Q$
0.0091542	5.189e-10	0.00020355	0.00012629	497.36	-0.027898
0.0098471	-8.1267e-10	1.2416e-05	-8.3125e-06	3446.5	-16.715
0.015703	8.8626e-08	1.0567e-05	2.8571e-09	6.2602	-0.0061524
0.018211	-9.0521e-10	0.00011344	2.3708e-06	163.75	0.087758
0.018981	-5.8727e-10	0.0047862	-0.0029898	1039.9	0.01454
0.022301	1.1839e-09	0.002295	-8.3951e-05	240.67	-0.010523
0.022708	7.0562e-08	0.0002196	7.4705e-08	15.254	-0.001928
0.026915	1.2391e-07	4.4516e-05	-2.4297e-09	88.781	0.011745
0.028067	1.8287e-05	1.1266e-06	-5.7813e-08	23.988	-1.4658
0.029667	1.029e-07	2.5957e-06	8.942e-07	732.03	2.5837
0.030703	3.8504e-05	2.1338e-07	-1.5521e-08	35.927	-0.065384

Table 1. Estimation results regarding frequencies (f), amplitudes (α) and Q-values.

6. REFERENCES

Kay, S. M. (1988). *Modern Spectral Estimation, Theory & Application.* Prentice-Hall. Englewood Cliffs, New Jersey.

Kung, S. Y. (1978). A new identification and model reduction algorithm via singular value decomposition. In: *Proc. 12th Asilomar Conference on Circuits, Systems and Computers, Pacific Grove, CA.* pp. 705–714.

McKelvey, T. and M. Viberg (2001). A robust frequency domain subspace algorithm for multicomponent harmonic retrieval. In: *Proc. of 35th Asilomar Conference on Signals, Systems and Computers.*

McKelvey, T., H. Akçay and L. Ljung (1996). Subspace-based multivariable system identification from frequency response data. *IEEE Trans. on Automatic Control* **41**(7), 960–979.

Roy, R. and T. Kailath (1989). ESPRIT - Estimation of signal parameters via rotational invariance techniques. *IEEE Trans. on Acoustics, Speech and Signal Processing* **37**(7), 984–995.

Stoica, P. and R. Moses (1997). *Introduction to Spectral Analysis.* Prentice-Hall. Englewood Cliffs, USA.

Taflove, A (1995). *Computational Electrodynamics: The Finite-Difference Time-Domain Method.* Norwood, MA: Artech House.

Verhaegen, M. and P. Dewilde (1992). Subspace model identification. Part I: The output-error state-space model identification class of algorithms. *Int. J. Control* **56**, 1187–1210.

Viberg, M. (1995). Subspace-based methods for the identification of linear time-invariant systems. *Automatica* **31**(12), 1835–1851.

Yee, K S (1966). Numerical solution of initial boundary value problems involving Maxwell's equations in isotropic media. *IEEE Trans. Antennas Propagat.* **AP-14**, 302–307.

IFAC Publications
www.elsevier.com/locate/ifac

APPLICATION OF A RECURSIVE SUBSPACE IDENTIFICATION ALGORITHM TO CHANGE DETECTION

Hiroshi Oku *

* *Osaka Institute of Technology*
5-16-1, Omiya, Asahi-ku, Osaka 535-8585, JAPAN

Abstract: This paper presents a new change detection method with the aid of subspace identification. The proposed method is based on monitoring a change in variance of a statistic generated by a recursive subspace identification algorithm. An asymptotic property of the statistic is presented. Without changes during sampling, it is shown that, under relevant assumptions, the statistic converges in probability to a stack of noise vectors multiplied from the left side by a Toeplitz matrix. A numerical example illustrates that the proposed method can detect changes in the dynamics of a system, without being disturbed by changes in the dynamics of an input signal which are not our concern.

Keywords: system identification, subspace methods, recursive algorithms, instrumental variable methods, fault detection, statistical analysis, multivariable systems.

1. INTRODUCTION

The problem considered in this paper is *to detect changes in the dynamics of a system under surveillance in real time, without being disturbed by changes in the dynamics of the input signal.* Such a situation as in the problem statement may be encountered, for example, on preprocessing of data sampled for system identification, more precisely, when we segment the data by each operating point of a (probably nonlinear) system for identification of local linear time-invariant models (Leith and Leithead, 2000).

Statistics-based change detection methods have widely been studied and there exists extensive literature. Basseville (1988) gives in-depth systematic surveys of this research area, and Basseville and Nikiforov (1993) provide thorough knowledge, from theory to applications, of change detection. Basseville *et al.* (2000) have integrated subspace-based identification technique into fault detection algorithms for vibration monitoring. Thus, the model structure employed in (Basseville *et al.*, 2000) has no control input signal. They have assumed that the order of a model is known.

In this paper, a new on-line change detection scheme using a recursive subspace identification algorithm is presented. The innovations form (Verhaegen, 1994) is employed as the model structure. It is not assumed to know the order of a model. The approach which we will take is to monitor a change in the variance of a statistic generated by a recursive subspace identification algorithm, which is used as a whitening filter. Such statistic as presented here is a relevant indicator to detection of changes in the dynamics of a system under surveillance, since its variance becomes significantly large when the system undergoes a change. In this case attention is given to the fact that the variance in the presence of changes should be regarded as unknown. To monitor changes in the variance, the geometric moving average (GMA) algorithm (Basseville and Nikiforov, 1993) integrated with the maximum likelihood estimation is applied for a decision function.

Asymptotic properties of subspace identification have extensively been investigated by, e.g., (Deistler *et al.*, 1995; Jansson and Wahlberg, 1996; Peternell *et al.*, 1996; Viberg *et al.*, 1997; Jansson and Wahlberg, 1998; Jansson, 2000; Bauer and Jansson, 2000). In this paper, however, an asymptotic property of the statistic used for the proposed change detection method is investigated. More precisely, without changes during sampling, it is shown that, under relevant assumptions, the statistic converges in probability to a stack of noise vector multiplied from the left side by a Toeplitz matrix, which consists of parameter matrices of the stochastic part of a system considered. This property is not violated by changes in the dynamics of the input signal. A numerical example illustrates that the statistic is not affected by them.

Throughout this paper, tr· denotes the trace operator while $\|\cdot\|_2$ denotes the matrix 2-norm(Golub and Loan, 1996).

2. PROBLEM FORMULATION

Let us consider a system modelled by a collection of linear, finite dimensional, discrete time, time-invariant, state-space, local models, each of which is described, at every operating point, by the following innovations form:

$$x_{t+1} = A_i x_t + B_i u_t + K_i e_t \quad (1a)$$

$$y_t = C_i x_t + D_i u_t + e_t, \quad (1b)$$

where $t \in \mathbb{Z}, y_t, e_t \in \mathbb{R}^l$, $u_t \in \mathbb{R}^m$, $x_t \in \mathbb{R}^n$, and the suffix of the coefficient matrices, i ($i = 1, 2, \cdots$), represents the corresponding operating point of the system. e_t is assumed to be the Gaussian white noise with zero mean and covariance matrix equal to $\sigma_e^2 I$, that is, $\mathbf{E}[e_t] = 0$ and $\mathbf{E}\left[e_s e_t^T\right] = \delta_{st}\sigma_e^2 I$, where $\mathbf{E}$ denotes expectation and δ_{st} the Kronecker's delta. The exogenous input u_t is assumed to be bounded, pseudo-stationary (Ljung, 1999), relevantly persistently excited and independent of e_t in a relevant sense, which will be defined later. The matrices A_i and $A_i - K_i C_i$ for $\forall i$ are assumed to be stable throughout this paper. For $\forall i$, (A_i, C_i) is assumed to be observable and $(A_i, [B_i \ K_i])$ to be reachable.

In order for the change detection scheme considered here to succeed, the following two assumptions are assumed. Firstly, every local model is different in its dynamics from each other. Secondly, abrupt transitions from one operating point to another occur occasionally. Note that we do not mind whether changes in the dynamics of the input signal may occur. However, these input changes are *not* our concern and should not be detected. The problem is *to detect changes in the dynamics (i.e., transition from one operating point to another) of a system under surveillance in real time, without being disturbed by changes in the dynamics of the input signal.*

3. BRIEF REVIEW OF RECURSIVE SUBSPACE IDENTIFICATION

In this and the next sections, let us assume that the system stays at an operating point. Let us omit the subscription i from the coefficient matrices for notational brevity. For the rest of the paper, capital calligraphic letters are reserved for denoting such structured matrices as Hankel and Toeplitz matrices and so on. For a sampled sequence of a signal $\{\cdots, w_0, w_1, \cdots, w_N, \cdots\}$, the Hankel matrices $\mathscr{W}_N$, $\mathscr{W}_{P,N}$ and the vector of the stack of finite successive samples $w_{\cdot}(\cdot)$ are respectively defined, with two integers α and β, by

$$\mathscr{W}_N := \begin{bmatrix} w_1 & \cdots & w_N \\ \vdots & & \vdots \\ w_\alpha & \cdots & w_{N+\alpha-1} \end{bmatrix} =: \begin{bmatrix} w_{\alpha}(1) & \cdots & w_{\alpha}(N) \end{bmatrix}, \quad (2)$$

$$\mathscr{W}_{P,N} := \begin{bmatrix} w_{1-\beta} & \cdots & w_{N-\beta} \\ \vdots & & \vdots \\ w_0 & \cdots & w_{N-1} \end{bmatrix} =: \begin{bmatrix} w_{\beta}(1-\beta) & \cdots & w_{\beta}(N-\beta) \end{bmatrix}. \quad (3)$$

It is assumed that $\alpha \geq n+1$ and $\beta \geq n$.

Definition 1. (Assumptions on exogenous input). Given a positive integer N_0, the exogenous input u_t is called *persistently exciting* if the following relations hold for $\forall N > N_0$ including infinity:

$$\frac{1}{N}\mathscr{U}_N \mathscr{U}_N^T > 0, \quad (4)$$

$$\frac{1}{N}\begin{bmatrix} \Phi_N \\ \mathscr{U}_N \end{bmatrix} \begin{bmatrix} \Phi_N^T & \mathscr{U}_N^T \end{bmatrix} > 0, \quad (5)$$

where $\Phi_N := \begin{bmatrix} \mathscr{U}_{P,N}^T & \mathscr{Y}_{P,N}^T \end{bmatrix}^T$.

The exogenous input u_t is called *uncorrelated* with the noise e_t if for $\forall \tau$

$$\frac{1}{N}\sum_{t=1}^{N-\tau} e_t u_{t+\tau}^T \longrightarrow 0 \quad (6)$$

as $N \to \infty$ *with probability* 1.

Note that the word "*uncorrelation*" is used here with slight abuse of terminology(Hannan and Deistler, 1988; Bauer and Jansson, 2000). Note also that, from (4), it is shown that for $\forall N > N_0$ including infinity there exist a positive number $M < \infty$ such that

$$\left\| \left(\frac{1}{N}\mathscr{U}_N \mathscr{U}_N^T \right)^{-1} \right\|_2 \leq M. \quad (7)$$

In this case, the exogenous input u_t will be called *inversely bounded.*

Taking it into account that $(A - KC)^\beta$ is negligible when a sufficiently large β is chosen, from (1) the relation between the input and output Hankel matrices can be given by (Jansson and Wahlberg, 1996)

$$\mathscr{Y}_N = \begin{bmatrix} \mathscr{O}\mathscr{L} & \mathscr{H} \end{bmatrix} \begin{bmatrix} \Phi_N \\ \mathscr{U}_N \end{bmatrix} + \mathscr{K}\mathscr{E}_N =: \Theta Z_N + \mathscr{K}\mathscr{E}_N, \quad (8)$$

where $\mathscr{O}$ is the so-called *extended observability matrix* defined by

$$\mathscr{O} := \left[C^T \cdots (CA^{\alpha-1})^T \right]^T, \qquad (9)$$

$\mathscr{L}$ is the reversed controllability matrix defined by $\mathscr{L} := \left[\mathscr{L}_u \ \mathscr{L}_y \right]$, where

$$\mathscr{L}_u := \left[(A-KC)^{\beta-1}(B-KD), \ \cdots, \ (B-KD) \right],$$
$$\mathscr{L}_y := \left[(A-KC)^{\beta-1}K, \ \cdots, \ K \right].$$

The lower-triangular Toeplitz matrices, $\mathscr{H}$ and $\mathscr{K}$, are respectively defined by

$$\mathscr{H} := \begin{bmatrix} D & & O \\ CB & D & \\ \vdots & & \ddots \\ CA^{s-2}B & \cdots & CB \ D \end{bmatrix}, \qquad (10)$$

$$\mathscr{K} := \begin{bmatrix} I & & O \\ CK & I & \\ \vdots & \ddots & \ddots \\ CA^{s-2}K & \cdots & CK \ I \end{bmatrix}. \qquad (11)$$

From (8), an estimate of $\Theta_\Phi := \mathscr{O}\mathscr{L}$, denoted by $\widehat{\Theta}_{\Phi,N}$, can be derived from the solution to the minimization with respect to Θ

$$\min_{\Theta} f(\Theta), \qquad (12)$$

where

$$f(\Theta) := \frac{1}{N}\mathrm{tr}\mathscr{K}\mathscr{E}_N(\mathscr{K}\mathscr{E}_N)^T \qquad (13)$$
$$= \frac{1}{N}\mathrm{tr}(\mathscr{Y}_N - \Theta Z_N)(\mathscr{Y}_N - \Theta Z_N)^T, \qquad (14)$$

and actually is given by

$$\widehat{\Theta}_{\Phi,N} = \frac{1}{N}\mathscr{Y}_N \Pi^{\perp}_{\mathscr{U}_N}\Phi_N^T \left(\frac{1}{N}\Phi_N \Pi^{\perp}_{\mathscr{U}_N}\Phi_N^T\right)^{-1} \qquad (15)$$

where

$$\Pi^{\perp}_{\mathscr{U}_N} := I - \frac{1}{N}\mathscr{U}_N^T \left(\frac{1}{N}\mathscr{U}_N\mathscr{U}_N^T\right)^{-1}\mathscr{U}_N. \qquad (16)$$

Essentially, subspace identification is implemented by the singular value decomposition of $\widehat{\Theta}_{\Phi,N}$ multiplied from both sides by relevant weighting matrices, followed by solving linear equations with respect to (A,B,C,D) (Van Overschee and De Moor, 1996). Therefore, the recursive update of the estimate $\widehat{\Theta}_{\Phi,N}$ has been the most important issue of recursive subspace identification (Verhaegen and Deprettere, 1991; Oku and Kimura, 1999; Oku *et al.*, 2001; Oku and Kimura, 2002).

When the pair of the data $(u_{N+\alpha}, y_{N+\alpha})$ is sampled at the instant $N+\alpha$, the parameter Θ_Φ is estimated in a recursive manner by the following equation with the help of the auxiliary equations omitted here and found in (Oku and Kimura, 1999; Oku *et al.*, 2001):

$$\widehat{\Theta}_{\Phi,N+1} = \widehat{\Theta}_{\Phi,N} - \xi_{N+1}(r_{N+1} + \widehat{\Theta}_{\Phi,N}q_{N+1})q_{N+1}^T\Psi_N^{-1}, \qquad (17)$$

where, with $\phi(N+1) := [u_\beta(N-\beta+1)^T \ y_\beta(N-\beta+1)^T]^T$,

$$\Psi_N := \Phi_N \Pi^{\perp}_{\mathscr{U}_N}\Phi_N^T, \qquad (18)$$

$$r_{N+1} := y_{\alpha}(N+1) - \frac{1}{N}\mathscr{Y}_N\mathscr{U}_N^T\left(\frac{1}{N}\mathscr{U}_N\mathscr{U}_N^T\right)^{-1} u_\alpha(N+1), \qquad (19)$$

$$q_{N+1} := \frac{1}{N}\Phi_N\mathscr{U}_N^T\left(\frac{1}{N}\mathscr{U}_N\mathscr{U}_N^T\right)^{-1} u_\alpha(N+1) - \phi(N+1), \qquad (20)$$

$$\zeta_{N+1} := \left(1 + \frac{1}{N}u_\alpha(N+1)^T\left(\frac{1}{N}\mathscr{U}_N\mathscr{U}_N^T\right)^{-1} u_\alpha(N+1)\right)^{-1},$$

$$\xi_{N+1} := \left(\frac{1}{\zeta_{N+1}} + \frac{1}{N}q_{N+1}^T\Psi_N^{-1}q_{N+1}\right)^{-1},$$

Note that the above recursive algorithm (17) can easily be integrated with an exponential forgetting factor (Oku and Kimura, 1999; Oku, 2000). The recursive algorithm with the exponential forgetting factor will be used for the numerical example in this paper.

4. ASYMPTOTIC PROPERTY OF A STATISTIC

If no change occurred in the dynamics of the system during sampling, the estimate $\widehat{\Theta}_{\Phi,N}$ would converge to the value which satisfies the equation (8) as N goes to infinity. In the case of no change of the system with a sufficient number of samples, the second term on the right hand side of (17) must be very small or driven only by the noise signal e_t. In this section, an asymptotic property of a statistic which is useful for change detection is presented. It is summarized by the following theorem:

Theorem 2. Let α and β be chosen sufficiently large but finite integers. Assume that the exogenous input u_t satisfies the aforementioned assumptions. Suppose that the system stays at an operating point, and that there exists $\Theta^* := \left[\Theta^*_\Phi \ \Theta^*_{\mathscr{U}}\right]$ such that for $\forall N > N_0$

$$\mathscr{Y}_N = \Theta^*_\Phi\Phi_N + \Theta^*_{\mathscr{U}}\mathscr{U}_N + \mathscr{K}\mathscr{E}_N, \qquad (21)$$
$$y_\alpha(N+1) = \Theta^*_\Phi\phi(N+1) + \Theta^*_{\mathscr{U}}u_\alpha(N+1) + \mathscr{K}e_\alpha(N+1). \qquad (22)$$

Define $e_\alpha(\infty) := \left[e_{1\infty}^T \cdots e_{\alpha\infty}^T\right]^T$, where $e_{i\infty}^T \in \mathbb{R}^l$, $i = 1, \cdots, \alpha$, are zero-mean Gaussian random vectors with the covariance matrix equal to $\mathbf{E}\left[e_{i\infty}e_{j\infty}^T\right] = \delta_{ij}\sigma_e^2 I$. Then,

$$r_{N+1} + \Theta^*_\Phi q_{N+1} \longrightarrow \mathscr{K}e_\alpha(\infty) \qquad (23)$$

in probability as N goes to infinity.

PROOF. From (19), (20),(21), (22), we have

$$r_{N+1} = -\Theta^*_\Phi q_{N+1} + \mathscr{K}e_\alpha(N+1) - \mathscr{K}\frac{1}{N}\mathscr{E}_N\mathscr{U}_N\left(\frac{1}{N}\mathscr{U}_N\mathscr{U}_N^T\right)^{-1} u_\alpha(N+1). \qquad (24)$$

Note that from the assumption of uncorrelation between u_t and e_t

$$\frac{1}{N}\mathscr{E}_N\mathscr{U}_N \longrightarrow 0 \tag{25}$$

with probability 1 as N goes to infinity. Therefore, for $\forall\varepsilon > 0$,

$$\begin{aligned}
&\lim_{N\to\infty} \mathbf{P}(\|r_{N+1} + \Theta_\Phi^* q_{N+1} - \mathscr{K} e_{\alpha(N+1)}\| \geq \varepsilon) \\
&= \lim_{N\to\infty} \mathbf{P}\left(\|\mathscr{K}\tfrac{1}{N}\mathscr{E}_N\mathscr{U}_N\left(\tfrac{1}{N}\mathscr{U}_N\mathscr{U}_N^T\right)^{-1} u_{\alpha(N+1)}\| \geq \varepsilon\right) \\
&\leq \lim_{N\to\infty} \mathbf{P}\left(\|\mathscr{K}\|_2 \cdot \|\tfrac{1}{N}\mathscr{E}_N\mathscr{U}_N\|_2 \cdot \|\left(\tfrac{1}{N}\mathscr{U}_N\mathscr{U}_N^T\right)^{-1}\|_2 \right.\\
&\qquad \left. \cdot\|u_{\alpha(N+1)}\| \geq \varepsilon\right) \\
&= 0,
\end{aligned}$$

due to boundedness, (25) and *inverse boundedness* w.r.t. u_t. Hence, this concludes the proof.

This implies that the recursive subspace identification algorithm has the potential ability for a whitening filter since from (2) and (11) the top block row of the vector $\mathscr{K} e_{\alpha(N+1)}$ equals e_{N+1}. Θ_Φ^* on the left hand side of (23) will be replaced by the estimate $\widehat{\Theta}_{\Phi,N}$ since it must be unknown.

5. CHANGE DETECTION USING GMA

Once the system undergoes an abrupt change, the recursive subspace identification algorithm (17) goes into a transient state and the theorem is not applicable for a while after the change. It is because the equality in (24) does not hold during the transient state. In the transient state, the statistics of the top block row of $r_{N+1} + \widehat{\Theta}_{\Phi,N} q_{N+1}$ are dominated by those of the output and input signals. If the output y_t and the noise e_t has a reasonable signal-to-noise ratio, a significant behavior can be observed in the top block row of $r_{N+1} + \widehat{\Theta}_{\Phi,N} q_{N+1}$ after the change. This notable property is expected to avoid false alarms. Moreover, it is empirically known that such behavior as mentioned before is irrelevant to changes in the dynamics of the input source. Hence, the discussion above explains the reason why the top block row of $r_{N+1} + \widehat{\Theta}_{\Phi,N} q_{N+1}$ generated recursively by the recursive subspace identification algorithm is a new promising test statistic for detection of abrupt changes in the dynamics of the system, without being disturbed by changes of the input signal.

For notational simplicity, let $\widehat{\varepsilon}_{N+1}$ denote the top block row of $r_{N+1} + \widehat{\Theta}_{\Phi,N} q_{N+1}$. Making use of the real-time observation of $\widehat{\varepsilon}_{N+1}$, an on-line change detection scheme is developed in this paper. The key point of the proposed change detection scheme is to decide *whether the covariance of the test statistic* $\{\widehat{\varepsilon}_t\}$ *is significantly larger than that of* $\{e_t\}$ *(i.e.,* $\sigma_0^2 I$*)* or not. In other words, the system concerned seems to stay an operating point if no significant difference in the covariance can be seen. Otherwise, it seems to jump from one operating point to another, and this is what we want to detect.

The change detection scheme tests between the two following hypotheses:

Hypotheses

$$\begin{cases} \mathbf{H}_0: & \text{The system stays at one operating point.} \\ \mathbf{H}_1: & \text{A change occurs.} \end{cases}$$

Note that, as long as the noise e_t retains the constant covariance $\sigma_0^2 I$, the sampling and test can continue sequentially.

From now on, let us consider the single output case, that is, $l = 1$ for ease of discussion. Note that it is straightforward to extend the discussion to the multiple output case. According to (Basseville and Nikiforov, 1993), the geometric moving average (GMA) algorithm in the case of a change in the variance *would* be given by

$$g_t = \sum_{i=0}^{\infty} \gamma_i s_{t-i}, \quad s_t = \ln\frac{\sigma_0}{\sigma_1} + \left(\frac{1}{\sigma_0^2} - \frac{1}{\sigma_1^2}\right)\frac{\widehat{\varepsilon}_t^2}{2}, \tag{26}$$

where $\gamma_i := \lambda(1-\lambda)^i$ for $0 < \lambda \leq 1$, σ_1^2 denotes the *variance of* $\widehat{\varepsilon}_t$ *in the transient state.* With $g_0 = 0$, g_t is rewritten in a recursive manner as

$$g_t = (1-\lambda)g_{t-1} + \lambda s_t. \tag{27}$$

However, σ_1^2 should be treated as an unknown variable since the knowledge of (A,B,C,D,K) is required in order to evaluate σ_1^2. To circumvent it, the maximum likelihood estimate of σ_1^2 is introduced to (26) and (27) as follows:

$$g_t = \sup_{\sigma_1}\left(\frac{\sigma_1^2 - \sigma_0^2}{\sigma_1^2\sigma_0^2}\tilde{g}_t + \ln\frac{\sigma_0}{\sigma_1}\right), \tag{28}$$

where

$$\tilde{g}_t = (1-\lambda)\tilde{g}_{t-1} + \lambda\frac{\widehat{\varepsilon}_t^2}{2} \quad \text{with } \tilde{g}_0 = 0. \tag{29}$$

Note that $\tilde{g}_t$ can be computed in a recursive manner without the knowledge of σ_1^2. It is easy to show that $\hat{\sigma}_1^2 = 2\tilde{g}_t$ attains the supremum in (28), and then, we have

$$g_t = \left(\frac{\tilde{g}_t}{\sigma_0^2} - \frac{1}{2}\right) + \ln\frac{\sigma_0}{\sqrt{2\tilde{g}_t}}. \tag{30}$$

Consequently, the decision rule d is given by

$$d = \begin{cases} 0 \text{ if } g_t < h; \ \mathbf{H}_0 \text{ is chosen,} \\ 1 \text{ if } g_t \geq h; \ \mathbf{H}_1 \text{ is chosen,} \end{cases} \tag{31}$$

where h is a conveniently chosen threshold.

6. NUMERICAL EXAMPLE

Consider the following ARX model with two operating points:

$$y_t = \frac{B(q^{-1})}{A_i(q^{-1})}u_t + \frac{1}{A_i(q^{-1})}e_t, \tag{32}$$

where q^{-1} denotes the backward shift operator, and for $i = 1,2$,

$$A_i(q) = 1 - (\sqrt{3}a_i + b_i)q^{-1} + (a_i^2 + \sqrt{3}a_i b_i)q^{-2} - a_i^2 b_i q^{-3}, \quad (33)$$

$$B(q) = 0.2q^{-1} + 0.08q^{-2}. \quad (34)$$

The pair of scalars (a_i, b_i), for $i = 1,2$, determines the location of the poles of the system. (a_1, b_1) corresponds to light crosses in Fig. 1, while (a_2, b_2) to dark ones.

The system abruptly changes from one operating point to the other at the sampling instants 2000, and vice versa at 5000. The dynamics of the input u_t changes at $k = 1500$, 3500 and 7000. For the time intervals, $[0, 1500]$ and $[3501, 7000]$, a zero-mean white sequence with unit variance is used as the input signal. For the intervals $[1501, 3500]$ and $[7001, 9000]$, the input equals the sum of a zero-mean white sequence with unit variance filtered with a 10th-order Butterworth filter (cutoff 0.6 times the Nyquist frequency) and a zero-mean white sequence with variance 0.01. The upper figure in Fig. 2 shows the profile of the input signal used here. The noise e_t has the normal distribution with zero mean and 0.05^2 of covariance. The signal-to-noise ratio with respect to the signal y_t and the noise e_t is around 23.5dB. The lower figure in Fig. 2 illustrates the output and noise used here. We take $\alpha = \beta = 8$ for the Hankel matrices, while $\lambda = 0.75$ for the GMA algorithm.

Fig. 3 illustrates the test statistic $\widehat{\varepsilon}_t$. For a while after 2000 or 5000, $\widehat{\varepsilon}_t$ has significant peaks, and otherwise it looks like a white noise with covariance of about 0.0028, which is very close to 0.05^2. This fact supports the theorem.

Fig. 4 illustrates the GMA algorithm with unknown σ_1^2. Significant peaks can be seen after 2000 and 5000, at which the system changes from one operating point to the other. On the other hand, no significant behavior can be seen around 1500, 3500 and 7000, at which the dynamics of the input signal changes. It demonstrates that the proposed change detection method has the ability to detect changes in the dynamics of the system without being disturbed by changes in the dynamics of the input signal.

Fig. 5 illustrates the maximum likelihood estimates of the unknown variance σ_1^2 at each sampling instant. Except for data around 2000 and 5000, the sample mean of the sequence of the maximum likelihood estimates of σ_1^2 is about 0.0028, close to σ_0^2, and the sample variance is about 8.6×10^{-6}.

7. CONCLUSION

A new change detection method integrated with a recursive subspace identification algorithm has been presented. An asymptotic property of a test statistic used in this paper has been investigated. A numerical example has illustrated that the test statistic $\widehat{\varepsilon}_t$ sensitively responds to a change in the dynamics of a system under surveillance and g_t gives significant peaks after the change occurs. The example has also demonstrated that $\widehat{\varepsilon}_t$ is not affected by a change in the input source.

Acknowledgement

The author is very grateful to Professors Tohru Katayama and Giorgio Picci for giving him the opportunity to take part in the invited session on subspace identification and applications.

REFERENCES

Basseville, M. (1988). Detecting Changes in Signals and Systems – A Survey. *Automatica* **24**(3), 309–326.

Basseville, M. and I. V. Nikiforov (1993). *Detection of Abrupt Changes*. Prentice Hall. New Jersey.

Basseville, M., M. Abdelghani and A. Benveniste (2000). Subspace-based fault detection algorithm for vibration monitoring. *Automatica* **36**(1), 101–109.

Bauer, D. and M. Jansson (2000). Analysis of the asymptotic properties of the MOESP type of subspace algorithms. *Automatica* **36**(4), 497–509.

Deistler, M., K. Peternell and W. Scherrer (1995). Consistency and relative efficiency of subspace methods. *Automatica* **31**(12), 1865–1875.

Golub, G. H. and C. F. Van Loan (1996). *Matrix Computations, 3rd Ed.*. The Johns Hopkis University Press. Baltimore.

Hannan, E. J. and M. Deistler (1988). *The Statistical Theory of Linear Systems*. John Wiley & Sons. New York.

Jansson, M. (2000). Asymptotic variance analysis of subspace identification methods. In: *Proceedings of IFAC System Identification*. Santa Barbara, California.

Jansson, M. and B. Wahlberg (1996). A linear regression approach to state-space subspace system identification. *Signal Processing* **52**, 103–129.

Jansson, M. and B. Wahlberg (1998). On consistency of subspace methods for system identification. *Automatica* **34**(12), 1507–1519.

Leith, D. J. and W. E. Leithead (2000). Survey of gain-scheduling analysis and design. *Int. J. Control* **73**(11), 1001–1025.

Ljung, L. (1999). *System Identification, 2nd Ed.*. Prentice Hall. New Jersey.

Oku, H. (2000). Sequential subspace state-space system identification and state estimation of unknown multivariable systems. PhD thesis. University of Tokyo.

Oku, H. and H. Kimura (1999). A recursive 4SID from the input-output point of view. *Asian J. Control* **1**(4), 258–269.

Oku, H. and H. Kimura (2002). Recursive 4SID algorithms using gradient type subspace tracking. *Automatica* **38**(6), 1035–1043.

Oku, H., G. Nijsse, M. Verhaegen and V. Verdult (2001). Change detection in the dynamics with recursive subspace identification. In: *Proceedings of the 40th CDC*. Orlando, Florida. pp. 2297–2302.

Peternell, K., W. Scherrer and M. Deistler (1996). Statistical analysis of novel subspace identification methods. *Signal Processing* **52**, 161–177.

Van Overschee, P. and B. De Moor (1996). *Subspace Identification for Linear Systems*. Kluwer Academic Publishers. Massachusetts.

Verhaegen, M. (1994). Identification of the deterministic part of MIMO state space models given in innovations form from input-output data. *Automatica* **30**(1), 61–74.

Verhaegen, M. and E. Deprettere (1991). A fast, recursive MIMO state space model identification algorithm. In: *Proceedings of the 30th CDC*. Brighton. pp. 1349–1354.

Viberg, M., B. Wahlberg and B. Ottersten (1997). Analysis of state space system identification methods based on instrumental variables and subspace fitting. *Automatica* **33**(9), 101–109.

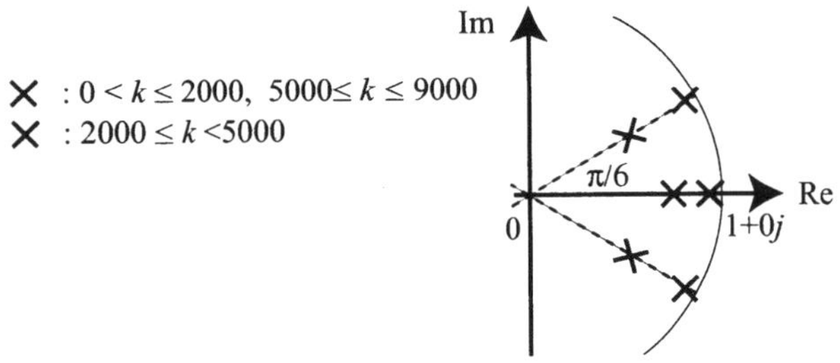

Fig. 1. Poles

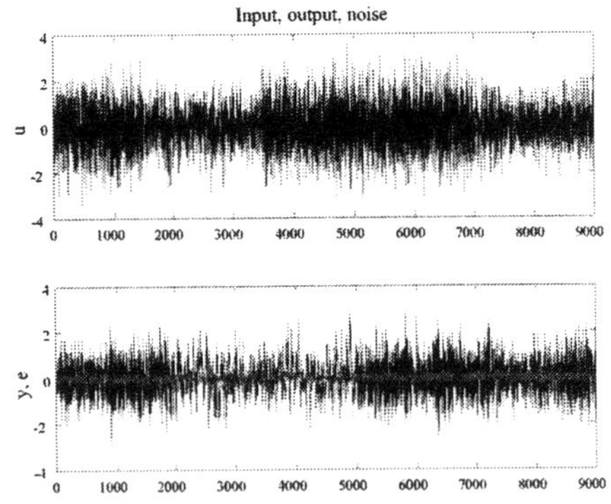

Fig. 2. Input, output and noise

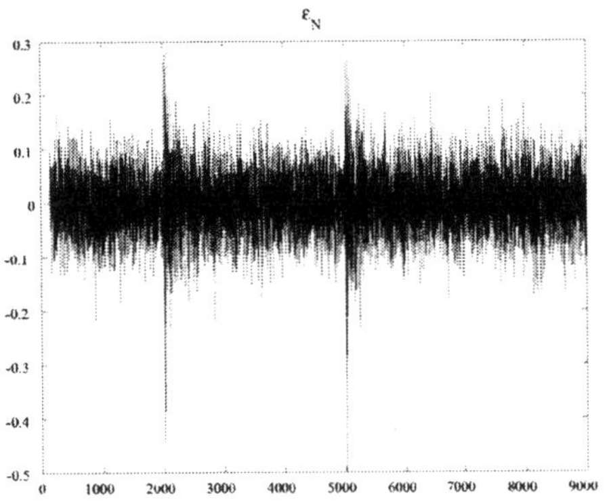

Fig. 3. Test signal $\widehat{\varepsilon}_t$

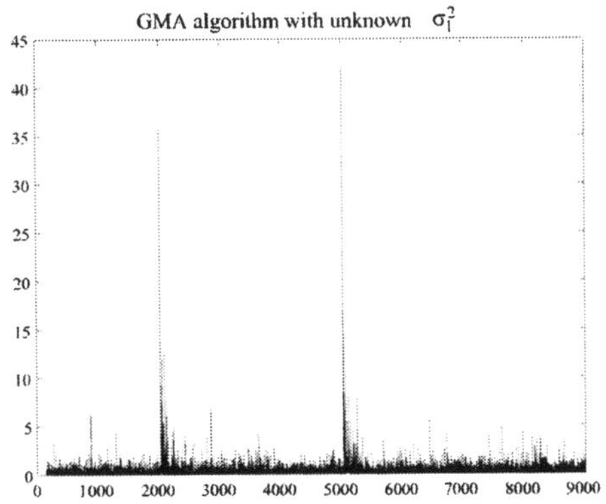

Fig. 4. GMA algorithm g_t

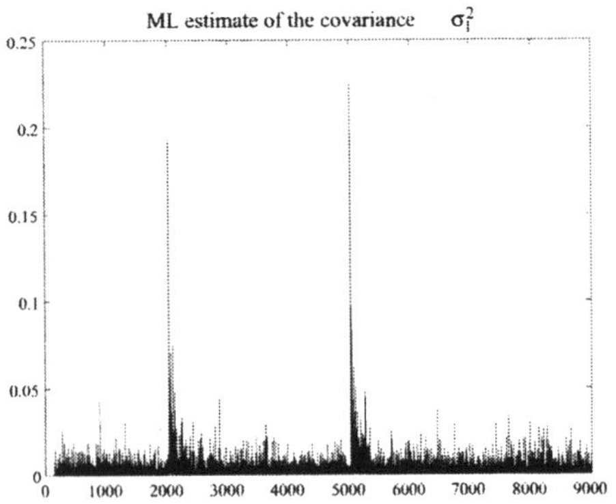

Fig. 5. Maximum likelihood estimate of σ_1^2

IFAC Publications
www.elsevier.com/locate/ifac

SUBSPACE-BASED MODAL IDENTIFICATION AND MONITORING OF LARGE STRUCTURES: A SCILAB TOOLBOX

Laurent Mevel *,[1] **Maurice Goursat** ** **Michèle Basseville** *,[2]
Albert Benveniste *,[1]

* IRISA, *Campus de Beaulieu, 35042 Rennes Cedex, F.-*
Firstname.Name@irisa.fr
** INRIA, *BP 105, Rocquencourt, 78153 Le Chesnay Cedex, F.-*
Maurice.Goursat@inria.fr

Abstract: Stochastic subspace-based structural identification and damage detection and localization methods are discussed, together with their implementation within the free INRIA software Scilab. Particular emphasis is put on the handling of high order models for large structures. *Copyright © 2003 IFAC*

Keywords: Subspace-based identification algorithms, large structures, vibration monitoring, statistical tests, damage detection, damage localization.

1. INTRODUCTION

Modal analysis of vibrating structures in operating conditions becomes a mastered technique for engineers in many different domains. Transportation systems (cars, airplanes, space rockets, trains...) are the most common and important industrial sector where such problems arise, for different purposes: fiability, comfort (for cars and airplanes), performance (maximize the payload of a rocket)... The monitoring of civil structures (high rise buildings, bridges...) is another domain subject to increasing interest.

We have been developing a complete identification and detection package, built within the Scilab development environment with a full GUI. The identification part can be used in a wide range of applications, from simple modal identification to fully automated modal monitoring. The detection part is built for damage localization and online monitoring of mechanical structures, under different hypotheses testing cases.

In section 2, we briefly recall some of the theoretical foundations of the identification procedure. The detection and localization issues are addressed in section 3. Implementation issues and experimental results are described in sections 4 and 5 for the identification and detection/localization parts, respectively. The former have been obtained for aircrafts, with both simulated and in-flight measured data; the latter for a laboratory test case of a civil structure.

2. STOCHASTIC SUBSPACE-BASED STRUCTURAL IDENTIFICATION

A major issue for in-operation structural vibration analysis, based on measurements from accelerometers or strain gauges, is to identify modes and modal shapes of mechanical systems subject to an uncontrolled, *unmeasured* and *nonstationary* excitation (Hermans and Van der Auweraer, 1999; Parloo *et al.*, 2001). Typical examples are offshore structures subject to swell, buildings subject to wind or earthquake, bridges subject to traffic, dams, wings subject to flutter in flight, and turbines subject to steam turbulence, friction in bearings, and imperfect balancing. Modeling modes and modal shapes through state space representations (Juang, 1994) shows that struc-

[1] Also with INRIA.
[2] Also with CNRS.

tural analysis is an important instance of identification of the eigenstructure of a linear multi-variable system.

2.1 *Output-only data*

It has been argued that structural model identification during normal operating conditions, with uncontrolled and unmeasured excitation, calls for the use of *output-only* identification methods (Peeters and de Roeck, 1999). Because of the nonstationarity of the unknown excitation, we advocate for the use of an output-only and covariance-driven subspace-based stochastic identification method (Basseville *et al.*, 2001). The difference between the covariance-driven form of subspace algorithms which we use and the usual data-driven form is minor, at least for eigenstructure identification. The covariance-driven form, however, turns out to be simple to handle with for merging data from different setups. The method deeply exploits the factorization property of the Hankel matrix of a linear dynamical system into the observability matrix times the controllability matrix. The state-transition matrix is obtained from the shift invariance property of the observability matrix. The modes and modeshapes are recovered through eigenstructure analysis of the state-transition matrix.

2.2 *Multi-patchs data*

This method has been extended to the handling of non-simultaneously measured data from multiple sensor setups (multi-patchs) involving both fixed (reference) and moving sensors (Mevel *et al.*, 2002a; Mevel, Goursat and Basseville, 2003a). This extension also deeply relies on the Hankel matrix factorization property mentioned above. The idea is to right-normalize the covariance matrices to make them looking as if they were obtained with the same excitation. One interesting computational feature of the resulting algorithm is that it mainly amounts to apply the above classical subspace identification algorithm to a Hankel matrix obtained by interleaving the block-columns of the Hankel matrices built on the reference sensors and the block-rows of the suitably normalized Hankel matrices built on the moving sensors.

2.3 *Robustness*

Both the classical and the multi-patchs subspace identification methods have been shown consistent and robust with respect to nonstationary excitation (Benveniste and Fuchs, 1985; Mevel *et al.*, 2002b).

2.4 *Input-output data*

In some applications, e.g. in aeronautics, input-output data are also available. The covariance-driven subspace identification method can be modified for handling such data. The idea (Guyader and Mevel, 2003) consists in defining orthogonalized output data as residuals of the projection of the outputs onto a sliding window of the inputs, and to apply the above covariance-driven subspace algorithm to these orthogonalized outputs.

2.5 *Handling high order systems*

Structural identification requires handling very large models. Actual structures possess of course a huge number of modes, as captured by the FE models (hundreds or thousands). More important, the mechanical engineer is frequently interested in recovering several tens of modes in a specified bandwidth, as a result of identification. Thus model reduction is applied in a non standard setting: from huge models, to large ones! It is our experience (and this conforms the experience of the mechanical engineering community) that statistical model order selection criteria (such as AIC, BIC,...) perform very poorly in this case.

Therefore a great effort has been devoted, in the toolbox, to address this issue. The technique consists in fusing, in some appropriate way, the results of identification for many different orders, including orders that are much larger than the expected number of modes. Typically, if several tens of modes are wished for identification, then identification up to an order of a few hundreds should be performed, as some relevant poles appear in the identified model, only with an order significantly exceeding twice the number of wished poles. The fusion of the results for these different models is performed in an assisted way, using a sophisticated GUI to manipulate so-called "waterfalls of modes". Without this post-processing, the basic algorithms have little efficiency. Overall, the resulting performance is impressive.

2.6 *Automated modal analysis in a monitoring approach*

The modal analysis approach has been embedded in a fully automated online monitoring procedure, which allows the monitoring of frequencies and damping factors over time. Full subspace and recursive versions have been implemented.

3. STOCHASTIC SUBSPACE-BASED DAMAGE DETECTION AND LOCALIZATION

The vibration monitoring problem is addressed as the double task of detecting damages seen as changes in the eigenstructure of a linear dynamic system, and localizing the detected damages within (a FEM of) the monitored structure.

3.1 *Damage detection*

Health monitoring techniques based on processing vibration measurements basically handle two types of characteristics: the *structural parameters* (mass, stiffness, flexibility, damping) and the *modal parameters* (modal frequencies, and associated damping values and modeshapes); see (Doebling *et al.*, 1996; Natke and Cempel, 1997) and references therein. A central question for monitoring is to compute *changes* in those characteristics and to assess their *significance*. For the *frequencies*, crucial issues are then: how to compute the changes, to assess that the changes are significant, to handle *correlations* among individual changes. Furthermore, it has been widely acknowledged that, whereas changes in frequencies bear useful information for damage *detection*, information on changes in (the curvature of) modeshapes is mandatory for performing damage *localization*. Then, similar issues arise, and assessing the significance of (usually small) changes in the modeshapes, and handling the (usually high) correlations among individual modeshape changes are still considered as opened questions (Natke and Cempel, 1997; Farrar, Doebling and Nix, 2001).

The proposed damage detection algorithm (Basseville, Abdelghani and Benveniste, 2000) aims at addressing the issues. It is based on the stochastic subspace-based covariance-driven identification method and on the statistical local approach to the design of detection algorithms (Basseville, 1998). Moreover no analytical model is handled at this stage. This algorithm basically: generates a residual from a parameter estimating function associated with the subspace identification algorithm; evaluates the residual from an asymptotic Gaussianity result; altogether computes a global χ^2-test, which performs a sensitivity analysis of the residuals to the damages, relatively to uncertainties and noises. This χ^2-test enjoys a useful invariance property (Basseville, 1999).

3.2 *Damage localization*

Damage localization consists in determining which part of the structure has been affected by the damage, more precisely which (groups of) elements of the structural parameters matrices (e.g. masses, stiffness coefficients,...) have changed. This problem is often addressed as an inverse estimation problem, based on an analytical model in the pre-damage stage and on modal identification in the post-damage stage. Typically, the deviations in the structural parameters corresponding to the observed deviation in the modal parameters are searched for using model updating techniques (Friswell, Mottershead and Ahmadian, 2001).

In the proposed approach, damage localization is stated as a detection problem instead (Basseville *et al.*, 1993; Basseville, Mevel and Goursat, 2003). Of course, an analytical model is handled at this stage. This problem is addressed by plugging sensitivities of the modes and modeshapes w.r.t. FEM parameters in the above setting. This provides us with directional tests, which perform the same type of fault-to-noise sensitivity analysis of the residual. These tests deliver damage diagnostics and localization information, without solving any inverse problem.

These damage detection and localization methods have proven useful in a number of simulated and real application examples (Basseville *et al.*, 1993; Abdelghani *et al.*, 1999; Hermans, Van der Auweraer and Mevel, 1999; Mevel, Hermans and Van der Auweraer, 1999; Mevel, Goursat and Basseville, 2003a; Mevel, Goursat and Basseville, 2003b). The design of the algorithms is fairly general, and this methodology has been applied in other application domains as well (Basseville, 1998).

4. EXPERIMENTAL ISSUES AND RESULTS: IDENTIFICATION

In this section, we report on identification results obtained on both in-flight measurements and new aircrafts simulated data, provided by AIRBUS France and Avions Marcel Dassault, within the Eureka project FLITE. The key issue is the very high number of eigenfrequencies.

Using the toolbox, it is easy to quickly scan the different values and to check the results by clicking on the neighboring points in the graphics, giving the subwindow with the corresponding modal parameters, see Fig. 1-2. We do the same work on data corresponding to in-flight measurements. Successive data sets are available. Thus we can examine the evolution of the modal characteristics with the modifications of the aircraft, for example the decreasing mass of the fuel in the tanks. Compare Fig. 3-4 with Fig. 5-6.

These examples illustrate two interesting features of the method: the powerful capability to separate very close eigenfrequencies and to follow the evolution of the damping during the flight.

An example of automated modal analysis is displayed on Fig. 7-8.

5. EXPERIMENTAL ISSUES AND RESULTS: DAMAGE LOCALIZATION

The steel-quake structure is used at the Joint Research Centre in Ispra (Italy) to test the performance of steel

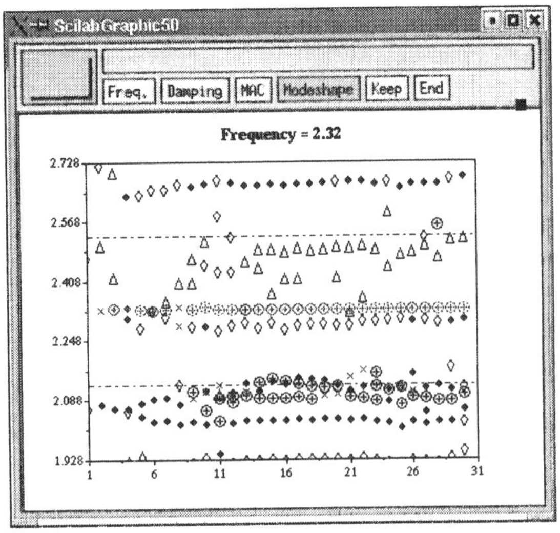

Figure 1. Stabilization diagram focussed on one frequency.

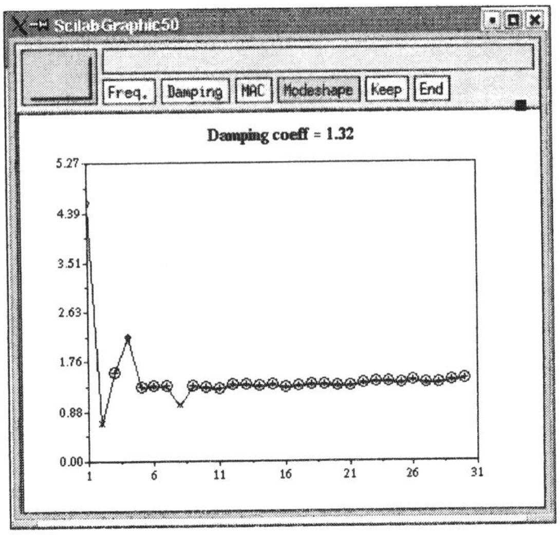

Figure 2. The corresponding damping values.

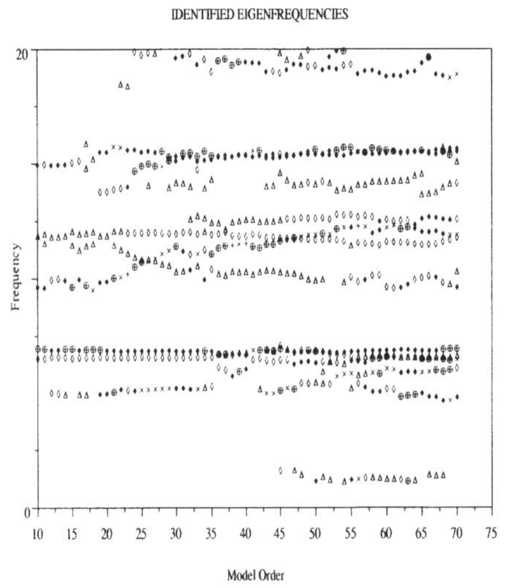

Figure 3. Flight beginning - Eigenfrequencies.

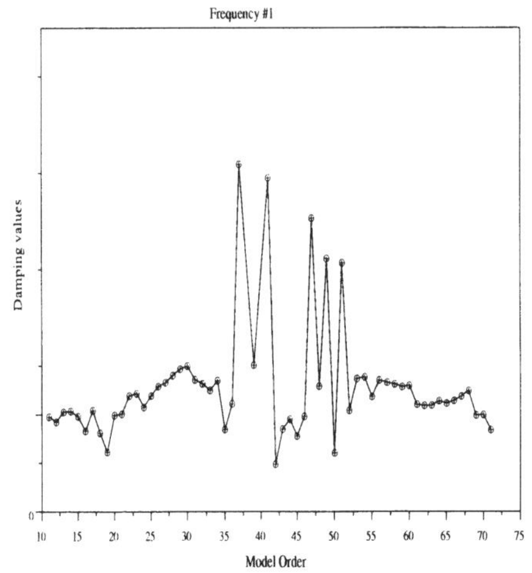

Figure 4. Flight beginning - Damping values for one frequency.

buildings during earthquakes. The structure is a two-storey frame, excited with the aid of an impact hammer. For each impact location, eight to ten experiments are performed, yielding data samples from the structure under different non stationary excitations. Fifteen accelerometers are used for recording the reaction of the structure to the hammer impacts.

We already reported successful damage localization results obtained on the bridge Z24, another COST F3 WG2 benchmark (Mevel, Goursat and Basseville, 2003b). In this section, we report damage localization results obtained on the steel-quake structure, using the new fully automated modal localization toolbox embedded in the Scilab Modal toolbox.

Damage localization has been performed on all sensors using a rough FE model of the steel-quake structure. This FE model has been obtained from the SAMCEF model used in COST F3, and updated using the Structural Dynamics Toolbox of E. Balmes [3]. Although we advocate that using an inaccurate FE model is not critical for our localization method which handles modal changes *directions*, too large differences between the FE model and the structure may lead to localization errors.

As a whole, the localization results obtained with this rough FE model have been satisfactory, as shown on Fig. 9. See (Basseville, Mevel and Goursat, 2003) for more details.

6. CONCLUSION

In that paper, we have described the features available in the Scilab Modal Toolbox and shown its application in identification and damage localization to some mechanical structures, benchmarks of past and present European networks. Example of automated modal analysis have been shown in (Mevel, Basseville and Benveniste, 2003). Examples of online damage detection test can be found in (Peeters et al., 2003). Flutter

[3] `http://www.sdtools.com/sdt/`

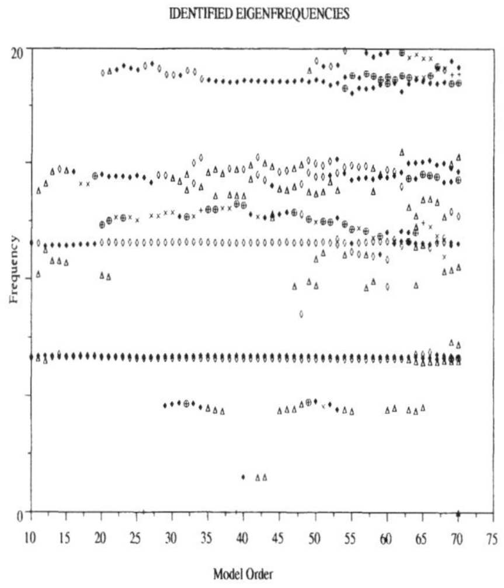

Figure 5. Less fuel in tanks - Eigenfrequencies.

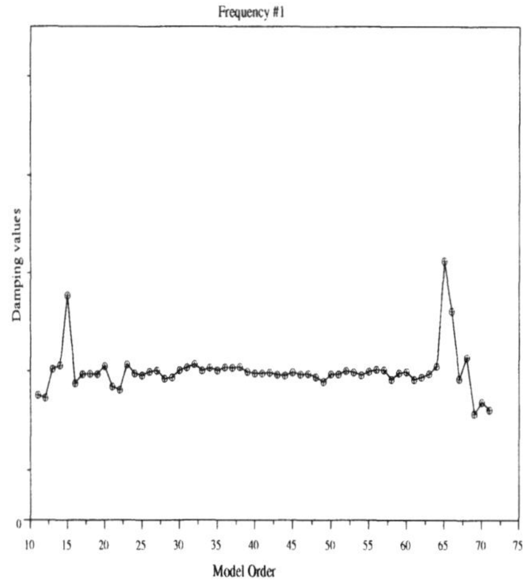

Figure 6. Less fuel in tanks - Damping values for one frequency.

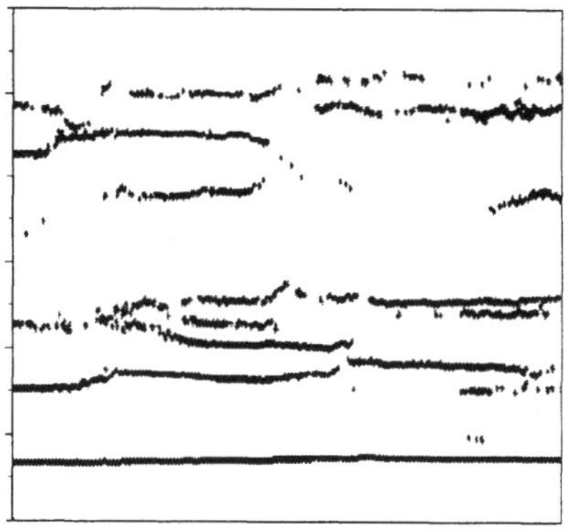

Figure 7. Automated identification: frequency vs time.

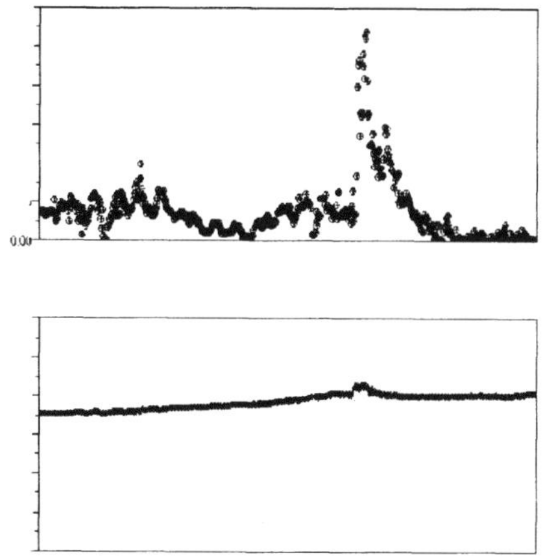

Figure 8. Zoom on one mode : frequency (bottom) and damping (top) vs time.

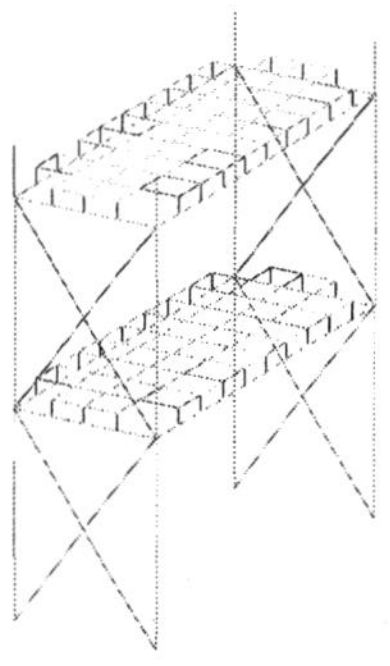

Figure 9. Damage localization result.

monitoring is also available in the toolbox and presented in (Mevel, Basseville and Benveniste, 2003). Damage localization theory and practices have been exposed in (Basseville, Mevel and Goursat, 2003).

The covariance-driven subspace algorithm has been implemented within two toolboxes: the IN-OP module of the LMS software CADA-X, and the modal analysis module of the free INRIA software Scilab. The damage detection and localization methods have been implemented within Scilab, and are currently integrated within LMS software environment.

ACKNOWLEDGMENTS

This work has been carried out within the framework of the Eureka projects no 1562 SINOPSYS (Model based **S**tructural monitoring using **in**-**op**eration **sys**tem identification) coordinated by LMS, Leuven, Belgium, and no 2419 FLITE (**Fli**ght **T**est **E**asy), coordinated by Sopemea, Velizy-Villacoublay, France. The contribution of Yann Veillard to the development of the Scilab toolbox is acknowledged.

7. REFERENCES

M. Abdelghani, M. Basseville, A. Benveniste, L. Mevel, E. Balmès, L. Hermans and H. Van der Auweraer (1999). Assessment of subspace fault detection algorithms on a realistic simulator-based example. *Proc. 17th Int. Modal Analysis Conf.*, Kissimmee, FL.

M. Basseville (1998). On-board component fault detection and isolation using the statistical local approach. *Automatica*, **34**(11), 1391–1416.

M. Basseville (1999). An invariance property of some subspace-based detection algorithms. *IEEE Trans. Signal Processing*, **SP-47**(12), 3398–3401.

M. Basseville, M. Abdelghani, A. Benveniste (1999). Subspace-based fault detection algorithms for vibration monitoring. *Automatica*, **36**(1), 101–109.

M. Basseville, L. Mevel, M. Goursat (2003). Statistical model-based damage detection and localization: subspace-based residuals and damage-to-noise sensitivity ratios. *Jal of Sound and Vibration*, to appear.

M. Basseville, A. Benveniste, M. Goursat, L. Hermans, L. Mevel, H. Van der Auweraer (2001). Output-only subspace-based structural identification: from theory to industrial testing practice. *ASME Jal Dynamic Systems Measurement and Control*, Special Issue on *Identification of Mechanical Systems*, **123**(4), 668–676.

M. Basseville, A. Benveniste, B. Gach-Devauchelle, M. Goursat, D. Bonnecase, P. Dorey, M. Prevosto, M. Olagnon (1993). Damage monitoring in vibration mechanics: issues in diagnostics and predictive maintenance. *Mechanical Systems and Signal Processing*, **7**(5), 401–423.

A. Benveniste and J.-J. Fuchs (1985). Single sample modal identification of a non-stationary stochastic process. *IEEE Trans. Automatic Control*, **AC-30**(1), 66–74.

S.W. Doebling, C.R. Farrar, M.B. Prime and D.W. Shevitz (1996). Damage identification and health monitoring of structural and mechanical systems from changes in their vibration characteristics: a literature review. *Los Alamos National Laboratory Report* LA-13070-MS.

C.R. Farrar, S.W. Doebling and D.A. Nix (2001). Vibration-based structural damage identification. *The Royal Society, Philosophical Transactions: Mathematical, Physical and Engineering Sciences*, **359**(1778), 131–150.

M.I. Friswell, J.E. Mottershead and H. Ahmadian (2001). Finite element model updating using experimental test data: parameterization and regularization. *Transactions of the Royal Society of London*, Series A, Special Issue on *Experimental Modal Analysis*, **359**, 169–186.

A. Guyader and L. Mevel (2003). Covariance driven subspace methods: input/output vs output-only. *Proc. 21st Int. Modal Analysis Conf.*, Kissimmee, FL.

L. Hermans and H. Van der Auweraer (1999). Modal testing and analysis of structures under operational conditions: industrial applications. *Mechanical Systems and Signal Processing*, **13**, 193–216.

L. Hermans, H. Van der Auweraer and L. Mevel (1999). Health monitoring and detection of a fatigue problem of a sports car. *Proc. 17th Int. Modal Analysis Conf.*, Kissimmee, FL.

J.N. Juang (1994). *Applied System Identification.* Prentice Hall, Englewood Cliffs, NJ.

L. Mevel, M. Basseville and A. Benveniste (2003). Statistical approach to flutter monitoring. *These Proceedings.*

L. Mevel, M. Goursat and M. Basseville (2003a). Stochastic subspace-based structural identification and damage detection - Application to the steel-quake benchmark. *Mechanical Systems and Signal Processing*, Special Issue on *COST F3 benchmarks*, **17**(1), 91–101.

L. Mevel, M. Goursat and M. Basseville (2003b). Stochastic subspace-based structural identification and damage detection and localization - Application to the Z24 bridge benchmark'. *Mechanical Systems and Signal Processing*, Special Issue on *COST F3 benchmarks*, **17**(1), 143–151.

L. Mevel, L. Hermans, H. Van der Auweraer (1999). Application of a subspace-based fault detection method to industrial structures. *Mechanical Systems and Signal Processing*, **13**(6), 823–838.

L. Mevel, M. Basseville, A. Benveniste, M. Goursat (2002a). Merging sensor data from multiple measurement setups for nonstationary subspace-based modal analysis. *Jal Sound and Vibration*, **249**(4), 719–741.

L. Mevel, A. Benveniste, M. Basseville, M. Goursat (2002b). Blind subspace-based eigenstructure identification under nonstationary excitation using moving sensors. *IEEE Trans. Signal Processing*, **SP-50**(1), 719–741.

H.G. Natke and C. Cempel (1997). *Model-Aided Diagnosis of Mechanical Systems: Fundamentals, Detection, Localization, Assessment.* Springer-Verlag.

E. Parloo, P. Verboven, P. Guillaume and M. Van Overmeire (2001). Maximum likelihood identification of modal parameters from nonstationary operational data. *Proc. 19th Int. Modal Analysis Conf.*, Kissimmee, FL, 425–431.

B. Peeters and G. de Roeck (1999). Reference-based stochastic subspace identification for output-only modal analysis. *Mechanical Systems and Signal Processing*, **13**, 855–878.

B. Peeters, L. Mevel, S. Vandanluit, P. Guillaume, M. Goursat, A. Vecchio, and H. Van der Auweraer (2003). Online vibration based crack detection durng fatigue testing. *Proc. 5th Int. Conf. on Damage Assessment of Structures.* July 2003.

IFAC Publications
www.elsevier.com/locate/ifac

IDENTIFYING POSITIVE REAL MODELS IN SUBSPACE IDENTIFICATION BY USING REGULARIZATION

Ivan Goethals *,1 **Tony Van Gestel** *,2 **Johan Suykens** *,3
Paul Van Dooren **,4 **Bart De Moor** *,5

* *Kasteelpark Arenberg 10, B3001 Heverlee-Leuven, Belgium*
** *Av. Georges Lemaître 4, B1348 Louvain-la-Neuve, Belgium*

Abstract: This paper deals with the lack of positive realness of identified models that may be encountered in many stochastic subspace identification procedures. Lack of positive realness is an often neglected, but important problem. Subspace identification algorithms fail to return a valid linear model if the so-called covariance model, which is obtained from an intermediate realization step in the subspace identification algorithm, is not positive real. The main contribution of this paper is to introduce a regularization approach to impose positive realness on the covariance model. It is shown that positive realness can be imposed by adding a regularization term to a least squares cost function appearing in the subspace identification procedure. *Copyright © 2003 IFAC*

Keywords: stochastic systems, subspace methods, robustness, regularization

1. INTRODUCTION

In this paper, we will consider stable systems and models of the form:

$$\begin{aligned} x_{k+1} &= Ax_k + w_k, \\ y_k &= Cx_k + v_k, \end{aligned} \tag{1}$$

with

$$\mathcal{E}\left\{\begin{bmatrix} w_p \\ v_p \end{bmatrix} \begin{bmatrix} w_q^T & v_q^T \end{bmatrix}\right\} = \begin{bmatrix} Q & S \\ S^T & R \end{bmatrix} \delta_{pq} \geq 0, \tag{2}$$

where $\mathcal{E}\{\cdot\}$ denotes the expected value operator and δ_{pq} the Kronecker delta. It is assumed that $\mathcal{E}\left\{\begin{bmatrix} w_p \\ v_p \end{bmatrix} x_k^T\right\} = 0, \forall p \geq k$. The elements of the vector $y_k \in \mathbb{R}^l$ are given observations at the discrete time index k of the l outputs of the system. The vector $x_k \in \mathbb{R}^n$ is the unknown state vector at time k. The unobserved process and measurement noise $w_k \in \mathbb{R}^n$ and $v_k \in \mathbb{R}^l$ are assumed to be white, zero mean, Gaussian with covariance matrices as given in (2). The system matrices A, C and the covariance matrices Q, S, and R have appropriate dimensions.

Stochastic subspace identification methods are ideally suited to identify models of the form (1). Typically, in a first step the measured output sequence $y_0, y_1, \ldots, y_{N-1}$ is stored in block Hankel matrices containing a user defined number of block rows i, and a certain number of columns j, so that $N = 2i + j - 1$, see (Van Overschee and De Moor, 1993)(Van Overschee and De Moor, 1996) for an extensive survey of this procedure. Kalman filter state sequences $\hat{X}_i \in \mathbb{R}^{n\times j}$ and $\hat{X}_{i+1} \in \mathbb{R}^{n\times j}$ of the system and an estimate of the system order $\hat{n}$ are then obtained by using geometric operations of subspaces spanned by the column and row vectors of these Hankel matrices.

In a second step, a so called covariance model $(\hat{A}, \hat{G}, \hat{C}, \hat{D})$, is estimated, where $\hat{G}$ is an estimate for the covariance matrix between states and observations $G = \mathcal{E}\{x_{k+1}y_k^T\}$ and $\hat{D} = \frac{\hat{\Lambda}_0}{2}$ is an estimate for $\frac{\Lambda_0}{2}$, with $\Lambda_m = \mathcal{E}\{y_{k+m}y_k^T\}$, $m \geq 0$ the output

[1] I. Goethals is a Research Assistant with the Fund for Scientific Research-Flanders (FWO-Vlaanderen).
[2] T. Van Gestel is a Postdoctoral Researcher with the FWO-Vlaanderen.
[3] J. Suykens is a Postdoctoral Researcher with the FWO-Vlaanderen.
[4] P. Van Dooren is Professor at the Catholic University of Louvain.
[5] B. De Moor is a full professor with the KULeuven.

covariance matrices. $\hat{A}$ and $\hat{C}$ are estimates for the matrices A and C in (1), which are obtained as the solution to a least squares problem:

$$(\hat{A}, \hat{C}) = \arg\min_{A,C} J_1(A, C), \quad (3)$$

with

$$J_1(A, C) = \left\| \begin{bmatrix} \hat{X}_{i+1} \\ Y_{i|i} \end{bmatrix} - \begin{bmatrix} A \\ C \end{bmatrix} \cdot \hat{X}_i \right\|_F^2, \quad (4)$$

where

$$Y_{i|i} = \begin{bmatrix} y_i \; y_{i+1} \; \dots \; y_{i+j-1} \end{bmatrix}. \quad (5)$$

Using the definitions for G and Λ_m above, one can derive that

$$\Lambda_m = CA^{m-1}G, \; \Lambda_{-m} = \Lambda_m^T, \quad m \geq 1. \quad (6)$$

Hence, the output covariances can be considered as Markov parameters of a deterministic linear time invariant system with system matrices (A, G, C, D).

From the estimated model $(\hat{A}, \hat{G}, \hat{C}, \hat{D})$, In a last step, a model is constructed in forward innovation form:

$$\begin{aligned} \hat{x}_{k+1} &= \hat{A}\hat{x}_k + \hat{K}e_k, \\ y_k &= \hat{C}\hat{x}_k + e_k, \end{aligned} \quad (7)$$

from which estimates of the error covariances of the system can be derived. The forward innovation model is obtained by first calculating the forward state covariance matrix $\hat{P} = \mathcal{E}\left\{\hat{x}_k\hat{x}_k^T\right\}$ of the covariance model through the solution of the forward algebraic Riccati equation:

$$\begin{aligned} \hat{P} = \hat{A}\hat{P}\hat{A}^T + (\hat{G} - \hat{A}\hat{P}\hat{C}^T)\cdot \\ (\hat{\Lambda}_0 - \hat{C}\hat{P}\hat{C}^T)^{-1}(\hat{G} - \hat{A}\hat{P}\hat{C}^T)^T, \end{aligned} \quad (8)$$

with the forward Kalman filter gain $\hat{K} = (\hat{G} - \hat{A}\hat{P}\hat{C}^T)(\hat{\Lambda}_0 - \hat{C}\hat{P}\hat{C}^T)^{-1}$. The resulting model matrices of the stochastic system are $(\hat{A}, \hat{K}, \hat{C}, I_l)$ and the covariance matrix $\mathcal{E}\left\{e_ke_k^T\right\}$ is given by $\hat{R} = \hat{\Lambda}_0 - \hat{C}\hat{P}\hat{C}^T$.

It is important to note here, that the forward innovation model can only be obtained if the forward algebraic Riccati equation (8) has a positive definite solution. It can be shown that this is the case if and only if the infinite sequence $\{\hat{\Lambda}_m\}_{m=0}^{\infty}$ with $\hat{\Lambda}_m = \hat{C}\hat{A}^{m-1}\hat{G}$, $m > 0$, and $\hat{\Lambda}_0 = \frac{1}{j}Y_{i|i}Y_{i|i}^T$, is a "valid" covariance sequence with positive definite Toeplitz matrix (Dahlén *et al.*, 1998; Faurre *et al.*, 1978). This is equivalent with the model $(\hat{A}, \hat{G}, \hat{C}, \hat{D})$, with $\hat{D} + \hat{D}^T = \hat{\Lambda}_0$, being positive real. Hence, if the positive realness property is not satisfied, no meaningful stochastic model will be obtained. This problem may appear in practical applications. The covariance model, for example, is built on a finite number of observed covariances. Even if these were exact ($j \rightarrow \infty$), the realization algorithm does not ensure that the infinite covariance sequence $\{\hat{\Lambda}_m\}_{m=0}^{\infty} = \hat{C}\hat{A}^{m-1}\hat{G}$, derived from the finite sequence $\{\hat{\Lambda}_m\}_{m=0}^{2i-1}$, is positive. Hence the choice of i has a direct influence on the possible occurence of positivity problems (Oono, 1981; Lindquist and Picci, 1996). Secondly, for j finite, the observed covariances are subject to statistical errors that may increase the probability for positive realness problems to occur. Finally the ability of $(\hat{A}, \hat{G}, \hat{C}, \hat{D})$ to model the observed covariance sequence is clearly dependent on the choice of the model order $\hat{n}$. The influence of the parameters i, j and $\hat{n}$ will be illustrated in section 3. For a further theoretical description, the reader is referred to (Lindquist and Picci, 1996).

In this paper we propose a solution to impose positive realness on a formerly identifed stochastic model by adding a regularization term that involves the system matrices $\hat{A}$ and $\hat{C}$, and we analyse its performance and compare it with already existing techniques.

2. IMPOSING POSITIVE REALNESS BY USING REGULARIZATION

2.1 *Main idea*

The estimation problem that we consider is the following: given matrices $\hat{X}_{i+1}, Y_{i|i}$ and $\hat{X}_i$ and given the estimates $\hat{G}$ and $\hat{\Lambda}_0$, estimate the model matrices $\hat{A}, \hat{C}$ such that the resulting model $\hat{A}, \hat{G}, \hat{C}, \hat{\Lambda}_0$ is positive real. To impose positive realness, we will add a regularization term to the cost function $J_1(A, C)$ from (3):

$$(\tilde{A}_c, \tilde{C}_c) = \arg\min_{A,C} J_1(A, C) + cJ_2(A, C), \quad (9)$$

with

$$J_2(A, C) = \text{Tr}\left(\begin{bmatrix} A \\ C \end{bmatrix} W \begin{bmatrix} A \\ C \end{bmatrix}^T\right), \quad (10)$$

where $c \geq 0$ is a positive real scalar and W a positive definite matrix of appropriate dimensions that satisfies $W - \hat{G}\hat{\Lambda}_0^{-1}\hat{G}^T > 0$. A similar regularization term $\text{Tr}\left(AWA^T\right)$, involving only the system matrix A was described in (Van Gestel *et al.*, 2001), and was shown to impose stability on a model. We will show that by adding the output matrix C to the regularization term, the model can not only be made stable, but also positive real, provided the regularization coefficient c is chosen sufficiently large.

By the choice of the regularization term $J_2(A, C)$, the optimal solution of the minimization problem is found as

$$\begin{aligned} \begin{bmatrix} \tilde{A}_c \\ \tilde{C}_c \end{bmatrix} &= \begin{bmatrix} \hat{X}_{i+1} \\ Y_{i|i} \end{bmatrix} \cdot \hat{X}_i^T \cdot \left[\hat{X}_i\hat{X}_i^T + cW\right]^{-1} & (11) \\ &= \begin{bmatrix} \hat{A} \\ \hat{C} \end{bmatrix} \hat{X}_i\hat{X}_i^T \left[\hat{X}_i\hat{X}_i^T + cW\right]^{-1}. & (12) \end{aligned}$$

From the optimality of the least squares estimate (11), it follows that for $c_1, c_2 \geq 0$:

$$\begin{aligned} J_1(\tilde{A}_{c_2}, \tilde{C}_{c_2}) + c_1 J_2(\tilde{A}_{c_2}, \tilde{C}_{c_2}) \\ \geq J_1(\tilde{A}_{c_1}, \tilde{C}_{c_1}) + c_1 J_2(\tilde{A}_{c_1}, \tilde{C}_{c_1}), \end{aligned} \quad (13)$$

$$J_1(\tilde{A}_{c_1}, \tilde{C}_{c_1}) + c_2 J_2(\tilde{A}_{c_1}, \tilde{C}_{c_1}) \geq J_1(\tilde{A}_{c_2}, \tilde{C}_{c_2}) + c_2 J_2(\tilde{A}_{c_2}, \tilde{C}_{c_2}), \quad (14)$$

where (14) can be rewritten as:

$$\begin{aligned} J_1(\tilde{A}_{c_1}, \tilde{C}_{c_1}) &+ c_1 J_2(\tilde{A}_{c_1}, \tilde{C}_{c_1}) \\ &+ (\Delta c) J_2(\tilde{A}_{c_1}, \tilde{C}_{c_1}) \\ \geq J_1(\tilde{A}_{c_2}, \tilde{C}_{c_2}) &+ c_1 J_2(\tilde{A}_{c_2}, \tilde{C}_{c_2}) \\ &+ (\Delta c) J_2(\tilde{A}_{c_2}, \tilde{C}_{c_2}), \quad (15) \end{aligned}$$

where $\Delta c = c_2 - c_1$. Combining (13) and (15) it is easily seen that the regularization term $J_2(\tilde{A}_c, \tilde{C}_c)$ is a non-increasing function of c.

2.2 *Choosing the regularization parameter*

2.2.1. *An upper-bound*

The following lemma, (Goethals *et al.*, 2003), states that positive realness can always be imposed, by using the regularization term introduced in (9), provided the regularization coefficient c is chosen sufficiently large.

Lemma 1. Let $\hat{G}$, $\hat{\Lambda}_0$ be given. Let $W = Q_W Q_W^T > 0$, $W - \hat{G}\hat{\Lambda}_0^{-1}\hat{G}^T > 0$, and define $\hat{\Sigma} = X_i X_i^T$, $P_0 = \hat{\Sigma}W^{-1}\hat{\Sigma} - \hat{\Sigma}\begin{bmatrix}\hat{A}^T & \hat{C}^T\end{bmatrix}\begin{bmatrix} W & \hat{G} \\ \hat{G}^T & \hat{\Lambda}_0 \end{bmatrix}\begin{bmatrix}\hat{A} \\ \hat{C}\end{bmatrix}\hat{\Sigma}$. Then there exists a c^* such that the system $\tilde{A}_c$, $\hat{G}$, $\tilde{C}_c$, $\hat{\Lambda}_0$, with $\tilde{A}_c$ and $\tilde{C}_c$ as in (11), is positive real for $c \geq c^*$, with $c^* = \max_{i|\theta_i \in \mathbb{R}^+} \theta_i$, and θ the set of generalized eigenvalues of the following eigenvalue problem:

$$\theta = \lambda\left(\begin{bmatrix} 0_{\hat{n}} & -I_{\hat{n}} \\ P_0 & 2\hat{\Sigma} \end{bmatrix}, -\begin{bmatrix} I_{\hat{n}} & 0_{\hat{n}} \\ 0_{\hat{n}} & W \end{bmatrix}\right). \quad (16)$$

Hence, provided the conditions of Lemma 1 are met, a positive real model is always obtained for $c \geq c^*$, and in particular $c = c^*$, with $c*$ as in Lemma 1. Furthermore, since any positive real model is necessarily stable, which follows immediately from the upper left part of the Schur complement of the Algebraic Riccati equation

$$\begin{bmatrix} P & G \\ G^T & D + D^T \end{bmatrix} - \begin{bmatrix} APA^T & APC^T \\ CPA^T & CPC^T \end{bmatrix} \geq 0, \quad (17)$$

stability is automatically guaranteed. However, c^* can be a too conservative estimate. In general it seems reasonable to keep the amount of regularization as low as possible. Hence, one should search for the smallest possible $c \leq c^*$ for which a positive real model is found.

2.2.2. *A lower bound*

A lower bound c_s for c can be found from a theorem presented in (Van Gestel *et al.*, 2001) that states that all eigenvalues of A_c can be made to lie within a closed disc with a given radius γ, provided $c \geq c_s = \max_{i|\vartheta_i \in \mathbb{R}^+} \vartheta_i$, where $\vartheta = \lambda\left(\begin{bmatrix} 0 & -I \\ P_0 & P_1 \end{bmatrix}, -\begin{bmatrix} I & 0 \\ 0 & P_2 \end{bmatrix}\right)$ is the set of eigenvalues of a Generalized Eigenvalue problem with $P_2 = -\gamma W \otimes \gamma W$, $P_1 = -\gamma W \otimes \hat{\Sigma} - \gamma\hat{\Sigma} \otimes W$, and $P_0 = \hat{A}\hat{\Sigma}\otimes\hat{A}\hat{\Sigma} - \gamma\hat{\Sigma}\otimes\gamma\hat{\Sigma}$. Furthermore c_s is shown to be the smallest regularization coefficient with this property. Hence, as shown in figure 1, a minimal c imposing positive realness will always satisfy $c_s \leq c \leq c^*$.

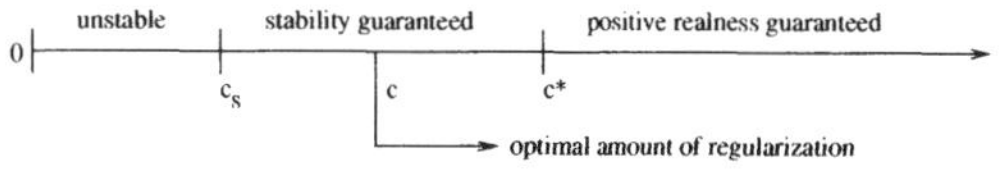

Fig. 1. Finding the optimal amount of regularization

When the realization $(\tilde{A}_{c_s}, \hat{G}, \tilde{C}_{c_s}, \hat{\Lambda}_0)$ is not yet positive real, i.e., $S_z(z) + S_z^T(z^{-1}) < 0$ for a certain $z = e^{j\theta}$, we can find a $c \geq c_s$ imposing positive realness, for instance by applying a bisection algorithm on the interval $c_s \leq c \leq c^*$.

Some alternative techniques have been reported in the literature in order to impose positive realness on a covariance model (Van Overschee and De Moor, 1996; Vaccaro and Vukina, 1993; Peternell, 1995; Marí *et al.*, 2000), many of whom are related to regularization principles. Apart from changing $\hat{A}$ and $\hat{C}$, regularization could also be applied to $\hat{G}$, $\hat{\Lambda}_0$, or a combination of both. A common problem for many of these alternatives is that they cannot be used if the covariance model is unstable. Apart from the technique proposed in this paper, which will be abbreviated as $\text{REG}_{\hat{A},\hat{C}}$ and of whom performance results will be given in the following sections, we will also discuss the performance of the following techniques:

- **SDP:** In (Marí *et al.*, 2000) a new identification scheme based was proposed, based on existing stochastic subspace methods and Semi Definite Programming (SDP). A stable $\tilde{A}$ is obtained by solving:

$$\begin{aligned} \min_{\tilde{A},\hat{P}} \quad & \|(\hat{A} - \tilde{A})\hat{P}\|_2 \\ \text{s.t.} \quad & \hat{P} > 0 \\ & \hat{P} - \tilde{A}\hat{P}\tilde{A}^T > 0. \end{aligned} \quad (18)$$

  Positive realness is thereafter imposed by solving a similar SDP-problem involving vectors of stacked covariance sequences. The performance of the SDP-technique was evaluated using software written by the authors and published on their website.

- **RES:** In (Van Overschee and De Moor, 1996) the residuals ρ_w and ρ_v of the least squares problem

$$\begin{bmatrix} \hat{X}_{i+1} \\ Y_{i|i} \end{bmatrix} = \begin{bmatrix} A \\ C \end{bmatrix}\hat{X}_i + \begin{bmatrix} \rho_w \\ \rho_v \end{bmatrix} \quad (19)$$

  are used to get estimates for Q, S and R that are guaranteed to be positive:

$$\begin{bmatrix} Q & S \\ S^T & R \end{bmatrix} = \mathcal{E}\left\{\begin{bmatrix} \rho_w \\ \rho_v \end{bmatrix}\begin{bmatrix}\rho_w^T & \rho_v^T\end{bmatrix}\right\}. \quad (20)$$

The algorithm leads to biased estimates, unless $i \to \infty$, and is only applicable to stable models.

- **REG$_{\hat{\Lambda}_0}$:** Regularization on $\hat{\Lambda}_0$, as proposed in (Peternell, 1995). From $S_z(z) = \hat{D} + \hat{C}(zI_n - \hat{A})^{-1}\hat{G}$ and $\hat{\Lambda}_0 = \hat{D} + \hat{D}^T$ it is easily seen that $S_z(z) + S_z^T(z^{-1})$ can always be made positive provided $\hat{\Lambda}_0$ is chosen large enough. The method only works for stable models.

- **REG$_{\hat{G}}$:** In (Vaccaro and Vukina, 1993) one starts from REG$_{\hat{\Lambda}_0}$ to make the spectrum positive and solve the Riccati equation (8) for $\hat{P}$. The new $\hat{\Lambda}_0$, which will be denoted as $\tilde{\Lambda}_0$ and $\hat{P}$ are used to obtain an adjusted $\hat{G}$, denoted as $\tilde{G}$, after which $\hat{\Lambda}_0$ is again set to its initial value.

$$\begin{aligned} \hat{P} &= \hat{A}\hat{P}\hat{A}^T + (\hat{G} - \hat{A}\hat{P}\hat{C}^T)\cdot \\ &\quad (\tilde{\Lambda}_0 - \hat{C}\hat{P}\hat{C}^T)^{-1}(\hat{G} - \hat{A}\hat{P}\hat{C}^T)^T \\ &= \hat{A}\hat{P}\hat{A}^T + (\tilde{G} - \hat{A}\hat{P}\hat{C}^T)\cdot \\ &\quad (\hat{\Lambda}_0 - \hat{C}\hat{P}\hat{C}^T)^{-1}(\tilde{G} - \hat{A}\hat{P}\hat{C}^T)^T \end{aligned} \quad (21)$$

The model $(\hat{A}, \tilde{G}, \hat{C}, \hat{D})$ can be shown to be positive real. The technique works on stable models only.

3. EMPIRICAL EVALUATION AND SIMULATION RESULTS

A known system was used to create output samples from Gaussian, zero mean, unit variance, white noise sequences. For each output sequence the stochastic subspace approach described in section 1 was used in combination with techniques to impose positive realness where necessary. The following system was used for the simulation:

$$H(z) = \frac{(z - 0.99e^{\pm 2j})(z - 0.98e^{\pm 1.4j})}{(z - 0.8e^{\pm 2.1j})(z - 0.8e^{\pm j})} \quad (22)$$

$$\cdot \frac{(z - 0.99e^{\pm 0.6j})(z \pm 0.9)}{(z - 0.8e^{\pm 1.7j})(z - 0.8e^{0.8j})} \quad (23)$$

The results of the simulations are reported in Table 1. The table contains the results of 4 different experiments, each with a different choice of the parameters $\hat{n}$ (order of the model), i (number of block-rows), and N (number of observations). For each experiment, 1000 noise-sequences were generated with the desired length N, and an equal number of covariance models were produced. The number of covariance models that needed corrections for stability and/or positive realness are reported. As unstable models are always non positive real, the latter number will always be greater than the former. Below this information, the performance of each technique on these non positive real models is given. The performance on all non-positive real, but stable models is given at the left hand side. The results for the unstable models are given at the right hand side. The performance measures d_∞, d_2, d_1 used in the tables are norms of the differences between the transfer functions of the simulated and the identified stochastic models:

$$d_p = ||H(z) - \hat{H}(z)||_p \quad (24)$$

with $p = 1, 2, \infty$ and $\hat{H}(x)$ is the identified model. Note that results for the techniques REG$_{\hat{G}}$, REG$_{\hat{\Lambda}_0}$, and RES on unstable models are sometimes available in the tables, even though it was stated earlier in this paper that these techniques do not work for unstable models. The reason is that for these techniques the regularization procedure described in (Van Gestel *et al.*, 2001) was used to impose stability on the covariance model prior to imposing positive realness. This to avoid ending up in a hard-failure mode during the experiments, and to maintain the possibility to compare the performance of all the techniques, even on unstable models. In some cases however, the experiments did return invalid results for some of the entries in the table (denoted by '-'), for instance if the total number of unstable models is zero. In this case, averaging over these models was impossible.

Performance of the techniques
Two techniques, RES and REG$_{\hat{A},\hat{C}}$ clearly outperform the others. For some experiments the former results in slightly better estimates, however problems with this method might occur as the system order is increased. To visualize this, in Figure 2 the estimated spectral densities for the fourth experiment ($\hat{n} = 10$, $i = 16$, $N = 500$) averaged over all 1000 runs (including the ones which did not need correction) are given, together with the spectral density of the original model and a 95% error region. Note the spikes in the average spectral density and its confidence bounds in many techniques, indicating bad performance on at least some of the 1000 sequences used for the experiment. Note also that in principle, without adaptation, the RES technique only works for stable models, a condition which is seldomly satisfied for non positive real models.

Influence of i, $\hat{n}$ and N
It is interesting to have a look into the influence of the parameters $\hat{n}$, i and N on the occurrence of positive realness problems. In Table 1, decreasing i from 16 to 12 clearly resulted in a much higher number of non positive real models. It is well known that when the modeling order $\hat{n}$ increases, the probability to obtain unstable models increases considerably (see also (Van Gestel *et al.*, 2001)). This can also be observed in the table. Finally, it is observed that for the example described in this paper the influence of N on the occurrence of positivity problems is relatively low compared to that of $\hat{n}$ and i.

4. CONCLUSIONS

Stochastic subspace methods for the identification of linear time-invariant systems are known to be asymptotically unbiased. However, for a finite amount of data, and depending on the choice of some used defined variables as the modeling order and the number of covariance lags used in the identification procedure, the procedure might break down due to positive realness problems. In this paper a regularization approach was proposed to impose positive realness on a formerly identified covariance model. It was shown that, if an adequate amount of regularization is used, a positive real model can always be obtained. The simulation results clearly indicate that this new approach yields better models than other existing techniques.

5. ACKNOWLEDGEMENTS

Our research is supported by grants from several funding agencies and sources:

Research Council KUL: Concerted Research Action GOA-Mefisto 666 (Mathematical Engineering), IDO (IOTA Oncology, Genetic networks), several PhD/postdoc & fellow grants;

Flemish Government: Fund for Scientific Research Flanders (several PhD/postdoc grants, projects G.0256.97 (subspace), G.0115.01 (bio-i and microarrays), G.0240.99 (multilinear algebra), G.0197.02 (power islands), G.0407.02 (support vector machines), research communities ICCoS, ANMMM), AWI (Bil. Int. Collaboration Hungary/ Poland), IWT (Soft4s (softsensors), STWW-Genprom (gene promotor prediction), GBOU-McKnow (Knowledge management algorithms), Eureka-Impact (MPC-control), Eureka-FLiTE (flutter modeling), several PhD grants);

Belgian Federal Government: DWTC (IUAP IV-02 (1996-2001) and IUAP V-10-29 (2002-2006): Dynamical Systems and Control: Computation, Identification & Modelling), Program Sustainable Development PODO-II (CP-TR-18: Sustainibility effects of Traffic Management Systems);

Direct contract research: Verhaert, Electrabel, Elia, Data4s, IPCOS;

6. REFERENCES

Dahlén, A., A. Lindquist and J. Mari (1998). Experimental evidence showing that stochastic subspace identification methods may fail. *Systems and Control Letters* **34**, 302–312.

Faurre, P., M. Clerget and F. German (1978). Opérateurs rationnels positifs, application à l'hyperstabilité et aux processus aléatoires. *Méthodes Mathématiques de l'Informatique*.

Goethals, I., T. Van Gestel, J. Suykens, P. Van Dooren and B. De Moor (2003). Identification of positive real models in subspace identification by using regularization, accepted for publication in. *IEEE, Transactions on Automatic Control*.

Lindquist, A. and G. Picci (1996). Canonical correlation analysis, approximate covariance extension, and identification of stationary time series. *Automatica* **32 no 5**, 209–233.

Marí, J., P. Stoica and T. McKelvey (2000). Vector arma estimation: A reliable subspace approach. *IEEE Transactions on Signal Processing* **SP-48**, 2092–2104.

Oono, Y. (1981). Introduction to pseudo-positive-real functions. *Proc. 1981 Int. Symp. Circuits and Systems* **Chicago**, 469–472.

Peternell, K. (1995). *Identification of linear dynamic systems by subspace and realization-based algorithms, PhD thesis*. T.U. Wien. Vienna.

Vaccaro, R.J. and T. Vukina (1993). A solution to the positivity problem in the state-space approach to modeling vector-valued time series. *Journal of Economic Dynamics and Control* **17**, 401–421.

Van Gestel, T., J. Suykens, P. Van Dooren and B. De Moor (2001). Identification of stable models in subspace identification by using regularization. *IEEE Transactions on Automatic Control* **46(9)**, 1416–1420.

Van Overschee, P. and B. De Moor (1993). Subspace algorithms for the stochastic identification problem. *Automatica* **29, no 3**, 649–660.

Van Overschee, P. and B. De Moor (1996). *Subspace Identification for Linear Systems: Theory - Implementation - Applications*. Kluwer Academic Publishers. Boston/London/Dordrecht.

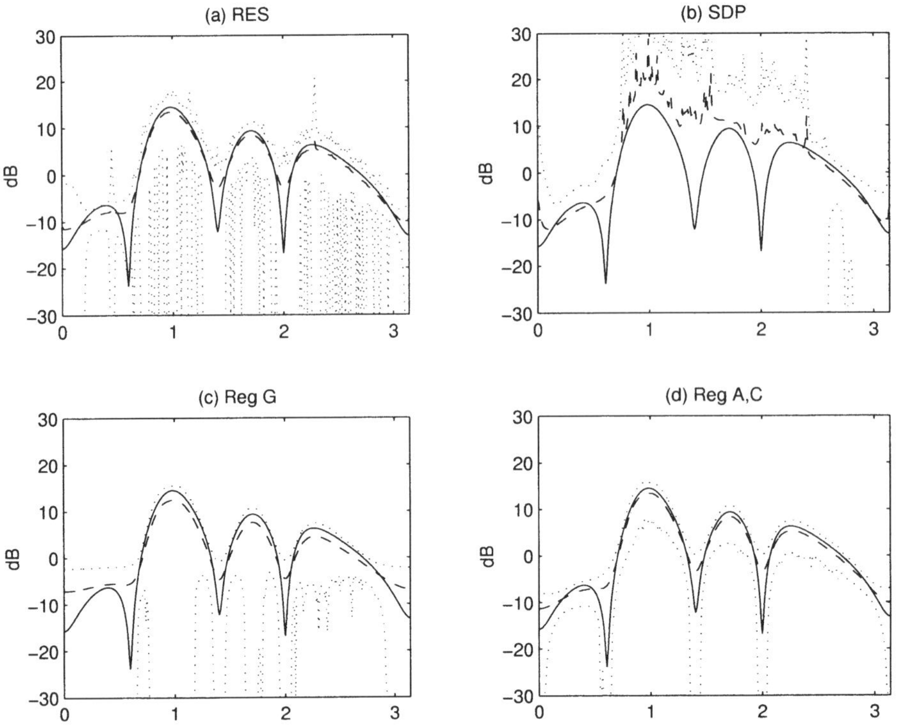

Fig. 2. Averaged Spectral density over 1000 runs for the simulation example ($H(z)$) with $\hat{n} = 10, i = 16, N = 500$ (dashed line) with 95% error region (dotted line). The solid line is the spectral density of the original model used for simulation. The numerical results are summarized in Table 1

Table 1. Performance for various techniques on the simulation example ($H(z)$). Results for $\text{REG}_{\hat{G}}$, $\text{REG}_{\hat{\Lambda}_0}$, RES on unstable models are emphasized to stress the fact that they cannot be obtained without making the model stable first

$n = 8, i = 16, N = 500$			Not positive real 528/1000				Unstable 0/ 1000			
	Stable models					Unstable models				
	$\text{REG}_{\hat{A},\hat{C}}$	$\text{REG}_{\hat{G}}$	$\text{REG}_{\hat{\Lambda}_0}$	RES	SDP	$\text{REG}_{\hat{A},\hat{C}}$	$\text{REG}_{\hat{G}}$	$\text{REG}_{\hat{\Lambda}_0}$	RES	SDP
Mean(d_∞)	1.6	2.05	2.24	1.46	9.54	-	-	-	-	-
Var(d_∞)	0.324	0.666	0.624	0.239	504	-	-	-	-	-
Mean(d_2)	0.571	0.695	0.771	0.549	1.93	-	-	-	-	-
Var(d_2)	0.0181	0.0573	0.0566	0.0146	3.41	-	-	-	-	-
Mean(d_1)	1.35	1.76	1.82	1.32	3.47	-	-	-	-	-
Var(d_1)	0.0813	0.459	0.31	0.0673	3.38	-	-	-	-	-
$n = 8, i = 12, N = 500$			Not positive real 794/1000				Unstable 4/ 1000			
	Stable models					Unstable models				
	$\text{REG}_{\hat{A},\hat{C}}$	$\text{REG}_{\hat{G}}$	$\text{REG}_{\hat{\Lambda}_0}$	RES	SDP	$\text{REG}_{\hat{A},\hat{C}}$	$\text{REG}_{\hat{G}}$	$\text{REG}_{\hat{\Lambda}_0}$	RES	SDP
Mean(d_∞)	1.55	2.19	2.42	1.48	3.59	2.48	*3.82*	*4.67*	*10.1*	8.01
Var(d_∞)	0.253	0.684	0.665	0.518	37.2	0.0523	*0.388*	*0.00468*	*15.1*	68.4
Mean(d_2)	0.577	0.75	0.846	0.549	1.12	1	-	-	*1.19*	1.62
Var(d_2)	0.0171	0.0662	0.0716	0.0159	0.274	0.00826	-	-	*0.035*	0.8
Mean(d_1)	1.37	1.87	2.02	1.29	2.54	2.47	*2.93*	*4.56*	*2.59*	3.16
Var(d_1)	0.0881	0.475	0.426	0.0662	0.843	0.0872	*0.222*	*0.000303*	*0.115*	1
$n = 8, i = 16, N = 1000$			Not positive real 544/1000				Unstable 1/ 1000			
	Stable models					Unstable models				
	$\text{REG}_{\hat{A},\hat{C}}$	$\text{REG}_{\hat{G}}$	$\text{REG}_{\hat{\Lambda}_0}$	RES	SDP	$\text{REG}_{\hat{A},\hat{C}}$	$\text{REG}_{\hat{G}}$	$\text{REG}_{\hat{\Lambda}_0}$	RES	SDP
Mean(d_∞)	1.15	1.58	1.75	1.05	8	1.46	*1.63*	*4.68*	*1.48*	41.9
Var(d_∞)	0.157	0.495	0.445	0.0977	1.84e+03	-	-	-	-	-
Mean(d_2)	0.413	0.533	0.602	0.418	1.48	0.55	-	-	0.607	8.05
Var(d_2)	0.00939	0.0456	0.0407	0.00594	4.41	-	-	-	-	-
Mean(d_1)	0.972	1.5	1.41	1.03	2.74	1.44	*1.54*	*4.54*	*1.63*	10.3
Var(d_1)	0.0431	0.716	0.208	0.0275	2.43	-	-	-	-	-
$n = 10, i = 16, N = 500$			Not positive real 727/1000				Unstable 182/ 1000			
	Stable models					Unstable models				
	$\text{REG}_{\hat{A},\hat{C}}$	$\text{REG}_{\hat{G}}$	$\text{REG}_{\hat{\Lambda}_0}$	RES	SDP	$\text{REG}_{\hat{A},\hat{C}}$	$\text{REG}_{\hat{G}}$	$\text{REG}_{\hat{\Lambda}_0}$	RES	SDP
Mean(d_∞)	1.7	2.45	2.63	2.19	18.3	2.48	*4.34*	*4.93*	*4.46*	15.1
Var(d_∞)	0.488	1.57	1.04	6.02	4e+03	2.31	*5.69*	*0.293*	*16*	3.76e+03
Mean(d_2)	0.579	0.784	0.889	0.591	2.46	0.696	-	-	*0.723*	2.06
Var(d_2)	0.0172	0.0907	0.117	0.0294	8.59	0.0298	-	-	*0.0623*	9.65
Mean(d_1)	1.36	1.98	2.12	1.35	4.08	1.65	*2.9*	*4.53*	*1.6*	3.47
Var(d_1)	0.0709	0.643	0.76	0.0655	6.7	0.174	*0.698*	*0.00353*	*0.168*	6.82

IFAC Publications
www.elsevier.com/locate/ifac

MODELING HUMAN GAITS WITH SUBTLETIES

Alessandro Bissacco * **Payam Saisan** ** **Stefano Soatto** *

* *Department of Computer Science, University of California, Los Angeles, CA 90095*
{bissacco,soatto}@cs.ucla.edu
** *Department of Electrical Engineering, University of California, Los Angeles, CA 90095* saisan@ee.ucla.edu

Abstract: We present a novel approach to modeling subtleties in human motion. We represent the trajectories of a certain number of salient features on the human body as the output of a dynamical system driven by an unknown stochastic input. We present several techniques for inferring model parameters and input signal distributions corresponding to different optimality criteria, and evaluate the corresponding models for accuracy and predictive power. In particular we exploit the higher order statistical information content in motion data to arrive at input signals with independent components and show that the human motion synthesized from non-Gaussian inputs capture best the subtle complexities of the motion data. *Copyright © 2003 IFAC*

Keywords: Image-based modeling and rendering, higher-order statistics, dynamic ICA, human motion modeling, animation

1. INTRODUCTION

Human gaits can convey rich and subtle information. By looking at a person walking from afar, we can sometimes tell whether she is happy, tired, or wounded. We can often make accurate guesses as to individual traits such as the rough age or gender, or even identify a known individual from her gaits. It is clear that such information is encoded not in each individual static pose, but in the *dynamics* of the moving body: in Johansson's experiments Johansson (1973) one cannot tell much from a single frame, but when the sequence is animated suddenly the event becomes easy to parse and identify [1].

Modeling the subtleties in human motion can play a crucial role in a number of applications ranging from security (recognizing individuals from their gaits) to entertainment and the arts (motion capture and synthesis). For the sake of concreteness, we take as our driving application image-based modeling for the purpose of motion synthesis. The idea is simple: we want to collect motion-capture data [2] for an individual, and from these data build a model that can then be used to generate (novel) synthetic motions, for instance of an animated character.

[1] Naturally there is a great deal of information in the photometry and geometry of the scene that can be conveyed in a single static frame. However, in this study we concentrate on the scene dynamics. Johansson's experiments are enlightening because they show that even after stripping a sequence of all of its pictorial content, the dynamics of moving dots still retains information a remarkable amount of information.

[2] In particular, trajectories of a collection of marker positions in space.

In order to do so, we define a gait as the output of a dynamical system driven by a stochastic process with an unknown distribution. While this was done in A. Bissacco and Soatto (2001) for the case of Linear Gaussian models, in this work we generalize the model to arbitrary input distributions. Learning a model then amounts to inferring the model dynamics as well as the input distribution. Unlike the Gaussian case, no closed-form exists, and even the optimality criterion is somewhat open to discussion. In this work we explore several alternatives including Gaussian mixtures, resulting in various mixture Kalman filter models for inference, to exponential classes of densities, resulting in a dynamic version of independent component analysis (ICA) Comon (1994).

The quality of the model inferred can be measured by the size of the residual, for instance the Kullbach-Leibler divergence or the L^2 norm. That measures how well the model fits the training data. However, more importantly for the driving application we are considering, one can simulate the model forward, and visually inspect whether the resulting simulation captures the "character" of the input set. More quantitatively, one can use a portion of the data to learn a model, use the model to predict future data, and then compare that to the real data acquired in subsequent times.

Before delving into the model, we discuss how our approach relates to the state of the art.

1.1 Relation to previous work and contribution of this paper

Modeling subtleties in human or animal motion has been subject of considerable attention lately. Bregler and coworkers Torresani *et al.* (2001) have proposed a variety of methods to model non-rigid motions in an attempt to capture subtleties. Such models are built as linear combinations of a collection of "key" poses, learned using principal component analysis from motion capture data. Similar ideas were used by Brand in his morphable models (see Brand (2001) and references therein). These approaches are also related to a linear-Gaussian model that is a special case of what we describe below. A. Bissacco and Soatto (2001) used a similar model for the purpose of recognition, and defined a metric on the space of model that was used for classification. To the best of our knowledge, dynamical systems with arbitrary input distributions have never been used to model human gaits before.

Local representation of motion based on optical flow has been exploited in Black (1999); Little and Boyd (1996), and view-based methods are proposed in Bobick (1996); Black (1996); Giese and Poggio (2000). Other approaches are based on principal component analysis Yacoob and Black (1999), parameterization of the motion on joint angles Campbell and Bobick (1995) and snake fitting Niyogi (1994). Estimation of motion from stereo Wren (1998) and multiple view systems Gavrila and Davis (1996) have also been investigated. In Bregler (1997) a mixed-state statistical model for the representation of motion has been proposed. In this Switching Linear Dynamic Model a stochastic finite-state automata at the highest level switches between local linear Gaussian models. Estimation and recognition is performed with Expectation-Maximization approaches using particle filters North *et al.* (2000); Black and Jepson (1998) or structured variational inference techniques Pavlovic *et al.* (2000).

Our models are discrete-time, continuous-state dynamical systems, and the action is coded in the dynamical model (i.e. the system parameters) as well as the input distribution. We assume that, using whatever method of the ones described above, either the joint angles or the marker trajectories are given to us. Our work, therefore, comes at a level of abstraction higher than typical motion detection and tracking algorithms. It uses track data to infer a global model of the dynamics of a human gait, using dynamical systems with unknown input distribution.

2. MODELING SUBTLETIES

We assume that we are given the trajectory of a number of distinctive feature points on the human body. For instance, when a motion capture system is used for data acquisition, these points correspond to the position of a number of markers placed ad-hoc on the human subject, for instance at her joints. We denote the position of each marker i ad the time instant t by $y_i(t) \in \mathbb{R}^3$. Alternatively, one may consider the joint angles corresponding to a skeletal model of the subject. In that case, one may call the joint angle i at time t $y_i(t) \in [0,\ C)$, where C_i is a constant that can be either π or 2π depending upon the joint. In any case, the dataset consists of a trajectory in some M-dimensional space; for simplicity we consider the first case where

$$y(t) \in \mathbb{R}^{3M},\ t \in [0,\ t_f] \tag{1}$$

However, the considerations developed here can be transposed to any other representation of the input data one chooses.

As we anticipated, we model $y(t)$ as the output of a dynamical system driven by a stochastic input. That is, we assume that at each instant of time there exists a vector $x(t) \in \mathbb{R}^N$, and suitable functions f and h such that

$$x(t+1) = f(x(t)) + v(t); \; x(0) = x_0 \quad (2)$$
$$y(t) = h(x(t)) + w(t) \quad (3)$$

where x_0 is an unknown but constant vector (the initial condition), and $v(t)$ and $w(t)$ are white, zero-mean stochastic processes. While we assume that $w(t)$ is normally distributed, $w(t) \sim \mathcal{N}(0, R)$, we allow $v(t)$ to be a sample from an unknown probability density q:

$$v(t) \overset{IID}{\sim} q(v). \quad (4)$$

Our goal, then, is to infer the input density q, the initial condition x_0, the order of the model N, the dynamics f and the output map h from a time series $\{y(t)\}_{t\in[0,\ t_f]}$. Naturally, there are many ways of doing so, depending on what inference criterion one chooses. In the following section we describe several criteria that we have evaluated, and the resulting learning algorithms. For the sake of simplicity we restrict our attention to linear dynamics $f(x) = Ax$ and linear output maps $h(x) = Cx$, shifting the emphasis of the modeling power to the input distribution q. Some of the techniques outlined below can be extended to arbitrary nonlinear models, although this is beyond the scope of this paper.

3. INFERENCE CRITERIA AND LEARNING ALGORITHMS

Given a sequence of output data $\{y(t)\}_{t\in[0,\ t_f]}$ generated by a model of the form

$$x(t+1) = Ax(t) + v(t) \quad (5)$$
$$y(t) = Cx(t) + w(t)$$
$$x(0) = x_0; \; v(t) \overset{IID}{\sim} q(v)$$

Our goal is that of finding the *optimal* estimates of the unknowns

$$\hat{A}, \hat{C}, \hat{q}(\cdot), \hat{x}(t). \quad (6)$$

We can choose several optimality criteria, depending on whether we seek for the maximum likelihood parameters, the maximum a-posteriori (given a prior distribution on the unknown), the parameters that result in a maximally independent estimated input sequence $\hat{v}(t)$, or the parameters that result in the minimum variance of the output error $y(t) - \hat{y}(t)$. There is no right or wrong criterion; all that one can do is to test several, and compare the results. As we show in the experimental section, comparison can be performed by verifying how well the model captures the training data (residuals), or how well the model can predict future data (prediction error). In addition, for the given application, we can simulate the model and visually inspect the results to verify whether subtleties have indeed been captured.

3.1 Linear Gaussian models and maximum likelihood

If we assume that the input distribution is Gaussian, $q(\cdot) \in \mathcal{N}(0, Q)$, then there are closed-form solution for the identification of the model parameters A, C, Q, R and the state sequence $x(t)$ that minimize the likelihood of the output error. In particular, subspace identification algorithms can be used for this purpose. Since the emphasis of this paper is on non-Gaussian modeling, we refer the reader to Overschee and Moor (1993) for details, and to the experimental section for experiments.

3.2 Dynam-ICA: dynamic independent component analysis

An optimality criterion for inference of model and input descriptions can be constructed by requiring the estimated input sequence $\hat{v}(t)$ to be a realization from a stochastic process that has maximally independent components. Independence can be expressed in terms of the mutual information among input components, which in turn can be written in terms of the Kullback-Leibler divergence.

This approach results in a semi-parametric statistical inference problem, where one has to simultaneously infer the (finite-dimensional) model parameters as well as the (infinite-dimensional) input distribution q. This is essentially an independent component analysis (ICA) problem.

In its conventional static form, ICA attempts to decompose a random vector into a linear combination of statistically independent components. If we call $y \in \mathbb{R}^m$ the random vector, then ICA looks for a matrix $C \in \mathbb{R}^{m\times n}$ with $n \leq m$ and a random vector $x \in \mathbb{R}^n$ with independent components, $p_{\mathbf{x}}(x_1, \ldots, x_n) = p_1(x_1) \ldots p_n(x_n)$ such that

$$y = Cx. \quad (7)$$

The unknowns C and p_i can be estimated by minimizing the mutual information $I(y||Cx) \doteq \int p_{\mathbf{y}} \log \frac{p_{\mathbf{y}}}{p_{Cx}} dy$, computed or approximated using a number of independent and identically distributed (IID) samples from $p_{\mathbf{y}}$: $y(1), \ldots, y(t) \overset{IID}{\sim} p_{\mathbf{y}}$. Typically the process y is assumed to be ergodic, and therefore a time series is used in lieu of a fair sample.

What we have here, however, is a dynamic ICA problem of separating independent components mixed by linear dynamical (state-space) system.

Let us rewrite the output of the model at time t:

$$y(t) = \left[CA^t,\ CA^{t-1}B, \ldots, CB\right] \begin{bmatrix} x(0) \\ v(0) \\ \vdots \\ v(t-1) \end{bmatrix} \doteq \tilde{\mathcal{C}}^t \tilde{\mathbf{V}} \quad (8)$$

and stack the observations $y(1), \ldots, y(t)$ into a vector $\mathbf{Y}^t$ to obtain

$$\mathbf{Y}^t = \tilde{\mathcal{C}}^t \tilde{\mathbf{V}}. \quad (9)$$

One may be tempted to invoke the independence of the components of $\mathbf{V}$ – based on the assumptions that $v(t)$ is white (time samples are independent) and has independent components – and use standard ICA to estimate the mixing matrix $\tilde{\mathcal{C}}^t$. This, however, does not work because it is not possible to use time realizations as independent samples of $\mathbf{Y}$ due to the initial condition $x(0)$.

In what follows, we make the simplifying assumption that the initial condition x_0 is zero. One may conjecture that if t is large enough and A is stable the effect of initial condition will wane; therefore, the assumption may not be as restrictive. Under this assumption, the problem of dynamic ICA can be posed as follows. Consider $\mathbf{Y}^t(k) = [y((k-1)t)^T, \ldots, y(kt-1)^T]^T$, and similarly for $\mathbf{V}^t(k)$. Furthermore, let $\mathcal{C}^t$ be the matrix obtaining by completing, in the sense of Toeplitz, the following matrix

$$\begin{bmatrix} CB & & & \\ CAB & CB & & \\ \vdots & \vdots & \ddots & \\ CA^{t-1}B & CA^{t-2}B & \ldots & CB \end{bmatrix}. \quad (10)$$

Then $\hat{A}, \hat{B}, \hat{C}$ can be found sub-optimally by first estimating the mixing matrix $\mathcal{C}^t$ having the particular structure above from a set of independent samples $\mathbf{Y}^t(1), \ldots, \mathbf{Y}^t(k)$ (notice that $\mathbf{Y}^t(i)$ and $\mathbf{Y}^t(j)$ do not share components $y(k)$):

$$\hat{\mathcal{C}}^t(A, B, C) = \arg\min_{\mathcal{C}^t} I(\mathbf{Y}^t(i) || \mathcal{C}^t \mathbf{V}^t(i)) \quad (11)$$

Under the assumption of stationarity, the model above can be thought of input-output form as $\mathbf{Y}^t(i) = \mathcal{C}^t \mathbf{V}^t(i)$. This is the familiar form of the blind source separation problem. We can reformulate this into an optimization problem where parameters of the model $\mathcal{C}^t_{A,B,C}$ can be learned using Amari's natural gradient flow Amari (1998). A detailed derivation of the parameter learning algorithm for this deconvolution problem is presented in Zhang. and Cichock (1998), where a second stage linear state-space demixing model is used to estimate the input signal with independent components.

In the next section we test this paradigm and learn the model parameters to recover the independent input signals deriving a dynamical system, namely the joint angles for human subjects during walking and running sequences.

We put the algorithm to test by synthesizing a new sequence given the dynamical system parameters of the system and non-Gaussian statistics. Once a model has been inferred, synthesis can be performed by simulating the model forward, that is by sampling an input from the distribution $\hat{q}$, and using it to compute the one-step increment of the state $\hat{x}(t+1)$ and hence the output trajectory $\hat{y}(t)$. In some cases, an explicit expression of the density q may be available. In some other cases, when the order of the model N is very high, sampling may be non trivial, and techniques such as Gibbs sampling may be necessary Geman and Geman (1984). One can apply the state-space demixing system described in Zhang. and Cichock (1998) to the training data to estimate the input distribution $\hat{q}$. The histogram of the values assumed by the elements y_i of $y(t)$ can be used as a discrete approximation $\hat{q}_i$ of the input density. Then synthesis can be easily performed by drawing samples from estimated distributions $\hat{q}_i$.

4. EXPERIMENTS

In this section we learn a model of the form (7) and the statistics of the input v for a number of human motion data sequences corresponding to different subjects' walking and running. The data entails the trajectories of four joint angles corresponding to shoulder, elbow, hips and knees. A typical sequence was about 48 to 78 frames, obtained by applying standard image-based tracking techniques (Bregler (1998)) to video streams taken with a 30Hz camera.

We obtained the dynamical system parameters and estimated the input sequence $v(t)$ with independent components. We then generated a random sample corresponding to the distributions obtained from $v(t)$ to form the input for synthesized sequences. Together with the estimates of the system matrices, A, B and C, we generated new sequences of joint angles. In our preliminary run of the experiments, we implemented a simplified version of the learning algorithm. That is, we first computed the system parameters, matrices A and C using known system identification techniques such as subspace ID and expectation maximization. We then used these to compute the innovations $e(t) = x(t+1) - Ax(t)$. With $e(t) = Bv(t)$ we learned the mixing matrix B using the standard ICA formulation and recovered maximally independent components, $v_i(t)$. We then sampled from the non-gaussian density function of each $v_i(t)$ to form new input sequences, $\hat{v}_i(t)$. Figure 1 shows the innovations $e(t)$ and the prediction error $e(t) - B\hat{v}(t)$ for one side of the body, the time progression of shoulder, elbow, hip and knee angles. As can be seen in the figures the best results were obtained with the infomax ICA.

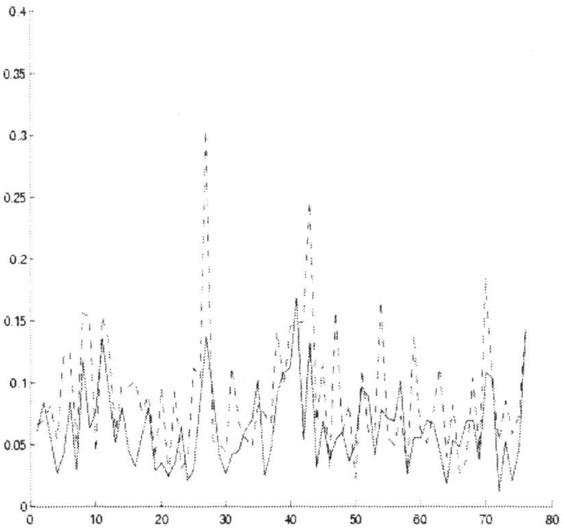

Fig. 1. *Norm of innovations* $\|e(t)\|$ *(solid line) and prediction errors* $\|e(t) - B\hat{v}(t)\|$ *(dashed line) normalized with respect the state* $\|x(t)\|$. *Horizontal axis is time, in frames.*

5. DISCUSSION

We described a novel approach for representing the process of human gaits, in this case walking and running, as the output of a linear dynamical system driven by a stochastic process with independent components. We presented a method for identifying a model from data when transient effects to due initial conditions are neglected. In particular we formulated the model identification process in an information theoretic framework and exploited higher order statistical information content in motion data to form input with independent components. Experimental results suggest the non-Gaussian input models capture best the complexity of the underlying process. We demonstrated that we could use this model to synthesize from it and generate novel instances of different styles of human walking and running motion sequences.

REFERENCES

A. Bissacco, A. Chiuso, Y. Ma and S. Soatto (2001). Recognition of human gaits. *In Proc. of the IEEE Intl. Conf. on Comp. Vision and Patt Recog. pages 401-417.*

Amari, S. (1998). Natural gradient works efficiently in learning. *Neural Computation, 10:251–276.*

Black, M. J. (1996). Eigentracking: robust matching and tracking of articulated objects.

Black, M. J. (1999). Explaining optical flow events with parameterized spatio-temporal models. *In Proc. of Conference on Computer Vision and Pattern Recognition, volume 1, pages 326–332.*

Black, M. J. and A. D. Jepson (1998). A probabilistic framework for matching temporal trajectories: Condensation-based recognition of gestures and expressions. *In Proc. of European Conference on Computer Vision, volume 1, pages 909-24.*

Bobick, A. F. (1996). Appearance-based representation of action.

Brand, M. (2001). Morphable 3d models from video. *In Proc. International Conference on Computer Vision and Pattern Recognition.*

Bregler, C. (1997). Learning and recognizing human dynamics in video sequences. *In Proc. of the Conference on Computer Vision and Pattern Recognition, pages 568–574.*

Bregler, C. (1998). Tracking people with twists and exponential maps. *In Proc. International Conference on Computer Vision and Pattern Recognition.*

Campbell, L. and A. Bobick (1995). Recognition of human body motion using phase space constraints. *In Proc. IEEE Conf. on Comp. Vision and Pattern Recogn. page 8.*

Comon, P. (1994). Independent component analysis, a new concept. *Signal Processing, 36:287–314.*

Gavrila, D. M. and L. S. Davis (1996). Tracking of humans in action: a 3-d model-based approach.

Geman, S. and D. Geman (1984). Stochastic relaxation, gibbs distributions, and the bayesian restoration of images. *In IEEE Transactions on Pattern Analysis and Machine Intelligence, PAMI-6, pp. 721–741.*

Giese, M. A. and T. Poggio (2000). Morphable models for the analysis and synthesis of complex motion patterns. *In International Journal of Computer Vision, volume 38(1), pages 1264–1274.*

Johansson, G. (1973). Visual perception of biological motion and a model for its analysis. *Perception and Psychophysics, 14(2):201–211.*

Little, J. J. and J. E. Boyd (1996). Recognizing people by their gait: the shape of motion.

Niyogi, A. A. (1994). Analyzing and recognizing walking figures in xyt. *In Proc. IEEE Conf. on Comp. Vision and Pattern Recogn., pages 469–474, Seattle, June.*

North, B., A. Blake, M. Isard and J. Rittscher (2000). Learning and classification of complex dynamics. *In IEEE Transaction on Pattern Analysis and Machine Intelligence, volume 22(9), pages 1016-34.*

Overschee, P. Van and B. De Moor (1993). Subspace algorithms for the stochastic identifica-

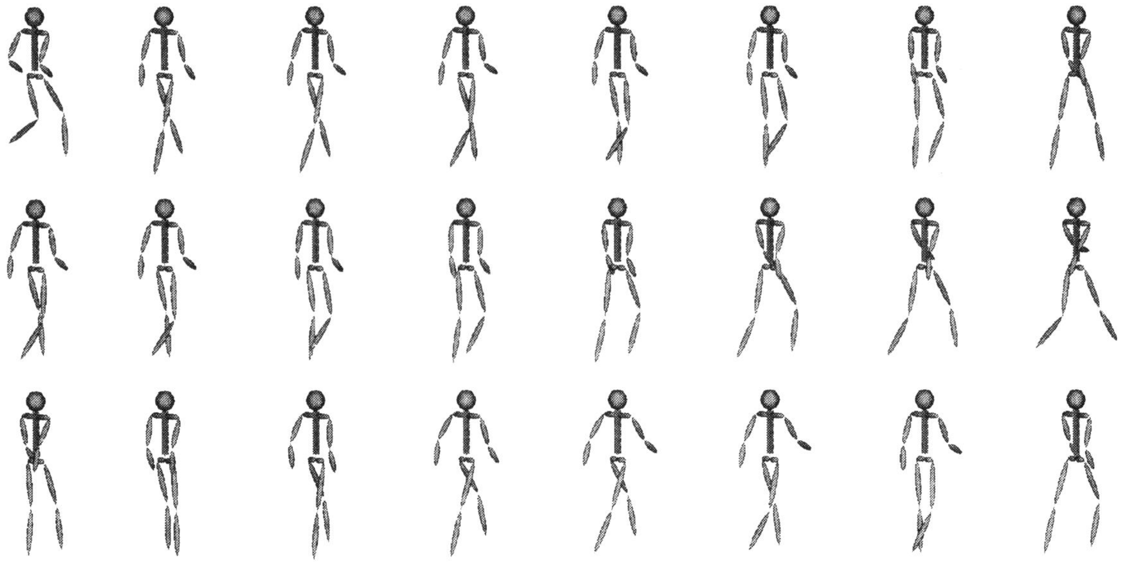

Fig. 2. *Motion sequences. Top row is the original walking data, second row is the synthesized sequence using Gaussian input and third row is the ICA based non-Gaussian input driven sequence.*

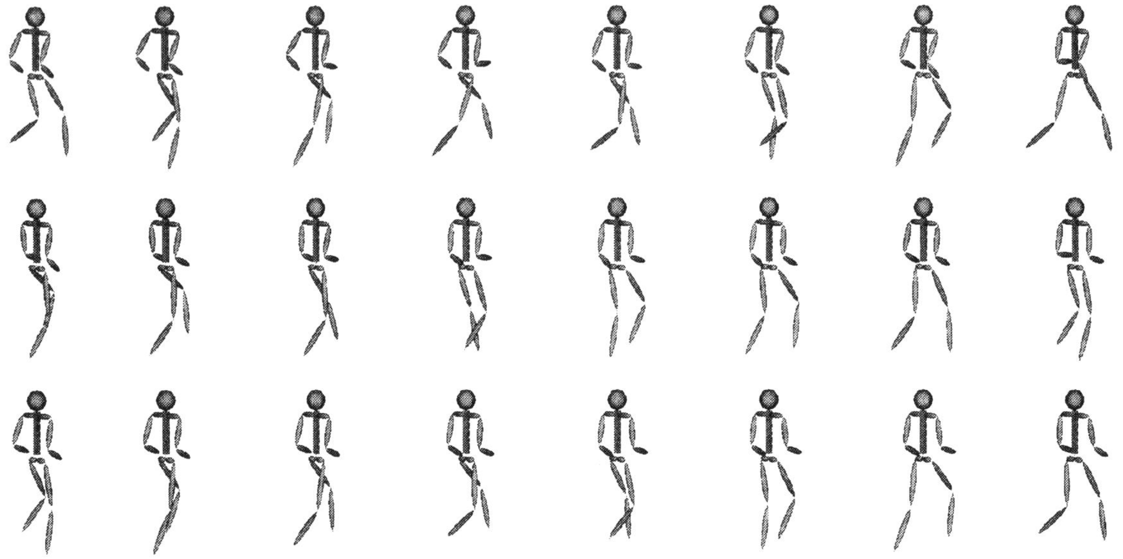

Fig. 3. *Motion sequences. Top row is the original running data, second row is the synthesized sequence using Gaussian input and third row is the ICA based non-Gaussian input driven sequence.*

tion problem. *Automatica, 29:649–660.*

Pavlovic, V., J. Rehg and J. MacCormick (2000). Impact of dynamic model learning on classification of human motio. *In Proc. International Conference on Computer Vision and Pattern Recognition.*

Torresani, L., D. Yang, G. Alexander and C. Bregler (2001). Tracking and modelling non-rigid objects with rank constraint. *In Proc. International Conference on Computer Vision and Pattern Recognition.*

Wren, C. (1998). Dynamic models of human motion.

Yacoob, Y. and M. J. Black (1999). Parameterized modeling and recognition of activities. *In Computer Vision and Image Understanding, volume 73(2), pages 232–247.*

Zhang., L. and A. Cichock (1998). Blind deconvolution of dynamical systems : A state-space approach. *Proceedings of the IEEE. Workshop on NNSP'98, 123–131.*

IFAC Publications
www.elsevier.com/locate/ifac

REDUCTION OF LARGE-SCALE GROUNDWATER FLOW MODELS VIA THE GALERKIN PROJECTION

P.T.M. Vermeulen * **A.W. Heemink** * **C.B.M Te Stroet** **

* *Department of Applied Mathematical Analysis, Faculty of Electrical Engineering, Mathematics and Computer Science, Delft University of Technology, P.O. Box* 5031, 2600 *GA Delft, the Netherlands*
** *Netherlands Institute of Applied Geoscience TNO, National Geological Survey, P.O. Box* 80015, 3508 *TA Utrecht, the Netherlands*

Abstract: In this paper we describe a reduced model structure that describes the hydraulic head $\mathbf{h}$ for groundwater flow models as a linear combination of a set of spatial patterns $\mathbf{P}$ with time-varying coefficients $\mathbf{r}$. We discuss a data-driven technique to extract patterns $\mathbf{P}$ (EOFs) that span a subspace of model results that captures most of the relevant information of the original model. We make use of the patterns to obtain a reduced dynamic model for the time-varying coefficients via a Galerkin Projection. This technique substitutes $\mathbf{h}$ within the PDE for groundwater flow by the reduced model structure $\mathbf{P}^{\mathrm{T}}\mathbf{r}$. We acquire a dynamic reduced model for $d\mathbf{r}/dt$ by multiplying the outcome with $\mathbf{P}^{\mathrm{T}}$. The vector dimension of $\mathbf{r}$ is often small compared to the original dimension of $\mathbf{h}$, and a model which operates within a lower dimension requires less computational time. The method has been evaluated for a realistic case, whereby we achieved a maximal reduction in computational time of ≈ 80. The reduced model has a promising prospect as its efficiency increases whenever the number of grid cells increases and the parameterization of the original model grows in complexity. *Copyright* © 2003 *IFAC*

Keywords: Groundwater, Empirical Orthogonal Functions (EOFs), Large-scale, Linear, Model reduction, Numerical methods

1. INTRODUCTION

The access and accuracy of digital data nowadays, allows the use of numerical groundwater models with a huge amount of grid cells (often $>$ 1 million). Such large transient models have long computational times which is especially a disadvantage whenever we need to evaluate the model for many different cases, for example to optimize a pumping strategy. The main idea of the here presented reduction method, is to seek a simpler model structure, that is able to represent the original model for an intended study. *Such a simple model structure describes the hydraulic head as a linear combination of spatial patterns and time-dependent coefficients.*

There are different approaches mentioned in literature to specify patterns, such as analytical-, system- and data-driven techniques. The advantage of *analytical-driven* patterns (Fourier functions, wavelets or polynomials) as described by Hooimeijer (2001), is that they are given a priori. A disadvantage is that they are non-economical regarding the amount of patterns required (Cazemier et al., 1998). More efficient is the *system-driven* technique of computing patterns as an eigenvalue decomposition of the system matrix differential equations. It can be computed efficiently with the Lanczos algorithm and has been done for the equation of groundwater flow (Dunbar and Woodbury, 1989). Unfortunately, in most realistic applications the system matrix does not contain a relatively few dominant eigenvalues and the possibilities for reduction are therefore limited. Finally, the *data-driven* pattern identification technique uses specific model results (empirical data) for which the eigenvalue decomposition produces a minimal set of patterns which span the data set optimally. In the different fields of science, this technique is also called Proper Orthogonal Decomposition (POD) or Empirical Orthogonal Functions (EOFs). The data-driven technique appeared to be most optimal, despite the extra computational time required to generate data with the original model.

To create a reduced model for the time-dependent coefficients, we applied a Galerkin Projection (Karhunen Loève Expansion). This technique substitutes a reduced model structure within a PDE of the original model, and projects the outcome onto the patterns. This technique is not new and most of the concepts and methods presented in this paper have been mathematically proven by Newman (1996). The technique has been applied in a wide range of scientific fields, such as turbulence and image processing (Sirovich, 1987), rapid thermal chemical vapor deposition (Adomaitis, 1995), in fluid dynamics for a lid-driven cavity with a rotating rod (Hoffman Jørgensen and Sørensen, 2000), in flow reactors governed by nonlinear equations (Park and Cho, 1996).

In this paper we describe the data-driven EOF technique and a Galerkin Projection (GP) for the PDE for groundwater flow. The acquired reduced model is implemented using the concepts and software of MODFLOW (McDonald and Harbaugh, 1988).

2. MATHEMATICAL FORMULATION FOR GROUNDWATER FLOW

Groundwater flow can be described in two dimensions by the PDE based upon Darcy's law and the equation of continuity:

$$\frac{\partial}{\partial x}\left(T_x\frac{\partial h}{\partial x}\right)+\frac{\partial}{\partial y}\left(T_y\frac{\partial h}{\partial y}\right)=q, \qquad (1)$$

where h represents a hydraulic head within the model (L); x, y are Cartesian coordinates (L); t is time (T); q is the discharge or recharge term (LT^{-1}); and T_x, T_y are the conductances in x,y-direction, respectively (L^2T^{-1}). For simplicity during further elaborations, we ignore the groundwater flow in z-direction. The q-term can be expressed as:

$$q=S\frac{\partial h}{\partial t}+T_s\left(h-h_{\mathrm{riv}}\right)+w, \qquad (2)$$

where S is the storage coefficient and defines the amount of groundwater which comes out of the storage due to the evolution of h in time t; T_s is the riverbed conductance (L^2T^{-1}); h_{riv} is the corresponding water level (L); and w is a constant discharge or recharge term (LT^{-1}).

3. REDUCED MODEL

3.1 Reduced Model Structure

The reduced model structure is based upon the assumption that h can be expressed as a linear combination of a set of spatially distributed patterns with time-varying coefficients. The approximated hydraulic head $\hat{h}$ becomes:

$$\hat{h}(\underline{x},t)=h_*(\underline{x})+\sum_{n=1}^{n_p}p_n(\underline{x})r_n(t), \qquad (3)$$

where $\underline{x}$ is a location in Euclidean space; p_n is a value for the nth pattern out of n_p (−); r_n is the nth time-dependent coefficient (L); and h_* is the reference hydraulic head (L). As often $n_p \ll n_m$ (number of grid cells), we save an enormous amount of computational time compared to the original model, as we only need to compute n_p time-dependent coefficients.

The h_* represents a steady-state solution of the original model with model impulses that remain constant during the entire simulation. Therefore it is not necessary to capture those *fixed impulses*, (variables with ^{f}) by patterns too. All other variable impulses are called *pattern impulses* (variables with ^{p}). The reduced model is capable of simulating a variation within the variables w^{p} and $h^{\mathrm{p}}_{\mathrm{riv}}$. Whenever $T_x, T_y, S, w^{\mathrm{f}}, T_s^{\mathrm{f}}$ and T_s^{p} are changed, we need to identify a new set of patterns.

3.2 Pattern Identification (EOFs)

To find EOFs, we create an empirical data set in which n_e evaluations of the original model are arranged in *snapshot*-vector (Sirovich, 1987), so: $\mathbf{H} = [\mathbf{h}_1, \mathbf{h}_2, ..., \mathbf{h}_{n_e}]$. We substract the mean vector $\bar{\mathbf{h}}$ (Hoffman Jørgensen and Sørensen, 2000) and we weight each snapshot vector to scale them (Newman, 1996). Finally, we substract $\mathbf{h}_*$ as the patterns are superpositioned upon the reference head, so each column n in $\mathbf{H}$ is processed as:

$$\mathbf{d}_n=||(\mathbf{h}_n-\bar{\mathbf{h}})-\mathbf{h}_*||^{-1}\left((\mathbf{h}_n-\bar{\mathbf{h}})-\mathbf{h}_*\right) \qquad (4)$$

We collect the centered vectors as columns in $\mathbf{D} = \left[\mathbf{d}_1, \mathbf{d}_2, ..., \mathbf{d}_{n_e}\right]$ and mathematically the patterns (EOFs) are the eigenvectors of the covariance matrix of $\mathbf{D}$. The corresponding eigenvalues provide a measure of relative "energy" associated with the pattern that can be interpreted as the relative amount of time that h spends along the corresponding pattern. To avoid the need of a long computational time to solve the eigenvalue problem for a very high-dimensional covariance matrix $\mathbf{C}_d = \mathbf{D}\mathbf{D}^{\mathrm{T}}$, we define a "reduced" covariance matrix $\mathbf{C}_s$ as $\mathbf{D}^{\mathrm{T}}\mathbf{D}$ (Krysl et al., 2000), for which we solve the eigenvalue problem. We transform the eigenvectors $\mathbf{G}$ of $\mathbf{C}_s$ with dimension n_e into eigenvectors $\mathbf{V}$

with dimension n_m by applying Golub and van Loan (1989):

$$\mathbf{V} = \mathbf{DG}\Lambda^{-\frac{1}{2}}, \tag{5}$$

Eventually, the EOFs can be found by normalizing the eigenvectors and collect them as columns within matrix $\mathbf{P} = [\mathbf{p}_1, \mathbf{p}_2, ..., \mathbf{p}_{n_p}]$. Herein, the importance of each pattern n is given by:

$$\varphi_n = \frac{\lambda_n}{\sum_{m=1}^{n_e} \lambda_m} \cdot 100\%. \tag{6}$$

The reduced model structure should be able to reproduce the original model for at least that amount of variance (φ^e) totally explained by the selected set of patterns (n_p). It appeared that $\varphi^e \geq 99\%$ to obtain reliable coefficients r from a reduced model.

3.3 Reduced Model via a Galerkin Projection

To compute the time-dependent coefficients r we define an ODE for a reduced model that describes dr/dt:

$$f(t)r + \underline{q} = \frac{dr}{dt} \tag{7}$$

wherein f is a linear function that determines how r evolutes with t, and $\underline{q}$ is a reduced forcing term. In this paper we use the Galerkin Projection (GP) to acquire a formulation for dr/dt. Therefore we substitute the reduced model structure (Eq. 3) within the original PDE (Eq. 1), so:

$$\frac{\partial}{\partial x}\left(T_x \sum_{n=1}^{n_p} \frac{\partial p_n}{\partial x} r_n\right) + \frac{\partial}{\partial y}\left(T_y \sum_{n=1}^{n_p} \frac{\partial p_n}{\partial y} r_n\right) = \underbrace{S\left(\sum_{n=1}^{n_p} p_n \frac{\partial r_n}{\partial t}\right)}_{1^{st} term} + \underbrace{T_s^{\mathrm{p}}\left(\sum_{n=1}^{n_p} p_n r_n - \left(h_{\mathrm{riv}}^{\mathrm{p}} - h_*\right)\right)}_{2^{nd} term} + \underbrace{T_s^{\mathrm{f}}\left(\sum_{n=1}^{n_p} p_n r_n - \left(h_{\mathrm{riv}}^{\mathrm{f}} - h_*\right)\right)}_{3^{rd} term} + w^{\mathrm{p}}. \tag{8}$$

It should be noticed that h_* has been left out from the substitution as it is independent of r. It is allowed for these type of confined linear models as h_* does not affect the computation of the second order differential and the first term in the RHS of the equation. The lack of h_* within the second term, is corrected by adjusting the surface water levels $h_{\mathrm{riv}}^{\mathrm{p}}$ by h_*.

Whenever we apply a finite-difference approximation for the pattern derivative of space, we obtain via

$$\left(\underbrace{\frac{\partial}{\partial x}\left(\mathbf{T}_x \frac{\partial \mathbf{P}}{\partial x}\right)}_{\mathbf{U}_x} + \underbrace{\frac{\partial}{\partial y}\left(\mathbf{T}_y \frac{\partial \mathbf{P}}{\partial y}\right)}_{\mathbf{U}_y}\right)\mathbf{r} = \mathbf{SP}\frac{\partial \mathbf{r}}{\partial t} + \underbrace{\mathbf{T}_s^{\mathrm{p}}\mathbf{P}}_{\mathbf{U}_s^{\mathrm{p}}}\mathbf{r} - \underbrace{\mathbf{T}_s^{\mathrm{p}}\left(\mathbf{h}_{\mathrm{riv}}^{\mathrm{p}} - \mathbf{h}_*\right)}_{\mathbf{f}_s^{\mathrm{p}}} + \underbrace{\mathbf{T}_s^{\mathrm{f}}\mathbf{P}}_{\mathbf{U}_s^{\mathrm{f}}}\mathbf{r} + \mathbf{w}^{\mathrm{p}} \tag{9}$$

a simplified state-space expression:

$$\underbrace{\left(\mathbf{U}_x + \mathbf{U}_y - \mathbf{U}_s^{\mathrm{p}} - \mathbf{U}_s^{\mathrm{f}}\right)}_{\mathbf{U}}\mathbf{r} = \mathbf{SP}\frac{d\mathbf{r}}{dt} + \underbrace{\mathbf{w}^{\mathrm{p}} - \mathbf{f}^{\mathrm{p}}}_{\mathbf{b}^{\mathrm{p}}}. \tag{10}$$

The assimilation of the third term (Eq. 8) needs some comment as it represents a *fixed impulse* that has been processed already within h_*. The impulse still relates to h and hence it will dewater more or less whenever h changes. The corresponding conductance T_s^{f} should be included in the reduced model as it becomes by $\mathbf{U}_s^{\mathrm{f}}$ eventually part of the LHS of the equation. The term $T_s^{\mathrm{f}}\left(h_{\mathrm{riv}}^{\mathrm{f}} - h_*\right)$ is skipped as anything should be removed from the RHS that is assimilated already by h_*.

The dimension of the only unknown vector $\mathbf{r}$ is n_p, though the dimension of all other matrices and vectors are n_m. This implies that there are more equations than necessary to compute $\mathbf{r}$. To eliminate the superfluous equations, we apply a projection with $\mathbf{P}^{\mathrm{T}}$:

$$\underbrace{(\mathbf{P}^{\mathrm{T}}\mathbf{U})}_{\mathbf{N}}\mathbf{r} = \underbrace{(\mathbf{P}^{\mathrm{T}}\mathbf{S}\mathbf{P}\overbrace{\mathbf{P}^{\mathrm{T}}\mathbf{P}}^{\mathbf{I}})}_{\mathbf{M}}\frac{d\mathbf{r}}{dt} + \mathbf{P}^{\mathrm{T}}\mathbf{b}^{\mathrm{p}}. \tag{11}$$

This is a ODE for a reduced model for groundwater flow that can be solved for t by applying an implicit Euler scheme. This result eventually in:

$$\boxed{\left(\mathbf{N} - \frac{1}{\Delta t}\mathbf{M}\right)\mathbf{r}^m = -\frac{1}{\Delta t}\mathbf{M}\mathbf{r}^{m-1} + \left(\mathbf{P}^{\mathrm{T}}\mathbf{b}^{\mathrm{p}}\right)^m.} \tag{12}$$

The reduced model consist of three time-independent matrices $\mathbf{N}$ and $\mathbf{M}$ — both with a low dimension $[n_p \times n_p]$ — and $\mathbf{P}^{\mathrm{T}}$ with dimension $[n_p \times n_m]$. During the actual simulation, the product $\mathbf{P}^{\mathrm{T}}\mathbf{b}^{\mathrm{p}}$ should be computed for each time-step m and after that it is easy to obtain $\mathbf{r}^m$ as all matrices and vectors in the equation are low dimensional.

4. REALISTIC CASE

4.1 Description of the Original Model

We use a supra-regional groundwater model to evaluate the performance and accuracy of the reduced model. The groundwater model is characterized by regional areas with free-floating water tables and regions with intense dewatering systems. The rate of the latter relates linearly to the difference between h and h_{riv}; hence, the model is strictly linear.

The model contains 61632 nodes, of which $n_m = 32949$ are active. All others are inactive (no-flow) due to an irregular boundary and/or several aquifers which thin out. The underground is modeled by 9 model layers that describe a complex system of successive high and extremely low permeabilities. Within each model layer several extraction wells are active.

The main purpose of the reduced model was to simulate the effects of change in recharge together with the impact of a change in rate for the extraction wells within the model layer $2, 3, ..., 9$ simultaneously. It resulted in $n_v = 9$ groups that we varied independently. We combined the stresses within each groups randomly and simulated the total effect for $n_t = 150$ time steps, each with $\Delta t = 10$ days.

4.2 Resulting Patterns

To compute appropriate patterns, we filled a data set with an ensemble of snapshot vectors. The chosen ensemble will influence the patterns and therefore the accuracy of the reduced model. On the other hand, it is a challenge to minimize the amount of snapshot vectors (n_e). Therefore, we compute a transient impulse-response for each group individually and collect snapshots for specific time steps that differ in h significantly. This is also advantageous from a numerical point of view, as it leads to a better posed eigenproblem (Cazemier et al., 1998). We collected 9 snapshots for the recharge group with: $\Delta t^m = 2\Delta t^{m-1}$ with $\Delta t^1 = 10$ days. For the well group we computed 4 snapshots with $\Delta t^1 = 10, \Delta t^2 = 10, \Delta t^3 = 50, \Delta t^4 = 250$ days. The most important issue herein is that Δt should be small initially, and may be increased as time elapses to obtain a set that characterizes the impulse-response optimally (Park and Cho, 1996). It resulted in a data set that contained $n_e = 42$ snapshot vectors, including a zero-vector. The computational effort was in fact less than a complete run with the original model, which needed $n_t = 150$ evaluations.

With the $n_e = 42$ snapshot vectors we computed the patterns as described in §3.2; the relative eigenvalues φ_n are given in Fig. 1.

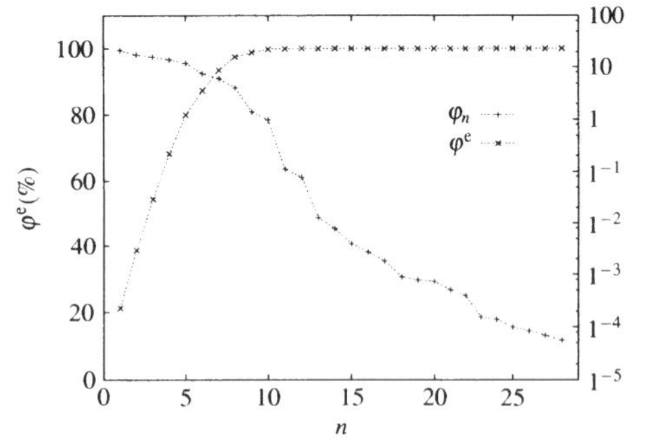

Fig. 1. Pattern number n versus the relative eigenvalues φ_n and the expected variance φ^e.

The maximal explained variance for a pattern ($\varphi_1 = 20.42\%$) declines rapidly ($\varphi_n \ll 1\%$; $n \in \{9, ..., 28\}$). Even though we constructed our 42 snapshots carefully, the contribution of the eigenvalues φ_n, $n \in \{29, ..., 42\}$ to the *expected* variance $\varphi^e(\%)$ is negligible. Apparently the spatial distribution of h is complex but its behaviour in time is not complex at all (Sirovich, 1987; Park and Cho, 1996; Hoffmann Jørgensen and Sørensen, 2000). The spatial structure of the major pattern $\mathbf{p}_1$ is given in Fig. 2. The amplitude within a major pattern ($\varphi_n > 1\%$) is related to those areas within the model that show a strong variation in h.

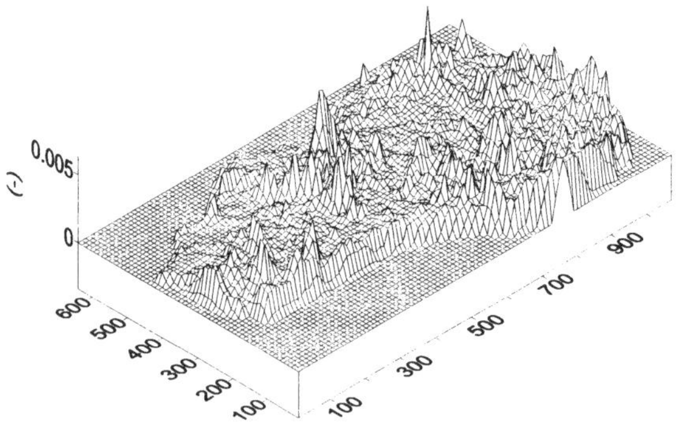

Fig. 2. Spatial structure of pattern $\mathbf{p}_1$ ($\varphi_1 = 20.42\%$) for model layer 1.

4.3 Performance of the Reduced Model

4.3.1. Initial Conditions The initial time-dependent coefficient $\mathbf{r}^0$ can be obtained by computing:

$$\mathbf{r}^0 = \mathbf{P}^\mathrm{T}(\mathbf{h}^0 - \mathbf{h}_*), \tag{13}$$

where $\mathbf{h}^0$ is the initial hydraulic head. Eq. 13 is only meaningful if it satisfies:

$$\mathbf{h}^0 - \mathbf{h}_* = \alpha_1 \mathbf{p}_1 + \alpha_2 \mathbf{p}_2 + ... + \alpha_{n_p} \mathbf{p}_{n_p}, \tag{14}$$

where $\alpha_1, ..., \alpha_{n_p}$ are coefficients.

4.3.2. Reconstruction of the Hydraulic Head Whenever we need the hydraulic head for a desired node i, j at a specific time-step m, we can reconstruct it with the reduced model structure and the computed time-dependent coefficients r_n^m as:

$$\hat{h}^m_{(i,j)} = h_{*(i,j)} + \sum_{n=1}^{n_p} p_{(i,j),n} r_n^m. \tag{15}$$

It is not neccessary to perform this reconstruction to proceed to the next time-step. This results in an enormeous time effort as we simulate $\Delta r/\Delta t$ instead of $\Delta h/\Delta t$.

4.3.3. Accuracy The accuracy of a reduced model is expressed as the relative mean absolute error (Van Overschee and De Moor, 1996):

$$\mathrm{MAE}^\mathrm{r} = 100 \cdot \frac{1}{n_t} \sum_{m=1}^{n_t} \frac{|\mathbf{h}^m - \hat{\mathbf{h}}^m|}{|\mathbf{h}^m - \mathbf{h}_*|} \quad [\%], \tag{16}$$

We have computed MAE^r over a total simulation time of $n_t = 150$, see Fig. 3. The strong reduction of MAE^r between $n_p = 9$ and $n_p = 10$ can be explained by the fact that we simulate a scenario with 9 independent groups (recharge and wells within 8 model layers, see

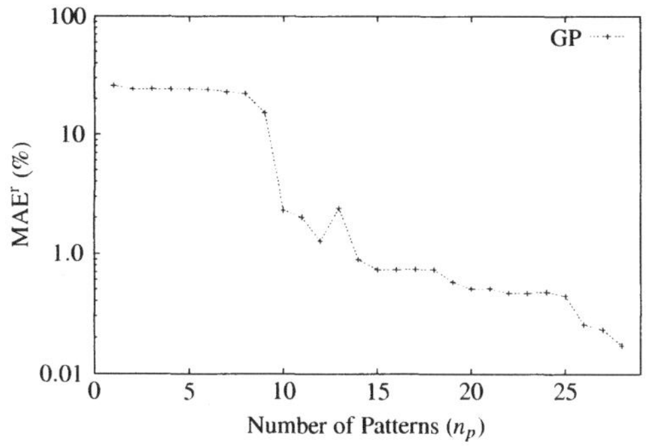

Fig. 3. Graph of n_p versus the error MAE^r over $n_t = 150$ time steps.

§4.1 last paragraph) and hence beyond that point, the accuracy gathers strength. It is strongly related to the purpose of the reduced model, but an error of MAE^r = 0.5% should be acceptable. To meet this criterion, we need at least $n_p = 20$ patterns, in other words, **we have reduced the original model, which operated within 32949 dimensions, to a model which operates within 20 dimensions only**. The cost of it in terms of model accuracy is negligible.

4.3.4. Piezometric heads With the selected $n_p = 20$ patterns we computed the mean hydraulic head for the original model as:

$$\bar{\mathbf{h}}_{\text{original}} = \frac{1}{n_t} \sum_{m=1}^{n_t} \mathbf{h}^m. \tag{17}$$

In Fig. 4 we have plotted the $\bar{\mathbf{h}}_{\text{original}}$ for model layer 1 over $n_t = 150$ time steps. We observe a complex distribution of h and whenever we plot the mean hydraulic head for the reduced model ($\bar{\mathbf{h}}_{\text{reduced}}$, Fig. 5), we observe that it is very well resembling $\bar{\mathbf{h}}_{\text{original}}$.

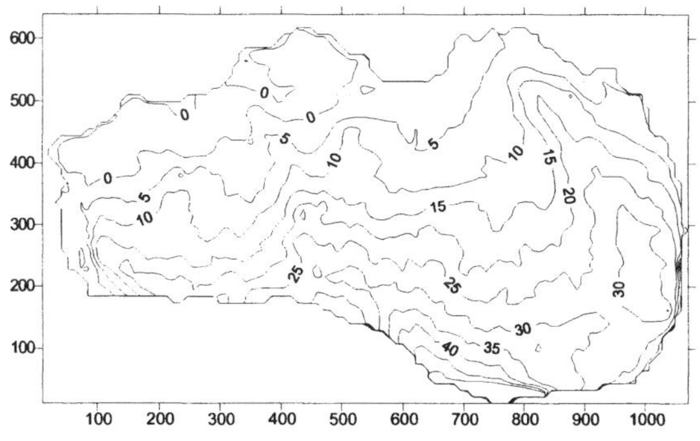

Fig. 4. $\bar{\mathbf{h}}_{\text{original}}$ for model layer 1.

Although MAE$^r \approx 0.5\%$ for the entire domain, throughout the model it can be more or less. In Fig. 6 we have plotted a time series for a specific grid cell that has an enlarged value MAE$^r \approx 6\%$. During most of the simulation time, $\hat{h}$ follows the behaviour of h accurately, but at several time-steps the difference is enlarged (e.g. $t = 350$; $\hat{h} - h \approx 0.15$ m). This temporary error is caused by the absence of some or all of the rejected patterns (φ_n ; $n \in \{21, ..., 28\}$). The particular rejected pattern affects this location locally but contributes negligible to the entire model. We could include them

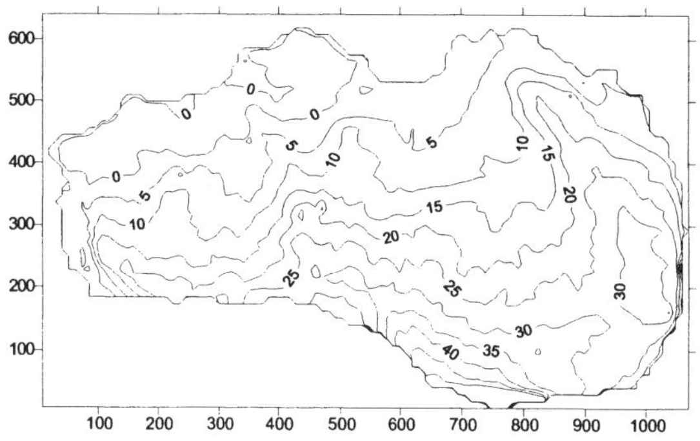

Fig. 5. $\bar{\mathbf{h}}_{\text{reduced}}$ for model layer 1.

to decrease the error. Most of the grid cells perform rather well and possess MAE$^r \ll 6\%$. In Fig. 7 we plotted a time series that possesses MAE$^r \approx 0.5\%$. Here, the reduced model simulates the behaviour of the model perfectly.

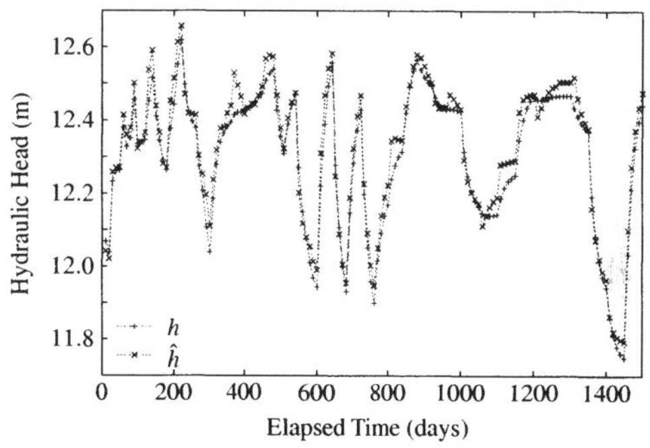

Fig. 6. Time series of h and $\hat{h}$ for a specific location in the model that has MAE$^r \approx 6\%$.

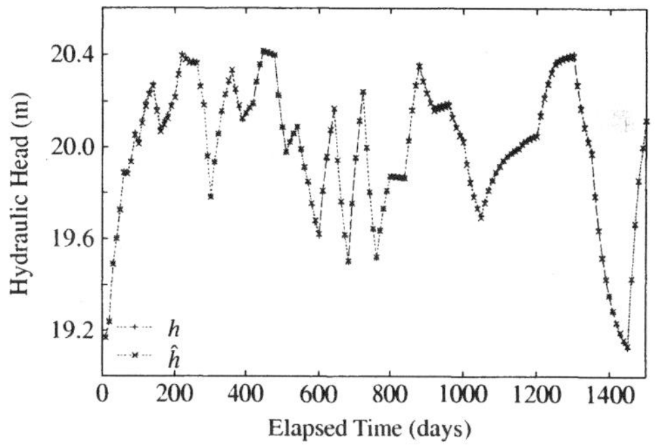

Fig. 7. Time series of h and $\hat{h}$ for a specific location in the model that has MAE$^r \approx 0.5\%$.

4.4 CPU Time Reduction

The key thought of a reduced model is to compute time-independent matrices in advance and to compute the others during the actual simulation.

4.4.1. Preparation Time We have plotted the preparation time t^p and it is, as expected, linearly related to n_p, see Fig. 8. For a configuration with $n_p = 20$ patterns, $t^p \approx 11$ sec. Combined with the time taken by the computation of the 42 snapshots (≈ 60 sec), it is still significantly less than the time taken by one simulation swith the original model ($\approx$ 225 sec). The t^{p} is remarkable reduced whenever we compare it to the t^{p} for a State-Space Projection as described by Vermeulen et al. (2002).

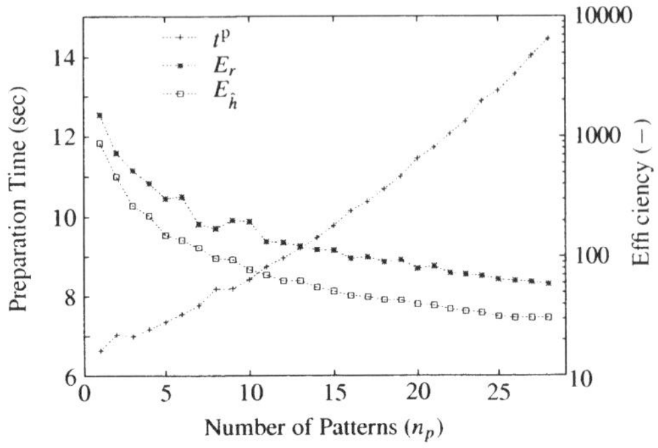

Fig. 8. Graph of n_p versus the preparation time t^{p} and the efficiency E_r and $E_{\hat{h}}$.

4.4.2. Simulation Efficiency We compute the reduction in computational time (E_r), as a fraction of the time taken by the PCG-solver of MODFLOW (t^s) to solve $\mathbf{h}^m$, and the time taken by the reduced model (t^s_{GP}) to solve $\mathbf{r}^m$. **With $\mathbf{n_p} = 20$ we obtained a maximal time efficiency of $\mathbf{E_r} = \mathbf{t^s}/\mathbf{t^s_{GP}} \approx= \mathbf{80}$**, see Fig. 8. Whenever we reconstruct $\hat{\mathbf{h}}$ for each time-step, the time efficiency is still considerable; $E_{\hat{h}} \approx 40$. In practice, $E_{\hat{h}}$ increases as we do not apply a reconstruction for each time-step and moreover, we do not reconstruct the entire vector $\hat{\mathbf{h}}$ (Eq. 3).

As the CPU time for MODFLOW tends to increase with a factor n_m^2, the CPU time for the reduced model increases linearly with n_p. Therefore the reduction model increases in efficiency with n_m.

5. CONCLUSIONS AND FUTURE DIRECTIONS

This paper described a method to develop a reduced model for a linear numerical groundwater flow model. We defined a reduced model structure that consisted of a linear combination of a set of spatial patterns $\mathbf{P}$ and time-varying coefficients $\mathbf{r}$. The patterns are called Empirical Orthogonal Functions (EOFs), and they span a subspace of model results that captures most of the relevant information of the original model. The time-varying coefficients were derived by substituting the hydraulic head $\mathbf{h}$ in the original PDE by the reduced model structure $\mathbf{P}^{\mathrm{T}}\mathbf{r}$ and project the outcome onto the EOFs. We created a reduced model with $n_p = 20$ paterns that explained $\varphi^{\mathrm{e}} = 99\%$ of the original model variance and it simulated a realistic case within an acceptable accuracy constraint (relative mean absolute error $< 0.5\%$). Specific for this case, we achieved a maximal time efficiency of $E_r \approx 80$. The reduced model has a promising prospect as its efficiency increases whenever the number of grid cells increases and the parameterization within the original model grows in complexity.

In future research we intend to focus on non-linear problems, such as unconfined aquifers and dewatering systems, such as drains. Experiences from other sciences showed that the reduced model is truly valuable in parameter estimation (Park and Cho, 1996) and in the computation of the Reduced Order Kalman Filters (Heemink et al., 2001). These topics have our interest also.

REFERENCES

Adomaitis RA. 1995. RTCVD Model Reduction: A Collocation on Empirical Eigenfunctions Approach. Technical Report T.R. 95-64. *Inst. Systems Research.*

Cazemier W, Verstappen RWCP, Veldman AEP. 1998. Proper orthogonal decomposition and low-dimensional models for driven cavity flows. *Physics of Fluids* **10**, 1685-1699.

Dunbar WS, Woodbury AD. 1989. Application of the Lanczos algorithm to the solution of the groundwater flow equation. *Water Resources Research* **25**(3), 551-558.

Golub G, Loan van A. 1989. *Matrix Computations.* John Hopkins University Press 2nd edition.

Heemink AW, Verlaan M, Segers AJ. 2001. Variance reduced ensemble Kalman filtering. *Mon. Weather Rev.* **129**(7), 1718-1728.

Hoffmann Jørgensen B, Sørensen JN. 2000. Proper Orthogonal Decomposition and low-dimensional modelling. *ERCOFTAC Bulletin* **46**, 44-51.

Hooimeijer MA. 2001. *Reduction of Complex Computational Models.* Printed by SIECA REPRO: Delft, The Netherlands.

Krysl P, Lall S, Marsden JE. 2000. Dimensional Model Reduction in Non-linear Finite Element Dynamics of Solids and Structures. *Int. J. Numer. Meth. Engng* **00**, 00-00.

McDonald MG, Harbaugh AW. 1988. *A Modular Three-Dimensional Finite-Difference Groundwater Flow Model.* U.S. Geological Survey, Open-File Report 83-875, Book 6, Chapter A1.

Newman AJ. 1996. Model reduction via the Karhunen Loève expansion Part I: An exposition. Technical Report T.R. 96-32. *Inst. Systems Research.*

Overschee Van P, Moor De B. 1996. *Subspace Identification for Linear Systems. Theory, Implementation, Applications.* Kluwer Academic Publishers.

Park HM, Cho CH. 1996. Low dimensional modeling of flow reactors. *Int. J. Heat and Mass Transfer* **39**(16), 3311-3323.

Sirovich L. 1987. Turbulence and the dynamics of coherent structures; Part I: Coherent Structures. *Quarterly Appl. Math.* **45**(3), 561-571.

Vermeulen PTM, Heemink AW, Stroet te CBM. 2002. Reduction of large-scale numerical groundwater flow models. Final Proc. XIV Int. Conf. on Computational Methods in Water Resources, Delft, the Netherlands **1**, 397-404.

IFAC Publications
www.elsevier.com/locate/ifac

MODEL REDUCTION FOR LARGE-SCALE LINEAR APPLICATIONS

Karen Willcox * **Alexandre Megretski** **

* *Department of Aeronautics and Astronautics, Massachusetts Institute of Technology, Cambridge, MA 02139*
** *Department of Electrical Engineering and Computer Science, Massachusetts Institute of Technology, Cambridge, MA 02139*

Abstract: Three model reduction methods are considered in the context of large scale fluid dynamic applications: the widely-used proper orthogonal decomposition, the Arnoldi method and a new Fourier method. The new method uses a Fourier expansion of the transfer function in discrete frequency to efficiently calculate reduced models with guaranteed stability and accuracy properties. Each method is described and then applied to the case of flow through a supersonic diffuser. The Fourier model reduction approach is found to be superior in all aspects; it is computationally more economical, preserves the stability of the original system, uses both input and output information to yield efficient models, and is valid over a wide range of frequencies. *Copyright © 2003 IFAC*

Keywords: Large-Scale, Model Reduction, Aerospace Systems, Computational Fluid Dynamics

1. INTRODUCTION

In the past decade, model reduction has become popular throughout the fluid dynamics community. While computational fluid dynamic (CFD) methods produce accurate models for problems of interest, their size and computational expense render them unsuitable for many applications. In particular, model reduction has been widely used for aeroelastic applications where the flow model must be coupled to a structural model (Dowell and Hall, 2001). While several different reduction techniques have been considered for CFD applications, including eigenmodes (Hall, 1994) and the Arnoldi method (Willcox *et al.*, 2002), the proper orthogonal decomposition (POD) is by far the most widely used in the fluid dynamics community (Holmes *et al.*, 1996).

First introduced in the context of turbulence (Lumley, 1967), the POD method of snapshots has been developed as a way to apply the technique to large-scale systems (Sirovich, 1987). This approach requires a set of flow solutions, or "snapshots" from a CFD simulation. These snapshots are then used to create a reduced-space basis, onto which the CFD governing equations are projected. While the POD basis is optimal in the sense that it minimizes the error between the snapshots and their projection in the reduced space, there are no guarantees as to the quality of the reduced-order model as an approximation of the original CFD system. In particular, one can not even guarantee that POD-based reduction of a stable CFD model will result in a stable reduced-order model.

Many effective, more rigorous reduction techniques have been developed in a controls context. The quality of a reduced-order system $\hat{G}$ as an approximation of the original system G is defined as the H-Infinity norm of the difference between their transfer functions:

$$\|\hat{G} - G\|_\infty = \sup_{\omega \in \mathbf{R}} |\hat{G}(j\omega) - G(j\omega)| \qquad (1)$$

No polynomial-time algorithm is known for determining the optimal reduced-order model, that is, one which minimizes the above norm. Algorithms such as Hankel model reduction (Adamjan *et al.*, 1971; Bettayeb *et al.*, 1980; Kung and Lin, 1981) and balanced truncation (Moore, 1981) have been widely used throughout the controls community to generate suboptimal reduced models with strong guarantees of quality. These algorithms can be performed in polynomial time; however, the computational requirements make them impractical for application to large systems such as those encountered in CFD applications. Several methods have been developed for computing approximations to the grammians for large systems, including the approximate subspace iteration (Baker *et al.*, 1996), least squares approximation (Scottedward Hodel, 1991) and Krylov subspace methods (Jaimoukha and Kasenally, 1994; Gudmundsson and Laub, 1994; Sorensen and Antoulas, 2002); however, these algorithms are complicated and computationally intensive.

In Willcox and Megretski (2003), a new technique is described for model reduction of large-scale systems. Fourier model reduction (FMR) uses an efficient iterative procedure to calculate Fourier coefficients of the transfer function in the discrete frequency domain. The resulting reduced-order models are guaranteed to be stable and the approximation error satisfies a known bound, which depends on the smoothness of the original transfer function.

In this paper, FMR is compared to the POD and Arnoldi methods for a CFD application of flow through a supersonic diffuser. In this example, reduced-order models are required in order to derive active control strategies. In the following section, each of the three algorithms, POD, Arnoldi, and FMR, are briefly outlined. Results are then presented for the chosen example and finally, conclusions are drawn.

2. MODEL REDUCTION FOR CFD

We consider the task of finding a low-order, state-space model

$$\hat{G}: \quad \frac{d}{dt}\hat{x}(t) = \hat{A}\hat{x}(t) + \hat{B}u(t), \qquad \hat{y}(t) = \hat{C}\hat{x}(t) + \hat{D}u(t) \tag{2}$$

which approximates well the given stable model

$$G: \quad E\frac{d}{dt}x(t) = Ax(t) + Bu(t), \qquad y(t) = Cx(t) + Du(t). \tag{3}$$

For the case of CFD applications, $x(t) \in \mathbf{R}^n$ is the state vector containing the n unknown perturbation flow quantities at each point in the computational grid, while $\hat{x}(t) \in \mathbf{R}^k$ is the kth-order reduced state vector. The vectors $u(t)$ and $y(t)$ contain the system inputs and outputs respectively. For simplicity, single-input, single-output systems will be considered here; however, all algorithms extend to the multiple-input, multiple-output case. The matrices E, A, B, C and D in (3) arise from the CFD formulation, and are evaluated at steady-state flow conditions. Typically, A and E are sparse matrices of very large dimension ($n > 10^4$). The descriptor matrix E is included for generality, and may contain some zero rows, which arise from implementation of flow boundary conditions. On solid walls, a condition is imposed on the flow velocity, while at farfield boundaries certain flow parameters are specified, depending on the nature of the boundary (inflow/outflow) and the local flow conditions (subsonic/supersonic). Although these prescribed quantities could be condensed out of (3) to obtain a smaller state-space system, such a manipulation is often complicated and can destroy the sparsity of the system. The more general form of the system is therefore considered.

2.1 *Proper Orthogonal Decomposition*

The POD has been widely used as a method of performing model reduction for large, CFD systems (Holmes *et al.*, 1996). The method of snapshots (Sirovich, 1987) is a way to construct a reduced-space basis using flow solutions or "snapshots". If x^i is the flow solution at a time t_i, then the POD basis vectors can be constructed as follows.

The correlation matrix R is first formed by computing the inner product between every pair of snapshots

$$R_{ik} = \frac{1}{m}\left(x^i, x^k\right), \tag{4}$$

where m is the number of snapshots and $\left(x^i, x^k\right)$ denotes the inner product between x^i and x^k. The eigenvalues λ_i and eigenvectors ψ^i of R are then computed. The jth POD basis vector, Φ_j, is given by a linear combination of snapshots

$$\Phi_j = \sum_{i=1}^{m} \psi_i^j x^i, \tag{5}$$

where ψ_i^j denotes the ith element of the jth eigenvector. The magnitude of the jth eigenvalue, λ_j, describes the relative importance of the jth POD basis vector.

Once the orthonormal set of POD basis vectors has been computed, the reduced-order model is obtained by projecting the CFD solution onto the reduced-space basis:

$$x(t) = \sum_{i=1}^{m} \hat{x}_i(t)\Phi_i. \tag{6}$$

Substituting this expression into the original system (3) and using orthogonality, we obtain the reduced-order system (2).

Often, the POD snapshots are obtained from a simulation of the CFD model. One issue with this approach is an appropriate choice of input to the simulation. This input choice is critical, since the resulting basis will capture only those dynamics present in the snapshot ensemble. This can be a problematic issue for many applications, such as flow control design, where the dynamics of the controlled and uncontrolled systems might differ significantly. An alternative approach is to apply the POD in the frequency domain (Kim, 1998). Rather than selecting a time-dependent input function, one selects a set of sample frequencies. The corresponding flow solutions can then be obtained by solving the frequency domain CFD equations

$$X(\omega) = [j\omega_i E - A]^{-1} B \tag{7}$$

where $u(t) = e^{j\omega_i t}$, $x(t) = Xe^{j\omega_i t}$, and ω_i is the ith sample frequency.

Frequency domain POD approaches typically yield better results; however, the computational cost of the method is high. The nth-order system given by (7) must be solved for each frequency selected. In a typical CFD application, a large number of frequency points are required to obtain satisfactory models. In particular, for three-dimensional applications, the cost of this approach is often prohibitive.

2.2 *Arnoldi Method*

The Arnoldi method is one of a set of moment-matching model reduction techniques. Consider a Taylor series expansion of the transfer function

$$G(s) = C\left[sE - A\right]^{-1} B + D \tag{8}$$

about the point $s = s_0$. We can write

$$G(s) = \sum_{j=0}^{\infty} m_j (s - s_0)^j, \tag{9}$$

where

$$m_0 = D + C\tilde{A}^{-1}B, \tag{10}$$
$$m_j = C\left(-\tilde{A}^{-1}E\right)^j \tilde{A}^{-1}B \; (j = 1, 2, \ldots) \tag{11}$$

and $\tilde{A} = (s_0 E - A)$. The coefficient m_j is known as the jth moment of $G(s)$ about $s = s_0$.

The Arnoldi method uses an efficient iterative process to generate a set of vectors that spans the kth-order Krylov subspace defined by

$$\mathcal{K}_k(\tilde{A}^{-1}E, \tilde{A}^{-1}B) = span\{\tilde{A}^{-1}B, (\tilde{A}^{-1}E)\tilde{A}^{-1}B, \ldots, (\tilde{A}^{-1}E)^{k-1}\tilde{A}^{-1}B\}, \tag{12}$$

The reduced-order model is obtained by projecting onto the kth-order Krylov subspace and matches k moments of the system transfer function about s_0. The flow solution $x(t)$ is represented by an expansion of the form (6), where Φ_i now represents the ith Arnoldi vector.

2.3 *Fourier Model Reduction*

A new method of model reduction for large-scale applications is described in Willcox and Megretski (2003). A transfer function expansion concept similar to that described for the Arnoldi method is used; however, the expansion is now performed in the discrete frequency domain. Using the identity

$$G(s) = g(z) = d + c(zI - a)^{-1}b \tag{13}$$
$$\text{for} \quad z = \frac{s + \omega_0}{s - \omega_0},$$

where

$$d = D + C(\omega_0 E - A)^{-1}B, \tag{14}$$
$$a = -(\omega_0 E + A)(\omega_0 E - A)^{-1}, \tag{15}$$
$$c = 2\omega_0 C(\omega_0 E - A)^{-1}, \tag{16}$$
$$b = -E(\omega_0 E - A)^{-1}B, \tag{17}$$

and ω_0 is some fixed positive real number, the transfer function $G(s)$ has the Fourier decomposition

$$G(s) = \sum_{j=0}^{\infty} G_j \left(\frac{s - \omega_0}{s + \omega_0}\right)^j, \tag{18}$$

where

$$G_0 = d, \quad G_j = ca^{j-1}b \; (j = 1, 2, \ldots). \tag{19}$$

The Fourier expansion converges exponentially for $|z| > \rho(a)$, where $\rho(a)$ denotes the spectral radius of a, defined as the maximal absolute value of its eigenvalues. The first m Fourier coefficients are easy to calculate using the efficient iterative process

$$G_j = ch_{j-1}, \; h_j = ah_{j-1} \; (j = 1, \ldots, m), \tag{20}$$
$$\text{where } h_0 = b,$$

which is expected to be "stable" since g is stable, i.e. $\rho(a) < 1$.

The calculated $m + 1$ Fourier coefficients are used to construct an mth-order discrete time reduced model, which can then be converted to an mth-order state-space system. An effective approach is to use the above iterative procedure to form the intermediate discrete time system

$$\hat{g}: \quad \hat{x}[t+1] = \hat{a}\hat{x}[t] + \hat{b}u[t],$$
$$\hat{y}[t] = \hat{c}\hat{x}[t] + \hat{d}u[t], \tag{21}$$

where

$$\hat{a} = \begin{bmatrix} 0 & 0 & 0 & 0 & \cdots \\ 1 & 0 & 0 & 0 & \cdots \\ 0 & 1 & 0 & 0 & \cdots \\ 0 & 0 & 1 & 0 & \cdots \\ 0 & 0 & 0 & \ddots & \end{bmatrix} \quad \hat{b} = \begin{bmatrix} 1 \\ 0 \\ 0 \\ 0 \\ \vdots \end{bmatrix}$$

$$\hat{c} = [G_1 \quad G_2 \quad \ldots \quad G_m] \qquad \hat{d} = G_0 \tag{22}$$

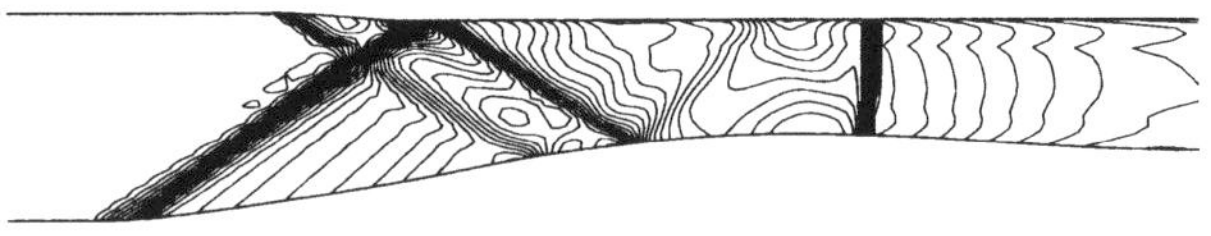

Fig. 1. Mach contours for steady flow through supersonic diffuser. Steady-state inflow Mach number is 2.2.

using several hundred states. This intermediate model can then be further reduced via balanced truncation. This second reduction step can be applied efficiently, since the Hankel matrix of the system (21) is known to be

$$\Gamma = \begin{bmatrix} G_1 & G_2 & G_3 & \dots & G_{m-1} & G_m \\ G_2 & G_3 & G_4 & \dots & G_m & 0 \\ G_3 & G_4 & G_5 & \dots & 0 & 0 \\ \vdots & \vdots & \vdots & & \vdots & \vdots \\ G_m & 0 & 0 & \dots & 0 & 0 \end{bmatrix}. \quad (23)$$

3. RESULTS

3.1 *Supersonic Diffuser*

Results will be presented for a CFD application of active control of a supersonic diffuser as shown in Figure 1. The diffuser operates at a nominal Mach number of 2.2; however, it is subject to perturbations in the incoming flow, which may be due (for example) to atmospheric variations. In nominal operation, there is a strong shock downstream of the diffuser throat, as can be seen from the Mach contours plotted in Figure 1. Incoming disturbances can cause the shock to move forward towards the throat. When the shock sits at the throat, the inlet is unstable, since any disturbance that moves the shock slightly upstream will cause it to move forward rapidly, leading to unstart of the inlet. This is extremely undesirable, since unstart results in a large loss of thrust. In order to prevent unstart from occurring, one option is to actively control the position of the shock. This control may be effected through flow bleeding upstream of the diffuser throat. In order to derive effective active control strategies, it is imperative to have low-order models which accurately capture the relevant dynamics.

The CFD formulation for this problem is described fully in Lassaux (2002). The governing equations considered are the two-dimensional Euler equations, which are linearized for unsteady flows. The CFD model considered here has 3078 grid points and 11,730 unknowns.

We consider the transfer function between bleed actuation and average Mach number at the throat. Bleed occurs through small slots located on the lower wall between 46% and 49% of the inlet overall length. Frequencies of interest lie in the range $f/f_0 = 0$ to $f/f_0 = 2$, where $f_0 = a_0/h$, a_0 is the freestream speed of sound and h is the height of the diffuser; a wider range will be plotted to gain further insight to the performance of the models.

Figure 2 shows the magnitude and phase of this transfer function as calculated by the CFD model and three reduced-order models each of size $k = 10$. The FMR model was calculated by using 201 Fourier coefficients (calculated at the cost of a single CFD matrix inversion) with $\omega_0 = 5$ to construct the Hankel matrix in (23). This 200th-order system was then further reduced to ten states using explicit balanced truncation. One might consider a similar approach using the Arnoldi method: first calculate 200 Arnoldi vectors (at the cost of a single CFD matrix inversion), and then further reduce the resulting system with balanced truncation. Although this two-step Arnoldi approach might be effective for some applications, for this case it could not be applied, since the model resulting from projection onto the 200th-order Arnoldi basis was unstable. The Arnoldi model was therefore constructed directly by projection onto the reduced-space basis spanned by the first ten Arnoldi vectors computed about $s_0 = 0$. Finally, the POD model was obtained by computing 41 snapshots at 21 equally-spaced frequencies from $f/f_0 = 0$ to $f/f_0 = 2$. This required the inversion of one real and 20 complex n^{th}-order matrices.

It can be seen from Figure 2 that the FMR model matches the CFD results well over the entire frequency range plotted, with a small discrepancy at higher frequencies. The Arnoldi model matches well for low frequencies, but shows considerable error for $f/f_0 > 1.3$. The POD model has some undesirable oscillations at low frequencies, and strictly is only valid over the frequency range sampled in the snapshot ensemble ($f/f_0 < 2$).

The performance of the POD and Arnoldi models can be improved by increasing the size of the reduced-order models. Figure 3 shows the results using 30 Arnoldi vectors and 15 POD basis vectors. The agreement at low frequencies is now very good for all models, but the POD and Arnoldi models still show discrepancy at higher frequencies. The POD model could be further improved by including more snapshots in the ensemble; however, each additional frequency considered requires an nth-order complex matrix inversion. The Arnoldi model could be further improved by increasing the size of the basis; however, this was found to result in unstable reduced-order models.

Not only is the accuracy of the FMR model in Figures 2 and 3 better than the other models, but also the tech-

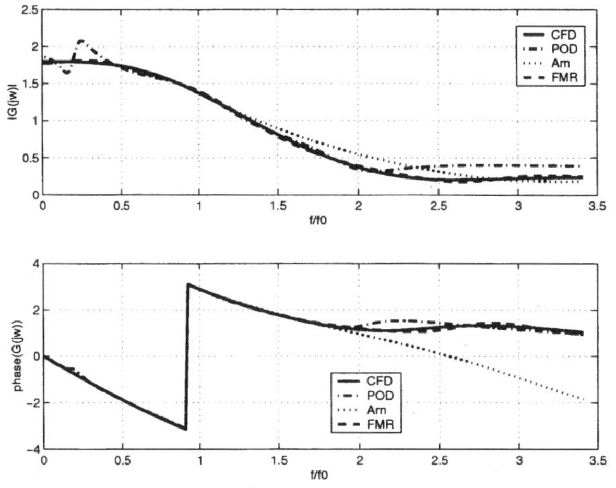

Fig. 2. Transfer function from bleed actuation to average throat Mach number for supersonic diffuser. Results from CFD model ($n = 11,730$) are compared to FMR, POD and Arnoldi models with $k = 10$ states.

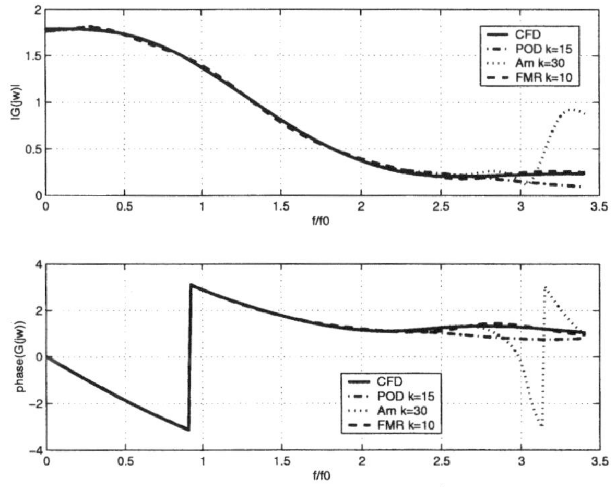

Fig. 3. Transfer function from bleed actuation to average throat Mach number for supersonic diffuser. Results from CFD model ($n = 11,730$) are compared to FMR with $k = 10$ states, POD with $k = 15$ states, and Arnoldi with $k = 30$ states.

nique compares favorably to the alternative methods in other aspects. One significant advantage of FMR is that, if the original CFD system is stable, the reduced-order models are also guaranteed to be stable. This is not the case for the Arnoldi and POD methods, and in practice, these techniques often result in unstable reduced models. For the supersonic diffuser considered here, this was observed for the Arnoldi model if too many basis vectors were used.

Secondly, the computational cost of deriving the models is of the same order for both the FMR and Arnoldi methods, that is, on the order of one n^{th}-order matrix inversion. The FMR requires some extra computation to transform from discrete to continuous time, plus the extra step of balanced truncation on the intermediate reduced model; however, the number of matrix solves typically dominates the reduction cost for large-scale applications. The computational cost of the POD method is much higher than for the other methods.

Finally, the FMR model yields more accurate results over a large frequency range, with fewer states. This is partly due to the fact that FMR considers both inputs and outputs in the reduction process, while Arnoldi and POD consider only system inputs and therefore yield inefficient models. Moreover, POD models are restricted to the frequency range contained within the snapshot ensemble. Arnoldi models are restricted to a frequency range close to that chosen for the Taylor series expansion, in this case $s_0 = 0$. The frequency range of the FMR model is controlled by the choice of the parameter ω_0 (see Willcox and Megretski (2003) for details). For this case, choosing $\omega_0 = 5$ enabled a large frequency range to be approximated accurately.

4. CONCLUSIONS

Three reduction methods have been compared for a large CFD system: the POD, Arnoldi method and Fourier model reduction. The most effective approach to reduction is found to be use of FMR to derive an intermediate Hankel matrix of order several hundred, followed by balanced truncation to obtain the final reduced-order model. While POD is the most commonly used method for reduction of CFD systems, it has several drawbacks. Firstly, the POD method is expensive, since a complex CFD system solve is required for each frequency considered; moreover, a large number of frequencies must be included if the reduced model is to yield accurate results. Secondly, the POD method only accounts for inputs when performing the reduction, and thus the resulting reduced models are often inefficient. Finally, no guarantees are available for the quality of the reduced-order model; in particular, the models may be unstable.

The Arnoldi method is better than the POD in terms of computational expense; however, this approach also does not consider outputs in the reduction process and is not guaranteed to produce a stable reduced-order model. This issue of stability was found to be a problem in the supersonic diffuser flow considered in this paper. The new FMR procedure addresses all these issues and provides an efficient, reliable method for model reduction of large scale linear systems.

5. REFERENCES

Adamjan, V.M., D.Z. Arov and M.G. Krein (1971). Analytic Properties of Schmidt Pairs for a Hankel Operator and the Generalized Schur-Takagi Problem. *Math. USSR Sbornik* **15**, 31–73.

Baker, M.L., D.L. Mingori and P.J. Goggins (1996). Approximate Subspace Iteration for Constructing Internally Balanced Reduced Order Models of Unsteady Aerodynamic Systems. AIAA Paper 96-1441.

Bettayeb, M., L.M. Silverman and M.G. Safonov (1980). Optimal Approximation of Continuous-Time Systems. *Proceedings of the 19th IEEE Conference on Decision and Control.*

Dowell, E.H. and K.C. Hall (2001). Modeling of fluid-structure interaction. *Annual Review of Fluid Mechanics* **33**, 445–90.

Gudmundsson, T. and A.J. Laub (1994). Approximate Solution of Large Sparse Lyapunov Equations. *IEEE Transactions on Automatic Control* **39**(5), 1110–1114.

Hall, K.C. (1994). Eigenanalysis of Unsteady Flows About Airfoils, Cascades and Wings. *AIAA Journal* **32**(12), 2426–2432.

Holmes, P., J.L. Lumley and G. Berkooz (1996). *Turbulence, Coherent Structures, Dynamical Systems and Symmetry*. Cambridge University Press. Cambridge, UK.

Jaimoukha, I.M. and E.M. Kasenally (1994). Krylov Subspace Methods for Solving Large Lyapunov Equations. *SIAM Journal of Numerical Analysis* **31**(1), 227–251.

Kim, T. (1998). Frequency-Domain Karhunen-Loeve Method and Its Application to Linear Dynamic Systems. *AIAA Journal* **36**(11), 2117–2123.

Kung, S-Y. and D.W. Lin (1981). Optimal Hankel-Norm Model Reductions: Multivariable Systems. *IEEE Transactions on Automatic Control* **AC-26**(1), 832–52.

Lassaux, G. (2002). High-Fidelity Reduced-Order Aerodynamic Models: Application to Active Control of Engine Inlets.. Master's thesis. Dept. of Aeronautics and Astronautics, MIT.

Lumley, J.L. (1967). The Structures of Inhomogeneous Turbulent Flow. *Atmospheric Turbulence and Radio Wave Propagation* pp. 166–178.

Moore, B.C. (1981). Principal Component Analysis in Linear Systems: Controllability, Observability, and Model Reduction. *IEEE Transactions on Automatic Control* **AC-26**(1), 17–31.

Scottedward Hodel, A. (1991). Least Squares Approximate Solution of the Lyapunov Equation. Proceedings of the 30th IEEE Conference on Decision and Control.

Sirovich, L. (1987). Turbulence and the Dynamics of Coherent Structures. Part 1 : Coherent Structures. *Quarterly of Applied Mathematics* **45**(3), 561–571.

Sorensen, D.C. and A.C. Antoulas (2002). The sylvester equation and approximate balanced reduction. Linear Algebra and Its Applications, Fourth Special Issue on Linear Systems and Control, edited by Blondel et al., 351-352, pp. 671-700.

Willcox, K. and A. Megretski (2003). Fourier Model Reduction for Large-Scale Applications in Computational Fluid Dynamics. To be presented at SIAM Conference on Applied Linear Algebra, Williamsburg, VA.

Willcox, K.E., J. Peraire and J. White (2002). An Arnoldi approach for generation of reduced-order models for turbomachinery. *Computers and Fluids* **31**(3), 369–89.

IFAC Publications
www.elsevier.com/locate/ifac

REDUCED ORDER MODELING OF AN INDUSTRIAL FEEDER MODEL

P. Astrid and S. Weiland *,[1] **A. Twerda** **

* *Department of Electrical Engineering, Eindhoven University of Technology, P.O. Box 513, 5600 MB Eindhoven, The Netherlands.*
Email: p.astrid, s.weiland@tue.nl
** *TNO Institute of Applied Physics*
P.O.Box 155, 2600 AD Delft, The Netherlands
E-mail: twerda@tpd.tno.nl

Abstract: Models of glass furnaces are described by a set of nonlinear partial differential equations which govern the mass, momentum and energy balances and a number of non-linear functions of the independent scalars which describe the dependent variables like viscosities and densities in a fluid. In this paper we consider the problem to simulate such a glass furnace and to substantially reduce the complexity of a glass feeder model while keeping its main dynamical properties. A Computational Fluid Dynamics (CFD) method is applied to simulate the dynamics of the physical and chemical processes that take place in the furnace. As a result of a fine discretization applied to the CFD, the resulting model is too large to be incorporated in control or optimization modules. In this paper, we present a reduced order modelling technique by proper orthogonal decomposition applied to an industrial glass feeder model. It is shown that with less than 0.1% of the states of the original model, the temperature deviations between the original and the reduced order model is negligible.

Keywords: CFD models, reduced order modeling, proper orthogonal decomposition

1. THE FEEDER MODEL

A glass feeder is part of a glass furnace which is located between the refiner and the glass exit point. The feeder is fed by an inflow glass (rate) while glass is pulled out of the feeder (with a certain pull rate) in its end-part. In the first part of the feeder, the glass is cooled very slowly and under very strict conditions so as to maintain small temperature gradients across the height and the width of the feeder. In the end-part of the feeder, the glass is not cooled anymore, but the temperature differences of the glass are kept to a minimum. For the control of the quality of the glass it is extremely important to precisely regulate the temperature within the feeder(R.Beerkens *et al.*, 1997).

[1] This work is partially supported by EET-REGLA project

Above the glass melt, the crown temperature is set to control the heat transfer process within the feeder. Sometimes, a stirrer is also placed to manipulate the heat transfer by convection. Heat transfer by conduction is also taking place by the feeder walls. Disturbances in the feeder may occur from the variation of pull rates, or the rate of inflow glass material.

Figure 1 shows a sketch of the glass feeder with the grid divisions of the numerical model. This is a simplified sketch of a feeder from the glass manufacturer REXAM in Dongen, the Netherlands. It has dimensions of $8.5\mathtt{m} \times 0.55\mathtt{m} \times 2\mathtt{m}$ in length, height, and width, respectively. This defines a Cartesian spatial volume $\mathbb{X} = [0, 8.5] \times [0, 0.55] \times [0, 2]$ whose coordinates are denoted by x (length), y (height) and z (width). The glass melt is placed at a height $y = 0.34\mathtt{m}$. The nominal pull rate for this feeder is 50 tons per day. The nominal crown temperature for this feeder is $1480\mathtt{K}$.

The nominal inlet temperature of the glass material is $1448.15\mathtt{K}$. In general, the glass flow in the feeder can be

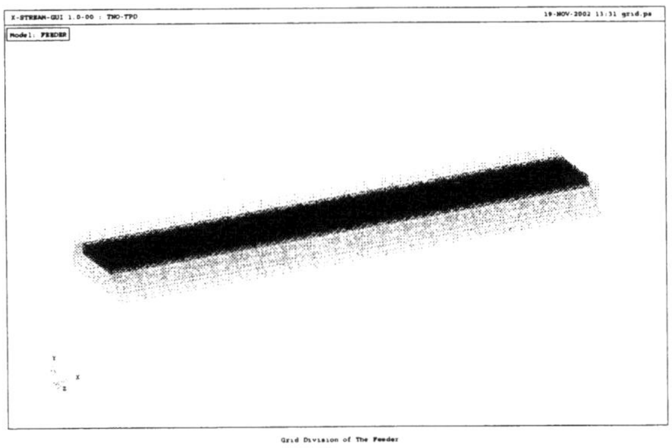

Fig. 1. Sketch of the Glass Feeder and the Grid Division, red blocks correspond to glass and yellow blocks to feeder walls

considered as an incompressible and laminar flow. The governing equations for the feeder are Navier-Stokes equations that describe the velocity field (u, v, w) in the x, y and z direction respectively as well as the pressure field p and the energy equations for the temperature field T (B.Bird *et al.*, 1960). The Navier-Stokes are solved for the glass media only, while the energy equations are solved for both heat transfer in the glass media and through the feeder walls. In this case, the physical parameters such as densities of both glass and walls are kept constant. The viscosity of glass μ is $50\mathtt{Ns/m^2}$, density ρ at $2400\mathtt{kg/m^3}$, thermal conductivity $80\mathtt{W/m.K}$, and specific heat c_p at $1160\mathtt{J/kg.K}$. For the walls, the density ρ is at $2000\mathtt{kg/m^3}$, thermal conductivity λ is $3.0\mathtt{W/m.K}$ and specific heat c_p is $1250\mathtt{J/kg.K}$.

The simulation is run to get the step response from initial conditions of steady velocity, pressure, and temperature fields at the nominal conditions (denoted by subscript $\mathtt{ss}$). The inflow rate is decreased to 10% and the crown temperature is increased to $20\mathtt{K}$ to $1500.15K$. The change of inflow rate and crown temperature modifies the Dirichlet boundary conditions velocity at x-direction u at $x = 0$ to $0.0032\mathtt{m/s}$ and the Dirichlet boundary conditions of T at $y = 0.55$ to $1500.15\mathtt{K}$.

Continuity Equation of glass

$$\frac{\partial \rho}{\partial t} + \frac{\partial \rho u}{\partial x} + \frac{\partial \rho v}{\partial v} + \frac{\partial \rho w}{\partial z} = 0 \qquad (1)$$

Momentum in x-direction of glass

$$\frac{\partial \rho u}{\partial t} + \frac{\partial \rho uu}{\partial x} + \frac{\partial \rho uv}{\partial y} + \frac{\partial \rho uw}{\partial z} = \frac{\partial}{\partial x}\left(2\mu\frac{\partial u}{\partial x}\right) +$$
$$+ \frac{\partial}{\partial y}\left(\mu\frac{\partial u}{\partial y}\right) + \frac{\partial}{\partial z}\left(\mu\frac{\partial u}{\partial z} + \mu\frac{\partial w}{\partial x}\right) - \frac{\partial p}{\partial x}$$
$$u(x,y,z)|_{t=0,x,y,z\neq\mathtt{boundaries}} = u_{ss}$$
$$u(x,y,z)|_{x=0} = 0.0032, \quad \frac{\partial u}{\partial x}|_{x=8.5} = 0$$
$$u(x,y,z)|_{y=0} = 0, \quad \frac{\partial u}{\partial n}|_{y=0.55} = 0$$
$$\frac{\partial u}{\partial n}|_{z=0.4,z=1.6} = 0 \qquad (2)$$

Momentum in y-direction of glass

$$\frac{\partial \rho v}{\partial t} + \frac{\partial \rho uv}{\partial x} + \frac{\partial \rho vv}{\partial y} + \frac{\partial \rho vw}{\partial z} = \frac{\partial}{\partial x}\left(\mu\frac{\partial u}{\partial y} + \mu\frac{\partial v}{\partial x}\right)$$
$$+ \left(2\mu\frac{\partial v}{\partial y}\right) + \frac{\partial}{\partial z}\left(\mu\frac{\partial v}{\partial z} + \mu\frac{\partial w}{\partial y}\right) - \frac{\partial p}{\partial y}$$
$$v(x,y,z)|_{t=0,x,y,z\neq\mathtt{boundaries}} = v_{ss}$$
$$v(x,y,z)|_{x=0} = 0, \quad \frac{\partial v}{\partial n}|_{x=8.5} = 0$$
$$v(x,y,z)|_{y=0.34} = 0, \quad v(x,y,z)|_{y=0.55} = 0$$
$$\frac{\partial v}{\partial n}|_{z=0.4,z=1.6} = 0 \qquad (3)$$

Momentum in z-direction

$$\frac{\partial \rho w}{\partial t} + \frac{\partial \rho uw}{\partial x} + \frac{\partial \rho vw}{\partial y} + \frac{\partial \rho ww}{\partial z} = \frac{\partial}{\partial x}\left(\mu\frac{\partial u}{\partial z} + \mu\frac{\partial w}{\partial x}\right)$$
$$+ \frac{\partial}{\partial y}\left(\mu\frac{\partial v}{\partial z} + \mu\frac{\partial w}{\partial y}\right) + \frac{\partial}{\partial z}\left(2\mu\frac{\partial w}{\partial z}\right) - \frac{\partial p}{\partial z}$$
$$w(x,y,z)|_{t=0,x,y,z\neq\mathtt{boundaries}} = w_{ss}$$
$$w(x,y,z)|_{x=0} = 0, \quad \frac{\partial w}{\partial n}|_{x=8.5} = 0$$
$$w(x,y,z)|_{y=0.34} = 0, \quad \frac{\partial w}{\partial n}|_{y=0.55} = 0$$
$$w(x,y,z)|_{z=0.4} = 0, \quad w(x,y,z)|_{z=1.6} = 0 \qquad (4)$$

Energy Equation, solved for walls and glass

$$\frac{\partial \rho c_p T}{\partial t} + \frac{\partial(\rho c_p u T)}{\partial x} + \frac{\partial(\rho c_p v T)}{\partial y} + \frac{\partial(\rho c_p w T)}{\partial z}$$
$$= \frac{\partial}{\partial x}\left(\lambda\frac{\partial T}{\partial x}\right) + \frac{\partial}{\partial y}\left(\lambda\frac{\partial T}{\partial y}\right) + \frac{\partial}{\partial z}\left(\lambda\frac{\partial T}{\partial z}\right) +$$
$$+ S(x,y,z)\, T(x,y,z)|_{t=0,x,y,z\neq\mathtt{boundaries}} = T_{ss}$$
$$T(x,y,z)|_{x=0} = 1448.15, \quad \frac{\partial T}{\partial n}|_{x=8.5} = 0$$
$$\frac{\partial T}{\partial n}|_{y=0} = 0, T(x,y,z)|_{y=0.55} = 1500.15$$
$$\frac{\partial T}{\partial n}|_{z=0} = 0, \frac{\partial T}{\partial n}|_{z=2} = 0 \qquad (5)$$

To solve the equations numerically over the spatial domain $\mathbb{X}$ and a finite time domain $\mathbb{T}$, the feeder is initially divided into $20 \times 15 \times 16$ grid cells in the spatial domain $\mathbb{X}$. The simulations and the model reduction are performed using the GTM-X software (TNO-TPD, 2000). Pressure equations don't exist explicitly, instead one has to solve the momentum equations iteratively to obtain the correct pressure field which fulfils the continuity constraint. To enhance the quality of the simulation also some grid points are placed in the wall of the feeder. Therefore, the total number of grid cells used in the simulation becomes 7128. The number of equations to be solved for all velocities, pressure, and temperature field are therefore $(3+1+1) \times 7128 = 35640$ equations.

2. MODEL REDUCTION STRATEGY

The model is described by a number of partial differential equations in real valued variable T (temperature, say,) over a (gridded) domain $\mathbb{X} \times \mathbb{T}$ of spatial and temporal coordinates.

Suppose that the differential equation has the general form

$$R(T(x,t)) = \frac{\partial T}{\partial t}(x,t) - D(T(x,t)) = 0 \quad (6)$$

with $D(\cdot)$ an operator that may involve spatial derivatives and other functions (P.Astrid *et al.*, 2002), $x \in \mathbb{X}$ and $t \in \mathbb{T}$.

Suppose that $\langle \cdot, \cdot \rangle$ defines an inner product on the space $\mathbb{X}$. We postulate that T can be expanded as a Fourier series (or spectral form)

$$T(x,t) = \sum_{i \in \mathbb{I}} a_i(t)\varphi_i(x) \qquad x \in \mathbb{X}, \quad t \in \mathbb{T}$$

where $\varphi_i(\cdot)$ is a set of orthonormal basis functions in $\mathbb{X}$, with i ranging over a countable index set $\mathbb{I} \subseteq \mathbb{N}$. An nth order approximation of (7) is then given by the truncated sequence

$$T_n(x,t) = \sum_{i=1}^{n} a_i(t)\varphi_i(x) \qquad x \in \mathbb{X}, \quad t \in \mathbb{T}. \quad (7)$$

The method of Proper Orthogonal Decompositions requires that the Galerkin projection of the residual $R(T_n(x,t))$ on the space spanned by the basis functions $\varphi_i(\cdot)$, $i = 1, \ldots, n$ vanishes, cf. (G.Karniadakis and S.Sherwin, 1999) (M.Kirby, 2001).

That is,

$$\langle R(T(x,t)), \varphi_i(x) \rangle = 0 \qquad i = 1, \ldots, n \quad (8)$$

for all $t \in \mathbb{T}$.

2.1 *Determination of time varying coefficients $a_i(t)$*

For every choice of orthonormal basis functions, the requirement (8) defines a finite number of constraints on the time varying coefficients $a_i(t)$, $i \in \mathbb{I}$. For linear operators R in (6), the condition (8) is equivalent to setting

$$\dot{a}_i(t) = \langle D\left(\sum_{i=1}^{n} a_i(t)\varphi_i(x)\right), \varphi_i(x) \rangle \quad (9)$$

That is, the time-varying coefficients $a_i(t)$ associated with the approximation (7) can be found by solving a system of n ordinary differential equations. The initial conditions of (9) is determined from:

$$a_i(0) = \langle T(x,0), \varphi_i(x) \rangle \quad (10)$$

2.2 *Determination of basis function $\varphi_i(x)$*

The basis functions $\varphi_i(x)$ with $i \in \mathbb{I}$ are the *modes* of the spatial dynamics of $T(x,t)$. A characteristic feature of the reduction method that we employ here is that the basis functions $\varphi_i(x)$, with $i \in \mathbb{I}$, are obtained from *data*. This data may be real data derived from the feeder, or simulated data obtained from a simulation on the full scale model. To determine $\varphi_i(x)$, we consider the data

$$\tilde{T}(x,t) \qquad x \in \mathbb{X}, t \in \mathbb{T}$$

where $\mathbb{X}$ is a gridded space of dimension K and $\mathbb{T}$ is also gridded and has dimension L. The data is stored in a matrix T_{snap} which is defined as $T_{\text{snap}} = (T_{\text{snap}})_{ij}$ with $i = 1, \ldots, K$ and $j = 1, \ldots, L$. That is,

$$T_{snap} = \left(\text{col}_i \tilde{T}(x_i, t_1) \ \ldots \ \text{col}_i \tilde{T}(x_i, t_L)\right)$$

We then calculate a singular value decomposition of T_{snap} as

$$T_{\text{snap}} = \Phi \Sigma \Psi^{\top} \qquad \Phi \in \mathbb{R}^{K \times K}, \quad \Psi \in \mathbb{R}^{L \times L} \text{ unitary}$$

where Σ is a $K \times L$ matrix which contains the singular values of T_{snap} in *nonincreasing order* on its main diagonal. Let

$$\Phi = (\varphi_1, \ldots, \varphi_K)$$

denote the column-partitioning of Φ and set, for all $i \in \mathbb{I}$, the basis functions

$$\varphi_i(x) = \langle \phi_i, x \rangle$$

where $x \in \mathbb{X}$. Note that the calculation of the first n ($n < K$) basis functions requires only the first n left singular vectors of T_{snap}, which will be a more efficient calculation for high dimensional spatial domains.

In the reduced model, only n dominant basis vectors $\varphi_i(\cdot)$ are taken which corresponds to the n largest singular value in Σ. With $n < K$ specified, the reduced model is described by the n ordinary differential equations (9) with $i = 1, \ldots n$, and with initial condition (10). Thus, starting from a model governed by PDE's, (typical CFD models), the application of a Proper Orthogonal Decomposition allows a reduction of the original model by collecting the most dominant spatial information in n basis vectors so as to obtain a reduced order model described by n ordinary differential equations.

3. RESULTS

For every variable in the feeder model, i.e. the velocities u, v and w in the x, y, z direction, the pressure field, and the temperature field, this reduction strategy has been applied. Snapshots are collected from the transient simulations of the feeder models when the crown temperature is increased 20°C and the inflow rate is decreased by 10%.

After application of singular value decomposition, n eigenvectors correspond to n largest eigenvalues are retained. These eigenvectors are used as the basis vectors in the reduced order calculation.

Figure 2 and Figure 3 show the results of the reduced temperature field in comparison with the original temperature field at $y = 0.42\text{m}$. The original temperature field is solved by $K = 7128$ equations and the reduced temperature field is obtained by applying the POD technique and taking $n = 6$ dominant modes which correspond to 6 largest eigenvalues of the temperature snapshots. It can be observed that the reduced order model approximates the temperature field of the original model closely.

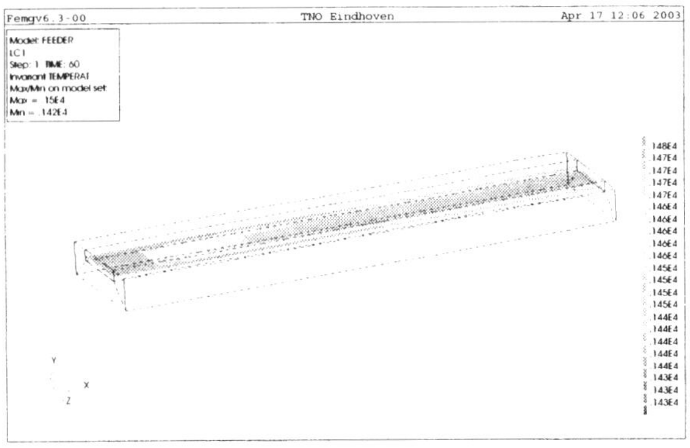

Fig. 2. Original Temperature Field at $y = 0.42\text{m}$

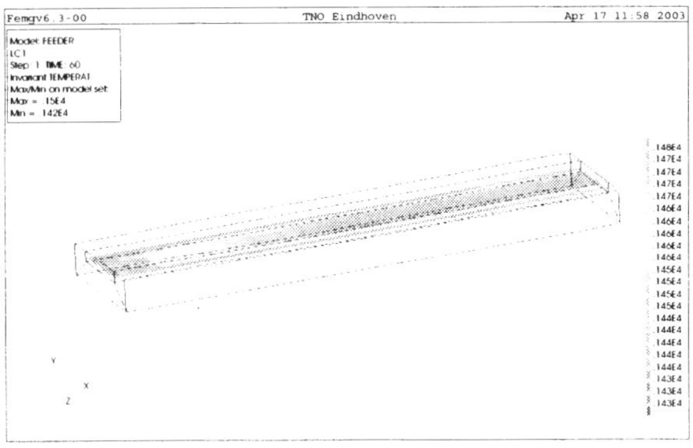

Fig. 3. Reduced Temperature Field at $y = 0.42\text{m}$

The original and reduced temperature distributions of the feeder walls are also shown in Figure 4 and Figure 5. Temperature distribution of the feeder walls determine the heat transfer process in the glass, thus it is important to have accurate estimations of this distribution.

Actually, the average of the absolute error (the absolute error of every time step until new steady state condition is achieved averaged for every grid point) is very small, it accounts about $0.6°C$ maximum. Figure. 6 shows this time-average absolute error plot for every grid point. Thus, even though the reduced model order is only 0.085% of the original model, the average absolute deviation is less than 0.05% with respect to

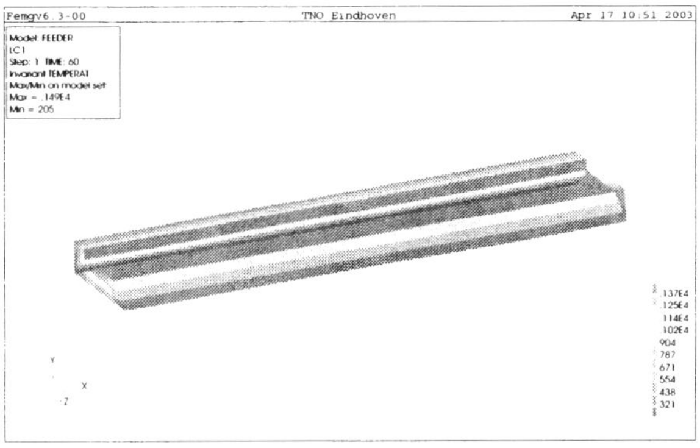

Fig. 4. Original Temperature Field of Feeder Walls

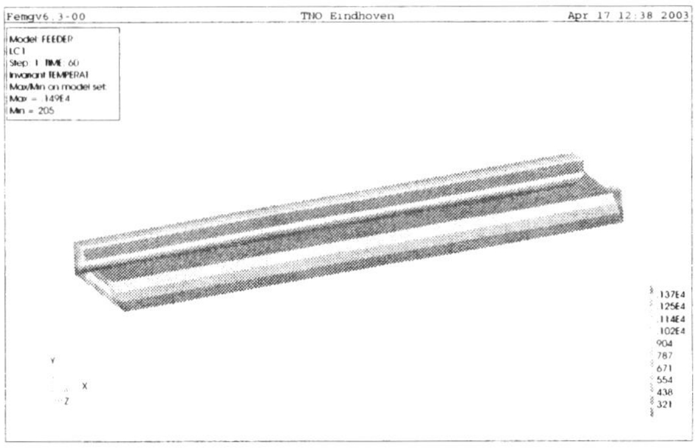

Fig. 5. Reduced Temperature Field of Feeder Walls

the temperature order of the glass melt, which lies in the range of 1400K.

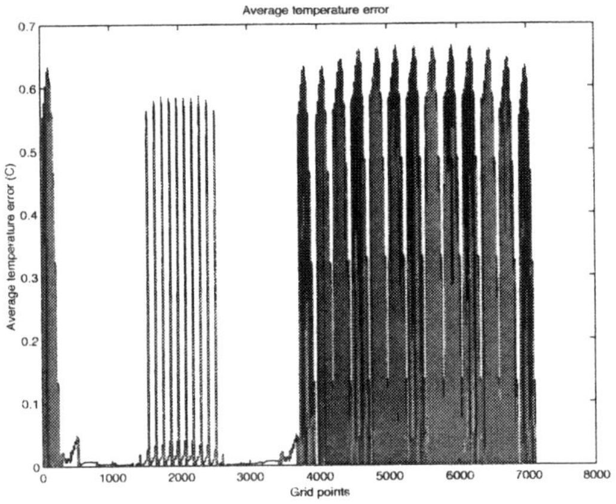

Fig. 6. Time-averaged absolute error of every grid point

The high accuracy of reduced order flow simulation is also appealing. Figure 7 and Figure 8 show how closed the velocity field is. The average absolute error for u is less than 3.10^{-6}m/s while the order of the velocity field is 10^{-3}. Hence, by retaining only 0.085% of the original model order, we obtain accuracy of 99.93%.

For this feeder, the dominant flow direction is in the x-direction. In real practice, to modify heat transfer by convection along the height and the width of the feeder, sometimes electronic boostings are applied. The presence of boostings will modify the flow pattern and increase its complexity since local turbulence will be generated.

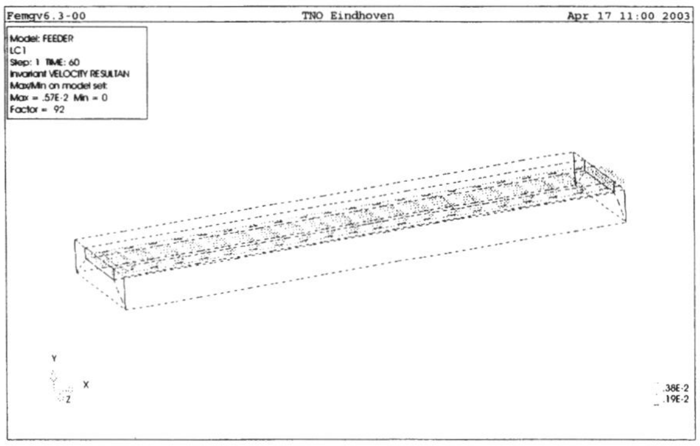

Fig. 7. Original Velocity Field at $y = 0.42\mathtt{m}$

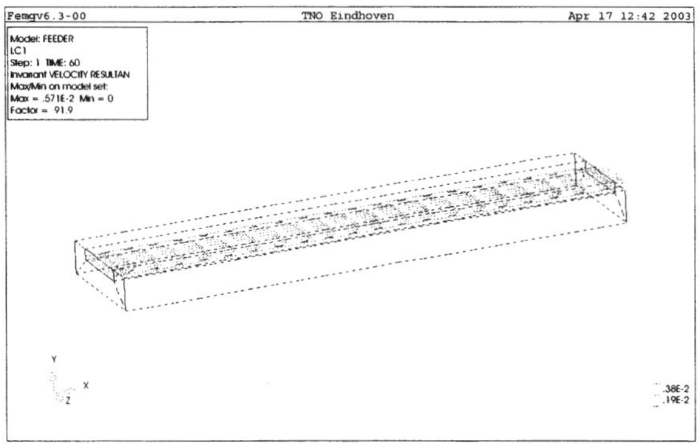

Fig. 8. Reduced Velocity Field at $y = 0.42\mathtt{m}$

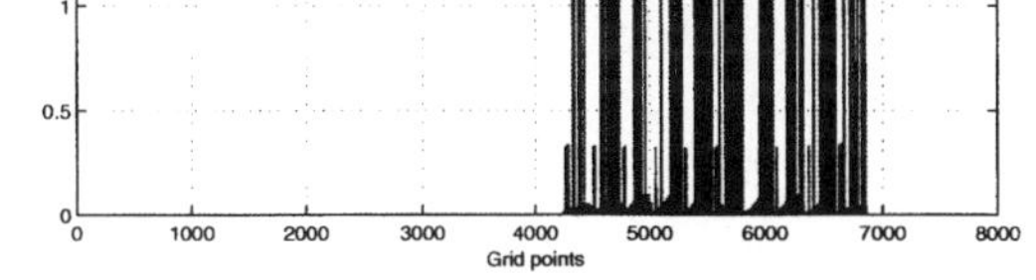

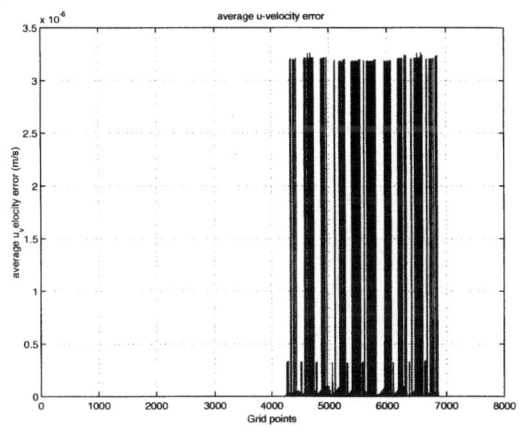

Fig. 9. Time-averaged absolute error of every grid point

One important issue in model reduction is how snapshots should be generated to obtain good results for a wide operating range. There is no general rule for this, but in principle one has to be able to identify which inputs are important for the "ergodicity" (P.Holmes *et al.*, 1998)of the systems.In the case of this feeder model, slight change of inflow rate affects the dynamics more than the slight change of the crown temperature. Hence it is important to have rate variations when generating snapshots in this case, possibly with higher sampling rate during the first transient simulations so that any change in flow dynamics are captured. However, when there is little possibility of drastic change of material throughput rates in real-life situations, there is no necessity to have extensive collection of snapshots.

4. CONCLUSION AND OUTLOOK

POD technique is a very efficient method to reduce high-fidelity models such as CFD models. With very few basic modes, the original dynamics can be described with negligible equations. The effectiveness of POD technique also relies on the quality of the generated snapshots, especially when solving flow equations.

In the near future, we will generate local turbulence by means of electronic boostings to observe the effectiveness of this reduced order modeling technique.

5. ACKNOWLEDGEMENTS

The authors greatly appreciate the assistance of scientific staff at TNO Institute of Applied Physics in Delft and Eindhoven that made this work possible.

6. REFERENCES

B.Bird, W.E.Stewart and E.N.Lightfoot (1960). *Transport Phenomena*. John Wiley and Sons. New York.

G.Karniadakis and S.Sherwin (1999). *Spectral/hp Element Method for CFD*. Oxford University Press. New York.

M.Kirby (2001). *Geometric Data Analysis*. Wiley Interscience.

P.Astrid, L.Huisman, S.Weiland and A.C.P.M.Backx (2002). Reduction and predictive controller design of a computational fluid dynamics model. *Proceedings of 41st IEEE Conference on Decision and Control*.

P.Holmes, J.L.Lumley and G.Berkooz (1998). *Turbulence, Coherent Structures, Dynamical Systems and Symmetry*. Cambridge University Press. Cambridge.

R.Beerkens, H.de Wall and F.Simonis (1997). *Handbook for Glass Technologists*. TNO-TPD Glass Technology. Eindhoven.

TNO-TPD (2000). *GTM-X User Manual*. TNO Institute of Applied Physics. Delft.

IFAC Publications
www.elsevier.com/locate/ifac

IDENTIFICATION OF THE TOPOLOGY OF A POWER SYSTEM NETWORK

Y. Hassaine * , **E. Walter** ** , **M. Dancre** *** , **B. Delourme** **** **and P. Panciatici** ****

* *Réseau de Transport d'Électricité and Laboratoire des signaux et systèmes, CNRS-Supélec-Université Paris-Sud.*
** *Laboratoire des signaux et systèmes, CNRS-Supélec-Université Paris-Sud.*
*** *Électricité de France R&D Department.*
**** *Réseau de Transport d'Électricité*

Abstract: This paper is devoted to the important practical problem of identifying the topology inside a suspect pocket of a power system network. Because the vector of the parameters to be estimated includes Boolean variables indicating the existence or absence of connection as well as real variables, this is a hybrid problem. It is solved here in a guaranteed way by ressorting to interval analysis and the notion of constraint propagation. The methodology is illustrated on the IEEE 14-bus network. *Copyright © 2003 IFAC*

Keywords: constraint propagation, electric power systems, hybrid identification.

1. INTRODUCTION

State estimation was introduced in the context of power systems in the early seventies (Schweppe *et al.*, 1970). Raw data are first collected and processed in order to build a reliable data base from which the state can be identified. The resulting state estimate is then used to perform security analysis and to take economic decisions by energy management systems operating in dispatching centers. This raw data processing is performed every few minutes. Most often, the topology of the network is assumed to be known, and the only variables to be identified are continuous electrical variables. Unfortunately, this may not be realistic, as there are many reasons why the actual topology of the network may be unknown. Such errors of topology may lead to a completely erroneous estimation of the continuous electrical variables. It is therefore of fundamental practical importance to address topology identification based on telemetered values. Several methods can be found in the literature for the identification of topology (Liu and Wu, 1989) and (Clement and Davis, 1988), and particularly noteworthy is the global vision proposed in (Monticelli *et al.*, 1998) under the name of *generalized state estimation.* Most of the methods available attempt to detect and correct the topology errors using state estimation residuals. Such an approach fails in the case of interacting data errors. Moreover, local optimization methods are used, and the results obtained are local solutions in the neighborhood of the starting point, without any guarantee of global validity.

In this paper, a new approach for topology identification is proposed, which provides guaranteed results in the sense that no solution can be lost. It is based on interval analysis and interval constraint propagation. Because all uncertain variables are described by intervals, it is not necessary to assume that probability density functions are available, as with Bayesian approaches.

The paper is organized as follows. In Section 2, the context of state estimation in electrical power networks is briefly presented, the equations connecting the observations to the state of the network are recalled, and topology problems are introduced. Two test-cases that will serve to illustrate the methodology proposed are described. In Section 3, the very few concepts of interval analysis and constraint propagation needed are presented. Finally, Section 4 reports the results obtained with this method on the two test-cases introduced in Section 2.

2. ESTIMATION IN POWER NETWORKS

This section starts by a description of the various types of data available and associated uncertainty. A classical state estimator used to estimate the continuous electrical variables based on these data is then presented. The limitations of this estimator when the topology is erroneous lead to attempting the identification of topology via the introduction of discrete state variables.

2.1 *Input data and their uncertainty*

There are three types of input data, namely the telemetered continuous data, the physical characteristics of branches (lines and transformers) and the telemetered discrete data (for instance transformer tape positions and the status of circuit breakers). The telemetered continuous data consist of voltage magnitudes, active or reactive branch flows and active or reactive injection flows. These quantities are measured at specific locations in the network. For the French national grid, about 5000 telemetered continuous data are collected every 2 mn. These data are corrupted by measurement noise, generaly taken as white, and may contain outliers due, e.g. , to sensor failures or asynchronous data readings. The physical characteristics of branches (such as resistance and reactance) are read from a data base. This data base may be doubtfull because of climatic variation or ageing of network elements. The telemetered discrete data include transformer tape positions and circuit breaker statuses. Transformer tape positions specify the transformer ratio, which can be set at several discrete values. The corresponding discrete data may be erroneous, for instance due to the asynchronicity of measurements. Finally, the circuit breakers connect or disconnect lines, generators and transformers of the network, so their status describes the network topology. For the French national grid, the statuses of about 10,000 breakers are updated. Asyncronicity may again lead to errors in these data and thus to a wrong topology. Two types of topology problems can be distinguished:

- *topology branch problem*: a branch may be disconnected or connected to another bus than originally thought,
- *topology bus problem*: a bus may be split into several buses or merged with another one.

In the next section, it will be assumed that the topology is correct, and that the only state variables to be estimated are the voltage magnitudes and phases at each bus of the network, which corresponds to a node of the associated graph.

2.2 *Identification of continuous variables*

The classical approach for state estimation in the context of power system networks is based on weighted least squares. The state is usually assumed to be static and thus equivalent to a vector of parameters to be identified. Others approaches based on robust estimators were implemented, such as LAV estimators (Singh and Alvarado, 1994) or Huber's M-estimators (Baldick *et al.*, 1997). Dynamic state estimation has also been considered via extended Kalman filtering (Mallieu *et al.*, 1986). Each of the underlying algorithms involves a knowledge-based model. This model is nonlinear and often linearized to facilitate solution.

In the context of very high voltage grids, active and reactive power flows are written as :

$$\begin{aligned} T_{ij}^{a} &= V_i V_j y_{ij} \rho_{ij} \rho_{ji} \sin(\theta_i - \theta_j - \delta_{ij}) \\ &\quad + V_i^2 \rho_{ij}^2 y_{ij} \sin \delta_{ij}, \end{aligned} \tag{1}$$

$$\begin{aligned} T_{ij}^{r} &= -V_i V_j y_{ij} \rho_{ij} \rho_{ji} \cos(\theta_i - \theta_j - \delta_{ij}) + \\ &\quad V_i^2 \rho_{ij}^2 (y_{ij} \cos \delta_{ij} - H_{ij}), \end{aligned} \tag{2}$$

where V_i and θ_i are the modulus and phase of the voltage at bus i, y_{ik} and $\delta_{ij} - \frac{\pi}{2}$ are the modulus and phase of the serial admittance between buses i and j, ρ_{ij} is the transformer ratio in the branch between buses i and j and H_{ij} is the shunt admittance of the quadripole between buses i and j. The nodal injection I_n at bus n is the sum of all productions and demands at this bus. The vector of measurements $\mathbf{z}$ can be expressed in terms of the state vector $\mathbf{x}$ of the system as:

$$\mathbf{z} = \mathbf{h}(\mathbf{x}) + \mathbf{e}. \tag{3}$$

$\mathbf{z}$ consists of measured voltages, active and reactive power flows and nodal injections, $\mathbf{x}$ consists of voltage moduli and phases, $\mathbf{e}$ is the error vector, and $\mathbf{h}$ is the vector of nonlinear measurement equations defined by (1) and (2). Equation (3) is most often solved iteratively using the least-squares approach

$$\begin{aligned} \hat{\mathbf{x}}_{\mathbf{k+1}} &= \hat{\mathbf{x}}_{\mathbf{k}} \\ &\quad -(\mathbf{H}(\hat{\mathbf{x}}_{\mathbf{k}})^{\mathrm{T}} \mathbf{W}^{-1} \mathbf{H}(\hat{\mathbf{x}}_{\mathbf{k}}))^{-1} \mathbf{H}(\hat{\mathbf{x}}_{\mathbf{k}})^{\mathrm{T}} \mathbf{W}^{-1} \mathbf{e}_{\mathbf{k}}, \end{aligned} \tag{4}$$

where $\hat{\mathbf{x}}_{\mathbf{k}}$ is the estimate of $\mathbf{x}$ at iteration k, $\mathbf{W} = E(\mathbf{e}\mathbf{e}^{\mathrm{T}})$ is the covariance matrix of the error vector (assumed known) and $\mathbf{H} = \frac{\partial \mathbf{h}}{\partial \mathbf{x}}$ is the Jacobian matrix of $\mathbf{h}$.

The state estimator (4) is well known not to be robust to outliers. A post-treatment must therefore be performed in order to detect and reject them. The method is based on residual analysis and requires highly redundant measurements. The measurement residual vector at $\hat{\mathbf{x}}$ is $\mathbf{r} = \mathbf{z} - \mathbf{h}(\hat{\mathbf{x}})$. If m is the number of measurements, the normalized residuals are :

$$e_i^N = (z_i - h_i(\hat{\mathbf{x}}))/W_{ii}, i = 1, ..., m. \quad (5)$$

In the absence of outliers the mean of the normalized residuals should be zero. The hypothesis of presence of outliers is rejected if $max_i|\mathbf{e}_i^N| < \gamma$, where γ is the detection threshold. Another method based on the χ^2-test and the $\mathbf{e}_i^N$-test has been shown to be effective in detecting and identifying outliers in the context of power systems (Van Cutsem, 1986).

2.3 *Limitations of the classical state estimator*

Even after robustification, the state estimator of the previous section cannot cope with errors in the topology of the network resulting from erroneously telemetered discrete variables. Contrary to outliers regarding the telemetered continuous data, which could be detected and eliminated based on residual analysis, the topological errors must be corrected for a reliable estimation of the continuous state variables to become possible.

This paper will deal with the identification of topology via the introduction of discrete state variables. The resulting identification problem is then hybrid, with both real and Boolean variables. The identification of transformer tap position could be addressed in a similar manner and should form the subject of future studies. It will be assumed that a suspect pocket (Monticelli *et al.*, 1998) has been located, for instance using the methods described in (Clement and Davis, 1988) and (Liu and Wu, 1989). A finite number of possible topologies inside this suspect pocket will be described using Boolean variables, and the identification of the proper topology will be performed by estimating the values of these Boolean variables. Two test cases TC_1 and TC_2 will be used to illustrate the methodology. Both are based on the classical IEEE 14-bus network (see Fig. 1). It is assumed that the part of the network below the transformers on Fig. 1 corresponds to a suspect pocket whose topology is doubtful. The power flows across these transformers are taken as belonging to a 20% interval around the measured value. The relative precision is 5% for voltage measurements and 20% for active and reactive power flows measurements. The prior domains for all non-measured variables are computed as follows. The power flow T_{ij} from bus i to bus j is assumed to belong to the interval $[-2y_{ij}V_n^2, 2y_{ij}V_n^2]$, where V_n is the nominal voltage magnitude, and y_{ij} is the modulus of the admittance of the branch connecting buses i and j, both given. Active injections are assumed to be within 20% of forecasted or measured value. No information is available about non-measured reactive injections, which implies that the prior domains for these variables are taken equal to $[-\infty, \infty]$. It is also assumed that the difference of the phases between the two extremities of any branch is smaller than 12°. Finally, nonmeasured voltages magnitudes are assumed to be within 15% of their nominal values.

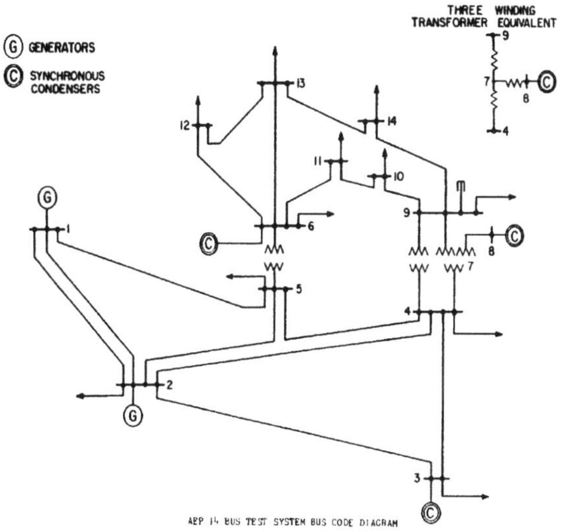

Fig. 1. IEEE 14-bus network

In Test-case TC_1, it is feared that the branch between buses 2 and 3 may actually be connecting buses 3 and 4, in parallel with an already existing branch. Decision on the topology should be based on voltage measured at buses 2 and 3 and on active and reactive power flows T_{34} and T_{12}. In Test-case TC_2, bus 2 may either actually correspond to a single bus or be split into buses 2_1 and 2_2. In the latter case, bus 2_1 is connected to buses 1 and 4, and bus 2_2 to buses 3 and 5. Decision should be based on active and reactive power flows T_{23} and T_{25}.

3. INTERVAL FORMULATION

3.1 *Basic concepts*

Interval analysis (see, e.g., (Moore, 1979) and (Jaulin *et al.*, 2001)), makes it possible to derive *guaranteed* numerical algorithms for the solution of mathematical problems. Guaranteed means here that no solution can be lost. Outer (and sometimes inner) approximations of the solution set can be obtained. This solution set is not required to be convex or even connected for the methodology to apply. All quantities are assumed to belong to intervals, which describe their degree of uncertainty. Deterministic quantities will be represented by degenerate intervals with zero width. Completely unknown quantities will be assumed to belong to the interval $[-\infty, \infty]$.

It is a simple matter to extend all classical operations on real numbers to intervals. Let $\underline{x}$ and $\overline{x}$ be the lower and upper bounds of the interval $[x]$. The addition of two intervals, for instance, is defined as follows

$$[x] + [y] = [\underline{x} + \underline{y}, \overline{x} + \overline{y}]. \quad (6)$$

Multiplication, division, union and intersection of two intervals can similarly be defined. All elementary functions operating on real numbers can also be extended to intervals. For instance $\exp([x]) = [\exp(\underline{x}), \exp(\overline{x})]$. Interval analysis can be used to compute intervals guaranteed to contain the possible values of all uncertain variables. Computation takes advantage of (i) prior information, (ii) the experimental data and (iii) the structure of the mathematical model under consideration.

Consider a vector $\mathbf{x}$ comprising n variables $x_i \in \mathbb{R}$, $i = 1, \ldots, n$. Assume that these variables are linked by m equality constraints

$$f_j(x_1, \ldots, x_n) = 0,\ j = 1, \ldots, m, \tag{7}$$

these constraints can be expressed in vector form as

$$\mathbf{f}(\mathbf{x}) = \mathbf{0}. \tag{8}$$

Each variable x_i belongs to some prior domain, which will here be assumed to be an interval for simplicity. Thus, the prior domain for $\mathbf{x}$ is a vector interval (or *box*), denoted by $[\mathbf{x}]$. Looking for the set $\mathscr{S}$ of all $\mathbf{x}$ in $[\mathbf{x}]$ such that (8) is satisfied is a *constraint satisfaction problem* (CSP), concisely denoted by

$$H : (\mathbf{f}(\mathbf{x}) = \mathbf{0}, \mathbf{x} \in [\mathbf{x}]). \tag{9}$$

Interval analysis may use a *contractor* to replace the prior box $[\mathbf{x}]$ by a smaller box $[\mathbf{x}']$ such that the solution set remains unchanged $\mathscr{S} \subset [\mathbf{x}'] \subset [\mathbf{x}]$. A possible method to build contractors is via interval constraint propagation.

3.2 *Interval constraint propagation*

To understand the basic idea of interval constraint propagation, consider the intervals variables $x \in [3,7]$, $y \in [0,5]$ and the boolean variable $\alpha \in \{0,1\}$. Assume that are linked by the constraint $\alpha x - y = 4$. This constraint can also be written as $x = (4+y)/\alpha$, $y = \alpha x - 4$ or $\alpha = (4+y)/x$. Each of these relations has been obtained by isolating one of the variables in the initial expression. The idea is to put α in the interval $[0,1]$ and try to eliminate the upper or the lower part of the initial interval. As soon as this is obtained, the value of α becomes certain and equal to the integer that remains in the contracted interval. Since $x \in [3,7]$, $y \in [0,5]$ and $\alpha \in [0,1]$ then $x \in (4 + [0,5])/[0,1] = [4,\infty]$. As $x = (4+y)/\alpha$ and $x \in [3,7]$, then $x \in [4,\infty] \cap [3,7] = [4,7]$. The same operations are performed on the other variables y and α, and the resulting contracted domains for x, y and α are $[4,7]$, $[0,3]$ and $[4/7,1]$ but because α is Boolean $\alpha = 1$.

The contractions above can be described by the following equations

$$\begin{cases} [x] = [x] \cap ((4+[y])/[\alpha]), \\ [y] = [y] \cap ([\alpha][x] - 4), \\ [\alpha] = [\alpha] \cap ((4+[y])/[x]). \end{cases} \tag{10}$$

The introduction of the constraint $\alpha(\alpha - 1) = 0$ can accelerate the convergence of α to the singleton 0 or 1. Constraint propagation may be applied on (10) as long as contraction is taking place.

3.3 *CSP for topology identification*

Interval analysis and the notion of CSP are now applied to topology identification as part of the generalized state estimation problem. It is assumed that a suspect pocket has been isolated and that the suspect branches and buses have been localized inside this suspect pocket. Let T_{ij}^a and T_{ij}^r be the active and reactive power flows from bus i to bus j defined in (1) and (2). They satisfy

$$\begin{cases} T_{ij}^a = V_iV_j(S_{ij}G_{ij}^c - G_{ij}^sC_{ij}) + V_i^2\tilde{G}_{ij}^s, \\ T_{ij}^r = -V_iV_j(S_{ij}G_{ij}^s + G_{ij}^cC_{ij}) + V_i^2(\tilde{G}_{ij}^c - \tilde{H}_{ij}), \end{cases} \tag{11}$$

with

$$\begin{cases} C_{ij} = \cos(\theta_i - \theta_j), \\ S_{ij} = \sin(\theta_i - \theta_j), \\ G_{ij}^s = y_{ij}\rho_{ij}\rho_{ji}\sin(\delta_{ij}), \\ G_{ij}^c = y_{ij}\rho_{ij}\rho_{ji}\cos(\delta_{ij}), \\ \tilde{G}_{ij}^s = y_{ij}\rho_{ij}^2\sin(\delta_{ij}), \\ \tilde{G}_{ij}^c = y_{ij}\rho_{ij}^2\cos(\delta_{ij}), \\ \tilde{H}_{ij} = \rho_{ij}^2H_{ij}. \end{cases} \tag{12}$$

Using, as indicated in (12), C_{ij} and S_{ij} as uncertain variables instead of θ_i and θ_j drastically simplifies interval computations, as the evaluation of trigonometric functions is thus avoided.

To take into account the properties of the sine and cosine functions, one must add the following constraints in the CSP

$$\begin{cases} S_{ij}^2 + C_{ij}^2 = 1, \\ C_{ij} = C_{ji}, \\ S_{ij} = -S_{ji}. \end{cases} \tag{13}$$

The most basic topology problems faced by the operators concern the *identification of branch topology*. In these problems, one of the branches originating at bus n ends at some unknown location to be identified from a set of possible buses. Each of the possible terminal buses n_i of a suspect branch l is associated with a binary status coefficient α_{li}. This coefficient is equal to zero if the bus n_i is not connected to the branch l, and to one if it is connected. Since the branch l is only connected to one terminal bus, only one of the α_{li} differs from zero. Let $\mathscr{I}_l$ be the set of the indices of all buses that may be connected with the bus n via the suspect branch l. A disconnected branch can be modeled by introducing a fictitious bus n_0 characterized by zero active injection, $T_{n0}^a = 0$, with $\alpha_{ln_0} = 1$. In this case, the index 0 will be introduced in $\mathscr{I}_l$. To guarantee that a unique α_{li} is equal to one, all the others being zero, the following constraints are included in the CSP

$$\sum_{i \in \mathscr{I}_l} \alpha_{li} = 1, \tag{14}$$

$$\forall (i,j) \in \mathscr{I}_l^2,\ i \neq j :\ \alpha_{li}\alpha_{lj} = 0. \tag{15}$$

With this notation, the power injection at bus n is

$$I_n = \sum_{j \in \mathscr{S}_n} T_{nj} + \sum_{l \in \mathscr{L}_n} \sum_{i \in \mathscr{I}_l} \alpha_{li}T_{ni}, \tag{16}$$

where $\mathscr{L}_n$ is the set of the indices of the suspect branches originating at bus n, and $\mathscr{S}_n$ is the set of the indices of the terminal buses of the reliable branches originating at bus n.

Fig. 2 describes the situation in the simple case $\mathscr{L}_n = \{1\}$, $\mathscr{I}_l = \{0,1,2,3,4\}$ and $\mathscr{S}_n=\{m\}$.

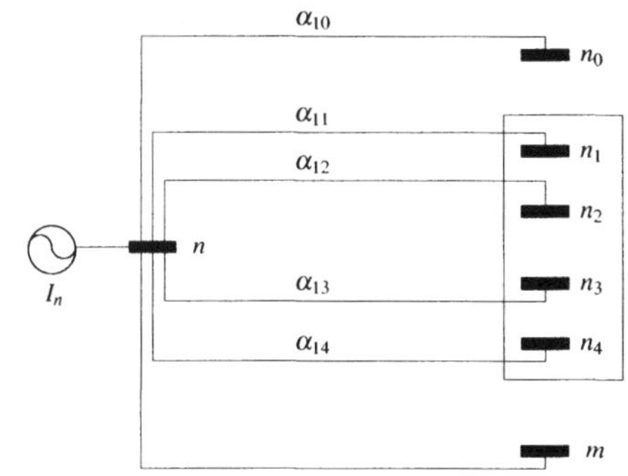

Fig. 2. Problem of branch topology; only one of the α_{li} is equal to one, all the others are zero

A more complex topology problem is the identification of bus topology, which consists of allocating buses correctly inside a substation. Let $\mathscr{K}$ be the set of the indices of the suspect substations. In the suspect substation indexed by $k \in \mathscr{K}$, the maximum number of buses is N_k. For the sake of simplicity, it will be assumed that only one branch topology is associated with the n-th possible bus topology, where $n = 1,...,N_k$. For each bus i ($i \leq n$) of an n-bus substation, the corresponding injection is denoted by I^k_{ni}, and the set of the coefficients of the connected branches is denoted by $\mathscr{M}^k_{ni}$. Again, the binary status coefficient α_{kn} takes the value one if the substation indexed by k involves n buses and zero otherwise. With this notation, the constraints can be written as

$$\alpha_{k1}(I^k_{11} - \sum_{m \in \mathscr{M}^k_{11}} T_m) = 0,$$

$$\begin{cases} \alpha_{k2}(I^k_{21} - \sum_{m \in \mathscr{M}^k_{21}} T_m) = 0, \\ \alpha_{k2}(I^k_{22} - \sum_{m \in \mathscr{M}^k_{22}} T_m) = 0, \end{cases}$$

$$\vdots$$

$$\begin{cases} \alpha_{kn}(I^k_{n1} - \sum_{m \in \mathscr{M}^k_{n1}} T_m) = 0, \\ \vdots \qquad \vdots \\ \alpha_{kn}(I^k_{ni} - \sum_{m \in \mathscr{M}^k_{ni}} T_m) = 0, \\ \vdots \qquad \vdots \\ \alpha_{kn}(I^k_{nn} - \sum_{m \in \mathscr{M}^k_{nn}} T_m) = 0, \end{cases} \tag{17}$$

$k \in \mathscr{K}, i \leq n \in \{1,...,N_k\}$. For any $n \in \{1,...,N_k\}$, the goal is to find the single α_{kn} that differs from zero.

The next two constraints are added to guarantee that the solution for the status coefficients is unique

$$\sum_{n=0}^{N_k} \alpha_{kn} = 1, \tag{18}$$

$$\forall (n,m) \in \{1,...,N_k\}^2,\ n \neq m: \ \alpha_{kn}\alpha_{km} = 0. \tag{19}$$

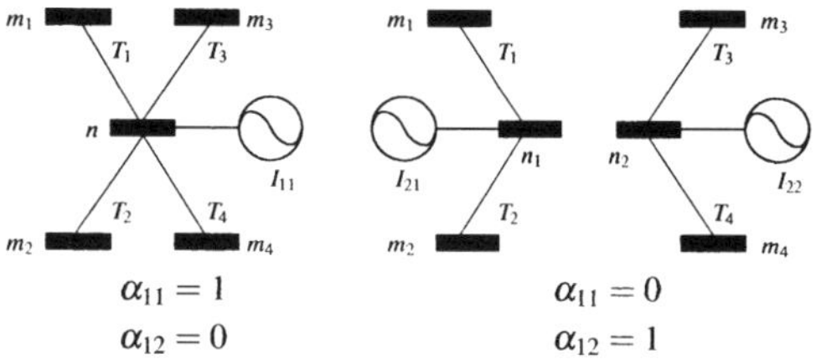

Fig. 3. Topology bus problem; the single bus n of the left diagram is split into the buses n_1 and n_2 of the right diagram

Fig. 3 illustrates a simple case where $\mathscr{K} = \{1\}$, $N_1 = 2$, $\mathscr{M}^1_{11} = \{1,2,3,4\}$, $\mathscr{M}^1_{21} = \{1,2\}$ and $\mathscr{M}^1_{22} = \{3,4\}$. Equation (17) then becomes

$$\begin{cases} \alpha_{11}(I_{11} - (T_1+T_2+T_3+T_4)) = 0, \\ \alpha_{12}(I_{21} - (T_1+T_2)) = 0, \\ \alpha_{12}(I_{22} - (T_3+T_4)) = 0. \end{cases} \tag{20}$$

In order to solve (20) using constraint propagation, and to avoid numerical instabilities, additional information regarding the injections and available in the data base must be taken into account. For this example, $I_{11} = I_{21} = I$ and $I_{22} = 0$, and (20) becomes

$$\begin{cases} I - (T_1+T_2) - \alpha_{11}(T_3+T_4) = 0, \\ \alpha_{12}(T_3+T_4) = 0. \end{cases} \tag{21}$$

For the identification of branch or bus topology, the model equations can be put in vector form as

$$[\mathbf{z}] = [h]([\mathbf{v}], [\alpha], [\mathbf{C}], [\mathbf{S}]), \tag{22}$$

where $[\mathbf{z}]$ is the interval vector of the power flows and injections, $[\mathbf{v}]$ is the interval vector of the voltage magnitude moduli, $[\mathbf{C}]$ and $[\mathbf{S}]$ are the interval matrices of the sines and cosines of the phase shifts and $[\alpha]$ is the interval matrix of status coefficients. The prior intervals for each status coefficient is $[0,1]$, and it is expected to be contracted into the degenerate interval $[0,0]$ or $[1,1]$ when solving the associated CSP.

The description of the two test cases in Section 2.3 indicates how the other prior intervals are computed. The approach advocated in this paper is to contract the prior domains for all variables until a decision is reached regarding the value of the Boolean status coefficients or no further contraction can be obtained.

4. SIMULATIONS AND RESULTS

This approach has been implemented using the interval solver *RealPaver* (Granvilliers, 2002). RealPaver contracts the prior domains for $[\mathbf{z}]$, $[\mathbf{v}]$, $[\alpha]$, $[\mathbf{C}]$ and $[\mathbf{S}]$, and generates a sequence of boxes containing the set $\mathscr{S}$ of all solutions to (22).

For TC_1, the branch indicated on Fig. 1 as connecting buses 2 and 3 may indeed connect these buses ($\alpha_1 = 1$)

or actually connect buses 3 and 4 in parallel with the existing branch structure of ($\alpha_2 = 1$). Again, the prior domain for each of the status variables α_i is taken as $[0,1]$ and the contracted domains for these status variables are as indicated in Table 1. The estimator concludes that the structure of Fig. 1 is correct, with a branch connecting buses 2 and 3.

For TC_2, the problem is deciding whether bus 2 is indeed a single bus ($\alpha_1 = 1$) or should be split into buses 2_1 and 2_2 as explained in Section 2.3 ($\alpha_2 = 1$). As usual, the prior domain for the status variables α_i is $[0,1]$. The contracted domains for these status variables are as indicated in Table 1, and the estimator thus concludes that bus 2 should not be split.

Table 1. Status coefficients

	Prior intervals	Results
TC_1	$\alpha_1 = [0,1]$ $\alpha_2 = [0,1]$	$\alpha_1 = 1$ $\alpha_2 = 0$
TC_2	$\alpha_1 = [0,1]$ $\alpha_2 = [0,1]$	$\alpha_1 = 1$ $\alpha_2 = 0$

To accelerate contraction and increase the accuracy of the results, it is interesting to introduce as many pertinent constraints as possible. For instance, the following constraint can be added $T_{ij}^a T_{ji}^a \leq 0$, to guarantee that the active power flows at the endpoints of a branch are of opposite signs. A more accurate constraint could be obtained by taking into account the maximum loss in each branch.

5. CONCLUSIONS

A new method has been proposed to identify the topology in a suspect pocket of a power system network, based on interval analysis and constraint propagation. Generalized state estimation is thus implemented in a guaranteed way (no topology consistent with the data can be lost). It should also be noted that the nonlinear model was handled as such, without linearization. As the results from the standard IEEE 14-bus network tend to show, the method is promising and makes it possible to identify the bus and branch topologies in suspect pockets. Other applications to the identification of discrete parameters in power system networks should be investigated in the near future. The same approach could also be applied to other types of hybrid systems.

6. REFERENCES

Baldick, R., K.A. Clements, Z.P. Dzigal and P.W. Davis (1997). Implementating nonquadratic objective functions for state estimation and bad data rejection. *IEEE Transactions on Power Systems* **12**, 376–382.

Clement, K.A. and P.W. Davis (1988). Detection and identification of topology errors in electric power systems. *IEEE Transactions on Power Systems* **3**, 1748–1753.

Granvilliers, L. (2002). www.sciences.univ-nantes.fr /info/perso/permanents/granvil/realpaver.

Jaulin, L., M. Kieffer, O. Didrit and E. Walter (2001). *Applied Interval Analysis*. Springer-Verlag. London.

Liu, W.H.E. and F.F. Wu (1989). Detection of topology errors by using state estimation. *IEEE Transactions on Power Systems* **4**, 176–183.

Mallieu, D., T. Van Cutsem, P. Rousseaux and M. Ribbens-Pavella (1986). Dynamic multilevel filtering for real-time estimation of electric power systems. *Control-Theory and Advanced Technology* **2**, 255–272.

Van Cutsem, T. (1986). *Estimation d'état hiérarchisée*. PhD dissertation, Liège.

Monticelli, A., O. Alsac, N. Vempati and B. Stott (1998). Generalized state estimation. *IEEE Transactions on Power Systems* **13**, 1069–1075.

Moore, R.E. (1979). *Methods and Applications of Interval Analysis*. SIAM. Philadelphia.

Schweppe, F.C., J. Wildes and D.B. Rom (1970). Power system static state estimation. *IEEE Transactions on Power Apparatus and Systems* pp. 120–135.

Singh, H. and F.L. Alvarado (1994). Network topology determination using least absolute value state estimation. *IEEE Power Engineering Society for Presentation at the IEEE/PES*.

IFAC Publications
www.elsevier.com/locate/ifac

LPV IDENTIFICATION OF A DIESEL ENGINE TORQUE MODEL

Xiukun Wei†, Luigi Del Re‡

Institute of Design and Control of Machatronical Systems
Johannes Kepler University, Linz, Austria.
Email:†wei.xiukun@jku.at, ‡Luigi.delre@jku.at

Abstract: Torque estimation has an increasing importance in the field of automotive control, as most engine controllers rely on torque estimation today. In this paper a new approach based on a linear parameter varying (LPV) model is proposed. It relies on a nonlinear subspace identification method and exploits the natural dependency of many engine processes on the rotational speed, but does not use a physical model, which is very time consuming and difficult to obtain. To this end, first a LPV state space method is introduced and then experimental results are presented which show that the model exhibits high accuracy. *Copyright © 2003 IFAC*

Keywords: Parameter Varying, Identification, Automotive, State Space, LPV.

INTRODUCTION

Today the torque control of automotive engines is the basis of every engine control structure and has a direct impact on every important aspect of automotive performance and passenger comfort (Jurgen, 1998). To use torque control, however, a measurement or at least an estimation of the instantaneous engine torque is required. As direct measurement is too expensive, several methods have been proposed to estimate the instantaneous torque, usually using some already available sensors, like rotational speed. However, these methods have serious disadvantages (Jurgen, 1998), as the precision of the estimation depends very strongly on the precision of the engine model parameters. These, in turn, are usually known only approximately, as important factors, like wear or production tolerances, cannot be known in advance, but strongly influence the model output. Therefore, it seems appropriate to look for a physically based model structure which can be easily identified from data. In our case, we consider the full path between accelerator pedal angle and torque, assuming that rotational speed is known, which is usually the case.

The nonlinearity of the engine prevents a simple description of the engine by classical linear methods, like the transfer function technique and state space method. Instead, observing that the time scale of the engine is fixed by the rotational speed, and that most parameters of the engine depends on it, a *linear parameter varying* (LPV) model seems a sensible compromise between the too large complexity of general nonlinear systems and the information loss associated with a linear model.

The research on LPV systems has received much interest in the recent years(Bamieh, 1999; Bamieh and Giarre, 2000). The dynamics of LPV systems is usually supposed to depend on exogenous parameters whose trajectories are unknown in advance before the plant is stabilized or synthesized but can be measured on-line by the controller. In the recent years the identification technique has achieved some progress(Verdult, 2001).

In this paper, we use this method to derive and identify a LPV diesel engine torque state space model. The method is tested using a production BMW 47D DI diesel engine equipped with exhaust gas recirculation(EGR) and a variable geometry turbocharger(VGT) on a dynamic test bench.

This paper is organized as follows. In section one, the LPV system model and its nonlinear subspace identification approach are reviewed. In the second part, an assumed LPV engine torque model is proposed from some prior knowledge and this model is used to define the LPV model structure for identification experiments. The experiment design and results are presented in the fourth part. In the last section, the LPV model structure is discussed and a new nonlinear model similar to a Hammerstein model is proposed. Concluding remarks close the paper.

1. SUBSPACE IDENTIFICATION METHOD OF LPV SYSTEMS

There are two main approaches for identification of LPV systems. One is the input-output method (Bamieh and Giarre, 2000), the other is based on state space method (Verdult and Verhaegen, 2000), the subspace nonlinear identification method that is applied in this paper. It will be reviewed in this section.

Considering a LPV plant represented in the state space by the following system equations:

$$x_{k+1} = A\begin{bmatrix} x_k \\ p_k \otimes x_k \end{bmatrix} + B\begin{bmatrix} u_k \\ p_k \otimes u_k \end{bmatrix} + w_k \quad (1)$$

$$y_k = C\begin{bmatrix} x_k \\ p_k \otimes x_k \end{bmatrix} + D\begin{bmatrix} u_k \\ p_k \otimes u_k \end{bmatrix} + v_k \quad (2)$$

where $\otimes$ denotes the Kronecker matrix product, $x_k \in R^n$ represent the unknown states, $u_k \in R^m$ are the inputs, $y_k \in R^l$ are the outputs, $p_k \in R^s$ are the exogenous input parameters that are measurable in real time, and $v_k \in R^l, w_k \in R^n$ are unknown white noise that are independent of p_k and u_k. The identification problem is how to estimate the system matrix A, B, C, D using finite data sequences of inputs and outputs under the assumption that not all states are measurable.

Until now there are no precise methods presented for this identification problem. The recent research is compiled in (Verdult, 2001). All these approaches are approximation methods and the main results are reviewed here.

$$P_k = [p_k, p_{k+1}, \cdots, p_{k+N-1}] \quad (3)$$
$$X_k = [x_k, x_{k+1}, \cdots, x_{k+N-1}] \quad (4)$$
$$U_k = [u_k, u_{k+1}, \cdots, u_{k+N-1}] \quad (5)$$
$$W_k = [w_k, w_{k+1}, \cdots, w_{k+N-1}] \quad (6)$$
$$Y_k = [y_k, y_{k+1}, \cdots, y_{k+N-1}] \quad (7)$$
$$V_k = [v_k, v_{k+1}, \cdots, v_{k+N-1}] \quad (8)$$

$$X_{j|j} = \begin{bmatrix} X_j \\ P_j \odot X_j \end{bmatrix} \quad (9)$$
$$X_{k+j|j} = \begin{bmatrix} X_{k+j-1|j} \\ P_{k+j} \odot X_{k+j-1|j} \end{bmatrix} \quad (10)$$
$$U_{j|j} = \begin{bmatrix} U_j \\ P_j \odot U_j \end{bmatrix} \quad (11)$$
$$U_{k+j|j} = \begin{bmatrix} U_{k+j} \\ P_{k+j} \odot U_{k+j} \\ U_{k+j-1|j} \\ P_{k+j} \odot U_{k+j-1|j} \end{bmatrix} \quad (12)$$

$$W_{j|j} = W_j \quad (13)$$
$$W_{k+j|j} = \begin{bmatrix} W_{k+j} \\ W_{k+j-1|j} \\ P_{k+j} \odot W_{k+j-1|j} \end{bmatrix} \quad (14)$$
$$Y_{j|j} = Y_j \quad (15)$$
$$Y_{k+j|j} = \begin{bmatrix} Y_{k+j} \\ Y_{k+j-1|j} \end{bmatrix} \quad (16)$$
$$V_{j|j} = V_j \quad (17)$$
$$V_{k+j|j} = \begin{bmatrix} V_{k+j} \\ V_{k+j-1|j} \end{bmatrix} \quad (18)$$

Lemma 1. Given the matrices(3)-(18).For the LPV system (1)-(2) it holds that

$$\begin{aligned} X_{k+j} &= \Delta_k^x X_{k+j-1|j} + \Delta_k^u U_{k+j-1|j} \\ &\quad + \Delta_k^w W_{k+j-1|j} \quad (19) \\ Y_{k+j|j} &= H_k^x X_{k+j-1|j} + H_k^u U_{k+j-1|j} \\ &\quad + H_k^w W_{k+j-1|j} + G_k^u U_{k+j} \\ &\quad + V_{k+j|j} \quad (20) \end{aligned}$$

The definitions of the coefficients in (19) and (20) can be found in (Verdult, 2001) and are omitted here for the sake of brevity.

Lemma 2. Let $H_k^x = \begin{bmatrix} \Gamma_k & \widetilde{H_k^x} \end{bmatrix}$, equation (20) is transformed into

$$\begin{aligned} Y_{k+j|j} &= \Gamma_k X_j + \widetilde{H_k^x}\left(P_{k+j-1|j} \odot X_j\right) \\ &\quad + H_k^u U_{k+j-1|j} + H_k^w W_{k+j-1|j} \\ &\quad + G_k^x U_{k+j} + V_{k+j|j} \quad (21) \end{aligned}$$

Algorithm 1. The subspace identification method of LPV systems based on the equation (21) needs several steps to be achieved, which are listed below:

1 *Approximate the X_j in equation (21). Different identification methods are derived from different approximation techniques.*

2 *Transform equation (21) to an approximation linear equation. Solve the approximation equation and get the approximate estimation of ΓX.*

3 *Using the subspace identification theory to retrieve the state estimation and system order using Singular Value Decomposition method.*
4 *Using the estimation value of state to recover the parameters of matrixes A, B, C and D.*
5 *Using the nonlinear optimization method to get a more precise model.*

More details about the LPV Subspace Identification Method which is applied in this paper can be found in (Verdult, 2001).

2. A DIESEL ENGINE LPV MODEL

In this section an assumed diesel engine LPV model is presented. It is based on some prior knowledge of engine structure, running characteristic as well as some physical laws.

The engine torque (T_e) is selected as the output variable and a nonlinear function of the accelerator pedal position (θ) is selected as the input variable. The engine system states and system order are estimated directly with the LPV subspace identification algorithm. The system parameters are assumed to be related with the engine speed from experience. So the engine speed is selected as the exogenous parameter($p_k = N_e$).

In a diesel engine, torque and power are controlled by the amount of fuel injected into the engine chamber (James, 1998). The fuel flow is controlled by a chain of physical and software components relaying the injectors to the accelerator pedal position(θ). This chain includes several limiters, *e.g.* smoke maps, which limits the maximum injection at a given engine speed. Therefore, the engine output torque is a dynamical nonlinear function relating the accelerate pedal angle(θ).

In this identification experiment the following static nonlinear function is supposed.

$$f(\theta) = k_1\theta^{\frac{1}{2}} + k_2\theta + k_3\theta^2 \quad (22)$$

where k_1, k_2, k_3 are three constant coefficients. The nonlinear function can also be extended to include some higher or lower order terms such as the cube of θ or the cube root of θ to get a more precise model. The following model structure is applied in the identification experiment.

$$\begin{aligned} x_{k+1} &= A_1x_k + A_2p_kx_k + B_1u_k \\ &\quad + B_2p_ku_k + w_k \quad (23) \\ y_k &= C_1x_k + C_2p_kx_k + D_1u_k \\ &\quad + D_2p_ku_k + v_k \quad (24) \end{aligned}$$

where $A_1, A_2 \in R^{n\times n}, B_1, B_2 \in R^{n\times m}, C_1, C_2 \in R^{l\times n}, D_1, D_2 \in R^{l\times m}$. The system order n is determined by the identification method using the singular value decomposition. The output number l equals 1 and only engine torque(T_e) is selected as an output. The input number m equals 3 for identification the engine model using the linear combination of θ , $\theta^{\frac{1}{2}}$ and θ^2, hence, u_k is choosed as $u_k = \left[\theta^{\frac{1}{2}}, \theta, \theta^2\right]^T$. Just as the foregoing proposed, the engine speed N_e is selected as p_k. The three constant coefficients k_1, k_2, k_3 are estimated along with other parameters. $[\cdot]^T$ denotes the transpose of matrix $[\cdot]$.

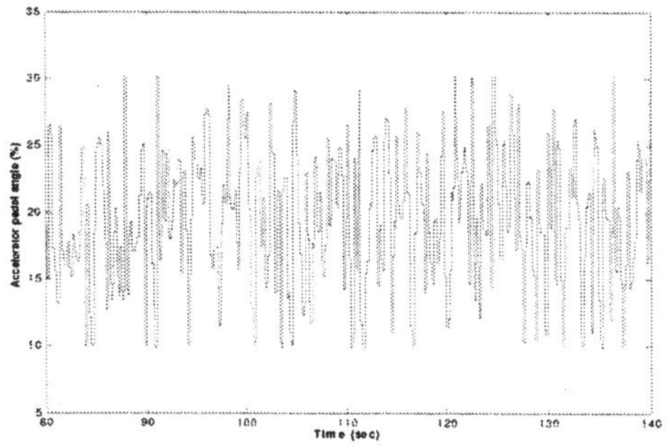

Fig. 1. The identification input signal: accelerator pedal angle sequence. This was used to estimate the model parameters under constant load 70 N.m

3. THE EXPERIMENT DESIGN AND RESULTS

The identification experiments data are generated on a dynamic diesel engine test bench with a BMW47D, 110KW, 280Nm turbocharged diesel engine. Two experiments are performed, one is on an operation point with constant load torque $T_l = 70N.m$ and the other is on the operation point with constant load torque $T_l = 120N.m$. In both cases, EGR is active. The sampled time constant is 0.02 second and $20,000$ samples are collected. About $3,000$ samples are used for the torque model estimation experiment and another $2,000$ samples are selected for model validation tests.

3.1 *The design of the input signals*

As for every identification method, the input excitation must be designed in such a way that it provides an appropriate spectrum and persistence excitation. To this end, the input signals are generated by filtering a band-limited white noise sequence and the output amplitude of the signals is limited by a limiter. Figure 1 shows a sequence of the accelerator pedal angle θ that is used for the torque model identification of the diesel engine. The load torque signal is generated by a load torque simulator which is kept constant at the operation point.

3.2 *The identification results*

Using the measured data a lot of off-line identification experiments were performed by trial-and-error approach to determine the structure of $f(\theta)$(at first only θ is used and then plus its square term θ^2 and then add its square root term $\theta^{\frac{1}{2}}$ at each trial). The best model structure of $f(\theta)$ is the one presented in the second section.The model quality is evaluated by three criterions. One quality of the output signal generated by the identified LPV model, is measured using the variance accounted for(VAF), which is defined as

$$VAF = \max\{1 - \frac{var\,(y_k - \widehat{y}_k)}{var\,(y_k)}\} \times 100\% \quad (25)$$

where $\widehat{y}_k$ denotes the estimated output signal, and $var(\cdot)$ denotes the variance of a quasi-stationary signal. Another quality is the maximum absolute value of output error(MOE). The fit(FIT) quality of the model is used which is defined as:

$$FIT = \frac{\|y_k - \widehat{y}_k\|}{\sqrt{N}} \quad (26)$$

where N is the length size of the data and $\|\cdot\|$ denotes the norm-2. This quality can express the space distance between real value y_k and estimated value $\widehat{y}_k$.

A linear time invariant state space model estimated through the linear subspace method N4SID(Overschee and Moor, 1994; Overschee and Moor, 1996)and the LPV model are identified for comparison. The performance functions of the models are shown in Table 1 for constant load torque $T_l = 70Nm$. At this operation point a second order(ORD) torque model is estimated using the LPV approach with nonlinear optimization and a tenth order model from N4SID is estimated. Table 2 is for the identification results with constant load torque $T_l = 120Nm$, at this operation point, a first order LPV model is used to compare with the tenth order model from N4SID method. Figure 2 to 5 are the identification and validation results under constant load torque $T_l = 70Nm$ via LPV and N4SID technique and figure 6 to 9 are the results under constant load torque $T_l = 120Nm$. Clearly LPV model is more precise than the N4SID linear model as well as a lower order. In all the figures only part of the curves for identification and validation are shown so that they can be clearly viewed.

Table 1 The qualities with load $T_l = 70Nm$

	Method	MOE	VAF	FIT	ORD
Estimation	*LPV*	7.50	95	1.75	2
Estimation	*N4SID*	8.91	84	3.08	10
Validation	*LPV*	18.31	91	2.39	2
Validation	*N4SID*	10.02	83	3.65	10

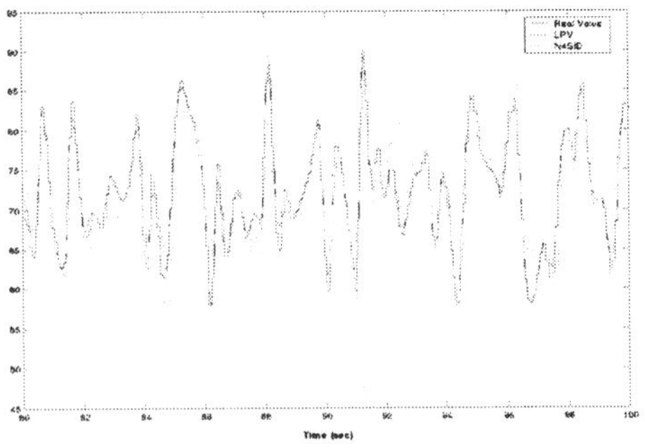

Fig. 2. The real torque curve and estimation torque curves using identification data via LPV and N4SID methods under constant load $T_l = 70$ Nm.

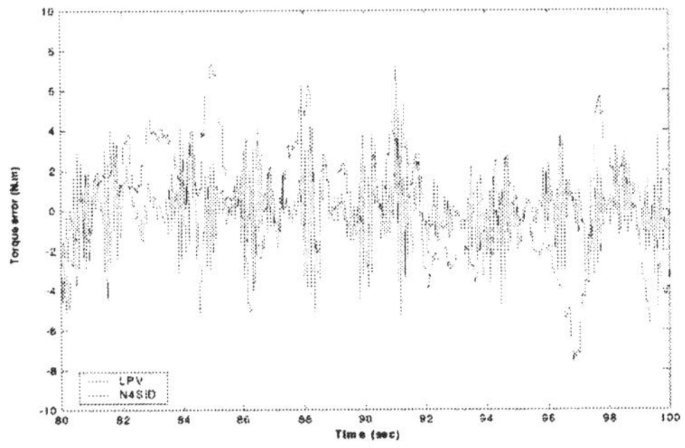

Fig. 3. The estimation error curves using identification data via LPV and N4SID methods with constant load torque $T_l = 70$ Nm.

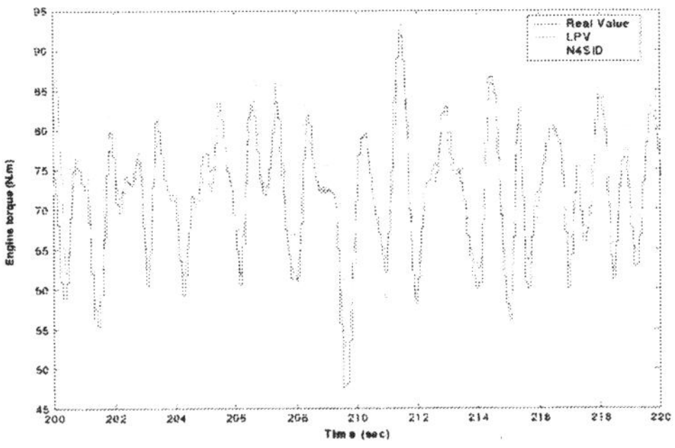

Fig. 4. The real torque curve and estimation torque curves using validation data via LPV and N4SID methods under constant load $T_l = 70$ Nm.

Table 2 The qualities with load $T_l = 120\ Nm$

	Method	MOE	VAF	FIT	ORD
Estimation	*LPV*	5.06	89	1.56	1
Estimation	*N4SID*	7.00	83	1.90	10
Validation	*LPV*	6.14	88	1.56	1
Validation	*N4SID*	7.56	85	1.80	10

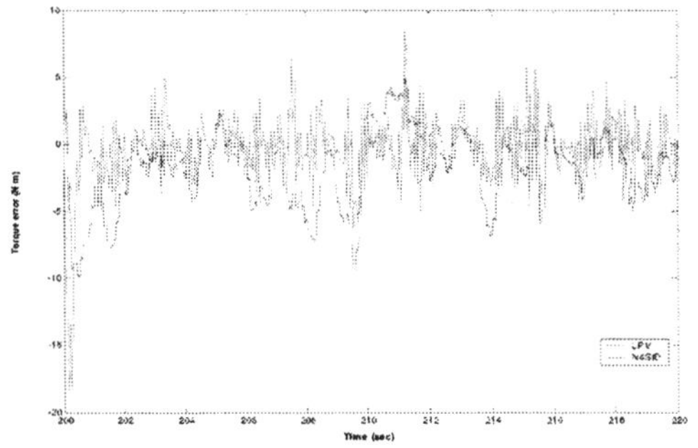

Fig. 5. The estimation error curves using validation data via LPV and N4SID methods with constant load torque $T_l = 70$ Nm.

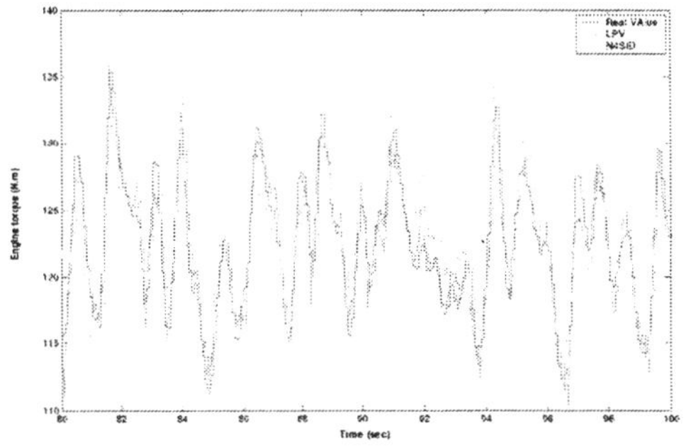

Fig. 6. The real torque curve and estimation torque curves using identification data via LPV and N4SID methods under constant load $T_l = 120$ Nm.

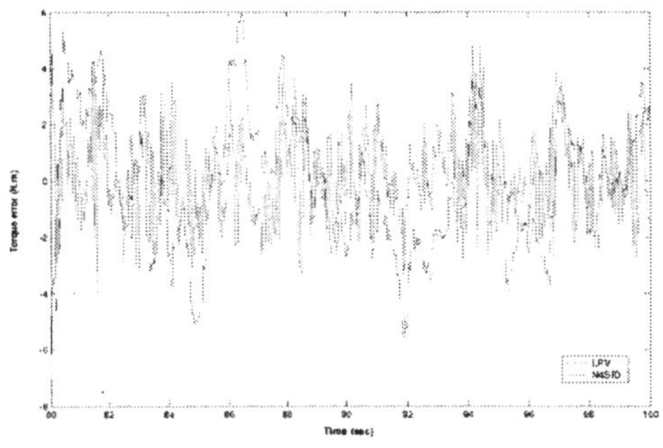

Fig. 7. The estimation error curves using identification data via LPV and N4SID methods with constant load torque $T_l = 120$ Nm.

4. CONCLUSION

From the results that are presented in the previous section, the LPV model has shown to be more precise than the linear model from N4SID method while retaining a significantly lower order. The LPV nonlinear torque model can be described as the following:

$$x_{k+1} = (A_1 + A_2 N_e)\, x_k + (B_1 + B_2 N_e)\, u_k \quad (27)$$

$$T_e = (C_1 + C_2 N_e)\, x_k + (D_1 + D_2 N_e)\, u_k \quad (28)$$

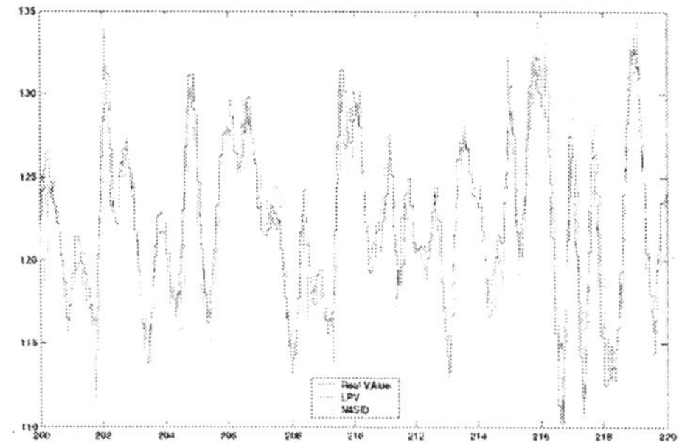

Fig. 8. The real torque curve and estimation torque curves using validation data via LPV and N4SID methods under constant load $T_l = 120$ Nm.

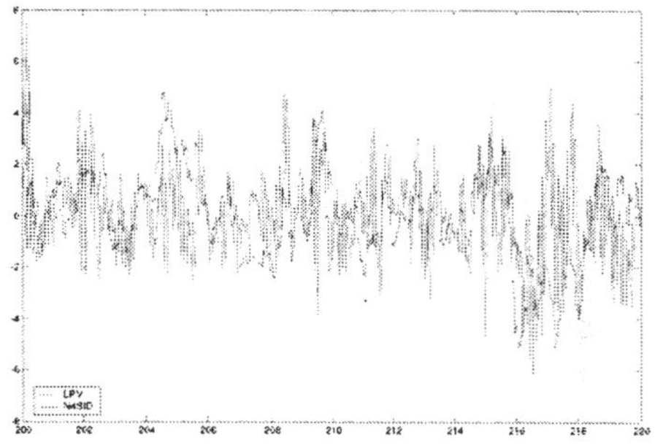

Fig. 9. The estimation error curves using validation data via LPV and N4SID methods with constant load torque $T_l = 120$ Nm.

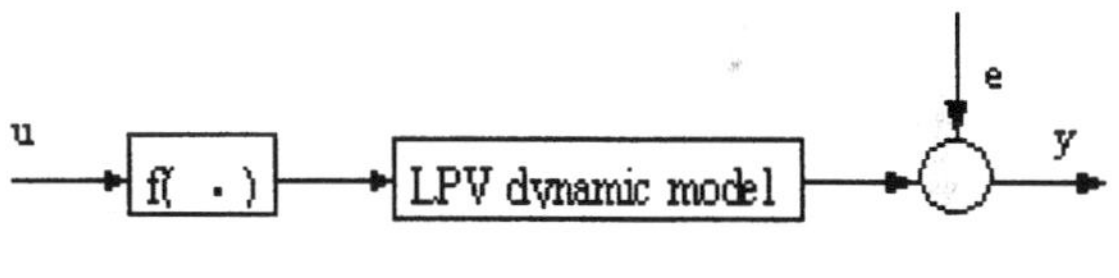

Fig. 10. A quasi-Hammerstein nonlinear Model

where A_1, A_2, B_1, B_2, C_1, C_2, D_1, D_2 are some constant matrices of appropriate dimension and the definition of other variables are the same as described in the first section. Therefore, the diesel torque model consists of a static nonlinear gain element cascaded with a LPV dynamic system similar to a Hammerstein model. This model structure is presented in figure 10.

The results obtained from the identification experiments are promising. However, the models are only identified in their stable operation points with a limited range of input and output. This model is valid only at these operation points and a global engine torque model is still not obtained because its dynamics is very complex. This will be another research target and further theory research and experiments are needed. These will include the global model identification and application to torque control and fault detection of diesel engine systems.

ACKNOWLEDGMENT

This work has been performed under the research grant 4.3 of the LCM Center of Competence in Mechatronics in Linz, whose support is gratefully acknowledged.

The authors would like to thank their co-workers Engelbert, G. and Peter, L. for their assistance to implement the experiments as well as Andreas, S. for his discussions about the identification experiment design and implementation.

5. REFERENCES

Bamieh, B. A. (1999). Identification of linear parameters varying models, *Procedings of the 38th conference on Decision and Control* pp. 1505–1510.

Bamieh, B. A. and Giarre, L. (2000). Identification for a general class of LPV models, *IFAC System Identification* .

James, S. E. (1998). *Modern Automotive Technology*, The goodheart-willcox company,inc., Tinley Park, Illinois.

Jurgen, R. K. (1998). *Electronic Engine Contol Technologies*, Society of Automotive Engineers, Inc., U.S.A.

Overschee, P. V. and Moor, B. D. (1994). N4SID: Subspace algorithms for the identification of combined deterministic-stochastic systems, *Automatica* **30**: 75–93.

Overschee, P. V. and Moor, B. D. (1996). *Subspace Identification for Linear Systems*, Kluwer Academic Publications.

Verdult, V. (2001). *Nonlinear System Identification(A State-Space Approach)*, PhD thesis, University of Twente, The Netherlands.

Verdult, V. and Verhaegen, M. (2000). Identification of multivariable linear parameter-varying systems based on subspace techniques, *Proceedings of the 39th IEEE Conference on Decision and Control* .

IFAC Publications
www.elsevier.com/locate/ifac

IDENTIFICATION AND CONTROL OF A PV-SUPPLIED SEPARATELY EXCITED DC MOTOR USING UNIVERSAL LEARNING NETWORKS

Ahmed Hussein * **Kotaro Hirasawa** ** **Jinglu Hu** **

* *Graduate School of Information Science and Electrical Engineering, Kyushu University 812-8581, 6-10-1 Hakozaki, Higashi-Ku, Fukuoka, JAPAN*
ahmed@cig.ees.kyushu-u.ac.jp
** *Graduate School of Information, Production and Systems, Waseda University 808-0135, 2-7 Hibikino, Wakamatsu-Ku, Kitakyushu, Fukuoka, JAPAN*

Abstract: This paper describes the use of Universal Learning Networks (ULNs) in the identification and control of a separately excited dc motor loaded with a centrifugal pump and fed from Photovoltaic (PV) generator via dc-dc buck-boost converter. The Universal Learning Network Identifier (ULNI) is trained offline using the forward propagation algorithm to emulate the dynamic behavior of the dc motor system. Then this identifier is used, instead of the motor system, for the online training of the Universal Learning Network Controller (ULNC). As a result, the motor speed can follow an arbitrarily selected reference signal. Furthermore, the overall system can operate at the Maximum Power Point (MPP) of the PV generator, which is the optimal operating point. The simulation results showed a good performance for the identifier and the controller as well. *Copyright © 2003 IFAC*

Keywords: Neural Networks, DC Motors, PV Generators, Power Converters

1. INTRODUCTION

Since the interests in the environmental issues are increasing and the conventional sources of energy are decreasing, renewable forms of energy such as solar energy are explored, where no noise, no pollution, no fuel and little maintenance are needed. Many researchers have investigated the steady state and transient performance of the dc motors (separately, series and shunt) when they are supplied directly from Photovoltaic (PV) sources (Appelbaum, 1986; Appelbaum and Sarma, 1989 and alghuwanem, 1992). It is revealed that the separately excited dc motor loaded with a centrifugal pump is the best candidate as far as better matching of the PV generator. Also to extract the maximum power from the PV source, a power converter must be connected between the PV source and the dc motor.

Many kinds of Neural Networks (NNs) have been devised and applied in different fields. The vast majority of NNs in use are those whose parameters are tuned by the gradient-based supervised learning. This category includes feedforward/multi-layer NNs, various types of recurrent NNs, radial basis function NNs and some networks with special architecture such as time delay NNs. Although it seems that these networks have different architecture and trained by distinguishable training algorithms, they can be unified in a single framework, with regard to their architecture and learning algorithms. Therefore, Universal Learning Networks (ULNs) have been proposed to provide a universal framework for the class of NNs and to unify their learning algorithms (Hirasawa, *et al.*, 2000) Also ULNs can be used in the stability analysis of any dynamical system by checking if their first and higher order derivatives converge to zero or not. Therefore this paper is considered as preliminary study that precedes the proposed stability analysis method that is not addressed here but will be discussed later in a separate paper.

This paper introduces two ULNs trained by the forward propagation algorithm as a new application in the field of DC motor drives. At first, the Universal Learning Network Identifier (ULNI) is trained offline to copy the dynamic behavior of the overall system. This ULNI is then used for the online training of the Universal Learning Network Controller (ULNC), which controls the converter duty ratio, to allow the motor speed to follow a specified reference signal and to extract the maximum power from the PV source.

This paper is organized as follows: In the next section, a brief introduction about ULNs and how they could be trained by the forward propagation algorithms is given. In section 3, a mathematical model for the proposed dc motor system is derived. The offline identification of the dc motor system using the ULNI and its simulation results are given in section 4. Then the online control algorithm for the motor system using the ULNC and its simulation results are investigated in section 5. Finally the major conclusion and some thoughts for the future research are given in section 6.

2. UNIVERSAL LEARNING NETWORKS

As said before, the purpose of ULNs is to provide a general framework for modeling and control of the complex systems that widely used in the real world. Also they can be used in the stability analysis of any nonlinear dynamical system by calculating their first and higher ordered derivatives.

2.1 Learning of ULNs

Based on the gradient method given in (1), learning of ULNs is realized by minimizing a criterion function, E.

$$\lambda \leftarrow \lambda - \eta \frac{\partial^{\dagger} E}{\partial \lambda}, \quad (1)$$

where λ is the network free parameter to be adjusted, η is the learning rate, and $\frac{\partial^{\dagger} E}{\partial \lambda}$ is the ordered derivative which means the net change of E caused by the change of λ while other variables being constant.

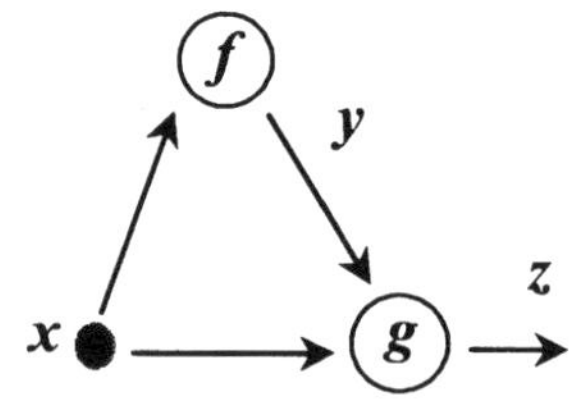

Fig. 1. Ordered and ordinary derivatives

To understand the difference between the ordered derivatives and the ordinary partial derivatives, let us consider the simple network shown in figure 1, where z is function of both x and y; while y is function in x:

$$y = f(x) \quad (2)$$

$$z = g(x, y) \quad (3)$$

The ordinary partial derivative $\partial z/\partial x$ is calculated without considering the relation between x and y as given in (4).

$$\frac{\partial z}{\partial x} = \frac{\partial g(x, y)}{\partial x} \quad (4)$$

On the other hand, the ordered derivative is calculated considering the indirect causal relationship as in (5).

$$\frac{\partial^{\dagger} z}{\partial x} = \frac{\partial g(x, f(x))}{\partial x} = \frac{\partial g(x, y)}{\partial y}|_{y=f(x)} \frac{\partial f(x)}{\partial x} + \frac{\partial g(x, y)}{\partial x}|_{y=f(x)} \quad (5)$$

Therefore, the ordered derivative considers both the direct and indirect paths.

2.2 Derivative Calculation

To calculate the ordered derivative $\frac{\partial^{\dagger} E}{\partial \lambda}$, both the direct and indirect relation between E and λ are considered. The direct relation is calculated by $\partial E/\partial \lambda$, and its value always equal zero for our case. But the indirect relation is calculated either by backward or forward propagation algorithms. Here the forward propagation algorithm is explained only.

To evaluate the indirect relation between E and λ, the node output (h_r), that directly influence E is considered as intermediate variable. Therefore, $\frac{\partial^{\dagger} E}{\partial \lambda}$ is calculated in (6).

$$\frac{\partial^{\dagger} E}{\partial \lambda} = \sum_{r \in J_o} \sum_{s \in T_o} [\frac{\partial E}{\partial h_r(s)} \frac{\partial^{\dagger} h_r(s)}{\partial \lambda}] + \frac{\partial E}{\partial \lambda} \quad (6)$$

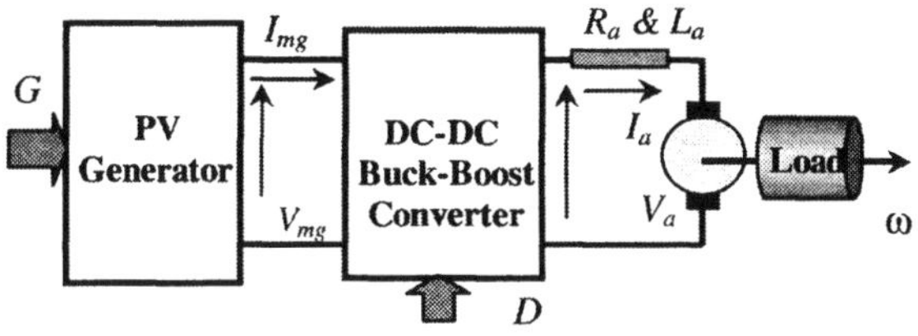

Fig. 2. The proposed dc motor system

where, J_o is the set of the suffixes of the nodes directly related to the evaluation of E. And T_o is the set of time instants at which E is evaluated. T_o is also referred as the updating window size instants.

Let us introduce the notation $P(j,t,\lambda)$ to represent the ordered derivative $\frac{\partial^\dagger h_j(t)}{\partial \lambda}$, which can be calculated using the forward propagation algorithm given in (7).

$$P(j,t,\lambda) = \sum_{i \in J_F(j)} [\frac{\partial h_j(t)}{\partial h_i(t-\tau_{ij})} P(i, t-\tau_{ij}, \lambda)] + \frac{\partial h_j(t)}{\partial \lambda} \quad (j \in J) \text{ and } (t \in T_o) \quad (7)$$

where, $J_F(j)$ is the set of suffixes of the nodes connected to node j. J is the set of suffixes of the total nodes. And τ_{ij} is the time delay of the branches from node i to node j. If the branch connecting nodes i and j has no time delay, then τ_{ij} is set to zero. The ordered derivative $\frac{\partial^\dagger E}{\partial \lambda}$ can be calculated using both (6) and (7) iteratively.

3. SYSTEM MODEL

The dc motor system consists of three main devices: PV generator, dc-dc converter and dc motor coupled to a centrifugal pump as shown in figure 2. Since this system is not installed yet, a mathematical model is needed to generate the input/output training data. In the following subsections a mathematical model for each device is developed and combined together to form the complete model used in the simulation studies.

3.1 PV Generator Model

The PV generator consists of solar cells connected in series and parallel fashion to provide the desired voltage and current required by the system. This PV generator exhibits a nonlinear insolation-dependent voltage-current characteristic governed by (8) and shown in figure 3.

$$V_g = \frac{1}{\Lambda_g} \ln(\frac{G \times I_{phg} + I_{og} - I_g}{I_{og}}) - I_g R_{sg} \quad (8)$$

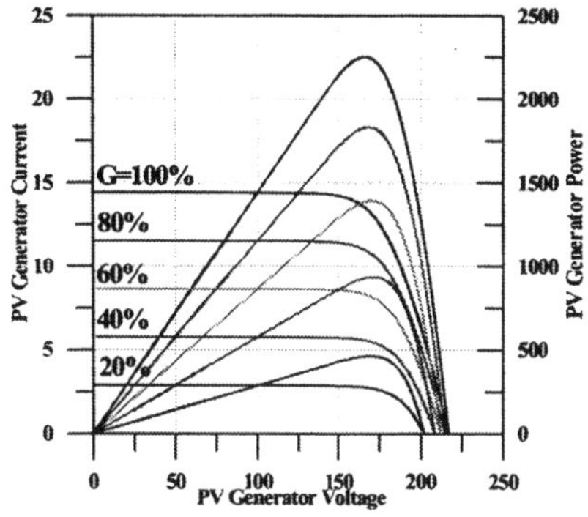

Fig. 3. PV generator characteristics

where V_g is the PV generator voltage; I_g is the PV generator current; $\Lambda_g = \frac{\Lambda}{N_s}$ is the PV generator constant; $\Lambda = \frac{q}{\varepsilon.Z.T}$, is the solar cell constant; $\varepsilon = 1.0$ is the completion factor; $T = 298.15°C$ is the absolute temperature; $q = 1.602 \times 10^{-19}$C is the electric charge; $Z = 1.38 \times 10^{-23}$J/K is the Boltzman constant; $N_s = 360$ is the series connected solar cells; $R_{sg} = Rs\frac{N_s}{N_p}$ is the PV generator series resistance; $R_s = 0.0152\Omega$ is the series resistance of one solar cell; $N_p = 3$ is the parallel paths; $I_{phg} = I_{ph} \times Np$ is the insolation-dependent photo current of the PV generator; $I_{ph} = 4.8A$ is the photo current of one solar cell; $I_{og} = I_o \times Np$ is the PV generator reverse saturation current; $I_o = 3.0797 \times 10^{-10}A$ is the reverse saturation current of one solar cell; G is the per unit solar insolation, and 1.0 per unit of $G = 1000W/m^2$.

From figure 3, at any value of G there is only one point at which the output power is maximum. This point is called the Maximum Power Point (MPP). Due to the high cost of the PV generators, it is recommended to operate at this MPP at all values of G. To locate this point, its corresponding voltage (V_{mg}) and current (I_{mg}) must be determined.

3.2 DC Motor Model

The dynamics of the separately excited dc motor and its load are represented by the following set of differential equations:

$$v_a(t) = R_a i_a(t) + L_a \frac{di_a(t)}{dt} + K\omega(t) \quad (9)$$

$$Ki_a(t) = A_1 + B\omega(t) + J_a \frac{d\omega(t)}{dt} + T_L(t) \quad (10)$$

$$T_L(t) = A_2 + \xi\omega^{1.8} \quad (11)$$

Where $V_a = 120volt$, is the motor voltage; $I_a = 9.2A$, is the motor current; $\omega = 1500rpm$, is the motor speed; $J_a = 0.02365Kg.m^2$ is the motor shaft inertia; $R_a = 1.5\Omega$, is the armature resistance; $L_a = 0.2H$, is the armature inductance; $K = 0.67609Nm.A^{-1}$, is the torque and back EMF constant; $A_1 = 0.2Nm$, is the motor friction; $A_2 = 0.3Nm$, is the load friction; $B = 0.002387Nm.s.rad^{-1}$, is the motor damping; $\xi = 0.00059Nm.s.rad^{-1}$, is the load torque constant.

Using a sampling time interval ΔT of 0.001s, and a first-order finite-difference approximation for the motor speed and current, the finite difference equation that governs the discrete-time dynamics of the dc motor is given in (12).

$$\begin{aligned}\omega(k) &= \alpha v_a(k-1) - \beta - \gamma\omega(k-1) - \theta\omega^{1.8}(k-1) \\ &\quad + \sigma\omega(k-2) + \psi\omega^{1.8}(k-2) \qquad (12)\end{aligned}$$

Where $\alpha, \beta, \gamma, \theta, \sigma$ and ψ are constants depend on ΔT and the motor/load parameters.

3.3 DC-DC Buck-Boost Converter Model

The most important parameter of the converter is its chopping ratio Y that depends on the duty ratio D through a nonlinear relation given in (13). Since $D(k)$ is changing with sampling instant k, the converter acts as a transformer whose turns ratio is varying. This converter is inserted between the PV generator and the dc motor to match their characteristics.

$$Y(k) = \frac{D(k)}{1 - D(k)} \qquad (13)$$

Assuming the converter is ideal, then its input and output powers are equal resulting in the following relation:

$$\frac{v_a(k)}{V_{mg}(k)} = \frac{I_{mg}(k)}{i_a(k)} = Y(k) \qquad (14)$$

4. OFFLINE SYSTEM IDENTIFICATION

The identification scheme of the dc motor system is shown in figure 4, where the ULNI is placed in parallel with the dc motor system model. The values $D(k), D(k-1), V_{mg}(k), \omega(k-1)$ and $\omega(k-2)$ are selected as its inputs, whereas the motor identified speed $\hat{\omega}(k)$ is the only output. Both $D(k)$ and $G(k)$ are generated randomly as shown in figure 5. Their values are assumed such that they cover the expected range of the real input signals. The identifier free parameters λ_i are updated every $T_o = 1000$ sampling instants based on the gradient method given in (15).

$$\lambda_i \leftarrow \lambda_i - \eta_i \frac{\partial^\dagger E_i}{\partial \lambda_i} + \mu_i \Delta\lambda_i, \qquad (15)$$

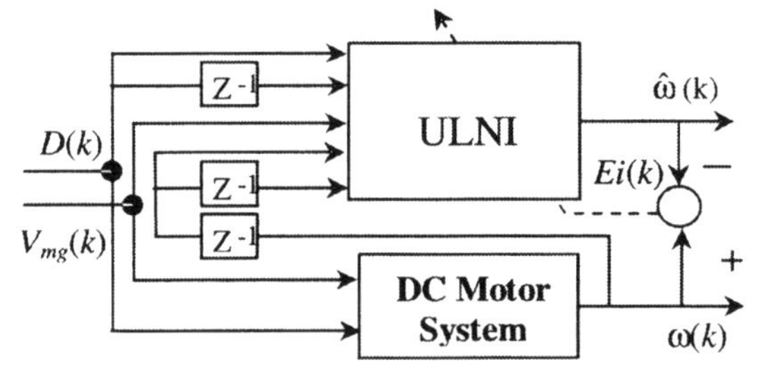

Fig. 4. DC motor system identification scheme

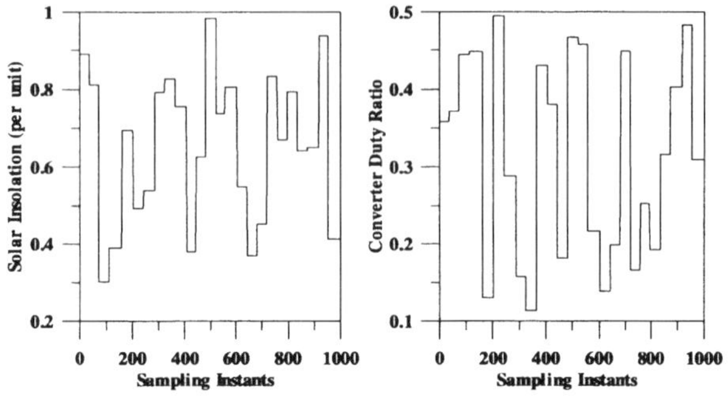

Fig. 5. Training data for the identifier network

Where η_i and μ_i are the learning rate and the momentum coefficient of the identifier network.

The main objective of the ULNI is to minimize the error signal E_i between $\hat{\omega}(s)$ and $\omega(s)$ given in (16)

$$E_i = \frac{1}{2} \sum_{s \in T_o} [\omega(s) - \hat{\omega}(s)]^2 \qquad (16)$$

As said before to calculate $\frac{\partial^\dagger E_i}{\partial \lambda_i}$, $\hat{\omega}(s)$ is considered as an intermediate variable. Therefore equation (6) can be rewritten as in (17)

$$\frac{\partial^\dagger E_i}{\partial \lambda_i} = \sum_{s \in T_o} [(\hat{\omega}(s) - \omega(s)) \frac{\partial^\dagger \hat{\omega}(s)}{\partial \lambda_i}] \qquad (17)$$

To calculate $\frac{\partial^\dagger \hat{\omega}(s)}{\partial \lambda_i}$, the ordered derivatives of all nodes that located before the identifier output nodes must be calculated first.

4.1 ULNI Simulation Results

The identification network used here is a feed forward network with architecture 5-8-1. The output layer node uses a sigmoidal activation function, while those of the hidden layer use a bipolar sigmoidal activation function. In the training procedure of the identifier network, there is no exact rule to select the number of the hidden nodes or the parameters η_i and μ_i. Usually the choice is based on the experience and the trial-and-error method. Therefore, the objective is to

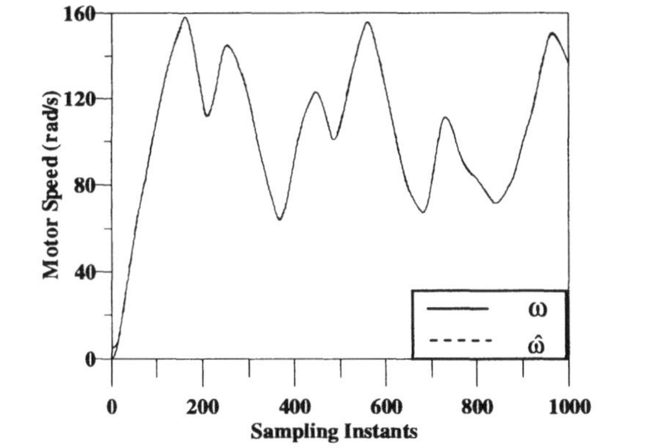

Fig. 6. Actual and identified motor speed at the end of identification process

find a combination of parameters, which gives an identification error with an acceptable tolerance. For $\eta_i = 1 \times 10^{-7}$, $\mu_i = 0.09$ and after 5×10^5 iterations, the final error E_i is found to be 0.03. At the end of identification process, $\hat{\omega}$ and ω are compared together as shown in figure 6. From this figure, it is found that there is a small difference between these two curves at the first few sampling instants. This may due to the starting characteristics of the dc motor. Otherwise the two curves are almost the same. Although the difference between the two curves is very small, this can't guarantee efficient identification algorithm. Consequently, exciting the dc motor with another set of $G(k)$ and $D(k)$ - as given in (18) and (19), respectively- must examine the identifier performance. During the examination procedure, $\hat{\omega}(k-1)$ and $\hat{\omega}(k-2)$ are used as inputs to the ULNI instead of $\omega(k-1)$ and $\omega(k-2)$. The examination results that are not shown here -for brevity- showed that both of $\hat{\omega}$ and ω are very close to each other indicating an accurate identification algorithm.

$$G(k) = 0.3 + 0.7\sin(\pi k \Delta T) \quad (18)$$

$$D(k) = 0.1 + 0.5\sin(\pi k \Delta T) \quad (19)$$

5. ONLINE SYSTEM CONTROL

The main objective of the Universal Learning Network controller (ULNC) is to continuously control the converter duty ratio. Consequently the dc motor can operate at the MPP of the PV generator. In addition to that, its speed can follow a reference signal (ω_{ref}), which assigned values correspond to the desired speed range as described by (20).

$$\omega_{ref}(k) = 110 + 15[\sin(2\pi k \Delta T) + \cos(2\pi k \Delta T)] \quad (20)$$

The control scheme of the proposed dc motor system is shown in figure 7, in which the dc motor system is replaced by the ULNI. The values

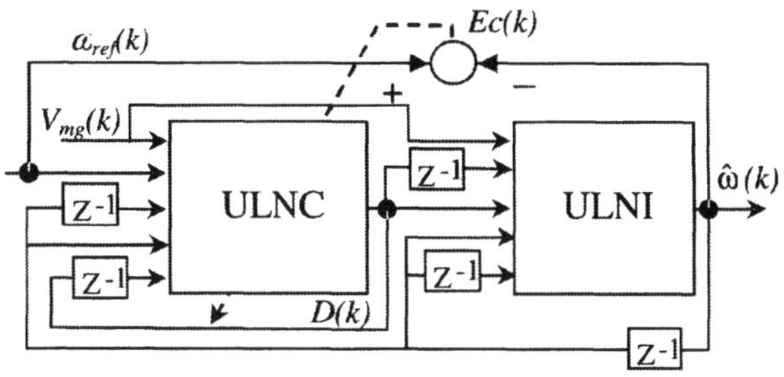

Fig. 7. DC motor system control scheme

$V_{mg}(k), \omega_{ref}(k), \hat{\omega}(k-1), \hat{\omega}(k-2)$ and $D(k-1)$ are selected as the inputs to the ULNC, while the converter duty ratio $D(k)$ is its only output. The ULNC free parameters λ_c are updated every window size $T_o = 10$ sampling instants based on the gradient method given in (21) in order to minimize the error $E_c(k)$ that given in (22).

$$\lambda_c \leftarrow \lambda_c - \eta_c \frac{\partial^{\dagger} E_c}{\partial \lambda_c} + \mu_c \Delta \lambda_c, \quad (21)$$

Where η_c and μ_c are the learning rate and the momentum coefficient of the controller network.

$$E_c = \frac{1}{2} \sum_{s \in T_o} [\omega_{ref}(s) - \hat{\omega}(s)]^2 \quad (22)$$

The ordered derivative, $\frac{\partial^{\dagger} E_c}{\partial \lambda_c}$, can be calculated by considering $\hat{\omega}(s)$ as intermediate variable as explained in (23)

$$\frac{\partial^{\dagger} E_c}{\partial \lambda_c} = \sum_{s \in T_o} [(\hat{\omega}(s) - \omega_{ref}(s)) \frac{\partial^{\dagger} \hat{\omega}(s)}{\partial \lambda_c}] \quad (23)$$

To calculate $\frac{\partial^{\dagger} \hat{\omega}(s)}{\partial \lambda_c}$, the ordered derivatives of all nodes that located before the identifier output nodes, including the controller nodes, must be calculated first.

The controller performance is investigated for different shapes of the reference signal and showed a good tracking capability. For brevity, the only results discussed here are those related to the per unit solar insolation $G(k)$ and the reference speed $\omega_{ref}(k)$ that given in (18) and (20), respectively.

5.1 ULNC Simulation Results

The network architecture, used as a controller, is similar to that used for identification, except that the number of hidden nodes is reduced to 5. Based on trial-and-error method, both of η_c and μ_c are adjusted to 1×10^{-4} and 0.17 respectively.

The controller learning curve, which indicates the change of the controller error, E_c, with the

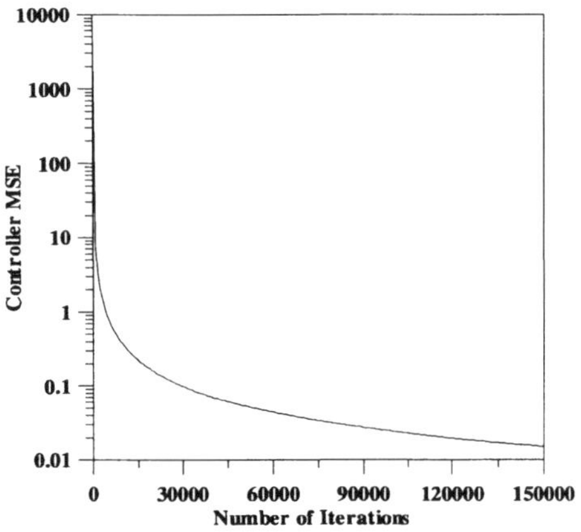

Fig. 8. Learning curve of the ULNC

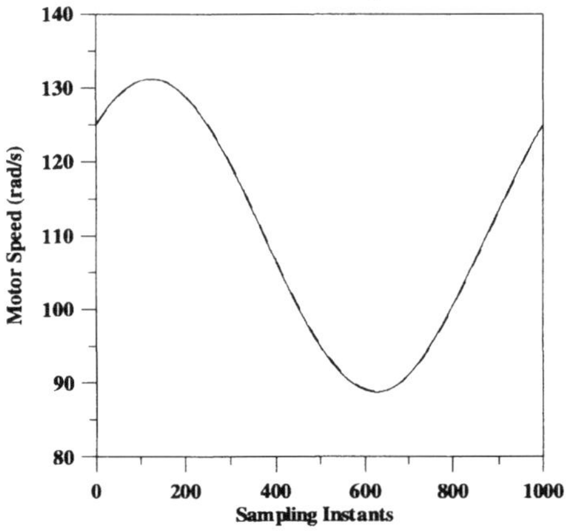

Fig. 9. Speed comparison at the end of operation of the ULNC

number of iterations, is shown in figure 8. From this figure, it is observed that after 15×10^4 iterations, an error less than 0.02 is obtained. At this point, the motor identified speed is compared to the reference signal as shown in figure 9. From which, it is found that the two curves are very close to each other indicating a good performance for the controller and the identifier as well. The variation of the controller output, $D(k)$, with time at the end of the controller operation is shown in figure 10. From this figure, the converter duty ratio $D(k)$ is changing smoothly depending on the shape of the reference signal.

6. CONCLUSION AND FUTURE RESEARCH

In this paper, a separately excited dc motor loaded with a centrifugal pump and fed from PV generator through a dc-dc buck-boost converter has been controlled successfully using a Universal Learning Network Controller (ULNC). Two universal learning networks are used. At first, the Universal Learning Network Identifier (ULNI) is trained offline to emulate the input/output behavior of the dc motor system. Then the ULNC is trained

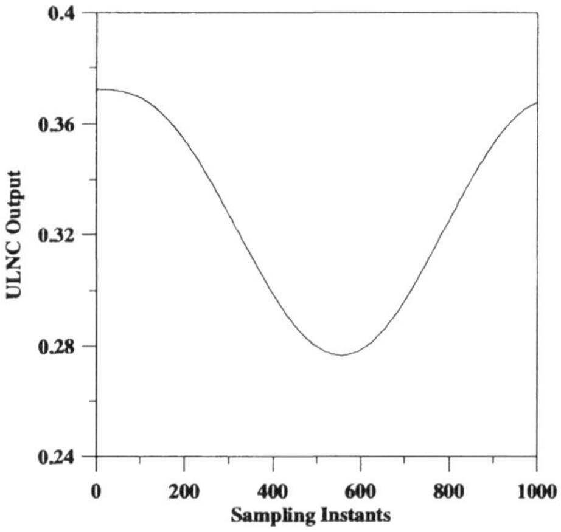

Fig. 10. Controller output at the end of operation

online to control the converter duty ratio to allow the motor speed to follow a reference signal and to extract the maximum power from the PV source. Both networks are trained using the forward propagation algorithm, which is the main property of the ULNs. The simulation results showed a good performance for the identifier and the controller as well. The overall system stability was not discussed in this paper even it is a unique point. But it will be discussed later using the higher-order derivatives of the ULNs. Also the dc motor system will be verified experimentally. Therefore the system identification and control algorithms will be designed based on a real data. So that, the mathematical model used here will not be used any more.

ACKNOWLEDGMENTS

This research was partly supported by the 21st Century COE Program.

REFERENCES

Appelbaum, J. (1986) Starting and steady state characteristics of dc motors powered by solar cell generators. *IEEE Trans. Energy Conversion*, **Vol. 1**, pp. 17-25.

Appelbaum, J. and M.S. Sarma, (1989) The operation of permanent magnet dc motors powered by a common source of solar cells. *IEEE Trans. Energy Conversion*, **Vol. 4**, pp. 635-642.

Alghuwainem, S.M. (1992) Steady state performance of dc motors supplied from photovoltaic generators with a step-up converter. *IEEE Trans. Energy Conversion*, **Vol. 7**, pp. 267-272.

Hirasawa, K., X. Wang, J. Murata, J. Hu, and C. Jin (2000). Universal learning network and its application to chaos control. *Neural Network*, **Vol. 13**, pp. 239-253.

IFAC Publications
www.elsevier.com/locate/ifac

VALIDATION OF STABILITY FOR AN INDUCTION MACHINE DRIVE USING EXPERIMENTS [1]

Henrik Mosskull *,** **Bo Wahlberg** * **Johann Galic** **

* *S3 - Automatic Control, KTH, SE-100 44 Stockholm, Sweden.*
(bo.wahlberg@s3.kth.se)
** *Bombardier Transportation, SE-721 73 Västerås, Sweden*

Abstract: The objective of this contribution is to estimate and validate the closed loop properties of a non-linear induction machine drive. In particular, it is of importance to be able to verify the stability margins of the system. First, a linear approximate transfer function model of the system is identified from experimental data. The remaining model errors are mainly due to non-linear dynamics and distortion. In a second step, the gain of the nonlinear error system is estimated. Robust stability can then be analyzed by combining the small gain theorem by a stability test of the linear model. The approach taken is inspired by the work (Schoukens *et al.*, 2002), on identification of the stability of feedback systems in the presence of nonlinear distortions. The problem at hand addresses several of the challenges of system identification for control. The system is highly nonlinear and dependent of the operating conditions. The stability problem is nontrivial, and has large practical implications. The approach is evaluated on a hardware/software simulation platform. *Copyright © 2003 IFAC*

Keywords: Identification of nonlinear system, identification for control, frequency domain system identification, application of identification, induction machine, stability.

1. INTRODUCTION

Bombardier Transportation is a world leader in the development and supply of complete rail vehicle transportation systems from large-scale urban transit systems to airport people movers. New developments in power converter technology and control technology have lead to higher levels of performance, increased safety and improved comfort. Control of these modern traction drives is a challenging problem, where the overall system stability is of most importance.

Our objective is to estimate and validate the closed loop properties of a non-linear induction drive from experimental data. The paper is organized as follows. In Section 2 it is outlined how to use system identification to estimate a linear approximate model of the system, together with bounds on the non-linear model errors. Section 3 gives an overview of robust stability using the small gain theorem. In Section 4, it is explained how to model the induction motor drive, so the standard framework of robust feedback control can be applied to analyze stability. Experimental results of the approach are given in Section 5, while Section 6 concludes the paper.

2. MODELLING AND IDENTIFICATION

The theory and practice of system identification of linear time invariant (LTI) models are well developed and advanced methods and tools are available. The problem gets more difficult when considering non-linear systems. When applying standard identification methods for estimating linear models to data from a nonlinear system, the best linear approximation in a certain sense is obtained. Consider a non-linear dynamical system with input $\{u(t)\}$ and corresponding

[1] This work was partially supported by The Swedish Science Foundation

output $\{y(t)\}$. The best linear model of this input-output relation, in the sense that it produces the same second order statistics, is given by the frequency response

$$G_0(e^{i\omega}) = \left[\frac{\Phi_{yu}(e^{i\omega})}{\Phi_{uu}(e^{i\omega})}\right]_{causal} \tag{1}$$

where subscript *causal* denotes the causal part, $\Phi_{uu}(e^{i\omega})$ is the power spectral density of the input and $\Phi_{yu}(e^{i\omega})$ is the cross power spectral density of the input and output signals. This linear approximation depends of course on the specific input signal. This is closely related to linearization using Taylor expansion of the nonlinear system, which often is done around a given operating point.

We will consider variations around a specified operating point $u(t) = u_0$ and $y(t) = y_0$. It is often easier to remove the DC-components before forming the model, i.e. to work with the signals $\Delta u(t) = u(t) - u_0$ and $\Delta y(t) = y(t) - y_0$. We will assume that the DC-components have been removed, but will still use the notations $u(t)$ and $y(t)$.

A common choice of input signal is $u(t) = Csin(\omega_0 t)$, i.e. a sinusoidal signal with frequency ω_0 and amplitude C. Let $y(t)$ be the corresponding output. From observations $\{y(t), u(t),\ t = 1,2\ldots N\}$, the (finite) Fourier transforms of the signals can be calculated as

$$U_N(e^{i\omega}) = \frac{1}{\sqrt{N}}\sum_{t=1}^{N} u(t)e^{-i\omega t}, \tag{2}$$

$$Y_N(e^{i\omega}) = \frac{1}{\sqrt{N}}\sum_{t=1}^{N} y(t)e^{-i\omega t}. \tag{3}$$

The Fourier transform of $u(t)$, is then approximately equal to $NC^2/4$ for $\omega = \omega_0$ and almost zero otherwise. The empirical transfer function estimate equals

$$\hat{G}_N(e^{i\omega_0}) = \frac{Y_N(e^{i\omega_0})}{U_N(e^{i\omega_0})}. \tag{4}$$

It is possible to show, (Ljung, 1987), that $\hat{G}_N(e^{i\omega_0})$ converges to $G_0(e^{i\omega_0})$, under certain assumptions on the system and the disturbances, as N tends to infinity. In this case $G_0(e^{i\omega_0})$ is the best linear frequency response approximation of the nonlinear system for a sinusoidal input signal with amplitude C and frequency ω_0 (around the operating point u_0).

A parametric time-domain model can be estimated using a prediction error identification method. This approach can be viewed as finding a parametric approximation of the frequency response $G_0(e^{i\omega})$. Consider the linear time-invariant output error model

$$y(t) = G(q,\theta)u(t) + e(t) \tag{5}$$

Here $\{y(t)\}$ $\{u(t)\}$ are the output and input signals, respectively, and $\{e(t)\}$ is additive noise. The transfer function $G(q,\theta)$ is a function of the shift operator $q^{-1}u(t) = u(t-1)$ and an unknown parameter vector θ.

Given observations $\{y(t), u(t), t = 1,2\ldots N\}$ from the underlying non-linear system, the prediction error estimate is given by

$$\hat{\theta}_N = \arg\min_{\theta}\sum_{t=1}^{N}|y(t) - G(q,\theta)u(t)|^2 \tag{6}$$

From (Ljung, 2001) it follows that

$$\hat{\theta}_N \to \arg\min_{\theta}\int_{-\pi}^{\pi}|G_0(e^{i\omega}) - G(e^{i\omega},\theta)|^2 d\omega, \tag{7}$$

i.e. the asymptotic parametric model is obtained by approximating G_0 in a frequency weighted sense.

Let us now represent the true system as

$$y(t) = G_0(q)u(t) + v(t), \tag{8}$$

where the additive error term $v(t)$ accounts for model errors as well as noise. Here $G_0(q)$ denotes the best linear model as discussed above. In case of a frequency domain model, the signal $G_0(q)u(t)$ can be calculated from the inverse Fourier transform of $G_o(e^{i\omega})U_N(e^{i\omega})$.

We will consider the following model error model

$$v(t) = v_{NL}(t) + e(t) \tag{9}$$

$$v_{NL}(t) = \tilde{g}(u),\ u = [u(t)\, u(t-1)\ldots u(0)]^T \tag{10}$$

$$||v_{NL}|| \leq \beta||u|| + \alpha, \quad \alpha,\beta \geq 0. \tag{11}$$

Here the L_2-norm is used, e.g.

$$||v_{NL}|| = \left[\sum_{t=1}^{\infty}[v_{NL}(t)]^2\right]^{1/2}. \tag{12}$$

The L_2-norm can also be calculated in the frequency domain using Parsevals relationship,

$$\sum_{t=1}^{N}[v_{NL}(t)]^2 = \sum_{k=1}^{N}|V_N(e^{i2\pi k/N})|^2, \tag{13}$$

where $V_N(e^{i\omega})$ is the Fourier-transform of $v_{NL}(t)$.

The error term has been decomposed into one non-linear input signal contribution $v_{NL}(t)$ and additive noise $e(t)$. The factor β is the gain of the non-linear system $\tilde{g}$. The constant α can be used to model offsets and external L_2 signals, see (Ljung, 2001). The L_2-gain of the nonlinear system $\tilde{g}$ is formally defined as

$$||\tilde{g}|| = \max_u \frac{||v_{NL}||}{||u||} \quad (14)$$

The problem is of course to find the input sequence which gives the maximal gain. For linear systems, the worst case input signal is a sinus and the problem simplifies to finding the frequency corresponding to maximum gain.

The gain of a static non-linearity can rather easily be estimated, in particular for single input single output systems. One way is to bound the observed data using piece-wise linear functions. The approach is very closely related to estimation of Wiener- and Hammarstein models. It is of course possible to estimate a model of the nonlinear system and then compute the gain of the estimate. Here model structures such as neural network, kernel models, local linear models, piecewise linear models can be used.

To conclude, the choice of input signal is most important when identifying a nonlinear system. It should resemble the input signal used in the application of the model. However, it is also important to obtain sufficient excitation to give a good estimate.

3. ROBUST STABILITY

Input-output L_2 stability of two interconnected systems, as illustrated in Figure 1 can now be studied using the small-gain theorem.

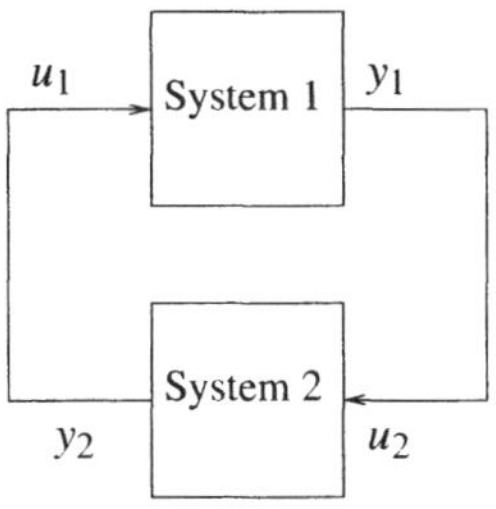

Fig. 1. Stability of two interconnected systems.

Suppose both systems are L_2 stable, i.e.

$$||y_1|| \leq \beta_1 ||u_1|| + \alpha_1 \quad (15)$$
$$||y_2|| \leq \beta_2 ||u_2|| + \alpha_2 \quad (16)$$

Then the feedback connection $u_1 = y_2$ and $u_2 = y_1$ is L_2 stable if $\beta_1\beta_2 < 1$, see (Khalil, 2002). This test is however quite conservative, since no phase information is used. However, by combining this result with stability tests for linear systems less conservative conditions can be obtained. An example is the circle criterion for static non-linearities.

Assume that the system (8) is controlled by a linear feedback regulator $F(q)$. The corresponding closed loop system is illustrated in Figure 2.

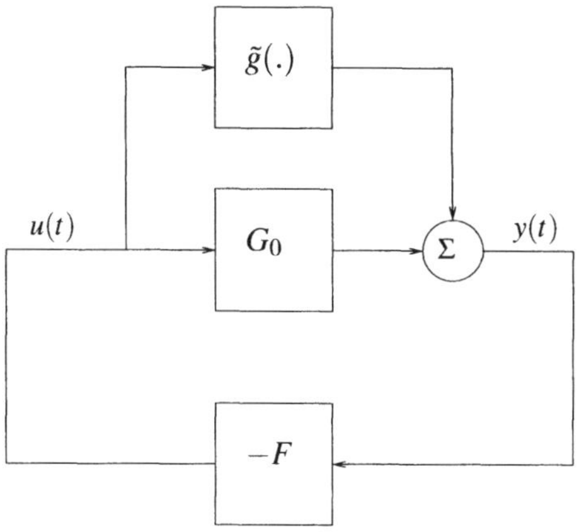

Fig. 2. Closed loop system.

The transfer function of the closed loop linear part of this system equals

$$\frac{-F(q)}{1 + G_0(q)F(q)}, \quad (17)$$

which means that the system can be re-represented as shown in Figure 3.

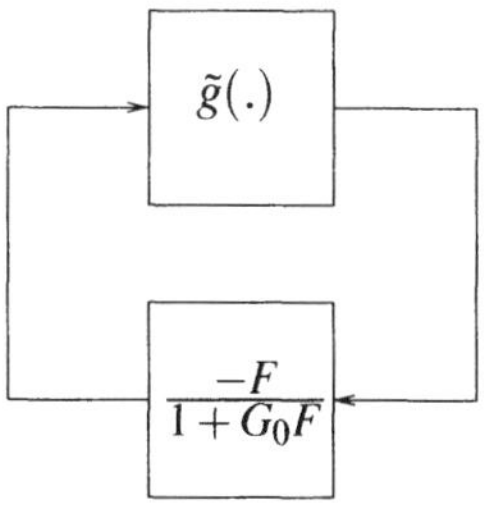

Fig. 3. Re-written closed loop system.

To start with, the closed loop linear system $F(q)/(1 + G_0(q)F(q))$ has to be stable. This is typically checked by plotting the Nyquist curve of $G_0(e^{i\omega})F(e^{i\omega})$. The small gain criterion can then be applied to analyze the stability of the closed loop system in Figure 3.

Assume that $F(q)/(1 + G_0(q)F(q))$ is stable, and that

$$\left| \frac{F(e^{i\omega})}{1 + G_0(e^{i\omega})F(e^{i\omega})} \right| < \beta_1, \quad \forall\omega \quad (18)$$
$$||\tilde{g}|| = \beta_2 \quad (19)$$

The feedback system in Figure 3 is input-output stable if $\beta_1\beta_2 < 1$.

A less conservative result can be obtained if it is possible to directly calculate the total open loop gain

$$v_{NL}(t) = \tilde{g}(u) \quad (20)$$
$$w(t) = -\frac{F(q)}{1 + G_0(q)F(q)} v_{NL}(t), \quad (21)$$

i.e. the gain from input u to the signal w. If

$$||w|| \leq \beta ||u|| + \alpha, \quad (22)$$

where $0 \leq \beta < 1$, the small-gain theorem implies that the closed loop system is stable. The advantage of this approach from a practical point of view is that it may be easier to estimate the maximum-gain input for this system, since it also involves the linear part.

We have outlined a discrete time theory. It is however trivial to modify all results to continuous time systems.

4. MODELLING OF INDUCTION MACHINE DRIVES

A typical drive with voltage source inverter, induction machine and RLC-network is shown in Figure 4.

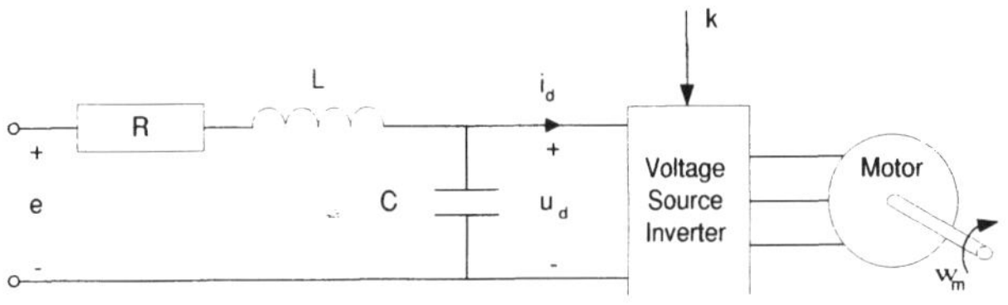

Fig. 4. Induction machine drive

The motors, usually one to four in parallel, are fed by a voltage source inverter converting a DC-voltage, $u_d(t)$, to three AC-voltages with variable amplitude and frequency. In order to suppress high frequency distortions generated by the inverter or the supply voltage $e(t)$, the drive has an RLC filter on the input. The resistance indicated in the figure represents resistance of the physical inductor and the cables. To minimize losses this resistance must be small, which implies that the filter is highly resonant.

The drive is controlled by the signal $k(t)$, which determines the state of each converter phase as a function of time. Furthermore, ω_M denotes the mechanical rotor speed. The electric RLC circuit can be described by the equations

$$u_d(t) = \frac{1}{C}\int_{t_0}^{t} i_C(\tau)d\tau + u_d(t_0) \tag{23}$$

$$i_L(t) = i_C(t) + i_d(t) \tag{24}$$

$$e(t) = Ri_L(t) + L\frac{d}{dt}i_L(t) + u_d(t) \tag{25}$$

with the obvious definitions of the currents i_L and i_C. Taking the Laplace transform of these equations gives

$$U_d(s) = Z_E(s)E(s) - Z_{DC}(s)I_d(s), \tag{26}$$

$$Z_E(s) = \frac{1}{LCs^2 + RCs + 1}, \tag{27}$$

$$Z_{DC}(s) = \frac{Ls + R}{LCs^2 + RCs + 1}. \tag{28}$$

The relation between the input voltage $u_d(t)$ to the voltage source inverter and the corresponding current $i_d(t)$ is determined by a quite complex non-linear dynamical system, which for example depends on the motor speed and load. Denote this input-output relation, called the input admittance, by

$$i_d(t) = f(u_d) \tag{29}$$

The key observation is now that this problem fits the stability framework outlined in Section 3. The equivalent feedback diagram is given in Figure 5.

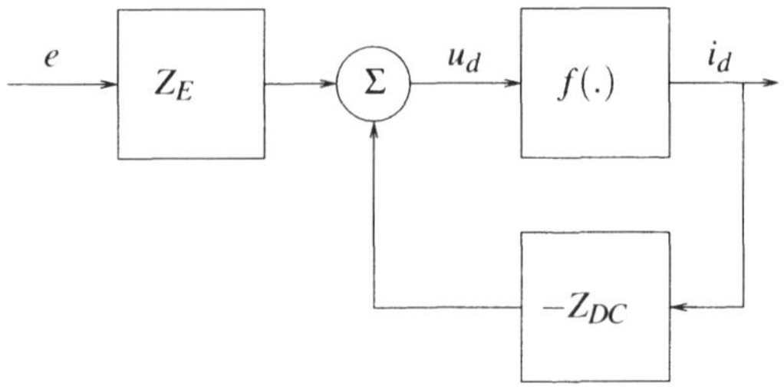

Fig. 5. Feedback representation

The impedance $Z_{DC}(s)$ will take the role of the controller, and the impedance $Z_E(s)$ will act as a pre-filter of the reference signal.

We have now represented the induction machine drive in a form which will allow stability and robustness analysis using an identified model of the input admittance and an estimation of the gain of the non-linear error system.

5. EXPERIMENTAL RESULTS

The aim of this chapter is to evaluate the stability margins of an induction motor drive through experiments. All experiments have been done on a hardware in the loop simulator at Bombardier Transportation, Sweden. An indication of the operating conditions are given by the following data:

DC-link capacitance	$C = 0.004$ F
DC-link inductance	$L = 0.005$ H
DC-link resistance	$R = 0.04$ Ohm
Nominal DC-link voltage	1700 V
Nominal motor speed	29 Hz

The resonance frequency of the input filter is given by $\omega_0 = 1/\sqrt{LC}$ and the damping factor by $\zeta = R\sqrt{C}/(2\sqrt{L})$. The damping factor for the system in the example will be around 0.02 and the resonance frequency equals 35.5 Hz.

The properties of the drive depend on the motor speed and torque load. Here we have evaluated the system for a number of operating points, for three load conditions at 100 different motor speeds. The motor speed has been varied between 0.2 pu (corresponds to 20 % of nominal speed) and 1.6 pu, at equidistant increments and the torque has been set to zero, maximum and minimum, respectively.

The linear transfer functions G_0 are estimated by Fourier analysis, see (4) and Section 6.2 in (Ljung, 1987). The corresponding frequency responses $G_0(e^{i\omega})$ for zero torque and for the specified values of motor speed, are given in Figure 6.

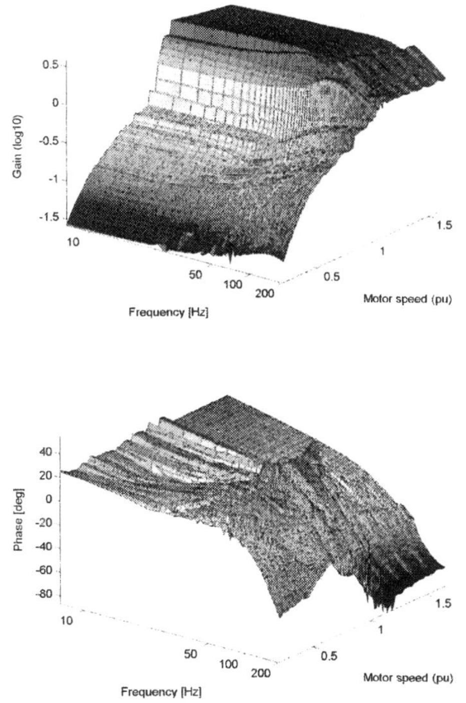

Fig. 6. Bode-plots of estimated frequency responses $G_0(e^{i\omega})$ as a function of frequency and motor speed at zero torque.

The Nyquist-plot of G_0Z_{DC} for zero torque, given Figure 7, shows that the linear closed loop system will be stable! The same will hold for the other operating points.

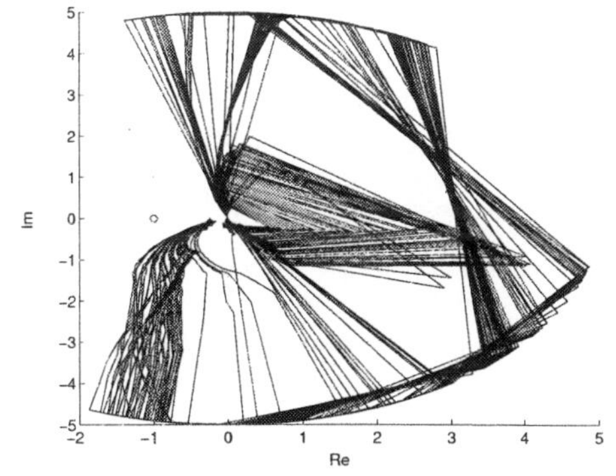

Fig. 7. Nyquist curves of G_0Z_{DC} for the operating points with zero torque. Observe that too large gains have been projected into the plot.

The model error is calculated from the in open-loop measured voltage and current

$$v_{NL}(t) = \tilde{g}(u_d) = i_d(t) - G_0(q)u_d(t) \qquad (30)$$

All calculations are, however, done in the frequency domain, since only the frequency response of G_0 is estimated. This error signal is then filtered by the linear transfer function

$$\frac{-F}{1+G_0F} = \frac{-Z_{DC}}{1+G_0Z_{DC}} \qquad (31)$$

to give the signal

$$w(t) = -\frac{Z_{DC}}{1+G_0Z_{DC}}v_{NL}(t). \qquad (32)$$

The gain from u_d to w is then estimated as

$$\hat{\beta} = \left[\frac{\sum_k |W_N(e^{i\omega_k})|^2}{\sum_k |[U_d]_N(e^{i\omega_k})|^2}\right]^{1/2}, \qquad (33)$$

where W_N and $[U_d]_N$ are the Fourier transforms of w and u_d, respectively. Here we have used Parseval's relationship (13) to calculate the gain in the frequency domain.

5.1 *Experimental Results*

A most difficult problem is to find the input signal which gives the maximum gain. By studying the magnitude plot of $Z_{DC}/(1+G_0Z_{DC})$ given in Figure 8 we can predict that the worst case input signal will have most of its energy around the resonant frequency of this transfer function. The choice of amplitude is important to obtain a good signal to noise ratio, but also to take the non-linear effects into account. The gain from u_d to w has been calculated for a large range of input signals. The conclusion is that a sinusoidal signal with frequency equals to the resonant frequency of $Z_{DC}/(1+G_0Z_{DC})$ gives a reasonable estimate of the maximum gain for the total non-linear system. This is in particular true in the more resonant low motor speed region, while a more wide band-signal gives slightly better results for the higher motor speed case.

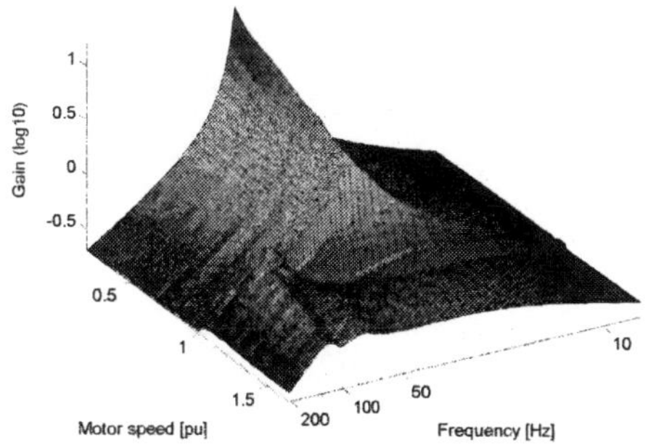

Fig. 8. Magnitude plot of $\frac{Z_{DC}}{1+G_0Z_{DC}}$ as a function of motor speed. The high gain peaks around the resonance frequency of the transfer function occur at low motor speed.

Figure 9 showes the experimental results of the estimated stability gains for zero torque as a function of motor speed.

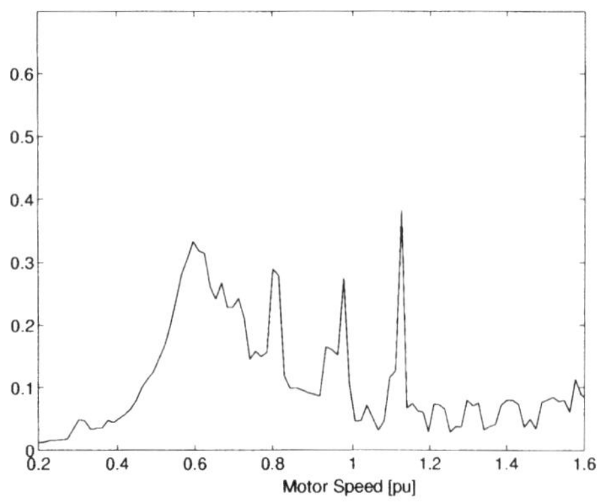

Fig. 9. Estimated stability gains $\hat{\beta}$ for zero torque as a function of motor speed.

Figures 10 and 11 give the experimental results for two other typical operations, namely maximal positive torque (driving) and maximal negative torque (braking).

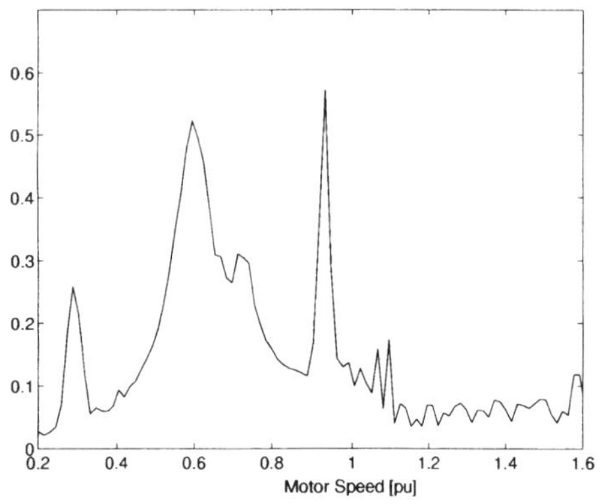

Fig. 10. Estimated stability gains $\hat{\beta}$ when driving as a function of motor speed.

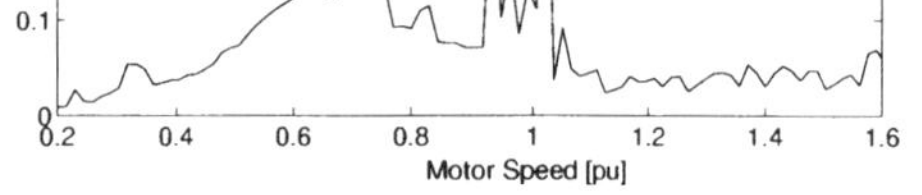

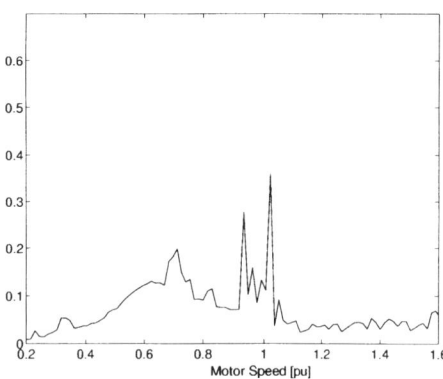

Fig. 11. Estimated stability gains $\hat{\beta}$ when breaking as a function of motor speed.

This result shows that the gains are well below one, and that we thus have quite good stability margin for most motor speeds. The results are more reliable for lower motor speeds. It should also be noticed that the controller of the motor has been carefully designed to optimize stability and performance.

6. CONCLUSION

Stabilization of induction motor drives with weak power supply is an important practical problem, and is for example discussed in (Bae *et al.*, 2001). The design of controllers for induction motors can be done using linearized model, see (Mosskull, 2002). Hence, it is of importance the to verify stability using experiments. Here we have studied how to validate stability margins using models obtained from frequency domain system identification together with simple estimates of the gain of the model error. Future work includes more advanced error models, to obtain less conservative stability results. One interesting approach is to use Integral Quadratic Constraints (IQC) to obtain better stability measures which also include phase information.

7. REFERENCES

Bae, B-H, B-H Cho and S-K Sul (2001). Damping control strategy for the vector controlled traction drive. In: *EPE*. Graz, Austria.

Khalil, H.K. (2002). *Nonlinear Systems, third edition.* Prentice Hall. New Jersey, USA.

Ljung, L. (1987). *System Identification: Theory for the User*. Prentice-Hall. Englewood Cliffs, NJ.

Ljung, Lennart (2001). Estimating linear time invariant models of non-linear time-varying systems. *European Journal of Control* **7**(2-3), 203–219. Semi-plenary presentation at the European Control Conference, Sept 2001.

Mosskull, H. (2003). DC-Link stabilization of an induction machine drive. In: *European Conference on Power Electronics and Applications*. Toulouse, France.

Schoukens, J, R. Pintelon and T. Dobrowiecki (2002). Identification of the stability of feedback systems in the presenence of nonlinear distorsion. In: *15th IFAC World Congress*. Barcelona, Spain.

IFAC Publications
www.elsevier.com/locate/ifac

AUTOMATIC STEERING CONTROL SYSTEM DESIGN UTILIZING A VISUAL FEEDBACK APPROACH

— System Identification and Control Experiments with a Radio-Controlled Car —

Shuichi Adachi[1), Tadahiro Fujihira[1) and Yukihiro Fujiwara[2)

1) Department of Electrical and Electronic Engineering, Utsunomiya University,
7-1-2 Yoto, Utsunomiya, 321-8585, Japan
e-mail : adachis@cc.utsunomiya-u.ac.jp
2) Honda R & D Co., Ltd / Utsunomiya University, Japan

Abstract: Advanced safety vehicles (ASV) assist drivers in avoiding traffic accidents, and research in all areas of automatic driving systems is necessary to further develop ASV technology. This paper focuses on a visual feedback approach in which an automatic driving system is obtained by recognizing the road trajectory using image information. The validity of this approach is examined by conducting experiments with a radio-controlled car. First, a practical image processing algorithm that recognizes white lines on the road is proposed. Next, a model of the radio-controlled car is built by system identification experiments and an automatic steering control system is designed based upon $\mathcal{H}_\infty$ control theory. Finally, the effectiveness of the proposed control system is examined via traveling experiments. *Copyright © 2003 IFAC*

Keywords: automatic steering, visual feedback, system identification, robust control.

1. INTRODUCTION

Intelligent Vehicle Highway Systems (IVHS) and Advanced Vehicle Control Systems (AVCS) have been investigated in order to prevent traffic accidents through intelligent sensing and control (Iguchi, 1996 and Miyazaki, 1997). In order to improve traffic accident prevention, it is necessary to both increase vehicle intelligence and to autonomously control vehicle mobility. There are, however, many obstacles to the realization of such functions, such as the recognition of the vehicles' running environments.

One way of recognizing a running environment is by using a camera (Dickmanns *et al*, 1988). A visual feedback approach, which employs the image information captured by a camera, has been studied before (Hutchinson *et al*, 1996 and Kovacs *et al*, 1998).

The feature value based method, a method that recognizes the shape of the road using feature values that are calculated from the camera's image information, has been proposed (Komori *et al*, 2000). This method has some advantages because it utilizes scalar-valued feature values instead of vector-valued position information. For example, the feature value based method doesn't need to transform the information from a two-dimensional image to a three-dimensional model, so less computational time is needed than that of the position based method. Komori's feature value based method has also been extended to a model containing uncertainty (Fujiwara *et al*, 2002). The effectiveness of these formulations was examined in numerical simulations, however, a real-car application was not conducted. This paper examines the effectiveness of the visual feedback approach for an automatic steering control system in a radio-controlled car.

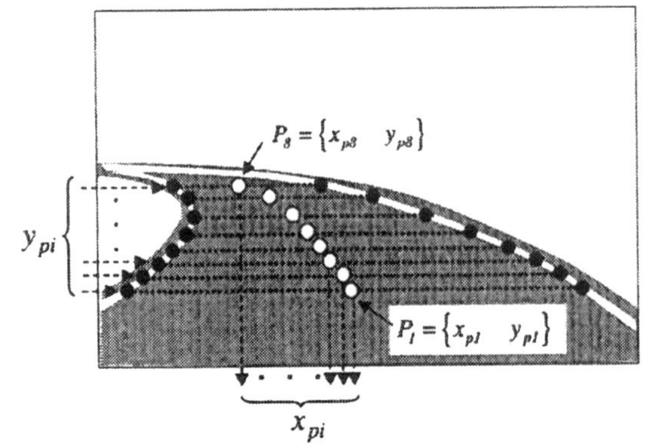

Figure 1: An example of a forward image and feature points P_i.

The paper is organized as follows. In Section 2, a practical image-processing algorithm is proposed. This algorithm calculates the feature value based upon the camera's image information. The mathematical model derived in Section 3 integrates the dynamics of the feature value and the vehicle dynamics of the radio-controlled car. An automatic steering control system based upon $\mathcal{H}_\infty$ control theory is designed in Section 4. The effectiveness of the designed system is then examined by traveling experiments in Section 5, and Section 6 concludes the paper.

2. ROAD SHAPE RECOGNITION USING IMAGE INFORMATION

2.1 Definition of feature value

An example of a forward image obtained by the camera is shown in Fig.1. In this figure, the filled circles, •, are extracted from positions on the white lines. Circles, ∘, are defined as feature points $P_i = [x_{pi} \ y_{pi}]^T \ (i = 1, 2, ..., 8)$. A feature value is then defined as

$$X_{sum} = \sum_{i=1}^{8} a_i x_{pi} \tag{1}$$

where a_i are weighting constants. These constants are design parameters.

2.2 Image processing algorithm

A practical algorithm to calculate the feature value based upon the image data is proposed below. The flowchart of this algorithm is shown in Fig.2, and each step is briefly illustrated using the image data shown in Fig.3.

In Step 1, an image frame is measured. Then, in Step 2, the initial image processing regions are de-

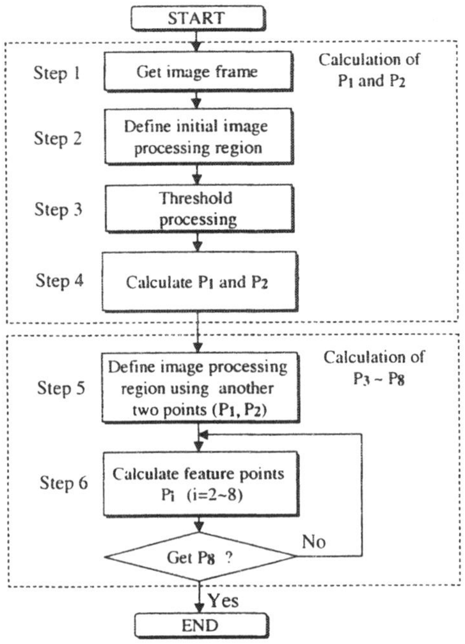

Figure 2: Flowchart for image processing.

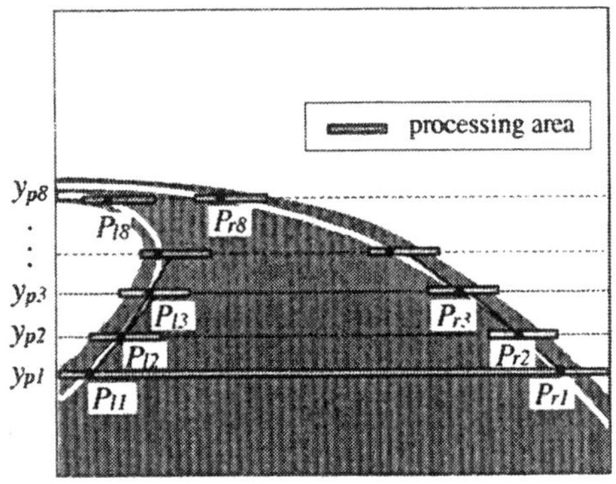

Figure 3: An example of image data.

fined in order to determine the points (P_{l1}, P_{r1}) and (P_{l2}, P_{r2}) shown in Fig.3. In Step 3, the white lines on the road are extracted by applying threshold processing to these regions, and points (P_{l1}, P_{r1}) and (P_{l2}, P_{r2}) on the road are determined. The center of each point, *i.e.*, $P_1 = \{x_{p1}, y_{p1}\}$ and $P_2 = \{x_{p2}, y_{p2}\}$ are calculated in Step 4.

Points P_3 to P_8 are determined in Steps 5 and 6. Step 5 restricts the image processing regions based upon the knowledge of (P_{l1}, P_{l2}) and (P_{r1}, P_{r2}). For example, P_{l3} is determined by searching around a restricted region defined by the area around where a line connecting P_{l1} and P_{l2} crosses the scanning line for y_{p3}. The remaining points are similarly determined. Step 6 then calculates the center points, *i.e.*, the feature points P_3 to P_8.

Figure 4: Radio-controlled car.

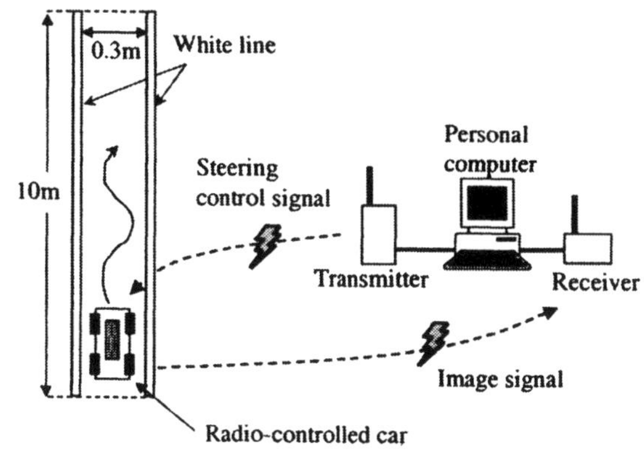

Figure 5: Configuration of the identification experiment.

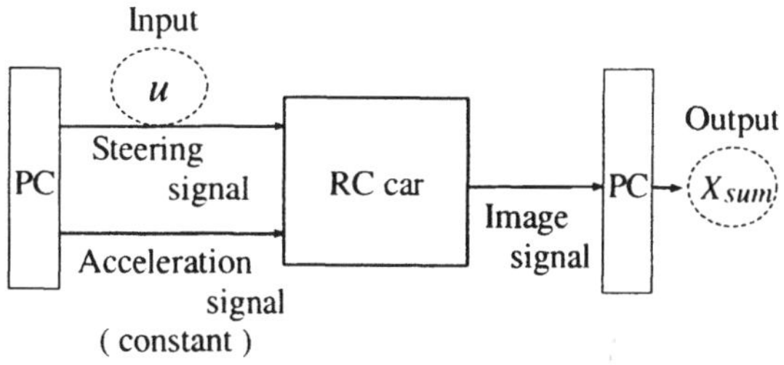

Figure 6: Signal flow of the identification experiment.

These steps are calculated in each frame. This algorithm is used to determine the feature points P_1 to P_8, and then Eq.(1) is used to calculate the feature values. The main advantage of the algorithm is that it restricts the image processing region, which shortens the calculation time and improves the robustness of the system.

3. MODELING OF RADIO-CONTROLLED CAR BY A SYSTEM IDENTIFICATION EXPERIMENT

3.1 Radio-controlled car

The radio-controlled car (RC car), shown in Fig.4, was used as a control object for the automatic steering control system based upon visual feedback. A CCD camera was installed at the center of the body of the RC car, which was a 1/10-scale model of a real car. System identification theory (*e.g.* Ljung, 1999) was used to create a model of the RC car because the precise values of the physical parameters, such as the center of gravity and the cornering forces, were unknown.

3.2 System identification experiment

The configuration of the system identification experiment is shown in Fig.5. In this experiment, the RC car was running between white lines 0.3m in width and 10m in length. The signal flow in this experiment is shown in Fig.6. The image information captured by the camera was transmitted to a personal computer (PC) by radio.

In this experiment, the input signal is the steering voltage signal u and the corresponding output signal is the feature value X_{sum}. The purpose of this identification, therefore, was to build a model from the steering signal to a feature value. It is noted that the RC car traveled at a constant speed.

A PRBS (pseudo random binary signal) with a 0.1s clock interval and 3V amplitude was employed as an input signal. The sampling interval was 1/30s, which was designated according to the camera's frame rate. The experiment ran for 20s, so 600 input-output data points were measured. These results are shown in Fig.7.

The least-squares method with the ARX model (ARX method), the prediction-error method with the ARMAX model, and spectral analysis were applied as identification methods. The identified gain characteristics are shown in Fig.8. From these plots it is evident that the methods all coincided well. Because of this, the most simple model, the ARX model, was selected as the identification model.

The identified 29th-order ARX model was then reduced to a 2nd-order model in consideration of pole/zero cancellation. A Bode diagram of the 2nd-order model is shown in Fig.9. The gain characteristics of the 2nd-order model shown in the upper figure (dashed line) matches that of the 29th-order model (dash-dotted line) quite well. On the other hand, it can be seen from the lower figure

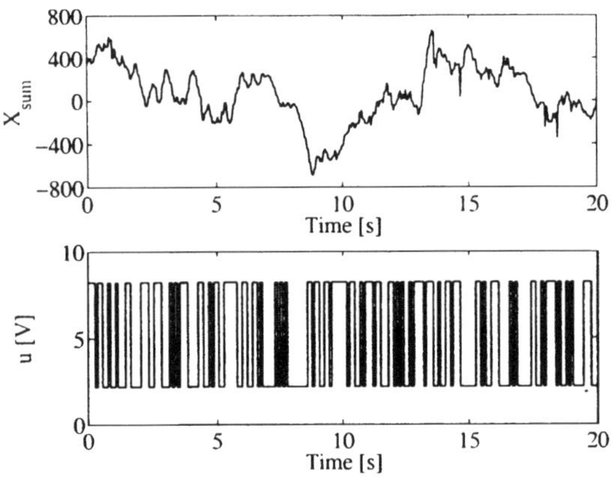

Figure 7: Input and output signals measured b the identification experiment.

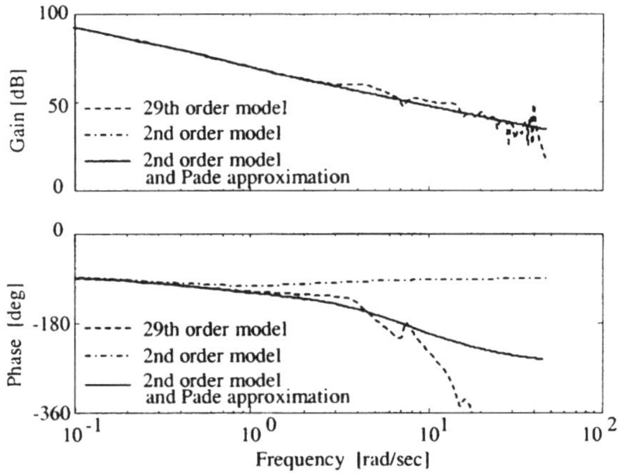

Figure 9: Bode diagram of the identified ARX models.

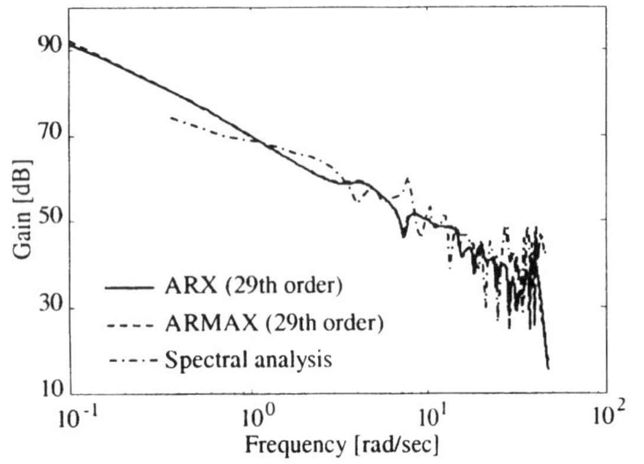

Figure 8: Gain characteristics for various identified models.

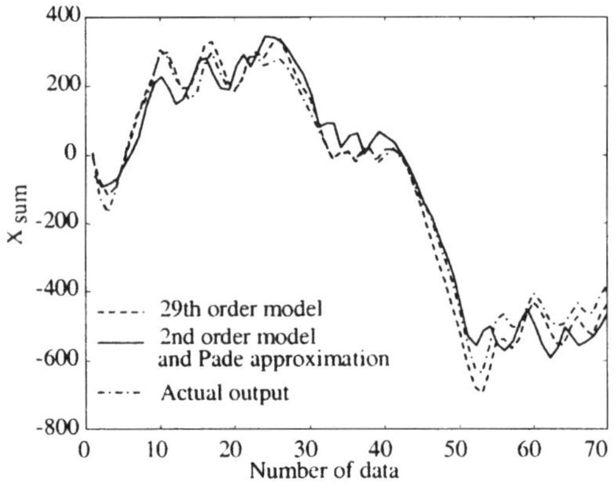

Figure 10: Comparison of time responses.

that there is a significant difference in phase characteristics between these two models. This characteristic signifies that the identification object has a dead-time due to the image processing system.

In order to account for the dead-time, a 1st-order Padé approximation is added to the 2nd-order model. As a result, a 3rd-order model

$$P_n(s) = \frac{-2372(s+1.268)(s-7.5)}{(s+0.7388)(s+0.009443)(s+7.5)} \qquad (2)$$

is obtained, which is depicted by the solid line in Fig.9. Eq.(2) is utilized as the nominal model in the control system design.

The phase characteristics of the model defined by Eq.(2) accords well with the 29th-order model below 1rad/s. The time responses of these models are plotted in Fig.10. It is clear that the model (2) plotted as the solid line agrees with the actual output.

4. STEERING CONTROL SYSTEM DESIGN BY H_∞ CONTROL

An automatic steering control system was designed based upon the nominal model identified in Eq.(2). The purpose of the control object is to reduce the sensitivity to changes in the shape of the road and to achieve a robust stability against modeling error. Therefore, the controller was designed by solving a mixed-sensitivity problem based upon $\mathcal{H}_\infty$ control theory.

The generalized plant, $\boldsymbol{G}$, is shown in Fig.11, where W_T is a frequency weighting function for the unmodeled dynamics and W_S is one for the sensitivity function.

The multiplicative uncertainty, Δ_m, can be evaluated based upon the difference between the identified 29th-order model and the nominal model, shown in Eq.(2). In order to account for Δ_m, W_T is chosen as

$$W_T(s) = \frac{12.73s+1}{70}. \qquad (3)$$

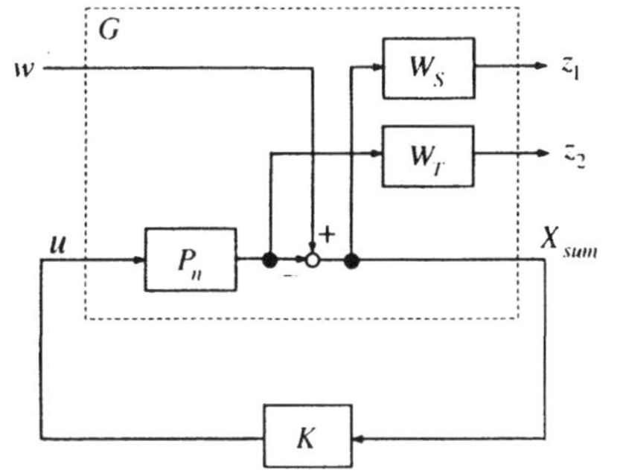

Figure 11: Block diagram of the generalized plant.

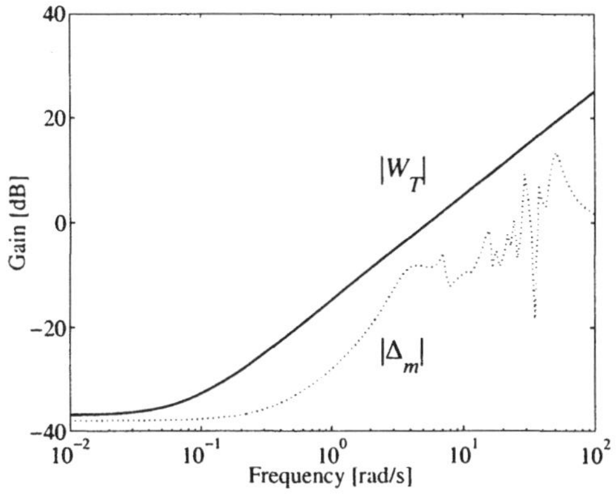

Figure 12: Gain characteristics of Δ_m and W_T.

The gain characteristics of Δ_m and W_T are shown in Fig.12.

On the other hand, $W_s(s)$ was chosen as a 1st-order low-pass filter:

$$W_s(s) = \frac{20}{3.98s + 0.001} \tag{4}$$

in consideration of the frequency characteristics for a road shape change.

The $\mathcal{H}_\infty$ mixed-sensitivity problem for the generalized plant $\boldsymbol{G}$ was solved in order to define a controller $K(s)$ such that

$$\left\| \begin{array}{c} W_S(s)S(s) \\ W_T(s)T(s) \end{array} \right\|_\infty < 1 \tag{5}$$

was satisfied for the closed-loop transfer function from the external input w to the controlled output z_1 and z_2.

As a result, the controller was determined to be

$$K(s) = \frac{0.1541(s+0.0094)(s+0.7388)(s+7.5)}{(s+1.268)(s+0.0001)(s+33.15 \pm j1.187)}. \tag{6}$$

This controller was then transformed into a discrete-time representation via a bilinear transformation with a sampling interval of 1/30s.

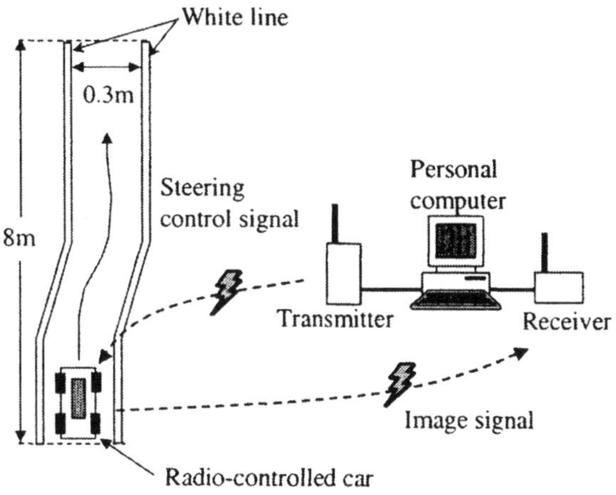

Figure 13: Configuration of control experiments.

5. CONTROL EXPERIMENT

5.1 Configuration of the control experiment

In order to examine the effectiveness of the proposed design method, traveling experiments using the RC car were conducted. The configuration of the experiment is shown in Fig.13. The traveling lane was 0.3m in width and 8m in length, and eventually makes a slight turn to the right.

The signal flow at each sampling time is summarized as follows.

1. The camera on the RC car captures the forward image.
2. The image data is transmitted to the PC by radio.
3. The feature value X_{sum} is calculated using the proposed algorithm explained in Section 2.2.
4. The control input, *i.e.* the steering signal, is calculated by the controller design method described in Section 4.
5. The signal is transmitted to the RC car by radio.

The purpose of the traveling experiment was to show that the RC car could run on the road and not deviate between the white lines.

5.2 Experimental results

Input (steering signal) and output (X_{sum}) data measured by the traveling experiment are shown in Fig.14. The center position of the steering wheel corresponds with 5.2V. Voltages above 5.2V

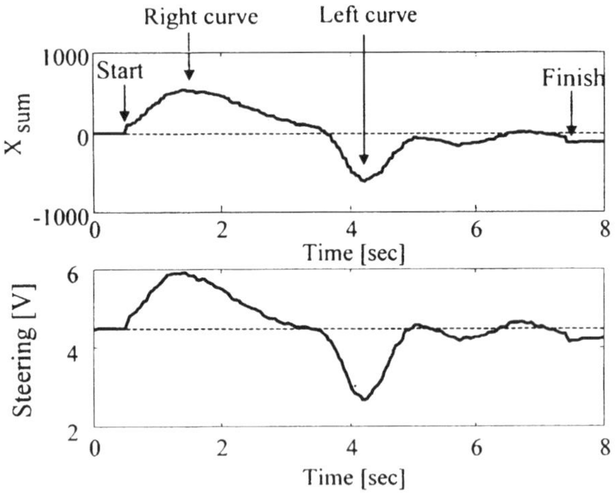

Figure 14: Input-output data measured by the control experiment.

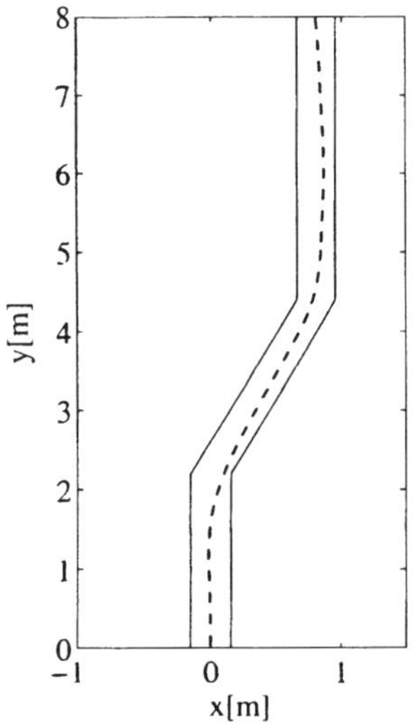

Figure 15: Traveling path of RC car.

signal a turn to the right, and voltages below 5.2V signal a turn to the left. From the figure, it can be seen that when the RC car turns to the right, X_{sum} increases. On the other hand, when the car turns to the left, X_{sum} decreases. X_{sum} becomes 0 for straight line traveling after 5s.

The traveling path of the RC car, shown in Fig.14, is depicted as a dashed line in Fig.15. It is clear that the RC car was able to remain between the white lines using the proposed control system.

6. CONCLUSIONS

In this paper, the effectiveness of an automatic steering control system utilizing visual feedback was examined through system identification and traveling experiments with an RC car. First, a practical image processing algorithm that recognizes the white lines on a road was proposed. Next, ARX modeling and Padé approximation were used in a system identification to determine both a nominal model and the corresponding multiplicative uncertainty for the automatic steering system design. The mixed-sensitivity problem was then solved, yielding an automatic steering control system. Finally, the traveling experiment was successfully conducted and the proposed design method was found to be effective.

Acknowledgement: The authors gratefully acknowledge Mr.Arai and Mr.Yoshii of Utsunomiya University for their cooperation on both the computer simulation and the experiments.

REFERENCES

Dickmanns,E.D. and V.Grefe (1988). Applications of dynamic monocular machine vision. *Machine Vision and Applications*, pp.214–261, Springer-Verlag New York Inc.

Fujiwara,Y., M.Yoshii and S.Adachi (2002). Automated steering control systems design for passenger vehicle in consideration of vehicle parameters' uncertainties (in Japanese). *Trans. of IEE of Japan*, Vol.122-D, No.6, pp.652–658.

Hutchinson,S. and P.I.Corke (1996). A tutorial on visual servo control. *IEEE Trans. on Robotics and Automation*, Vol.12, No.5, pp.651–670.

Iguchi,M. (1997). Background of advanced safety vehicle (ASV) development program (in Japanese). *The Journal of SICE*, Vol.36, No.3, pp.163–164.

Komori,T., Y.Fujiwara, M.Fujita and K.Uchida (2000). Automated driving by visual servoing (in Japanese). *Trans. of IEE of Japan*, Vol.120-C, No.4, pp.501–506.

Kovacs, G., J.Bokor, L.Palkovics, L.Gianone, A.Semsey, and P.Szell (1998). Lane-departure detection and control system for commercial vehicles. *IEEE International Conference on Intelligent Vehicles*, Stuttgart, Germany, pp.46–50.

Ljung, L. (1999). *System Identification - Theory for the User* (2nd Edition). Englewood Cliffs, NJ: Prentice Hall PTR.

Miyazaki,T. (1997). Plan for promotion of the development of advanced safety vehicle (ASV) (in Japanese). *The Journal of SICE*, Vol.36, No.3, pp.165–167.

IFAC Publications
www.elsevier.com/locate/ifac

Application of the RBF-type ARX Modelling and Control to Gas Turbine Combined Cycle SCR Systems

Y. Toyoda*, H. Peng, T. Ozaki**, K. Nakano*** and H. Shioya******

**: Niihama National College of Technology, toyoda@mec.niihama-nct.ac.jp*
***: The Institute of Statistical Mathematics, peng@ism.ac.jp, ozaki@ism.ac.jp*
****: The University of Electro-Communications, nakano@ee.uec.ac.jp*
*****: Bailey Japan Co., Ltd., hshioya@bailey.co.jp*

Abstract: Most electric power companies in Japan are installing the SCR system for NOx emissions. While these systems are effective, control and maintenance problems often result in excess reagent ammonia (NH_3) usage and NH_3 slip. The delay in the NOx signal from the stack causes inaccurate NH_3 injection resulting in poor NOx control. This paper discusses significant aspects of the technical and economic evaluation by the desktop simulation using data from the historical database in the GTCC plant computers. The significant merits of the proposed predictive control is that the multi-step ahead predictor is formulized as a nonlinear version based on the RBF-type ARX models. The RBF-type ARX model-based predictive control is one of solutions for the SCR for the GTCC optimization. *Copyright © 2003 IFAC*

Keywords: nonlinear process, GTCC, SCR system, RBF-type ARX model, nonlinear model-based predictive control, energy saving.

1. INTRODUCTION

In Japan, GTCC (Gas Turbine-Steam Turbine Combined Cycle) power plants are increasing more and more for their higher efficiency. Especially, the natural gas-firing GTCC power plant is most effective in terms of thermal efficiency and lower CO_2 energy source. Most power plants install the SCR (Selected Catalytic Reduction) systems for NOx emissions. This process is a typical nonlinear system due to the chemical reaction. While the SCR systems are effective at achieving the required NOx levels, control and maintenance problems often result in higher NH_3 usage than necessary . The delay of measured NOx signal from an analyzer causes inaccuracies in NH3 injection resulting in poor control performance. Furthermore, excessive NH_3 leads to NH_3 slip and damages the catalysts. This paper discusses significant aspects of the technical and economical evaluation about the desktop simulation using the RBF-type ARX model-based predictive control. The SCR system configuration installed at GTCC is as shown in Fig.1. In this paper, NOx at SCR inlet is called "INOx", and NOx at SCR outlet called "SNOx" for convenience. Where "HRSG" means the Heat Recovery Steam Generators.

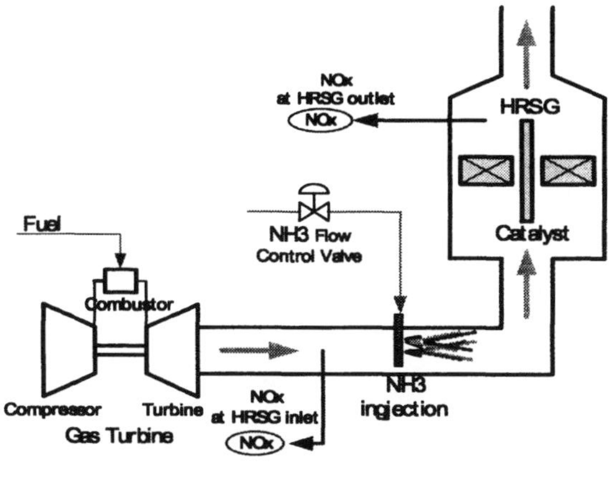

Fig.1. The SCR System in the GTCC

2. SYSTEM MODELING

2.1 The RBF-type ARX Modelling Approach

In practical process control problems, there is a class of nonlinear processes that is described by poles dependent on various operating conditions. Such industrial processes seem to be linear within relatively smaller change of operating conditions, on the other hand they show nonlinear dynamics under wide range of operating conditions. In order to address these issues, desired dynamic models should include the linear dynamics in the nonlinear dynamics. In order to cope with these issues, we propose the RBF-type ARX model (Toyoda, 1997, Peng, 2000) which is capable of describing the system nonlinearity within the whole operating range.

$$y_t = \Phi_0(X_{t-1}) + \sum_{i=1}^{q_y} \Phi_{y,i}(X_{t-1}) \cdot y_{t-i} + \sum_{i=1}^{q_u} \Phi_{u,i}(X_{t-i}) \cdot u_{t-i} + \sum_{i=1}^{q_d} \Phi_{d,i}(X_{t-i}) \cdot d_{t-i} + e_t \quad (1)$$

$$\Phi_0 = c_0 + \sum_{k=1}^{m} c_k^0 \exp\left(-\lambda_k \cdot \|X_{t-1} - Z_k\|^2\right)$$

$$\Phi_{j,i} = c_{i,0}^y + \sum_{k=1}^{m} c_{i,k}^y \exp\left(-\lambda_k \cdot \|X_{t-1} - Z_k\|^2\right), \quad j=y,u,d \quad (2)$$

where, $d \subset R^l$ is the disturbance, X_t is the key process variables for describing the nonlinear phenomena, Φ_0 and $\Phi_{j,i}$ are the constants, λ_k is the scaling factor, and Z_k is the center parameter of the Gaussian. u and y are the control and output sequence of the process and e is a zero-mean white noise.

2.2 Model Input/ Output Selection

In order to model the SCR process, data were collected from the historical database in the GTCC plant computers. Two relevant variables have been selected as follows:

- NH3 flow rate setpoint
- NOx emissions at the SCR inlet

The ambient temperatures around the SCR system have been neglected because of their slight variations during the operations.

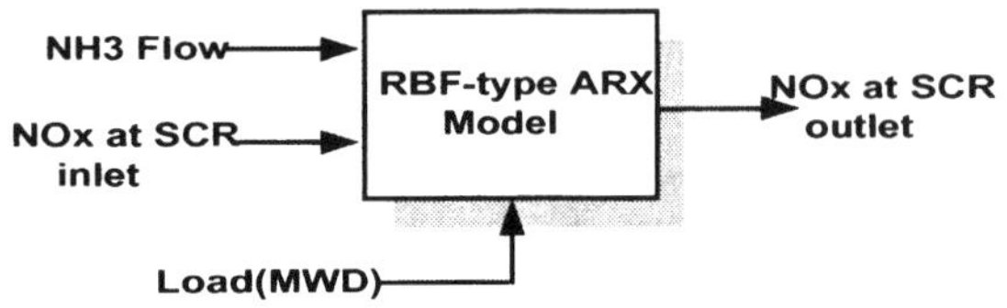

Fig.2. Model inputs/ outputs

Modelling process is done in the following steps.

- Determining the manipulated variables and controlled variables.
- Determining the sampling period corresponding to the controlled plant bandwidth.
- Collecting data at the specified sampling period from the historical database in the commercial operations.
- Estimating the both linear and nonlinear parameters of the RBF-type ARX models.
- Evaluating the pole assignment of the RBF-type ARX models by desktop simulation for checking their stability.
- Evaluating the models by desktop simulation using the data-sets for training.

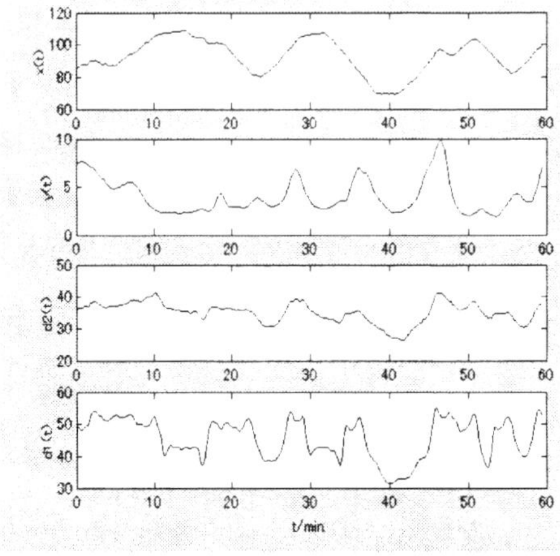

Fig.3. Example of the training data-set:
x(t) : Time series plot of MWD
y(t) : Time series plot of SNOx
d1(t) : Time series plot of INOx
d2(t) : Time series plot of NH3 injection

In Fig.3, INOx and SNOx mean NOx emissions at the SCR inlet and at the SCR exit respectively.

2.3 Model Parameter Estimation

We use general optimization algorithm for estimation of all parameters of the RBF-type ARX models. In this paper, the Levenberg-Marquardt Method (LMM) and the Least Squares Method (LSM) are presented for the parameter estimation. The point of our method is to divide all parameters into the linear part and the nonlinear part and use the aforementioned LMM and the LSM alternately. Namely, the search in nonlinear subspace uses the LMM and the search in linear subspace uses the LSM respectively (Peng, et al., 2001).

For the RBF-type ARX models, the parameters of the linear part are given as

$$\theta_L = \{c_i^0, c_{j,i}^y, c_{k,i}^u, c_{k,i}^d \mid i = 0,1,\cdots;m, j = 1,\cdots;q_y; k = 1,2,\cdots;q_u\}$$

$$\theta_L \in \Re^{(m+1)\times(q_y+q_u+q_d+1)} \quad (3\text{-a})$$

and the parameters of the nonlinear part as

$$\theta_N \equiv \left[\lambda_1, \lambda_2, \cdots, \lambda_p, Z_1^T, Z_2^T, \cdots, Z_p^T\right]^T$$
$$\theta_N \in \Re^{p+p\times n_x}$$
(3-b)

This procedure is as follows:
Select the initial values θ_N^0 using the empirical equation (4).

$$\lambda_k^0 = -\log \varepsilon_k \Big/ \max\left\{\left\|X(t-1) - Z_k^0\right\|^2\right\} \quad (4)$$
$$\varepsilon_k \in [0.1, 0.0001]$$

We can estimate θ_L^0 using the LSM under fixed θ_N^0 as follows:

Initial parameters: θ_N^0, θ_L^0

Z_k^0: subset of the state vector X(t-1) by choosing randomly

θ_L^0 is determined using the LSM

$$\theta_L^0 = \left[\sum_{t=\tau}^{M} \overline{\Phi}_{t,0} \overline{\Phi}_{t,0}^T\right]^{-1} \sum_{t=\tau}^{M} \overline{\Phi}_{t,0} \cdot \overline{y}(t) \quad (5)$$

where $\overline{\Phi}_{t,0} = \Phi(\theta_N^0, X(t-1))$

$$X(t-1) = \begin{bmatrix} y(t), y(t-1), \cdots; y(t-q_y), u(t), u(t-1), \cdots; u(t-q_u), \\ d(t), d(t-1), \cdots; d(t-q_d) \end{bmatrix} \quad (6)$$

is the measured data-sets, τ is the time delay in $X(t-1)$, and M is the length of the data.
Then we use the objective function as

$$V(\theta_N, \theta_L) \equiv \left\|F(\theta_N, \theta_L)\right\|^2$$

$$F(\theta_N, \theta_L) \equiv \begin{bmatrix} f(\theta_N, \theta_L, \overline{X}(\tau)) - \overline{y}(\tau+1) \\ f(\theta_N, \theta_L, \overline{X}(\tau+1)) - \overline{y}(\tau+2) \\ \vdots \\ f(\theta_N, \theta_L, \overline{X}(M-1)) - \overline{y}(M) \end{bmatrix} \quad (7)$$

The k is iteration steps, then gradient matrix of $F(\theta_N^k, \theta_L^k)$ with respect to θ_N is as

$$\nabla f(\theta_N^k) \equiv \left\{\partial F(\theta_N^k, \theta_L^k) / \partial \theta_N^k\right\} \quad (8)$$

The nonlinear parameters is obtained from

$$\theta_N^{k+1} = \theta_N^k + \beta_{k+1} \cdot d_{k+1} \quad (9)$$

where d_k is the search direction vector and calculated the following equation:

$$\left\{\nabla f(\theta_N^k) \nabla f(\theta_N^k)^T + \gamma_k I\right\} \cdot d_{k+1} = -\nabla f(\theta_N^k) F(\theta_N^k, \theta_L^k) \quad (10)$$

Besides, β_{k+1} is the step size so as to minimize F using the line search procedure. Fig.4 shows the residual and its histogram using the estimated parameters of the RBF-type ARX models for the training data-sets.

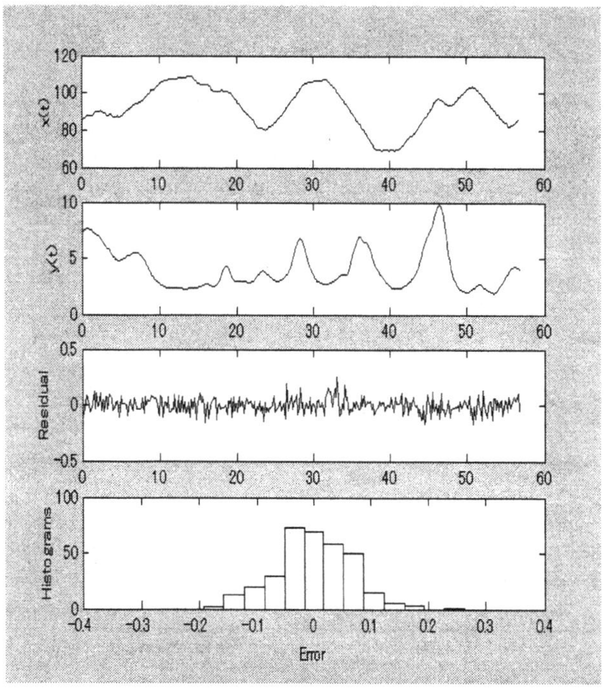

Fig.4. The residual and the error histogram using the RBF-type ARX models
MWD: Time series plot of the unit load
SNOx: Time series plot of the NOx at the SCR exit
Residula: Time series plot of the prediction error
Histogram: Prediction error histogram

3. DESIRED ADVANCED CONTROL

The use of advanced control technique like our proposed control method is based on a analysis about the conventional PID control technologies by evaluating the economical and environmental benefits. We can see that the desired advanced control is realized using the real time optimization under constraints on controlled variables and on manipulated variables.

3.1 Poles Dependent On the Load Signal

Fig.5 shows the estimated ARX model poles continuously changed in a manner which is dependent on the operating-point state. We can verify the effectiveness of the proposed RBF-type ARX models for a class of nonlinearity of the SCR processes.

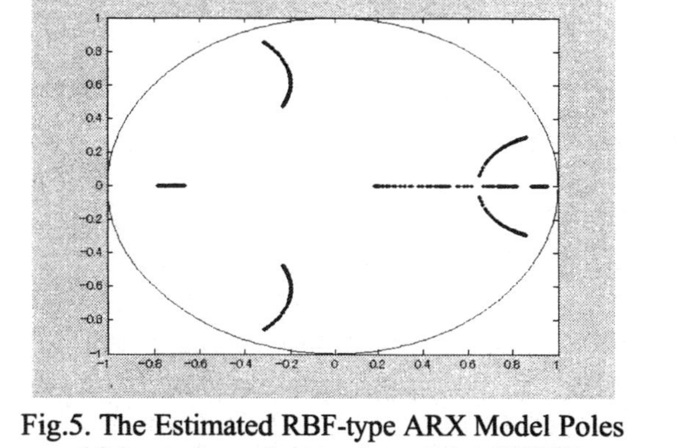

Fig.5. The Estimated RBF-type ARX Model Poles which are dependent on the unit load MWD

3.2 Nonlinear Model-Based Predictive Control using the RBF-type ARX model

The nonlinear model-based predictive control applications use a real-time control programs that combines optimization and dynamic control. The proposed predictive control system is composed of the following parts: a process model using the RBF-type ARX models, a control model based on these models and a dynamic optimizer. We can find a sequence for manipulated variables *u(k)* that minimizes the quadratic objective function over the model prediction horizon "*N*" and over the controller horizon "*Nu*" (Clarke et al., 1987).

$$\min_{U(t)} \tilde{J} = \frac{1}{2} U(t)^T \left(G_t^T G_t + R\right) \cdot U(t) + \{Y_0(t) - Y_r(t)\}^T G_t U(t) \quad (11)$$

$$\text{subject to} \begin{bmatrix} G_t \\ -G_t \end{bmatrix} \cdot U(t) \le \begin{bmatrix} Y_{max} - Y_0(t) \\ -Y_{min} + Y_0(t) \end{bmatrix} \quad (12)$$

$$U_{min} \le U \le U_{max}, \Delta U_{min} \le \Delta U \le \Delta U_{max} \quad (13)$$

Predictive output is described as below.

$$\hat{Y}(t) = G_t U(t) + Y_0(t) \quad (14)$$

where,

$$\hat{Y}(t) = \left[\hat{y}(t+1|t), \hat{y}(t+2|t), \cdots \hat{y}(t+N|t)\right]^T$$
$$Y_0(t) = \left[y_0(t+1|t), y_0(t+2|t), \cdots y_0(t+N|t)\right]^T$$
$$U(t) = \left[u(t), u(t+1), \cdots u(t+N_u-1)\right]^T$$
$$Y_r(t) = \left[y_r(t+1), y_r(t+2), \cdots y_r(t+N)\right]^T \quad (15)$$

Here *$Y_r(t)$* is the desired output sequence.

$$G_t = \begin{bmatrix} g_{t,0} & 0 & \cdots & 0 \\ g_{t,1} & g_{t,0} & \cdots & \vdots \\ \vdots & \vdots & \cdots & \vdots \\ g_{t,N-1} & g_{t,N-2} & \cdots & \sum_{i=0}^{N-N_u} g_{t,i} \end{bmatrix} \quad (16)$$

$$G_{t,j}^T\left(q^{-1}\right) = g_{t,0} + g_{t,1}q^{-1} + \cdots + g_{t,j-1}q^{-j+1} \quad (17)$$

The RBF-type ARX model is described as

$$A_t\left(q^{-1}\right)y(t) = a_{0t} + B_t\left(q^{-1}\right)\{U_{pid}(t-1) + U_{npc}(t-1)\} + D_{1t}\left(q^{-1}\right)v_1(t-1) + D_{2t}\left(q^{-1}\right)v_2(t-1) + \xi(t) \quad (18)$$

where U_{pid} and U_{npc} are the control inputs from the existing PID controller and from the NPC, respectively and v_1 and v_2 are disturbances. Therefore, the *j*-th step-ahead optimal predictive output is derived using the following equations:

$$\hat{y}(t+j|t) = G_{t,j}^T\left(q^{-1}\right) \cdot u(t+j-1) + y_0(t+j|t)$$
$$y_0(t+j|t) = E_{t,j}^T(1) \cdot a_{0,t} + F_{t,j}\left(q^{-1}\right) \cdot y(t-1) + H_{t,j}\left(q^{-1}\right) \cdot u(t-1) + E^T{}_{t,j}\left(q^{-1}\right) D_t\left(q^{-1}\right) \cdot v(t+j-1) \quad (19)$$

where the polynomials *E, F, G* and *H* are the solutions of two Diophantine equations:

$$1 = E_{t,j}^T\left(q^{-1}\right) \cdot A_t\left(q^{-1}\right) + q^{-j} F_{t,j}\left(q^{-1}\right)$$
$$E_{t,j}^T\left(q^{-1}\right) \cdot B_t\left(q^{-1}\right) = G_{t,j}^T\left(q^{-1}\right) + q^{-j} H_{t,j}\left(q^{-1}\right) \quad (20)$$

The Quadratic Programming algorithm available in MATLAB Toolbox is applied for the above optimization. Fig.6 shows the configuration of the proposed predictive control system for the SCR system (Peng, et al., 2002).

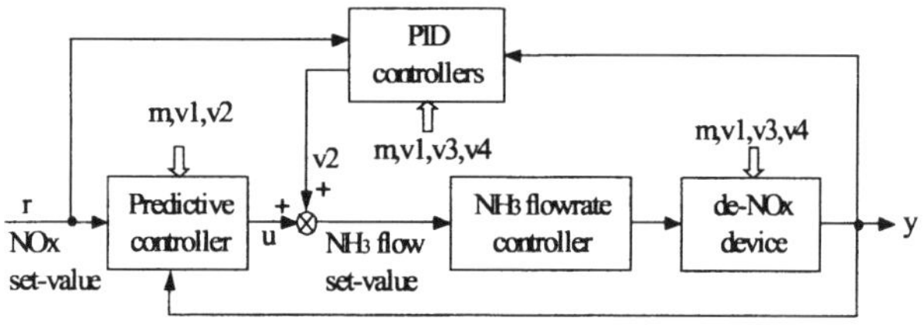

Fig.6. Configuration of the proposed predictive control system

Fig.7 illustrates the example of the control performance improvement between the proposed predictive control based on the RBF-type ARX models (NPC) with *N*=20, *Nu*=5 and the existing PID control. Conventional SCR control system relies on a simple feed-forward loop, developed using both the relationship between Nox and MWD and the chemical equation, therefore the poor control performance results in higher reagent NH3 usage. On the other hand, we can see that the proposed control system can keep the SNOx concentration *y(t)* around 6ppm resulting in 12% less NH3 usage.

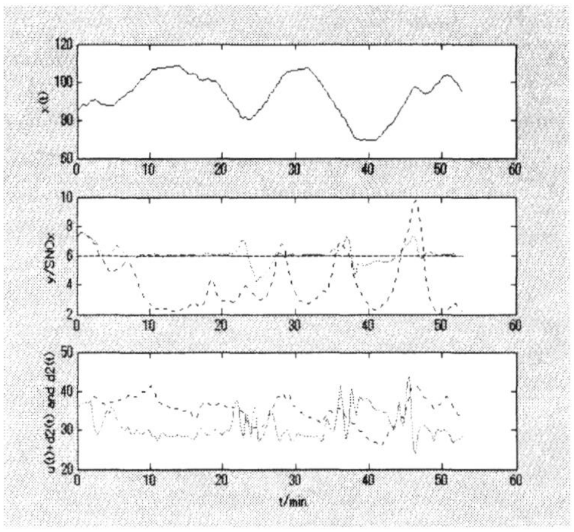

Fig.7. Control Performance
The RBF-type ARX model-based predictive control (solid) vs PID control (dotted)
x(t) : Time series plot of the unit load
SNOx: Time series plot of the NOx at the SCR exit
u(t)+d2(t) and d2(t): Time series plot of the Upid(t)+Unpc(t) and the Unpc(t)

Fig.8 shows the comparison of predicted and measured data for the training data-sets. The figure also shows the histogram of predictive error and model performance. We can clearly find the effectiveness of the proposed modelling approach.

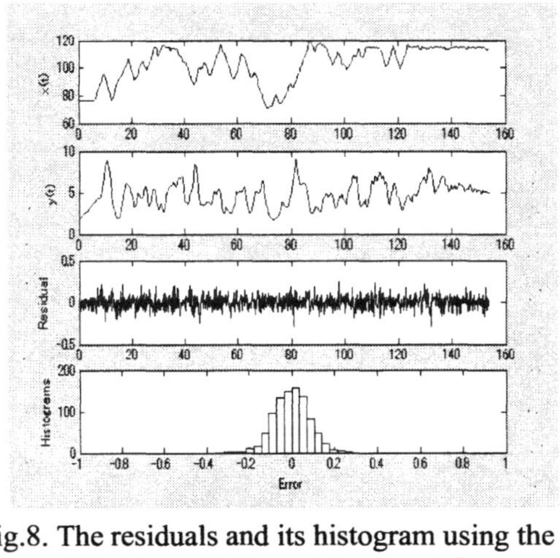

Fig.8. The residuals and its histogram using the RBF-type ARX model estimated
x(t): Time series plot of the unit load
y(t): Time series plot of the SNOx
Histogram: Prediction error histogram

Fig.9 shows the control quality improvement for validation. The figures shows the unit load MWD, the comparison of manipulated variable U_{pid}, $U_{pid} + U_{npc}$, and manipulated variable U_{pid} respectively. During this time period the GTCC plant was operated from 80MW to 120MW. Using the NPC, the average value of the control deviation of SNOx is achieved 1/80 and the variance of SNOx deviation is achieved 1/9.

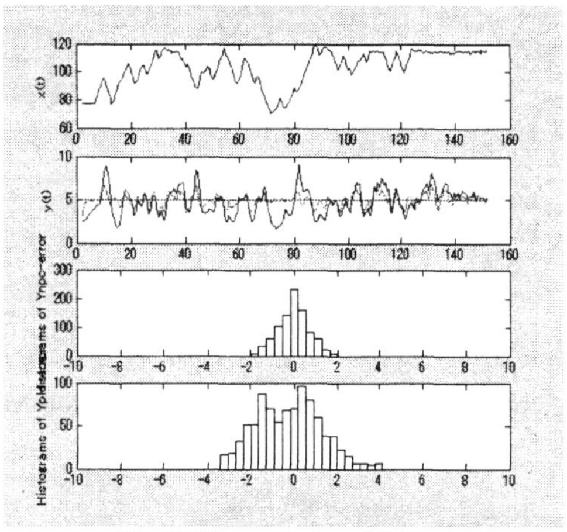

Fig.9. Control error histogram
The RBF-type ARX model-based predictive control (red) vs PID control (black)
x(t): Time series plot of the unit load
y(t): Time series plot of the SNOx (NPC vs PID)
Histogram: Control error histogram using the predictive control and using the PID controller

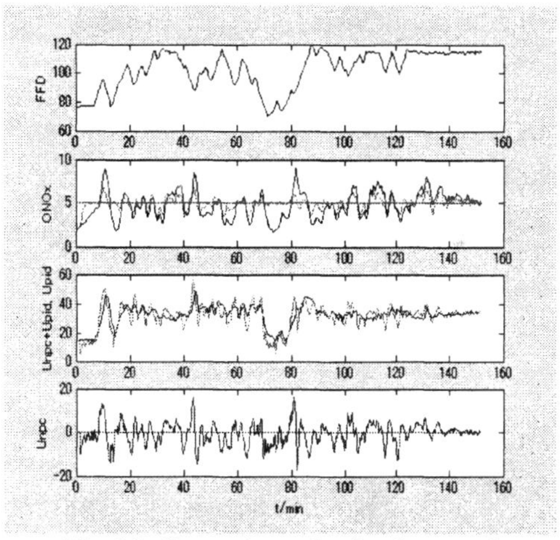

Fig.10. Control results for the test data
The RBF-type ARX model-based predictive control (red) vs PID control (black)
Set-value of y(t): 5 ppm
x(t): Time series plot of the unit load
y(t): Time series plot of the SNOx
Histogram: Control error histogram using the predictive control and using the PID controller

As shown in Fig.11, the proposed predictive control system NPC will be installed on a computer that is designed to connect the data highway in the plant and will operate in both the predictive control mode and the PID control mode.

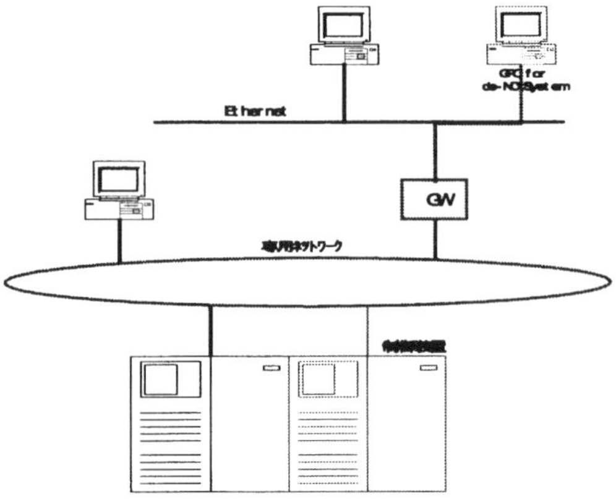

Fig.11 The proposed NPC system in the GTCC plant computer systems

4. CONCLUSIONS

For a class of nonlinear processes whose static and dynamic characteristics are dependent upon exogenous variables, the multi-step predictive control based on the RBF-type ARX models was reported. The significant merit of the proposed predictive control for the nonlinear process is that the multi step-ahead predictor is formulized as a nonlinear version of future control based on the RBF-type ARX models. The RBF-type ARX model-based predictive control is one of solutions for GTCC SCR optimization and control. The benefits of the proposed solution are qualified in both economic and environmental terms. In the desktop simulation, we found that the average value of the control deviation of SNOx is achieved 1/80 and the variance of SNOx deviation is achieved 1/9 by the proposed NPC.

REFERENCES

Toyoda, Y., Oda, K., and Ozaki, T.: The nonlinear system identification method for advanced control of the fossil power plants, Proceedings of 11th IFAC Symposium on System Identification, Fukuoka, Japan, 8/11 (1997)

Peng, H., Ozaki, T., Toyoda, Y., and Oda, K.: Exponential ARX model based generalized predictive control for thermal power plants, Proceedings of System Identification(SYSID2000), California, USA, 645/650 (2000)

Peng, H., Ozaki, T., Toyoda, Y.: A new estimation method for thermal power plants, Proceedings of IFAC Conference on New Technologies for Computer Control, 602/607 (2001)

Clarke, D.W., Mohtadi, C., and Tuffs, P.S.: Generalized predictive control-Part 1. The Basic Algorithm, Automatics, Vol.23, No.2, 137/148 (1987)

Clarke, D.W, Mohtadi, C., and Tuffs, P.S.: Generalized predictive control-Part 2. Extensions and Interpretations, Automatics, Vol.23, No.2, 149/160 (1987)

Peng, H., Ozaki, T., Toyoda, Y.: Nonlinear predictive control based on a global model identified off-line, Proceedings of the American Control Conference, 4197/4202 (2002)

IFAC Publications
www.elsevier.com/locate/ifac

AUTOMATIC TIME SERIES IDENTIFICATION SPECTRAL ANALYSIS with MATLAB TOOLBOX ARMASA

Piet M.T. Broersen,

Signals and Systems Group, Department of Applied Physics
Delft University of Technology

Abstract: ARMASA provides a new automatic spectral estimator for random data. For stationary stochastic observations, time series identification gives a better accuracy in spectral estimation than what can be obtained by FFT analysis with windowed and tapered periodograms. The parameters of the time series model accurately represent the spectral density and the covariance function of the data. The increased computational speed gives the possibility to compute hundreds of models and to select only one. The three linear time series model types are: autoregressive (AR), moving average (MA) and the combined ARMA models. The ARMAsel algorithm computes models of the three types for a large number of candidate model orders. The computer first selects the best order for each of the model types separately. Then, a single type is selected from those three models by looking for the smallest prediction error. That selected model includes precisely the statistically significant details that are present in the data, and no more. *Copyright © 2003 IFAC*

Keywords: ARMA model, autoregressive model, covariance estimation, moving average model, order selection, parametric model, spectral estimation.

1. INTRODUCTION

The FFT algorithm of Cooley and Tuckey (1965) had a great impact on the practice of spectral estimation. The reduced computer effort enabled the routine analysis of extensive sets of data with windowed and tapered periodograms. Until about the year 2000, the computational demands for time series analysis were too heavy for use in data with unknown model type and model order. For that reason, analysis with periodograms has been the main practical tool for spectral analysis for such a long time. However, the vast literature on windowing and tapering did not solve the problem that the optimum solution for type and width of the spectral window requires the knowledge of the exact spectral density (Priestley, 1981). It cannot be deduced with sufficient reliability from the estimated spectrum of measured data.

Time series analysis is a new perspective. Time series models can be subdivided into three model types: AR or Autoregressive models, MA or Moving Average models and the combined ARMA models (Priestley, 1981). Theoretically, any stationary stochastic process with a continuous spectral density can be expressed as an unique AR(∞) or MA(∞) process In practice, finite order MA or AR models for those infinite order processes are accurate enough, because the true parameters of higher orders decrease rapidly for most processes. Broersen (2002) showed that a time series model is the best spectral estimate for stochastic data. Recently, the increased speed of computers has created the possibility to evaluate a large number of candidate time series models. From a single selected model, the covariance and the spectral density are computed. The accuracy of the parametric spectrum is typically better than the best of all possible periodogram estimates and the Cramér-Rao lower bound is often approached in simulations.

For AR estimation, Burg's (1967) algorithm is preferred as compromise between bias and variance. Akaike (1974) solved order selection for AR models with an asymptotic criterion. Broersen (2000a) used finite sample equivalents, which are necessary if model orders higher than one tenth of the sample size are considered. Order selection can be interpreted as looking for the lowest order p where the extrapolated

AR(p) correlation function is not significantly different from the data correlation at larger distance. That property is verified up to the highest order that is considered for AR order selection, so taking high candidate orders for AR selection diminishes the probability of missing significant details. For MA and ARMA models, robust algorithms exist which never have problems with convergence and always find stationary and/or invertible models. Durbin's methods (1959, 1960) are good for a safe, robust and practical solution of the MA and ARMA estimation. They use the parameters of a long AR model as the input for MA estimation. MA Models estimated with Durbin's methods are always invertible; ARMA models are also invertible and stationary. This is an important advantage for estimation in finite samples. Recently, an automatic selection algorithm for the model type has been proposed; the single selected AR, MA or ARMA model is denoted the ARMAsel model. This paper describes the robust algorithms that have been used in the identification of the model order and type in ARMAsel that is part of the ARMASA Matlab toolbox of Broersen (2001) for automatic spectral analysis with time series models, and is available in the public domain. Broersen, *et al.*, (2002) also develop a time series algorithm for missing data or for irregularly sampled data.

2. ROBUST ALGORITHMS

Time series have three different linear model types, autoregressive or AR, moving average or MA and combined ARMA. An ARMA(p,q) process can be written as (Priestley, 1981):

$$x_n + a_1 x_{n-1} + \cdots + a_p x_{n-p} = \varepsilon_n + b_1 \varepsilon_{n-1} + \cdots + b_q \varepsilon_{n-q}\,, \tag{1}$$

where ε_n is a purely random process of independent identically distributed stochastic variables with zero mean and variance σ_ε^2. This process is purely AR for q=0 and purely MA for p=0. The roots of the AR parameter polynomial

$$A(z) = 1 + a_1 z^{-1} + \cdots + a_p z^{-p} \tag{2}$$

are denoted the poles of the ARMA(p,q) process. The roots of $B(z)$, defined likewise, are the zeros. Processes and models are called stationary if all poles are strictly within the unit circle and they are invertible if all zeros are within the unit circle. A shorthand notation for an ARMA process is:

$$A(z)x_n = B(z)\varepsilon_n \tag{3}$$

where $z^{-1}x_n = x_{n-1}$. The only assumption is that data are supposed to represent a stationary stochastic process. The power spectrum $h(\omega)$ of the ARMA(p,q) model is completely determined by the parameters in (1) and the innovation variance σ_ε^2 and is given by:

$$h(\omega) = \frac{\sigma_\varepsilon^2}{2\pi}\left|1 + \sum_{i=1}^{q} b_i \,\mathrm{e}^{-j\omega i}\right|^2 \Big/ \left|1 + \sum_{i=1}^{p} a_i \,\mathrm{e}^{-j\omega i}\right|^2 . \tag{4}$$

Direct methods to derive the covariance from the parameters are available in the ARMASA toolbox. The transform of an estimated AR(L) model $\hat{A}_L(z)$ into a positive semi-definite covariance function $\hat{R}(k)$, $k=0,\ldots,L$ or from $\hat{R}(k)$ to $\hat{A}(z)$ is made with the Yule-Walker equations (Kay and Marple, 1981):

$$\sum_{i=0}^{L} \hat{a}_i \hat{R}(m-i) = 0, \qquad m = 1, 2, \cdots, L. \\ \textit{with} \qquad \hat{R}(-|k|) = \hat{R}(k),\ k > 0 \tag{5}$$

The same equation can be used to extrapolate the covariance function beyond L, by using (5) for index $m > L$. It is worth noticing that (5) can also be used for the true parameters and the true covariance, for which it has been derived originally. Also programs to compute $R(k)$ for ARMA processes are available.

2.1 Model Accuracy

The prediction error as a quality measure allows a mutual comparison of AR, MA and ARMA models. It is defined as the squared error of prediction when applying the estimated model to another independent realization of the same true process. The Model Error ME (Broersen, 1998) is a scaled version of the prediction error PE. With $\hat{A}(z)$ and $\hat{B}(z)$ denoting the estimated model, the ME is defined as:

$$ME = ME\left(\frac{\hat{B}}{\hat{A}}, \frac{B}{A}\right) = N\left(\frac{PE}{\sigma_\varepsilon^2} - 1\right). \tag{6}$$

The expectation of the ME for unbiased estimated models is independent of N and equals the number of estimated parameters. Broersen (1998) derived an efficient expression for the computation of the ME in the time domain. PE can be written as var(η_n), where η_n is defined as a new ARMA process derived from the true and the estimated parameter polynomials as

$$\eta_n = \frac{\hat{A}(z)}{\hat{B}(z)} x_n = \frac{\hat{A}(z)B(z)}{\hat{B}(z)A(z)} \varepsilon_n\,, \tag{7}$$

where (3) has been used to substitute x_n. It can be proved that var(η_n) ≥ var(ε_n). The ME is suited to describe the model quality for different sample sizes and for the three different model types, hence also with $\hat{A}(z)$ or $\hat{B}(z)$ equal to 1.

It is well known that the best fitting AR(r) model to an arbitrary process is just the solution of the Yule-Walker relations in (5) of order r to equate the AR(r) covariance to the first r covariances of the process. The same solution is found by writing the arbitrary process as a long AR process and to use the first r reflection coefficients to determine the AR(r) model. This formulation can be given formally with the Model Error ME. Hence, the best AR(r) model of an arbitrary ARMA process can be written as:

$$\hat{A}_r(z) = \arg\min_{\hat{A}_r(z)} \left\{ ME\left(\frac{1}{\tilde{A}_r(z)}, \frac{B(z)}{A(z)}\right)\right\},\ \forall r\,. \tag{8}$$

This can be solved easily with the Yule-Walker relations (3) applied to the covariance function generated by the right-hand side ARMA process *B/A*. Some rules for computations with the ME are given:

$$ME\left(\frac{\hat{B}}{\hat{A}},\frac{B}{A}\right)=ME\left(\frac{1}{1},\frac{\hat{A}B}{A\hat{B}}\right)=ME\left(\frac{1}{\hat{A}},\frac{B}{A\hat{B}}\right) \quad (9)$$

and

$$ME\left(\frac{\hat{B}}{1},\frac{\hat{A}B}{A}\right)=ME\left(\frac{A}{\hat{A}B},\frac{1}{\hat{B}}\right)\approx ME\left(\frac{1}{\hat{B}},\frac{A}{\hat{A}B}\right). \quad (10)$$

The final step in (10), interchanging the sequence of the arguments in ME, is only a close approximation for well fitting models, where var(η_n) is about $\{1+O(1/N)\}$var(ε_n), only slightly greater than var(ε_n). Nevertheless, it will often be necessary in the sequel to apply this interchange: the first representation of (10), fitting a MA model to the right-hand side ARMA process has no easy or computationally attractive solution for $\hat{B}(z)$ that minimizes the ME. The solution for $\hat{B}(z)$ would be highly non-linear. The last representation has an easy computable minimizer for $\hat{B}(z)$, for given $\hat{A}(z)$, $A(z)$ and $B(z)$. That is the solution for which the parameters of $\hat{B}(z)$, considered as an AR process generate the same correlation as the ARMA process $A(z)/\hat{A}(z)\,B(z)$.

2.2 AR Estimation

This model type is the backbone of time series analysis in practice. The algorithm of Burg (1967) for AR parameter estimation recursively estimates *L* reflection coefficients k_i from consecutive equidistant data. Reflection coefficients are used to recursively determine the parameters $\hat{A}_q(z)$ of all model orders *q* between 1 and *L*, with the Levinson-Durbin formulas (Kay and Marple, 1981):

$$\begin{aligned} &\hat{a}_1^1 = k_1 \\ &\hat{a}_i^q = \hat{a}_i^{q-1} + k_q \hat{a}_{q-i}^{q-1}, \qquad 1 \le i < q \\ &\hat{a}_q^q = k_q \qquad 1 \le q \le L \end{aligned} \quad (11)$$

It is sure that the model will be stationary, with all roots of each model $\hat{A}_q(z)$ within the unit circle. Let *RES(p)* denote the variance of the residuals after estimating the AR(*p*) model. The order of AR models is selected with a finite sample order selection criterion *CIC(p)* of Broersen (2000a), defined as:

$$CIC(p) = \ln\{RES(p)\} + \max\left[\prod_{i=1}^{p}\frac{1+1/(N+1-i)}{1-1/(N+1-i)}-1\ ,\ 3\sum_{i=1}^{p}\frac{1}{N+1-i}\right]. \quad (12)$$

The coefficients $1/(N+1-i)$ in this criterion are the finite sample approximations for the variances of estimated reflection coefficients of order *i*. The asymptotical theoretical value of that variance is $1/N$, but the number of contributions to the Burg solution is only $N+1-i$ for order *i*. The order with the minimum of *CIC(p)* is selected. The highest candidate order *L* can be chosen as high as *N*-1 without numerical complications, but generally $L=N/2$ is taken as an upper boundary for accurate estimation. For large data sets, *L* is taken as 1000 to limit the computing time. However, especially if an order will be selected close to that chosen maximum candidate order, it is advisable to verify whether a relaxation of the experimenters restrictions on the maximum order *L* has influence on the selected order.

2.3 MA Estimation

Durbin's (1959) MA method estimates all zeros inside the unit circle and fulfills all requirements to be useful under all circumstances. Theoretically, a MA(q) model is exactly equivalent with an AR(∞) model, by using $B_q(z)=1/A_\infty(z)$. Durbin's method uses the *estimated* parameters of a long AR model $A^{long}(z)$ to approximate the MA model. Of course, the order of $A^{long}(z)$ has to be finite in estimation. The true MA and the approximating AR representation are:

$$\begin{aligned} &x_n = B_q(z)\varepsilon_n \\ &A^{long}(z)x_n \approx \varepsilon_n \ . \end{aligned} \quad (13)$$

Minimizing the difference between two models can be done with the ME. With (10), the obvious estimator for the MA parameters based on (13) is:

$$\hat{B}_q(z) = \arg\min_{\tilde{B}_q(z)}\left\{ME\left(\frac{1}{\tilde{B}_q(z)},\frac{A^{long}(z)}{1}\right)\right\},\ \forall q. \quad (14)$$

For a long time, Durbin's method had a poor reputation in the literature. Mostly, the best AR order for AR prediction of the observations has been used as the length for $A^{long}(z)$, in theory and in practice. Indeed, the accuracy of Durbin's method is poor then. However, Broersen (2000b) showed that the performance of Durbin's algorithm can be improved considerably by selecting a new and much higher order *K* for the intermediate AR model $A^{long}(z)$. The theoretical selection principle for *K* is that the new order is characterized by the most accurate parameter estimates. In practice, that order *K* is chosen with a sliding window algorithm as twice the AR order *p'* which is selected for the data with minimal *CIC(p')* plus the number *q'* of MA parameters that is to be estimated: $K=2p'+q'$. If the calculated *K* is greater than *L*, the AR(*L*) model is used for $A^{long}(z)$, with the highest candidate AR order. MA(q') models are generally estimated from $q'=1$ until $q'=N/5$. The MA order *q* is selected with the asymptotical selection criterion *GIC(q,3)* that is defined as (Broersen, 2002):

$$GIC(q,3) = \ln\{RES(q)\} + 3q/N. \quad (15)$$

Here *RES(q)* is computed as the variance of the filtered signal $x_n/\hat{B}_q(z)$. This asymptotically based order selection criterion is generally a slight improvement with respect to the AIC criterion of Akaike (1974) that is equal to *GIC(p,2)* with penalty factor 2 in (15). The penalty factor, α in $GIC(q,\alpha)$, handles the selection compromise between bias in underfit models, which have too few parameters included, and variance due to overfit models with too many estimated parameters (Broersen, 2000a). The compromise $\alpha = 3$ for the penalty is based on asymptotic arguments. It has a good performance in simulations for finite samples and on practical data, better than $\alpha = 2$ or $\alpha = 4$.

2.4 ARMA Estimation

ARMA models are most difficult to estimate. A new problem arises in splitting the dynamics in AR and MA parts. No general theoretical principle to derive efficient ARMA estimators can be given for finite data samples and many ad hoc solutions can be studied, each with a different performance when applied to various data. ARMA models can be computed with Durbin's (1960) two methods. The first ARMA method of Durbin uses both reconstructed residuals and previous measured observations as regressors in a least squares solution.

$$\begin{aligned} A(z)x_n &= B(z)\varepsilon_n \\ A^{long}(z)x_n &= \hat{\varepsilon}_n \end{aligned} \quad . \qquad (16)$$

The residuals $\hat{\varepsilon}_n$ are reconstructed by using a truncated estimated long AR model $A^{long}(z)$. If the order of $A^{long}(z)$ would be infinite, the reconstruction of $\hat{\varepsilon}_n$ could be perfect theoretically, but for the chosen finite length $K=3p'+9$, the accuracy of the reconstruction is reasonable. Parameters are estimated in Durbin's first method by minimizing

$$V_{ARMA(p,q)} = \sum_{n=\max(p,q)}^{N} \left\{ \begin{matrix} x_n + \hat{a}_1 x_{n-1} + \cdots + \hat{a}_p x_{n-p} \\ -\hat{\varepsilon}_n - \hat{b}_1\hat{\varepsilon}_{n-1} - \cdots - \hat{b}_q\hat{\varepsilon}_{n-q} \end{matrix} \right\}^2 . \qquad (17)$$

This initial ARMA solution is not efficient, and not guaranteed to be stable and invertible. That problem is solved in Durbin's (1960) second ARMA method. Using the AR parameters $\hat{A}_p(z)$ of Durbin's first ARMA method (17) as initial conditions, updates of the MA parameters $\hat{B}_q(z)$ and afterwards of the AR parameters can be made. For updating the MA parameters, an adaptation the MA method is used. With (10), (14) and (16), it can be seen that

$$\hat{B}_q(z) = \arg\min_{\tilde{B}_q(z)} \left\{ ME\left(\frac{1}{\tilde{B}_q(z)}, \frac{A^{long}(z)}{\hat{A}_p(z)} \right) \right\}, \ \forall q \ . \qquad (18)$$

With the long AR model divided by the initially estimated $\hat{A}_p(z)$ from (17), the improved MA parameters $\hat{B}_q(z)$ are estimated from the quotient $A^{long}(z)/\hat{A}_p(z)$, like in (14) from $A^{long}(z)$. Broersen (2000b) described a sliding window choice for the order K of the intermediate AR model $A^{long}(z)$ for ARMA(p,q) models. It is three times the AR order as selected with $CIC(p)$ plus the number of parameters that has to be estimated: $K=3p'+p+q$. Again, the accuracy of the second method of Durbin is very much dependent on the proper choice of the order of $A^{long}(z)$. The poor reputation of Durbin's method in the literature is mainly caused by using a wrong order for $A^{long}(z)$. Having obtained this new estimate $\hat{B}_q(z)$ for the MA parameters, the initial estimate $\hat{A}_p(z)$, obtained with (17) is updated by:

$$\hat{A}_p(z) = \arg\min_{\tilde{A}_p(z)} \left\{ ME\left(\frac{1}{\tilde{A}_p(z)}, \frac{1}{A^{long}(z)\hat{B}_q(z)} \right) \right\}, \ \forall q \ . \qquad (19)$$

Improved parameters estimates for the parameters $\hat{A}_p(z)$ are then found by equating the covariance function of $\hat{A}_p(z)$ to the first p terms of the covariance generated by the right-hand side of (19). The standard ARMAsel implementation calculates ARMA(r,r-1) models for $r = 1,\ldots,N/10$. The residual variance is computed as the variance of the filtered signal $\{\hat{A}_r(z)/\hat{B}_{r-1}(z)\}x_n$. The criterion $GIC(q,3)$ of (15) is also used for the selection of the $2r$-1 parameters of the ARMA(r,r-1) model, with $GIC(2r-1,3)$. The limitation to ARMA(r,r-1) models instead of to arbitrary orders for the AR and the MA part has two reasons. The first is that ARMA models require more computation time than AR or MA models. Given a certain amount of computation time, it might be more informative to compute ARMA(r,r-1) models of order r from 1 to 100 than all possible ARMA(i,j) models with i and j both limited to a maximum of 10 or 11. The second reason is that the quality of order selection depends on the number of competing models. Generally, the final selection quality, measured in ME, becomes worse according as the number of close competitors among the candidates increases. The limitation to only ARMA(r,r-1) models limits the number of close competitors, because ARMA(r+$1,r$-1) and ARMA(r,r) are not considered as candidates. This has a favorable influence on the quality of the order selection.

Broersen and de Waele (2002) developed reduced statistics software that uses only a long AR model of the data for MA and ARMA estimation, not the data themselves. The initial conditions $\hat{A}_p(z)$ can be found from that long AR model and also order selection can be carried out by a comparison of estimated (AR)MA models to the long AR model. This software is available from the author, upon request. Also time series software for missing data will be developed.

3. SELECTION OF THE MODEL TYPE

If the structure of stationary stochastic data is unknown, it will be a question whether AR, MA or ARMA will give the best fitting model with the smallest number of estimated parameters. It is advisable to use all three types of time series models and to select between them. Due to the differences in computation algorithms, it is not advised to consider AR(p) as a subclass of ARMA(p,q) with q=0 and likewise p=0 for MA(q) models. The selection of the model type would have been straightforward if the same order selection criterion could have been used for three model types individually. The same order selection criterion would then also be used for selection of the model type. However, the estimation of MA and ARMA models with a robust algorithm required the use of a high order intermediate AR model. Such high AR orders can only be selected properly with a special criterion like $CIC(p)$ in (12) that is adapted to Burg's AR estimates and that cannot be used for MA and ARMA. Another argument to separate AR from MA and ARMA is that the residual variance is more or less minimized during AR estimation with Burg's algorithm, but it is computed afterwards in MA and ARMA estimation. Hence, the behavior of order selection criteria, especially the selection bias in the residual variance, which is caused by selecting the smallest residual variance from a number of candidates, is different for AR on one hand and MA and ARMA on the other hand. Therefore, different order selection criteria, $CIC(p)$ of (12) and GIC(q,3) of (15), have been recommended. The finite sample behavior of AR estimation means that the influence of extra estimated parameters on the residual variance depends on the model order, especially for orders greater than N/10. Therefore, the penalty has to depend on the model order in $CIC(p)$. This property of residuals prevents the selection of the model type with a single asymptotic selection criterion. $GIC(q,3)$ has a constant penalty for additional parameters, independent of the order. This criterion could perhaps be used to choose between the selected MA(q) and ARMA(r,r-1) models, but a new principle is necessary to include the AR(p) model in the automatic selection of the model type.

The choice between the best AR, MA and ARMA models can only be made with an objective statistical criterion if the best model of each individual type is computed and selected with an algorithm that depends only on the data, not on subjective choices of the experimenter. This means that all models must be obtained without constraints on the roots, without mirroring of roots, without elimination of undesirable model choices and without influence of the maximum candidate model order, provided that this order is taken high enough. Examples of algorithms and order selection criteria that fulfill the requirements have been used in the ARMAsel algorithm; they have been described in the previous section. Others can replace them, as long as only statistics is involved in parameter estimation and order selection. The quality of models is measured by their ability to predict future observations: the prediction error PE or the model error ME in (6). In simulations, knowledge of the true process is used to obtain an efficient estimate for PE and ME. In practice, an estimate for PE could also be obtained if an independent realization of the same process is available, which had not been used for estimation of the parameters. Substitution of the different models in that new realization would reveal the best predicting model type with the smallest PE, the squared error of prediction. In practice, however, all available data are preferably used for the estimation to obtain the best possible accuracy. No independent data are left unused to evaluate the prediction error with sufficient reliability. Only the residual variance is known as an accuracy measure.

Theoretically, a relation between the prediction error and the residual variance has been established for AR processes. The asymptotic relation was the basis of the first order selection criterion used in time series: the final prediction error FPE of Akaike (1970). Broersen (2000a) developed a finite sample expression for this AR relation. A limitation of those relations is that they are only strictly valid for model orders above the true process order. However, this limitation is in practice not important for selection. It may be expected that the three selected models of the three types will contain all statistically significant parameters. Parameters of higher orders are small; otherwise the higher order would have been selected. Small parameters will cause a bias contribution that will generally be negligible. Therefore, it is justified to apply in practice the theoretical properties, which are valid above the true process order, to all model orders above a properly selected model order. For a measured value of the residual variance $RES(p)$ with Burg's AR method, the conditional expectation of the squared prediction error for the selected AR(p) model is found with the finite sample theory as:

$$PE(p) = RES(p) \prod_{i=1}^{p} \frac{1+1/(N+1-i)}{1-1/(N+1-i)} . \tag{20}$$

Like in $CIC(p)$ of (12), the empirical variance coefficients $1/(N+1-i)$ are used in $PE(p)$.

The expectations of the prediction error for MA(q) and ARMA(r,r-1) models will be based on the asymptotic theory. The asymptotic theory is more applicable for MA models than for AR models in finite samples. It turns out that the accuracy of the asymptotic theory is very good for MA orders q until N/8, it is acceptable for orders until N/4 and it becomes worse for orders q=N/2, where the error is 80%. Such high orders, however, are seldom or never selected in practical data. The usual highest MA order considered in ARMAsel is N/5 and for ARMA models the usual limit is N/10. Therefore, there is no reason to introduce a refined finite sample formula for MA and ARMA prediction. The asymptotic formula for the conditional expectation squared error of prediction error for a given estimate of $RES(m)$ is:

$$PE(m) = RES(m) \frac{1+m/N}{1-m/N} , \tag{21}$$

where m denotes the number of estimated parameters.

From the three previously selected models AR(p), MA(q) and ARMA(r,r-1), the type with the smallest estimate of the prediction error PE with (20) for AR or (21) for MA and ARMA is chosen. This gives a single time series model with the ARMAsel algorithm, with type and order automatically selected on purely statistical arguments. This concludes the automatic identification of time series models. After the computation of hundreds of models, a single one is selected automatically and all others are forgotten.

4. SIMULATIONS

It is not possible to evaluate the quality of this automatic identification algorithm with theoretical means, mainly because the asymptotical maximum likelihood theory is not accurate enough in finite samples of time series. The final quality is influenced by a great number of choices in the algorithms and by the order selection criteria. It is not interesting to obtain a good performance for only certain types of data. It is much more important that the performance on measured data is never poor, and that it approaches the Cramér-Rao lower bound in simulations with processes without very high peaks or very deep valleys in the spectral density. High spectral peaks indicate a long covariance function. The length of the measured data must always be longer than the significant covariance length to find proper parameter estimates. Numerous simulations have been carried out with the separate AR, MA and ARMA algorithms, with and without order selection, before the automatic identification was developed.

The program demo_armasa is included in the ARMASA toolbox. Most possibilities and auxiliary programs are demonstrated there, to facilitate the use of the estimated models. The demo gives examples of the computation of spectra and covariance. It shows how to obtain an indication of the accuracy of all computed models, also of those, which have not been selected. This is at the same time an indication whether it is important to use only the selected model or how close competitors of another order or type are.

Results of spectral densities computed from models selected with ARMAsel have been compared to spectral estimates from tapered and windowed periodograms. As a periodogram can be seen as a MA process, the comparison can be made with the ME. The results have never given examples in which the best of many windowed periodogram estimates was of the same quality as the automatically selected time series model of ARMAsel.

5. CONCLUDING REMARKS

The ARMAsel algorithm estimates *N/2* AR models, *N/5* MA models and *N/10* ARMA models from *N* measured observations. Then it selects automatically one model. First, order selection is applied separately to the three model types AR, MA and ARMA. The expected prediction errors of the selected AR(p), MA(q) and ARMA($r,r-1$) models are used for selection of the model type. This new algorithm has a good performance for small samples as well as for a very large number of observations. It requires data as input and delivers without any interference of the user the parameters of the single selected time series model as output. The accuracy of the model with selected type and order is only slightly worse than the accuracy of the model that could be estimated if type and order would a priori be known. For small samples, the accuracy may even be better.

REFERENCES

Akaike, H. (1970). Statistical Predictor Identification. *Ann. Inst. Statist. Math.*, **22**, pp. 203-217.

Akaike, H. (1974). A new look at the statistical model identification., *IEEE Trans. Autom. Control*, **AC-19**, pp. 716-723.

Broersen, P.M.T. (1998). The quality of models for ARMA processes. *IEEE Trans. on Signal Process.*, **46**, pp. 1749-1752.

Broersen, P.M.T. (2000a). Finite sample criteria for autoregressive order selection. *IEEE Trans. on Signal Process.*, **48**, pp. 3550-3558.

Broersen, P.M.T. (2000b). Autoregressive model orders for Durbin's MA and ARMA estimators. *IEEE Trans. on Signal Process.*, **48**, pp. 2454-2457.

Broersen, P.M.T. (2001). ARMASA toolbox, *http: //www.tn.tudelft.nl/mmr/downloads/default.htm.*

Broersen, P.M.T. (2002). Automatic spectral analysis with time series models. *IEEE Trans. on Instrumentation and Measurement*, **51**, pp. 211-216.

Broersen, P.M.T. and S. de Waele (2002). Selection of order and type of time series models estimated from reduced statistics. *Proc. 19th IEEE/IMTC Conf.*, Anchorage, Alaska, pp. 1309-1314.

Broersen, P.M.T., R. Bos and S. de Waele (2002). Spectral analysis of irregularly sampled data with autoregressive models. *Proc. 15th IFAC World Conference*, paper 651, 6 pp.

Burg, J. P., (1967). Maximum entropy spectral analysis. *Proc 37th Meeting Soc. Of Exploration Geophysicists,* 6 pp.

Cooley J.W. and J.W. Tuckey (1965). An algorithm for the machine calculation of complex Fourier series, *Mathematics of computation*, **19**, pp. 297-301.

Durbin, J. (1959). Efficient estimation of parameters in moving average models. *Biometrika*, **46**, pp. 306-316.

Durbin, J. (1960). The Fitting of time series models. *Revue Inst. Int. de Stat.*, **28**, pp. 233-243.

Kay, S.M. and S.L. Marple (1981) Spectrum analysis - a modern perspective. *Proc. IEEE*, **69**, pp. 1380-1419.

Priestley, M.B. (1981). *Spectral Analysis and Time* Series. London, U.K.: Academic Press.

The ARMASA toolbox for Matlab is available at:
http://www.tn.tudelft.nl/mmr/downloads/default.htm
and it can also be found under spectral analysis at
www.mathworks.com/matlabcentral/fileexchange/

IFAC Publications
www.elsevier.com/locate/ifac

MULTI-EDIP – AN INTERACTIVE SOFTWARE PACKAGE FOR PROCESS IDENTIFICATION

Jerzy Kasprzyk

Institute of Automatic Control, Silesian University of Technology, Gliwice, Poland

Abstract: An interactive intelligent software environment *MULTI-EDIP* for computer aided signal and system identification is presented. Motivation for the development of intelligent software for process identification is discussed. A summary of services offered by *MULTI-EDIP* and its main features are presented, particularly the problem of expert advice in model structure determination is highlighted. An example of intelligent support in electro-acoustic plant identification for active noise control application is described. *Copyright © 2003 IFAC*

Keywords: Interactive programs, Process identification, Parameter estimation, Validation, MIMO, Models, Active noise control.

1. INTRODUCTION

One of the fundamental activities in science and technology is modelling. It can be defined as the attempt of representing processes we try to understand by means of models, which we already comprehend. The main goal of identification is to determine mathematical models describing the behaviour of processes based on prior knowledge and measurements collected – if possible – in specially designed identification experiments. These models don't attempt to depict the internal mechanism of the system – they just aggregate it into a convenient and parsimonious representation. Convenient means that the models have simple and standard forms, useful for analysis and applications. Parsimonious means that the number of parameters is kept as small as possible without unduly sacrificing the models explanatory and predictive power.

Process identification is nowadays done in areas as diverse as engineering including aeronautics, acoustics and vibration analysis, bioengineering, chemical engineering, control engineering, electronics, mechanical engineering, signal processing, technical diagnostic, telecommunication; hard sciences including physics and chemistry; life sciences including biology, ecology, agriculture, horticulture and medicine, earth sciences including geology, hydrology and seismography, social sciences including economics and sociology.

Process identification may be looked upon as a generalized measurement technique, providing users with hardware and software necessary to transform raw measurement data into mathematical models representing, in a comprehensive form, the essential features of systems or signal behaviour. It constitutes a unique blend of mathematical methods, manifold programming techniques and practical experience.

It is generally accepted that some form of computer support is a necessary prerequisite for successful process identification, as the result of identification depends on a considerable extent upon the availability of a user-friendly system giving access to robust and high-quality numerical identifications software, some data base tools and some tools for producing graphical outputs. Therefore in early 1990-ties some software tools for interactive intelligent identification were proposed, e.g. ESPION (Haest, *et al.*, 1990), SEXI (Gentil *et al.*, 1990), EFPI (Niederliński, *et al.*, 1991), ISID (Overschee, *et al.*, 1994)

It is also generally known that for successful process identification some special and hard to get expertise is necessary, an expertise which is elusive and difficult to define but nevertheless very real. Thus it is obvious to postulate that computer support for identification should go beyond standard number crunching and data base services (as made available e.g. by appropriate identification toolboxes) and deliver those services that may be expected from true identification experts. To use a catchword: computer supported identification should provide some intelligent services.

1.1 Intelligent process identification

The word *intelligence* appears more and more often in a control and identification related context, without being precisely defined. The default definition implied by most of what is being published seems to be that some control or identification algorithm or software is regarded as intelligent if it is based upon (or contains) paradigms which somewhat resemble human inference paradigms (Passino, 1993). E.g. a piece of software using recursive least squares is hardly ever thought to be intelligent, but if it is extended by a front-end containing a rule-based knowledge base which extends the software scope so as to allow e.g. automatic order determination, one somehow succumbs to a temptation of calling it by a more impressive name. Some other paradigms, provided they seem to be remotely related to processes presumably going on in living systems (lets say fuzzy logic, neural processing, genetic algorithms), are also used as excuses for calling a particular control or identification contraption intelligent. This usage seems to contradict a long established pattern of utilising the word intelligent in everyday and not-so-everyday speech. This pattern can be summarised as follows:

1. The adjective intelligent conveys both the meaning of reasonableness or soundness (which could be regarded as defining some good but standard action) as well as the meanings of brilliancy, cleverness and brightness (which are used to describe something that is above standard and somehow surprising). It seems that all those using the word intelligent in a control– or identification–related context are referring rather to something to be above standard. Therefore we could safely in what follows confine the discussion to the second meaning of intelligent.
2. The adjective intelligent is granted to some action as a result of judgement that might be called expert acceptance. This judgement is passed by experts, which enjoy the reputation of having some standard of what constitutes intelligent behaviour in similar actions.
3. The fact of calling some action intelligent does not seem to depend on the actor arriving at the action with the help of some particular philosophy of acting, which guarantees intelligence of actions. Intelligence of action is generally and universally judged by its fruits, not by the method the action was determined. This method is anyway very seldom known.
4. The nature of intelligence seems to be rather transitory, it seems to belong to a receding horizon: what was considered to be intelligent at some particular instant of time might well be considered routine, obvious or standard at some later instant of time. E.g. to those steam-engine operators which were Watt's contemporaries, the steam engine speed controller might appear to be a highly intelligent piece of hardware, whereas to contemporary control engineers it is just a dull standard.

The above arguments make it easier to accept the following rather subjective definition of intelligence: *a performance is considered intelligent, if experts comparing it with what is standard in their domain of expertise, consider it as such.*

The aim of this paper is to present the *Multivariate Systems and Signals Analyser MULTI-EDIP* – a tool, which not only provides basic number-crunching and data base services, but using some expert know-how, intelligently supports the user in consecutive steps of process identification: from designing an experiment to verifying the model.

2. A SUMMARY OF MULTI-EDIP FUNCTIONS

MULTI-EDIP is the latest chain in a long development project. It started in 1980-ties at the Institute of Automatic Control; Silesian University of Technology, under the name *EDIP* (*Expert for Process Identification*), see Niederliński, *et al.* (1991, 1994). It has been continued during the years 1994-97 as *MULTI-EDIP for DOS* and is being pursued up to now as *MULTI-EDIP for Windows*. The basic assumptions formulated and implemented in *EDIP* for SISO systems and scalar signals, have been extended in *MULTI-EDIP* for MIMO systems and vector signals.

On the basis of configuration parameters declared by the user, *MULTI-EDIP* offers: data generation, data preparation and model identification together with model validation.

2.1 Data generation

The only source of data used by *MULTI-EDIP* are *data files* containing the data generated in any way. However, system offers also two possibilities for data generation:

1. The associated application *SIMULATOR MULTI-EDIP* allows to *simulate* a broad spectrum of deterministic and stochastic time-series, scalar or vector, as well as multidimensional dynamic systems. The simulation mode gives the user plenty of opportunities to test all identification procedures.
2. The associated application *PAIO MULTI-EDIP* allows to perform *real-world identification experiments* using specialised precision analogue input-output units. These units provide up to four analogue-digital inputs to connect with measurement transducers and up to four digital-analogue outputs to excite the system under consideration.

No matter what mechanism is used for data generation, the user is urged to generate for each experiment two data sets: an *estimation data set*, used for model identification, and a *validation data set*, used for model validation.

2.2 Data preparation

Raw data, i.e. data collected from some identification experiment, are not likely to be suitable for immediate processing by some identification algorithms because of their possible deficiencies or the need to enhance some features of particular interest. In the data preparation mode, *MULTI-EDIP* offers the following services:

- data checking (removing outliers, calculating histograms or statistical parameters, testing time-invariability, etc.);
- data editing (decimation, interpolation, choice of a subsequence of interest);
- data preprocessing (filtering the sample sequence, normalization, data scaling, removing averages, polynomial trends or periodical components, integrating, differencing, etc.).

2.3 Model identification

MULTI-EDIP provides support for identification of time-series models, scalar or vector, such as:

- stationary or time-varying stochastic parametric models (*AR, MA, ARMA* and theirs integrated versions as well as the neural counterpart *NAR*);
- deterministic models (models of polynomial trends, discrete spectra models);
- nonparametric models – correlation and frequency-domain models, stationary and time varying (auto- and cross-correlation functions, cepstrum, power spectral density, and synch spectrum).

The next class of models supported by *MULTI-EDIP* consists of stationary and time-varying models of systems, SISO or MIMO, such as:

- parametric models (*ARX, ARMAX, FIR, OE, BJ* and neural counterparts *NARX* and *NFIR*.);
- nonparametric models – correlation and frequency domain models (frequency transfer functions, coherence functions, pulse response, power spectral density of disturbances, etc.).

The basic estimation methods used by *MULTI-EDIP* are:

- for parametric models: ordinary least squares, recursive least squares with exponential data discounting, recursive prediction error method, recursive pseudolinear regression, instrumental variable, backpropagation, Levenberg-Marquardt;
- for nonparametric models: correlation and classical spectral estimation methods, parametric methods for discrete and continuous spectra identification.

For spectral analysis of time-varying processes there are two ways of processing the data:

- direct methods, based on classical FFT – consecutive models are estimated for a time window moving across the data;
- indirect methods – parametric model with assumed structure are identified using a recursive algorithm with exponential data discounting (so called forgetting factor) and used to compute a spectral model.

2.4 Model validation

Model validation is the process of establishing or refuting the soundness of a particular model. This validation is usually done with a special data set, the so called *validation data set*, to be generated together with the basic *estimation data set*. The essence of validation is to subject an obtained model to some tests. *MULTI-EDIP* offers the following validation tests:

- visual tests (simulation, one-step ahead prediction, step- and pulse- responses, frequency transfer functions);
- basic tests (whiteness of prediction error and conditioning number of data matrix);
- additional tests (conditioning number tests for input- and output- correlation matrices, correlation between inputs and prediction error, pole-zero cancellation tests);
- comparison tests (loss function, information criteria as *AIC, BIC*);
- tests of model sensitivity on randomness of data.

MULTI-EDIP is asking by default for a validation data set for any of these tests. However this may be overridden by the user, who may apply the estimation data set for the same purposes

3. MAIN PROPERTIES OF MULTI-EDIP

3.1 System philosophy

At basis of the system construction lies an assumption that process identification and model analysis/verification should be done on the highest possible level of abstraction, i.e. on the symbolic level. Therefore, the user is dealing only with basic concepts of process identification, as e.g.: input-output models, time series models, model classes (polynomial, spectral, neural), types and structures, estimation procedures, model validation and model checking tests. He is not dealing with bits, bytes, instructions, syntax, semantics, algorithms, etc. *MULTI-EDIP* frees the user from doing any computer programming by providing full control of all functions and services through a system of windows and pull-down menus.

The main motivation for this assumption came from the exigencies of teaching process identification: students performing laboratory identification experiments should work on exactly the same conceptual level as the one used in exposing them to the main ideas of identification.

Intelligent user support should be provided by:

- combining the model to be identified with the most appropriate identification method;
- offering expert advice for model structure selection;
- providing a set of default values for some important parameters;
- checking the correctness of some parameters declared by the user;
- proposing a set of verification procedures.

Roughly speaking, the basic philosophy of *MULTI-EDIP* is anti*MATLAB*. Of course, the presented design concept involves an important trade-off between user friendliness and tool flexibility. The features built into *MULTI-EDIP* enable even less knowledgeable users to obtain satisfactory results and understand them. However, in *MULTI-EDIP* more advanced users can not create their own algorithms of data processing, modify implemented methods or identify some unusual models as they can using *MATLAB* identification toolboxes.

Nevertheless the spectrum of basic and intelligent services is broad enough to support typical *MULTI-EDIP* users, e.g. academic teachers running courses on process identification and engineers implementing advanced control systems.

The main techniques used to implement some of the services are

1. Default values and default procedures, suggested to users as they proceed using *MULTI-EDIP*. Needles to say, all the default suggestions may be overridden by users who think they know better or who just want to experiment in order to gain deeper knowledge of the techniques made available.
2. Checking correctness and reasonableness of parameters declared by users or procedures initiated by the user. Values that don't pass the check may be used further only by a separate users decision. Procedures which are not basically sound produce a warning, that may be overridden by the user.
3. A hypertext context sensitive help system, providing comments, suggestions and explanation of what has been obtained or what should be done next.
4. A simulation mode, allowing the user to test any identification method first on a set of simulated data derived from a model of precisely defined structure with precisely known parameters. This service is crucial for building users confidence in the tool and preparing them to use it effectively.

3.2 User interface

The *main window* of *MULTI-EDIP* is a multi-document interface (MDI). This window is an area where other children-windows may appear. They display many kinds of plots, like time- and frequency-domain plots, histograms, polar and Nyquist plots.

The results of parameter model estimation, the effects of model validation tests, information about data preparation and operations done on files are presented in a *report window*. During the session, the contents of this window can be edited, saved in a file or printed.

Identified nonparametric models, results of some model validation tests as well as time-domain plots of prepared data are presented as diagrams. Each kind of diagram allows some actions such as reading of coordinates of any point in the diagram, change of scale (logarithmic, linear), zooming of a chosen part of the diagram, data carrying between windows, comparison with results saved in a file or with a mean value of a group of files, saving results in a file and printing the diagram.

For time-varying models, the results are shown as three-dimensional diagrams, that present changes in time for identified models. In such a window it is possible to plot and move a cross-section plane, to change a point of watching on a diagram, to show a cross-section in a two-dimensional window and to print a diagram.

3.3 Model structure determination

To support parametric model identification the following approach is suggested and partially performed automatically:

1. Start from a model structure that is consistent with prior knowledge.
2. Estimate model parameters using suitable algorithm according to estimation criterion.
3. Validate the estimated model.
4. If the result is not good enough, try another model structure and repeat from step 2, until suitable model is obtained.

This identification loop can be repeated a number of times. Its crucial part is handling the search of "optimal" structure, i.e. the structure for which the model is the best one according to the chosen criteria. It is especially important when there is no reason to assume a model structure, e.g. when a linear model is identified to approximate a dynamic behaviour of a non-linear system around the set point. The natural way to do this is an "exhaustive search", which means that models for all structures within the assumed bounds ought to be estimated. This approach assures that the "global optimum" will be achieved, but due to a huge number of possible model structures it is time-consuming. Therefore, a second method is implemented, based on the author's experience. In this method, the search through the structure space is driven by a set of heuristic rules that attempt to reach a (sub) optimal point (the best structure) in a small number of steps (Kasprzyk, 1997).

4. EXAMPLE OF INTELLIGENT SUPPORT IN ACOUSTIC PLANT IDENTIFICATION

MULTI-EDIP proved useful in many different applications. One of them is development of active noise control (ANC) systems, where a typical problem is identification of electro-acoustic plants for feedforward control systems. In the example considered models of so called secondary and feedback paths are necessary to design an adaptive feedforward active noise control system creating a local 3-dimensional zone of quiet surrounding a single (error) microphone in a reverberant enclosure, see Błażej and Ogonowski (2001). This enclosure is disturbed by a noise that should be reduced using two control loudspeakers.

The structure of control system implies that before the ANC is activated, transfer functions between the outputs of digital compensators (that drive control loudspeakers) and the signal acquired from error microphone should be estimated (Elliot, 1999). Identification of acoustic plants is not a trivial task because they are not typical low pass systems as other industrial plants and due to reverberations their frequency characteristics are very irregular, with a significant number of valleys and peaks.

Identified plant is of a TISO type (Two-Input-Single-Output) representing D/A converters, reconstruction filters, amplifiers, control loudspeakers, acoustic space between these loudspeakers and the error microphone, anti-aliasing filter and A/D converter. It was shown (Figwer, *et al.*, 1993) that specially designed excitation in the form of a vector multisine signal minimises the input-generated identification error and enables to properly estimate addendum to the overall system output due to the fact that all inputs are orthogonal and hence partial outputs are mutually orthogonal as well. Therefore for identification experiment a bivariate multisine excitation (Niederliński and Figwer, 1995) with a period $N=2048$ samples was generated and used to drive the control loudspeakers. The sampling frequency was 500 Hz and cut-off frequency of both filters (anti-aliasing and reconstruction) was fixed to 150 Hz.

Figwer (1998) discusses the problem of nonparametric identification for presented acoustic plant by means of correlation analysis. Here the results of model structure selection in parametric model identification are presented.

Since every channel consists of the same type elements (D/A converter, reconstruction filter, amplifier etc.) it was assumed that relationships between inputs $u_1(i)$ and $u_2(i)$ and output $y(i)$ in presented TISO plant can be described by an *ARX* model:

$$A(z^{-1})y(i) = \sum_{j=1}^{2} z^{-dj} B_j(z^{-1})u_j(i) + e(i) \qquad (1)$$

where: z^{-1} is the backward shift operator, $u_j(i)$ and $y(i)$ are i-th samples of j-th (j=1,2) input and output respectively, d_j denotes the pure delay in the j-th control path, $A(z^{-1})$ and $B_j(z^{-1})$ are polynomials of the operator z^{-1} with orders dA and dB_j respectively, $e(i)$ is a white noise modelling disturbances acting on the output signal.

4.1 Expert advice in model structure selection

A model structure is specified by a set of integer numbers defining polynomial orders and delays. In this case the model structure can be defined as:

$$([d_1, d_2], dA, [dB_1, dB_2]). \qquad (2)$$

The problem of structure determination is specially difficult for acoustic plants, where it is very hard to guess the proper orders and delays. Simulation and real-data identification experiments carried on by the author showed, that model quality and computing time depend on the proper choice of a starting point, so a heuristic algorithm of searching a (sub) optimal structure was implemented in *MULTI-EDIP*. In this case it may be summarised as follows:

1. For each pair – input u_j (j=1,2) and output y – search the optimal structure of a SISO model.

2. As the starting point in TISO system identification, for consecutive j-th control channel take orders and delay obtained for j-th SISO model, but for common polynomial (i.e. A) choose the greatest order of corresponding polynomial. Then search (by varying orders or delays) such a TISO model structure that minimises the chosen criterion (e.g. *BIC*).

As a rule, an overparametrization (i.e. too high orders) occurs while SISO model identification is performed for a TISO system. It is rather obvious, because the model attempts to explain the overall input-output behaviour only by means of the one input-output relation. Therefore, handling the search of polynomial orders in the second step is usually done by decreasing the polynomial orders.

Step 1 consists of the following stages:

1.1. For consecutive delays within the admissible set identify SISO models with the lowest admissible orders and choose the delay that minimises cost function.

1.2. For the chosen delay identify SISO models for consecutive polynomial orders within the admissible set and choose the structure that minimises the chosen criterion.

1.3. For selected structure try to vary orders of consecutive polynomials as well as the delay until a minimum of the criterion is reached

The complexity of the identified electro-acoustic plant implies that identified model should be of a high order, so models having polynomial orders within the range 21 – 40 were tested. The lower bound resulted from filter orders and prior experiments while the higher bound stem from an assumption that not more than 20 orders should be tested. Similarly, taking into account the dimensions of the plant and delays caused by the filters, it was assumed that delays may vary within the range 4 – 12.

The results of consecutive steps in model structure selection using *BIC* criterion are as follows:

Identification of the first channel. In the first step the delay $d_1 = 7$ was chosen. Next models with orders $dA_1 = dB_1 = 21,\ldots,40$ were tested and a model with order 34 was determined. Finally a model with the structure (9,34,34) was chosen.

Identification of the second channel. In the first step the delay $d_2 = 7$ was determined. Next models with orders $dA_2 = dB_2 = 21,\ldots,40$ were tested and a model with order 21 was determined. Finally a model with the structure (10,19,17) was chosen.

Identification of the entire TISO model. Starting from the structure ([9,10],34,[34,17]) and varying consecutive delays and orders, a final model with the structure ([9,10],34,[37,17]) was found.

In the procedure presented above only 31 SISO models for the first channel, 35 SISO models for the second channel and 43 TISO models were estimated and tested, while in an "exhaustive search" more than 600000 TISO models ought to be estimated and tested. However, the proposed method couldn't guarantee that the model adequate to a global optimum of chosen criterion was obtained.

It should be notice that *MULTI-EDIP* explores orders and delays also beyond the assumed bounds. It is done when such violation allows to obtain a model with a lower value of the criterion than the value reached at the bounds.

CONCLUSIONS

It seems that automation of a knowledge-intensive task like process identification is feasible. However, it proved to be rather difficult, software-wise demanding and generally time-consuming. Thanks to extensive testing of *MULTI-EDIP* by groups of students working on laboratory projects and Ph.D. students working on theses on active noise control, many simple and sophisticated errors have been discovered and removed. Nevertheless, some new ideas still wait to be tested.

ACKNOWLEDGMENT

The partial financial support of this research by the State Committee of Scientific Research (KBN) under the grant 4 T11A 010 22 is gratefully acknowledged.

REFERENCES

Błażej, M. and Z. Ogonowski (2001). Adaptive 3-D Zone of Quiet in an Enclosure Using LMS-Type Algorithms. *Proceedings of the 7th IEEE Int. Conference MMAR 2001,* **2**, Międzyzdroje, 805-810.

Elliot, S. J. (1999). Down with Noise. Practical control systems for combating audible noise show up in aerospace, general aviation, and military roles. *IEEE Spectrum*.

Figwer, J. (1998). Two-Input System Identification Using Bivariate Multisine Excitation. *Proceedings of the 5th IEEE Int. Conference MMAR 1998,* **2**, Międzyzdroje, 571-576.

Figwer, J., A. Niederliński and J. Kasprzyk (1993). A New Approach to the Identification of Linear Discrete-Time MISO Systems. *Archives of Control Sciences,* **2**, 223-239.

Gentil, S., A. Barraud, and K. Szafnicki (1990). SEXI: An Expert Identification Package. *Automatica*, **26**, 803-809.

Haest, M., G. Bastin, M. Gevers and V. Wertz (1990). ESPION: an Expert System for System Identification. *Automatica*, **26**, 85-95.

Kasprzyk, J. (1997). Model Structure Determination in Parametric Model Identification, *Systems Science*, **23**, 89-95.

Niederliński, A., Kasprzyk, J., Figwer, J. (1991). EFPI - An Integrated Software Environment for System- and Signal Identification. *Proceedings of the 9th IFAC/IFORS Symposium on Identification and System Parameter Estimation*, Budapest, 567-572.

Niederliński, A., J. Kasprzyk and J. Figwer (1994). Computer Aided Process Identification. In: *Computer Aided Control System Design*. (M. Brdyś and K. Malinowski, (Ed.)), 73-98. World Scientific Publishing Co. Pte. Ltd., Singapore.

Niederliński, A. and J. Figwer (1995). Using the DFT to Synthesize Bivariate Orthogonal White Noise Series. *IEEE Transactions on Signal Processing*, **3**, 749-758.

Overschee, P. van, B. De Moor, H. Aling, R. Kosut and S. Boyd (1994). ISID II. A Fully Interactive Identification Module for Xmath. *Preprints of 10th IFAC Symposium on System Identification*, **4**, Copenhagen, 1-2.

Passino, K. M. (1993) Bridging the Gap between Conventional and Intelligent Control. *IEEE Control Systems*, June 1993, 12-18.

IFAC Publications
www.elsevier.com/locate/ifac

KALMTOOL
FOR USE WITH MATLAB

Magnus Nørgaard *, **Niels Kjølstad Poulsen** *** and **Ole Ravn** **

* *Widex A/S, Ny Vestergaardsvej 25, DK-3500 Værløse, Denmark*
** *Ørsted•DTU, building 326, Technical University of Denmark,DK-2800 Kgs. Lyngby, Denmark*
*** *Informatics and Mathematical Modelling, building 321, Technical University of Denmark, DK-2800 Kgs. Lyngby, Denmark*

Abstract: The KALMTOOL toolbox is a set of MATLAB tools for state estimation for nonlinear systems. The toolbox contains functions for extended Kalman filtering as well as for two new filters called the DD1 filter and the DD2 filter. The toolbox specifically addresses the problem of not having observations available at all sampling instants. All functions are available as m-functions but for faster (much faster!) execution, the DD1 and DD2 filters are also available as C Mex files for MATLAB under Windows and Linux. The toolbox requires MATLAB 6. No additional toolboxes are required. *Copyright © 2003 IFAC*

Keywords: Kalman filter, State estimation, non-linear systems, Extended Kalman filters.

1. INTRODUCTION

In this paper a newly developed toolbox, Kalmtool, will be described. The purpose of the toolbox is to make implementations of the new DD1 and DD2 filters available for solving nonlinear state estimation problems and to enable a comparison with a conventional method like the extended Kalman filter.

If you are familiar with the EKF, you will quickly find out about the advantage of the new filters. First of all, they are easier to use. Secondly, you can expect a similar (DD1) or better (DD2) performance than with the EKF. On the downside, the new filers tend to require more computations than the EKF.

During our work we found that the extended Kalman filter was somewhat inconvenient to use in some of our applications. A small modification of the application sometimes had serious implications on the EKF implementation. Moreover, it was often difficult to implement. Our problem was that the EKF requires a linearization of the system model. Sometimes this is easy to find but sometimes it can be pretty hard. In any case, it makes things inflexible. If a small change is made in the model, one has to work out a new set of derivatives. This is particularly inconvenient in model calibration where certain model parameters are temporarily included in the state vector and estimated simultaneously with the actual states.

So far, three publications related to these filters are available. A short introduction can be found in (Nørgaard *et al.*, 1999), an expanded version in (Nørgaard *et al.*, 2000) and the most thorough description in (Nørgaard *et al.*, 1998).

2. BACKGROUND

In order to use the filter routines you need the following prerequisites:

A state space model in the form:

$$\begin{aligned} x_{k+1} &= f(x_k, u_k, v_k) \\ y_k &= g(x_k, w_k) \end{aligned}$$

The noise covariance matrices

$$\begin{aligned} Q &= \mathbf{E}\left\{v_k v_k^T\right\} \\ R &= \mathbf{E}\left\{w_k w_k^T\right\} \end{aligned}$$

Initial estimates of state and covariance matrix

$$\begin{aligned} \bar{x}_0 &= \mathbf{E}\left\{x_0\right\} \\ P_0 &= \mathbf{E}\left\{(x_0 - \bar{x}_0)(x_0 - \bar{x}_0)^T\right\} \end{aligned}$$

Input-output data sets

$$\begin{aligned} y &= \left\{y_0, y_1, y_2, \ldots\right\} \\ u &= \left\{u_0, u_1, u_2, \ldots\right\} \end{aligned}$$

For application of the extended Kalman filter you must also derive the linearized state and observation equations:

$$x_{k+1} \approx f(\hat{x}_k, u_k, \bar{v}_k) + A(k)(x_k - \hat{x}_k) + F(k)(v_k - \bar{v}_k)$$
$$y_k \approx g(\bar{x}_k, \bar{w}_k) + C(k)(x_k - \bar{x}_k) + G(k)(w_k - \bar{w}_k)$$

where

$$A(k) = \left.\frac{\partial f(x, u_k, \bar{v}_k)}{\partial x}\right|_{x=\hat{x}_k} \quad F(k) = \left.\frac{\partial f(\hat{x}_k, u_k, v)}{\partial v}\right|_{v=\bar{v}_k}$$
$$C(k) = \left.\frac{\partial g(x, \bar{w}_k)}{\partial x}\right|_{x=\bar{x}_k} \quad G(k) = \left.\frac{\partial g(\bar{x}_k, w)}{\partial w}\right|_{w=\bar{w}_k}.$$

When you have collected all this information you must specify the model in MATLAB functions. These functions must conform to a particular structure, which is discussed below

3. WRITING THE EQUATIONS IN M-FUNCTIONS

If you wish to run the DD1 or DD2 filter you must write two functions. One should contain the state equation, and the other should contain the output equation. If you are working with more than one observation stream, you must write an output equation function for each stream. This application will not be covered any further here, but more details can be found in the manual accompanying the toolbox. If you are going to use the EKF, it is necessary to write an additional function that specifies the linearization of the two equations.

3.1 *The state equation*

As an example, let us implement the state equation in a function called **myxfunc**. The necessary components are shown below (for a nonsense system!).

```
function xout=myxfunc(x,u,v)

% Make variables static
persistent mypar1 mypar2;

% Check if variables should be initialized
if nargin==1,
   mypar1 = x(1)*0.5 + x(2);
   mypar2 = 75*x(3);
   return
end

% A priori update of states
xout    = zeros(3,1);

xout(1) = x(1) + mypar2*cos(x(2)+u(1)*v(1));
xout(2) = x(3) + mypar2*cos(x(2)+u(2)*v(2));
xout(3) = mypar1*x(1) + v(3);
```

Dissection of the function The header must always look like this:

function xout=myxfunc(x,u,v)

The function and variable names are unimportant, but the function must always take 3 arguments and return one output. The arguments, which should be (column) vectors, are the current state estimate, control input, and process noise (in that order). The function should output the *a priori* state update. Argument 2 and 3 *must* be present even if there are no inputs or process noise.

By using the *persistent* declaration, it is possible to maintain parameters from one function call to another. This is convenient as one can initialize certain parameters before the filtering.

There should always be an initialization section in the function. This must take the form

```
if nargin==1,
   .. do initialization stuff
   return
end
```

Before the actual filtering is performed, *myxfunc* will be invoked as *myxfunc(opt.init)*. Thus, by specifying parameter initializations in the variable *opt.init*, these parameters are passed to the function through the argument *x*. *opt.init* is a struct, which can contain various optional parameters. More details are provided later on. The initialization section must be included even if there are no such initializations. In this case it should just contain the *return* statement.

The last part of the function is the actual state update. If *x* is not also used as the return variable, make sure that the returned variable is a column vector.

3.2 *The output equation*

The output equation is written in an m-function with a similar format (again containing nonsense):

```
function y=myyfunc(x,w)
% Make variables static
persistent mypar3

% Check if variables should be initialized
if nargin==1
   mypar3=x(4);
   return
end

% Calculate output estimate
y = mypar3*x.*x+w;
```

The function should take two arguments: the state vector and the measurement noise vector. Apart from that, the function has the same structure as *xfunc*.

3.3 *Linearization of the equations (EKF only)*

When using the EKF it is necessary to write a separate function that contains the linearization of the two equations. The function must have the following format (with still more pedagogical nonsense)

```
function [M,N]=mylinfunc(x,u,vw,flag)
% Make variables static
persistent mypar1 mypar2;
```

```
% Check if variables should be initialized
if nargin==1,
   mypar1 = x(1)*0.5 + x(2);
   mypar2 = 75*x(3);
   A0 = diag([1 1]);
   F0 = zeros(2,2);
   C0 = [1 0];
   G0 = [];
   return
end

% Linearize state equation
if flag==0,
  M = A0;
  M(1,2)=mypar1*(sin(x(1)*u(1))+x(2)*u(2));

  N = F0;
  N(1,1) = mypar2/x(2);
  N(2,2) = x(1)*x(2);

% Linearize output equation
elseif flag==1,
   M=C0;
   N=G0;
end
```

Dissection of the function The header must always look like this:

function [M,N]=my_linfunc(x,u,vw,flag)

The function and variable names are unimportant, but the function must always take 4 input arguments and return 2 outputs. The arguments, three (column) vectors and an integer, are the current state estimate, control input, process noise or measurement noise, and a flag (in that order). If *flag=0* then *vw* is process noise and the function should linearize the state equation and return the matrix *A* in *M* and the matrix *F* in *N*. If *flag=1* then *vw* is measurement noise and the function should linearize the output equation and return the *C* matrix in *M* and the *G* matrix in *N*.

Argument 2 and 3 *must* be present even if there are no inputs or no process/measurement noise.

4. RUNNING THE FILTERS

When input and observation data are available, and the appropriate m-functions have been written, it is straightforward to run the filters. The DD2 filter is invoked in the following way:

```
[xhat,Smat]=dd2('myxfunc','myyfunc', ...
                x0,P0,Q,R,u,y,tidx);
```

u is a matrix of inputs to the system. The first row contains the inputs applied at time *k=0*, the second row contains the inputs at time *k=1*, etc. If the system has no inputs, the empty matrix *[ ]* is passed.

y is a matrix of observations and *tidx* is a vector with time stamps of the observations in *y*. The first row in *y* contains the observations acquired at the sample number specified in the first element of *tidx*, the second row in *y* contains the observations acquired at the sample number specified in the second element of *tidx*, etc. If observations are available at all sampling instants, we simply have that *tidx=0:size(u,1)*.

The state estimates are returned in the matrix *xhat*. The first row in *xhat* is the estimate at time *k=0*. If there is no observation update at that time, it will simply be the argument *x0* that was passed to the filter function upon the call.

The matrix *Smat* is a matrix containing coefficients of the Cholesky factors of the estimation error covariance matrices. The format of this matrix will be further explained in the following section.

The DD1 filter is called in the exact same way. For the extended Kalman filter it is necessary to include the name of your m-function that performs the linearizations:

```
[xhat,Pmat]=ekf('myxfunc','myyfunc', ...
 'mylinfunc',x0,P0,Q,R,u,y,tidx)
```

The coefficients of the covariance matrices (and not their Cholesky factor as was the case above) are returned in the matrix *Pmat*. The format of this matrix will also be explained in the following section.

If the noise processes have mean values different from 0, these can of course be subtracted at the appropriate places in the functions *myxfunc* and *myyfunc*. They can also be passed to the filter functions through the optional data structure 'optpar'. For example,

```
[xhat,Smat]=dd2('myxfunc','myyfunc', ...
          x0,P0,Q,R,u,y,tidx,optpar);
```

4.1 *Optional parameters*

In most applications, the state space model is less complicated than shown in the introduction. F.ex, different types of linear terms are common:

$$\begin{aligned}
x_{k+1} &= Ax_k + f(u_k) + Fv_k \\
x_{k+1} &= Ax_k + f(u_k, v_k) \\
x_{k+1} &= f(x_k, u_k) + Fv_k \\
y_k &= Cx_k + Gw_k \\
y_k &= Cx_k + g(w_k) \\
y_k &= g(x_k) + Gw_k
\end{aligned}$$

By including one or more og the matrices $\{A, F, C, G\}$ in the optional structure variable *optpar*, the filtering functions will take advantage of the equation structure and therefore run faster. As example of including one of the matrices in structure, let

```
opt.C = 5*eye(3);
```

If the noise enters linearly, the noise contribution can be left out of the m-functions described above. This is **not** the case for terms like Ax_k and Cx_k. This is because the information provided through *opt* is for use only in the various covariance updates. *optpar* can also include initialization parameters for the user-written m-functions. This is done through the *optpar.init* element. Upon call of the filter functions, *optpar.init* (which can be either a vector or a matrix) will be passed as the first argument to the user's functions in the initialization process prior to the actual filtering, e.g., *myxfunc(opt.init)*

5. ANALYZING THE RESULTS

The performance of the filters can be evaluated with the function *kalmeval*. This function calculates the output estimates and compares these to the actual observations. Additionally, it plots the state estimates along with three times their (estimated) standard deviations. For the DD2 filter, the function is called as follows:

```
[yhat,RMS]=kalmeval('dd2','myyfunc', ...
                    R,xhat,Smat,y,tidx,optpar)
```

and similarly for the DD1 filter. *RMS* is the RMS error between observations and predictions.

To evaluate the extended Kalman filter, the function is called as follows (notice that *Pmat* is passed instead of *Smat*):

```
[yhat,RMS]=kalmeval('ekf','myyfunc', ...
                    R,xhat,Pmat,y,tidx,optpar)
```

If you would like to take a closer look at the covariance estimates, four functions are available to accommodate this. Two of the functions are used together with the DD1 and DD2 filters. These filters works on Cholesky factors of the covariance matrices and not on the covariance matrices themselves. If the covariance matrix is denoted P, the Cholesky factor, S, is an upper triangular matrix with the property

$$P = SS^T$$

For sample no. k, the elements of $S(k)$ are stored in row $k+1$ of the matrix, *Smat*:

$$S(k) = \begin{bmatrix} S_{11} & S_{12} & S_{13} \\ 0 & S_{22} & S_{23} \\ 0 & 0 & S_{33} \end{bmatrix}$$

$$\rightarrow \quad Smat = \begin{bmatrix} & \vdots & & & \vdots & \\ S_{11} & S_{12} & S_{13} & S_{22} & S_{23} & S_{33} \\ & \vdots & & & \vdots & \end{bmatrix}$$

With the function *smat2cov* it is possible to extract the covariance matrix at a specific sample number:

```
P = smat2cov(Smat,k+1)
```

extracts the covariance matrix at time k (recall that the estimates for k=0 are stored in the first row).

The function *smat2var* uses *Smat* to calculate the variance of each state estimate, i.e., the diagonal of the covariance matrix:

```
V = smat2var(Smat)
```

The first column contains the variance estimates for the first state, etc.

To extract the covariances estimated by the extended Kalman filter, two similar functions are available. *mat2cov* takes *Pmat* as input and returns the covariance matrix at a specified sample number. The format of *Pmat* is shown below. If $P(k)$ is the covariance matrix at time k then row $k+1$ of *Pmat* is organized as follows:

$$P(k) = \begin{bmatrix} P_{11} & P_{12} & P_{13} \\ P_{12} & P_{22} & P_{23} \\ P_{13} & P_{23} & P_{33} \end{bmatrix}$$

$$\rightarrow \quad Pmat = \begin{bmatrix} & \vdots & & & \vdots & \\ P_{11} & P_{12} & P_{13} & P_{22} & P_{23} & P_{33} \\ & \vdots & & & \vdots & \end{bmatrix}$$

$P(k)$ is extracted by

```
P = mat2cov(Pmat,k+1)
```

The function *mat2var* works the same way as *smat2var* except that it takes *Pmat* as input.

6. WORKING WITH THE MEX-FILES

With a little extra effort one can experience a tremendous increase in execution speed by using the MEX alternatives to the m-functions. This alternative is available only for the DD1 and DD2 filters as the increase in execution speed is particularly pronounced for these.

Whereas before one had to write m-functions containing state and output equations, in the MEX case it is necessary to write similar functions in C. As a prototyping stage it is always a good idea to start out in MATLAB. When things are working here you can "translate" the m-functions to C and use the MEX functions.

The MEX files operate on a special matrix format for storage of vectors and matrices. In the C code a "matrix" will be pointer to a data structure. A new matrix is declared and memory allocated by:

```
matrix *M;
M = mmake(rows,columns);
```

The data structure contains three variables: number of rows (*M->row*), number of columns (*M->col*) and a pointer to an array that contains pointers to the memory locations where each row is stored (the content of *M->mat[0]* points to the first element in the matrix). A number of functions are available to operate on the matrices.

7. STEP 1: WRITING STATE AND OUTPUT EQUATIONS IN C

Like in the MATLAB case you must write separate functions for state equation and observation equation(s). Let us first take a look at the format of the state equation function.

7.1 *State Equation*

```
/* Function prototype */
 int myxfunc(matrix*, matrix*, matrix*,
                              matrix*, int);

/* The state equation function */
 int myxfunc(matrix *xbar, matrix *xhat,
         matrix *u, matrix *v, int flag)
 {
   /* Variable declarations */
   int a, b, c;
   double d, e, f;
   static int h, i, j;
   static matrix *M, *N;

   /* Initializations */
   if (flag == -1){
      ... Initialize static variables
      return 0;

   /* Clean up */
   else if (flag == -2){
      ... free matrices that you might
          have allocated
      return 0;
   }

   /* Normal call of function */
   else{
      ... Perform state update
          (insert in xbar)
      return 0;
   }
}
```

Dissection of the function Except for the function and variable names, which can be arbitrary, the function call must have exactly the above format. *xbar* is the *a priori* state estimate; i.e., the output of the function. *xhat* is the previous state estimate, *u* is the input (it must be included in the argument list, but it does not have to be used), and *v* is the process noise. All four arguments to the function are column vectors; i.e., matrices with one column.

The last argument, *flag*, is used for specifying whether the function is called in "initialization mode", in "clean up mode", or in "filtering mode". The modes are explained below:

Initialization mode
The function is called once in this mode prior to the filtering. It is included because often it is useful to remember certain parameters, vectors, matrices, etc, from one call to another rather than having to recalculate them at every sample. In particular, this section is used for vector/matrix allocations (call of the *mmake* function). As *mmake* performs a dynamic memory allocation this is not something one should carry out at every sample. Variables that are to be remembered from call to call should be declared "static".

If you wish to pass certain parameters to the function upon filtering (in contrast to hard coding all the information), the filter function has an optional argument, which can be used for passing such parameters. This argument must be a matrix, and it will be passed to the function through the first argument (*xbar*) during initialization.

Clean up mode
The function will be called in this mode just before termination of the filter function. If you have dynamically allocated memory for matrices or arrays, this is the place to free the memory. Matrices allocated with *mmake* are deallocated with the *mfree* command.

Filtering mode
In this section of the function, the actual state update is placed. The updated states are placed in the first vector in the argument list (*xbar*).

7.2 *Observation Equation*

The observation equation has a similar format:

```
/* Function prototype */
 int myyfunc(matrix*, matrix*, matrix*, int);

/* The state equation function */
 int myyfunc(matrix *ybar, matrix *xbar,
                matrix *w, int flag)
 {
   /* Variable declarations */

   /* Initializations */
   if (flag == -1){
      ... Initialize static variables
      return 0;

   /* Clean up */
   else if (flag == -2){
      ... free matrices that you might
          have allocated
      return 0;
   }

   /* Normal call of function */
   else{
      ... Calculate output estimate
          (insert in ybar)
      return 0;
   }
}
```

For applications with multiple observation streams the output function must be augmented to handle this. This is different from the MATLAB case where a separate function was written for each output stream. A more thorough explanation is provided in the toolbox manual.

```
   /* Normal call of function */
   else if (flag == 0){
      ... Calculate output estimate for stream 1
      return 0;
   }
   else if (flag == 1){
      ... Calculate output estimate for stream 2
      return 0;
   }
```

8. STEP 2: MODIFYING THE TEMPLATE AND COMPILE THE FUNCTION

All the components must now be put together and compiled. First place the functions described above in the same file (place both prototype declarations in the top). Secondly, select a filter function. Two templates files are available: *dd1c.c* and *dd2c.c* (or *dd1mc.c* and *dd2mc.c* for multiple observation streams). Make a copy of the template corresponding to the filter function you wish to use and give it a new name, such as *myddfilter.c*. Now open the file in an editor. In the top of the file there are three "define" statements: *KALMFILE*, *XFUNC*, and *YFUNC*. After *KALMFILE* you write the name of the file containing your functions, after *XFUNC* you write the name of the state equation function, and after *YFUNC* you write the name of the output equation function:

```
#define KALMFILE "myfile.c"
#define XFUNC myxfunc
#define YFUNC myyfunc
```

The file is compiled by issuing the MATLAB command **mex**. The compilation depends on the platform you are working on. Under MS WINDOWS. MATLAB is now bundled with the LCC compiler. For Matlab under Linux, the GCC compiler is assumed:

```
% PC/Windows, lcc
   >> mex myddfilter.c kalmlblcc.obj
% PC/Linux
   >> mex myddfilter.c kalmlblx.o
```

9. STEP 3: CALLING THE FILTER ROUTINES

The MEX function is now ready to be called from MATLAB. The call is *almost* the same as when you call the m-function counterpart. The main difference is that you should not call the function with the covariance matrices, but with a root, S, for which $P = SS^T$ (not necessarily a Cholesky factor).

```
>> Sv = chol(Q)';
>> Sw = sqrtm(R);
>> [v,d] = eig(P0);
>> Sx0   = real(v*sqrt(d));

>> [xhat,Smat]=myddfilter(x0,Sx0,Sv, ...
                        Sw,u,y,tidx,optidx);
```

For illustration purposes three different ways to factorize the covariance matrices have been shown above. You can use any which one you prefer.

To evaluate the result of the filtering you can use the MATLAB functions described previously.

10. REFERENCE

Filter Functions

dd1	DD1 filter.
dd1m	DD1 filter for systems with multiple observation streams.
dd2	DD2 filter.
dd2m	DD2 filter for systems with multiple observation streams.
ekf	Extended Kalman filter.
ekfm	Extended Kalman filter for systems with multiple observation streams.

MEX files

dd1c	C-Mex counterpart to the 'dd1' function.
dd1mc	C-Mex counterpart to the 'dd1m' function.
dd2c	C-Mex counterpart to the 'dd2' function.
dd2mc	C-Mex counterpart to the 'dd2m' function.
kalmlb**	Object file that must be linked with the mex files.
xytest	Test C-functions before the filtering is performed.

Utilities

kalmeval	Evaluate filter performance.
mat2cov	Extract covariance matrix from vector containing the upper triangular elements.
mat2var	Extract variance estimates from matrix containing covariance estimates.
smat2cov	restore covariance matrix from vector of Cholesky factor elements.
smat2var	Calculate variance estimate for each state from the Cholesky factored covariance matrices. covariance estimates.
triag	Triangularization with Householder transformation.

Demonstrations

agvdemo	Position and orientation estimation and calibration of an AGV.
falldemo	Falling body example (a continuous-time example).
demomex	Generates MEX files to speed up the demonstrations.

CONCLUSION

In this paper we have presented the toolbox KALMTOOL which a set of MATLAB tools for state estimation for nonlinear systems. It contains functions for extended Kalman filtering as well as for the two new filters the DD1 filter and the DD2 filter. The toolbox can be downloaded from

www.iau.dtu.dk/research/control/kalmtool.html

11. REFERENCES

Nørgaard, M., N.K. Poulsen and O. Ravn (1998). Advances in derivative-free state estimation for nonlinear systems. Technical Report IMM-REP-1998-15. Department of Mathematical Modelling, Technical University of Denmark.

Nørgaard, M., N.K. Poulsen and O. Ravn (1999). Easy and accurate state estimation for nonlinear systems. In: *14th IFAC World Conference in Beijing, China*. pp. 343–348.

Nørgaard, M., N.K. Poulsen and O. Ravn (2000). New developments in state estimation for nonlinear systems. *Automatica* **36**(11), 1627–1638.

IFAC Publications
www.elsevier.com/locate/ifac

THE ADAPT$_X$ SOFTWARE FOR AUTOMATED AND REAL-TIME MULTIVARIABLE SYSTEM IDENTIFICATION

Wallace E. Larimore

Adaptics, Inc, 1717 Briar Ridge Road, McLean, VA 22101
Phone: 703 532-0062, Fax: 703 536-3319, Email: larimore@adaptics.com

Abstract. Recent developments in the ADAPT$_X$ software for automated system identification are discussed. Unknown delays in systems occur frequently in industrial processes and other time series data. Such delays can considerably increase the state order and the number of estimated parameters. New methods and software are discussed that give fast, numerically stable and accurate determination of the multivariable delay structure that is then used for efficient subspace modeling. A second innovation is the development of a leaps and bounds type algorithm for the determination of the significant inputs that influence the outputs. The automated system identification capability has been extended for real-time use with the dSPACE data aquisition hardware. This is of considerable interest in automotive and aerospace applications. Finally, a Java graphical user interface is under development.

Key Words: Automated system identification, Real-time, Delay determination, Input selection.

1. INTRODUCTION

The ADAPT$_X$ software for automated multivariable system identification has been in use now for over a decade with a number of major achievements discussed in detail in Larimore (1999, 2000). In this paper, first an overview is given of the ADAPT$_X$ approach to system identification and how it differs from conventional system identification in general and the available subspace system identification software in particular. Then, in following sections new features in the ADAPT$_X$ software that have been or soon will be released are discussed.

A general overview of ADAPT$_X$ is given in Larimore (2000) including the approach to completely automatic computation, the theory of canonical variate analysis (CVA) upon which it is based, the computational steps in the software, and several of the major applications to which it has been applied. A more detailed tutorial discussion of the theory is contained in Larimore (1999). In the next section, an overview of the ADAPT$_X$ approach is given that emphasizes some of the unique aspects of the software.

In recent years, several enhancements to ADAPT$_X$ have been developed, tested on industrial applications, and released or will be release for general use:

- DelaySelect. Automatic identification of the multivariable delay structure of a system
- StrucSelect. Automatic identification of the significant sets of system inputs, and the inputs of little or no influence
- Real-Time Interface. A real-time interface to data aquisition hardware for use in on-line and real-time system identification
- Java GUI. A Java Graphical User Interface for interfacing with the various hardware and software platforms now using ADAPT$_X$.

These developments greatly extend the applicability of automated multivariable system identification to multivariable systems with large delays relative to the sample period, systems with some inputs having little or no effect of the outputs, and real-time systems. In such real-time systems, identification may need to be done in real time, for system monitoring or adaptation through on-line control design. In the sections below, the new features of ADAPT$_X$ are discussed. The ADAPT$_X$ software is available running under Matlab on Windows and UNIX platforms. It is also available as a C++ library for Windows and UNIX platforms.

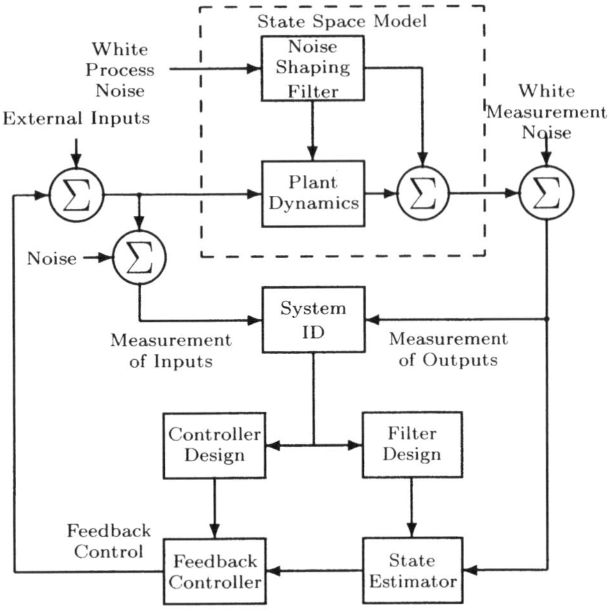

Figure 1: Structure of System Identification and Adaptive Control

2. OVERVIEW OF ADAPT$_X$

In this section, an overview of the approach of ADAPT$_X$ to system identification is given. Earily development of the CVA method are discussed in Larimore (1983, 1990a). Fig. 1 gives a schematic of the ADAPT$_X$ system identification procedure in the context of closed-loop control and state estimation. This approach differs from the other available software in providing automatic identification of complex, large scale, systems using completely stable and fast computational methods:

- Automation of the choice of number of lags of past and future, choice of state order, determination of delay structure, and choice of system input variables.
- Near optimal or very good performance of these automatic choices has been confirmed by theory (Larimore, 1999; Bauer, 2000; Bauer and Ljung, 2002;) and simulation (Larimore, 1996; Juricek et al., 2001; Deistler et al., 1996; Bauer et al., 1999; Ljung, 2003; Hunter, 1999).
- Completely stable computation that is very accurate, often to machine precision, and has been applied to large scale systems involving hundreds of states and thousands of parameters with completely stable computation (Hunter, 1999)
- Fast computational algoriths suitable for real-time implementation (Peloubet et al., 1990), and even a version of the algorithm implemented on a massively parallel systolic array (Martin, 1990).
- Real-time interface to the dSPACE data aquisition hardware for use in on-line systems for applications such as automotive and aerospace.
- Monitoring for system changes or faults and fault diagnosis (Larimore, 1997; Wang et al., 1997; Juricek, 2002)

The original version of ADAPT$_X$ released a decade ago included the completely automatic identification of a system from multivariable input and output data. This included the selection of the number of lags of the past and future as well as the model state order. A forerunner of ADAPT$_X$ was used in on-line identification of aircraft wing flutter discussed below. A basic vision for the CVA method has been the possibility of making multivariable system identification completely automatic and computationally stable, and yet retain the computational accuracy of near optimal performance.

The development of theory over the last decade has come a long way in explaining the outstanding behavior of the CVA procedure and its observed near optimal behavior in may cases, but there are still some open questions to be resolved. The likelihood function can be expressed in terms of the future predicted from the past, and this expression is exact except for end effects that become insignificant with increasing sample size (Larimore, 1997, 1999). The CVA procedure provides optimal prediction of the future from the past under a specified rank constraint, namely the state order. Bauer (2000) considers a closely related expression in the asymptotic efficiency of CVA (which he calls cca). Bauer and Ljung (2002) study the choice of weighting matrices in 'Larimore' type subspace algorithms with no inputs that are basically a generalized singular value decomposition (SVD). They show that the optimal weighting corresponds to the CVA weighting.

There have been a number of simulation demonstrations of subspace algorithms. In several cases, CVA was shown to exhibit maximum likelihood efficiency (optimal behavior) as the sample size became quite large (10^4) (Larimore, 1996; Deistler et al., 1996). Cases were also found demonstrating that the default weighting in the N4SID algorithm were far inferior to the CVA weighting. Even when the CVA weighting was used in the N4SID software (Juricek et al., 2001), the result was still substantially worse. Ljung (2003) confirmed this and found an improvement in the computation that makes N4SID more comparable with ADAPT$_X$.

The computational methods used in ADAPT$_X$ are based on highly stable and accurate methods, in some cases not previously kown in the literature (Larimore, 1990b). The CVA method is completely transparent to any scaling of the multivariable data. In fact it can be shown that CVA is the only scaling in subspace algorithms that has this property. As an aside, a maximum likelihood algorithm must have this property. As a result, CVA is appropriate for use in large scale systems that may involve the estimation of thousands of parameters. Identification of large scale systems is not feasible using numerical optimization methods such as maximum likelihood, also called prediction error methods.

Table 1: Computational Requirements of Subspace Algorithms

ALGORITHM	FLOP COUNT	STORAGE
N4SID, MOESP	$O(MN^2 + M^2N + M^3)$	$O(N^2 + MN)$
CVA (ADAPTx)	$O(MN + M^3)$	$O(N + M^2)$
FSD	$O(MN + M^2)$	$O(N + M^2)$

ADAPT_X has been applied to a number of large scale multivariable systems, and has resulted in some major improvements in accuracy and automation of the system identification. Hunter (1999) gives an extensive comparison of CVA and the Eigensystem Realization Algorithm (ERA) of Juang and Pappas (1985) in simulations of vibration in a lithographic stage. The structure was a finite element model of a lithographic stage including actuator and measurement dynamics with a total of 84 states, 3 input actuators, and 3 output sensors that involved estimating more than 700 parameters.

On the other hand, fast computational methods have been implemented in ADAPT_X that are considerably faster than the N4SID algorithm and are comparable to recently developed fast computational algorithms for subspace methods (Cho et al., 1994). The Fast Subspace Decomposition (FSD) algorithm exploits the block shift structure of the various matrices. These methods, however, have potential problems of numerical instability (Cho et al., 1994). The ADAPTx implementation of the CVA procedure in Larimore (2001) takes advantage of the structure to reduce the computation, but uses only numerically stable SVD computations.

Table 1 gives a comparison of the computation and storage requirements for the various algorithms, where N is the data length or sample size and M is the number of lags of the past and future used in the computation. The main difference between the fast algorithms CVA and FSD and the other two is the term N in the fast algorithms whereas the others have term N^2. Thus the N4SID and MOESP algorithms require about a factor of N more computation, which can become very large for large N, and they require a factor of around N more memory. This can make a considerable difference. For example, for a sample size $N =$ 10,000 and the number of past lags $M = 100$, N4SID and MOESP require approximately 100 GFLOP and 1 GBYTES, whereas CVA and FSD require about 10 MFLOP and 100 KBYTES. Usually to apply any of the algorithms requires enough data to obtain accurate estimates of the parameters. In such case, it is necessary that $M^2 \leq N$, and the CVA algorithm is then nearly as fast as the FSD algorithm.

The CVA method has even been demonstrated on massively parallel systolic arrays of processors. Larimore and Luk (1988) and Martin (1990) show a considerable speedup in the computation is potentially possible using parallel algorithms on a systolic array of processors where the speedup is proportional to the number of processors. This has an enormous potential given the recent development by IBM of processor arrays of nearly 1000 1 GHz processors per computer, each the size of a PC, with the potential to interconnect up to 1000 such computers in an array of 10^6 processors. Since there is no known limit to the applicability of the CVA method, this could make possible the real-time implementation of very large scale systems.

ADAPT_X now has available a real-time interface to the dSPACE data aquisition hardware. This makes possible the real-time and on-line aquisition of data, identification of the system model, and use of the system model in on-line control design or system monitoring. An example of such a system occurred very earily in the development of CVA. This involved a wind tunnel test in 1986 of on-line adaptive control of unstable aircraft wing flutter using CVA system identification and linear quadratic gaussian (LQG) control design (Peloubet et al., 1990). The system had 2 inputs, 6 outputs, and substantial wind tunnel turbulence. This example illustrates the use of a single system identification and control design procedure to successfully identify over 100,000 multivariable systems with up to 30 states for a wide range of system dynamics and structural configurations. There were no failures of the algorithm that was operating completely automatically with reidentification and on-line control design performed once per second during a 3 week period.

There have recently been a number of studies of process monitoring and fault detection trying to isolate and use a low dimensional subspace where the process is under normal operation. The chemometric methods of partial least squares and principal component analysis have been used to try to identify such a subspace. These methods have been mostly developed and applied for static regression problems and not dynamical systems. The CVA method provides a precise and rigorous approach to identifying the low dimensional state space that contains the process behavior apart from random observation noise. The use of CVA in process monitoring has also been recently discussed by Negiz and Cinar (1997).

A precise test of hypothesis for change of the system verses no change was developed in Larimore (1997) using the fitting of dynamic system models using CVA. Since this test is a generalized likelihood ratio test, it is a most powerful test of the hypothesis of change verses no change for all possible changes of the system. This test has been applied to the detection of additive disturbances and changes in the dynamics in simulations of a continuous stirred tank reactor in Wang et al. (1997). The disturbances included ramps, steps, and sine waves. The CVA procedure was found to be much more sensitive for detection of disturbances than standard Shewhart charts or more recent principal component analysis (PCA) methods. In some cases, CVA could detect disturbances 500 times smaller in amplitude than Shewhart charts or PCA.

Juricek (2002, and forthcomming papers) has recently done extensive analysis of the CVA method for process monitoring. A very significant development is the use of CVA for computing local statistics for detecting and isolation process changes and faults.

The following sections discuss the recent new developments and enhancements to the ADAPT$_X$ software including the delay structure selection, input structure selection, the real-time interface, and the Java interface.

3. MULTIVARIABLE DELAY MODELING

In this section the procedure for determination of the delay structure and using it in the subspace modeling is discussed. A detailed discussion of the method is in Larimore (2003) including several examples. The method includes a fast order-recursive algorithm that is computationally stable and accurate to machine precision even for colinear data or singular covariance matrices (Larimore, 1990b, 2002) and Zhou and Zhu (2002). Applications of this delay estimation and modeling procedure are given in Juricek (2002).

The basic idea is to determine the delay structure between all pairs of inputs and outputs as well as any delayed internal feedback between pairs of present output components and past output components. The delay structure is determined by fitting ARX models as discussed below. Once the delay structure is determined, additional synthetic inputs are introduced in the subspace model to be identified and the subspace state space model is identified. This can result in a considerable reduction of the order of the identified state space model and number of estimated parameters resulting in a considerable inprovement in the identified model accuracy.

A very powerful method for determining the delay structure is by fitting autoregressive models with inputs (ARX) to the data with various delays present in the inputs and outputs of the process. Consider the fitting of ARX models for the observed inputs u_t and outputs y_t of the form

$$y(t) = \sum_{s=1}^{p} \alpha(s) y(t-s) + \sum_{s=0}^{p} \beta(s) u(t-s) + e(t) \quad (1)$$

where p and q are respectively the AR and X orders. The functions $\alpha(s)$ and $\beta(s)$ are respectively the pulse response functions from the past outputs to the present outputs and from the past and present inputs to the present outputs. The objective is to determine the delay times of these pulse response functions. For a matrix pulse response function $h(s)$, the delay from the $j-th$ input $u_j(t)$ to the $i-th$ output is $k\Delta t$ where Δt is the sample time if

$$h_{ij}(s) = 0 \quad \text{for} \quad 0 \geq s < k; \quad h_{ij}(k) \neq 0. \qquad (2)$$

For an pulse response function $h(s)$, define the delay matrix $\Delta(h(s))$ as the matrix of nonnegative integers containing the above delay times for all pairs (i,j). In particular, define $\Delta(\beta(s))$ as the *input-output* delay. By replacing past inputs $u(t-s)$ with past outputs $y(t-s)$ in the above argument, the *internal feedback* delay matrix $\Delta(\alpha(s))$ is defined.

An internal feedback delay will often be manifest in a different way. Suppose that the system is a low order autoregression with no delay so that there is significant correlation between the last few outputs and the present output. Suppose that in addition to this structure, there is a pure delay of 12 of the output that is feed back internally in the system. In this case the determination of such a delay is more involved but as will be seen below can be done in a similar way.

Any $n \times m$ matrix pulse response function $h(s)$ containing delays can be written as a delay structure from n inputs to mn possibly delayed variables that cascade into a $n \times mn$ pulse response function with no delays. Depending of the particular delay structure of a system, the delay structure may involve far less than mn variables. The basic idea exploited is to determine the delay structure of the autoregressive model. Then only the delayed variables are used that contribute to a net reduction of the number of estimated parameters and a corresponding decrease in the prediction error of the corresponding model. There are a number of situations where this strategy will significantly improve the accuracy of the identified model. If the delay times are long relative to the state order of the system, then a considerable improvement will result. Also if only a few of the input-output pairs involve long delays, then a substantial improvement can result.

The first step in the computation is to determine the delay matrices $\Delta(\alpha_s)$ and $\Delta(\beta_s)$ from the observational data. This can be accomplished by the following steps:

- Choose Maximum ARX Order. Fit ARX models using the order recursive procedure described by Larimore (1990b) that is computationally stable and well conditioned even for collinear and illconditioned data. This procedure adds lags one at a time up to some chosen upper bound on ARX order. Compute the AIC for all order ARX models fitted and choose the ARX order corresponding to the minimum AIC. This defines the optimal order for ARX modeling and is used in the CVA computations for the length of the past and future.
- Choose Optimal Input-output Delays. Now for each output component $y_i(t)$, consider the multiinput single output ARX model. Consider the order recursive procedure where the terms associated with input j are removed from the ARX model starting with lags $s = 0$, then $s = 1$, $s = 2$, ... This is the equivalent of setting the coefficients $\beta_{ij}(s) = 0$ successively for the sets $\{s = 0\}$, $\{s = 0, 1\}$, $\{s = 0, 1, 2\}$.... For each such model, the AIC measure of model fit is computed and the model with the minimum AIC is the optimal delay for input j to output i.
- Choose Optimal Autoregressive Delays. Repeat the above step but consider lags of the output $y_j(s)$ instead of lags of the inputs $u_j(s)$ to determine the presence of pure autoregressive delays. In addition, where there is no pure delay, look for a delayed response after the initial impulse response becomes insignificant. This is done by successively removing autoregressive lags after the initial response dies out until a signifi-

cant response is encountered at a longer lag. This is directly related to time series seasonal models as discussed below.

- Fit Subspace Models with Synthetic Delayed Inputs. The above procedure will flag the potentially important delays in inputs and internal output feedback of the system. These can then be used to construct the synthetic delayed inputs to use in subspace modeling. There may be several combinations of such synthetic delayed inputs to try particularly if some of these are of low statistical significance. These alternative subspace models can be compared using the AIC to determine the more plausable models.

The above procedure has both computation and statistical advantages. The procedure is computationally efficient and numerically stable. Precise tests of hypotheses are performed to determine the optimal multivariable delay structure. Significant differential delays from one input to various outputs is handled by introducing delayed versions of that input as additional synthetic inputs to the system.

4. INPUT SELECTION

A major problem in modeling multivariable systems is the selection of the significant inputs to the system. Often, some of the inputs have little or no influence on the outputs, so that including them in the model will reduce the accuracy of the mdel. The problem then is to determine sets of inputs that have a significant effect on the outputs. The problem is closely related to the subset regression problem, but includes the issues of modeling dynamic process and time series.

The approach taken is essentially that of the regression by leaps and bounds (RLEAP) (Furnival, 1974). The major issue in these type of techniques is to efficiently search the very large number of possible combinations of input variables to find the more significant sets and quantatively evaluate the model quality. The procedure RLEAP is to essentially determine which variables are essential so that when they are deleted the model quality is so poor that the deletion of even irrelevant variables will not lead to enough improvement. In this way large branches are pruned from the search tree of possible models. For examples (Furnival, 1974), for 50 variables with $2^{50} = 10^5$ possible models, only several hundred remain after such pruning.

In the StrucSelect procedure, the Akaike information Criterion (AIC) is used for the measure of model quality. This has a fundamental interpretation as the measure of model approximation based on the basic notion of sufficiency in statistics (Larimore, 1996). The AIC is computed for all plausable models identified by RLEAP.

An example of the use of StrucSelect is given in Eserin (1999) for an industrial boiler for steam and power generation in a paper mill. The boiler drum collects return water and is a reserve for the boiler operation. The drum water level is the modeled system output that must be maintained within some tolerances. There were 13 inputs to the system that were considered for modeling.

The StrucSelect procedure was applied to the data. Of the $2^{13} = 8096$ possible combinations of input variable, the StructSelect procedure ended up evaluating several hundred models that were contenders involving between 8 and 12 inputs.

The implementation has two aspects, (1) evaluation of autoregressive approximations using the order-recursive methods discussed in the Delay Estimation, and (2) evaluation of the state space models fitted using the CVA computation. Of course the state space method is much more accurate, but for a very large number of inputs, it may be of interest to use the ARX order-recursive methods which is extremely fast since each ARX model fitting requires only a very small increment of computation in moving through the tree search of models.

5. INTERFACE FOR REAL-TIME IDENTIFICATION

The objective of the Real-time Interface was to seamlessly integrate dSPACE real-time data acquisition hardware and Adaptics system identification software in the MATLAB environment. In the future, other data aquisition environments will also be included.

The benefit is that it creates a unified environment in which experimental input-output data can be captured and processed using ADAPTx to generate MIMO system models for system analysis and controller design. Thus, systems can be analyzed and controllers designed/implemented while the plant is in operation, resulting in significant reductions in design time and the ability to deal with variations in system dynamics. In addition, controllers that are designed using the identified model can be directly implemented on the dSPACE system.

The initial release of the interface is in the form of a MATLAB m-function that utilizes dSPACE's MLIB MATLAB interface library to capture input-output data from dSPACE real-time hardware and save it as a .mat file formatted for ADAPTx. The user specifies the variables to be captured, sampling period and data capture length. ADAPTx then processes the data in the .mat file based on standard user inputs (max no. of lags, bias and feedthrough).

In future extensions, this interface will be further enhanced to include a GUI that allows the user to select input/output variables, sampling period, capture length, data pre-processing options, system identification options, and also controller design parameters.

Current and past Realtime applications include automotive engine control and monitoring, automotive power train, wind tunnel testing (Peloubet et al, 1990), and photolithography (Hunter, 1999, a simulation).

6. JAVA GRAPHICAL USER INTERFACE

The development of a graphical user interface (GUI) in Java is in progress that will allow use of a common interface across platforms. Platforms covered will include Windows, UNIX, and Lunix, and the software envirnments of Matlab, C++, and real-time envirnments such as dSPACE.

7. REFERENCES

Bauer, M., and L. Ljung (2002. Some facts about the choice of the weighting matrices in Larimore type of subspace algorithms. *Automatica* **38**(5), 763-774.

Bauer, D., M. Deistler and W. Scherrer (1999). Consistency and asymptotic normality of some subspace algorithms for systems without observed inputs. *Automatica* **35**(10), 1243-1254.

Bauer, M. (2000) Asymptotic efficiency of cca subspace methods in the case of no exogenous inputs. Technical Report LiTH-ISY-R-2262. Department of Eletrical Engineering, Linkoping University. SE-581 83 Linkoping, Sweden.

Cho, Y.M., G. Xu and T Kailath (1994). Fast identification of state space models via explotation of displacement structure. *Trans. Automatic Control* **39**(10).

Deistler, M., K. Peternell and W. Scherrer (1996). Consistency and relative efficiency of subspace methods. *Automatica* **31** pp. 1865-75.

Eserin, P.K.N., (1999). Applications of canonical variate analysis to the dynamic modeling and control of drum level in industrial boilers. *Proc. 1999 American Control Conference* pp. 1163-1166.

Furnival, G.M., R.C. Wilson, Jr. (1974). Regression by leaps and bounds. *Technometrics*, **16**, pp. 499-511.

Hunter, N.F., (1999). Comparing CVA and ERA in transfer function measurements of lithography applications. *Proc. 1999 American Control Conference*, pp. 1171-1175.

Juang, J. and R. Pappas (1985). An eigensystem realization algorithm (ERA) for modal parameter identification and model reduction. *J. Guidance, Control and Dynamics*, **8**(5), pp. 620-627.

Juricek, B. 2002. *Multivariable Statistical Methods for System Identification and Process Monitoring.* PhD dissertation, University of California, Santa Barbara, Department of Chemical Engineering.

Juricek, B.C., D.E. Seborg, and W.E. Larimore (2001). Identification of the Tennessee Eastman challenge process with subspace methods. *Control Engineering Practice*, **9**(12), pp. 1337-1351.

Juricek, B.C., D.E. Seborg, and W.E. Larimore (2002). Identification of multivariable, linear, dynamic models: comparison of regression and subspace techniques. *Ind. Eng. Chem. Res.*, **41**, pp. 2185-2203.

Larimore, W.E. (1983). System identification, reduced-order filtering and modeling via canonical variate analysis. *Proc. 1983 American Control Conference*, H.S. Rao and T. Dorato, Eds., pp. 445-51.

Larimore, W.E. (1990a). Canonical variate analysis for system identification, filtering, and adaptive control. *Proc. 29th IEEE Conference on Decision and Control*, **1**, pp. 635-9.

Larimore, W.E. (1990b). Order-recursive factorization of the pseudoinverse of a covariance matrix. *IEEE Trans. of Automatic Control*, **35**, pp. 1299-1303.

Larimore, W.E. (1996). Statistical optimality and canonical variate analysis system identification. *Signal Processing*, **52**, pp. 131-144.

Larimore, W.E. (1997). Optimal reduced rank modeling, prediction, monitoring, and control using canonical variate analysis. *Preprints IFAC Advanced Control of Chemical Processes*, pp. 61-6.

Larimore, W.E. (1999). Automated multivariable system identification and industrial applications. *Proc. American Control Conference*, pp. 1148-62.

Larimore, W.E. (2000). The ADAPTx software for automated multivariable system identification. *Proc. IFAC Symposium on System Identification 2000.*

Larimore, W.E. (2001), *ADAPT$_X$ Automated System Identification Software Users Manual*, Adaptics, Inc, 40 Fairchild Drive, Reading, MA 01867.

Larimore, W.E. (2002). Reply to 'Comment on 'Order-recursive factorization of the pseudoinverse of a covariance matrix.' ' *IEEE Trans. Automat. Contr.* **47** pp. 1953-7.

Larimore, W.E. (2003). Inferring multivariable delay and seasonal structure for subspace models. To appear in *Proc. IFAC Symposium on System Identification 2003.* Rotterdam, The Netherlands.

Larimore, W.E., and F.T. Luk (1988). System identification and control using SVDs on systolic arrays. *SPIE Symposium on Innovative Science and Technology, Proc. of Conference on High Speed Computing*, **880**.

Ljung, L. (2003). Aspects and experiences of user choices in subspace identification methods. To appear in *Proc. IFAC Symposium on System Identification 2003.* Rotterdam, The Netherlands.

Martin, D. (1990). Canonical variate analysis of large-scale systems on a transputer systolic array. *Proc. 29th IEEE Conference on Decision and Control*, **1**, pp. 596-603.

Negiz, A, and A. Cinar (1997). PLS, balanced, and canonical variate realization techniques for identifying VARMA models in state space. *Chemometrics and Intelligent Laboratory Systems*, **38**, pp. 209-221.

Peloubet, R.P., R.L. Haller, R.M Bolding (1990). On-line adaptive control of unstable aircraft wing flutter. *Proc. 29th IEEE Conference on Decision and Control*, **1**, pp. 643-51.

Van Overschee, P. and B. DeMoor (1966). *Subspace Identification of Linear Systems: Theory, Iimplementation, Applications.* Kluwer Academic Publishers.

Wang, Y., D.E. Seborg and Larimore (1997). Process monitoring based on canonical variate analysis. *IFAC 1997 Int. Symp. on Advanced Control of Chemical Processes*, Banff, Canada.

Zhou, J., and Zhu, Y. (2002). Note on 'Order-recursive factorization of the pseudoinverse of a covariance matrix'. *IEEE Trans. Automat. Contr.* **47**, pp. 1952.

IFAC Publications
www.elsevier.com/locate/ifac

FREQUENCY DOMAIN SYSTEM IDENTIFICATION TOOLBOX FOR MATLAB: AUTOMATIC PROCESSING – FROM DATA TO MODELS

István Kollár[*], Rik Pintelon[**], Yves Rolain[**], Johan Schoukens[**], and Gyula Simon[*]

[*]*Budapest University of Technology and Economics*
Department of Measurement and Information Systems
H-1521 Budapest, Magyar tudósok krt. 2. Hungary
Fax: +36 1 463-4112, email:[kollar,simon]@mit.bme.hu
[**]*Vrije Universiteit Brussel, Dienst ELEC, Belgium*
Pleinlaan 2, B-1050 Brussel, Belgium, fax: +32 2 629-2850
Email: [Johan.Schoukens,Rik.Pintelon,Yves.Rolain]@vub.ac.be

Abstract: The Frequency Domain System Identification Toolbox for MATLAB is an effective tool for the identification of linear dynamic system models from measured data. Since the use of advanced system identification methods often requires a lot of programming work, the attention can be deflected from the modelling issues. Therefore, a Graphical User Interface (GUI) was developed which allows the experimenter to visually follow and control the data processing and modelling steps.

However, for many users it is still desirable to obtain good models with as few decisions to be made as it is possible. Therefore, automatic processing steps have been added to the GUI. Identification can be done now in this toolbox with a minimum of user interactions in the graphics windows, and a reasonable model is returned, ready for control or physical analysis. *Copyright © 2003 IFAC*

Keywords: system identification, frequency domain, automatic data processing, model order selection, Fourier analysis, transfer function, graphical user interface, GUI, MATLAB.

1 INTRODUCTION

System identification is a very difficult task. It needs a lot of insight into the physics of the phenomenon under test, and into the modelling process. The noisy nature of measured signals demands appropriate procedures for experiment design, measurement, data preprocessing, modelling, parameter estimation and model validation. In system identification theoretical and numerical considerations need to be followed, and for many users this is already prohibitive in the related methods.

These aspects were clear for us even after the first release of the Frequency Domain System Identification Toolbox in 1994. We made several steps to narrow the gap between the available functions and the users, as described below.

2 BRINGING THE PROGRAMS CLOSER TO THE USERS

If we want to let people use a program, we must make it as simple to use as it is possible. There are different steps we can make.

2.1 Simple function calls with default values of arguments

The first obstacle for a user is the requirement to give all the possible options in each routine. In general purpose programs many options are provided to control the type of iteration, stop criteria, data reduction, handling of messages and informative plots, etc. The choice among these settings requires good knowledge of the theory, and significant experience. However, in many cases, reasonable choices can be made by utilising some properties of the data, or sacrificing a part of the speed of the program for reliable convergence. Therefore, many function arguments can be given default values, sometimes in dependence on the data.

2.2 Data structures

An eminent difficulty in the use of general purpose programs is the handling of several interdependent aspect of the data. A measurement does not only consist of the measured samples, but also of the sampling frequency, circumstances like periodic or random excitation, frequency contents, amplitude units, names of measurement channels, date, etc. If these are all to be given, function calls will usually get badly organized, difficult to check and debug, and frightening for the user. The solution for this problem was the use of complex data types: *cells* and *structures*, later *objects*. These are since several years available in Matlab. All function heads were carefully redesigned to be able to use as simply as it was possible. Moreover, some function calls were made dependent on the object type in the arguments. However, despite the efforts, we regularly got questions concerning the use of individual functions. It was apparent that what we did was not enough for the users to forget programming, and focus at the modelling task.

2.3 Graphical User Interface

The most drastic step towards user-friendliness was the introduction of a Graphical User Interface (GUI). This has been described in previous papers (Kollár et al, 1999; Kollár et al, 2000). The main window of the GUI can be seen in Fig. 1.

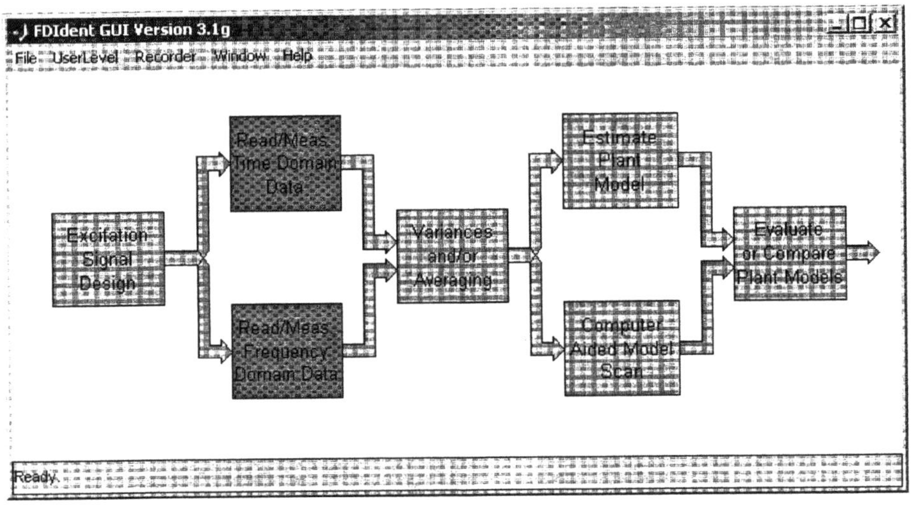

Fig. 1. Starting window of the Graphical User Interface

The basic idea is that opening any block, the user has access to processing tools (controls), and by proper selections the procedure can be directed to properly run.

Even in this intuitively clear tool, the users has several alternatives, that is, there are several decisions to make. This is sometimes still frustrating, and not always necessary. Therefore, already in the first design we made two decisions:

- Almost all parameters have default values, set properly for the given data. This not only allows the user to accept the computer-suggested settings, but also paves the way for automatic runs.
- Three different user levels allow to make a global choice among "Automatic", "Interactive" and "Advanced" processing. Those who do not want to bother with many decisions can run through the processing steps with default values.

3 NOVELTY NOW: AUTOMATIC PROCESSING

As we saw above, the direction of development is towards an easy-to-use tool. We try to let the computer program overtake as much as it is possible from the from the user. This proved to be the right direction, but we still had to pose several questions where the user needed to consider and answer. Therefore, the most logical step was to localize such places, and make the whole process as automatic as it is possible. The ultimate goal is to have a tool which absorbs the data, and returns verified results. Here are the points where we could automate, and overtake the task of decisions from the user.

3.1 Automatic Assembly of Data Structures (Data Objects)

Providing consistent and properly described data to an algorithm is a real pain. Despite of different helps and descriptions, creation of a proper full object was not easy for somebody who wanted to focus on identification rather than on technical details of data import. A set of data import windows made it possible to consistently generate objects from the descriptors of the data and of the experiment.

3.2 Reliable Preprocessing of Periodic Data

When measuring and processing periodic signals, there are two "catches". One is that it can often happen that sampling and the signal frequencies are not coherent, despite of true efforts of the user; second is that it is often cumbersome to provide the set of frequencies of the signal components. An automatic tool has been developed which determines the period length, strong components, and variance of the measurement noise (Schoukens et al, 2002). This allows the user to simply plug the measured signals into the GUI, and watch how the estimated Fourier coefficients and variances come out.

3.3 Automatic Model Order Selection

Determining the proper model order is a very difficult task. Different criteria exist (Akaike, Rissanen, etc.), but they only work if nonlinearity errors are negligible compared to observation noise. Maybe the most popular solution is to scan the reasonable numerator/denominator order combinations, and select on the basis of some user-defined criteria.

Analysis of the system equations in the proper mathematical basis allows to use a systematic approach: small singular values in the singular value decomposition tell how much overmodelled is the system, and careful "peeling" of the models allows to reach the proper minimum order model (Simon et al, 2000).

4 QUESTIONS STILL NEEDED TO BE POSED

The dream of a user is an intelligent machine which takes the data, and returns good results while the user drinks coffee. This sounds good, and *can almost be achieved.* In a typical "Automatic" session, the user needs to give only the following answers:

1) Bring the data into Matlab, and tell the GUI import window where are these. There are 2 possibilities: time domain measurements and frequency function measurements.
   Time domain measurements (assuming periodic excitations):
   - input and output time series
   - sampling time
   - antialias filters on/off (more precisely, whether lowpass filtered input/output measurements are made (bandpass experiment), or the input signal is computer generated, and the output is simply sampled (ZOH experiment))
   - stationary/transient state

   Optional parameters may also be given, but these are not obligatory.

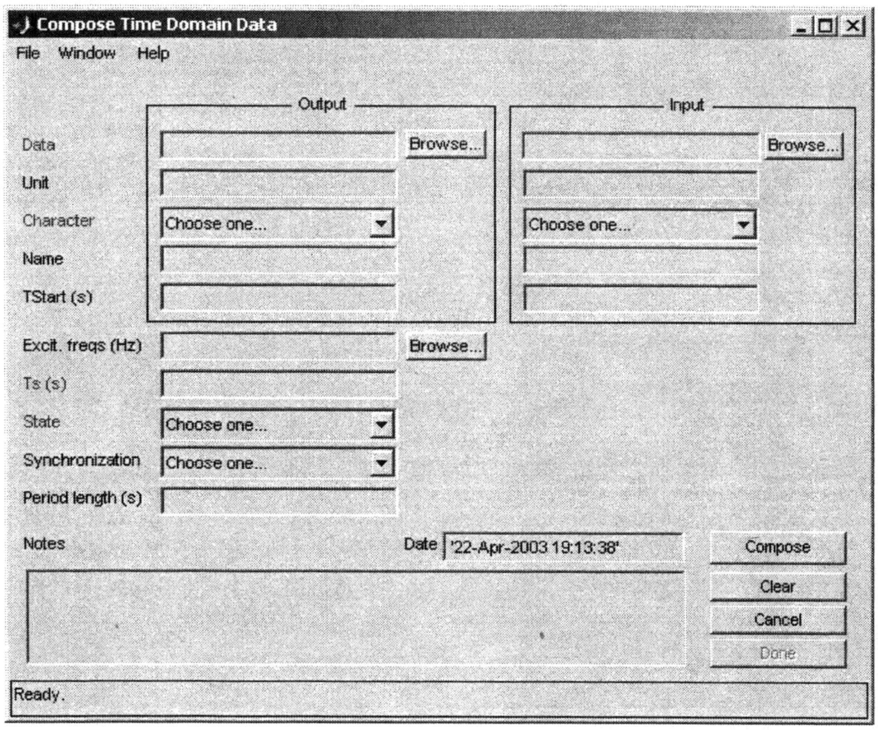

Fig. 2 Composition of time domain data: the necessary fields (see above) are shown in the real life in red

Frequency function measurements, typically:

- FRF magnitude and phase, at which frequencies
- antialias filters on/off (more precisely, whether lowpass filtered input/output measurements are made (bandpass experiment), or the input signal is computer generated, and the output is simply sampled (ZOH experiment))

Optional parameters may also be given, but these are not obligatory.

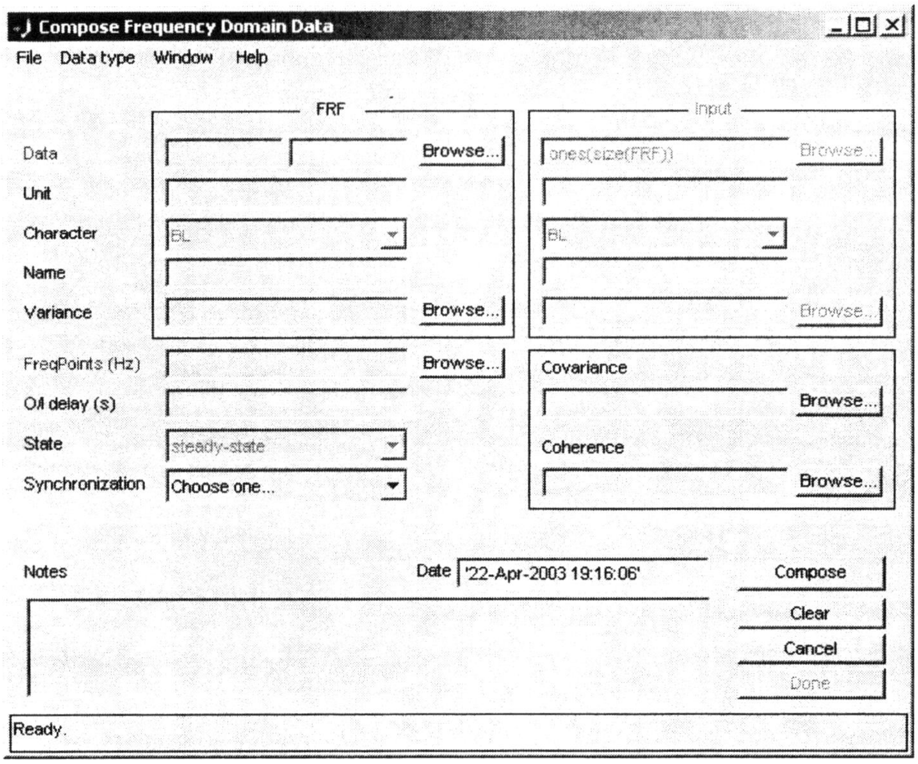

Fig. 3 Composition of frequency domain data: the necessary fields (see above) are shown in the real life in red

2) Choose between *s*-domain and *z*-domain (continuous-time or discrete-time models)

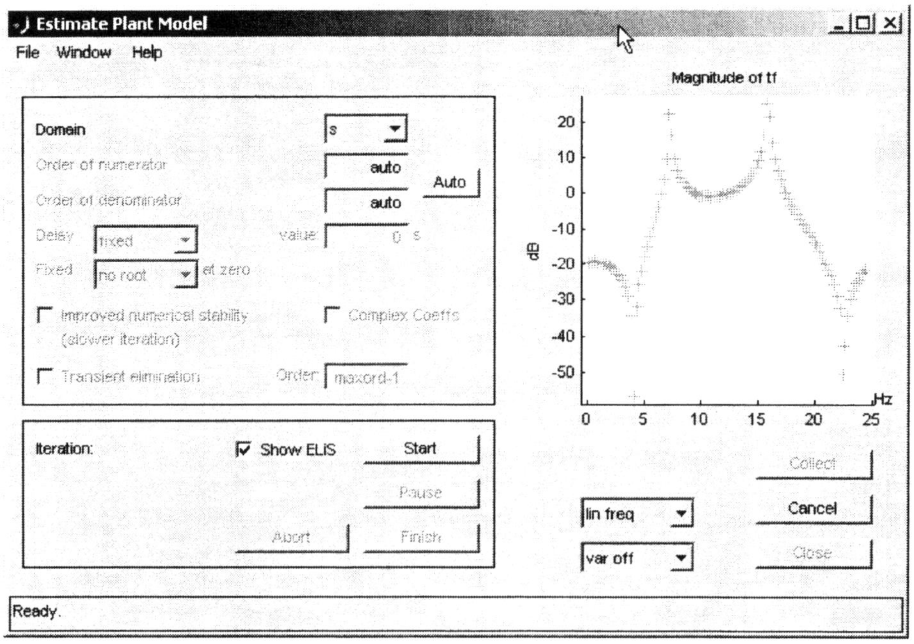

Fig. 4 Almost all controls are inactive in the "Estimate Plant Model" window: the user need only to select the domain, and press "Start". The "auto" strings in the order windows mean that the best orders will be determined automatically.

The domain could also be fixed beforehand, but in many cases at least its choice is required, and this is not a complicated question to answer at all.

5. CONCLUSIONS

Automatic processing is already possible in identification, using careful program design and recently developed algorithms. An example for frequency domain identification is available in (FDIDENT, 1999-2003). The described automatisms allow the user to forget programming pains, and focus at identification itself.

REFERENCES

Balogh L., I. Kollár, and G. Gueret, "Variance of Fourier coefficients calculated from overlapped signal segments for system identification," *Instrumentation and Measurement Technology Conference, 2002. IMTC/2002. Proceedings of the 19th IEEE*, Vol. 1, pp. 1065-1070.

FDIDENT (1999-2003), *Frequency Domain System Identification Toolbox Developers' Page*. `http://elec.vub.ac.be/fdident/`

Kollár, I., R. Pintelon G. Román, G. Simon and J. Schoukens (1999), Graphical User Interface, Objects, and Improved Numerical Stability – New Developments in the Frequency Domain System Identification Toolbox. *Electronic publication.* `http://www.mit.bme.hu/~kollar/...` `...topics/fdidgui/fdident-paper.html`

Kollár, I., J. Schoukens, R. Pintelon, G. Simon and G. Román (2000), "Extension for the Frequency Domain System Identification Toolbox: Graphical User Interface, Objects, Improved Numerical Stability." *Preprints of the IFAC Symposium on System Identification, SYSID 2000, 21-23 June 2000, Santa Barbara, CA, USA. CD-ROM.*

Pintelon, R., and J. Schoukens (2001). *System Identification - A Frequency Domain Approach.* IEEE Press, Piscataway, NJ.

Schoukens, J. and R. Pintelon (1991). *Identification of Linear Systems: A Practical Guideline to Accurate Modeling.* Pergamon Press, Oxford.

Schoukens, J., Y. Rolain, Simon G., and R. Pintelon (2002) "Fully automated spectral analysis of periodic signals," *Instrumentation and Measurement Technology Conference, 2002. IMTC/2002. Proceedings of the 19th IEEE,* Vol. 1, pp. 299-302.

Simon, G., J. Schoukens, and Y. Rolain (2000), "Automatic Model Selection for Linear Time-Invariant Systems," Proceedings of the12th IFAC Symposium on System Identification, SYSID 2000, Santa Barbara, CA, USA, 21-23 June 2000, Vol. I., pp. 379-384.
Extended electronic version: Simon, G., J. Schoukens, and Y. Rolain, "Automatic Model Selection for Linear Time-Invariant Systems – Practical Issues" `http://www.mit.bme.hu/~simon/...` `...publications/automodel.ps.zip`

IFAC Publications
www.elsevier.com/locate/ifac

PREDICTION ALGORITHMS: COMPLEXITY, CONCENTRATION AND CONVEXITY

Peter L. Bartlett

Division of Computer Science and Department of Statistics
University of California, Berkeley
bartlett@eecs.berkeley.edu

Abstract: In this paper, we review two families of algorithms used to estimate large-scale statistical models for prediction problems, kernel methods and boosting algorithms. We focus on the computational and statistical properties of prediction algorithms of this kind. Convexity plays an important role for these algorithms, since they exploit the computational advantages of convex optimization procedures. However, in addition to its computational advantages, the use of convexity in these methods also confers some attractive statistical properties. We present some recent results that show the advantages of convexity for estimation rates, the rates at which the prediction accuracies approach their optimal values. In addition, we present results that quantify the cost of using a convex loss function in place of the real loss function of interest. *Copyright © 2003 IFAC*

Keywords: prediction problems; machine learning; statistical analysis; convex optimization

1. INTRODUCTION

We consider several recent machine learning methods that have proven successful in a variety of multivariate prediction problems. These methods include support vector machines and other kernel methods (Boser *et al.*, 1992; Cortes and Vapnik, 1995; Cristianini and Shawe-Taylor, 2000; Schölkopf and Smola, 2002), and boosting algorithms, such as AdaBoost (Freund and Schapire, 1997). Because these methods involve large-scale statistical models, computational efficiency is an important consideration, and the development of these methods has been influenced by the availability of efficient algorithms for convex optimization. At the heart of each of these methods is the optimization of some convex criterion over a convex set. In addition to its computational advantages, the use of convexity in models and prediction methods has statistical consequences. In this paper we examine some of these consequences.

Consider a multivariate prediction problem in which a covariate $X \in \mathcal{X}$ is associated with a response variable $Y \in \mathcal{Y}$. We quantify the cost of an incorrect prediction using a loss function, $l : \mathcal{Y}^2 \to \mathbb{R}^+$, so that $l(\hat{y}, y)$ is the loss incurred when $\hat{y}$ is predicted but the correct response is y. The aim is to choose a function $f : \mathcal{X} \to \mathcal{Y}$, typically from a particular class $\mathcal{F}$ of functions, so that the expectation of the loss, $\mathbf{E}l(f(X), Y)$, is small, where the expectation is with respect to some probability distribution over $\mathcal{X} \times \mathcal{Y}$. We assume that we do not know this probability distribution, but we get information about it in the form of (X, Y) pairs.

Three key issues arise in prediction problems of this kind:

approximation error — which depends on how well suited the model $\mathcal{F}$ is to the problem,

estimation error — which arises in estimating $f \in \mathcal{F}$ from limited data, and

computational complexity — since the choice of f usually involves a nontrivial optimization problem.

In examining these three issues, we shall focus on the central role played by the convexity of the loss function l, and convexity of the class of functions $\mathcal{F}$.

Our motivating examples of large-scale statistical models $\mathcal{F}$ arise from kernel methods, where $\mathcal{F}$ is a subset of a reproducing kernel Hilbert space of functions on $\mathcal{X}$, and boosting algorithms, where $\mathcal{F}$ is a subset of the linear span of a set of simpler functions. Section 2 describes these algorithms and presents some of their properties.

Both of these methods involve an infinite dimensional convex function class $\mathcal{F}$. As a result, the statistical issue is crucial, and careful regularization is essential to choose an appropriate model complexity, so as to balance the representational power of the model with the statistical cost of estimating it from data. Section 3 presents recent estimation error bounds that measure the complexity of a model in terms of so-called *Rademacher averages*. These results motivate regularization methods that are used to choose an appropriate model complexity.

Both boosting and kernel methods choose a function from a convex class to minimize a convex criterion, even when the performance criterion of interest involves the expectation of a non-convex loss function. Section 4 presents estimation error bounds for convex problems such as these. It turns out that the estimation error decreases to zero more rapidly with the sample size in these cases. These error bounds use a more refined notion of model complexity: a local version of the Rademacher averages. A converse result is also presented, which shows that, in some cases, these faster rates cannot be achieved when the class $\mathcal{F}$ is not convex.

Section 5 considers the effect of convexity on the approximation error. In many practical problems, such as pattern classification problems, the loss function of interest is not convex, but a convex upper bound is used as a surrogate, and it is important to understand how the former behaves when the latter is minimized.

Section 6 discusses the implications of these results for system identification, and presents some directions for future work.

2. PREDICTION ALGORITHMS

In this section we describe two families of prediction algorithms: kernel methods and boosting methods. We consider the following situation: we have a number of independent, identically distributed pairs,

$$(X, Y), (X_1, Y_1), \ldots, (X_n, Y_n)$$

from the product space, $\mathcal{X} \times \mathcal{Y}$, where $\mathcal{Y}$ is a bounded subset of $\mathbb{R}$. (We shall consider extensions to dependent data in Section 6.) We have a bounded loss function, $l : \mathcal{Y}^2 \to [0, 1]$, and a prediction algorithm aims to choose a function $f : \mathcal{X} \to \mathbb{R}$ so that the risk, $R(f) = \mathbf{E}l(Y, f(X))$, is small. The algorithms that we consider choose a function f with a small empirical risk,

$$\mathbf{E}_n l(f(X), Y),$$

where we use $\mathbf{E}_n$ to denote the expectation under the empirical distribution, or sample average,

$$\mathbf{E}_n g(X, Y) = \frac{1}{n} \sum_{i=1}^{n} g(X_i, Y_i).$$

Often the class of functions from which f is chosen is so large that it is always possible to obtain a function with a small empirical risk. In such cases, it is essential to use some kind of regularization, to ensure that the empirical risk is a good approximation of the risk. For instance, if we define a norm on the space of functions, we might choose f to minimize the regularized empirical risk,

$$\mathbf{E}_n l(f(X), Y) + C\|f\|,$$

where C is a regularization coefficient that determines the trade-off between the complexity of f and its performance on the training data.

2.1 Kernel methods

Kernel methods use functions from a subset $\mathcal{F}$ of a *reproducing kernel Hilbert space* (RKHS). For more details on kernel methods, and especially support vector machines (Boser *et al.*, 1992; Cortes and Vapnik, 1995), a popular example of these methods, see (Cristianini and Shawe-Taylor, 2000; Schölkopf and Smola, 2002; Vapnik, 1998).

Recall that a RKHS $\mathcal{H}$ of real functions on $\mathcal{X}$ is a Hilbert space for which there is a reproducing kernel, $k : \mathcal{X}^2 \to \mathbb{R}$, which satisfies the following two properties:

(1) k has the reproducing property: for all $x \in \mathcal{X}$,

$$\langle k(x, \cdot), f \rangle = f(x)$$

(2) k spans $\mathcal{H}$, that is,

$$\text{span}\left(\{k(x, \cdot) : x \in \mathcal{X}\}\right)$$

is dense in H.

Equivalently, we can define a reproducing kernel Hilbert space H as the completion of the span of a symmetric function k that satisfies a certain positive definiteness property. Thus, it is possible to work with a large, even infinite-dimensional, space of functions using only linear combinations of functions of the form $x \mapsto k(x', x)$.

Kernel methods typically choose a function $f \in \mathcal{H}$ so as to minimize a combination of the sample average of a loss function and a regularization term involving the squared norm, $\|f\|^2_{\mathcal{H}}$, of the function in the RKHS. For example, the kernel ridge regression approach is to minimize

$$\frac{1}{n}\sum_{i=1}^{n}(f(x_i) - y_i)^2 + C\|f\|^2_{\mathcal{H}},$$

where C is a positive regularization coefficient. The following theorem shows that for any regularization term that is an increasing function of the norm, the minimizer has an expansion in terms of the data.

Theorem 1. Fix a kernel k and let $\mathcal{H}$ be the corresponding RKHS. For any function $L : \mathbb{R}^n \to \mathbb{R}$ and any increasing function $\Omega : \mathbb{R} \to \mathbb{R}$, define an objective function,

$$J(f) = L(f(x_1), \ldots, f(x_n)) + \Omega\left(\|f\|^2_{\mathcal{H}}\right). \quad (1)$$

If $J(f^*) = \min_{f \in \mathcal{H}} J(f)$, then for some real coefficients $\alpha_1, \ldots, \alpha_n$, we can write f as

$$f(\cdot) = \sum_{i=1}^{n} \alpha_i k(x_i, \cdot). \quad (2)$$

The proof is straightforward: for any $f \in \mathcal{H}$, let f_0 be the projection of f on to the subspace spanned by the $k(x_i, \cdot)$. Then the reproducing property implies

$$f(x_i) = \langle f, k(x_i, \cdot)\rangle = \langle f_0, k(x_i, \cdot)\rangle = f_0(x_i),$$

which, with Pythagorus' theorem, implies that

$$\begin{aligned} J(f) &= L(f_0(x_1), \ldots, f_0(x_n)) \\ &\quad + \Omega\left(\|f_0\|^2_{\mathcal{H}} + \|f - f_0\|^2_{\mathcal{H}}\right) \\ &\geq J(f_0), \end{aligned}$$

with equality iff f lies in the span of the $k(x_i, \cdot)$. See (Kimeldorf and Wahba, 1971) for the first result of this form and (Schölkopf and Smola, 2002) for a review of subsequent refinements.

This result implies that an objective function that is of this form (a norm-regularized sample average of a loss function) can be expressed entirely in terms of the *Gram matrix* of inner products, $K \in \mathbb{R}^{n \times n}$, with

$$K_{ij} = k(x_i, x_j).$$

Specifically, for any f of the form (2), we have

$$\begin{aligned} (f(x_1), \ldots, f(x_n))' &= K\alpha, \\ \|f\|^2_{\mathcal{H}} &= \alpha' K \alpha. \end{aligned}$$

where $\alpha = (\alpha_1, \ldots, \alpha_n)' \in \mathbb{R}^n$. Thus, for the optimization of a $J : \mathcal{H} \to \mathbb{R}$ of the form (1), it suffices to consider the dual optimization problem

$$J^*(\alpha) = L(K\alpha) + \Omega(\alpha' K \alpha).$$

This representation has a number of advantages. For typical choices of L and Ω, it is a finite-dimensional convex optimization problem, often a quadratic program. For many loss functions, the solution is sparse, that is, the number of non-zero α_i is small, which has advantages for the efficiency of evaluating the function f. It is only necessary to compute k on the finite sample, which avoids the need to work explicitly with the inner product space $\mathcal{H}$, which may be large. On the other hand, there can be advantages to using the primal form of the optimization problem (1); see, for example, (Suykens *et al.*, 2002).

The major design decision in algorithms of this kind is the choice of the kernel k. For vector data, a common choice is the gaussian radial basis kernel,

$$k(u, v) = \exp\left(-\|u - v\|^2/\sigma^2\right),$$

is commonly used. In this case, the corresponding RKHS is infinite dimensional. Often, information about the problem domain can be used to choose an inner product that is more appropriate for the problem. Kernel methods have been successfully applied to document discrimination using kernels that compare word frequency counts (Joachims, 1998). In this case, the dimensionality of the RKHS is not infinite, but it is very large; it is the number of distinct words. More sophisticated kernels have been defined for text and other sequence data, such as genome sequences (Watkins, 2000), for images (Romdhani *et al.*, 2001), and for data for which an accurate generative model can be estimated (Tsuda *et al.*, 2002).

2.2 Boosting algorithms

Boosting algorithms make predictions using functions from the linear span of a fixed class of functions. The first boosting algorithm arose in a constructive proof of the following surprising result about two-class pattern classification (Schapire, 1990).

Theorem 2. Consider a class $\mathcal{G}$ of functions mapping from $\mathcal{X}$ to $\{\pm 1\}$, a sequence

$$(x_1, y_1), \ldots, (x_n, y_n) \in \mathcal{X} \times \{\pm 1\}$$

where $y_i = f^*(x_i)$ for some $f^* \in \mathcal{G}$, and the discrete loss, $l(\hat{y}, y)$, defined as the indicator function of $\hat{y} \neq y$. If there is a polynomial-time algorithm (we call it the *base learning algorithm*) that can choose an $f \in \mathcal{G}$ that has empirical risk slightly better than random guessing,

$$\mathbf{E}_n l(f(X), Y) \leq \frac{1}{2} - \gamma,$$

then there is a polynomial time 'boosting' algorithm that can choose a g that has empirical risk smaller than ϵ.

The boosting algorithm from the proof of this result repeatedly calls the learning algorithm for $\mathcal{F}$, and forms a thresholded linear combination of the functions that it returns. Although this result makes strong assumptions about the process generating the data, when a particular boosting algorithm—AdaBoost (Freund and Schapire, 1997)—was tried in practice, it gave significant performance improvements over the learning algorithm for the class $\mathcal{G}$ of basis functions (Drucker and Cortes, 1996; Freund and Schapire, 1996). Typical basis function classes used in practice include decision trees, regression trees, and neural networks.

We can understand algorithms such as AdaBoost as a particular example of the greedy convex optimization approach described in Figure 1. In this algorithm, $\mathcal{G}$ is a class of $[-1, 1]$-valued functions on $\mathcal{X}$ (the basis functions from Theorem 2) and $A : \text{span}(\mathcal{G}) \to \mathbb{R}$ is a convex objective function. Figure 2 shows an analogous algorithm for convex combinations, that is, functions from the convex hull of $\mathcal{G}$.

Fig. 1. Greedy boosting algorithm for linear combinations.

set $f_0 = 0$
for $t = 1, \ldots, T$
  choose $\alpha_t \in \Lambda \subseteq \mathbb{R}$ and $g_t \in \mathcal{G}$ to minimize

$$A(f_{t-1} + \alpha_t g_t).$$

  set $f_t = f_{t-1} + \alpha_t g_t$
return f_T.

Fig. 2. Greedy boosting algorithm for convex combinations.

set $f_0 = 0$
for $t = 1, \ldots, T$
  choose $\alpha_t \in [0, 1]$ and $g_t \in \mathcal{G}$ to minimize

$$A((1 - \alpha_t)f_{t-1} + \alpha_t g_t).$$

  set $f_t = (1 - \alpha_t)f_{t-1} + \alpha_t g_t$
return f_T.

The Adaboost algorithm is a special case of the algorithm of Figure 1, where the objective function A is defined by

$$A(f) = \mathbf{E}_n \exp(-Yf(X)),$$

where $Y \in \{\pm 1\}$ is the binary response variable. This approach has been extended to a variety of related objective functions, such as logistic regression (Friedman *et al.*, 2000). With the freedom of choosing an element f of the (possibly infinite-dimensional) space of functions spanned by $\mathcal{G}$, regularization is essential. In some cases, allowing the algorithm to run for only a small number of iterations is a successful regularization strategy. This is the approach typically used with Adaboost. With some assumptions on the size of the steps α_t taken by the algorithm, we can view early stopping as constraining the one norm of the parameters. Of course, restricting the functions f_t to be convex combinations, as in the algorithm of Figure 2, constrains this norm more directly.

For a quadratic objective function,

$$A(f) = \mathbf{E}_n(Y - f(X))^2,$$

the algorithm of Figure 2 is equivalent to the greedy optimization approach proposed in (Jones, 1997) and applied to neural network parameter optimization in (Barron, 1993; Lee *et al.*, 1996). Indeed, a neural network with a one norm constraint on its parameters is an example of the class of functions used by the algorithm of Figure 2. Related algorithms have been studied in (Friedman *et al.*, 2000; Mason *et al.*, 2000; Zhang, 2003; Rätsch *et al.*, 2001).

The following theorem, from (Zhang, 2003), shows that the boosting algorithm of Figure 2 converges to an optimal solution.

Theorem 3. Suppose that, for some finite M, the convex loss function $l : \mathcal{Y}^2 \to \mathbb{R}$ is such that for all $y' \in \mathcal{Y}$, $\partial^2/\partial y^2 l(y, y') \leq M$. Let $A(f) = \mathbf{E}_n l(f(X), Y)$. Then the algorithm of Figure 2 produces a function f_T satisfying

$$A(f_T) \leq \inf_{f \in \text{co}(\mathcal{G})} A(f) + \frac{2M}{(T+2)}.$$

Thus, if the empirical risk, $\mathbf{E}_n l_g$, can be efficiently optimized over the class $\mathcal{G}$, then the empirical risk can be efficiently optimized over the convex hull of $\mathcal{G}$.

There are analogous results for the convergence rates of the boosting algorithms of Figure 1 (Zhang and Yu, 2003).

3. COMPLEXITY, CONCENTRATION, AND ESTIMATION ERROR

Suppose a prediction algorithm produces a function f from some class $\mathcal{F}$, with the aim of minimizing the risk, $\mathbf{E}l(Y, f(X))$. The optimal choice is an $f \in \mathcal{F}$ that achieves the minimal risk,

$$\inf_{f\in\mathcal{F}} \mathbf{E}l(Y, f(X)).$$

Since we only receive information about the stochastic process that generates the data from the finite sample, our choice $\hat{f} \in \mathcal{F}$ suffers an estimation error,

$$\mathbf{E}l(Y, \hat{f}(X)) - \inf_{f\in\mathcal{F}} \mathbf{E}l(Y, f(X)).$$

In this section, we present bounds on the estimation error for prediction algorithms. Such bounds motivate regularization schemes, since they show how the estimation error depends on the complexity of the function class. For instance, we shall see that the regularization term that is commonly used in kernel methods, involving the RKHS norm, can be motivated by upper bounds on the estimation error. It is possible to show that a bound on estimation error can be used to define a *complexity regularization scheme*, that is, an algorithm for choosing the model complexity by penalizing complex models. For instance, Theorems 5 and 7 below show that, for any λ, with high probability over the sample, every f in an RKHS $\mathcal{H}$ satisfying $\|f\|_{\mathcal{H}} \leq \lambda$ has

$$\mathbf{E}f \leq \mathbf{E}_n f + \lambda c_n,$$

where c_n is a constant that depends on n, $\mathcal{H}$, and properties of the probability distribution P. This motivates a complexity regularization scheme in which we choose $f \in \mathcal{H}$ to minimize the upper bound,

$$\mathbf{E}_n f + c_n \|f\|_{\mathcal{H}}.$$

This has the form of the objective function in Theorem 1. It is possible to give performance guarantees for (slightly modified versions of) these complexity regularization schemes; see (Bartlett *et al.*, 2002) for details.

The bounds in this section are based on results describing the convergence of sample averages to expectations, uniformly over the class $\mathcal{F}$. We use a measure of complexity of a function class known as the *Rademacher averages*.

Definition 4. Suppose that $\mathcal{F}$ is a class of real-valued functions defined on $\mathcal{X}$, and P is a probability distribution on $\mathcal{X}$. Then the *Rademacher averages of $\mathcal{F}$* are

$$R_n(\mathcal{F}) = \mathbf{E} \sup_{f\in\mathcal{F}} \frac{1}{n} \sum_{i=1}^{n} \epsilon_i f(X_i).$$

where $X_1, \ldots, X_n \in \mathcal{X}$ are independently chosen according to P, and $\epsilon_1, \ldots, \epsilon_n$ are independent uniform $\{\pm 1\}$-valued random variables (Rademacher random variables). The *empirical Rademacher averages of $\mathcal{F}$* are

$$\hat{R}_n(\mathcal{F}) = \mathbf{E}\left[\sup_{f\in\mathcal{F}} \frac{1}{n} \sum_{i=1}^{n} \epsilon_i f(X_i) \middle| X_1, \ldots, X_n \right].$$

The Rademacher averages are a natural notion of the complexity of the class $\mathcal{F}$ with respect to the probability distribution P. The ϵ_i define a random vector in the n-cube. If $R_n(\mathcal{F})$ is large, then for many such random vectors, there is a function that defines a vector $(f(X_1), \ldots, f(X_n))$ with large inner product with the random direction. In that case, we should not expect a small empirical average of an $f \in \mathcal{F}$ to imply a small expectation. The converse is also true.

Theorem 5. For $\mathcal{F} \subseteq [0,1]^{\mathcal{X}}$ and $\delta > 0$, with probability at least $1 - \delta$ over i.i.d. sequences $X_1, \ldots, X_n \in \mathcal{X}$, we have

$$\sup_{f\in\mathcal{F}} \left(\mathbf{E}f - \frac{1}{n} \sum_{i=1}^{n} f(X_i) \right) \leq 2R_n(\mathcal{F}) + \sqrt{\frac{\log 1/\delta}{2n}}.$$

Thus, if $\hat{f} \in \mathcal{F}$ minimizes $\mathbf{E}_n f$, then with probability at least $1 - \delta$,

$$\mathbf{E}\hat{f} \leq \inf_{f\in\mathcal{F}} \mathbf{E}f + 2R_n(\mathcal{F}) + c\sqrt{\frac{\log 1/\delta}{n}}.$$

The proof is in two steps. The first step uses the bounded differences concentration inequality (McDiarmid, 1989), to show that

$$Z = \sup_{f\in\mathcal{F}} \left(\mathbf{E}f - \frac{1}{n} \sum_{i=1}^{n} f(X_i) \right)$$

is with high probability close to its expectation. (The conditions of this concentration result require that the random variable Z changes only slightly whenever an X_i changes. This is clearly satisfied here, because of the boundedness of the range of functions in $\mathcal{F}$.) The second step uses an easy symmetrization argument to show that

$$\mathbf{E}Z \leq 2R_n(\mathcal{F}).$$

(There is also a partial converse to this result, which shows that the inequality cannot be loose unless the class $\mathcal{F}$ is small.)

It is easy to see that we can accurately estimate $R_n(\mathcal{F})$ from the data: it is the expectation of a random variable that also satisfies the conditions of the bounded differences concentration inequality. This implies that, with probability at least $1 - \delta$ over the random X_i and ϵ_i,

$$\mathbf{E}\hat{f} \leq \inf_{f\in\mathcal{F}} \mathbf{E}f + \sup_{f\in\mathcal{F}} \frac{2}{n} \sum_{i=1}^{n} \epsilon_i f(X_i) + c\sqrt{\frac{\log 1/\delta}{n}}$$

for some universal constant c.

In our application, we will apply these results to the loss functions $l_f : \mathcal{X} \times \mathcal{Y} \to \mathbb{R}$ defined by

$$l_f(x, y) = l(f(x), y).$$

Suppose that the function $\alpha \mapsto l(\alpha, y)$ is L-Lipschitz for all y, that is, there is an L such that for all $y \in \mathcal{Y}$ and all α_1, α_2 in the range of functions in $\mathcal{F}$, we have

$$|l(\alpha_1, y) - l(\alpha_2, y)| \leq L|\alpha_1 - \alpha_2|.$$

Then we can relate the Rademacher averages for the class of loss functions,

$$l_{\mathcal{F}} = \{l_f : f \in \mathcal{F}\},$$

to the Rademacher averages for the class $\mathcal{F}$. Indeed, the Ledoux-Talagrand contraction inequality (Ledoux and Talagrand, 1991) implies that

$$R_n(l_{\mathcal{F}}) \leq L R_n(\mathcal{F}).$$

The Rademacher averages can be bounded in terms of other notions of complexity, such as covering numbers and combinatorial dimensions (see, for example, (Anthony and Bartlett, 1999; Vidyasagar, 1997)), which allows us to use Theorem 5 to obtain many other error estimates, in most cases in an improved form; see, for example, (Bartlett *et al.*, 2002; Bartlett and Mendelson, 2002; Koltchinskii and Panchenko, 2002; Mendelson, 2002*b*; Dudley, 1999).

For the prediction algorithms described in Section 2, we can compute the Rademacher averages directly. The following theorem is straightforward. It shows that the Rademacher averages of the scaled convex hull of a class $\mathcal{G}$ are equal to the scaled Rademacher averages of $\mathcal{G}$.

Theorem 6. For a class $\mathcal{F} \subseteq \mathbb{R}^{\mathcal{X}}$,

(1) For $\lambda \in \mathbb{R}$, if $\lambda\mathcal{F} = \{\lambda f : f \in \mathcal{F}\}$, then $R_n(\lambda\mathcal{F}) = |\lambda| R_n(\mathcal{F})$.
(2) $R_n(\text{co}(\mathcal{F})) = R_n(\mathcal{F})$.
(3) Hence, for $\mathcal{G} = \lambda\,\text{co}(\mathcal{F})$),

$$R_n(\mathcal{G}) = |\lambda| R_n(\mathcal{F}).$$

For kernel classes, the empirical Rademacher averages can be estimated by the sum of the eigenvalues of the Gram matrix, which is readily computable from the data. There is a similar result for the Rademacher averages, in terms of the eigenvalues of a certain integral operator.

Theorem 7. For a kernel k and the corresponding RKHS $\mathcal{H}$, define $\mathcal{F} = \{f \in \mathcal{H} : \|f\|_{\mathcal{H}} \leq \lambda\}$. Then we have

$$\hat{R}_n(\mathcal{F}) \leq \lambda \frac{\sqrt{\text{trace}(K)}}{n} = \lambda \sqrt{\frac{\bar{R}^2}{n}},$$

where $K_{ij} = k(X_i, X_j)$ is the $n \times n$ Gram matrix, and so its trace is n times the average squared norm of the point evaluation functionals in $\mathcal{H}$,

$$\bar{R}^2 = \frac{1}{n} \sum_{i=1}^{n} \|k(X_i, \cdot)\|_{\mathcal{H}}^2.$$

Furthermore,

$$R_n(\mathcal{F}) \leq \lambda \sqrt{\frac{\sum_{j=1}^{\infty} \theta_j}{n}},$$

where $\theta_1, \theta_2, \ldots$ are the non-increasing eigenvalues of the integral operator $T_k : L_2(P) \to L_2(P)$ defined by

$$T_k f = \int_{\mathcal{X}} k(\cdot, x) f(x) dP(x).$$

We can estimate the Rademacher averages for certain convex neural network classes. For instance, consider the class $\mathcal{F}_B$ of two layer networks defined on $\mathcal{X} = \mathbb{R}^d$, with a bound on the one-norm of the parameters, that is, functions in $\mathcal{F}_B$ are of the form

$$x \mapsto \sum_{i=1}^{k} w_i \sigma\left(v_i^T x\right),$$

for any $k \geq 1$, $w \in \mathbb{R}^k$, and $v_i \in \mathbb{R}^d$ satisfying $\|w\|_1 \leq B$ and $\|v_i\|_1 \leq B$. Here, $\sigma : \mathbb{R} \to [-1, 1]$ is a monotonic nonlinearity. This class is infinite-dimensional. However, we can write

$$\mathcal{F}_B = B\,\text{co}\left(\sigma\left(B\,\text{co}(\mathcal{G})\right)\right),$$

where $\mathcal{G} = \{(x_1, \ldots, x_d) \mapsto x_j : 1 \leq j \leq d\}$ and we use $\sigma(A)$ to denote $\{x \mapsto \sigma(f(x)) : f \in A\}$. Thus, using the properties of the Rademacher averages we have already encountered, together with an estimate for the Rademacher averages for a finite set (Massart, 2000), we have the following result

Theorem 8. For the neural network class $\mathcal{F}_B$ defined above,

$$R_n(\mathcal{F}_B) \leq B^2 \sqrt{\frac{2 \log d}{n}},$$

where B is the bound on the parameters and d is the dimensionality of the input space.

4. CONVEXITY AND ESTIMATION ERROR

The approach in the previous section relied on convergence of empirical averages to expectations, uniformly over a class $\mathcal{F}$. This uniform convergence property suffices to ensure that the expected loss (risk) of the empirical risk minimizer $\hat{f}$ is small. However, it is not necessary for the deviations between empirical averages and expectations to be small uniformly over the class. In particular, if the worst-case difference between empirical averages and expectations occurs far from $\hat{f}$, these estimates might be too conservative. Indeed, in many problems it is possible to do better. Rather than the loss class $l_{\mathcal{F}}$, we consider the class of excess loss,

$$\mathcal{L}_{\mathcal{F}} = \{l_f - l_{f^*} : f \in \mathcal{F}\}$$
$$= \{(x,y) \mapsto l(f(x),y) - l(f^*(x),y) : f \in \mathcal{F}\},$$

where $f^* \in \mathcal{F}$ is the minimizer of the expected loss, $f^* = \arg\min_{f\in\mathcal{F}} \mathbf{E}l_f$. Then if $\mathcal{F}$ is convex and l is strictly convex, we can obtain faster convergence rates. The crucial property of the loss function is its *modulus of convexity.*

Definition 9. (Modulus of convexity). For a convex function $f : \mathbb{R} \to \mathbb{R}$, the *modulus of convexity* of f is the function $\delta : [0,\infty) \to [0,\infty]$ satisfying

$$\delta(\epsilon) = \inf\left\{\frac{f(x_1)+f(x_2)}{2} - f\left(\frac{x_1+x_2}{2}\right) : x_1, x_2 \in S,\ |x_1, x_2| \geq \epsilon\right\}.$$

In the following theorem, we use a more refined measure of the complexity of the excess loss class $\mathcal{L}_{\mathcal{F}}$, which we call the *local Rademacher averages.* These are the Rademacher averages of a small subset,

$$R_n\left(\{g \in \mathcal{L}_{\mathcal{F}} : \mathbf{E}g \leq \epsilon\}\right).$$

Theorem 10. There is a constant K for which the following holds. Consider a loss function $l : \mathcal{Y}^2 \to [0,B]$. Suppose that, for all y, the function $l(\cdot,y)$ is L-Lipschitz and strictly convex, with modulus of convexity $\delta(\epsilon) \geq c\epsilon^r$. Define $\beta = \min(1, 2/r)$. Fix a convex class $\mathcal{F}$ of real functions on $\mathcal{X}$. For i.i.d. data $(X_1,Y_1),\ldots,(X_n,Y_n)$, let $\hat{f} \in \mathcal{F}$ be the minimizer of the empirical risk, $\mathbf{E}_n l(f(X),Y)$. Then with probability at least $1-e^{-x}$,

$$\mathbf{E}l_{\hat{f}} \leq \mathbf{E}l_{f^*} + \epsilon,$$

where

$$\epsilon = K\max\left\{\epsilon^*, \left(\frac{c_r L^2 x}{n}\right)^{1/(2-\beta)}, \frac{BLx}{n}\right\},$$
$$\epsilon^* \geq R_n\left(\{g \in \mathcal{L}_{\mathcal{F}} : \mathbf{E}g \leq \epsilon^*\}\right),$$
$$c_r = \begin{cases} (2c)^{-2/r} & \text{if } r \geq 2, \\ (2c)^{-1}B^{2-r} & \text{otherwise.}\end{cases}$$

The theorem is from (Bartlett *et al.*, 2003*b*). It simplifies and generalizes results from (Mendelson, 2002*b*), which built on ideas of (Lee *et al.*, 1996). The proof is based on the observation that for a strictly convex loss the excess loss function has well-behaved variance: if l is roughly quadratic, then for any probability distribution we have

$$\mathbf{E}\left(l_f - l_{f^*}\right)^2 \leq c\mathbf{E}\left(l_f - l_{f^*}\right),$$

for some universal constant c. Having control of the variance implies that, even though the fluctuations of $\mathbf{E}_n l_f$ might be large, the fluctuations of $\mathbf{E}_n l_f$ and $\mathbf{E}_n l_{f^*}$ are strongly correlated when f is not too far from f^*, as it is when f is the empirical minimizer. This fact, together with some recent concentration inequalities (see (Bousquet, 2002) and the references cited there), implies the theorem.

For the prediction algorithms described in Section 2, we can also compute the local Rademacher averages. In the following theorem, we consider the example of scaled convex hulls, a class used by boosting algorithms. The assumption that $\mathcal{G}$ is finite dimensional can be replaced by an assumption on its covering numbers or a suitable combinatorial dimension; see (Bartlett *et al.*, 2003*b*) for details.

Theorem 11. For the class $\mathcal{F} = \lambda\,\mathrm{co}(\mathcal{G}))$, where $\mathcal{G}$ is a subset of a d-dimensional subspace, and a loss function l for which the function $l(\cdot,y)$ is uniformly Lipschitz and strictly convex, with modulus of convexity $\delta(\epsilon) \geq c\epsilon^2$, we have

$$R_n\left(\{g \in \mathcal{L}_{\mathcal{F}} : \mathbf{E}g \leq \epsilon\}\right)$$
$$\leq c_1\lambda\left(n^{-(d+2)/(2d+2)} + n^{-1/2}\epsilon^{1/(d+2)}\right).$$

The result involving Rademacher averages of the full class, Theorem 6, cannot give rates better than $n^{-1/2}$, whereas the rate above is always faster, particularly if d is small.

Similarly, for kernel classes we get improved rates. This theorem combines results from (Bartlett *et al.*, 2003*a*) and (Mendelson, 2002*a*).

Theorem 12. For a kernel k and the corresponding RKHS $\mathcal{H}$, define $\mathcal{F} = \{f \in \mathcal{H} : \|f\|_{\mathcal{H}} \leq \lambda\}$. Suppose l is a loss function for which the function $l(\cdot,y)$ is uniformly Lipschitz and strictly convex, with modulus of convexity $\delta(\epsilon) \geq c\epsilon^2$. Then we have

$$R_n\left(\{g \in \mathcal{L}_{\mathcal{F}} : \mathbf{E}g \leq \epsilon\}\right)$$
$$\leq c_1\lambda\left(\frac{1}{n}\sum_{j=1}^{\infty}\min\{c_2\epsilon, \theta_i\}\right)^{1/2}$$

where $\theta_1, \theta_2, \ldots$ are the non-increasing eigenvalues of the integral operator $T_k : L_2(P) \to L_2(P)$ defined by

$$T_k f = \int_{\mathcal{X}} k(\cdot,x)f(x)dP(x).$$

Since the trace of T_k is always at least as large as the sum in this expression, the bound is no worse than that involving the Rademacher averages of the full class, Theorem 7. If the eigenvalues decrease rapidly, the estimation error rate we obtain by exploiting the convexity information can be considerably better. For instance, if they decrease exponentially quickly and the loss function has a quadratic modulus of convexity, then Theorems 10 and 12 imply that, with high probability,

$$\mathbf{E}l_{\hat{f}} \leq \inf_{f:\|f\|\leq\lambda} \mathbf{E}l_f + \lambda^2 c_n,$$

where c_n is a constant that depends on the sample size n and the eigenvalues of T_k. Bounds of this kind motivate a complexity regularization scheme in which we choose $f \in \mathcal{H}$ to minimize

$$\mathbf{E}_n f + c_n \|f\|_H^2.$$

Again, this has the form of the objective function in Theorem 1.

In the previous section, an easy concentration argument showed that we could estimate the Rademacher averages of the full class from the empirical Rademacher averages. For the local Rademacher averages considered here, something similar is also possible. This is more complicated, in part because the function class

$$\{l_f - l_{f^*} : f \in \mathcal{F}, \mathbf{E}\left(l_f - l_{f^*}\right) \leq \epsilon\}$$

is defined in terms of the risk minimizer f^* and the probability distribution, which are both unknown. It turns out that for a convex class $\mathcal{F}$ and strictly convex loss l, we can replace these local Rademacher averages with the following empirical local Rademacher averages.

$$\hat{\xi}_{\mathcal{F}}(\epsilon) = \hat{R}_n\left(\left\{f \in \mathcal{F} : \mathbf{E}_n\left(f - \hat{f}\right)^2 \leq c\epsilon\right\}\right),$$

where $\hat{f}$ is the minimizer of the empirical risk, $\mathbf{E}_n l_f$. In the case of kernel classes, this allows us to replace the eigenvalues of the integral operator T_k with the eigenvalues of the Gram matrix, since

$$\hat{\xi}_{\mathcal{F}}(\epsilon) \leq c\left(\frac{1}{n}\sum_{i=1}^{n} \min\left\{\frac{\lambda_i}{n}, \epsilon\right\}\right)^{1/2}.$$

where $\lambda_1, \cdots, \lambda_n$ are the eigenvalues of the Gram matrix $K_{ij} = k(X_i, X_j)$. See (Bartlett *et al.*, 2003*a*) for details.

These results show that, if the loss function is strictly convex and the class $\mathcal{F}$ is convex, then the estimation error decreases rapidly to zero as a function of the sample size n, in some cases considerably faster than $n^{-1/2}$. There is a converse result for the condition on the convexity of the class $\mathcal{F}$. The following theorem, from (Lee *et al.*, 1998), shows that for a nonconvex class with quadratic loss, the minimax convergence rate of the estimation error

$$\mathbf{E}l(\hat{f}(X), Y) - \inf_{f\in\mathcal{F}} \mathbf{E}l(f(X), Y)$$

cannot be faster than $n^{-1/2}$.

Theorem 13. Fix a function class $\mathcal{F}$ and the quadratic loss function, $l(\hat{y}, y) = (\hat{y} - y)^2$. For $\epsilon, \delta, B > 0$, define $n(\epsilon, \delta, B)$ as the smallest integer satisfying the following condition. For every sample size $n \geq n(\epsilon, \delta, B)$,

$$\inf_{\hat{f}} \sup_{P} P^n \left\{D_n : \mathbf{E}l(\hat{f}(D_n, X), Y) > \inf_{f\in\mathcal{F}} \mathbf{E}l(f(X), Y) + \epsilon\right\} \leq \delta,$$

where the infimum is over prediction algorithms $\hat{f}$ that take as input a random X and an i.i.d. data sequence $D_n = (X_1, Y_1, \ldots, X_n, Y_n)$ chosen according to the probability distribution P, and the supremum is over all probability distributions P on $\mathcal{X} \times [-B, B]$.

If, for some probability distribution $P_{\mathcal{X}}$ on $\mathcal{X}$, the closure of $\mathcal{F}$ in $L_2(P_{\mathcal{X}})$ is not convex, then for sufficiently large B, there is a constant c for which

$$n(\epsilon, \delta, B) \geq c\frac{\ln(1/\delta)}{\epsilon^2}.$$

5. CONVEXITY AND APPROXIMATION

It is clear from the statistical and computational results of the previous sections that if we have a convex loss function and a nonconvex function class $\mathcal{F}$, then working with the convex hull of $\mathcal{F}$ does not add significant computational cost, and can reduce the estimation error. Clearly, it increases the function class, so it can only help the approximation properties. That is, if the loss function is convex, it often makes sense to ensure that the function class is convex.

Is the same true of the loss function? That is, if the function class is convex, should we replace a non-convex loss function with a convex one? In some cases, this gives significant computational advantages. One important example arises in pattern classification problems. Consider two-class pattern classification, which is a prediction problem of the kind we have been considering where the response variable Y takes values in the binary set $\{\pm 1\}$ and the loss function is the indicator function of $\text{sign}(f(x)) \neq y$. For most interesting function classes, the problem of approximately minimizing risk in a minimax setting is provably hard (Arora *et al.*, 1997, for example). However, if we replace this nonconvex loss function by a surrogate of the form $\tilde{l}(f(x), y) = \phi(yf(x))$, where ϕ is convex, the optimization problem can often be solved efficiently. This is the approach used by support vector machines (kernel methods for pattern classification), AdaBoost, and logistic regression. In such cases, it is important to understand the impact of this choice on the statistical properties of the method. In particular, we are interested in how the risk,

$$\begin{aligned} R(f) &= \Pr(Y \neq \text{sign}(f(X))) \\ &= \mathbf{E}l(f(X), Y), \end{aligned}$$

is related to the ϕ-risk,

$$R_\phi(f) = \mathbf{E}\phi(Yf(X)),$$

since the latter is the quantity that these methods minimize. The following theorem, from (Bartlett *et al.*, 2003*b*), considers the excess risk, that is, how much the risk of a function exceeds the optimal or Bayes risk,

$$R^* = \inf_f \Pr(Y \neq \text{sign}(f(X))),$$

where the infimum is over all measurable f, and the excess ϕ-risk, that is, how much the ϕ-risk exceeds the optimal ϕ-risk,

$$R^*_\phi = \inf_g \mathbf{E}\phi(Yg(X)).$$

We first define a functional transform of the loss function.

Definition 14. We define the *ψ-transform* of a loss function as follows. Given $\phi : \mathbb{R} \to [0,\infty)$, define the function $\psi : [0,1] \to [0,\infty)$ by $\psi = \tilde{\psi}^{**}$, where

$$\tilde{\psi}(\theta) = \inf_{\alpha:\alpha\leq 0}\left(\frac{1+\theta}{2}\phi(\alpha) + \frac{1-\theta}{2}\phi(-\alpha)\right) - \inf_{\alpha\in\mathbb{R}}\left(\frac{1+\theta}{2}\phi(\alpha) + \frac{1-\theta}{2}\phi(-\alpha)\right).$$

and $g^{**} : [0,1] \to \mathbb{R}$ is the Fenchel-Legendre biconjugate of $g : [0,1] \to \mathbb{R}$, which is defined by

$$\text{epi } g^{**} = \overline{\text{co}} \text{ epi } g.$$

Here, $\overline{\text{co}}\ S$ is the closure of the convex hull of the set S, and epi g is the epigraph of the function g, that is, the set $\{(x,t) : x \in [0,1], g(x) \leq t\}$.

Theorem 15. For any nonnegative loss function ϕ defined on $\mathbb{R}$, any measurable $f : \mathcal{X} \to \mathbb{R}$ and any probability distribution on $\mathcal{X} \times \{\pm 1\}$, we have

$$\psi(R(f) - R^*) \leq R_\phi(f) - R^*_\phi.$$

This bound shows that minimizing the ϕ-risk leads to minimal risk provided that the nonnegative function ψ is zero only at zero. It turns out that for convex ϕ this is the case if and only if ϕ is differentiable at zero and has $\phi'(0) < 0$. It can be shown that the bound cannot be improved anywhere, in general (Bartlett *et al.*, 2003*b*). Thus, it quantifies the cost of replacing the loss function of interest with a convex surrogate. It would be interesting to extend this result to other nonconvex loss functions; the definition of ψ is specific to the discrete loss l.

6. SYSTEM IDENTIFICATION, DEPENDENT DATA, AND OPEN PROBLEMS

We can express certain system identification problems as prediction problems of the kind studied here. Define $\mathcal{X}$ as the space of input sequences and $\mathcal{Y}$ as the space of system outputs (for instance, $\mathcal{Y} \subseteq \mathbb{R}$). Consider a class $\mathcal{F}$ of functions mapping from $\mathcal{X}$ to $\mathcal{Y}$, the family of models that we wish to identify.

Most of the results presented in this paper are based on the assumption that the pairs

$$(X_1, Y_1), \ldots, (X_n, Y_n), (X, Y)$$

are independent and identically distributed. In a system identification problem, this is not a reasonable assumption: the input sequences are of the form $X_t = (u_t, u_{t-1}, u_{t-2}, \ldots)$ for some infinite sequence (u_t). However, if the sequences (u_t) and (y_t) and the model class $\mathcal{F}$ satisfy some reasonable conditions, it is not difficult to extend the results in this paper to this setting. In particular, suppose that

(1) (u_t) and (y_t) are stationary stochastic processes satisfying a suitable mixing condition, so that (X_t, Y_t) and (X_{t+s}, Y_{t+s}) become 'closer to independent' in some sense as s increases, and
(2) models in $\mathcal{F}$ uniformly satisfy a fading memory property, which implies that for all input sequences,

$$f(u_t, u_{t-1}, u_{t-2}, \ldots)$$

and

$$f(u_t, \ldots, u_{t-s})$$

become close as s increases.

Then it is possible to relate the behavior of an empirical risk minimization algorithm in this setting to its behavior with i.i.d. data. In this way, many of the results presented above, including Theorems 5 and 10, can be extended to the system identification setting. In particular, this would show how rapidly the expected loss of the empirical minimizer converges to the minimal value on the model class $\mathcal{F}$. For details on this and related approaches see, for example, (Nobel and Dembo, 1993; Yu, 1994; Weyer, 2000; Vidyasagar and Karandikar, 2001; Karandikar and Vidyasagar, 2002)

On the other hand, it is not as simple to extend data-dependent error estimates to this setting, such as the bounds in terms of the empirical Rademacher averages, and their consequences in terms of eigenvalues of the kernel matrices. These results are appealing because they use the data to measure the complexity of the function class. It would be interesting to develop versions of these results that apply to mixing stochastic processes.

REFERENCES

Anthony, Martin and Peter L. Bartlett (1999). *Neural Network Learning: Theoretical Foun-*

dations. Cambridge University Press. Cambridge, UK.

Arora, S., L. Babai, J. Stern and Z. Sweedyk (1997). The hardness of approximate optima in lattices, codes, and systems of linear equations. *Journal of Computer and System Sciences* **54**, 317–331.

Barron, Andrew R. (1993). Universal approximation bounds for superpositions of a sigmoidal function. *IEEE Transactions on Information Theory* **39**, 930–945.

Bartlett, P. L. and S. Mendelson (2002). Rademacher and Gaussian complexities: Risk bounds and structural results. *Journal of Machine Learning Research* **3**, 463–482.

Bartlett, P. L., O. Bousquet and S. Mendelson (2003*a*). Local Rademacher complexities. Technical report. University of California at Berkeley.

Bartlett, P. L., S. Boucheron and G. Lugosi (2002). Model selection and error estimation. *Machine Learning* **48**, 85–113.

Bartlett, P., M. I. Jordan and J. D. McAuliffe (2003*b*). Convexity, classification, and risk bounds. Technical Report 638. Department of Statistics, University of California at Berkeley.

Boser, B. E., I. M. Guyon and V. N. Vapnik (1992). A training algorithm for optimal margin classifiers. In: *Proceedings of the 5th Annual Workshop on Computational Learning Theory.* ACM Press, New York, NY. pp. 144–152.

Bousquet, O. (2002). A Bennett concentration inequality and its application to suprema of empirical processes. *C. R. Acad. Sci. Paris Ser. I* **334**, 495–500.

Cortes, C. and V.N. Vapnik (1995). Support-vector networks. *Machine Learning* **20**, 273–297.

Cristianini, Nello and John Shawe-Taylor (2000). *An Introduction to Support Vector Methods.* Cambridge University Press.

Drucker, Harris and Corinna Cortes (1996). Boosting decision trees. In: *Advances in Neural Information Processing Systems.* Vol. 8. pp. 479–485.

Dudley, R. M. (1999). *Uniform Central Limit Theorems.* Cambridge University Press.

Freund, Y. and R.E. Schapire (1997). A decision-theoretic generalization of on-line learning and an application to boosting. *Journal of Computer and System Sciences* **55**(1), 119–139.

Freund, Yoav and Robert E. Schapire (1996). Experiments with a new boosting algorithm. In: *Proceedings of the 13th International Conference on Machine Learning.* Morgan Kaufmann. pp. 148–156.

Friedman, J., T. Hastie and R. Tibshirani (2000). Additive logistic regression: a statistical view of boosting. *Annals of Statistics* **28**, 337–374.

Joachims, T. (1998). Text categorization with support vector machines: Learning with many relevant features. In: *Proceedings of the European Conference on Machine Learning.* Springer. pp. 137–142.

Jones, Lee K. (1997). The computational intractability of training sigmoidal neural networks. *IEEE Transactions on Information Theory* **43**(1), 167–173.

Karandikar, Rajeeva L. and M. Vidyasagar (2002). Rates of uniform convergence of empirical means with mixing processes. *Statistics and Probability Letters* **58**(3), 297–307.

Kimeldorf, George S. and Grace Wahba (1971). Some results on Tchebycheffian spline functions. *Journal of Mathematical Analysis and Applications* **33**(1), 82–95.

Koltchinskii, V. and D. Panchenko (2002). Empirical margin distributions and bounding the generalization error of combined classifiers. *Annals of Statistics* **30**(1), 1–50.

Ledoux, M. and M. Talagrand (1991). *Probability in Banach Spaces: Isoperimetry and Processes.* Springer.

Lee, W. S., P. L. Bartlett and R. C. Williamson (1996). Efficient agnostic learning of neural networks with bounded fan-in. *IEEE Transactions on Information Theory* **42**(6), 2118–2132.

Lee, Wee Sun, Peter L. Bartlett and Robert C. Williamson (1998). The importance of convexity in learning with squared loss. *IEEE Transactions on Information Theory* **44**(5), 1974–1980.

Mason, L., J. Baxter, P.L. Bartlett and M. Frean (2000). Functional gradient techniques for combining hypotheses. In: *Advances in Large Margin Classifiers* (A.J. Smola, P.L. Bartlett, B. Schölkopf and D. Schuurmans, Eds.). pp. 221–247. MIT Press. Cambridge, MA.

Massart, P. (2000). Some applications of concentration inequalities to statistics. *Annales de la Faculté des Sciences de Toulouse* **IX**, 245–303.

McDiarmid, C. (1989). On the method of bounded differences. In: *Surveys in Combinatorics 1989.* pp. 148–188. Cambridge University Press.

Mendelson, S. (2002*a*). Geometric parameters of kernel machines. In: *Proceedings of the 15th Annual Conference on Computational Learning Theory.* pp. 29–43.

Mendelson, S. (2002*b*). Improving the sample complexity using global data. *IEEE Transactions on Information Theory* **48**(7), 1977–1991.

Nobel, A. B. and A. Dembo (1993). A note on uniform laws of averages for dependent processes. *Statistics and Probability Letters* **17**, 169–172.

Rätsch, G., S. Mika and M. K. Warmuth (2001). On the convergence of leveraging. Technical Report NC-TR-01-098. NeuroCOLT2, Royal Holloway College.

Romdhani, S., B. Schölkopf, P. Torr and A. Blake (2001). Computationally efficient face detection. In: *Proceedings of the International Conference on Computer Vision*. pp. 695–700.

Schapire, Robert E. (1990). The strength of weak learnability. *Machine Learning* **5**(2), 197–227.

Schölkopf, B. and A. Smola (2002). *Learning with Kernels*. MIT Press. Cambridge, MA.

Suykens, J. A. K., T. Van Gestel, J. De Brabanter, B. De Moor and J. Vandewalle (2002). *Least Squares Support Vector Machines*. World Scientific.

Tsuda, K., T. Kin and K. Asai (2002). Marginalized kernels for biological sequences. *Bioinformatics* **18(Suppl. 1)**, S268–S275.

Vapnik, V. N. (1998). *Statistical Learning Theory*. Wiley and Sons. New York.

Vidyasagar, M. (1997). *A Theory of Learning and Generalization, with Applications to Neural Networks and Control Systems*. Springer. New York.

Vidyasagar, M. and R. L. Karandikar (2001). A learning theory approach to system identification and stochastic adaptive control. In: *IFAC Symposium on Adaptation and Learning*.

Watkins, C. (2000). Dynamic alignment kernels. In: *Advances in Large Margin Classifiers* (A.J. Smola, P.L. Bartlett, B. Schölkopf and D. Schuurmans, Eds.). pp. 39–50. MIT Press. Cambridge, MA.

Weyer, E. (2000). Finite sample properties of system identification of ARX models under mixing conditions. *Automatica* **36**(9), 1291–1299.

Yu, B. (1994). Rates of convergence for empirical processes of stationary mixing sequences. *Annals of Probability* **22**, 94–116.

Zhang, T. (2003). Sequential greedy approximation for certain convex optimization problems. *IEEE Transactions on Information Theory* **49**(3), 682–691.

Zhang, T. and B. Yu (2003). Boosting with early stopping: convergence and consistency. Technical Report 635. Department of Statistics, UC Berkeley.

IFAC Publications
www.elsevier.com/locate/ifac

Grey–Box Model Calibrator & Validator

Torsten Bohlin, Dept. Signals, Sensors & Systems, KTH, Sweden
Alf J Isaksson, ABB Corporate Research, Sweden

Abstract

This contribution presents a tool for designing 'grey–box' models with applications primarily to continuous industrial production processes. It is the result of more than ten years development at the Automatic Control group at KTH. This has involved development of basic theory for 'grey–box' identification and a number of case studies of full–scale industrial production processes, as well as construction of the MoCaVa software tool (Model Calibrator & Validator) with user interface adapted to the task of identification under partial structure and input uncertainty. The program uses Matlab. It is available from http://mocava.s3.kth.se free of charge.

Introduction

Commercially available tools for making mathematical models of dynamic processes are of two kinds, with different demands on the user. On one hand there are *Modelling* tools, generally associated with simulation software (*e.g.* Dymola, http://www.dynasim.se/www/Publications.pdf), which require the user to provide complete specification of the equations governing the process, either expressed as statements written in some modelling language (*e.g.* Modelica, Tiller, 2001), or by picking and connecting components from a library. This alternative may be supported by combining the modelling tools with tools for parameter optimization (*e.g.* HQP, http://sourceforge.net/projects/hqp). Call this 'white–box' identification.

On the other hand there are 'black–box' system identification tools (*e.g.* Matlab System Identification Tool Box), which requires the user to accept one of the generic model structures (*e.g.* linear) and then to determine which tools to use in the particular case, and in what order, as well as the values of a number of design parameters (*e.g.* order numbers, weighting factors, etc.). Finally, the user must interpret the resulting model, which is expressed in a form that primarily is not adapted to the physical object. Unless the model is to be used directly for design of feedback control, there is some further translation to do.

Generally, the user has two sources of information on which to base the model making: *prior knowledge* and *response data*. 'White–box' identification uses mainly one source and 'black–box' identification the other. The strength of 'white–box' identification is that it will allow the user to exploit invariant prior knowledge. Its weakness is its inability to cope with the unknown and with random effects in the object and its environment. The latter is the strength of 'black–box' identification based on statistical methods, but also means that the reproducibility of its results may be in doubt. In essence, 'black–box' identification produces 'data descriptions', and repeating the experiment may well produce a much different model. This may or may not be serious, depending on what the model is to be used for.

When system identification tools are applied to industrial processes prior knowledge is typically *uncertain* and *incomplete*, the effects of unknown input ('disturbances') are *not negligible*, and it is desirable to have *reproducibility* of the model, for instance for the monitoring of unmeasured variables, for feed–forward control, or for long–range prediction of variables with much delayed responses to control action. For such cases both classes of admissible models (free but deterministic *vs* stochastic but given) are too restricted.

The idea of Grey–box identification is therefore to combine the strengths of the two approaches in order to reduce the effects of their weaknesses. This requires a widening of the class of admissible models to include some class of user–written structures with both deterministic and stochastic input. The parameter estimation problems associated with this can generally be solved by adopting the Maximum Likelihood principle, and some tools have been developed to support this, *e.g.* CYPROS (CAMO A/S, 1987), MATRIXx (Gupta *et al*, 1993), IdKit (Bohlin and Graebe, 1994), CTSM (Kristensen, Melgaard and Madsen, 2002). Even if this allows the user to make models in some cases where the other approaches fail, it also makes the tools more difficult to use.

Now, each tool implements the solution of a System Identification Problem: Given a parametric class of models, find the member that fits given experiment data with the minimum loss according to a given criterion (Ljung, 1987). Sometimes "given" means that prerequisites are build into the software, sometimes that they are expected as input to the software.

When one is faced with a given *object* instead, and possibly with a given *purpose* for the model, it is certainly not obvious how to get the answers to the questions posed by identification software, or to other questions that seem relevant, like the following: How can I make use of what I *know* in advance? How much of my prior knowledge is *useful*, or even *correct* when used in the particular environment? What do I do with the *disturbances* I cannot eliminate? Are my *experiment data* sufficient and relevant? How do I know when the model is *good* enough?

In all cases the problems of setting up tasks for the software to solve and interpreting the results are crucial for the model making to succeed. It is therefore important that developers of modelling and identification software do what they can to facilitate the answering. It is not necessarily a desirable ambition to make the software more 'automatic' by demanding less from the user. He/she is still responsible for the quality of the result, and any input that a user is able to provide, but is not asked for, may be a waste of information and reduce quality. A better goal is therefore to make the software demand its input in a *form* that the user can supply more easily.

Secondly, user input (both prior knowledge and experiment data) is often uncertain, irrelevant, contradictory, or even false. A second goal for the software designer is therefore to provide tools for appraising the user's input. Admittedly, any software must have something "given", but it makes a difference whether the software wants *assumptions*, taken for facts, or *hypotheses*, that will be subject to tests.

The rather long introduction has finally arrived at the problem treated in this paper. In summary, the user of identification tool kits must contribute two things: *i*) The sequence in which to apply the tools and *ii*) specifications pertaining to the particular case. The sequel will present some support to the user in dealing with these tasks. The contents emphasize what MoCaVa can and cannot do rather that how it is constructed and how it operates. The latter would require more space to be informative, and is left to references (Bohlin, 2001). Thus, the first sections narrate the reasoning behind the design of MoCaVa, in order to illuminate the circumstances for its use. This is followed by a brief overview of its main features. The paper ends with a briefly commented list of case studies and other applications, mainly to industrial processes, in order to support our claim that MoCaVa is not a 'toy'.

The grey–box identification procedure

Considering the diversity of a user's prior information, originating in a variety of more or less reliable sources, it is very unlikely that one would be able to formulate, much less solve, a mathematical problem that, given prior input and data, would produce a 'best' model according to a given criterion (and thus be able retain the usual definition of the identification problem).

However, it is possible to conceive *multistep* procedures for generating a model that satisfies many of the demands one may have on it, and taking the user's input into account. The steps in these procedures will require the solution of less demanding subproblems, like fitting to data, and testing whether one model is significantly better than another. The literature offers principles and ideas for solving many of the subproblems, and a number of those have been compiled into a systematic procedure for grey–box identification (Bohlin, 1991a, 1994a). One of the procedures has also been implemented as a User's Shell (IKUS) to IdKit, Linux version (Bohlin, 1993). However, its principles are general and can be implemented with other tool boxes that are general enough and open enough.

In essence, IKUS processes sets of model structures with increasing complexity. It uses the SFI rule (Stepwise Forward Inclusion, Leontaritis and Billings, 1987) to decide in which direction to expand a model structure, the Maximum Likelihood principle to find its parameters, and the maximum–power Likelihood–Ratio test to falsify wrong attempts of expansion. Thus, IKUS works by updating two sets of model structures: the 'tentative', which is the best so far, and the 'alternatives', some of which may or may not be better. In this way it will find the simplest adequate structure and its most likely parameter values.

The IKUS shell provides support to the user's first task of determining the sequence in which to apply the tools. Notice that it does not mean a fixed procedure for running an identification session. The steps in a session still depend on the outcome of the previous steps, and the sequel still needs to be controlled by the user. What the software can contribute are interpretations of partial results, a much reduced menu for the next logical step, and often proposals of what to do next. Grey–box identification is necessarily interactive and occasionally requires further input. The latter does in fact make the task easier, since the user does not have to specify all in advance.

The IdKit tool box

In addition to the IKUS shell MoCaVa builds on IdKit, which is a set of programs for solving subproblems, mainly fitting and falsification of model structures. IdKit processes given models of the form

$$dx(t)/dt = G[x(t), w(t), u(t), t, p]$$
$$z(t) = H[x(t), w(t), u(t), t, p]$$
$$y(t_k) = Cz(t_k) + Fe(k)$$

where u is known input, x is continuous state, z is the continuous response, and y is the output of sampling sensors. $w(t)$ is a Wiener process, $e(k)$ are 'white noise' sequences, p are known or unknown parameters, t is continuous time, and k a sampling counter.

The *fitting* of subsets of the parameters p is done by minimizing the loss function Q

$$Find\ \bar{p} = \arg\min_p\ Q(p, d, \overline{\mathcal{M}} | p \in \overline{\mathcal{P}})$$

where d is the data sample, $\overline{\mathcal{M}}$ the tentative model class, and $\overline{\mathcal{P}}$ is the free parameter space of the tentative model structure $\overline{\mathcal{M}}(\overline{\mathcal{P}})$.

The *falsification* is done using the criterion

$$Reject\ \overline{\mathcal{P}}\ if\ Q(\bar{p}, d, \overline{\mathcal{M}}) - Q(p, d, \mathcal{M}) > \chi^2_n/2$$

for some $p \in \mathcal{P}$, where $\mathcal{P} \supset \overline{\mathcal{P}}$ is the free parameter space of the wider alternative structure $\mathcal{M}(\mathcal{P})$, $\mathcal{M} \supseteq \overline{\mathcal{M}}$ is the possibly wider model class, and χ^2_n is the chi–square variable for a given risk of accepting an unnecessarily complex model structure with n parameters too many.

Notice that the same loss function Q appears in both criteria. And since Q has a given form derived from quite general optimality criteria (Bohlin, 1993), this eliminates two conventional user's tasks, *viz.* those of selecting a loss function (including the specification of weighting factors) and of selecting a test criterion. The common loss function is the negative logarithm of the asymptotic likelihood function for long data samples. It is a (known) function of the model's prediction errors and their covariances.

Purpose and motivation of MoCaVa

MoCaVa has been conceived for the following scenario: Suppose a production process is to be described by a dynamic model for simulation or other purposes. A number of submodels (or first principles or heuristic relations) for parts of the process are available as prior information, developed under more or less well controlled conditions. However, when the submodels are assembled into a model for the integrated process, all their inputs and outputs are no longer controlled or measured, and the environment is no longer the same as when the submodels were developed. In addition, unmodelled phenomena and unmeasurable input (disturbances) may effect the responses substantially. It is not known which of the submodels that are needed for a satisfactory model, or whether there will still remain unexplained phenomena in the data, when all prior information has been used. It is in the nature of things that prior model information is more or less precise, reliable, and relevant. It may even be false.

The combination of the IdKit tool box and the IKUS shell has the potential of answering the questions posed by the scenario and in the introduction, provided the user enters the adequate input. What remains for MoCaVa to do is to make this easier.

Now, designing general support for the user's second task of entering prior knowledge is obviously more difficult than that solved by IKUS. In principle, it would require no less than making tools for translating general user's prior knowledge into the specific input accepted by the particular tool box. What can reasonably be done is creating a number of *forms* for specifying 'user prior knowledge', acceptable to a class of users with particular background and education. It should then be possible to design a 'user interface', or 'front end' suiting the particular class of users.

Modelling

Out of the three basic subtasks, *viz. modelling, fitting,* and *falsification,* the last two are case–independent enough to have satisfactory general solutions. The key issue is the modelling. It must be easy to create the expanding sequences of case–dependent model structures that the 'grey–box' identification procedure requires.

Obviously, the modelling support in MoCaVa is limited in generality by what IdKit can handle. It is limited further by what is *a priori* imagined will be possible for the user to handle. The model set used in MoCaVa is therefore structured further to adapt to common properties of industrial production processes, in particular to *transport* processes. The latter may be characterized as systems of separate units, each one accepting flows of commodities from one or more preceding units, changing their properties, and feeding the product to one or more following units. Since there is an obvious cause–and–effect relationship between the input and output variables of the units, state–vector models (defined by assignment statements) are convenient to use in those cases.

Secondly, the operation of an individual unit is generally a result of interaction between particular physical phenomena (at least first principles are generally expressed in this way). Also the different phenomena may be described by submodels.

A third common characteristic of production processes is that the operation of some units may be affected by the operations of other, *control* units. Instead of flows (mass or energy), they produce information input to the affected unit, but are still describable by the same type of submodel.

In order to satisfy the requirements MoCaVa is able to administer the creation of submodels, and to connect them into systems.

A second restriction built into the model classes used in MoCaVa is also motivated by common properties of industrial production systems: Model components are structured as the following four–boxes system

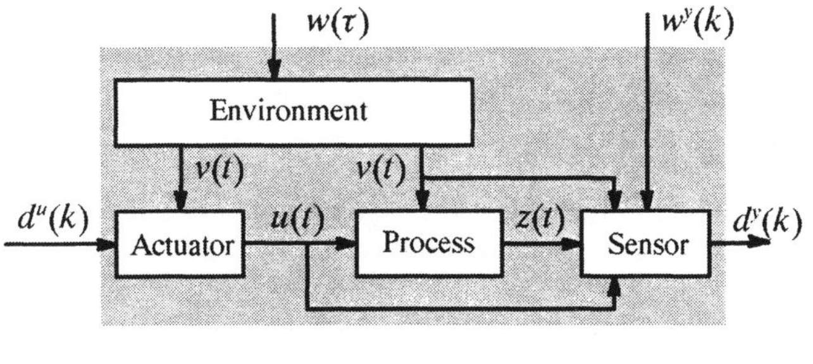

The point is to relieve the user from the intricacies of writing the hybrid and stochastic models describing the Environment, Actuator, and Sensor models. Being less particular to the case than the Process proper, they allow MoCaVa to offer menus of black boxes for the user to pick from. This necessarily means reduced flexibility.

The tentative models to be fitted and tested consist of systems of connected components of the form

$$[s(t), z(t), dx(t)/dt] = G[x(t), v(t), u(t), t, p]$$

were v is unknown input from a stochastic 'environment, and s are 'signal output' used to connect components.

Basically, components are defined once, while the tentative structures to be tested are created by *activating* a selection of them. This allows the easy creation of a large number of alternatives from relatively few components, and thus reduces the modelling chores.

The principle of connecting a component is to give its signal outputs the same names as the parameters they are to replace. This is done when the component is defined. Thus, components may be connected or disconnected very conveniently (at the appropriate check points in the user interaction) by clicking the mouse in a table of created components. The motivation for this design is twofold: *i*) Regardless of whether a component will be used to describe a physical unit producing input to another unit, or to describe a particular physical phenomenon inside the unit, the user knows *a priori* which parameter it is to replace. *ii*) In grey–box identification with uncertain structure it is more important to be able to change the structure and complexity, than to have flexibility in the creation of components.

The disadvantage of requiring each component to be an 'instant', with prespecified connections, is compensated in the design of MoCaVa–2 in two ways, both to facilitate reuse of model components: *i*) A 'copy' option, allowing the user to reproduce a component and change the connecting variable names in a displayed table. *ii*) A 'call' option, allowing the user to include calls to user–defined or library functions.

A basic tenet in the modelling in MoCaVa is that causality between variables is prior information. The rationale for this is the focus on industrial transport processes, not all being describable by deterministic functional relations. In processes with unknown input causality is important prior knowledge, eliminating some model structures from being considered that would otherwise agree with observed covariation in the data (Bohlin, 1991a). Since the list of components is built up gradually, each component added only when test results have indicated that the model so far is inadequate, it is normally easy to see what is cause and what is effect, as well as where to connect an amendment.

The tentative models are all in the form of ODE–systems with sampled data. PDE–models are not supported explicitly, but can be treated with the help of a user–provided discretization routine (Funkquist, 1995). ADE–models are not supported explicitly. However, explicit algebraic equations can be modelled using fast dynamic states, and resolved using the stiff ODE solver in IdKit. For mathematical reasons models must be continuously linearizable with respect to states, unknown parameters, and random input.

Calibration and validation

MoCaVa consists of two procedures MoCa (for Model Calibration) and MoVa (for Model Validation). Both use the same tool box for solving the subproblems, and implement general rules to provide support to the problems of setting up a sequence of tasks for the tool box and interpreting the results.

MoCa resolves the *calibration* task by recursively fitting and testing a series of model structures, starting with the simplest and most reliable submodels. The structure is then expanded by a process of 'pruning and cultivation': submodels that do not contribute significantly to reducing the overall loss are eliminated from consideration, while those that do contribute are candidates for further refinement. The roles of the two 'actors' in the interactive process is that MoCa does the elimination and also suggests the most likely of a limited number of alternative refinements, while the user suggests the alternatives to be considered and also accepts or overrules MoCa's suggestions. In this way a user will be able to exercise his/her modelling skill and knowledge about the physical process in order to reduce the number of tests that would be needed otherwise. The design of MoCa is based on the belief that, although it may be hard to know the right model structure in advance, an engineer is usually good at amending a model, when there is evidence pointing to where it fails.

MoVa supports the *validation* task in a more heuristic manner: Basically, it is a tool box for various kinds of simulation designed to test the performance and robustness of the calibrated model for selected purposes: *parameter estimation*, *prediction*, *monitoring*, or *control*, with or without feedback, with or without control adaptation, and for various prediction ranges.

The MoVa user's shell administers a procedure of reduction of the calibrated model, as long as the result is still good enough for a number of selected purposes. This also solves the problem of appraising the experiment data: Only if even the unreduced model is not good enough for the purpose, will one need more or better data.

Both validation and initial model generation is also supported by a separate simulation package.

Features of MoCaVa

MoCaVa is implemented under Microsoft Windows and Matlab. It uses the IKUS user's shell (written as M–files) to control the execution of the tasks in IdKit (written in C), and uses a special front end (written as M–files) to return partial results and ask for user input. All are open source files.

Scope

MoCaVa handles systems of non–linear, deterministic state–vector models with deterministic and stochastic input. It accepts noisy and irregular data.

MoCaVa administers easy modification of model structures in the form of systems of connected components.

MoCaVa separates the submodels (components) into the four blocks of 'sensor', 'process proper', 'actuator', and 'environment', in the double interest of increasing computing efficiency and facilitating the modelling. The user explicitly enters hypothetical relations only between the continuous–time input and output of the process proper. The remaining three blocks involve relations between continuous–time variables and discrete–time data and random variables, and are modelled by selecting from a library. Thus, disturbances are modelled by allocating their sources to the 'environment', and picking a model of the source from among a few standard types. Random disturbances are of real and cannot be avoided, but this is a way to evade the subtleties of the mathematics of stochastic variables. As importantly, it prevents a user from writing stochastic models that IdKit cannot integrate.

MoCaVa allows feedback between components. This may or may not create 'algebraic loops', which are resolved automatically, in principle by treating them as dynamic loops with infinitely short response times. In this way the order of causality will not be violated, and no formula manipulation will be needed. Hence, the original form of the model structure is retained during the processing, which facilitates debugging and generally helps the user to understand what is going on.

Modelling options

The modelling in MoCaVa is based on writing assignment statements. The form is a restricted set of Matlab M–statements (including `for`, `if` and `else` statements, as well as elementary mathematical functions and calls to other user–written dynamical functions).

Reuse of models is facilitated in two ways: *i*) A 'copy' option creates an identical component with a new name, and causes a window to be opened for changing the names of its input and output arguments. *ii*) The user may also define sub–functions with generic argument names (and in this way create a user's library), which are accessed by calling statements in the component definitions.

MoCaVa supports 'import' of model structures from Dymola. They must be of 'all continuous' type, that is, hybrid models cannot be processed, and neither can models containing serious discontinuities like 'dry friction' and 'dead–zone'. The names and object–oriented structure of the variables and parameters used in Dymola are also imported. This allows MoCaVa to use the Dymola model (`dsmodel.c`) as a 'computing engine', and means that only input parameters that need to be estimated, and only output variables that need to be fitted to data also need to be connected for processing by MoCaVa.

Expanding and refining model structures

MoCaVa builds larger models from a current, inadequate model in two possible ways:
• Expanding a model by appending components modelling separate process units (with sensors, actuators, and environment)
• Refining a model by augmenting components modelling internal physical phenomena.
Refinement is done gradually and not before necessary.

A component is connected automatically as soon as it has been activated (which is done my clicking in a menu of defined components). This facilitates the processing of alternatives, which is a clear advantage in the calibration of uncertain model structures.

MoCaVa produces a block–diagram of the system of currently activated components. It is not processed by MoCaVa, but serves as confirmation that the expanded structure is in accordance with the user's intentions.

Results

MoCaVa produces validated models in two respects:
1) The end product of the 'calibration' process is the simplest model (within the framework given by the user) that is not contradicted by data with a given confidence.

2) The result of the 'validation' process is a number of measures indicating whether or not a number of reduced models will be good enough for a number of given purposes. The options are briefly: *i*) Reproducibility (determines whether adaptation is needed or not). *ii*) Short–range or long–range prediction (determines whether feedback control is needed or not). *iii*) Prediction of measured and unmeasured variables (determines whether a sensor is needed or not). *iv*) Parameter accuracy (determines whether more data is needed or not).

Other features

MoCaVa comes with a HTML manual containing step–wise instructions and hints on how to run the calibration and validation sessions efficiently.

MoCaVa has also a run–time help function as support for the decisions and specifications the user has to provide for the interactive identification process to be efficient.

MoCaVa is open to the user, in the sense that physical variables are displayed and tentative sub–models retain their forms during the processing. This facilitates debugging of the user's components. Since this rules out automatic formula manipulation and reduction, efficient computing must be achieved by other means:
• Compiled C–routines are used (instead of M–statements) for the CPU–demanding tasks. (This also simplifies the assembling of sub–models into a system, since the C–linker provides the services needed.)
• Three routines supervise the on–line updating of the model's dynamics (= small–signal transition matrices), in order to take advantage of the fact that transition matrices of large models composed of many components tend to be sparse in three ways: They do not change at all times, many entries are zero, and all entries do not depend on all parameters.

MoCaVa consists of a number of independently executable script files for the various tasks that have to be handled in a calibration session, but employs the IKUS user's shell to run the tasks according to a procedure of proper calibration. The shell guarantees that scripts are executed in a theoretically correct order and input arguments are set properly. The shell also serves to administer a full log of the outcome of and status at each step in the session. Since the latter can be rather long in realistic cases, and the session status involves quite a number of variables, whose values depend on the history of the session, a user is recommended not to run the scripts directly, even though this is possible to do.

MoCaVa allows the user to suspend the session at any point, in order to return to Matlab, or to the operating system, or even to log out. He/she can then resume the session from the same point, or go back a selected number of steps. To allow this MoCaVa features a 'pilot', which shows a map of the calibration procedure and points at the current position. This is particularly useful, if the user wants to step back and try another decision.

The user's contribution

In summary, a user of MoCaVa must be able to do the following:
• Define one or more submodels in the form of state equations.
• Classify the input variables in the equations as either known input, disturbances, parameters, constants, or time.
• Assign sources of the known input, such as other models, or else selecting a rule for interpolating between sampled data input.
• Specify the type of disturbance input, *i.e.* pick a model from the library.
• Specify various attributes of some of the variables, such as dimensions, scales, nominal values, and ranges.
• Determine causality between components, *i.e.* the order of execution.
• Suggest one or more alternative submodels to augment, when the current tentative model structure has been rejected.
• Specify one or more purposes, if that is other than finding the simplest model that agrees with available data and prior knowledge.

A user does *not* have to specify:
• All submodels, in particular not those of disturbances,
• Which of the known submodels that are needed,
• The values of all parameters,
• The number of parameters to calibrate,
• A fitting criterion,
• A falsification criterion,
• A purpose, provided the simplest model that satisfies data and prior knowledge will suffice.

Case studies and applications

The purpose of the case studies has been to test the theory and the software being developed in parallel, and to see what problems may appear in trying to apply the theory to practice. The cases were selected as suitable to grey–box modelling, *i.e.* partial prior knowledge, substantial random (= unknown) input, and contaminated data. In most cases physical data from either pilot plants or full–scale production units have been analyzed using predecessors of MoCaVa. The following is a list of case studies done within the grey–box identification program at KTH.

Baker's yeast production. This is a biotechnical process. Data were collected during production of 'mother yeast' using a pilot plant at Jästbolaget in Sweden, normally used for that purpose and for experiments (Fan, 1990; Fan and Bohlin, 1989).

Steel rinsing. This is a process in industrial steel production. Data were collected from a full–scale production unit at the Domnarvet plant in Sweden. One of the four control variables was varied experimentally for the purpose of identification. (Sohlberg, 1991, 1993; Bohlin, 1991b, 1994b)

Mould level control in continuous casting. This is a process in industrial steel production. Data were collected from a full–scale production unit (Graebe, Elseley, and Goodwin, 1992).

Continuous pulp digester. This is an industrial pulp production process. Data were collected during experiments on a full–scale production unit at the SCA Wifsta–Östrand plant in Sweden. It is the largest process treated successfully so far: Five coupled nonlinear PDE with three internal stochastic disturbances where converted into a seventy–statesODE system, before being processed by IdKit (Funkquist, 1994, 1995).

Cement milling. This is a subprocess in industrial cement production. Data were collected during experiments on a full–scale production unit at Lafarge Canada, Richmond (Havelange, 1995).

Recovery boiler. This is an industrial process in pulp production, recovering chemicals and energy from the pulp digesting process. The main physical process is combustion. Data were collected during normal production at the Husum and Värö plants in Sweden (Bohlin, internal report, unpublished)

Bending stiffness of cardboard. This is the main quality variable of an industrial card–board manufacting process. Data were collected during normal production at the Frövi plant in Sweden (Bohlin, 1996; Pettersson, 1989)

EEG signals. This is a physiological process; the signals are obtained by placing electrodes on the human scalp. In essence, some hypotheses about the source of the electrical processes in the brain producing the signals were tested using the method in MoCaVa. Simulated data were used, in addition to physical data recorded at the University of Houston, Texas (Markusson and Bohlin, 1997, Markusson, 2002).

Drive train identification: In an on–going project (see other paper in this invited session) mechanical parameters are estimated for drive systems of rolling mills, paper machines etc. MoCaVa was used for rapid prototyping of the semi–physical parameter estimation. The end product, however, is written as a Matlab software directly dedicated to the drive train application.

Table 1 lists the number of variables involved in the case studies in order to give a measure of model size.

Table 1. Sizes of case models: Number of variables.

Case	*in*	*out*	*states*	*dist.*	*par.*
Baker's yeast	2	4	4	1	25
Steel rinsing	4	5	5	6	30
Pulp digester	2	3	70	3	25
Cement milling	2	2	2	0	6
Recovery boiler	7	2	1	1	6
Bending stiffness	16	2	8	2	20
EEG signals	0	2	19	2	9

References

Bohlin, T. (2001). *A Grey–BoxProcess Identification Tool: Theory and Practice.* Technical Report IR–S3–REG–0103, Royal Inst. Technology, Stockholm, Sweden.

Bohlin, T. (1996). Modelling the bending stiffness of card-board at the Frövifors plant. Report IR–S3–REG–9608, Royal Institute of Technology, Stockholm, Sweden.

Bohlin, T. (1994a). Derivation of a 'designer's guide' for interactive 'grey–box' identification of nonlinear stochastic objects. *Int, J. Control,* **5**, 1505–1524.

Bohlin, T. (1994b). A case study of Grey Box identification. *Automatica,* **30**, 307–318.

Bohlin, T. (1993). A Designer's Guide for Grey–BoxIdentification of Nonlinear Dynamic Systems with Random Disturbances. *IFAC 1993 World Congress,* Sydney, Australia, July 1993.

Bohlin, T. (1991a). *Interactive System Identification. Prospects and Pitfalls.* Springer–Verlag, Berlin, 1991.

Bohlin, T. (1991b). *Grey–Box Identification: A Case Study.* Report TRITA–REG 91/1, Automatic Control, Royal Institute of Technology, Stockholm, Sweden.

Bohlin, T. and S. Graebe (1994). Issues in nonlinear stochastic grey–box identification. *Int. J. Adaptive Control and Signal Processing.* **9** 465–490.

Fan, P. (1990). *Modelling and Control of a Fermentation Process.* Doctoral Thesis, TRITA–REG 90/4, Automatic Control, Royal Institute of Technology, Stockholm, Sweden.

Fan, P. and T. Bohlin (1989). Modelling, Estimation and Control of Baker's Yeast Production. *IEEE International Conference on Control and Application,* Jerusalem, Israel.

Camo A/S (1987). *Modelling & Identification.* Trondheim, Norway.

Funkquist, J. (1995). *Modelling and Identification of a Distributed Parameter Process: The Continuous Digester.* Doctoral thesis, TRITA–REG–9504, Automatic Control, Royal Institute of Technology, Stockholm, Sweden.

Funkquist, J. (1994). On Modeling and Identification of a Continuous Pulp Digester. SYSID94, Copenhagen, Denmark.

Graebe, S. F., G. Elsley, and G. C. Goodwin (1992). *Nonlinear identification and control of mold level oscillations in continuous bloom casting.* Report EE9204, Centre for Industrial Control Science, University of Newcastle, Australia.

Gupta, K. and D. Groshans and S. P. Houtchens (1993). $\mathrm{MATRIX_X}$. In D. A. Linkens (Ed.), *CAD for Control Systems.* Dekker, New York, 1993.

Havelange, O. (1995). *Grey Box Modelling of a Cement Milling Circuit.* Report IR–RT–EX–9510, Automatic Control, Royal Institute of Technology, Stockholm, Sweden.

Kristensen, N.R., H. Melgaard, and H. Madsen (2002). *CTSM 2.1 – User's Guide.* Technical University of Denmark, Lyngby, Denmark.

Leontaritis, I. J. and S. A. Billings (1987). Model selection and validation methods for non–linear systems. *Int. J. Control,* **45**, 311–341.

Ljung, L. (1987). *System Identification. Theory for the User.* Prentice–Hall, Englewood Cliffs, USA.

Markusson, O. (2002). *Model and System Inversion with Applications to Nonlinear System Identification and Control.* Doctoral thesis, Royal Institute of Technology, Stockholm, Sweden.

Markusson, O. and T. Bohlin (1997). *Identification of a nonlinear EEG–generating model.* IEEE Workshop on Nonlinear Signal and Image Processing, Michigan, USA. Report IR–S3–REG–9712, Automatic Control, Royal Institute of Technology, Stockholm, Sweden.

Pettersson, J. (1998). *On Model Based Estimation of Quality Variables for Paper Manufacturing.* Licentiate thesis, TRITA–S3–REG–9804, Royal Institute of Technology, Stockholm, Sweden.

Sohlberg, B. (1993). *Supervision and Control of a Steel Strip Rinsing Process.* Doctoral thesis, TRITA–REG–9302, Automatic Control, Royal Institute of Technology, Stockholm, Sweden.

Sohlberg, B. (1991). Computer aided modelling of a rinsing process, *IMACS Symposium MTCS,* Casablanca, May 1991.

Tiller, M. (2001). *Introduction to Physical Modeling with Modelica.* Kluwer Academic Publisher, Boston, 2001.

IFAC Publications
www.elsevier.com/locate/ifac

INITIALIZATION OF PHYSICAL PARAMETER ESTIMATES

Pablo A. Parrilo * and **Lennart Ljung** **

* *Automatic Control Laboratory, ETH-Zentrum, Zürich, Switzerland, email:* parrilo@aut.ee.ethz.ch
** *Division of Automatic Control, Linköping University, SE-58183, Linköping, Sweden, email:* ljung@isy.liu.se

Abstract: Grey box models of dynamical systems contain designated parameters with physical interpretation to be estimated from input-output data. This often gives distinct advantages over black-box models in terms of fewer parameters to estimate and hence better statistical accuracy. The basic theory for how this can be done is well established. The main practical obstacle may however be how the search for the estimates should be initialized. In this contribution we review the difficulties and point to a possibility to use semidefinite programming and a sum-of-squares formulation to achieve guaranteed consistent initial values for the physical parameters. *Copyright © 2003 IFAC*

1. GREY-BOX MODELS

A linear dynamic model of Grey-box character has the form

$$\begin{aligned} \dot{x} &= A(\theta)x + B(\theta)u + w \\ y &= C(\theta)x + e \end{aligned} \tag{1}$$

Here y and u are the measured output and input signals, x are the non-measured states, and w and e are non-measurable disturbances. The matrices A, B, C are matrices with partly known entries, but with some unknown parameters, denoted by θ. The problem is to estimate these parameters from sampled-data measurements of y and u.

In this paper we shall assume that there exists a true parameter vector θ_0 such that the data have been generated by (1) for this value:

$$\begin{aligned} \dot{x} &= A(\theta_0)x + B(\theta_0)u + w \\ y &= C(\theta_0)x + e \end{aligned} \tag{2}$$

The standard solution the linear grey-box problem is to form the Kalman predictor for the sampled data: Equation (1) is transformed to discrete time, in line with the intersample behavior of the input, and then the corresponding Kalman predictor is computed. This will require some assumptions about the noises w, e. These assumptions may or may not be parameterized by θ. The predictor then takes the form

$$\begin{aligned} \hat{x}(t+T) =& F(\theta)x(t) + G(\theta)u(t) \\ &+ K(\theta)(y(t) - C(\theta)\hat{x}(t)) \\ \hat{y}(t|\theta) =& C(\theta)\hat{x}(t) \end{aligned}$$

Here T is the sampling interval, and F, G, K are computed from A, B, C, T and the noise assumptions in a well known way. For example, if w is assumed to be 0, then $K = 0$, which gives the *Output error case.*

The parameter vector θ is then estimated by minimizing

$$V_N(\theta) = \sum_{t=1}^{N} \|y(t) - \hat{y}(t|\theta)\|^2 \tag{3}$$

with respect to θ. This is the *Prediction error method,* closely related to the *Maximum likelihood method.*

The globally minimizing value to (3) is the estimate, denoted by $\hat{\theta}_N$. We would like this estimator to be *consistent* i.e. that $\hat{\theta}_N$ converges to the true parameter value θ_0 as $N \to \infty$.

In general, the minimization must be done by iterative search of the type

$$\hat{\theta}_N^{(i+1)} = \hat{\theta}_N^{(i)} + \mu R_N^{(i)} V_N'(\hat{\theta}_N^{(i)}) \qquad (4)$$

See, e.g., (Ljung 1999) for a comprehensive treatment of all this.

The iterative search for the parameter estimate by minimization of (3) is inherent in all Maximum likelihood and prediction error methods for model structures that are not linear regressions (i.e. $\hat{y}(t|\theta)$ is a non-linear function of θ).

We shall in this article only treat the case that the matrices $A(\theta), B(\theta)$ and $C(\theta)$ are such that each entry is either known or a component of θ. Moreover, no links between the parameters in the different matrices are allowed. This corresponds exactly to the case of the model object `idss` with `SSParameterization = Structured` in the SYSTEM IDENTIFICATION TOOLBOX, (Ljung 2000).

The general case, where $A(\theta), B(\theta), C(\theta)$ are arbitrary functions of θ are treated using the model object `idgrey` in the Toolbox, but that case is not treated in this contribution.

Let us remark, that if there are p outputs, m inputs and n states, a maximum of $(p+m)n$ parameters associated with A, B, C can be estimated. This is the number of parameters in an identifiable canonical representation of the system. So, if the dimension of θ is larger than this number, the individual components of the parameter vector cannot be identified, and one can normally as well use a black-box model parameterization. In other words, the process knowledge represented by such a parameterization of (1) has no information value regarding the process dynamics.

2. INITIALIZATION PROBLEMS

The search method (4) requires an initial parameter value $\hat{\theta}_N^{(0)}$ where the search is initialized. Depending on the quality of this initial guess, the iterate $\hat{\theta}_N^{(i)}$ may, as $i \to \infty$, wander off to infinity, to a local minimum of V_N, get stuck, or converge to the desired value $\hat{\theta}_N$, i.e. the global minimum of V_N. In other words, we need to find a $\hat{\theta}_N^{(0)}$ that lies in the *domain of attraction* of the global minimum. This is what the current contribution is about.

Is this a pressing problem?

We have generated a number of random, stable systems (using `rss` in MATLAB) and simulated them with noise-free data. We have then randomly picked a certain number of parameters in these models that are to be considered as unknown, to be estimated. We have then minimized V_N 100 times starting at 100 different randomly chosen initial values of θ (but only such values that give stable models), and by Monte Carlo runs estimated the probability of ending up in the correct parameter values.

The results of these simulations are shown in figures 1 – 5.

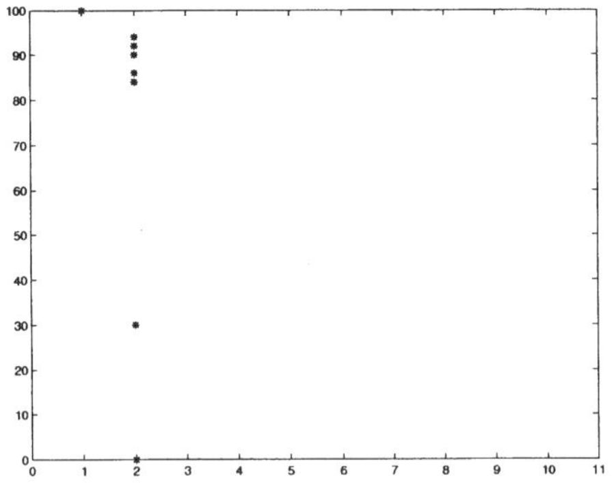

Fig. 1. The result of Monte Carlo runs on first order systems. Each star corresponds to a randomly generated system of order 1 with the number of unknown parameters given by the x-axis. The y-value is the success-rate in % for converging to the correct parameters from a random, stable initial model. This number was determined from 100 random initializations. There are 10 stars in each column, each corresponding to a certain system with given unknown parameters.

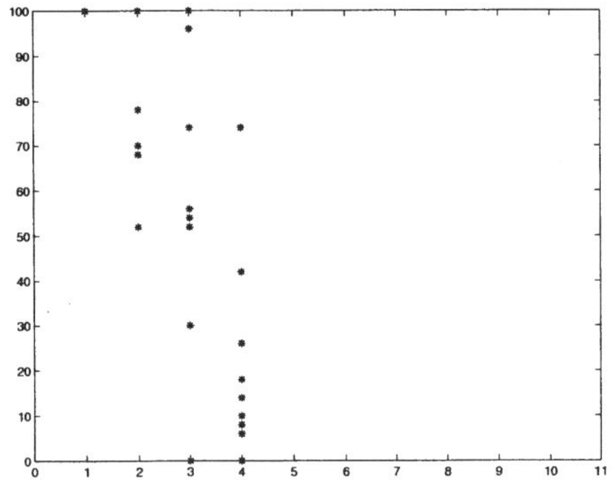

Fig. 2. As figure 1, but for second order systems.

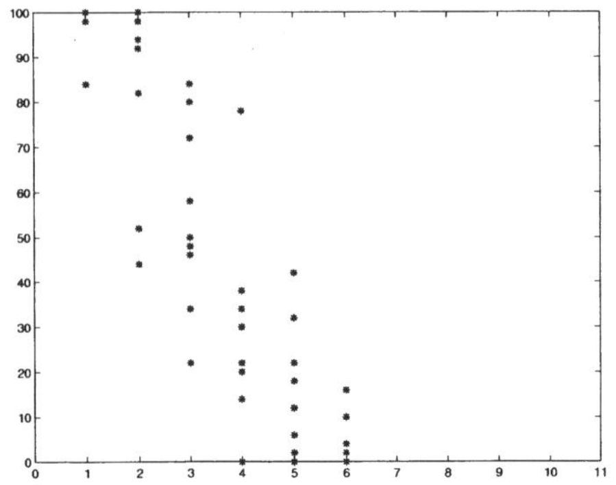

Fig. 3. As figure 1, but for third order systems.

The figures show that initialization in structured state-space models may be a highly non-trivial problem. For example, Figure 5 shows that for

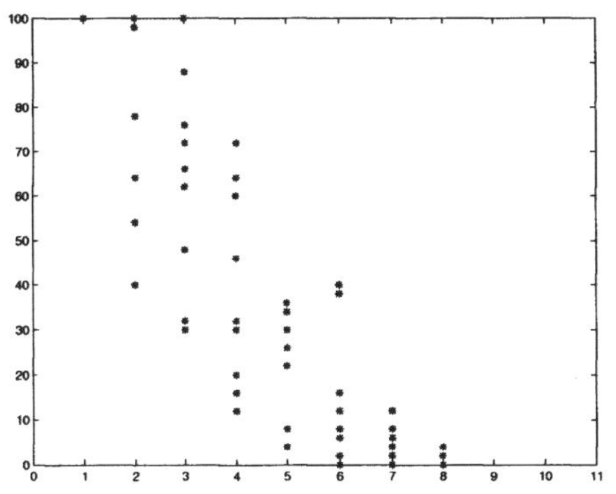

Fig. 4. As figure 1, but for fourth order systems.

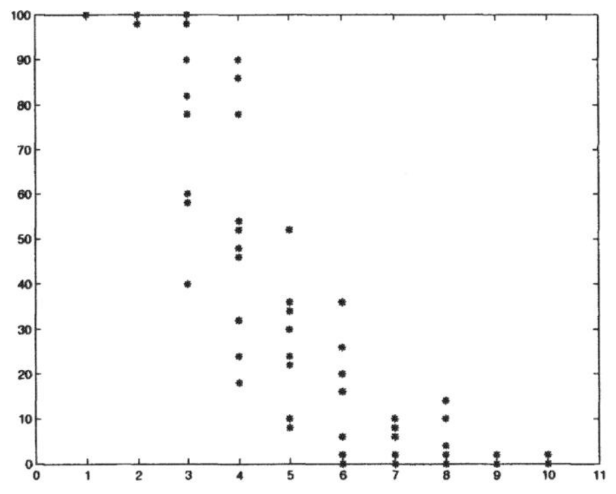

Fig. 5. As figure 1, but for fifth order systems.

most 5th order system with 10 unknown parameters the probability of convergence to the correct parameter vector from a random initial value is less than 2%. Some remedies that have been suggested include

- Use physical insight for the initialization. Since the parameters have physical significance, it should be possible to have good initial guesses. This is no doubt a very sensible approach. However, the domain of attraction of the global minimum may be small, and prior knowledge may not be sufficient to find it.
- Use global minimization techniques, such as simulated annealing or the genetic algorithm. Conceptually, these can be thought of as (more or less thoughtful) random restarts. The figures which in some cases show 0% success-rate over 100 random initializations give an idea of how difficult this may be.
- Use algebraic tools of different kinds. For example, in (Ljung and Glad 1994) it has been shown that any globally identifiable model structure can be rearranged (at the price of mis-handling the noise structure) to a linear regression. This regression could be used to estimate the initial parameter values using least squares. However, the computational burden in using Ritt's algorithm for this could be high.

What we seek is a procedure that, asymptotically, as $N \to \infty$ gives an initial estimate $\hat{\theta}_N^{(0)}$ that is guaranteed to lie in the domain of attraction of the global estimate $\hat{\theta}_N$. Since the size of the domain of attraction is not known, this means that this initial estimate $\hat{\theta}_N^{(0)}$ must itself be a consistent estimate, i.e. $\hat{\theta}_N^{(0)} \to \theta_0$ as $N \to \infty$. The initial estimate is then converted to an estimate $\hat{\theta}_N$ that is both consistent and efficient (having optimal variance properties) by the minimization scheme (4).

3. USING A BLACK-BOX MODEL FOR INITIALIZATION

What we need is consequently a consistent estimator for θ in (1), that does not depend on initial guesses. A natural question is if we can employ consistent, non-iterative methods for black-box model for this purpose. A black-box state-space model is given by

$$\dot{x} = Ax + Bu$$
$$y = Cx$$

without any particular structure for the matrices A, B and C.

So called *subspace methods*, e.g. (Larimore 1983), (Van Overschee and DeMoor 1996), (Ljung 1999), Chapter 10, have the advantage of producing generically consistent black-box estimates of state space systems, without any iterative search.

Thus a subspace method provides a consistent scheme to produce this system from data, as the data length tends to infinity. In other words, after conversion to continuous time, we will in the limit obtain state space matrices A_0, B_0, C_0 that realize the same input-output properties as (2).

That means that there exists an invertible matrix T, such that

$$TA(\theta_0) = A_0T$$
$$TB(\theta_0) = B_0$$
$$C(\theta_0) = C_0T$$

An obvious question is whether such estimates can be used to find an initial estimate $\hat{\theta}_0$ for the minimization of (3).

In some cases this may be straightforward. For example, if the numerators and denominators of

$$G(s,\theta) = C(\theta)(sI - A(\theta))^{-1}B(\theta)$$

turn out to be linear in the components of θ, then we can identify coefficients in this expression with

$$G_0(s) = C_0(sI - A_0)^{-1}B_0$$

In general, though, identifying such coefficients leads to a set of non-linear equations in θ. To solve these numerically may require iterative procedures, and then we are back to the problem of finding a good initial estimate.

4. BASIC SET-UP

The basic approach, as explained in (Xie and Ljung 2002), is to reformulate the question as a polynomial optimization problem, and attempt to solve:

$$(\hat{\theta}, \hat{T}) = \operatorname*{argmin}_{\theta, T} h(\theta, T) \tag{5a}$$

$$h(\theta, T) = \|T \cdot A - A_0(\theta) \cdot T\|_F^2 + \|T \cdot B - B_0(\theta)\|_F^2 + \|C - C_0(\theta)T\|_F^2 \tag{5b}$$

Here $\|\cdot\|_F^2$ denotes the Frobenius norm, i.e. the sum of all squared matrix elements. Under the stated assumptions, the function $h(\theta, T)$ is a multivariate polynomial in the estimates θ and the entries of the matrix T. Our approach to this minimization problem, as outlined in (Parrilo 2000), (Parrilo and Sturmfels 2003), will use sum of squares decompositions to provide a lower bound on the optimal value. If certain additional conditions on the numerical solution are satisfied, we will also obtain the optimal parameter values, corresponding to the global minimizer.

Optimization and sums of squares. We can obtain a lower bound the optimal value of $h(\theta, T)$ by finding the largest real number λ for which $h(\theta, T) - \lambda$ is a sum of squares of polynomials. This can be done by solving a semidefinite program.

We explain the generalities of the method next, referring the reader to the cited works for the details and further applications. The discussion is followed by some specific comments on how to exploit the particular algebraic structure of the polynomial that appears in our problem.

To find a sum of squares decomposition of a multivariate polynomial, we try to express $h(\theta, T) - \lambda$ as a quadratic form in a properly chosen set of variables $\mathbf{z}$, i.e.,

$$h(\theta, T) - \lambda = \mathbf{z}^* Q \mathbf{z}. \tag{6}$$

The choice of auxiliary variables $\mathbf{z}$ will depend on the structure of the polynomial $h(\theta, T)$: for instance, for a dense polynomial of degree $2d$, we would choose as the z_i all the monomials of degree d. In general, the variables z_i will not be algebraically independent, and therefore some quadratic relations (or *syzygies*) will exist among them. This implies the existence of an affine subspace of matrices Q for which (6) holds. This subspace will contain a positive semidefinite matrix if and only if the polynomial has a sum of squares representation. By the geometric interpretation of SDP, finding the maximum value of λ for which (6) holds is equivalent to the solution of a semidefinite program.

Example 1. Consider the nonconvex quartic polynomial in two variables $F(\theta, t)$ described below, and define $z_1 := 1, z_2 := \theta, z_3 := t, z_4 = t^2$:

$$F(\theta, t) - \lambda = \theta^2 - 4\theta t + 3t^4 - t^2 - \theta + 5 - \lambda$$

$$= \begin{bmatrix} 1 \\ \theta \\ t \\ t^2 \end{bmatrix}^T \begin{bmatrix} q_{11} & q_{12} & q_{13} & q_{14} \\ q_{12} & q_{22} & q_{23} & q_{24} \\ q_{13} & q_{23} & q_{33} & q_{34} \\ q_{14} & q_{24} & q_{34} & q_{44} \end{bmatrix} \begin{bmatrix} 1 \\ \theta \\ t \\ t^2 \end{bmatrix}.$$

In order for these two expressions to be identical, the following linear equalities should hold:

$$\begin{aligned} q_{11} &= 5 - \lambda, & 2q_{12} &= -1, & 2q_{13} &= 0 \\ q_{33} + 2q_{14} &= -1, & 2q_{23} &= -4, & 2q_{24} &= 0, \\ q_{22} &= 1, & 2q_{34} &= 0, & q_{44} &= 3. \end{aligned}$$

We want to maximize λ subject to these linear equations, with Q being positive semidefinite. The optimal λ, Q satisfying all these constraints can be found using semidefinite programming. The optimal solution in this case is given by $\lambda = 0.75$ and

$$Q = \begin{bmatrix} 4.25 & -0.5 & 0 & -3 \\ -0.5 & 1 & -2 & 0 \\ 0 & -2 & 5 & 0 \\ -3 & 0 & 0 & 3 \end{bmatrix}.$$

Factorizing $Q = L^T L$, we have the sum of squares decomposition:

$$\begin{aligned} F(\theta, t) &= 0.75 + (0.866025t - 0.866025t^2)^2 \\ &+ (0.970143\theta - 2.06155t - 0.363803t^2)^2 \\ &+ (2.06155 - 0.242536\theta - 1.45521t^2)^2. \end{aligned}$$

We have proved thus the lower bound $F \geq 0.75$. From the dual solution of the SDP, since it has rank one, we directly obtain the global minimizer $(\theta, t) = (2.5, 1)$. ◇

Exploiting structure. To achieve efficiency and reliability in the numerical implementation, it is important to exploit as much structure as possible in the polynomial to be minimized. For this, we notice that our objective function $h(\theta, T)$ in (5b) is a quartic polynomial in the variables θ_i and T_{jk}. However, since the original equations are bilinear, this implies that the polynomial $h(\theta, T)$ is actually *biquadratic*, having degree two in each of the $\theta_i T_{jk}$ (in other words, no quartic terms in any single variable appear).

As a consequence, in the formulation of the SDP instead of using *all* the monomials of degree less than or equal to 2 in T_{ij}, p_i — a total of $\binom{n^2+m+2}{2}$, only those in $\{1, \theta_i, T_{jk}, \theta_i \cdot T_{jk}\}$ will be needed, with a smaller total of $(1 + n^2)(1 +$

m). For instance, if $n = 4, m = 5$, the first number is 253, the second being only 102. In practice however, due to the internal structure of the parameterization, an even more reduced number of parameters may be actually needed. We illustrate these simplifications in Example 2. An important practical remark is that there may be a difference in computational terms between posing the problem as an optimization over T or over T^{-1}, since the structure of the polynomial may be simpler in one case versus the other. We also notice that even though nonnegative polynomial are not necessarily sum of squares in the general case, in this specific application the objective function has exactly this form.

Example 2. The physically-based parameterization is given by:

$$A_0 = \begin{bmatrix} 0 & 1 \\ 0 & \theta_1 \end{bmatrix}, \quad B_0 = \begin{bmatrix} 0 \\ \theta_2 \end{bmatrix}, \quad C_0 = \begin{bmatrix} 1 & 0 \end{bmatrix}. \tag{7}$$

The original matrices, estimated using subspace methods, are:

$$A = \begin{bmatrix} -4.125 & -0.111 \\ -1.7815 & -0.114 \end{bmatrix} \quad B = \begin{bmatrix} 1.3068 \\ 0.4189 \end{bmatrix}$$
$$C = \begin{bmatrix} 0.5518 & -1.669 \end{bmatrix}.$$

Since there are two uncertain parameters, and the state dimension is two, this corresponds in our notation to the case $n = 2, m = 2$. The procedure described earlier would generate an SDP with matrices of size equal to 15×15. However, due to the structure of the matrices in (7), only some of the bilinear terms appear. As a consequence, it is possible to further reduce the problem to an 8×8 semidefinite program indexed by the variables:

$$\theta_1, \theta_2, T_{11}, T_{12}, T_{21}, T_{22}, \theta_1 T_{21}, \theta_1 T_{22}.$$

After solving the corresponding SDPs, we obtain the optimal parameters and similarity transform:

$$\theta_1 = -4.1871, \quad \theta_2 = 0.9791, \tag{8}$$

$$T = \begin{bmatrix} 0.5405 & -1.6657 \\ 0.7407 & 0.0263 \end{bmatrix}. \tag{9}$$

◇

Results. The method presented works fairly well in small problems. When there are too many parameters and/or states (and here "too many" is around 20-24 unknowns), the SDP/SOS method slows down considerably due to the large size of the matrices. This is partly a consequence of the use of standard interior-point solvers for solving the corresponding SDPs.

A negative feature of the approach described is that sometimes it may produce answers that, while mathematically correct and optimal, are not useful in terms of the original problem. This can be traced back to the choice of objective function in (5b). Even if we find the globally optimal matrix T that minimizes it (as our method usually does), this may not produce (in theory and/or practice) a "good" coordinate transformation. The reason for this is that we are not weighting enough the possibility of T being close to singularity. To give a concrete example, for the instance analyzed in (Xie and Ljung 2002), an "optimal" set of parameters θ can be obtained: $[0, 0, 0.1639, 0]$. However, the corresponding matrix T achieves a nearly zero objective value, but is very close to singularity with the estimated parameters being far off the "true" ones.

A possibility we are exploring is to modify the objective function given in (5b) by an alternative expression that penalizes directly a measure of the deviation between the models. Natural choices in this regard are system norms such as the 2- or ∞-norm of the difference between the corresponding transfer functions, i.e., $\|G(s,\theta) - G_0(s)\|$. In the 2-norm case, for instance, this is a rational function of the unknown parameters. As mentioned earlier, in general this will give rise to a nonconvex problem, but one that we can also attempt to globally minimize using SOS/SDP techniques.

A clear advantage of this approach is that we get rid of all the T_{ij}, ending up with a much smaller number of decision variables (equal to the number of parameters θ_i), though the resulting polynomials will in general have higher degree. In this framework, it would also be possible to add additional constraints on the parameters, for instance to restrict them to a range where they are physically meaningful.

5. CONCLUSIONS

We have pointed to one way of initializing the (prediction error) search for physical parameters in structured linear dynamical models. It gives a solution the problem of consistent initializations and of local minima of the criterion function. The SOS problem gives a consistent estimate of the parameters, as more and more data become available and thus the estimated A, B, C come close, within a similarity transformation to the true system matrices. This consistent estimate will thus approach the domain of attraction of the true parameter values, and the estimate will then have all the asymptotically optimal properties of ML and prediction error estimates.

While the methods as presented work satisfactorily for small problems, further research is needed towards obtaining improved problem formulations, as well as enhancing the computational efficiency of the SOS-based approach.

6. REFERENCES

Larimore, W. E. (1983). System identification, reduced order filtering and modelling via canonical variate analysis. In: *Proc 1983 American Control Conference*. San Francisco.

Ljung, L. (1999). *System Identification - Theory for the User*. 2nd ed.. Prentice-Hall. Upper Saddle River, N.J.

Ljung, L. (2000). *System Identification Toolbox for use with* MATLAB. *Version 5.*. 5th ed.. The MathWorks, Inc. Natick, MA.

Ljung, L. and T. Glad (1994). On global identifiability of arbitrary model parameterizations. *Automatica* **30**(2), pp 265–276.

Parrilo, P. A. (2000). Structured semidefinite programs and semi-algebraic geometry methods in robustness and optimization. PhD thesis. California Institute of Technology. Available at `http://www.cds.caltech.edu/~pablo/`.

Parrilo, P. A. and B. Sturmfels (2003). Minimizing polynomial functions. In: *Algorithmic and Quantitative Real Algebraic Geometry* (S. Basu and L. González-Vega, Eds.). Vol. 60 of *DIMACS Series in Discrete Mathematics and Theoretical Computer Science*. Preprint available from `arXiv:math.OC/0103170`.

Van Overschee, P. and B. DeMoor (1996). *Subspace Identification of Linear Systems: Theory, Implementation, Applications*. Kluwer Academic Publishers.

Xie, L. L. and L. Ljung (2002). Estimate physical parameters by black box modeling. In: *Proc. 21st Chinese Control Conference*. Hangzhou, China. pp. 673–677.

IFAC Publications
www.elsevier.com/locate/ifac

PARAMETER ESTIMATION IN LINEAR DIFFERENTIAL-ALGEBRAIC EQUATIONS [1]

Markus Gerdin, Torkel Glad and Lennart Ljung

Division of Automatic Control
Department of Electrical Engineering
Linköpings universitet, SE-581 83 Linköping, Sweden
{gerdin,torkel,ljung}@isy.liu.se

Abstract: This paper describes how parameter estimation can be performed in linear DAE systems. Both time domain and frequency domain identification are examined. The results are exemplified on a small system. A potential application for the algorithms is to make parameter estimation in models generated by a modeling language like Modelica. *Copyright © 2003 IFAC*

Keywords: Descriptor systems, Implicit systems, Singular systems, Parameter estimation, Identification, Parameter identification.

1. INTRODUCTION

In recent years so-called object-oriented modeling languages have become increasingly popular. Examples of such languages are Omola, Dymola and Modelica, (Mattsson *et al.*, 1998; Tiller, 2001). Modeling languages of this type make it possible to build models by connecting submodels in a manner that parallels the physical construction. A consequence of this viewpoint is that it is usually not possible to specify in advance what variables are inputs or outputs from a given submodel. A further consequence of this is that the resulting model is not in state space form. Instead the model is a collection of equations, some of which contain derivatives and some of which are static relations. A model of this form is sometimes referred to as a DAE (*differential algebraic equations*) model or a *descriptor model*. It can be noted that these models are a special case of the so-called behavioral models discussed in, e.g., (Polderman and Willems, 1998). When constructing models of this form it is common that some parameters have unknown values and have to be determined from identification experiments. It is thus of interest to examine how identification can be performed for models of this form. In the present paper some issues of principle will be addressed and a concrete example will illustrate the conclusions.

When a model has been defined in an object oriented language, one can mark the parameters that are to be identified, mark the inputs and outputs that are used in an identification experiment and give the corresponding data files. Ideally this should be enough, so that then the transformation into a form suitable for identification should be automatic.

Here only linear models will be considered. The object oriented languages discussed above are equation oriented. This means that all models consist of systems of equations (or descriptions that can be converted to systems of equations). The equations relate the physical variables and possibly first order derivatives of them. This means that a model can be described in the form

$$E\dot{\xi}(t) = J\xi(t) + Ku(t) \tag{1a}$$
$$y(t) = L\xi(t) \tag{1b}$$

where ξ is a vector of physical variables, u is the input and y is the measured output. The matrices E, J, K and L contain physical parameters that describe the system. Some of these parameters are those that are to be estimated from an identification experiment. The model above is not in state space form but forms a DAE.

[1] Supported by the Swedish Research Council and by the Foundation for Strategic Research (SSF).

If a system is modeled by just writing down the equations describing its different parts, the matrix E is typically non-invertible so that the extraction of a state space description is not completely straightforward. The goal of the present paper is to discuss how this model can be transformed to a form where well-known identification techniques can be directly applied. In particular we will discuss what the implications are for the noise models.

Previously, not very much work has been done on parameter identification in DAE systems. One of the works on the nonlinear case is (Schittkowski, 2002).

2. DAE SYSTEMS

The model structure that will be examined is thus (1). It will be assumed that $\det(sE-J)$ is not zero for all s. If the determinant were zero for all s, it can be realized by Laplace transforming the equations that $\xi(t)$ will not be uniquely determined by $u(t)$ so this is a reasonable assumption. Given this assumption, it is possible to transform (1) to a form which resembles a normal state space description. Transformations have been suggested in the literature, see for example (Dai, 1989; Brenan *et al.*, 1996) and references in (Verghese *et al.*, 1981). Here the one derived below will be used:

Lemma 1. If $\det(sE-J) \not\equiv 0$, the solution of (1) can be described by the *normal form*

$$\dot{w}_1(t) = Aw_1(t) + B_1u(t) \tag{2a}$$

$$w_2(t) = \sum_{i=0}^{m-1}(-N)^i D_1 u^{(i)}(t) \tag{2b}$$

$$\xi(t) = T^{-1}\begin{bmatrix} w_1(t) \\ w_2(t) \end{bmatrix} \tag{2c}$$

$$y(t) = LT^{-1}\begin{bmatrix} w_1(t) \\ w_2(t) \end{bmatrix}. \tag{2d}$$

PROOF. It follows e.g. from (Brenan *et al.*, 1996) that there exists invertible matrices P and T such that the transformation $PET\dot{\xi} = PJT\xi + PKu$ gives the form (dependence on t is here omitted in the notation)

$$\begin{bmatrix} I & 0 \\ 0 & N \end{bmatrix}\begin{bmatrix} \dot{w}_1 \\ \dot{w}_2 \end{bmatrix} + \begin{bmatrix} -A & 0 \\ 0 & I \end{bmatrix}\begin{bmatrix} w_1 \\ w_2 \end{bmatrix} = \begin{bmatrix} B_1 \\ D_1 \end{bmatrix} u \tag{3a}$$

$$\xi = T^{-1}\begin{bmatrix} w_1 \\ w_2 \end{bmatrix} \tag{3b}$$

where N is nilpotent, that is $N^m = 0$ for some m. Now, if $N = 0$ then $w_2(t) = D_1u(t)$ and the proof is done. If $N \neq 0$ the second row of (3a) is multiplied with N to get

$$N^2\dot{w}_2(t) + Nw_2(t) = ND_1u(t). \tag{4}$$

Then (4) is differentiated and the second row of (3a) is inserted. This gives

$$w_2(t) = D_1u(t) - ND_1\dot{u}(t) + N^2\ddot{w}_2(t). \tag{5}$$

If $N^2 = 0$ the proof is done, otherwise the process is continued until $N^m = 0$ (this is true for some m as N is nilpotent). The resulting expression is then

$$w_2(t) = \sum_{i=0}^{m-1}(-N)^i D_1 u^{(i)}(t) \tag{6}$$

and the proof is complete. □

This result is consistent with the result in (Rosenbrock, 1970) that all linear systems are equivalent to a state space description, possibly with derivatives of the input added to the output. Note that the transformation in Lemma 1 may not be suitable for systems where the matrices E, J, K and L change abruptly, see further (Verghese *et al.*, 1981).

3. DISCRETIZATION

When doing system identification, the measured data that is to be used to identify the parameters is usually available as sampled data in the time domain, so in this section it will be examined what the exact discrete time description for a linear DAE system is under certain assumptions on the input.

It is a standard result, e.g., (Åström and Wittenmark, 1984), that the solution of a continuous-time state space system can be described by a discrete-time state space system if the input is piecewise constant, or in general is completely determined by the values in the sample points. In the DAE case the input cannot be assumed to be piecewise constant as the system according to Lemma 1 may depend on derivatives of the input. Instead it will be assumed that the $m-1$:th derivative of the input is piecewise constant and that the system is on the form (2) which is possible according to Lemma 1. This will make it possible to use the result for normal state space systems, and the first step of the derivation is to define a vector with the input and some of its derivatives:

$$w_3(t) = \begin{bmatrix} u(t)^T & \dot{u}(t)^T & \dots & u^{(m-2)}(t)^T \end{bmatrix}^T \tag{7}$$

This leads to the relation (8). (In the rest of this section dependence on the time t will be omitted if the dependence is clear.)

$$\dot{w}_3 = \underbrace{\begin{bmatrix} 0 & I & \dots & 0 \\ \vdots & \vdots & \ddots & \vdots \\ 0 & 0 & \dots & I \\ 0 & 0 & \dots & 0 \end{bmatrix}}_{\Psi_1} w_3 + \underbrace{\begin{bmatrix} 0 \\ \vdots \\ 0 \\ I \end{bmatrix}}_{\Psi_2} u^{(m-1)} \tag{8}$$

Rewriting (2) using w_3 gives

$$\dot{w}_1 = Aw_1 + \underbrace{\begin{bmatrix} B_1 & 0 & \dots & 0 \end{bmatrix}}_{\Psi_3} w_3 \tag{9a}$$

$$w_2 = \underbrace{\left[I \ (-N) \ \dots \ (-N)^{m-2} \right] D_1}_{\Psi_4} w_3$$

$$+ \underbrace{(-N)^{m-1} D_1}_{\Psi_5} u^{(m-1)} \quad (9b)$$

$$\dot{w}_3 = \Psi_1 w_3 + \Psi_2 u^{(m-1)} \quad (9c)$$

$$y = LT^{-1} \begin{bmatrix} w_1 \\ w_2 \end{bmatrix} \quad (9d)$$

If w_2 is eliminated, these equations can be seen as a normal state space equation with $u^{(m-1)}$ as the input:

$$\begin{bmatrix} \dot{w}_1 \\ \dot{w}_3 \end{bmatrix} = \underbrace{\begin{bmatrix} A & \Psi_3 \\ 0 & \Psi_1 \end{bmatrix}}_{\tilde{A}} \begin{bmatrix} w_1 \\ w_3 \end{bmatrix} + \underbrace{\begin{bmatrix} 0 \\ \Psi_2 \end{bmatrix}}_{\tilde{B}} u^{(m-1)} \quad (10a)$$

$$y = \underbrace{LT^{-1} \begin{bmatrix} I & 0 \\ 0 & \Psi_4 \end{bmatrix}}_{\tilde{C}} \begin{bmatrix} w_1 \\ w_3 \end{bmatrix} + \underbrace{LT^{-1} \begin{bmatrix} 0 \\ \Psi_5 \end{bmatrix}}_{\tilde{D}} u^{(m-1)}$$

(10b)

Defining $x = \left[w_1^T \ w_3^T \right]^T$ this can be written in the compact form

$$\dot{x} = \tilde{A}x + \tilde{B}u^{(m-1)} \quad (11a)$$

$$y = \tilde{C}x + \tilde{D}u^{(m-1)}. \quad (11b)$$

Now, if it is assumed that $u^{(m-1)}$ is piecewise constant, the result for discretization of state space systems can be applied to (11) to give an exact discrete time description of the original time DAE system (1). The above discussion proves the following result:

Result 1. Consider the DAE system (1) with the normal form (2). If $u^{(m-1)}(t)$ is constant for $T_s k \le t < T_s k + T_s$ for constant T_s and $k = 0, 1, 2, \dots$, then $y(T_s k)$ is exactly described by the discrete time state space system

$$x(T_s k + T_s) = \Phi x(T_s k) + \Gamma u^{(m-1)}(T_s k) \quad (12a)$$

$$y(T_s k) = \tilde{C}x(T_s k) + \tilde{D}u^{(m-1)}(T_s k). \quad (12b)$$

where

$$\Phi = e^{\tilde{A}T_s}, \Gamma = \int_0^{T_s} e^{\tilde{A}\tau} d\tau \tilde{B}. \quad (13)$$

Note that there are other assumptions on the behavior of $u^{(m-1)}$ between the sample points which also makes it possible to calculate an exact discrete time description, for example that it is piecewise linear.

4. NOISE MODEL

Systems are often affected by other factors than the known inputs. These factors can from a modeling perspective be seen as unmeasured inputs. It is also possible that the measurement $y(t)$ is corrupted by noise. These effects can be included by adding a *noise model*.

In the case of linear DAE systems a noise model can be added according to (14), where $v_1(t)$ represents the unmodeled inputs and $v_2(t)$ represents the measurement noise. M is a constant matrix.

$$E\dot{\xi}(t) = J\xi(t) + Ku(t) + Mv_1(t) \quad (14a)$$

$$y(t) = L\xi(t) + v_2(t) \quad (14b)$$

(14) can also be written as

$$E\dot{\xi}(t) = J\xi(t) + \left[K \ M \right] \begin{bmatrix} u(t) \\ v_1(t) \end{bmatrix} \quad (15a)$$

$$y(t) = L\xi(t) + v_2(t). \quad (15b)$$

Under certain assumptions, it is possible to make a transformation to a state space system as in the case without noise model. The first step to achieve this is to apply Lemma 1 to (15) get that the solution can be described by the normal form (in the rest of this section dependence of t will be omitted from the notation if the dependence is clear)

$$\dot{w}_1 = Aw_1 + B_1 u + B_2 v_1 \quad (16a)$$

$$w_2 = \sum_{i=0}^{m-1} (-N)^i D_1 u^{(i)} + \sum_{i=0}^{m-1} (-N)^i D_2 v_1^{(i)} \quad (16b)$$

$$y = LT^{-1} \begin{bmatrix} w_1 \\ w_2 \end{bmatrix} + v_2. \quad (16c)$$

A normal assumption when E is the identity matrix (i.e. in the case of a normal state space description) is that $v_1(t)$ and $v_2(t)$ are white noise signals. Sometimes it is also assumed that they have a Gaussian distribution. The reason is usually that the state estimation problem then is easy (the solution is obtained by the Kalman filter). To estimate the states of (14) it is desirable to make the same assumption here. Time continuous white noise signals are however delicate mathematical objects, for example the integral and derivative are not well defined in the normal sense. See further (Åström, 1970). Therefore one has to consider carefully what (14) means if $v_1(t)$ and $v_2(t)$ are assumed to be white noise signals. Furthermore derivatives of $v_1(t)$ cannot be allowed to occur in (16), that is it must be required that

$$ND_2 = 0. \quad (17)$$

Note that (17) is related to the so called *impulse controllability*, see for example (Dai, 1989), with respect to $v_1(t)$, but this viewpoint is not further pursued here.

Now, if it is assumed that the matrix M in (14) is such that (17) is fulfilled (see (Schön *et al.*, 2003) for a discussion on the relation between (17) and M), (16) can be written as

$$\dot{w}_1 = Aw_1 + B_1 u + B_2 v_1 \quad (18a)$$

$$w_2 = \sum_{i=0}^{m-1} (-N)^i D_1 u^{(i)} + D_2 v_1 \quad (18b)$$

$$y = LT^{-1} \begin{bmatrix} w_1 \\ w_2 \end{bmatrix} + v_2. \quad (18c)$$

Now (18) is to be transformed to a state space description with $u^{(m-1)}$ as the input using the same method as in section 3. The vector w_3 is thus defined according to (7), which gives the description

$$\dot{w}_1 = Aw_1 + \Psi_3 w_3 + B_2 v_1 \tag{19a}$$

$$w_2 = \Psi_4 w_3 + \Psi_5 u^{(m-1)} + D_2 v_1 \tag{19b}$$

$$\dot{w}_3 = \Psi_1 w_3 + \Psi_2 u^{(m-1)} \tag{19c}$$

$$y = LT^{-1} \begin{bmatrix} w_1 \\ w_2 \end{bmatrix} + v_2. \tag{19d}$$

Eliminating $w_2(t)$ and letting $x = \begin{bmatrix} w_1^T & w_3^T \end{bmatrix}^T$ gives

$$\dot{x} = \tilde{A}x + \tilde{B}u^{(m-1)} + \underbrace{\begin{bmatrix} B_2 \\ 0 \end{bmatrix}}_{\tilde{B}_2} v_1 \tag{20a}$$

$$y = \tilde{C}x + \tilde{D}u^{(m-1)} + \underbrace{LT^{-1} \begin{bmatrix} 0 \\ D_2 \end{bmatrix}}_{\tilde{N}} v_1 + v_2 \tag{20b}$$

which can be written as

$$\dot{x} = \tilde{A}x + \tilde{B}u^{(m-1)} + \tilde{B}_2 v_1 \tag{21a}$$

$$y = \tilde{C}x + \tilde{D}u^{(m-1)} + \begin{bmatrix} \tilde{N} & I \end{bmatrix} \begin{bmatrix} v_1 \\ v_2 \end{bmatrix}. \tag{21b}$$

It is thus possible to describe the solutions of a DAE system with noise model (14) with a state space system under the assumption (17) that $N D_2 = 0$. Note however, that in the state space model, the noise at the output equation is related to the noise at the state equation through the $v_1(t)$ term. This relation can however be eliminated if the stronger assumption that $D_2 = 0$ is made. Then $\tilde{N} = 0$ so the state space description simplifies to

$$\dot{x} = \tilde{A}x + \tilde{B}u^{(m-1)} + \tilde{B}_2 v_1 \tag{22a}$$

$$y = \tilde{C}x + \tilde{D}u^{(m-1)} + v_2. \tag{22b}$$

Here, the noise at the state and output equations are related only if v_1 and v_2 are.

4.1 *Discretization*

Also when a noise model is included, it is interesting to examine what the discrete time description of the DAE system is. If the noise is assumed to be piecewise constant, Result 1 can be used to obtain a discrete time representation. It can also be assumed that $v_1(t)$ and $v_2(t)$ is *white noise* in (21) and (22). For this case formulas which can transform the state space descriptions (21) and (22) to discrete time are available in for example (Åström, 1970) and are not repeated here.

4.2 *A shortcut*

It may be a substantial work to examine what the corresponding state space and discrete time descriptions of (14) is. If there is no prior knowledge about the noise model, and it is desired to identify it as a black box model, it might be better to add the noise model after the conversion to the state space form (11), or even after discretization of the system. Then the noise model does not need to be converted and the assumption (17) does not have to be considered.

5. KALMAN FILTERING

It is now established that it under some assumptions is possible to transfer a continuous time DAE system to a discrete time state space system which gives an equivalent description of the output at the sampling instants. This opens up the possibility to use a discrete time *Kalman filter* to estimate the states and make predictions. To be concrete, assume that a DAE system has been converted to the state space system (23). It is then straightforward to implement a discrete time invariant Kalman filter, e.g., (Anderson and Moore, 1979).

$$x(t+1) = Ax(t) + Bu(t) + Nv_1(t) \tag{23a}$$

$$y(t) = Cx(t) + Du(t) + v_2(t). \tag{23b}$$

Note that a discretized DAE system just is a special case of this model with proper time scaling and redefinitions. For example $u^{(m-1)}(t)$ is renamed to $u(t)$.

It can be noted that the approach to Kalman filtering for DAE models suggested in this section is related to the approach for discrete time descriptor systems which is presented in (Dai, 1987).

6. TIME DOMAIN IDENTIFICATION

As it has been shown that the DAE model under some assumptions can be described by a discrete time state space system, a large amount of theory for system identification is available, e.g., (Ljung, 1999). The probably most common method available is the prediction error method with a quadratic criterion. This method is based on using the Kalman filter as a one-step ahead predictor and then minimizing the prediction error. It is also possible to do maximum likelihood estimation if $v_1(t)$ and $v_2(t)$ are assumed to have a Gaussian distribution. Then the Kalman filter is used to calculate the probability distribution of the output, and the parameters are estimated by maximizing the probability of the measured output.

7. FREQUENCY DOMAIN IDENTIFICATION

The work which has been done until now has been based on transforming the DAE system to the so called normal form. Although the normal form always exists if $\det(sE - J) \not\equiv 0$, there may be problems to calculate it, for example one might encounter numerical difficulties. The state space form which the normal form produces might also not always be well suited for estimation purposes, especially if the input has to be differentiated numerically.

A way around these problems is identification in the frequency domain in continuous time, where the model should be specified by transfer functions according to (24).

$$Y(\omega) = G(i\omega)U(\omega) + H(i\omega)V_1(\omega) + V_2(\omega). \quad (24)$$

A linear DAE system with a noise model (14) can be transformed to the form (24) under the usual assumption that $\det(sE - J) \not\equiv 0$. The only difference from the transfer functions of a state space system is that G and H may not be proper in the general case. Therefore $u(t)$ must be differentiable sufficiently many times, which of course is as many times as it must be differentiated in the normal form. Furthermore, if $v_1(t)$ is assumed to be white noise, the M matrix must be such that H is proper. To transform (14) to the frequency domain the equations are Fourier transformed to get

$$Ei\omega\xi(\omega) = J\xi(\omega) + KU(\omega) + MV_1(\omega) \quad (25a)$$
$$Y(\omega) = L\xi(\omega) + V_2(\omega). \quad (25b)$$

The desired description (26) is then achieved by eliminating $\xi(\omega)$.

$$Y(\omega) = \underbrace{L(i\omega E - J)^{-1}K}_{G(i\omega)}U(\omega) + \underbrace{L(i\omega E - J)^{-1}M}_{H(i\omega)}V_1(\omega) + V_2(\omega) \quad (26)$$

Now, when it is established that a DAE system can be described by transfer functions, all that has to be done is to plug the transfer function into any identification algorithm for the frequency domain. Books which treat identification in the frequency domain are (Ljung, 1999; Pintelon and Schoukens, 2001). One possible selection of identification method is to minimize the criterion

$$\sum_{k=1}^{N} |Y(\omega_k) - G(i\omega_k)U(\omega_k)|^2 \cdot \frac{1}{|H(i\omega_k)|^2} \quad (27)$$

with respect to the unknown parameters.

Note that the Fourier transforms of the input and output signals are needed, but these can be computed from time domain data.

8. AN EXAMPLE

In this section the algorithms examined earlier in the paper are exemplified on a physical system. The system setup examined is a model of a generator, see Figure 1. Dependence on time is omitted in this section.

The model can be described as follows: The input is the angle θ_{in} on the left axis. This axis is connected to a rotating mass with inertia J which is rotated the angle θ and rotates with the angular velocity ω. The torque acting on the left side of the mass is M_1 and the torque on the right side is M_2. The mass is then

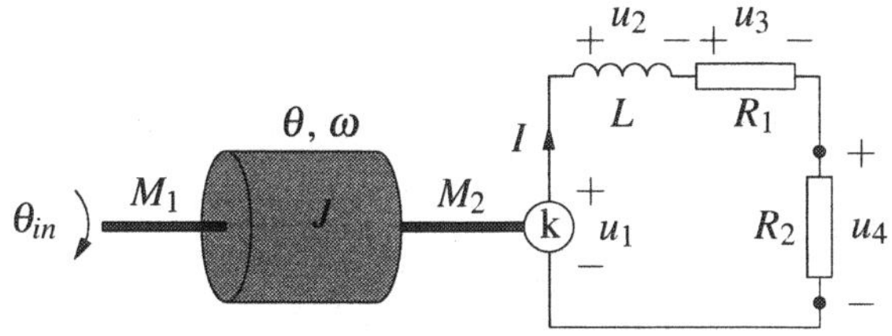

Fig. 1. A model of a generator.

connected to a second axis which is connected to the actual generator. The variables describing the second axis and the electrical quantities are then assumed to depend on each other according to $M_2 = kI$ and $u_1 = k\omega$ for some constant k. The rest of the electrical circuit consists of two resistors and one inductor. The measured output is the voltage u_4.

Note that it here is easy to spot that too many variables are included in the model if the torques are not interesting, as the mass then will have no impact on the system. This is to show that the identification algorithm treats this automatically. (In larger models it may be more difficult to spot that some variables are unnecessary.)

The system is now modeled in an object-oriented manner by just writing down the equations describing the different parts and the ways they are connected:

$\theta_{in} = \theta$	$\dot{\theta} = \omega$	$J\dot{\omega} = M_1 + M_2$
$M_2 = kI$	$u_1 = k\omega$	$u_2 = L\dot{I}$
$u_3 = R_1 I$	$u_4 = R_2 I$	$u_1 = u_2 + u_3 + u_4$

To apply the identification methods examined in Section 2, these equations must be written on the form (1):

$$\begin{pmatrix} 0&0&0&0&0&0&0&0&0 \\ 0&0&0&1&0&0&0&0&0 \\ 0&0&J&0&0&0&0&0&0 \\ 0&0&0&0&0&0&0&0&0 \\ 0&0&0&0&0&0&0&0&0 \\ 0&0&0&0&L&0&0&0&0 \\ 0&0&0&0&0&0&0&0&0 \\ 0&0&0&0&0&0&0&0&0 \\ 0&0&0&0&0&0&0&0&0 \end{pmatrix} \begin{pmatrix} \dot{M}_1 \\ \dot{M}_2 \\ \dot{\omega} \\ \dot{\theta} \\ \dot{I} \\ \dot{u}_1 \\ \dot{u}_2 \\ \dot{u}_3 \\ \dot{u}_4 \end{pmatrix} +$$

$$\begin{pmatrix} 0&0&0&1&0&0&0&0&0 \\ 0&0&-1&0&0&0&0&0&0 \\ -1&-1&0&0&0&0&0&0&0 \\ 0&1&0&0&-k&0&0&0&0 \\ 0&0&-k&0&0&1&0&0&0 \\ 0&0&0&0&0&0&-1&0&0 \\ 0&0&0&0&-R_1&0&0&1&0 \\ 0&0&0&0&-R_2&0&0&0&1 \\ 0&0&0&0&0&1&-1&-1&-1 \end{pmatrix} \begin{pmatrix} M_1 \\ M_2 \\ \omega \\ \theta \\ I \\ u_1 \\ u_2 \\ u_3 \\ u_4 \end{pmatrix} = \begin{pmatrix} 1 \\ 0 \\ 0 \\ 0 \\ 0 \\ 0 \\ 0 \\ 0 \\ 0 \end{pmatrix} \theta_{in}$$

$$y = \begin{pmatrix} 0&0&0&0&0&0&0&0&1 \end{pmatrix} \begin{pmatrix} M_1 \\ M_2 \\ \omega \\ \theta \\ I \\ u_1 \\ u_2 \\ u_3 \\ u_4 \end{pmatrix}$$

8.1 *Time domain identification*

The first step when identifying in the time domain is to transform the model to the form (11) (here, no noise model is added before the transformation). In an automated identification procedure, this transformation is probably best made numerically for each value of the parameters that the identification algorithm tries during the minimization. Stable numerical algorithms for making this computation is a subject for future research.

In this simple case, it is however possible to make the transformation to a state space system by algebraic manipulations. If it is assumed that $L \neq 0$, the transformed system is

$$\dot{x} = -\frac{R_1 + R_2}{L}x + \frac{k^2(R_1 + R_2)}{L^2}\theta_{in} \tag{28a}$$

$$y = -\frac{R_1}{k}x + \frac{R_1 k}{L}\theta_{in}. \tag{28b}$$

If it is assumed that $L = 0$, the transformed system is

$$y = \frac{R_2 k}{R_1 + R_2}\dot{\theta}_{in}. \tag{29}$$

In both of these cases, the transformation leads to a description which can be used in standard identification algorithms for the time domain.

8.2 *Frequency domain identification*

To do identification in the frequency domain, the transfer function for the model must be calculated. This can be done using equation (26) (it is assumed that $M = 0$ in this equation, i.e., an output error model is selected). The result is

$$G(s) = \frac{R_2 ks}{Ls + R_1 + R_2}. \tag{30}$$

This transfer function can be used in standard identification algorithms for the frequency domain.

9. CONCLUSIONS

This paper has treated how the parameter identification problem can be solved for linear DAE systems. In the time domain, a prediction error method or maximum likelihood identification can be implemented by constructing a Kalman filter. This requires that the descriptor system is converted to a special form. Although this form can always be calculated in theory under mild assumptions, it may be difficult to find in practice. Therefore it might be better to do identification in the frequency domain in some cases.

Future studies on the subject include to examine if there are numerically robust algorithms to find the normal form and to examine how identification can be done for nonlinear DAE models.

10. REFERENCES

Anderson, B.D.O. and J.B. Moore (1979). *Optimal Filtering*. Information and System Sciences Series. Prentice-Hall inc. Englewood Cliffs, New Jersey.

Åström, K.J. (1970). *Introduction to Stochastic Control Theory*. Mathematics in Science and Engineering. Academic Press. New York and London.

Åström, K.J. and B. Wittenmark (1984). *Computer Controlled Systems, Theory and Design*. Information and System Sciences Series. Prentice-Hall. Englewood Cliffs, N.J.

Brenan, K.E., S.L. Campbell and L.R. Petzold (1996). *Numerical Solution of Initial-Value Problems in Differential-Algebraic Equations*. Classics In Applied Mathematics. SIAM. Philadelphia.

Dai, L. (1987). State estimation schemes for singular systems. In: *Preprints of the 10th IFAC World Congress, Munich, Germany*. Vol. 9. pp. 211–215.

Dai, L. (1989). *Singular Control Systems*. Lecture Notes in Control and Information Sciences. Springer-Verlag. Berlin, New York.

Ljung, L. (1999). *System Identification - Theory for the User*. Information and System Sciences Series. 2. ed. Prentice Hall PTR. Upper Saddle River, N.J.

Mattsson, S.E., H. Elmqvist and M. Otter (1998). Physical system modeling with Modelica. *Control Eng. Pract.* **6**, 501–510.

Pintelon, R. and J. Schoukens (2001). *System Identification: A frequency domain approach*. IEEE Press. New York.

Polderman, J.W. and J.C. Willems (1998). *Introduction to Mathematical Systems Theory: a behavioral approach*. Number 26 in: *Texts in Applied Mathematics*. Springer-Verlag. New York.

Rosenbrock, H. H. (1970). *State-space and Multivariable Theory*. John Wiley & Sons, Inc., New York.

Schittkowski, K. (2002). *Numerical Data Fitting in Dynamical Systems*. Kluwer Academic Publishers. Dordrecht.

Schön, T., M. Gerdin, T. Glad and F. Gustafsson (2003). A modeling and filtering framework for linear implicit systems. Technical Report LiTH-ISY-R-2498. Dept. of Electr. Eng., Linköpings universitet. www.control.isy.liu.se/publications/.

Tiller, M. (2001). *Introduction to Physical Modeling with Modelica*. Kluwer. Boston, Mass.

Verghese, G.C., B.C. Lévy and T. Kailath (1981). A generalized state-space for singular systems. *IEEE T. Automat. Contr.* **AC-26**(4), 811–831.

IFAC Publications
www.elsevier.com/locate/ifac

MODEL VALIDATION IN NON-LINEAR CONTINUOUS-DISCRETE GREY-BOX MODELS

Jan Holst * Erik Lindström * Henrik Madsen ** Henrik Aalborg Nielsen **

* *Centre for Mathematical Sciences Lund Institute of Technology Box 118, Se-221 00 Lund, Sweden.*
** *Informatics and Mathematical Modeling Technical University of Denmark Dk-2800 Kongens Lyngby, Denmark*

Abstract: The validation part of the modeling procedure is used in the iterative process of model building and as an acceptance test of the obtained model. This paper presents a number of tools for validating non-linear continuous-discrete time stochastic state space models. For equidistantly sampled data a minor transformation of the one-step prediction errors is appropriate and the tools applied can also be used to suggest extensions of the model. Alternatively, we may regard the diffusion term as unmodelled systematic variation and in this case use the estimated diffusion as a tool to suggest possible sources of such variation. Finally, in the general case we may apply a transformation, based on numerical solution of the Fokker-Planck equation, in order to obtain standardized residuals which under the model are i.i.d. Gaussian. Standard tests can then be applied to these residuals. *Copyright © 2003 IFAC*

Keywords: Grey-box models, validation, stochastic differential equations, lag-dependent functions, Fokker-Planck equation

1. INTRODUCTION

We study validation procedures for a general class of multivariate stochastic systems. The generic model for such systems is specified using a continuous-discrete time stochastic state space formulation as

$$\begin{aligned} dX(t) &= \mu(t,u(t),X(t),\theta)dt + \\ &\quad \sigma(t,u(t),X(t),\theta)dW(t), \\ Y_k &= h(t_k,u(t_k),X(t_k),\theta) + e_k. \end{aligned} \quad (1)$$

In this model, the drift is denoted $\mu(t,u(t),X(t),\theta) : [0,T] \times \mathbb{R}^l \times \mathbb{R}^n \times \mathbb{R}^d \mapsto \mathbb{R}^n$, the diffusion is $\sigma(t,u(t),X(t),\theta) : [0,T] \times \mathbb{R}^l \times \mathbb{R}^n \times \mathbb{R}^d \mapsto \mathbb{R}^{n\times m}$ and $dW(t)$ is a m-dimensional Wiener process, i.e. $X(t)$ is a n-dimensional system of stochastic differential equations, driven by a m-dimensional vector of Wiener processes. The stochastic integrals are interpreted in the sense of Itô, cf. (Kloeden and Platen, 1992) or (Madsen and Holst, 2000). It is assumed that the system may be influenced by some known external signal $u(t) \in \mathbb{R}^l$, often called input signals. Furthermore, the observations are obtained through the mapping $h(t_k,u(t_k),X(t_k),\theta) : [0,T] \times \mathbb{R}^l \times \mathbb{R}^n \times \mathbb{R}^d \mapsto \mathbb{R}^p$, and the p-dimensional vector of observations, Y_k, is only observed at $t_k, k = 1,2,\ldots$, which not nescessarily are equidistant. The observation error, e_k, is assumed to be Gaussian, independent and having the covariance matrix R_e.

Models of the type (1) are applicable for grey-box modelling since they allow a priory knowledge about the structure of e.g. the differential equations governing the system to be build directly into the model. Estimation of the d-dimensional parameter vector θ can be performed using a maximum likelihood method, where it is assumed that the transition probabilities can be approximized by conditional Gaussian densities. In this case the likelihood function can be evaluated by using iterated Extended Kalman filters. We have used CTSM [1], see (Madsen and Melgaard, 1991; Kristensen *et al.*, 2002), to estimate the parameters. An overview of methods for parameter estimation of discretely observed stochastic differential equations is given in (Nielsen *et al.*, 2000)

As part of inference methodology in systems with this model formulation, it is not uncommon to assume that the diffusion term is independent of the state of the system as many systems can be formulated (or transformed using Itô's formula as described in (Madsen and Holst, 2000) or (Nygaard and Madsen, 2001)) to a system where the diffusion is independent on the state. It should however be noted that the physical interpretation of the states in the original formulation of the system might disappear as a consequence of such a transformation. Furthermore, the diffusion term being part of the description of possible unmodelled dynamics may be used as indicator for extensions in case of failing validation tests. Thus, its desirable also to be able to handle the state dependent diffusion term, and in this paper both state dependent and state independent diffusions are studied.

Our approach to validation aims at identifying characteristics which, under the hypothesis that the model is adequate, can be verified based on data. Other approaches to validation which we do not consider here include quantification of the prediction performance. Also, for grey-box models, a comparison of estimated coefficients with values known from the literature or otherwere will often be appropriate.

Model validation is often used in an iterative model building process. To this end it is desirable that the tools used for validation also indicates relevant extensions of the model. Examples include the structure of serial dependence in specially constructed residual series and the cross dependence between these series and the input.

In Section 2 we discuss the use of one-step prediction errors or a simple transformation of these. Section 3 describes how the estimated diffusion term can be used to pinpoint potential areas of model improvement. The presented methodology may often be applicable in situations where the validation has proven the model to be inadequate. An advanced method, based on numerical solution of the Fokker-Planck equation for constructing standardized residuals is described in Section 4. This method is applicable to non-equidistantly sampled non-linear systems as well.

2. ANALYSIS OF ONE-STEP PREDICTION ERRORS

The validation procedure may be based on a sequence of residuals, computed as the difference between the measured system outputs and the one-step predictions, i.e.

$$r'(t_k) = Y_k - E[Y_k|\mathscr{Y}^{k-1}], \qquad (2)$$

where E denotes expectation and $\mathscr{Y}^{k-1}$ describes the available information at time t_{k-1} on which the non-linear prediction of the output signal as well as the estimates of the unknown structure and parameters is based.

An exact evaluation of the one-step transition probabilities (or conditional densities) can be found using successive prediction based on Kolmogorov's forward equation and updating using Bayes' rule. This approach, which is computationally demanding, allows for a very general validation technique and is further discussed in Section 4. In this Section however, an alternative, and computationally less demanding validation method is presented. Since the SDE's in (1) are driven by a Wiener process, and since increments of a Wiener process are Gaussian, it is reasonable to assume, under some regularity conditions, that the conditional densities can be well approximated by Gaussian densities. This means that filter methods (like the Extended Kalman Filter) can be used to calculated the conditional mean and covariance of the one-step prediction. The standard deviation of the one-step prediction error can be used to normalize the sequence of residuals, and the normalized residuals will be noted r_k in the following. Note that for linear, time-invariant equidistantly sampled systems the normalization is not required.

In the remaining part of this section it is assumed that the signals are sampled equidistantly. Under this condition, residual analysis deals with the analysis of normalized estimates of the one-step prediction errors, called r_k above. If the model is correct r_k will have properties similar to that of observations over a finite time span from a multivariate white noise process with constant covariance matrix. Below the term "residuals" will be used for "normalized residuals". Generally, since Grey-box Models in principle are inherently non-linear it is desirable for the methods to be able to detect and estimate non-linear dependencies.

It is important to plot each of the residual series and judge the plots for non-stationarity of the series and non-constancy of the variance. For linear system models as given by (1) the residuals as defined above are Gaussian if the model is correct and this assumption can be tested formally, see e.g. (Holst *et al.*, 1993). However, for non-linear models the residuals need not be Gaussian even if the model is correct. A simple graphical check of the residual distribution could be obtained by simulating a new dataset using the estimated model. Based on re-estimation of the model *simulated residuals* can be calculated. Given that the model is correct a QQ-plot of the simulated residuals against the actual residuals should be grouped around the line of identity. However, deviations from the line of identity may also be caused by serial dependence

[1] The program is available at http://www.imm.dtu.dk/ctsm

of the actual residuals and this should be the main focus of the residual analysis, which can be split in (i) the serial dependence of each of the residuals series, (ii) the cross-dependence between each of the residual series, and (iii) the cross-dependence between the individual residual series and the inputs. For linear dependence (correlation) a number of methods exists, see (Brockwell and Davis, 1987; Holst *et al.*, 1993).

Lag Dependence

Some of the most important methods are based on the Sample Auto Correlation Function (SACF) for investigation of serial correlation and the Sample Cross Correlation Function (SCCF) together with prewhitening (Brockwell and Davis, 1987) for investigation of correlation between series. For non-linear system models it is reasonable to require that the tools used for validating the fit of the model is able to detect non-linear dependencies. For this purpose the linear tools mentioned above are not adequate, in fact non-linear deterministic processes exists for which the SACF will be close to zero for non-zero lags. An example of such a process is $x_k = 4x_{k-1}(1-x_{k-1})$ with an initial value of 4/5, see also (Granger, 1983).

It is possible to generalize the SACF to allow detection of non-linear dependencies. Let $f_\ell(r) = E[r_k \mid r_{k-\ell} = r]$ be the conditional mean of the residual process. It is readily shown that except for minor differences in the denominators, the estimate of the autocorrelation function in lag ℓ is equal to the estimate of the correlation coefficient between r_k and $r_{k-\ell}$ using the residuals $\{r_1, \ldots, r_N\}$. The estimates are asymptotically equivalent. Furthermore, exploiting the fact that correlation and R-squared values for linear models are closely linked allows the SACF in lag ℓ to be expressed as $\mathrm{sign}(\hat{f}_\ell(r_{max}) - \hat{f}_\ell(r_{min}))\sqrt{R^2_{0(\ell)}}$, where $\hat{f}_\ell(r)$ and $R^2_{0(\ell)}$ are obtained by fitting a straight line by least squares to the ℓ-lagged scatter plot, r_{min} and r_{max} are the minimum and maximum of the residuals. The generalization which allows the detection of non-linear dependencies is based on estimating $f_\ell(r)$ using a smoother and calculating the value of $R^2_{0(\ell)}$ for this smooth. Thus, the Lag Dependence Function in lag ℓ, $LDF(\ell)$, is calculated as

$$LDF(\ell) = \mathrm{sign}\left(\hat{f}_\ell(r_{max})) - \hat{f}_\ell(r_{max})\right) \sqrt{(R^2_{0(\ell)})_+} \quad (3)$$

The subscript "+" indicates truncation of negative values, which might occur for some lags when the smoother is highly inadequate, see (Nielsen and Madsen, 2001) for details. $LDF(\ell)$ is defined for any smoother or parametric model of the conditional mean, including kernel estimates, local regression, smoothing splines, regression splines, and neural networks. For local linear regression the $SACF(\ell)$ is obtained as the limit of $LDF(\ell)$ when the bandwidth (and the number of residuals) tends to infinity. In (Nielsen and Madsen, 2001) it is shown how bootstrapping can be used to obtain confidence intervals of $LDF(\ell)$ under the assumption that the residuals i.i.d.

By example, it is shown in (Nielsen and Madsen, 2001) that the cost paid to be able to detect non-linearities is reduced power against linear alternatives. For this reason the method should be used as a complement to the SACF and a variety of smoothers should be used.

As an example consider the discrete-time process $\{X_t\}$ defined by $X_t = e_t + 2\cos e_{t-1}$, where $\{e_t\}$ is a standard Gaussian white noise process. Figure 1 shows SACF and LDF based on $N = 1000$ simulated observations from $\{X_t\}$. On SACF an approximate 95% confidence interval $(\pm 2/\sqrt{N})$ for white noise is shown. LDF is calculated using a local linear smoother together with a 50% nearest neighbour bandwidth. The confidence interval for white noise is standard normal intervals (Nielsen and Madsen, 2001) with a coverage of 95% based on 200 bootstrap replicates. The calculations are performed using the S-PLUS library LDF[2]. It is seen that only LDF is capable at detecting the non-linear serial dependence. Also, note that the approximate 95% confidence intervals for LDF is wider than for SACF.

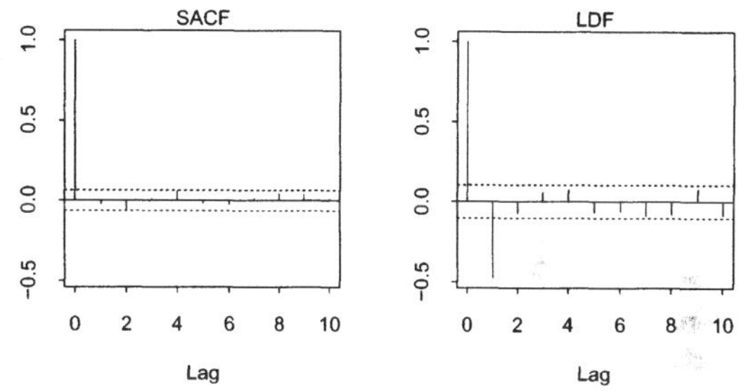

Fig. 1. SACF and LDF for 1000 observations from a simulated non-linear process.

Furthermore in (Nielsen and Madsen, 2001), a quantity termed the Non-linear Lag Dependence Function (NLDF) is defined. NLDF can be used to investigate and test if a strictly non-linear serial dependence is present in the series under consideration. Also, for investigating the lag-structure of serial dependence they suggests a generalization of the Sample Partial Auto Correlation Function.

Another method for detecting lag dependency based on kernel estimates is found in (Hjellvik and Tjøstheim, 1996). Also, methods based on the mutual information coefficient and Kendall's τ exist, see (Granger and Lin, 1994).

[2] The package can be downloaded from http://www.imm.dtu.dk/~han/software.html

Cross Dependence

As noted in (Nielsen and Madsen, 2001) the SCCF can be generalized in a similar way, but prewhitening is not possible for non-linear models. For this reason the generalization is dependent on one of the series not exhibiting serial dependence. The analysis of the residuals might indicate that these are independent or for designed experiments the input might be white noise. Also when none of these requirements are true it will often be relevant to investigate the dependence between the input series and the residuals because this may point towards necessary extensions of the model. One possible way to accomplish this will be to fit FIR-type non-linear additive models of increasing order and compare the R-squared values obtained. Also the method described in Section 3 will be applicable.

3. USING THE ESTIMATED DIFFUSION TERM

One of the benefits of using the model (1) in grey-box modelling is due to the decomposition of the prediction error into system error and measurement error. Since the system error accounts for models approximations the estimated elements of the diffusion term facilitates an evaluation of an identified system. More specifically the estimated elements of the diffusion term pinpoints model deficiencies and provides means to uncover the structural origin of these deficiencies.

If the diffusion term has significiant elements, these elements points at *potential* areas for model improvement. Obviously, if the identified diffusion term is insignificiant, then the system error is negligible and hence it can be concluded that the model provides an adequate description of the observed dynamics of the system.

Consider for instance the parameters, θ in (1), which are assumed to be constant. However, if one of the parameters, say θ_i, actually depends on either a state, an input variable or the time, then the model needs to be extended, and possibly with a non-linear or non-stationary term. Also functional relations involving unmeasured variables can be uncovered, and in general the presence of a significiant (and large) diffusion term indicates that the corresponding element of the drift term might be incorrect.

In (Kristensen *et al.*, 2003) a systematic framework is suggested for detecting such unknown functional relations and for uncovering their structural origin. The framework, which is based on an interplay between stochastic differential equation modeling, statistical tests and non-parametric modeling will be briefly described in the following.

Assume that at some stage of the modelling cycle the parameters of a continuous-discrete time stochastic state space model like (1) is estimated, and that an inspection of the diffusion terms leads to the suspicion that some of the inherent phenomena described by the corresponding drift term are incorrectly modelled.

More specifically, it is assumed that the model maker is able to suggest a specific phenomena, $v(t)$ for further investigation. In many cases $v(t)$ is equal to a parameter which is suspected for not being constant – as described previously.

Consider now the extended model

$$\begin{aligned} dX^*(t) &= \mu^*(t,u(t),X^*(t),\theta^*)dt + \\ &\quad \sigma^*(t,u(t),\theta^*)dW^*(t) \\ Y_k &= h(t_k,u(t_k),X^*(t_k),\theta^*) + e_k \end{aligned} \tag{4}$$

where $X^*(t) = (X^T(t)\ v(t))^T$, $\sigma^* \in \mathbb{R}^{(n+1)\times(m+1)}$ and $W^*(t)$ is now a $m+1$-dimensional standard Wiener process. σ^* is composed of the original σ, but with an added term corresponding to $dv(t)$ and the other new elements are zero. Finally,

$$\mu^*(t,u(t),X^*(t),\theta^*) = \begin{pmatrix} \mu(t,u(t),X(t),\theta^*) \\ 0 \end{pmatrix} \tag{5}$$

Hence, any variation in $v(t)$ must be explained by the corresponding element of the expanded diffusion term, which in turn means that if the parameters of this element are significiant, this indicates that $v(t)$ is not constant. Correspondingly, if the extended drift term is now correct then a significiant decrease of the previously significant diffusion terms is seen.

The next step is now to uncover the structural origin of the found model deficiencies, i.e. a model for $v(t)$. The procedure for doing this is based on a combination of state filtering and the ability of non-parametric regression methods to provide visualizable estimates of the unknown functional relations.

The state space formulation of the extended model facilitates state estimates, $\hat{v}(t|t)$ of the phenomena of interest. If CTSM is used for parameter estimation, the state estimates are directly available. A plot of these state estimates against $\hat{X}(t|t)$ and $u(t)$ for all the available time points followed by non-parametric regressions are now providing estimates of the possible functional relations between $v(t)$ and the state and input variables. By considering the non-parametric relations physical insights, or trial and error, is subsequently used to provide a parametric description of $v(t)$, and hence an improved model is obtained.

4. A TRANSFORMATION METHOD

If the SDE's in (1) is driven by vector Brownian motion and the process is sampled often and equidistantly, then we know that the approximation by Gaussian densities often gives reasonable results. In this section we address the problems that occur when these conditions are not fulfilled.

Previous work on validation of non-Gaussian models with non-equidistant sampling is most often assuming the non-linear model to be univariate, see (Aït-Sahalia, 1996), (Aït-Sahalia, 2002), (Bak *et al.*, 1999) or (Lindström, 2003). The most restrictive of these is (Aït-Sahalia, 1996), where the data generating process is required to be strictly stationary. The test is constructed by comparing the (non parametrically estimated) stationary distribution to the stationary parametric distribution. This approach has several limitations. (Bak *et al.*, 1999) estimates the drift term and diffusion term non parametrically to find an appropriate model, and suggests a lack-of-fit test for validation of models based on stochastic differential equations.

The basic idea of model validation in this paper as a whole is to work with standardized residuals. Hence, also in the non-Gaussian and the non-equidistantly-sampled case we aim at finding a sequence of transformations giving standardized, identically distributed residuals. Such a transformation starts by solving the Fokker-Planck equation numerically to calculate the (cumulative) transition probability density. This is then used to transform all innovations to approximative independent, uniformly distributed random variables testing dependence and distributional properties for the transformed innovations, which is almost equivalent to testing standardized residuals. A detailed treatment of this transformation technique for validation is found in (Lindström, 2003). It may be generalized to several dimensions.

The transformation

Thus assume that we can calculate the cumulative transition probability function for the p-vector Y_k, $F_{Y_k}(y) = P(Y_{1,t_k} \leq y_1, \ldots, Y_{p,t_k} \leq y_p)$. In fact, the transition probability density for Y_k can be calculated by calculating the transition probability density for X_t using the Fokker-Planck equation, transforming $X(t)$ to $h(t, X(t))$ and convolving the distribution with the distribution of the measurement noise. Alternatively Monte Carlo simulations can be used to calculate the cumulative distribution function. This is the approach in (Pedersen, 1994).

The process $R_k = F_{Y_k}(Y_k)$ may be tested for distributional properties and dependence, it should ideally be composed of independent and uniformly distributed random variables. It can also be transformed to an approximative Gaussian process by applying the inverse method, $Z_i = \Phi^{-1}(R_i)$, $\Phi(\cdot)$ is the standard Gaussian distribution function. All previous tests can then be applied to these new standardized residuals.

Moreover, the residuals are stochastically independent implying that no dependence to either lagged values of the process or to any external signal should be statistically significant (corrected for multiple testing of hypothesis). Still, any remaining dependence will indicate possible improvements of the model.

An Example

The Cox-Ingersoll-Ross model is commonly used for describing the term structure of interest rate, cf. (Cox *et al.*, 1985). The model is given by the differential equation

$$\begin{aligned} dX(t) &= \alpha(\beta - X(t))dt + \sigma\sqrt{X(t)}dW(t) \\ Y_k &= X(t_k) \end{aligned}, \qquad (6)$$

where $\alpha > 0$, $\beta > 0$ and $\sigma > 0$. The process is stationary, non-negative and varying around β. The data used was simulated by means of a Milstein scheme, see (Kloeden and Platen, 1992), using equidistant observations. The distance between the observations is $\Delta t = 1$ and the parameters used were chosen as $\alpha = 0.17, \beta = 0.05$ and $\sigma = 0.07$ i.e.

$$dX(t) = 0.17(0.05 - X(t))dt + 0.07\sqrt{X(t)}dW(t).$$

The simulated process is presented in Figure 2. The process is skewed and does never reach zero. The distribution is compared to a Gaussian distribution in the QQ-plot and dependence is also examined.

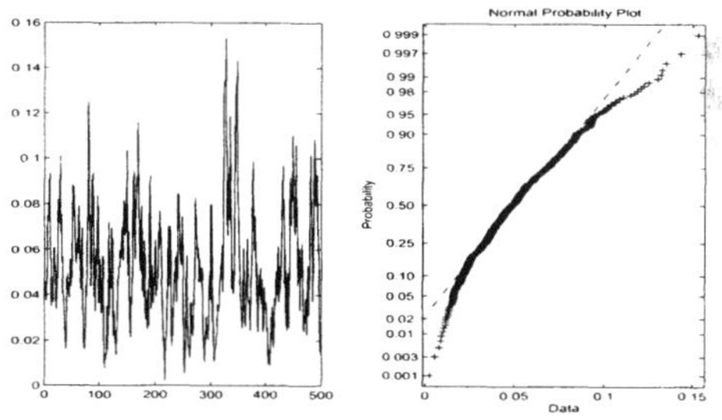

Fig. 2. A realization of the Cox-Ingersoll-Ross process (left) and the corresponding QQ-plot for the realization (right).

After transforming the data using the calculated distribution function and the inverse of the standard Gaussian distribution function, we get an empirical distribution function for the residuals. A graphical analysis of the residuals is given in Figure 3. The residuals are Gaussian and they show no linear or quadratic dependence.

5. CONCLUSION

This paper presents tools and concepts which may be used for validation of models based on non-linear continuous-discrete stochastic differential equations. This class of models is one of the most promising tools for Grey-Box modeling where a priori knowledge and observations are used in combination to obtain a model of the system under consideration.

Our approach to validation is that under the model we identify characteristics which must be present and we apply appropriate tools for identifying deviations from these characteristics. Ideally, when the model under consideration is not appropriate, these tools should also indicate how the model should be modified.

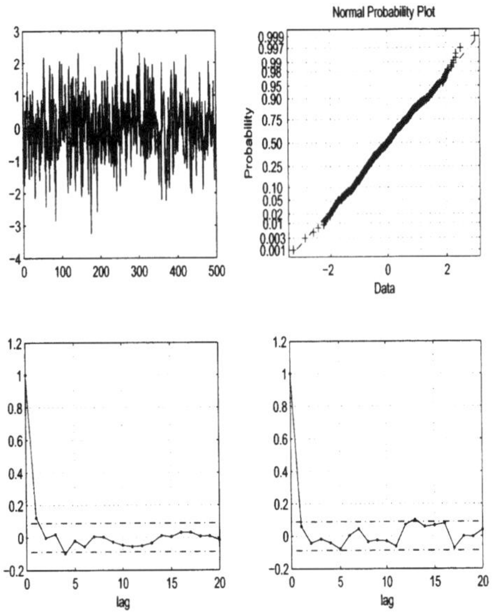

Fig. 3. The Gaussian residuals calculated from the Cox-Ingersoll-Ross process (top left) and the corresponding QQ-plot for the realization (top right). The autocorrelation function for the process (bottom left) and for the squared process (bottom right) are compared to the 95 % confidence intervals.

Depending on the specific model structure and on the sampling plan different approaches to validation are required. For samples equidistant in time it is often possible to work with normalized one-step prediction errors. For this setup we present tools which can be used to detect non-linear dependencies.

When the input processes do not have properties similar to that of white noise processes detecting non-linear cross dependencies is complicated by the fact that prewhitening is inappropriate. In this case it will often be appropriate to regard the diffusion term as unmodelled systematic variation and use the estimated diffusion as a tool to suggest possible sources of such variation. The approach just mentioned is also appropriate when the samples are not equidistant in time.

Finally, for the general case of non-equidistantly sampled systems modelled by non-linear models with state-dependent diffusion a transformation is presented. As a result of this transformation, under the model, i.i.d. Gaussian residuals are obtained, and standard tests can be applied to these residuals.

REFERENCES

Aït-Sahalia, Y. (1996). Testing Continuous-Time Models of the Spot Interest Rate. *The Review of Financial Studies* **9**(2), 385–426.

Aït-Sahalia, Y. (2002). Maximum Likelihood Estimation of Discretely Sampled Diffusions: A Closed–Form Approximation Approach. *Econometrica* **70**, 223–262.

Bak, J., H. Madsen and H. Aa. Nielsen (1999). Goodness of fit of stochastic differential equations. In: *(in Danish) Symposium i Anvendt Statistik* (Peter Lind and Anders Holm, Eds.).

Brockwell, P. J. and R. A. Davis (1987). *Time series: theory and methods*. Springer-Verlag Inc.

Cox, J. C., J. E. Ingersoll and S. A. Ross (1985). A theory of the term structure of interest rates. *Econometrica* **53**, 385–407.

Granger, C. and J.-L. Lin (1994). Using the mutual information coefficient to identify lags in non-linear models. *Journal of Time Series Analysis* **15**(4), 371–384.

Granger, C. W. J. (1983). Forecasting white noise. In: *Applied Time Series Analysis of Economic Data* (Arnold Zellner, Ed.). pp. 308–314. U.S. Department of Commerce. Washington.

Hjellvik, V. and D. Tjøstheim (1996). Nonparametric statistics for testing of linearity and serial independence. *Journal of Nonparametric Statistics* **6**, 223–251.

Holst, J., U. Holst, H. Madsen and H. Melgaard (1993). Validation of grey box models. *Adaptive Systems in Control and Signal Processing 1992 and IFAC Symposia Series* (8), 53–60.

Kloeden, P. E. and E. Platen (1992). *Numerical Solution of Stochastic Differential Equations*. Springer-Verlag.

Kristensen, N. R., H. Madsen and H. Melgaard (2002). *CTSM 2.1*. Technical University of Denmark. IMM, Building 321, DTU, 2800 Lyngby.

Kristensen, N. R., H. Madsen and S. B. Jørgensen (2003). A method for systematic improvements of grey-box process models. Submitted.

Lindström, E. (2003). Goodness of Fit Tests for Diffusion Processes. Division of Mathematical Statistics, Centre of Mathematical Sciences, Lund University.

Madsen, H. and H. Melgaard (1991). The mathematical and numerical methods used in CTLSM. Technical Report 7. Department of Mathematical Modelling, Technical University of Denmark. Lyngby, Denmark.

Madsen, H. and J. Holst (2000). *Modelling Non-Linear and Non-stationary Time Series*. Informatics and Mathematical Modelling, The Technical University of Denmark, Kongens Lyngby, Denmark.

Nielsen, H. Aa. and H. Madsen (2001). A generalization of some classical time series tools. *Computational Statistics and Data Analysis* **37**(1), 13–31.

Nielsen, J.N., H. Madsen and P.C. Young (2000). Parameter estimation in stochastic differential equations; an overview. *Annual Reviews in Control* **24**, 83–94.

Nygaard, J. and H. Madsen (2001). Applying the EKF to stochastic differential equations with level effects. *Automatica* **37**, 107–112.

Pedersen, A. R. (1994). Uniform residuals for discretely observed diffusion processes. Technical Report 292. Department of Theoretical Statistics, University of Aarhus.

IFAC Publications
www.elsevier.com/locate/ifac

IDENTIFICATION OF MECHANICAL PARAMETERS IN DRIVE TRAIN SYSTEMS

Alf J. Isaksson[1], Rickard Lindkvist and Xiaojing Zhang
ABB, Corporate Research, SE-721 78 Västerås, Sweden

Mattias Nordin[1] and Mats Tallfors
ABB, Automation Technologies, SE-721 67 Västerås, Sweden

Abstract: The aim of this paper is to study the estimation of mechanical parameters for drive trains. The task is to fit a two-mass model with backlash to experimental data. The suggested approach is to use a procedure based on three dedicated experiments: 1) One experiment with a sequence of setpoint steps while maintaining the controller in automatic. This enables estimation of the static gain yielding the total inertia damping and static friction. 2) Then an experiment tailored to guaranteeing that no gap openings are encountered, thus enabling estimation of the other mechanical parameters. Black-box identification of a general third order transfer function is used to find initial parameters, followed by fine tuning using a physical model. 3) Finally yet another experiment that with certainty contains gap openings. Use this data set to estimate the gap size, while keeping the other model parameters fixed.

This procedure has been developed and tested successfully on numerous data sets simulated by the drive dynamics analyzer (DDA) program. The methods have then been implemented as part of DDA, and verified on real data from a test rig as well as on industrial data from both rolling mills and paper industry. *Copyright © 2003 IFAC*

Keywords: Grey-box identification, Black-box identification, Drive train, Backlash

1. INTRODUCTION

ABB Automation Technologies, Västerås, already have a Matlab based program (Drive Dynamics Analyzer -- DDA) for simulation and speed controller tuning of mechanical drive systems. Typical applications include rolling mills, paper machines, large fan systems, etc.

Today the modelling is dependent on the collection of mechanical data of motor as well as load from data sheet provided by the suppliers. This is often a cumbersome and time-consuming effort. Moreover, the gear-boxes normally have significant gear play, which introduces backlash into the drive system. This backlash increases with wear and usually cannot be accurately determined without performing experiments.

Hence, the gathering of mechanical data (including gap size) has been identified as a bottleneck in the tuning procedure. Therefore it has been suggested to estimate the mechanical parameters based on a fit to experimental data instead.

In a collaboration between ABB Corporate Research and ABB Automation Technologies new routines have been developed (implemented as part of DDA) for the identification of parameters in two-mass models.

2. BACKGROUND

Many industrial drives consist of motor, flexible shafts, gear-boxes and load, which form a multi-mass system. Backlash is introduced due to existence of elements like gear-boxes and flexible couplings. In this case, the system model becomes non-linear.

[1] For questions or comments please use email addresses *Alf.Isaksson@se.abb.com* or *Mattias.C.Nordin@se.abb.com*

The working hypothesis of this project is that in many cases the model can be simplified to a two-mass system where the first mass represents the motor, the second mass represents the load, and the shaft is considered as mass and inertial free. Such drive train system serves as a model for rolling mill, paper machine, large fan, etc.
The main task is then to estimate the unknown parameters of the two-mass model by making a numerical fit to experiment data.

3. TWO-MASS LINEAR MODEL

In drive train systems, backlash introduces a non-linearity which makes the mathematical model much more difficult to estimate. The following linear model of an uncertain two-mass system without backlash was presented in (Nordin, 2000):

$$J_m \dot{\omega}_m = -c_m \omega_m - T_s + T_m - T_0 \quad (1)$$

$$J_l \dot{\omega}_l = -c_l \omega_l + T_s \quad (2)$$

$$\dot{\theta}_d = \omega_m - \omega_l \quad (3)$$

with transmitted shaft torque

$$T_s = k_s \theta_d + c_s (\omega_m - \omega_l) \quad (4)$$

where the system input is T_m, the motor torque, and the measured output is ω_m, the motor speed.
The other two signals in the model are $\theta_d = \theta_m - \theta_l$, the angle difference between motor and load , and ω_l, the load speed.

The parameters to estimate are:

T_0 the idle torque, Nm
J_m the motor moment of inertia, kgm^2
c_m the viscous motor friction, Nm/(rad/s)
J_l the load moment of inertia, kgm^2
c_l the viscous load friction, Nm/(rad/s)
k_s the shaft elasticity, Nm/rad
c_s the shaft damping, Nm/(rad/s)

The suggested approach is to fit parameters directly in the model (1)-(4) by solving the system of differential equations and forming a least-squares criterion by comparing the computed motor speed to the measured one.

There are, however, at least two things that complicate matters. Firstly, we need to somehow obtain initial estimates of the seven parameters listed above. Secondly, when there is significant backlash in the system we need to introduce a model for this as well and estimate a gap parameter.

In an attempt to circumvent these two complications an approach based on two dedicated dynamic experiments was tested:

1) Perform a tailored experiment that guarantees that no gap openings are encountered. Then use black-box identification of a general third order transfer function to find initial mechanical parameters.
2) Fine tune the mechanical parameters on the same data using a physical model.
3) Perform another experiment that with certainty contains gap openings. Use this data set to estimate the gap size, while keeping the other model parameters fixed.

These steps are described in some more detail in the subsequent sections.

4. BLACK-BOX IDENTIFICATION

Since the gap free two-mass model (1)-(4) in the previous section is linear, a natural idea is to study its transfer function.
The uncompensated transfer function of the open loop, $P_{um}(s)$, from T_m to ω_m is given by:

$$P_{um}(s) = \frac{J_l s^2 + (c_l + c_s) s + k_s}{d(s)} \quad (5)$$

with

$$d(s) = J_m J_l s^3 + (J_l (c_m + c_s) + J_m (c_l + c_s)) s^2 + ((J_l + J_m) k_s + c_m c_l + c_m c_s + c_l c_s) s + (c_m + c_l) k_s$$

A natural approach to finding the initial parameter estimates would be by estimating a general third-order discrete-time transfer function and convert it to a general continuous-time one in the form

$$G(s) = \frac{b_2 s^2 + b_1 s + b_0}{s^3 + a_2 s^2 + a_1 s + a_0} \quad (6)$$

Comparison of equations (5) and (6) then gives the following system of equations:

$$b_2 = \frac{1}{J_m} \quad (7)$$

$$b_1 = \frac{c_l + c_s}{J_m J_l} \quad (8)$$

$$b_0 = \frac{k_s}{J_m J_l} \quad (9)$$

$$a_2 = \frac{c_m + c_s}{J_m} + \frac{c_l + c_s}{J_l} \quad (10)$$

$$a_1 = \frac{(k_s (J_m + J_l) + c_m c_l + c_m c_s + c_s c_l)}{J_m J_l} \quad (11)$$

$$a_0 = \frac{k_s (c_m + c_l)}{J_m J_l} \quad (12)$$

The idea is then to use estimates of the a- and b-parameters to solve this system of equations for J_m, J_l, c_m, c_l, k_s and c_s.

The system of equations can indeed be solved in, for example, Maple using so-called Groebner bases. However, it turns out that for typical values of the mechanical parameters the solution is very sensitive. In particular the inertia dampings c_l and c_m are extremely sensitive to small variations in the estimated discrete-time parameters. The reason is that they are small in comparison to the other parameters, and basically only influence the static gain

$$\frac{b_0}{a_0} = \frac{1}{c_m + c_l}$$

The static gain is, however, very difficult to estimate from dynamic experiments, since it is so small that for short data records the system acts almost like an integrating system.

Unfortunately the bad accuracy of the inertia dampings also influences the estimate of the other mechanical parameters to such an extent that it was found that a method based on solving the complete system of equations is not feasible.

Instead an alternative approach was taken, neglecting c_l and c_m altogether for the initialization. Then the system (5) has a pure integration, i.e. it has the form

$$G(s) = \frac{b_2 s^2 + b_1 s + b_0}{s(s^2 + a_2 s + a_1)} \tag{13}$$

where

$$b_2 = \frac{1}{J_m} \tag{14}$$

$$b_1 = \frac{c_s}{J_m J_l} \tag{15}$$

$$b_0 = \frac{k_s}{J_m J_l} \tag{16}$$

$$a_2 = \frac{c_s (J_m + J_l)}{J_m J_l} \tag{17}$$

$$a_1 = \frac{k_s (J_m + J_l)}{J_m J_l} \tag{18}$$

Notice that in equations (14) to (18) there are now five equations but only four unknown parameters. Several alternatives are therefore possible (including solving the over determined system of equations in a least-squares sense). We have tried them all and in the end settled for using equations (14), (16), (17) and (18). Hence the equation (15) is excluded in the solution.

It now remains to find initial estimates for the inertia dampings c_l and c_m, since they were excluded in the black-box identification. Notice also that since mean values are typically subtracted from the signals before black-box identification no estimate of the idle torque T_0 was obtained either.

To solve these two remaining problems we suggest introduction of yet another dedicated experiment described in the next section.

5. IDENTIFICATION OF STATIC GAIN

As indicated above the goal with performing a static experiment is to estimate the sum of viscous motor friction and load friction, as well as the offset created by the idle torque.

The model (1)-(3) implies the following static relationship

$$\overline{T}_m = (c_m + c_l)\overline{\omega}_m + T_0$$

where $\overline{\omega}_m$ is static motor speed (rad/s) and $\overline{T}_m$ is static motor torque (Nm). Hence, the aim of the static experiment is to find the slope $c = c_m + c_l$ and offset T_0 of a straight line.

Therefore an experiment is conducted with the speed controller in automatic, where a sequence of setpoint steps are made spanning the operating range. The torque and speed signals are then run through a method described in Cao and Rhinehart (1995) that automatically finds data points where the system is in steady-state.

The data in all sufficiently long steady-state intervals are then averaged to form one observation of the straight line. Given several such observations an over determined system of equations is formed, which is easily solved in a least-squares sense by Matlab.

Since only the total damping can be estimated, some further assumption has to be made before initializing the individual inertia dampings. We have chosen as a first guess to distribute the damping relative to the size of the corresponding inertia, i.e.

$$c_m = \frac{cJ_m}{J_m + J_l}; \quad c_l = \frac{cJ_l}{J_m + J_l}.$$

6. GREY-BOX IDENTIFICATION

Once initial estimates of the mechanical parameters have been obtained using the procedures described in Sections 4 and 5, fine tuning is made by fitting the mechanical parameters directly, using a least-squares criterion

$$V = \sum_k (\omega_m(k) - \hat{\omega}_m(k,\theta))^2$$

where $\omega_m(k)$ denotes the (discrete-time) measurement of motor speed and $\hat{\omega}_m(k,\theta)$ denotes the motor speed simulated through the model (1)-(3) at the same time instants, using the current mechanical parameter values here denoted by the vector θ.

In a pre-study the general purpose Matlab based program MoCaVa -- (Bohlin and Isaksson, 2003) -- was used to verify the feasibility of the approach. For the final tool a modified version of the Matlab toolbox Diffpar -- (Edsberg and Wedin , 1993, 1995), (Edsberg and Wikström, 1995) -- has instead been used to develop a graphical user interface dedicated to the drive train problem.

7. BACKLASH ESTIMATION

Once the linear mechanical parameters have been obtained, another experiment is made that is guaranteed to open up the gap.

The model (1)-(4) is modified by replacing (4) with the following backlash model:

$$z = \theta_d + c_s(\omega_m - \omega_l)$$

$$T_s = \frac{1}{2} k_s (z + \sqrt{\delta + z^2}) \qquad if\ z = \geq -\alpha$$

$$T_s = \frac{1}{2} k_s (z + 2\alpha - \sqrt{\delta + (z + 2\alpha)^2}) \quad if\ z < -\alpha$$

This corresponds to a dead-zone model, corrected for the effect of damping, see Nordin (2000). Furthermore a smoothness parameter δ has been introduced to give a continuous derivative. Unfortunately this instead causes a discontinuity of T_s at $z = -\alpha$. However, if a very small value of δ is used, this does not seem to cause any numerical problems.

In the current implementation the linear parameters are fixed and only the gap parameter α is allowed to vary. Since obtaining an initial estimate for α is very difficult, the program instead initializes the search at several user chosen starting values. Once again prototyping was made using MoCaVa, while the end product is built on Diffpar.

8. SUMMARY AND SOFTWARE TOOL

At this point a summary of the developed approach may be appropriate. The results in Sections 3-7 describe a procedure based on three dedicated experiments using four identification methods that should be run sequentially:

1. A steady-state experiment with controller in automatic, yielding estimates of $c_m + c_l$ and T_0.
2. A dynamic backlash free experiment followed by
   - initial estimation by black-box identification using Matlab's System Identification Toolbox
   - fine tuning using Diffpar and the mechanical model (1)-(3), yielding estimates of J_m, J_l, c_m, c_l, k_s and c_s.
3. A dynamic experiment with guaranteed backlash to estimate the gap parameter α, keeping all other parameters fixed.

A user interface has been created that reflects this four step procedure, where the second and third step share the same data. In each step the user can choose whether to use the results from the previous step, or to manually insert values. Thus making it possible to, for example, identify mechanical parameters from dynamic data even if a steady-state experiment is not available.

The main window of this identification tool is shown in Figure 5.

9. REAL DATA EXAMPLE

Real data has been collected on a test rig at ABB Automation Technologies, as well is from both rolling mills and paper mills. The results so far have been very encouraging. We will here show the results from one set of data from the test rig, consisting of one AC motor on the drive side, and a DC motor as load. The system should be possible to model very well using a two-mass model. From data sheets the motor inertia is given as $J_m = 0.40$, the load inertia $J_l = 5.6$ and shaft stiffness $k_s = 3300$ (with negligible shaft inertia). The test rig also has a rubber belt coupling, which has a similar effect as a gear play.

First the steady-state experiment is shown in Figure 1, where in the lower plot the automatically found steady-state intervals are marked. Obviously there is very little damping (i.e. the line has almost zero slope), while there is significant static friction resulting in a non-zero idle torque. These facts are also reflected in the obtained parameter estimates $c = 0.00184$ and $T_0 = 15.14$.

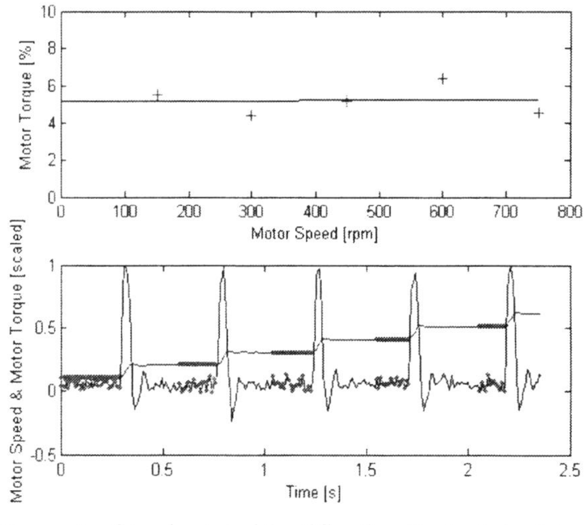

Figure 1. Steady state identification for test rig data.

In Figures 2 and 3 the data and model fit for the backlash free experiment are shown. Figure 2 shows the fit after initial black-box identification and Figure 3 the fine tuned fit after linear grey-box estimation.

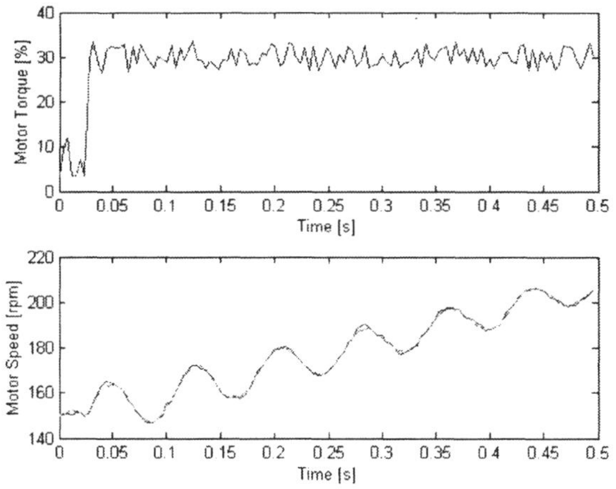

Figure 2. Data and model fit of black-box identification for test rig data (data solid, simulation dashed).

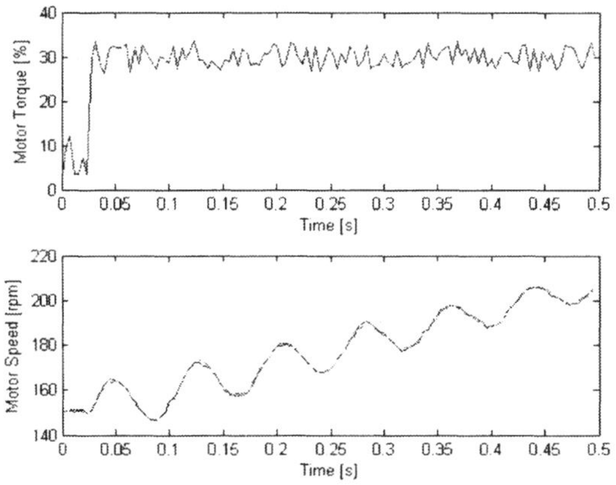

Figure 3. Data and model fit for fine tuning on backlash free test rig data (data solid, simulation dashed).

The estimation results for the mechanical parameters are summarized in Table 1. Notice that even though from visual inspection the fit looks very good already after the initial black-box estimation, the fine tuning in fact reduces the loss function by almost 25 %. According to the "true" values given above both the load inertia and shaft stiffness estimates are quite acceptable, whereas the motor inertia is estimated about 50 % too large. In this case, however, it is probably fair to claim that the estimated parameters better represent the real system than the ones obtained from data sheet.

Table 1. Summary of estimation results for backlash free data.

	Black-box	Grey-box
J_m	0.601	0.587
J_l	5.554	5.179
c_m	0.000180	0.000176
c_l	0.0017	0.0017
k_s	3445	3310
c_s	3.324	3.115
Resonance freq.	12.68	12.60
Loss function	0.00856	0.00656

Finally, an attempt was made to estimate also the gap parameter using the third set of experiment data (see Figure 4) yielding the result $\alpha = 4.3$ degrees. Obviously the gap model is not able to describe the more damped behaviour of the belt coupling, but at least manages to quite well capture the correct resonance frequency.

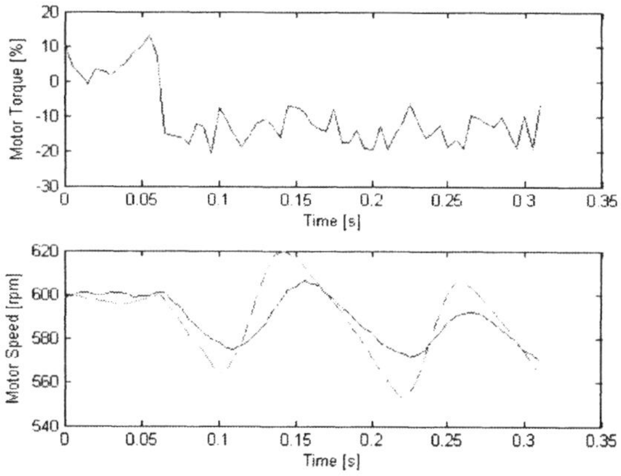

Figure 4. Data with backlash and corresponding model fit after gap identification for test rig data (data solid, simulation dashed).

10. CONCLUSIONS

This paper has described a way to identify mechanical parameters in drive train systems. The methods have been implemented in Matlab as part of the Drive Dymamics Analyzer (DDA), and this identification extension to the tool will be released in April 2003. As mentioned above the current program is restricted to two-mass models. A possible future extension is to be able to indicate whether using three or more masses would lead to a significantly better fit.

REFERENCES

Bohlin T. and A.J. Isaksson (2003). Grey-Box Model Calibrator & Validator. SYSID 2003. This invited session.

Cao S. and R.R. Rhinehart (1995). An efficient method of on-line identification of steady state. Journal of Process Control. Vol. 5, No. 6, pp. 363-374.

Edsberg L. and P.-Å. Wedin (1993). Diffpar: A toolbox for parameter estimation. Technical report, Dept. Numerical Analysis, KTH, TRITA-NA-9308.

Edsberg L. and P.Å. Wedin (1995). Numerical tools for parameter estimation in ODE-systems. Optimization Methods and Software, Vol. 6, pp. 193-217.

Edsberg L. and G. Wikström (1995). Toolbox for Parameter Estimation and Simulation in Dynamic Systems with Application to Chemical Kinetics, inProceedings of the Nordic Matlab Conference, pp 78-83.

Nordin M. (2000). "*Nonlinear Backlash Compensation for Speed Controlled Elastic Systems*", Doctoral Thesis, Department of Mathematics, KTH, Stockholm.

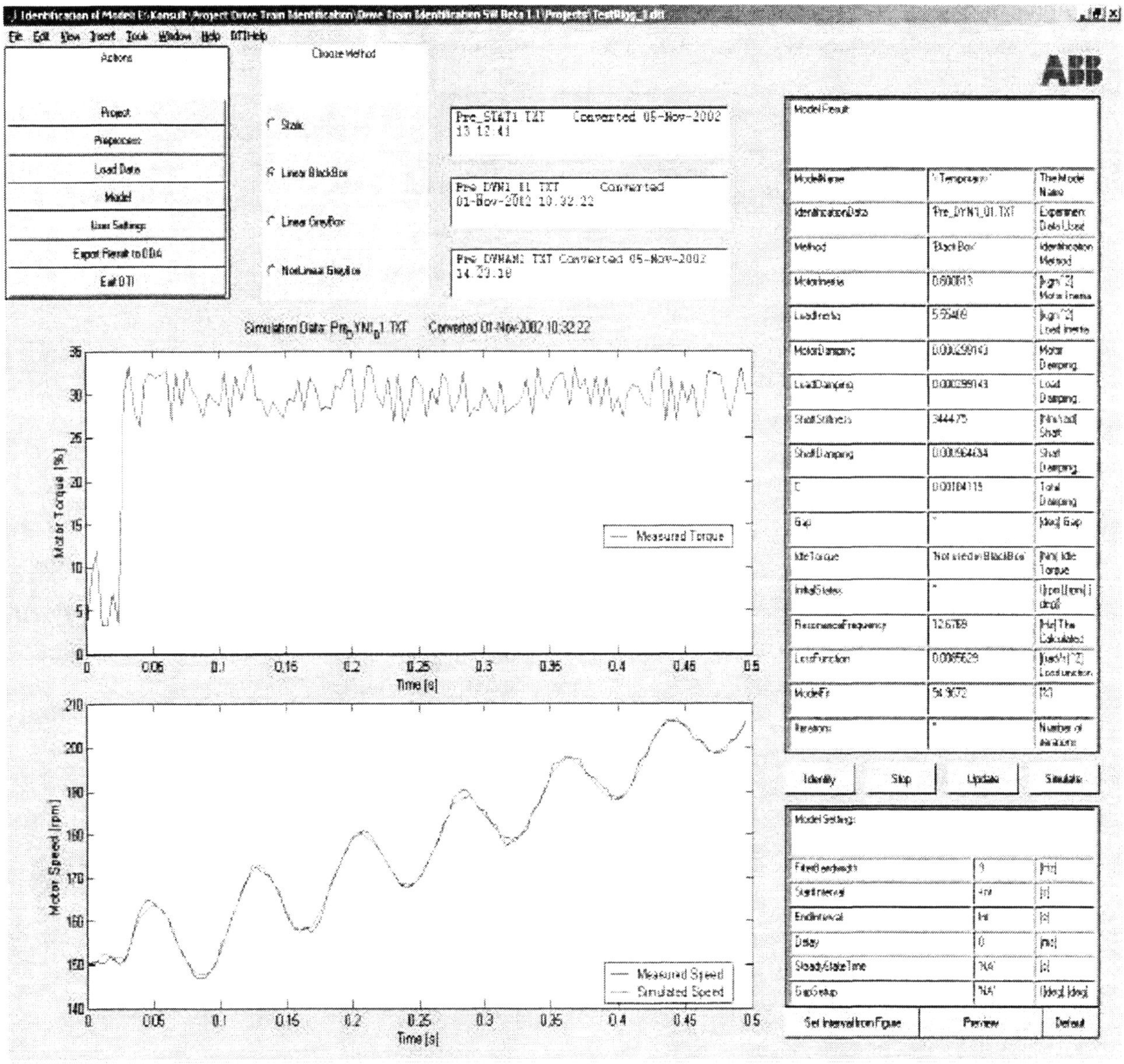

Figure 5. Main window of the Drive Train Identification Tool.

IFAC Publications
www.elsevier.com/locate/ifac

IDENTIFICATION AND MODEL PREDICTIVE CONTROL OF A pH NEUTRALIZATION PROCESS BASED ON LINEAR AND WIENER MODELS

Juan C. Gómez *,[1] **Arthur Jutan** **

* *Department of Chemical and Biochemical Engineering*
The University of Western Ontario
London, Ontario, Canada N6A 5B9
e-mail: jgomez@uwo.ca
** *e-mail:* ajutan@uwo.ca

Abstract: In this paper, a comparison between linear model-based and Wiener model-based Identification and Predictive Control of a pH neutralization process is presented. Input-Output data from a nonlinear first principles simulation model of the pH neutralization process are used for subspace-based identification of black-box linear and Wiener-type models. The proposed nonlinear subspace method has the advantage that it delivers a Wiener model in a format which is suitable for its use in a standard linear model-based predictive control scheme. The identified models are used as the internal models in a model predictive controller which is used to control the nonlinear white-box simulation model. Simulation results show that the Wiener-based model predictive controller outperforms the one based on the linear model. *Copyright © 2003 IFAC*

Keywords: Subspace Identification Methods, Wiener Model, Model Predictive Control, Process Control, pH Neutralization Process.

1. INTRODUCTION

The control of pH processes is a problem frequently encountered in the chemical process and biotechnological industries. It has been recognized as a challenging problem due to the time-varying and nonlinear characteristics of the pH processes. This is particularly true when control has to be achieved in the neutral range (pH between 6 and 8) when only strong acids and strong bases are present. Several techniques have been proposed in the literature for the control of pH processes, most of them resorting to nonlinear adaptive schemes (see for instance (Henson and Seborg, 1994), (Henson and Seborg, 1997), (Narayanan *et al.*, 1997)).

Some research has also been conducted on Model Predictive Control (MPC) of pH processes. Model predictive control refers to a class of control algorithms in which a dynamic process model is used to predict and optimize process performance (see for instance the book (Maciejowski, 2002)). MPC has been used in industry for more than thirty years, and has become an industry standard mainly do to its intrinsic capability for dealing with constraints [2] and with multivariable systems. Most commercially available MPC technologies are based on a linear model of the process. For processes which are highly nonlinear, the performance of an MPC based on a linear model can be poor. This has motivated the development of Nonlinear Model Predictive Control (NMPC), where a more accurate (nonlinear) model of the plant is used for prediction and optimization (see for instance (Henson, 1998) for

[1] On leave from the Laboratory for System Dynamics and Signal Processing, FCEIA, Universidad Nacional de Rosario, Riobamba 245 Bis, 2000 Rosario, Argentina. Author to whom all correspondence should be addressed.

[2] Often, a most efficient (and profitable) operation requires the process to work at or near such constraints.

a survey on the current status and future directions on Nonlinear Model Predictive Control).

Many of the current NMPC schemes are based on physical models of the process. However, in many cases such models are difficult to derive, and often not available at all. In these cases it makes sense to use a nonlinear empirical model, identified from input-output measurements. Some works where this approach has been followed are for instance: (Pottmann and Seborg, 1997) where a nonlinear predictive control scheme based on radial basis functions models is proposed, and (Fruzzetti *et al.*, 1997) where the NMPC is based on a Hammerstein model. In all these works the paradigmatic application has been pH neutralization processes.

From an identification point of view, pH processes have often been considered in the literature as having a Wiener structure (see for instance (Kalafatis *et al.*, 1995)) consisting in the cascade connection of a linear time-invariant (LTI) system followed by a static (memoryless) nonlinearity. In this structure, the linear block represents the mixing dynamics of the reagent streams in the stirring tank reactor (CSTR), while the static nonlinearity represents the nonlinear *titration curve* which gives the pH of the effluent solution as a function of the chemical components. Several methods have been proposed in the literature for the identification of Wiener models (see for instance (Greblicki, 1994), (Wigren, 1993), for some classical identification methods for Wiener models). More recently, some research interest has been focused on extending linear subspace identification methods (*c.f.*, (Van Overschee and de Moor, 1994), (Verhaegen, 1994), (Larimore, 1990)) for this class of nonlinear models (see (Westwick and Verhaegen, 1996), (Verhaegen and Westwick, 1996), (Gómez and Baeyens, 2002) for subspace identification of Hammerstein and Wiener models).

In this paper, subspace methods are used to identify a linear model and a Wiener model of the pH neutralization process, for their use in a Model Predictive Controller. The CVA[3] method by Larimore is employed to identify a linear model, while the method proposed in (Gómez and Baeyens, 2002) is used to identify a Wiener model of the process. Furthermore, a comparison between the performance of MPC based on the identified linear model and on the identified Wiener model of the pH neutralization process is carried out.

2. SUBSPACE-BASED WIENER MODEL IDENTIFICATION

In this section, the subspace-based Wiener model identification method originally introduced in (Gómez and Baeyens, 2002) is reviewed.

[3] CVA: Canonical Variate Analysis (Larimore, 1990).

2.1 *Problem Formulation*

A (multivariable) Wiener model is schematically depicted in figure 1. The model consists of the cascade connection of an LTI system followed by a zero-memory nonlinear element with input-output characteristic given by $\mathbf{N}(\cdot)$. The LTI subsystem has a state-space representation of the form

$$x_{k+1} = Ax_k + Bu_k + \omega_k, \quad (1)$$
$$v_k = Cx_k + Du_k + \vartheta_k, \quad (2)$$

where A, B, C and D, are the system matrices of appropriate dimensions, and where $x_k \in \mathbb{R}^n$, $v_k \in \mathbb{R}^m$, $u_k \in \mathbb{R}^p$, and $\vartheta_k \in \mathbb{R}^m$, represent the LTI system state, output, input, and process noise vectors at time k, respectively.

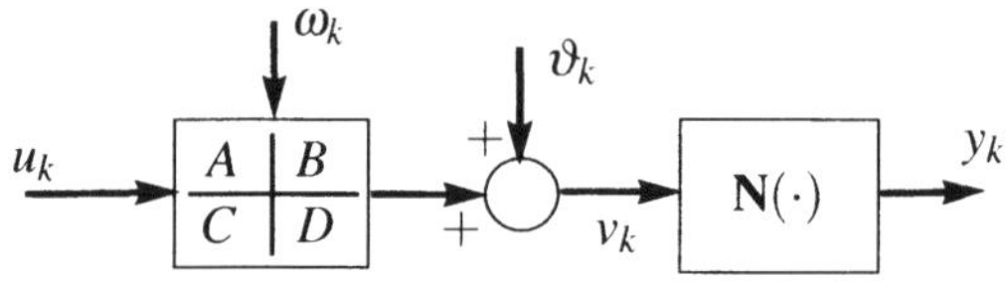

Fig. 1. Multivariable Wiener Model.

It will be assumed that the static nonlinear function $\mathbf{N}(\cdot) : \mathbb{R}^m \to \mathbb{R}^m$ is invertible, and that its inverse $\mathbf{N}^{-1}(\cdot)$ can be described as a linear combination of basis functions in the form

$$\mathbf{N}^{-1}(y_k) = \sum_{i=1}^{r} \alpha_i g_i(y_k) \quad (3)$$

where $g_i(\cdot) : \mathbb{R}^m \to \mathbb{R}^m$, $(i = 1, \cdots, r)$, are the assumed known nonlinear basis functions, and $\alpha_i \in \mathbb{R}^{m \times m}$, $(i = 1, \cdots, r)$, are unknown matrix parameters. Typically, the basis functions are polynomials[4], but they can also be basis functions generated by translations and dilations of a *mother* function (*e.g.*, Wavelets, or Radial Basis Functions).

With this representation for the static nonlinearity, equation (2) can be written as

$$\alpha Y_k \triangleq \sum_{i=1}^{r} \alpha_i g_i(y_k) = Cx_k + Du_k + \vartheta_k, \quad (4)$$

where $\alpha \triangleq [\alpha_1, \cdots, \alpha_r]$, $Y_k \triangleq \left[g_1^T(y_k), \cdots, g_r^T(y_k)\right]^T$. The Wiener model can then be described as

$$x_{k+1} = Ax_k + Bu_k + \omega_k, \quad (5)$$
$$Y_k = \widetilde{C}x_k + \widetilde{D}u_k + \widetilde{\vartheta}_k, \quad (6)$$

with $\widetilde{C} \triangleq \alpha^{\dagger}C$, $\widetilde{D} \triangleq \alpha^{\dagger}D$, $\widetilde{\vartheta}_k \triangleq \alpha^{\dagger}\vartheta_k$, and where $\alpha^{\dagger}$ stands for the left pseudoinverse of α. It is clear from equations (5)-(6) that the parametrization (1),(2), (3) is not unique, since any parameter matrices βC, βD and $\alpha^{\dagger}\beta^{-1}$, for some nonsingular matrix $\beta \in$

[4] Any smooth function in an interval can be represented with arbitrary accuracy by a polynomial of suffi ciently high order.

$\mathbb{R}^{m\times m}$, provide the same description (5)-(6). To obtain a one-to-one parametrization, *i.e.* for the system to be identifiable, additional constraints must be imposed on the parameters. A technique that can be used to obtain uniqueness is to normalize the parameter matrices $\alpha^\dagger$, that is to assume for instance that $\|\alpha^\dagger\|_2 = 1$. Under this assumption the parametrization (1),(2), (3) is unique.

Equations (5)-(6) can be interpreted as a state-space realization of an LTI system whose output Y_k is a filtered (by the assumed known basis functions $g_i(\cdot)$) version of the original output y_k. It is then clear that any available subspace identification algorithm (such as the N4SID algorithm by Van Overschee and de Moor (Van Overschee and de Moor, 1994), the MOESP algorithm by Verhaegen (Verhaegen, 1994), or the CVA algorithm by Larimore (Larimore, 1990), can be employed to obtain estimates of the system matrices $A, B, \widetilde{C}$, and $\widetilde{D}$ from input-output data.

2.2 *Subspace Identification Algorithm*

Given the estimates $\widehat{A}, \widehat{B}$, $\widehat{\widetilde{C}}$ and $\widehat{\widetilde{D}}$ of the matrices $A, B, \widetilde{C}$, and $\widetilde{D}$, respectively, the problem is how to compute estimates of the matrices C, D and α. Matrices $\widetilde{C}$ and $\widetilde{D}$ can be expressed in a combined form as

$$\left[\widetilde{C}\widetilde{D}\right] = \alpha^\dagger \left[CD\right]. \tag{7}$$

It is clear that the best (in the mean squares sense) estimates of matrices C, D and α are such that

$$\left(\widehat{C}, \widehat{D}, \widehat{\alpha}^\dagger\right) = \underset{C,D,\alpha^\dagger}{\arg\min}\left\{\left\|\left[\widehat{\widetilde{C}}\ \widehat{\widetilde{D}}\right] - \alpha^\dagger \left[C\ D\right]\right\|_2^2\right\} \tag{8}$$

The solution to this minimization problem is provided by the SVD of the matrix $\left[\widehat{\widetilde{C}}\ \widehat{\widetilde{D}}\right]$. The result is summarized in the following Theorem.

Theorem 2.1. Let $\left[\widehat{\widetilde{C}}\ \widehat{\widetilde{D}}\right] \in \mathbb{R}^{mr\times(n+p)}$ have rank $s > m$, and let the *economy-size* SVD of $\left[\widehat{\widetilde{C}}\ \widehat{\widetilde{D}}\right]$ be given by

$$\left[\widehat{\widetilde{C}}\ \widehat{\widetilde{D}}\right] = U_s\Sigma_s V_s^T \triangleq \left[U_1\ U_2\right]\begin{bmatrix}\Sigma_1 & 0\\ 0 & \Sigma_2\end{bmatrix}\begin{bmatrix}V_1^T\\ V_2^T\end{bmatrix} \tag{9}$$

where Σ_s is a diagonal matrix containing the s nonzero singular values $(\sigma_i, i = 1,\cdots,s)$ of $\left[\widehat{\widetilde{C}}\ \widehat{\widetilde{D}}\right]$ in nonincreasing order, and where the unitary matrices $U_s = \left[u_1\ u_2\ \cdots\ u_s\right] \in \mathbb{R}^{mr\times s}$ and $V_s = \left[v_1\ v_2\ \cdots\ v_s\right] \in \mathbb{R}^{(n+p)\times s}$ contain the corresponding left and right singular vectors, respectively. The partition of the SVD in (9) is such that the following dimensions for the matrices Σ_1, U_1 and V_1 hold, $\Sigma_1 \in \mathbb{R}^{m\times m}$, $U_1 \in \mathbb{R}^{mr\times m}$, and $V_1 \in \mathbb{R}^{(n+p)\times m}$.

Then, the matrices $\widehat{\alpha}^\dagger \in \mathbb{R}^{mr\times m}$, $\widehat{C} \in \mathbb{R}^{m\times n}$, and $\widehat{D} \in \mathbb{R}^{m\times p}$ that minimize the norm

$$\left\|\left[\widehat{\widetilde{C}}\ \widehat{\widetilde{D}}\right] - \alpha^\dagger \left[C\ D\right]\right\|_2^2,$$

are given by

$$\left(\widehat{\alpha}^\dagger, \left[\widehat{C}\ \widehat{D}\right]\right) = \left(U_1, \Sigma_1 V_1^T\right), \tag{10}$$

and the approximation error is given by

$$\left\|\left[\widehat{\widetilde{C}}\ \widehat{\widetilde{D}}\right] - \widehat{\alpha}^\dagger \left[\widehat{C}\ \widehat{D}\right]\right\|_2^2 = \sigma_{m+1}^2. \tag{11}$$

Proof: See (Gómez and Baeyens, 2002). ∎

Based on this result, the Subspace Identification Algorithm for the Wiener model can be summarized as follows

Algorithm 2.1.

Step 1: Compute estimates $\left(\widehat{A}, \widehat{B}, \widehat{\widetilde{C}}, \widehat{\widetilde{D}}\right)$ of the systems matrices $\left(A, B, \widetilde{C}, \widetilde{D}\right)$ in (5)-(6) using any available subspace algorithm for LTI systems.

Step 2: Compute the *economy size* SVD of $\left[\widehat{\widetilde{C}}\ \widehat{\widetilde{D}}\right]$ as in Theorem 2.1 , and the partition of this decomposition as in equation (9).

Step 3: Compute the estimates of the parameter matrices C, D and α as $\left[\widehat{C}\ \widehat{D}\right] = \Sigma_1 V_1^T$, and $\widehat{\alpha} = U_1^\dagger$, respectively, with U_1, V_1 and Σ_1 defined as in Theorem 2.1. □

3. PROCESS DESCRIPTION

The pH neutralization process consists of an acid (HNO_3) stream, a base ($NaOH$) stream, and a buffer ($NaHCO_3$) stream that are mixed in a constant volume (V) stirring tank. The process is schematically depicted in figure 2, and corresponds to a bench-scale plant at UCSB (see (Henson and Seborg, 1994), (Henson and Seborg, 1997)).

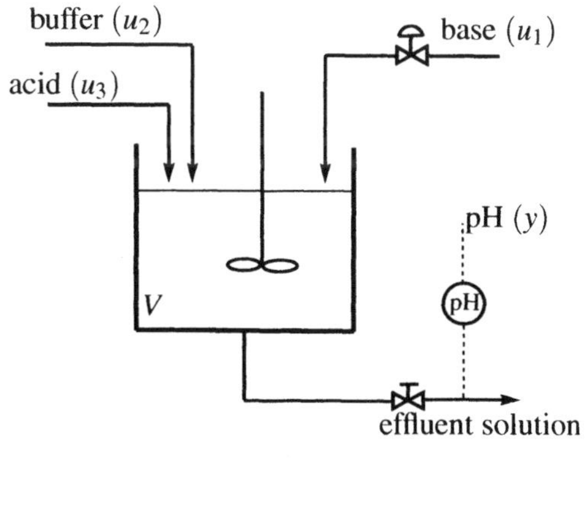

Fig. 2. Schematic representation of the pH neutralization process.

The inputs to the system are the base (volumetric) flow rate (u_1), the buffer flow rate (u_2), and the acid flow rate (u_3), while the output (y) is the pH of the effluent solution. The acid flow rate (u_3), as well as the volume (V) of the tank are assumed to be constant.

Usually, the objective is to control the pH of the effluent solution by manipulating the base flow rate, despite the variations of the unmeasured buffer flow rate.

A simulation model, based on first principles, was presented in (Henson and Seborg, 1994) introducing two *reaction invariants* (one related to a charge balance and the other to a balance on the carbonate ion) for each inlet stream. The reaction invariants will be denoted here as (W_{a1}, W_{b1}), (W_{a2}, W_{b2}), (W_{a3}, W_{b3}) and (W_a, W_b), for the base stream, the buffer stream, the acid stream, and the effluent solution, respectively. The model is highly nonlinear due to the implicit output equation, known as *titration curve* (equation (13)). The dynamic model for the reaction invariants of the effluent solution (W_a, W_b), in state-space form, is given by (Henson and Seborg, 1994), (Henson and Seborg, 1997):

$$\dot{x} = f(x) + g(x)u_1 + p(x)u_2, \quad (12)$$

$$h(x, y) = 0, \quad (13)$$

where

$$x \triangleq [x_1, x_2]^T = [W_a, W_b]^T, \quad (14)$$

$$f(x) = \left[\frac{u_3}{V}(W_{a3} - x_1), \frac{u_3}{V}(W_{b3} - x_2)\right]^T, \quad (15)$$

$$g(x) = \left[\frac{1}{V}(W_{a1} - x_1), \frac{1}{V}(W_{b1} - x_2)\right]^T, \quad (16)$$

$$p(x) = \left[\frac{1}{V}(W_{a2} - x_1), \frac{1}{V}(W_{b2} - x_2)\right]^T, \quad (17)$$

$$h(x, y) = x_1 + 10^{y-14} - 10^{-y} + x_2 \frac{1 + 2 \times 10^{y-pK_2}}{1 + 10^{pK_1 - y} + 10^{y-pK_2}} \quad (18)$$

Here, the parameters pK_1 and pK_2 are the first and second disassociation constants of the weak acid H_2CO_3. The nominal operating conditions of the system are given in (Henson and Seborg, 1994), (Henson and Seborg, 1997), and they are reproduced in Table 1 for the sake of completeness.

Table 1. Nominal Operating Conditions

$u_3 = 16.60$ ml/s	$u_2 = 0.55$ ml/s
$u_1 = 15.55$ ml/s	$V = 2900$ ml
$W_{a1} = -3.05 \times 10^{-3}$ M	$W_{a2} = -3 \times 10^{-2}$ M
$W_{a3} = 3 \times 10^{-3}$ M	$W_a = -4.32 \times 10^{-4}$ M
$W_{b1} = 5 \times 10^{-5}$ M	$W_{b2} = 3 \times 10^{-2}$ M
$W_{b3} = 0$ M	$W_b = 5.28 \times 10^{-4}$ M
$pK_1 = 6.35$	$pK_2 = 10.25$
$y = 7.0$	

4. PROCESS IDENTIFICATION

In this section, the nonlinear white box model (12)-(13) of the process is used to generate input-output data for the (black-box) identification of a Wiener model and a linear model of the process. The nonlinear subspace identification method of Section 2 is used to estimate a Wiener model, while the CVA subspace algorithm (Larimore, 1990) is used for the estimation of a linear state-space model of the process.

For the purposes of identification, the model (12)-(13) was excited with band limited white noise around the nominal value of the base flow rate, keeping the buffer flow rate and the acid flow rate constant in their nominal values. The output of the system was corrupted with additive Gaussian white noise with zero mean and standard deviation $\sigma = 0.001$, in order to simulate the more realistic situation of having measurement noise. The power of the output noise was chosen so that not to obscure the focus of the paper which is the comparison of the performance of the identified Wiener and linear models in the context of model predictive control.

The first one thousand data were used for the estimation of the models, while the following six hundred data were used for validation. The estimation and validation input-output data are represented in figure 3.

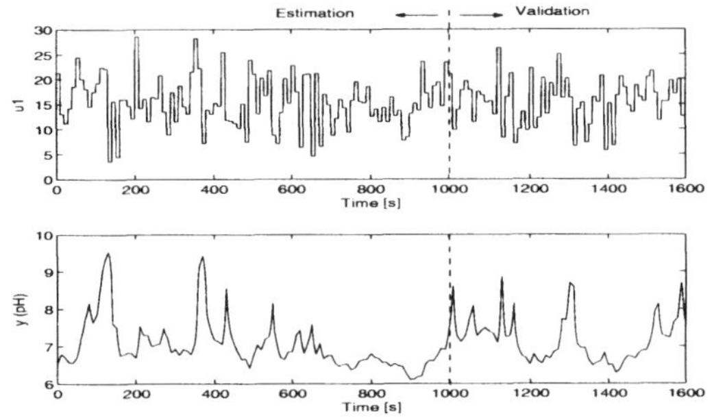

Fig. 3. Estimation (first 1000 points) and Validation (remaining 600 points) Input-Output Data.

In order to determine the model order of both the Wiener and the linear models, identification experiments were performed for 10 different realizations of the measurement noise, and for model orders in the range from 2 to 10. Only the stability of the identified models was considered, and the results are summarized in Table 2. There, the letter 'S' indicates that stable models for the 10 different realizations of the noise were obtained, while the letter 'U' indicates that at least one of the models for the ten different realizations, was unstable (actually, an average of 6 out of 10 were unstable in these cases). The default option ('Prediction') for the 'focus' property of the n4sid algorithm in version 5 of the System Identification Toolbox for use with Matlab (hereafter referred as SIT) was employed for the first step of the nonlinear identification algorithm, as well as for the identification of the linear model. This option means that the models are determined by minimizing the prediction errors, and corresponds to the optimal weighting from a statistical variance point of view (Ljung, 2000).

In this sense Wiener models perform better than linear models, since they remain stable for a wider range of model orders. Based on these results, and in order to

Table 2. Model Order and Stability ('S': stable , 'U': unstable)

Mod. order	2	3	4	5	6	7	8	9	10
Wiener	S	S	S	S	S	S	S	S	S
Linear	S	S	U	U	U	U	U	U	U

be able to make a comparison, a model order equal to 3 was selected using Akaike's criterion.

The input-output data in figure 3 were used to identify a Wiener model using the algorithm presented in Section 2. The CVA algorithm by Larimore (Larimore, 1990) as implemented by Ljung in SIT (Ljung, 2000), was used in the first step of the algorithm. A third order model, with transfer function

$$\widehat{G}(z) = \frac{0.0251z^2 - 0.0488z + 0.0237}{z^3 - 2.9210z^2 + 2.8435z - 0.9226}$$

was estimated for the linear block in the Wiener model. On the other hand, the following third order polynomial,

$$\widehat{\mathbf{N}}^{-1}(y_k) = 0.8918y_k^3 + 0.4065y_k^2 + 0.1988y_k$$

was estimated for the inverse of the nonlinear block in the Wiener model. The estimated nonlinear characteristic is represented in the left plot of figure 4. The true and estimated output (for both the estimation and the validation data) are represented in the right plot of figure 4, where a good agreement between them can be observed.

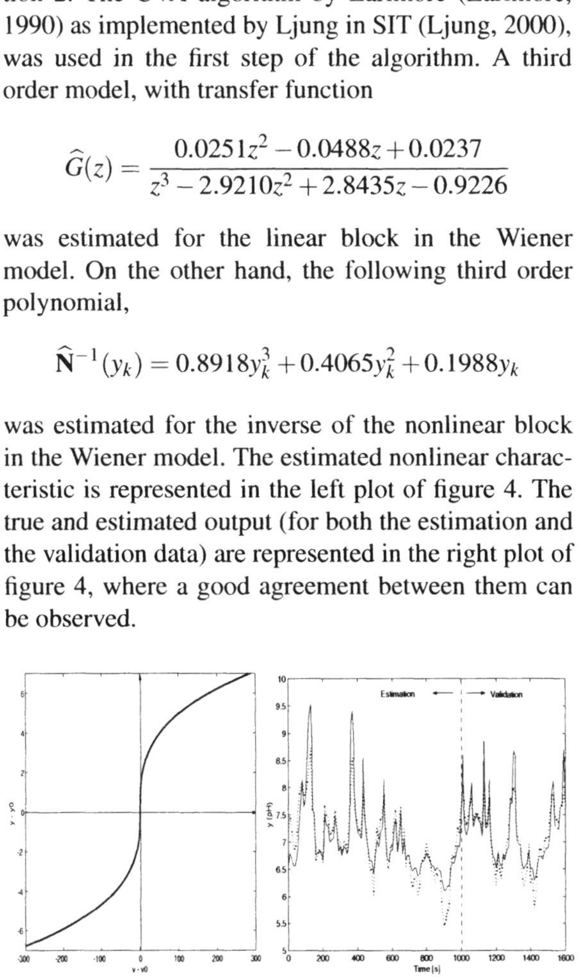

Fig. 4. Left Plot:Estimated Nonlinear Characteristic. Right Plot: True (solid line) and Estimated (dotted line) Output (Estimation and Validation Data).

The same data were used to estimate a third order linear model using the CVA algorithm by Larimore (Larimore, 1990), as implemented in SIT (Ljung, 2000). The estimated transfer function is

$$\widehat{G}_{lin}(z) = \frac{0.0084z^2 - 0.0133z + 0.0054}{z^3 - 2.5515z^2 + 2.1610z - 0.6085}$$

For the purposes of comparison between the Wiener and the linear models, three different performance criteria were considered, *viz.*, the Mean Square Error[5] (MSE), the Best Fit[6] (FIT) and the Variance Accounted For[7] (VAF) criterion (Ljung, 1999). The results are summarized in Table 3. It can be seen that, for this example, the prediction capability of the Wiener model is (slightly) better than that of the linear model.

Table 3. Error Comparison between Wiener and Linear Models

	MSE	FIT	VAF
Wiener	0.0728	56.8456	81.5129
Linear	0.0832	53.8870	78.7940

5. MODEL PREDICTIVE CONTROL

In this section, both the Wiener and the linear models identified in Section 4, are used for constrained model predictive control of the pH neutralization process. Saturation constraints in the manipulated variable are imposed to take into account the minimum and the maximum aperture of the valve regulating the base flow rate. For both cases (Wiener Model-based MPC and Linear model-based MPC), a lower limit of 0 ml/s and an upper limit of 30 ml/s were chosen for this variable, so that the nominal operating condition is approximately in the middle of the range (the nominal value of the base flow rate being 15.55 ml/s). The prediction horizon was chosen equal to 20, while the number of control moves was chosen equal to 5.

Simulations were performed using the nlmpcsim function in the MPC Toolbox for use with Matlab (Morari and Ricker, 1994). The function allows the simulation of closed-loop systems with saturation constraints on the manipulated variables using linear models in the *step* format for nonlinear plants represented as SIMULINK S-functions. There was no need to modify the nlmpcsim function for the simulation of the Wiener model based MPC. If the reader goes back to the identification method presented in Section 2, she will realize that in the first step of the algorithm, an equivalent linear model of the plant is constructed

[5] The MSE is defined as

$$MSE = \frac{1}{N}\sum_{k=1}^{N}(y_k - \widehat{y}_k)^2,$$

where y_k denotes the real output, $\widehat{y}_k$ denotes the output of the model, and N is the number of validation data.

[6] The Best Fit is defined as

$$FIT = \left(1 - \frac{\|Y - Y_v\|}{\|Y - y_{mean}\|}\right) \times 100,$$

where Y is a vector containing the output of the model when simulated with the validation input data, Y_v is a vector with the validation output data, and y_{mean} is the mean value of y.

[7] The Variance Accounted For is defined as

$$VAF = \max\left\{1 - \frac{\mathrm{Var}\{y - \widehat{y}\}}{\mathrm{Var}\{y\}}, 0\right\} \times 100\%,$$

where $y = \{y_k\}_{k=1}^{N}$ denotes the real output sequence, $\widehat{y} = \{\widehat{y}_k\}_{k=1}^{N}$ denotes the model output sequence, and $\mathrm{Var}\{\cdot\}$ denotes the variance of a quasi-stationary signal.

whose outputs are the actual outputs of the plant filtered by the (nonlinear and known) basis functions used to describe the inverse of the static nonlinearity. It is this equivalent linear model that can be used in the nlmpcsim function. This, of course, is an additional advantage of the proposed identification method, since it delivers a Wiener model in a format than can be used directly in a standard linear model-based MPC.

The simulation results corresponding to the linear model-based predictive control of the process are shown in the left plot of figure 5, while those corresponding to the Wiener model-based predictive control are shown in the right plot of the same figure. It can be observed that the Wiener model-based MPC performs better than the linear model-based one, when the operating region is far from the nominal operating conditions (*i.e.*, pH equal to 7).

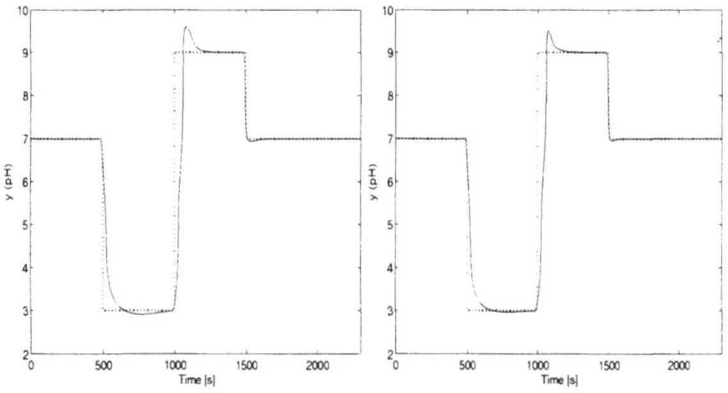

Fig. 5. Setpoint (dotted line) and Output (solid line) with MPC based on linear model (left plot), and Wiener model (right plot).

6. CONCLUDING REMARKS

In this paper, a comparison between linear model-based and Wiener model-based Identification and Predictive Control of a pH neutralization process has been presented. Input-Output data from a nonlinear first principles simulation model of the process were used for subspace-based identification of black-box linear and Wiener-type models. Simulation results showed that the identified Wiener models remained stable over a much wider range of model orders, in comparison with the identified linear models. In addition, the proposed nonlinear subspace method delivers a Wiener model in a format that can be used directly in a standard linear model-based predictive control. The identified models were used as the internal models in a model predictive control of the nonlinear white-box simulation model. Simulation results showed that, for the considered application, the Wiener-based model predictive controller outperformed the one based on the linear model, particularly when the system was operating away from the nominal operating conditions.

7. REFERENCES

Fruzzetti, K., A. Palazoglu and K. McDonald (1997). Nonlinear model predictive control using Hammerstein models. *J. of Process Control* **7**(1), 31–41.

Gómez, J. C. and E. Baeyens (2002). Subspace identification of multivariable Hammerstein and Wiener models. In: *Proc. of the 15th. IFAC World Congress*. Barcelona, Spain. pp. 2849–2854.

Greblicki, W. (1994). Nonparametric identification of Wiener systems by orthogonal series. *IEEE Trans. on Autom. Control* **39**(10), 2077–2086.

Henson, M. (1998). Nonlinear model predictive control: current status and future directions. *Comput. Chem. Engng.* **23**, 187–202.

Henson, M. and D. Seborg (1994). Adaptive nonlinear control of a pH neutralization process. *IEEE Trans. on Control Syst. Tech.* **2**(3), 169–182.

Henson, M. and D. Seborg (1997). *Nonlinear Process Control*. Chap. 4: "Feedback Linearizing Control". Prentice Hall. N.J.

Kalafatis, A., N. Arifin, L. Wang and W. Cluett (1995). A new approach to the identification of pH processes based on the Wiener model. *Chem. Engng. Sci.* **50**(23), 3693–3701.

Larimore, W. (1990). Canonical variate analysis in identification, filtering, and adaptive control. In: *Proc. of the 29th CDC*. Hawaii. pp. 596–604.

Ljung, L. (1999). *System Identification: Theory for the User*. 2nd ed.. Prentice-Hall, Inc.. New Jersey.

Ljung, L. (2000). *System Identification Toolbox, User's Guide, ver. 5*. The MathWorks, Inc.. Natick, MA.

Maciejowski, J. (2002). *Predictive Control with Constraints*. Prentice Hall.

Morari, M. and N. Ricker (1994). *Model Predictive Control Toolbox - For use with Matlab, User's Guide*. The MathWorks, Inc.. Natick, MA.

Narayanan, N., P. Krishnaswamy and G. Rangaiah (1997). An adaptive internal model control strategy for pH neutralization. *Chem. Engng. Sci.* **52**(18), 3067–3074.

Pottmann, M. and D. Seborg (1997). A nonlinear predictive control strategy based on radial basis functions models. *Comput. Chem. Engng.* **21**(9), 965–980.

Van Overschee, P. and B. de Moor (1994). N4SID: Subspace algorithms for the identification of combined deterministic-stochastic systems. *Automatica* **30**(1), 75–93.

Verhaegen, M. (1994). Identification of the deterministic part of MIMO state space models given in innovations form from input-ouput data. *Automatica* **30**(1), 61–74.

Verhaegen, M. and D. Westwick (1996). Identifying MIMO Hammerstein systems in the context of subspace model identification methods. *Int. J. Control* **63**(2), 331–349.

Westwick, D. and M. Verhaegen (1996). Identifying MIMO Wiener systems using subspace model identification methods. *Signal Processing* **52**, 235–258.

Wigren, T. (1993). Recursive prediction error identification using the nonlinear Wiener model. *Automatica* **29**(4), 1011–1025.

IFAC Publications
www.elsevier.com/locate/ifac

LOCAL MODELLING OF NONLINEAR DYNAMIC SYSTEMS USING DIRECT WEIGHT OPTIMIZATION

Jacob Roll * **Alexander Nazin** ** **Lennart Ljung** *

* *Div. of Automatic Control, Linköping University, SE-58183 Linköping, Sweden, e-mail:* `roll, ljung@isy.liu.se`
** *Institute of Control Sciences, Profsoyuznaya str., 65, 117997 Moscow, Russia, e-mail:* `nazine@ipu.rssi.ru`

Abstract: Local models and methods construct function estimates or predictions from observations in a local neighborhood of the point of interest. The bandwidth, i.e., how large the local neighborhood should be, is often determined based on asymptotic analysis. In this paper, an alternative, non-asymptotic approach that minimizes a uniform upper bound on the mean square error for a linear estimate is used. It is shown that the estimator is obtained from a quadratic program, that an automatic bandwidth selection is obtained, and that the approach can be seen as a local version of fitting affine models to data. Finally, the approach is applied to two benchmark systems. *Copyright © 2003 IFAC*

Keywords: Non-parametric identification, Function approximation, Minimax techniques, Quadratic programming, Nonlinear systems, Mean-square error, Local structures

1. INTRODUCTION: LOCAL MODELS

Non-linear black box models of dynamical systems have long been of central interest in system identification, see, e.g., the survey (Sjöberg *et al.*, 1995). In the control community mostly models of function expansion type have been applied, like Artificial Neural Network (ANN) models, wavelets, and (neuro-)fuzzy models. In statistics, the interest has been focused on various local methods, like kernel methods, (Nadaraya, 1964), local polynomial approaches, (Fan and Gijbels, 1996) and trees, (Breiman *et al.*, 1984).

A *local* model or method builds the function estimate or prediction from observations in a local neighborhood of the point of interest. Also most function expansion methods are of this character: A radial basis neural network is built up from basis functions with local support, and the standard sigmoidal (one hidden layer feed-forward) network is local around certain hyperplanes in the regressor space.

No matter what type of local modelling approach is taken, the central problem is the *bandwidth* question:

- How big should the local neighborhood be?

This is a variant of the classical bias/variance tradeoff. Intuitively, it is clear that the answer must depend on three items:

(1) How many data are available (and how are they distributed)?
(2) How smooth is the function surface (supposed to be)?
(3) How much noise is there in the observations?

This problem has been studied extensively in the statistical literature, and there are several solutions based on asymptotic (in the number of observations) analysis (Fan and Gijbels, 1996; Cleveland and Loader, 1994).

In this paper, another solution that is not based on the asymptotic behavior of the estimates is proposed. Based on a smoothness measure and noise variance, a uniform upper bound is computed of the mean square error (MSE) of a linear estimate, as a function of the estimator parameters. This upper bound is then minimized. It turns out that this problem can be reformulated as a quadratic programming (QP) problem, which can be solved efficiently. It also turns out that this solution has many of the key features of the asymptotically optimal estimators, but for finite number of observations it produces better guaranteed error bounds. The method will be called a *direct weight optimization* (DWO) approach, since the weights of the linear estimator are directly optimized, without any reference to kernels or parametric models. An early result in this direction was given in (Sacks and Ylvisaker, 1978).

The basic idea was presented earlier in (Roll *et al.*, 2002) for approximation of univariate functions, where examples also showed that it is sometimes superior to an asymptotically based local linear estimator. In this paper, the approach is extended to multivariate functions and put into a system identification context.

The paper is organized as follows: In Section 2, the problem is stated, and a very brief overview of some of the existing local approaches is given. The new approach is described in Section 3, and some of its most important properties are derived in Section 4. In Section 5 some examples are given of how the DWO approach can be used for prediction. Some extensions are described in Section 6 and conclusions are given in Section 7.

2. PROBLEM STATEMENT

Consider the problem of estimating the value $f(\varphi_0)$ of an unknown multivariate function $f : \mathbf{R}^n \to \mathbf{R}$ at a given point φ_0, given a set of input-output pairs $\{(\varphi_k, y_k)\}_{k=1}^N$, coming from the relation

$$y_k = f(\varphi_k) + e_k \tag{1}$$

Assume that the function f is continuously differentiable, and that there is a Lipschitz constant L such that

$$\|\nabla f(\varphi_1) - \nabla f(\varphi_2)\| \le L\|\varphi_1 - \varphi_2\| \tag{2}$$

where $\|\cdot\|$ is the Euclidean norm. Denote the class of functions satisfying these assumptions by $\mathscr{F}_L$.

The noise terms e_k are independent random variables with $Ee_k = 0$ and $Ee_k^2 = \sigma_k^2$, and should be independent of the regression variables φ_k. Both L and σ_k are assumed to be positive constants, given a priori. The notation

$$\tilde{\varphi}_k = \varphi_k - \varphi_0 \tag{3}$$

and $X = (\varphi_1, \dots, \varphi_N)$ will also be used.

The given problem arises, e.g., when predicting outputs from a system of NARX structure (Sjöberg *et al.*, 1995), where the system dynamics is known to satisfy (2). In this case, φ_k is composed of previous inputs $u_{k-\tau}$ and outputs $y_{k-\tau}$, i.e.,

$$\varphi_k = (y_{k-1} \; \cdots \; y_{k-n_a} \; u_{k-1} \; \cdots \; u_{k-n_b})^T \tag{4}$$

This fits in especially well within the "Model-On-Demand" (MOD) framework (see Stenman (1999)).

When the regression vectors depend on old values of y, though, this means that φ_i and e_j will not be independent for all values of i, j anymore (this is also noticed in (Stenman, 1999)). In most practical cases, however, this will only have small effects. The implications are further discussed in (Roll, 2003). Note also that for NFIR models such problems will not occur.

A common approach in statistics for estimation problems of this kind is to use a linear estimator in the form

$$\hat{f}(\varphi_0) = \sum_{k=1}^N w_k y_k \tag{5}$$

where $\hat{f}(\varphi_0)$ is our estimate of $f(\varphi_0)$. The problem then reduces to finding good weights w_k, which give reasonably small bias and variance of the estimate over the class $\mathscr{F}_L$. A common measure of the performance of the estimator is the *mean squared error* (MSE) (Fan and Gijbels, 1996) defined by

$$MSE(\hat{f}, \varphi_0) = E[(\hat{f}(\varphi_0) - f(\varphi_0))^2 | X] \tag{6}$$

Often one tries to minimize the worst-case MSE, i.e., achieve the *linear minimax risk*, defined by

$$R(X, L) = \inf_{\hat{f}(\varphi_0)} \sup_{f \in \mathscr{F}_L} E[(\hat{f}(\varphi_0) - f(\varphi_0))^2 | X] \tag{7}$$

where the infimum is taken over all linear estimators.

A classic family of methods are the *kernel methods*, where a *kernel function* K is used to determine the weights. An example is the Nadaraya-Watson estimator (Nadaraya, 1964; Watson, 1964). Common choices of kernels are the Gaussian kernel and the *spherical Epanechnikov kernel*

$$K(u) = C(1 - \|u\|^2)_+ \tag{8}$$

where $(\cdot)_+ = \max\{\cdot, 0\}$ and C is a normalization constant. The width of the kernel is specified by introducing a *bandwidth matrix* H, and using the kernel $K_H(u) = \det(H)^{-1} K(H^{-1}u)$ when calculating the weights. In the setting of this paper (i.e., where all that is known is that f satisfies (2)), it is sufficient to consider $H = hI$.

Another popular approach is the *local polynomial modelling* approach (Fan and Gijbels, 1996), where the estimator is determined by locally fitting a polynomial to the given data. A weighted least-squares problem is obtained, which for a first-order polynomial takes the form

$$\hat{\beta} = \arg\min_{\beta} \sum_{k=1}^N K_H(\tilde{\varphi}_k) \left(y_k - (\beta_0 + \beta_1^T \tilde{\varphi}_k)\right)^2 \tag{9}$$

Here, $\hat{f}(\varphi_0) = \hat{\beta}_0$, and the weights w_k in (5) are thus implicitly determined. The bandwidth h can be determined to asymptotically achieve the linear minimax risk (see Fan and Gijbels (1996) for details).

The different methods given in this section for choosing the weights w_k in the linear estimator (5) were all justified using asymptotic arguments, as $N \to \infty$. However, in reality only a finite number of data is given. Furthermore, these data may be sparsely and non-uniformly distributed. This might deteriorate the performance of the estimation methods.

In the following section, a non-asymptotic approach for determining the weights will be presented, based on a uniform (over $\mathscr{F}_L$) upper bound on the MSE (and hence on the linear minimax risk).

3. A DIRECT WEIGHT OPTIMIZATION APPROACH

As in the previous section, a linear estimator in the form (5) will be considered. For the bias to be bounded over $\mathscr{F}_L$, the following requirements on the weights w_k are needed (see Roll (2003)):

$$\sum_{k=1}^{N} w_k = 1 \tag{10a}$$

$$\sum_{k=1}^{N} w_k \tilde{\varphi}_k = 0 \tag{10b}$$

Under these restrictions, any linear function is estimated with zero bias. Note also that the local linear estimator described in the last section will automatically satisfy these requirements.

From (2), one can obtain

$$f(\varphi) = f(\varphi_0) + \nabla^T f(\varphi_0)\tilde{\varphi} + \frac{c(\varphi)}{2}\|\tilde{\varphi}\|^2 \tag{11}$$

for some $c(\varphi)$, $|c(\varphi)| \leq L$. Using all these equations, one gets

$$\hat{f}(\varphi_0) = f(\varphi_0) + \sum_{k=1}^{N} w_k \frac{c(\varphi_k)}{2}\|\tilde{\varphi}_k\|^2 + \sum_{k=1}^{N} w_k e_k$$

and an upper bound on the MSE (over $\mathscr{F}_L$)

$$E[(\hat{f}(\varphi_0) - f(\varphi_0))^2|X] \leq \frac{L^2}{4}\left(\sum_{k=1}^{N} \|\tilde{\varphi}_k\|^2 |w_k|\right)^2 + \sum_{k=1}^{N} w_k^2 \sigma_k^2 \tag{12}$$

Note that this bound is tight and attained by a paraboloid with the Hessian $\nabla^2 f(\varphi) = L \cdot I$ if the weights w_k are non-negative.

It is now natural to minimize the upper bound in (12). Hence, the values of w_k are chosen to minimize the following convex optimization problem:

$$\begin{aligned} \min_{w_k} \quad & \frac{L^2}{4}\left(\sum_{k=1}^{N} \|\tilde{\varphi}_k\|^2 |w_k|\right)^2 + \sum_{k=1}^{N} w_k^2 \sigma_k^2 \\ \text{subj. to} \quad & \sum_{k=1}^{N} w_k = 1 \\ & \sum_{k=1}^{N} w_k \tilde{\varphi}_k = 0 \end{aligned} \tag{13}$$

This problem is equivalent to the following QP:

$$\begin{aligned} \min_{w_k, s_k} \quad & \frac{L^2}{4}\left(\sum_{k=1}^{N} \|\tilde{\varphi}_k\|^2 s_k\right)^2 + \sum_{k=1}^{N} s_k^2 \sigma_k^2 \\ \text{subj. to} \quad & s_k \geq w_k \\ & s_k \geq -w_k \\ & \sum_{k=1}^{N} w_k = 1 \\ & \sum_{k=1}^{N} w_k \tilde{\varphi}_k = 0 \end{aligned} \tag{14}$$

The equivalence between (13) and (14) can be shown as follows: Given a feasible solution $w = (w_1, \ldots, w_N)$ to (13), a feasible solution to (14) with the same value of the objective function can easily be obtained by using the same w and $s_k = |w_k|$. Hence (14) is a relaxation of (13), and it suffices to show that when the optimal value of (14) is reached, then $s_k = |w_k|$ for all $k = 1, \ldots, N$. Suppose, e.g., that $s_1 > |w_1|$. Then, without changing any other variables, the value of the objective function can be reduced by decreasing s_1. This can be seen by observing that the coefficient before s_1 in the first sum of the objective function is non-negative, and the coefficient before s_1^2 in the second sum is positive, so decreasing s_1 will decrease the objective function. Therefore, when the objective function will reach its minimum, then $s_k = |w_k|$, and the equivalence is shown.

It should be pointed out, that the fact that the upper bound in (12) is tight for non-negative weights w_k does not necessarily mean that minimizing (14) yields the weights that minimize the worst-case MSE, even if the resulting weights are positive. The reason is that a subset of the weights that really minimize the worst-case MSE may be negative, and so the upper bound (12) is not tight for these weights. However, preliminary experimental results show that the solution of (14) gives an upper bound which is mostly within a few percents from the optimal worst-case MSE.

Solving the QP (14) can be done very efficiently using standard solvers, e.g., CPLEX (ILOG, Inc., 2000).

4. SOME BASIC PROPERTIES OF THE DWO APPROACH

Since the DWO approach minimizes an upper bound on the linear minimax risk, one would expect that the weights w_k would asymptotically converge to the weights of the local linear estimator using the Epanechnikov kernel with an asymptotically optimal bandwidth (Fan and Gijbels, 1996). This is also the case under certain assumptions, as the following theorem shows. For a proof of the result, see Roll (2003).

Theorem 4.1. Consider the problem of estimating an unknown univariate function $f : \mathbf{R} \to \mathbf{R}$, $f \in \mathscr{F}_L$ at a given internal point $\varphi_0 \in (-1/2, 1/2)$ under a *fixed, equidistant design model* as follows

$$\varphi_k = \frac{k-1}{N-1} - \frac{1}{2}, \quad k = 1, \ldots, N \tag{15}$$

and with $\sigma_k = \sigma > 0$ for all $k = 1, \ldots, N$. Let

$$w^{opt} = (w_1^{opt}, \ldots, w_N^{opt})$$

be the minimum point of the upper bound function

$$U(w) = \left(\frac{L}{2}\sum_{k=1}^{N} |w_k||\tilde{\varphi}_k|^2\right)^2 + \sigma^2 \sum_{k=1}^{N} w_k^2 \qquad (16)$$

subject to the constraints (10). Then asymptotically, as $N \to \infty$,

$$w_k^{opt} \approx \frac{3}{4} C_N \left(1 - \left(\frac{\tilde{\varphi}_k}{h_N}\right)^2\right)_+, \quad k = 1, \ldots, N \quad (17)$$

where

$$C_N \asymp \frac{1}{Nh_N}, \quad h_N \asymp \left(\frac{15\sigma^2}{L^2 N}\right)^{1/5} \quad \text{as } N \to \infty \quad (18)$$

Hence, the optimal weights (17) approximately coincide with related asymptotically optimal weights and bandwidth of the local polynomial estimator for the worst case function in $\mathscr{F}_L$.

Here $a_N \asymp b_N$ means that $a_N / b_N \to 1$ as $N \to \infty$ for two real sequences (a_N) and (b_N).

Remark. When the data are symmetrically distributed around φ_0, e.g., when $\varphi_0 = 0$, the relation (17) will hold exactly also for finite N. In other words, the non-zero weights will lie along a parabola given by (17), with an asymptotic bandwidth given by (18).

Another interesting feature of the DWO approach is that in most cases, the weights w_k corresponding to φ_k lying beyond a certain distance from φ_0 will be zero. This can be regarded as an *automatic bandwidth selection*, which means that the user does not have to bother about how many of the samples should be included in the estimator. In fact, the following theorem holds (see also (Sacks and Ylvisaker, 1978) for a similar theorem in a slightly different setting):

Theorem 4.2. Suppose that the problem (14) is feasible, and $\sigma_k = \sigma > 0$ for all $k = 1, \ldots, N$. Then there exist $\mu_1 \in \mathbf{R}$, $\mu_2 \in \mathbf{R}^n$, and $\mu_3 \in \mathbf{R}$, $\mu_3 \geq 0$, such that for an optimal solution (s^{opt}, w^{opt}), it holds that

$$w_k^{opt} = \qquad (19)$$
$$\begin{cases} \mu_1 + \mu_2^T \tilde{\varphi}_k - \mu_3 \|\tilde{\varphi}_k\|^2, & \mu_3 \|\tilde{\varphi}_k\|^2 \leq \mu_1 + \mu_2^T \tilde{\varphi}_k \\ 0, & |\mu_1 + \mu_2^T \tilde{\varphi}_k| \leq \mu_3 \|\tilde{\varphi}_k\|^2 \\ \mu_1 + \mu_2^T \tilde{\varphi}_k + \mu_3 \|\tilde{\varphi}_k\|^2, & \mu_1 + \mu_2^T \tilde{\varphi}_k \leq -\mu_3 \|\tilde{\varphi}_k\|^2 \end{cases}$$

Proof: See (Roll, 2003).

Figure 1 shows, for the univariate case, the principal shape of the curve along which the weights w_k^{opt} are placed. When $\mu_3 \|\tilde{\varphi}_k\|^2 \leq \mu_1 + \mu_2^T \tilde{\varphi}_k$ (which in the figure corresponds to the dashed line being above the upper dash-dotted parabola), the weights will be positive. When $\mu_1 + \mu_2^T \tilde{\varphi}_k \leq -\mu_3 \|\tilde{\varphi}_k\|^2$ (the dashed line is below the lower dash-dotted parabola), the weights are negative, and otherwise they are zero.

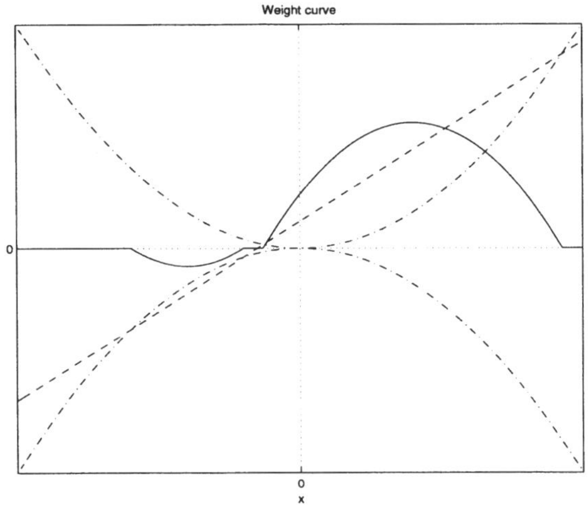

Fig. 1. Principal shape of the weight curve (solid curve). The dash-dotted parabolas are $\pm\mu_3\tilde{\varphi}^2$, and the dashed line is $\mu_1 + \mu_2^T \tilde{\varphi}$. (The weight curve is scaled by a factor 4 to make the figure more clear.)

The last result opens up for a possible reduction of the computational complexity: Since many of the weights w_k will be zero, one can already beforehand exclude data that will most likely correspond to zero weights, thus making the QP (14) considerably smaller. Having solved (14), one can easily check whether or not the excluded weights really should be zero, by checking if the excluded data points satisfy $|\mu_1 + \mu_2^T \tilde{\varphi}_k| \leq \mu_3 \|\tilde{\varphi}_k\|^2$ (the middle case of (19)). For more details, see (Roll, 2003).

A third property to note is that, when $L = 0$, the optimal weights from (13) equal the weights implicitly obtained from (9), with $K_H(\tilde{\varphi}_k)$ replaced by $1/\sigma_k^2$, just as one would expect. Since the latter approach gives a global affine model of ARX type (which is independent of φ_0), the weights obtained from (13) with $L = 0$ can also be seen as corresponding to an underlying global affine model. Hence, the proposed DWO approach can be interpreted as an alternative local version of fitting an "affine ARX model" to the data. The property is given in the following theorem:

Theorem 4.3. For $L = 0$, (13) gives the same weights as a local linear model given by (9) with $K_H(\tilde{\varphi}_k) = 1/\sigma_k^2$.

Proof: With $K_H(\tilde{\varphi}_k) = 1/\sigma_k^2$ and

$$\beta = \begin{pmatrix} \beta_0 \\ \beta_1 \end{pmatrix}, \quad \Phi = \begin{pmatrix} 1 & \ldots & 1 \\ \varphi_1 & \ldots & \varphi_N \end{pmatrix},$$
$$Y = \begin{pmatrix} y_1 \\ \vdots \\ y_N \end{pmatrix}, \quad Q = \begin{pmatrix} 1/\sigma_1 & & 0 \\ & \ddots & \\ 0 & & 1/\sigma_N \end{pmatrix}$$

$\hat{\beta}_0$ obtained from (9) can be written

$$\hat{\beta}_0 = \mathbf{e}_1^T (\Phi Q^2 \Phi^T)^{-1} \Phi Q^2 Y \qquad (20)$$

where $\mathbf{e}_1$ is the first vector of the standard basis. This corresponds to the weights

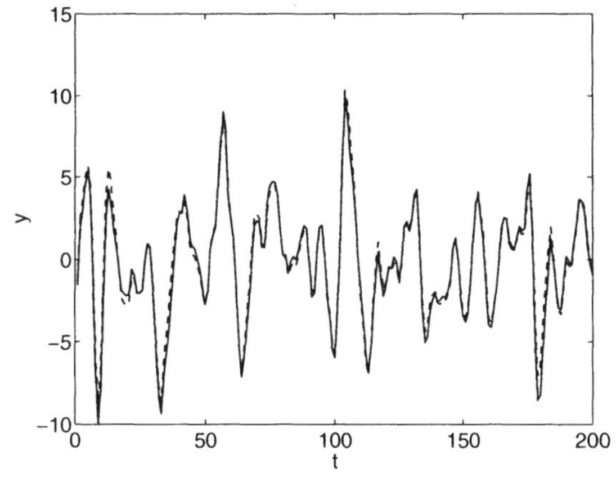

Fig. 2. Simulated (solid) and true (dashed) output for system (24) with $L_0 = 0$, modelled using the DWO approach with $L = 0.01$.

$$w^{LLM} = Q^2\Phi^T(\Phi Q^2\Phi^T)^{-1}\mathbf{e}_1 \tag{21}$$

On the other hand, if $L = 0$, (13) can be formulated as

$$\begin{aligned} \min_{w} \quad & w^T Q^{-2} w \\ \text{subj. to} \quad & \Phi w = \mathbf{e}_1 \end{aligned} \tag{22}$$

or, letting $v = Q^{-1}w$,

$$\begin{aligned} \min_{v} \quad & v^T v \\ \text{subj. to} \quad & \Phi Q v = \mathbf{e}_1 \end{aligned}$$

But this is nothing else than finding the least-norm solution of a system of linear equations, which is easily done using the pseudoinverse, i.e., $v = (\Phi Q)^{\dagger}\mathbf{e}_1$. Hence

$$w = Qv = Q^2\Phi^T(\Phi Q^2\Phi^T)^{-1}\mathbf{e}_1 \tag{23}$$

which are exactly the same weights as (21).

5. EXAMPLES

Example 5.1. To give an example of the DWO approach, it was applied to an extended version of the linear so-called Åström system (Ljung, 1999):

$$\begin{aligned} y(t) = & \qquad\qquad\qquad\qquad\qquad\qquad (24) \\ & 1.5y(t-1) - 0.7y(t-2) + u(t-1) + 0.5u(t-2) \\ & + \alpha + L_0(\cos y(t-1) + 0.5u^2(t-1)) + e(t) \end{aligned}$$

First α and L_0 were set to zero. To get estimation data, $u(t)$ and $e(t)$ were both selected as random Gaussian sequences of length 500, with unit variance. As validation data, 200 samples of noise-free data were selected, with $u(t)$ generated in the same way as for the estimation data. The simulated output for $L = 0.01$ is shown in Figure 2. As can be seen, the simulated output follows the true output well (84.9% fit). For $L = 0$, the result is the same as fitting an affine model using a least-squares criterion, according to Theorem 4.3, giving 85.9% fit. As comparison, a linear ARX model was also estimated, and performed slightly better compared to the other approaches (91.8% fit), as expected, since the true system was linear.

Choosing $L_0 = L = 1$, $\alpha = 1$ and using an estimation data sequence of 50000 samples (generated as above), yielded a fit of 52.4% for the simulated output, as compared to 41.8% for a linear ARX model (estimated after removing means from the data). The corresponding numbers for one-step-ahead predictions were 94.8% and 77.5%, respectively.

Example 5.2. As another example, a nonlinear benchmark system proposed by (Narendra and Li, 1996) is considered. The system is defined in state-space form by

$$\begin{aligned} x_1(t+1) &= \left(\frac{x_1(t)}{1+x_1^2(t)} + 1\right)\sin x_2(t) \\ x_2(t+1) &= x_2(t)\cos x_2(t) + x_1(t)e^{-\frac{x_1^2(t)+x_2^2(t)}{8}} \\ &\quad + \frac{u^3(t)}{1+u^2(t)+0.5\cos(x_1(t)+x_2(t))} \\ y(t) &= \frac{x_1(t)}{1+0.5\sin x_2(t)} + \frac{x_2(t)}{1+0.5\sin x_1(t)} + e(t) \end{aligned} \tag{25}$$

The noise term $e(t)$ is added in accordance with (Stenman, 1999) and has a variance of 0.1. The states are assumed not to be measurable, and following the discussion in (Stenman, 1999), a NARX331 structure is used to model the system, i.e.,

$$\varphi(t) =$$
$$(y(t-1)\; y(t-2)\; y(t-3)\; u(t-1)\; u(t-2)\; u(t-3))^T$$

As estimation data, $N = 50000$ samples were generated using a uniformly distributed random input $u(t) \in [-2.5, 2.5]$. For validation the input signal

$$u(t) = \sin\frac{2\pi t}{10} + \sin\frac{2\pi t}{25}, \quad t = 1,\ldots,200$$

was used. Figure 3 shows the simulated output when L was chosen to be 0.1. The results are reasonable (49.7% fit), although it should be noted that the Lipschitz constant is not known a priori, and is chosen ad hoc to be constant over the entire state-space. In fact, since the real system is not of NARX structure, there might not even exist a such a Lipschitz constant. Therefore, combining the approach with a local estimation of L using an algorithm similar to the bandwidth selection methods in, e.g., (Stenman, 1999), would probably improve the results.

6. EXTENSIONS

6.1 *Prior knowledge*

One may think of the constraints (10) as *ad hoc*, although reasonable. These constraints can be removed if instead a restricted family of functions is considered with some prior knowledge of the function value and its derivative:

$$|f(\varphi_0) - a| \le A, \qquad |f'(\varphi_0) - b| \le B \tag{26}$$

and formulate the corresponding min-max MSE problem. It can then be shown, see (Roll, 2003), that for

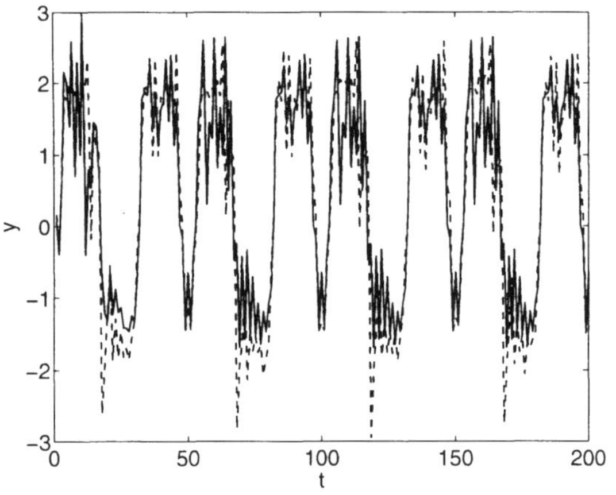

Fig. 3. Simulated (solid) and true (dashed) output for system (25), modelled using the DWO approach with $L = 0.1$.

sufficiently large A and B, the solution will automatically be subject to the constraints (10).

Moreover, if a reasonable linear estimate of $\nabla f(\varphi_0)$ is formed and plugged in as a known value (with or without error), the resulting estimator will obey (10b) (see Roll (2003) for the univariate case).

6.2 *More smoothness*

The criterion that has been used assumes the function to be once continuously differentiable with a bound on the Lipschitz constant of the derivative. It is quite easy to extend this to any degree of assumed differentiability. The min-max problem will still be reducible to a QP problem, (Roll, 2003).

7. CONCLUSIONS

The proposed non-asymptotic DWO approach to local modelling has a number of interesting features:

- The problem is phrased without any reference to bandwidth. The formulation offers the possibility to use all observations. Theorem 4.2 however shows that there is a bandwidth feature even for a finite number of measurements: Observations outside a certain band carry weights that are exactly zero.
- Although the DWO approach does not give strictly better estimates (in the MSE sense) than, say, the local polynomial approach in all cases, the important point is that the delivered guaranteed MSE bound is better than what other approaches can offer. In practice it is of course only this guaranteed bound that can be used for confidence intervals etc, since the actual MSE depends on the unknown function.
- As is illustrated in (Roll *et al.*, 2002), the improvement over asymptotically optimal estimates is more pronounced (naturally enough) for fewer data, and more non-uniformly distributed observation points φ_k. For applications to higher regressor dimensions and dynamical systems, this is a very valuable property.

Finally, one may ask how to find the "input values" L and σ_k^2 to the algorithms. This is the same problem as for the kernel methods and the local polynomial approach, and has been extensively studied in the statistical literature. In (Fan and Gijbels, 1996) related ideas can be found.

The proposed approach can be used as an alternative to building non-linear black-box models in an MOD fashion and applied to, for example, model predictive control. See (Stenman, 1999) for such ideas.

8. REFERENCES

Breiman, L., J. H. Friedman, R. A. Olshen and C. J. Stone (1984). *Classification and Regression Trees*. Wadsworth & Brooks.

Cleveland, W. S. and C. Loader (1994). Smoothing by local regression: Principles and methods. Technical report. AT&T Bell Laboratories. 600 Mountain Avenue, Murray Hill, NJ 07974, USA.

Fan, J. and I. Gijbels (1996). *Local Polynomial Modelling and Its Applications*. Chapman & Hall.

ILOG, Inc. (2000). *CPLEX 7.0 User's Manual*. Gentilly, France.

Ljung, L. (1999). *System Identification: Theory for the User*. 2nd ed.. Prentice-Hall.

Nadaraya, E. (1964). On estimating regression. *Theory of Probability and its Applications* **10**, 186–190.

Narendra, K. S. and S.-M. Li (1996). Neural networks in control systems. In: *Mathematical Perspectives on Neural Networks* (P. Smolensky, M. C. Mozer and D. E. Rumelhart, Eds.). Chap. 11, pp. 347–394. Lawrence Erlbaum Associates.

Roll, J. (2003). Local and Piecewise Affine Approaches to System Identification. PhD thesis. Department of Electrical Engineering, Linköping University. SE-581 83 Linköping, Sweden.

Roll, J., A. Nazin and L. Ljung (2002). A non-asymptotic approach to local modelling. In: *The 41st IEEE Conference on Decision and Control*. pp. 638–643.

Sacks, J. and D. Ylvisaker (1978). Linear estimation for approximately linear models. *The Annals of Statistics* **6**(5), 1122–1137.

Sjöberg, J., Q. Zhang, L. Ljung, A. Benveniste, B. Delyon, P. Y. Glorennec, H. Hjalmarsson and A. Juditsky (1995). Nonlinear black-box modeling in system identification: a unified overview. *Automatica* **31**(12), 1691–1724.

Stenman, A. (1999). Model on Demand: Algorithms, Analysis and Applications. PhD thesis. Department of Electrical Engineering, Linköping University. SE-581 83 Linköping, Sweden.

Watson, G. (1964). Smooth regression analysis. *Sankhyā* **A**(26), 359–372.

IFAC Publications
www.elsevier.com/locate/ifac

OPTIMALITY IN SM IDENTIFICATION OF NONLINEAR SYSTEMS

Mario Milanese and Carlo Novara

Dipartimento di Automatica e Informatica, Politecnico di Torino
Corso Duca degli Abruzzi 24, 10129 Torino, Italy
Email: milanese@polito.it, novara@polito.it

Abstract: In the paper the problem of identifying discrete time nonlinear systems from finite and noise-corrupted measurements is considered. Most methods in the literature are based on the estimation of a model within a finitely parametrized model class describing the functional form of involved nonlinearities. A key problem in these methods is the proper choice of the model class, typically realized by a search, from the simplest to more complex ones (linear, bilinear, polynomial, neural networks, etc.). In this paper an alternative approach, based on a Set Membership framework is presented, not requiring assumptions on the functional form of the regression function describing the relations between measured input and output, but assuming only some information on its regularity, given by bounds on its gradient. In this way, the problem of considering approximate functional forms is circumvented. Moreover, noise is assumed to be bounded, in contrast with statistical methods, which rely on assumptions such as stationarity, ergodicity, uncorrelation, type of distribution, etc., whose validity may be difficult to be reliably tested and is lost in presence of approximate modeling. Necessary and sufficient conditions for assumptions validation are given. An optimal interval estimate of the regression function is obtained, providing its uncertainty range for any assigned regressor values. The set estimate allows to derive an optimal identification algorithm, giving estimates with minimal guaranteed L_p error on the overall domain of the regressors. The properties of the optimal estimate are investigated and its worst-case L_p identification error is evaluated. *Copyright © 2003 IFAC*

Keywords: Nonlinear Identification, Model quality, Set Membership Identification, Optimal Algorithm.

1. INTRODUCTION

Consider a nonlinear dynamic system of the form:

$$y_{t+1} = f_o(w_t) \qquad (1)$$

$w_t = [y_t \dots y_{t-n_y+1}\ u_t^1 \dots u_{t-n_1+1}^1 \dots u_t^m \dots u_{t-n_m+1}^m]^T$ where $y_t, u_t^1, \dots, u_t^m \in \Re$, $f_o : \Re^n \to \Re$, $n = n_y + \sum_{i=1}^m n_i$.

Suppose that the function f_o is not known, but a set of noise corrupted measurements of y_t and w_t, $t = 1, 2, \dots, T$ is available, and it is of interest to make an inference on the system (e.g. identification, prediction, smoothing, filtering, control design, decision making, fault detection, etc.).

The inference is described by the operator $I(f_o, W_T) : L_p(W) \times \Re^{n \times T} \longrightarrow \mathcal{I}$ where $W_T \doteq [w_1, w_2, \dots, w_T] \in \Re^{n \times T}$, W is a bounded set in $\Re^n$, $L_p(W)$ is the space of functions f for which $|f(w)|$ is Lebesgue-integrable over W, equipped with the norm $\|f\|_p \doteq \left[\int_W |f(w)|^p dw\right]^{\frac{1}{p}}$, $p \in [1, \infty)$ and $\|f\|_\infty \doteq$ess-sup$_{w \in W} |f(w)|$, and $\mathcal{I}$ is a normed space called inference space.

Clearly, being f_o and W_T unknown, the exact inference $I(f_o, W_T)$ cannot be derived and the usual approach is to obtain from data some estimate $\widehat{f}$ of f_o, and $\widehat{W}_T$ of W_T and to make the approximate inference $I\left(\widehat{f}, \widehat{W}_T\right)$. Then, two basic problems arise: i) evaluate the inference error measured as $e\left(\widehat{f}, \widehat{W}_T\right) \doteq \|\, I(f_o, W_T) - I\left(\widehat{f}, \widehat{W}_T\right) \|_{\mathcal{I}}$, where $\|\cdot\|_{\mathcal{I}}$ is a suitable norm; ii) find estimates $\widehat{f}$ and $\widehat{W}_T$ minimizing $e\left(\widehat{f}, \widehat{W}_T\right)$.

In this paper the focus is on the case that the desired inference is identification of f_o, i.e. $I(f_o, W_T) = f_o$. The case that desired inference is prediction has been considered in (Novara and Milanese, 2001; Milanese and Novara, 2002*a*).

In the identification problem here investigated, the aim is to find an estimate $\widehat{f}$ of f_o, giving small, possibly minimal, identification error $\left\| f_o - \widehat{f} \right\|_p$. However, this error is not known and indeed, since data are finite and noise corrupted, no reliable estimate on the identification error can be derived if no information is available on f_o and on noise. The information on f_o is typically given by considering that it belongs to some subset K of functions $f : L_p(W) \to \Re$. In some cases, the knowledge of the laws governing the system (mechanical, economical, biological, etc.) generating the data, may allow to have reliable information on its structure. In many other situations, due to the fact that the laws are too complex or not sufficiently known, this is not possible or not convenient and the usual approach is to consider that f_o belongs to a finitely parametrized set of functions $K \doteq \{f(p), \, p \in \Re^q\}$. Then, measured data are used to derive an estimate $\widehat{p}$ of p and $f(\widehat{p})$ is used to obtain inferences. Basic to this approach is the proper choice of the set of functions $f(p)$, typically realized by some search on different functional forms, starting from the simplest ones, such as linear models, and moving to more complex ones, such as piecewise linear, bilinear, polynomial, neural networks, etc. (Haber and Unbehauen, 1990; Sjöberg *et al.*, 1995; Narenda and Mukhopadhyay, 1997; Isermann *et al.*, 1997). This search may be quite time consuming, and in any case leads to approximate model structures only. The evaluation of the effects of such approximation on inference errors appears at present to be a largely open problem.

In this paper an alternative approach is taken, formulating the problem in a Set Membership (SM) framework, used in linear systems identification to deal with modeling errors, see e.g. (Milanese and Tempo, 1985; Milanese and Vicino, 1991; Milanese *et al.*, 1996). The SM approach does not assume to know the functional form of f_o, but uses only some information on its regularity, given by bounds on the gradient of f_o. In this way, the problem of considering approximate functional forms of f_o is circumvented. Moreover, the method assumes only that the noise is bounded, in contrast with statistical approaches, which rely on assumptions such as stationarity, ergodicity, uncorrelation, type of distribution, etc. The validity of these assumptions may be difficult to be reliably tested in many applications and is lost in presence of approximate modeling.

The proposed approach has strong connections with method used in approximation theory for approximating multivariable functions with bounded derivatives, from the knowledge of a finite number of their values (see (Traub *et al.*, 1988; Novak, 1988; Pinkus, 1985; Wasilkowski and Woźniakowski, 2001) and the references therein). In this literature, noise free measurements are typically assumed, and weaker optimality concepts are considered than the one of the present paper (see the remark at the end of next section for a more specific discussion).

The paper is organized as follows. In section 2 the identification problem is formulated in a Set Membership framework, defining the type of considered assumptions, the guaranteed identification error and optimality concept. In section 3, necessary and sufficient conditions for assumptions validation are given and it is shown how they can be used for assessing the variables appearing in the assumptions. Moreover, an optimal interval estimate of $f_o(w)$ for any $w \in W$ is derived. In section 4, an optimal point estimate of f_o, having minimal guaranteed L_p identification error, is obtained and its properties are investigated. In section 5 two variations of the method are proposed, which may give significant improvement, allowing to adapt to properties of data, such as variable gradient bounds and quite different magnitude of gradient components.

2. THE NONLINEAR SM APPROACH

Consider that a set of noise corrupted data $\widetilde{Y}_T = [\widetilde{y}_{t+1}, \; t = 1,..,T]$ and $\widetilde{W}_T = [\widetilde{w}_t, \; t = 1,..,T]$ generated by (1) is available. Then:

$$\widetilde{y}_{t+1} = f_o(\widetilde{w}_t) + d_t, \; t = 1,..,T \tag{2}$$

where the term d_t accounts for the fact y_{t+1} and w_t are not exactly known.

The aim is to derive an estimate $\hat{f}$ of f_o from available measurements $(\widetilde{Y}_T, \widetilde{W}_T)$, i.e. $\hat{f} = \phi(\widetilde{Y}_T, \widetilde{W}_T)$. The operator ϕ, called identification algorithm, should be chosen to give small (possibly minimal) error $e(\hat{f}) = ||f_o - \hat{f}||_p$. This error is not known, since from available data it is only known that $f_o \in F(\widetilde{Y}_T, \widetilde{W}_T)$, the set of all f that can have generated the data. This set, even in case of exact measurements, is unbounded, since the mapping generating data from given f is not injective. Then, whatever algorithm ϕ is chosen, no information on the identification error can be derived, unless some assumptions are made on the function f_o and the noise d. The typical approach in the literature is to assume a finitely parametrized functional form for f_o (linear, bilinear, neural network, etc.) and statistical models on the noise (Haber and Unbehauen, 1990; Sjöberg *et al.*, 1995; Narenda and Mukhopadhyay, 1997; Isermann *et al.*, 1997). In the present SM-IBC approach, different and somewhat weaker assumptions are taken, not requiring the selection of a functional form for f_o, but related to its rate of variation. Moreover, the noise sequence $D_T = [d_1, d_2, ..., d_T]$ is only supposed to be bounded.

Prior assumptions on f_o:
$f_o \in K \doteq \{f \in C^1(W) : \|f'(w)\| \leq \gamma, \; \forall w \in W\}$
Prior assumptions on noise:
$D_T \in \mathcal{D} \doteq \{[d_1, ..., d_T] : |d_t| \leq \varepsilon_t, \; t = 1, 2, ..., T\}$ ■

Here, $f'(w)$ denotes the gradient of $f(w)$ and $\|x\| \doteq \sqrt{\sum_{i=1}^{n} x_i^2}$ is the Euclidean norm.

A key role in this Set Membership framework is played by the Feasible Systems Set, often called "unfalsified systems set", i.e. the set of all systems consistent with prior information and measured data.

Definition 1. Feasible Systems Set
The Feasible Systems Set FSS_T is:

$$FSS_T \doteq \{f \in K : |\tilde{y}_{t+1} - f(\widetilde{w}_t)| \leq \varepsilon_t, t{=}1{,}2{,}...{,}T\} \quad (3)$$

■

The Feasible Systems Set FSS_T summarizes all the information on the mechanism generating the data that is available up to time T. If prior assumptions are "true", then $f_o \in FSS_T$, an important property for evaluating the accuracy of inferences that can be done on the system. In particular, it follows that $f_o(w)$ is bounded as:

$$\underline{f}(w) \leq f_o(w) \leq \overline{f}(w), \forall w \in W \quad (4)$$

$$\begin{aligned} \overline{f}(w) &= \sup_{f \in FSS_T} f(w) \\ \underline{f}(w) &= \inf_{f \in FSSF_T} f(w) \end{aligned} \quad (5)$$

Provided that prior assumptions hold, $\overline{f}(w)$ and $\underline{f}(w)$ are the tightest upper and lower bounds of $f_o(w)$ and are called *optimal bounds.*

As typical in any identification theory, the problem of checking the validity of prior assumptions arises. The only thing that can be actually done is to check if prior assumptions are invalidated by data, evaluating if no system exists consistent with data and assumptions, i.e. if FSS_T is empty. Indeed the fact that the priors are consistent with the present data, i.e. $FSS_T \neq \emptyset$, does not exclude that they may be not consistent with future data. However, it is usual to introduce the concept of prior assumption validation as follows.

Definition 2. Validation of prior assumptions
Prior assumptions are considered validated if:

$$FSS_T \neq \emptyset$$

■

In the following, the FSS_T is assumed to be non-empty. If empty, the prior assumptions on the system and the noise are invalidated by data and have to be suitably modified to give a non-empty FSS_T.

An identification algorithm ϕ is an operator mapping all available information about function f_o, noise d, data $(\widetilde{Y}_T, \widetilde{W}_T)$ until time T, summarized by FSS_T, into an estimate $\hat{f} \in L_p(W)$ of function f_o:

$$\phi(FSS_T) = \hat{f} \simeq f_o$$

The related L_p error is:

$$e(\hat{f}) = e(\phi(FSS_T)) = \left\|f_o - \hat{f}\right\|_p$$

This error cannot be exactly computed, since it is only known that $f_o \in FSS_T$, but its tightest bound is given by $e(\hat{f}) \leq \sup_{f \in FSS_T} \left\|f - \hat{f}\right\|_p$.

This motivates the following definition of the identification error, often indicated as local worst-case or guaranteed error.

Definition 3. (Local) *identification error*
The (local) identification error of $\hat{f} = \phi(FSS_T)$ is:

$$E[\phi(FSS_T)] = E(\hat{f}) \doteq \sup_{f \in FSS_T} \left\|f - \hat{f}\right\|_p$$

■

Looking for algorithms that minimize the identification error, leads to the following optimality concepts.

Definition 4. (Locally) *optimal algorithm*
An algorithm ϕ^* is called (locally) optimal if:

$$E[\phi^*(FSS_T)] = \inf_{\phi} E[\phi(FSS_T)] =$$
$$= \inf_{\hat{f}} \sup_{f \in FSS_T} \left\|f - \hat{f}\right\|_p = r_I$$

■

The quantity r_I, called (local) *radius of information*, gives the minimal identification error that can be guaranteed by any estimate based on the available information up to time T.

Remark

Substituting $\tilde{y}_{t+1}$ in (3) with its expression (2), it is clear that FSS_T depends on f_o and the actual realization of noise sequence $D_T = [d_1, d_2, ..., d_T]$. Thus, the (local) identification error depends on f_o and D_t, i.e.:

$$E\left(\hat{f}\right) = E\left(\hat{f}, f_o, D_t\right)$$

A global identification error of given algorithms ϕ is often considered, defined as:

$$E^g(\phi) \doteq \sup_{f_o \in K,\ D_t \in \mathcal{D}} E(\phi(FSS_T), f_o, D_t)$$

An algorithm ϕ^g is called globally optimal if:

$$E^g(\phi^g) = \inf_{\phi} E^g(\phi)$$

This is the optimality concept usually investigated in the approximation theory literature ((Traub *et al.*, 1988; Novak, 1988; Pinkus, 1985; Wasilkowski and Woźniakowski, 2001)). Note that a (locally) optimal algorithm ϕ^* is globally optimal, but ϕ^g is not in general locally optimal. Thus, the (local) optimality concept investigated in this paper is stronger than the global optimality concept investigated in the above cited literature. As just noted before, for given algorithm, the tightest bound on $||f_o - \phi(FSS_T)||$ is given by $E(\phi(FSS_T), f_o, D_t)$. The algorithm ϕ^* minimizes $E(\phi(FSS_T), f_o, D_t)$ for any f_o and D_T, while ϕ^g minimizes it only for worst case $f \in K$ and noise sequence in B.

3. ASSUMPTIONS VALIDATION AND OPTIMAL BOUNDS EVALUATION

Necessary and sufficient condition for checking the assumptions validity are now given. Let us define the functions:

$$f_u(w) \doteq \min_{t=1,\ldots,T} \left(\overline{h}_t + \gamma \|w - \widetilde{w}_t\|\right)$$
$$f_l(w) \doteq \max_{t=1,\ldots,T} \left(\underline{h}_t - \gamma \|w - \widetilde{w}_t\|\right) \quad (6)$$

where $\overline{h}_t \doteq \widetilde{y}_{t+1} + \varepsilon_t$ and $\underline{h}_t \doteq \widetilde{y}_{t+1} - \varepsilon_t$.

Theorem 1.

i) $f_u(\widetilde{w}_t) \geq \underline{h}_t,\ t = 1, 2, ..., T$
is necessary condition for prior assumptions to be validated.
ii) $f_u(\widetilde{w}_t) > \underline{h}_t,\ t = 1, 2, ..., T$
is sufficient condition for prior assumptions to be validated.
Proof. See (Milanese and Novara, 2002*b*). ■

Note that there is essentially no "gap" between the necessary and sufficient conditions, since condition $\overline{f}(\widetilde{w}_t) \geq \underline{h}_t + \delta,\ t = 1, 2, ..., T$ is sufficient for any $\delta > 0$ arbitrarily small and necessary for $\delta = 0$. In the paper it is assumed that the sufficient condition holds. If not, values of the constants appearing in the assumptions on function f_o and on noise d_t are accordingly changed. The validation Theorem 1 can be used for assessing the values of such constants such that sufficient condition holds (see (Novara and Milanese, 2001; Milanese and Novara, 2002*a*)).
The function f_u and f_l allow also to solve the problem of finding the optimal interval estimate of $f_o(w)$ for given $w \in W$. In fact, from (4) and (5) it follows that the smallest interval guaranteed to include $f_o(w)$, is given by the interval $[\underline{f}(w), \overline{f}(w)]$, where $\underline{f}(w)$ and $\overline{f}(w)$ are the optimal bounds. The following theorem shows that the optimal bounds are actually given by f_u and f_l.

Theorem 2.

The functions f_u and f_l given in (6) are optimal bounds i.e.:

$$\overline{f}(w) = \min_{t=1,\ldots,T} \left(\overline{h}_t + \gamma \|w - \widetilde{w}_t\|\right) \doteq f_u(w)$$
$$\underline{f}(w) = \max_{t=1,\ldots,T} \left(\underline{h}_t - \gamma \|w - \widetilde{w}_t\|\right) \doteq f_l(w)$$

Proof. See (Milanese and Novara, 2002*b*). ■

In order to investigate the properties of the optimal bounds $\underline{f}$ and $\overline{f}$ and of the optimal identification algorithm derived in the next section, the concept of Hyperbolic Voronoi Diagram (HVD) introduced in (Novara and Milanese, 2000), is now given. The Hyperbolic Voronoi Diagrams can be defined as follows.
Consider the set of points:

$$\widetilde{W}_T \doteq [\widetilde{w}_1, \widetilde{w}_2, ..., \widetilde{w}_T]$$

and a $T \times T$ antisymmetric matrix η. Then define:

- The $(n-1)$-dimensional hyperbola $H_{t\tau}$:
$H_{t\tau} \doteq \{w : \|w - \widetilde{w}_t\| - \|w - \widetilde{w}_\tau\| = \eta_{t\tau},\ t \neq \tau\}$
- The n-dimensional regions $S_{t\tau}$ containing $\widetilde{w}_t$:
$S_{t\tau} \doteq \{w : \|w - \widetilde{w}_t\| - \|w - \widetilde{w}_\tau\| < \eta_{\tau t},\ t \neq \tau\}$
- The hyperbolic cell C_t:
$C_t \doteq \bigcap_{\tau \neq t} S_{t\tau}$

The intersections between the surfaces $H_{t\tau}$ generate other cells of dimension d, with $0 \leq d \leq n-1$ called d-faces. The cells C_t are called n-faces while the 0-faces are also called vertices.

Definition 5.

The Hyperbolic Voronoi Diagram $V\left(\widetilde{W}_T, \eta\right)$ is defined as the set of all d-faces, $0 \leq d \leq n$. ■

If $\eta_{t\tau} = 0, \forall t, \tau$, all hyperbola $H_{t\tau}$ degenerate into hyperplanes and the definitions become the ones of standard Voronoi diagrams (Edelsbrunner, 1987).
The next result shows that $\underline{f}$ and $\overline{f}$ are differentiable almost everywhere (a.e.) in W. For given $\underline{f}$ and $\overline{f}$, consider the HVD $\overline{V} \doteq V\left(\widetilde{W}_T, \overline{\eta}\right)$ and $\underline{V} \doteq V\left(\widetilde{W}_T, \underline{\eta}\right)$, where $\overline{\eta}_{\tau t} = \left(\overline{h}_\tau - \overline{h}_t\right)/\gamma$, $\underline{\eta}_{\tau t} = \left(\underline{h}_t - \underline{h}_\tau\right)/\gamma$. Let $\overline{V}^d$ and $\underline{V}^d$ be the sets of the d-faces of the HVD $\overline{V}$ and $\underline{V}$ respectively, with $d < n$. Let co$\underline{M}$ and co$\overline{M}$ be the complements in W of the sets $\underline{M} \doteq \cup_{d<n} \underline{V}^d \cup \widetilde{W}_T$ and $\overline{M} \doteq \cup_{d<n} \overline{V}^d \cup \widetilde{W}_T$, i.e. $\underline{M} \cup$ co$\underline{M} = W$ and $\overline{M} \cup$ co$\overline{M} = W$. Note that $\underline{M}$ and $\overline{M}$ are sets of zero measure in $\Re^n$. In fact, $\widetilde{W}_T$ is a set composed by a finite number of points and $\overline{V}^d$, $\underline{V}^d$ are composed by a finite number of d-dimensional sets, with $d < n$.

Theorem 3.

The functions $\underline{f}$ and $\overline{f}$ are Lipschitz-continuous on W. Moreover:
i) $\underline{f}$ is differentiable $\forall w \in$ co$\underline{M}$ and:
$\left\|\underline{f}'(w)\right\| = \gamma,\ \forall w \in$ *co*$\underline{M}$.
ii) $\overline{f}$ is differentiable $\forall w \in$ co$\overline{M}$ and:
$\left\|\overline{f}'(w)\right\| = \gamma,\ \forall w \in$ *co*$\overline{M}$.

Proof. See (Milanese and Novara, 2002*b*). ■

4. OPTIMAL ALGORITHM AND ESTIMATE

Let the function f_c be defined as:

$$f_c(w) \doteq \frac{1}{2}\left[\underline{f}(w) + \overline{f}(w)\right]$$

where $\underline{f}(w)$ and $\overline{f}(w)$ are given in Theorem 2.
We will show that the algorithm $\phi_c(FSS_T) = f_c$ is optimal for any L_p norm. In order to prove this property, we need some preliminary results about $\underline{f}$, $\overline{f}$ and f_c. At first, it is shown that f_c is Lipschitz - continuous and almost everywhere differentiable.
Let $\overline{V}^d$ and $\underline{V}^d$ be the sets of the d-faces of $\overline{V}$ and $\underline{V}$ respectively, with $d < n$. Let co$M$ the complement in W of the set $M \doteq \cup_{d<n}\left(\overline{V}^d \cup \underline{V}^d\right) \cup \widetilde{W}_T$, i.e. $M \cup$ co$M = W$. Note that M is a set of zero measure in $\Re^n$. In fact, $\widetilde{W}_T$ is a set composed

by a finite number of points and $\overline{V}^d$, $\underline{V}^d$ are sets composed by a finite number of d-dimensional sets, with $d < n$.

Theorem 4.

i) The function f_c is Lipschitz-continuous on W.
ii) $f_c(w)$ is differentiable $\forall w \in coM$ and:
$\|f_c'(w)\|_2 \leq \gamma$, $\forall w \in \text{co}M$.

Proof. See (Milanese and Novara, 2002*b*). ∎

We need also the following technical Lemma.

Lemma 1. Let $\overline{FSS_T}$ be the closure of FSS_T with respect to (wrt) norm $\|f\|_S = \|f\|_\infty + \|f'\|_2$. Then, $\underline{f}, \overline{f}, f_c \in \overline{FSS_T}$.

Proof. See (Milanese and Novara, 2002*b*). ∎

Now the main result of this section can be proven.

Theorem 5.

For any $L_p(W)$ norm, with $p \in [1, \infty]$:
i) The identification algorithm $\phi_c(FSS_T) = f_c$ is optimal
ii) $E(f_c) = \frac{1}{2}\left\|\overline{f} - \underline{f}\right\|_p = r_I = \inf_\phi E[\phi(FSS_T)]$

Proof.
From Theorems 5 and 6 it follows that $\overline{f}, \underline{f}$ and f_c are bounded on W which is bounded. Then, $\overline{f}, \underline{f}, f_c \in L_p(W)$. The diameter of FSS_T is:
$d(FSS_T) \leq [\int_W |\sup_{f_1 \in FSS_T} f_1(w) +$
$- \inf_{f_2 \in FSS_T} f_2(w)|^p dw]^{\frac{1}{p}} = \|\overline{f} - \underline{f}\|_p$. Then:

$$d(FSS_T) \leq \|\overline{f} - \underline{f}\|_p \qquad (7)$$

From Lemma 1, since FSS_T is dense in $\overline{FSS_T}$ wrt $\|\cdot\|_S$, it follows that $\forall \delta > 0$, $\exists \overline{f}_\delta, \underline{f}_\delta \in FSS_T$ such that $\|\overline{f} - \overline{f}_\delta\|_\infty < \delta$ and $\|\underline{f} - \underline{f}_\delta\|_\infty < \delta$. Then:

$$\begin{aligned} \|\overline{f}_\delta - \underline{f}_\delta\|_p \geq \|\overline{f} - \underline{f}\|_p - \|\overline{f}_\delta - \overline{f}\|_p + \\ -\|\underline{f} - \underline{f}_\delta\|_p \geq \|\overline{f} - \underline{f}\|_p - 2\delta\Omega_p \end{aligned} \qquad (8)$$

where $\Omega_p = [\int_W dw]^{\frac{1}{p}}$. Since δ can be taken as small as desired, (7) and (8) implies that:

$$d(FSS_T) = \|\overline{f} - \underline{f}\|_p \qquad (9)$$

On the other hand:

$$\begin{aligned} E(f_c) &\leq [\int_W \sup_{f \in FSS_T} |f(w) - f_c(w)|^p dw]^{\frac{1}{p}} \\ &= \frac{1}{2}\|\overline{f} - \underline{f}\|_p = \frac{1}{2}d(FSS_T) \end{aligned} \qquad (10)$$

From Definition 3 and 4 we have:

$$r_I = r(FSS_T) \leq E(f_c) \qquad (11)$$

Then, from (10), (11) and the well known relations between radius and diameter of a set, it follows that $E(f_c) = r_I = \frac{1}{2}\|\overline{f} - \underline{f}\|_p$, thus proving claims i) and ii). ∎

Note that the optimal estimate $f_c$is a Chebicheff center of FSS_T, i.e.:

$$\sup_{\tilde{f} \in FSS_T} \left\|\tilde{f} - f_c\right\|_p = \inf_f \sup_{\tilde{f} \in FSS_T} \left\|\tilde{f} - f\right\|_p$$

but it does not belong to FSS_T,since it is not differentiable everywhere. However, functions belonging to FSS_T approximating f_c in L_p norm with arbitrary precision can be found, as stated in the following result, an immediate consequence of Lemma 1.

Theorem 6.

$\forall \delta > 0$, $\exists\ f_\delta \in FSS_T$ such that $\|f_\delta - f_c\|_p < \delta$.

Proof. See (Milanese and Novara, 2002*b*). ∎

5. LOCAL ASSUMPTIONS AND REGRESSORS SCALING

5.1 *Local assumptions*

So far a global bound on $\|f_o'(w)\|$ over all W is assumed. However, a local approach can be taken in order to obtain improvements in identification accuracy, e.g. by assuming different bounds γ_k on suitable partitions W_k of W. This is similar to what done in identification of piece-wise linear model, where partitions W_k are looked for, over which $f_o(w)$ can be considered approximately linear, i.e. $f_o'(w) \simeq$ const., $\forall w \in W_k$, (see e.g. (Stenman *et al.*, 1996; Zheng and Kimura, 2001; Sontag, 1981; Ferrari-Trecate *et al.*, 2001)). However, finding such partitions may be not an easy task. A very simple alternative approach allowing to use local assumptions on f_o, is based on the evaluation of a function f_a approximating f_o(using any desired method, e.g. the SM one proposed in this paper assuming global gradient bounds) and on the application of the method described in this paper to the residue function, defined as:

$$f_\Delta(w) \doteq f_o(w) - f_a(w)$$

using the set of values:

$$\Delta y_{t+1} = \tilde{y}_{t+1} - f_a(\tilde{w}_t), \ t = 1, 2, ..., T$$

The global bound $\|f_\Delta'(w)\| = \|f_o'(w) - f_a'(w)\| \leq \gamma_\Delta$ on the residue function f_Δ implies the local bound $\|f_a'(w)\| - \gamma_\Delta \leq \|f_o'(w)\| \leq \|f_a'(w)\| + \gamma_\Delta$ for function f_o.

5.2 *Regressors scaling*

The problem of selecting suitable scalings of regressors is now investigated, in order to adapt to the properties of data. Suitable scaling may result very important when the gradient components have quite different magnitudes (Stenman *et al.*, 1996; Wasilkowski and Woźniakowski, 2001). The problem is posed as follows. Some estimates of the quantities $\mu_i = \max_{w \in W} \left|\frac{\partial f_o(w)}{\partial w_i}\right|$, $i = 1, 2, ..., n$ can be derived (e.g. from a neural approximation of f_o or directly from data). Here, w_i denotes the i-the component of vector $w \in W \subseteq \Re^n$. These estimates support the evidence that: $f_o \in K_\infty^\mu \doteq \{f \in C^1, \|f'(w)\|_\infty^\mu \leq \gamma, \forall w \in W\}$ where $\|x\|_\infty^\mu \doteq \max_{i=1,...,n} |x_i| \mu_i^{-1}$, $\mu_i > 0$ denotes the weighted ℓ_∞ norm.

Then, $f_o \in K_\infty^\mu$ could be used as prior assumption on the unknown function f_o. Unfortunately, dealing with such a type of prior appears not easy, and weighted l_2 bounds on the gradient are used, of the form:

$f_o \in K_\infty^\nu \doteq \{f \in C^1(W), \|f'(w)\|_2^\nu \leq \gamma, \forall w \in W\}$ where $\|x\|_2^\nu \doteq \sqrt{\sum_{i=1}^n \nu_i x_i^2}$.

Outer approximations $K_2^\nu \supseteq K_\infty^\mu$ can be looked for, by suitably choosing ν. Since $K_2^\nu \supseteq K_\infty^\mu \Leftrightarrow B_2^\nu \supseteq B_\infty^\mu$, where $B_2^\nu \doteq \{x \in R^n : \|x\|_2^\nu \leq 1\}$, $B_\infty^\mu \doteq \{x \in R^n : \|x\|_\infty^\mu \leq 1\}$, the problem is equivalent to look, in the n-dimensional gradient space, for outer approximations of the weighted ℓ_∞ ball B_∞^μ with a weighted ℓ_2 ball B_2^ν. By taking the ratio of the volumes of the two balls as measure of approximation goodness, minimal volume outer approximation is optimal. The following lemma shows how the optimal solution can be obtained.

Lemma 2. The optimal (minimal volume) outer approximation B_2^ν of B_∞^μ is given by:

$$\nu_i = (n\mu_i^2)^{-1}, \; i = 1, ..., n$$

■

Let us define the scaled regressors:

$$v_i = \frac{1}{\sqrt{\nu_i}} w_i, \; i = 1, 2, ..., n$$

and, with a slight abuse of notation, denote $f(w) = f(w_i, .., w_i) = f(\sqrt{\nu_1} v_i, .., \sqrt{\nu_n} v_i) = f(v)$. Then $\frac{\partial f(v)}{\partial v_i} = \frac{\partial f(w)}{\partial w_i} \sqrt{\nu_i}$ and $\|f'(w)\|_2^\nu =$ $= \sqrt{\sum_{i=1}^n \nu_i \left(\frac{\partial f(w)}{\partial w_i}\right)^2} = \sqrt{\sum_{i=1}^n \left(\frac{\partial f(v)}{\partial v_i}\right)^2} = \|f'(v)\|$. Thus, considering the scaled regressors v, a bound on the euclidean norm of gradient is obtained. The results presented in the previous sections can be directly applied by substituting regressors w with scaled regressors v.

6. REFERENCES

Edelsbrunner, H. (1987). *Algorithms in Combinatorial Geometry.* Springer-Verlag. Berlin.

Ferrari-Trecate, G., M. Muselli, D. Liberati and M. Morari (2001). A clustering technique for the identification of piecewise affine systems. *A. Sangiovanni-Vincentelli and M.D. Di Benedetto, Eds., Hybrid Systems: Computation and Control, Lecture Notes in Computer Science. Springer Verlag.*

Haber, R. and H. Unbehauen (1990). Structure identification of nonlinear dynamic systems-a survey on input/output approaches. *Automatica* **26**, 651–677.

Isermann, R., S. Ernst and O. Nelles (1997). Identification with dynamic neural networks -architectures, comparisons, applications-. In: *Sysid 97.* Vol. 3. pp. 997–1022.

Milanese, M. and A. Vicino (1991). Optimal algorithms estimation theory for dynamic systems with set membership uncertainty: an overview. *Automatica* **27**, 997–1009.

Milanese, M. and C. Novara (2002*a*). Nonlinear Set Membership prediction of river flow. In: *Proc. of the 41th IEEE Conference on Decision and Control.* Las Vegas, Nevada.

Milanese, M. and C. Novara (2002*b*). Set Membership identification of nonlinear systems. *Politecnico di Torino Internal Report.*

Milanese, M. and R. Tempo (1985). Optimal algorithms theory for robust estimation and prediction. *IEEE Transaction on Automatic Control* **30**, 730–738.

Milanese, M., J. Norton, H Piet Lahanier and E. Walter (1996). *Bounding Approaches to System Identification.* Plenum Press.

Narenda, K. S. and S. Mukhopadhyay (1997). Neural networks for system identification. In: *Sysid 97.* Vol. 2. pp. 763–770.

Novak, E. (1988). *Deterministic and Stochastic Error Bounds in Numerical Analysis.* Vol. 1349. Springer-Verlag. Berlin.

Novara, C. and M. Milanese (2000). Set Membership identification of nonlinear systems. In: *Proc. of the 39th IEEE Conference on Decision and Control.* Sydney, AU. pp. 2831–2836.

Novara, C. and M. Milanese (2001). Set Membership prediction of nonlinear time series. In: *Proc. of the 40th IEEE Conference on Decision and Control.* Orlando, FL. pp. 2131–2136.

Pinkus, A. (1985). *n-Widths in Approximation Theory.* Springer-Verlag. Berlin.

Sjöberg, J., Q. Zhang, L. Ljung, A. Benveniste, B.Delyon, P. Glorennec, H. Hjalmarsson and A. Juditsky (1995). Nonlinear black-box modeling in system identification: a unified overview. *Automatica* **31**, 1691–1723.

Sontag, E. D. (1981). Nonlinear regulation. The piecewise linear approach. *IEEE Trans. Automatic Control* **26**, 346–357.

Stenman, A., F. Gustafsson and Ljung (1996). Just in time models for dynamical systems. In: *Proc. of the 35th IEEE Conference on Decision and Control.* Kobe, Japan. pp. 1115–1120.

Traub, J. F., G. W. Wasilkowski and H. Woźniakowski (1988). *Information-Based Complexity.* Academic Press, Inc.

Wasilkowski, G. W. and H. Woźniakowski (2001). Complexity of weighted approximation over R^d. *Journal of Complexity* (17), 722–740.

Zheng, Q. and H. Kimura (2001). Just in time modeling for function prediction and its applications. *Asian Journal of Control, Vol. 3, No. 1,* pp. 35–44.

IFAC Publications
www.elsevier.com/locate/ifac

A SUBOPTIMAL BOOTSTRAP METHOD FOR STRUCTURE DETECTION OF NONLINEAR OUTPUT-ERROR MODELS

Sunil L. Kukreja *

* *Division of Automatic Control, Department of Electrical Engineering, Linköpings universitet, SE-581 83 Linköping, Sweden*
Fax: +46-13-282622, Email: sunil@isy.liu.se

Abstract: Identification of nonlinear systems involves estimating unknown parameters and structure detection, selection of a subset of candidate terms that best describe the observed output. For nonlinear systems simple output additive noise can generate multiplicative terms between the input, output and noise. The terms associated with noise need to be modeled to obtain unbiased parameter estimates, significantly increasing the number of candidate terms to be estimated and considered. In special cases, it may be possible to use an output error (OE) model structure and the instrumental variable (IV) estimator to obtain unbiased parameters without modeling the noise. This significantly reduces the dimensionality of the structure computation problem. In this paper we propose a suboptimal bootstrap structure detection algorithm. The applicability of this suboptimal bootstrap structure detection (SOBSD) technique for nonlinear polynomial models was evaluated by estimating the structure of a simple NARMAX model and comparing the results to the t-test. *Copyright © 2003 IFAC*

Keywords: NARMAX, Nonlinear Systems, Structure Detection, System Identification

1. INTRODUCTION

The NARMAX (Nonlinear AutoRegressive, Moving Average eXogenous) structure is a general parametric form for modeling nonlinear systems (Leontaritis and Billings, 1985). This structure describes both the stochastic and deterministic components of nonlinear systems. Many nonlinear systems are a special case of the general NARMAX structure (Chen and Billings, 1989). In this paper, we focus on a special class of NARMAX models; nonlinear polynomial models. The polynomial NARMAX structure models the input-output relationship as a nonlinear difference equation of the form

$$z(n) = f^l[z(n-1),\cdots,z(n-n_y),u(n),\cdots, u(n-n_u),e(n-1),\cdots,e(n-n_e)]+e(n). \quad (1)$$

f^l denotes a nonlinear mapping, u is the controlled or exogenous input, z is the measured output, and e is the uncontrolled input or innovation. This nonlinear mapping may include a variety of nonlinear terms, such as terms raised to an integer power, products of present and past inputs, past outputs, or cross-terms. This system description encompasses many forms of nonlinear difference equations that are linear-in-the-parameters. Thus, we do not have problems with local minima.

For nonlinear systems, output additive noise can produce multiplicative terms between input, output and itself. To compute unbiased parameter estimates a noise model needs to be estimated. As a result, the number of candidate terms can become large for even moderately complex models making structure detection difficult. For example, consider a model of order: $O=[n_u=4,\ n_y=4,\ n_e=4,\ l=2]$. A model of this order has $p=105$ candidate terms (Billings and Voon, 1984).

To limit the number of terms to be considered an output error (OE) formulation can be used, where it may not be necessary not postulate a noise model (Ljung, 1999). The instrumental variable (IV) estimator can be used to compute an unbiased parameter set for an OE model structure without modeling the noise. The IV method provides unbiased results when the noise terms are represented within the NARMAX model as a linear expansion (Billings and Voon, 1984). This special case of the general NARMAX model (1) can be written as

$$z(n) = f^l[u(n), \cdots, u(n-n_u)] + [z(n-1), \cdots, z(n-n_y), e(n-1), \cdots, e(n-n_e)] + e(n). \quad (2)$$

The class of models that give linear output and error terms include blocked structured N-L models (static nonlinearity followed by a causal, linear, time-invariant, dynamic system) such as Hammerstein models, bilinear models, etc.

We define the model order, for this model set, as $O = [n_u\, n_y\, l]$ where n_u is the maximum input lag, n_y the maximum output lag, and l the maximum nonlinearity order. The maximum number of candidate terms in a model (2) with n_y and n_u dynamic terms and lth order nonlinearity is

$$p = n_y + \sum_{i=1}^{l} p_i + 1; \quad (3)$$

$$p_i = \frac{p_{i-1}(n_u + i)}{i}, \quad p_0 = 1.$$

In practical identification experiments there is often some *a priori* knowledge regarding the type of structures that may be realistic to consider. Therefore, in this paper, we assume the (i) system order, O, is known and (ii) process to be identified lies within the model class. Consider again the previous example with order, $O = [n_u = 4, n_y = 4, n_e = 4, l = 2]$. If there is *a priori* knowledge suggesting the model lies within the model class (2), the order may be redifined as $O = [n_u = 4, n_y = 4, l = 2]$ and only $p = 55$ terms need be considered; a significant reduction.

Recently, Kukreja *et al.* (1999) demonstrated that bootstrap is a useful tool for structure detection of nonlinear models. The bootstrap is a numerical procedure for estimating parameter statistics that requires few assumptions (Efron, 1979). The conditions needed to apply bootstrap to regression analysis are quite mild; namely, that the errors be independent, identically distributed, and have zero-mean.

In this paper, we show that our suboptimal bootstrap method yields good results for structure detection in the presence of additive output noise with 0 dB signal-to-noise ratio (SNR).

The organization of this paper is as follows. Section 2 reviews some commonly used techniques for assessing parameter significance while the instrumental variables estimator is summarized in §3. Bootstrap sampling theory and its application to structure detection is discussed in §4. Our bootstrap structure detection algorithm is presented in §5 while §6 illustrates the results of our algorithm on a simulated system. Section 7 provides a discussion of our findings and §8 summarizes the conclusions of our study.

2. STRUCTURE DETECTION

The structure detection problem is that of selecting a subset of candidate terms that best describes the output. Several methods for NARMAX structure detection have been proposed including hypothesis testing of differences between means via the t-test exhaustive search and analysis of variance (ANOVA)

The t-test in combination with regression analysis is sometimes referred to as a form of hypothesis testing by computing the differences between means (Ostrom, 1990). This procedure assumes that an accurate estimate of parameter variance is available. This assumption is violated for over-parameterized models and, therefore, may lead to inaccurate estimates of system structure (Kukreja *et al.*, 1999)

Exhaustive search for significant regressors has been used in the statistical community for many years (Seber, 1977). This approach searches over every possible subset of the full model. The best model is deemed as the one that minimizes some criterion such as AIC or root mean-square error (Ljung, 1999). This technique is known to fail even with few candidate terms and small amplitude noise due to local minima (Lind, 2001).

Recently, analysis of variance (ANOVA) has been applied to structure detection of nonlinear dynamic models (Lind, 2001). ANOVA computes structure via hypothesis tests of the residual sum of squares and F-distribution (Miller, 1997; Montgomery, 1991). This technique is sensitive to grouping data into various levels (Lind, 2001).

3. PARAMETER ESTIMATION

NARMAX models provide a concise system representation; however, any noise on the output enters the model as product terms with the system input and output. Consider the system shown in fig. 1. The system

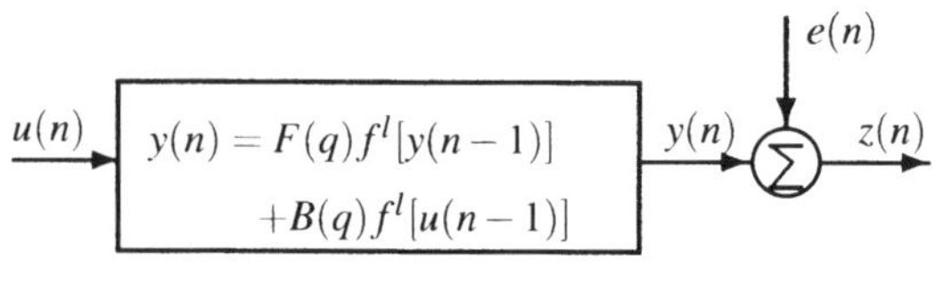

Fig. 1. Output error (OE) system configuration. $u(n)$: input. $y(n)$: true system output. $e(n)$: Gaussian, zero-mean, noise. $z(n)$: measured output.

output is assumed to be a function of both current and

past inputs and past outputs. The noise term $e(n)$ is assumed to be a stationary, zero-mean, random process, uncorrelated with itself and the input.

A least squares formulation for this system is

$$Z = \Psi_{zu}\theta + e \tag{4}$$

where Z is a $N \times 1$ vector of measured outputs, Ψ_{zu} is a nonsingular $N \times p$ matrix of regressors, based on input-output only, θ is a $p \times 1$ vector of unknown parameters, and e is a $N \times 1$ vector of an inaccessible zero-mean white-noise process. This ordinary least-squares formulation assumes that Ψ_{zu} is deterministic. However, since Ψ_{zu} is a function of the system inputs and measured outputs Ψ_{zu} is not deterministic. As a result applying ordinary least-squares directly will give biased parameter estimates since e (4) will not be zero-mean (Ljung, 1999; Seber, 1977).

This makes identification of NARMAX models complicated since a NARMAX system description gives a model in which the noise can be correlated with true system output, input and itself. To obtain unbiased parameters other estimation techniques based on least-squares must be used. One such method, used for NARMAX models, is instrumental variables (IV). IV addresses the bias problem by using an instrument matrix that is correlated with the regressors but uncorrelated with the noise.

The IV method is based on selecting an instrument matrix V which satisfies the conditions

$$\lim_{N\to\infty} \frac{1}{N} V^T \Psi_{zu} = R; \quad R \text{ is nonsingular} \tag{5}$$

$$\lim_{N\to\infty} \frac{1}{N} V^T (Z - \Psi_{zu}\theta) = 0.$$

This ensures that the estimate $\hat{\theta} = (V^T \Psi_{zu})^{-1} V^T Z$ is unbiased since the instrument matrix is not correlated with the errors (Söderström and Stoica, 1989; Söderström and Stoica, 1983).

In general, the instrumental variable algorithm will yield biased estimates for nonlinear systems. Instrumental variables can only be applied to nonlinear systems if certain properties of the system noise are satisfied (see example in (Billings and Voon, 1984) on pp. 608-9). The IV method gives unbiased results when the noise terms are represented within the NARMAX model as a linear expansion and, therefore, it always satisfies the conditions of (5). In this paper, we assume that the system(s) under test is/are represented by a linear expansion of outputs (and hence its errors) making IV an appropriate choice of estimator.

One way to satisfy (5) is to define V^T to have the same structure as Ψ_{zu}^T but with z replaced by $\hat{y}$ (Söderström and Stoica, 1983; Ljung, 1999). The columns of V^T associated with u are unchanged since it is assumed the input is measured with negligible error. In IV, the NARMAX formulation of (2) is redefined into a OE model by replacing z with $\hat{y}$; making it a deterministic least-squares problem. However, it is difficult to form an uncorrelated instrument matrix and consistency cannot always be guaranteed (Stoica and Söderström, 1981; Ljung, 1999). For this reason, we consider the IV method a *suboptimal* estimator.

4. THE BOOTSTRAP

The bootstrap is a powerful numerical technique for computing parameter statistics in situations where conventional techniques will likely fail. The conditions needed for bootstrap methods to give accurate results are quite mild; namely, that the errors be independent and identically distributed (i.i.d.) with zero mean. The bootstrap achieves with a computer what the user would do in practice if it were possible: Monte-Carlo analysis. With bootstrap, observations are randomly reassigned, and estimates recomputed. These assignments and recomputations are done a "large" number of times and treated as repeated experiments.

4.1 Bootstrap Sampling Theory

The bootstrap method assumes that observed data arise from an unknown distribution, F, which has given, say, the observed data $x = (x_1, x_2, \cdots, x_N)$. From the observed data a statistic of interest, $\hat{\theta} = s(x)$, is computed where "s" represents some estimator of the statistic (Efron and Tibshirani, 1993). With bootstrap, the empirical distribution $\hat{F}$ gives bootstrap samples $x^* = (x_1^*, x_2^*, \cdots, x_n^*)$ by random sampling, from which bootstrap replications of the statistic of interest, $\hat{\theta}^* = s(x^*)$ are calculated. The empirical distribution function $\hat{F}$ is defined to be the discrete distribution that puts probability $1/N$ on each value $x_i; i = 1, 2, \cdots, N$.

4.2 The Bootstrap Operation

Given $z_1, \ldots, z_N$, let $\hat{\varepsilon}_1^*, \ldots, \hat{\varepsilon}_N^*$ be conditionally independent, with common distribution $\hat{F}_N$. Let $\hat{Z}^* = \Psi^* \hat{\theta} + \hat{\varepsilon}^*$. Informally, $\hat{\varepsilon}^*$ is obtained by resampling the residuals, $\hat{\varepsilon}$, and $\hat{Z}^*$ is generated from the data, using the regression model with $\hat{\theta}$ as the vector of parameters and $\hat{F}_N$ as the distribution of the residuals.

Next, consider giving the "starred" data $(\Psi^*, \hat{Z}^*)$ to another experimenter and asking him or her to estimate $\hat{\theta}$. The instrumental variable estimate is simply $\hat{\theta}^* = (V^{T*} \Psi^*)^{-1} V^* \hat{Z}^*$. The bootstrap principle is that the distribution of $(\hat{\theta}^* - \hat{\theta})$, which can be computed directly from the data approximates the distribution of $(\hat{\theta} - \theta)$.

Bickel and Freedman (1982) studied the linear regression model where the number of data points N and parameters p were both large. For the full p-dimensional distribution of the least-squares estimates the bootstrap distribution will converge to the true unknown distribution as (Bickel and Freedman, 1982)

$$\gamma = \frac{p^2}{N} \to 0. \tag{6}$$

Initially, p cannot change, consequently the accuracy of the bootstrap estimate is determined by the data length, N, available for estimation. For structure detection, to obtain consistent results the relevant criterion for bootstrap convergence is given by (6).

5. APPLICATION TO STRUCTURE DETECTION

Application of bootstrap to structure detection involves two steps: (i) computing a series of parameter replications, in which bootstrap data is generated to compute new bootstrap parameter estimates, and (ii) forming percentile intervals for hypothesis testing, where the significance of the parameters is determined. Bootstrap data is formed by first estimating the residuals of the identified model; these residuals are then resampled with replacement, centered and then added to the predicted output to generate bootstrap replications of the output. B bootstrap data sets are generated to estimate B bootstrap parameter replications.

Significance of the parameters is determined by forming percentile intervals. The estimates from B parameter replications are ranked in increasing order and the $B \cdot \alpha$th and $B \cdot (1-\alpha)$th values in the ordered list of the B replications are used as an upper and lower bound for the parameter deviation with an αth and $(1-\alpha)$th level of significance, respectively (Efron and Tibshirani, 1993). The significance of each parameter is determined by checking if 0 lies in its interval: if so, the parameter is rejected. This leads to the following algorithm to detect structure of linear-in-the-parameter models.

5.1 SOBSD Algorithm

(1) Compute an initial estimate of the unknown parameter vector as: $\hat{\theta} = (V^T \Psi_{zu})^{-1} V^T Z$.
(2) Estimate the residuals as: $\hat{\varepsilon} = Z - \hat{Z}$.
(3) Generate B bootstrap data sets as $\hat{Z}^* = \Psi^* \hat{\theta} + \hat{\varepsilon}^*$.
(4) Compute B bootstrap parameter replications as: $\hat{\theta}^* = (V^{T*} \Psi^*)^{-1} V^* \hat{Z}^*$.
(5) Form percentile intervals for each parameter by ranking estimates from the B parameter replications in increasing order.
(6) Compute the upper and lower bounds of each parameter's confidence interval for a desired α level of significance.
(7) Determine if zero lies in the interval of each parameter.
(8) If zero lies in the interval for any parameter remove it from the regression.
(9) Compute a new estimate of the parameter vector and residuals as in 1 & 2.
(10) Go to 3 until convergence.

6. SIMULATIONS

The efficacy of the SOBSD algorithm was assessed using Monte-Carlo simulations of a nonlinear system. The inputs were uniformly distributed, white, zero-mean, random sequences with unit variance. Two hundred Monte-Carlo simulations were generated in which each input-output realization was unique, and had a unique Gaussian, white, zero-mean, noise sequence added to the output, with 0 dB SNR. The IV estimator was used to compute system parameters and the instrument matrix, V, was formed with measured outputs replaced by predicted outputs.

6.1 Convergence Analysis of Bootstrap

In this study we assessed how small $\gamma = p^2/N$ must be to achieve a high probability of true model structure selection using the IV estimator. In addition, we evaluated whether increasing the number of bootstrap replications could compensate for insufficient data length.

For identification, the data length was determined as $N = \left\lceil \frac{p^2}{\gamma} \right\rceil$. γ was decreased in increments 0.01 in the interval $\gamma = [0.2, 0.01]$. An initial estimate of the system parameters was computed and B bootstrap replications were generated to assess the distribution of each parameter. For each γ (or N) the identification was restarted for increasing B in the interval $B = [50, 300]$ in increments of 50. For every γ and B, each parameter was tested for significance at the 95% confidence level. The SOBSD routine's performance was compared with the t-test, at the same confidence level.

For each input-output realization, we classified the structure detection result into one of three categories:

(1) Exact Model: A model which contains only true system terms,
(2) Over-modeled: A model with all its true system terms plus spurious parameters and
(3) Under-modeled: A model without all its true system terms. An under-modeled system may contain spurious terms as well.

6.2 Second-Order NARMAX Model

We studied the second-order system:

$$y(n) = 0.4[u(n-1) + u^2(n-1)] + 0.8y(n-1) \quad (7)$$

which is of order $O = [1, 1, 2]$ and has 7 candidate terms (3). The true system is described by only three parameters, two lagged inputs and one lagged output. This system was studied because it has a small number of candidate terms.

Fig. 2 shows the results for this model. The left panel shows the rate a which our SOBSD algorithm selected the exact model, over-modeled and under-modeled for decreasing γ (increasing N) and increasing B. The right panel shows the rate a which the t-test selected the exact model, over-modeled and under-modeled for decreasing γ.

The SOBSD method had a 2-0.5% rate of under-modeling for $\gamma = 0.2 - 0.1$ and $\forall B$ levels. The rate of

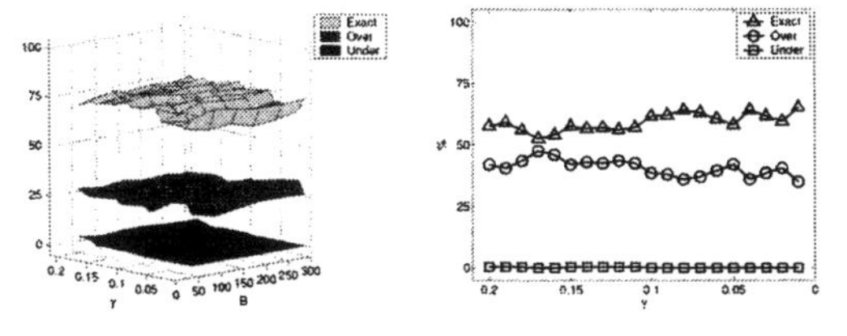

Fig. 2. Predicted structure of a second-order NARMAX model. Left panel: Selection rate of SOBSD algorithm for decreasing γ and increasing B. X-axis: B. Y-axis: γ. Z-axis: Percent selection. Right Panel: Selection rate of the t-test for decreasing γ. X-axis: γ. Y-axis: Percent selection.

under-modeling was constant at 0% $\forall B$ and $\gamma < 0.1$. The rate of over-modeling was approximately constant $\forall \gamma$ and B, with a maximum of 27% at $\gamma = 0.2$ and a minimum of 25% at $\gamma = 0.01$. The rate of selecting the exact model was approximately constant for all B but increased with smaller γ, with a minimum of 70% at $\gamma = 0.2$ and a maximum of 75% at $\gamma = 0.01$. The t-test had a 0.5% rate of under-modeling for $\gamma = 0.2 - 0.1$ and 0% for $\gamma < 0.1$. The rate of over-modeling was 42-35% for $\gamma = 0.2 - 0.01$ and the rate of selecting the exact model was 56-65% for $\gamma = 0.2 - 0.01$.

For the second-order NARMAX model, the t-test and SOBSD, had a low rate of selecting an under-modeled model. However, performance of the two techniques diverged for the rate of selecting an over-modeled model and exact model. The SOBSD method selected an over-modeled model with a maximum rate of 27% while the t-test selected an over-modeled model with a maximum rate of 42%. The rate of selecting the exact model peaked at 75% for SOBSD and only 65% for the t-test. Clearly, for this second-order NARMAX model SOBSD outperformed the t-test.

6.3 Parameter Statistics

To assess why the accuracy of our SOBSD technique is better than the t-test a Monte-Carlo analysis was performed. Statistics of each parameter was computed for the second-order NARMAX model (7).

The asymptotic means and variance were estimated using Monte-Carlo simulations consisting of 1,000 runs. For the t-test, estimates were computed from a single realization using standard regression techniques (Ljung, 1999). For the SOBSD method, estimates were also computed from a single realization and $B = 100$ bootstrap realizations. Parameter means and variance were first calculated for the full model. The simulation paradigm was the same as described previously.

Assuming a nonlinear model with a linear expansion of output terms, model (7) is fully described as

$$y(n) = \theta_0 + \theta_1 u(n) + \boldsymbol{\theta_2 u(n-1)} + \theta_3 u^2(n) + \theta_4 u(n)u(n-1) + \boldsymbol{\theta_5 u^2(n-1)} + \boldsymbol{\theta_6 y(n-1)} \quad (8)$$

where the true system parameters and regressors are shown in **bold**.

Fig. 3 shows the Monte-Carlo, regression and bootstrap mean and variance estimate for each parameter. In each panel, the square (□) represents the Monte-Carlo estimate, the triangle (△) represents regression estimate and the circle (○) represents the bootstrap estimate. The horizontal dashed line in each panel is the true parameter value.

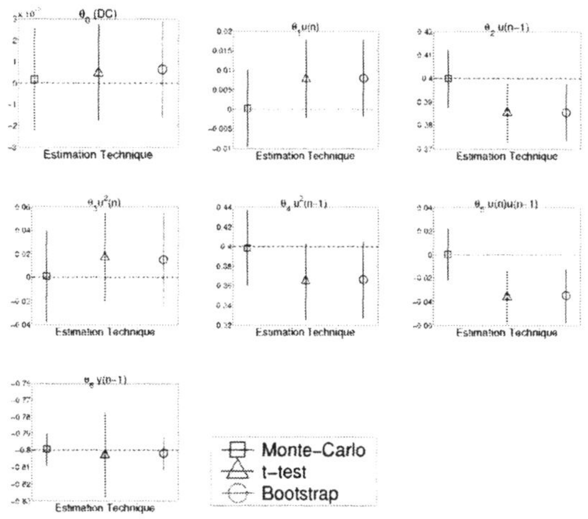

Fig. 3. Standard deviation about the mean of spurious and true terms for a second-order NARMAX model (7) when the full model is postulated. Abscissa: Estimation Technique. Ordinate: Mean parameter value. Dashed Line: True parameter value.

The plots show the standard deviation about the mean for each parameter. These are initial estimates for the full model, before spurious parameters were removed from the regression to obtain an improved estimate. The statistics calculated using bootstrap were closer to the Monte-Carlo estimate, while those associated with the regression method were not as close to the Monte-Carlo estimate. This result was not surprising since prior to the removal of any parameters the model was over-parameterized; giving an inaccurate estimate of the residuals and, therefore, yielding biased estimates of parameter statistics.

Next, estimates of parameter statistics were recomputed after removing spurious terms. Parameter statistics calculated for the exact or true model are plotted in fig. 4. The results show that the standard deviation computed using bootstrap were closer to the Monte-Carlo estimate than those based on regression techniques. It is surprising to see that the regression estimates were different from the Monte-Carlo even when the exact structure was used. This deviation may be because the regression estimates were based on a single realization. For a different realization it may be possible to compute more accurate estimates based on regression techniques. This result gives some insight as to why regression methods alone perform poorly

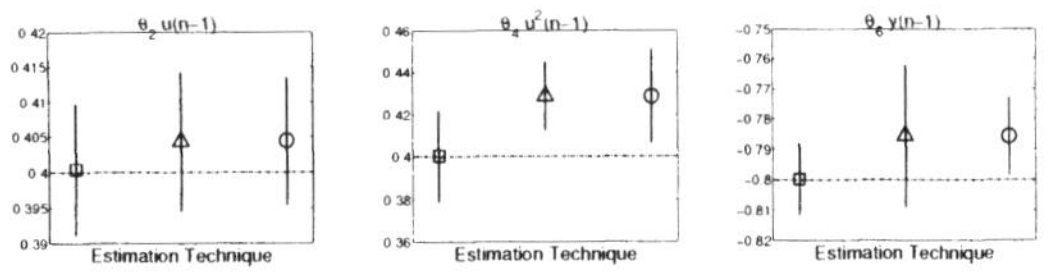

Fig. 4. Standard deviation about the mean of true terms for a second-order NARMAX model (7) when the exact model is postulated. Abscissa: Estimation Technique. Ordinate: Mean parameter value. Dashed Line: True parameter value.

when applied to structure detection; since they may provide poor estimates of parameter statistics.

7. DISCUSSION

7.1 Second and Third-Order Models

For the simple nonlinear model studied here, results demonstrated that the t-test and SOBSD had a low rate of selecting an under-modeled model but performance diverged for the rate of selecting an over-modeled model and exact model. Note in fig. 2, the plot is not "smooth". This non-smooth appearance is because we only used 200 Monte-Carlo simulations for these studies. If the number of Monte-Carlo simulations are increased the plots should become smooth.

7.2 Decreasing γ Versus Increasing B

Convergence analysis of $\gamma \rightarrow 0.01$ and $B \rightarrow 300$ empirically demonstrated that as data length is increased the rate of exact model selection will likely increase asymptotically to 100%; as $\gamma \rightarrow 0$ (Bickel and Freedman, 1982; Ljung, 1999). The results also demonstrated that increasing B did not compensate for insufficient data length and, therefore, it is computationally inefficient to select large B, $B \gtrsim 100$.

This study suggests that as $\gamma \rightarrow 0$ ($N \rightarrow \infty$) our SOBSD algorithm will, at worst, provide a model that is slightly over-parameterized. We consider an over-parameterized model better than an under-modeled model since it is not possible, with our backwards elimination approach, to re-enter a parameter into the model once it has been removed. An over-parameterized model which still contains its true parameters is clearly more useful than one which has dropped a true parameter. Moreover, our results are given for poor SNR conditions (i.e., 0 dB SNR) therefore they should be widely applicable under most experimental conditions.

8. CONCLUSIONS

This result demonstrates that the SOBSD algorithm may be a robust method for detecting the structure of Hammerstein structure models and resistant to noise. Using bootstrap, it is possible to compute good estimates of parameter statistics resulting in accurate estimates of model structure. Therefore, the SOBSD algorithm is useful for structure detection.

9. ACKNOWLEDGMENTS

Supported by grants from Vetenskapsrådet, The Swedish Research Council.

Dedicated in loving memory of Margherita Rapagna (25 August 1968 – 20 May 2002).

REFERENCES

Bickel, P.J. and D.A. Freedman (1982). Bootstrapping regression models with many parameters. Technical Report 7. Department of Statistics, University of California (Berkeley). Berkeley, California.

Billings, S.A. and W.S.F. Voon (1984). Least squares parameter estimation algorithms for non-linear systems. *Int.J. System Sci.* **15**(6), 601–615.

Chen, S. and S.A. Billings (1989). Representations of non-linear systems: the NARMAX model. *Int. J. Control* **49**(3), 1013–1032.

Efron, B. (1979). Computer and the theory of statistics: Thinking the unthinkable. *SIAM Review* **21**(4), 460–480.

Efron, B. and R.J. Tibshirani (1993). *An Introduction to the Bootstrap*. first ed.. Chapman & Hall. New York.

Kukreja, S.L., H.L. Galiana and R.E. Kearney (1999). Structure detection of NARMAX models using bootstrap methods. In: *Proceedings of the 38th IEEE Conference on Decision and Control*. Vol. 38. Phoenix, Arizona. pp. 1071–1076.

Leontaritis, I.J. and S.A. Billings (1985). Input-output parametric models for non-linear systems part I: deterministic non-linear systems. *Int. J. Control* **41**(2), 303–328.

Lind, I. (2001). Regressor selection in system identification using ANOVA. Technical Report Licentiate Thesis no. 921. Department of Electrical Engineering, Linköping University. SE-581 83 Linköping, Sweden.

Ljung, L. (1999). *System Identification: Theory for the User*. second ed.. Prentice Hall, Inc.. Englewood Cliffs, New Jersey.

Miller, Jr., R.G. (1997). *Beyond ANOVA*. Chapman and Hall. London.

Montgomery, D.C. (1991). *Design and Analysis of Experiments*. 3rd ed.. John Wiley & Sons. New York.

Ostrom, C.W. (1990). *Time Series Analysis: Regression Techniques*. second ed.. Sage Publications. Newberry Park, Calif.

Seber, G.A.F. (1977). *Linear Regression Analysis*. first ed.. John Wiley & Sons. New York.

Söderström, T. and P. Stoica (1983). *Instrumental Variable Methods for System Identification*. Springer-Verlag.

Söderström, T. and P. Stoica (1989). *System Identification*. Prentice-Hall International.

Stoica, P. and T. Söderström (1981). Asymptotic behavior of some bootstrap estimators. *Int. J. Control* **33**(3), 433–454.

IFAC Publications
www.elsevier.com/locate/ifac

IDENTIFICATION OF NONLINEAR PARAMETRICALLY VARYING MODELS USING SEPARABLE LEAST SQUARES

F. Previdi * **M. Lovera** **

* *Dipartimento di Ingegneria Gestionale e dell'Informazione,*
Universita' di Bergamo
Via Marconi 5, 24044 Dalmine (BG), Italy
Phone: + 39 035 2052.353; Fax: +39 035 562779
Email: previdi@unibg.it
** *Dipartimento di Elettronica e Informazione, Politecnico di Milano*
P.za Leonardo da Vinci 32, 20133 Milano, Italy
Phone: + 39 02 23993592; Fax: +39 02 23993412
Email: lovera@elet.polimi.it

Abstract: The aim of this paper is to propose a novel identification aalgorithm based on separable least squares ideas, for a class of nonlinear, possibly parameter-varying, input/output models. These models are given in the form of a Linear Fractional Transformation (LFT) where the "forward" part is represented by a conventional linear regression and the "feedback" part is given by a nonlinear map which can take into account scheduling variables available for measurement. The nonlinear part of the model can be parameterised according to various paradigms, like, e.g., Neural Network (NN) or NARX. *Copyright © 2003 IFAC*

Keywords: Nonlinear models, Identification algorithms, Least squares method, Parametric variation, Parameter estimation.

1. INTRODUCTION

The literature on non linear system identification provides advanced tools for the estimation of a wide variety of model classes (see for example (Bittanti and Picci, 1996; Chen and Billings, 1989; Sjoberg *et al.*, 1995; Murray-Smith and Johansen, 1997)). A particularly important issue within the field of nonlinear identification is represented by the problem of identifying suitable dynamic models for nonlinear plants which are led to operate in a number of different equilibria. In this case, it would be desirable to have model classes in which conventional input variables and scheduling variables (i.e., variables defining the operating point of the plant) enter the model in distinct ways. (Skeppstedt *et al.*, 1992; Johansen and Foss, 1995). Linear Parametrically Varying (LPV) models have been recently proposed as a way of dealing with this kind of problems. These models are linear ones, in which a vector of scheduling variables enters the system matrices in an affine or linear fractional way (Lee and Poolla, 1996; Lovera *et al.*, 1996; Bamieh and Giarré, 1999). Note that LPV model structures are routinely used in the formulation of robust control problems for uncertain systems (Apkarian and Gahinet, 1995) and for the design of gain-scheduling nonlinear controllers (Hunt and Johansen, 1997; Previdi, 1999; Shamma and Athans, 1990). Recently, a novel class of NonLinear, Parameter-Varying (NLPV) models suitable for system identification purposes was proposed in (Previdi and Lovera, 1999; Previdi and Lovera, 2003). NLPV models combine the advantages of a parametrically varying structure and the generality of the NARMAX (*Nonlinear AutoRegressive eXogenous Moving Average*) class and they are given in the form of a discrete-time Linear Fractional Transformation (LFT) where the "forward" part is represented by a conventional linear regression and the "feedback" part is given by a nonlinear dynamic map which can take into account scheduling variables available for measurement. In (Previdi and Lovera, 2003) the nonlinear part of the model was parameterised by a Neural Network (NN) and an identification procedure based on the iterative estimation of the parameters of the linear part and those of the nonlinear part was proposed. Also, it was shown that a strong analogy between this model class and the well known class of local model networks (LMNs), which can be exploited for control applications for the proposed model class. However, by inspection of the prediction error minimisation (PEM) cost function for the considered model class, one can recognize that the associated optimisation problem is in the form of so-called *separable least squares* (SLS), (see the classical paper (Golub and Pereyra, 1973)), i.e., the parameters of the linear part and those of the nonlinear parts can be estimated separately. Separability in least squares problems has been exploited in system identification work in a number of previous studies. In particular, in (Bruls *et al.*, 1999) the problem of identifying linear state space models is dealt with, while, more recently, the extension of the approach to bilinear models was presented in (Verdult and Verhaegen, 2001; Verdult, 2002). Separability turns out to be even more useful when applied to input/output models such as the NLPV class, and makes it possible to formulate a particularly efficient NLPV identification algorithm, which can be easily adapted to deal with various types of parameterisations for the nonlinear part of the model, like NN and Nonlinear Autoregressive models with eXogenous input (NARX) models. The aim of this paper is to present this SLS identification algorithm for models of the NLPV class, and to discuss the main properties of such an identification procedure, in the light of the theory of SLS optimisation.

2. THE NLPV MODEL CLASS

In the following, discrete-time SISO (*Single Input Single Output*) systems are considered. The symbol $y(t)$ indicates the value of the variable y at the (discrete) time step $t \in Z^+$. So, $u(t) \in \mathbb{R}$ and $y(t) \in \mathbb{R}$, for all $t = 0, 1, \ldots$. I_n is the identity matrix of dimension n. Finally, if A is a generic matrix, A' is its transpose and $A^\dagger$ its pseudoinverse.

The conventional LPV model structure can be represented by the interconnection of a linear time-invariant system with inputs u, z and outputs y and w and the block $z = \Delta w$, where Δ is formed from the elements of the time-varying, measurable parameter vector $\delta \in \mathbb{R}^r$ according to

$$\Delta = \text{blockdiag}(\delta_1 I_{n_1}, \delta_2 I_{n_2}, \ldots, \delta_r I_{n_r}). \quad (1)$$

The key point in such representations for dynamic systems is the *separation* between "conventional" control inputs (u) and other input variables (δ) which play the role of scheduling variables, i.e., they determine how the dynamics of the model change in time, according to, e.g., changes in the operating point of the system.

It is clear that the above described LFT structure gives rise to an LPV model in which the parameters are *rational* functions of the elements of vector δ. In order to take into account more complex parametric structures in the model class, a generalisation of the LPV-LFT model class is considered in this work: while the "forward" part of the LFT retains the LTI form, the "feedback" part is now modelled with a neural network, which is also fed with the vector δ as input. The main advantages of this model class are the following:

(1) When δ is nonzero, one obtains a parametrically varying model class which is far more general than the LPV-LFT one.
(2) When $\delta = 0$, it is possible to show that this model class is strongly related to the class of local model networks, although with some interesting differences (see (Previdi and Lovera, 2003)).

Formally, the NLPV model class can be described by the model structure

$$\begin{aligned} y(t) &= a_1(t)y(t-1) + \cdots + a_{n_a}(t)y(t-n_a) + \\ &+ b_1(t)u(t-1) + \cdots + b_{n_b}(t)u(t-n_b) + e(t), \end{aligned} \quad (2)$$

where n_a is the maximum output lag and n_b is the maximum input lag in the model. Note that the parameters $a_i(t)$, $i = 1, \ldots, n_a$ and $b_j(t)$, $j = 1, \ldots, n_b$ are also time-varying. A delay of n_k time steps can be introduced by forcing to zero the coefficients $b_j(t)$ for $j = 1, \ldots, n_k - 1$. The term $e(t)$ incorporates modelling error and disturbance effects.

Eq. (2) can be also seen as a particular case of the general model family

$$y(t) = g(\varphi(t), t) + e(t) \quad (3)$$

where $\varphi(t) \in \mathbb{R}^{n_a+n_b}$, for all t, is the model regressor and $g(\cdot, t) : \mathbb{R}^{n_a+n_b} \to \mathbb{R}$, for all t, is a generic time-varying function. The regressor vector is simply the collection of the past measurements of input and output used to predict the current output value, i.e.

$$\varphi(t) = [y(t-1), \ldots, y(t-n_a), u(t-1), \ldots, u(t-n_b)]'.$$

However, in this work a particular representation of the previous Eq. (3) is considered, i.e., the linear regression with time varying parameters of Eq. (2), which can be also written as

$$y(t) = \vartheta'(t)\varphi(t) + e(t) \quad (4)$$

where $\vartheta(t) \in \mathbb{R}^{n_a+n_b}$, for all t, is a time-varying model parameter vector, i.e.

$$\vartheta(t) = [a_1(t), \ldots, a_{n_a}(t), b_1(t), \ldots, b_{n_b}(t)]' \quad (5)$$

Now, suppose that the model parameters of Eq. (5) are time-varying according to the following pseudo-affine transformations:

$$a_i(t) = a_i^1 + a_i^2 z(t), \quad i = 1, \ldots, n_a \quad (6)$$

$$b_j(t) = b_j^1 + b_j^2 z(t), \quad j = 1, \ldots, n_b \tag{7}$$

where $z(t) \in \mathbb{R}$, for all t, is a scheduling variable that is assumed to be the output of a generic time-invariant model

$$z(t,\eta) = f(\psi(t),\eta) \tag{8}$$

where $\eta \in \mathbb{R}^m$ is a vector of parameters and $f : \mathbb{R}^{n_y+n_u+n_\delta} \to \mathbb{R}$ is a nonlinear dynamic mapping between the vector $\psi(t) \in \mathbb{R}^{n_y+n_u+n_\delta+1}$, for all t, and the output z. The vector $\psi(t)$ is the regressor of the scheduling model, i.e.

$$\psi(t) = [y(t-1), \ldots, y(t-n_y), u(t-1), \ldots, u(t-n_u), \delta(t-1), \ldots, \delta(t-n_\delta)]'.$$

It is worth noting that $\psi(t)$ is not only a collection of past measurements of input and output, but a new variable δ appears. δ is an exogenous signal carrying additional information on the current operating regime of the system: in this framework, such kind of information can be easily incorporated in the model, simply by enlarging the scheduling regressor vector $\psi(t)$. The parameterisation of the static nonlinear mapping $f(\cdot)$ will be discussed in Section 5. Finally, some additional assumptions should be made, concerning the structure of the regressors $\varphi(t)$ and $\psi(t)$. In fact, it is sensible that the model should obtain information about the operating conditions without using more information than those available and actually used to model the plant dynamics. As the output $z(t)$ of the nonlinear part of the model is only devoted to give an estimate of the current operating conditions of the plant at time k, it will be assumed in the following that the model regressor $\varphi(t)$ and the scheduling regressor $\psi(t)$ are such that $\max\{n_y, n_u\} \leqslant \max\{n_a, n_b\}$.

3. FORMULATION OF THE IDENTIFICATION PROBLEM

The NLPV model of Equations from (4) to (2) can be written as

$$y(t) = \theta' \begin{pmatrix} \varphi(t) \\ \varphi(t)z(t,\eta) \end{pmatrix} \tag{9}$$

where $\theta \in \mathbb{R}^{2n_a+2n_b}$ is

$$\theta = [a_1^1, \ldots, a_{n_a}^1, b_1^1, \ldots, b_{n_b}^1, a_1^2, \ldots, a_{n_a}^2, b_1^2, \ldots, b_{n_b}^2]' \tag{10}$$

and

$$\varphi(t) = [y(t-1), \ldots, y(t-n_a), u(t-1), \ldots, u(t-n_b)]'.$$

The problem is to find an estimate $(\hat{\theta}, \hat{\eta})$ of the parameters of the model

$$\hat{y}(t) = \sum_{j=1}^{n} \theta_j \phi_j(t,\eta) \tag{11}$$

where $n = 2n_a + 2n_b$, θ_j indicates the jth component of the vector θ and $\phi_j(t,\eta)$ is the jth component of the vector $\phi(t,\eta) \in \mathbb{R}^{2n_a+2n_b}$ defined as follows, consistently with Eqs. (9)

$$\begin{aligned} \phi(t,\eta) = [&y(t-1), \cdots, y(t-n_a), \\ &u(t-1), \cdots, u(t-n_b), \\ &y(t-1)z(t,\eta), \cdots, y(t-n_a)z(t,\eta), \\ &u(t-1)z(t,\eta), \cdots, u(t-n_b)z(t,\eta)]'. \end{aligned} \tag{12}$$

The parameters must be estimated by minimization of the cost function

$$J(\theta,\eta) = \sum_{i=1}^{N} \left(y_i - \sum_{j=1}^{n} \theta_j \phi_j(t_i,\eta) \right), \tag{13}$$

which can be easily rewritten as

$$J(\theta,\eta) = \|\mathbf{y} - \Phi(\eta)\theta\|_2^2 \tag{14}$$

where

$$\Phi(\eta) = \begin{pmatrix} \phi'(t_1,\eta) \\ \phi'(t_2,\eta) \\ \vdots \\ \phi'(t_N,\eta) \end{pmatrix}. \tag{15}$$

While the problem of minimising (14) can be certainly faced as a general non linear optimisation problem, significant improvements in the estimation procedure can be obtained by exploiting the particular structure of the cost function. Indeed, Equations (11)-(14) define a separable least squares problem, the structure of which has been first investigated in (Golub and Pereyra, 1973). A summary of the main properties of SLS problems is given in the following Section.

4. ON SEPARABLE LEAST SQUARES PROBLEMS

Consider the problem of estimating the parameters a, p of the regression model

$$J_1(a,p) = \|\mathbf{y} - \Phi(p)\mathbf{g}(a)\|_2^2. \tag{16}$$

where a is a s-dimensional vector, p is a m-dimensional vector, $\mathbf{g}(a)$ is a n-dimensional vector function and $\Phi(p)$ is a $(N \times n)$-dimensional matrix function. The approach proposed by Golub and Pereyra for the minimisation of J_1 is based on the following two assumptions (the meaning of which will be made clear in the following):

Assumption 1 The system of nonlinear equations $\mathbf{g}(a) = b$ has at least one solution for any vector $b \in R^n$.

Assumption 2 The matrix $\Phi(p)$ has constant rank $r \leq \min(N,n)$ for $p \in \Omega \subset R^m$, Ω being an open set containing the desired solution.

In order to obtain a separation between the estimation of a and p, define the following modified cost function (depending only on the parameters p), called Variable Projection Functional

$$J_2(p) = \|\mathbf{y} - \Phi(p)\Phi^{\dagger}(p)\mathbf{y}\|_2^2. \tag{17}$$

The idea behind the separable least squares approach to the problem is the following: minimize

$$J_1(a,p) = \|\mathbf{y} - \Phi(p)\mathbf{g}(a)\|_2^2$$

with respect to the new parameter vector $b = \mathbf{g}(a)$. In this way, a LS problem is obtained:

$$J_1(b,p) = \|\mathbf{y} - \Phi(p)b\|_2^2$$

the solution of which, for any p, is given by

$$\hat{b} = \Phi^{\dagger}(p)\mathbf{y}.$$

Substituting $\hat{b}$ in $J_1(b,p)$, Eq. (17) is easily obtained whose minimization is an estimate $\hat{p}$. Finally, an estimate $\hat{a}$ is obtained by solving $\Phi^{\dagger}(\hat{p})\mathbf{y} = \mathbf{g}(a)$, thanks to Assumption 1. The above procedure is guaranteed to minimize the original cost function J_1 thanks to the following result:

Theorem 1. (Golub-Pereyra). Let J_1 and J_2 be defined as above. Then under Assumptions 1 and 2 the following holds:

(1) If $\hat{p}$ is a critical point (or a global minimizer in Ω) of $J_2(p)$ and $\hat{a}$ satisfies $g(\hat{a}) = \Phi^{\dagger}(\hat{p})\mathbf{y}$ then $(\hat{a},\hat{p})$ is a critical point (or a global minimizer for $p \in \Omega$) of $J_1(a,p)$ and $J_1(\hat{a},\hat{p}) = J_2(\hat{p})$.

(2) If $(\hat{a},\hat{p})$ is a critical point (or a global minimizer for $p \in \Omega$) of $J_1(a,p)$ then $\hat{p}$ is a critical point (or a global minimizer in Ω) of $J_2(p)$ and $J_2(\hat{p}) = J_1(\hat{a},\hat{p})$. Furthermore, if there is an unique $\hat{a}$ among the minimizing pairs of J_1, then $\hat{a}$ must satisfy $g(\hat{a}) = \Phi^{\dagger}(\hat{p})\mathbf{y}$.

5. SLS IDENTIFICATION OF NLPV MODELS

In order to apply the results described in Section 4 to the NLPV identification problem, we must first check the validity of the assumptions in this framework. Assumption 1 is trivially verified (with the notation of Section 4 we have that $b = a$). As for Assumption 2, since matrix $\Phi(\eta)$ is constructed from the available input/output data, it corresponds to an identifiability condition for the model class. Actually, $\Phi(\eta)$ depends also on the estimated parameter vector $\hat{\eta}$. However, in practice, the verification of Assumption 2 must be performed after Step 1 in the identification algorithm, with a given estimate $\hat{\eta}$ of the parameter vector.

On the basis of this discussion, the identification algorithm can be described as the following two steps procedure:

Step 1 Compute an estimate $\hat{\eta}$ of η as

$$\hat{\eta} = \arg\min_{\eta} \|P_{\Phi}^{\perp}(\eta)\mathbf{y}\|_2^2 \tag{18}$$

where $P_{\Phi}^{\perp}(\eta) = I - \Phi(\eta)\Phi^{\dagger}(\eta)$.

Step 2 Compute

$$\hat{\theta} = \Phi^{\dagger}(\hat{\eta})\mathbf{y}. \tag{19}$$

In order to estimate $\hat{\eta}$ by solving Eq. (18), an iterative procedure is needed. In Golub-Pereyra there is an example of application to the Gauss-Netwon-Marquardt (GNM) algorithm, summarized in the following. The GNM $(i+1)th$ iteration is

$$\hat{\eta}_{i+1} = \hat{\eta}_i - K_i^{\dagger} r_i \tag{20}$$

where

$$K_i = \begin{pmatrix} \mathbf{D}\left(P_{\Phi}^{\perp}(\hat{\eta}_i)\mathbf{y}\right) \\ \hline \nu_i I_n \end{pmatrix}, \quad r_i = \begin{pmatrix} P_{\Phi}^{\perp}(\hat{\eta}_i)\mathbf{y} \\ \hline 0_n \end{pmatrix} \tag{21}$$

with ν_i for $i = 0, \ldots$, a sequence of non–negative auxiliary parameters and where $\mathbf{D}$ indicates the Frechet derivative operator. Note that the Frechet derivative of a projection operator $P_A(\mathbf{x})$ can be written as

$$\begin{aligned} &\mathbf{D}(P_A(\mathbf{x})) \equiv \mathbf{D}\left(A(\mathbf{x})A^{\dagger}(\mathbf{x})\right) = \\ &= P_A^{\perp}(\mathbf{x})\mathbf{D}(A(\mathbf{x}))A^{\dagger}(\mathbf{x}) + \left(P_A^{\perp}(\mathbf{x})\mathbf{D}\left(A(\mathbf{x})\right)A^{\dagger}(\mathbf{x})\right)', \end{aligned} \tag{22}$$

and (22) holds provided that $A(\mathbf{x})$ has ***constant rank*** on an open set in the $\mathbf{x}$ parameters space. Clearly this includes the particular case of a full rank matrix A.

Remark 1. Observe that, by defining $h_i = \hat{\eta}_{i+1} - \hat{\eta}_i$ we can rewrite Eq. (20) as $h_i = -K_i^{\dagger} r_i$ which is simply the LS solution of the equation $K_i h_i = -r_i$. So, depending on the condition of the matrix K_i different methods can be used to calculate the increments h_i.

So, the identification algorithm can be described as follows: choose an initial value of the parameters η_0, $\nu_0 > 0$ at iteration step $i = 0$. Then, at each i repeat the following steps:

1 Compute $P_{\Phi}^{\perp}(\eta_i)\mathbf{y}$
2 Compute $\mathbf{D}\left(P_{\Phi}^{\perp}(\hat{\eta}_i)\mathbf{y}\right)$
3 Build K_i and r_i
4 Compute $h_i = -K_i^{\dagger} r_i$ (or solve $K_i h_i = -r_i$)
5 Update the estimate $\hat{\eta}_{i+1} = \hat{\eta}_i + h_i$
6 Increase or decrease the parameter ν_i, using a multiplicative factor Λ_i (according to some criterion) $\nu_{i+1} = \Lambda_i \nu_i$

In order to complete the definition of the algorithm, we need

$$\begin{aligned} \mathbf{D}\left(P_{\Phi}^{\perp}(\hat{\eta}_i)\mathbf{y}\right) &= -P_{\Phi}^{\perp}(\hat{\eta}_i)\mathbf{D}\left(\Phi(\hat{\eta}_i)\right)\Phi^{\dagger}(\hat{\eta}_i)\mathbf{y} + \\ &\quad - \left(P_{\Phi}^{\perp}(\hat{\eta}_i)\mathbf{D}\left(\Phi(\hat{\eta}_i)\right)\Phi^{\dagger}(\hat{\eta}_i)\right)'\mathbf{y}. \end{aligned} \tag{23}$$

In Eq. (23) we must compute $\mathbf{D}(\Phi(\eta))$ which is a tensor with dimension $k \times (N \times n)$, i.e. is a set of k matrices with dimension $N \times n$ each one containing the partial derivatives of the elements of Φ with respect to one of the component of the parameter vector η. Now, suppose that in Eq. (8) $z(t,\eta)$ is the output of a multilayer feedforward neural network, i.e.

$$f(\psi(t),\eta) \equiv \sigma\left(\sum_{h=1}^{n_h} \alpha_h \sigma\left(\sum_{j=1}^{m} w_{hj}\psi_j(t) + \beta_h\right)\right)$$

where n_h is the number of units in the hidden layer; α_h for $h = 1, \ldots, n_h$ are the parameters connecting the hidden layer to the network output; w_{hj} are the elements of the matrix $W \in \mathbb{R}^{n_h} \times \mathbb{R}^m$ with $m = n_y + n_u + n_\delta + 1$ built with the weights connecting the network input to the hidden layer; $\beta_h \in \mathbb{R}$ are the bias parameters; $\sigma(\cdot)$ is a nonlinear sigmoidal scalar function: in this paper, the hyperbolic tangent function has been used. So, the neural network realizes a function $f : \mathbb{R}^{n_y+n_u+n_\delta+1} \rightarrow [-1, 1]$. Obviously, the parameter vector $\eta \in \mathbb{R}^k$, with $k = (m+2)n_h$, is easily obtained by collecting the neural network parameters, i.e.

$$\eta = [\alpha_1, \ldots, \alpha_{n_h}, \beta_1, \ldots, \beta_{n_h}, w_{11}, \ldots, w_{n_h, n_y+n_u+n_\delta+1}].$$

In this case, the k elements of $\mathbf{D}(\Phi(\eta))$ can be explicitly computed as follows. First of all, for $l = 1, \ldots, n_h$, let's define:

$$\Sigma_l(t_i, \eta) = \left(1 - \sigma^2\left(\sum_{h=1}^{n_h} \alpha_h \sigma\left(\sum_{j=1}^{m} w_{hj} \psi_j(t) + \beta_h\right)\right)\right) \sigma\left(\sum_{j=1}^{m} w_{lj} \psi_j(t) + \beta_l\right)$$

$$\Xi_l(t_i, \eta) = \alpha_l \left(1 - \sigma^2\left(\sum_{h=1}^{n_h} \alpha_h \sigma\left(\sum_{j=1}^{m} w_{hj} \psi_j(t) + \beta_h\right)\right)\right) \left(1 - \sigma^2\left(\sum_{j=1}^{m} w_{lj} \psi_j(t) + \beta_l\right)\right)$$

Then, the elements of $\mathbf{D}_l(\Phi(\eta))$ are given by

$$\begin{pmatrix} 0 & \ldots & 0 & y(t_1-1)\Sigma_l(t_1,\eta) & \ldots & u(t_1-n_b)\Sigma_l(t_1,\eta) \\ 0 & \ldots & 0 & y(t_2-1)\Sigma_l(t_2,\eta) & \ldots & u(t_2-n_b)\Sigma_l(t_2,\eta) \\ \vdots & \ddots & \vdots & \vdots & \ddots & \vdots \\ 0 & \ldots & 0 & y(t_N-1)\Sigma_l(t_N,\eta) & \ldots & u(t_N-n_b)\Sigma_l(t_N,\eta) \end{pmatrix}$$

for $l = 1, \ldots, n_h$, and by

$$\begin{pmatrix} 0 & \ldots & 0 & y(t_1-1)\Xi_{l-n_h}(t_1,\eta) & \ldots & u(t_1-n_b)\Xi_{l-n_h}(t_1,\eta) \\ 0 & \ldots & 0 & y(t_2-1)\Xi_{l-n_h}(t_2,\eta) & \ldots & u(t_2-n_b)\Xi_{l-n_h}(t_2,\eta) \\ \vdots & \ddots & \vdots & \vdots & \ddots & \vdots \\ 0 & \ldots & 0 & y(t_N-1)\Xi_{l-n_h}(t_N,\eta) & \ldots & u(t_N-n_b)\Xi_{l-n_h}(t_N,\eta) \end{pmatrix}$$

for $l = n_h + 1, \ldots, 2n_h$. The remaining elements of the tensor, for $l = 2n_h + 1, \ldots, (m+2)n_h$, can be divided into n_h sub–vectors, each of them with m elements. Being r the index labeling each sub–vector, so $r = 1, \ldots, n_h$, and s the index labeling the elements in the current sub–vector, so $s = 1, \ldots, m$, we can write the generic lth elements follows, where $l = (r-1)m + s + 2n_h$

$$\begin{pmatrix} 0 & \ldots & 0 & y(t_1-1)\Xi_r(t_1,\eta)\psi_s(t_1) & \ldots & u(t_1-n_b)\Xi_r(t_1,\eta)\psi_s(t_1) \\ 0 & \ldots & 0 & y(t_2-1)\Xi_r(t_2,\eta)\psi_s(t_1) & \ldots & u(t_2-n_b)\Xi_r(t_2,\eta)\psi_s(t_1) \\ \vdots & \ddots & \vdots & \vdots & \ddots & \vdots \\ 0 & \ldots & 0 & y(t_N-1)\Xi_r(t_N,\eta)\psi_s(t_1) & \ldots & u(t_N-n_b)\Xi_r(t_N,\eta)\psi_s(t_1) \end{pmatrix}.$$

6. SIMULATION EXAMPLES

Two sets of I/O data of $N = 1000$ samples each have been generated using a system belonging to the model class used for identification, so that no structural identification issues must be considered. So, the system and the model have the following structure: $n_a = 2$, $n_b = 2$, $n_k = 1$, $n_y = 2$, $n_u = 2$, $n_k = 1$, $n_h = 4$. At this stage, no exogeneous δ signal has been considered. The first data set has been used for parameter estimation, the second data set has been used for validation. In Fig. 1 it is shown the plot of the cost function $MSE = \frac{1}{N}J(\theta, \eta)$ where $J(\theta, \eta)$ is defined in Eq. (13), during the estimation procedure. The minimum value for the cost function during estimation is $MSE = 0.0084$. In Fig. 2 a plot of 100 samples drawn from the validation data set is compared with the simulated output. Following (Billings and Zhu, 1994),

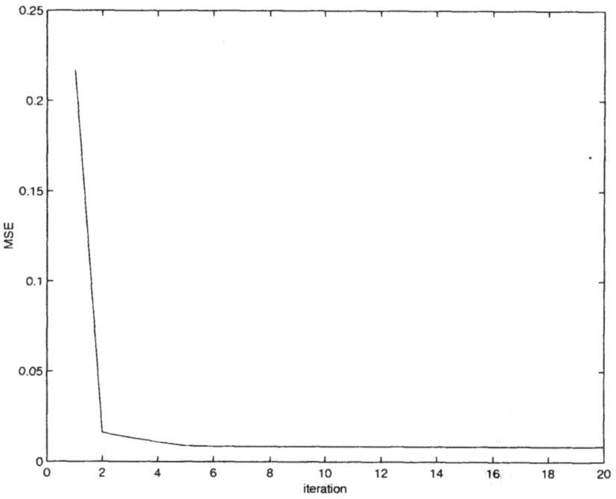

Fig. 1. Mean Square Error during estimation.

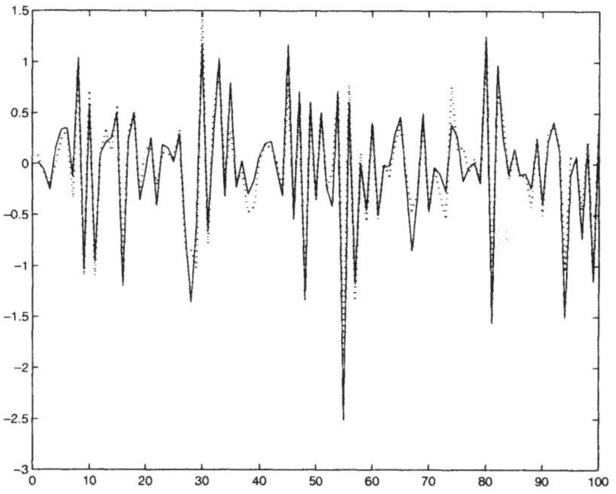

Fig. 2. Comparison between a subset of the validation output (solid line) and the output simulated using the estimated model.

the residuals are analysed using high order cross correlation functions, estimated from the available data. The following estimator has been used for two finite realizations of length N of two stochastic processes x_1 and x_2 with mean values $\bar{x}_1$ and $\bar{x}_2$ respectively:

$$\hat{\phi}_{x_1 x_2} = \frac{\sum_{t=1}^{N-\tau}(x_1(t+\tau) - \bar{x}_1)(x_2(t) - \bar{x}_2)}{(\sum_{t=1}^{N}(x_1(t) - \bar{x}_1)^2)^{1/2}(\sum_{t=1}^{N}(x_2(t) - \bar{x}_2)^2)^{1/2}}$$

Specifically, using this estimator, the following six statistical hypotheses have been tested at 95% confidence level: $\hat{\phi}_{eu} = 0$, $\hat{\phi}_{e^2u^2} = 0$, $\hat{\phi}_{eu^2} = 0$, $\hat{\phi}_{e\beta} = 0$, $\hat{\phi}_{u^2\alpha} = 0$ $\forall\tau$, and

$$\hat{\phi}_{e^2\alpha} = \begin{cases} k, & \tau = 0 \\ 0, & otherwise \end{cases}$$

where $\alpha(t)=y(t)e(t)$, $\beta=u(t)e(t), t=1,\ldots,N$, with $e(t)=y(t)-\hat{y}(t)$ the simulation residual. Figure 3 shows that the estimated model can be considered acceptable.

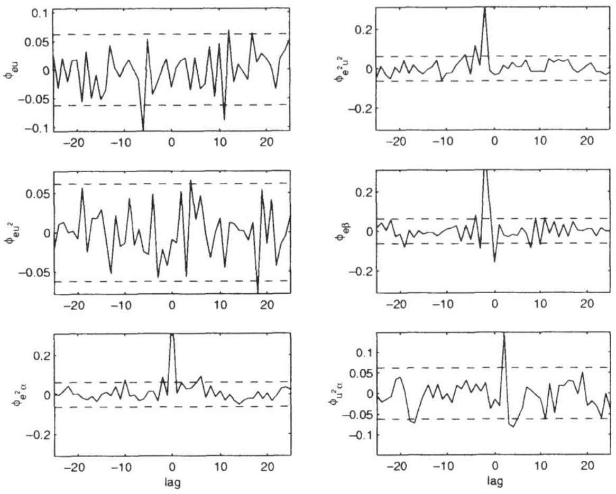

Fig. 3. Plots of high order cross correlation functions with 95% confidence level.

7. CONCLUDING REMARKS

In this work, a class of models is proposed for nonlinear and parameter varying system identification purposes, in which the parameters vary according to a nonlinear dynamic law. An identification procedure based on Separable Least Squares ideas has been worked out, which provides an accurate and efficient solution to the identification problem.

Acknowledgements Paper supported by the MIUR project "Innovative Techniques for Identification and Adaptive Control of Industrial systems".

8. REFERENCES

Apkarian, P. and P. Gahinet (1995). A Convex Characterization of Gain-Scheduling H_∞ Controllers. *IEEE Trans. Automatic Control* **40**, 853–864.

Bamieh, B. and L. Giarré (1999). Identification of linear parameter varying models. In: *Proc. of the Conference on Decision and Control*. Phoenix, AZ, USA.

Billings, S. A. and Q. M. Zhu (1994). Nonlinear model validation using correlation tests. *International Journal of Control* **44**, 235–244.

Bittanti, S. and Picci, G., Eds.) (1996). *Identification, adaptation and learning - The science of learning models from data*. Springer-Verlag, Berlin.

Bruls, J., C.T. Chou, B.R.J. Haverkamp and M. Verhaegen (1999). Linear and non-linear system identification using separable least-squares. *European Journal of Control* **5**(1), 116–128.

Chen, S. and S. A. Billings (1989). Representations of non-linear systems: the NARMAX model. *Int. J. of Control* **49**, 1013–1032.

Golub, G. and V. Pereyra (1973). The differentiation of pseudoinverses and nonlinear least squares problems whose variables separate. *SIAM Journal of Numerical Analysis* **10**, 413–432.

Hunt, K. J. and T. A. Johansen (1997). Design and Analysis of Gain-Scheduled Control using Local Controller Networks. *Int. Journal of Control* **66**, 619–651.

Johansen, T. A. and B. A. Foss (1995). Identification of Non-Linear System Structure and Parameters using Regime Decomposition. *Automatica* **31**, 321–326.

Lee, L. H. and K. Poolla (1996). Identification of Linear Parameter-Varying Systems via LFT. In: *Proc. of the 35th Conference on Decision and Control*. Japan. pp. 1545–1550.

Lovera, M., M. Verhaegen and C. T. Chou (1996). State space identification of MIMO linear parameter varying models. In: *Proc. of the International Conference on Mathematical Theory of Networks and Systems*. Padova, Italy.

Murray-Smith, R. and T. A. Johansen (1997). *Multiple Model Approaches to Modeling and Control*. Taylor and Francis, London.

Previdi, F. (1999). *Identification and Control using Local Linear Models*. PhD Thesis. Politecnico di Milano, Italy.

Previdi, F. and M. Lovera (1999). Identification of a class of linear models with nonlinearly varying parameters. In: *Proc. of the European Control Conference*. Karlsruhe, Germany.

Previdi, F. and M. Lovera (2003). Identification of a class of nonlinear parametrically varying models. *International Journal on Adaptive Control and Signal Processing* **17**, 33–50.

Shamma, J. S. and M. Athans (1990). Analysis of Gain-Scheduled Control for Nonlinear Plants. *IEEE Trans. Automatic Control* **35**, 898–907.

Sjoberg, J., Q. Zhang, L. Ljung, A. Benveniste, B. Delyon, P. Glorennec, H. Hjalmarsson and A. Juditsky (1995). Nonlinear Black-box Modeling in System Identification: a Unified Overview. *Automatica* **31**, 1691–1724.

Skeppstedt, A., L. Ljung and M. Millnert (1992). Construction of composite models from observed data. *Int. J. of Control* **55**, 141–152.

Verdult, V. (2002). Nonlinear system identification: a state space approach. PhD thesis. University of Twente.

Verdult, V. and M. Verhaegen (2001). Identification of multivariable bilinear state space systems based on subspace techniques and separable least squares optimization. *Int. J. of Control* **74**, 1824–1836.

IFAC Publications
www.elsevier.com/locate/ifac

MODELING AND IDENTIFICATION OF RATE-INDEPENDENT HYSTERESIS USING A SEMILINEAR DUHEM MODEL [1]

JinHyoung Oh and Dennis S. Bernstein

Department of Aerospace Engineering,
The University of Michigan, Ann Arbor, MI 48109-2140

Abstract: In this paper we consider a semilinear Duhem model. The input-output map of the model is rate-independent, thus yielding persistent phase shift (that is, hysteresis) at arbitrarily low frequency. For the semilinear Duhem model we reparameterize the response in terms of the control input, and we provide sufficient conditions for convergence to a hysteresis map. A constrained least squares method is developed to identify the hysteresis map using the semilinear Duhem model. *Copyright © 2003 IFAC*

Keywords: Hysteresis, Modelling, Nonlinearity, Identification

1. INTRODUCTION

Hysteresis is a widespread phenomenon in many engineering areas. Although there is no precise definition of hysteresis, we adopt the intuitive notion that hysteresis is effectively *DC phase shift*, that is, phase shift that persists as the frequency content of the input signal approaches DC. Consequently, hysteresis is an inherently nonlinear phenomenon since the phase shift of linear systems always approaches zero degrees as the input frequency decreases.

To illustrate this point of view, consider the mechanism shown in Figure 1. The equation of motion is given by

$$m\ddot{q}(t) + c\dot{q}(t) + kd_w\left(q(t) - r(t)\right) = 0, \qquad (1)$$

where $d_w(z)$ is the deadzone function with width w. Because of the deadzone at the attachment point, there is a phase shift between $r(t)$ and $q(t)$. The presence of hysteresis is not obvious during dynamic operation, since the phase shift is a consequence of both the gap and the dynamics.

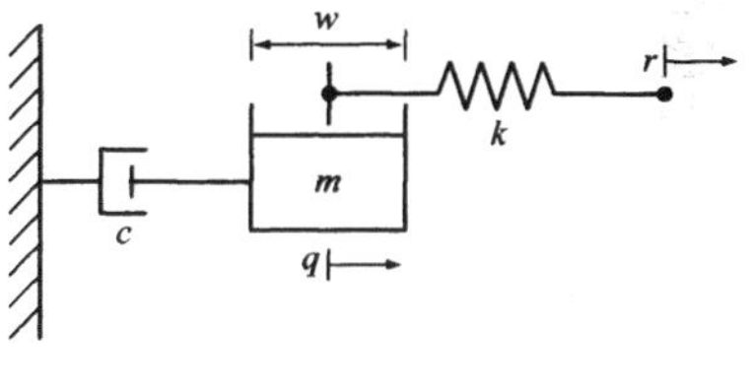

Fig. 1. Mass-dashpot-spring system with deadzone.

However, Figure 2 reveals that the phase shift persists near DC, that is, at asymptotically low frequency.

Alternatively, consider the relationship between the magnetic field $H(t)$ and the magnetic flux $B(t)$ of ferromagnetically soft materials of the isoperm type (Coleman and Hodgdon, 1986)

$$\dot{B}(t) = \alpha|\dot{H}(t)|[bH(t) - B(t)] + a\dot{H}(t). \qquad (2)$$

Figure 3 shows the relationship between $H(t)$ and $B(t)$. The presence of phase shift for low frequency inputs indicates that this system is hysteretic.

Although the examples discussed above are both hysteretic, the response of the mass-spring system depends on the input frequency, or more generally, is affected by the time dependence of the input.

[1] This research was supported in part by the National Science Foundation under grant ECS–0225799.

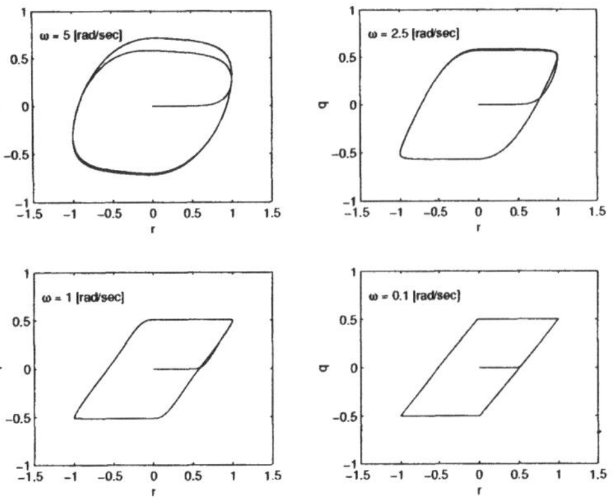

Fig. 2. Input-output map for the mass-dashpot-spring system with deadzone.

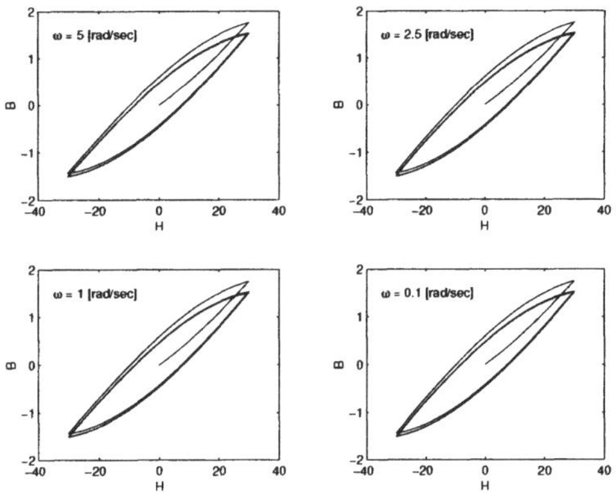

Fig. 3. Input-output map for the model of ferromagnetically soft materials of the isoperm type.

However, the ferromagnetic material model has the same input-output response for all frequencies and types of input.

One of the most successful hysteresis models is the Preisach model(Macki *et al.*, 1993, and references therein). Preisach models are frequency independent and thus they are inherently kinematic. However, Preisach models are computationally demanding, requiring gridding of the plane.

In contrast, the examples discussed above are finite dimensional, and thus they suggest alternatives to the Preisach model. In fact, both of these examples illustrate hysteresis models that have been studied in the literature. In particular, the mass-spring example is suggested in Kransnosel'skii and Pokrovskii (1980), p. 93, as an approximation to a hysteron model, while the ferromagnetic material model is a Duhem model, a class of hysteresis models extensively studied in Chua and Bass (1972).

The purpose of this paper is to extend the existing analysis of Duhem models in order to understand their properties for modeling and identification.

2. GENERALIZED DUHEM MODEL

Consider the single input-single output *generalized Duhem model* given by

$$\dot{x}(t) = f(x(t), u(t))g(\dot{u}(t)), \tag{3}$$
$$y(t) = h(x(t), u(t)), \quad x(0) = x_0,\ t \geq 0, \tag{4}$$

where $x \in \mathbb{R}^n$, $y \in \mathbb{R}$, $u \in \mathbb{R}$, $f : \mathbb{R}^n \times \mathbb{R} \to \mathbb{R}^{n\times r}$, and $g : \mathbb{R} \to \mathbb{R}^r$. We assume that the solution to (3) exists and is unique on all finite intervals. The following definition will be useful.

Definition 1. The generalized Duhem model (3), (4) is *time-scale invariant* if, for all $x(t)$ and $u(t)$ satisfying (3), all initial conditions x_0, and all $\alpha > 0$, it follows that $x_\alpha(t) \triangleq x(\alpha t)$ and $u_\alpha(t) \triangleq u(\alpha t)$ also satisfy (3).

The following definition is needed to characterize time-scale invariant generalized Duhem models.

Definition 2. The function g is *positively homogeneous* if $g(\alpha v) = \alpha g(v)$ for all $\alpha \geq 0$ and $v \in \mathbb{R}$.

The following result generalizes Property 9 of Chua and Bass (1972).

Proposition 1. Assume that g is positively homogeneous. Then the generalized Duhem model (3), (4) is time-scale invariant.

The time-scale invariant generalized Duhem model has several alternative representations. The following lemma is needed for further discussion.

Lemma 1. Assume g is positively homogeneous. Then there exist $h_+ \in \mathbb{R}^r$ and $h_- \in \mathbb{R}^r$ such that

$$g(v) = \begin{cases} vh_+, & v \geq 0, \\ vh_-, & v < 0. \end{cases} \tag{5}$$

Assume g is positively homogeneous. Then (3) and (4) can be rewritten as

$$\dot{x}(t) = \begin{cases} f_+(x(t), u(t))\,\dot{u}(t), & \dot{u}(t) \geq 0, \\ f_-(x(t), u(t))\,\dot{u}(t), & \dot{u}(t) < 0, \end{cases} \tag{6}$$
$$y(t) = h(x(t), u(t)),\ x(0) = x_0,\ t \geq 0, \tag{7}$$

where $f_+(x(t), u(t)) \triangleq f(x(t), u(t))h_+$ and $f_-(x(t), u(t)) \triangleq f(x(t), u(t))h_-$. Note that (6) can be viewed as a switching system with respect to the sign of $\dot{u}(t)$. Next define $\dot{u}_+(t) \triangleq \max\{0, \dot{u}(t)\}$, and $\dot{u}_-(t) \triangleq \min\{0, \dot{u}(t)\}$. Then (6) can be written as

$$\dot{x}(t) = [\dot{u}_+(t)I_n\ \dot{u}_-(t)I_n]\begin{bmatrix} f_+(x(t), u(t)) \\ f_-(x(t), u(t)) \end{bmatrix}, \tag{8}$$
$$y(t) = h(x(t), u(t)), \quad x(0) = x_0,\ t \geq 0. \tag{9}$$

which is the *classical Duhem model* (Macki *et al.*, 1993).

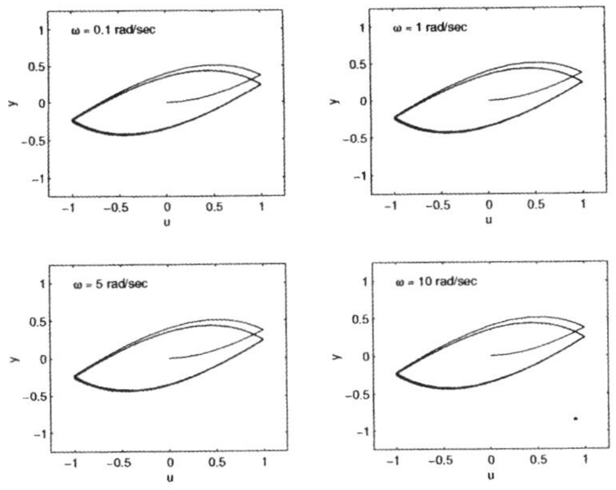

Fig. 4. Input-output maps of Example 1 with $g(\dot{u}) = |\dot{u}|$ and $u(t) = \sin t$.

Example 1. Consider the generalized Duhem model

$$\dot{x}(t) = (-x(t) + u(t))g(\dot{u}(t)), \quad (10)$$
$$y(t) = x(t), \quad x(0) = 0,\ t \geq 0,. \quad (11)$$

First, let $g(\dot{u}) = |\dot{u}|$. Since $|\dot{u}|$ is positively homogeneous, Proposition 1 implies that (10), (11) is time-scale invariant and the input-output maps of the model with different input frequencies are identical as shown in Figure 4. Next let $g(\dot{u}) = \dot{u}^2$, which is not positively homogeneous. Figure 5 shows that the input-output map of (10) and (11) depends on input frequency.

3. REPARAMETERIZATION OF THE TIME-SCALE INVARIANT GENERALIZED DUHEM MODEL

Consider the time-scale invariant generalized Duhem model (6), (7), where $u(t)$ is piecewise monotonic. Suppose $\dot{u}(t) \neq 0$. Then dividing both sides of (6) by $\dot{u}(t)$ yields

$$\frac{\mathrm{d}x(t)}{\mathrm{d}u(t)} = \begin{cases} f_+(x(t), u(t)), & \dot{u}(t) > 0, \\ f_-(x(t), u(t)), & \dot{u}(t) < 0. \end{cases} \quad (12)$$

Now suppose $\dot{u}(t) = 0$. Then (6) becomes $\dot{x}(t) = 0$ and thus $x(t)$ is constant. Therefore, the time-scale invariant generalized Duhem model (6), (7) can be reparameterized with u considered as the independent variable. Let $\hat{x}(u)$ and $\hat{y}(u)$ be the reparameterized variables of (6) and (7) such that $\hat{x}(u) \triangleq x(u(t))$ and $\hat{y}(u) \triangleq y(u(t))$. Then the reparameterized time-scale invariant generalized Duhem model becomes

$$\frac{\mathrm{d}\hat{x}(u)}{\mathrm{d}u} = \begin{cases} f_+(\hat{x}(u), u), & \text{if } u \text{ increases}, \\ f_-(\hat{x}(u), u), & \text{if } u \text{ decreases}, \end{cases} \quad (13)$$
$$\hat{y}(u) = h(\hat{x}(u), u), \quad \hat{x}(u_0) = x_0, \quad (14)$$

where $u_0 = u(0)$. Note that (13) and (14) can be viewed as a time-varying dynamical system with nonmonotonic time.

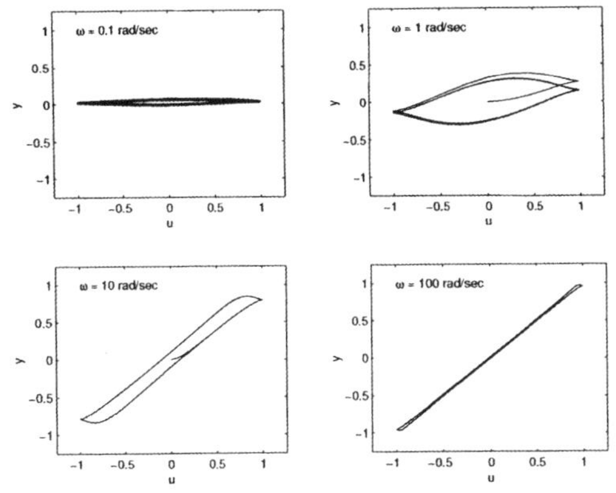

Fig. 5. Input-output maps of Example 1 with $g(\dot{u}) = \dot{u}^2$ and $u(t) = \sin t$.

Example 2. Reconsider Example 1 with two different inputs, namely, sinusoidal and triangle inputs with same period and amplitude. When $g(\dot{u}) = |\dot{u}|$, the input-output maps under different inputs are identical. This shows that the input-output map is not affected by the time dependence of u. However, when $g(\dot{u}) = \dot{u}^2$, the input-output maps under different inputs are different as shown in Figure 6.

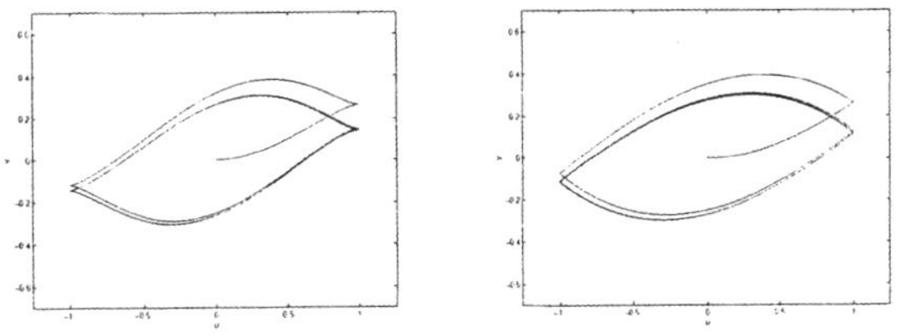

(a) (b)

Fig. 6. Input-output maps of Example 2 with $g(\dot{u}) = \dot{u}^2$ under (a) sinusoidal input, and (b) triangle input.

4. SEMILINEAR DUHEM MODEL

As a specialization of (8) and (9), in this section we consider the *semilinear Duhem model*

$$\dot{x}(t) = [\dot{u}_+(t)I_n \ \dot{u}_-(t)I_n]\times \left(\begin{bmatrix} A_+ \\ A_- \end{bmatrix} x(t) + \begin{bmatrix} B_+ \\ B_- \end{bmatrix} u(t) + \begin{bmatrix} E_+ \\ E_- \end{bmatrix} \right), \quad (15)$$
$$y(t) = Cx(t) + Du(t), \quad x(0) = x_0,\ t \geq 0, \quad (16)$$

where $A_+ \in \mathbb{R}^{n\times n}$, $A_- \in \mathbb{R}^{n\times n}$, $B_+ \in \mathbb{R}^n$, $B_- \in \mathbb{R}^n$, $E_+ \in \mathbb{R}^n$, $E_- \in \mathbb{R}^n$, $C \in \mathbb{R}^{1\times n}$, and $D \in \mathbb{R}$. Note that (15), (16) are a time-scale invariant generalized Duhem model of the form (8), (9) with $f_+(x(t), u(t)) = A_+x(t) + B_+u(t) + E_+$, $f_-(x(t), u(t)) = A_-x(t) + B_-u(t) + E_-$, and $h(x(t), u(t)) = Cx(t) + Du(t)$. We exclude pathological inputs by assuming that $u(t)$ is piecewise monotonic. Reparameterizing (15) and (16) in terms of u yields

$$\frac{\mathrm{d}\hat{x}(u)}{\mathrm{d}u} = \begin{cases} A_+\hat{x}(u) + B_+u + E_+, & \text{if } u \uparrow, \\ A_-\hat{x}(u) + B_-u + E_-, & \text{if } u \downarrow, \end{cases} \quad (17)$$

$$\hat{y}(u) = C\hat{x}(u) + Du, \quad \hat{x}(u_0) = x_0, \tag{18}$$

where $u_0 = u(0)$. For the following lemma, A^{D} denote the Drazin generalized inverse of A and let $\operatorname{ind} A$ denote the index number of A (p. 122, Campbell and C. D. Meyer, 1979).

Lemma 2. Let $r_+ = \operatorname{ind} A_+$ and $r_- = \operatorname{ind} A_-$. Then the forward-time ramp response of (17) is given by $\hat{x}(u) = e^{A_+(u-u_0)}x_0 + \mathcal{X}_+(u,u_0) + \mathcal{Y}_+(u-u_0) - \mathcal{Z}_+(u,u_0)$, $u \geq u_0$, and the backward-time ramp response of (17) is given by $\hat{x}(u) = e^{A_-(u-u_0)}x_0 + \mathcal{X}_-(u,u_0) + \mathcal{Y}_-(u-u_0) - \mathcal{Z}_-(u,u_0)$, $u \leq u_0$, where

$$\mathcal{X}_+(u,u_0) \triangleq \left(I - A_+A_+^{\mathrm{D}}\right)\sum_{k=1}^{r_+}\frac{u+ku_0}{(k+1)!}(u-u_0)^k A_+^{k-1}B_+,$$

$$\mathcal{X}_-(u,u_0) \triangleq \left(I - A_-A_-^{\mathrm{D}}\right)\sum_{k=1}^{r_-}\frac{u+ku_0}{(k+1)!}(u-u_0)^k A_-^{k-1}B_-$$

$$\mathcal{Y}_+(u) \triangleq \left(I - A_+A_+^{\mathrm{D}}\right)\sum_{k=1}^{r_+}\frac{1}{k!}u^k A_+^{k-1}E_+,$$

$$\mathcal{Y}_-(u) \triangleq \left(I - A_-A_-^{\mathrm{D}}\right)\sum_{k=1}^{r_-}\frac{1}{k!}u^k A_-^{k-1}E_-,$$

$$\begin{aligned}\mathcal{Z}_+(u,u_0) \triangleq{}& A_+^{\mathrm{D}}(uI - u_0e^{A_+(u-u_0)})B_+ + A_+^{2\mathrm{D}}\times\\ &(I - e^{A_+(u-u_0)})B_+ + A_+^{\mathrm{D}}(I - e^{A_+(u-u_0)})E_+,\\ \mathcal{Z}_-(u,u_0) \triangleq{}& A_-^{\mathrm{D}}(uI - u_0e^{A_-(u-u_0)})B_- + A_-^{2\mathrm{D}}\times\\ &(I - e^{A_-(u-u_0)})B_- + A_-^{\mathrm{D}}(I - e^{A_-(u-u_0)})E_-,\end{aligned}$$

Let $\rho(A)$ denote the spectral radius of $A \in \mathbb{R}^{n\times n}$. We now state the main result on the existence of hysteretic maps of the semilinear Duhem model.

Theorem 1. Let $u(t)$ and $y(t)$ satisfy the semilinear Duhem model (15), (16). Suppose $u(t)$ is piecewise monotonic and periodic with period T and has exactly one local maximum $u_{\max}$ in $[0,T)$ and exactly one local minimum $u_{\min}$ in $[0,T)$. Furthermore, let $\beta \triangleq u_{\max} - u_{\min}$ and assume that $\rho\left(e^{\beta A_+}e^{-\beta A_-}\right) < 1$. Then the input-output map of $u(t)$ and $y(t)$ converges to the closed curve in $\mathbb{R}^2$ given for $u \in [u_{\min}, u_{\max}]$ by

$$\begin{aligned}\hat{y}_+(u) ={}& Ce^{A_+(u-u_{\min})}\hat{x}_+ + C\mathcal{X}_+(u,u_{\min})\\ &+ C\mathcal{Y}_+(u-u_{\min}) - C\mathcal{Z}_+(u,u_{\min}) + Du,\\ \hat{y}_-(u) ={}& Ce^{A_-(u-u_{\max})}\hat{x}_- + C\mathcal{X}_-(u,u_{\max})\\ &+ C\mathcal{Y}_-(u-u_{\max}) - C\mathcal{Z}_-(u,u_{\max}) + Du,\end{aligned}$$

where

$$\hat{x}_+ \triangleq \left(I - e^{-\beta A_-}e^{\beta A_+}\right)^{-1}\left(e^{-\beta A_-}\mathcal{W}_+ + \mathcal{W}_-\right),$$

$$\hat{x}_- \triangleq \left(I - e^{\beta A_+}e^{-\beta A_-}\right)^{-1}\left(e^{\beta A_+}\mathcal{W}_- + \mathcal{W}_+\right),$$

$$\mathcal{W}_+ \triangleq \mathcal{X}_+(u_{\max},u_{\min}) + \mathcal{Y}_+(\beta) - \mathcal{Z}_+(u_{\max},u_{\min}),$$

$$\mathcal{W}_- \triangleq \mathcal{X}_-(u_{\min},u_{\max}) + \mathcal{Y}_-(-\beta) - \mathcal{Z}_+(u_{\max},u_{\min}).$$

As a special case of (15), (16), consider the semilinear Duhem model

$$\dot{x}(t) = \left[\dot{u}_+(t)I \;\; \dot{u}_-(t)I\right] \times \left(\begin{bmatrix}h_+A\\h_-A\end{bmatrix}x(t) + \begin{bmatrix}h_+B\\h_-B\end{bmatrix}u(t) + \begin{bmatrix}h_+E\\h_-E\end{bmatrix}\right), \tag{19}$$

$$y(t) = Cx(t) + Du(t), \quad x(0) = x_0, \; t \geq 0, \tag{20}$$

where $A \in \mathbb{R}^{n\times n}$, $B \in \mathbb{R}^n$, $E \in \mathbb{R}^n$, $h_+ \in \mathbb{R}$, $h_- \in \mathbb{R}$, $C \in \mathbb{R}^{1\times n}$, and $D \in \mathbb{R}$. The following result is the specialization of Theorem 1 to (19), (20).

Corollary 1. Let $u(t)$ and $y(t)$ satisfy the semilinear Duhem model (19), (20). Suppose A is asymptotically stable, and $u(t)$ is piecewise monotonic and periodic with period T and has exactly one local maximum $u_{\max}$ in $[0,T)$ and exactly one local minimum $u_{\min}$ in $[0,T)$. Furthermore, assume $h_- < h_+$. Then the input-output map of $u(t)$ and $y(t)$ converges to the closed curve in $\mathbb{R}^2$ given for $u \in [u_{\min}, u_{\max}]$ by

$$\begin{aligned}\hat{y}_+(u) ={}& Ce^{h_+A(u-u_{\min})}\hat{x}_+ + C\mathcal{V}_+(u) + Du,\\ \hat{y}_-(u) ={}& Ce^{h_-A(u-u_{\max})}\hat{x}_- + C\mathcal{V}_-(u) + Du,\end{aligned}$$

where

$$\hat{x}_+ \triangleq (I - e^{\beta(h_+-h_-)A})^{-1}(e^{-\beta h_-A}\mathcal{V}_+(u_{\max}) + \mathcal{V}_-(u_{\min})),$$

$$\hat{x}_- \triangleq (I - e^{\beta(h_+-h_-)A})^{-1}(e^{\beta h_+A}\mathcal{V}_-(u_{\min}) + \mathcal{V}_+(u_{\max})),$$

$$\begin{aligned}\mathcal{V}_+(u) \triangleq{}& A^{-1}(uI - u_{\min}e^{h_+A(u-u_{\min})}B + h_+^{-1}A^{-2}\times\\ &(I - e^{h_+A(u-u_{\min})})B - A^{-1}(I - e^{h_+A(u-u_{\min})})E,\\ \mathcal{V}_-(u) \triangleq{}& A^{-1}(uI - u_{\max}e^{h_-A(u-u_{\max})}B + h_-^{-1}A^{-2}\times\\ &(I - e^{h_-A(u-u_{\max})})B - A^{-1}(I - e^{h_-A(u-u_{\max})})E,\end{aligned}$$

Example 3. Consider the semilinear Duhem model (19), (20) with

$$A = \begin{bmatrix}-1 & 4\\-4 & -1\end{bmatrix}, \; B = \begin{bmatrix}0\\1\end{bmatrix}, \; C = \begin{bmatrix}0 & 1\end{bmatrix},$$
$$x_0 = \begin{bmatrix}0.15 & 0.15\end{bmatrix}^{\mathrm{T}}.$$

Suppose $u(t) = \sin t$, $t \geq 0$ and let $h_+ = 1$ and $h_- = -1$. Then since A is asymptotically stable and $h_- < h_+$, Corollary 1 implies that the input-output map of $u(t)$ and $y(t)$ converges to a closed curve as $t \to \infty$. Indeed, $\rho(e^{2h_+A}e^{-2h_-A}) = 0.0183 < 1$, and the closed curve is shown in Figure 7a. Now let $h_+ = 1$ and $h_- = 1.1$. Then $h_- > h_+$ and the input-output map of $u(t)$ and $y(t)$ does not converge, as shown in Figure 7b. In this case, $\rho(e^{2h_+A}e^{-2h_-A}) = 1.2214 > 1$.

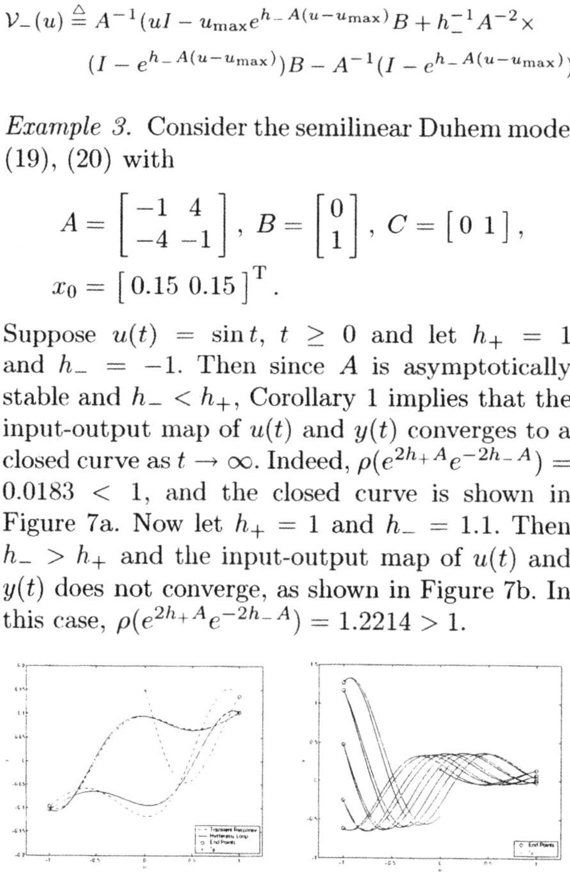

(a) (b)

Fig. 7. Input-output maps of Example 3 with $h_+ = 1$ and (a) $h_- = -1$, and (b) $h_- = 1.1$.

5. IDENTIFICATION OF THE SEMILINEAR DUHEM MODEL

In this section we develop an identification method based on the semilinear Duhem model. Specifically, consider input-output curves $\hat{y}_+(u)$ and $\hat{y}_-(u)$, $u \in [u_{\min}, u_{\max}]$, that form a closed curve in $\mathbb{R}^2$. Then the semilinear Duhem model identification problem is to find an order n and matrices $A_+ \in \mathbb{R}^{n\times n}$, $A_- \in \mathbb{R}^{n\times n}$, $B_+ \in \mathbb{R}^n$, $B_- \in \mathbb{R}^n$, and $C \in \mathbb{R}^{1\times n}$ such that, $\hat{y}_+(u)$, $\hat{y}_-(u)$ satisfy (17), (18) in steady state with $E_+ = E_- = 0$ and $D = 0$. Furthermore, to guarantee convergence to the hysteresis map, we require that the stability condition $\rho(e^{\beta A_+}e^{-\beta A_-}) < 1$ be satisfied.

The semilinear Duhem model identification problem is equivalent to identifying two linear systems whose forward-time ramp response and backward-time ramp response coincide with $\hat{y}_+(u)$ and $\hat{y}_-(u)$, respectively under the stability condition. Note that the independent variable of the linear system (17) is nonmonotonic, since u is increasing for $\hat{y}_+(u)$ and decreasing for $\hat{y}_-(u)$. To avoid backward-in-time identification, we introduce a monotonically increasing independent variable, $\hat{u} \in [u_{\min}, 2u_{\max} - u_{\min}]$. Since the time-scale invariance property of the semilinear Duhem model renders the input-output map unaffected by the time dependence on u, we reparameterize $\hat{y}_+(u)$ and $\hat{y}_-(u)$ in terms of $\hat{u}$. Specifically, we 'flip over' $\hat{y}_-(u)$ as shown Figure 8 and define

$$\hat{y}(\hat{u}) \triangleq \begin{cases} \hat{y}_+(\hat{u}), & u_{\min} \le \hat{u} < u_{\max}, \\ \hat{y}_-(u_{\max} + u_{\min} - \hat{u}), & u_{\max} \le \hat{u} \le 2u_{\max} - u_{\min}, \end{cases}$$

$$u(\hat{u}) \triangleq \begin{cases} \hat{u}, & u_{\min} \le \hat{u} < u_{\max}, \\ \hat{u}_{\max} + u_{\min} - \hat{u}, & u_{\max} \le \hat{u} \le 2u_{\max} - u_{\min}. \end{cases}$$

Then the identification problem is to find system matrices associated with the input $\hat{u}(\hat{u})$ and the output $\hat{y}(\hat{u})$, $u \in [u_{\min}, 2u_{\max} - u_{\min}]$.

Now, let y_k and u_k, $k = 0, \ldots, 2l-1$, be $2l$ measurements taken from $\hat{y}(\hat{u})$ and $u(\hat{u})$, respectively. Then we determine system matrices $\hat{A}_+$, $\hat{A}_-$, $\hat{B}_+$, $\hat{B}_-$, and $\hat{C}$ to approximately satisfy the *discrete-time semilinear Duhem model*

$$x_{k+1} = \begin{cases} \hat{A}_+ x_k + \hat{B}_+ u_k, & k = 0, \ldots, l-1, \\ \hat{A}_- x_k + \hat{B}_- u_k, & k = l, \ldots, 2l-1, \end{cases} \tag{21}$$

$$y_k = \hat{C} x_k, \tag{22}$$

Note that the stability condition for (21), (22) is $\rho(A_+^l A_-^l) < 1$.

Finding system matrices is nontrivial, since the input u_k is not persistently exciting. Nevertheless, we use the nonminimal input/state/output representation approach. Suppose u_k and y_k satisfy the m-dimensional ARX model

$$y_{k+1} = -\alpha_1^+ y_k - \cdots - \alpha_m^+ y_{k-m+1} + \beta_1^+ u_k + \cdots + \beta_m^+ u_{k-m+1}, \tag{23}$$

for $k = 0, \ldots, l-1$, where $\alpha_i^+ \in \mathbb{R}$ and $\beta_i^+ \in \mathbb{R}$, $i = 1, \ldots, m$, are system parameters. Then (23) has a nonminimal state space representation of order $2m - 1^{\text{th}}$ given by

$$x_{k+1} = \hat{A}_+ x_k + \hat{B}_+ u_k, \quad k = 0, \ldots, l-1, \tag{24}$$

$$y_k = \hat{C} x_k, \tag{25}$$

where

$$x_k \triangleq \begin{bmatrix} y_k & \cdots & y_{k-m+1} & u_{k-1} & \cdots & u_{k-m+1} \end{bmatrix}^{\text{T}},$$

$$\hat{A}_+ \triangleq \begin{bmatrix} -\alpha_1^+ & \cdots & -\alpha_{m-1}^+ & -\alpha_m^+ & \beta_2^+ & \cdots & \beta_{m-1}^+ & \beta_m^+ \\ 1 & \cdots & 0 & 0 & 0 & \cdots & \cdots & 0 \\ \vdots & \ddots & \vdots & \vdots & \vdots & & & \vdots \\ 0 & \cdots & 1 & 0 & 0 & \cdots & \cdots & 0 \\ 0 & \cdots & 0 & 0 & 0 & \cdots & 0 & 0 \\ 0 & \cdots & 0 & 0 & 1 & \cdots & 0 & 0 \\ \vdots & & \vdots & \vdots & \vdots & \ddots & \vdots & \vdots \\ 0 & \cdots & 0 & 0 & 0 & \cdots & 1 & 0 \end{bmatrix},$$

$$\hat{B}_+ \triangleq \begin{bmatrix} \beta_1^+ & 0 & \cdots & 0 \end{bmatrix}^{\text{T}}, \quad \hat{C} \triangleq \begin{bmatrix} 1 & 0 & \cdots & 0 \end{bmatrix}.$$

We define $\hat{A}_- \in \mathbb{R}^{(2m-1)\times(2m-1)}$ and $\hat{B}_- \in \mathbb{R}^{2m-1}$ for $k = l, \ldots, 2l-1$ analogously. Note that the initial state x_0 for (24) involves data taken from the backward system of (21), namely, $y_{2l-1}, \ldots, y_{2l-m+1}$, and analogously for x_l. This ensures continuity of the state over $[u_{\min}, 2u_{\max} - u_{\min}]$.

Since x_k consists of inputs and outputs, we can immediately determine the state x_k for all $k = 0, \ldots, 2l-1$. Now, define $\Phi_{1_+} \in \mathbb{R}^{(2m-1)\times l}$, $\Phi_{2_+} \in \mathbb{R}^{(2m-1)\times l}$, and $U_{1_+} \in \mathbb{R}^{1\times l}$ as

$$\Phi_{1_+} \triangleq \begin{bmatrix} x_0 & x_1 & \cdots & x_{l-1} \end{bmatrix}, \quad \Phi_{2_+} \triangleq \begin{bmatrix} x_1 & x_2 \cdots & x_l \end{bmatrix},$$

$$U_{1_+} \triangleq \begin{bmatrix} u_0 & u_1 & \cdots & u_{l-1} \end{bmatrix}.$$

Then $\hat{A}_+$, $\hat{B}_+$, $\hat{A}_-$ and $\hat{B}_-$ are determined from minimizing

$$\left\| \Phi_{2_+} - [\hat{A}_+ \ \hat{B}_+] \begin{bmatrix} \Phi_{1_+} \\ U_{1_+} \end{bmatrix} \right\|_{\text{F}} + \left\| \Phi_{2_-} - [\hat{A}_- \ \hat{B}_-] \begin{bmatrix} \Phi_{1_-} \\ U_{1_-} \end{bmatrix} \right\|_{\text{F}}, \tag{26}$$

subject to

$$\rho(A_+^l A_-^l) < 1. \tag{27}$$

Since the stability condition (27) is not convex, we use the fact that if $\bar{\sigma}(A) < 1$, where $\bar{\sigma}(A)$ is the maximum singular value of A, then $\rho(A) < 1$ and thus $\rho(A^l) < 1$. Then from the submultiplicative property of the maximum singular value, we have the alternative stability condition

$$\bar{\sigma}(A_+) < 1 \text{ and } \bar{\sigma}(A_-) < 1. \tag{28}$$

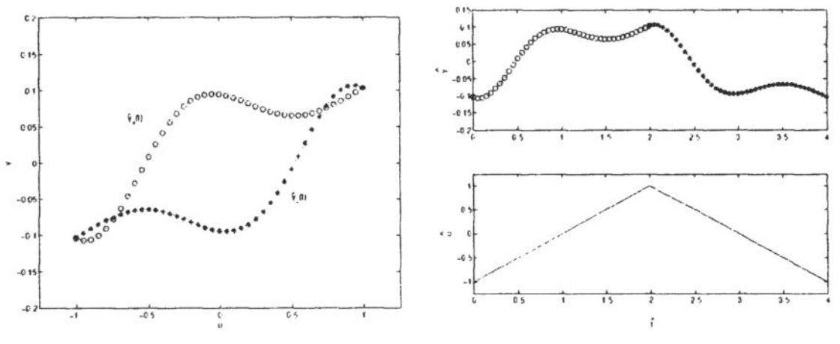

(a) (b)

Fig. 8. The given input-output curves $\hat{y}_+(u)$ and $\hat{y}_-(u)$ in (a) are considered as the input-output map of $\hat{y}(\hat{u})$ and $u(\hat{u})$ in (b).

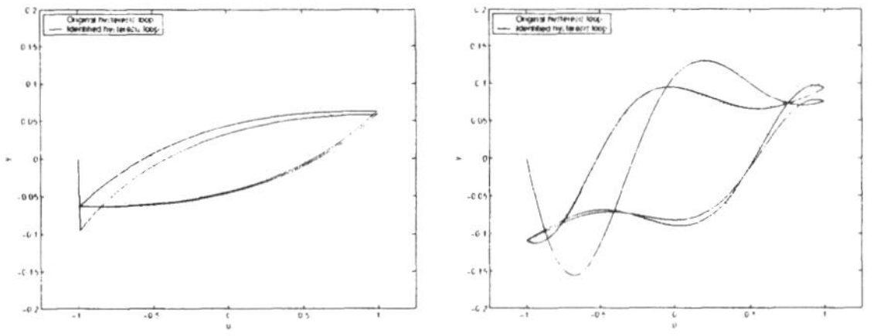

(a) (b)

Fig. 9. The identification of $\hat{y}_+(u)$, $\hat{y}_-(u)$ from Example 3 with (a) $p = 1$ and (b) $p = 7$.

Therefore, the original constrained least squares problem (26), (27) can be rewritten as two separate least squares problems with a convex constraint

$$\min_{A_+, B_+} \left\| \Phi_{2_+} - \left[\hat{A}_+ \ \hat{B}_+ \right] \begin{bmatrix} \Phi_{1_+} \\ U_{1_+} \end{bmatrix} \right\|_F, \quad \text{subject to } \bar{\sigma}(A_+) < 1, \tag{29}$$

and

$$\min_{A_-, B_-} \left\| \Phi_{2_-} - \left[\hat{A}_- \ \hat{B}_- \right] \begin{bmatrix} \Phi_{1_-} \\ U_{1_-} \end{bmatrix} \right\|_F, \quad \text{subject to } \bar{\sigma}(A_-) < 1. \tag{30}$$

The constrained least squares problems (29) and (30) are quadratic programming problem with positive-semi-definite constraints. Then we can find $\hat{A}_+$, $\hat{A}_-$, $\hat{B}_+$, and $\hat{B}_-$ by minimizing a linear function over symmetric cones. For details of a similar algorithm, see Lacy and Bernstein (2003).

Remark 1. The alternative stability condition (28) is conservative and may result an overly constrained solution. To avoid the conservatism, we consider $\bar{\sigma}(A_+) < p$ and $\bar{\sigma}(A_-) < p$, where $p \geq 1$. The condition $\rho(A_+^l A_-^l) < 1$ is checked a posteriori to verify stability.

Example 4. Suppose $\hat{y}_+(u)$ and $\hat{y}_-(u)$ are given as Figure 9 from Example 3. The identification is performed with $m = 2$, and thus the identified system is of order 3. Figure 9a shows the input-output map of the identified system with $p = 1$. Although $\rho(A_+^l A_-^l) = 0.0262$ and thus the stability condition is met, the least squares cost is 0.0553 and the input-output map poorly fits $\hat{y}_+(u)$ and $\hat{y}_-(u)$. Then the upper bound $p = 1$ is increased to $p = 7$, where $\rho(A_+^l A_-^l) = 0.0134$. The input-output map fits the original hysteresis better as shown in Figure 9b, and the least squares cost is 5.12×10^{-9}.

Example 5. Consider a dynamic stall model of an oscillating 2-D airfoil. The dashed line of Figure 10 represents dynamic stall between the angle of attack and the lift coefficient of an airplane (Carr *et al.*, 1977). The identification is performed with $m = 20$, hence the identified system is of order 39. Figure 10a shows the input-output map of the identified system and it does not fit the stall loop

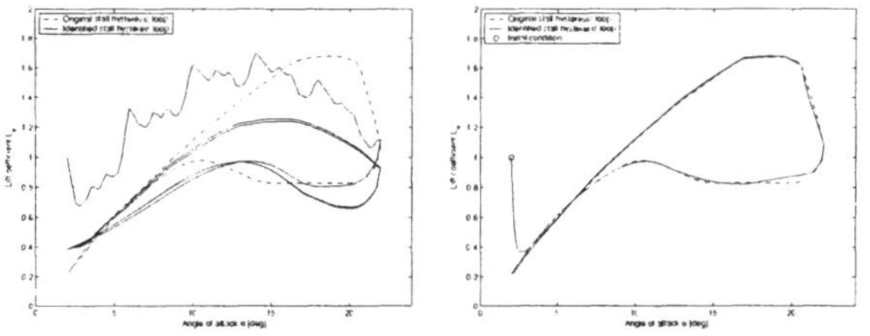

(a) (b)

Fig. 10. The identification of the dynamic stall model (a) without additional subintervals and (b) with 10 subintervals.

well. To improve the fit, we divide $[u_{\min}, u_{\max}]$ into 10 subintervals and identify the system matrices for each subintervals. Hence the model is a switching system which switches according to the subinterval of u. Figure 10b shows the input-output map of the identified system.

6. Conclusion

In this paper we introduced the rate-independent semilinear Duhem model. The analysis of this model was facilitated by a reparameterization in terms of the control input. By analyzing the iterated ramp response, we obtained sufficient conditions for convergence to a hysteresis map. Finally the semilinear Duhem model provided the basis for a least squares identification method based on a convex optimization algorithm.

REFERENCES

Campbell, S. L. and Jr. C. D. Meyer (1979). *Generalized Inverse of Linear Transformations.* Pitman Publishing Ltd.. London.

Carr, L. W., K. W. McAlister and W. J. McCroskey (1977). Analysis of the development of dynamic stall based on oscillating airfoil measurements. Technical Report TN D-8382. NASA.

Chua, L. O. and S. C. Bass (1972). A generalized hysteresis model. *IEEE Trans. Circuit Theory* **19**(1), 36–48.

Coleman, B. D. and M. L. Hodgdon (1986). A constitutive relation for rate-independent hysteresis in ferromagnetically soft materials. *Int. J. Eng. Sci.* **24**, 897–919.

Kransnosel'skii, M. A. and A. V. Pokrovskii (1980). *Systems with Hysteresis.* Springer-Verlag. New York, NY.

Lacy, S. S. and D. S. Bernstein (2003). Subspace identification with guaranteed stability using constained optimization. In: *Proc. Amer. Contr. Conf.* To be appeared.

Macki, J. W., P. Nistri and P. Zecca (1993). Mathematical models of hysteresis. *SIAM Review* **35**(1), 94–123.

IFAC Publications
www.elsevier.com/locate/ifac

LEAST SQUARES HARMONIC SIGNAL ANALYSIS USING PERIODIC ORBITS OF ODEs

Torbjörn Wigren, Emad Abd-Elrady and Torsten Söderström

Systems and Control, Department of Information Technology,
Uppsala University, PO Box 337, SE-751 05, Uppsala, SWEDEN.
torbjorn.wigren@it.uu.se, emad.abdelrady@it.uu.se, torsten.soderstrom@it.uu.se

Abstract: The idea of the paper is to model harmonic signals by means of second order nonlinear ordinary differential equations (ODEs). This is motivated by the well known theoretical results on the existence of periodic orbits for nonlinear second order ODEs. The right hand side functions of the ODE are the estimated quantities, this being accomplished by the use of a polynomial parameterisation. A least squares estimation algorithm is then derived. The methodology reduces the number of parameters needed in cases where the signal generation can be accurately described by the suggested model, thereby enhancing estimation performance. *Copyright © 2003 IFAC*

Keywords: Identification, Nonlinear systems, Periodic motion, Phase plane.

1. INTRODUCTION

The modelling of signals with harmonic spectra has widespread applications. Examples include vibration analysis and overtone analysis in power networks as well as the measurement of linearity in electronic power amplifiers and other devices.

The field of harmonic signal analysis has been extensively studied. The periodogram method in combination with the fast Fourier transform (FFT) forms a baseline against which other methods can be evaluated, see (Stoica and Moses, 1997). A large number of references (Kumaresan *et al.* ,1982; Li and Stoica,1996; Rife and Boorstyn, 1976; Stoica and Moses, 1997) discuss estimation algorithms applicable to harmonic signal analysis. This includes ARMA model based methods, the nonlinear least squares method, the high order Yule-Walker method, the Pisarenko method, the MUSIC method, the ESPRIT method, the APES method as well as more direct parametric approaches. Further extensions can be found in (Händel and Tichavsky, 1994; Stoica and Nehorai, 1989). The harmonic relation between the frequencies of the spectral components is explicitly exploited e.g. in the comb filter approach of Nehorai and Porat (1986) and in the recursive modelling of static nonlinearities and a fundamental frequency as in (Wigren and Händel, 1996; Abd-Elrady, 2001; Abd-Elrady, 2002).

The present paper is inspired by one possible model for the generation of periodic signals, namely nonlinear ordinary differential equations (ODEs). There is a rich theory on the subject as outlined in e.g. (Khalil, 1996). The focus here will be on periodic orbits and their properties. Some of the strongest results of the theory concern ODEs with two state variables. The reason is that closed periodic orbits in R^2 that do not intersect themselves divide the space into one part interior to the orbit and one part exterior to the orbit. The mathematical consequence is that there are several powerful theorems on the existence of periodic solutions to ODEs in R^2, and hence it seems to be advantageous to base estimation algorithms on second order ODEs. The specific signal model of the paper is therefore obtained by introducing a polynomial parameterisation of the right hand side of a general second order ODE, and by defining the harmonic signal to be modelled as a function of the states of this ODE. A least squares algorithm is then derived. The performance of the algorithm is illustrated in a simulation study.

What are the advantages of the approach taken? First, many systems that generate harmonic signals are best described by nonlinear ODEs. Examples include tunnel diodes, pendulums, biological predator-prey systems and radio frequency synthesisers, see (Khalil, 1996). Many of these systems are described by second order ODEs with polynomial right hand sides. It can therefore be expected that there are good opportunities to obtain highly accurate models by estimating only a few parameters. The parsimony principle, e.g. (Söderström and Stoica, 1989), suggests that the achievable accuracy would be improved by the proposed methods, as compared e.g. to the periodogram and other methods that do not impose the same amount of prior information on the solution. Even if the data generating the ODE model would have a dimension higher than two, it could be expected that addition of parameters to the right hand side of the ODE would still allow for accurate modelling, compensating for the order discrepancy.

The paper is organised as follows. Section 2 introduces the details on the model. Section 3 discusses the algorithm, while section 4 presents a simulation study. Conclusions appear in section 5. In order to organise the conditions imposed in the paper, these are ordered C1), C2), etc.

2. MODEL AND PARAMETERISATION

2.1 Measurements And Modelled Signals

The starting point is the *discrete time* measured signal $z(t)$, where

$$z(t) = y(t) + e(t). \qquad (1)$$

Here $y(t)$ is the *continuous time* signal to be modelled and $e(t)$ is the discrete time measurement noise. It is assumed here that $y(t)$ is periodic, *i.e.*

C1) $y(t+T) = y(t)$, $\forall t \in R$, $0 < T < \infty$.

Furthermore, $e(t)$ is assumed to be zero mean Gaussian white noise, *i.e.*

C2) $e(t) \in N(0, \sigma^2)$, $E[e(t)e(t+kT_s)] = \delta_{k,0}\sigma^2$.

T_S denotes the sampling period.

2.2 Model Structures

As stated above, the main idea of the paper is to model the generation of the signal $y(t)$ by means of an unknown parameter vector $\boldsymbol{\theta}$ and an ordinary differential equation of order two, as shown in (2), (3)

$$\boldsymbol{\theta} = \begin{pmatrix} \boldsymbol{\theta}_1^T & \boldsymbol{\theta}_2^T \end{pmatrix}^T \qquad (2)$$

$$\begin{pmatrix} \dfrac{dx_1(t)}{dt} \\ \dfrac{dx_2(t)}{dt} \end{pmatrix} = \begin{pmatrix} f_1(x_1(t), x_2(t), \boldsymbol{\theta}_1) \\ f_2(x_1(t), x_2(t), \boldsymbol{\theta}_2) \end{pmatrix}$$
$$z(t) = \begin{pmatrix} c_1 & c_2 \end{pmatrix} \begin{pmatrix} x_1(t) \\ x_2(t) \end{pmatrix} + e(t). \qquad (3)$$

In (3) $\begin{pmatrix} x_1(t) & x_2(t) \end{pmatrix}^T$ is the state vector and $\begin{pmatrix} c_1 & c_2 \end{pmatrix}$ contains the *selected* output weighting factors.

At this point it is highly relevant to pose the question whether the model structure (3) may be too general. It is stressed that this issue requires further study. However, as will be shown, there may be certain advantages with (3).

In (Wigren *et. al*, 2002), it is assumed that the second order ODE

$$\frac{d^2 y}{dt^2} = f_2\left(y, \frac{dy}{dt}, \boldsymbol{\theta}_2\right) \qquad (4)$$

generates the harmonic *signal that is measured.* This fact allows the state variable and model selection

$$\begin{pmatrix} x_1 \\ x_2 \end{pmatrix} = \begin{pmatrix} y \\ dy/dt \end{pmatrix}. \qquad (5)$$

$$\begin{pmatrix} \dfrac{dx_1(t)}{dt} \\ \dfrac{dx_2(t)}{dt} \end{pmatrix} = \begin{pmatrix} x_2(t) \\ f_2(x_1(t), x_2(t), \boldsymbol{\theta}_2) \end{pmatrix}.$$
$$z(t) = \begin{pmatrix} 1 & 0 \end{pmatrix} \begin{pmatrix} x_1(t) \\ x_2(t) \end{pmatrix} + e(t) \qquad (6)$$

The model (6) depends only on the parameters of the second right hand side function of (3), a fact that should be advantageous from a computational and performance point of view. The obvious question is then - are there other facts that could motivate the use of (3) instead of (6)?

One key to the answer to this question is the output equations of (3) and (6). Physically, it may very well be the case that the system is governed by (4), but the generated periodic signal may be a nonlinear function of y and dy/dt. It could also be the case that this function is not well known and/or noninvertible, and the question is then what can be done in such situations? One idea would be to reuse the output relation of (6) and try and find a representation of the (state space) ODE such that the first state equals the sought output. In order to study the implications of this idea, the following system is studied

$$\frac{d^2y}{dt^2} = f\left(y, \frac{dy}{dt}, \boldsymbol{\theta}\right)$$
$$z = g\left(y, \frac{dy}{dt}\right) \quad (7)$$

With the states chosen as

$$\begin{pmatrix} x_1 \\ x_2 \end{pmatrix} = \begin{pmatrix} z \\ dy/dt \end{pmatrix}, \quad (8)$$

it follows by differentiation that

$$\frac{dx_1}{dt} = \frac{\partial g(y, x_2)}{\partial y} x_2 + \frac{\partial g(y, x_2)}{\partial x_2} f(y, x_2, \boldsymbol{\theta})$$
$$\frac{dx_2}{dt} = f(y, x_2, \boldsymbol{\theta}). \quad (9)$$

It remains to find an expression for y. This follows from (7) and (8), since under mild regularity conditions (Khalil, 1996) the inverse function theorem allows at least a local solution of the output equation with respect to y, *i.e.*

$$x_1 = g(y, x_2) \Rightarrow y = h(x_1, x_2). \quad (10)$$

When (10) is inserted in (9), it is clear that the result requires and fits into the structure (3), with the weights of the output equation being equal to those of (6). The conclusion is that the model structure (3) may extend the validity of the proposed harmonic signal modelling approach, beyond that of (6). Despite this fact, the treatment will be based on the model (6) in this paper.

2.3 Parameterisation

A natural approach is now to expand the right hand side of the second *state* equation of (6) in terms of known basis functions $\left\{b_k(x_1(t), x_2(t))\right\}_{k=0}^{\infty}$, modelling the right hand side as a truncated superposition of these functions. In case of a polynomial model, a suitable parameterisation is

$$f_2(x_1(t), x_2(t), \boldsymbol{\theta}_2) = \sum_{l=0}^{L_2} \sum_{m=0}^{M_2} \theta_{2,l,m} x_1^l(t) x_2^m(t) \quad (11)$$

$$\boldsymbol{\theta}_2 = \begin{pmatrix} \theta_{2,0,0} & \ldots & \theta_{2,0,M_2} & \ldots & \theta_{2,L_2,0} & \ldots & \theta_{2,L_2,M_2} \end{pmatrix}^T \quad (12)$$

Remark 1: No scale factor problems are expected with this parameterization. The reason is that the transformation

$$y' = ky \quad (13)$$

transforms (4) to

$$\frac{d^2y'}{dt^2} = kf_2\left(\frac{1}{k}y', \frac{1}{k}\frac{dy'}{dt}, \boldsymbol{\theta}_2\right). \quad (14)$$

Since the polynomial model is a general function expansion, the function of (14) can be modelled equally well as the function of (4).

2.4 Discretization

In order to formulate complete discrete time models, the continuous time ODE model (6) needs to be discretized. This is done by exploiting an Euler forward numerical integration scheme. For simplicity, the discretization interval is selected to be equal to the sampling period T_S resulting in

$$x_1(t + T_S) = x_1(t) + T_S x_2(t)$$
$$x_2(t + T_S) = x_2(t) + T_S \sum_{l=0}^{L_2} \sum_{m=0}^{M_2} \theta_{2,l,m} x_1^l(t) x_2^m(t) \quad (15)$$

3. ALGORITHM

A number of different algorithms can now be derived from the model of the previous section. Kalman and extended Kalman filter algorithms based on a similar model are e.g. described in (Wigren *et al.*, 2002).

In order to formulate a least squares algorithm, the model must be formulated in linear regression form. A complication is then that both states cannot be measured and hence one additional signal needs to be generated from the measured signal. In this paper this is accomplished by applying a differentiating filter to the measured signal. The signals (which are here the same as the modelled states plus noise) are selected so that the measured signal equals the first state of the state model plus noise and so that the second state equals the *differentiated* measured signal plus (differentiated) noise, cf. (6). The differentiation is performed by a discrete time differentiating filter described by

$$\mathbf{x}_d(t + T_S) = \mathbf{K}(T_S)\mathbf{x}_d(t) + \mathbf{L}(T_S)z(t)$$
$$\left(dz(t)/dt\right)_{est} = \mathbf{M}^T(T_S)\mathbf{x}_d(t) \quad (16)$$

Here the subscript $(\ )_{est}$ is used to highlight the fact that the signal is constructed from a measured signal. See (Carlsson, 1989) for differentiating filter details.

To proceed, note that (15) results in the model

$$x_1(t + T_S) - x_1(t) = T_S x_2(t)$$

$$x_2(t+T_S) - x_2(t) = \boldsymbol{\varphi}_2^T(x_1(t), x_2(t))\boldsymbol{\theta}_2 . \qquad (17)$$

$$\frac{1}{T_S}\boldsymbol{\varphi}_2^T(x_1(t), x_2(t)) =$$
$$\begin{pmatrix}1 & \dots & x_2^{M_2}(t) & \dots & x_1^{L_2}(t) & \dots & x_1^{L_2}(t)x_2^{M_2}(t)\end{pmatrix} \qquad (18)$$

$$\boldsymbol{\theta}_2 =$$
$$\begin{pmatrix}\theta_{2,0,0} & \dots & \theta_{2,0,M_2} & \dots & \theta_{2,L_2,0} & \dots & \theta_{2,L_2,M_2}\end{pmatrix}^T \qquad (19)$$

By replacing the states of (17) with estimates based on measured data, *i.e.* by replacing $x_1(t)$ with $z(t)$ and $x_2(t)$ with $(dz(t)/dt)_{est}$, the second state equation of (15) results in

$$\Delta z(t+T_S) = (dz(t+T_S)/dt)_{est} - (dz(t)/dt)_{est} =$$
$$\boldsymbol{\varphi}_2^T\left(z(t), (dz(t)/dt)_{est}\right)\boldsymbol{\theta}_2 + \varepsilon_2(t+T_S) \qquad (20)$$

The first state equation of (17) is not needed. The expression (20) follows by performing a Taylor series expansion around $\begin{pmatrix}z(t) & (dz(t)/dt)_{est}\end{pmatrix}^T$. Note that when this is done, sums of noise samples that are delayed with a varying number of sampling periods appear in the combined error of each state equation. The combined regression error, $\varepsilon_2(t+T_S)$, can therefore not be expected to be white. This means that the least squares estimator will, in the end, be biased. However, for a high signal to noise ratio the accuracy could still be relatively good. Secondly, the least squares solution, although biased, may be useful as an initial value for more elaborate nonlinear search methods.

Now, by defining the vectors and matrices (it is assumed that data is available at times $t_1, \dots, t_1 + (N-1)T_S$)

$$\Delta\mathbf{Z}(t_1, N, T_S) = \left(\Delta z^T(t_1+T_S) \ \dots \ \Delta z^T(t_1+NT_S)\right)^T \qquad (21)$$

$$\boldsymbol{\Phi}^T(t_1, N, T_S) = \left(\boldsymbol{\varphi}\left(z(t_1), (dz(t_1)/dt)_{est}\right) \ \dots \ \boldsymbol{\varphi}\left(z(t_1+(N-1)T_S), (dz(t_1+(N-1)T_S)/dt)_{est}\right)\right) \qquad (22)$$

the following estimate results, see (Söderström and Stoica, 1989)

$$\left(\boldsymbol{\Phi}^T(t_1, N, T_S)\boldsymbol{\Phi}(t_1, N, T_S)\right)\hat{\boldsymbol{\theta}}^{LS}(t_1, N, T_S) = \boldsymbol{\Phi}^T(t_1, N, T_S)\ \Delta\mathbf{Z}(t_1, N, T_S). \qquad (23)$$

Note that it is in general not numerically sound to solve the least squares problem via the normal equations (23). Rather, a formulation as an over-determined system of equations should be used, see (Söderström and Stoica, 1989) for details.

4. NUMERICAL EXAMPLES

The Van der Pol oscillator (Khalil, 1996) was selected as the underlying system in example 1 – example 3 of the simulation study. In these examples, the oscillator was described by

$$\begin{pmatrix}\frac{dx_1}{dt} \\ \frac{dx_2}{dt}\end{pmatrix} = \begin{pmatrix}x_2 \\ -x_1 + 2(1-x_1^2)x_2\end{pmatrix}. \qquad (24)$$

The Matlab™ routine *ode45* was used to solve (24). The initial states of (24) were selected as $\begin{pmatrix}x_1(0) & x_2(0)\end{pmatrix}^T = \begin{pmatrix}0 & 1\end{pmatrix}^T$. The measured signal was in all examples selected as the first state with white Gaussian noise added. The differentiated signal was obtained by applying a simple first order difference (Euler forward approximation). All results below are based on data runs of length $N = 10^4$.

Example 1. The example illustrates the effect of the sampling period T_S. The noise level was selected so that an RMS SNR of 60 dB was obtained. The estimated model used third degree polynomials. The algorithm was then run for different sampling periods. As a measure of performance,

$$V = \left\|\hat{\boldsymbol{\theta}}^{LS}(t_1, N, T_S) - \boldsymbol{\theta}_2^0\right\|_2 \qquad (25)$$

was computed and plotted as a function of T_S in Fig. 1. In (25), $\boldsymbol{\theta}_2^0$ denotes the true parameter vector.

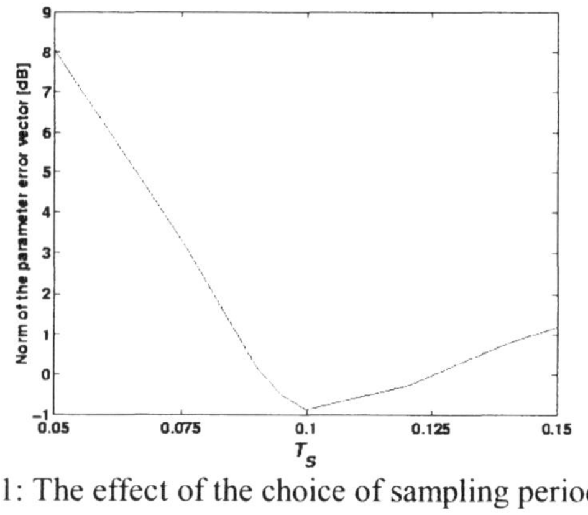

Fig. 1: The effect of the choice of sampling period.

The minimum of the plot can be explained as follows. For small values of T_S, the numerical differentiation results in a large noise amplification. The error caused by the regression vector then becomes dominant. For large values of T_S the most pronounced effect is instead modelling errors (bias) introduced by the Euler forward numerical discretization scheme. The former error decreases with T_S while the latter error in general increases with T_S, and hence an optimal T_S may exist.

Example 2. The example illustrates the effect of the signal to noise ratio (SNR) on the performance of the algorithm. The setup was the same as in example 1, with the exceptions that the SNR was varied and that $L_2 = M_2 = 2$. The sampling periods were selected to be $T_S = 0.075$ s and $T_S = 0.2$ s. Fig. 3 and Fig. 4 were generated using $T_S = 0.075$ s.

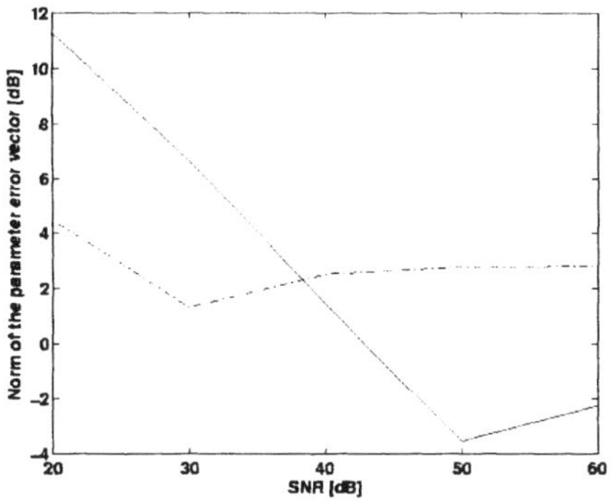

Fig. 2: The variation of the performance with the SNR for $T_S = 0.2$ s (dashed) and $T_S = 0.075$ s (solid).

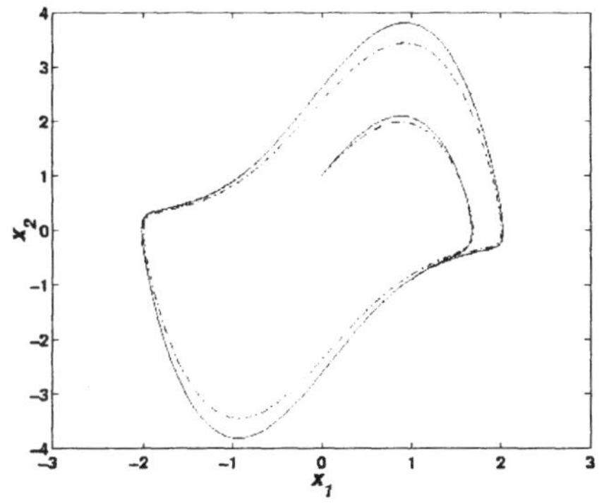

Fig. 3: Phase plane plot obtained with SNR=50 dB. System (solid) and identified model (dashed).

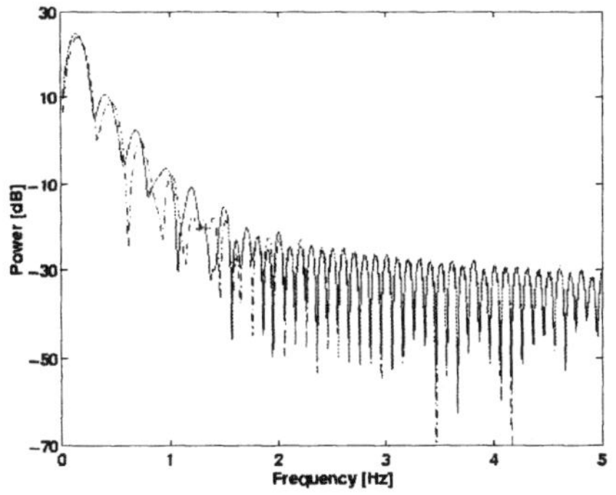

Fig. 4: Periodogram obtained with SNR=50 dB. System (solid) and identified model (dashed).

The plots of the performance measure (25) in Fig. 2 are consistent with the explanation given in response to Fig. 1. As a further illustration a phase plane plot and a periodogram appear in Fig. 3 and Fig. 4, respectively.

Example 3. The example illustrates the effect of over-parameterisation on the performance of the algorithm. Simulations were run for the polynomial degrees 2, 3, 4 and 5 using SNR=60 dB and $T_S = 0.1$ s. The performance as expressed by (25) is illustrated in Fig. 5. The result indicates that the algorithm can cope with degrees that are incorrect with at least one unit.

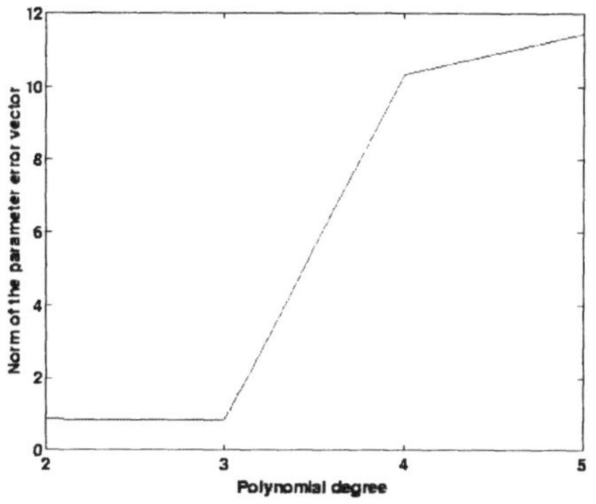

Fig. 5: The effect of the model order on the performance.

Example 4. This example illustrates the effect of under- modelling, by identification of the un-damped physical pendulum system,

$$\frac{d^2y}{dt^2} = -\frac{g}{l}\sin(y) \tag{26}$$

Here $g = 9.81\ m/s^2$ and $l = 1.00\ m$. The sampling time was selected as $T_S = 0.01$ s. The system was initiated at rest with a 90 deg angle with the vertical. All simulations used SNR=50 dB. The system was identified with polynomial degrees ranging from 2 to 7 as illustrated in Fig. 6 - Fig. 8.

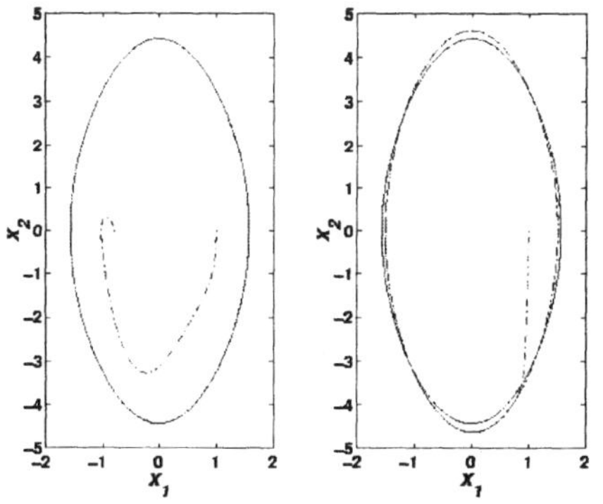

Fig. 6: Phase plane plot for polynomial degrees 2 (left) and 3 (right). System (solid) and identified model (dashed).

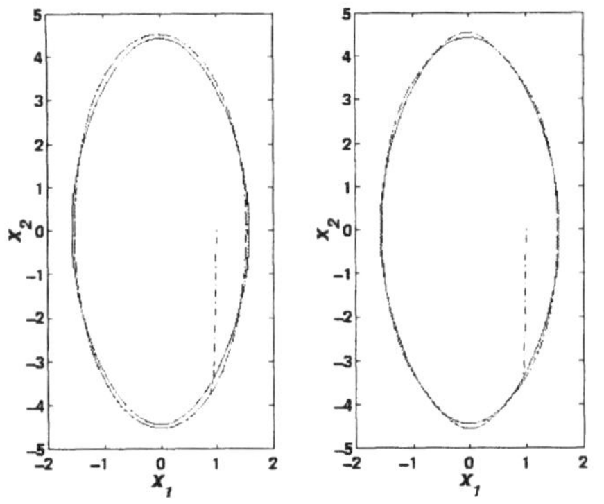

Fig. 7: Phase plane plot for polynomial degrees 4 (left) and 5 (right). System (solid) and identified model (dashed).

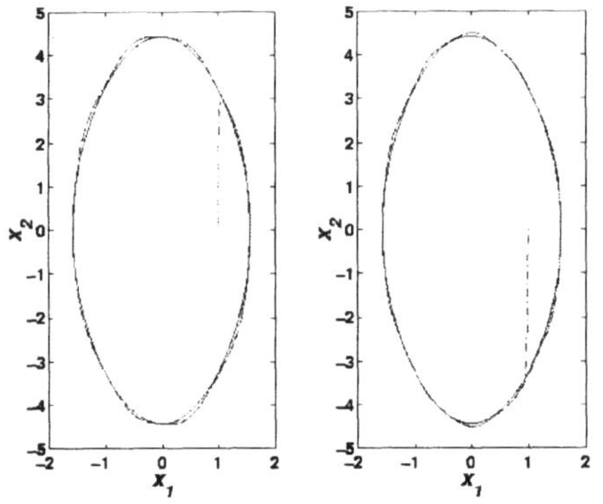

Fig. 8: Phase plane plot for polynomial degrees 6 (left) and 7 (right). System (solid) and identified model (dashed).

The polynomial degrees 3, 4, 5, 6 and 7 generated accurate and stable periodic orbits, a result consistent with the Taylor series expansion of $\sin(x)$.

5. CONCLUSIONS

The main idea of this paper is to model periodic signals as being generated by second order nonlinear ODEs. A linear in the parameters polynomial model was used in order to derive a least squares algorithm. This algorithm was tested in a simulation study and was shown to work well for high signal to noise ratios. Robustness with respect to over- and under-modelling was also demonstrated. The sampling period was found to be an important tuning parameter. Further research is needed to extend the operating region towards lower signal to noise ratios. Topics for future research also include treatment of alternative model structures and parameterisation strategies, as well as a theoretical analysis of the performance of the algorithm.

ACKNOWLEDGEMENT

The work of this paper was supported by the Swedish Research Council for Engineering Sciences under contract 98-564.

REFERENCES

Abd-Elrady, E. (2001). Study of a nonlinear recursive method for harmonic signal modeling. *20th IASTED Int. Conf. on Modeling, Identification and Control,* **2**, Innsbruck, Austria.

Abd-Elrady, E. (2002). An adaptive grid point RPEM algorithm for harmonic signal modeling. *15th IFAC World Congress,* Barcelona, Spain.

Carlsson, B. (1989). Digital Differentiating Filters and Model Based Fault Detection. *PhD-thesis, Acta Univ. Ups., Uppsala Dissertations From the Faculty of Science 28,* Uppsala, Sweden.

Händel, P. and P. Tichavsky (1994). Adaptive estimation for periodic signal enhancement and tracking. *Int. J. Adaptive Control, Signal Processing,* **8**, 447-456.

Khalil, H. K. (1996). *Nonlinear Systems, 2nd Ed.* Prentice Hall, Upper Saddle River, NJ.

Kumaresan, R. and D. W. Tufts (1982). Estimating the parameters of exponentially damped sinusoids and pole-zero modeling in noise. *IEEE Trans. Acoust. Speech, Signal Processing,* **ASSP-30**, 833-840.

Li, J. and P. Stoica (1996). An adaptive filtering approach to spectral estimation and SAR imaging. *IEEE Trans. Signal Processing,* **44**, 1469-1484.

Nehorai, A. and B. Porat (1986). Adaptive comb filtering for harmonic signal enhancement. *IEEE Trans. Acoust., Speech, Signal Processing*, **ASSP-34**, 1124-1138.

Rife, D. C. and R. R. Boorstyn (1976). Multiple tone parameter estimation from discrete-time observations. *Bell Syst. Tech. J.,* **55**, 1389-1410.

Stoica, P. and A. Nehorai (1989). Statistical analysis of two nonlinear estimators of sine wave parameters in the colored noise case. *Circ. Syst. Signal Proc.*, **8**, 3-15.

Stoica, P. and R. Moses (1997). *Introduction To Spectral Analysis.* Prentice Hall, Upper Saddle River, NJ.

Söderström, T. and P. Stoica (1989). *System Identification.* Prentice Hall Int., Hemel Hempstead, UK.

Wigren, T. and P. Händel (1996). Harmonic signal modeling using adaptive nonlinear function estimation. *IEEE International Conference on Acoustics, Speech and Signal Processing,* Atlanta, GA, USA.

Wigren, T., Abd-Elrady, E. and T. Söderström (2002). Harmonic signal analysis with Kalman filters using periodic orbits of nonlinear ODEs. Submitted.